# Introduction to TRAFFIC ENGINEERING

**R Srinivasa Kumar**
*BE (Civil Engineering),*
*ME (Highway and Transportation Engineering),*
*PhD (Transportation Engineering)*

**Universities Press**

**INTRODUCTION TO TRAFFIC ENGINEERING**

**UNIVERSITIES PRESS (INDIA) PRIVATE LIMITED**

*Registered Office*
3-6-747/1/A & 3-6-754/1, Himayatnagar, Hyderabad 500 029, Telangana, India
info@universitiespress.com; www.universitiespress.com

*Distributed by*
**Orient Blackswan Private Limited**

*Registered Office*
3-6-752 Himayatnagar, Hyderabad 500 029, Telangana, India

*Other Offices*
Bengaluru, Bhopal, Chennai, Guwahati, Hyderabad, Jaipur, Kolkata, Lucknow
Mumbai, New Delhi, Noida, Patna, Vijayawada

501440

ISBN 978-93-86235-47-3



*Typeset in* Times New Roman 10 points *by*
MacroTex Solutions, Chennai 600 091, Tamil Nadu, India

*Printed at*
Yash Printographics, Noida

*Published by*
Universities Press (India) Private Limited
3-6-747/1/A & 3-6-754/1, Himayatnagar, Hyderabad 500 029, Telangana, India

# Preface

Traffic Engineering is a subject that plays an important role in the life of anyone who uses the world's vast network of roads – as a pedestrian, or passenger or driver of any mode of vehicle. Our daily experience of travelling from one place to another can be made safe and simple when adequate facilities are planned, designed and implemented based on traffic engineering principles.

The objective of this book is to present contemporary knowledge on traffic engineering, and the related topics of road infrastructure and facilities, to undergraduate and postgraduate students, and to researchers and practicing engineers. This book also aims to guide students appearing for the professional engineering exams conducted by different state governments and in national-level talent search tests, such as GATE and IES.

*Introduction to Traffic Engineering* provides an in-depth discussion on a wide range of topics, organised in 22 chapters. Chapter 1 introduces the reader to the broad scope of road traffic engineering and provides an overview of the applicable surveys. Road user and vehicle characteristics are discussed in Chapter 2, while highway geometric design principles are described in Chapter 3. Chapters 4 to 9 examine in detail the different types of traffic data collection methods/studies and their analysis methods. Chapter 10 elaborates on fundamental principles and derivations of traffic flow characteristics and their parameters. Chapters 11 to 15 introduce the different types of intersections and their basic design guidelines. Parking studies are dealt with in Chapter 16, while road markings, traffic signs and street lighting are discussed in Chapters 17, 18 and 19, respectively. Chapters 20 and 21 provide detailed information about the investigation of road traffic accidents, road traffic studies, reconstruction of accidents, and statistical analysis of accident data. The final chapter explores the future of traffic engineering in the form of intelligent transport systems, with case studies describing the VICS system of Japan, USA and other countries.

An exhaustive list of references has been provided at the end of the book, along with an appendix on the roadway lighting evaluation forms.

This book contains more than 500 figures/illustrations/photographs that enable the reader to visualise the fundamental concepts explained in the chapters. Numerous solved examples, particularly those related to the Indian Roads Congress (IRC), Bureau of Indian Standards (BIS) and American Association of State Highway and Transportation Officials (AASHTO), help to strengthen and expand the student's understanding of these standards. Each chapter ends with a self-evaluation section that comprises multiple choice questions (MCQs) and review/exercise questions.

I welcome suggestions and ideas from the readers to further improve this textbook and enable me to achieve my broad objective of dissemination of technical knowledge in India among students, teachers and practicing professionals.

**R Srinivasa Kumar**

# Acknowledgements

During the preparation of this textbook, I drew liberally from many publications (cited in the references at the end of the book) and was in contact with several authors and organisations, as acknowledged below.

In addition, a special thank you is due to the following agencies and their representatives who generously supported and shared information critical for the development of this book.

- Federal Highway Administration (FHWA), Washington, DC, USA
  - Lydia C Conrad, HOP, Office of Operations, Federal Highway Administration (FHWA), Washington, DC, USA
- American Association of State Highway and Transportation Officials (AASHTO), Washington, DC, USA
- Institute of Transportation Engineers (ITE), Washington, DC, USA
- Oregon Department of Transportation, Salem, Oregon, USA (www.oregon.gov/odot/hwy/engservices)
  - Dave Joe Polly, Senior Standards Engineer, Oregon Department of Transportation (ODOT), Salem, Oregon, USA
  - Sheila Lyons, Pedestrian and Bicycle Program Manager, Oregon Department of Transportation (ODOT), Salem, Oregon, USA
- Nancy Brickman, Oregon State University, Corvallis, Oregon, USA (http://cce.oregonstate.edu/traffic-safety-workshops)
- Washington State Department of Transportation (WSDOT), Washington, DC, USA (www.wsdot.wa.gov)
  - Chris Schroedel, Policy Research and Publications Engineer, Washington State Department of Transportation (WSDOT), Development Division, Washington, DC, USA
  - Kimberly Colburn, HQ Customer Service, Washington State Department of Transportation (WSDOT), Olympia Headquarters, USA
- Transportation Research Board of National Scientific Council, Washington, DC, USA
- Australian Road Research Board (ARRB), Australia (www.austroads.com.au)
  - Elaena Gardner, Communications Manager, Austroads, Australia
- National Cooperative Highway Research Program (NCHRP), Washington, DC, USA
- National Academy of Sciences (NAS), Washington, DC, USA
  - Maria Crawford, Editor, National Academy of Sciences (NAS), Washington, DC, USA
- United States Access Board, Washington, DC, USA (www.access-board.gov)
- Unified Traffic and Transportation Infrastructure (Planning & Engineering) Centre, New Delhi, India
- Japan Traffic Management Technology Association, Tokyo, Japan
  - Masahiro Ono, President, Research & Development Department, Japan Traffic Management Technology Association, Tokyo, Japan

    - Yoshimi Tanaka, Deputy Manager, Research & Development Department, Japan Traffic Management Technology Association, Tokyo, Japan
- Georgia Department of Transportation, Atlanta, Georgia, USA
- JAMAR Technologies Inc., Pennsylvania, USA (www.jamartech.com)
    - Mike Overholt, JAMAR Technologies Inc., Pennsylvania, USA
- Harald Kolodziej, General Manager, TCS People Counting, New South Wales, Australia (www.tcsforsurveys.com.au)
- Mikros Systems Pty Ltd., Pretoria, South Africa (www.mikros.co.za)
    - Rob Sik, Mikros Systems Pty Ltd., Pretoria, South Africa
- Essae Digitronics Pvt, Ltd., Chennai, India (www.essae.in)
    - Samuel Praveen Wood, Manager, Transient Weighing Division (TWD), Essae Digitronics Pvt, Ltd., Chennai, India
- Kistler Group, Winterthur, Switzerland (www.kistler.com)
    - Christoph Gilgen, Product Manager - SBF Road & Traffic Kistler Instrumente AG, Switzerland
- Traffic and Parking Control Co. Inc., Milwaukee, Wisconsin, USA (www.tapconet.com)
    - Andrew Bergholz, Vice President of Sales, Traffic and Parking Control Co. Inc., Milwaukee, Wisconsin, USA
- Xtralis Pty Ltd., Dublin, Republic of Ireland (www.xtralis.com)
    - Ian Ehrenberg, Senior Vice President and General Manager, Xtralis Pty Ltd., Dublin, Ireland
- SmarTek Systems Inc., Woodbridge, Virgina, USA (www.smarteksys.com)
    - Greg Pieper, SmarTek Systems Inc., Woodbridge, Virgina, USA
- Vyara Tiles Pvt. Ltd., Vyara, Gujarat, India (http://vyaratiles.in)
    - Mili Jain, Vyara Tiles Pvt. Ltd., Vyara, Gujarat, India
- Standards for Highways, UK (www.standardsforhighways.co.uk)
- Nebraska Department of Roads, Lincoln, Nebraska, USA
- California Department of Transportation, Sacramento, California, USA
- National Roads Authority, Dublin, Ireland (www.dttas.ie)
- Colorado Department of Transportation (CDOT), Denver, Colorado, USA (www.codot.gov)
    - Tina Littleton, Customer Service Representative, Colorado Department of Transportation (CDOT), Denver, Colorado, USA
- American Traffic Safety Services Association (ATSSA), Fredericksburg, Virginia, USA (www.atssa.com)
- Minnesota Department of Transportation (MnDOT), St. Paul, Minnesota, USA
    - Sue Zarling, Office of Traffic, Security and Technology, Traffic Electrical Systems Engineer, Minnesota Department of Transportation (MnDOT), St. Paul, Minnesota, USA
- South Dakota Department of Transportation (SDDOT), Pierre, South Dakota, USA (www.sddot.com)
    - Mark A Leiferman,Chief Road Design Engineer, South Dakota Department of Transportation (SDDOT), Pierre, South Dakota, USA
- New Jersey Department of Transportation (NJDOT), Trenton, New Jersey, USA
    - Carl Kneidinger, New Jersey Department of Transportation (NJDOT) Director of Permits, Electrical and Claims (PEC), Trenton, New Jersey, USA
- Alberta Transportation, Government of Alberta, Canada
    - Des Williamson, Executive Director, Technical Standards Branch, Alberta Transportation, Government of Alberta, Canada

- Molly Wagner, Director of Communications, Dewberry (www.dewberry.com)
- Ohio Department of Transportation, Columbus, Ohio, USA
  - Duane Soisson, Traffic Control Design Engineer, Office of Roadway Engineering, Ohio Department of Transportation, Columbus, Ohio, USA
- City of Saint John, New Brunswick, Canada (www.saintjohn.ca)
  - Timothy D O'Reilly, Deputy Commissioner, Parks and Public Spaces, Traffic Engineer, City of Saint John, New Brunswick, Canada
- Land Transport Authority, Development & Building Control Division, Singapore (www.lta.gov.sg)
  - Low L C Daisy, Deputy Director, Development & Building Control, Singapore
- Zehntner GmbH Testing Instruments, Sissach, Switzerland (www.zehntner.com)
  - Vanessa Mantella, Zehntner GmbH Testing Instruments, Sissach, Switzerland
- DELTA, Denmark (http://roadsensors.madebydelta.com)
  - Kjeld Aabye, DELTA, Denmark
- Munro Instruments, Harlow, UK (www.munroinstruments.co.uk)
  - BS Pressland, Munro Instruments, Harlow, UK
- Findlay Irvine Ltd, Midlothian, Scotland, UK (www.findlayirvine.com)
  - Alister Mille, Findlay Irvine Ltd, Midlothian, Scotland, UK
- East Nippon Expressway Company Limited, Tokyo, Japan (www.e-nexco.co.jp)
  - Kumada, Manager of Nexco-East International Department, East Nippon Expressway Company Limited, Tokyo, Japan
- Central Nippon Expressway Company Limited, Nagoya, Japan (http://global.c-nexco.co.jp)
  - Mitsuho O`Toole, International Business Department, Central Nippon Expressway Company Limited, Nagoya, Japan
- Metro Infrasys, New Delhi, India (www.metroinfrasys.com)
  - Sachin Bhatia, Metro Infrasys, New Delhi, India
- International Road Dynamics Inc., Saskatoon, Saskatchewan, Canada (www.irdinc.com)
  - Kerry Mitchell, Technical Design Co-ordinator, Saskatoon, Saskatchewan, Canada
- Schuh & Co. GmbH, Germering, Germany (www.schuhco.de)
  - Rainer Heer, Schuh & Co. GmbH, Germering, Germany
- VICS Centre, Tokyo, Japan (www.vics.or.jp)
- Mettler-Toledo India Pvt. Ltd., Mumbai, India (www.mt.com)
  - Brian Howell, Product Manager – MTWT, Mettler-Toledo India Pvt. Ltd., Mumbai, India
- Moxa Americas Inc., Brea, California, USA (www.moxa.com)
  - Oliver Wang, Regional Marketing Manager, Moxa Americas Inc., Brea, California
- Prof. Dusan Teodorovic, Transportation Infrastructure and Systems Engineering, Virginia Tech, Blacksburg, Virginia, USA
- Prof. Antonio A. Trani, Transportation Infrastructure and Systems Engineering, Virginia Tech, Blacksburg, Virginia, USA
- Prof. V Bhikshma, Head, Department of Civil Engineering, Osmania University, Hyderabad, India

To accomplish the task of bringing out this book, I am indebted to Dr. D Venkateshwar Rao, Professor of Civil Engineering (retired). He has been a source of inspiration and has guided me at every stage of the redrafting of the text.

A special thank you is due to Mr. Madhu Reddy of Universities Press Pvt Ltd, and his team – Mr. Thomas Mathew Rajesh, Mr. Kallol Das, Ms. Madhavi Sethupathi and Mr. Janardhan Reddy.

I am grateful to the Vice-Chancellor and Registrar (Osmania University), Principal, Vice-Principal and Dean (University College of Engineering), Head (Department of Civil Engineering) and all the faculty members, including non-teaching staff, of the Department of Civil Engineering, University College of Engineering, Osmania University, Hyderabad, for their encouragement and support.

I also thank my students Rohith Mane, Soma Prashant and Sindhu Chouhan for assisting me in the preparation of some parts of the manuscript.

Finally, I would like to express my gratitude to my family, including Mrs. Rangoju Saritha, Ms. Rangoju Gayatrisai, Mr. Rangoju Vivek Dhir, Mrs. Rangoju Parvathamma and Mrs. NVL Narasamma.

I have referred to many published articles on different topics. It is difficult to quote all of them. However, I thank them profusely.

Although efforts were made to check and improve the text while writing, editing and proofreading, there may still be a few omissions and errors. I would be very grateful for any positive and constructive suggestions from the readers for further improvements to the text.

**R Srinivasa Kumar**

# Contents

*To all the researchers who contributed to the development of various methods and models which are being used in Traffic Engineering to solve traffic-related problems*

CHAPTER

# 1 Introduction to Road Traffic Engineering

## 1.0 INTRODUCTION

**Traffic** is defined as pedestrians, bicyclists, ridden or herded animals, all types of motorised vehicles, trams and other conveyances, either singly or together, using any highway for the purpose of travel (FHWA, 2003).

**Traffic Engineering** is a part of Transportation Engineering that uses engineering techniques to promote safety and efficiency in the *orderly movement* of all road users (i.e., vehicular traffic and pedestrians). The objective of traffic engineering is to provide efficient and safe flow of traffic in all weather conditions on different terrains that have well-designed roads, geometric components and their appurtenances.

The focus is on the *functional aspects* relating to road geometric components with traffic regulatory devices, and their management. These traffic components include road geometrics, traffic signs, traffic lights, traffic signals, road surface markings, delineators, traffic flow parameters, driver characteristics, standards and policies on traffic flow and urban planning issues, work zone safety measures, detour plans, development of counter measures to reduce road accidents, intelligent transportation systems and related appurtenances with their infrastructure facilities.

## 1.1 TRAFFIC PROBLEMS

As the volume of traffic keeps on increasing, the resulting traffic demand of the present and future years should be adequately accommodated on existing as well as new highway systems – a challenging task for traffic engineers and transport planners. Provision of such timely traffic facilities is key to the success of any traffic operation and control system. The following possibilities are considered as the broad scope of road traffic engineering and may be explored to ensure systematic development of traffic facilities and infrastructure needs:

- Traffic control and road management measures to cater to present and future needs
- Planning of new roadway and highway development projects to meet the forecasted traffic needs
- Allocation of budgetary and non-budgetary funds, with appropriate public–private partnership (PPP) models, for sustainable highway development
- Enforcement, control, management, administration and safety measures of traffic, including vehicles and road users, for healthy and eco-friendly traffic flow
- Establishment of dedicated and non-profit traffic research institutions at local, regional and national levels for updating, disseminating and revising the knowledge on related technological aspects

## 1.2 REASONS FOR COLLECTING DATA

Traffic data in different forms is collected for the following reasons:

- Managing the existing roadway physical system with replaced or repaired facilities or with the anticipated schedule of improvements.
- Upgrading of any existing road to a National Highway or widening of the road to a two- or multi-lane roadway.
- Determining the rate of road distress due to traffic volume and weather conditions. This data is used for the pavement management system and for assessing the future needs of the budget.
- Investigating trends in traffic growth over time so as to forecast future transportation needs in a study area.
- Understanding the needs and choices of the public and industry based on the nature of travel demand, based on choice of mode.
- Assessing traffic delays and determining the level of service of existing traffic facilities such as at signalised intersections.
- Calibrating basic relationships or parameters such as perception verses reaction time, discharge headways at signalised road junctions, and values of headway and spacing.
- Assessing the effectiveness of improvements using before-and-after studies of traffic data. For example, the number of road accidents is often compared and justified with the investment.
- Assessing the impact on traffic due to new facilities – a measure that is widely practiced in various municipalities before going on to provide no-objection certificates based on the criteria for parking, congestion, air pollution, sound pollution and future scope of developments.
- Evaluating an existing facility or system performance by periodically conducting studies to determine the quantity and quality of accessibility, mobility and ease of service to the public.

## 1.3 TYPES OF STUDIES

For the reasons and purposes mentioned above, the following types of studies are conducted in any selected area:

- **Traffic volume studies**: This is the most basic traffic study and the data collected is used for many purposes.
- **Speed studies**: The speed of vehicles is measured for the implementation of safe operation of traffic at all locations, times and in any climate.
- **Travel-time studies**: Journey speed with delays along any selected road segment is used to measure the quality of traffic flow service.
- **Delay studies**: These are considered a part of travel-time studies and deal with factors that the user finds particularly annoying or which cause delay/discomfort.
- **Density studies**: These studies are mostly carried out by indirect measurement.
- **Accident studies**: These involve investigation of accident characteristics, causal factors, specific locations, grey/black spots and preventive measures.
- **Parking studies**: These involve the collection of data about inventory, parking supply/demand and required parking facilities.
- **Goods movement and transit studies**: These involve the collection of data about truck volume, truck loading/unloading facilities and transit systems.

A list of traffic surveys carried out across the world for the improvement/development of road projects are presented in Table 1.1.

**Table 1.1** Summary of the main types of traffic surveys, with intended output

| *Survey type* | *Required information/ data* | *Methods of survey** | *Output* |
|---|---|---|---|
| Road inventory | Road and network geometric characteristics | • Observation<br>• Survey of India maps<br>• Google maps | Geometric details, land use and road traffic furniture |
| Traffic volume | Traffic demand | • Manual counts<br>• Automatic counts<br>▪ Intrusive<br>▪ Non-intrusive | Pedestrian and vehicular traffic volume, flows on links and across junctions, annual average daily traffic (AADT), traffic growth rate, peak hour factor (PHF) and other traffic factors |
| Spot speed | Speed control, safety measures and vehicle speed performance on road links | • Manual measurement by short base method<br>• Radar measurement | $85^{th}$ percentile speed, space mean speed, median speed, modal speed, pace and per cent values in pace |
| Road network speed and delay | Route performance | • Floating car method | Average and running speed on routes/ networks, delay and congestion at points |
| Road junction speed and delay | Junction performance | • Stopped vehicle count<br>• Elevated observer method | Approach delay, total delay, distribution of delay by turning movements and causes of delay |
| Saturation flow | Capacity of junction and design of signal time with phases | • Flow profile method<br>• Saturated period count | Saturation flow and capacity of junction/lane |
| Origin and destination | Traffic demand forecasting | • Registration number methods<br>• Post card method<br>• Home interview | Trip matrix, route choice and through traffic travel time |
| Parking use | Parking supply and demand for space | • Parking patrol survey | Types of parking in use, space usage for parking and occupancy time |
| Accident report | Accident reduction/ preventive measures | • Field investigations<br>• Statistical data collection | Causes and reasons for accidents, block spots |

* Only a few methods are listed here. A detailed discussion is presented in subsequent chapters.

### 1.3.1 Frequency of Traffic Data Collection

Traffic surveys are conducted as needed, with specific requests, special projects and input for the design of various transportation facilities as the main reasons for collecting data. However, any one of the following may be considered as suitable/guideline:

- Hourly
- Daily
- Weekly
- Monthly
- Quarterly
- Yearly
- Every two years
- Every four years
- Every five years
- As and when required by a project in demand

These road traffic studies and their objective details are briefly described in subsequent chapters.

## EXERCISES

### Multiple Choice Questions

1. Match List-I (Study) with List-II (Purpose) and select the correct answer using the codes given below the lists:

   **List-I**

   A. Secondary road system — 1. Demographic study
   B. Economic studies — 2. Traffic volume
   C. Engineering studies — 3. State highways
   D. Road use studies — 4. Topographic study

   **List-II**

   | | A | B | C | D |
   |---|---|---|---|---|
   | (a) | 1 | 3 | 2 | 4 |
   | (b) | 1 | 3 | 4 | 2 |
   | (c) | 3 | 1 | 4 | 2 |
   | (d) | 3 | 1 | 2 | 4 |

2. Which set of traffic studies is necessary for the functional design and capacity design of highways?
   (a) Road condition and safety studies
   (b) Speed and accident studies
   (c) Volume and speed studies
   (d) Axle load and volume studies

### Answers

1. (c) 2. (c)

### Questions

1. Define traffic engineering with reference to road transportation.
2. What are the five possibilities that may be explored to ensure the systematic development of traffic facilities and infrastructure needs?
3. State the reasons for collecting traffic data.
4. State the type of traffic studies conducted with their intended purpose.
5. What is the frequency of traffic data collection? Write your comments on the frequency of conducting studies with reference to project needs?

CHAPTER

# 2 Road User and Vehicle Characteristics

## 2.0 INTRODUCTION

The main objective of the design of traffic facilities is to ensure orderly and systematic movement of traffic within safety and economic considerations. To achieve this objective, there should be coordination among the components involved. The **main components** of traffic engineering are listed below:

- Drivers
- Pedestrians
- Vehicles
- Roads
- Traffic control devices
- The general environment

The characteristics of each of these components affect the performance of any transportation system, and the traffic engineer should have an understanding of the local/regional problems and a sound knowledge of how these factors interact with each other.

We will now examine in detail the characteristics of these components and their relevance with respect to the performance of traffic operations.

## 2.1 ROAD USER

Road users can be broadly classified into three types: drivers, pedestrians and passengers (Fig. 2.1). Regardless of their type, they come together to form a complex human element – one that displays large variations in performance while interacting with the transport environment. This is of crucial concern to traffic engineers in their quest for optimal design of facilities.

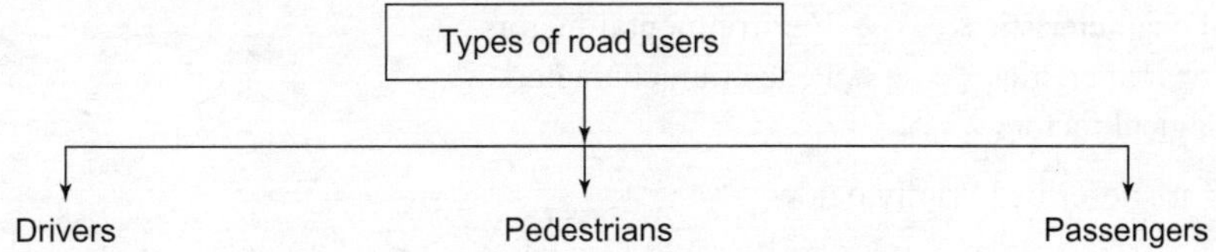


**Fig. 2.1** Types of road users (humans)

For example, there will be a large variation in performance between an average pedestrian and an elderly person while crossing a street. Thus, the design engineer should taken into account the different types of people – children, the elderly and the physically challenged – as well as the different types of vehicles, good and average drivers and the prevailing climatic conditions, while designing the transport environment.

According to a survey (www.visualexpert.com), over 90% of road accidents are caused fully or in part by human error. Of these, 90% of accidents are caused by perception error and the remaining 10%

by response error. Most of these errors are due to perception and attention failure or due to inadequate highway visibility conditions. Therefore, knowledge of human performance and their capabilities and behavioural characteristics in traffic flow operations is a vital input for the design of the various geometric elements of roadways.

The internal and external factors that affect general driver behaviour are presented in Fig. 2.2.

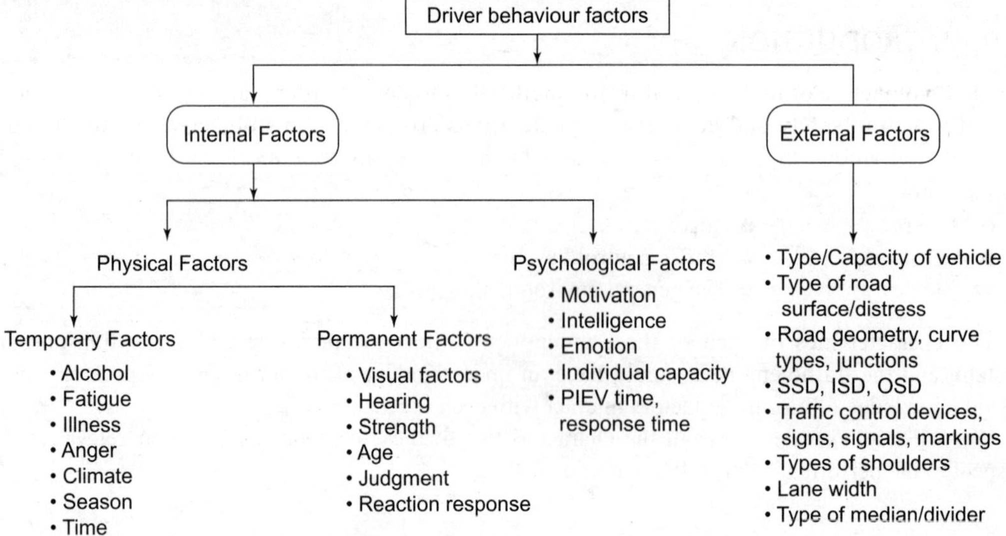


**Fig. 2.2** Behaviour factors affecting driver performance

### 2.1.1 Road User Characteristics

The factors affecting road user characteristics may broadly be classified under five heads:

- Physical characteristics
- Mental characteristics
- Psychological factors
- Environmental factors
- Pedestrian characteristics

These factors are described briefly below.

**Physical characteristics:** The physical characteristics that are required of a road user are vision, sense of touch, hearing and strength.

- **Vision**: Visual factors affect the ability of a road user to detect objects, receive information, and process and identify objects. Vision depends on the following abilities of the eyes.
    - **Visual acuity or clearness of vision**: It is the ability of the eyes to see the fine details of an object under average illumination conditions. Visual acuity is the one of the important factors in the placement of traffic information devices such as roadside signs. Visual acuity is of two types: static and dynamic. Static acuity relates to stationary objects while dynamic acuity relates to moving objects. Clarity of vision depends on the cone angle of vision (Fig. 2.3) and the speed of the driver under the given conditions of illumination. Some cone angles of vision and their corresponding clarity of vision level are presented in Table 2.1.

**Table 2.1** Cone angles and clarity of vision

| *Cone angle of vision* | *Clarity of vision* |
|---|---|
| 3–5° | Very clear vision; legend can be read within this cone |
| 5–7° | Clear and good vision |
| 10–12° | Satisfactory vision; object shapes/colour can be identified |
| >12° | Blurred vision starts; movement of objects can be detected |

- **Peripheral vision**: This is the field of view beyond the cone of clearest vision (Fig. 2.3). In this field of view, an individual can only see objects without clarity of details/colour. The cone of vision can provide a warning to the driver and act as a safety measure. In static conditions, the cone angle of peripheral vision ranges between 120° and 160°; the cone angle reduces with increase in speed. For example, the peripheral vision cone angle may reduce from 100° to 40° as the driving speed increases from 30 km/hr to 90 km/hr.

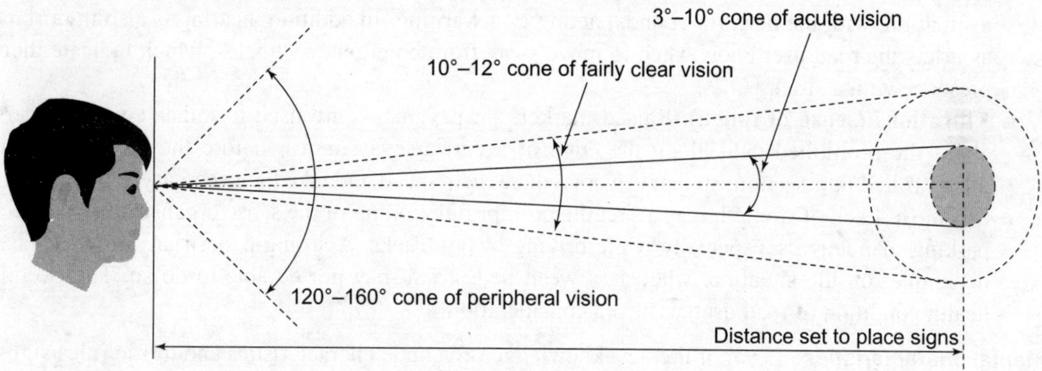


**Fig. 2.3** Field of vision

- **Eye movement and visual concentration**: The movement of the eyes and the time taken to focus on different objects may be categorised as shown in Table 2.2.

**Table 2.2** Eye movement and time lapsed

| *Eye movement* | *Time taken* |
|---|---|
| Fixing of eyes on objects | 0.10–0.30 seconds |
| Jumping of eyes | 0.12–0.23 seconds |
| Shifting of eyes (converging and diverging) | 0.30–0.50 seconds |

- **Colour vision**: As we know, different colours have different wavelengths, and thus, relative visibility. At very low levels of illumination, colour sensation may disappear altogether. Colour blindness is another factor which may be due to defective vision.

- **Glare vision and recovery**: Glare vision results in decrease in the seeing ability of a driver and can thus cause discomfort to the driver. It depends on the eye's adaptability to changes in the intensity of light. Glare recovery is the time duration in which a driver recovers from the effect of glare, just after passing a light source. It may be as high as 6 seconds when moving from light to dark conditions and as low as 3 seconds when moving from dark to light conditions. However, elderly drivers may experience a longer recovery time. The effect of glare on drivers/pedestrians can be minimised by reducing the brightness of lights and by increasing the height of the light source.
- **Perception of space**: Dynamic judgment and assessment of space and dimensions with time depends on:
  - Convergence of eye, binocular accommodation
  - Straining to see through dust, fog or smoke
  - Memory based on past experience of such situations/places

  These capabilities are useful for assessing the traffic flow situation, and also while overtaking, turning, lane changing, traversing adjacent to islands and intersections/interchanges/roundabouts where large visual angles are required to judge the desired driving path.

- **Hearing**: Sound is the most important factor by which drivers and pedestrians move away from a situation, with the horn (or sounds) acting as a warning. In addition, hearing is also important as it lets the road user know when to move away from emergency vehicles, which indicate their presence with a siren.
- **Vibrations (sense of touch)**: Raised markers on pavements are used to indicate/warn drivers about the prohibited path along the edge of a carriageway or just before the crossing of an uncontrolled intersection, so that drivers reduce their speed and take the necessary action.
- **Strength**: Lack of strength may affect the prompt functioning of the sense organs and also affect parking manoeuvres, especially while driving heavy trucks. A strong pedestrian can walk fast, depending on the situation, whereas a weak pedestrian may not be able to do so. The overall health condition of road users is also a consideration.

**Mental characteristics:** Level of literacy, knowledge of vehicle characteristics and traffic rules/signs, driving/walking skills and the ability to understand the surroundings based on intelligent analysis of each situation on the road can influence the performance of the road user.

**Psychological/Behavioural factors:** Factors such as attentiveness, impatience, anxiety, anger, fear and the attitude of the road user may affect traffic performance to a great extent. The willingness to allow other drivers to overtake and to give way to other vehicles and pedestrians are some of the behavioural factors of road users in a traffic stream.

**Environmental factors:** Rain, fog, snow, mist, dust, sun radiation, cold weather, wind intensity, night visibility and surface water drainage conditions also affect the performance of road users.

**Pedestrian characteristics:** Pedestrians cross the streets at mid-block locations (i.e., across a straight section length of a road) and at intersections (see Fig. 4.15). Walking speed is the most critical input parameter in the design of pedestrian crossings at intersections. The intersections may be signalised or un-signalised. The 85$^{th}$ percentile of walking speed is used for the design of signal timings at signalised intersections. The gap acceptance and time-gap accepting/rejecting behaviour of pedestrians is considered in the design of un-signalised intersections. A study conducted on the behaviour of pedestrians at mid-block crossings on urban roads in India found that the temporal and spatial 85$^{th}$ percentile gap values varied from 5.0 sec to 5.8 sec, and 82 m to 95 m, respectively, for different combinations of variables (Pawar and Patil, 2015). The factors affecting the gap acceptance behaviour of pedestrians comprise the speed of the approaching vehicles, width of the street, frequency distribution of gaps in the traffic stream, waiting time, age, gender and others.

As large variations exist in the walking speed of pedestrians with reference to age, gender, physical fitness and location, the 85th percentile speed of the sampled pedestrian walking speed in the locality is used as the pedestrian walking speed for design purposes. In the absence of such data, a walking speed of 1.2 m/sec may be assumed for the design of timing signals and location planning of off-street parking facilities which are adjacent to the central business district (CBD). The 15th percentile speed of 0.95 m/sec (Fig. 2.4) is used for the design of pedestrian crossing facilities in India (IRC:103-2012).

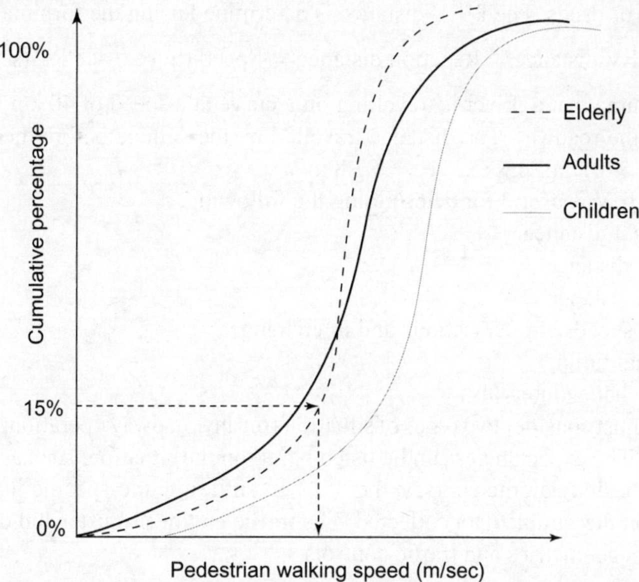


**Fig. 2.4** Typical presentation of sampled pedestrian crossing speed data

## 2.1.2 Reaction Process

**Perception–Reaction time (PRT):** It is the time taken by a driver to initiate a physical response on detection of an event. The driver performs four distinct processes in sequence: detection, identification, decision and response. The factors affecting PRT are age, fatigue level, complexity of the situation and the presence of alcohol/drugs in the bloodstream of the road user. The complexity of the reaction depends on the situation which, in turn, may depend on vehicle defects, traffic flow conditions, road geometrics, time of day and weather conditions.

The average PRT value commonly suggested for sight distance design purposes is 2.5 seconds (based on the 90th percentile criterion by the American Association of State Highway and Transportation Officials or AASHTO (2004)) and the PRT value for driver response to signal timing purpose is 1.0 seconds (based on the 85th percentile criterion by the Institute of Transportation Engineers or ITE (2001)).The 85th percentile represents the value that 85% of the population can meet or exceed.

**Perception, identification, emotion and volition (PIEV):** The time required for this process depends on the complexity of the task, and the level of expectancy and variability in the driver's capabilities. The PIEV process is as follows:

1. **Perception**: Stimulus of observation takes place while perceiving the situation ahead.
2. **Identification**: Identification and understanding of the stimulus with speed alternatives develops.
3. **Emotion**: A decision/judgment is made about the best course of action.
4. **Volition**: Reaction of the decision.

The PIEV time depends on the driver's physical and mental capabilities, fatigue level and the influence of alcohol or drugs. The PIEV distance is determined using the formula

$$\text{PIEV distance} = \text{Reaction distance} = \text{Speed (m/sec)} \times \text{PRT (sec)} \tag{2.1}$$

For example, assume that a driver is travelling on a curve at a speed of 40 km/hr and suddenly sees an animal crossing the roadway. The distance travelled by the vehicle before the driver's foot presses the brake is $(0.278 \times 40)$ m/sec $\times$ 2.5 sec = 27.8 m.

The concept of PIEV is useful for determining the following:

- Overtaking sight distance
- Stopping sight distance
- Decision sight distance
- Safe approach speeds at intersections and interchanges
- Design of signal timing
- Traffic signal change intervals

*Note*: The designer must consider two types of situations on any roadway operation, namely, the expected and the unexpected. The expectancies can be listed based on the situation, such as continuity, temporal and event. Average design parameters (say, the $85^{th}$ percentile) or site-specific input parameter values are selected based on the sample data collected. The traffic engineer must avoid designing unexpected events into roadways geometrics and traffic controls.

**Design driver:** Following are the indicative and average driver characteristics/capabilities chosen by the traffic designer, and such a driver is referred to as the design driver.

- PRT = 2.5 seconds (AASHTO)
- Height of the driver's eye level above the road surface = 1.2 m
- Comfortable deceleration rate = 3.0 m/sec$^2$ (ITE) or 3.4 m/sec$^2$ (AASHTO)
- Allowable jerk = 0.7 m/sec$^3$; standard values of lateral jerk range from 0.3 m/s$^3$ to 0.9 m/s$^3$ (AASHTO)

  *Note*: The rate of change of acceleration, i.e., the third derivative of displacement, is known as jerk.

The critical gap value should be maintained between 4.0 seconds and 7.5 seconds, depending on the complexity of the manoeuvre, speed of opposing streams and the situation.

## 2.1.3 Rate of Change of the Visual Angle

As a vehicle passes by an object on the side of the road (say, a signboard), the visual angle ($\theta$) of the driver changes (Fig. 2.5). This means that as the $L$ value decreases with vehicle movement along the road, the $\theta$ value of the driver vision angle increases.

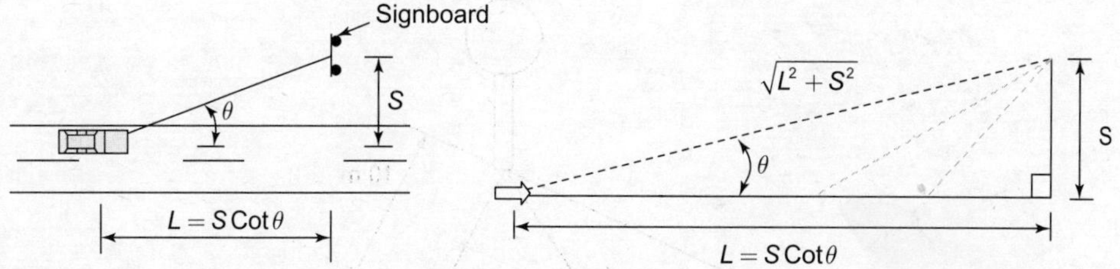


**Fig. 2.5** Change of visual angle with driver position

The rate of change of visual angle ($d\theta$) of the driver due to the lateral movement of his vehicle over a distance ($L$) can be determined as given below (Michaels and Cozan, 1963):

Let $S$ be the lateral distance of an object, measured perpendicular to the path traced by the vehicle; $L$ be the longitudinal distance, measured along the path traced by the vehicle; $\theta$ be the visual angle; and $v$ be the vehicle speed. From the free body diagram (Fig. 2.5)

$$\tan\theta = \frac{S}{L}$$

Rewriting the above, we get

$$L = S\,\text{Cot}\,\theta$$

Differentiating the above with respect to time, $t$

$$\frac{dL}{dt} = \frac{d}{dt}(S\,\text{Cot}\,\theta) = S\,\text{Cosec}^2\theta$$

(–ve sign is neglected, which indicates the diminishing value of $L$)

We know that $\dfrac{dL}{dt}$ = Vehicle velocity = v .

$$v = S\left(\frac{\sqrt{L^2+S^2}}{S}\right)^2$$

$$\text{Rate of change of visual angle} = \frac{d\theta}{dt} = \left(\frac{vS}{L^2+S^2}\right) \quad (2.2)$$

**Example 2.1**

A car is travelling at a speed of 70 km/hr. The driver, at an instantaneous distance of 90 m, first observes the lateral displacement of a sign post, placed 10 m to the left of the driver's path (Fig. 2.6). With this site situation and travel condition, what would be the distance along the road at which the driver first observes the displacement of the sign post, if it is placed at an offset distance of 15 m?

**Answer:** Vehicle velocity = 70 km/hr × 0.278 = 19.46 m/sec

Take the given values as, $S$ = 10 m and $L$ = 90 m. By Eq. 2.2, the rate of change of visual angle is

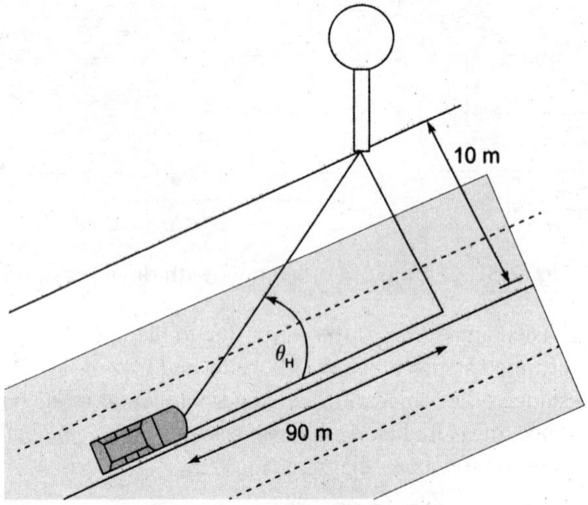


**Fig. 2.6** Example problem 2.1

$$= \frac{d\theta}{dt} = \left(\frac{vS}{L^2 + S^2}\right) = \left(\frac{19.46 \times 10}{90^2 + 10^2}\right) = \left(\frac{194.6}{8200}\right)$$

If the above rate of change of visual angle is assumed to be constant, let us now take $S = 15$ m and find $L$. Equating both the rates of change of visual angle, we get

$$\left(\frac{19.46 \times 15}{L^2 + 15^2}\right) = \left(\frac{194.6}{8200}\right)$$

$$L^2 = \left(\frac{8200 \times 19.46 \times 15}{194.6}\right) - 15^2$$

The driver must be cautious at an earlier distance $(L) = 109.88$ m.

## Example 2.2

A vehicle is travelling at a speed of 70 km/hr and the driver, at an instantaneous distance of 150 m, first observes the lateral displacement of a sign post, placed 5 m to the left of the driver's path (Fig. 2.7). With this site situation and travel condition, at what later distance should the sign be placed if the driver first views the displacement at double the longitudinal distance?

**Answer:** Vehicle velocity = 70 km/hr × 0.278 = 19.46 m/sec
Take the given values as, $S = 5$ m and $L = 150$ m. By Eq. 2.2, the rate of change of visual angle

$$= \frac{d\theta}{dt} = \left(\frac{vS}{L^2 + S^2}\right) = \left(\frac{19.46 \times 5}{150^2 + 5^2}\right)$$

If the above rate of change of visual angle is assumed to be constant, let us now take $L = 2 \times 150$ m $= 300$ m, and find $S$. Equating both the rates of change of visual angle, we get

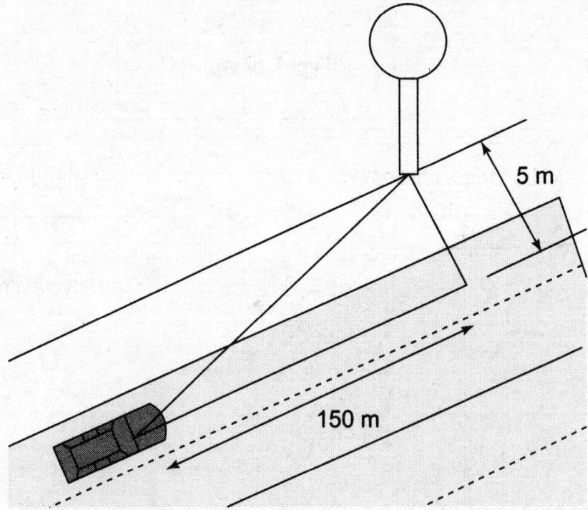


**Fig. 2.7** Example problem 2.2

$$\left(\frac{19.46 \times 5}{150^2 + 5^2}\right) = \left(\frac{19.46 \times S}{300^2 + S^2}\right)$$

$$300^2 + S^2 = 4505 \times S \tag{2.3}$$

$$S^2 - 4505\,S + 300^2 = 0$$

The above algebraic equation is in the form $ax^2 + bx + c = 0$. Its factors can be determined as $\dfrac{-b \pm \sqrt{b^2 - 4ac}}{2a}$

From Eq. 2.3, let us substitute $a = 1$, $b = -4505$ and $c = 300^2$

$$= \frac{-(-4505) \pm \sqrt{(-4505)^2 - (4 \times 1 \times 300^2)}}{2 \times 1}$$

$$= \frac{(4505) + \sqrt{(-4505)^2 - (4 \times 1 \times 300^2)}}{2 \times 1} = 4484.93\,\text{m}$$

or

$$= \frac{(4505) - \sqrt{(-4505)^2 - (4 \times 1 \times 300^2)}}{2 \times 1} = 20.067\,\text{m}$$

The sign should be placed at an offset distance of ($S$) $\approx 20$ m.

## 2.2 VEHICLE

The vehicles that operate on roadways can be broadly classified into two types: motorised and non-motorised (Fig. 2.8).

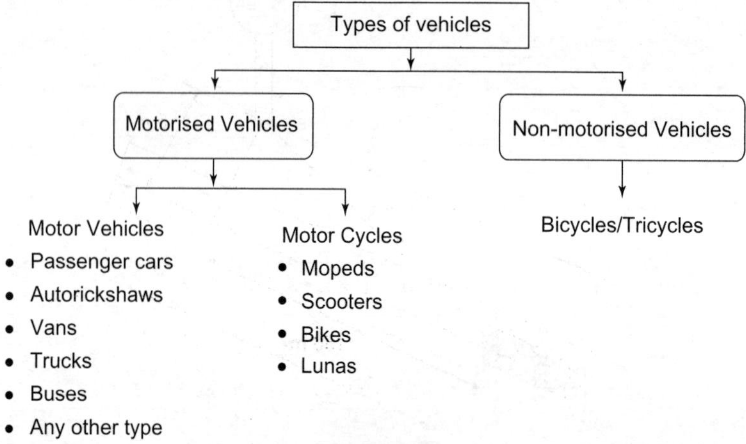


**Fig. 2.8** Classification of vehicles

### 2.2.1 Vehicle Characteristics

Traffic engineers should have knowledge of vehicle performance characteristics because the geometric design of roadways must take into account and accommodate such vehicle-related inputs. The road design should cater to the needs of present and future anticipated vehicles.

The geometric design of highways is carried out by considering three types of characteristics: static, kinematic and dynamic. The physical and performance features of each are discussed below.

**Static characteristics:** The physical features and maximum permissible loads for different types of vehicles in India (IRC: 3-1983) are given in Table 2.3.

**Table 2.3** Maximum permissible gross weight of vehicles in India

| *Vehicle class* | *Vehicle type* | *Maximum gross weight (tonnes)* |
|---|---|---|
| Type 2 | | Both single tyres: 12 |
| | | Front single tyre; rear dual tyre: 16.2 |
| Type 3 | | 24 |
| Type 2-S1 | | 26.4 |
| Type 2-S2 | | 34.2 |

*(Cont'd)*

**Table 2.3** (Continued)

| *Vehicle class* | *Vehicle type* | *Maximum gross weight (tonnes)* |
|---|---|---|
| Type 3-S1 |  | 34.2 |
| Type 3-S2 |  | 42 |
| Type 2-2 | 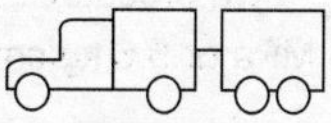 | 36.6 |
| Type 3-2 |  | 44.4 |
| Type 2-3 | 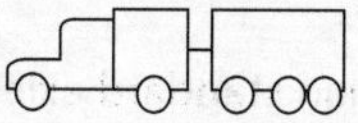 | |
| Type 3-3 |  | 52.2 |

**Vehicle dimensions:** The maximum values of width, height, wheel-base distance, overhang and length of vehicles influence the selection of design criteria such as limiting of lane width, kerb/curve radii with extra widening of carriageway, operating speeds over a road-section length, vertical clearance under a bridge, tunnel and under pass, provision of standard size of parking bays with turning radius and other such dimensions. The details of such factors and the permissible values applicable according to IRC specifications are presented in *Chapter 3*.

**Factors relating to axle loads and vehicle gross loads:** The input values relating to axle load, configuration of axle with wheels, and gross weight of vehicles are useful for pavement thickness design purposes.

- **Standard axle loads**: Generally, the load carried by any one truck is not the same as that carried by an axle. Each axle load will impart a certain amount of damage or distress on the pavement. The degree of distress caused by different loads of axle will increase as the magnitude of load and repetitions increases. Under mixed traffic conditions, repetitions of different axles having different loads plying on a road will not indicate any meaningful value relating to how much damage has been caused to the pavement due to their combined action. Different axle loads will cause different degrees of damage. Therefore, it is customary to convert repetitions of axles having different loads to an equivalent standard axle load. The configuration of the Indian standard single axle with dual wheel assembly is presented in Fig. 2.9.

  According to the Indian practice (IRC: 37-2012), standard axle load is prescribed for different axle load configurations, as follows:

  - Single axle with single wheel on either side: 65 kN
  - Single axle with dual wheels on either side: 80 kN
  - Tandem axle with dual wheels on either side: 148 kN
  - Tridem axles with dual wheels on either side: 224 kN

  IRC charts (IRC: 37-2012, IRC: SP: 72-2014, IRC: 58-2015 and IRC: SP: 62-2014) have been prepared for a specified number of repetitions of such equivalent standard axle loads on the

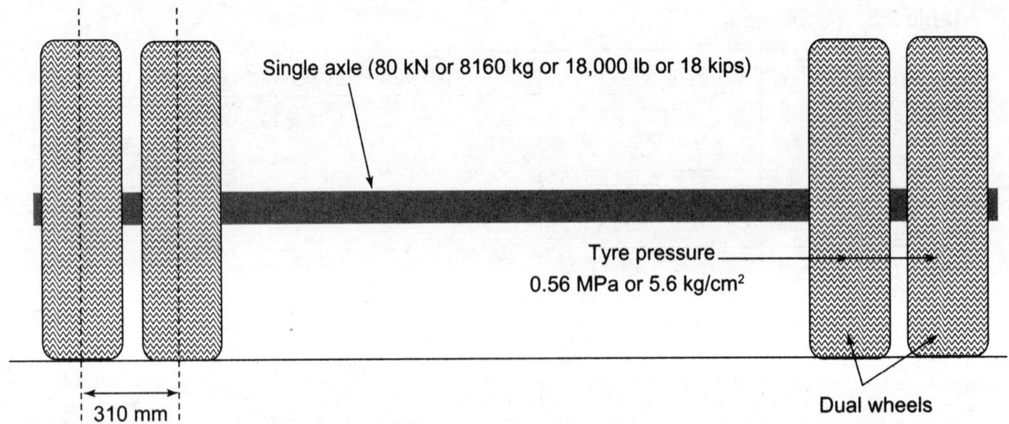


**Fig. 2.9** Details of the Indian standard axle

pavement layers. The remaining life of in-service pavements is also incorporated in the charts and specified in terms of the number of repetitions of the standard axle.

- **Configuration of axle with wheels**: The total weight of a vehicle is carried by its axles. The load on the axles is transferred to the wheels and this load is ultimately transferred on to the surface of the pavement in contact with the tyres.

  To keep wheel load–induced stresses on pavements within allowable limits, the total vehicle load is distributed onto wider areas of pavement by using more axles and wheels. This is why more axles and wheels are fitted onto heavy load–carrying trucks. Figure 2.10 shows the different types of configurations of axles and wheels used in the manufacture of road vehicles.

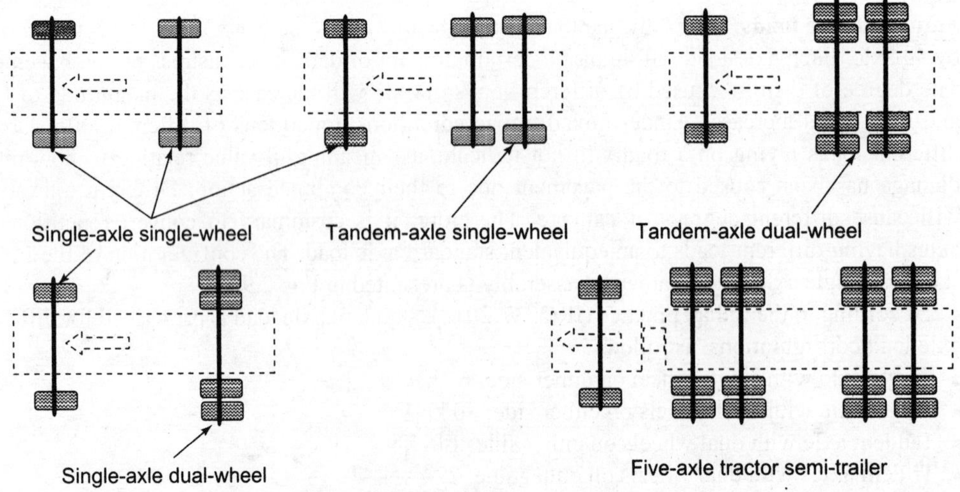


**Fig. 2.10** Different axle and wheel configurations

- **Legal axle load limit**: To safeguard pavements against overloading, different agencies (government or any competent authority) in different countries have prescribed the maximum allowable limits of axle load, depending on their method of compaction, materials used, considerations relating to mix design as well as pavement crust thickness. The maximum allowable axle load limit is referred to as the legal axle load limit. This means that no axle of a vehicle should carry more than the legal axle load. The details of the legal axle load limits adopted in India (IRC: 37-2012) are given in Table 2.4.

**Table 2.4** Legal axle load (tonnes) limits adopted in India

| *Single axle* | *Tandem axle* | *Tridem axle* |
|---|---|---|
| 10.2 | 19.0 | 24.0 |

**Kinematic characteristics**

- **Speed limit**: Variability in the performance of drivers is the main factor in the operational functionality of any road facility. To take this variability into account, the speed limit of any road section is often fixed as the 85$^{th}$ percentile. This indicates that at least 85% of the user population is safely covered (Fig. 2.11).

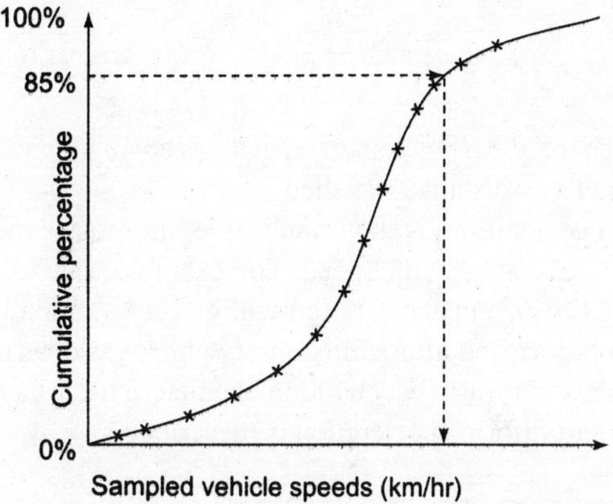


**Fig. 2.11** Typical presentation of traffic speed data

- **Braking performance**: The braking performance and deceleration characteristics of vehicles are very important factors in the safety evaluation of vehicles with roadway geometric design considerations. The time and distance travelled by vehicles after application of the brake is an important design consideration, which depends on the characteristics of the vehicle and the surface characteristics of the pavement. The distance to decelerate a vehicle can be determined by

$$\text{Breaking distance}\left(\text{BD}\right) = \frac{u^2 - v^2}{f \pm G} \tag{2.4}$$

where $u$ = initial, i.e., the operating speed of the vehicle (m/sec); $v$ = final speed of the vehicle (m/sec); $f$ = coefficient of longitudinal friction of pavement surface; and $G$ = grade (decimal value, use the + sign for upgrade and – sign for downgrade).

The influence of braking performance is considered for the design of the following geometric elements of a roadway:

- Safe stopping sight distance, including at intersection triangles
- Safe clearance between vehicles, between Amber and Red
- Placement of signs along a roadway

PRT is, by default, considered for all of the above cases.

- **Acceleration characteristics:** The acceleration characteristics of vehicles are not only useful for accident reconstruction/investigation purposes but also for estimating the time required to cross an intersection area at signalised intersections, overtaking distance and for gap acceptance purposes.

  The acceleration of vehicles depends on the horsepower (hp) and speed of the vehicle and also on grades. Acceleration can be of two types: uniform and non-uniform. The basic assumption made in the case of uniform acceleration is that the acceleration of the vehicle remains constant throughout the travelled distance with time. Following are the widely used equations of physics applicable to uniform acceleration.

$$v = u + at \tag{2.5}$$

$$x = ut + \tfrac{1}{2} at^2 \tag{2.6}$$

$$x = \frac{v^2 - u^2}{2a} \tag{2.7}$$

where $u$ = initial speed of the vehicle; $v$ = final speed of the vehicle; $a$ = acceleration or retardation; $t$ = time; and $x$ = distance travelled.

However, uniform acceleration is not applicable in most practical situations. The rate of acceleration of vehicles varies with speed. For example, the value of acceleration will be maximum when speed is zero, and acceleration will be zero at maximum speed. For this reason, the rate of acceleration before and after collision of vehicles is determined by tests. The method of determination of the non-uniform acceleration characteristics of a vehicle is explained below.

Let a vehicle with non-uniform acceleration be described as

$$\text{Rate of change of velocity} = \text{Acceleration} = \frac{dv}{dt} = p - qv \tag{2.8}$$

where $v$ = vehicle speed and $p$ and $q$ are constants. Rewriting Eq. 2.8, we get

$$dt = \left(\frac{1}{p - qv}\right) dv \tag{2.9}$$

$$\int_0^t dt = \int_u^v \left(\frac{1}{p - qv}\right) dv$$

Integrating the above expression with a time interval of $t = 0$ seconds, the initial speed = $u$. When $t = t$, the speed will be final, say $v$. If speed = 0, the acceleration ($a$) of the vehicle will be maximum, which is equal to $p$ (i.e., $a = p$). If acceleration = 0, the vehicle speed will be maximum, which is equal to $p/q$. (i.e., $v = p/q$).

Integrating Eq. 2.9, we get

$$t = -\frac{1}{q}\left(\log(p - qv)\right)_u^v$$

$$t = -\frac{1}{q}\left(\log(p - qv) - \log(p - qu)\right)$$

$$t = -\frac{1}{q}\left|\log(\frac{p - qv}{p - qu})\right| \tag{2.9a}$$

$$\log(\frac{p - qv}{p - qu}) = -qt$$

$$\frac{p - qv}{p - qu} = e^{-qt}$$

$$p - qv = e^{-qt}\left(p - qu\right) = pe^{-qt} - que^{-qt}$$

$$p - qv - pe^{-qt} + que^{-qt} = 0$$

Dividing by $q$,

$$\frac{p}{q} - v - \frac{p}{q}e^{-qt} + ue^{-qt} = 0$$

$$\frac{p}{q}\left(1 - e^{-qt}\right) - v + ue^{-qt} = 0$$

$$v = \frac{p}{q}\left(1 - e^{-qt}\right) + ue^{-qt} \tag{2.10}$$

The above equation describes the speed of the vehicle at any time $t$.

Substituting Eq. 2.10 in Eq. 2.8, we obtain the relationship between acceleration and time.

$$\frac{dv}{dt} = p - qv = p - q\left(\frac{p}{q}\left(1 - e^{-qt}\right) + ue^{-qt}\right)$$

$$\frac{dv}{dt} = p - q\left|\frac{p}{q}\left(1 - e^{-qt}\right) + ue^{-qt}\right| = pe^{-qt} - que^{-qt}$$

$$\frac{dv}{dt} = \left(p - qu\right)e^{-qt} \tag{2.11}$$

It is known that $v = \dfrac{dx}{dt}$. Now, substituting Eq. 2.10 in the above equation, we get

$$\frac{dx}{dt} = \frac{p}{q}\left(1 - e^{-qt}\right) + ue^{-qt}$$

By integration, we obtain the following equation as a relationship between distance and time

$$x = \frac{p}{q}t - \frac{p}{q^2}\left(1 - e^{-qt}\right) + \frac{u}{q}(1 - e^{-qt}) \tag{2.12}$$

**Example 2.3**

A car travelling at 20.0 m/sec skids to a stop in 2.5 seconds. Determine the skidding distance of the car (assume uniform acceleration).

**Answer:**

$$d = (v_i + v_f)/2 \times t = (20.0 \text{ m/sec} + 0 \text{ m/sec})/2 \times 2.5 \text{ sec}$$

$$d = (10.0 \text{ m/sec}) \times 2.5 \text{ sec} = 25.5 \text{ m}$$

**Example 2.4**

The following data is collected to evaluate the travel characteristics of exclusive bus lanes in a city:

- Distance between the two bus stops considered for the study = 3.2 km
- Design speed of the bus lane = 80 km/hr
- Maximum acceleration rate of the bus = 1.5 m/sec$^2$
- Maximum deceleration rate of the bus = 1.2 m/sec$^2$
- Maximum running speed = 50 km/hr

Determine the acceleration distance, deceleration distance, distance travelled at a constant speed and the space mean speed between the bus stops.

**Answer:** Let us assign and convert the given values as:
$u$ = Initial speed = 0.0 km/hr = 0.0 m/sec; $v$ = Running speed = 50 km/hr = 13.88 m/sec
$L_{Total}$ = Distance between the two bus stops = 3.2 km = 3200 m
$a_{acceleration}$ = Maximum acceleration rate = 1.5 m/sec$^2$
$a_{deceleration}$ = Maximum deceleration rate = 1.2 m/sec$^2$

Based on the fundamental concepts of physics, the required parameters are calculated as follows.

Distance required by the bus to accelerate from stop condition to attain a speed of 50 km/hr:

$$L_{acceleration} = \left(\frac{v^2 - u^2}{2 \times a_{acceleration}}\right) = \left(\frac{13.88^2 - 0^2}{2 \times 1.5}\right) = 64.2\,\text{m}$$

Distance required by the bus to decelerate from a speed of 50 km/hr to the stop condition:

$$L_{deceleration} = \left(\frac{v^2 - u^2}{2 \times a_{deceleration}}\right) = \left(\frac{0^2 - 13.88^2}{2 \times (-1.2)}\right) = 80.2\,\text{m}$$

The distance travelled at a constant speed of 50 km/hr between the bus stops:

$$L_{50\,km/hr} = L_{Total} - \left(L_{acceleration} + L_{deceleration}\right)$$

$$L_{50\,km/hr} = 3200 - (64.2 + 80.2) = 3055.6\,\text{m}$$

Let the time required between these three conditions of travel be calculated as:

$$t_{\text{Total}} = t_{\text{acceleration}} + t_{\text{50 km/hr}} + t_{\text{deceleration}}$$

$$t_{acceleration} = \left(\frac{v-u}{a_{acceleration}}\right) = \left(\frac{13.88-0}{1.5}\right) = 9.25\,\text{sec}$$

$$t_{deceleration} = \left(\frac{u-v}{a_{deceleration}}\right) = \left(\frac{0-13.88}{-1.2}\right) = 11.56\,\text{sec}$$

$$t_{\text{50 km/hr}} = \left(\frac{L_{50}\,\text{km}}{50\,\text{km/hr}}\right) = \left(\frac{3055.6\,\text{m}}{13.88\,\text{m/sec}}\right) = 220.14\,\text{sec}$$

$$t_{\text{Total}} = 9.25\text{ sec} + 220.14\text{ sec} + 11.56\text{ sec} = 240.95\text{ sec}$$

$$\text{Time mean speed of the bus between the two stations} = \frac{3200\,\text{m}}{240.95\,\text{sec}} = 13.28\text{ m/sec} = 47.7\text{ km/hr.}$$

**Example 2.5**

A sign is required to be placed in advance of a horizontal curve of a highway. The highway has an 85th percentile speed of 81 km/hr. The vehicles travel at a constant speed of 72 km/hr during PRT and they travel on the curve at the advised speed of 36 km/hr. Assume the average value of PRT of the drivers to be 2.5 seconds and the average value of deceleration of the vehicles as 3.0 m/sec$^2$. What is the minimum distance at which the directional sign should be placed in advance of the horizontal curve, if the sign legibility distance is 75 m?

**Answer:** Convert the given speeds to the unit of m/sec:
$u$ = Initial speed = 81 km/hr = 22.5 m/sec; $v$ = Final speed = 36 km/hr = 10 m/sec
$t$ = Perception–reaction time = 2.5 sec; $d$ = Deceleration value of vehicles = 3.0 m/sec$^2$
Let, $S_d$ = Total distance travelled from the moment when the sign is observed
$S_d$ = Perception–reaction distance at speed $u$ + Breaking distance needed to decelerate to speed $v$

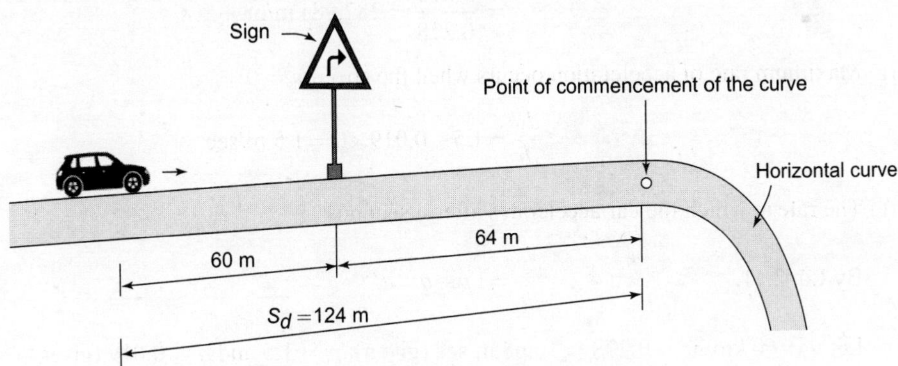


Fig. 2.12 Example problem 2.4

$$S_d = (u \times t) + \left( \frac{u^2 - v^2}{2 \times d} \right)$$

Substituting the given values, we get

$$S_d = (22.5 \times 2.5) + \left( \frac{22.5^2 - 10^2}{2 \times 3.0} \right) = (56.25\,\text{m}) + (67.70\ \text{m})$$

$$S_d = 123.95 \approx 124\,\text{m}$$

The vehicles will travel 56.25 m before braking, and then travel 67.70 m while decelerating to the speed of 36 km/hr.

The minimum distance at which the sign should be placed in advance of the curve = 124 – 60 m = 64 m.

**Example 2.6**

A car is travelling at 60 km/hr and the driver decides to overtake a vehicle by pressing the accelerator. The accelerating behaviour of the car is described by $\frac{dv}{dt} = 1.5 - 0.019v$, where $v$ is the speed in m/sec and $t$ is the time in seconds. Determine

(i) The maximum speed of the car
(ii) The maximum rate of acceleration
(iii) The rate at which the car accelerates after 4 seconds
(iv) The time the car takes to reach a speed of 100 km/hr

**Answer:**

(i) The speed of the car will be maximum when its acceleration = 0

$$\frac{dv}{dt} = 1.5 - 0.019v = 0$$

$$v = 1.5/0.019 = 78.94\ \text{m/sec}$$

$$= \frac{78.74}{0.278} = 282.98\,\text{km/hr}$$

(ii) Maximum rate of acceleration occurs when the speed $v = 0$

$$\frac{dv}{dt} = 1.5 - 0.019 \times 0 = 1.5\ \text{m/sec}^2$$

(iii) The rate at which the car accelerates after 4 seconds

By Eq. 2.11, $$\frac{dv}{dt} = (p - qu)e^{-qt}$$

Let $u$ = 60 km/hr × 0.278 = 16.68 m/sec (given); $p$ = 1.5 and $q$ = 0.019 (given)

$$\text{Acceleration just after 4 sec} = (p - qu)e^{-qt}$$

$$= (1.5 - 0.019 \times 16.68)e^{-0.019 \times 4}$$

$$= (1.5 - 0.019 \times 16.68) \times 0.9268 = 1.0964 \text{ m/sec}^2$$

(iv) What is the time the car takes to reach a speed of 100 km/hr from 60 km/hr?
Let $v$ = 100 km/hr × 0.278 = 27.8 m/sec (given).

By Eq. 2.9a,
$$t = -\frac{1}{q}\left(\log\left(\frac{p - qv}{p - qu}\right)\right)$$

$$t = -\frac{1}{0.019}\left(\log\left(\frac{1.5 - 0.019 \times 27.8}{1.5 - 0.019 \times 16.68}\right)\right)$$

$$t = -\frac{1}{0.019}(\log(0.8214)) = -\frac{1}{0.019}(-0.085)$$

$$= 4.4736 \text{ sec.}$$

**Dynamic characteristics:** Transient pavement response defines the response of a pavement to a moving tyre load input. The effects of operating vehicle speed, vehicle mass inertia, suspension and the damping system are considered for modelling/designing vehicle components for vehicle pavement interaction (Fig. 2.13).

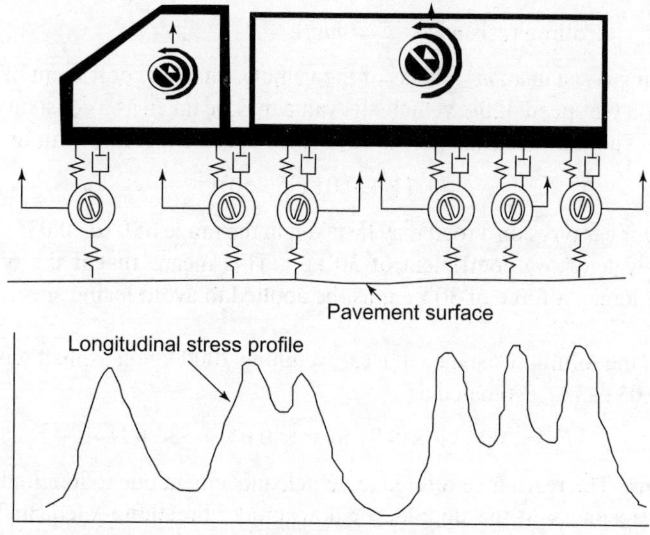


**Fig. 2.13** Typical model of vehicle attributed with load–mass damper system (based on Roebuck et al., 2012)

Vehicle dynamics, indicating the influence of each component on pavement performance, is obtained by analytical, experimental and computer-based simulations (Todd and Kulakowski, 1989). However, implementation of such complex theoretical models with validation by costly experimentation, augmented by computer simulation, may be suitably explored, depending on the precision and need. Modelling of vehicle components with vibration analysis that fits into the vehicle–pavement interaction system and freight damage (Steyn et al., 2012) is out of the scope of this textbook.

**Other characteristics:** A few factors relating to vehicle characteristics that influence the design elements of roadways are discussed below (Bennett and Greenwood, 2001).

- **Inertial force**: It is the force that acts on a vehicle in motion. It is defined by Newton's law

$$F = m \times a \tag{2.13}$$

where $F$ = inertial force; $m$ = mass of the vehicle; and $a$ = effective acceleration.

- **Air resistance**: Resistance to vehicle motion depends on several factors; however, the main factors may be listed as direction of wind, speed of the vehicle, density of air and front face area of the vehicle. Air resistance ($F_a$) as skin friction of the vehicle may be determined using the formula

$$F_a = 0.5 \times Cd \times A \times \rho \times v^2 \tag{2.14}$$

where $F_a$ = aerodynamic force opposing the motion of the vehicle (N); $Cd$ = aerodynamic drag coefficient; $A$ = projected frontal face area of the vehicle (m$^2$); and $V$ = velocity of the vehicle (m/sec).

The average $Cd$ value may be taken as 0.32 for passenger cars, 0.5–1.5 for trucks and 0.6–0.7 for buses.

- **Rolling resistance:** When wheels roll over the road surface, surface irregularities and roughness may cause deformation of tyres. The resulting shocks and impact hinder the rolling motion of the wheels. Rolling resistance depends on the surface characteristics and speed of the vehicle. The equation used for determining rolling resistance is

$$\text{Rolling resistance, } F_r = mgf \tag{2.15}$$

where $F_r$ = rolling resistance; $m$ = mass of the vehicle; and $f$ = coefficient of rolling resistance, which depends on the speed of the vehicle. Its value may be taken as a constant ($f_o$) up to the speed ($V$) of 50 km/hr. The following formula is used if the speed exceeds 50 km/hr

$$f = f_o\,(1 + 0.01\,(V - 50)) \tag{2.16}$$

The $f_o$ value for car tyres on a road may be taken in the range of 0.01–0.035. Thus, a coefficient of 0.030 is equivalent to a coefficient of 30 kg/t. This means that if the tyre is supporting a wheel-load of 1 tonne, a force of 30 kg must be applied to avoid losing speed because of rolling resistance.

For example, the rolling resistance of a car weighing 2000 kg on asphalt with a rolling friction coefficient of 0.03 can be estimated as

$$F_r = 2000 \text{ kg} \times 9.81 \text{ m/s}^2 \times 0.03 = 588.6 \text{ N}$$

- **Grade resistance**: The resistance offered to vehicle movement due to longitudinal slope depends on the profile steepness in the upgrade or downgrade direction. Adequate tractive energy of vehicles is required to overcome such grade resistance, and it is determined by

$$F_g = \frac{mGg}{100} \tag{2.17}$$

where $F_g$ = grade resistance (N); $m$ = mass of the vehicle (kg); $G$ = magnitude of grade (%); and $g$ = acceleration due to gravity (9.81 m/sec$^2$).

The approximate value of the total power required by a vehicle to overcome these three resisting forces is calculated using the formula

Total horsepower required = Horsepower required against (Grade resistance, $F_g$ + Rolling resistance, $F_r$ + Air resistance, $F_a$)

These calculations are useful for planning and designing road geometric facilities and are also applicable for traffic regulations relating to speed, including acceleration/deceleration at a turning, and merging and diverging stretches.

## 2.3 ROAD CHARACTERISTICS

An overview of the road characteristics that influence vehicle performance is given below.

### 2.3.1 Surface Characteristics

Data about traffic volume and the composition of traffic will dictate the traffic engineer's choice of the type of pavement and the number of lanes required. The intensity of roughness, skidding characteristics and type/severity of distress of the road surface will influence the intensity of noise generation, speed reduction, traffic congestion locations, vehicle operation cost and vehicle maintenance cost, including the service life of tyres. More details on these topics can be obtained from *Pavement Evaluation and Maintenance Management*, published by the same author with Universities Press.

### 2.3.2 Lighting

The lighting characteristics of vehicle headlamps are useful for determining the minimum length of sag curves, and a few important standards are given below (IRC: SP: 23-1993).

- Height of headlamp above the road surface = 0.75
- Height of the driver's eye level above the road surface = 1.2 m
- Angle of spread of headlight or light cone focus angle = 1°
- Minimum height of the obstacle above the road surface = 0.15 m

Details pertaining to these factors, which are useful for providing safe sight distance along vertical curves, and the related design of highways, are presented in *Chapter 3*.

The intensity of illumination on a roadway will influence traffic speed and pedestrian flow rate characteristics, including road-safety considerations. Providing adequate lighting in terms of intensity of flux from street lights is one of the essential factors considered during night time, cloudy and underpass/tunnel operations of traffic. The light-reflecting characteristics of a pavement surface during the day, at night and in wet weather conditions will also influence the operational safety of vehicles. The factors relating to the illumination of roadways and their effect on vehicle performance are discussed in *Chapter 19*.

## 2.4 TRAFFIC CONTROL DEVICES

Information about traffic volume and the composition of traffic will dictate the choice of devices/mechanism to be adopted for safe operation and efficient flow of traffic. Data about traffic volume in the morning and evening peak hours is useful for traffic management purposes, such as for the adoption of tidal flow technique. Various types of traffic control devices, including signals, signs, road markings and intelligent transportation systems, are discussed in detail in subsequent chapters of this book.

## 2.5 THE GENERAL ENVIRONMENT

The construction of roads and the traffic flow through a built-up environment as well as through rural areas will have a definite impact on the overall quality of life of people residing in that area. Some of the adverse effects of such construction are listed below:

- Air and noise pollution
- Congestion to pedestrian flow and to regular traffic movement in an area due to the impact of construction or heavy traffic volume
- Land development impacts on the socio-economic health of local residents
- Poor visibility due to dust and suspended dust particles in the air
- Vibrations due to the movement of heavy and fast-moving vehicles
- Road accidents due to possible interactions between vehicles and between vehicles and pedestrians or stationary objects
- Pollution to local water bodies due to oil spills from automobiles
- Effect on the growth of greenery in the area adjacent to the road

These elements need to be considered before and after implementation of any road project so that the living environment can be preserved for the sustainable development of any society.

## EXERCISES

### Multiple Choice Questions

1. What is the standard percentile value of driver characteristics used for the design of roadway geometric facilities?
   (a) 85$^{th}$ percentile (b) 90$^{th}$ percentile (c) 95$^{th}$ percentile (d) 80$^{th}$ percentile
2. What is the cone angle range of fairly clear vision?
   (a) 5–10° (b) 3–5° (c) 10–12° (d) 12–15°
3. What is the rolling resistance of a car weighing 1000 kg on an asphalt road surface with a rolling friction coefficient of 0.03. Take acceleration due to gravity as 10 m/sec$^2$.
   (a) 3333.33 N (b) 300 N (c) 333333.33 N (d) 3 N
   (*Hint*: $F_r$ = 1000 kg × 10 m/s$^2$ × 0.03 = 300 N)
4. What is the standard percentile value taken for the design of pedestrian crossing facilities in India?
   (a) 15$^{th}$ percentile (b) 90$^{th}$ percentile (c) 95$^{th}$ percentile (d) 85$^{th}$ percentile
5. A driver travelling on a forest road at a speed of 40 km/hr observes a deer sleeping on the road. If the driver applies his brakes immediately without any PRT, and starts decelerating at 4 m/sec$^2$, what is the distance the car will travel before stopping?
   (a) 16.65 m (b) 33.36 m (c) 16.71 m (d) 50.01 m
   (*Hint*: Speed = 40 km/hr × 0.278 = 11.12 m/sec

$$t_{deceleration} = \left(\frac{u-v}{a_{deceleration}}\right) = \left(\frac{11.12-0}{-3.7}\right) = 3.0\,\text{sec}$$

   The distance travelled during this 3.0 sec = (11.12 × 3.0) + (½ × −3.7 × 3$^2$) = 33.36 − 16.65 = 16.71 m)

6. What is the maximum width of vehicles permitted in India?
   (a) 1.75 m (b) 2.4 m (c) 2.5 m (d) 1.8 m
7. For normal application, no vehicle other than a double-decker bus shall have a height exceeding __________ and a height exceeding 4.2 m when carrying ISO series 1 freight containers. Double-decker buses may, however, have a height not exceeding 4.75 m.
   (a) 3.5 m (b) 3.8 m (c) 4.2 m (d) 4.75 m
8. No vehicle other than a double-decker bus shall have a height exceeding __________ when carrying ISO series 1 freight containers.
   (a) 3.5 m (b) 3.8 m (c) 4.2 m (d) 4.75 m
9. What is the maximum height permitted for double-decker buses?
   (a) 5.5 m (b) 3.8 m (c) 4.2 m (d) 4.75 m
10. What is the maximum overall length of a single-unit truck, exclusive of the front and rear bumpers, and having two or more axles?
    (a) 10 m (b) 11 m (c) 16.5 m (d) 19.5 m
11. What is the maximum overall length of a single-unit bus, exclusive of the front and rear bumpers, and having two or more axles?
    (a) 12 m (b) 11 m (c) 16.5 m (d) 19.5 m
12. The maximum overall length of a truck-tractor semi-trailer combination, exclusive of the front and rear bumpers, shall be __________.
    (a) 11 m (b) 12 m (c) 16 m (d) 18 m
13. According to IRC: 3, the maximum overall length of a truck-trailer combination, exclusive of front and rear bumpers, shall be __________.
    (a) 12 m (b) 16 m (c) 18 m (d) 23 m
14. According to IRC: 3, the total gross weight imposed on the highway by a single axle fitted with dual wheels shall not exceed __________ tonnes.
    (a) 6 (b) 8 (c) 10.2 (d) 14.7
15. According to IRC: 3, the total gross weight imposed on the highway by a single axle with single wheels shall not exceed __________ tonnes.
    (a) 6 (b) 8 (c) 10.2 (d) 14.7
16. According to IRC: 3, the total gross weight imposed on the highway by two axles in tandem articulated from a common attachment to the vehicle or individually attached to the vehicles with standard spacing between the axles, shall not exceed __________ tonnes.
    (a) 8 (b) 10.2 (c) 18 (d) 19.5
17. According to IRC: 3, the maximum permissible gross weight of a Type 2 vehicle having both single tyres is __________ tonnes.
    (a) 12 (b) 16.2 (c) 24 (d) 26.4
18. According to IRC: 3, the maximum permissible gross weight of a Type 2 vehicle having a single tyre front axle and a dual tyre rear axle is __________ tonnes.
    (a) 12 (b) 16.2 (c) 24 (d) 26.4
19. According to IRC: 3, the maximum permissible gross weight of a Type 2 vehicle having single tyres on both of its axles is __________ tonnes.
    (a) 12 (b) 16.2 (c) 24 (d) 26.4

20. According to IRC: 3, the maximum permissible gross weight of a Type 2-S2 vehicle is ________ tonnes.
    (a) 24 (b) 26.4 (c) 34.2 (d) 42
21. According to IRC: 3, the maximum permissible gross weight of a Type 3-3 vehicle having single tyres on both of its axles is ________ tonnes.
    (a) 52.2 (b) 44.4 (c) 36.6 (d) 45.0
22. According to IRC: 3, what is the standard space provided between the axles of a tandem two-axle configuration?
    (a) Less than 1.2 and greater than 0.5 m
    (b) Not less than 1.2 and greater than 0.5 m
    (c) Not less than 1.2 m but not more than 2.5 m
    (d) Not less than 1.75 m but not more than 2.5 m
23. The figure below indicates what type of vehicle according to the IRC: 3 standard? 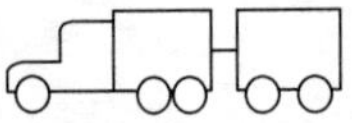
    (a) Vehicle Type 2-2 (b) Vehicle Type 3-2 (c) Vehicle Type 2-3 (d) Vehicle Type 3-3
24. The figure below indicates what type of vehicle according to the IRC: 3 standard? 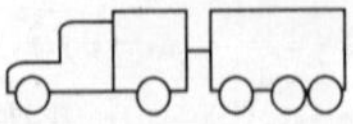
    (a) Vehicle Type 2-2 (b) Vehicle Type 3-2 (c) Vehicle Type 2-3 (d) Vehicle Type 3-3
25. The figure below indicates what type of vehicle according to the IRC: 3 standard? 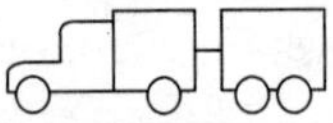
    (a) Vehicle Type 2-2 (b) Vehicle Type 3-2 (c) Vehicle Type 2-3 (d) Vehicle Type 3-3
26. The figure below indicates what type of vehicle according to the IRC: 3 standard? 
    (a) Vehicle Type 2-2 (b) Vehicle Type 3-2 (c) Vehicle Type 2-3 (d) Vehicle Type 3-3

## Answers

1. (a) 2. (b) 3. (b) 4. (a) 5. (c) 6. (c) 7. (b) 8. (c) 9. (d) 10. (b) 11. (a)
12. (c) 13. (c) 14. (c) 15. (a) 16. (c) 17. (a) 18. (b) 19. (a) 20. (c) 21. (a) 22. (c)
23. (b) 24. (c) 25. (a) 26. (d)

## Questions

1. State the components of traffic engineering, along with the characteristics affecting the performance of any traffic flow system.
2. Classify and list the behaviour factors affecting driver performance.
3. Classify road users and list the factors affecting each type.
4. Write an overview of the physical characteristics of road users and explain their relevance to the geometric design of roads and traffic flow operations.

5. Explain the visual characteristics of road users relating to (a) visual acuity, (b) peripheral vision, (c) eye movement and visual concentration, (d) colour vision, (e) glare effect and (f) perception space.
6. Write an overview of pedestrian characteristics and explain their relevance to the geometric design of roads and traffic flow operations.
7. Explain the PIEV theory. What are the factors affecting PRT?
8. Define design driver and explain the design driver characteristics.
9. Derive the expression for determining the rate of change of the visual angle with driver position.
10. A driver is driving a vehicle at a speed of 80 km/hr and he first observes a sign post which was placed at an instantaneous distance of 100 m. The offset distance of the sign post is 12 m, to the left of the driver's path. Determine what would be the distance along the road at which the driver will first observe the displacement of the sign post, if it is placed at an offset distance of 15 m?
11. A car is travelling at a speed of 110 km/hr and the driver first observes, at an instantaneous distance of 120 m, the lateral displacement of a sign post, placed 9 m to the left of the driver's path. Determine at what later distance the sign post should be placed if the driver starts displacing it at half the longitudinal distance?
12. Write the details about the maximum permissible gross weight of vehicles in India, with appropriate sketches of the vehicles for each case.
13. Classify and state the vehicle types.
14. Explain the factors relating to axle load and vehicle gross load that affect the performance of vehicles on the road.
15. Define standard axle and legal axle load and give the values of different types of standard axle and legal axle loads.
16. State the concept of speed limit and draw a sketch.
17. Write the factors and explain how braking performance and the deceleration characteristics of vehicles are used for the safety evaluation of vehicles.
18. Write the details of the acceleration characteristics of vehicles used for the safety evaluation of vehicles.
19. Derive an expression for determining the non-uniform acceleration characteristics of a vehicle, with standard notations.
20. A car is travelling at a speed of 100 km/hr and the driver decides to overtake a vehicle by pressing the accelerator. The accelerating behaviour of the car is described by

$$\frac{dv}{dt} = 1.4 - 0.02v$$

where $v$ is the speed in m/sec and $t$ is the time in seconds. Determine

(a) The maximum speed of the car
(b) The maximum rate of acceleration
(c) The rate at which the car accelerates after 5 seconds
(d) The time the car takes to reach a speed of 120 km/hr

21. Explain the following vehicle characteristics that influence the design elements of a roadway:
(a) Inertial force (b) Rolling resistance (c) Air resistance (d) Grade resistance
22. Write an overview of road surface characteristics and explain their relevance to traffic flow operations.
23. Give details of the lighting characteristics of vehicle headlamp lighting useful for determining the minimum length of sag curves, with a few important standards.

CHAPTER

# 3 Highway Geometric Design

## 3.0 INTRODUCTION

Developing a highway geometric design involves determining the physical dimensions of the various geometric elements necessary for a highway, based on scientific and practical requirements. A geometric design is deemed to be excellent if it allows the safe, smooth and economic movement of pedestrians and vehicles – both motorised and non-motorised. If the layout of different geometric elements and their dimensions are defective, it poses serious problems which may be impossible to rectify at a later stage due to developments along the roadside (Losh, 1932). Therefore, the need for good geometric design becomes more pressing. Appropriate specifications and the recommendations of various standard institutions should be suitably implemented in the design.

**Good highway geometric design** should take into account the following aspects:

- Cross-sectional elements
- Vehicle dimensions and permissible axle loads
- Sight distances
- Elements in horizontal alignment
- Elements in vertical alignment
- Intersection layout
- Cross-drainage works, namely, causeways, culverts, bridges, etc.
- Accommodation of important utility services within the right of way. These utility services may include sewers, drainage and water pipelines, cables, transmission towers, telephones lines and gas pipelines and those carrying combustible materials (IRC: 98-1997)
- Location of new road–rail level crossings (IRC: 39-1986)
- Noise and light barriers
- Minimum width of land to be acquired, keeping in view future developments
- In hilly areas, alternative routes must be provided in case of road blockages due to land slide, rock fall, etc. (IRC: 52-2001)
- Road safety measures
- Other issues such as markings, lighting, channelisation, flyovers, road arboriculture and roads for military use to cater to special vehicles

The **guidelines** to be considered while planning the geometric design of highways are as follows:

- The design should, at all times and in all situations and environments, be safe, economical and compatible.
- Vehicle dimensions such as length, width and height should be taken into consideration.
- Suitable provisions should be made to accommodate future increased traffic intensity and for developments in the locality along the road network.
- The design should be consistent and conform to the standard specifications of IRC or other international institutions (e.g., American Association of State Highway and Transportation Officials (AASHTO)).
- Factors relating to humans and vehicles should be suitably considered.

- Minimum total costs, which include construction, maintenance, operational and user costs, should be calculated (often termed **life cycle cost analysis**).
- The design should be comprehensive and complete, with drainage systems, signs, markings, arboriculture, medians, footpaths, signals and lighting systems.

The design details of the various geometric elements of a highway are discussed below.

## 3.1 VEHICLE DIMENSIONS

Vehicle dimensions such as length, width, height and gross weight play an important role in the design of various geometric elements of highways (Table 3.1).

**Table 3.1** Vehicle parameters useful for designing geometric elements of highways

| *Vehicle parameter* | *Description* |
|---|---|
| Width | Width and number of lanes or lane capacity<br>Width of shoulder and design of signals<br>Parking area, cross-drainage works<br>Lateral placement of vehicles |
| Height | Location of electric and other cables<br>Flyovers and underpasses |
| Length | Bridges and underpasses<br>Highway junctions, hill roads, curves and hairpin bends<br>Overtaking zones, highway capacity |
| Engine capacity | Design speed, skid resistance and braking distance<br>Gradient, grade compensation and length of vertical curve<br>Superelevation on curves<br>Safety measures at salient locations |

Table 3.2 lists the standard vehicle dimensions according to the IRC (86-1983; 3-1983).

**Table 3.2** Standard vehicle parameters useful for designing geometric elements of highways

| *Vehicle parameter* | *Maximum allowable value (m)* |
|---|---|
| Width | 2.44 |
| Height | For single deck: 3.81<br>For double deck: 4.72 |
| Length† | With two axles: 10.67<br>More than two axles: 12.19<br>Tractor with semi-trailer: 15.24<br>Tractor with trailer: 18.29 |

†Maximum allowable length of vehicles is 31 m in the USA and 25 m in Canada (Wang and Cartmell, 1998).

The standard weights of axles and their configurations are discussed in *Chapter 2*.

## 3.2 TERRAIN CLASSIFICATION

The geometric design of highways is based on the topography of the terrain (general nature of the ground slope) on which the highway is to be laid and the corresponding design speeds specified. The cost of construction and maintenance of highways depends mainly on the type of the terrain. For example, the cost of road construction in a mountainous area is higher than that on flat or rolling ground. The cross slope of a terrain may be classified into four types. A typical terrain classification (IRC: 73-1980) along with the general ground slopes encountered is presented in Fig. 3.1.

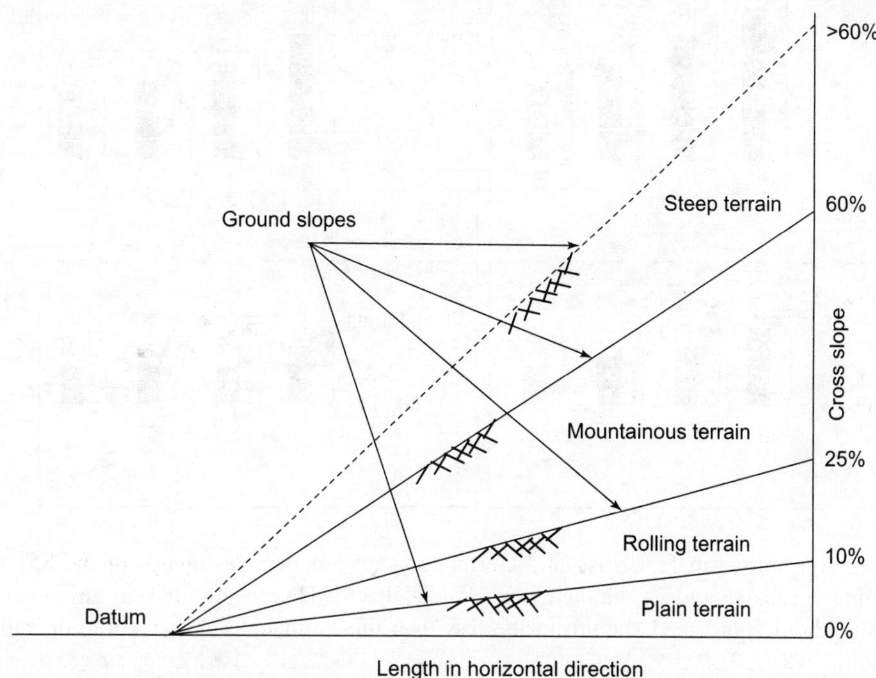


**Fig. 3.1** Terrain classification based on cross slope

## 3.3 DESIGN SPEED

Design speed may be defined as the maximum average safe operating speed of vehicles that is either specified or maintained by most drivers travelling through all the features of the roadway. It is one of the prime parameters that dictate the overall geometry of highways in plain as well as hilly areas.

The dimensions and layout of most geometric clcments are derived from design speed considerations. Field observations are made to correlate different speeds with the location of various highway geometric elements. Based on the field observations and scientific analysis, IRC, AASHTO and other institutions have recommended certain minimum acceptable geometric standards.

The design speeds prescribed by IRC (Table 3.3) depend on the type of cross slope of the country, the terrain classification and the importance of the highway. According to IRC (73-1980), the ruling design speed is generally considered a design criterion and taken as a primary input parameter for the design of various geometric elements of highways. The minimum design speed is considered if the ruling design speed cannot be adopted for practical reasons and other considerations. Appropriate design speeds may be selected based on site conditions, feasibility, economics of construction and traffic operations. However, care should be taken to maintain uniform vehicular speed throughout the length of the road, even when terrain conditions differ.

**Table 3.3** Graphical presentation of design speeds (km/hour), a typical example (NH = National Highway; SH = State Highway; MDR = Major District Road; ODR = Other District Road; VR = Village Road) (Modified from IRC: 73-1980 and IRC: SP: 48-1998)

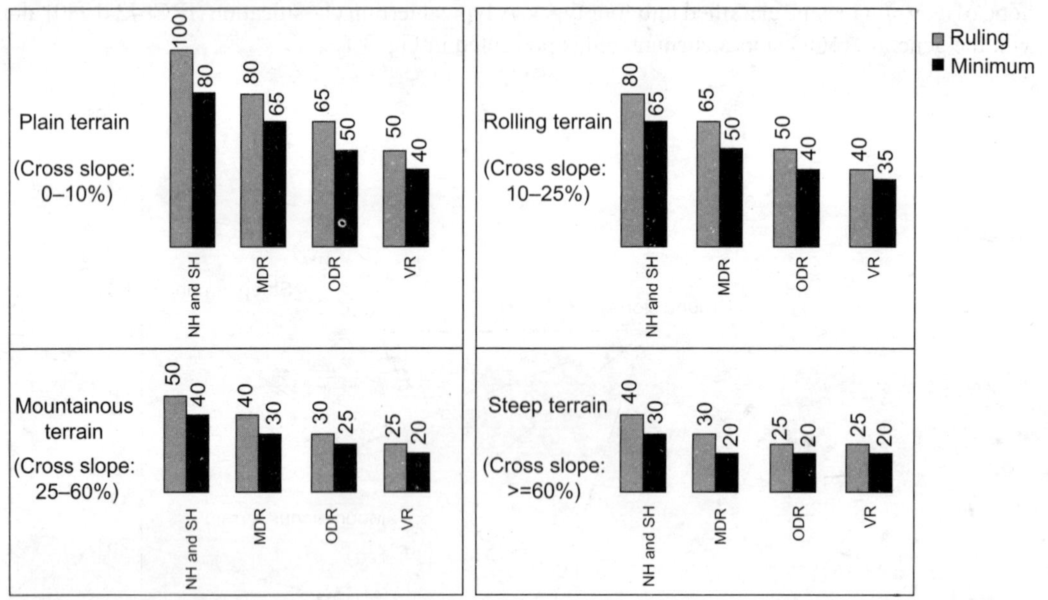


If the design must satisfy British and German design practices, estimates of the 85th percentile of operating speeds should be adopted (Kanellaidis, 1996). This means that on any given segment of roadway, the design speed should not be less than the estimated 85th percentile operating speed (Nicholson, 1998).

To design horizontal curves, the operating speed (at the 85th percentile) is generally used as a parameter for controlling the speed (Bonneson, 1999). If the proposed design speed is to be assumed, it should be a logical one, determined primarily by considering the natural terrain, land use and highway classification (AASHTO, 1990). However, the design of highway geometric elements based on assumed design speed is no longer considered adequate, especially for the lower design speed of highways (Hirsh et al., 1986).

## 3.4 CROSS-SECTIONAL ELEMENTS

Cross-sectional elements include all the visible physical features of a highway (Fig. 3.2). Some of the elements important in highway geometric design are briefly discussed in the following sections.

For more details on the cross-sections of urban roads, please refer to IRC: 69-1977.

### 3.4.1 Boundary Lines

Boundary lines can be classified as road boundaries, building lines and control lines. Each boundary line serves a particular purpose (Fig. 3.2). No construction should be made within the limits of the building

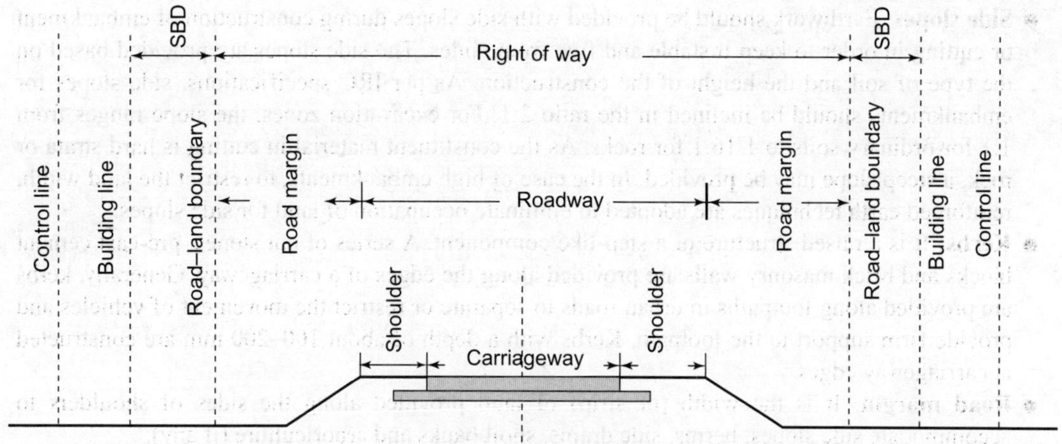


**Fig. 3.2** Salient cross-sectional details of a highway

lines, to ensure uniformity of road width. Within the control lines, the government has limited control over the construction of structures. Based on the terrain and type of road to be laid, IRC: SP: 15-1974 recommends standard overall widths between these respective lines. These widths of land should be specified for new highway alignments to help in land acquisition. They should also be monitored to prevent illegal developments along the highways. The land width between the road boundary and the building line is called the **setback distance** (SBD). The standard values of SBD recommended for national and state highways are 3–6 m in plain and rolling terrain and 3–5 m for the remaining classes in all terrains (IRC: 73-1980).

## 3.4.2 Right of Way or Road–Land Width

Right of way is the width of land acquired to accommodate the width of the carriageway, the shoulders and road margins. It is generally acquired for at least five decades of the service life of a highway. Necessary provision has to be made for future road-widening or anticipated developments along the alignment. This is because land acquisition, especially in built-up or urban areas, is very difficult and leads to legal problems and payment of a heavy compensation (Losh, 1932).

The required right of way for different terrain and road classes as recommended by the IRC (73-1980) is furnished in Table 3.4.

**Table 3.4** Desirable right of way (m) for different road classes in plain and rolling terrain, a typical example (Modified from IRC: 73-1980)

| *Road class* | *Open areas (m)* | *Built-up areas (m)* |
|---|---|---|
| NH and SH | 30–60 (normal: 45) | 30–60 (normal: 30) |
| MDR | 25–30 (normal: 25) | 15–25 (normal: 20) |
| ODR | 15–25 (normal: 15) | 15–20 (normal: 15) |
| VR | 12–18 (normal: 42) | 10–15 (normal: 10) |

- **Side slopes**: Earthwork should be provided with side slopes during construction of embankment or cutting in order to keep it stable and free from slides. The side slopes are provided based on the type of soil and the height of the construction. As per IRC specifications, side slopes for embankments should be inclined in the ratio 2:1. For excavation zones, the slope ranges from 1:1 for ordinary soils to 1/16:1 for rocks. As the constituent material in cutting is hard strata or rock, a steep slope may be provided. In the case of high embankments, to restrict the land width, reinforced earth techniques are adopted to eliminate occupation of land for side slopes.
- **Kerbs:** It is a raised structure or a step-like component. A series of cut stones, pre-cast cement blocks and brick masonry walls are provided along the edges of a carriageway. Generally, kerbs are provided along footpaths in urban roads to separate or restrict the movement of vehicles and provide firm support to the footpath. Kerbs with a depth of about 100–200 mm are constructed at carriageway edges.
- **Road margin**: It is the width (or strip) of land provided along the sides of shoulders to accommodate side slopes, berms, side drains, spoil banks and arboriculture (if any).
- **Side drains**: Side drains are provided within the width of the road margin, on either side of the alignment, to drain off surface water to a safe distance. Suitable longitudinal gradients are provided according to the terrain. Side drains are designed to carry away storm water and also to control the groundwater table (GWT) at a safe depth below the sub-grade by preventing capillary rise.

### 3.4.3 Carriageway Width and Roadway Width

The carriageway (CW) is the main width of the road responsible for providing a strong, stable and smooth riding surface. It is designed based on traffic volume, design life, lane capacity, vehicle dimensions and the clearance to be maintained between adjacent vehicles and the road edge. The IRC (73-1980) specifications regarding the width of carriageways and roadways, based on the number of lanes, are given in Tables 3.5 and 3.6, respectively.

**Table 3.5** Width of carriageway

| *Lane* | *Width of carriageway (m)* |
|---|---|
| Single lane | 3.75 |
| Single lane (less important or village roads) | 3.0 |
| Double lanes (without raised kerbs) | 7.0 |
| Double lanes (with raised kerbs) | 7.5 |
| Double lanes of less important roads | 5.5 |
| Multi-lane | 3.5 per lane |

### 3.4.4 Shoulder

Shoulder is the width of paved surface laid along both sides of the carriageway to facilitate overtaking, temporary parking, walking and for emergency halting. Shoulders provide a lateral edge support to the carriageway along its length, a level surface during overtaking operations and facilitate draining of the surface water to the side drains. The width of the shoulder should be equal to half the difference between the widths of the roadway and the carriageway.

**Table 3.6** Width of roadway or formation

| *Road class* | | *Roadway width (m) for terrain** | |
|---|---|---|---|
| | | *Plain and rolling* | *Mountainous and steep* |
| NH and SH | Single lane | 12.0 | 6.25 |
| | Double lane | 12.0 | 8.8 |
| MDR | Single lane | 9.0 | 4.75 |
| | Double lane | 9.0 | |
| ODR | Single lane | 7.5 | 4.75 |
| | Double lane | 9.0 | |
| VR (Single lane) | | 7.5 | 4.0 |

* These values may vary subject to site conditions as recommended by IRC: 73-1980.

Depending on requirements, shoulders are paved using earthen surfaces, a mechanically compacted gravel–soil mix or stabilised soil using bituminous materials or cement or lime. In India, the outward minimum slope of shoulders for earthen surfaces is 0.5% steeper than the pavement camber (which should be a minimum of 3%). For paved shoulders, a suitable value is selected from Table 3.7. On superelevated pavement sections, the same value of cross fall is provided as that of the pavement (IRC: 73-1980).

### 3.4.5 Camber

The cross slope or slope in the transverse direction provided to the carriageway surface on either side of its centre line is called a camber or cross fall (Fig. 3.3). With reference to the level of the carriageway edges, the middle portion of the carriageway surface is raised. This cross fall or transverse slope of the carriageway surface facilitates the quick drain-off of rain water on to the shoulders and prevents stagnation on the carriageway.

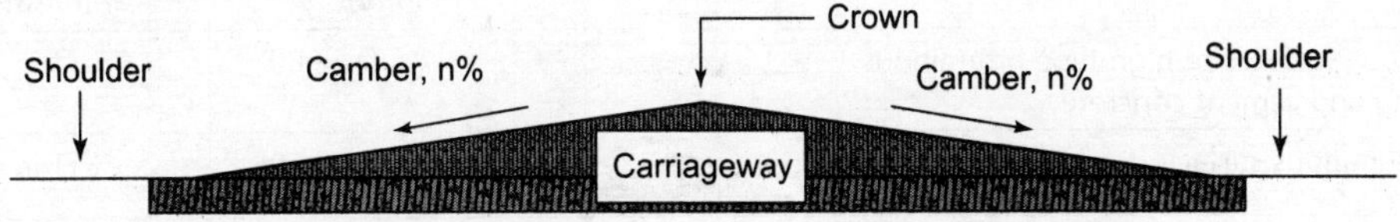


**Fig. 3.3** Details of transverse slope or camber

Sufficient camber should be provided for safe surface drainage. The value for the camber should be carefully specified based on rainfall intensity, type of materials used for construction, required smoothness and imperviousness of the carriageway. Generally, more camber is provided for unpaved or earthen surfaces than for paved surfaces. This is because paved surfaces are smooth and do not allow rain water to percolate beneath the layers. At the same time, care should be taken not to provide more camber than is required for unpaved surfaces. It may create erosion or rutting or potholes.

The classification of camber based on the shape of the sloped surface is given in Fig. 3.4. Table 3.7 shows the maximum allowable camber for different surfaces of pavements recommended by the IRC (73-1980).

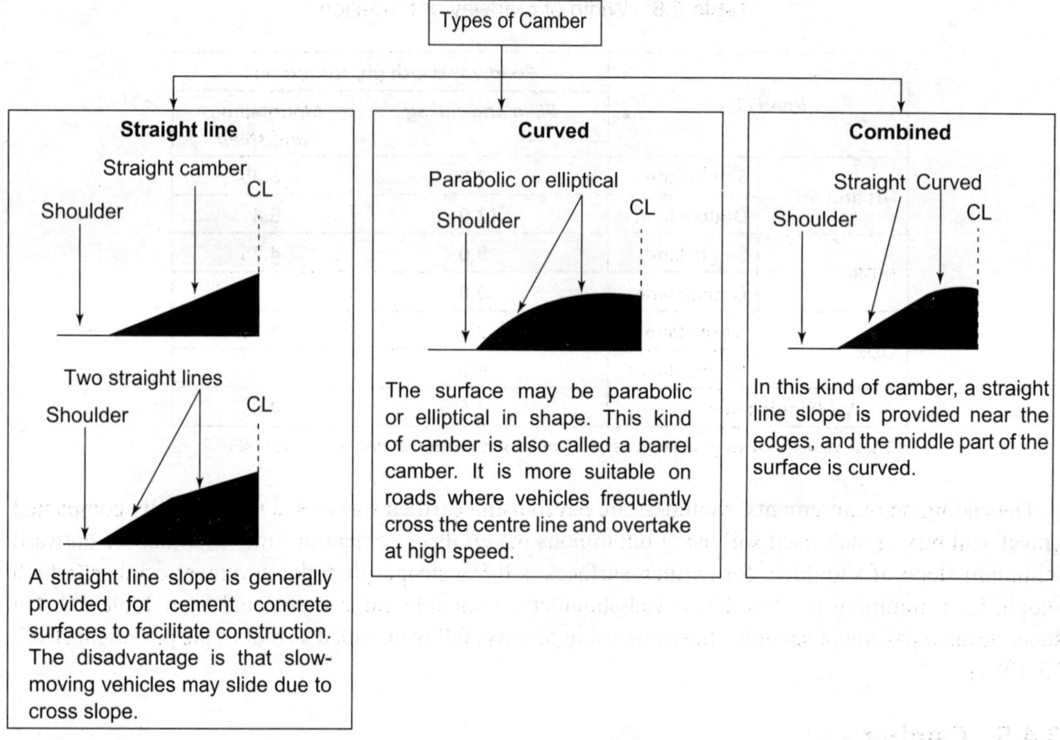


**Fig. 3.4** Types of camber

**Table 3.7** Values of camber for different surfaces of pavements, a typical example

(Modified from IRC: 73-1980)

| *Type of pavement surface* | *Symbolic profile shape* | *Range of camber* | |
|---|---|---|---|
| | | *For light rainfall intensity* | *For heavy rainfall intensity* |
| Impervious surfaces of high-type bituminous surfacing and cement concrete | | 1.7% (1 in 60) | 2.0% (1 in 50) |
| Thin bituminous surfacing | | 2.0% (1 in 50) | 2.5% (1 in 40) |
| Water-bound macadam (WBM) and gravel | | 2.5% (1 in 40) | 3.0% (1 in 33) |
| Earthen road | | 3.0% (1 in 33) | 4.0% (1 in 25) |

If the camber is provided with a parabolic curved surface, the following relationship may be used to determine the offsets from the centre line at the crown point of a carriageway (Fig. 3.4(a)).

$$\frac{O_i}{(X_i)^2} = \frac{h}{\left(\frac{\text{CW}}{2}\right)^2} \tag{3.0}$$

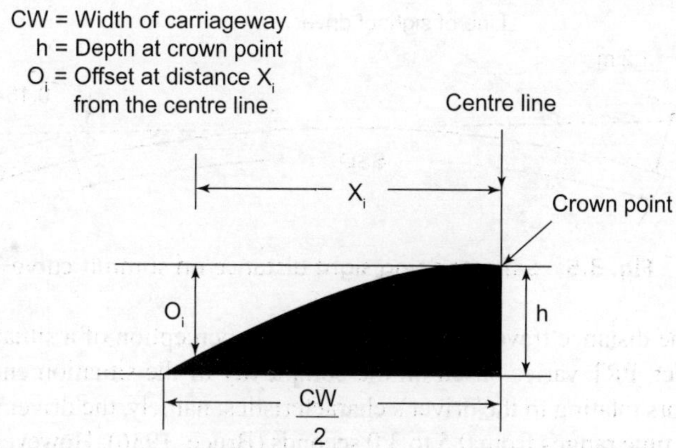


**Fig. 3.4(a)** Details of the offset for a parabolic camber surface

## 3.5 SIGHT DISTANCE

Sight distance is the maximum distance visible to the driver along the centre line of a highway. The desirable length of sight distance depends on many factors; e.g., characteristics of the driver and vehicles, design speed, superelevation, horizontal and vertical curvature. Driver characteristics include physical and perception capabilities to react to eventualities while driving. Vehicle characteristics consist of efficiency of the braking system, engine capacity and dimensions (Table 3.2). Sight distance is classified as follows:

- Safe stopping sight distance (SSD) or non-passing sight distance
- Overtaking sight distance (OSD) or passing sight distance
- Intersection sight distance (ISD)

### 3.5.1 Safe Stopping Sight Distance (SSD)

The minimum distance travelled by a vehicle after the driver perceives an obstruction (distance travelled during perception time), acts on his observation (lag distance), applies the brakes and the vehicle stops (braking distance) is called the safe stopping sight distance. The distance and definition varies depending on the terrain on which the highway is laid. For example, in the case of a summit curve, according to IRC, SSD may be defined as the safe minimum visible straight distance of a driver along the centre line of a highway when the height of the driver's eye and an object in the opposite direction above the road surface is 1.2 m and 0.15 m, respectively (Fig. 3.5) (IRC: 66-1976). A more detailed discussion on curves can be found in Section 3.11.2.

**Perception–Reaction Time (PRT):** It is the time required for a driver, while driving the vehicle, to observe, notice, assess and mentally analyse a given situation – the driver will then decide whether to react, apply the brakes or steer away. In other words, while driving, it is the minimum time required for a driver to react instantly and exercise control. PRT may be explained by the PIEV theory.

**PIEV theory**: According to this theory, a driver **p**erceives a situation on the road, analyses it using his **i**ntelligence, **e**motionally feels it and reacts with **v**olition.

$$\text{Stopping sight distance (SSD)} = \text{Lag distance (LD)} + \text{Braking distance (BD)} \quad (3.1a)$$

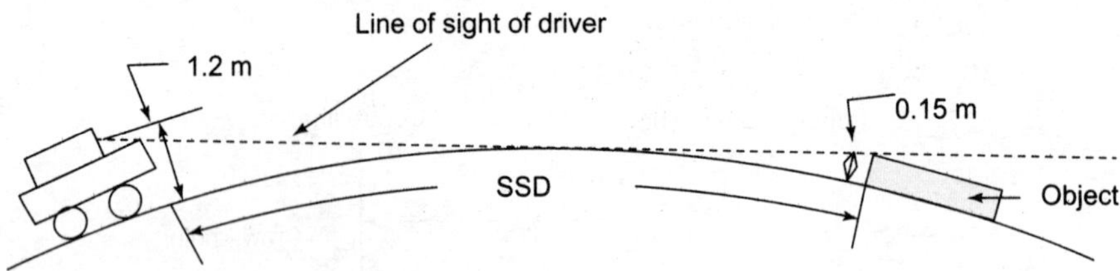


Fig. 3.5 Safe stopping sight distance on summit curve

**Lag distance**: It is the distance travelled by a vehicle from perception of a situation to the application of brakes by the driver. PRT varies based on the complexity of the situation encountered (AASHTO, 1990) and many factors relating to the driver's characteristics, namely, the driver's age, intelligence and physical fitness. This time ranges from 0.5 to 3.0 seconds (Bruce, 1936). However, for design purposes, the total time required from perception to reaction (say, application of brakes) is taken to be 2.5 seconds (IRC: 66-1976; AASHTO, 1994).

**Braking distance**: It is the distance travelled by a vehicle just after the application of brakes, either to bring the vehicle to a complete halt or to steer it away. A vehicle moves even after the application of brakes since the kinetic energy of the vehicle cannot be instantly reduced to zero. Following are the factors affecting braking distance:

- Velocity of the travelling vehicle and its inertial force
- Brake efficiency of the vehicle and the type of braking system
- Surface friction or road surface characteristics
- Longitudinal slope or gradient of a road

According to IRC, the value of longitudinal friction factor ($f_s$) varies from 0.35 at 100 kmph to 0.40 at 20 kmph.

We will now calculate SSD on different terrains.

**Case 1:** On a level road surface (Fig. 3.6)

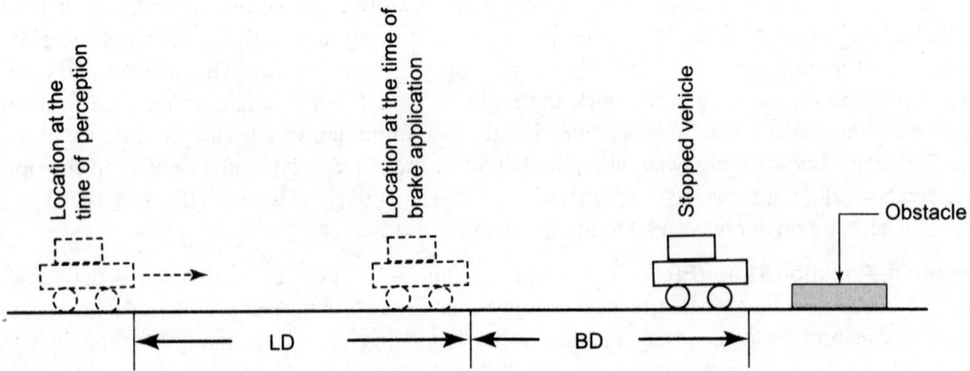


Fig. 3.6 Safe stopping sight distance on a plain road

$$\text{Lag distance (LD)} = v \times t$$

**Braking distance (BD):** It may be derived by equating the kinetic energy of the moving vehicle to the work done by the frictional force (F) of the road surface to stop the vehicle.

$$\frac{1}{2}mv^2 = f_s WL$$

$$\frac{W}{2g}v^2 = f_s WL \qquad \left(\because\ W = mg;\ m = \frac{W}{g}\right)$$

$$L = \frac{v^2}{2g\,f_s}$$

$$\text{SSD} = vt + L$$

$$\text{SSD} = vt + \frac{v^2}{2g\,f_s} \tag{3.1b}$$

where $L$ = braking distance (m); $v$ = speed of the vehicle (m/s); $f_s$ = coefficient of surface friction (0.35 to 0.40 for a speed range of 100 to 20 kmph); and $g$ = acceleration due to gravity (9.81 m/s$^2$).

**Case 2:** On a sloped road surface (Fig. 3.7)

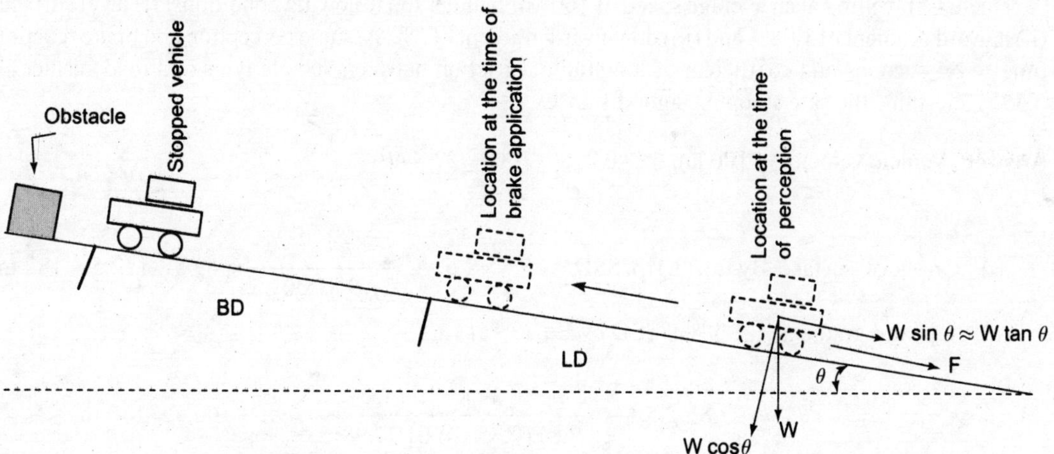


**Fig. 3.7** Stopping sight distance on a sloped road

$$\text{Lag distance (LD)} = v\,t$$

Braking distance (BD), with reference to Fig. 3.7

$$\frac{1}{2}mv^2 = f_s WL + WL\tan\theta$$

$$\frac{W}{2g}v^2 = f_s WL + WL\tan\theta$$

$$\frac{v^2}{2g} = L(f_s + \tan\theta)$$

$$L = \frac{v^2}{2g(f_s + \tan\theta)} = \frac{v^2}{2g(f_s + 0.01n)}$$

[since $\tan\theta = +0.01\,n$, if the ascending gradient is taken as $+\,n\%$ or 1 in $n$]

$$L = \frac{v^2}{2g(f_s - 0.01n)}$$

[since $\tan\theta = -0.01\,n$, if the descending gradient is taken as $-\,n\%$ or 1 in $n$]

$$\text{SSD} = (LD + BD) = vt + L$$

$$\therefore \quad \text{SSD} = vt + \frac{v^2}{2g(f_s \pm 0.01\,n)} \tag{3.1c}$$

**Example 3.1**

A vehicle is travelling at an average speed of 100 km/h under the following conditions: (i) level surface, (ii) upward gradient of 1.98% and (iii) downward gradient of 2%. Assume perception and brake reaction time = 2.5 seconds and coefficient of longitudinal friction between vehicle tyres and road surface = 0.35. Determine the safe stopping sight distances.

**Answer**: Vehicle velocity = 100 km/h = 0.278 × 100 = 27.8 m/s

(i) On a level surface: By Eq. (3.1b), $\text{SSD} = 27.8 \times 2.5 + \dfrac{27.8^2}{2 \times 9.81 \times 0.35} = 69.5 + 112.5 = 182$ m

(ii) Upward gradient = 1.98%: SSD by Eq. (3.1c)

$$= 27.8 \times 2.5 + \frac{27.8^2}{2 \times 9.81\,(0.35 + 0.01 \times 1.98)} = 69.5 + 106.5 = 176 \text{ m}$$

(iii) Downward gradient of 2%: $\text{SSD} = 27.8 \times 2.5 + \dfrac{27.8^2}{2 \times 9.81\,(0.35 - 0.01 \times 2)} = 69.5 + 119.5$

$$= 189\,\text{m}$$

### 3.5.2 Overtaking Sight Distance (OSD) or Passing Sight Distance (PSD)

Under mixed traffic conditions, different vehicles travel at different speeds. However, for comfortable travel, vehicles travelling at design speed frequently need to overtake slow-moving vehicles. During this process, the fast-moving vehicle enters into the opposite lane for some distance until it overtakes the slow-moving vehicle and re-enters the original lane. For safe overtaking, the driver should be provided a sufficiently long sight distance ahead in order to avoid collision with vehicles coming in the opposite direction.

Overtaking sight distance may be defined as the minimum distance on an undivided highway required by a vehicle/driver to overtake a slow-moving vehicle that is travelling in the same direction safely against approaching traffic in the opposite direction. The minimum required OSD for different design speeds and vehicles have been obtained by AASHTO (1994) based on extensive field studies (Prisk, 1941) and empirical formulas (AASHTO, 1990).

The following standard conditions are assumed (AASHTO, 1994). The average speed differential between the impeding and overtaking vehicles passing at design speed is assumed to be 16 km/h. That means, if an overtaking vehicle is travelling at 100 km/h, the impeding vehicle is assumed to be travelling at 100 – 16 = 84 km/h. The maximum allowable difference of 16 km/h between the design and operating speeds is taken as a widely used assumption in highway design in order to achieve operational consistency and improve driving comfort and safety (Kanellaidis, 1991).

Let three vehicles travel on a two-lane undivided straight road, as shown in Fig. 3.8. The fast-moving vehicle F is travelling at design speed and the slow-moving vehicle S is travelling at a uniform speed slower than F. The third vehicle O is travelling at design speed in the opposite direction on the opposite lane.

$L_1$ = Distance travelled during a PRT of $t_1$ seconds and during the initial acceleration to the point of encroachment on the right lane. The perception time is the duration of time required for the driver of vehicle F to make the decision to overtake the slow-moving vehicle S.

$L_2$ = Total distance travelled by vehicle F from the time it overtakes S, enters the opposite (right) lane, accelerates while travelling along the side of vehicle S for some distance and re-enters the left lane after finishing overtaking.

$L_3$ = Distance between vehicle F and vehicle O at the end of its manoeuvre. It usually ranges from 30 to 90 m (AASHTO, 1994).

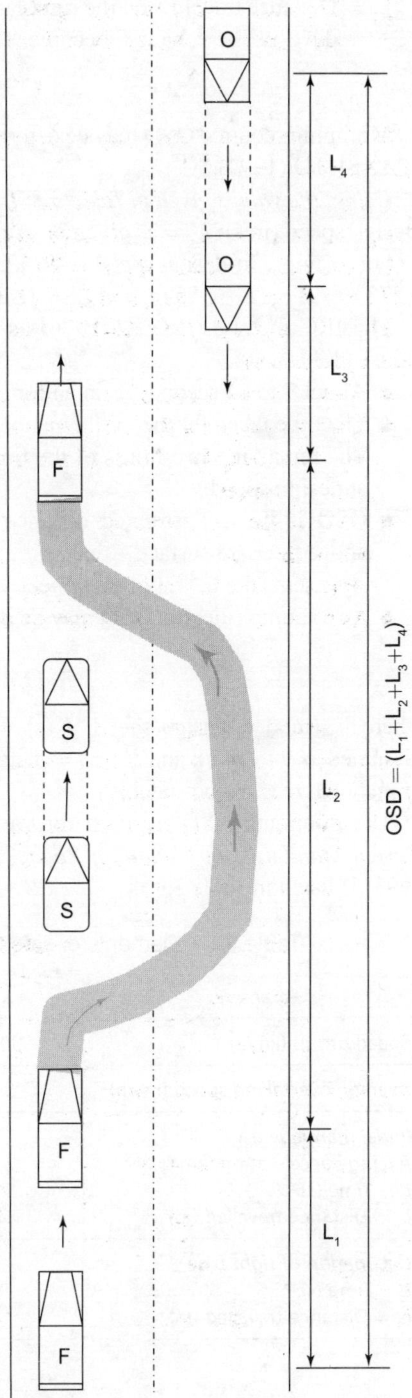


Fig. 3.8 Details of overtaking on a two-lane undivided carriageway

$L_4$ = Distance travelled on the right lane by the opposing vehicle O at design speed for two-thirds of the time the vehicle F occupies the right lane, or equal to $2L_2/3$.

$$OSD = L_1 + L_2 + L_3 + L_4 \tag{3.2}$$

Minimum required OSD may be determined using standard values (Table 3.9) or empirical equations of AASHTO (1994).

*Using the values given in Table 3.8*: $L_1 = t_1$ (s) × Speed of the impeding vehicle (m/s); $L_2 = t_2$ (s) × Design speed (m/s); $L_3$ = as given or suitably assumed; and $L_4 = 2/3^{rd}$ of $L_2$.

For example, if design speed = 70 km/h, $L_1 = 4.0 \times (70 - 16) \times 0.278 = 60$ m; $L_2 = 10 \times 70 \times 0.278 = 195$ m; $L_3 = 55$ m; and $L_4 = (2/3) \times 195 = 130$, OSD = 60 + 195 + 55 + 130 = 440 m.

The IRC method (IRC: 66-1976) assumes the following conditions and recommends safe OSD values (Table 3.9):

- About 9 to 14 seconds is taken for completion of an overtaking manoeuvre.
- The time taken for the initial manoeuvre is about one-third the total time required for overtaking. The remaining two-thirds of the time is accounted for by the opposing vehicle that is travelling at design speed.
- OSD is the unobstructed distance between two points on the carriageway which are 1.2 m above the road surface – one point represents the position of the driver's eye and the other point represents the height of the object.
- As a thumb rule, the OSD may be determined as

$$OSD = (5/3) \times (T\,V_d) + L_3$$

where $V_d$ (m/s) = design speed; $T$ (s) = total time required to complete the overtaking manoeuvre (assumed as 9 to 14 s); and $L_3$ (m) = distance between the overtaking and impeding vehicles at the end of manoeuvre (assume suitably).

The minimum OSD required for various design speeds according to the *Manual of Geometric Design Standards for Canadian Roads*, Transportation Association of Canada (TAC) (Sparks et al., 1993) is listed in Table 3.9.

**Table 3.8** Elements of safe OSD on undivided two-lane two-way carriageways

| *Parameter* | *Value* | | | |
|---|---|---|---|---|
| Speed group (km/h) | 50–65 | 66–80 | 81–95 | 96–110 |
| Average overtaking speed (km/h) | 56.2 | 70.0 | 84.5 | 99.8 |
| *Initial manoeuvre*<br>Average acceleration (km/h/s)†<br>$t_1$ = Time (s)†<br>$L_1$ = Distance travelled (m) | <br>2.25<br>3.6<br>45 | <br>2.30<br>4.0<br>65 | <br>2.37<br>4.3<br>90 | <br>2.41<br>4.5<br>110 |
| *Occupation of right lane*<br>$t_2$ = Time (s)†<br>$L_2$ = Distance travelled (m) | <br>9.3<br>145 | <br>10.0<br>195 | <br>10.7<br>250 | <br>11.3<br>315 |

(Cont'd)

**Table 3.8** Continued

| *Parameter* | *Value* | | | |
|---|---|---|---|---|
| *Clearance length*<br>$L_3$ = Distance travelled (m)[†] | 30 | 55 | 75 | 90 |
| *Opposing vehicle*<br>$L_4$ = Distance travelled (m) | 95 | 130 | 165 | 210 |
| OSD = $(L_1 + L_2 + L_3 + L_4)$ = | 315 | 445 | 580 | 725 |

[†] For consistent speed relation, the observed values have been adjusted slightly.
Adapted from Table III-4 in *Policy on Geometric Design of Highways and Streets*, 1994, by the American Association of State Highway and Transportation Officials (AASHTO), Washington, DC, USA.

**Table 3.9** Overtaking sight distance

| *Speed (km/h)* | *Safe OSD (m) as per Indian Roads Congress (from IRC: 66-1976)* | *Minimum OSD‡ (m) as per the Manual of Geometric Design Standards for Canadian Roads, Transportation Association of Canada (TAC) (from Sparks et al., 1993)* |
|---|---|---|
| 40 | 165 | |
| 50 | 235 | 340 |
| 60 | 300 | 420 |
| 65 | 340 | |
| 70 | | 480 |
| 80 | 470 | 560 |
| 90 | | 620 |
| 100 | 640 | 680 |
| 110 | | 740 |
| 120 | | 800 |
| 130 | | 860 |

[‡] TAC recommends that the operating speed be 10 km/h less than the highway design speed

The following factors/salient features are considered to be helpful in **determining the minimum OSD** (Wang and Cartmell, 1998):

- The longer the passing and/or impeding (slow-moving) vehicle, the longer the OSD. It has been shown that an increase in the length of the impeding vehicle has the same affect on OSD as that of an increase in the length of the passing vehicle.
- The greater the acceleration of the overtaking vehicle, the smaller the OSD. This means that if the overtaking vehicle has an engine of greater capacity, it requires a shorter OSD.
- If the speed of the impeding vehicle is greater, the required OSD is much larger and extra care should be taken.

- As the speed of the opposing vehicle increases, there is a linear increase in OSD.
- It was observed by another study (Sparks et al., 1993) that the values of OSD are significantly ($\approx$ 40%) greater when overtaking a long truck (23–25 m) than when overtaking a car (4.5 m) at an operating speed of 100 km/h. It was also noted that whether a truck is 23 m or 25 m long, it does not have a significant impact on the OSD requirements.
- The differential speed between the overtaken and the overtaking vehicles has the greatest influence on determining the required OSD (Sparks et al., 1993).

Apart from the above cited references, there are other sources where the reader may obtain information on the determination of OSD according to generalised versions of models (Lieberman, 1982; Glennon, 1988) and models based on probability (Sparks et al., 1993).

### 3.5.3 Intermediate Sight Distance (ISD)

At approaches to intersections between the main highway and intermediate or single or unimportant roads, intermediate sight distance (ISD) is provided. ISD is also used when provision of OSD is not possible due to economic reasons or space limitations at the site. According to Indian practices, ISD is taken as two times the value of the safe stopping sight distance (SSD). It is measured between two points on the carriageway which are 1.2 m above the road surface – one point represents the position of the driver's eye and the other point represents the height of the object (IRC: 66-1976). For practical reasons, ISDs are not provided on summits and horizontal curves. If there is no space for SSD, restrictive pavement markings (IRC: 35-1970) or 'No Overtaking' signboards should be installed at appropriate places along the road.

### 3.5.4 Intersection Sight Distance (ISSD)

Estimation of intermediate sight distance becomes more crucial at intersections where no traffic control devices are installed. At such locations, minimum sight distance (xz) should be provided with a suitable margin or limit beyond which buildings should not be constructed, to avoid obstruction of vision along the xy line (Fig. 3.9). The area covered by the xyz triangle is called the sight triangle. As stated before, there should not be any obstructions to vision in this area. The sight distance (xz) is a function of the time gap ($t_g$) needed for a stopped vehicle on a minor road to turn or pass through a major road (Harwood et al., 2000). At point B, a total stop board is provided for vehicles approaching the junction from the minor road. The time-gap acceptance method is specified by AASHTO (2001) for determination of ISSD based on the assumption that (i) the time gap is independent of the approach speed on the major road, (ii) the intersection angle is 90° and (iii) both the major and the minor roads are straight and have no curvature in the horizontal and vertical planes. The required sight distance (xz) is calculated by the following relationship.

$$\text{Sight distance (xz)} = v \times t_g \tag{3.3}$$

where $v$ = the design speed on the major road (m/s) and $t_g$ = the time gap (s) required for a vehicle on the minor road to proceed. For different types of vehicles, the value of $t_g$ may be obtained from *A Policy on the Geometric Design of Highways and Streets* (AASHTO, 2001).

Adequate sight distance should be provided at intersections to enable drivers to view vehicles approaching and decide whether to stop, reduce speed or cross the main road smoothly without collision (in case of un-controlled intersections). According to Indian practices (IRC: 66-1976), the minimum visibility distance (yz) along the minor road is taken as 15 m and the average time gap to travel the

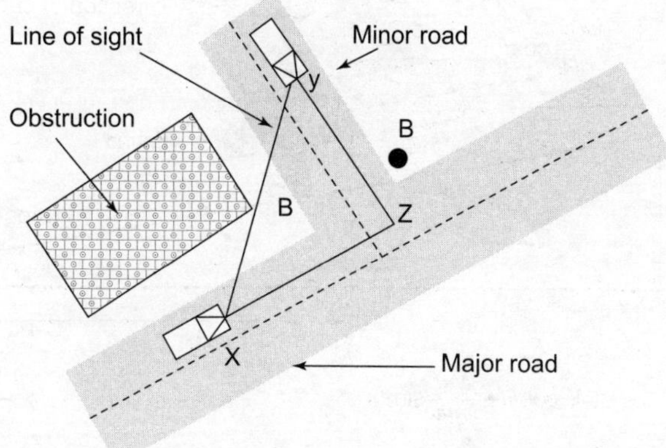


Fig. 3.9 Details of intermediate sight distance

distance (xz) along the major road is assumed to be 8 seconds. For example, for a vehicle standing on a minor road 15 m before a major road, the minimum visibility distance required along the major road is equal to $v \times 8$, where $v$ is the design speed in m/s on the major road.

### 3.5.5 Sight Distance at Medians

In the case of median crossing, vehicles generally stop adjacent to the median and cross or make a U-turn when sufficient distance is available from the opposite on-coming vehicle (B) (Fig. 3.10). For crossing a median, the minimum required sight distance (SD) may be determined from the relationship, SD = $v \times T$, where $v$ = design speed of the vehicle on the opposite lane (m/s); $T = t_p + t_c$; $t_p$ = perception time (s) and $t_c$ = time (s) required to cross or perform a U-turn of a stopped vehicle adjacent to the median.

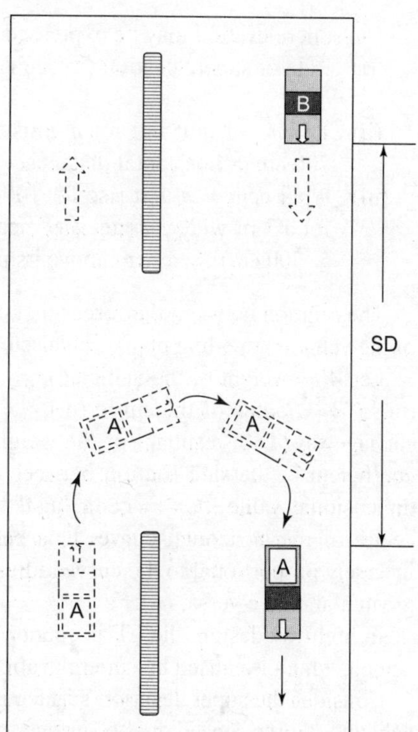


Fig. 3.10 Sight distance at a median turning

## 3.6 SUPERELEVATION

If a vehicle is travelling along a horizontal curve, there is a chance that it may fall, either by overturning or by skidding, due to centrifugal force. The intensity of the centrifugal force depends on the radius of the curve, and the weight and velocity of the travelling vehicle. To counteract the centrifugal force acting on the vehicle while travelling on the horizontal curve, the outer edge of the carriageway surface is elevated with respect to the inner side. This elevated outer edge of the carriageway or the provision of a slope in the transverse direction of the carriageway along a horizontal curve is called **superelevation** ($e$) or cant or banking. The different forces acting on a vehicle travelling on a curved path with respect to its centre of gravity (cg) are presented in Fig. 3.11.

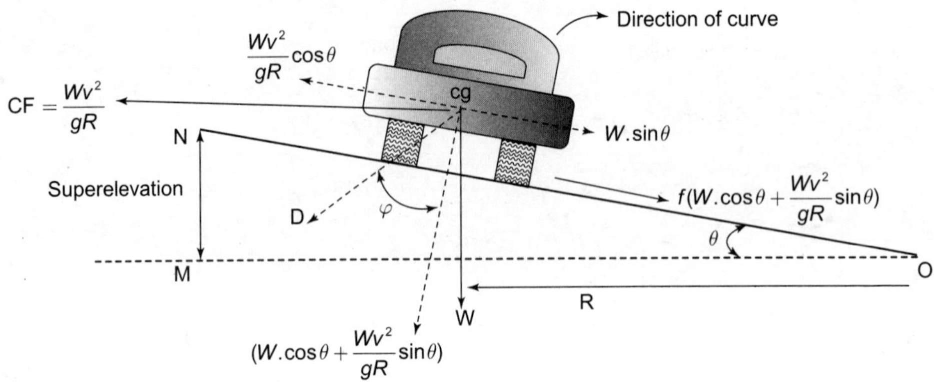


**Fig. 3.11** Forces acting on a vehicle travelling on a circular curve provided with superelevation (not to scale)

The superelevation may be expressed as given below.

(i) Rate of superelevation (in decimal form) $= e = \tan\theta = \frac{\text{MN}}{\text{ON}}$.

(ii) 1 in $n$ = 1 unit rise for $n$ units of horizontal distance; for example, 1 in 700 = 1 cm rise for 700 cm of horizontal distance.

(iii) $n$ per cent = $n$ unit rise for 100 units of horizontal distance; for example, 7% superelevation for a 7-m-wide carriageway means that the outer edge of the carriageway is raised by [(7/100) × 700 cm)] = 49 cm above its inner edge.

The relation between the forces that cause friction across the carriageway and the centrifugal force of the vehicle travelling on a circular curve provided with superelevation is given below.

Let $W$ = weight of the vehicle; $g$ = acceleration due to gravity (m/s$^2$); $v$ = velocity of the vehicle (m/s); $R$ = radius of the curve (m); $\theta$ = angle of the carriageway surface or transverse slope of the carriageway; $D$ = resultant of the weight of the vehicle and the centrifugal force; and $f = \tan(\phi)$ = coefficient of the side friction between the tyres and the road surface or side friction factor (a non-dimensional value). CF = centrifugal force which pushes the travelling vehicle outwards from the centre of the horizontal curve. It is perpendicular to the force of gravity. The centrifugal force is inversely proportional to the curve radius. As the radius of the curve becomes smaller, the CF becomes greater, and vice versa.

In highway design, the CF is counteracted by side friction and the component of the weight of the vehicle which is caused by superelevation (Awadallah, 2005).

Consider the superelevation surface in Fig. 3.11. CF = Horizontal component of the weight of the vehicle + Force caused due to surface friction of the carriageway against the sliding

$$\frac{Wv^2}{gR}\cos\theta = W\sin\theta + f(W\cos\theta + \frac{Wv^2}{gR}\sin\theta)$$

$$\frac{v^2}{gR}\cos\theta = \sin\theta + f\cos\theta + f\frac{v^2}{gR}\sin\theta$$

$$\frac{v^2}{gR}(\cos\theta - f\sin\theta) = \sin\theta + f\cos\theta$$

Dividing both sides by $\cos\theta$

$$\frac{v^2}{gR}(1 - f\tan\theta) = \tan\theta + f$$

$$\frac{v^2}{gR}(1 - f\times e) = e + f \quad (\text{since } \tan\theta = e)$$

$$\frac{v^2}{gR}(1 - 0) = e + f$$

(since $f \times e \approx 0$ when the maximum allowable values of $f$ and $e$ are taken)

$$e + f = \frac{v^2}{gR} \tag{3.4}$$

From the above relationship (Eq. 3.4), for a selected design speed, the minimum required radius of the horizontal curve can be determined by substituting the maximum values of superelevation and side friction factor.

### 3.6.1 Design of Superelevation

According to the IRC, the centrifugal force produced due to 75% of the design speed is taken care of by superelevation and the remaining 25% is counteracted by the surface friction of the carriageway. Based on the above assumption, the rate of superelevation ($e$) is designed using the following formula (IRC: 73-1980).

$$e = \frac{(0.75\,v)^2}{gR} \tag{3.5}$$

According to Indian (IRC: 73-1980) and US guidelines (AASHTO, 1984), design speeds are used for planning the superelevation, whereas in Germany, operating speeds are used (Kanellaidis, 1991). Operating speeds are usually expressed as the 85$^{th}$ percentile of passenger cars in free-flow conditions (Lamm and Choueiri, 1987).

### 3.6.2 Design Rate of Superelevation

Five maximum superelevation rates (12, 10, 8, 6 and 4%) are considered according to US guidelines (AASHTO, 1984). However, a maximum superelevation rate ($e_{max}$) of 8% is taken as the US mean representative value and it is suggested that 4% or 6% be taken for urban areas. Equation (3.4) is used to determine the minimum radius of the curve. According to German guidelines, a 7% rate in general and an 8% rate for exceptional conditions are considered (Kanellaidis, 1991). In Australia and New Zealand, $e_{max}$ is generally taken as 0.1% (Nicholson, 1998). In India, the permissible $e_{max}$ values recommended by IRC (IRC: 73-1980 and IRC: 86-1983) are given below.

- In plain and rolling terrain, including snowbound areas : 7% (1 in 16)
- In hilly areas not bounded by snow : 10% (1 in 10)
- In urban areas with frequent intersections : 4% (1 in 25)

When the superelevation determined from Eq. (3.5) is less than the road camber, the normal camber should be provided along the curved section; the superelevation need not be provided (IRC: 73-1980).

The above differences in the limiting values of superelevation may be because of variability in the speed behaviour of drivers, climatic conditions (warm or cold), carriageway characteristics and terrain conditions prevailing in different countries. Lower values of $e_{max}$ are suitable in cold and snowfall areas. This is because heavy vehicles travelling on curves even at low speeds tend to slide down due to the low coefficients of side friction offered by snowy surfaces (Awadallah, 2005).

**Example 3.2**

Determine the curve radii beyond which no superelevation is provided when the design speed is 70 km/h and the camber is 2%.

**Answer**: $v = 0.278 \times 70 = 19.46$ m/s. Using Eq. (3.5), $e = \dfrac{(0.75\, v)^2}{gR}$. Substitute the $e$ value as equal to camber value.

$$R \geq \frac{(0.75\, v)^2}{g \times (\text{camber})} = \frac{(0.75 \times 19.46)^2}{9.81 \times \left(\dfrac{2}{100}\right)} = 1085.7\,\text{m}$$

∴ For the given conditions, no superelevation is required when the radius is greater than 1086 m.

### 3.6.3 Design Side Friction Factor

The optimal design side friction factor ($f_{max}$) is one that assures safe driving on a horizontal curve having minimum permissible radius, maximum superelevation and design speed (AASHTO, 2001). $f_{max}$ is generally used to define the minimum radius available for the design of the horizontal curve. The friction factor depends on many parameters and can vary spatially and temporally (prevailing weather conditions). Variations in the available values of friction and the friction demanded by different vehicles travelling at different speeds lead to variations in the margin of safety (Nicholson, 1998). For these reasons, $f_{max}$ should be carefully selected. For various design speeds, different values of $f_{max}$ recommended by several highway agencies are shown in Fig. 3.12 (Bonneson, 1999).

As per AASHTO (2001), for the design of low-speed facilities (where the design speed is taken as $\leq$70 km/h), the value of the side friction factor varies from 0.16 to 0.31 – when a minimum design speed of 30 km/h and a maximum design speed of 70 km/h is taken, respectively. In a study, it was observed that on the horizontal curves, drivers accept a higher side friction factor with higher speed reduction (Bonneson, 1999). This means that when there is an increase in speed reduction, there is greater demand for side friction. Therefore, for the design of low-speed facilities such as rural roads and urban streets, appropriate values of side friction factor should be taken to avoid exceeding the intended design speed on curves (Gattis et al., 2005).

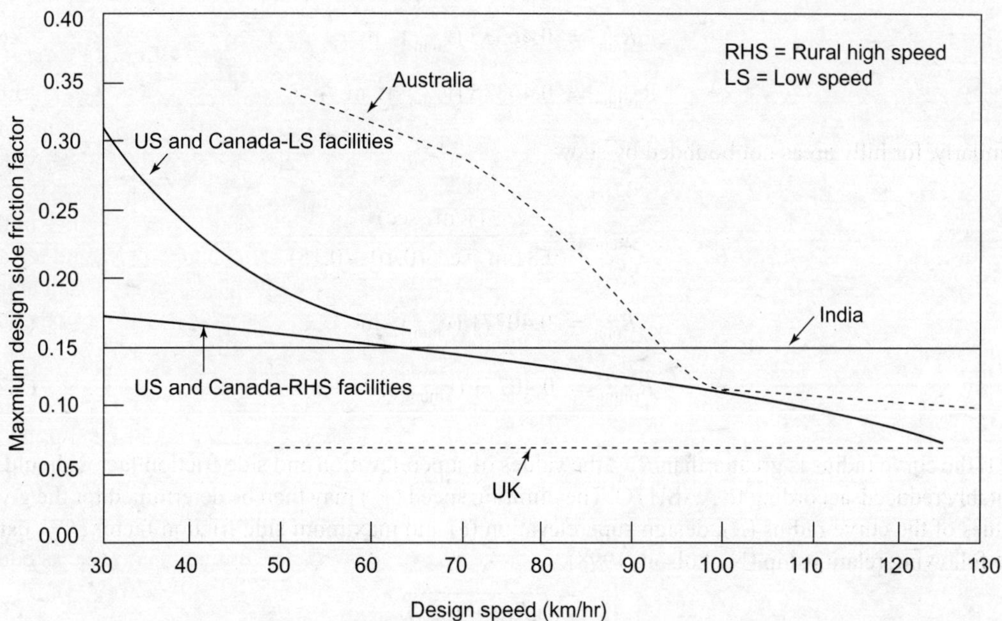


**Fig. 3.12** Maximum design side friction factors recommended by several highway agencies (developed from Bonneson, 1999)

In India, the value of $f_{max}$ recommended for all design speeds is 0.15 (IRC:73-1980). From Fig. 3.12, it can be observed that UK and India are the exceptions, where the values are insensitive to speed; however, the values of $f_{max}$ vary with speed.

## 3.6.4 Analysis of Curve Radii

Ruling and minimum curve radii can be determined using the ruling and minimum design speeds, respectively. Depending on the type of terrain and design speed (Table 3.3), suitable equations may be used (either (3.6) or (3.7), given below). The IRC has recommended radii beyond which no superelevation is required and $R_{min}$ for different terrain conditions (IRC:73-1980).

We have
$$R = \frac{v^2}{g\,(e + f)}$$

The above equation may be expressed as follows.

$$R_{min} = \frac{v^2}{g\ (e_{max} + f_{max})} \tag{3.6a}$$

For plain, rolling terrian and hilly areas, including snowbound locations

$$R_{min} = \frac{(v\ \mathrm{m/sec})^2}{9.81\ \mathrm{m/sec^2}(0.07 + 0.15)}$$

$$R_{min} = 0.46335\ (v_{min})^2\ \text{m} \tag{3.6b}$$

$$R_{ruling} = 0.46335\ (v_{ruling})^2\ \text{m} \tag{3.6c}$$

Similarly, for hilly areas not bounded by snow

$$R_{min} = \frac{(v\ \text{m/sec})^2}{9.81\ \text{m/sec}^2(0.10+0.15)}$$

$$R_{min} = 0.40771\ (v_{min})^2\ \text{m} \tag{3.7a}$$

$$R_{ruling} = 0.40771\ (v_{ruling})^2\ \text{m} \tag{3.7b}$$

If the curve radius is greater than $R_{min}$, the values of superelevation and side friction factor should be suitably reduced, according to AASHTO. The limiting speed ($V_L$) may then be determined for the given values of the curve radius ($R$), design superelevation ($e$) and maximum side friction factor ($f_{max}$) using the following relationship (Nicholson, 1998).

$$V_L = \sqrt{gR(e+f_{max})} \tag{3.8}$$

The difference between limiting speed ($V_L$) and design speed ($v$) can be considered as the design margin of safety. It is also called the **design speed margin of safety (DSMS)**. Variations in safety margin arise on horizontal curves mainly due to variations in available friction and the friction demanded by drivers. When we fix the design criteria for horizontal curves, the maximum allowable coefficient of side friction ($f_{max}$) is generally considered because at this critical condition, overturning/or sliding of vehicles is sufficient to cause discomfort to drivers and thereby restrict higher speeds (Dabbour et al., 2004).

Values of variance and mean of DSMS are determined for a given section of highway consisting of horizontal curves. If the variance of DSMS is small, it indicates that the design consistency is good. If the mean value is large, it indicates that the available safety is better (Nicholson, 1998; Easa, 2003).

**Conditions assumed**

(i) When $f = 0$, $e = v^2/gR$. This means that the total centrifugal force is counteracted by superelevation only because it is assumed that there will be no frictional force. By providing this calculated value of superelevation ($e$) at design speed, the pressure distribution under both wheels will be equal. This condition is termed **equilibrium cant** (Fig. 3.13).

(ii) If $e < v^2/gR$, the total centrifugal force is not counteracted by the provided superelevation. It is assumed that frictional forces will come into play to compensate the same.

(iii) If $e = 0, f = v^2/gR$. Here, it is assumed that surface frictional forces are entirely responsible for counteracting the total centrifugal force. If the $f$ value provided is less than the calculated value, the vehicle may slide down.

**Example 3.3**

For a proposed national highway (NH), determine the minimum required value of the radius when the design speed = 100 km/h, $f_{max}$ = 0.15 and $e_{max}$ = 0.07.

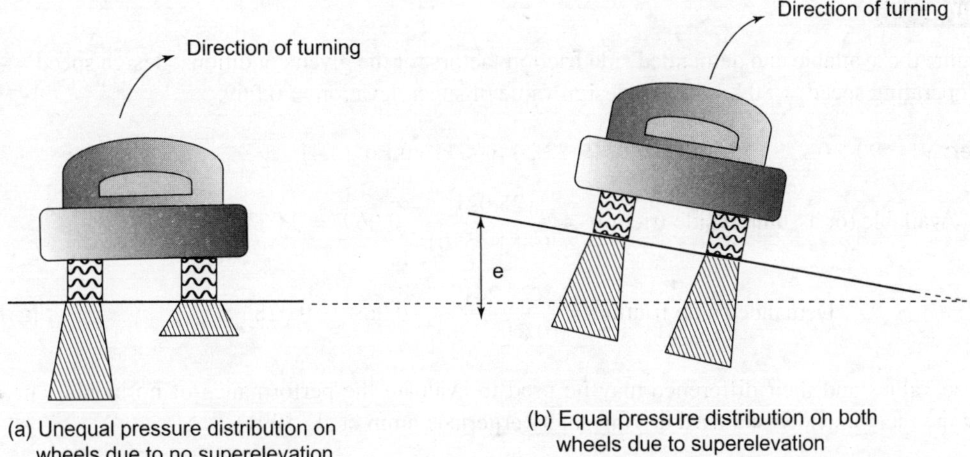


**Fig. 3.13** Variation in wheel-load distribution on the road surface at curves

**Answer**: Let $v = 100 \times 0.278 = 27.8$ m/s. The minimum radius required for the curve corresponding to $V$ is determined by Eq. (3.6b).

$$R_{min} = 0.46335\,(v)^2 = 0.46335\,(27.8)^2 = 358\,\text{m}$$

**Example 3.4**

Determine the design speed margin of safety (DSMS) when the design speed = 90 km/h and the horizontal circular curve radius = 520 m. Assume suitable required values.

**Answer**: Assume $f_{max} = 0.15$ and $e_{max} = 0.07$. Let $v = 90 \times 0.278 = 25.02$ m/s. From Eq. (3.5)

$$e = \frac{(0.75 \times 25.02)^2}{(9.81 \times 520)} = 0.069 < e_{max};\ \text{provide } e = 6.9\%$$

From Eq. (3.4), the actual required value of $f$ is determined as

$$f = \frac{v^2}{gR} - e = \frac{(25.02)^2}{9.81 \times 520} - 0.069 = 0.0537$$

$$\text{Friction utilisation ratio} = 100 \times f/f_{max} = 100 \times 0.0537/\,0.15 = 35.8\%$$

It is determined that the required side friction is less than the assumed value of maximum side friction. Therefore, the design is safe.

The corresponding limiting speed may be determined by Eq. (3.8).

$$V_L = \sqrt{gR(e + f_{max})} = \sqrt{9.81 \times 520(0.069 + 0.15)} = 33.42\ \text{m/s} = 120.32\ \text{km/h}$$

Therefore, the design speed margin of safety (DSMS) = $V_L$ – design speed ($v$)

$$= 120.23 - 90 = 30.32\ \text{km/h}$$

**Example 3.5**

Determine the available and demanded side friction factors for the given conditions: design speed = 90 km/h; operating speed = 80 km/h; and design value of superelevation = 0.069.

**Answer:** $v_d = 90 \times 0.278 = 25.02$ m/s; $v_o = 22.24$ m/s. From Eq. (3.4)

$$\text{Available (or assumed) side friction} = \frac{(25.02)^2}{(9.81 \times 520)} - 0.069 = 0.0537$$

$$\text{Demanded side friction} = \frac{(22.24)^2}{(9.81 \times 520)} - 0.069 = 0.028$$

These values and their difference may be used to evaluate the performance of horizontal curves against the recommended standard driving safety criteria (Lamm et al., 1991).

**Example 3.6**

A national highway (NH) passes through plain, rolling and hilly areas. According to IRC guidelines, the design superelevation for the given condition is:

(i) For plain terrain: Ruling design speed = 80 kmph; horizontal curve radius = 215 m.
(ii) For rolling terrain: Ruling design speed = 80 kmph; horizontal curve radius = 300 m.
(iii) For hilly areas: Ruling design speed = 100 kmph; horizontal curve radius = 137 m.

**Answer:** Assume $e_{max} = 0.07$ and $f_{max} = 0.15$.

(i) *For plain terrain*:

$v = 80 \times 0.278$ kmph = 22.24 m/s. From Eq. (3.5)

From Eq. (3.5) $$e = \frac{(0.75 \times 22.24)^2}{(9.81 \times 215)} = 0.1319 > e_{max}$$

Therefore, provide $e = e_{max}$ and check the value of $f_{max}$.

From Eq. (3.4) $$\text{Available } f = \frac{v^2}{gR} - e_{max} = \frac{22.24^2}{(9.81 \times 215)} - 0.07 = 0.164 > f_{max}$$

Therefore, determine maximum allowable speed when $e = e_{max}$ and $f = f_{max}$ values are taken

$$V_L = \sqrt{9.81 \times 215(0.07 + 0.15)} = 21.54 \text{ m/s} = 77.48 \text{ kmph}$$

(ii) *For rolling terrain*:

$v = 80 \times 0.278$ kmph = 22.24 m/s.

From Eq. (3.5) $$e = \frac{(0.75 \times 22.24)^2}{(9.81 \times 300)} = 0.094 > e_{max}$$

Therefore, take $e = e_{max} = 0.07$ and check the value of $f_{max}$.

From Eq. (3.4) Available $f = \frac{v^2}{gR} - e = \frac{22.24^2}{(9.81 \times 300)} - 0.07 = 0.098 < f_{max}$

Provide $e = 0.07$ for the given speed.

(iii) *For hilly areas*:

$$v = 100 \times 0.278 \text{ kmph} = 27.8 \text{ m/s}.$$

From Eq. (3.5) $e = \frac{(0.75 \times 27.8)^2}{(9.81 \times 137)} = 0.080 < e_{max}$ where $e_{max} = 0.1$.

Therefore, provide superelevation rate $= 0.08 \times 100 = 8\%$.

**Example 3.7**

For achieving the best surface drainage conditions, economy in construction and environmental compatibility with surroundings, it was proposed to provide negative superelevation on a horizontal curve of a multi-lane highway. Determine the minimum radius of the horizontal curve given the following conditions: operating speed of vehicles = 60 km/h, $e_{max} = -2.4\ \%$ and $f_{max} = 0.08$.

**Answer:** $v = 60 \times 0.278 = 16.68$ m/s; $e_{max} = -2.4\ \% = -0.024$ and $f_{max} = 0.08$. From Eq. (3.6a)

$$R_{min} = \frac{16.68^2}{9.81 \times (-0.024 + 0.08)} = 506.4 \text{ m}$$

### 3.6.5 Attaining Superelevation

Superelevation may be attained by adopting any one of the methods given in Table 3.10. The choice of method depends on the situation encountered.

A reference line or axis of roadway along the inner edge or outer edge or centre line is considered the datum line or axis of rotation. About the axis of rotation, the respective points of the roadway surface are revolved in two stages.

In the first stage, the outer half of the cambered roadway surface is gradually rotated to the horizontal (i.e., normal cross slope = 0) and the inner half of the roadway surface is kept at the normal camber. This means that the normal camber on the outer half of the roadway is removed (Hinkle, 1930) and gradually made horizontal over a length of road. This is what is termed a **tangent run out** ($L_{TR}$). In the second stage, the horizontally placed outer half of the roadway surface is gradually rotated about the axis of rotation till it matches the slope of the inner half of the roadway surface (at Q in Fig. 3.14). Further, the whole inclined roadway surface is revolved about the axis of rotation till full superelevation is attained at the point when the circular curve begins. The curve between the end of the tangent run out and the beginning of the circular curve is called the **spiral curve** (or transition curve) and its length is termed the **superelevation run off** or length of run off ($L_R$). This length is determined based on the superelevation rate and design speed. Details of how superelevation is attained by rotating the roadway surface about the centre line of a highway with a combined curve are explained below.

**Table 3.10** Suitability of different methods for attaining superelevation

| *Method of attaining superelevation** | *Remarks* |
|---|---|
| Rotation of roadway surface about its inner edge<br>Inner edge taken as axis of rotation | • Facilitates drainage when the inner edge level of the roadway is required to be unaltered<br>• Provides uniform appearance of the superelevated curve<br>• Used more often in earth-filling operations<br>• Centre line of the carriageway may have to be shifted along the length of the curve<br>• Suitable for multi-lane highways having medians and ramps with baseline |
| Rotation of roadway surface about its outer edge<br>Outer edge taken as axis of rotation | • Used more often in earth-cutting operations<br>• Care should be taken to drain off the accumulated rain water at the inner portion of the curve<br>• Centre line of the carriageway may have to be shifted along the length of the curve<br>• Suitable for multi-lane highways having medians and ramps with baseline |
| Rotation of a roadway surface about its centre line (as illustrated in Fig. 3.14 (b)) | • Most frequently adopted method, since the least distortion occurs to the construction centre line of the roadway. Thus, the vehicle maintains stability while manoeuvering the curve. |

*Reference: IRC: 73-1980 and IRC:86-1983.

From Fig. 3.14, it is observed that the superelevation provided is gradually increased to its full value in stages – (i) point $T_1$ to point P and (ii) point P to junction point A of transition and the circular curves. The full value of superelevation is provided on the circular curve portion from point A to junction point B of the circular and transition curves. Similarly, on the down slope, the full value of superelevation (at B) is deceased to zero from points B to Q', P' and $T_2$.

Let us consider the combined curve (Fig. 3.14), where a circular curve is augmented with similar spiral curves at either ends. A normal camber of C (in per cent) is provided up to the point $T_1$. The width of the carriageway is termed $W$ (in metres). The maximum allowable longitudinal slope is $G$ (in per cent).

The relationships required for superelevation design with reference to Fig. 3.14 (b) may be obtained as as follows:

From the fundamental principles of rectangular hyperbola (Taiganidis and Kanellaidis, 1991)

$$\frac{L_{TR}}{L_R} = \frac{e_{T1}}{\text{full } e} = \frac{Y_{T1}}{Y_A} \tag{3.9}$$

where $e_{T1}$ and $e_A$ are the values of superelevation (%) having depths $Y_{T1}$ and $Y_A$ at points $T_1$ and A, respectively. For this condition, $e_{T1} = C_n$ (normal camber in %) and $e_A$= full $e$ (full superelevation in %).

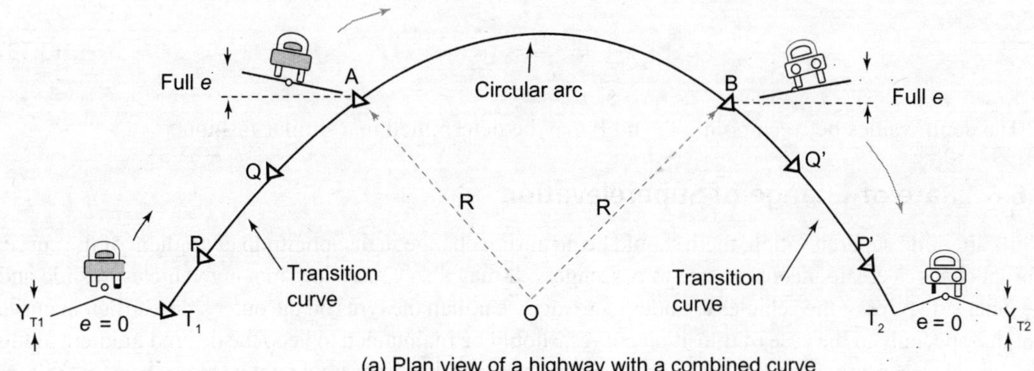


(a) Plan view of a highway with a combined curve

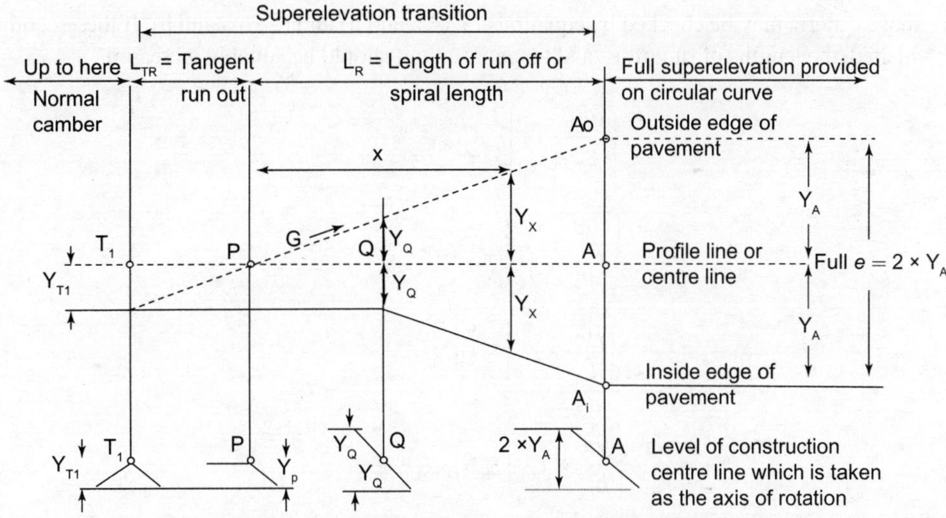


(b) Profile of the roadway along the highway as given in (a).

**Fig. 3.14** Superelevation design by rotating the roadway about the centre line (not to scale)

$$\text{Length of superelevation transition } (L_{ST}) = L_{TR} + L_R \tag{3.10}$$

$$\text{Depth, } Y_{T1} = Y_p = Y_Q = \frac{W}{2} \times \frac{C_n(\%)}{100} \tag{3.11}$$

$$\text{Depth, } Y_A = \frac{W}{2} \times \frac{\text{full } e\ (\%)}{100} \tag{3.12}$$

From any horizontal distance $x$ units from the point P, the superelevated depth $Y_x$ can be determined based on the principle of similar triangles by considering the triangle $PAA_0$.

$$\frac{Y_x}{Y_A} = \frac{x}{L_R} \tag{3.13}$$

The depth values between points $T_1$ and P may be determined in a similar fashion.

### 3.6.6 Rate of Change of Superelevation

While attaining superelevation, there should be no abrupt change in the longitudinal gradient of the curve. Abrupt changes create inconvenience to passengers. It may also cause slow-moving vehicles to slide and the inner side tyres of the vehicles to undergo greater wear than the tyres on the outer side. Sufficient curve length, especially in the case of transition curves, should be maintained to keep the desired gradient along the curves. According to the IRC, the rate of change of superelevation should not be more than 1 in 150 ($\approx$ 0.67 %) for plain and rolling terrain ($G_{\text{P\&R}}$) and 1 in 60 ($\approx$ 1.67 %) for hilly or mountainous terrain ($G_{\text{H\&M}}$). The above criteria may be checked by equations (3.14 a and b) and (3.15 a and b). If these conditions are not satisfied, the lengths of run off or/and tangent run out should be suitably increased.

$$\left(\frac{100}{L_R} \times Y_A\right) \leq G_{\text{P\&R}}\ \% \tag{3.14a}$$

$$\left(\frac{100}{L_{TR}} \times Y_{T1}\right) \leq G_{\text{P\&R}}\ \% \tag{3.14b}$$

$$\left(\frac{100}{L_R} \times Y_A\right) \leq G_{\text{H\&M}}\ \% \tag{3.15a}$$

$$\left(\frac{100}{L_{TR}} \times Y_{T1}\right) \leq G_{\text{H\&M}}\ \% \tag{3.15b}$$

where $G_{\text{p\&R}}$ and $G_{\text{H\&M}}$ are longitudinal grades in plain and rolling, and hilly and mountainous terrain, respectively.

According to normal practice, the length of the transition curve is made longer than the length of the run off. However, the length of the superelevation transition ($L_{ST}$) should be long enough to make the pavement rotation from the normal crown to the full superelevated rate imperceptible to drivers (Medina and Tarko, 2007).

Where no transition curve is provided, the superelevation ($e$) should be attained gradually over a length – two-thirds of the value of superelevation should be attained on a straight section before the beginning of the circular curve and the remaining one-third value should be attained on the curve (AASHTO and IRC: 52-2001). For better speed operating conditions near circular curves, 70–90% of the run off length should be placed over the tangent length (AASHTO, 2004). To view superelevation designs for multi-lane divided highways, AASHTO recommendations may be referred.

**Example 3.8**

Determine the minimum length of the tangent run out, when camber = 2%, rate of design superelevation = 7% and length of spiral = 90 m.

**Answer**: From Eq. (3.9), let us assume

Length of spiral = Length of superelevation run off ($L_R$)

$$\frac{L_{TR}}{90\,\text{m}} = \frac{2\,\%}{7\,\%}; \quad \therefore L_{TR} = 25.7\,\text{m}$$

**Example 3.9**

Determine the values of superelevation at salient locations along a combined curve with reference to Fig. 3.15 and check for maximum allowable longitudinal gradient according to the IRC guidelines. The pavement consists of an undivided dual lane in rolling terrain and has a width of 7.0 m. The other conditions are the same as given in the above problem.

**Answer**: From the above problem, we get $C = 2\%$; full $e = 7\%$; $L_R = 90$ m and $L_{TR} = 25.7$ m. Assume that full superelevation should be provided throughout the length of the circular curve. The notations followed will be as given in Fig. 3.14 (a) and (b).
From Eq. (3.10)

Length of superelevation transition ($L_{ST}$) = 25.7 + 90 = 115.7 m.

From Eq. (3.11) $\quad$ Depth $Y_{T1} = Y_p = Y_Q = \frac{7.0}{2} \times \frac{2\,\%}{100} = 0.07\,\text{m} = 7\,\text{cm}$

From Eq. (3.12) $\quad$ Depth $Y_A = \frac{7.0}{2} \times \frac{7\,\%}{100} = 0.245\,\text{m} = 24.5$ cm

As an example, let us take a point at a horizontal distance of 50 m from the point P. The depth $Y_{@50m}$ can be determined by Eq. (3.13).

$$\frac{Y_{@50m}}{0.245\ \text{m}} = \frac{50\ \text{m}}{90\ \text{m}}; \quad Y_{@50m} = 0.136\,\text{m} = 13.6\ \text{cm}$$

Similarly, other depths at required intervals may be determined. Check for maximum allowable longitudinal gradient limit. As the highway is in rolling terrain, the value of $G_{P\&R} \approx 0.67\%$. From Eq. (3.14 a and b)

$$\left(\frac{100}{90\ \text{m}} \times 0.245\ \text{m}\right) \le G_{P\&R}; \qquad 0.272\,\% \le G_{P\&R};\ or$$

$$\left(\frac{100}{25.7\ \text{m}} \times 0.07\ \text{m}\right) \le G_{P\&R}: \quad 0.272 \le G_{P\&R}$$

For the given conditions, the calculated longitudinal gradient value is less than the maximum allowable limit. Hence, the design is safe.

## 3.7 WIDENING OF CARRIAGEWAY ON CURVES

There are many reasons why carriageways are widened on curves:

- When a four-wheeled vehicle turns along a horizontal curve, the rear wheels do not pass exactly over the same path traced by the front wheels because the rigid-based axle assembly of the rear wheels tracks along the inner side of the curve. Therefore, the carriageway must have some additional width on curves to accommodate this deviated path of the rear wheels.

- As a safe margin on curves, drivers try to keep an offset distance on both sides of the vehicles. This extra space is maintained while travelling on curves – the extent of the extra space depends on the driver's psychological characteristics.

Let us consider the right-angled triangle ABC in Fig. 3.15. The interior angle at B is 90º.

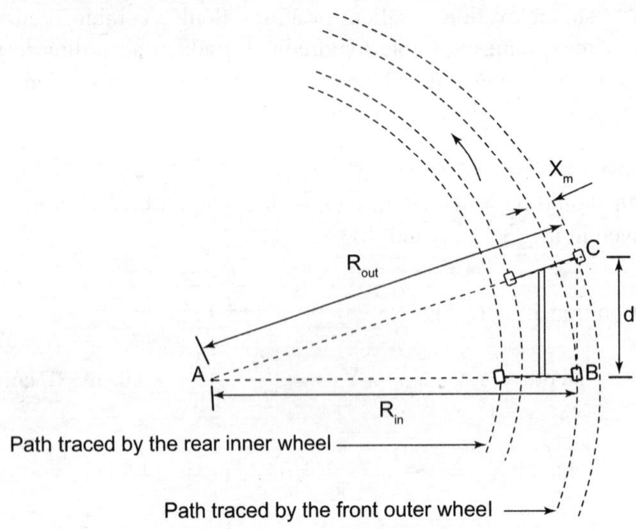


**Fig. 3.15** Widening of carriageway on curves

Let $d \approx$ BC = wheel base length (m); $R_{out}$ = AC = radius of path traced by the outer front wheel (m); $R_{in}$ = AB = radius of the path traced by the outer rear wheel (m); $X_m = R_{out} - R_{in}$ = mechanical widening of carriageway (m); and $R_c$ = radius of the curve with respect to the centre line of the carriageway or mean curve radius (m).

From triangle ABC

$$AB^2 = AC^2 - BC^2$$

$$R_{in}^2 = R_{out}^2 - d^2$$

$$(R_{out} - X_m)^2 = R_{out}^2 - d^2$$

If $R_{out} = R_{in} = R_c$, and we simplify the above equation, $X_m$ = mechanical widening = $0.5d^2/R_c$ (approximately). If the number of traffic lanes is $n$, the above equation can be rewritten as

$$X_m = 0.5nd^2/R_c \tag{3.16}$$

In addition to the mechanical widening of curves, it is practical to increase the value of widening to allow margin space during steering and overtaking operations. This extra widening of the carriageway on curves is termed **psychological widening** ($X_p$). Based on field studies and experience, the IRC has suggested the following formula for determining $X_p$:

$$X_p = 0.36\, v/\sqrt{R} \tag{3.17}$$

where $v$ = design speed in m/s.

Total widening, $X_e$ (in meters) = Mechanical widening ($X_m$) + Psychological widening ($X_p$)

$$X_e = 0.5nd^2/R_c + 0.36\, v/\sqrt{R_c} \tag{3.18}$$

According to the IRC: 73 guidelines, extra widening of curves is not required if the curve of the radius is longer than 60 m and 300 m for single and double-lane carriageways, respectively. In the case of hill roads with sharp radii of curves, additional width may be provided along the inner side of the curve. The extra width should be equally provided on both sides of the carriageway. The full width of extra widening is attained gradually – starting from zero at the beginning of the curve and reaching its full value in the middle of the curve; from there it is gradually reduced to zero at the end of the curve.

## 3.8 HORIZONTAL CURVES

Horizontal curves are used to divert highway alignments from the straight route to any other path. The length and type of curve will be decided based on the location of the site and the horizontal alignment requirements. Horizontal curves are classified based on their geometry as follows:

- Simple circular curves
- Compound curves and multi-compound (three-centred) curves
- Reverse curves
- Transition curves
- Mixed curves

The first three types of curves originate from the fundamental geometric principles of circular arcs. The simple circular arc consists of an arc with a constant finite radius throughout its length. The typical geometry of a circular curve is shown in Fig. 3.16.

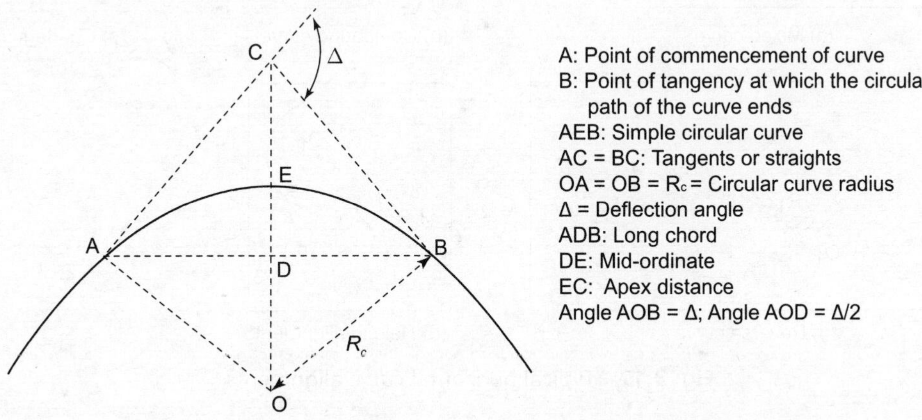


**Fig. 3.16** Elements of a simple circular curve

The well-known standard relationships for determining various geometric elements of circular curves are given below:

$$\text{Tangent length} = \text{AC} = \text{BC} = R_c \tan\left(\frac{\Delta}{2}\right) \tag{3.19}$$

$$\text{Length of long chord} = \text{ADB} = 2R_c \sin\left(\frac{\Delta}{2}\right) \tag{3.20}$$

$$\text{Mid-ordinate} = \text{DE} = R_c\left(1 - \cos\left(\frac{\Delta}{2}\right)\right) \tag{3.21}$$

$$\text{Apex distance} = \text{CE} = R_c\left(\sec\left(\frac{\Delta}{2}\right) - 1\right) \tag{3.22}$$

$$\text{Length of circular curve} = \text{AEB} = \frac{\pi R_c \Delta}{180^\circ} \tag{3.23}$$

Circular curves are generally adopted for slow-moving traffic whereas compound curves are used where deviation of alignment is inevitable. A compound curve consists of a combination of two or more circular arcs having different radii, joined together in the same direction. Their curve centres fall on the same side of the curves.

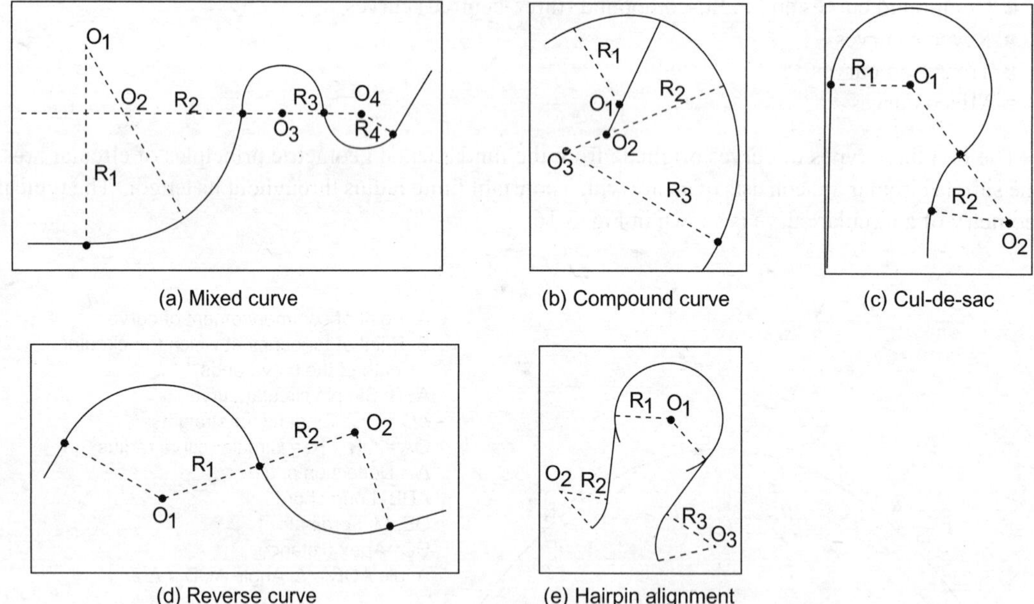


**Fig. 3.17** Typical horizontal curve alignments

In the case of reverse curves, the circular arcs of different (or same) radii are connected with their centres lying in opposite directions (Fig. 3.17 (d)).

The use of reverse curves in highway alignment is favourable for the alignment of mainline or freeway interchange ramps, where there is minimal need for acquiring land (Dabbour et al., 2004). In the case of sharp curves, a sudden change in direction results in arduous steering where the available sight distance is also insufficient. To satisfy the demand for high driving speeds and automatic drives, transition spiral curves are used in the alignment of high-speed highways such as national highways and expressways.

### 3.8.1 Sight Distance on Horizontal Curves

When a vehicle travels on a horizontal curve, the driver's line of vision is limited by obstructions, such as buildings or any immovable structure, situated at the inside of the curve. Therefore, for horizontal curves, as a safe geometric design criterion, the sight distance provided along the curve should always be longer than the actual distance required for stopping a vehicle or for performing further action. The sight distance availability is governed by the following parameters (Easa, 1993):

- Geometric features of the horizontal curve, such as *Rc*, Δ, etc.
- Location of the obstacle causing vision obstruction
- Location of the driver with respect to the obstacle and an approaching object considered on the curve

The sight distance also depends on the design speed, characteristics of the vehicle, the driver and the complexity of the location (AASHTO). When an obstacle is located at the mid-point of the long chord of a circular curve on a single-lane carriageway, clear visibility in the lateral direction is obtained by providing a setback distance (DE) between the centre line of the carriageway and the point of obstruction (Fig. 3.18). IRC: 52-2001 has recommended different setback distances for different values of sight distance (as per Eq. (3.1a)) at different parameters. For the simple case considered in Fig. 3.18, the required setback distance (DE) may be determined by Eq. (3.21).

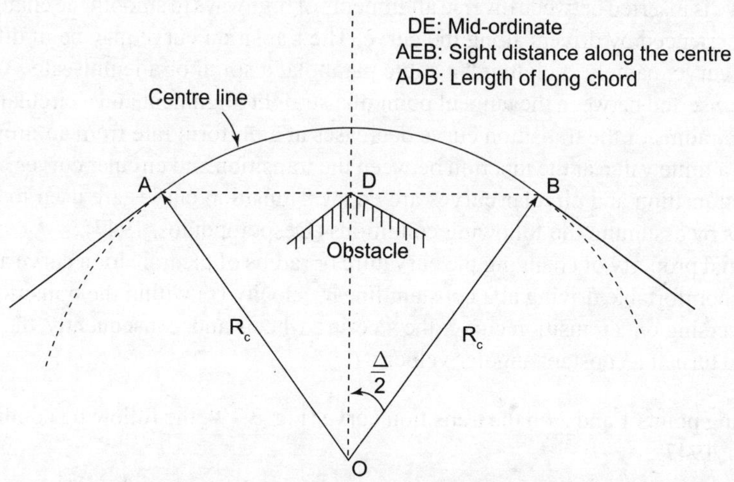


**Fig. 3.18** Sight distance along a circular curve

The available sight distances along circular curves for different locations of obstacles may be evaluated according to the AASHTO geometric design guidelines. Available and minimum sight distances on circular curves may also be obtained by exact solutions (Easa, 1991). Sight distances at intersections which are controlled by a stop sign may be determined using Design Aids – software which help to design and evaluate transportation systems (Easa et al., 2004).

Minimum sight distances on compound horizontal curves having different radii and for obstacles at different positions may be determined using relationships and design tables available in literature (Easa, 1993). In extreme cases, it is preferable to change the alignment, so as to avoid obstructions altogether.

**Example 3.10**

Determine the required setback distance when a two-lane–wide carriageway road has a curve radius of 550 m along its centre line and minimum stopping sight distance of 180 m.

**Answer**: With reference to Fig. 3.18, the given values are interpreted as follows: $R_c = 550$ m; sight distance $S$ = Curve length (AEB) = 180 m; EB = $R_c \times \Delta/2$.

Therefore $\Delta/2 = \text{EB} / R_c = 90/550 = 0.1636$ radians $\times$ (180°/π) = 9.373°

By Eq. (3.21), the required setback distance from the centre line of the road

$$= \text{DE} = 550\left(1-\cos\left(9.373\right)\right) = 7.34\,\text{m}$$

Otherwise, the following equation may be used to calculate the setback distance from the centre line of the road.

$$\text{DE} = \text{R}_c\left(1-\cos\left(\frac{28.647\times S}{\text{R}_c}\right)\right) = 550\left(1-\cos\left(\frac{28.647\times 180}{550}\right)\right) = 7.34\,\text{m}$$

### 3.8.2 Transition Curve

A transition curve is inserted between diverse alignments of highways to smooth the change in centripetal acceleration experienced by drivers along the curve. The transition curve may be in different forms of a mathematical curve, namely, a clothoid, a cubic parabola, a spiral or a lemniscate. As an example, a transition curve inserted between the tangent point of a straight alignment and a circular curve is shown in Fig. 3.19. The radius of the transition curve decreases at a uniform rate from an infinite value at the tangent point to a finite value at the junction between the transition and circular curves. At this junction, the radii of the transition and circular curves are equal. Transition curves are used to align the centre line of highways by assuming the following conditions (Despotopoulos, 1947).

- The essential property of changing the curvature or radius of the transition curve at a uniform rate facilitates comfortable driving at a constant linear velocity ($v$) within the transition.
- While traversing on a transition curve, the steering wheels and, consequently, the front wheels of the vehicle turn at a constant angular velocity ($\omega$).

By considering points 1 and 2 on the transition curve (Fig. 3.19), the following conditions hold good (Despotopoulos, 1947).

$$R_1 \times \omega \times t_1 = R_2 \times \omega \times t_2 = \text{constant; and } R_1 \times \frac{L_1}{v}$$

$$= R_2 \times \frac{L_2}{v} = \text{constant} \qquad (3.24\text{ a, b})$$

where $R_1$ and $R_2$ are the radii of curvature at points 1 and 2, respectively. $L_1$ and $L_2$ are the lengths from the beginning of the transition curve corresponding to times $t_1$ and $t_2$. However, the above relationships do not satisfy the given condition at the beginning point (or tangent point) of the transition curve.

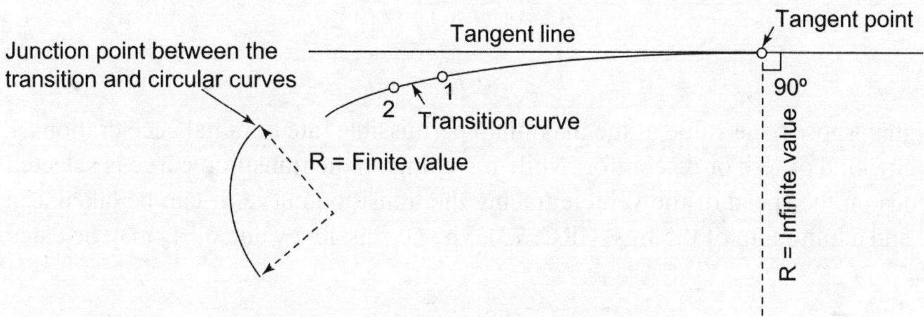


**Fig. 3.19** A typical combined curve

Transition curves with spirals are preferred in highway alignments for the following reasons (Craus and Polus, 1977; Lipičnik, 1998).

- Superelevation on the curved path can be increased at a uniform rate, which is the same as the rate of decrease of radius of the transition curve.
- They help avoid the sudden impact of centrifugal forces by providing a smooth change consistent with the vehicle's path.
- The centrifugal force is the reciprocal of the change in alignment curvature. When the speed of vehicles is assumed as constant, non-uniform change in the magnitude of the side pressure takes place due to abrupt changes in curvature. The differential side pressure (or thrust) has a detrimental effect on the surface of the road as well as on vehicle performance. In such alignment locations, transition curves should be inserted suitably.
- They facilitate gradual change in the turning angle of steering wheels.
- They provide safe and comfortable travel on curves, even at high speeds.

Following are the possible (or favourable) cases where transition curves can be inserted in the alignment of highways (Lipičnik, 1998):

- Between two straight parts of an alignment
- Between circular curves aligned in the same and apposite directions having the same or different radii
- Between the alignment of a straight part and a circular curve

## 3.8.3 Length of Transition Curve

Two considerations are important in calculating the required length of a transition curve: the rate of change of radial acceleration and the rate of change of superelevation. Based on these, the length of the transition curve can be calculated using two methods. The longer length of transition curve is recommended.

**By maximum allowable rate of change of radial acceleration:** Assume that the design speed and rate of increase of centripetal acceleration are constant (Greenstein, 1976). Let the maximum radial acceleration at the end of the transition curve (i.e., at the beginning point of the circular curve) be equal to $v^2/R_c$, where $v$ is the velocity of the vehicle in m/s. The radial acceleration will be zero at the tangent point when $R$ takes an infinite value.

$$\text{Time taken to travel on a length of transition curve (s)} = t = Lt/v$$

$$\text{Rate of change of radial acceleration } (A_R) = \text{Radial acceleration / Time taken to travel the distance}$$

$$A_R = (v^2/R_c) / (Lt/v)$$

Therefore $$L_t \geq v^3/(R_c \times A_R) \tag{3.25}$$

For design purposes, the value of the maximum permissible rate of radial acceleration ($A_R$) that does not impart any kind of jerk or discomfort while travelling on the transition curve is selected. The value of $A_R$ depends on the speed of the vehicle tracing the transition curve. It can be taken as a maximum of 0.8 m/s$^3$ and a minimum of 0.5 m/s$^3$ (IRC: 73). The permissible value of $A_R$ may be calculated using the formula

$$A_R = 80/(75 + 3.6\,v) \tag{3.26}$$

**By maximum allowable rate of change of superelevation:** To determine the minimum required length of the transition curve ($L_t$) according to the guidelines of IRC: 73 and as discussed in Section 3.6.6, the following relationships may be used

$$L_t \geq \left(\frac{100}{G_{P\&R}\ \%} \times Y_A\right) \quad 3.27)$$

$$L_t \geq \left(\frac{100}{G_{H\&M}\ \%} \times Y_A\right) \tag{3.28}$$

These notations carry the same meaning as explained in Section 3.6.6. They hold good when the superelevation is attained by rotating the surface of the pavement about its centre line.

**Example 3.11**

A horizontal curve of 270 m radius is to be designed for a speed of 65 kmph. The road is a two-lane road (7 m) and the superelevation is provided by rotating the pavement surface about its crown. The rate of introduction of superelevation is 1 in 150. If the maximum wheel base length is 6 m, calculate the required minimum length of the transition curve.

**Answer:** Design speed, $v = 65$ kmph $= 65 \times 0.278 = 18.07$ m/s

*Criterion 1:* By rate of change of acceleration (Eqs (3.25) and (3.26)),

$$L_t = \frac{v^3}{A_R \times R_c}$$

where $$A_R = \frac{80}{75+3.6\,v} = \frac{80}{(75+3.6\times 18.07)} = 0.571$$

$$\text{Take } A_R = 0.571 \quad (\because A_{R\min} = 0.5 \text{ and } A_{R\max} = 0.8)$$

$$L_t = \frac{18.07^3}{0.571\times 270} = 38.27 \text{ m}$$

Therefore, take $L_t = 39$ m.

*Criterion 2:* Based on the rate of change of superelevation, by Eq. (3.5)

Superelevation
$$e = \frac{(0.75\,v)^2}{g\,R_C} = \frac{(0.75 \times 18.07)^2}{9.81 \times 270}$$
$$= 0.0693 \text{ m} < e_{max} = 0.07 \text{ m}$$

$\therefore$ Provide full $e = 6.93\%$

Widening of carriageway on curve: From Eq. (3.18)

$$X_e = \frac{0.5 \times 2 \times 6^2}{270} + \frac{0.36 \times 18.05}{\sqrt{270}}$$
$$= 0.133 + 0.395 = 0.528 \text{ m}$$

Now, the width of carriageway, $W = W + X_e = 7.0 + 0.528 = 7.528\,\text{m}$

From Eq. (3.12) Depth, $Y_A = \frac{W}{2} \times \frac{\text{full } e\ (\%)}{100}$

$$= \frac{7.528}{2} \times \frac{6.93\ \%}{100} = 0.260\,\text{m}$$

From Eq. (2.27) $L_t \geq \left(\frac{100}{G_{P\&R}\ \%} \times Y_A\right)$

$$G_{P\&R}\ \% = \frac{1}{150} \times 100 = 0.67\%\ \text{(given)}$$

$$= \left(\frac{100}{0.67\ \%} \times 0.260\right) = 38.8 \text{ m}; \quad \therefore L_t \geq 39 \text{ m}$$

Therefore, provide final length of transition curve ($L_t$) = 40 m

***Note:*** Always select the higher value from the above criteria.

### 3.8.4 Elements of the Transition Curve

Generally, a spiral curve is the ideal transition curve because its radius of curvature changes inversely as the curve length increases from the tangent point. Figure 3.20 presents the basic elements of a combined curve having a circular curve adjoined by transition curves at either end.

Let $L_t$ = length of the transition curve; $L_c$ = length of the circular curve; P and P'= tangent points of transition curves; C and C' = junction points of transition and circular curves; $R$ = radius of the circular curve and radius of the transition curve at points C and C'; $X_Q$ and $X_T$ = distance along the tangent from point P; $Y_c$ = perpendicular ordinate from point T to point C; $\alpha$ = spiral angle with the line PI by a tangent drawn at C; $\theta$ = angle subtended by the circular arc with radius $R_{c;}$ $r$ = radius of the transition curve at a given point; and $\ell_t$ = any curved length of transition curve measured from origin P, having radius *r*.

From the fundamental principle of a clothoid spiral $r \propto \frac{1}{\ell_t}$.

It can also be written as $\quad r \times \ell_t = R_c \times \ell_t = C$, where $C$ is a constant (3.29)

The rectangular coordinates of the transition curve at $\ell_t$ measured from origin P may be written mathematically as (Fig. 3.20)

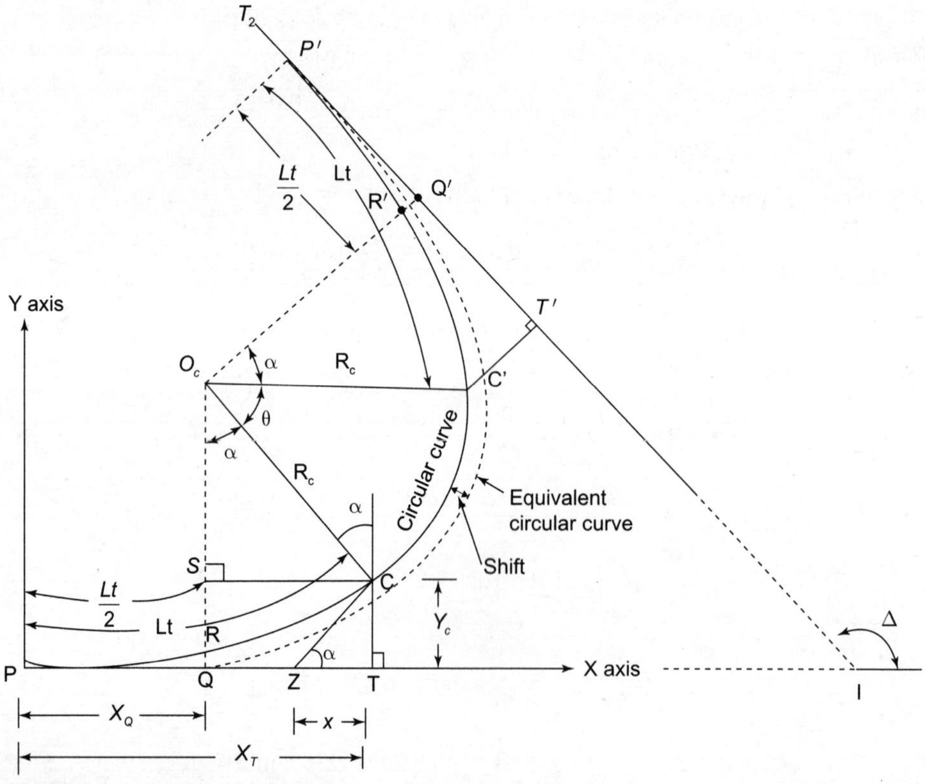


**Fig. 3.20** Elements of a horizontal combined curve having transition and circular curves (Elaborated based on Roy, 1992)

$$X = \ell_t - \frac{\ell_t^5}{40C^2} + \frac{\ell_t^9}{3456C^4} - \frac{\ell_t^{13}}{599040C^6} + \ldots \tag{3.30}$$

$$Y = \frac{\ell_t^3}{6C} - \frac{\ell_t^7}{336C^3} + \frac{\ell_t^{11}}{42240C^5} - \frac{\ell_t^{15}}{9676800C^7} + \ldots \tag{3.31}$$

Using the above coordinates, various elements of transition curves, as given in Fig. 3.20, may be determined approximately from the following relationships.

$$X_T = L_t - \frac{L_t^3}{40\,R_c^2} \tag{3.32}$$

$$X_Q = \frac{L_t}{2} - \frac{L_t^3}{1280\,R_c^2} \tag{3.33}$$

$$Y_C = \frac{L_t^2}{6\,R_c} - \frac{L_t^4}{336\,R_c^3} \tag{3.34}$$

where $X_T$ and $X_Q$ are the distances measured along the tangent line from the origin P to points T and Q, respectively, and $Y_c$ is the perpendicular distance between the tangent lines from points T and point C.

The other parameters of the transition circular curve may be determined based on the concept of equivalent circular curve, as given below (Roy, 1992).

From triangle $SCO_c$ $$\alpha = \sin^{-1}\left(\frac{SC}{CO_c}\right) = \sin^{-1}\left(\frac{QT}{R_c}\right)$$

$$\alpha = \sin^{-1}\left(\frac{X_T - X_Q}{R_c}\right) \tag{3.35}$$

$$\text{The distance, } QO_c = QS + SO_c = Y_C + R_c\cos(\alpha) \tag{3.36}$$

$$\therefore \quad \text{Shift} = QO_c - R_c \tag{3.37}$$

From triangle ZTC $$\text{Distance, } ZT = x = Y_C\cot(\alpha) \tag{3.38}$$

$$\text{Tangent distance, } ZC = Y_C\,\text{cosec}(\alpha) \tag{3.39}$$

$$L_c = \frac{\pi R_c \theta}{180^\circ} \tag{3.40}$$

$$\text{Total length of combined curve} = 2 \times L_t + L_c \tag{3.41}$$

$$\text{Total length of tangent, PI} = PQ + QI$$

$$PI = X_Q + (QO_c) \times \tan\left(\frac{\Delta}{2}\right) \tag{3.42}$$

$$\text{where } \Delta = 2\alpha + \theta \tag{3.43}$$

The design of highway transition curves can also be carried out using methods based on the unit curve concept (Rao, 1995) and by other methods (García and Conesa, 1997; Lipičnik, 1998).

The design of two- and three-centred compound curves, with and without transition spirals at the points of change of curvature, can be found in literature (AASHTO, 2001; Roy, 1994).

A generalised solution to determine various elements of hairpin alignments having transition curves is also available (García and Conesa, 1992).

**Example 3.12**

Determine the required length of the transition curve and calculate the salient elements of a combined curve consisting of a circular curve joined with a transition curve to either end. The required curve design inputs are as follows.
Road class = National Highway; Design speed = 50 kmph
Terrain and climate = Snowbound mountainous terrain; Road type = Undivided two lanes, 7.0 m wide
Radius of the horizontal curve ($R_c$) = 160 m

Angle of deflection between two tangents of the proposed combined curve ($\Delta$) = 64°
Chainage at the point of intersection (I) = 500.00 m
Permissible rate of change of superelevation = 1 in 60 ($G_{H\&M} \approx 1.67\%$)
The superelevation is attained by the rotation of the pavement surface about its centre line. Take the maximum permissible superelevation as 7%.

**Answer:** This problem is solved in two parts. The first part consists of calculating the length of the transition curve using two criteria. In the second part, elements of the combined curve are determined. The notations used here are with reference to Fig. 3.20.

$$\text{Design speed} = 50 \text{ kmph} = 50 \times 0.278 = 13.9 \text{ m/s}$$

**Part I:**

*Criterion 1*: By rate of change of radial acceleration and Eqs (3.25) and (3.26)

$$\text{Take } A_R = 0.639 \quad (\because A_{R\min} = 0.5 \text{ and } A_{R\max} = 0.8)$$

$$L_t = \frac{13.9^3}{0.639 \times 160} = \frac{2685.619}{102.24} = 26.27 \text{ m.}$$

Therefore, take $L_t = 27$ m.

$$L_t = \frac{v^3}{A_R \times R_c}$$

where

$$A_R = \frac{80}{75 + 3.6\,v} = \frac{80}{(75 + 3.6 \times 13.9)} = 0.639$$

*Criterion 2*: Based on the rate of change of superelevation by Eq. (3.5)

$$\text{Superelevation, } e = \frac{(0.75\,v)^2}{g\,R_c} = \frac{(0.75 \times 13.9)^2}{9.81 \times 160}$$

$$= 0.0692 \text{ m} < e_{\max} = 0.07 \text{ m}$$

$$\therefore \text{Provide, full } e = 6.92\%$$

Widening of carriageway on curve: From Eq. (3.18)

$$X_T = \frac{0.5 \times 2 \times 6^2}{160} + \frac{0.36 \times 13.9}{\sqrt{160}}$$

$$= 0.225 + 0.395 = 0.62 \text{ m}$$

Now, the width of the carriageway, $W = W + X_T = 7.0 + 0.62 = 7.62$ m

From Eq. (3.12) $\quad \text{Depth, } Y_A = \dfrac{W}{2} \times \dfrac{\text{full } e\ (\%)}{100} = \dfrac{7.62}{2} \times \dfrac{6.92\%}{100} = 0.263 \text{ m}$

From Eq. (3.23), $L_t \geq \left(\frac{100}{G_{H\&M}\ \%} \times Y_A\right)$

$$G_{P\&R}\ \% = \frac{1}{60} \times 100 = 1.67\ \%\ \text{(given)}$$

$$L_t \geq \left(\frac{100}{1.67\ \%} \times 0.263\right) = 15.7\ \text{m}; \qquad \therefore L_t \geq 15.7\ \text{m}$$

***Note***: Select the higher value from the above criteria.

Therefore, provide final length of transition curve $(L_t) = 27$ m.

**Part II:** Various geometric elements of the combined curve consisting of a circular curve joined with transition curves at either end are determined below. The notations used here are the same as those used in Fig. 3.20.

From Eq. (3.33)
$$X_Q = \frac{L_t}{2} - \frac{L_t^3}{1280\, R_c^2} = X_Q$$
$$= \frac{27}{2} - \frac{27^3}{1280 \times 27^2} = 13.4994\ \text{m}$$

From Eq. (3.32)
$$X_T = 27 - \frac{27^3}{40 \times 160^2} = 26.9808\ \text{m}$$

From Eq. (3.34)
$$Y_C = \frac{L_t^2}{6\, R_c} - \frac{L_t^4}{336\, R_c^3}$$
$$= \frac{27^2}{6 \times 160} - \frac{27^4}{336 \times 160^3} = 0.75899\ \text{m}$$

From Eq. (3.35)
$$\alpha = \sin^{-1}\left(\frac{X_T - X_Q}{R_c}\right)$$
$$= \sin^{-1}\left(\frac{26.9808 - 13.4994}{160}\right) = 4.8334°$$

From Eq. (3.36) Distance, $QO_c = QS + SO_c = 0.75899 + 160\cos(4.8334°)$
$$= 160.1900\ \text{m}$$

From Eq. (3.7) $\text{Shift} = QO_c - R_c = 160.1900 - 160 = 0.1900\ \text{m}$

From Eq. (3.42), Tangent length of combined curve (PI)

$$= X_Q + (QO_c) \times \tan\left(\frac{\Delta}{2}\right)$$

$$= 13.4994 + (160.1900)\ \tan\left(\frac{64^\circ}{2}\right) = 113.5972 \text{ m}$$

From Eq. (3.43) $\theta = \Delta - 2\alpha = 64^\circ - 2 \times 4.8334^\circ = 54.3332^\circ$

From Eq. (3.40) Length of circular curve $= L_c = \frac{\pi R_c \theta}{180^\circ} = \frac{\pi \times 160 \times 54.3332^\circ}{180^\circ} = 151.7269$ m

From Eq. (3.41) Length of combined curve $= 2 \times L_t + L_c$

$= 2 \times 27 + 151.7269 = 205.7269$ m

**Chainage at salient points on the combined curve:**

Chainage at point I = 500.00 m (given)

Chainage at point P = Chainage at I – PI = 500.00 – 113.5972 = 386.4028 m

Chainage at junction point (C), at the end of the transition curve and the beginning of the circular curve

= Chainage at P + $L_t$ = 386.4028 + 27 = 413.4028 m

Chainage at junction point (C$^1$), at the end point of the circular curve and the beginning of the second transition curve = Chainage at C$^1$ + $L_c$ = 413.4028 + 151.7269 = 565.1297 m

Chainage at point, P$^1$ = Chainage at C$^1$ + $L_t$ = 565.1297 + 27 = 592.1297 m

## 3.9 LONGITUDINAL GRADIENT

Generally, the natural ground consists of undulations or uneven surfaces. To make a level surface, it becomes necessary to fill up or cut the uneven ground surface according to the existing ground profile. Depending on the type of terrain, the magnitude of ground rise or fall is observed. This rate of rise or fall of ground along its length is termed **longitudinal slope** or gradient or grade. Profiles of natural ground and desired pavement surfaces are shown in Fig. 3.21.

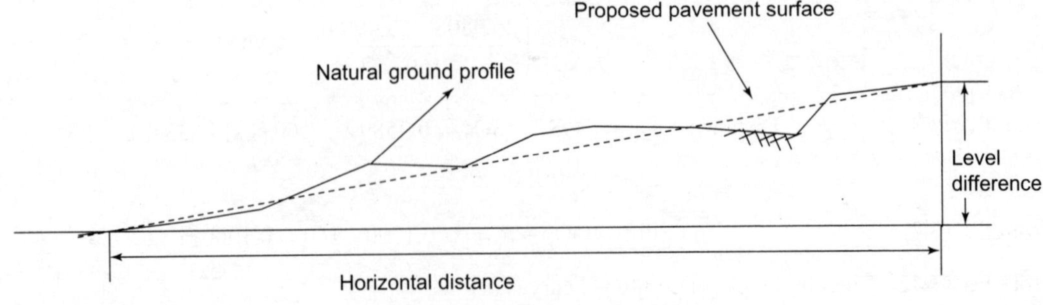


**Fig. 3.21** Longitudinal gradient

The gradient can be expressed in the following forms

1. In percentage: Grade (%) $= \dfrac{\text{Level difference}}{\text{Horizontal distance}} \times 100$
2. 1 in N: 1 vertical in $N$ horizontal distance
3. Vertical : Horizontal

**Factors affecting gradient**

- Existing ground profile and surrounding floor levels of developed areas, especially with reference to surface and sub-surface drainage
- Levels of control, mandatory and obligatory points along the desired alignment
- Type and importance of the highway
- Design speed and type of vehicles anticipated, i.e., bullock carts or motor vehicles
- Operating cost of vehicles and energy saving
- Cost of construction and practical problems with the surrounding locality
- Excessive excavation and high embankments
- Natural sub-soil, water table and artesian conditions
- Earthquake conditions

**Types of gradients**

- *Ruling gradient*: It is the maximum allowable grade or longitudinal slope considered by a designer during alignment of a highway in the vertical plane. This gradient is also termed the design gradient. The magnitude of the ruling gradient depends on the type of terrain, design speed, curves and altitude.
- *Limiting gradient:* This gradient is slightly steeper than the ruling gradient. Limiting gradient is provided when the natural ground profile is slightly steeper than the ruling gradient. It is necessary to adopt a steeper gradient, to keep the earthwork operation at minimum. The limiting gradient is generally adopted to connect plain and rolling terrains or to reach hilly areas.
- *Exception gradient:* This gradient is adopted exceptionally where the ground profile frequently changes its slope and needs an abrupt or a sudden rise or fall in vertical alignment. Exceptional gradient is still steeper in slope than the limiting gradient. Providing this gradient for a long distance at a time is inconvenient during ascent and more tractive effort is required from the vehicle to overcome the frictional and gravitational forces. Hence, exceptional gradient is provided for a short distance not exceeding 100 m in length at a stretch. Successive sections should be separated by flatter gradients for a distance of at least 100 m (IRC: 52-2001). These gradients are used in steep hilly areas.
- *Minimum gradient:* To drain off surface or rain water in side drains, it is necessary to provide a certain amount of longitudinal slope. A minimum amount of gradient is provided based on the surface characteristics of drains, intensity of rainfall, soil type, type of terrain and existing average floor level of the developed area. A minimum gradient of 0.5% and 1.0% is provided for lined and unlined side drains, respectively (IRC: 73-1980).

The values of the various gradients based on the terrain type are presented in Table 3.11 (IRC: 73-1980).

**Table 3.11** Graphical presentation of road gradients, a typical example (Modified from IRC: 66-1976)

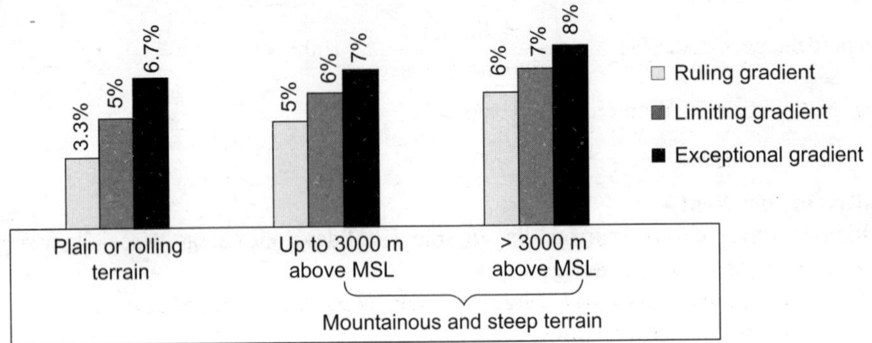


## 3.10 COMPENSATION OF GRADE ON HORIZONTAL CURVES

To facilitate easy driving on horizontal curves in hilly areas, a certain amount of grade is deducted from the design gradient value. Gradients steeper than 4% should be eased. The grade compensation value is determined using the following formula (IRC: 73-1980).

$$\text{Grade compensation} = (30 + R_c)/\, R_c \tag{3.44}$$

$$\text{The maximum allowable value of grade compensation} = 75/\, R_c \tag{3.45}$$

The method of determination of grade compensation is explained below.

**Example 3.13**

A ruling gradient of 1 in 18 is to be provided for a hill road having a horizontal curve radius of 70 m. What is the value of grade compensation?

**Answer:** Design gradient = (1/18) ×100 = 5.55 % (given)

Radius of horizontal curve, $R_c = 70$ m (given)

By Eq. (3.44) Grade compensation $= (30 + R_c)/\, R_c = (30 + 70)/\, 70 = 1.428\%$

From Eq. (3.45)

Maximum allowable grade compensation $= 75/\, R_c = 75/\, 70 = 1.071\%$

Now, select the least value from the above.

Grade compensation value = 1.07%. Therefore, net grade = 5.55 – 1.071 = 4.479%

## 3.11 VERTICAL CURVES

Proper design of vertical alignment of highways is important to ensure adequate sight distance and to provide uniform change in grade. This will facilitate a safe, smooth and comfortable ride over the road. Circular and parabolic curves are used in the vertical alignment of highways. Parabolic curves are preferred as the rate of change of slope along its length is constant and the sight distance available is more than that available for a circular curve having equal length (Fig. 3.22). Calculation of reduced levels along the parabolic curve is simple and easy to set out. Therefore, vertical axis parabolas are generally used for the design of vertical alignment of highways.

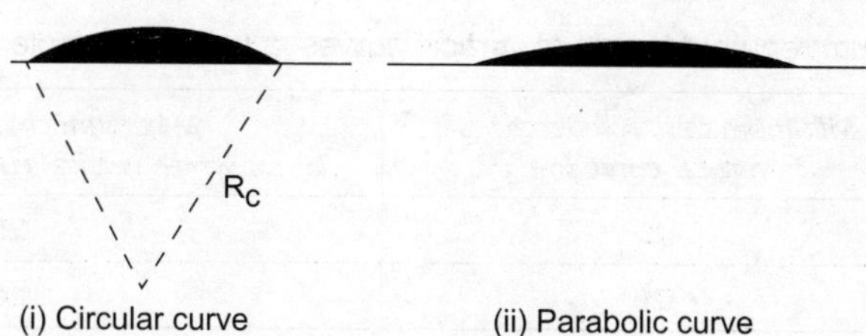


**Fig. 3.22** Availability of sight distance on a circular and on a parabolic vertical curve

### 3.11.1 Types of Vertical Curves

Based on the nature and intensity of steepness of adjacent grade lines, two types of vertical curves are formed (Fig. 3.23): summit (or crest) curves and sag (or valley) curves. An intersection point of adjacent grade lines is termed the **vertical point of intersection** (I). When the elevation (or reduced level, RL) of the point I is higher than or equal to any point on the vertical curve, such curves are termed **summit curves**. When the elevation of the point I is lower than or equal to any point on the vertical curve, such curves are termed **sag curves**.

The point of curvature (which is at the curve beginning) and the point of tangency (which is at the curve end point) are termed $T_1$ and $T_2$, respectively. The nature of the gradient ($g$%) in terms of positive and negative signs indicates the upward and downward slopes, respectively, of the ground. $G$ is the angle of deviation of adjacent grade lines which is equal to the algebraic difference of the tangent grades $|g_2 - g_1|$ and its value is expressed in decimal form. The above notations are used in the following paragraphs.

### 3.11.2 Length of Vertical Curves

It is the horizontal distance ($L$) measured between points $T_1$ and $T_2$ with reference to the provision of minimum sight distance ($S$) along the centre line of a highway to facilitate safe and efficient operations of vehicles at minimum cost of construction. Hence, determination of the length of the vertical curve is significant while designing the alignment of highways. In the case of lower grade change, vertical curves with minimum required length (IRC: 52-2001) should be provided for aesthetic reasons (Table 3.12).

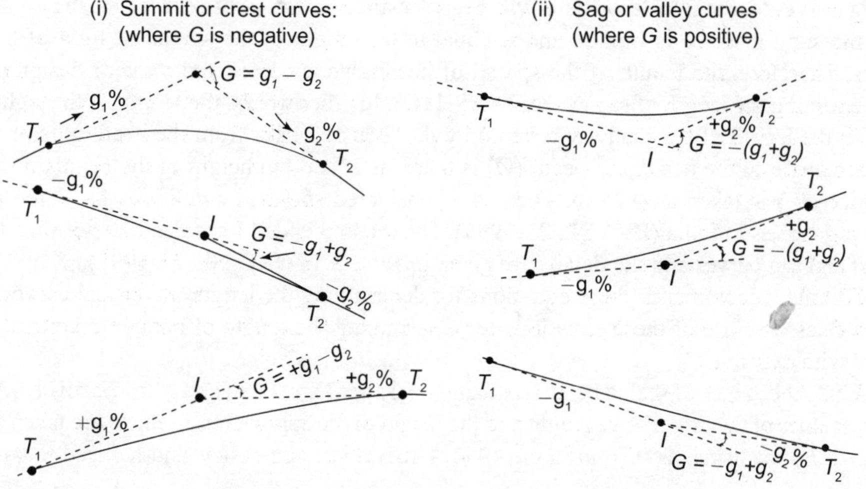


**Fig. 3.23** Possible cases of forming vertical road curves based on adjacent grade lines

**Table 3.12** Minimum required length of vertical curves, a typical example (Modified from IRC: 52-2001)

| *Design speed (m/sec)* | *Minimum desirable length of vertical curve (m)* | *Maximum change in grade (G) for which vertical curve is not required (%)* |
|---|---|---|
| ≤9.72 | 15 | 1.5 |
| 13.88 | 20 | 1.0 |

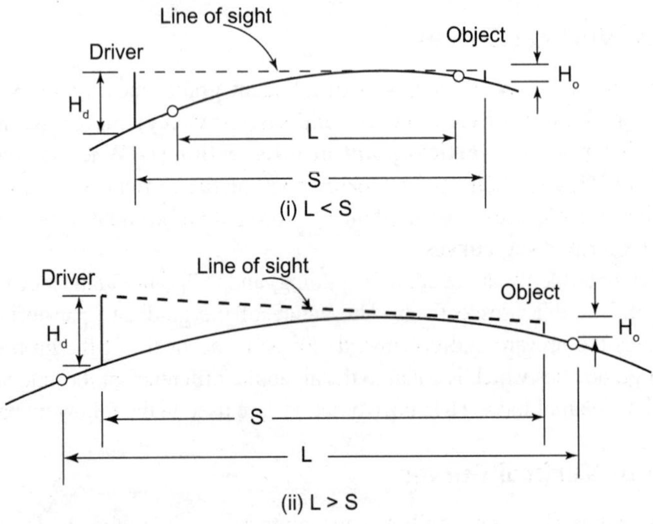


**Fig. 3.24** Design criteria for summit (or crest) curves

Sight distance ($S$) is a major design parameter used for determining the length of the summit (or crest) curves. The main reason is that the driver's visibility (or sight distance) is restricted by the high point of the curve.

For sag curves, during the day there will be no restriction for sight (except during snowfall) but at night, the sight distance is limited and is equal to the length of the spread of light of the vehicle headlights. Therefore, the length of the spread of headlights is considered a major design parameter while determining the length of sag curves (AASHTO). In other words, the length of the sight distance depends on the length of the roadway covered by the focussed light from the headlights of a vehicle. The upward angle of the headlamp beam (Ø) is taken as 1° and the height of the headlamp above the road surface ($H_h$) is taken as 0.75 m. These are considered standard conditions for determining the length of sag curves in India (IRC: SP: 23-1993). The recommended formulae to determine the length of summit and sag curves and the details are given in Table 3.13 (IRC: SP: 23-1993).

AASHTO also recommends these equations for determining the length of vertical curves, but with different values. The use of these equations for determining the length of parabolic vertical curves is illustrated with examples.

In the USA, the values of sealed beam headlamp ($H_h$) and Ø are 0.6 m and 1°, respectively (AASHTO, 2004). The values of the driver's eye height and the height of the approaching vehicle are taken as 1.08 m. The value of $H_h$ was taken as 0.76 m in the 1954 AASHTO design policy, and reduced to 0.6 m in the

**Table 3.13** Formulae to determine the length of summit and sag curves

<table>
<tr><th>Sl.</th><th>Equation for minimum curve length</th><th>For Indian conditions</th></tr>
<tr><td colspan="2">For summit curves (Eqs (3.46) and (3.47)); refer to Figs 3.24 and 3.25</td><td rowspan="6">$H_d$ = Height of the driver's eye level above the road surface<br>$H_o$ = Height of the obstacle above the road surface<br>If $S$ = ISD = OSD, $H_d = H_o$ = 1.2 m<br>If $S$ = SSD, $H_d$ =1.2 m, $H_o$ = 0.15 m<br>$G = |g2 - g1|$, angle of deviation in decimal value<br>$H_h$ = 0.75 m, height of headlight above the road surface<br>Ø = 1° = Angle of spread of headlight or light cone–focussed angle<br>The values of $S$ may be obtained from design speed relationships or taken according to IRC guidelines.</td></tr>
<tr><td>1</td><td>If $L>S$; $L = \dfrac{|G|S^2}{2(\sqrt{H_d}+\sqrt{H_o})^2}$</td></tr>
<tr><td>2</td><td>If $L<S$; $L = 2S - \dfrac{2(\sqrt{H_d}+\sqrt{H_o})^2}{|G|}$</td></tr>
<tr><td colspan="2">For sag curves (Eqs (3.48) and (3.49))</td></tr>
<tr><td>3</td><td>If L>S; $L = \dfrac{|G|S^2}{2(H_h + S\ \tan\theta)}$</td></tr>
<tr><td>4</td><td>If L<S; $L = 2S - \dfrac{2(H_h + S\ \tan\theta)}{|G|}$</td></tr>
</table>

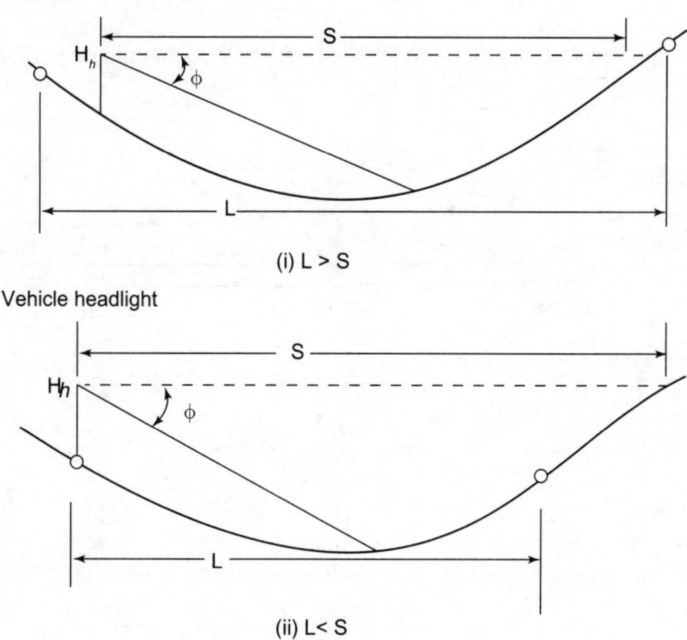


**Fig. 3.25** Design criteria for sag curves

1965 AASHTO design policy; it has remained the same since then. This reduction had the effect of lengthening sag curves by about 3–10% at different design speeds. In view of the latest developments in headlamp technology, a study in the USA (Hawkins, 2007) revealed that when $H_h$ = 0.6 m, it is more appropriate to take the Ø value as 0.75° rather than 1°. The effect would lengthen sag curves by 15–30% at different design speeds.

Truck and cab-over-engine tractor driver's eye height ($H_d$) may be taken as up to 2.75 m and the object height ($H_o$) may be taken as 0.15 or 0.46 m in North America (Easa, 1992).

### 3.11.3 Sight Distance on Sag Curves Passing Under an Overhead Structure

If a sag curve passes under an overhead structure, the structure may restrict the sight distance. The following equations may be used to find the exact solution of the minimum sight distance ($S$) on a sag curve with a centred overpass.

- According to Hickerson (1964), when $S < L$, the minimum sight distance ($S$) is determined based on

$$\frac{S}{L} = \frac{O_o + c - \left(\dfrac{H_d + H_o}{2}\right)}{2 \times O_o} \tag{3.50}$$

The similar triangles principle is used for the development of the above relationship.

- According to Easa (1992), when $S > L$ (Fig. 3.26)

$$S = \frac{1}{/G/}\left[\sqrt{(c + O_o - H_d)} + \sqrt{(c + O_o - H_o)}\right]^2 \tag{3.51}$$

where $c$ = vertical clearance; $O_o$ = as given by Eq. (3.54c); the remaining parameters are as given in Fig. 3.26.

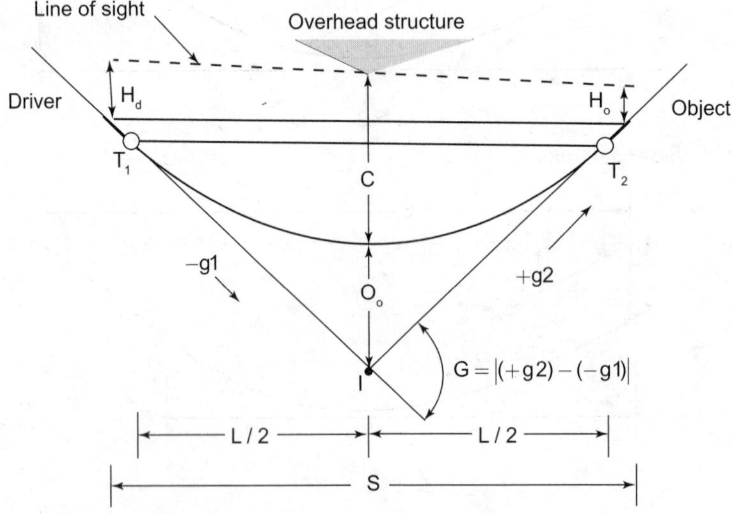


**Fig. 3.26** Sight distance on sag curves when passing under an overhead structure; condition $S > L$ (modified based on Hickerson (1964) and Easa (1992))

For normal Indian conditions, the minimum value of vertical clearance ($c$) at underpasses is taken as 5 m. To accommodate double-decker buses, this value should be increased to 5.5 m (IRC: 54-1974).

According to American practice, the minimum vertical clearance for highways with trucks is taken as 4.42 m (AASHTO).

Normally, sag vertical curves are designed based on the headlight criterion while minimum required sight distance is checked when the sag curve passes under an overhead structure (AASHTO, 2004). Determination of the minimum sight distance under the overpass is useful for improving existing conditions to facilitate better operational and cost-effective measures (Neuman and Glennon, 1984).

Thus far, we have discussed vertical curves which are symmetrical and consisting of a constant rate of change of grade. Unsymmetrical vertical curves will be formed when the lengths of tangents are unequal ($T_1I \neq IT_2$) and the curve consists of two arcs having unequal length with two different rates of change of grade connected by a common tangent point. The design of such unsymmetrical vertical curves may be based on sight distance and drainage requirements (Easa and Hassan, 1998).

Fig. 3.26 (a) A sag vertical road curve passing under a railway bridge

**Example 3.14**

A vertical summit curve is formed when an ascending gradient of 1 in 30 meets a descending gradient of 1 in 40. The curve is to be designed to provide OSD for a design speed of 80 km/h. Calculate the suitable length of the summit curve. Assume suitable data. (JNTU, APR/MAY 2005)

**Answer:** Ascending gradient $(+g1)$ = 1 in 30 = (1/30) × 100 = 3.33%

Descending gradient $(-g2)$ = 1 in 40 = (1/40) × 100 = 2.50%

Angle of deviation of adjacent grades

$G = (-g2) - (g1) = -2.5 - 3.33 = -5.833\% = -0.05833$; $|G| = 0.05833$

Assume Indian standard conditions. For the design speed of 80 km/h, assume minimum OSD value as $S = 470$ m (Table 3.9). Assume $L < S$. If $S$ = OSD, $H_d = H_o = 1.2$ m. From Eq. (3.47)

$$L = 2S - \frac{2\,(\sqrt{H_d} + \sqrt{H_o})^2}{|G|} = 2(470) - \frac{2(\sqrt{1.2} + \sqrt{1.2})^2}{0.05833} = 940 - \frac{2(4.8)}{0.05833} = 775.42 \text{ m}$$

Therefore, our assumption of $L < S$ is not correct, since $S = 470$ m. Now, reassume $L > S$ (or $S < L$). From Eq. (3.46)

$$L = \frac{|G|S^2}{2(\sqrt{H_d} + \sqrt{H_o})^2} = \frac{0.05833 \times (470)^2}{2(\sqrt{1.2} + \sqrt{1.2})^2} = \frac{12885.097}{2 \times 4.8} = 1342.19 \text{ m}$$

< 470 m (assumed value)
Therefore, the assumption of $L > S$ is correct. Hence, provide length of the curve = 1343 m.

**Example 3.15**

A summit curve is formed when an ascending gradient of 1 in 20 meets an ascending gradient of 1 in 80. If the value of SSD = 110 m, calculate the total length of summit curve.

**Answer:** Ascending gradient (+g1) = 1 in 20 = (1/20) × 100 = 5.00% (given)

Ascending gradient (+g2) = 1 in 80 = (1/80) × 100 = 1.25% (given)

$$G = g2 - (g1) = 1.25 - 5.00 = -3.75\% = -0.0375$$

$$|G| = 0.0375$$

Assume, SSD = $S$ = 110 m (given)

Length of the curve = $L$; Height of the driver's eye level above the road surface, $H_d = 1.2$ m; Height of an object above road surface, $H_o = 0.15$ m. Assume $L < S$. By Eq. (3.47)

$$L = 2S - \frac{2(\sqrt{H_d} + \sqrt{H_o})^2}{|G|} = 2 \times 110 - \frac{2(\sqrt{1.2} + \sqrt{0.15})^2}{0.0375}$$

$$= 220 - \frac{2(2.1985)}{0.0375} = 102.74 \text{ m}$$

Our assumption of $L < S$ is correct, since $S = 110$ m. Hence, provide $L = 103$ m.

**Example 3.16**

A valley curve is formed by a descending gradient of 1 in 30 meeting an ascending gradient of 1 in 50. If the design speed is 80 kmph, calculate the total length of the valley curve.

**Answer:** Descending gradient (–g1) = –1 in 30 = – (1/30) × 100 = –3.33%

Ascending gradient (+g2) = 1 in 50 = (1/50) × 100 = 2.00%

$$G = (g2) - (-g1) = 2.00 + 3.33 = 5.833\% = 0.05833;\ |G| = 0.05833$$

For a design speed of 80 km/h, stopping sight distance (SSD) may be assumed as $S = 120$ m (or may be determined by Eq. (3.1)). Assume $L > S$. By Eq. (3.48)

$$L = \frac{|G|S^2}{2(H_h + S\tan\phi)} = \frac{0.05833 \times 120^2}{2(0.75 + 120 \times \tan(1^\circ))}$$

$$= \frac{839.952}{2(0.75 + 120 \times 0.017455)} = 147.64 \text{ m}$$

Therefore, our assumption of $L > S$ is correct, since $S = 120$ m. Hence, provide $L = 148$ m.

**Example 3.17**

Determine the minimum required sight distance when a sag curve is passing under an overhead structure. The conditions given are:

(i) $S < L$; $|G| = 0.05833$; $L = 148$ m; $H_d = 1.2$ m; $H_o = 0.15$ m; and vertical clearance at midway of the sag curve under an overhead structure, $c = 4.42$ m.

(ii) $S > L$; $|G| = 0.0389$; $L = 74.2$ m; $H_d = 2.75$ m; $H_o = 0.15$ m; and vertical clearance between carriageway and an overhead structure, $c = 4.42$ m.

**Answer:** (i) When $S < L$: From Eq. (3.59*c*)

$$O_o = \frac{|G|}{8} \times L = \frac{0.05833}{8} \times 148 = 1.079 \text{ m}$$

By Eq. (3.50)
$$S = L\left[\frac{O_o + c - \left(\frac{H_d + H_o}{2}\right)}{2 \times O_o}\right] = 148\left[\frac{1.079 + 4.42 - \left(\frac{1.2 + 0.15}{2}\right)}{2 \times 1.079}\right] = 330.8 \text{ m}$$

(ii) When $S > L$:

$$O_o = \frac{0.0389}{8} \times 74.2 = 0.3608 \text{ m}$$

By Eq. (3.51)

$$S = \frac{1}{/G/}\left[\sqrt{(c + O_o - H_d)} + \sqrt{(c + O_o - H_o)}\right]^2$$

$$= \frac{1}{0.0389}\left[\sqrt{(4.42 + 0.3608 - 2.75)} + \sqrt{(4.42 + 0.3608 - 0.15)}\right]^2 = 328.9 \text{ m}$$

## 3.12 BUS BAY

The traffic in India is heterogeneous in character and composition, consisting of slow-moving animal-drawn vehicles and fast-moving motor vehicles of different sizes. For smooth flow of vehicles, the slow-moving ones should be segregated suitably and traffic in urban areas regulated. To avoid congestion at bus stops, bus bays of suitable geometry, which will segregate slow- and fast-moving traffic and facilitate safe passage of pedestrians (Fig. 3.27), should be constructed. Standard procedures for segregation and regulation of traffic in urban areas are given in detail in IRC: 70-1977.

## 3.13 ROAD HUMPS AND BUMPS

These are round-shaped elevated/protruded surfaces formed across roadways as physical devices to reduce/deter/control the speed of vehicles on minor or secondary/tertiary and unimportant un-controlled

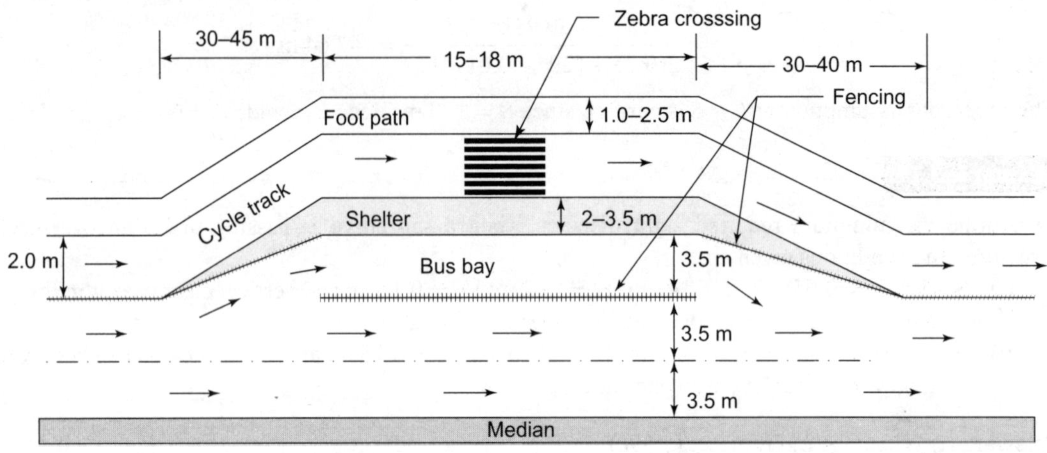


**Fig. 3.27** An example of a typical layout of a bus bay (Modified from IRC: 70-1977)

roads/junctions and also in accident-prone segments, residential streets, schools and manned and unmanned level crossings. Based on the cross-section, they are classified as road bumps and road humps (Fig. 3.28).

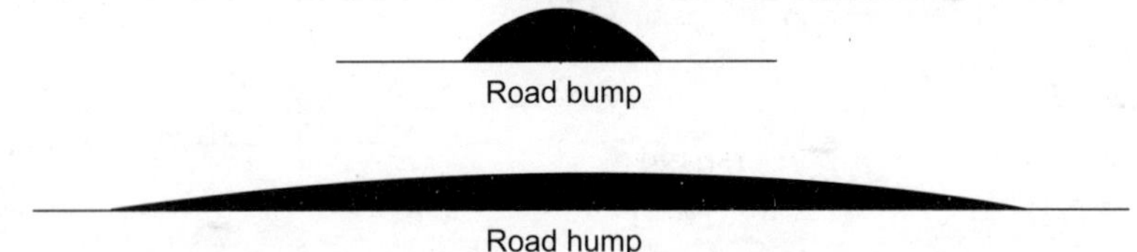


**Fig. 3.28** Typical cross-sections of road bump and road hump

Road humps are generally preferred because they are flatter than road bumps (Chadda and Cross, 1985). They are used in India and several other countries like the UK, USA, Australia and Finland (Jarvis, 1980; Clement, 1983; Fwa and Tan, 1992). The design elements of a road hump consist of (i) dimensions design and (ii) layout design, which deals with spacing and the number of humps to be installed. The road humps should be designed to facilitate smooth, safe and convenient control of the speed of vehicles and also the safety of pedestrians. Proper drainage arrangement should be made at road humps to inhibit/prevent the formation of puddles.

The geometric design of road humps may be carried out based on the findings of field experiments carried out by Fwa and Tan (1992). They observed that the ratio of the cross-sectional area ($A$) to width ($W$) of the road hump is significant for controlling hump cross-speed. The relationship is

$$\frac{1}{V_{85}} = 0.0212 + 0.4072\left(A/W\right) \tag{3.65}$$

where $V_{85}$ = desired design 85$^{th}$ percentile of the hump-crossing speed (km/h); ($A/W$) = area-to-width ratio (hump cross-section area ($A$) may be calculated according to the assumed surface profile) (Fig. 3.29).

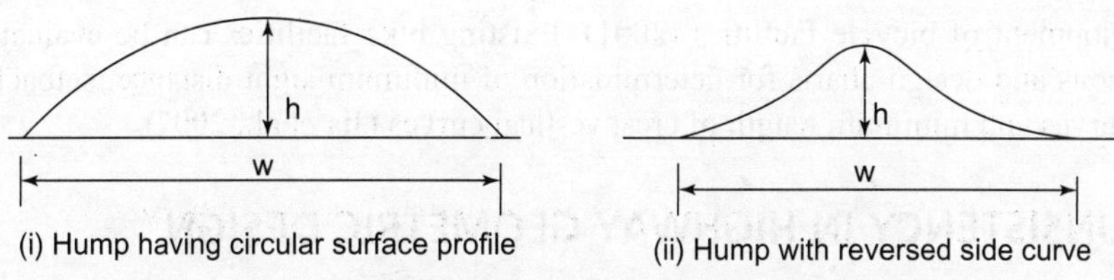


(i) Hump having circular surface profile (ii) Hump with reversed side curve

**Fig. 3.29** Typical surface profiles of road humps (from Fwa and Tan, 1992)

The above relationship is useful for designing hump geometry for a selected hump crossing vehicle speed. According to the TRRL Report (No: LR1017), hump width may vary from 2.44 m to 3.67 m and its height from 7.6 cm to 15.2 cm (Fwa and Tan, 1992).

The geometric design of road humps, in different countries, is given in Fig. 3.30. Drivers should be warned of the presence of a speed breaker, at least 40 m in advance. This can be done by installing visual aids according to the prescribed standards (IRC: 67-1977 and IRC: 99-1988).

It is good engineering practice to impose speed restrictions on traffic by installing mandatory speed limit signs, flash beacons and road markings of various types. It should be noted that speed breakers in the form of road bumps/humps are not intended to control the speed of vehicles as a substitute for mandatory, warning or informatory signs. They are merely provided to draw the attention of drivers to possible hazards ahead. Use of speed breakers on important and major intercity roads outside urban areas is not recommended by the IRC (IRC: 99-1988).

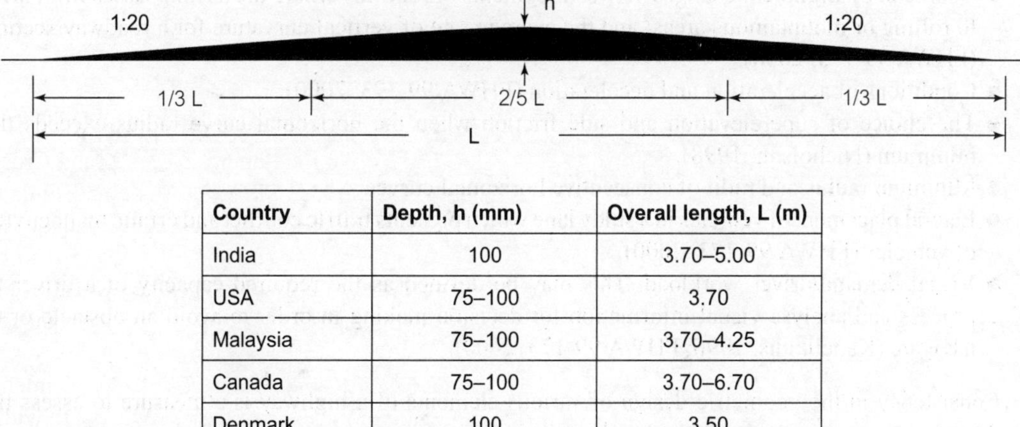


| Country | Depth, h (mm) | Overall length, L (m) |
|---|---|---|
| India | 100 | 3.70–5.00 |
| USA | 75–100 | 3.70 |
| Malaysia | 75–100 | 3.70–4.25 |
| Canada | 75–100 | 3.70–6.70 |
| Denmark | 100 | 3.50 |

**Fig. 3.30** Sectional details of hump speed breakers in different countries

## 3.14 GEOMETRIC DESIGN OF BIKE PATHS

Bike path facilities are useful for bicyclists, non-motorised vehicles and, very rarely, for certain kinds of motorised vehicles. The guidelines for bike path geometric design are described in the AASHTO guide for the development of bicycle facilities (2001). Existing bike facilities can be evaluated according to the equations and design charts for determination of minimum sight distance, setback distance on horizontal curves and minimum length of crest vertical curves (Jia et al., 2007).

## 3.15 CONSISTENCY IN HIGHWAY GEOMETRIC DESIGN

The design of highways and streets should be sound enough to fulfill the safety requirements by providing consistency of alignment in the horizontal and vertical planes, harmonisation between design speed and operating speed and human factors (Kanellaidis, 1996). The following factors are considered to detect design inconsistencies and promote highway safety:

- Consistency of geometric design can be evaluated by considering the following major categories (Gibreel et al., 1999):
  - (i) *Speed*: Operating speed and design speed
  - (ii) *Safety*: Influence of $e$ and $f$ on the stability of vehicles travelling on curves
  - (iii) *Performance*: Driver's workload and influence of various aesthetic aspects relating to highway geometric elements.
- Discrepancy between design and operating speeds and speed reduction between successive geometric features (Ottesen and Krammes, 2000).
- Consistency in speed measurement along highway segments (FHWA 99-171, 2000), consideration of operating speeds (Lamm et al., 1986) and differential operating speed are used to determine horizontal inconsistency (Leisch and Leisch, 1977).
- Difference between the estimated 85$^{th}$ percentile speeds between two consecutive highway sections (Kanellaidis, 1996).
- Character of alignment changes between segments of roadway where the terrain passes from level to rolling or mountainous areas; and the average rate of vertical curvature for a roadway section (FHWA 99-173, 2000).
- Conditions of acceleration and deceleration (FHWA 99-173, 2000).
- The choice of superelevation and side friction when the horizontal curve radius exceeds the minimum (Nicholson, 1998).
- Minimum radius and radii of consecutive horizontal curves.
- Lateral placement of vehicles in a study lane which presents traffic conflict and erratic manoeuvres of vehicles (FHWA 99-173, 2000).
- Visual demand/driver workload: This may be defined as the required capacity of a driver to process and analyse visual information for decision making in order to avoid an obstacle or to navigate (Kanellaidis, 1996; FHWA 99-173, 2000).

Consistency in the geometric design of various elements of a highway is a measure to assess the quality of alignment in the horizontal and vertical planes of new projects as well as on existing roads. The following relationships may be used to evaluate the design consistency of highways based on vehicle speeds (Lamm et al., 1988).

- Good design: $(V_{85} - V_d) \leq 10$ km/h
- Fair design: 10 km/h $< (V_{85} - V_d) \leq 20$ km/h
- Poor design: $(V_{85} - V_d) > 20$ km/h

where $V_{85}$ and $V_d$ are the operating and design speeds, respectively.

As a simple check, speed evaluation studies can be carried out on existing highways to provide more harmony with driver behaviour. For more details, readers may refer to the cited references.

## EXERCISES

### MULTIPLE CHOICE QUESTIONS

1. Transition curve is provided in horizontal alignment
   (a) To increase the radius of curvature
   (b) To introduce the superelevation gradually
   (c) To counteract the centripetal forces gradually
   (d) To prevent the vehicle from skidding laterally on curves
2. If the design speed is 100 km/hr, perception time of the driver is 2.5 seconds and coefficient of longitudinal friction is 0.35, the safe stopping sight distance will be ________.
   (a) 69.5 m (b) 41.8 m (c) 111.3 m (d) 182 m

   (*Hint*: On level surface: $SSD = 0.278 \times 100 \times 2.5 + \dfrac{(0.278 \times 100)^2}{2 \times 9.81 \times 0.35} = 182.044\ \text{m}$)
3. Given the safe stopping sight distance (SSD) as 182 m, the height of the driver's eye as 1.2 m, the height of the obstacle as 15 cm and the angle of deviation of adjacent grade lines as 0.08, the required length of the summit parabolic curve is ________.
   (a) 602.65 m (b) 309 m (c) 364 m (d) 546 m

   (*Hint*: Assume, L > S; $L = \dfrac{|G|S^2}{2(\sqrt{H_d} + \sqrt{Ho})^2} = \dfrac{0.08 \times 182^2}{2(\sqrt{1.2} + \sqrt{0.15})^2} = 602.65\ \text{m}$)

   120 m < 602.65; Our assumption is wrong.

   If L < S; $L = 2S - \dfrac{2(\sqrt{H_d} + \sqrt{Ho})^2}{|G|} = 2 \times 182 - \dfrac{2(\sqrt{1.2} + \sqrt{0.15})^2}{|0.08|} = 309.0368\ \text{m}$
4. The important factor considered in the design of summit curves on highways is
   (a) Comfort to passenger
   (b) Angle of spread of the headlight
   (c) Safe stopping sight distance
   (d) Upward gradient along the curve
5. Superelevation required on a horizontal circular curve of 200 m radius for a design speed of 70 km/hr with a coefficient of friction 0.15 is ________.
   (a) 6.5% (b) 10.5% (c) 7% (d) 4%

   (*Hint*: $v = 0.278 \times 70$ kmph = 19,.46 m/s. By Eq. 3.5,

   $$e = \frac{(0.75 \times 19.46)^2}{(9.81 \times 200)} = 0.108 = 10.5\% > e_{max} = 7\%$$

   Check the $f_{max}$ value. $f = \dfrac{v^2}{gR} - e = \dfrac{19.46^2}{(9.81 \times 200)} - 0.07 = 0.123 < f_{max} = 0.15$

   Hence, provide, $e = 0.07 = 7\%$ for the given speed)
6. If the superelevation provided on a horizontal curve of a hilly road is 1 in 10 and the coefficient friction is 0.14. The minimum radius of the horizontal curve on which the vehicles can travel safely at a design speed of 60 km/hr in metres is ________.

(a) 295 m (b) 354 m (c) 236 m (d) 118 m

(*Hint*: $e+f=\frac{v^2}{gR}$; $(\frac{1}{10})+0.14=\frac{(0.278\times 60)^2}{9.81\times R}$; $R=118.17$ m)

7. On highway stretches, where the required overtaking sight distance (OSD) cannot be provided, it is necessary to provide at least the length equal to
   (a) Three times the safe stopping sight distance (b) Intermediate sight distance
   (c) Safe stopping sight distance (d) Five times the stopping sight distance
8. The ideal form of the summit curve is
   (a) Circular curve (b) Parabola (c) Spiral (d) Lemniscate
9. The ideal form of the transition curve is
   (a) Circular curve (b) Parabola (c) Spiral (d) Clothoid
10. The objective of providing a transition curve on the approach to a horizontal curve is
    (a) To provide superelevation (b) To minimise the length of the horizontal curve
    (c) To simplify the laying out and construction procedure for the horizontal curve
    (d) To reduce sudden jerks to the allowed limit
11. The available sight distance is determined for sag or valley curves based on
    (a) Design speed (b) Height of the obstacle
    (c) Height of the driver's eye (d) Night visibility conditions
12. The value of the camber recommended for cement concrete roads in areas of light rainfall is
    (a) 1 in 60 (b) 1 in 50 (c) 1 in 40 (d) 1 in 33
13. The value of the camber recommended for bituminous surfaces in areas of light rainfall is
    (a) 1 in 60 (b) 1 in 50 (c) 1 in 40 (d) 1 in 33
14. The value of the camber recommended for earthen road surfaces in areas of heavy rainfall is
    (a) 1 in 50 (b) 1 in 40 (c) 1 in 33 (d) 1 in 25
15. The value of superelevation required for a vehicle travelling at a speed of 55 km/hr on a curve radius of 110 m, on a surface with coefficient of friction of 0.15 is
    (a) 0.07 (b) 0.10 (c) 0.66 (d) 0.04
16. If a descending gradient of 1 in 66 meets another ascending gradient of 1 in 33, then deviation angle is
    (a) 1/22 (b) 1/66 (c) 1/33 (d) 0.5%
17. A parabolic is used to connect a 2.8% upgrade with a 1.8% downgrade. The highest point on the summit curve is at a distance of ________ (measured horizontally from the point of curvature). The horizontal distance between the point of curvature and the point of tangency = 300 m.
    (a) 200.33 m (b) 182.6 m (c) 150.0 m (d) 233.3 m

(*Hint*: $G = g_2 - (g_1) = -1.8 - (2.8) = -4.6$ %; Therefore, $|G| = 0.046$ $\forall = \frac{|G|}{L} = \frac{0.046}{300}$

Distance of highest point on the curve from $T_1$: $x = \frac{g_1}{\forall} = \frac{0.028}{\left(\frac{0.046}{300}\right)} = 182.6$ m)

18. The width of a carriageway for a single lane is recommended to be
 (a) 3.0 m (b) 3.5 m (c) 3.75 m (d) 4.0 m
19. The width of carriageway for a double lane divided road is recommended to be
 (a) 5.5 m (b) 7.5 m (c) 7.0 m (d) 10.0 m
20. The width of a carriageway for an intermediate lane is recommended to be
 (a) 3.5 m (b) 3.75 m (c) 5.0 m (d) 5.5 m
21. The width of a carriageway for a double-lane undivided road is recommended to be
 (a) 5.5 m (b) 7.5 m (c) 7.0 m (d) 10.0 m
22. Stopping sight distance is the minimum distance available on a highway, which is the
 (a) Shortest distance of sufficient length to stop the vehicle without collision
 (b) Longest distance of sufficient length to stop the vehicle without collision
 (c) Distance of sufficient length along the centre line of the roadway to stop the vehicle without collision
 (d) Shortest sight distance of sufficient length across the curve to stop the vehicle without collision
23. The design value of lateral friction coefficient of highways is
 (a) 0.07 (b) 0.15 (c) 0.35 (d) 0.40
24. Camber on a carriageway is provided to take care of
 (a) Centrifugal forces (b) Lane alignment
 (c) Drainage (d) Segregate the traffic flow by direction
25. A parabolic is used to connect a 2.8% upgrade with 1.8% downgrade. What is the length of vertical curve if the rate of change of grade is 0.2% per 20 m chain?.
 (a) 200.33 m (b) 360 m (c) 100 m (d) 460 m
 (*Hint*: $G = g_2 - (g_1) = -1.8 - (2.8) = -4.6$ %; Length of curve = |G|/Permissible rate of change of grade = 4.6%/(0.2%/20 m) = 460 m)
26. Stopping sight distance and the coefficient of frictional coefficient are
 (a) Directly proportional to each other (b) Dependent on superelevation and design speed
 (c) Inversely proportional to each other (d) Dependent on pavement surface characteristics
27. A vehicle is moving at a speed of 85 km/hr on a road having 0.6% upward gradient. The driver applies brakes when he sees an obstruction ahead. His reaction time is 2.5 sec, assuming that the coefficient of longitudinal friction between the pavement and the tyre is 0.35. What is the distance he will traverse before the vehicle finally stops.
 (a) 141.8 m (b) 139.0 m (c) 218.9 m (d) None of these

 (*Hint*: Upward gradient $= 0.6\,\% = 0.006$; $SSD = 0.278 \times 85 \times 2.5 + \dfrac{0.278 \times 85^2}{2 \times 9.81(0.35 + 0.006)}$;
 $= 69.5 + 106.5 = 139.01$ mm)
28. A traffic stream is moving at an average speed of 80 km/hr with an average headway of 2.8 sec. The longitudinal distance between the consecutive vehicles is
 (a) 28.57 m (b) 62.27 m (c) 127.62 m (d) None of these
 (*Hint*: Speed = $0.278 \times 80 = 22.24$ m/sec; gap = $22.24 \times 2.8 = 62.27$ m)

29. A traffic stream is moving at an average speed of 100 km/hr with an average space headway of 27.8 m. The time headway between the consecutive vehicles is
(a) 0.1 sec (b) 0.2 sec (c) 1.0 sec (d) None of these
(*Hint*: Speed = 0.278x100 = 27.8 m/sec; time-gap = 27.8m/27.8m/sec = 1 sec)

30. For a pavement with a camber of 2% and design speed of 50 km/hr, the minimum radius beyond which no superelevation is required is
(a) 115.8 m (b) 553.9 m (c) 660.5 m (d) None of these

(*Hint*: $v = 0.278 \times 50 = 13.9$ m/s; $e = \dfrac{(0.75\,v)^2}{gR}$ and substitute the $e$ value as equal to camber value. $R \geq \dfrac{(0.75\,v)^2}{g \times (camber)} = \dfrac{(0.75 \times 13.9)^2}{9.81 \times \left(\dfrac{2}{100}\right)} = 553.9$ m)

31. What is the braking distance if the speed of a vehicle is 100 km/hr and the coefficient of longitudinal friction is 0.35?
(a) 112.5 m (b) 182 m (c) 55.6 m (d) 69.5 m

(*Hint*: $v = 0.278 \times 100 = 27.8$ m/s; Breaking Distance $= \dfrac{v^2}{2gf} = \dfrac{27.8^2}{2 \times 9.81 \times 0.35} = 112.5$ m)

32. A vehicle moving at 49 km/hr on an ascending gradient of road has to come to a stop before collision with an obstacle. The ratio of lag distance to braking distance = 4:3. Assume total reaction time of the driver as 2.5 sec and the coefficient of longitudinal friction as 0.35. Then, the value of ascending gradient is
(a) 2.02% (b) 1.01% (c) 3.03% (d) None of these

(*Hint*: $\dfrac{\text{Lag distance}}{\text{Break distance}} = \dfrac{vt}{\left(\dfrac{v^2}{2g(f+n)}\right)} = \dfrac{4}{3};\ \dfrac{2g(f+n)t}{v} = \dfrac{4}{3};\ n = \dfrac{4v}{3 \times (2gt)} - f$

$= \dfrac{4 \times 13.622}{3 \times (2 \times 9.81 \times 2.5)} - 0.35 = 0.0202 = 2.02\%$)

33. The total width of a carriageway required for a three-lane national highway at a horizontal curve of 330 m radius, considering a wheel base of 12.2 m and design speed of 100 km/hr is
(a) 1.25 m (b) 8.25 m (c) 11.25 m (d) 11.75 m

(*Hint*: Widening of carriageway on curve: From Eq. 3.18, $X_e = \dfrac{0.5 \times 3 \times 12.2^2}{330} + \dfrac{0.38 \times 27.8}{\sqrt{330}} = 1.258$ m.

Now, the width of carriageway, $W = W + X_e = 3 \times 3.5 + 1.258 = 11.758$ m)

34. While designing a hill road with a ruling gradient of 1 in 20, a sharp horizontal curve radius of 60 m is encountered. The compensated gradient as per IRC is
(a) 1.2% (b) 1.5% (c) 3.5% (d) 3.8%

(*Hint*: Ruling gradient = (1/20) × 100 = 5%;

The grade compensation value is determined using the following formula (IRC:73-1980)

Grade compensation = (30 + $R_c$)/ $R_c$ = (30+60)/60 = 1.5%

The maximum allowable value of grade compensation = 75/ $R_c$ = 75/60 = 1.25%

Select highest value, as grade compensation = 1.5%; Compensated gradient = 5% – 1.5% = 3.5%)

35. The stopping sight distance required on a two-way road when the design speed is 75 km/hr, reaction time of the driver is 2.5 sec and the coefficient of friction between the pavement and the tyre is 0.35 is

(a) 115.4 m (b) 230.8 m (c) 346.2 m (d) None of these

(*Hint*: $SSD = 0.278 \times 75 \times 2.5 + \dfrac{(0.278 \times 75)^2}{2 \times 9.81(0.35)} = 115.4$ mm

As traffic is two-way, the SSD = $2 \times SSD = 2 \times 115.4 = 230.8$ m)

36. The required setback distance when a two-lane–wide carriageway road has a curve radius of 250 m which is measured along its centre line and minimum stopping sight distance of 180 m is

(a) 8.0 m (b) 16.0 m (c) 24.0 m (d) 32.0 m

(*Hint*: Setback distance = $DE = R_c\left(1 - \text{Cos}\left(\dfrac{28.647 \times S}{R_c}\right)\right) = 250\left(1 - \text{Cos}\left(\dfrac{28.647 \times 180}{250}\right)\right) = 16.02$ m)

37. A summit curve joins two gradients of +2.8% and –2% for an intermediate sight distance (ISD) of 240 m. The curve length (which is greater than the ISD) to be provided as per IRC standards is

(a) 280 m (b) 288 m (c) 628.7 m (d) 388.3 m

(*Hint*: If L > S; $L = \dfrac{|G|S^2}{2(\sqrt{H_d} + \sqrt{Ho})^2} = \dfrac{0.048 \times 240^2}{2(\sqrt{1.2} + \sqrt{1.2})^2} = 288$ m)

38. It was found that a vehicle having a wheel base of 6 m is permitted to take a turn on a horizontal curve and the observed off-tracking of the outer wheel path is 0.09 m. What would be the value of widening of the two-lane carriageway required, if the speed of vehicle is given as 65 km/hr?

(a) 0.57 m (b) 0.18 m (c) 0.66 m (d) 0.75 m

(*Hint*: Design speed, $v$ = 65 kmph = 0.278 × 65 = 18.07 m/s; Off-tracking, $0.09 = \dfrac{0.5 \times 6^2}{R_c}$; Rc = 200 m; Widening of carriageway on curve: From Eq. 3.18,

$$X_e = \frac{0.5 \times 2 \times 6^2}{200} + \frac{0.38 \times 18.05}{\sqrt{200}} = 0.665 \text{ m})$$

39. The following shape of the camber is generally provided and best suited for cement concrete pavements.

(a) Straight line (b) Parabolic (c) Elliptical (d) Barrel

40. The following shape of the camber is more suitable on roads where vehicles frequently cross the centre line and overtake at high speed.

(a) Straight line (b) Parabolic (c) Sloped line (d) Straight and curved

41. What is the value of offset for a parabolic camber surface at a distance of 1.0 m from the centre line of a carriageway crown, if the given values of carriageway width is 3.75 m and depth at the crown point is 40 mm.

(a) 10.5 mm (b) 11.30 mm (c) 3.26 mm (d) 5.66 mm

($Hint$: $\frac{O_i}{(1.0)^2} = \frac{0.04}{\left(\frac{3.75}{2}\right)^2} == 0.0113 \text{ m} = 11.3 \text{ mm}$)

42. Coefficient of friction will be greater if the pavement surface is

(a) Rough and dry (b) Rough and wet (c) Smooth and dry (d) Smooth and wet

43. Compared to a level surface, on an ascending gradient, the stopping sight distance is

(a) Less (b) More

(c) Dependent on the speed (d) Dependent on the sag curve

44. The minimum length of the overtaking zone as per IRC recommendation is

(a) Overtaking sight distance (b) Two times the overtaking sight distance

(c) Three times the overtaking sight distance (d) Five times the overtaking sight distance

45. The safe stopping sight distance for a design speed of 60 km/hr for two-lane two-way traffic is ________. Assume coefficient of friction = 0.37 and the driver perception time = 2.5 sec.

(a) 80 m (b) 160 m (c) 240 m (d) 320 m

($Hint$: $SSD = 0.278 \times 60 \times 2.5 + \frac{(0.278 \times 60)^2}{2 \times 9.81(0.37)} = 80 \text{ mm}$)

46. The safe stopping sight distance for a design speed of 60 km/hr for two-lane one-way traffic is ________. Assume coefficient of friction = 0.37 and the driver perception time = 2.5 sec.

(a) 80 m (b) 160 m (c) 240 m (d) 320 m

($Hint$: $SSD = 0.278 \times 60 \times 2.5 + \frac{(0.278 \times 60)^2}{2 \times 9.81(0.37)} = 80 \text{ mm}$; Stopping sight distance when there Two-way traffic with single lane = 2 × SSD = 160 m)

47. The minimum sight distance required to avoid a head-on collision of two cars approaching from opposite directions is ________. The stopping sight distance for the first car and the second car is 153.6 m and 82.2 m, respectively.

(a) 153.6 m (b) 82.2 m (c) 71.4 m (d) 235.8 m

(*Hint*: Minimum sight distance required to avoid a head-on collision of two cars approaching from the opposite directions = 153.6 + 82.2 = 235.8 m)

48. The ruling design speed provided on national highways in rolling terrain as per IRC recommendations is

(a) 100 km/hr (b) 80 km/hr (c) 40 km/hr (d) 65 km/hr

49. The ruling design speed provided on national highways in mountainous terrain as per IRC recommendations is
(a) 100 km/hr (b) 80 km/hr (c) 50 km/hr (d) 65 km/hr
50. The ruling design speed provided on national highways in steep terrain as per IRC recommendations is
(a) 100 km/hr (b) 80 km/hr (c) 50 km/hr (d) 40 km/hr
51. The ruling design speed provided on village roads in plain terrain as per IRC recommendations is
(a) 50 km/hr (b) 40 km/hr (c) 35 km/hr (d) 20 km/hr
52. As per IRC, the centrifugal force produced due to two-thirds of the design speed is taken care of by the superelevation and the remaining one-third is counteracted by the surface friction of the carriageway. Based on this assumption, the rate of superelevation is given by
(a) $\frac{v^2}{2gR}$ (b) $\frac{0.75(v)^2}{2gR}$ (c) $\frac{v^2}{gR}$ (d) $\frac{(0.75v)^2}{2gR}$
53. As per IRC, the following equation is used to calculate the equilibrium superelevation
(a) $\frac{v^2}{2gR}$ (b) $\frac{0.75(v)^2}{2gR}$ (c) $\frac{v^2}{gR}$ (d) $\frac{(0.75v)^2}{gR}$
54. Find the incorrect statement, where $f$ = coefficient of friction, $e$ = superelevation, $v$ = design speed, $g$ = acceleration due to gravity and $R$ = radius of the circular curve.
(a) If $f = 0$, $e = v^2/gR$, the total centrifugal force is counteracted by superelevation only
(b) If $e < v^2/gR$, frictional forces will come into play to compensate the same.
(c) If $e = 0$, $f = v^2/gR$, surface frictional forces are entirely responsible for counteracting the total centrifugal force.
(d) If $f = 0$, $e = 0$, the total centrifugal force is partially counteracted by superelevation
55. In flat terrain, the advantage of balancing the earthwork is attained if the superelevation is provided by
(a) Rotating the inner edge (b) Rotating the outer edge
(c) Rotating about the centre line (d) Widening the carriageway
56. In plain terrain, drainage problems arise if the superelevation is provided by
(a) Rotating the inner edge (b) Rotating about the crown
(c) Crown shifted outward (d) Widening the carriageway
57. The following method of providing superelevation is preferred on very flat terrain in high rainfall areas when the road is not provided on embankment.
(a) Rotating the inner edge (b) Rotating the outer edge
(c) Rotating about the centre line (d) Widening the carriageway
58. The mechanical extra widening required for a 7.5 m divided carriageway on a horizontal curve radius R is given by ________ where $l$ = length of wheel base of vehicle.
(a) $\frac{l^2}{2R}$ (b) $\frac{l^2}{R}$ (c) $\frac{l^2}{3R}$ (d) $\frac{4l^2}{2R}$

59. The design gradient that is most widely considered for the geometric design of roads is
(a) Ruling gradient (b) Limiting gradient
(c) Minimum gradient (d) Exceptional gradient

60. As per IRC, the maximum width of a vehicle in India is
(a) 2.30 m (b) 2.44 m (c) 2.5 m (d) 3.0 m

61. The relationship between the length ($\ell$) and radius ($r$) of an ideal transition curve is given by
(a) $\ell_t \propto \frac{1}{r^2}$ (b) $r \propto \frac{1}{\ell_t}$ (c) $r \propto \frac{1}{\ell_t^2}$ (d) $r \propto \ell_t^2$

62. In the limiting case of overturning while traversing a simple circular curve by a vehicle with a uniform speed, the pressure distribution will be
(a) Outer wheels will be maximum (b) Inner wheels will be maximum
(c) Equal pressure on both the wheels (d) Double the pressure on the inner wheel

63. The setback distance is measured between these two lines
(a) Control line and building line (b) Control line and road boundary line
(c) Building line and crown (d) Building line and road land boundary line

64. The lag distance as per IRC, if the design speed is 60 km/hr, is
(a) 41.7 m (b) 82.2 m (c) 164.4 m (d) 247.6 m
(*Hint*: Lag distance = 0.278 × 60 × 2.5 sec = 41.7 m)

65. For design of overtaking sight distance, an average speed differential between impeding and overtaking vehicle passing at design peed is assumed to be
(a) 100 km/h (b) 0 km/h (c) 16 km/h (d) 30 km/h

66. The value of maximum permissible rate of radial acceleration (m/sec$^3$) is selected on the transition curve when the design speed of the vehicle is ________ km/hr
(a) $\frac{80}{75+V}$ (b) $\frac{75}{80+V}$ (c) $\frac{85}{75+V}$ (d) $\frac{80}{70+V}$

67. While designing road geometrics, the most important factors are
A. Superelevation B. Design speed C. Terrain type D. Land cost
(a) A and B (b) B and C (c) C and D (d) D and A

68. In the case of equilibrium cant, the pressure distribution on the inner and outer wheels would be ________ if the design speed = 60 km/hr and radius of the horizontal curve = 100 m.
(a) $\frac{6}{127}$ (b) $\frac{36}{127}$ (c) $\frac{6}{225}$ (d) $\frac{36}{225}$

69. The road camber is given in the following figure.

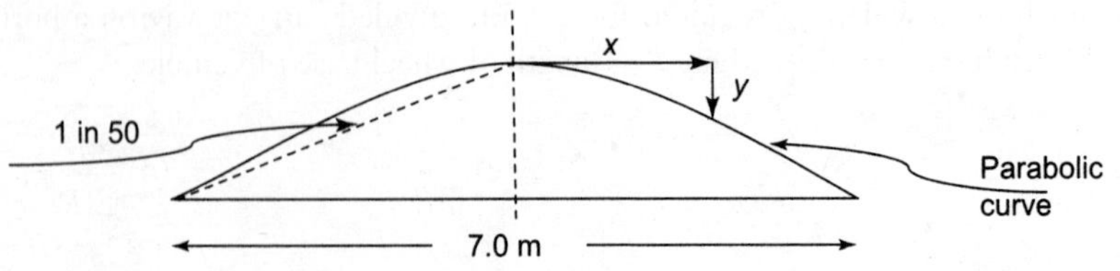

x
y
1 in 50
Parabolic curve
7.0 m

For designing this camber, the equation to be used is

(a) $y = \frac{x^2}{175}$ (b) $y = 2x^2$ (c) $y = \frac{x^2}{350}$ (d) $y = \frac{x^2}{225}$

(*Hint*: Formula used, $y = \frac{2x^2}{nW} = \frac{2x^2}{50 \times 7} = \frac{2x^2}{350} = \frac{2x^2}{175}$)

## Answers

| | | | | | | | | | |
|---|---|---|---|---|---|---|---|---|---|
| 1. (b) | 2. (d) | 3. (b) | 4. (c) | 5. (c) | 6. (d) | 7. (b) | 8. (b) | 9. (c) | 10. (d) |
| 11. (c) | 12. (a) | 13. (b) | 14. (d) | 15. (c) | 16. (a) | 17. (b) | 18. (c) | 19. (b) | 20. (d) |
| 21. (c) | 22. (c) | 23. (b) | 24. (c) | 25. (d) | 26. (c) | 27. (b) | 28. (b) | 29. (c) | 30. (b) |
| 31. (a) | 32. (a) | 33. (d) | 34. (c) | 35. (b) | 36. (b) | 37. (b) | 38. (c) | 39. (a) | 40. (b) |
| 41. (b) | 42. (a) | 43. (a) | 44. (c) | 45. (a) | 46. (b) | 47. (d) | 48. (b) | 49. (c) | 50. (d) |
| 51. (a) | 52. (d) | 53. (c) | 54. (d) | 55. (c) | 56. (a) | 57. (c) | 58. (b) | 59. (a) | 60. (b) |
| 61. (b) | 62. (a) | 63. (d) | 64. (a) | 65. (c) | 66. (a) | 67. (b) | 68. (b) | 69. (a) | |

## Questions

1. State the vehicle parameters useful for the geometric design of highways, with the relevant design element.
2. State the factors to be considered during the geometric design of highways.
3. Define terrain. What are the different terrains according to IRC: 73?
4. What is the significance of design speed in the geometric design of highways?
5. What is the significance of operating speed in the geometric design of highways?
6. Distinguish between design speed and operating speed.
7. Draw a neat sketch showing the salient cross-sectional details of a highway.
8. Define: (i) control line; (ii) building line; (iii) road land boundary line; (iv) setback distance (SBD); (v) right of way and roadway.
9. Define camber. Why is camber provided to road surfaces?
10. What are the different types of cambers based on the surface profile?
11. State the standard specifications of camber used for the construction of different types of roads based on the materials used.
12. State the formula used for determining a camber profile provided with a parabolic curved surface.
13. Define and state the different types of sight distances provided in the geometric design of highways with reference to the height of the driver's eye level and an object on the road.
14. Derive the formulae for determining the minimum sight distance in the following cases: (i) on a level road surface; (ii) on a sloped road surface.
15. Define overtaking sight distance (OSD or PSD) using a neat sketch with reference to Indian practices.
16. Define OSD. Explain the standard procedure of overtaking according to AASHTO using a neat sketch.
17. State the conditions assumed for determining OSD according to IRC: 66.

18. State the conditions assumed for determining OSD according to AASHTO.
19. State the factors affecting OSD and their influence on minimum OSD.
20. Define intermediate sight distance. How is it determined?
21. Define intersection sight distance using a neat sketch and state the design criterion according to IRC: 66.
22. State the assumption made by AASHTO for determining intersection sight distance.
23. Distinguish between intermediate sight distance and intersection sight distance using diagrams
24. How is the sight distance determined in medians?
25. Define superelevation and derive an expression based on radius of a curve, design speed, side friction and acceleration due to gravity.
26. What are the factors that affect superelevation?
27. State the assumptions and formula recommended by IRC: 73 for the determination of the rate of superelevation.
28. What are the standard values of superelevation adopted in different countries for different conditions?
29. Define the significance of side friction factor for various design parameters related to horizontal curves.
30. What are the factors affecting side friction factor?
31. Determine the minimum radius of a horizontal curve and restriction of speed recommended by IRC: 73.
32. What do you understand by design speed margin of safety (DSMS) and negative superelevation?
33. Define equilibrium cant using a neat sketch.
34. State and explain the different methods of attaining superelevation.
35. Describe the suitability of the different methods of attaining superelevation.
36. Formulate the relationships used for the design of superelevation by revolving the roadway about its centre line.
37. State the limiting values of the rate of change of superelevation for (i) plain and rolling terrain; (ii) hilly and mountainous terrain.
38. State the reasons for extra-widening of the carriageway on horizontal curves?
39. Derive an expression for determining the extra width of carriageway on curves. Also write the formula used for determining the total widening of curves.
40. Draw typical forms of different horizontal curves used for the centre line alignment of highways.
41. State the assumptions made for the design of transition curves for the geometric design of highways.
42. State the possible cases of insertion of transition curves in the centre line alignment of highways.
43. Why are transition curves with spirals preferred in highway alignment?
44. How is the length of a transition curve determined?
45. Derive the formulae for determination of the salient elements of a typical combined curve which consists of a circular arc connected with a transition curve on either side. Draw a neat sketch showing the salient features.

46. Derive expressions with neat sketches for the determination of salient elements of (i) twin centred compound circular curves and (iii) three-centred compound circular curves.
47. Define and describe the different types of longitudinal gradients according to IRC: SP23.
48. What are the factors affecting the selection of appropriate longitudinal gradients in highways?
49. Why is grade compensation necessary for the design of highway curves in hilly areas? What is the formula recommended by IRC: 73?
50. Define summit (or crest) and sag curves with reference to the angle of deviation of adjacent grade lines. Draw figures wherever applicable.
51. State when the design of vertical curves is not required with reference to the maximum change in grade and design speed.
52. State the design criteria, according to (i) IRC: SP-23 and (ii) AASHTO standards, used for determining the minimum length of vertical curves with reference to the height of the driver's eye level, the minimum depth of the object and the vehicle's headlight. Draw figures showing the salient details.
53. Write the formulae used to determine the minimum sight distance on a sag curve passing under an overhead structure. Draw figures showing salient details.
54. Distinguish between a road hump and a road bump.
55. Why is the use of speed breakers not favoured by the IRC? State a few good engineering practices to control speed on roads.
56. What do you understand by inconsistency in highway geometric design? State the factors to be considered to evaluate the consistency of the design geometric elements of a highway.
57. State the speed relationships between operating speed and design speed for evaluating the design consistency of highways.

CHAPTER

# 4 Traffic Volume Studies

## 4.0 INTRODUCTION

Traffic volume data is considered the most important input parameter of traffic streams. It is useful for ascertaining the importance of a route with respect to other routes, based on hourly variation, seasonal variation of traffic flow rate, distribution of traffic flow in a road network and its composition by vehicle type. Most of the Municipal Corporations in India, including the Town and Country Planning Department, will usually conduct detailed traffic and pedestrian surveys for preparing their Master Plans.

Traffic data with the associated vehicle, driver and pedestrian factors and their interrelationship with reference to road geometrics must be understood for proper road design and planning. We shall now discuss the basic definitions, types of traffic, methods/techniques adopted for traffic volume count and related aspects.

## 4.1 OBJECTIVES OF TRAFFIC VOLUME STUDIES

Traffic flow data is collected for planning, developing and managing road networks for the following reasons:

- To ascertain the necessity of road-widening and planning of new road network development
- To determine the general needs of structural and functional improvement, i.e., for overlay thickness design and thin overlays to meet roughness reduction requirements
- To ascertain the efficiency of the road network by comparing current traffic flow parameters with the level of service or with the calculated capacity
- To design signal systems, in particular, coordinated signal systems based on the directional movement of traffic at road junctions
- To enable prioritisation of maintenance work based on traffic volume, fund allocation for improvement schemes, and designation of a road as a national highway and expressway with their related facilities
- To assess the economic benefits arising from road improvements
- To estimate the needs of parking facilities
- To estimate the possible demands of traffic produced due to planning of new roads in a new town and the associated land use
- To assess the impact of traffic due to proposed new shopping malls or any such traffic-attracting and -generating service stations
- To determine traffic warrants against traffic control and traffic management techniques, such as no-entry, one-way, stop signs, etc.
- To project the future traffic growth for a given period and for deciding the optimal time for maintenance activities
- To classify roads/road networks on their functional basis
- To relate road accidents with traffic volumes and predominant vehicles

- To evaluate the performance of pavements based on traffic volume variability with seasonal and weather factors
- To establish the economic and social implications of the design of infrastructure facilities and, as part of their feasibility studies, of all development projects country-wide or region-wide
- To use weigh-in-motion (WIM) data relating to total tonnage moved on roads to estimate the freight travelling on each roadway. Such input is useful for determining the costs of congestion and the resulting benefits due to new construction and proposed operating strategies (FHWA, 2012).

## 4.2 TYPES OF TRAFFIC COUNTS

As a broad classification, the traffic flow on roads may be categorised as vehicular traffic and pedestrian traffic (Fig. 4.1). The summary of the data obtained from the respective traffic volume counts is used for specific projects or for the overall development/improvement of a road transportation facility.

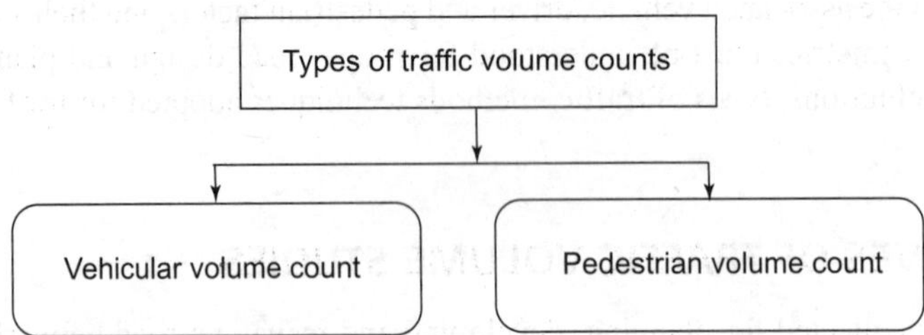


**Fig. 4.1** Types of traffic volume counts

### 4.2.1 Vehicular Traffic Volume (or Flow) Survey

Traffic volume (*VOL*) or flow (*Q*) is defined as the number of vehicles (*n*) passing across a section of a road, or a lane (Fig. 4.2), or a direction of a two-way highway, in a specified time interval (*t*), say, 1 hour or 15 minutes or 24 hours. In general, the volume (*Q*) is expressed as vehicles/hour (veh/h) or veh/day or veh/yr. This concept is also applicable for the counting of non-motorised vehicles.

$$Q = VOL = \frac{n}{t} \tag{4.1}$$

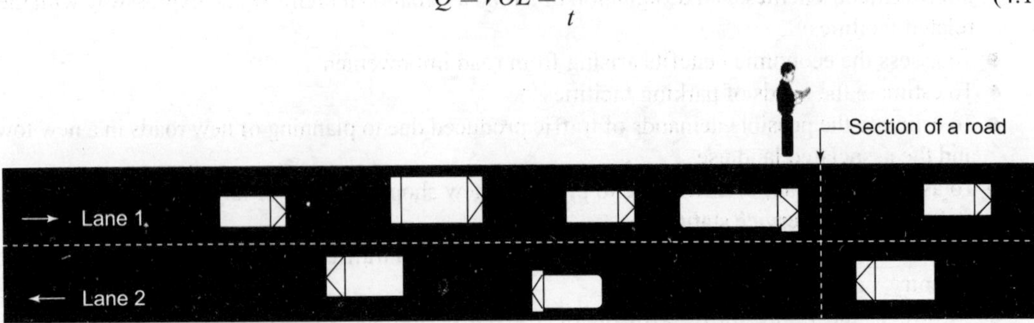


**Fig. 4.2** Traffic volume measurement: Vehicles crossing over a section of a road per lane

The passing of vehicle flow across a road section is random. For this reason, fluctuations in traffic volume are observed on any section of road with time. These fluctuations in flow can be observed with reference to seasons, days or months. On holidays and on Sundays, the traffic volume is observed to be less than the traffic volume observed on any full working day (except on Saturdays). The maximum traffic volume observed in an hour of a day is termed **peak hour traffic,** which is usually up to 10% of the total daily traffic flow, or 2–3 times the average hourly volume observed on that day.

The morning peak hour and the evening peak hour traffic indicate the maximum flow rate on a particular road section and provide a better measure of the operating conditions. The identified peak hour indicates intense activity, especially in urban areas, and the data is used for capacity building of the required infrastructure.

## 4.2.2 Pedestrian Traffic Volume Survey

The number of persons passing across a section of a road in a specified period of time is often reported as **pedestrian traffic volume**. In general, pedestrian traffic volume is expressed as ped/min, ped/hr, ped/day or ped/yr. It is known that pedestrian traffic volume changes depending on several factors, such as time of day, holidays, accidents, occasional events, road closure and weather conditions (Fig. 4.3).

**Fig. 4.3** Views of pedestrian traffic

## 4.3 INTRODUCTION TO PEDESTRIAN VOLUME STUDIES

Pedestrian surveys focus on pedestrian (i.e., a person on foot) movement in various places such as in central business districts (CBDs), road junctions and along roads. Problems at junctions with reference to pedestrian needs on specific issues and during peak hours are some of the important concerns of pedestrian studies. Major planning decisions related to transport facilities and the associated land use requires a proper understanding of the flow patterns of foot traffic.

## 4.4 OBJECTIVES OF PEDESTRIAN VOLUME STUDIES

Some of the important objectives of conducting pedestrian surveys (ASPO, 1965) are given below:

- To ensure the mobility and safety of pedestrians
- To minimise conflict with vehicular traffic movement
- To enhance the enjoyment of walking within urban localities
- To compare the level of pedestrian activity before and after developments
- To benchmark the occupancy of pedestrians in a public space or shopping mall or building in each hour of the day (if required, with their age/gender composition)
- To correlate pedestrian traffic patterns with pedestrian generators relating to land use, street parking facilities and transit maps indicating bus stops and railway stations
- To compare the resources of passing trade, say in a shopping centre, to that in competing centres
- To determine the hierarchy of streets in an urban locality used for pedestrian movement

## 4.5 USE CRITERIA FOR PEDESTRIAN VOLUME STUDIES

Pedestrian count data is used for several applications, based on three main criteria:

- **Mobility**: Pedestrian movement with reference to (i) conflicts with vehicular traffic at street intersections, mid-block locations and turnings; (ii) congestion on sidewalks and (iii) availability of routes/paths that approach various land-use applications such as business, shopping, workplace, transit points and recreational places.
- **Safety**: Enhancement of the safety of all types of road users by lessening *conflict* with vehicular traffic movement is the main concern under this category. A study conducted in Maine, USA revealed that pedestrian risk correlates more strongly with pedestrian volume than with vehicle volume (Gårder, 2004). Hence, it is important to plan and design facilities/amenities for pedestrians (for example, walkways/sidewalks, crosswalks, kerb ramps (excluding flares), landings, access ramps, pedestrian overpasses/underpasses, steps, elevators, platform lifts, shelters, toilets, plants and sitting benches).
- **Pleasure**: Comfort, pleasure and encouragement to pedestrians is provided by constructing various necessary facilities/sidewalk amenities such as canopies, waiting stations, adjacent artwork, road markings, pedestrian signals, plants and seating benches under shelters at the road-side.

## 4.6 TERMINOLOGY

- **Pedestrian:** Any person (adult or child) afoot, including people in wheelchairs or other such mobility devices. People who walk with bicycles are also included in this category.

- **Footpath:** A portion of the right-of-way of a road used as a pathway for movement of pedestrian traffic.
- **Street crossing:** The place where roads cross at grade (i.e., at the same level), and the roadway area is used for the movement of all types of vehicles and pedestrians.
- **Crosswalks:** A marked area on the carriageway which is provided for pedestrians to cross at street crossings or mid-block locations.
- **Shy-away distance** (Fig. 4.4): Pedestrians tend to keep a minimum distance from walls (15–45 cm), fences (30 cm), obstacles such as trees, benches and light poles (10–40 cm), opposite flows (60 cm), other pedestrians (27.5 cm), platform edges (80 cm) and in curves (15 cm), depending on the wall condition and their freedom of movement (Stucki et al., 2003; Daamen, 2004).

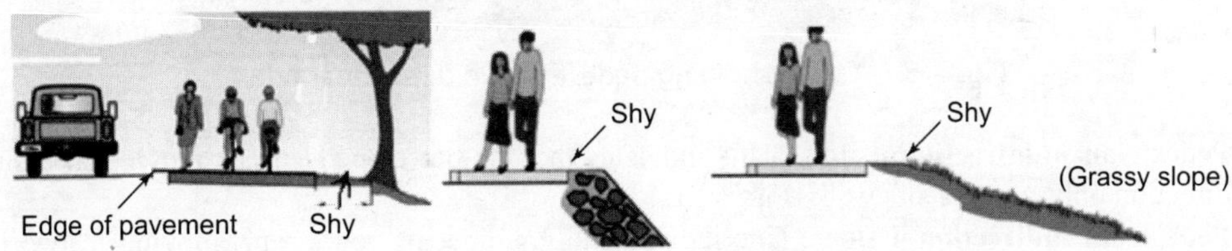


**Fig. 4.4** Shy-away distance

(*Source*: ODOT, 2011. *Courtesy*: Oregon Department of Transportation (ODOT), Oregon, USA)

- **Pedestrian speed:** The average pedestrian walking speed, expressed in m/sec. In general, walking speeds at mid-block locations are higher than at intersections. A walking speed of 1.2 m/sec may be assumed for the design of pedestrian crossing signals (TRB, 2000).
- **Pedestrian start-up time**: The time taken by pedestrians, after waiting for the signal, to start walking to cross the road. In other words, it is the duration of time from the start of the walking signal to the moment that the pedestrian actually steps off the kerb and starts to cross the road. In a study, the average start-up time was found to be 2.5 seconds for older pedestrians and 1.9 seconds for younger pedestrians (Dewar, 2002). However, 3.0 seconds may be taken as the start-up time for the design of signal timing at pedestrian crossings (TRB, 2000).
- **Pedestrian flow rate:** The number of pedestrians passing a section line on a road per unit time, expressed as p/min or p/15 min or p/hr. A section line is considered perpendicular to the direction of the pedestrian path.
- **Pedestrian density:** The average number of pedestrians per unit area within a walkway or queue length, expressed as p/m$^2$. Sometimes, the pedestrian flow rate per unit width of walkway is also expressed as p/min/m width.
- **Pedestrian space:** The average space provided for each pedestrian on a walkway, expressed as m$^2$/p. This is an inverse value of density and this unit is used more widely for analysing and designing pedestrian amenities/facilities.
- **Pedestrian platoon:** It represents the number of pedestrians walking together in group while crossing a road intersection or walking on a walkway. This parameter is useful for the design of pedestrian signals at intersections.

Fig. 4.5 A view showing pedestrian unidirectional flow

- **Pedestrian unidirectional flow:** This indicates the number of pedestrians moving in a single direction only, along a sidewalk (Fig. 4.5).
- **Pedestrian bidirectional flow:** This indicates the number of pedestrians moving in two (i.e., usually opposite) directions (Fig. 4.6).

Fig. 4.6 A view showing pedestrian bidirectional flow

- **Cross flow:** The flow that takes place roughly perpendicular to and which crosses another pedestrian stream.
- **Pedestrian circulation path** (Figs 4.7 and 4.8): This path includes independent pathways that are provided as walkways, sideways and shared bicycle pathways with other types of pedestrian facilities. Such paths may be provided adjacent to streets/highways, which may be separated from buffer/post-mounted objects/obstacles (Fig. 4.9).
- **Pedestrian access routes** (Figs 4.7 and 4.8): There is a need to continuously connect pedestrian circulating paths with adjacent or nearby pedestrian facilities. These are the connecting pathways/routes provided between pedestrian facilities and regular walkways. The pedestrian facilities may be listed as any one or a combination of steps, elevators, coffee-shops, kerb ramps, foot overbridges, etc.

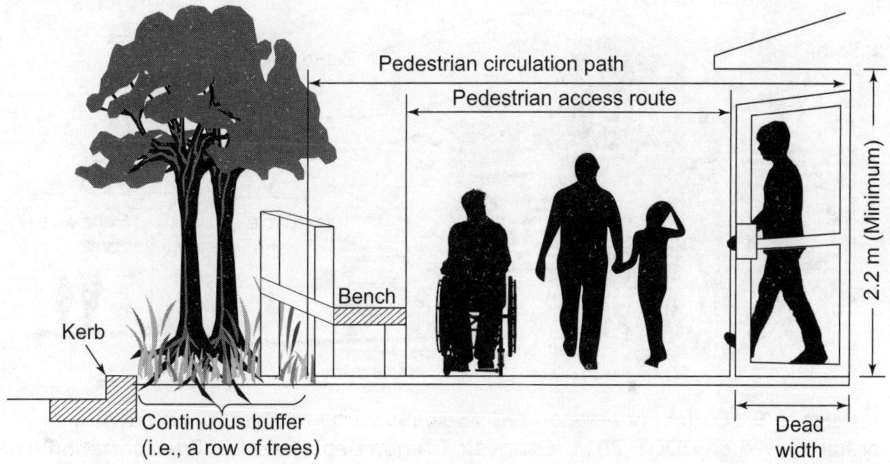


**Fig. 4.7** Details of a pedestrian walkway zone adjacent to a continuous buffer (modified and elaborated, based on www.wsdot.wa.gov)

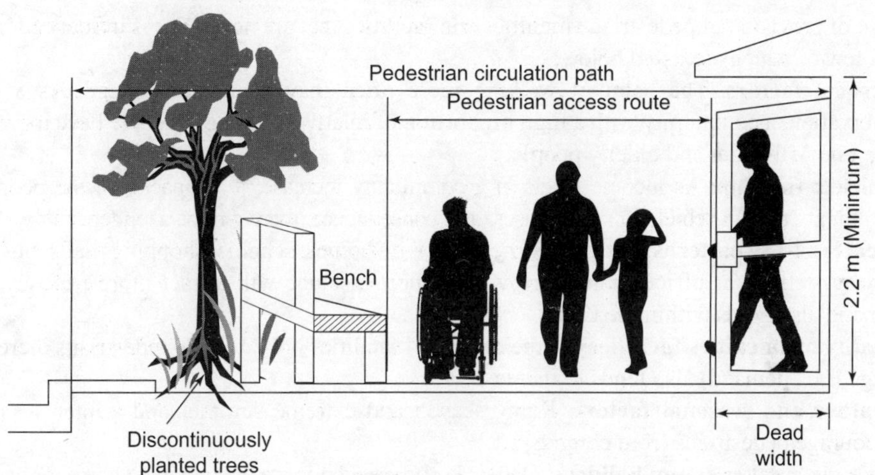


**Fig. 4.8** Details of a pedestrian walkway zone with a discontinuous buffer (modified and elaborated, based on www.wsdot.wa.gov)

## 4.7 TYPES OF PEDESTRIAN TRANSPORT NETWORKS

Pedestrian transport networks may be of the following types:

- Footpaths/Footways/Walkways
- Elevated walkways/Foot overbridges (FOBs)
- Subways
- Stairs/Ramps
- Escalators
- Travelators

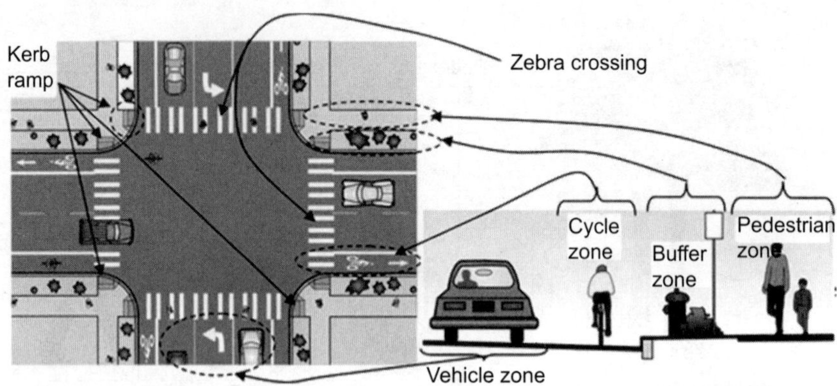


**Fig. 4.9** Buffer zone–separated sidewalks with kerb ramps and crossings (elaborated, based on ODOT, 2011; *Courtesy*: Oregon Department of Transportation (ODOT), Oregon, USA)

## 4.8 FACTORS AFFECTING PEDESTRIAN DEMAND

The extent of provision of pedestrian amenities and facilities that are necessary is influenced by several factors, a few of which are listed below:

- **Personal factors**: The habit of walking more often may be related to age. As a common observation, a community with a high proportion of relatively younger people have more walkers than one with older and elderly people.
- **Vehicle ownership**: As income levels in a community increase, the capacity of the people in the community to own vehicles also increases; as a consequence, there will be a tendency to walk less.
- **Local land-use patterns**: The places/precincts with homes, schools, shopping malls, bus stations, railway stations or offices which are within short distance will attract more pedestrians. This demand also holds within the CBD.
- **Quality of amenities/facilities**: As the quality of facilities provided for pedestrians increases, the usage (i.e., demand) also tends to increase.
- **Weather and seasonal factors**: Rainy season and extreme summer and winter seasons may discourage pedestrians from coming out.
- **Day time, weekends and holidays**: Day time and weekends are the two most commonly observed time periods during which pedestrians may walk around for recreational or holiday or other purposes.
- **Special occasions**: Special events in certain localities may attract more pedestrians, such as temples/churches/mosques/gurudwaras during festivals, melas and road-shows.

## 4.9 PEDESTRIAN DATA COLLECTION METHODS

The pedestrian count is a measure of the volume and direction of pedestrians passing (with their gender and age details) during a time period by location. Pedestrian volume counts are widely expressed as p/min, p/hr or p/day. Apart from the count, additional parameters such as flow characteristics, delays and impedance due to obstacles also need to be determined as part of data collection (Fig. 4.10).

The counting of pedestrians passing across a given section of a road can be carried out by different methods, and they may be broadly categorised as manual counting methods and automatic counting

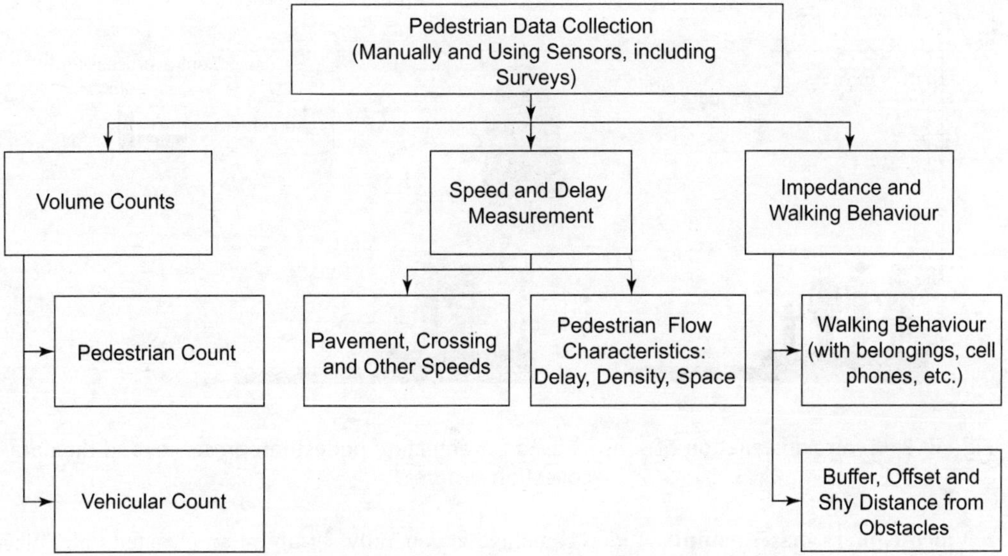


**Fig. 4.10** Pedestrian data collection methodology

methods. In general, pedestrian counts are performed in good weather conditions so as to obtain realistic pedestrian use data for a locality. These methods are briefly explained below (Fig. 4.11).

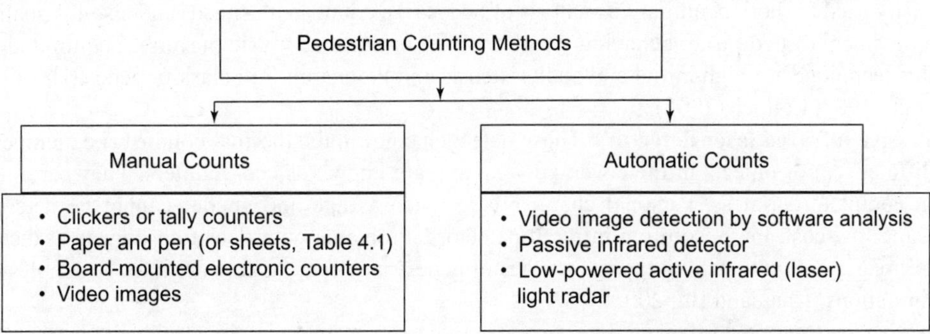


**Fig. 4.11** Pedestrian counting methods and techniques

### 4.9.1 Manual Counting Techniques

The pedestrians passing across any road section are manually counted and recorded with time. Tally counters, paper and pen or video cameras are used for this purpose (see Figs 5.2 and 5.3). The enumerators should be trained in advance and the manual collection of data for long periods may not be feasible for economic reasons.

### 4.9.2 Automatic Counting Techniques

The following techniques are widely used for the automatic counting of pedestrians (Fig. 4.12):

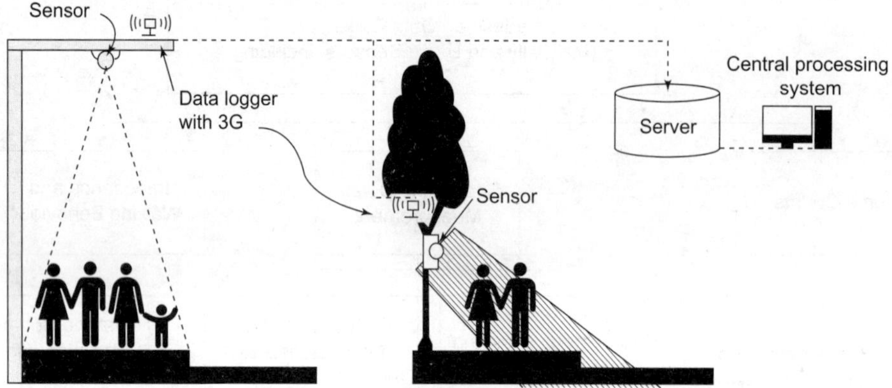


Fig. 4.12 Typical installation of sensors used for detecting pedestrian presence, and the data collection system

- **Video camera–based counts:** Video cameras are generally set up on an elevated pole, placed about 1.5–4 m above the floor. A view area of 10 m × 4 m may be considered as a study location with a combination of video camera and lens (Daamen, 2004).
  As the system records only images from the cameras, hardware and software tools are used to analyse the images and enable people-counting with specified characteristics. Image processing algorithms with the associated microprocessors built into the video camera can also be used, and the data is communicated by a wireless system. The data collected by this technique can be easily used for both quantitative analysis of pedestrian count (with speed) and also for qualitative assessment of pedestrian behaviour (with mobiles or portable music players). The limitations of this technique are high initial and analysis costs and the inability to classify pedestrians (Turvey et al., 1987; Lefloch, 2007).
- **Passive infrared laser detectors**: These detect changes in the thermal contrast (i.e., temperature differences) of objects in the coverage area, and are known as eco-counters. They can function in complete darkness. Personal characteristics such as age and gender cannot be determined using low-cost, low-resolution infrared detectors. Problems with detecting objects at the edges of images and in matching the trajectories of pedestrians over an area are a few examples of the limitations (Chan and Bu, 2005).
- **Active low-powered infrared detectors**: These detect any passing objects which obstruct the ray beam and count such obstructions as pedestrians.

As a common observation, pedestrian volumes increase in the day and decrease at night. Such behaviour with a specific pattern depends on the type of land use, time of day and trip purposes. Seasonal variations do exist in a year with respect to trips. For example, trips in the summer may be categorised as recreational trips.

## 4.10 LOCATION OF COUNTING STATIONS

The locations most commonly considered for pedestrian volume counts are given below (Fig. 4.13). These locations are used with reference to the intended pedestrian at crossing signals, foot overbridges and railings.

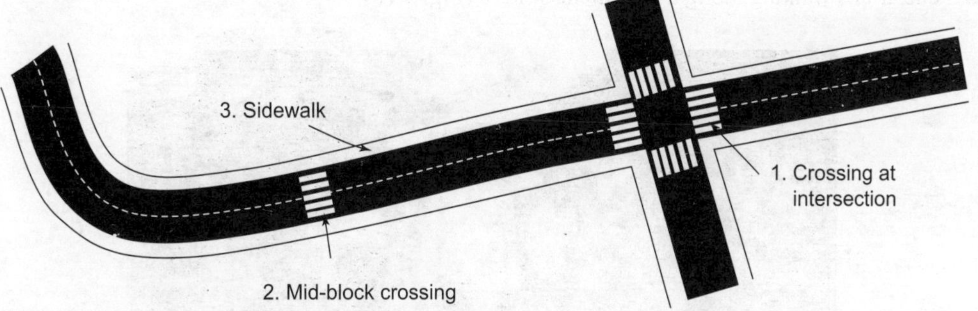


**Fig. 4.13** Commonly considered locations for pedestrian volume count on streets

- Pedestrians crossing an intersection (Fig. 4.14)

**Fig. 4.14** A view showing pedestrians crossing an intersection

- Pedestrians crossing a mid-block location on a road (Fig. 4.15)

**Fig. 4.15** A view showing pedestrians crossing a mid-block location on a road

- Pedestrians walking along a pavement or kerb (Fig. 4.16)

Fig. 4.16 A view showing pedestrians walking along a kerb

According to Fruin's observations, a critical condition exits at *corners* and pedestrian crossings at *intersections* not at *mid-block* locations (Fig. 4.17). Multi-directional pedestrian flow and queuing of pedestrians takes place at the corners of kerbs at junctions and opposite flow may take place between the corners while pedestrians mingle and interact with turning vehicles. For these reasons, the crossing speeds at mid-block locations are *greater* than the speeds observed at intersections (Bloomberg and Burden, 2006).

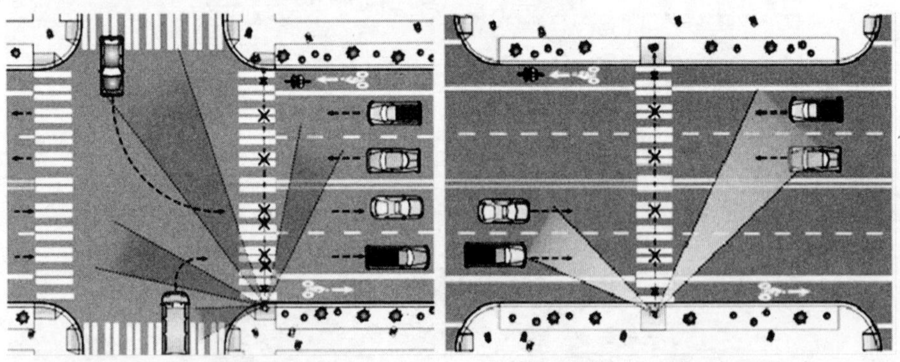

(a) Crossings at an intersection (b) Crossings at an mid-block location

Fig. 4.17 The number of conflicts is greater at intersections than at mid-block locations
(*Source*: ODOT, 2011; *Courtesy*: Oregon Department of Transportation, Oregon, USA)

## 4.11 PEDESTRIAN VOLUME CHARACTERISTICS

Pedestrian volume characteristics are essential for planning, designing and operating facilities/amenities at walkways, malls, intersections, shopping centres and parks. The pedestrian data collected is classified into different groups and their trends are analysed to plan/design the required/suitable facility. Trend analysis is carried out to determine the growth rate of pedestrian flow with time, and the time may be expressed as weekly, monthly, seasonal or yearly. Such a classification of pedestrians with their trend is commonly carried out based on the pedestrian volume characteristics. The following pedestrian volume characteristics may be used for the intended application (Cameron, 1976).

- Gender
- Shoppers
- Students
- Commuters
- Age
- Employees
- Visitors
- Mixed

## 4.12 DESIGN VOLUME

Peak hour pedestrian volume is commonly used as the maximum flow rate per hour and considered as design volume. Design volume is used for studying and designing the pedestrian-related traffic facilities in a locality.

The value of maximum rate of pedestrian flow *within the peak hour* can be estimated by using peak hour factor (PHF). Details of the calculation of PHF, with solved example problems, are presented in *Chapter 5*. A typical analysis of pedestrian traffic data and the calculation of various factors used to estimate equivalent volume counts from the short counts are presented below.

## 4.13 PRESENTATION OF PEDESTRIAN VOLUME DATA

Pedestrian volume data can be presented in different forms. Tabular columns and pie charts are the most conventional; they list the pedestrian volume totals (with age and gender) as hourly or half-hourly durations for each count location, differentiating the volumes by the direction of flow. In addition, the data can also be presented on maps showing the direction and volume data over a thickness scale. Such types of maps can also be used for representing the pedestrian traffic flow data obtained from origin and destination (OD) surveys.

A few standard forms of pedestrian data are presented in Figs 4.18–4.23.

**Table 4.1** A typical summary of pedestrian volume in a study area (Time: 8:30 am – 7:30 pm; Date: dd:mm:yy)

| *Street name* | *Number of pedestrians* | | | | | |
|---|---|---|---|---|---|---|
| | *Male* | | *Female* | | *Total* | |
| | *Number* | *Per cent* | *Number* | *Per cent* | *Number* | *Per cent* |
| Khairathabad | 789 | 19.4 | 344 | 15.3 | 1133 | 17.9 |
| Lakidikaphool | 1043 | 25.7 | 556 | 24.7 | 1599 | 25.3 |
| Assembly | 986 | 24.3 | 456 | 20.3 | 1442 | 22.8 |
| Nampally | 458 | 11.3 | 345 | 15.3 | 803 | 12.7 |
| Afjalgunj | 789 | 19.4 | 547 | 24.3 | 1336 | 21.2 |
| **Total** | 4065 | 100 | 2248 | 100 | 6313 | 100 |

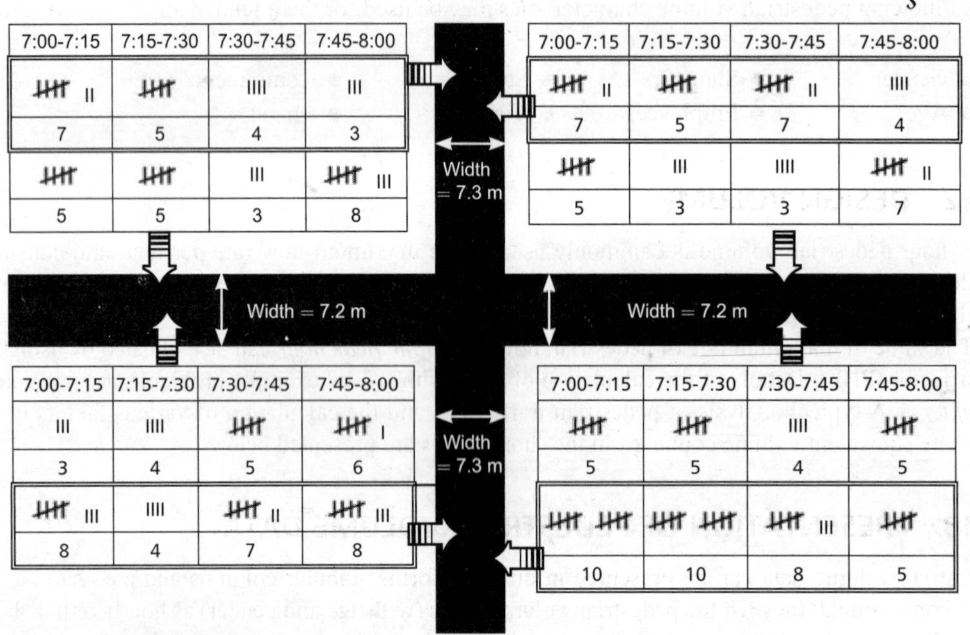


Summary of Pedestrian Movement

| | *Direction of pedestrian crossing* | | | | |
|---|---|---|---|---|---|
| *Time* | *North* | *South* | *East* | *West* | *Total* |
| *7:00-7:15* | 14 | 18 | 10 | 8 | 50 |
| *7:15-7:30* | 10 | 14 | 8 | 9 | 41 |
| *7:30-7:45* | 11 | 15 | 7 | 8 | 41 |
| *7:45-8:00* | 7 | 13 | 12 | 14 | 46 |
| Total | 42 | 60 | 37 | 39 | 178 / hour |

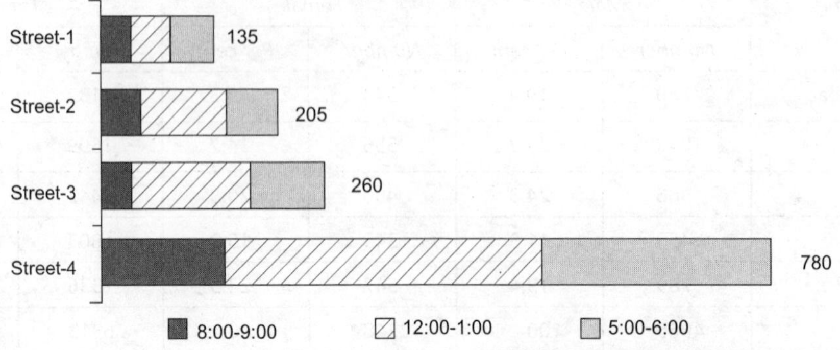


**Fig. 4.18** A typical presentation: "January 20XX" week-day pedestrian counts

**Table 4.2** A typical pedestrian volume sheet

| LOCATION: *City Park* | |
|---|---|
| CITY: *Hyderabad* | |
| TIME FROM: *7:00 AM* To *8:00 AM* | Date: *12th January 20XX* |
| INTERSECTION NAME: *RTC Jn.* | OBSERVER NAME: *Mr. John* |
| RAISED MEDIANS ON BOTH ROADS: | *YES*/ NO |
| WEATHER | *Sunny* |

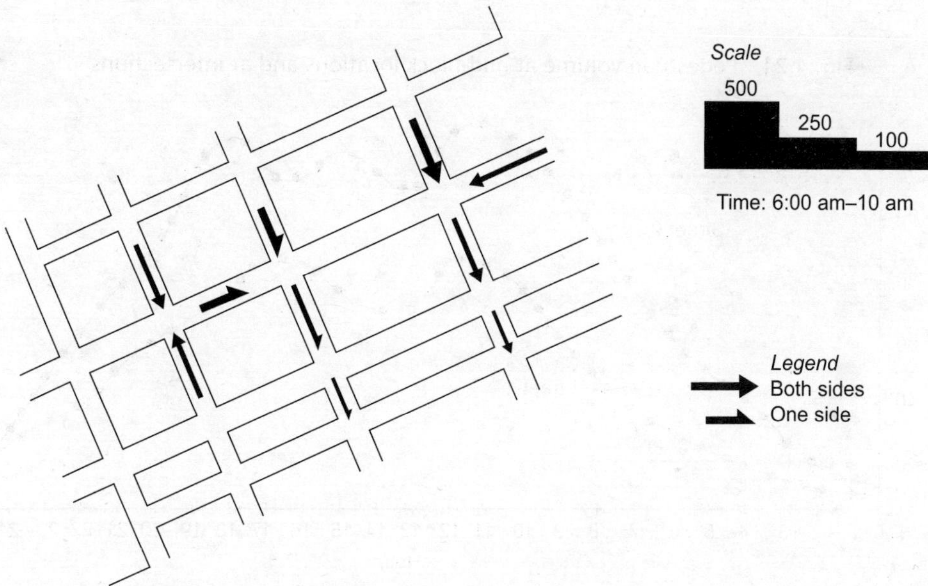


**Fig. 4.19** Volume and direction of pedestrian movement in a study area

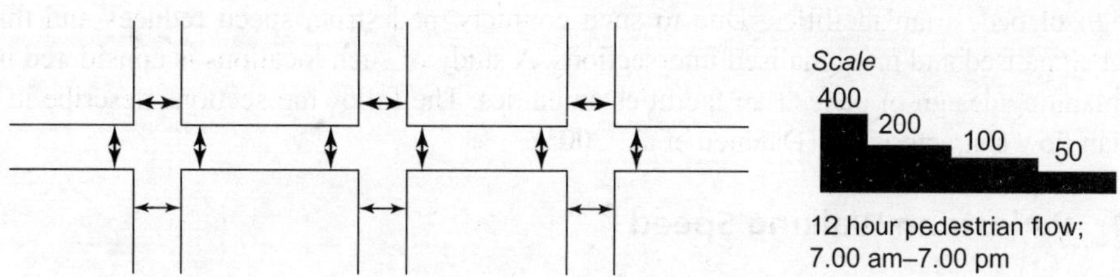


**Fig. 4.20** Pedestrian movement at intersections

## 4.14 PEDESTRIAN FLOW CHARACTERISTICS

Pedestrian flow characteristics such as speed, volume, rate of flow and density are similar to vehicular flow characteristics. In addition, with these parameters, pedestrian movement/crossing with reference to conflicts with vehicular movement at different locations is a subject of concern while planning

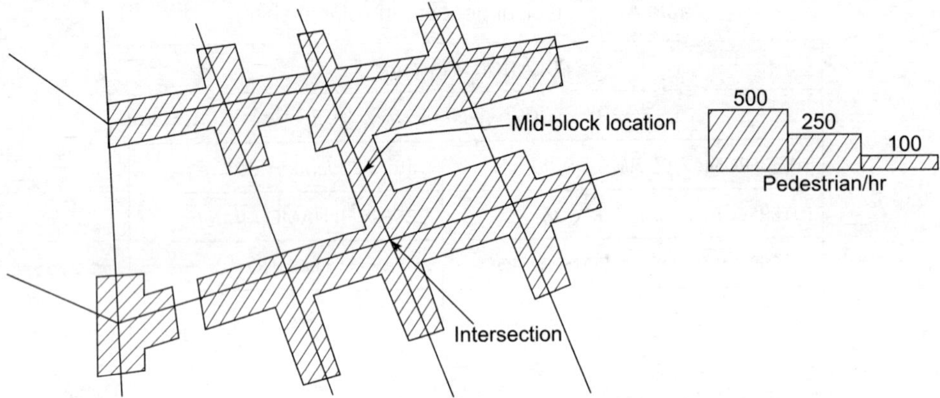


Fig. 4.21 Pedestrian volume at mid-block locations and at intersections

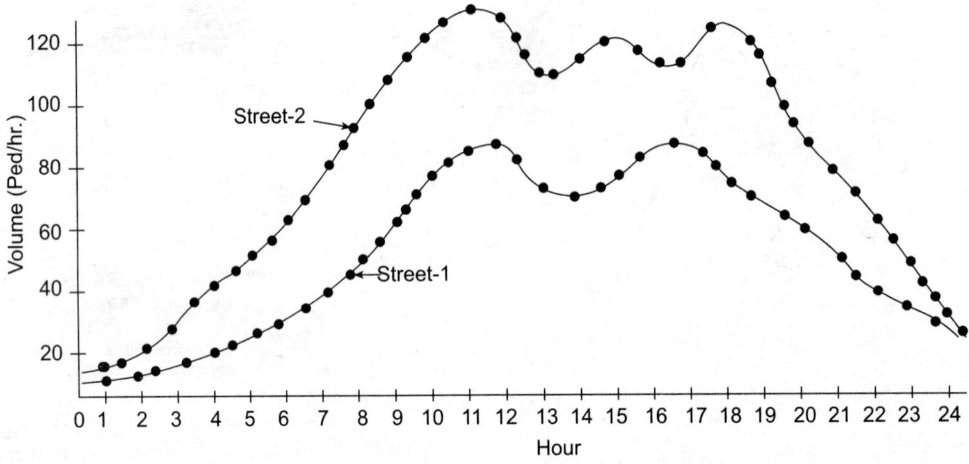


Fig. 4.22 An illustration of a typical pedestrian traffic flow pattern in a 24-hour period

the design of pedestrian facilities. Due to such conflicts, pedestrian speed reduces and this causes delays at signalised and un-signalised intersections. A study of such locations is considered important during planning/design of pedestrian facilities/amenities. The following sections describe in brief the pedestrian flow characteristics (Daamen et al., 2005).

### 4.14.1 Pedestrian Walking Speed

The walking speed is one of the primary characteristics required to evaluate pedestrian facilities. Walking speed (i.e., free speed in uncongested corridors) depends on several factors such as age, gender, height, direction, weather conditions, temperature, gradient, purpose of walking, infrastructure type and configuration. These factors may be categorised into three groups: (a) pedestrian characteristics, (b) movement conditions and (c) characteristics of infrastructure.

The walking speed of men is 10.9% higher than that of women. However, a high proportion of elderly (>60 years of age) pedestrians and children (<12 years of age) considerably reduce the speed of pedestrians walking in a group (Weidmann, 1993).

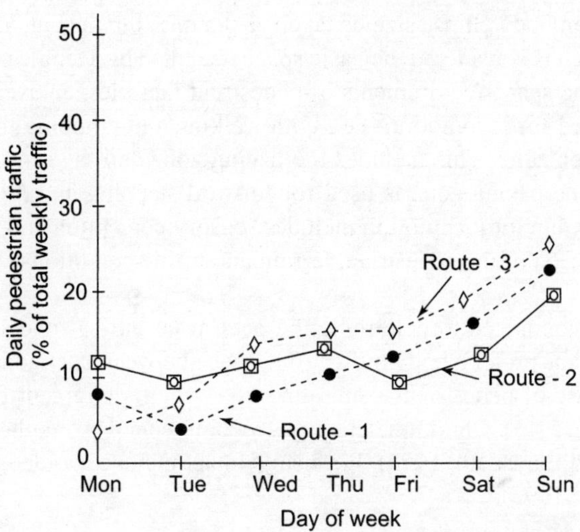


**Fig. 4.23** A typical Day of Week flow pattern of pedestrian traffic on a few streets

The speed of individual pedestrians appears to follow a normal distribution (Fig. 4.24). Several studies were conducted and observed that, pedestrian walking speeds range between 0.6 and 1.3 m/sec, with an average speed of 1.02 m/sec. The walking speed may go up to 1.6 m/s in the UK and USA. The average speed of pedestrians observed in different countries around the world may be taken as 1.34 m/sec, with a standard deviation of 0.37 m/sec (Daamen, 2004). The design speeds considered in India (IRC: 103-2012) are given below:

- 15th percentile speed used for the design of pedestrian crossing facilities $\approx$ 0.95 m/sec
- Average crossing speed of older pedestrians $\approx$ 0.79 m/sec
- Average crossing speed of younger/middle-aged pedestrians $\approx$ 0.98 m/sec
- Mean walking speed of physically challenged persons and users of walkers is 0.6 m/sec (Dewar, 2002).

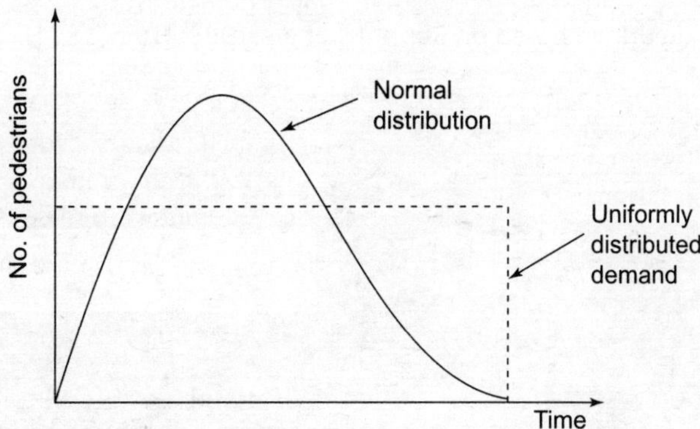


**Fig. 4.24** Disribution of pedestrian demand over time (schematic diagram)

### 4.14.2 Pedestrian Walking Space

The average space occupied by a pedestrian while walking is considered as one of the important components used to determine the minimum space requirement. Based on field studies, it was suggested that the average pedestrian body ellipse size be taken as 0.5 m × 0.6 m (Fig. 4.25 (a)). The total area is calculated as 0.3 m$^2$, which is considered the basic space occupied by a single pedestrian in the standing position. To determine the space requirements of pedestrian facilities, an average area of 0.75 m$^2$ per walking pedestrian is used for locomotion (i.e., both walking and climbing the stairs), and this space is termed the **body buffer zone**. This includes the pacing zone and the sensory zone (Fig. 4.25 (b)). The space required for the pacing zone is used for forward stepping and foot placement. The space required for the sensory zone for locomotion includes sensory perception and reaction, which depends on gender, height, age, speed of the pedestrian, terrain and traffic conditions (Fruin, 1971; Rouphail et al., 1998).

Several studies were conducted to determine the pedestrian buffer zone space requirements. One study found that pedestrians prefer a body buffer zone space between 0.27 and 0.84 m$^2$ (Pushkarev and Zupan, 1975)). In the case of pedestrians using umbrellas, the requirement for the body buffer zone space is >0.80 m$^2$ (Fig. 4.25 (c)). Optimum density, as recommended by another study, was between 1.3 and 1.8 p/m$^2$ (Virkler and Elayadath, 1994). Physical contact may be avoided at a density between 3.0 and 3.5 p/m$^2$ (Weidmann, 1993).

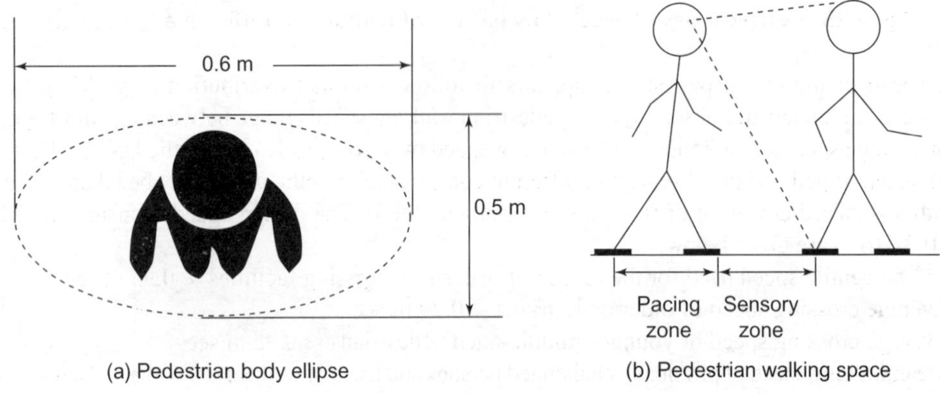


**Fig. 4.25** Pedestrian space characteristics
(Redrawn based on Rouphail et al., 1998; Fruin, 1971)

**Fig. 4.25(c)** Pedestrian body buffer zone space with an umbrella

Ultimately, the concept of buffer zone can be used for determining the inter-person spacing of pedestrians (Lee, 2005). For example, the separation space maintained by male and female subjects; and differences in the acceptance of closer spacing and higher levels of physical contact among eastern societies and western communities.

The following fundamental equation may be used to determine the width of a pedestrian facility (Dines and Brown 2001):

$$\text{Width} = \left( \frac{\text{Volume of pedestrians, } \dfrac{\text{p}}{\text{sec}}}{\left(\text{Mean density of pedestrians, } \dfrac{\text{p}}{\text{m}^2}\right) \times \left(\text{Speed of pedestrians, } \dfrac{\text{m}}{\text{sec}}\right)} \right) \tag{4.2}$$

### 4.14.3 Pedestrian Start-up Time

Pedestrians come with a wide range of characteristics and behaviours. As a common observation, start-up times were significantly longer for older pedestrians (range: 3.6 seconds for males to 3.95 seconds for females) than younger pedestrians (range: 2.76 seconds for males to 3.31 seconds for females). However, designers consider the start-up time value as 3.75 seconds and 3.0 seconds for older and younger pedestrians, respectively (TranSafety Inc., 1997). It must be noted that people with children often walk at slower speeds and such speeds may be obtained by conducting site-specific pedestrian studies.

### 4.14.4 Pedestrian Capacity

It is expressed as the number of pedestrians passing per minute per metre width of footpath. Pedestrian capacity depends on the average physical size and level of acceptance of the pedestrians to walk with closer spacing. The capacity of pedestrians (as one-directional flow on walkways in transit stations) observed by a study conducted in the UK (Daly et al., 1991) was 86 p/min/m, and in Hong Kong (Lam and Cheung, 2000) was 92 p/min/m. The following equation may be used to estimate the capacity of a walkway in other places.

$$\text{Capacity of walkway} = \text{Capacity per unit width of walkway} \times \text{Effective width of walkway} \tag{4.3}$$

### 4.14.5 Pedestrian Speed–Density (V–K) Relationship

The fundamental relationships between speed and density (i.e., volume) of pedestrian flow are analogous to vehicular flow. As the density of pedestrians increases, the space available for pedestrian passing decreases, and consequently, the pedestrian speed decreases as people try to avoid brushing against others and prevent repeated sudden stoppages. A typical trend of the relationship between the speed and density of pedestrians is presented in Fig. 4.26.

A typical trend of the speed of individual pedestrians walking at different locations is presented in Fig. 4.27. The speeds may be considered as slow, fast or average at the study location.

The following straight line relationship between the speed and the density of pedestrians is found to be widely known (Fruin, 1971a; Lam et al., 1995; Sarkar and Janardhan, 1997).

$$V = V_f \left(1 - \frac{k}{k_j}\right) \tag{4.4}$$

where $V$ = speed; $V_f$ = free flow speed; $K$ = density; and $K_j$ = jam density.

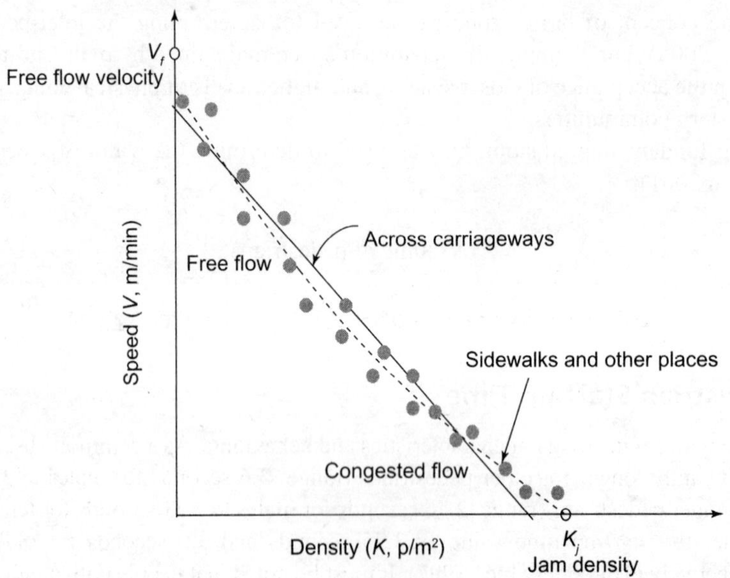


**Fig. 4.26** Relationship between speed and density (schematic diagram)

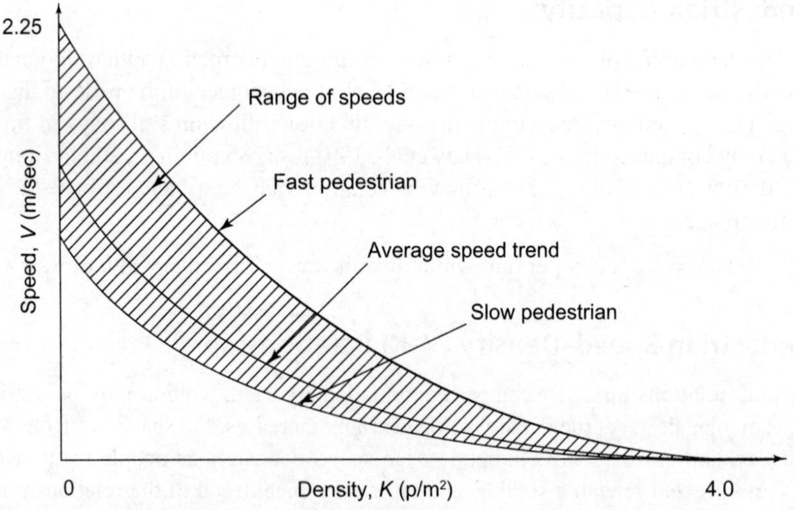


**Fig. 4.27** Typical trend of pedestrian speed observed under different walking conditions (schematic diagram redrawn based on Daamen, 2004)

Jam density ($K_j$) on a level surface was estimated by Fruin (1971) as 4.09 p/m$^2$, which refers to the space available per pedestrian – about 0.24 m$^2$. In general, the following observed values may be taken for the design of pedestrians facilities (Daamen, 2004):

- Jam density value may be taken as between 4.0 and 5.5 p/m$^2$.
- Flow of pedestrians appears to be unimpeded (i.e., not slowed or prevented) when the density is less than 0.5 p/m$^2$.

- Maximum flow or capacity flow may be taken between 1.1 and 1.67 p/m/sec (i.e., between 66 and 100 p/m/min). In Asian countries, it may vary between 1.48 p/m/sec and 1.53 p/m/sec (Sarkar and Janardhan, 1997; Tanariboon et al., 1986). In American and European countries, it may vary between 1.0 p/m/sec and 1.29 p/m/s (O'Flaherty and Parkinson, 1972; Virkler and Elayadath, 1994; Weidmann, 1993). Such observed densities ranging between low and high are found based on the conditions and situations. For example, low capacities are found near schools, whereas high capacities are found in railway or bus stations.

In addition to the straight line models,the double S-bend curve model (Weidmann, 1993), the two-regime model (Virkler and Elayadath, 1994) and the exponential walking speed models for bidirectional and multi-directional flow (Algadhi et al., 2001) are also in use in other parts of the globe.

### 4.14.6 Pedestrian Flow–Density Relationship

As flow increases, the density of pedestrians also increases up to a maximum value and then declines (Fig. 4.28). The relationship between flow, density and the speed of pedestrians is analogous to that of vehicular traffic, and the equation is as follows (Daamen, 2004):

$$\text{Pedestrian flow } (Q) = \text{Pedestrian density } (K) \times \text{Pedestrian speed } (V) \qquad (4.5)$$

where $Q$ = pedestrian flow rate, i.e., expressed as the number of pedestrians passing a cross-section of a road per minute per metre width (p/min/m); $K$ = pedestrian density, pedestrians per metre square (p/m$^2$); and $V$ = pedestrian speed, in metres per minute (m/min). It is the space mean speed and is calculated as the average speed of pedestrians present in an area at a study moment (i.e., usually expressed as metres per second).

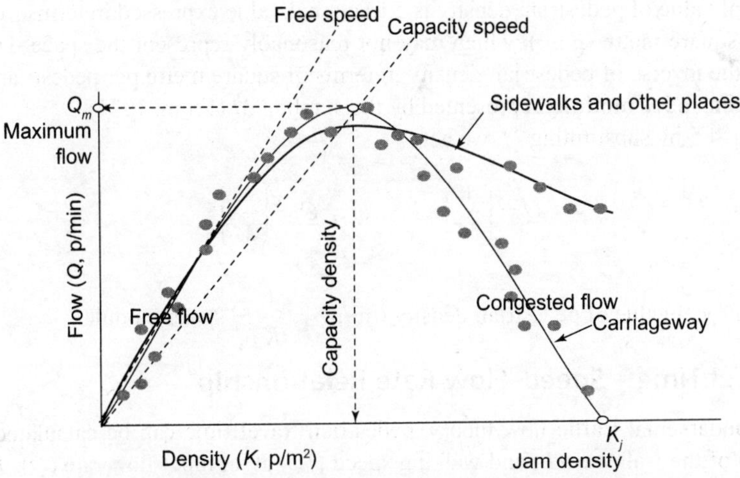


**Fig. 4.28** Relationship between flow and density (schematic diagram based on Daamen, 2004)

Maximum flow indicates the capacity and it can be represented as $Q_m$ (Fig. 4.28). The regions with density less than capacity are termed free flow regions, and the particular flow condition must be considered during the design of pedestrian facilities. It has been observed under a free flow condition that, as the width of the walkway increases, the values of flow rate and density also increase (Fig. 4.29). As the width of the walkway increases, a similar form of the flow–density relationship is assumed. It must also be noted that maximum flow cannot be achieved at the maximum walking speed (Daamen, 2004).

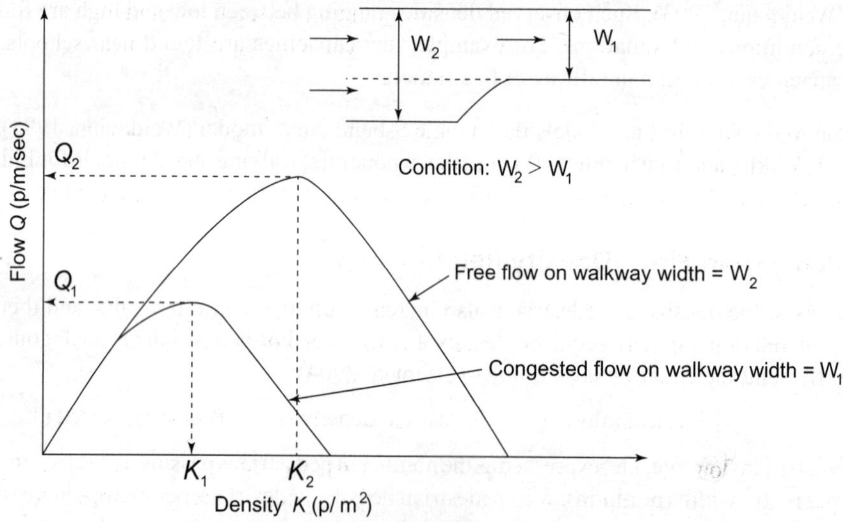


**Fig. 4.29** Typical effect of walkway width on the flow–density relationship (schematic diagram based on Daamen, 2004)

The numerical value of pedestrian density is a fractional value expressed in terms of the number of pedestrians per square metre (p/m$^2$), which may not reasonably represent the space. Hence, its value is expressed as the inverse of pedestrian density in terms of square metre per pedestrian (m$^2$/p), and is termed **pedestrian space module**, represented by the symbol, $M$ (Fruin, 1971).

Rewriting Eq. 4.5 by substituting $M$, we get

$$Q = \left(\frac{1}{M}\right) \times V \quad \text{or} \quad Q = \left(\frac{V}{M}\right) \tag{4.6}$$

where $M$ = reciprocal value of pedestrian density (m$^2$/p) $= \dfrac{1}{K}$ = space module.

### 4.14.7 Travel Time - Speed–Flow Rate Relationship

According to fundamental traffic flow theory, pedestrian travel time can be calculated based on the values of length of the walkway ($L$) and walking speed ($V_Q$) at an input flow rate ($Q$). The total travel time ($T$) for *each pedestrian* walking on a stationary walkway (Fig. 4.30) can be calculated using the following equation. The total duration of time comprises the free flow regime (Fruin, 1971) and the congestion regime (Daamen, 2004).

Total travel time ($T$) on a walkway = Free flow walking time on the walkway ($T_1$) + Passing delay on the walkway due to congestion ($T_2$) (4.7)

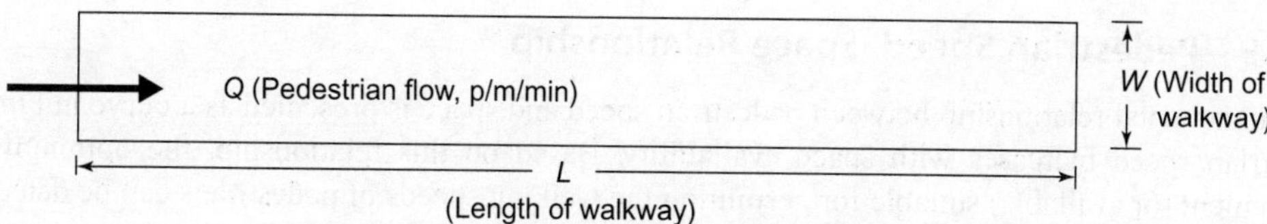


**Fig. 4.30** Schematic description of a stationary free flow condition (i.e., a non-movable plain walkway space or corridor) (after Fruin, 1984)

where $T_1$ = free flow walking time $= \frac{L_f}{V_Q}$; $T_2$ = walking time in congestion due to restricted conditions $= \frac{L_c}{V_C}$; $L_f$ = length of free flow space on a walkway; $V_Q$ = free flow speed observed on the walkway at an input rate of low $Q$; $L_c$ = length of congested space on the walkway; and $V_c$ = congested speed observed on the walkway at an input flow rate $Q$.

The congestion may be due to a revolving door, reduced width of a pathway, an obstructed pathway or due to queuing. For example, if the congested speed at a revolving door is 0.45 m/sec, the resulting congested speed ($V_c$) over a length ($L_c$) of 1.0 m would be equal to 2.22 seconds.

This model can be used to predict pedestrian traffic flow and to validate any walking model (Daamen, 2004).

## 4.14.8 Pedestrian Speed–Flow Relationship

The fundamental relationship between pedestrian speed and flow is presented as a curve in Fig. 4.31. Like vehicular movement, pedestrian speed increases with reduced flow, and vice versa. It is also similar in that the speed of a pedestrian on a walkway decreases with limited space, which is due to increased flow.

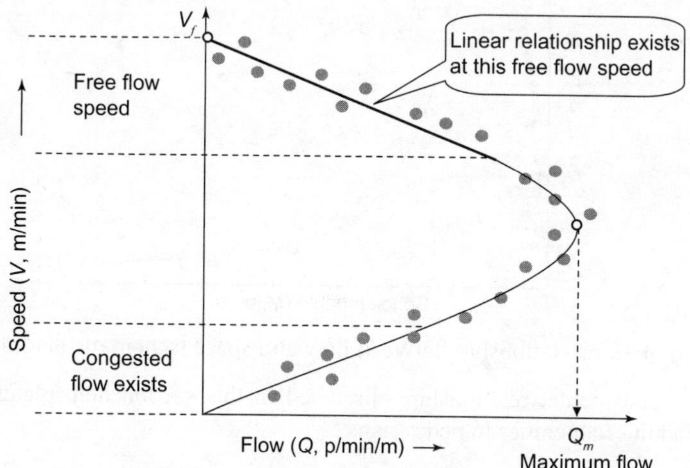


**Fig. 4.31** Relationship between speed and flow (schematic diagram)

### 4.14.9 Pedestrian Speed–Space Relationship

The fundamental relationship between pedestrian speed and space is presented as a curve in Fig. 4.32. Pedestrian speed increases with space availability. Based on this relationship, the optimum space requirement (or available) suitable for permitting the walking speeds of pedestrians can be determined (or evaluated) by conducting site-specific field studies. Such findings from surveys can reveal facts about the existing level of service (LOS) and the desired LOS.

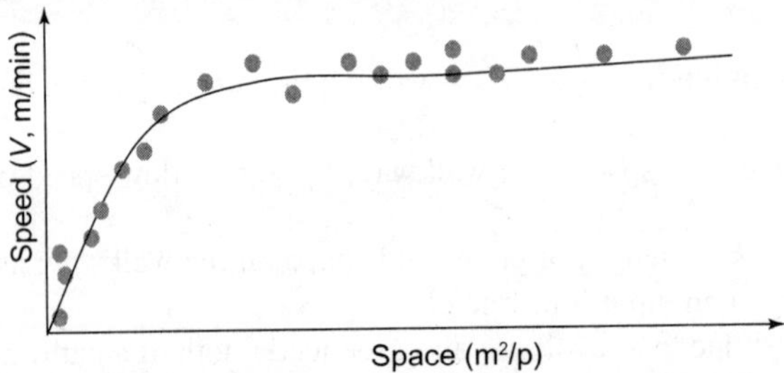


**Fig. 4.32** Relationship between speed and space (schematic diagram)

### 4.14.10 Pedestrian Flow–Space Relationship

The fundamental relationship between space and flow that was observed by researchers is presented in Fig. 4.33. Based on field studies, the optimum space per pedestrian can be determined at maximum flow condition, as occurs at a site. The average space required per person may vary between 0.4 and 0.9 $m^2/p$.

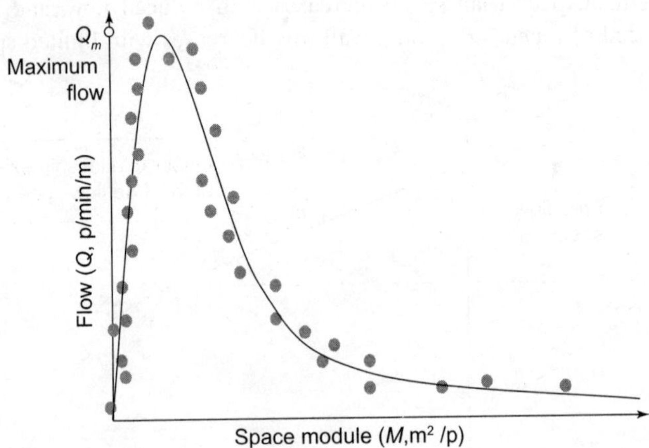


**Fig. 4.33** Relationship between flow and space (schematic diagram)

The basic and fundamental relationships discussed in this section help planners and designers provide suitable facilities/amenities to pedestrians.

## 4.15 LEVEL OF SERVICE (LOS)

The design of new pedestrian walkways/pathways and the performance evaluation of existing pedestrian infrastructure must be carried out by considering *comfort* and *safety*. The comfort of pedestrians is often described by the concept of level of service (LOS). The criteria adopted to judge the comfort are based on freedom of speed choice, frequency of forced hindrance in speed due to bidirectional or multi-directional or crossing or passing movements of pedestrians, the influence of obstacles and constraints to movement of pedestrians against brushing each other, particularly with the opposite gender. A broad categorisation of the factors affecting the quality of pedestrian facilities is presented in Fig. 4.34.

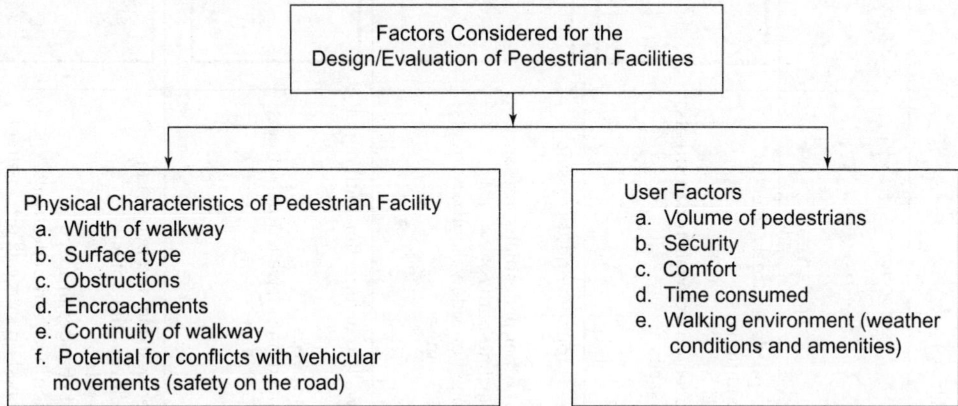


**Fig. 4.34** Factors affecting the quality of service of pedestrian facilities

## 4.15.1 LOS Criteria for Uninterrupted Flow Pedestrian Facility

Density and flow rate are the main parameters considered for the quantitative level of the service, of which there are six types: LOS-*A*, *B*, *C*, *D*, *E* and *F*, with *A* representing the *best* and *F* representing the *worst* walking conditions. The pedestrian flow conditions with respect to these six levels of service are presented in Table. 4.3. This method of determination of pedestrian LOS is widely known as Fruin's flow rate method (Bloomberg and Burden, 2006).

With reference to IRC: 103-2012, the following are the details used to determine the pedestrian level of service (relating to stationary and level walkways) using (a) pedestrian space and (b) pedestrian flow rate, as described by Fruin (1971) and TRB (2000).

(a)

$$\text{Pedestrian space}, M\left(\text{m}^2/\text{p}\right) = \frac{1}{\text{Density}(k)} \tag{4.8}$$

The density ($k$) of pedestrians in an area determines the LOS and it can be calculated using the following equation.

$$\text{Density}(k) = \frac{\text{Number of pedestrians in a study area of a walkway}(N)}{\text{Effective surface area of the walkway}(A)} \tag{4.9}$$

In general, the density value ($k$) is obtained at the time of maximum number (i.e., based on reasonable expectancy) of pedestrians observed in a study area. The effective walkway area ($A$) can be calculated using the effective width ($W_{\text{Effective}}$) of the walkway and the length of the walkway ($L$). The following equation is used to calculate the value of $W_{\text{Effective}}$ (Fig. 4.35):

**Table 4.3** Illustration of pedestrian level of service on walkways and sideways
(Courtesy: Indian Roads Congress, New Delhi; Transportation Research Board of National Scientific Council, Washington, DC; Georgia Department of Transportation, USA.)

| *LOS* | *Pedestrian space ($m^2$/p)* | | *Flow rate (p/m/min)* | | *Description* | *Schematic presentation (GDOT, 2003)* |
|---|---|---|---|---|---|---|
| | *IRC: 103-2012* | *TRB (2000)* | *IRC: 103-2012* | *TRB (2000)*[†] | | |
| LOS-A | >4.9 | >5.60<br>(>3.2, Fruin, 1971) | ≤12 | ≤16.0<br>(≤23.0, Fruin, 1971) | * Walking speeds are freely selected<br>* Ideal conditions<br>* No conflicts<br>* Travel speed on sidewalk urban streets: >1.33 m/sec | |
| LOS-B | >3.3–4.9 | 3.70–5.60<br>(2.3–3.2, Fruin, 1971) | ≤12–15 | 16.0–23.0<br>(23–33, Fruin, 1971) | * Free speeds found without crossing conflicts<br>* Aware of other pedestrians' presence while selecting paths<br>* Travel speed on sidewalk urban streets: >1.17–1.33 m/sec | |
| LOS-C | >1.9–3.3 | 2.20–3.70<br>(1.4–2.3, Fruin, 1971) | ≤15–21 | 23.0–33.0<br>(33–49, Fruin, 1971) | * Free to walk in unidirection with minor conflicts<br>* Reduced speeds<br>* Travel speed on sidewalk urban streets: >1.00–1.17 m/sec | |
| LOS-D | >1.3–1.9 | 1.40–2.20<br>(0.9–1.4, Fruin, 1971) | ≤21–27 | 33.0–49.0<br>(49–66, Fruin, 1971) | * Restricted walking and bypassing<br>* Reverse flow with conflicts<br>* Slow speeds<br>* Frequent changes in speed<br>* Interaction between pedestrians<br>* Travel speed on sidewalk urban streets: >0.83–1.00 m/sec | |

| LOS | Pedestrian space ($m^2/p$) | | Flow rate (p/m/min) | | Description | Schematic presentation (GDOT, 2003) |
|---|---|---|---|---|---|---|
| | IRC: 103-2012 | TRB (2000) | IRC: 103-2012 | TRB (2000)† | | |
| LOS-E | >0.6–1.3 | 0.75–1.40<br>(0.5–0.9, Fruin, 1971) | ≤27–45 | 49.0–75.0<br>(66–82, Fruin, 1971) | * Restricted speeds<br>* Move forward by shuffling<br>* Extreme difficulty in crossing/bypassing<br>* Stoppage and interrupted flow<br>* Travel speed on sidewalk urban streets: ≥0.58–0.83 m/sec | |
| LOS-F | ≤0.6 | ≤ 0.75<br>(≤ 0.5, Fruin, 1971) | Varies | Varies<br>(Variable to 82, Fruin, 1971) | * Severely restricted speeds<br>* Shuffling to move<br>* Unavoidable contact<br>* Sporadic, unstable flow<br>* Queued pedestrians<br>* Travel speed on sidewalk urban streets: <0.58 m/sec | |

† The TRB recommended flow rates are greater than the flow rates recommended by the IRC. This may be due to the higher walking speeds found in USA than in India. The clear space between adjacent pedestrians maintained in Asian countries is smaller than that observed in Western countries.

*Note*: The above OL criteria (i.e., in terms of flow rates and space) are not applicable in the case of platoon flow of pedestrians, as observed commonly in bus transit terminals, airports and railway stations. Platoon-adjusted LOC criteria values and conditions can be obtained from TRB (2000).

$$\text{Effective width of the walkway } (W_{\text{Effective}}) = \text{Total width of the walkway } (W_T) - \text{Sum of widths of obstacles and shy-away distances } (W_o) \tag{4.10}$$

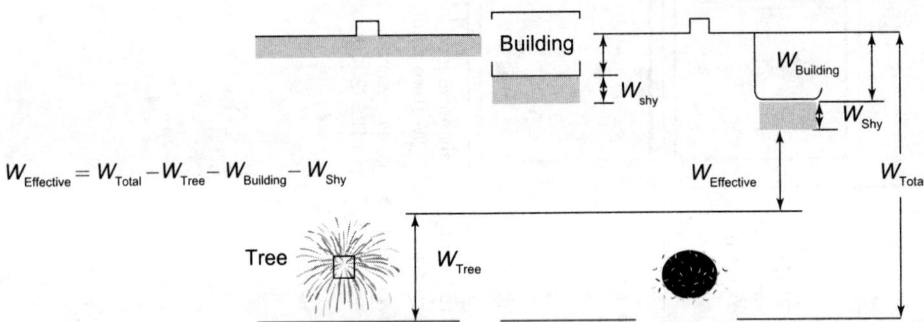


**Fig. 4.35** Typical example: Calculation of effective width ($W_o$) of the walkway

The pedestrian flow rate can be determined from the following equation (TRB, 2000):

(b)
$$\text{Pedestrian flow rate}\left(\frac{\frac{p}{\min}}{m}\right) = \frac{\text{Peak 15-minute pedestrian volume}}{15 \times W_{\text{Effective}}} \tag{4.11}$$

## 4.15.2 LOS Criteria for Stairways and Queuing

The LOS criteria for stairways and queuing as recommended by the TRB (2000) are presented in Table 4.4.

**Table 4.4** Partial details showing the LOS criteria for stairways and queuing areas (*Source*: TRB, 2000; *Courtesy*: Transportation Research Board of National Scientific Council, Washington, DC, USA.)

| LOS | *For stairs (TRB, 2000)* | | | *Queuing areas (Fruin, 1971)* | |
|---|---|---|---|---|---|
| | *Pedestrian space ($m^2/p$)* | *Flow rate (p/m/min)* | *Average horizontal speed (m/sec)* | *Pedestrian space ($m^2/p$)* | *Interperson space (m)* |
| A | ≤ 1.9 | ≤16 | >0.53 | ≤1.20 | ≥1.2 |
| B | 1.9–1.6 | 16–20 | | 1.20–0.90 | 1.1–1.2 |
| C | 1.6–1.1 | 20–26 | 0.48–0.53 | 0.90–0.70 | 0.9–1.1 |
| D | 1.1–0.70 | 26–36 | 0.42–0.48 | 0.70–0.30 | 0.6–0.9 |
| E | 0.70–0.5 | 36–49 | 0.40–0.42 | 0.30–0.2 | ≤0.6 |
| F | ≥0.5 | Variable | ≤0.40 | ≥0.2 | Touching |

The LOS criteria for pedestrian crossflows and pedestrian–bicycle shared two-way paths may be obtained from TRB (2000).

### 4.15.3 LOS Criteria for Interrupted Flow Pedestrian Facility

The interrupted flow condition is commonly referred to for two conditions: at signalised intersections and at un-signalised intersections. The walking time, crossing phenomenon and waiting time for signal change or for crossing the road are important factors considered for determining the LOS. The average delay experienced by pedestrians at these two locations is used to determine the LOS (Table 4.5).

**Table 4.5** LOS criteria for interrupted flow conditions (Source: TRB 2000, Courtesy: Transportation Research Board of National Scientific Council, Washington, DC, USA.)

| *LOS* | *At signalised intersections* | *At un-signalised intersections* | *Risk of acceptance* |
|---|---|---|---|
| | *Pedestrian delay*† *(sec/p)* | *Pedestrian delay (sec/p)* | |
| A | 0–10 | 0–5 | Low |
| B | ≥10–20 | ≥5–10 | |
| C | >20–30 | >10–20 | Moderate |
| D | >30–40 | >20–30‡ | |
| E | >40–60 | >30–45 | High |
| F | >60 | >45 | Very High |

† Reflects low to moderate conflicting vehicle volumes
‡ Impatience begins after 30 seconds of delay (Mahmassani et al., 1993)

It has been noted that the average delay of pedestrians observed at signalised intersection crossings is not constrained by the capacity of the pedestrians, even when flow rate reaches 83.3 p/min (TRB, 1998). However, pedestrian overloaded street corners (as temporary holding areas) and pedestrian crossing queues (as on circulation space) can affect vehicular flow and result in a demand for increased pedestrian Green time.

The formulae used to determine pedestrian space–time relationships can be obtained from TRB (2000). The following equation is recommended by TRB (2000) for calculating the average delay of pedestrians based on signal timing (in seconds):

$$\text{Average pedestrian delay} = \frac{(\text{Cycle length} - \text{Effecive green time for pedestrians})^2}{2 \times \text{Cycle length}} \qquad (4.12)$$

## 4.16 TIME–SPACE (TS) ANALYSIS METHOD

The LOS criteria and associated pedestrian traffic flow equations presented in the above sections are applicable when pedestrian movement is continuous, without any delays or waiting time. In such cases, pedestrian flow is assumed to be analogous to hydraulic flow in channels. In areas where pedestrian flow is *not* uniform but is discontinuous due to delays and queuing, an alternative analysis technique, widely known as the time–space (TS) method, is used. The requirement of circulation space for walking passengers and a holding place for standing is determined using this concept. For example, delay and waiting of pedestrian flow is common at places such as signalised intersections, on traffic islands before crossing and while queuing at a ticket counter (Maria and Braaksma, 1986).

Therefore, the TS analysis method is used for evaluating/checking the adequacy of pedestrian spaces, areas, flared controlled areas, platforms, kerb islands or street corners where both waiting and walking can occur simultaneously. Fruin and Benz's (1984) concept of TS is presented below.

$$\text{Time–Space}_{\text{Required}} = \sum_{i=1}^{n} P_i \times S_i \times T_i$$

where $P_i$ = number of people involved in the activity; $S_i$ = space required per person for the activity; and $T_i$ = time required for the activity.

The above value of time–space required is compared with the time–space available. The time–space available is the product of the available area (in m²) and the time it is available in (as analysis period, in minutes). A comparison between the required and available space per pedestrian is used for evaluating a pedestrian facility based on the LOS standards.

The following fundamental relationships derived from the above concepts are used for evaluating/determining pedestrian space requirements at these critical spaces (Benz, 1986).

$$\text{Time–Space supply (TSS)} \approx \text{Time–Space demand (TSD)} \tag{4.13}$$

$$\text{Average pedestrian space requirement, } M\left(\text{m}^2/\text{p}\right) = \frac{\text{TSS}}{\text{TSD}} \tag{4.14}$$

$$\text{TSS} = \text{Time period of the analysis period (min)} \times \text{Effective area of space (m}^2\text{)}$$

$$\text{TSD} = \text{No. of the pedestrians occupying the space} \times \text{Time of occupation of space (min)}$$

*Note 1*: As a common practice, the value of the time period of analysis is considered as a peak 15-minute period of analysis.

*Note 2*: For the determination of time–space (TS) availability at a street kerb corner on a signalised intersection (Fig. 4.36), the following expression is recommended by TRB (2000):

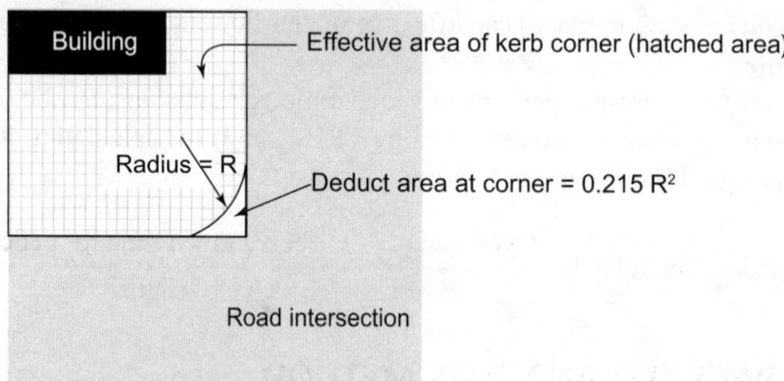


**Fig. 4.36** Details of deduct area at the kerb corner to determine the effective area of the kerb island (redrawn based on TRB, 2000)

$$TS_{\text{Available}} = (\text{Signal cycle length in minutes}) \times (\text{Effective area of kerb corner} - 0.215 \times \text{R}^2) \tag{4.15}$$

## 4.17 DESIGN GUIDELINES

**Width of the pavement:** The required width of the pavement depends on the pedestrian flow rate and LOS. The minimum effective walkway width is 1500 mm for a pedestrian flow rate per hour of not greater than 600, with LOS-A. The width may be 1220 mm for a pedestrian flow rate per hour not greater than 600, with LOS-B through E.

The minimum clear width of walkway provided should be 1800 mm, with a right-of-way ≥10.0 m. This clear width can accommodate two wheelchairs passing side by side (i.e., 900 mm + 900 mm). The design flow rate (p/hr) in both directions expected with a clear width of 1800 mm is 1350 and 1890 for LOS-B and LOS-C, respectively. Other widths (Fig. 4.37) and the pedestrian flow rate with their LOS may be obtained from IRC: 103-2012.

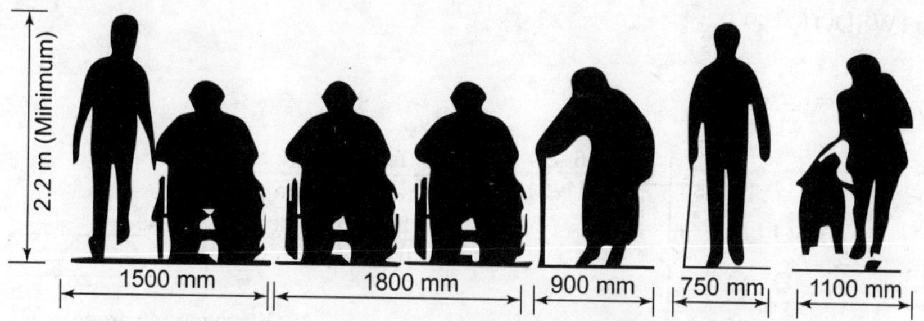


**Fig. 4.37** Typical dimensions of pedestrian width and clear height

Based on the type of land use, the minimum required dimensions of clear width pavement as recommended by UTTIPEC (2009) are as follows:

- Residential areas: 1.8 m (minimum)
- Commercial/Mixed use area: 2.5 m
- Commercial nodes: 4.00 m

**Fig. 4.38** Kerb height

**Kerb height:** The maximum kerb height (top of paving), including at bus stops, shall not exceed 150 mm (Fig. 4.38). However, on arterial roads, the kerb height is restricted to 100 mm (UTTIPEC, 2009).

**Longitudinal gradient:** The preferred walking gradient for generally accessible routes is between 0% and 3%. If the gradient is between 3% and 8%, a level area of 15 m should be provided for every 300 m of ramp length. The ramp gradient provided, intended to be used by the physically challenged (with wheelchairs) and those with strollers and bicycles, should not exceed 8.33% (i.e., 1:12) (Fig. 4.39). However, ramp slopes with up to 10% gradient may be built for short distances, when the street space available is inadequate.

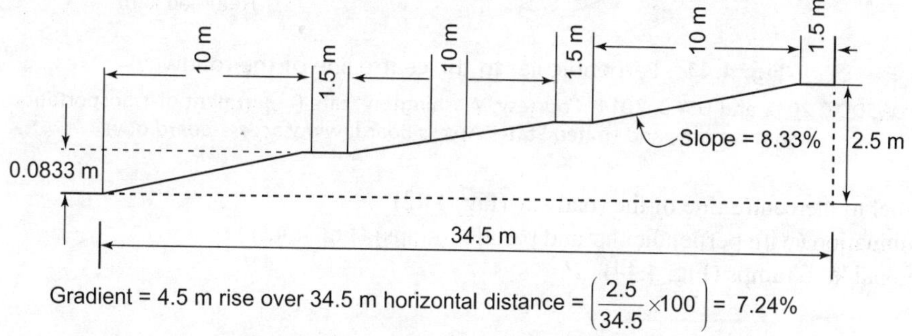


Gradient = 4.5 m rise over 34.5 m horizontal distance = $\left(\frac{2.5}{34.5}\times 100\right)$ = 7.24%

**Fig. 4.39** Gradient guidelines for access ramps and separated walkways

**Pavement kerb ramps:** Pavement kerb ramps provide crossing accessibility at intersections, mid-block locations, building entrances, transit platforms, bus stops and other areas where elevated walkways are edged with kerbing. In another words, the kerb ramps provide an accessible connection from the roadway surface to the raised sidewalk. The kerb ramps have a detectable warning surface to help pedestrians with disabilities (sight-impaired and physically-impaired) identify the street crossing location (Fig. 4.40). The transition between the surface of the pavement and the kerb should be smooth, without a lip (WSDOT, 1997)

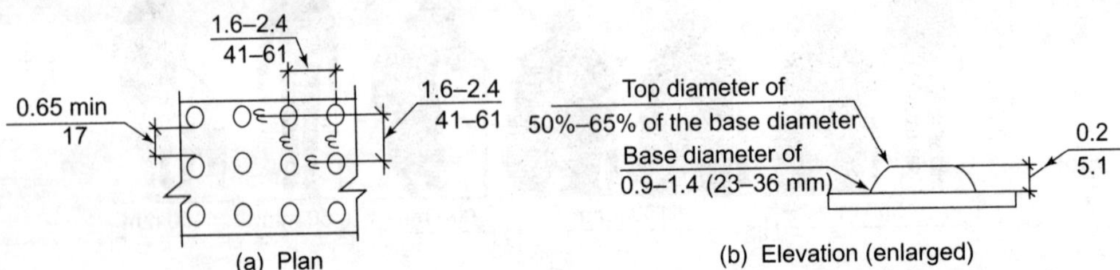


**Fig. 4.40** Size and spacing of truncated domes used as a detectable warning surface

(*Source*: USAB, 2011; www.access-board.gov/guidelines-and-standards/...)

The kerb ramps are oriented in each direction of pedestrian travel within the width of the pedestrian crossing they serve. Based on the requirements, kerb ramps are oriented as follows:

- Perpendicular to the centre line of the roadway (Fig. 4.41)

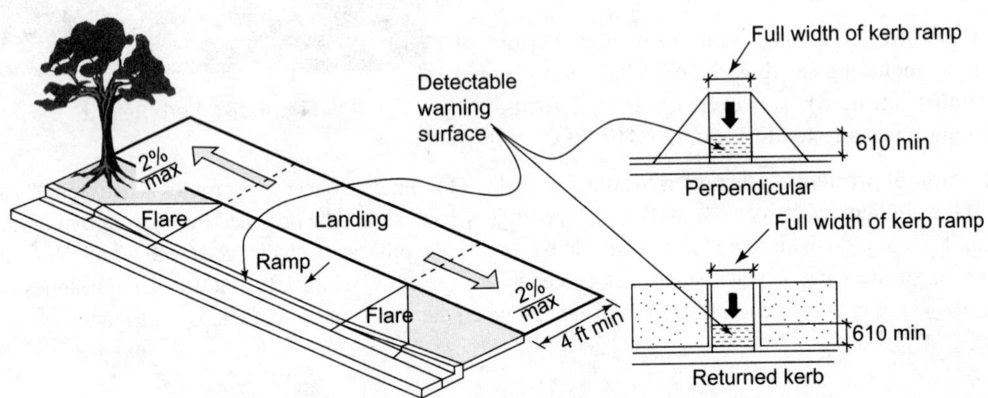


**Fig. 4.41** Perpendicular to the centre line of the roadway

(*Source*: WSDOT, 2012 and USAB, 2011. *Courtesy*: Washington State Department of Transportation, www.wsdot.wa.gov, and United States Access Board, www.access-board.gov)

- Parallel to the centre line of the roadway (Fig. 4.42)
- Combination (with perpendicular and parallel ramps) (Fig. 4.43)
- Diagonal kerb ramps (Fig. 4.44)

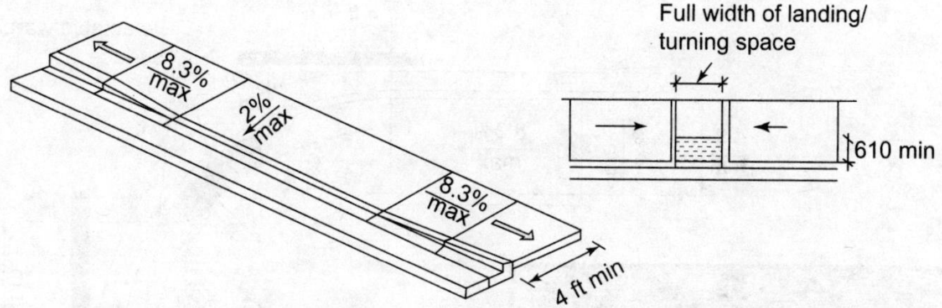


**Fig. 4.42** Parallel to the centre line of the roadway

(*Source*: WSDOT, 2012 and USAB, 2011. *Courtesy*: Washington State Department of Transportation, www.wsdot.wa.gov and United States Access Board, www.access-board.gov)

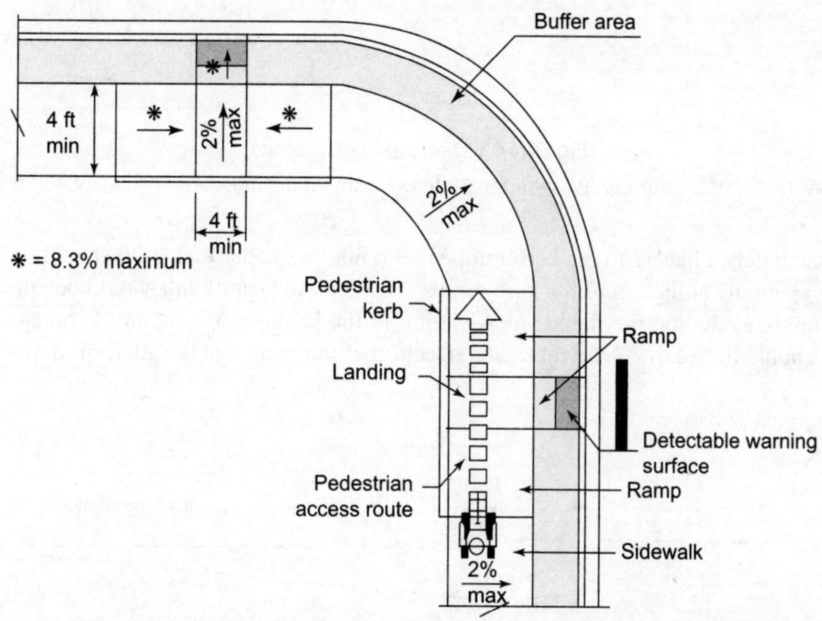


**Fig. 4.43** Combination (with perpendicular and parallel ramps)

(*Source*: WSDOT, 2012. *Courtesy*: Washington State Department of Transportation, www.wsdot.wa.gov )

The kerb with the orientations mentioned above are provided with drainage considerations, so that wheelchairs, strollers and bicycles can be eased across the kerb and the roadway, and vice versa.

The limiting value of the kerb ramp running slope should be taken as 7.5% (against a maximum of 8.3%) and the kerb ramp length must be limited to 4.5 m. The minimum desirable ramp width is 1219 mm. The limiting value of the cross slope kerb ramp (i.e., the flare measured perpendicular to the direction of travel) should be 1.5% (2% maximum). Counter slopes of adjoining gutters and road

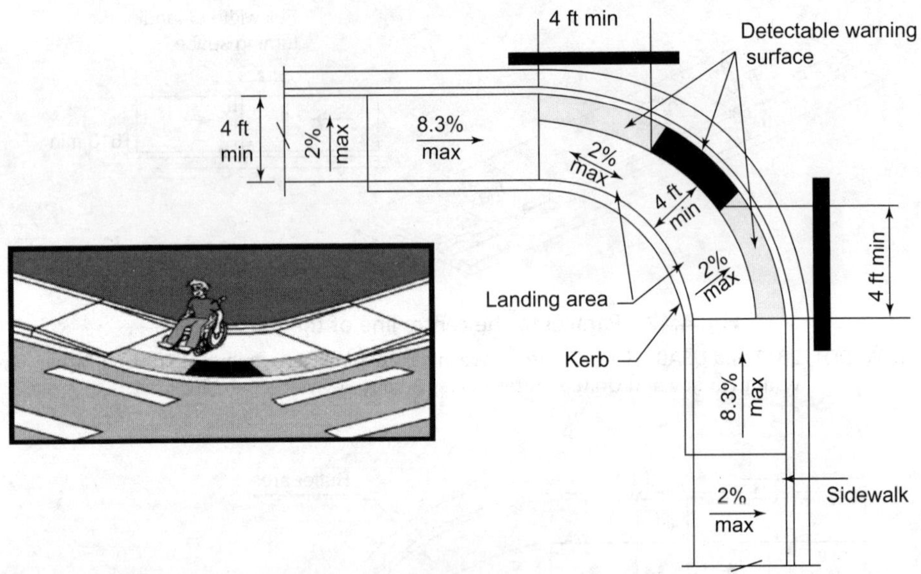


**Fig. 4.44** Diagonal kerb ramps
(*Source*: WSDOT, 2012. *Courtesy*: Washington State Department of Transportation, www.wsdot.wa.gov)

surfaces immediately adjacent to the kerb ramp should not be steeper than 1:20 (Fig. 4.45). Landings shall be provided only at the top of the kerb ramps. The landing clear width should be at least as wide as the kerb ramp, excluding the flared sides leading to the landing. As a common practice, the level landing size should be 1.2 m× 1.2 m at the intersection of the ramp and the adjacent flares.

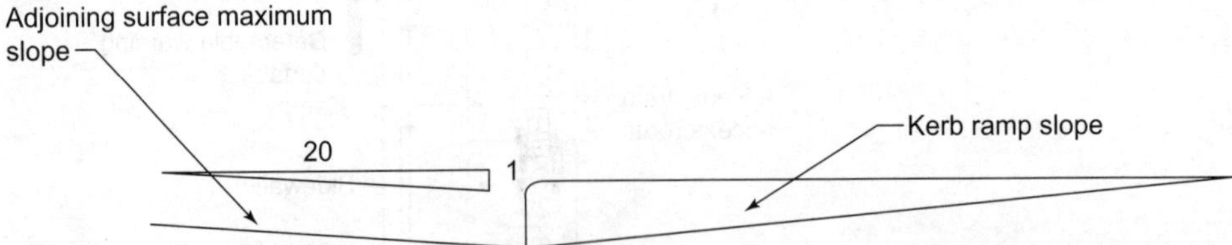


**Fig. 4.45** Counter slope of surfaces adjacent to kerb ramps
(*Source*: USAB, 2011. *Courtesy*: United States Access Board, USA)

**Stairways:** A combination of steps and landing area is provided for a ramp slope ≥8.33% (Fig. 4.46). In such cases, the minimum length of the landing = (Tread + 3 × Average pacing distance). As a common practice, the value of the risers is fixed as low as 130 mm and the treads as wide as 360 mm (Fig. 4.47). Stairway handrail height should be >1100 mm. The minimum stairway width for two-way movement is 760 mm + 760 mm. The minimum preferred width may be 1830 mm.

**Guard rails:** As a safety hazard, adequate height (>1.2 m) of mild steel (MS) guard rails should be provided on pedestrian walking areas next to any significant drop or against vehicular flow or with ramps/stairways (Fig. 4.48).

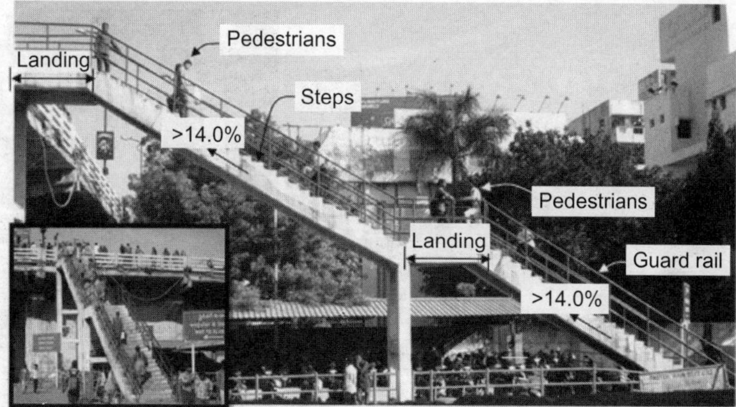


Fig. 4.46 A view of steps and landing (with step flight slope ≥14.0%)

**Access control:** Bollards or boulevards of standard dimension may be installed to restrict the entry of vehicles into a pedestrian pathway/facility (Fig. 4.49).

**Walking distance:** The most widely accepted walking distance, not exceeding 0.5 k, to transit stops/amenities may be considered while planning pedestrian facilities.

**Design period:** A 15-minute time period is used as the design period. Peak 15-minute volume (Peak hour factors × Peak hour volume) is used as the design pedestrian flow rate.

Fig. 4.47 Step dimensions

Fig. 4.48 A view of a walkway illustrated with effective width and guard rail height

Fig. 4.49 A view of bollards, restricting the entry of vehicles

**Pedestrian signal time gap:** Pedestrian signals are designed based on the time gap (in seconds) required for crossing a road. The following equation may be used to calculate the minimum time gap (TRB, 2000):

$$\text{Minimum time gap} = (W/V_p) + \text{Start-up time} + \text{Time gap between pedestrians} \times (N-1) \quad (4.16)$$

where $W$ = width of the roadway crossing (m) and $V_p$ = average walking speed of pedestrians while crossing (m/sec).

**Grade separated pedestrian crossing facilities:** Pedestrian road crossings are of three types (Fig. 4.50). At-grade pedestrian crossings are the shortest and most convenient type. However, it may not be feasible to provide them for the following reasons (IRC: 103-2012):

- Where insertion of an exclusive pedestrian phase signal increases the total traffic cycle time beyond 120 seconds.
- Uninterrupted flow is not permitted on urban major arterial roads and high-speed roads such as national highways and expressways.
- Where there is non-visibility of crossing/accessing pedestrians across an intersection.

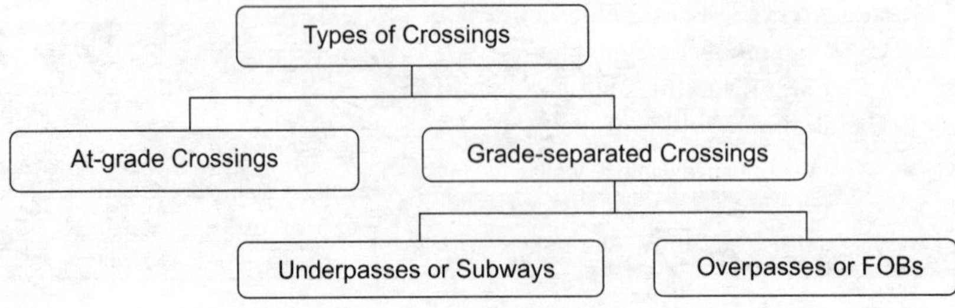


Fig. 4.50 Types of pedestrian crossings

- Occurrence of vehicular delay during pedestrian crossing and also based on the recorded evidence of vehicle–pedestrian collisions/crashes.
  The optional provisions with reference to their prioritisation in the order of ease and comfort to pedestrians and to the cost of construction (IRC: 103-2012) are as follows:
- **Hump subways** (Fig. 4.51): A clear raised height of 1.75 m with reference to the ground level is recommended. The depth of the hump should be 1.5 m. The ramp slope may be restricted to 8.33% to facilitate the physically challenged. Bicycles may also be taken through this hump subway, as the depth of the hump is low enough to lift the bicycle over the ramp/steps. Minimum width should be 4.8 m and minimum visibility of at least 50 lux should be provided across the two ends.

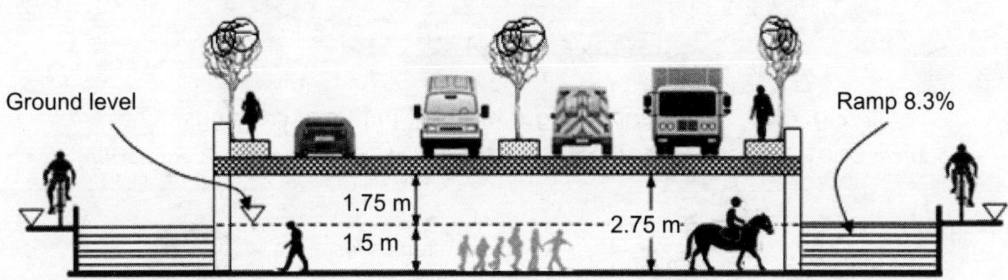


**Fig. 4.51** Sectional elevation of a hump subway (modified and elaborated based on IRC: 103)

- **Full subways** (Fig. 4.52): A total clear depth of 2.75 m should be provided under the ground level. The ramp slopes and use of this type of subway are similar to hump subways. Provision should be made while planning to ensure natural lighting/ventilation.

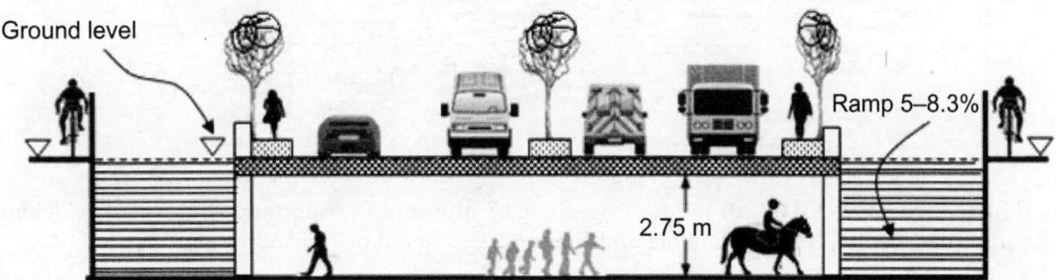


**Fig. 4.52** Sectional elevation of a full subway (modified and elaborated based on IRC: 103)

- **Foot overbridges** (FOB)/**Footbridge/Pedestrian overpass/Overhead crossings** (Fig. 4.53): These are bridges commonly designed for pedestrians and, in some cases, for animal traffic, horse riders and cyclists. FOB facilities are commonly designed as stairways in most urban localities (Fig. 4.54). The minimum width of the FOB should be 1800 mm. The ramp slope or step flight slope should preferably be 5%, with appropriate landings (Fig. 4.55).

The following guidelines recommended by the Unified Traffic and Transportation Infrastructure (Planning & Engineering) Centre, Delhi Development Authority, New Delhi, may be adopted where the roadway width is $\leq$30 m (UTTIPEC, 2009):

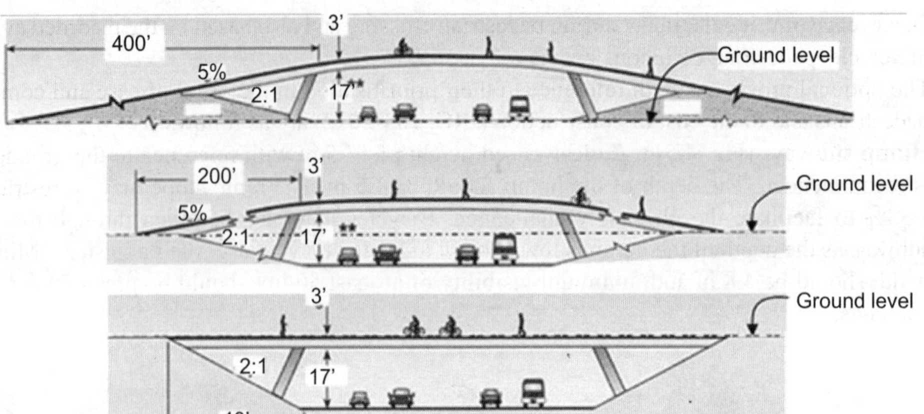


**Fig. 4.53** Different configurations of path overcrossings

(*Source*: ODOT, 2011. Courtesy: Oregon Department of Transportation, Oregon, USA)

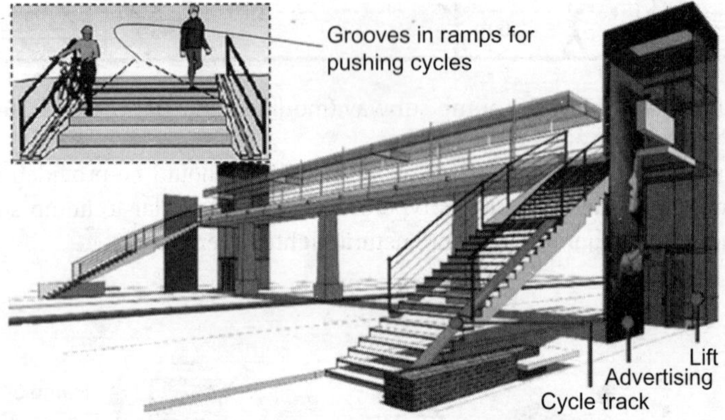


**Fig. 4.54** Suggested stairway (FOB) design

(*Source*: UTTIPEC, 2009 and ODOT, 2011. *Courtesy*: Unified Traffic and Transportation Infrastructure (Planning & Engineering) Centre, New Delhi, India and Oregon Department of Transportation, Oregon, USA)

**Fig. 4.55** View of a stairway (FOB)

- **Location criteria**: Mid-block locations in shopping areas, schools, key civic areas, residential areas, transit/bus stops and similar land-use pedestrian attractions
- **Universal accessibility** with a combination of either staircase + ramp or staircase + elevator
- **Minimum width of FOB walkway**: 2.50–3.00 m (FOB only for pedestrians); 3.50 m (pedestrians and cyclists)
- **Minimum width of staircase**: 2.5 m
- **Vertical clearance**: 5.5 m above roadways; the FOB should span the entire carriageway
- **Slope**: 8% on footbridge ramps and 5% with landings
- **Minimum size of elevator:** 1400 × 1400 mm
- **Cycle elevators** should be provided at every alternate FOB with a minimum size of 1400 × 2000 mm.
- CCTV cameras should be provided with full coverage of all types of pedestrian facilities for security purposes.

**Continuity in flow:** There should not be any abrupt/sudden break in the path of the pedestrian walkway, particularly with vehicular movement or due to obstruction or footpath encroachments. Pedestrian planners must accommodate this aspect during design auditing against the safety and continuity of the pedestrian facilities.

**Location of mid-block crossings:** Mid-block road crossings should be provided at every bus/train stop and important pedestrian facilities. As a common guideline (IRC: 102), the spacing interval may range 80–250 m in residential areas and 80–150 m in commercial areas.

**Pedestrian facilities in work zones:** Temporary installations during construction activities on the roadside require special care as they are a safety concern. The design details may be obtained from any standard manual (IRC: 103-2012; UTTIPEC, 2009; MUTCD, 2012).

**Floor surface friction:** The desirable value of the coefficient of surface friction of pedestrian flooring in public transit areas is ≥0.5.

**Traffic signage, markings and signals:** Examples of the signage and markings/indicators used to facilitate pedestrian flow are given below:

- Pedestrian crossing markings (Fig. 4.56)
- Flashers in the pavement
- Overhead/Elevated signs
- Blinking/Animated indicators of pedestrian movement

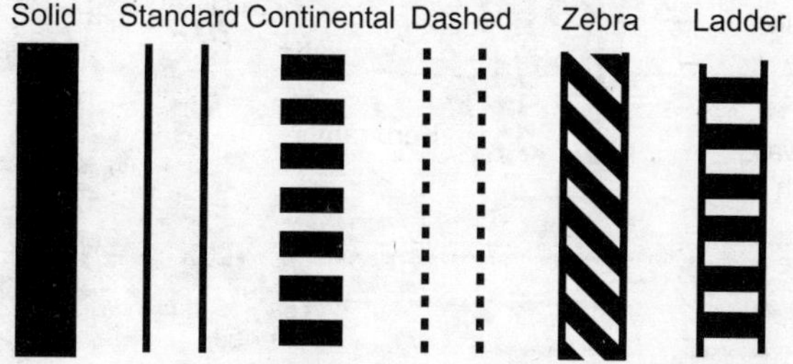


**Fig. 4.56 Types of pedestrian crossing marking patterns**

(*Source*: FHWA: RD-01-102; Zegeer et al., 2002. *Courtesy*: Federal Highway Administration (FHWA), US Department of Transportation, USA)

- Static signboards indicating cross flow of school children
- Pedestrian signals (approximately 20 seconds) synchronised with the nearest traffic signals (Fig. 4.57)
- Audible signals (beeps when signal is Green)

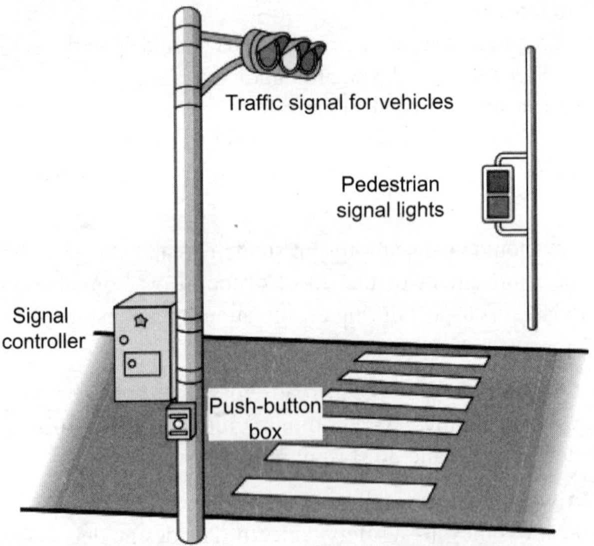


**Fig. 4.57(a)** Push-button signal controller

(*Source*: IUTR, 1998. *Courtesy*: Japan Traffic Management Technology Association, Japan)

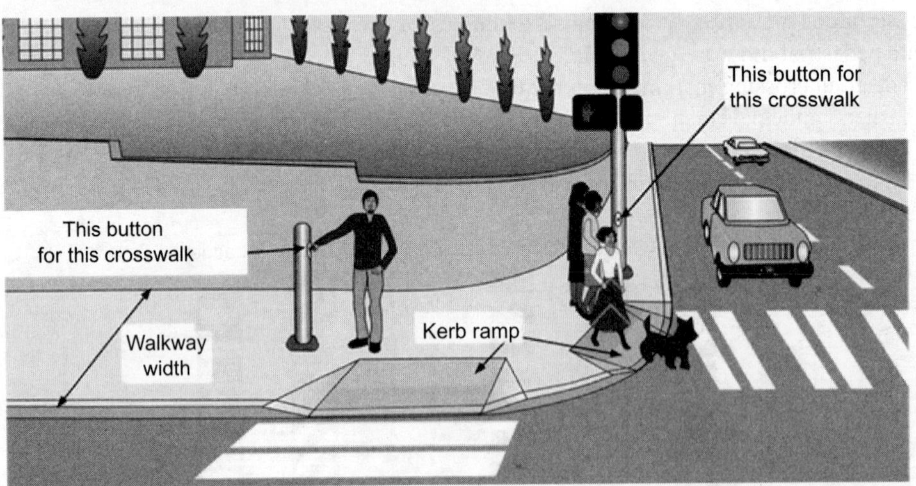


**Fig. 4.57(b)** Proper placement of pedestrian push-button

(Modified based on ODOT, 2011. *Courtesy*: Oregon Department of Transportation (ODOT), Oregon, USA)

## 4.18 SOLVED PROBLEMS

**Example 4.1**

Determine the proper width of a walkway required to accommodate 1000 pedestrians in 10 minutes. The walkway is located adjacent to a building compound wall on one side and a telephone crew box in the walk on the other side. Assume standard flow rate of 35 p/m/min.

**Answer:** Let us determine the minimum required width of the walkway as

$$\text{Width of the walkway} = \left(\frac{N}{T \times Q_{std}}\right) + \text{(Offset distance or buffer on one side)} + \text{(Keep-away distance due to impediments or obstacles)}$$

where $N$ = number of pedestrians anticipated to walk in the peak period (p/m/min) = 1000 p (given); $T$ = duration of the peak period (min) = 10 minutes (given); $Q_{\text{std}}$ = standard flow rate (p/m/min during the peak period) = 35 p/m/min (given, but we can also assume this value based on LOS criteria A/B/C).

Let us assume that

Offset distance on one side due to a compound wall = 0.45 m. Keep-away distance due to telephone crew box = 0.65 m. On substitution of values in the above equation, we get

$$\begin{aligned}\text{Width of the walkway} &= \left(\frac{1000}{10 \times 35}\right) + (0.45\text{ m}) + (0.65\text{ m})\\ &= (2.85) + (0.45\text{ m}) + (0.65\text{ m})\\ &= 3.95\text{ m, say } 4.0\text{ m (Fig. 4.58)}\end{aligned}$$

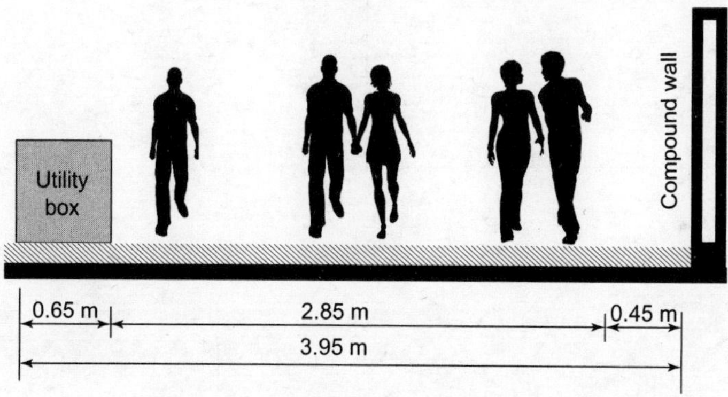


**Fig. 4.58** Proper width of the walkway

**Example 4.2**

Determine the LOS for average conditions for the following observations:

Pedestrian peak flow rate = 1000 pedestrians per 15 minutes. Total pavement width = 4.0 m. Kerb on one side and tree on other side of the walkway. No other obstructions.

**Answer:** Let us assume that the shy-away distance for the kerb = 0.5 m and for the tree = 0.6 m. The effective width of the walkway

$$(W_{Effective}) = 4.0 - 0.5 - 0.6 = 2.9 \text{ m}$$

The following TRB equation may be used to determine the pedestrian flow rate:

$$\text{Pedestrian flow rate}\left(\frac{\frac{p}{\min}}{m}\right) = \frac{\text{Peak 15-minute pedestrian volume}}{15 \times W_{\text{Effective}}}$$

$$= \frac{1000}{15 \times 2.9} = 22.98$$

Based on flow rate = 22.98 p/min/m, it is found from Table 4.3 that the LOS = B. That means that the above pavement is expected to be operated at LOS-B for average conditions.

**Example 4.3**

Determine the radius of the kerbs ($R_1$ and $R_2$) for the following widths of roadways with reference to Fig. 4.59: Roadway-1 width ($W_1$) = 9 m; Roadway-2 width ($W_2$) = 8 m; Roadway-3 width ($W_3$) = 10 m.

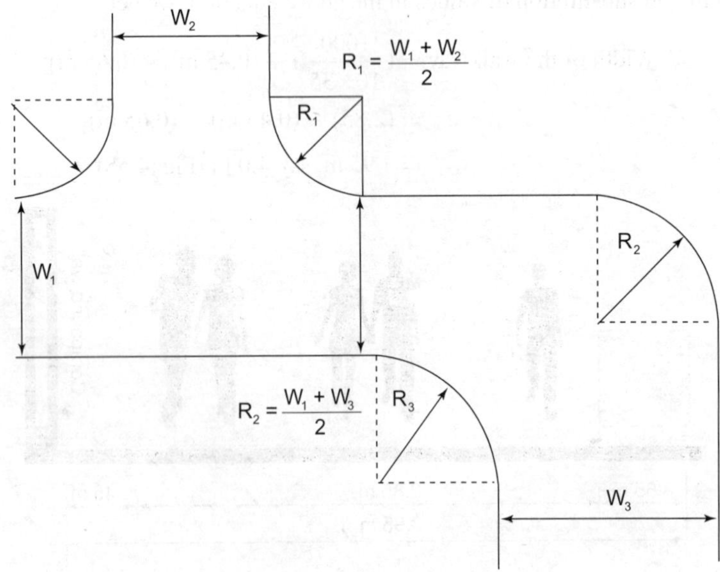


**Fig. 4.59** Determining the kerb radius

**Answer:** The calculation of kerb radii is as follows:

$$\text{Radius of the kerb} - 1: \; R_1 = \frac{9+8}{2} = 8.5 \, \text{m}$$

$$\text{Radius of the kerb} - 2: \; R_2 = \frac{9+10}{2} = 9.5 \, \text{m}$$

**Example 4.4**

Determine time delay of a pedestrian crossing at a signalised intersection (without pedestrian signals) for the following given signal timings:
Number of signal phases = 2. Cycle length = 90 seconds with a 4.0-second change interval. Major Street (15 m width): Phase-1 Green time = 50.0 seconds. Minor Street (10 m width): Phase-2 Green time = 32.0 seconds. The TRB (2000) recommended formula may be used.

**Answer:** The following TRB-recommended formula is used:

$$\text{Average pedestrian delay} = \frac{(\text{Cycle length} - \text{Effecive Green time for pedestrians})^2}{2 \times \text{Cycle length}}$$

$$\text{Average pedestrian delay on Major Street} = \frac{(90-32)^2}{2 \times 90} = 18.6\,\text{sec}$$

From Table 4.3: LOS on Major Street = B

$$\text{Average pedestrian delay on Minor Street} = \frac{(90-50)^2}{2 \times 90} = 8.8\,\text{sec}$$

From Table 4.3: LOS on Minor Street = A

**Example 4.5**

Determine pedestrian flow per metre of effective corridor width for the following observed values: Average normal walking speed of pedestrians = 0.9 m/sec. Average pedestrian space as LOS-C = 2.0 m²/p.

**Answer:** $\text{Volume in pedestrians per metre width of walkway per sec.} = \frac{\text{Average speed of pedestrians}}{\text{Average area per pedestrian}}$

$$\text{Volume in pedestrians per metre width of walkway per sec.} = \frac{0.9}{2.0} = 0.45\ \text{p/m/sec.}$$

$$\text{Volume in pedestrians per metre width of walkway per minute} = 27\ \text{p/m/min.}$$

*Note:* This method of analysis is applicable only where continuous flow takes place. The time–space analysis method may be adopted in case of delays, waiting and queuing exits in the study areas.

**Example 4.6**

Calculate the average area per pedestrian ($M$) required during a 15-minute period for the following given field conditions:
Number of passengers expected to cross a road at an intersection = 1000 during a peak 15 minutes. Area of ticket selling counter space = 180 m². Average walk time through this space = 12 seconds. Waiting time for the half of the passengers in queue line to purchase tickets = 10 seconds.

**Answer:** The space required per person is calculated by the formula given below

$$\text{Average pedestrian space required during the peak15 minute period, } M\left(\text{m}^2/\text{p}\right)=\frac{\text{TSS}}{\text{TSD}}$$

TSS = Time period of the analysis period (min) × Effective area of the space ($m^2$)

TSS = 15 minutes × 180 $m^2$ = 2700 $m^2$-min

TSD = No. of pedestrians occupying the space × Time of occupation of space (min)

$$\text{Time of occupation of space (sec)} = \{(\text{Average walk time}) + (\text{Average wait time})\}$$

$$\text{Time of occupation of space (sec)} = \left\{(12)+\left(\frac{1}{2}\right)\times(10)\right\} = 12+5 = 17 \text{ sec}$$

$$\text{Time of occupation of space (minutes)} = \frac{17}{60} = 0.2833 \text{ min}$$

$$\text{TDS} = 1000\times 0.283 = 283.3 \text{ p} - \text{min}$$

$$\text{Average}\frac{\text{space}}{\text{p}}\text{ required during the peak 15 minutes} = \frac{2700}{283.3} = 9.53 \text{ m}^2/\text{p}$$

$$\text{Average}\frac{\text{space}}{\text{p}}\text{ required during the peak15 minutes for ticket selling desk} = 9.53 \text{ m}^2/\text{p}$$

**Example 4.7**

Determine the width of a corridor 35 m long, with an estimated flow of 200 people per minute. The observed average walking speed = 0.95 m/sec. The desirable LOS is C, with pedestrian space required = 2.6 $m^2$/p. Take the time of the analysis period ($T$) as 1 minute. The predicted occupancy time of the pedestrians for purchasing tickets during the analysis period = 30 seconds.

**Answer:** The corridor width required can be calculated by adopting the time–space analysis method
Time Space Supply (TSS) ≈ Time Space Demand (TSD)
Let the corridor width (i.e., effective width) = $W$

$$\text{TSS} = 2.6 \text{ m}^2\text{per pedestrian}\times 200 \text{ No. of pedestrians}\times\left(\frac{30 \text{ sec}}{60}\right)$$

$$\text{TSD} = \{(1 \text{ minute})\times(W\times 35 \text{ m})\}$$

On substitution, we get

$$\left\{2.6 \text{ m}^2\text{per pedestrian}\times 200 \text{ pedestrians}\times\left(\frac{30 \text{ sec}}{60}\right)\right\} \approx \{(1 \text{ minute})\times(W\times 35 \text{ m})\}$$

Rewriting the above expression

$$\text{W} \approx \frac{2.6\frac{\text{m}^2}{\text{p}}\times 200 \text{ No.}\times\left(\frac{30}{60}\right)}{(1 \text{ minute})\times(35 \text{ m})} \approx \frac{2.6\frac{\text{m}^2}{\text{p}}\times 200\text{No.}\times 0.5 \text{ minute}}{(1 \text{ minute})\times(35 \text{ m})} = \frac{260}{35} = 7.42 \text{ m}$$

**Example 4.8**

Calculate the time gap required for pedestrians to cross at an intersection for the following intersections: Width of the roadway = 10.0 m. Average walking speed of pedestrians = 0.92 m/sec. Design value of start-up time of elderly pedestrians = 3.2 seconds. Time gap between pedestrians = 2.2 seconds. Expected number of rows of of pedestrians just after a film show = 6 (i.e., 5 in each row).

**Answer:** The following TRB (2000) recommended equation may be used to calculate the time gap requirement for the design pedestrian signal

Minimum time gap = $(W \div V_p)$ + Start-up time + Time gap between pedestrians × $(N - 1)$

Minimum time gap = $(10.0 \div 0.92) + 3.2 + [2.2 \times (6 - 1)] = (10.86) + 3.2 + [11]$

= 25.06 seconds

**Example 4.9**

A new bus shelter is required to provide a passenger loading area to protect the commuters. The peak passenger volume expected to use the shelter is 100. Determine the area required for the shelter.

**Answer:** Several studies were conducted to determine the pedestrian buffer zone space requirements. Based on field studies, it was suggested that the average pedestrian body ellipse size be taken as 0.5 m × 0.6 m, which works out to 0.3 $m^2$ per person. This value is considered the basic space occupied by a single pedestrian in the standing position.

The basic area required as standing position requirement = 0.3 $m^2$ per person × 100 = 33 $m^2$

Based on the optimum density criteria (Virkler and Elayadath, 1994), 1.3 and 1.8 p/$m^2$.
Area required for the passenger shelter should be in the range:
Minimum: 100/1.8 = 55 $m^2$; Maximum: 100/1.3 = 76 $m^2$

## EXERCISES

### MULTIPLE CHOICE QUESTIONS

1. The marked area on a carriageway which is provided for pedestrians crossing at street crossings or mid-block locations is
   (a) Footpath (b) Pedestrian crossing (c) Shy-away distance (d) Cross flow
2. The path that includes independent pathways that are provided as walkways, sideways, and shared bicycle pathways with other types of pedestrian facilities is
   (a) Footpath (b) Pedestrian circulation path (c) Shy-away distance (d) Pedestrian access path
3. The connecting pathways/routes provided between pedestrian facilities and regular walkways is
   (a) Footpath (b) Pedestrian circulation path (c) Shy-away distance (d) Pedestrian access path
4. The unit used to represent the measure of pedestrian space, where *p* represents pedestrians and *m* represents metre is
   (a) m (b) p/$m^2$ (c) $m^2$/p (d) m/p
5. What is the generally adopted mounting height of video cameras used for pedestrian volume counts?
   (a) 1.5–4.0 m (b) 0.5–1.2 m (c) >4.0 m (d) 0.1–2.0 m

6. The average walking speed of pedestrians considered for the design of pedestrian signals is _________ m/sec.
   (a) 0.9 (b) 1.0 (c) 1.1 (d) 1.2
7. As per IRC: 103, the 15th percentile speed used for the design of pedestrian crossing facilities is
   (a) 0.95 m/sec (b) 1.00 m/sec (c) 1.2 m/sec (d) 1.02 m/sec
8. As per IRC: 103, the average crossing speed of older pedestrians is
   (a) 0.79 m/sec (b) 1.00 m/sec (c) 1.2 m/sec (d) 1.02 m/sec
9. As per IRC: 103, the average crossing speed of younger/middle-aged pedestrians is
   (a) 0.98 m/sec (b) 1.00 m/sec (c) 1.2 m/sec (d) 1.02 m/sec
10. The standard dimensions of a pedestrian body ellipse are
    (a ) 0.6 m × 0.5 m (b) 0.6 m × 0.6 m (c) 0.7 m × 0.7 m (d) 0.7 m × 0.6 m
11. The standard dimensions of a pedestrian body buffer zone are
    (a) 0.3 m² (b) 0.75 m² (c) 0.7 m × 0.7 m (d) 0.7 m × 0.6 m
12. What is the width of the pathway required if the volume of pedestrians = 3 p/sec, mean density of pedestrians = 3 p/m² and speed of pedestrians = 1 m/sec?
    (a) 1 m (b) 2 m (c) 6 m (d) 9 m

$$\left(\textit{Hint}\text{: Width} = \left(\frac{\text{Volume of pedestrians}, \frac{\text{p}}{\text{sec}}}{\left(\text{Mean density of pedestrians}, \frac{\text{p}}{\text{m}^2}\right) \times \left(\text{Speed of pedestrians}, \frac{\text{m}}{\text{sec}}\right)}\right) = \left(\frac{3, \frac{\text{p}}{\text{sec}}}{\left(3, \frac{\text{p}}{\text{m}^2}\right) \times \left(1, \frac{\text{m}}{\text{sec}}\right)}\right)\right)$$

13. What is the capacity of a walkway (p/min/m), when the capacity per metre width of the walkway = 8 per minute and effective width of walkway = 10 m?
    (a) 8 (b) 80 (c) 1.25 (d) 800
    (*Hint*: Capacity of walkway = Capacity per unit width of walkway × Effective width of walkway
    Capacity of walkway = 8 × 10 = 80 p/min/m)
14. As per IRC: 103, for LOS-A, the minimum required pedestrian space (m²/p) is
    (a) 5.0 (b) 3.3 (c) 1.9 (d) 1.3
15. The minimum effective walkway width (mm) required to facilitate a pedestrian flow rate per hour not greater than 600 with LOS-A is
    (a) 1500 (b) 1220 (c) 600 (d)1600
16. As per IRC: 103, in case of hump subways, the required minimum good visibility to be provided across the two ends should be at least _________ lux.
    (a) 30 (b) 40 (c) 50 (d) 25
17. As per IRC: 103, to facilitate people using wheelchairs, the ramp slope may be restricted to
    (a) 8.33% (b) 3.33% (c) 0.33% (d) 5%
18. As per IRC: 103, for full subways, the total clear depth provided under ground level is _________ m.
    (a) 2.75 (b) 3.0 (c) 5.4 (d) 5.6
19. As per IRC: 103, the minimum width of the foot overbridge should be _________ mm.
    (a)1000 (b) 1200 (c) 1500 (d) 1800

20. As a common guideline of IRC: 103, the spacing interval of mid-block crossings in residential areas may range
    (a) 80–250 m (b) 80–150 m (c) 100–250 m (d) 100–300 m
21. As a common guideline of IRC: 102, the spacing interval of mid-block crossings in commercial areas may range
    (a) 80–250 m (b) 80–150 m (c) 100–250 m (d) 100–300 m
22. As per IRC: 103, the total clear depth provided under ground level required for a full subway is ________ m.
    (a) 1.75 (b) 2.75 (c) 1.5 (d) 5.4
23. As per IRC: 103, it may not always be feasible to provide at-grade pedestrian crossings for the following reason:
    (a) Where insertion of an exclusive pedestrian phase signal increases the total traffic cycle time beyond 120 seconds
    (b) Interrupted flow is not permitted on urban major arterial roads and high-speed roads such as national highways and expressways
    (c) Where there is inadequate visibility of crossing/accessing pedestrians across an intersection
    (d) Non-occurrence of vehicular delay during pedestrian crossing and also based on the recorded evidence of vehicle–pedestrian collisions/crashes
24. As per IRC: 103, for LOS-B, the design flow rate (pedestrian/hr) in both directions expected with a clear width of 1800 mm is
    (a) 1000 (b) 500 (c) 1350 (d) 1890
25. As per IRC: 103, for LOS-C, the design flow rate (pedestrian/hr) in both directions expected with a clear width of 1800 mm is
    (a) 1000 (b) 500 (c) 1350 (d) 1890

## Answers

1. (b) 2. (b) 3. (d) 4. (c) 5. (a) 6. (d) 7. (a) 8. (a) 9. (a) 10. (a)
11. (b) 12. (a) 13. (b) 14. (b) 15. (a) 16. (c) 17. (a) 18. (a) 19. (d) 20. (a)
21. (b) 22. (b) 23. (a) 24. (c) 25. (d)

## Questions

1. What are the objectives of conducting traffic volume studies?
2. Write the broad classification of traffic studies and briefly explain each.
3. Write a short note on vehicular and pedestrian traffic studies.
4. What data is collected by pedestrian traffic volume studies?
5. What are the objectives of conducting pedestrian volume studies?
6. What is the significance of conducting pedestrian traffic volume studies for urban infrastructure development?
7. Write the use criteria for pedestrian studies.
8. Define the following:

(a) Pedestrian (b) Footpath (c) Street crossing (d) Pedestrian crossing (e) Pedestrian density (f) Pedestrian platoon (g) Pedestrian unidirectional flow (h) Pedestrian bidirectional flow

9. Define the standard applicable specifications/practices and illustrate/describe the following with schematic diagrams:

   (a) Shy-away distance (b) Pedestrian circulation path (c) Pedestrian access routes

10. List and explain the factors affecting pedestrian demand.
11. Draw a flow chart showing details of the methodology adopted for pedestrian data collection.
12. Write the broad classification of pedestrian counting methods and the techniques adopted.
13. Explain the manual and automatic methods of pedestrian counting and the techniques.
14. Compare the manual and automatic methods of pedestrian counting with their suitability and limitations.
15. What are the critical locations selected for pedestrian volume counts? Write the suitability and limitations of each location.
16. State the parameters used to represent pedestrian volume characteristics.
17. Draw a typical sketch that represents the pedestrian volume data.
18. Draw a typical sketch showing that the number of conflicts at an intersection are greater than those at a mid-block section of a road.
19. Draw a sketch showing the different methods of installing sensors used for pedestrian data collection with the data collection system.
20. Draw a typical sketch of labelled buffer zone separated sidewalks with kerb ramps and crossings.
21. Draw a sketch showing the lateral dimensions related to dead width, pedestrian circulation path, pedestrian access route and discontinuously planted trees.
22. Draw and explain a sketch showing the differences between a pedestrian walkway zone adjacent to a continuous buffer and a discontinuous buffer.
23. List and explain the pedestrian flow characteristics using sketches showing the standard curves.
24. Explain the distribution of pedestrian demand over time relating to walking speed.
25. What are the factors affecting pedestrian walking speed? Also write the design speeds recommended by IRC: 103.
26. What are the characteristics of a standard pedestrian? Also write the details related to body buffer zone space, average pedestrian body ellipse size and optimum densities for normal and desirable cases where physical contact must be avoided.
27. Write the formula used to determine the width of a pedestrian facility, with usual notations.
28. Define pedestrian start-up time. State its significance in the design of pedestrian facilities. Also write the commonly adopted values of start-up time for older and younger pedestrians (write separately for male and female).
29. Define pedestrian capacity. State its significance in the design of pedestrian facilities.
30. Explain how pedestrian capacity is represented and estimated. Also write the most commonly observed value.
31. Draw a curve sketch and explain the relationship between pedestrian speed and density.
32. Explain the typical trend of pedestrian speeds observed under different walking conditions.

33. Write the standard observed values of jam density, flow of pedestrians and capacity flow which are taken for the design of pedestrian facilities.
34. Draw a curve sketch and explain the relationship between pedestrian flow and density.
35. Describe the typical effect of walkway width on the flow–density relationship.
36. Explain the relationship between pedestrian travel time, speed and the flow rate.
37. Describe a stationary free flow condition and state where it is applicable.
38. Draw a curve sketch and explain the relationship of pedestrian speed–flow.
39. Draw a curve sketch and explain the relationship of pedestrian speed–space.
40. Draw a curve sketch and explain the relationship of pedestrian flow–space.
41. What are the factors affecting the quality of service of a pedestrian facility?
42. Define the Level of Service related to pedestrian facilities.
43. Explain the concept of LOS criteria for uninterrupted flow pedestrian facility.
44. Explain the details of determining pedestrian LOS as per IRC:103 using (a) pedestrian space and (b) pedestrian flow rate.
45. Explain how average pedestrian delay is determined based on the signal timing.
46. Explain the time–space analysis method used for evaluating/checking adequacy of pedestrian facilities.
47. Write the IRC: 103 criteria for pedestrian LOS on walkways and sideways.
48. Write the most commonly adopted guidelines on the following components related to pedestrian facilities:

    (a) Width of sidewalk (b) Kerb height (c) Longitudinal gradient (d) Sidewalk kerb ramp (e) Stairway
49. Write the design guidelines for grade-separated pedestrian crossing facilities.
50. Explain the concept of pedestrian signal time gap and state how it is calculated.
51. Describe with sketches how traffic signage, markings and signals are used to facilitate pedestrian flow.

CHAPTER

# 5 Traffic Count Techniques and Analysis of Traffic Volume Data

## 5.0 TYPES OF TRAFFIC COUNTING METHODS

Counting of traffic flow across a given section of road can be carried out using different methods, which may be broadly categorised as **traditional** (using intrusive devices) and **non-traditional** (using non-intrusive devices) (Fig. 5.1). These methods are explained briefly below.

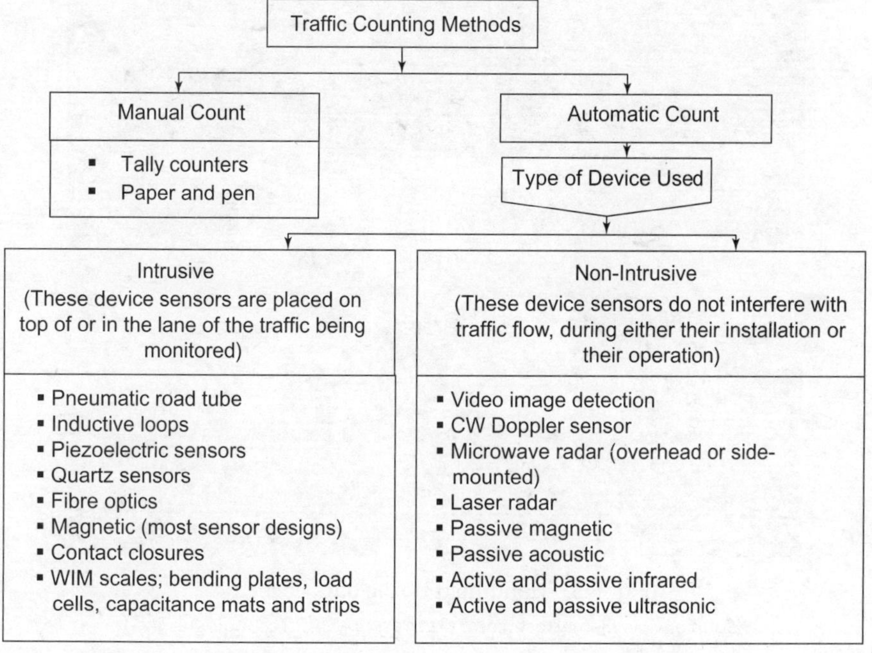


**Fig. 5.1** Traffic counting methods and the types of devices used (modified based on Skszek, 2001)

## 5.1 MANUAL TRAFFIC COUNT

This is the most commonly adopted method for collecting traffic flow data in less developed and least developed countries, where human resources are available at low cost. A group of people is assigned to count vehicles by visual observation and record the traffic as it passes a section of a lane or at each turning road at intersections. Such traffic enumerator teams are normally set up at various locations throughout the road network, at set intervals. Tally board counters or electronic counting boards may be used to count traffic flow (Figs 5.2 and 5.3). The recorded data is later classified and summarised according to the time duration, vehicle type and direction of flow.

**Fig. 5.2** Different types of tally counters
(*Source*: www.tcsforsurveys.com.au. *Courtesy*: Mr. Harald Kolodziej of TCS People Counting)

**Fig. 5.3** Hand-held traffic data collector
(*Source*: www.jamartech.com. *Courtesy*: JAMAR Technologies Inc.)

The duration of the traffic count is determined based on the end use of the data, and it is decided on at the commencement of the survey. As the task of manual counting is laborious, the survey teams need to be supervised and managed properly to ensure efficient and reliable collection of traffic data.

## 5.2 AUTOMATIC TRAFFIC COUNT (ATC)

Traffic data is collected using automatic detectors that are designed to continuously record traffic flow by vehicle class for discrete time periods (e.g., 1 minute, 15 minutes, hourly, weekly, monthly and yearly). The detection system essentially comprises traffic flow sensors and certain types of devices or installations that indicate the presence or passage of vehicles on a predetermined section of a road. The performance of these detectors is affected by the following factors:

- Sensor type
- Mounting type and location
- Sensor viewing angle (if applicable)
- Maintenance of the sensor system
- Intended application (i.e., the parameter(s) to be measured)
- Vehicle mix
- Configuration of roads
- Traffic congestion/jam

The intrusive sensors may be placed in one of the following positions (Klein et al., 2006*a*):

- Embedded in the pavement of the carriageway
- Embedded in the subgrade of the pavement
- Taped or attached to the surface of the carriageway, for temporary counting purposes

### 5.2.1 Intrusive Sensors

Details of intrusive (i.e., contact) sensors are given below.

**Pneumatic tube ATC:** This comprises two rubber pneumatic tube axle sensors connected to a mechanical counting system. The rubber tubes are installed on top of the surface of the carriageway, across the lane (Fig. 5.4). As the axle of the vehicle passes over the tubes, the resulting compression sends a burst of air to an air switch, which can be installed in any type of traffic counting device. The detected axle counts are converted to count, speed and vehicle class, depending on how the road tube configuration is structured, based on their axle base separation.

**Fig. 5.4** Road tube traffic data (volume, class, speed and gap) recorder
(*Source*: www.jamartech.com. *Courtesy*: JAMAR Technologies Inc.)

Pneumatic tube ATC is especially suitable for short-duration surveys. Axle detectors are capable of providing precise speeds and wheel positions over the entire vehicle spectrum, from motorcycles to heavy vehicles. The use of four-channel tubes can enhance the capacity to detect different vehicle axles that pass over the tubes across adjacent lanes at the same time (Fig. 5.5). Pneumatic tubes are cheap, accurate and reliable and can also be operated effectively over a huge range of environmental conditions.

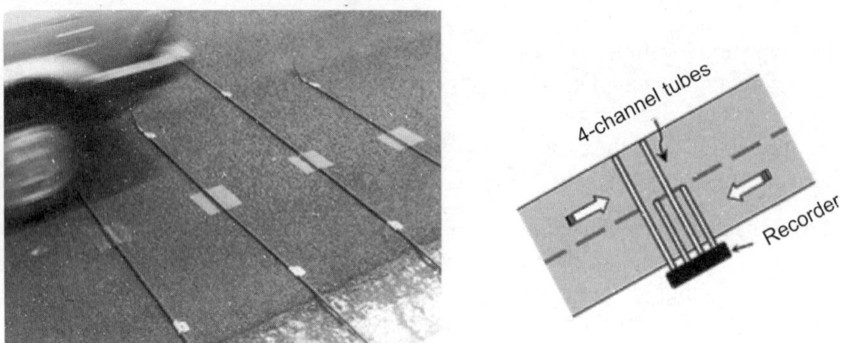


**Fig. 5.5** Parallel rubber pneumatic round tubes placed on top of the road, across the lane
(*Source*: www.tcsforsurveys.com.au. *Courtesy*: Mr. Harald Kolodziej of TCS People Counting)

The life of the pneumatic tube dictates the quality and accuracy of the counting system. This depends on the damage caused to the tubes by the repetitive thrust of traffic flow.

**Inductive loops:** These are used as axle count sensors. When suitably configured, inductive loops are capable of counting traffic, measuring the speed and length of vehicles (long or short) based on their axle configuration, the occupation rate and the time between vehicles. It must be noted that inductive loops *cannot* measure the inter-axle distance, an axle's weight and the vehicle's gross weight. The initial cost is low but maintenance (or replacement of the tubes due to wear and tear) is more than that of any other type of traffic counter. Inductive loops are insensitive to inclement weather conditions such as rain, snow, fog and heavy wind.

Permanent loops of any configuration/shape can be installed on roads with epoxy (slot cut size of 35 mm depth and 8 mm width) or be secured on the road surface with adhesive tape. Inductive loops degrade with deterioration of the pavement surface, period of use and weather.

The parameters of traffic flow that can be measured depend on the loop configuration installed on the section of road. For example, in the case of a single-lane road, a single-loop configuration (Fig. 5.6 (a)) is capable of only counting the number of vehicles that cross the loop. In the case of a double-loop configuration (Fig. 5.6 (b)), the loops can measure the number of vehicles, speed, length of vehicles, occupation rate and category of vehicle (i.e., short or long). In the case of lane straddling of vehicles between adjacent lanes on an undivided two-lane two-way carriageway, vehicles can be detected by installing the loops abreast of each other and spacing the adjacent loops within a minimum distance of 500 mm (Fig. 5.7). This reduces double counting of vehicles on sites where lane discipline is a problem.

**Fig. 5.6** Simple configurations of inductive loop at a traffic count station

**Magnetometers:** Based on the type of technology used, these are of two types: **two-axis fluxgate magnetometers** and **induction or search coil magnetometers**. These can be used as direct replacement for inductive loops and installed quickly into a single 3/8$^{th}$ inch saw cut. They are designed to reduce

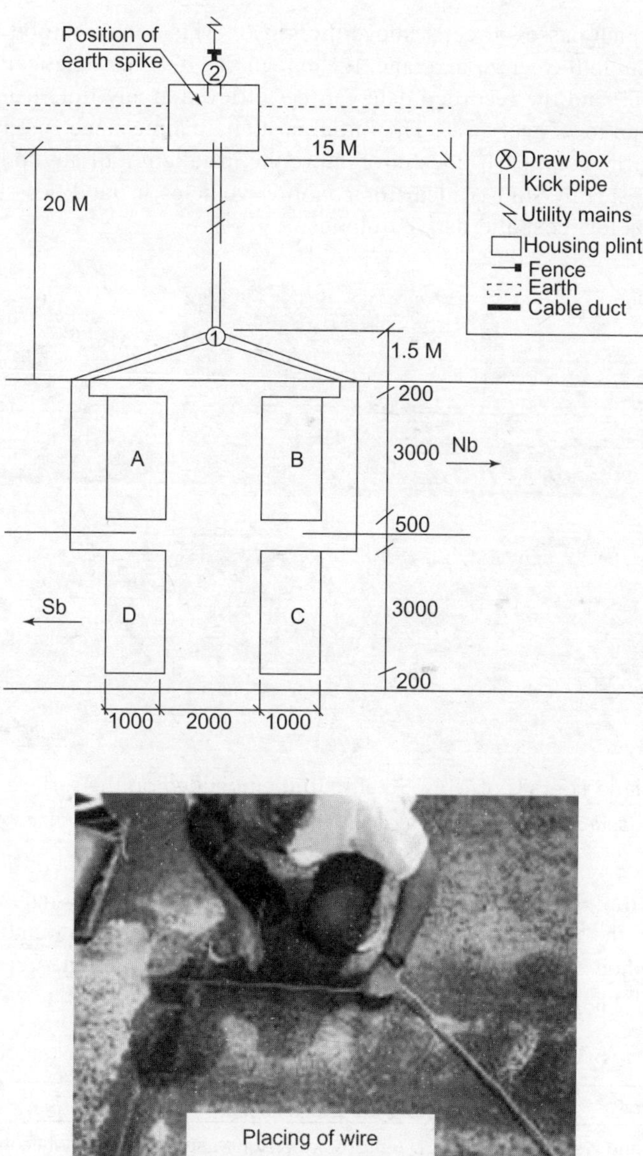


**Fig. 5.7** A standard method of installing inductive loops on the surface of a dual-lane carriageway (*Source*: www.mikros.co.za. *Courtesy*: Mikros Systems Pvt. Ltd.)

the effects of temperature fluctuations and destabilising magnetic fields. Magnetometers are useful for vehicle presence, counting, classification and speed.

As an example, the TRAX Stealth Stud (Fig. 5.8) from Jamar Technologies Inc. uses magnetometer technology for continuous counting of passing vehicles and can record volume data at one-minute intervals. It can detect any type of vehicle that includes a part made up of a metal chassis and which passes directly above the sensor. This device can be installed in a five-inch-diameter hole cut in the centre of a lane. It can recharge itself automatically using an in-built integrated solar panel, and the recorded data can be retrieved at any time using a wireless USB radio module, from up to 35 feet away. The initial cost and maintenance cost is low compared to any other type of traffic counter. It does not require the installation of any loop/tube or roadside cabinet. The TRAX Stealth Stud is ideal for counting vehicles in facilities such as parks/retail organisations and is a low-cost alternative to loops.

**Fig. 5.8** Installation details of TRAX Stealth Stud embedded on the surface of a pavement (*Source*: www.jamartech.com. *Courtesy*: JAMAR Technologies Inc.)

**Weigh in Motion (WIM):** Trucks or axle loads can be weighed in two ways: static weighing and weigh in motion (WIM). In the static weighing method, vehicles are stopped and weighed. However, in the weigh-in-motion method, vehicles are weighed dynamically, while in motion. A comparison of the two methods is given in Table 5.1.

**Table 5.1** Differences between static weighing and weigh in motion

| *Static weighing method* | *Weigh in motion (WIM)* |
|---|---|
| Vehicles are stopped and weighed. | Vehicles are weighed automatically, while in motion, without disturbing the driver. |
| Accurate weight measurement | Weight measurements may be influenced by parameters related to vehicle speed, suspension system of the vehicle, tyre pressure, acceleration and deceleration of the vehicle and dynamic forces produced due to pavement roughness, wind velocity, etc. |
| Takes more time and interrupts free flow of traffic; may pose problems related to safety. | There is no such interruption to flow of traffic; can weigh high volumes of traffic. |

(Continued)

**Table 5.1** (Continued)

| *Static weighing method* | *Weigh in motion (WIM)* |
|---|---|
| Other information such as body type of vehicles, loading type, etc. can be physically ascertained. | Collection of such information is not possible by automated WIM equipment. |
| Fewer number of vehicles can be measured. Only selected vehicles are weighed. There is a need for more personnel, time and space to weigh all the vehicles. | More vehicles can be measured. There is better coverage of vehicles as it is automatic. |
| Lower installation and maintenance costs | High installation and maintenance costs |
| The weigh pads can be installed at any location, as a temporary measurement station. | The WIM equipment can be installed at fixed locations, as a permanent measurement station. |

In the static method, the axle load of a vehicle is weighed using portable weights or load pads (Fig. 5.9) or a weighing platform. Only axle loads $\geq 3$ tonnes are taken into account for the design of pavements since the damage caused to the pavement by axles weighing $<3$ tonnes is negligible. However, for traffic analysis, vehicles of any axle weight are considered for the classification of vehicles based on their axle spacing and vehicle count.

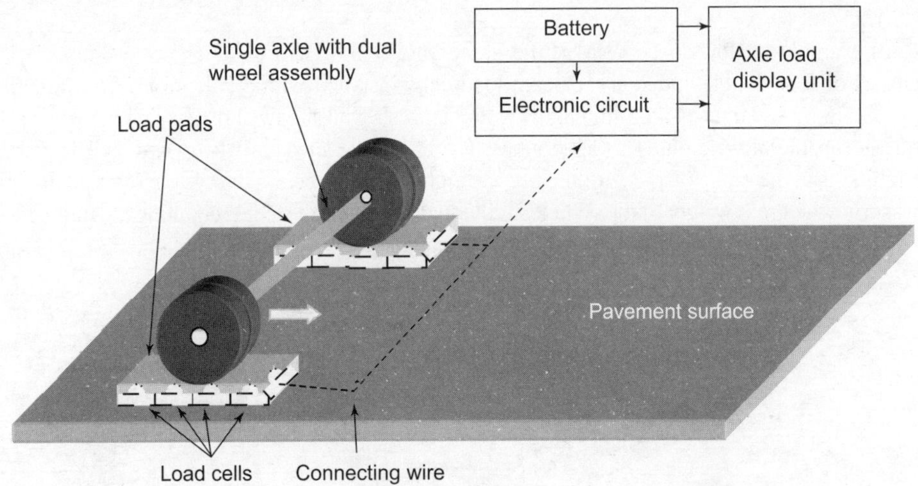


**Fig. 5.9** Axle load measurement using load pads (schematic diagram)

WIM technology has been used in traffic monitoring, weight enforcement, pavement design protocols, pavement management systems and transport safety for the last five decades (Jiang et al., 2009). The WIM system can weigh vehicles at highway speeds without requiring the vehicle to slow down, stop and get back to speed. Historically, WIM vehicle scales have been developed for weigh station 'ramp sorters', where overweight trucks are detected and then sorted for further verification on to the static scales. With the advent of technological developments, WIM applications are now being used for in-bound and out-bound weighing at seaports, border ports of entry, toll roads and at truck terminals where their weights are checked for compliance with reference to ASTM standard 1318-02 or IRC: 3.

The WIM system consists of a weighing sensor with associated electronics for integrating, recording and communicating the data to a control station by wire or by an internet-based system. Other uses of the WIM system are given below:

- Recording of vehicle weight data, which is used in analysis and pavement research projects. For example, to calculate the vehicle damage factor (VDF) and equivalent single-axle load (ESAL) repetitions.
- Pre-screening of trucks and commercial vehicles as part of vehicle enforcement operations.
- The weight information is used to calculate tolls on toll roads, bridges, expressways and tunnels.

Based on the speed of vehicles, the WIM system may be categorised as low speed (8–24 km/hr) or high speed (64–112 km/hr). High-speed WIMs are used for ramp sorting and for regular traffic counts. Low-speed WIMs are used in seaports, harbours and truck terminals.

The following sensors are used for WIM and their reliability depends on their frequency of calibration with the static weighing scale. Their initial cost and maintenance (or replacement of pads) will be higher. Any one of the following sensors alone or in combination with inductive loops are commonly used for the precise measurement of traffic counts with their classification, number of axles in a vehicle, axle load, gross weight, speed, date and time of travel. The associated electronic system with in-built software is capable of integrating such traffic data on a network level and producing user-customised summary reports.

- **Load cells WIM:** Load cells are used in weigh pads. Based on the type of electronic processor connected to the weigh pads, these scales can be used as slow-speed WIM or high-speed WIM.

  An example of slow-speed WIM is the Essae Digitronics Pvt Ltd weigh scale (Fig. 5.10), which consists of four double-ended shear beam load cells installed with a galvanised iron (GI) platform (size: 3275 mm width × 845 mm length) which is flush with the level of the pavement surface. This scale is capable of weighing 1 to 120 tonnes. The allowable speed is 1 to 15 km/hr. The accuracy of weighing is in the range of 1–3% of the gross weight of the vehicle. This system of scale is capable of measuring the gross weight of the vehicle, individual weight of the axles, counting the number of axles in the vehicle, and the speed of the vehicle with date and time of stamping.

**Fig. 5.10** Components, installation and use of slow-speed WIM
(*Source*: www.essae.in. *Courtesy*: TWD of Essae Digitronics Pvt, Ltd.)

  An example of a high-speed WIM is the Metler Toledo Inc. virtual WIM recommended for commercial vehicle enforcement. It consists of a set of two weighing scale platforms configured so that they are staggered and flush with the pavement surface (Fig. 5.11). Each of these two platforms (6 feet wide and 30 inches long) consists of four load cells (capacity 7.25 tonnes each) comprising a shear beam design with strain gauges. These two platforms weigh the left and right wheels independently. These scales are capable of weighing a 22.8-tonne single axle and have a minimum overload capacity of 150%. The allowable speed is 1–145 km/hr. The accuracy of weighing is in the range of 1% with static weighing. This system of scale is configured with an intrusive loop and video camera, and capable of producing vehicle count, gross weight of the vehicle, individual weight of axles, spacing between axles, number of axles in a vehicle, and speed of a vehicle with date and time of stamping. This system can be operated as an unmanned automated weighing system with internet access.

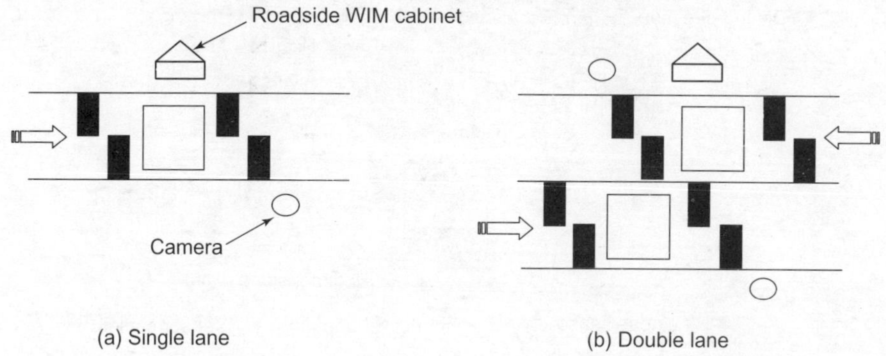


**Fig. 5.11** Best practices for the configuration of intrusive loop with staggered WIM scale platforms (FHWA, 2012 and www.mtwim.com)

- **Bending plate WIM:** The weigh pad with its borders is installed in a pavement and placed in the travel lane perpendicular to the direction of traffic flow. The plate of the weigh pad consists of high-strength steel with wire strain gauges running through two slots cut on its bottom. These wire strain gauges are bridged to a Wheatstone bridge and have electronic components used for temperature compensation. The operating speed on these pads can be between 5 and 200 km/h. An example of this type of WIM is the IRD-PAT Bending Plate™ WIM system.
- **Piezoelectric WIM:** Based on the type of material used, these sensors can be ceramic, polymer or quartz. They are different in shape, dimension, cost and sensitivity to environmental conditions. Strips of suitable length (i.e., across each lane) are installed directly into cut slots (about 1" × 0.7" or about 2" D × 3" W), and made flush with the surface level with a small amount of resin. This is used for permanent installations.

  An example of a piezo-quartz sensor is the Lineas® quartz crystal force piezoelectric sensor, developed by the Kistler Group, which is capable of weighing vehicles travelling at a speed of 5–200 km/h (Figs 5.12, 5.13 and 5.14). The Lineas® quartz piezoelectric sensor consists of an aluminium alloy profile, in the middle of which, quartz crystal discs are fixed at every 50 mm under pre-loading. The sensor is protected from its sides by special elastic material by which only vertical force measurement is sensed without absorbing any type of lateral force. The sensor strips (each of length 1.5–2.0 m) are positioned into a slot cut across the full width of a lane or road and installed flush with its surface with epoxy adhesive. As a wheel load rolls over the Lineas® sensors, the resulting vertical force is transferred on to the quartz crystals, with virtually no deformation in the sensor. The piezo-quartz discs yield an electrical charge proportionate to the force applied, and the electrical signals are amplified proportionately and converted to load units based on the calibration factor.

  A study (Jiang et al., 2009) conducted with test conditions revealed that piezoelectric quartz sensor technology has the best weight measurement accuracy, is non-sensitive to temperature

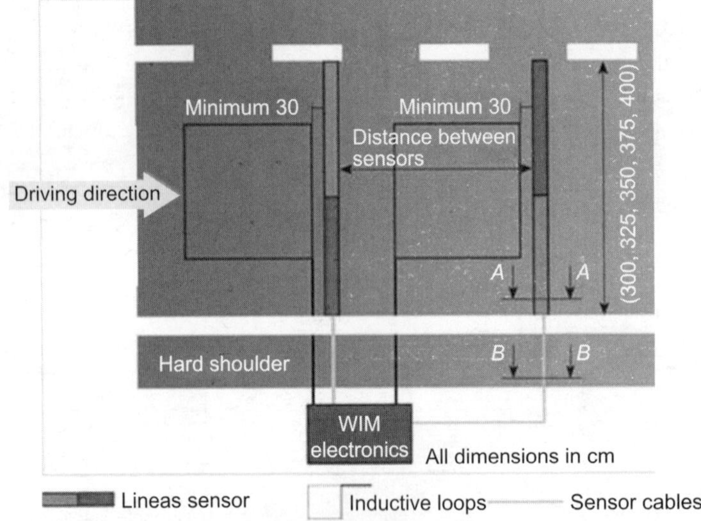


(a) A typical WIM system: two rows of two Lineas® WIM sensors (four sensors in total)

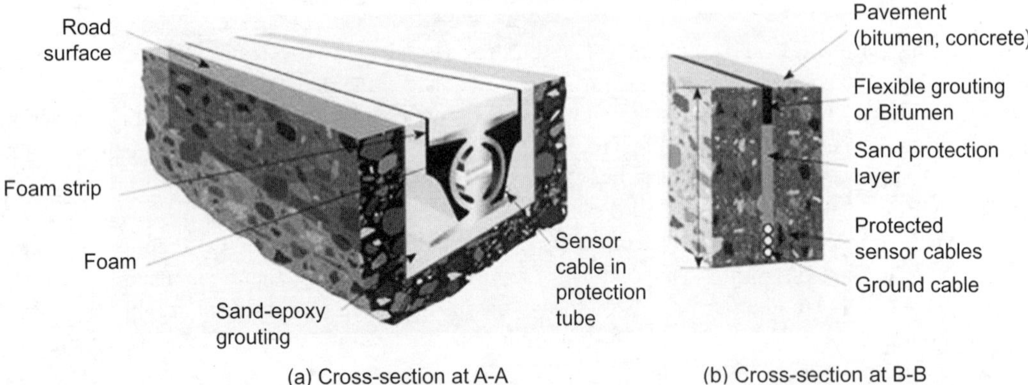


(a) Cross-section at A-A (b) Cross-section at B-B

**Fig. 5.12** Details of a WIM system Using Kistler Lineas® piezo-quartz crystal sensors (*Source*: www.kistler.com. *Courtesy*: Kistler Group, Switzerland)

(a) Cut slot (b) Mounting of sensor (c) Grouting (d) Sensor setting in slot

**Fig. 5.13** Installation of the Kistler Lineas® piezo-quartz crystal sensor on pavement surface (*Source*: www.kistler.com. *Courtesy*: Kistler Group, Switzerland)

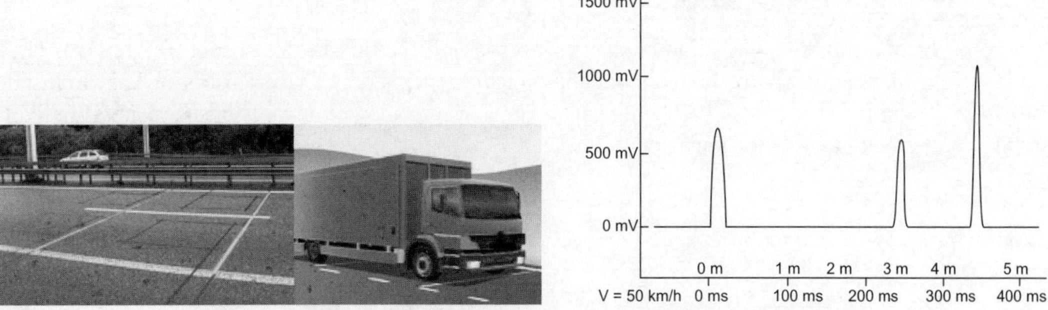


**Fig. 5.14** Typical view of the Kistler Lineas® piezo-quartz crystal sensor WIM system in use, with output
(*Source*: www.kistler.com. *Courtesy*: Kistler Group, Switzerland)

changes and provides good overall performance at a comparatively higher cost among the three types of sensors. However, the conventional piezoelectric sensors provide the lowest accuracy with the lowest cost among the other sensors used as WIM systems.

The measurable parameters of traffic flow depend on the type of sensors used and the loop configuration installed on a section of road. A few combinations of inductive loops and WIM sensors are described in Fig. 5.15.

| Sensor configuration | Remarks |
|---|---|
| | ▪ Standard recommended configuration<br>▪ Dual loop plus single (full width) piezo axle sensor<br>▪ Classification is based on axle spacing profiles<br>▪ Up to 8 lanes can be monitored. Information on individual vehicles and/or time sliced summaries is available |
| | ▪ Dual loop per lane<br>▪ Classification is based on inductive profile plus vehicle length<br>▪ Up to 8 lanes can be monitored. Information on individual vehicles and/or time sliced summaries is available |
| | ▪ Dual loop plus dual bending plate weigh sensor and a piezo axle detector<br>▪ In addition, up to three off-scale sensors can be accommodated to enchance weigh results<br>▪ Used for HSWIM information. Classification is based on axle spacing profiles. Piezo detector is used to detect light<br>▪ Up to 8 lanes can be monitored. Information on individual vehicles and/or time sliced summaries is available |
| | ▪ Dual loop plus two rows of Lineas® weigh sensors.<br>▪ Used for HSWIM information. Classification is based on axle spacing profiles<br>▪ Up to 4 lanes can be monitored. Information on individual vehicles and/or time sliced summaries is available |

**Fig. 5.15** Typical configurations of WIM sensors and their combinations
(*Source*: www.mikros.co.za. *Courtesy*: Mikros Systems Pty Ltd.)

### 5.2.2 Non-intrusive Sensors

The types of non-intrusive (i.e., non-contact) sensors are discussed in this section (Fig. 5.16). Their initial cost is moderate to high and recurring maintenance cost is low. These sensors are easy to install, safe to use and more suitable for collecting data in areas where snowploughs or dust sweepers are likely to disturb the road tubes/loops/intrusions.

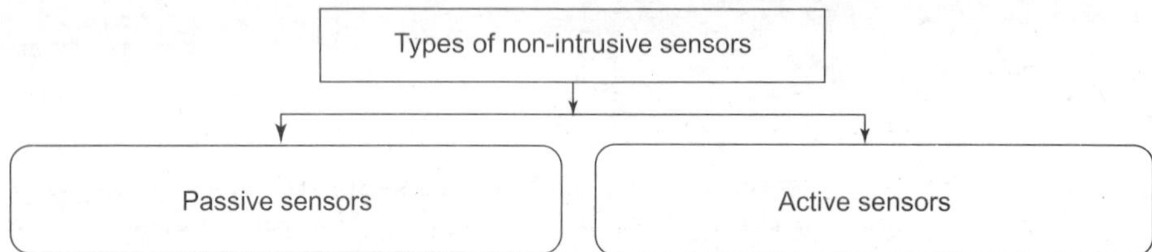


**Fig. 5.16** Categorisation of non-intrusive sensors based on the technology used

**Mounting of non-intrusive sensors:** Non-intrusive sensors may be installed over the roadway in any one of the following ways (Klein et al., 2006*a*):

- **Side fire**: Here, the sensor is mounted on an elevated object along the roadway, and offset from the nearest lane by some distance. The installation is simple, and easy to access and maintain. Based on the direction of sensor view, they may be (i) focused inclined downwards on to the lanes (Figs 5.17 and 5.18) or (ii) focused horizontally across the lanes to be monitored. The former position is also used for wide-area surveillance while the latter is applicable in case of laser radar sensors. The drawback of this position is that in the case of multi-lane roadways,

**Fig. 5.17** Simple configuration of in-lane detection event sensors
(*Source*: www.tapconet.com. *Courtesy*: Traffic and Parking Control Co. Inc.)

**Fig. 5.18** Remote traffic microwave sensor
(*Source*: www.tapconet.com. *Courtesy*: Traffic and Parking Control Co. Inc.)

the traffic using the farthest lane away from the location of the sensor may be obscured from the side fire sensors by vehicles travelling in the lane closer to the sensor – termed occlusion.

- **Overhead**: Here, the sensor is mounted on an elevated structure across the traffic lane, i.e, above the traffic flow (Fig. 5.19). The occlusion problem can be eliminated by this type of mounting. It was observed that overhead mounted sensors could detect more accurately than side fire sensors using the same technology. However, lane closure may be required for the installation and maintenance of these sensors, to avoid material being dropped during such activities.

**Fig. 5.19** Typical installation of overhead mounted sensors used for detecting vehicle on lanes. A combination detector, called a tri-detector, which uses a combination of Doppler, ultrasound and passive infrared in a single unit, is also shown in the top-right corner.
(*Source*: www.xtralis.com. *Courtesy*: Xtralis Pty Ltd.)

### Passive sensors

**Passive acoustic sensors:** These sensors are capable of detecting vehicles on multilanes at all speeds from free flow to stop and go, in all weather and visibility conditions. This type of sensor listens to the noise produced by vehicles and records it. An example of a passive acoustic sensor is the SmarTek Systems Acoustic Sensor (SAS-1), which is described briefly below.

Vehicle-generated acoustic signals (noise from engines, fans, belts, tyres) leave their source (the vehicle to be detected) and arrive at the acoustic sensor, which is installed on an elevated highway structure with proper alignment (Fig. 5.20). This type of side-mounted sensor is said to be in the side fire position (Fig. 5.21). The acoustic signal that travels from each vehicle from the respective lane will arrive at the sensor with a different signal level and at a different wave front angle (i.e., arrival angle). The system is capable of receiving signals from 91 such directions. As a vehicle moves or flows by the sensor, it produces acoustic images or blobs of high-intensity signal. The sensor is electronically equipped to convert the received signal wave front into digital form for further processing by the in-built software. This sensor is capable of providing vehicle counts with class and time of flow, on each lane.

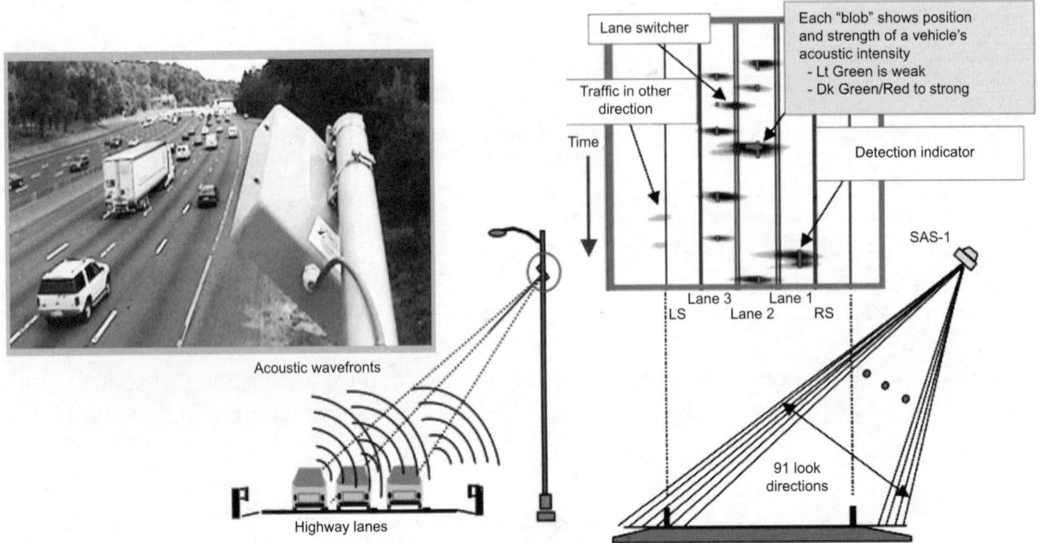


**Fig. 5.20** SmarTek Systems Acoustic Sensor (SAS-1) and the formation of its acoustic image (*Source*: www.smarteksys.com. *Courtesy*: SmarTek Systems™, Inc.)

**Microwave radar detectors:** These detectors emit focused high-frequency signals (range: 1–30 GHz) within a predefined band width (λ: 1–30 cm). If a vehicle moves into or through these signal-focused areas, the emitted signals reflect back to the detector in a shorter time period which is less than the normal reflection case. Due to Doppler shift between the emitted and received frequency of signals, the reflected incidence is detected as a vehicle presence with its direction, speed, gap during vehicle flow in multiple lanes and ultimately, the count. It has been noted that continuous wave Doppler sensors cannot detect stopped vehicles (Klein et al., 2006).

An example of this is the Radar Recorder (Fig. 5.22) from Jamar Technologies Inc, which uses microwave (24.4 GHz) technology for continuous detection of traffic vehicles, and measures their count, speed (range 2–130 mi/hr) and classification based on their length. It can be mounted on any utility pole or tree in the side fire position. The radar range is up to 400 ft. It can measure traffic data of up to two lanes of bidirectional traffic flow (Fig. 5.23) at any user-defined interval of 1, 5, 15, 30 or 60 minutes. The automatically recorded traffic data can be viewed in real time and also downloaded by direct connection to the unit or by wireless Bluetooth option from a vehicle, in any weather condition. The analysed summary data can also be used to determine the best time for enforcement, best location for enforcement and evaluate speeds before and after enforcement.

This device cannot used for traffic data collection at intersections, where vehicles are likely to park and at pedestrian crossings and where radar view is obstructed.

**Passive infrared detectors:** Passive sensors transmit no energy of their own (i.e., no artificial light source is used to detect vehicles or pedestrians). These detectors are capable of sensing the slightest changes in thermal radiation contrasts against the background as field of view. The light energy emitted from the vehicle, road surface and other objects in the field of view or energy emitted by the atmosphere and reflected by the vehicles, road surface and other objects are detected by these sensors (Fig. 5.24). The portion of the transmitted energy reflected from the vehicles is captured by the sensor receiver and it is focused by an optical system onto an infrared sensitive-plate mounted at its focal plane. The plate material converts the received energy into electrical signals using an attached real-time signal processing unit for detecting the presence of a vehicle.

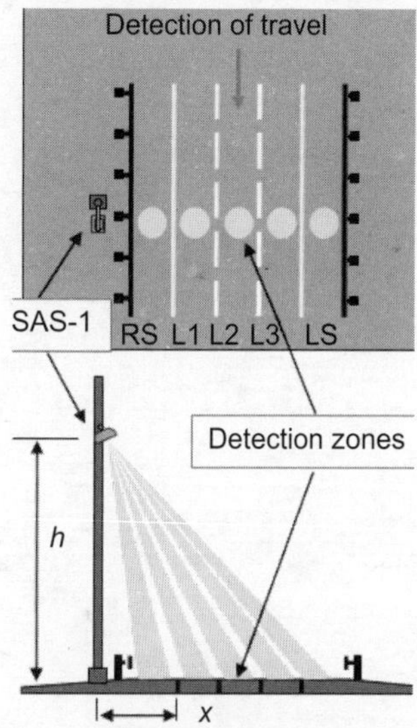


**Fig. 5.21** Side mounting for multi-lane monitoring
(*Source*: www.smarteksys.com. *Courtesy*: SmarTek Systems™, Inc.)

**Fig. 5.22** Radar recorder
(*Source*: www.jamartech.com. *Courtesy*: JAMAR Technologies Inc.)

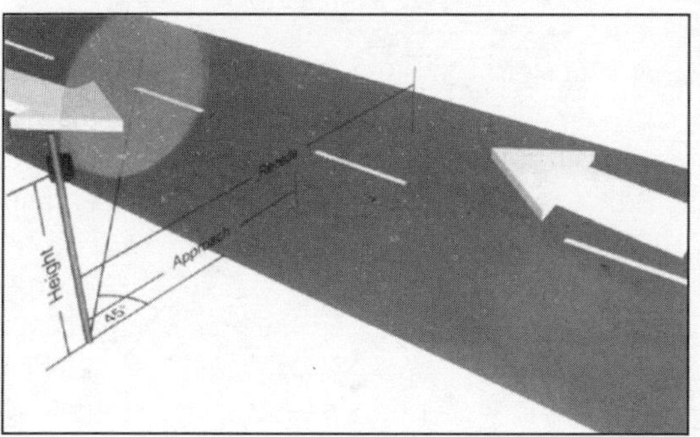

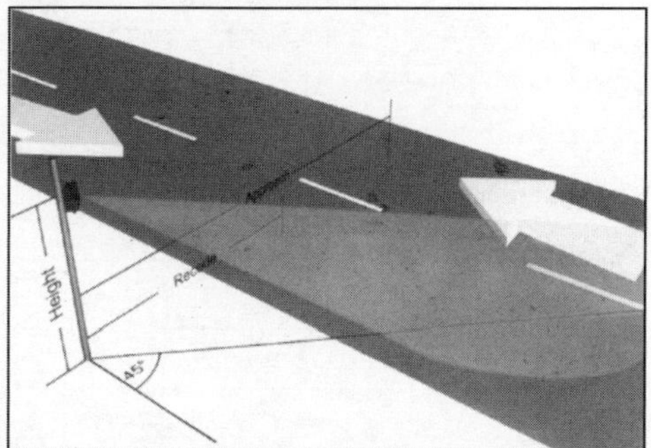


**Fig. 5.23** The working principle of the Radar Recorder
(*Source*: www.jamartech.com. *Courtesy*: JAMAR Technologies Inc.)

| Technologies | Ultrasonic, passive infrared | Radar | Passive infrared |
|---|---|---|---|
| | 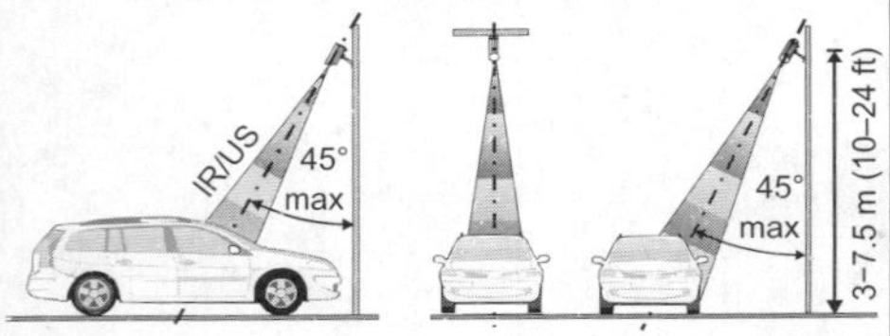 IR/US 45° max; 45° max; 3–7.5 m (10–24 ft) | 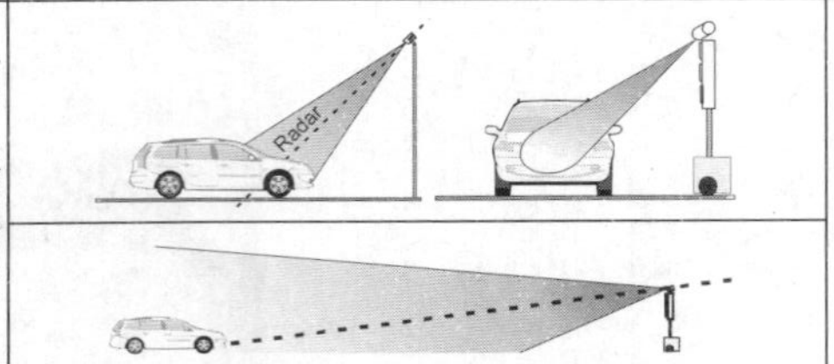 Radar | 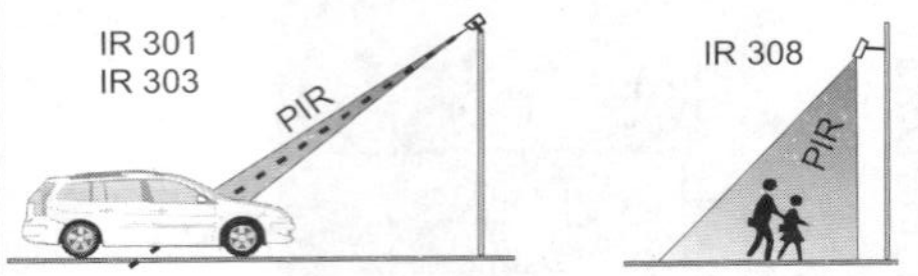 IR 301<br>IR 303 PIR; IR 308 PIR |
| Application | Free moving, heavy or stationary traffic overland and in cities up to 100 km/h | Free moving, heavy or stationary traffic overland and in cities up to 250 km/h | |
| Features | Traffic lights control via presence recognition<br>Vehicle counting<br>Real presence recognition of stationary objects<br>Height dependent classification | Traffic lights control via movement recognition can be set to detect aproaching or all vehicles adjustable speed and sensitivity threshold | Traffic lights control via temperature contrast recognition lane selective vehicle versions volumetric pedestrian version |
| Speed Range | 1–100 km/h | 4–250 km/h or 8–250 km/h adjustable | 0–250 km/h |
| Range | 1 m–8 m | 1 m–45 m | 1 m–50 m |

| Technologies | Radar, ultrasonic, passive infrared | | Radar, static and dynamic passive infrared | | Static and dynamic passive infrared | |
|---|---|---|---|---|---|---|
| | 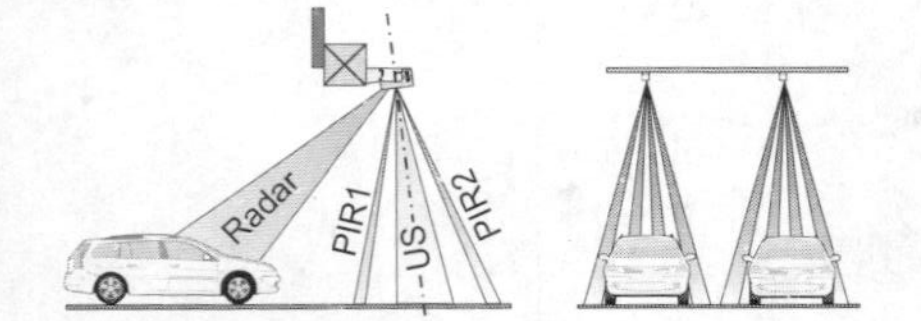 Radar PIR1 US PIR2 | | 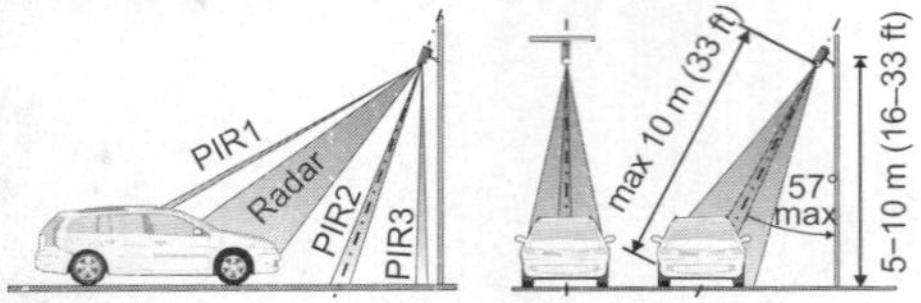 PIR1 Radar PIR2 PIR3; max 10 m (33 ft) 57° max 5–10 m (16–33 ft) | | 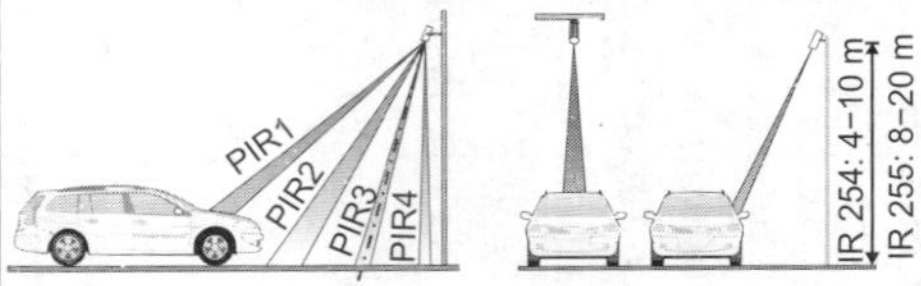 PIR1 PIR2 PIR3 PIR4; IR 254: 4–10 m IR 255: 8–20 m | |
| Application | Multilane highways, up to 250 km/h | | Moving, heavy or standing traffic, overland or in cities, up to 250 km/h | | | |
| Data Output via RS485/ Additional Functions | TLS Class (2 - 8 + 1)<br>Length<br>Speed<br>Occupation time<br>Distance to earlier vehicle | "Wrong-Way-Driver" alarm<br>Traffic jam warning<br>Internal error notes<br>Backfire operation with low power mode | TLS Class (2)<br>Length<br>Speed<br>Occupation time<br>Distance to earlier vehicle | "Wrong-Way-Driver" alarm<br>Traffic jam warning<br>Internal error notes<br>Low power mode supporting solar powering | Speed<br>Length | "Wrong-Way-Driver" alarm<br>Traffic jam warning<br>Internal error notes<br>Very low power consumption |
| Accuracy Range | Up to 100 km/h: ± 3 km/h<br>Above 100 km/h: ± 3% | | Up to 100 km/h: ± 3 km/h<br>Above 100 km/h: ± 3% | | Type. ±10% (20 events average) | |
| Speed Range | 0–250 km/h | | 0–250 km/h | | 0–250 km/h | |

**Fig. 5.24** ASIM traffic detectors

(*Source*: www.xtralis.com. *Courtesy*: Xtralis Pty Ltd.)

Infrared sensors can installed at a road-side post in the side fire position. These detectors can be used for multiple applications such as pedestrian count, vehicular traffic count, speed of vehicles, classification of vehicles by length, presence detection of vehicles in parking lots, detection of pedestrians at pedestrian crossings, transmission of traffic information to motorists, queue detection and the required extension of Green time at signalised intersections.

The performance of detection of vehicles by passive sensors may be affected by heavy rain, snow and dense fog. Some models may not be recommended for vehicle-presence detection (Klein et al., 2006).

#### Active sensors

**Active infrared detector or laser radar:** Active infrared sensors illuminate detection zones with low-power infrared energy which is transmitted by laser diodes operating in the near-infrared region of the electromagnetic spectrum at 0.85 mm. A portion of the transmitted energy is reflected by the vehicles back towards the sensor receiver. The emitted energy is captured by the sensor and focused by an optical system onto an infrared-sensitive plate mounted at its focal plane. The plate material converts the received energy into electrical signals using an attached real-time signal processing unit for detecting the presence of vehicles.

Laser radar detectors are an example of this case, as the energy transmitted as a laser beam is within the infrared spectrum (Fig. 5.24). Vehicle height, vehicle position, and speed with classification can be accurately measured by transmitting multiple laser beams across the width of the lane. Multiple lane operations are possible. The performance of these sensors may be affected by dense fog ($\leq$6 m) and unclear lens (Klein et al., 2006).

**Ultrasonic detectors:** These detectors emit acoustic signal bursts at high frequency, beyond the audible range of humans and most animals. If a vehicle moves into or through the signal-focused area, the emitted signals reflect back to the detector at a shorter time period, which is less than the normal case. The distance to the surface of the vehicle is determined from the travel time of the reflected signal (Fig. 5.24). As a result, the presence of standing vehicles can be detected and monitored with their class more precisely than with the other technologies such as infrared or radar. These detectors are also capable of multiple lane operations and detecting over-height vehicles (Klein et al., 2006).

**Video image processor–based detectors:** The camera's image is focused/adjusted to view a particular section of a road and its field view is termed the detection zone. Up to four or five such lanes of traffic flow can be set as the detection zone from each camera installed at a location. Such cameras, with their associated electronics, are capable of recording video signals. Their digital signal processor uses image processing algorithms which are part of their in-built software to detect vehicle presence at all speeds in all weather conditions, at surface street intersections or on a section of multi-lane road. The configured cameras can also detect motion and speed and classify the vehicles in the respective lanes. The data obtained from this automatic visual detection system is integrated with a WIM system at traffic count stations and used to report on overweight vehicle violations. For this purpose, licence plate detection techniques are also used to analyse the videographs.

An example of this is the SmarTek Systems Inc. video monitoring system which can automatically acquire and display traffic data (Fig. 5.25), including environmental condition data (when a temperature sensor is installed) at a remote site. The data acquired from multiple cameras installed at different remote road sites can be visualised as lane-by-lane traffic flow from a central monitoring station,

**Fig. 5.25** Video camera junction box and a typical video setup display
(*Source*: www.smarteksys.com. *Courtesy*: SmarTek Systems™, Inc.)

with reference to the geo-spatial map (Fig. 5.26). The Windows software associated with each camera provides the traffic count with vehicle class, speed, date and time stamp data. Real-time traffic data can also be obtained from the remote site using the internet.

**Detectors that use combined technologies:** The technologies mentioned above can be used in any combination (i.e., double or triple (Fig. 5.26)) just as a single detector can be used to obtain an optimised solution for a variety of applications such as traffic data acquisition, stop bar presence, time gap recognition, presence detection of vehicles and for actuated intersection control.

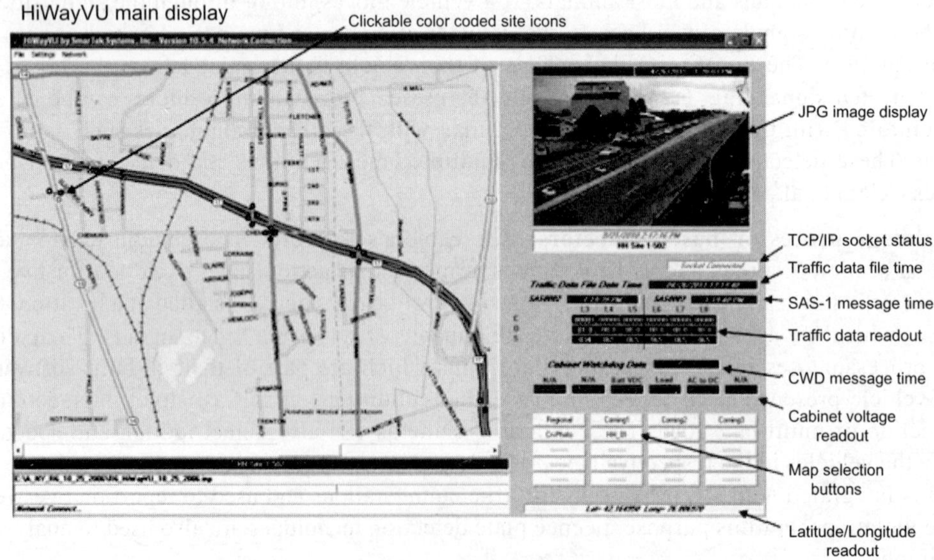


**Fig. 5.26** Highway traffic monitoring system showing various types of information available from a road-side cabinet with a map overview as a user interface
(*Source*: www.smarteksys.com. *Courtesy*: SmarTek Systems™, Inc.)

## 5.3 FACTORS AFFECTING THE SELECTION OF TECHNOLOGY

In addition to the above basic functionalities of intrusive and non-intrusive equipment, the following factors need to be considered while selecting the appropriate technology to be used to monitor motorised, non-motorised and mixed traffic (FHWA, 2012).

- The cost components relating to initial cost, installation cost, calibration cost, operating cost, maintenance cost and its salvage value
- The type of traffic to be monitored, e.g., motorised, non-motorised or mixed; urban or rural
- The number of lanes and the number of devices used for each lane
- Environmental conditions relating to temperature
- Accuracy of the traffic data collected
- Purpose and period of data collection: temporary or permanent
- Type of power source required: AC, DC or solar panel
- Method of data transfer/communication: wired or wireless or by using the internet
- Suitability of the proprietary algorithms that analyse/process the traffic data, e.g., vehicle detection based on the axle spacing permitted in a state or country or vehicle occupancy at an intersection
- Ability of the sensor and its software to provide output data that seamlessly integrates with the needs of a central traffic processing system
- Capability of the vendor to supply future maintenance/upgradation/warranty needs of the sensor components and software
- Pavement condition and period of maintenance, in case of intrusive sensors and WIM sensors
- Availability of technicians and engineers for installation and repair, and also for any agency-specific needs such as change in the output format of reports or change in analysis that meets the local government regulations
- Risk of theft or vandalism, i.e., action involving deliberate destruction/damage

## 5.4 DURATION OF TRAFFIC COUNT

Traffic volume surveys should be conducted preferably on a long-term basis, every day (24 hours a day, 365 days a year); and this is possible only using automatic counting equipment at permanent counting stations – often referred as **continuous count.** For short-duration counts, the traffic volume is counted annually; at 2–5-year intervals; 24 hours a day, seven days a week or three days a week; or 24 hours a day or a month of a year, depending on the importance of the road. The traffic surveys should preferably be conducted throughout the year by establishing permanent count stations. Otherwise, a short-term survey (usually, for seven days, 24 hours, 48 hours or 72 hours; consisting of Saturday, Sunday and Monday) may be considered a temporary count or coverage traffic count (IRC: 9).

A successful data collection programme should account for traffic variability, and the analysed data should be used to identify changes in the traffic patterns as they occur over time. Such traffic data is essentially used to support the Highway/Road Department/Ministry activities in planning, financing, engineering economics, design, environmental analysis, maintenance management and safety. Every country and every state has its own requirements for traffic data collection and budgeting/funding priorities that are applicable within geographical limitations and organisational constraints. These differences result in the Highway/Road Department/Ministry selecting/recommending different equipment for traffic data collection, adoptiong different data collection plans and emphasizing different forms/formats of data reporting (FHWA, 2013; Liggett et al., 2009).

## 5.5 TRAFFIC ATTRIBUTES

The type of devices and technology selected will depend on the type of traffic data to be collected. Sensor-recorded data is usually analysed with the proprietary algorithms of the device producers. For this purpose, suitable equipment/devices should be selected based on traffic data categorisation.
Traffic data may be categorised as follows:

**Non-motorised traffic**

- Bicycle
- Tri-cycle
- Bullock cart
- Horse-drawn vehicle
- Pedestrian

**Motorised traffic**

- Motor cycle and scooter
- Auto-rickshaw and jugaad (a motor-driven trolley used to transport people)
- Car, taxi, jeep and mini-van
- Bus
- Truck and multi-unit truck (with reference to multiple axles)

**Attributes of traffic flow**

- Volume
- Vehicle classification
- Speed
- Axle spacing and their number count per commercial vehicle
- Axle and gross weight of vehicle
- Gap between vehicles
- Lane occupancy

## 5.6 PRESENTATION OF TRAFFIC VOLUME DATA

The traffic data collected according to time, day and vehicle type are classified and the desired parameters such as ADT or AADT are computed. Summarised traffic data can be presented as tables and graphs showing the hourly, daily, weekly, monthly, seasonal and yearly variations in traffic flow, at a specified location or lane. Such integrated data from different sections of roads will provide useful and comprehensive information about the traffic volume on any road network.

### 5.6.1 Variations in Traffic Flow

Traffic flow varies spatially and temporally. Traffic flow data with reference to vehicle class, volume, speed, load capacity, direction and time in an area describe the current traffic condition, and such data can also be used to determine future trends in traffic growth. The different time scales that may be considered to study variations in traffic flow at a place are given below.

**Time of day or hourly variation:** It is well known that the use of a road changes in the course of a 24-hour day. In most locations, traffic volume increases in the day and decreases at night (Fig. 5.27). This is particularly applicable in urban localities, where traffic flow is maximum in an hour during the day and the night, often termed peak hour, with the traffic volume termed peak hour traffic (seen as two humps in Fig. 5.27). There may be two peak hours, i.e., one during the day and the other at night.

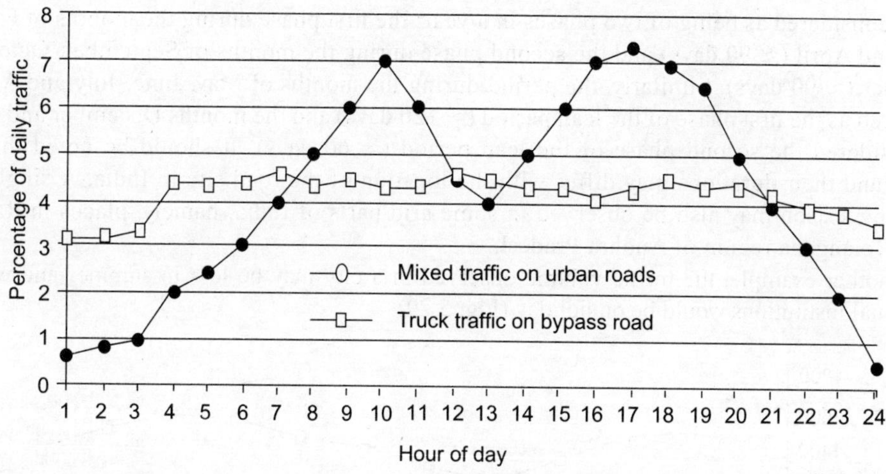


**Fig. 5.27** Typical trend of variation in traffic flow in a 24-hour day in an urban area

In case of routine truck traffic and commercial vehicular traffic, there may not be much variation during the day and at night, as most such trips are made as long-haul movements for commercial purposes, on business days. In case of regions where mineral resources are available, loaded trucks (classified as five-axle tractor and semi-trailer) move in one direction and unloaded or empty trucks (classified as three-axle single unit) move in the opposite direction. These trends of variations in traffic flow may change subject to site/locality-specific conditions, vehicle type and their directional traffic movements.

**Day of week (DoW) or daily variation:** DoW patterns differ in large part with reference to rural and urban locations, vehicle type and the business needs of the locality. A typical trend of traffic volume in a day of a week in an urban locality is shown in Fig. 5.28.

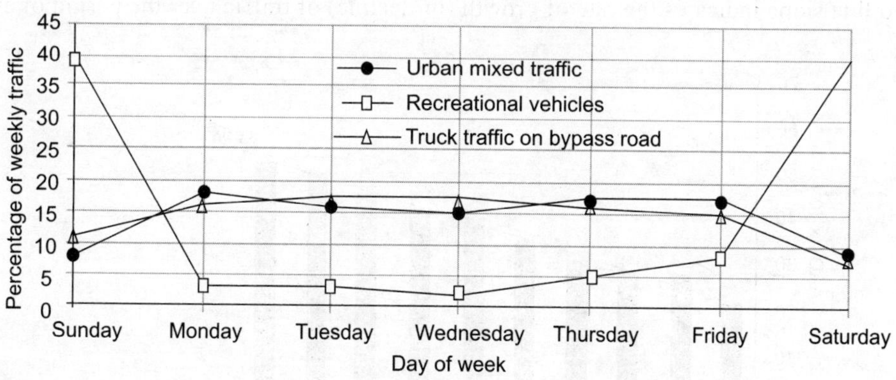


**Fig. 5.28** Typical trend of variation in traffic flow in a day of a week in an urban area

**Monthly or seasonal variation:** Seasonal variations in traffic flow may be observed from mean monthly values obtained over a period of a year. For example, peak harvesting and lean seasons may be observed in agriculture areas (Fig. 5.29). Generally, in India, the peak harvesting season may be considered as being of two phases in a year: the first phase during the months of February, March and April ($\approx$ 90 days) and the second phase during the months of September, October and November ($\approx$ 90 days). Similarly, the period during the months of May, June, July and August is considered as the first phase of the lean period ($\approx$ 120 days) and the months December and January are considered the second phase of the lean period ($\approx$ 60 days). It should be noted that these seasons and their durations may differ with the location of the road site in India. A single-phase harvesting season may also be observed in some arid parts of India, namely, places in Rajasthan and the Telangana region of Andhra Pradesh.

As another example, the traffic volume observed in a city may be less in summer and winter as educational institutions would be on holiday (Fig. 5.29).

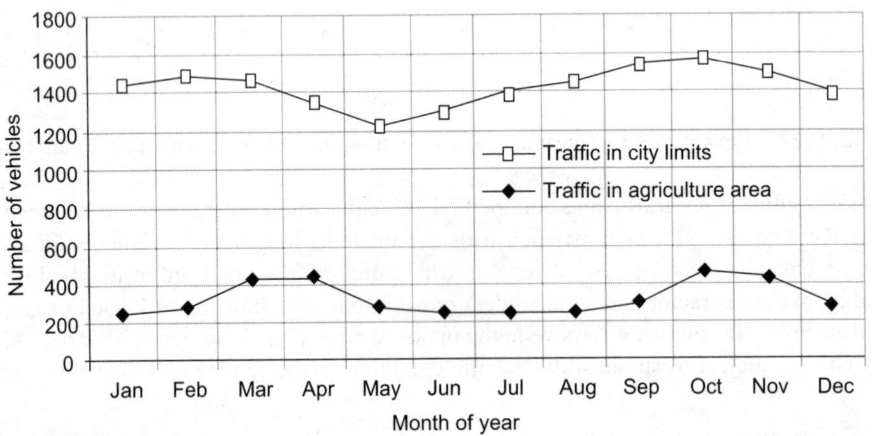


**Fig. 5.29** Typical trend of seasonal variation in traffic flow in a year in a locality

**Yearly variation:-** The yearly variation in traffic flow on a section of road can be plotted (Fig. 5.30); the trend line slope indicates the rate of growth (or decline) of traffic over the period of analysis.

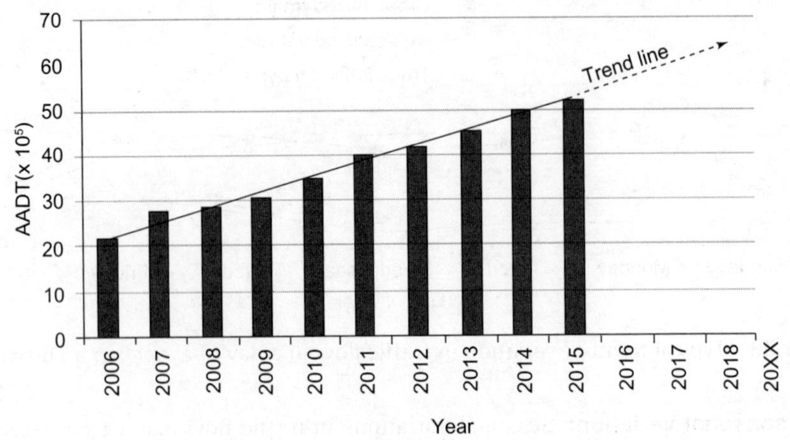


**Fig. 5.30** Typical trend of yearly variations in traffic growth

### 5.6.2 Summary Tables

A summary of the traffic volume data such as ADT, peak hour, non-peak hour, weekly average, monthly average, AADT and seasonal average may be presented in a tabular form (Table 5.2).

**Table 5.2** Typical form of a summary table

**Location of the site:..........Day & Date:............Direction of flow.......Weather:...................**

| *Time period Vehicle Type-1* | | *Number of vehicles with each class of vehicle* | | | | | |
|---|---|---|---|---|---|---|---|
| | | *Vehicle Type-2* | ..... | ... | ... | *Vehicle Type-n* | |
| Any one of the duration in | Hours | | | | | | |
| | Days | | | | | | |
| | Months | | | | | | |
| | Years | | | | | | |
| Individual Totals = | | | | | | | |
| | | | | | Grand Total= | | |

### 5.6.3 Graphic Summary of Flow at Intersections

Identification and interpretation of traffic manoeuvres are considered key elements in characterising the traffic flow and are considered for the design of various geometric elements. Vehicle manoeuvres based on their direction of movement or a combination of movements are categorised as shown in Fig. 5.31.

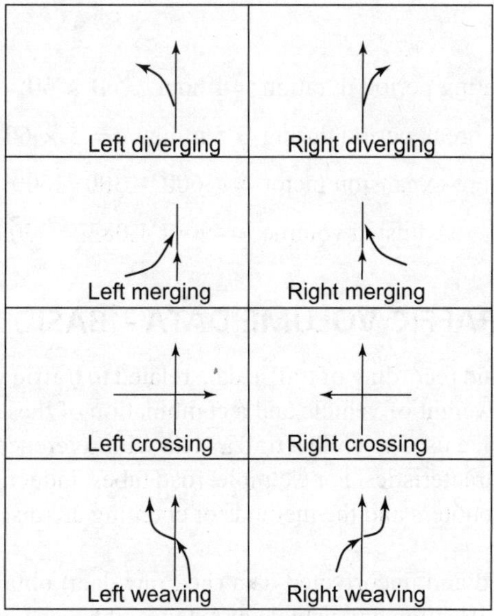


**Fig. 5.31** Types of traffic manoeuvres

The traffic manoeuvre data collected from such studies can be used to eliminate or minimise potentially dangerous manoeuvres of vehicular conflict before they become critical and to carry out accident investigations related to lane change manoeuvres and tyre trajectories on the road.

The count of traffic flow with manoeuvres through an intersection can be presented by denoting with arrows attached with the respective count boxes (Fig. 5.32). This type of presentation is useful for the design of intersection signal timings, accident possibility, queue length and for solving intersection-related problems.

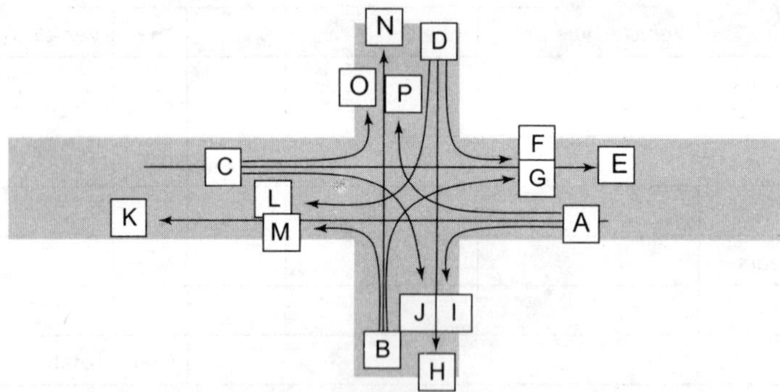


**Fig. 5.32** Typical presentation of traffic flow at intersections

**Example 5.1**

The manual count method was adopted at an isolated signalised intersection that has a cycle length of 100 seconds. The observers each record the counts for 1 hour and break for 5 minutes and then begin the next series of counts. The total vehicle count averages 490 vehicles in an hour of counting for one of the approach roads. What is the adjusted vehicle volume for this approach road movement?

**Answer:**

$$\text{Total counting period duration} = 1 \text{ hour} = 60 \times 60 = 3600 \text{ sec}$$

$$\text{Total break period during} = 5 \text{ minutes} = 5 \times 60 = 300 \text{ sec}$$

$$\text{Count expansion factor} = (3600 + 300)/3600 = 1.083$$

$$\text{Adjusted volume} = 490 \times 1.083 = 530.67 \approx 531 \text{ vehicles}$$

## 5.7 ANALYSIS OF TRAFFIC VOLUME DATA – BASIC TERMINOLOGY

- **Count**: The collection and recording of traffic data related to traffic volume, classification, speed, number of axles, gross weight of vehicle and a combination of these characteristics.
- **Counter**: A type of device used to collect traffic data with reference to any one of the above or a combination of such characteristics. For example, road tubes, induction loops, digital cameras and radar. Details of these counters and the methods of counting are discussed in detail in subsequent paragraphs.
- **Base data**: The unedited and uncorrected data (i.e., raw data) obtained directly from the traffic count with vehicle classification and vehicle or axle weights.

- **Base count:** A traffic count that has not been adjusted for the effects of trucks or seasons using axle factors and seasonal factors, respectively. The seasonal factors may be determined based on the day of the week or month of the year.
- **Daily truck volume (DTV)**: The total value of truck volume counted on a highway segment in a day.
- **Vehicle distance travelled (VDT)**: The distance travelled by all vehicles for a given period. Its value is typically computed for a geographic region. This parameter can be expressed as vehicle kilometres travelled or vehicle miles travelled.
- **Peak season:** A duration that represents thirteen consecutive weeks (i.e., 13 × 7 = 91 days) of the year with the highest traffic volume (FDOT, 2012).

## 5.8 TRAFFIC VOLUME DATA PRODUCTS

Traffic count data is used to calculate different parameters that are intended to represent the trend of traffic flow during a period of study time and also for the design and improvement of the transport facilities in a study area. The most widely used parameters that can be derived from traffic count data are described below.

### 5.8.1 Monthly Average Days of the Week Traffic (MADWT)

It is an estimate of the mean traffic volume for each day of the week, over a period of one month. It is the sum of all the traffic for each day of the week (Monday through Sunday) in a month, divided by the number of occurrences of that day in that month (AASHTO, 2004).

$$MADWT_{ij} = \frac{1}{n}\sum_{1}^{n} VOL_{ijn} \tag{5.1}$$

where $MADWT$ = monthly average day of the week traffic volume for the day of week $i$ (i.e., Monday = 1 through Sunday = 7) and the month $j$ (i.e., January = 1 through December = 12); $VOL_{ijn}$ = daily traffic volume for the $n^{th}$ occurrence of the day of the week $i^{th}$ in that month $j$; and $n$ = number of occurrences of day of the week $i$ in the month $j$ (usually, between 1 and 5, depending on the missing day data).

An example of the $MADWT_{ij}$ factor for the $i^{th}$ day of the week and $j^{th}$ month of the year are presented in Table 5.3.

**Table 5.3** $MADWT_{ij}$ factors (an 84-cell matrix)

| $j^{th}$ month of the year | $i^{th}$ day of the week | | | | | | |
|---|---|---|---|---|---|---|---|
| | *Monday* | *Tuesday* | *Wednesday* | *Thursday* | *Friday* | *Saturday* | *Sunday* |
| January | MADWT 1,1 | MADWT 1, 2 | MADWT 1,3 | MADWT 1, 4 | MADWT 1,5 | MADWT 1, 6 | MADWT 1, 7 |
| February | MADWT 2,1 | MADWT 2, 2 | MADWT 2,3 | MADWT 2, 4 | MADWT 2,5 | MADWT 2,6 | MADWT 2,7 |
| March | MADWT 3,1 | MADWT 3, 2 | MADWT 3,3 | MADWT 3, 4 | MADWT 3,5 | MADWT 3,6 | MADWT 3,7 |
| April | MADWT 4,1 | MADWT 4, 2 | MADWT 4,3 | MADWT 4, 4 | MADWT 4,5 | MADWT 4,6 | MADWT 4,7 |
| May | MADWT 5,1 | MADWT 5, 2 | MADWT 5,3 | MADWT 5, 4 | MADWT 5,5 | MADWT 5,6 | MADWT 5,7 |

(Continued)

**Table 5.3** (Continued)

| *j*th *month of the year* | *i*th *day of the week* | | | | | | |
|---|---|---|---|---|---|---|---|
| | *Monday* | *Tuesday* | *Wednesday* | *Thursday* | *Friday* | *Saturday* | *Sunday* |
| June | MADWT 6,1 | MADWT 6, 2 | MADWT 6,3 | MADWT 6, 4 | MADWT 6,5 | MADWT 6, 6 | MADWT 6, 7 |
| July | MADWT 7,1 | MADWT 7, 2 | MADWT 7,3 | MADWT 7, 4 | MADWT 7,5 | MADWT 7,6 | MADWT 7,7 |
| August | MADWT 8,1 | MADWT 8, 2 | MADWT 8,3 | MADWT 8, 4 | MADWT 8,5 | MADWT 8,6 | MADWT 8,7 |
| September | MADWT 9,1 | MADWT 9, 2 | MADWT 9,3 | MADWT 9, 4 | MADWT 9,5 | MADWT 9,6 | MADWT 9,7 |
| October | MADWT10,1 | MADWT10,2 | MADWT10,3 | MADWT10,4 | MADWT10,5 | MADWT10,6 | MADWT 10,7 |
| November | MADWT11,1 | MADWT11,2 | MADWT11,3 | MADWT11,4 | MADWT11,5 | MADWT11,6 | MADWT 11,7 |
| December | MADWT12,1 | MADWT12,2 | MADWT12,3 | MADWT12,4 | MADWT12,5 | MADWT12,6 | MADWT 12,7 |

### 5.8.2 Monthly Average Daily Traffic (MADT)

It is an estimate of the mean traffic volume for a month and can be calculated as the sum of the monthly average days of the week divided by seven. The seven comprises Monday through Sunday. The data should have at least one Monday, Tuesday, Wednesday, Thursday, Friday, Saturday and Sunday. In the absence of a MADW for each day of the week, the MADT may be calculated as the sum of monthly average of available days of a week divided by the number of those days of the week (AASHTO, 2004).

$$MADT_j = \frac{1}{7}\sum_{i=1}^{7} MADWT_{ij} \tag{5.2}$$

where $MADWT$ = monthly average day of the week traffic volume for the day of the week $i$ and the month $j$ (i.e., January = 1 through December = 12).

MADT reporting provides a month-by-month indication of the change in traffic trends, and the data can be used to monitor the effects on traffic flow due to specific events such as storms, fuel price shifts, economy, etc. (Wen, 2013).

### 5.8.3 Monthly Average Weekday Daily Traffic (MAWDT)

The MADT for Monday, Tuesday, Wednesday, Thursday and Friday are summed and then divided by five, as shown below, to obtain the MAWDT:

$$MAWDT_j = \frac{1}{5}\sum_{i=1}^{5} MADWT_{ij} \tag{5.3}$$

where $MADWT$ = monthly average day of the week traffic volume for the day of the week $i$ and the month $j$. (Monday = 1, Tuesday = 2, Wednesday = 3, Thursday = 4, Friday = 5, Saturday = 6, Sunday = 7).

### 5.8.4 Monthly Average Weekend Daily Traffic (MAWET)

The MADT for Saturday and Sunday are summed and then divided by two, as shown below:

$$MAWET_j = \frac{1}{2}\sum_{i=6}^{7} MADWT_{ij} \tag{5.4}$$

where *MADWT* = monthly average day of the week traffic volume for the day of the week *i* and the month *j*. Take Saturday = 6 and Sunday = 7.

### 5.8.5 Average Annual Weekday Traffic (AAWT)

The average of 24-hour traffic volume observed on all the weeks in a year (i.e., 365 days), at a location is considered. It is expressed as the total number of vehicles passing a section of a road on the week days in a year divided by 260. This volume may be estimated by adjusting a short-term count with weekly and monthly factors (AASHTO, 2004).

$$AAWT = \left( \frac{\text{Total volume of traffic on week days in a year}}{260\,\text{weeks}} \right) \tag{5.5}$$

### 5.8.6 Average Weekday Traffic (AWT)

The average of 24-hour traffic volume observed on a few weeks over a period of less than a year (say *N* number of weeks) at a location is considered. The traffic on a section of a road may be measured during the *N* number of weeks, which may be considered with reference to a season, festival days, months or few weeks.

$$AWT = \left( \frac{\text{Total volume of traffic in } N \text{ number of weeks}}{N \text{ number of weeks}} \right) \tag{5.6}$$

The relationship between AADT and ADT is analogous to the one between AAWT and AWT.

### 5.8.7 Annual Average Daily Truck Traffic (AADTT)

It is the ratio between the total truck traffic on a highway section for one year and the number of days in the year.

### 5.8.8 Annual Average Weekday Traffic (AAWDT)

The average computed from traffic volume measured during on weekdays (i.e., Monday, Tuesday, Wednesday, Thursday and Friday) at continuous monitored sites.

$$AAWDT = \frac{1}{12} \sum_{j=1}^{12} MAWDT_j \tag{5.7}$$

There *MAWDT* = monthly average weekday traffic volume for the month *j*.

### 5.8.9 Annual Average Day of Week Traffic (AADWT)

$$\text{AADWT} = \frac{1}{n} \sum_{j=1}^{n} MADWT_{ij} \tag{5.8}$$

where *MADWT* = monthly average day of week traffic; *i* = day of the week, *j* = month of the year and *n* = number of months for which MADWT values are available.

### 5.8.10 Average Annual Daily Traffic (AADT)

The average of 24-hour traffic volume observed over a full year (i.e., 365 days) at a location is considered. It is an estimate of the mean daily traffic on a road segment on all the days of the week (i.e., Monday through Sunday) over a period of one year. It is the most widely and frequently used traffic data statistic. AADT data is based on site-specific counts. It is the rate flow that indicates how busy a road is. It is expressed as the total number of vehicles passing a section of a road in a year divided by 365. It is a simple average volume of 365 days, except for leap years which have 366 days.

$$AADT = \left(\frac{\text{Total of average daily volume of traffic in a year}}{365\,\text{days}}\right) \tag{5.9}$$

or

$$AADT = \frac{\sum_{i=1}^{365} VOL_{24\,hour}}{365} \tag{5.10}$$

or

$$AADT = \frac{1}{n}\sum_{i=1}^{n} AADWT \tag{5.11}$$

where $AADWT$ = annual average day of week traffic, $i$ = day of the week and $n$ = number of days of the week for which AADWT values are available.

or

$$AADT = \frac{1}{12}\sum_{j=1}^{12} MADT_j \tag{5.12}$$

where $MADT$ = monthly average daily traffic and $j$ = month of the year.

The AADT can be calculated from MADWT, as given below. This AASHTO method is also known as the **average of averages method**. It accounts for the *missing days count* by giving equal weight to each day of the week in each month (Wang and Teng, 2004).

$$AADT = \frac{1}{84}\left[\sum_{j=1}^{12}\left(\sum_{i=1}^{7} MADWT_{ij}\right)\right] \tag{5.13}$$

where $MADWT$ = monthly average days of the week traffic, $i$ = day of the week (1 through 7) and $j$ = month of the year (1 through 12).

Alternatively, AADT can also be calculated by adjusting the *short-duration traffic volume* raw count, which reduces the effects of temporal bias, seasonal variation, equipment used and duration of the count. The respective correction factors are used for this purpose. In general, a 48-hour axle count may be converted to AADT using the following equation, as recommended by FHWA (2012):

$$AADT_{ij} = VOL_{ij} \times M_j \times DOW_j \times ACF_i \times G_j \tag{5.14}$$

where $AADT_{ij}$ = annual average daily traffic at location $i$ of factor group $j$; $VOL_{ij}$ = 48-hour axle volume at location $i$ of factor group $j$ (in case of a seven-day or any-other-duration count, the value of the average count is substituted and the parameter $DOW_j$ removed); $M_j$ = the monthly or seasonal correction factor for factor $j$; $DOW_j$ = the day of week factor for factor group $j$ (as applicable); $ACF_i$ = the axle correction factor for location $i$ (this parameter is removed if the inductive loop detector is used instead of a conventional pneumatic axle tube); and $G_j$ = the growth factor for factor group $j$ (this parameter is applicable if the count was only taken in a year other than the year for which AADT is being estimated).

It has been noted that the AADT estimated from the short-duration count may be associated with more error than that calculated from a continuous 365-day count (Wang and Teng, 2004). Details of these correction factors and their definitions are discussed in different sections of this chapter.

### 5.8.11 Annual Average Weekend Traffic (AAWET)

It is the average value of the monthly average weekend traffic volume over a year, given as

$$AAWET = \frac{1}{12}\sum_{j=1}^{12} MAWET_j \tag{5.15}$$

where $MAWET$ = monthly average weekend traffic volume for $j$ months.

### 5.8.12 Annual Vehicle Distance Travelled (AVDT)

It is a measure of the number of kilometres that vehicles are driven in one year. This value may be computed as

$$\text{AVDT} = \text{Daily vehicle distance travelled (DVDT)} \times \text{No. of days in a year} \tag{5.16}$$

$$\text{Section DVDT} = \text{Section AADT} \times \text{Section length} \tag{5.17}$$

$$\text{Section length} = \text{Distance between beginning and ending km-stones in a road class}$$

### 5.8.13 Vehicle Miles Travelled (VMT)

$$\text{VMT} = \text{AADT} \times \text{Length of the road segment in miles} \tag{5.18}$$

The values of VMTs in a state or locality may be summed and used as an indicator of the amount of traffic in that state or locality. The following formula may be used to calculate the VMT in a study area (Wang and Teng, 2004).

$$\begin{aligned} VMT = {} & \sum_{k=1}^{P}\left(AADT_k^{count\ station\ 1} \times L_k^{Segment\ 1}\right) + \left(AADT_k^{count\ station\ 2} \times L_k^{Segment\ 2}\right) \\ & + \sum_{k=1}^{Q}\left(AADT_k^{count\ station\ 1} \times L_k^{Segment\ 1}\right) + \left(AADT_k^{count\ station\ 2} \times L_k^{Segment\ 2}\right). \\ & + \sum_{k=1}^{R}\left(AADT_k^{count\ station\ 1} \times L_k^{Segment\ 1}\right) + \left(AADT_k^{count\ station\ 2} \times L_k^{Segment\ 2}\right). \\ & + \sum_{k=1}^{S}(\ ) + (\ ) + (\ ) + \sum_{k=1}^{T}(\ ) + (\ ) + (\ )\ldots \end{aligned} \tag{5.19}$$

where $P$ = number of segments or sections considered on road 1; $Q$ = number of segments or sections considered on road 2; $R$ = number of segments or sections considered on road 3; $S$, $T$ and...= number of segments or sections considered on roads 4, 5...n; $AADT$ = annual average daily traffic, in each direction on each segment; and $L$ = applicable length of the road segment in each direction.

### 5.8.14 Average Daily Traffic (ADT)

The average 24-hour traffic volume observed over a period of less than a year (say, $N$ number of days) at a location is considered. It is obtained by a short-duration traffic count, and the data is based on site-specific counts. The traffic on a section of road measured during the $N$ number of days may be considered with reference to a season, festival days, months, a week or say 3 days. The calculated value is reported as the ADT for any such period of the year. For this reason, the ADT is not considered as the standard measure of vehicular traffic flow on a section of road and is not the basis for most decisions related to transport projects.

$$ADT = \left(\frac{\text{Total daily volume of traffic in } N \text{ number of days}}{N \text{ number of days}}\right) = \text{Estimated } AADT \tag{5.20}$$

For example, if 750 vehicles were observed on a section of road over three days, the ADT = (750/3) = 250.

This method of calculating ADT is a simple average of all the days measured. The disadvantage of this method is that the missing data can cause inaccuracies in the determination of true AADT. The resulting bias in the AADT will be large when an unequal number of week day or weekend days are missing or a considerable number of days is missing from monthly or yearly traffic volume data. To overcome this effect, AASHTO adopted a different approach for calculating AADT. In this approach, first the average monthly days of the week are computed and these 84 values (i.e., 7 weeks × 12 months = 84) are then averaged to yield the AADT. The AASHTO (1992) recommended equation is as follows:

$$AADT = \frac{1}{7}\sum_{i=1}^{7}\left\{\frac{1}{12}\sum_{j=1}^{12}\left(\frac{1}{n}\sum_{k=1}^{n} VOL_{ijk}\right)\right\} \tag{5.21}$$

where $VOL$ = daily traffic for day $k$, of week $i$ and month $j$; $i$ = day of the week; $j$ = month of the year; $k$ = 1, when the day is the first occurrence of that day of the week in a month; 4 when it is the fourth day of the week; and $n$ = number of days of that day of the week in month $j$.

The AASHTO method explicitly accounts for missing data by weighting each day of the week the same and each month the same, regardless of how many days are actually present or missing within that category, provided there are between 1 and 5 records for each day of the week in each month (FHWA, 2013). The equation, average the measured values in the sequence of day of the month and month of the day, thus maintain the balance between week days and weekend days.

Alternatively, AADT can be calculated from ADT by multiplying it with the seasonal correction factor (SCF), axle correction factor (ACF) and annual growth factor (Wang, 2012).

$$AADT = ADT \times SCF \times ACF \times AGF \tag{5.22}$$

It must be noted that the previous year's factors are applied to the current year's data.

### 5.8.15 Axle Correction Factor (ACF)

It is a multiplication number used to adjust vehicle axle sensor-based data for WIM-recorded vehicles with more than two axles. ACF can be calculated by dividing the total number of axles based on automatic vehicle classification data by the total number of vehicles counted. It represents the average number of axles per vehicle.

In the case of a single sensor (i.e., pneumatic tube detector) used for counting the number of axles that pass a section of a lane, ACF is multiplied with the total axle count to obtain an estimate of vehicle volume. It is an indirect way of determining vehicle volume count from the total axle count. The following equation may be used to determine the correction factor (TRL, 1993):

$$\text{Correction factor} = \frac{\text{Number of vehicles observed}}{\left(\frac{1}{2} \times \text{Number of axles on those vehicles}\right)} \quad (5.23).$$

The above correction factor is calculated for a specific site (or for an area by averaging such factors at each site) based on a short sample count, with both peak and off-peak data.

### 5.8.16 Seasonal Correction Factor (SCF) or Seasonal Adjustment Factor (SAF)

It is used to adjust all short-duration counts that reflect seasonal changes in traffic volume. SCF is used to adjust short-duration counts influenced by seasonal variations; similar traffic characteristics can be obtained from permanent traffic counts.

The seasonal adjustment factor for each day of the week and month of the year combination can be calculated from the following equation:

$$F_{ij} = \frac{AADT}{MADWT_{ij}} \quad (5.24)$$

where $F_{ij}$ = seasonal adjustment factor for the $i^{th}$ day of the week and $j^{th}$ month of the year and $MADWT_{ij}$ = monthly average day of week traffic volume for the $i^{th}$ day of the week and $j^{th}$ month of the year.

An example of a seasonal adjustment factor ($F_{ij}$) for the $i^{th}$ day of the week and the $j^{th}$ month of the year are presented in Table 5.4.

**Table 5.4** Seasonal and day of week factors ($F_{ij}$)

| *$j^{th}$ month of the year* | *$i^{th}$ day of the week* | | | | | | |
|---|---|---|---|---|---|---|---|
| | *Monday* | *Tuesday* | *Wednesday* | *Thursday* | *Friday* | *Saturday* | *Sunday* |
| January | F1,1 | F1, 2 | F1,3 | F1, 4 | F1,5 | F1,6 | F1,7 |
| February | F2,1 | F2, 2 | F2,3 | F2, 4 | F2,5 | F2,6 | F2,7 |
| March | F3,1 | F3, 2 | F3,3 | F3, 4 | F3,5 | F3,6 | F3,7 |
| April | F4,1 | F4, 2 | F4,3 | F4, 4 | F4,5 | F4,6 | F4,7 |
| May | F5,1 | F5, 2 | F5,3 | F5, 4 | F5,5 | F5,6 | F5,7 |
| June | F6,1 | F6, 2 | F6,3 | F6, 4 | F6,5 | F6,6 | F6,7 |
| July | F7,1 | F7, 2 | F7,3 | F7, 4 | F7,5 | F7,6 | F7,7 |
| August | F8,1 | F8, 2 | F8,3 | F8, 4 | F8,5 | F8,6 | F8,7 |
| September | F9,1 | F9, 2 | F9,3 | F9, 4 | F9,5 | F9,6 | F9,7 |
| October | F10,1 | F10, 2 | F10,3 | F10, 4 | F10,5 | F10,6 | F10,7 |
| November | F11,1 | F11, 2 | F11,3 | F11, 4 | F11,5 | F11,6 | F11,7 |
| December | F12,1 | F12, 2 | F12,3 | F12, 4 | F12,5 | F12,6 | F12,7 |

The above factors are useful for calculating AADT based on short-duration count or coverage traffic count. For example, the following formula demonstrates the calculation of AADT for a short-duration count segment, where two consecutive daily traffic volumes are collected (Wang and Teng, 2004).

$$AADT = \frac{1}{2}\left[\left(F_{i,j} \times VOL_{i,j}\right) + \left(F_{i+1,j} \times VOL_{i+1,j}\right)\right] \tag{5.25}$$

Similarly, in the case of three consecutive daily traffic volumes

$$AADT = \frac{1}{3}\left[\left(F_{i,j} \times VOL_{i,j}\right) + \left(F_{i+1,j} \times VOL_{i+1,j}\right) + \left(F_{i+2,j} \times VOL_{i+2,j}\right)\right] \tag{5.26}$$

where $AADT$ = annual average daily traffic estimated from short-duration traffic count; $i$ = seven days in a week (1, 2, 3, 4, 5, 6 and 7); $j$ = twelve months in the year (1, 2, 3, 4, 5, 6, 7, 8, 9, 10, 11 and 12); $F_{ij}$ = applicable seasonal and day of week factor; and $VOL_{i,j}$ = daily traffic volume collected by the short-duration count.

### 5.8.17 Seasonal Average Daily Traffic (SADT)

It is a duration comprising those full months that contain 80% of the annual traffic volume observed at a station. The months used for calculation of SADT may not be consecutive because a place may have more than one season. To determine SADT, the respective MADT is used to compute the traffic volume for each month and the monthly volume is ranked from highest to lowest. Once the computed highest cumulative per cent of monthly volume exceeds 80% of the annual traffic volume, the respective MADTs are averaged and reported as SADT (FHWA, 2005)

### 5.8.18 Average Summer Daily Traffic (ASDT)

The traffic data collecting methods used for determining AADT and ASDT are the same; however, ASDT data is collected in summer only. ASDT is calculated as an *arithmetic mean* of the traffic counted in the respective period, like AADT. It is used to represent the average daily traffic flow over a summer. The value of ASDT calculated at a given location may not be equal to the value of AADT. The ASDT value may be more or less than the value of AADT, which demonstrates seasonal variations in the use of such roads in a locality (Božić et al., 2012).

### 5.8.19 Seasonal Average Daily Traffic Adjustment Factors (SADTAF)

These are used to estimate the SADT based on short-duration counts that reflect the seasonal changes in traffic volume.

The SADTAF for each day of the week based on short-duration counts can be calculated using the equation

$$SADTAF_{ij} = \frac{SADT}{MADWT_{ij}} \tag{5.27}$$

where $SADTAF_{ij}$ = seasonal average daily traffic adjustment factor for the $i^{th}$ day of the week and $j^{th}$ month of the year and $MADWT_{ij}$ = monthly average day of week traffic volume for the $i^{th}$ day of the week and $j^{th}$ month of the year.

### 5.8.20 Peak Hour Factor (PHF)

It is the ratio between the average traffic volume during the peak hour period to four times the average volume during the peak 15-minute period.

PHF represent traffic flow variation within the peak hour. The maximum value of PHF is 1.00 and the minimum value is 0.25. The minimum value would occur if the total volume observed in the peak hour is found in only one of the 15-minute intervals. Typically, the PHF value will fall in the range of 0.70 to 0.98. The higher the value of PHF, the lower the degree of variation in volume during peak hour, and vice versa. As the value of PHF is $\leq 0.50$, it represents a non-homogeneous traffic flow condition within the peak hour. The following expression may be used to calculate PHF:

$$PHF = \left( \frac{\text{Average volume during peak hour}}{4 \times \text{Average volume during peak 15-minute period}} \right) \tag{5.28}$$

For example, the traffic volume data obtained in a peak hour at a location is given below. Calculate the PHF.

| *Time duration within an hour* | *Traffic volume (number of vehicles)* |
|---|---|
| 10:00–10:15 | 950 |
| 10:15–10:30 | 1010 |
| 10:30–10:45 | 1050 |
| 10:45–11:00 | 990 |
| Total = | 4000 |

$$PHF = \left( \frac{4000}{4 \times 1050} \right) = 0.95$$

### 5.8.21 Design Hour (DH)

It is an hour with a traffic volume that is used for designing geometric elements and the traffic control devices of a road facility.

### 5.8.22 Design Year (DY)

It is a year in future for which the traffic facility is designed. It is usually 20 or 30 years from the opening or current year of a roadway.

### 5.8.23 Design Hour Volume (DHV)

It is the traffic volume that is estimated to use a highway segment during the design hour (DH) of the design year. It is an adopted volume of traffic which is obtained below peak and at $n^{th}$ highest hour (Fig. 5.33). Usually, the $30^{th}$ highest hourly volume is adopted as the DHV because the design of facilities for maximum volume of traffic observed in a year is uneconomical. It means this hourly volume will be exceeded only 29 times in a year.

As a recent practice, the particular hour which falls within the range of the knee of the curve (Fig. 5.33), where the slope of the curve changes most rapidly, is used as the design hour. As a general trend, the curve flattens more gradually in a rural road than in an urban facility (Liggett et al., 2009).

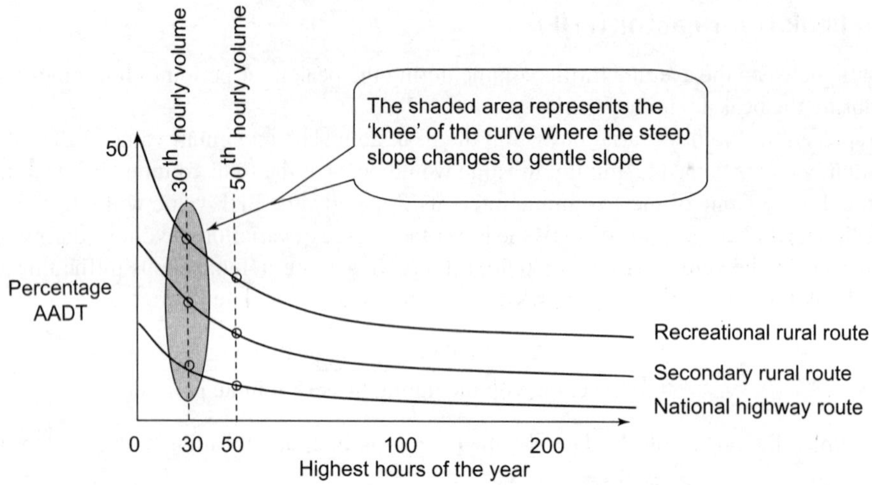


**Fig. 5.33** Typical relationships between hourly volume and AADT

DHV can be obtained using the following equation:

$$DHV = AADT \times K \textit{ for the design year} \tag{5.29}$$

For example, it can be observed from Fig. 5.33 that the variations in hourly volume up to the 50th highest hourly volumes in a year are less on a national highway than on a recreational rural route, where seasonal variations are predominant.

## 5.8.24 K-Factor or Design Hour Factor

It is the ratio between traffic volume in the study hour and AADT. The K-factor represents the proportion of AADT that is expected to occur in the design hour. The routes that are subjected to occasional extreme traffic volumes generally exhibit the highest K-factors. The K-factor is useful for corridor traffic forecasting and project traffic forecasting projects because it defines the volume of traffic that the road is designed to handle (FDOT, 2012).

Design hour volume (DHV) may be considered as the traffic volume in the study hour, and the calculated factor is termed the design hour factor-K. The relationship between DHV, AADT and K is expressed below

$$K = DHV \div AADT \tag{5.30}$$

## 5.8.25 Design Hour Truck (DHT)

It is the percent value of trucks estimated to use a highway segment during the design hour (DH) of the design year.

## 5.8.26 Design Hour Factor

It is the proportion of AADT that occurs during the design hour (TRB, 2000).

### 5.8.27 Directional Distribution (D) or Directional Split of Traffic

D is the percentage of traffic flow in one direction of a roadway during design hour. D is an important factor in the analysis of highway capacity and LOS. The number of lanes required to cater to the traffic to provide the desired LOS is assessed with reference to D.

### 5.8.28 Directional Design Hour Volume (DDHV)

Peak directional distribution may occur in a single direction during one hour in the morning and the other in the evening. The DDHV for the design year is used for the geometric design of highway facilities. The following equations may used for estimation of DDHV (FDOT, 2012):

$$DDHV \text{ (in peak direction)} = AADT \times D \times K \tag{5.31}$$

$$DDHV \text{ (in opposite direction)} = AADT \times (1 - D) \times K \tag{5.32}$$

## 5.9 SOLVED PROBLEMS

**Example 5.2**

The following data is collected from a continuous automatic vehicle counter and classifier. The monthly average daily traffic for the total volume and the average daily truck traffic data are summarised below.

| *Month* | *MADT* | *DOW* | *AADMT* |
|---|---|---|---|
| January | 53535 | Monday | 459 |
| February | 51172 | Tuesday | 479 |
| March | 57149 | Wednesday | 490 |
| April | 57677 | Thursday | 502 |
| May | 58340 | Friday | 505 |
| June | 59200 | Saturday | 450 |
| July | 58250 | Sunday | 415 |
| August | 59222 | | |
| September | 57240 | | |
| October | 58267 | | |
| November | 55582 | | |
| December | 51761 | | |

Determine (a) AADT, (b) seasonal (monthly) correction factors and (c) the truck day of week (DOW) factors.

In a related truck traffic survey, a short-term monitoring site on the same route 10 km to the north, two-day counts were taken in September and the summary of the data is given below.

| *Date* | *ADMT* | *ADT* |
|---|---|---|
| February, 11 (Tuesday) | 487 | 50761 |
| February, (Wednesday) | 503 | 51231 |

Correctly estimate the annual average daily truck traffic from the above traffic data by adjusting the respective seasonal correction factor and DOW factor.

**Answer:** The AADT and seasonal or monthly factors are calculated below:

| *Month* | *MADT* | *Seasonal or monthly factor* |
|---|---|---|
| [1] | [2] | [3] |
| January | 53535 | (56450 ÷ 53535) = 1.05 |
| February | 51172 | (56450 ÷ 51172) = 1.10 |
| March | 57149 | (56450 ÷ 57149) = 0.99 |
| April | 57677 | (56450 ÷ 57677) = 0.98 |
| May | 58340 | (56450 ÷ 58340) = 0.97 |
| June | 59200 | (56450 ÷ 59200) = 0.95 |
| July | 58250 | (56450 ÷ 58250) = 0.97 |
| August | 59222 | (56450 ÷ 59222) = 0.95 |
| September | 57240 | (56450 ÷ 57240) = 0.99 |
| October | 58267 | (56450 ÷ 58267) = 0.97 |
| November | 55582 | (56450 ÷ 55582) = 1.02 |
| December | 51761 | (56450÷ 51761) = 1.09 |
| Total | 677395 | - |
| **AADT** | **(677395/12) = 56450** | **(56450 ÷ 56450) = 1.00** |

The truck traffic DOW correction factors for each day of the week for the year are calculated below:

| *DOW* | *AADTT* | *Truck traffic DOW factor* |
|---|---|---|
| Monday | 459 | (471 ÷ 459) = 1.03 |
| Tuesday | 479 | (471 ÷ 479) = 0.98 |
| Wednesday | 490 | (471 ÷ 490) = 0.96 |
| Thursday | 502 | (471 ÷ 502) = 0.94 |
| Friday | 505 | (471 ÷ 505) = 0.93 |
| Saturday | 450 | (471 ÷ 450) = 1.05 |
| Sunday | 415 | (471 ÷ 415) = 1.14 |
| Total | 3300 | - |
| **Annual ADTT** | **(3300 ÷ 7) = 471** | **(471 ÷ 471) = 1.00** |

The given short count of truck class is averaged as follows:

| *Date* | *ADTT* |
|---|---|
| February, 11 (Tuesday) | 487 |
| February, 12 (Wednesday) | 503 |
| TOTAL | 990 |
| **Average** | **(990 ÷ 2) = 495** |

The above given short count of truck class is factored in for seasonality and for DOW, as Seasonal or monthly factor for February = 1.10; DOW factor for Tuesday = 0.98; DOW factor for Wednesday = 0.96

Estimate of AADTT for the month of February on Tuesday of week = 487 × 1.10 × 0.98 = 524.986
Estimate of AADTT for the month of February on Wednesday of week = 503 × 1.10 × 0.96 = 531.168
The above two AADTT estimates are averaged to provide the correct estimate of AADMT = (524.986 + 531.168) ÷ 2 = (1056.154) ÷ 2 ≈ 528

*Note:* Other classes of vehicles would also need to be adjusted as shown above.

**Example 5.3**

The details of vehicle classification count conducted manually (as a short-duration count) on a road section are given below.

| *Vehicle class* | *Vehicle type* | *Daily vehicle volume* | *Average number of axles per vehicle* | *Total number of axles* |
|---|---|---|---|---|
| Type 2 |  | 220 | 2 | 440 |
| Type 3 | 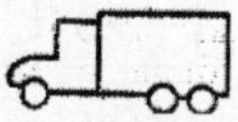 | 1968 | 3 | 5904 |
| Type 2-S1 |  | 347 | 3 | 1041 |
| Type 2-S2 |  | 208 | 4 | 832 |
| Type 3-S1 |  | 215 | 4 | 860 |
| Type 3-S2 |  | 167 | 5 | 835 |
| Type 2-2 |  | 94 | 4 | 376 |
| Type 3-2 |  | 10 | 5 | 50 |
| Type 2-3 |  | 27 | 5 | 135 |
| Type 3-3 |  | 11 | 6 | 66 |
| | Total Volume = | 3267 | Total Number of Axles | 10539 |

Determine (a) axle correction factor (ACF) from the above survey data and (b) number of vehicles that passed over near the WIM station if the number of axles counted is 5548.

**Answer:** **(a)** Axle correction factor (ACF) = Total number of axles ÷ Total volume =10539 ÷ 3267 = 3.225895 axles per vehicle.

**(b)** Number of vehicles that passed over near a WIM station = Number of axles counted ÷ ACF = 5548 ÷ 3.225895 = 1719.833 ≈ 1720 vehicles.

## Example 5.4

The values of mean average daily truck traffic (MADTT) of 10 truck types (IRC) with their observed temporal patterns over a 12-month period are given below.

| MADTT | *Vehicle class as per IRC* | | | | | | | | | |
|---|---|---|---|---|---|---|---|---|---|---|
| | *Type 2* | *Type 3* | *Type 2-S1* | *Type 2-S2* | *Type 3-S1* | *Type 3-S2* | *Type 2-2* | *Type 3-2* | *Type 2-3* | *Type 3-3* |
| **January** | 602 | 2590 | 1621 | 415 | 290 | 328 | 148 | 67 | 73 | 57 |
| **February** | 623 | 2610 | 1637 | 451 | 310 | 448 | 158 | 50 | 63 | 43 |
| **March** | 645 | 2630 | 1653 | 458 | 317 | 412 | 153 | 33 | 97 | 28 |
| **April** | 670 | 2849 | 1804 | 489 | 310 | 471 | 151 | 81 | 90 | 70 |
| **May** | 690 | 3019 | 1920 | 503 | 366 | 340 | 210 | 59 | 83 | 51 |
| **June** | 729 | 3267 | 2092 | 542 | 389 | 346 | 202 | 85 | 97 | 73 |
| **July** | 771 | 3476 | 2238 | 590 | 367 | 369 | 146 | 107 | 70 | 92 |
| **August** | 819 | 3678 | 2380 | 632 | 419 | 351 | 184 | 42 | 56 | 36 |
| **September** | 846 | 3590 | 2325 | 660 | 430 | 172 | 162 | 88 | 39 | 75 |
| **October** | 790 | 3456 | 2227 | 612 | 390 | 221 | 201 | 16 | 38 | 13 |
| **November** | 691 | 3178 | 1960 | 511 | 379 | 161 | 212 | 24 | 56 | 20 |
| **December** | 598 | 3167 | 1817 | 458 | 321 | 265 | 169 | 25 | 68 | 21 |
| **Total** | **8474** | **37510** | **23673** | **6321** | **4288** | **3885** | **2096** | **676** | **830** | **579** |

Determine the truck monthly factors that reflect their travel pattern throughout the year.

**Answer:** Monthly adjustment factors (MAF) for each class of vehicle as per the FHWA (2012) may be calculated using the following equation.

$$\text{MAF for the } i^{\text{th}} \text{ class of vehicle and month } j = (\text{MADTT}_{ij} \div \Sigma \text{MADTT}_{i}) \times 12$$

The given values are substituted in the above equation:
Monthly adjustment factor (MAF) for *Type 2* class of vehicle and month *January* = (602 ÷ 8474) × 12 = 0.85
Monthly adjustment factor (MAF) for *Type 3-S1* class of vehicle and month *May* = (366 ÷ 4288) × 12 = 1.02
The MAFs for the each class of vehicle are given below:

| MADTT | Vehicle class (ith) | | | | | | | | | |
|---|---|---|---|---|---|---|---|---|---|---|
| | Type 2 | Type 3 | Type 2-S1 | Type 2-S2 | Type 3-S1 | Type 3-S2 | Type 2-2 | Type 3-2 | Type 2-3 | Type 3-3 |
| **January** | 0.85 | 0.83 | 0.82 | 0.79 | 0.81 | 1.01 | 0.85 | 1.19 | 1.05 | 1.19 |
| **February** | 0.88 | 0.83 | 0.83 | 0.86 | 0.87 | 1.39 | 0.90 | 0.88 | 0.92 | 0.88 |
| **March** | 0.91 | 0.84 | 0.84 | 0.87 | 0.89 | 1.27 | 0.88 | 0.58 | 1.40 | 0.58 |
| **April** | 0.95 | 0.91 | 0.91 | 0.93 | 0.87 | 1.45 | 0.86 | 1.44 | 1.31 | 1.44 |
| **May** | 0.98 | 0.97 | 0.97 | 0.95 | 1.02 | 1.05 | 1.20 | 1.05 | 1.20 | 1.05 |
| **June** | 1.03 | 1.05 | 1.06 | 1.03 | 1.09 | 1.07 | 1.16 | 1.51 | 1.40 | 1.51 |
| **July** | 1.09 | 1.11 | 1.13 | 1.12 | 1.03 | 1.14 | 0.84 | 1.91 | 1.02 | 1.91 |
| **August** | 1.16 | 1.18 | 1.21 | 1.20 | 1.17 | 1.08 | 1.05 | 0.74 | 0.81 | 0.74 |
| **September** | 1.20 | 1.15 | 1.18 | 1.25 | 1.20 | 0.53 | 0.93 | 1.56 | 0.57 | 1.56 |
| **October** | 1.12 | 1.11 | 1.13 | 1.16 | 1.09 | 0.68 | 1.15 | 0.28 | 0.54 | 0.28 |
| **November** | 0.98 | 1.02 | 0.99 | 0.97 | 1.06 | 0.50 | 1.21 | 0.42 | 0.81 | 0.42 |
| **December** | 0.85 | 1.01 | 0.92 | 0.87 | 0.90 | 0.82 | 0.97 | 0.44 | 0.98 | 0.44 |
| **Total** | **12.00** | **12.00** | **12.00** | **12.00** | **12.00** | **12.00** | **12.00** | **12.00** | **12.00** | **12.00** |

**Example 5.5**

A typical example of the computation of truck hourly distribution factors on hourly basis is shown below. Column [3] data pertains to a 24-hour truck traffic count obtained on a section of road.

| Start time | End time | Hourly distribution of truck traffic | |
|---|---|---|---|
| | | Hourly AADTT | % |
| [1] | [2] | [3] | [4] |
| 00:00 | 01:00 | 10 | (10 ÷ 1427)× 100 = 0.70 |
| 01:00 | 02:00 | 12 | (12 ÷ 1427)× 100 = 0.84 |
| 02:00 | 03:00 | 14 | (14 ÷ 1427)× 100 = 0.98 |
| 03:00 | 04:00 | 15 | (15 ÷ 1427)× 100 = 1.05 |
| 04:00 | 05:00 | 24 | (24 ÷ 1427)× 100 = 1.68 |
| 05:00 | 06:00 | 39 | (39 ÷ 1427)× 100 = 2.73 |
| 06:00 | 07:00 | 49 | (49 ÷ 1427)× 100 = 3.43 |
| 07:00 | 08:00 | 71 | (71 ÷ 1427)× 100 = 4.98 |
| 08:00 | 09:00 | 81 | (81 ÷ 1427)× 100 = 5.68 |
| 09:00 | 10:00 | 80 | (80 ÷ 1427)× 100 = 5.61 |
| 10:00 | 11:00 | 82 | (82 ÷ 1427)× 100 = 5.75 |
| 11:00 | 12:00 | 93 | (93 ÷ 1427)× 100 = 6.52 |
| 12:00 | 13:00 | 102 | (102 ÷ 1427)× 100 = 7.15 |

(Continued)

**Table** (Continued)

| Start time | End time | Hourly distribution of truck traffic | |
|---|---|---|---|
| | | Hourly AADTT | % |
| 13:00 | 14:00 | 112 | $(112 \div 1427)\times 100 = 7.85$ |
| 14:00 | 15:00 | 98 | $(98 \div 1427)\times 100 = 6.87$ |
| 15:00 | 16:00 | 95 | $(95 \div 1427)\times 100 = 6.66$ |
| 16:00 | 17:00 | 89 | $(89 \div 1427)\times 100 = 6.24$ |
| 17:00 | 18:00 | 88 | $(88 \div 1427)\times 100 = 6.17$ |
| 18:00 | 19:00 | 71 | $(71 \div 1427)\times 100 = 4.98$ |
| 19:00 | 20:00 | 63 | $(63 \div 1427)\times 100 = 4.41$ |
| 20:00 | 21:00 | 58 | $(58 \div 1427)\times 100 = 4.06$ |
| 21:00 | 22:00 | 47 | $(47 \div 1427)\times 100 = 3.29$ |
| 22:00 | 23:00 | 21 | $(21 \div 1427)\times 100 = 1.47$ |
| 23:00 | 24:00 | 13 | $(13 \div 1427)\times 100 = 0.91$ |
| Total = | | 1427 | 100.00 % |

**Answer:** The values in Column [4] are the truck distribution hourly factors for each hour of the day.

**Example 5.6**

A summary of the details of traffic volume count carried out in a study area are given below.

| Road No. | Direction of traffic flow | Name of the count station | AADT | Segment length (mile) |
|---|---|---|---|---|
| Road-1 | North–South | CS-1-N-S | 22260 | 0.94 |
| | South–North | CS-1-S–N | 24369 | 0.94 |
| Road-1 | East–West | CS-2-E–W | 25511 | 1.24 |
| | West–East | CS-2-W–E | 21911 | 1.24 |
| Road-2 | NE–SW | CS-3- NE–SW | 61345 | 2.56 |
| | SW–NE | CS-3-SW–NE | 71128 | 2.56 |
| Road-3 | NW–SE | CS-4- NW–SE | 33476 | 0.89 |
| | SE–NW | CS-4- SE–NW | 37737 | 0.89 |

Calculate the VMT in the study area.

**Answer:** On substituting the given values in Eq. 5.19, we get

$$VMT = \sum_{k=1}^{Road-1} (22260\times0.94)+(24369\times0.94)+(25511\times1.24)+(21911\times1.24)$$

$$+\sum_{k=1}^{Road-2} (61345\times2.56)+(71128\times2.56)+\sum_{k=1}^{Road-3} (33476\times0.89)+(37737\times0.89)$$

$$VMT = \sum_{k=1}^{Road-1} (20924.4) + (22906.86) + (31633.64)$$

$$+ (27169.64) + \sum_{k=1}^{Road-2} (157043.2) + (182087.7).$$

$$+ \sum_{k=1}^{Road-3} (29793.64) + (33585.93)$$

$$VMT = \sum_{k=1}^{Road-1} 102634.5 + \sum_{k=1}^{Road-2} 339130.9 + \sum_{k=1}^{Road-3} 63379.57$$

$$\text{VMT} = 505145 \text{ vehicle-miles}$$

**Example 5.7**

Determine the months that are part of the seasonal average daily traffic (SADT) and calculate the value of SADT from the following summary of traffic count:

| *Month* | *Number of days* | *MADT* |
|---|---|---|
| January | 31 | 201 |
| February | 28 | 209 |
| March | 31 | 345 |
| April | 30 | 983 |
| May | 31 | 1273 |
| June | 30 | 1979 |
| July | 31 | 1950 |
| August | 31 | 1483 |
| September | 30 | 1339 |
| October | 31 | 934 |
| November | 30 | 299 |
| December | 31 | 254 |

**Answer:** Following are the details of the computation of seasonal average daily traffic (SADT) (FHWA, 2013):

$$\text{Column [4]} = \text{Column [2]} \times \text{Column [3]}$$

$$\text{Column [5]} = (\text{Column [4]} \div 343492) \times 100$$

| *Month* | *Days* | *MADT* | *Monthly volume* | *Per cent of year* | *Ranking by highest MADT volume* | *Cumulative percentile based onrRanking* | *Is MADT part of SADT?* |
|---|---|---|---|---|---|---|---|
| [1] | [2] | [3] | [4] = [2]×[3] | [5] | [6] | [7] | [8] |
| January | 31 | 201 | 6231 | 1.81 % | 11th | | No |
| February | 28 | 209 | 5852 | 1.70 % | 12th | | No |
| March | 31 | 345 | 10695 | 3.11 % | 8th | | No |
| April | 30 | 983 | 29490 | 8.59 % | 6th | 80.03 % | Yes |
| May | 31 | 1273 | 39463 | 11.49 % | 5th | 71.44 % | Yes |
| June | 30 | 1979 | 59370 | 17.28 % | 2nd | 34.88 % | Yes |
| July | 31 | 1950 | 60450 | 17.60 % | 1st | 17.60 % | Yes |
| August | 31 | 1483 | 45973 | 13.38 % | 3rd | 48.26 % | Yes |
| September | 30 | 1339 | 40170 | 11.69 % | 4th | 59.95 % | Yes |
| October | 31 | 934 | 28954 | 8.43 % | 7th | | No |
| November | 30 | 299 | 8970 | 2.61 % | 9th | | No |
| December | 31 | 254 | 7874 | 2.29 % | 10th | | No |
| **Total of annual volume =** | | | **343492** | **100.00%** | | | No |

The months that are part of SADT = April through September (i.e., 6 months)

| *Rank* | *MADT* |
|---|---|
| 6th | 983 |
| 5th | 1273 |
| 2nd | 1979 |
| 1st | 1950 |
| 3rd | 1483 |
| 4th | 1339 |
| **Total=** | **9007** |

Total value of MADT that is part of SADT:

SADT = (Total of contributing MADT) ÷ (No. of months that contribute more than 80% of total annual traffic volume)

$$\text{SADT} = (9007) \div (6 \text{ months}) = 1501$$

**Example 5.8**

The summary of average day of week traffic observed at a site A in each month of the previous year is given below.

| ADT of | Jan | Feb | Mar | Apr | May | Jun | Jul | Aug | Sep | Oct | Nov | Dec |
|---|---|---|---|---|---|---|---|---|---|---|---|---|
| Sun | 2563 | 4260 | 4973 | 4109 | 4265 | 4756 | 4598 | 4986 | 3897 | 3986 | 3289 | 2976 |
| Mon | 2670 | 2980 | 5983 | 3989 | 3467 | 4098 | 4256 | 4754 | 3976 | 3845 | 3487 | 3087 |
| Tue | 2779 | 3105 | 6191 | 4129 | 3987 | 4123 | 3986 | 4276 | 4189 | 3587 | 3198 | 2987 |
| Wed | 2898 | 3584 | 6278 | 4267 | 4108 | 3986 | 4127 | 4169 | 3998 | 3587 | 3456 | 2875 |
| Thu | 2681 | 3763 | 6343 | 4387 | 4099 | 4278 | 3967 | 4245 | 3867 | 3498 | 3612 | 2755 |
| Fri | 2451 | 3988 | 5769 | 3987 | 4156 | 4376 | 4125 | 3987 | 3764 | 3187 | 3289 | 2834 |
| Sat | 2374 | 4684 | 6105 | 4387 | 4379 | 4387 | 4512 | 4378 | 3687 | 3687 | 3067 | 2798 |

Determine (a) MADT, (b) AADT, (c) ratio between MADT and AADT and (d) ratio between each day of the week and MADT (i.e., seasonal adjustment factors).

As a part of current year traffic volume surveys, a temporary radar sensor has been installed on a connecting road segment located near the previous-year count-station (i.e., site-B) and traffic flow in the month of March for 48 hours on Thursday and Friday was recorded as 3015 and 2905, respectively. Estimate the AADT at site B based on the factors ($F_{ij}$) calculated from the previous year's traffic data at site A. The FHWA procedure of estimation may be adopted.

**Answer:** The details of the analysis pertaining to site A (with previous year count) are given below. The ADT values of each day of the week are summed and averaged to determine the respective MADT values. The AADT value is calculated as

$$AADT = \frac{1}{12}\sum_{k-1}^{12} MADT_k = \frac{47282}{12} = 3940.$$

The ratio between MADT and AADT and the respective day of week factors of site A the each month are calculated and presented below.

| | Jan | Feb | Mar | Apr | May | Jun | Jul | Aug | Sep | Oct | Nov | Dec |
|---|---|---|---|---|---|---|---|---|---|---|---|---|
| The 7-day sum | 18416 | 26364 | 41642 | 29255 | 28461 | 30004 | 29571 | 30795 | 27378 | 25377 | 23398 | 20312 |
| MADT (Average of 7-day sum) | 2631 | 3766 | 5949 | 4179 | 4066 | 4286 | 4224 | 4399 | 3911 | 3625 | 3343 | 2902 |
| AADT | 3940 | 3940 | 3940 | 3940 | 3940 | 3940 | 3940 | 3940 | 3940 | 3940 | 3940 | 3940 |
| MADT/AADT | 0.67 | 0.96 | 1.51 | 1.06 | 1.03 | 1.09 | 1.07 | 1.12 | 0.99 | 0.92 | 0.85 | 0.74 |
| SUN/MADT | 0.97 | 1.13 | 0.84 | 0.98 | 1.05 | 1.11 | 1.09 | 1.13 | 1.00 | 1.10 | 0.98 | 1.03 |
| MON/MADT | 1.01 | 0.79 | 1.01 | 0.95 | 0.85 | 0.96 | 1.01 | 1.08 | 1.02 | 1.06 | 1.04 | 1.06 |
| TUE/MADT | 1.06 | 0.82 | 1.04 | 0.99 | 0.98 | 0.96 | 0.94 | 0.97 | 1.07 | 0.99 | 0.96 | 1.03 |
| WED/MADT | 1.10 | 0.95 | 1.06 | 1.02 | 1.01 | 0.93 | 0.98 | 0.95 | 1.02 | 0.99 | 1.03 | 0.99 |
| THU/MADT | 1.02 | 1.00 | 1.07 | 1.05 | 1.01 | 1.00 | 0.94 | 0.96 | 0.99 | 0.96 | 1.08 | 0.95 |
| FRI/MADT | 0.93 | 1.06 | 0.97 | 0.95 | 1.02 | 1.02 | 0.98 | 0.91 | 0.96 | 0.88 | 0.98 | 0.98 |
| SAT/MADT | 0.90 | 1.24 | 1.03 | 1.05 | 1.08 | 1.02 | 1.07 | 1.00 | 0.94 | 1.02 | 0.92 | 0.96 |

The details of the analysis pertaining to site B (for current year estimate based on previous year traffic factors) are as follows:

The seasonal adjustment factors ($F_{ij}$) for the $i^{th}$ day of week and $j^{th}$ month of the year are obtained from the table:

Seasonal adjustment factor for Thursday (5th day of week) and March (3rd month) = $F_{5,3}$ = 0.97

Seasonal adjustment factor for Friday (6th day of week) and March (3rd month) = $F_{6,3}$ = 1.03

Average value of Thursday ($F_{5,3}$) and Friday ($F_{6,3}$) = $F_{average}$ = (1.07 + 0.97)/2 = 1.02

Therefore, it is found that for traffic monitoring during Friday and Saturday in the previous year, the average daily traffic was 1.02 times the March average daily traffic of the current year.

$$\text{Average daily traffic in the current year} = (3015 + 2905)/1.02 = (5920)/1.02 \approx 5804$$

From the above table, the ratio (March MADT)/AADT = 1.51

Therefore, estimated AADT at site B = 5804 × 1.51 ≈ 8764

Cumulative annual traffic at site B = 5804 × 365 ≈ 2118460.

*Note:* A similar procedure may be adopted with any duration for short-term traffic counts. However, it is not clear if the results obtained will be associated with unwanted error in such estimates.

**Example 5.9**

Determine the lane unitisation factors (LUF) for the four approach roads with the following details of traffic volume:

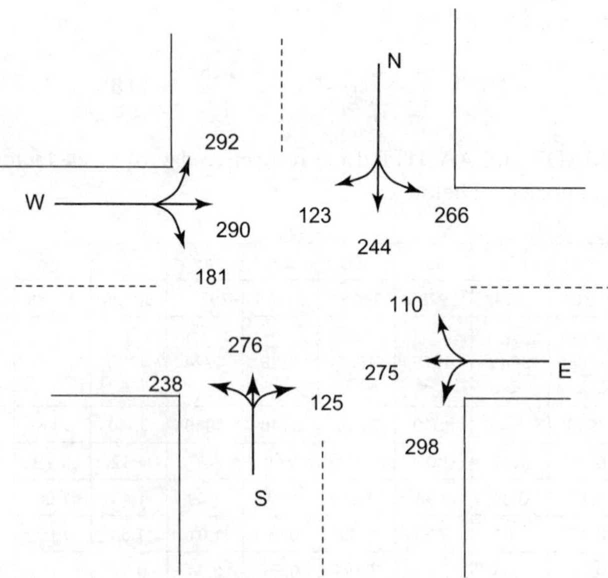


**Fig. 5.34** Directional distribution of traffic volume, veh/hr

**Answer:**

$$\text{LUF of an approach} = \frac{\text{Sum of approach volume}}{\text{No of lanes} \times \text{Highest observed volume in the approach}}$$

Using the above equation, we can calculate the LUF for the respective approaches, as follows:

$$\text{LUF of N approach} = \frac{(123+244+266)}{3\times 266} =$$

$$\text{LUF of E approach} = \frac{(110+275+298)}{3\times 298} =$$

$$\text{LUF of S approach} = \frac{(238+276+125)}{3\times 276} =$$

$$\text{LUF of W approach} = \frac{(292+290+181)}{3\times 292} =$$

# EXERCISES

## MULTIPLE CHOICE QUESTIONS

1. Which of the following is not used for automatic traffic count?
   (a) Pneumatic tube (b) Tally counter
   (c) Fibre optics (d) Passive acoustic
2. Which of the following is not a device that belongs to a non-intrusive type of equipment?
   (a) Magnetic device (b) Laser radar
   (c) Video detection (d) Passive magnetic
3. Which of the following is not a device that belongs to the intrusive type of equipment?
   (a) Contact closures (b) WIM scales
   (c) Magnetic sensor (d) Passive ultrasonic
4. Intrusive sensors may not be placed in one of the following positions:
   (a) Embedded in the pavement of the carriageway
   (b) Embedded in the subgrade of the pavement
   (c) Taped or attached to the surface of the carriageway, for temporary counting purposes
   (d) Mounted over the carriageway
5. In the case of a single-lane road, a single-induction-loop configuration is capable of counting the
   (a) Number vehicles that cross the loop (b) Both speed and the number of vehicles
   (c) Length of vehicles (d) Occupation rate and category of vehicle
6. In the case of lane straddling of vehicles in adjacent lanes on an undivided two-lane two-way carriageway, such vehicles can be detected by configuring the induction loops installed abreast of each other and spacing the adjacent loops at a minimum distance of
   (a) 200 mm (b) 300 mm (c) 400 mm (d) 500 mm
7. These devices are useful for vehicle presence, counting, classification and speed:
   (a) Magnetometers (b) WIM scales (c) Tally counters (d) Induction loops
8. Match List-I (methods of traffic volume counts) with List-II (equipment used) and select the correct answer using the codes given below the lists:

| List-I | List-II |
|---|---|
| A. Manual count | 1. Multiple pen recorder |
| B. Combination of manual and mechanical methods | 2. Piezoelectric sensors |
| C. Intrusive type | 3. Tally counters |
| D. Non-intrusive type | 4. Passive ultrasonic |

| | A | B | C | D |
|---|---|---|---|---|
| (a) | 1 | 3 | 4 | 2 |
| (b) | 3 | 1 | 4 | 2 |
| (c) | 1 | 3 | 2 | 4 |
| (d) | 3 | 1 | 2 | 4 |

9. Match List-I (type of study) with List-II (data collected) and select the correct answer using the codes given below the lists:

| List-I | List-II |
|---|---|
| A. Public transport inventory study | 1. Incident study |
| B. Public transport safety study | 2. Available routing and scheduling |
| C. Public transport usage study | 3. Vehicle - km, profit per km, cost per km |
| D. Public transport operators study | 4. Passenger - km, walking time, waiting time |

| | A | B | C | D |
|---|---|---|---|---|
| (a) | 2 | 1 | 4 | 3 |
| (b) | 1 | 2 | 3 | 4 |
| (c) | 2 | 1 | 3 | 4 |
| (d) | 3 | 1 | 4 | 2 |

10. Match List-I (type of survey) with List-II (method/instrument used) and select the correct answer using the codes given below the lists:

| List-I | List-II |
|---|---|
| A. Pedestrian volume study | 1. Automatic vehicle counter and intrusive loops |
| B. Speed and delay study | 2. Fixed time interview photographs |
| C. Origin and destination study | 3. Microwave radar |
| D. Multiple character studies | 4. Infrared sensors |
| | 5. Tag on car method |

| | A | B | C | D |
|---|---|---|---|---|
| (a) | 2 | 5 | 4 | 3 |
| (b) | 3 | 1 | 5 | 2 |
| (c) | 1 | 3 | 2 | 4 |
| (d) | 4 | 2 | 5 | 1 |

11. The attribute that belongs to non-motorised traffic is
    (a) Bus (b) Jugaad (c) Auto-rickshaw (d) Tri-cycle
12. The attribute that belongs to motorised traffic is
    (a) Jugaad (b) Bicycle (c) Tri-cycle (d) Pedestrian
13. Two loop detectors were spaced 1000 m apart. A truck was observed with the start time of the pulse of entry detector as 14:01:56 and at the exit detector as 14:03:16. What is the speed of the truck?
    (a) 56.2 km/hr (b) 37.5 km/hr (c) 45 km/hr (d) 24.6 km/hr
    (*Hint:* Distance = 1000 m; Time: Hours = 14 −14 =0; Minutes = 03 − 01 = 2 = 60 seconds; Seconds = 16 + 4 = 20 seconds; Speed = Distance/Time = 1000/80 sec = 12.5 m/sec = 45 km/ hr)
14. An induction loop sensor system indicates that 4400 axles passed over its sensor. It is known that 20% of the vehicles were three-axle trucks. How many two-axle cars actually passed over the sensor?
    (a) 2000 (b) 1572 (c) 880 (d) 3520
    (*Hint:* Let $(3 \times 0.2)N + (2 \times 0.8)N = 4400$, where $N$ = total vehicles = (4400/2.2) = 2000.)
15. A single-lane road is operated as a two-way lane. It has an average peak flow rate of cars of 50 PCU per 15-minute time interval. The directional distribution of traffic flow is 55:45. What is the heaviest one-way density of flow at an average speed of 40 km/hr?
    (a) 0.36 (b) 2.75 (c) 5.00 (d) 0.2
    (*Hint:* Flow rate /hr (Q) = Heaviest directional factor × Two-way flow
    = 0.55 × (50 PCU/(15/60) hr) = 110 PCU/hr; Density (K) = Q/V = (110P CU/hr)/(40 km/hr) = 2.75 PCU/km)
16. The following traffic volumes are observed during peak hour in an ***east***–west highway: Eastbound: 4500 vehicles; Westbound: 1500 vehicles. What is the peak hour directional distribution factor?
    (a) 3.00 (b) 0.33 (c) 0.25 (d) 0.75
    (*Hint:* Directional distribution factor = Peak direction volume/Total volume
    = 4500/(4500 + 1500) = 0.75)
17. A radar gun is used to collect the speed data of vehicles. Which of the following methods is used to determine the mean travel speed of the traffic based on the collected data?
    (a) Time mean speed method (b) Space mean speed method
    (c) Capacity and delay method (d) Moving car method
18. Which of the following factors is not required for the determination of space mean speed?
    (a) Length of the road section (b) Width of the road section
    (c) Travel time (d) Number of vehicles
19. A traffic count study was conducted at a location for a one-year period and a summary of the data is given below:
    Total monthly traffic = 164,2500 vehicles
    Sum of average daily traffic for the peak traffic months of June, July and August = 15000 vehicles

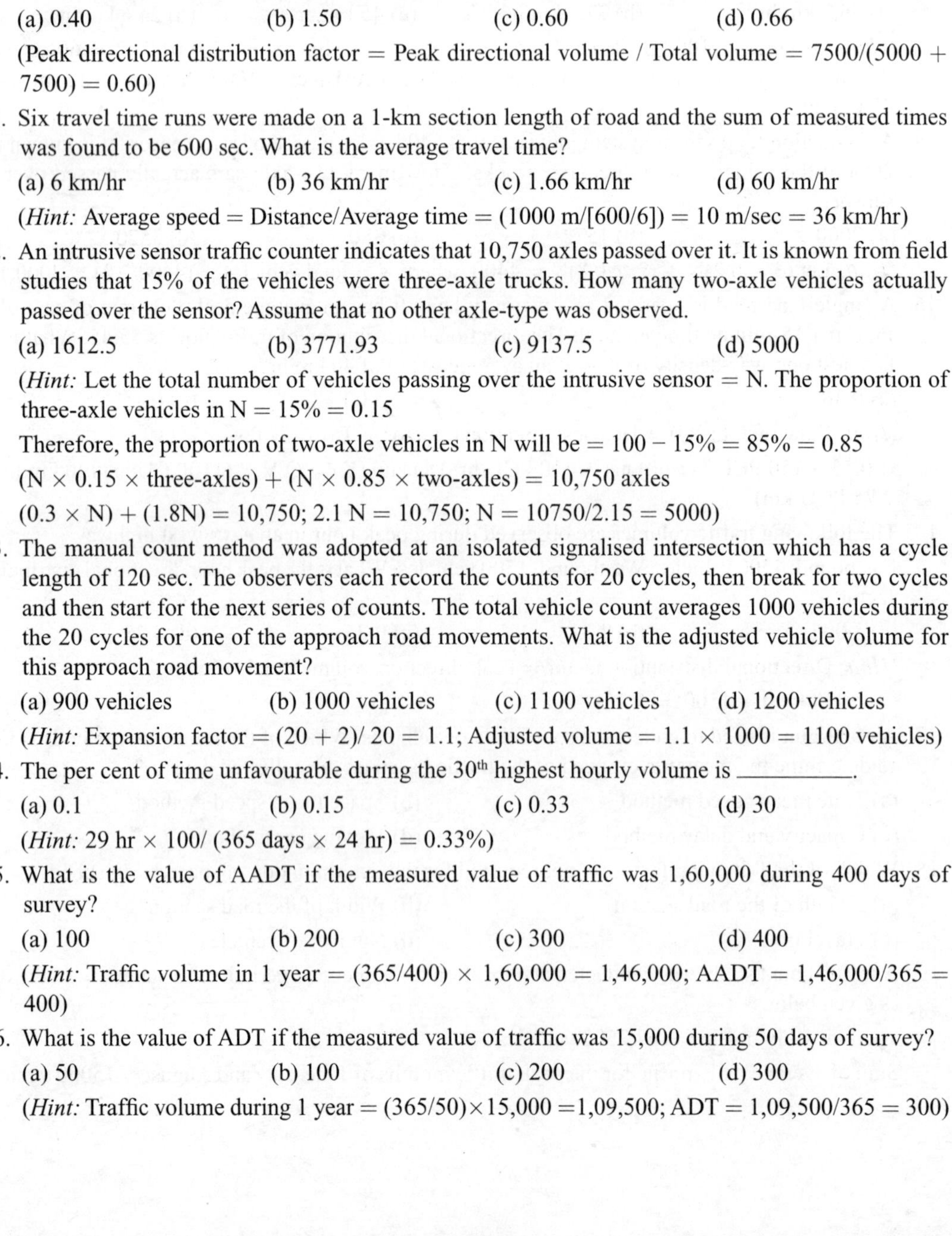

What is the seasonal factor for June through August?

(a) 0.9 (b) 0.3 (c) 1.11 (d) 3.0

(*Hint:* AADT = 1642500/365 = 4500 veh/day

Average daily traffic for June through August = 15000/3 = 5000 veh/day

Seasonal factor = 5000/4500 = 1.11)

20 What is the peak hour directional distribution factor if the peak hour observed traffic volume in the eastbound lane is 5000 vehicle and the westbound lane is 7500?

(a) 0.40 (b) 1.50 (c) 0.60 (d) 0.66

(Peak directional distribution factor = Peak directional volume / Total volume = 7500/(5000 + 7500) = 0.60)

21. Six travel time runs were made on a 1-km section length of road and the sum of measured times was found to be 600 sec. What is the average travel time?

(a) 6 km/hr (b) 36 km/hr (c) 1.66 km/hr (d) 60 km/hr

(*Hint:* Average speed = Distance/Average time = (1000 m/[600/6]) = 10 m/sec = 36 km/hr)

22. An intrusive sensor traffic counter indicates that 10,750 axles passed over it. It is known from field studies that 15% of the vehicles were three-axle trucks. How many two-axle vehicles actually passed over the sensor? Assume that no other axle-type was observed.

(a) 1612.5 (b) 3771.93 (c) 9137.5 (d) 5000

(*Hint:* Let the total number of vehicles passing over the intrusive sensor = N. The proportion of three-axle vehicles in N = 15% = 0.15

Therefore, the proportion of two-axle vehicles in N will be = 100 − 15% = 85% = 0.85

(N × 0.15 × three-axles) + (N × 0.85 × two-axles) = 10,750 axles

(0.3 × N) + (1.8N) = 10,750; 2.1 N = 10,750; N = 10750/2.15 = 5000)

23. The manual count method was adopted at an isolated signalised intersection which has a cycle length of 120 sec. The observers each record the counts for 20 cycles, then break for two cycles and then start for the next series of counts. The total vehicle count averages 1000 vehicles during the 20 cycles for one of the approach road movements. What is the adjusted vehicle volume for this approach road movement?

(a) 900 vehicles (b) 1000 vehicles (c) 1100 vehicles (d) 1200 vehicles

(*Hint:* Expansion factor = (20 + 2)/ 20 = 1.1; Adjusted volume = 1.1 × 1000 = 1100 vehicles)

24. The per cent of time unfavourable during the 30$^{th}$ highest hourly volume is ________.

(a) 0.1 (b) 0.15 (c) 0.33 (d) 30

(*Hint:* 29 hr × 100/ (365 days × 24 hr) = 0.33%)

25. What is the value of AADT if the measured value of traffic was 1,60,000 during 400 days of survey?

(a) 100 (b) 200 (c) 300 (d) 400

(*Hint:* Traffic volume in 1 year = (365/400) × 1,60,000 = 1,46,000; AADT = 1,46,000/365 = 400)

26. What is the value of ADT if the measured value of traffic was 15,000 during 50 days of survey?

(a) 50 (b) 100 (c) 200 (d) 300

(*Hint:* Traffic volume during 1 year = (365/50)×15,000 =1,09,500; ADT = 1,09,500/365 = 300)

27. What is the minimum value of a 15-minute peak hour factor on a section of road?
(a) 0.1 (b) 0.25 (c) 0.33 (d) 1.0
28. What is the normal range of a 15-minute peak hour factor on a section of a road?
(a) 0.7–0.98 (b) 0.2–1.0 (c) 0.5–0.33 (d) 0.0–1.0
29. What is the maximum value of a 15-minute peak hour factor on a section of a road?
(a) 10.0 (b) 0.25 (c) 0.33 (d) 1.0
30. If the total seven-day traffic volume observed during a week is 3500 and the traffic volume for Tuesday is 600, the monthly expansion factor for Tuesday would be
(a) 0.83 (b) 1.0 (c) 1.20 (d) 1.15
(*Hint:* MEF = $7 \times 600/3500 = 1.20$)

## Answers

| | | | | | | | | | |
|---|---|---|---|---|---|---|---|---|---|
| 1. (b) | 2. (a) | 3. (d) | 4. (d) | 5. (a) | 6. (d) | 7. (a) | 8. (d) | 9. (a) | 10. (d) |
| 11. (d) | 12. (a) | 13. (c) | 14. (a) | 15. (b) | 16. (d) | 17. (b) | 18. (b) | 19. (c) | 20. (c) |
| 21. (b) | 22. (d) | 23. (c) | 24. (c) | 25. (d) | 26. (d) | 27. (b) | 28. (a) | 29. (d) | 30. (c) |

## Questions

1. Write the details of the classification of traffic counting methods and the types of devices used.
2. Write how manual traffic counting is carried out using handheld devices and explain its suitability and limitations.
3. Compare the manual and automatic methods of traffic counts.
4. What are the automatic methods of traffic counts? State the factors affecting the performance of these detectors.
5. State the possible positions of installation of intrusive sensors.
6. State the possible positions of installation of non-intrusive sensors.
7. Write the classification details of intrusive sensors used for traffic counting. Explain any two methods in detail.
8. Explain the traffic counting method using pneumatic tubes. State its advantages and limitations.
9. Explain the traffic counting method using inductive loops. State its advantages and limitations.
10. Draw sketches showing the standard methods of installation of intrusive loops on the surface of single and dual-lanes with single/double loops. Compare and differentiate these two methods of installation.
11. What are the two magnetometers widely used for traffic counting? Also write their suitability and limitations.
12. Compare the static and WIM methods. State how WIM can be used for traffic counting.
13. Draw a typical schematic diagram showing the details of measuring axle loads using load pads. Also state its suitability and limitations.
14. What is WIM technology? State how it is advantageous over the static method.
15. List the primary purposes of the WIM system used on roadways.
16. Write an overview of the applicability of the load-cell type WIM system.

17. What are the best practices of the configuration of intrusive loops with staggered WIM scale platforms used for single lane and double lane?
18. Write an overview of the applicability of the bending plate WIM system for traffic counting surveys.
19. Write an overview of the applicability of the piezoelectric WIM system used for traffic counting, with sketches.
20. Compare the bending plate and piezoelectric WIM systems.
21. Draw sketches showing the different types of combinations of inductive loops with weight sensors and their measurable parameters of traffic flow.
22. Describe the following methods of installation of intrusive sensors with combinations:
    (a) Dual loop
    (b) Dual loop plus two rows of Lineas® weight sensors
    (c) Dual loop plus single full-width piezo-axle sensor
23. Draw a sketch showing details of dual loop plus bending plate weight sensor and a piezo-axle detector and also their measurable parameters of traffic flow.
24. Define and classify the non-intrusive sensors.
25. List the disadvantages of intrusive sensors.
26. What are the different types of mounting methods of non-intrusive sensors? List the drawbacks and advantages of each type.
27. Write the basic working principles of passive acoustic sensors used for traffic surveys. Also draw a sketch and describe with the details.
28. List the advantages and disadvantages of mounting sensors in the side fire position and overhead position.
29. State the basic working principles of microwave radar sensors used for traffic volume surveys. Also draw a sketch and describe the details.
30. Write the details of active infrared detector and passive infrared detector used for traffic volume surveys. Also list their specific applications.
31. Write a short note on ultrasonic detectors used for traffic volume surveys.
32. Explain how the video image processor–based detectors are useful for conducting traffic volume surveys. State their drawbacks and advantages.
33. What are the factors affecting the selection of appropriate technologies used for conducting traffic volume surveys.
34. Write a short note on the duration of traffic count. Explain the possible alternatives with the technology used.
35. Define (a) passive detectors and (b) active detectors used for traffic counts.
36. State the differences between passive detectors and active detectors and also write about the technology used for each case.
37. Define traffic attributes and explain their significance in conducting traffic volume surveys.
38. List the different forms of presentation of traffic volume data. Explain their applicability with reference to their applications and uses.

39. Draw typical sketches with details of:
    (a) Typical trend of variation in traffic flow during a 24-hour day, in an urban area
    (b) Typical trend of variation in traffic flow in a day of week, in an urban area
    (c) Typical trend of seasonal variation in traffic flow in a year, in a locality.
    (d) Typical trend of yearly variation in traffic growth
    (e) Types of traffic manoeuvres
    (f) Typical presentation of traffic flow at intersections
40. Define the following terms: (a) Base data (b) Base count (c) Vehicle distance travelled
41. Define and write the basic expressions that represent the following traffic volume data products:
    (a) Average Annual Daily Traffic
    (b) Monthly Average Days of the Week Traffic
    (c) Monthly Average Daily Traffic
    (d) Monthly Average Weekday Daily Traffic
    (e) Monthly Average Weekend Daily Traffic
    (f) Average Annual Weekday Traffic
    (g) Average Weekday Traffic
    (h) Annual Vehicle Distance Travelled
    (i) Vehicle Miles Travelled
42. State the difference between AADT and ADT.
43. Define Peak Hour Factor and state its maximum and minimum values.
44. Define ACF and state its significance in traffic counting.
45. Define Seasonal Correction Factor and state its use in AADT computations.
46. Define Seasonal Average Daily Traffic and state its importance in traffic volume analysis.
47. Write a short note on Average Summer Daily Traffic.
48. Define and write how Seasonal Average Daily Traffic Adjustment Factors are useful for traffic volume analysis.
49. Define these terms: (a) Design hour (b) Design year (c) Design hour volume
50. Define and design hour volume. Explain the relationship between AADT and the highest traffic volume observed during those years.
51. What is K-factor and state its applications.
52. Define DDHV and state its relationship with the K-factor used in traffic count analysis.
53. Comment on the traffic flow conditions for the following traffic volume data collected on a section of rural road during morning peak hour and evening peak hour:

| | *Cars* | *Buses* | *Trucks* | *Pedal Cycle* |
|---|---|---|---|---|
| Morning Peak Hour Traffic | 238 | 89 | 102 | 493 |
| Evening Peak Hour Traffic | 301 | 95 | 98 | 488 |

54. The following is the traffic volume data obtained in a peak hour at a location. Calculate the PHF.

| *Time duration within an hour* | *Traffic volume (number of vehicles)* |
|---|---|
| 18:00–18:15 | 959 |
| 18:15–18:30 | 1019 |
| 18:30–18:45 | 1010 |
| 18:45–19:00 | 989 |

55. The following are the details of vehicle classification count conducted manually (as a short-duration count) on a road section.

| *Vehicle Class* | *Vehicle Type* | *Daily Vehicle Volume* | *Average Number of Axles per Vehicle* | *Total Number of Axles* |
|---|---|---|---|---|
| Type 2 |  | 210 | 2 | 410 |
| Type 3 |  | 1908 | 3 | 4104 |
| Type 2-S1 |  | 340 | 3 | 1101 |
| Type 2-S2 |  | 108 | 4 | 812 |
| Type 3-S1 |  | 115 | 4 | 810 |
| Type 3-S2 |  | 267 | 5 | 805 |
| Type 2-2 |  | 44 | 4 | 276 |
| Type 3-2 | 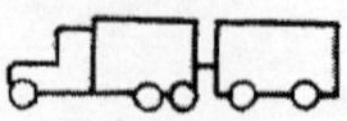 | 11 | 5 | 70 |
| Type 2-3 |  | 24 | 5 | 115 |
| Type 3-3 | 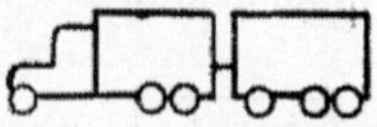 | 9 | 6 | 96 |

Determine (a) Axle correction factor (ACF) from the above survey data (b) Number of vehicles that passed over near the WIM station if the number of axles counted is 5102.

56. Following are the values of mean average daily truck traffic (MADTT) of 10 truck types (IRC) with their observed temporal patterns over a 12-month period.

| *MADTT* | *Vehicle Class as per IRC* | | | |
|---|---|---|---|---|
| | *Type 2* | *Type 3* | *Type 2-S1* | *Type 2-S2* |
| January | 512 | 2191 | 1121 | 405 |
| February | 521 | 2411 | 1137 | 401 |
| March | 515 | 2531 | 1153 | 411 |
| April | 579 | 2511 | 1204 | 469 |
| May | 593 | 3110 | 1220 | 553 |
| June | 619 | 3261 | 1992 | 532 |
| July | 670 | 3116 | 2038 | 580 |
| August | 709 | 3188 | 2180 | 612 |
| September | 706 | 3100 | 2225 | 610 |
| October | 699 | 3116 | 2027 | 602 |
| November | 599 | 3118 | 1860 | 501 |
| December | 518 | 3217 | 1711 | 408 |

Determine the truck monthly factors that reflect their travel pattern throughout the year.

57. Following is a summary of the details of traffic volume count carried out in a study area:

| *Road No.* | *Direction of traffic flow* | *Name of the Count Station* | *AADT* | *Segment length (mile)* |
|---|---|---|---|---|
| Road-A | N–S | CS-1-N–S | 21860 | 1.94 |
| | S–N | CS-1-S–N | 20569 | 1.94 |
| Road-B | E–W | CS-2-E–W | 24111 | 2.24 |
| | W–E | CS-2-W–E | 20811 | 2.24 |
| Road-C | NE–SW | CS-3- NE–SW | 60745 | 3.56 |
| | SW–NE | CS-3-SW–NE | 70828 | 3.56 |
| Road-D | NW–SE | CS-4- NW–SE | 31676 | 1.89 |
| | SE–NW | CS-4- SE–NW | 34137 | 1.89 |

Calculate the VMT in the study area.

58. Determine the months that are part of Seasonal Average Daily Traffic (SADT) and calculate the value of SADT from the following summary of traffic count:

| *Month* | *Number of Days* | *MADT* |
|---|---|---|
| January | 31 | 231 |
| February | 28 | 219 |
| March | 31 | 315 |
| April | 30 | 883 |
| May | 31 | 1073 |

(Continued)

**Table** (Continued)

| *Month* | *Number of Days* | *MADT* |
|---|---|---|
| June | 30 | 1179 |
| July | 31 | 1450 |
| August | 31 | 1583 |
| September | 30 | 1139 |
| October | 31 | 831 |
| November | 30 | 269 |
| December | 31 | 230 |

59. Determine the Lane Utilisation Factors (LUF) for the four approach roads with the following details of traffic volume:

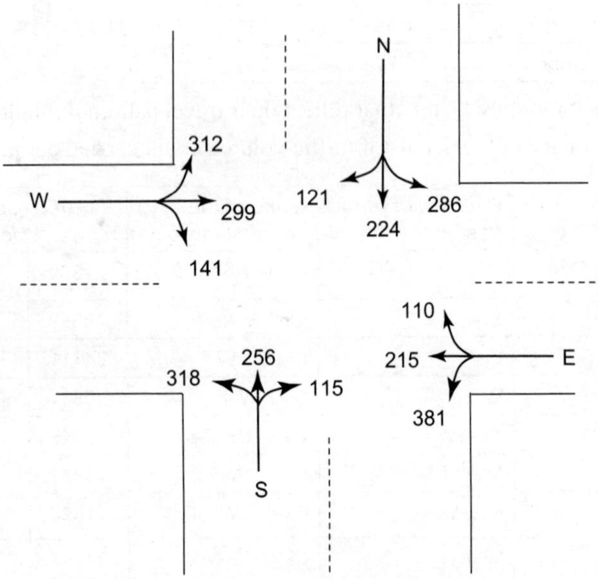

N
W
E
S
312
299
141
121
224
286
110
215
381
256
318
115

CHAPTER

# 6 Spot Speed Studies and Characteristics

## 6.0 INTRODUCTION TO SPEED STUDIES

Speed is the rate of motion of a vehicle and is expressed as the distance travelled in metres per unit time. Speed or velocity (V) is usually denoted in terms of kilometres per hour (km/h) or metres per second (m/s). The speed of vehicles is variable and is subject to *time* and *space*. The different types of speeds used to represent these variations are discussed below.

## 6.1 SPOT SPEED STUDY

Spot speed is the *instantaneous speed* of a vehicle measured over a specific distance (often <100 m) along a road. Spot speed can be measured using the following methods:

- An enoscope and a stopwatch timing procedure (Fig. 6.1)
- Intrusive sensors (see *Chapter 4*)
- A radar speedometer (see *Chapter 5*)
- Time-lapse photographic or videographic methods (see *Chapters 4* and *5*)
- A graphic recorder

The basic principle involved in these methods is to measure the time taken by a vehicle to travel between any two predetermined markings made along the road. The spot speed is then determined by dividing the measured distance between the markings by the elapsed time measured. It is expressed in terms of metres per second (m/s) or kilometres per hour (km/h). The distance between such markings along a carriageway may be generally taken as a plain and straight length between 30 and 100 m. According to a TRL (1993) guideline, the following distances may be considered with the average speed of traffic:

| Average speed of traffic (km/hr) | Distance between the markings (m) |
|---|---|
| <40 | 25 |
| 40–65 | 50 |
| >65 | 75 |

### 6.1.1 Factors Affecting Spot Speed

- Driver characteristics relating to age, sex and physical, psychological and mental capability.
- Vehicle characteristics relating to year of make, type/capacity of engine, body condition, weight of the vehicle and the manufacturer.
- Roadway characteristics relating to the type of terrain, grade with length, type of surface, curvature, curves, number of lanes with or without dividers, shoulder width and junctions.
- Operating conditions relating to composition of traffic, time of travel, headway, access control, encroachments on the right of way, speed restrictions, sight distance, day of the week and time of travel.
- Environmental factors relating to weather conditions and the season.

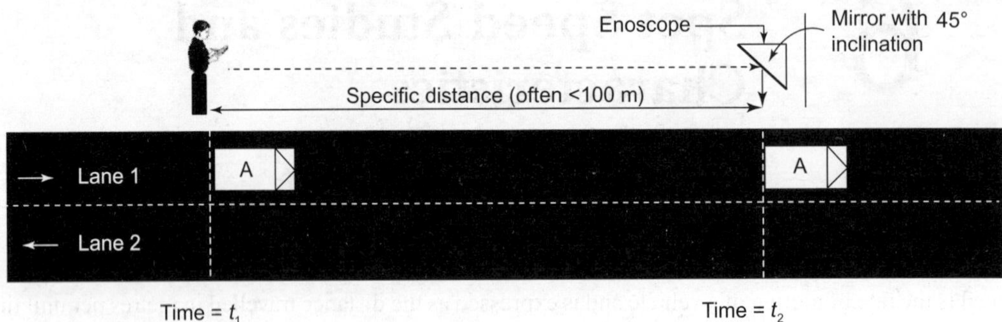


**Fig. 6.1** Concept of measurement of spot speed of a vehicle, using an enoscope

Spot speed studies are generally conducted in good weather and in normal traffic flow conditions.

### 6.1.2 Applications of Spot Speed Data

- Spot speed studies are used to estimate the distribution of the speeds of vehicles that are travelling in a stream of traffic on a particular length of a highway.
- Spot speed is used as a performance indicator of specific geometric components of a road, such as horizontal curves and vertical curves, to establish speed limits/trends, safe operating speed on different stretches of road including on curves, demarcation of speed zone boundaries, intersection sight distance, location of signboards and their visibility, design of signals, delay modelling and for accident analysis as well as before-and-after studies.

## 6.2 OTHER DEFINITIONS OF SPEED

The terminology used to characterise the speed of traffic using different parameters is explained below.

- **Operating speed**: If the design is according to British and German design practices, estimates of 85$^{th}$ percentile operating speeds should be adopted (Kanellaidis, 1996). This means that on any given segment of roadway, the design speed should not be less than the estimated 85$^{th}$ percentile operating speed (Nicholson, 1998). To design horizontal curves, the operating speed at the 85$^{th}$ percentile is generally used as a parameter for controlling the speed (Bonneson, 1999). If the proposed design speed is to be assumed, it should be a logical one, determined primarily by considering the natural terrain, land use and highway classification (AASHTO, 1990).
- **Project/Design speed**: Design speed may be defined as the maximum average safe operating speed of vehicles that is either specified or maintained by most drivers travelling through all the features of a roadway. It is one of the prime parameters that dictate the overall geometry of highways in plain as well as hilly areas. According to the IRC (IRC: 73-1980), the ruling design speed is generally considered as a design criterion and taken as a primary input parameter for the design of various geometric elements of highways.

  This value of speed is used to design geometric road elements such as the radius of curves and superelevation on curves. Further, the geometric design of highways is based on the topography of the terrain (i.e., general nature of the ground slope) and the corresponding design speeds specified.

- **Journey/Travel speed**: It is the average speed of a vehicle travelling between any two points, taking into account all possible delays in the course of the journey. The following expression can be used to calculate the journey speed.

$$\text{Journey speed} = \left(\frac{\text{Distance travelled}}{\text{Total time spent to cover the distance, including delays}}\right).$$

Minimum travel time is considered as the criterion for selecting the route, and hence, this parameter is very crucial in the evaluation of a highway system (NCUT, 1958). A smaller numerical difference between journey speed and running speed on a road indicates that travel conditions are satisfactory and that delays are marginal.

- **Running speed**: It is the average speed of a vehicle travelling between any two points without considering any delays in the course of the journey. The following expression can be used to calculate the running speed.

$$\text{Running speed} = \left(\frac{\text{Distance travelled}}{\text{Journey time} - \text{Time for delays and stoppages}}\right).$$

*Note*: The quality of traffic flow is best judged on the basis of the journey speed and running speed of vehicles. The ratio between running speed and journey speed describes the quality of flow. The higher the ratio, the better the quality of flow, and vice versa (Jalihal et al., 2005).

- **Time mean speed**: It is the *arithmetic mean speed* of all instantaneous vehicle speeds at a specified location on a road length. The time mean speed can be calculated as follows:

Time mean speed $$V_{mean} = \frac{\sum_{i=1}^{n} V_i}{n} \quad 6.1$$

where $V_i$ = individual velocities of different vehicles in a group and $n$ = total number of vehicles considered in the group.

In the case of grouped vehicular data, the following equation is used:

$$V_{mean} = \frac{\sum_{i=1}^{n} f_i \times V_i}{\sum_{i=1}^{n} f_i} \quad 6.2$$

where $f_i$ = frequency of the $i^{th}$ vehicle.

- **Space mean speed**: It is the mean travel speed of all the vehicles traversing a specified section-length of a road, over a specified time period. The space mean speed can be calculated as follows:

$$\text{Space mean speed} = \left(\frac{\text{Distance}}{\text{Average of the time spent to travel the distance}}\right)$$

$$\text{Space mean speed} = \frac{L}{\frac{\sum_{i=1}^{n} t_i}{n}} = \frac{nL}{\sum_{i=1}^{n} t_i} \quad 6.3$$

where $L$ = distance covered by the group of vehicles considered; $t_i$ = individual time spent by the $i^{th}$ vehicle to cover the distance $L$; and $n$ = total number of vehicles considered in the group.

Space mean speed is considered the harmonic mean of vehicles and is useful in traffic applications. It is always less than the time mean speed.

- **85th percentile speed**: It is the speed of vehicles that is only exceeded by 15% of vehicles. This speed is usually 20% higher than the average speed of the vehicles.
- **Median speed**: It is the speed that represents the middle value among a series of speed values arranged in ascending order. It is the 50th percentile speed which demarcates the point at which one half of the values of speed will be above median speed and the remaining half will be below it.
- **Modal speed**: It is the value of speed that is found to have the highest frequency of occurrence. In other words, it is the speed observed most frequently at a study site.
- **Pace**: It is an interval or range of speed; usually a 10-mi/hr interval, found in the greatest number of speed observations.

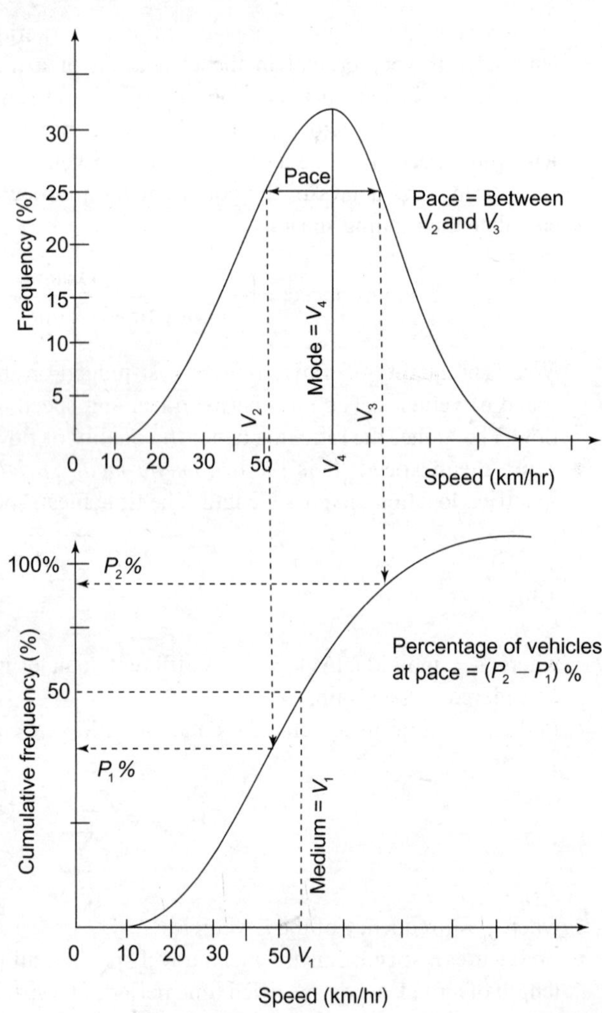


**Fig. 6.2** A typical presentation of spot speed observations

Once the data is collected, the descriptive statistics of speed that are distributed over a study length of a road need to be computed. The most commonly used descriptive statistics of speed are space mean speed, median speed, modal speed, $i^{th}$ percentile speed, pace, standard deviation and variance.

A typical presentation of spot speed observations by frequency curve and cumulative frequency curve are plotted in Fig. 6.2. The typical values of median, mode, pace and the per cent of vehicles at pace are shown on the respective curves.

## 6.2.1 Sample Size

The minimum number of spot speeds required for a degree of statistical accuracy can be determined using Eq. (8.4) (see *Chapter 8*).

## 6.3 SOLVED PROBLEMS

**Example 6.1**

This example describes the procedure of calculating time mean speed and space mean speed from the following data, obtained from a field survey.

**Answer:**

| *Vehicle number* | *Time (in seconds) taken to travel the distance L= 50 m* | *Speed (km/h)* |
|---|---|---|
| 1 | 4 | = 3.6 × (50 m/4 s) = 45 |
| 2 | 3 | = 3.6 × (50 m/3 s) = 60 |
| 3 | 4 | = 3.6 × (50 m/4 s) = 45 |
| 4 | 4 | = 3.6 × (50 m/4 s) = 45 |
| 5 | 5 | = 3.6 × (50 m/5 s) = 36 |
| 6 | 6 | = 3.6 × (50 m/6 s) = 30 |
| 7 | 3 | = 3.6 × (50 m/3 s) = 60 |
| 8 | 5 | = 3.6 × (50 m/5 s) = 36 |
| 9 | 6 | = 3.6 × (50 m/6 s) = 30 |
| 10 | 6 | = 3.6 × (50 m/6 s) = 30 |
| *n* = 10 | Total = 46 seconds | Total = 417 km/h |

$$\text{Time mean speed} = \frac{\sum_{i=1}^{n} V_i}{n} = \frac{417}{10} = 41.7 \text{ km/h}$$

$$\text{Time mean speed} = \frac{nL}{\sum_{i=1}^{n} t_i} = \frac{10 \times 50 \text{ m}}{46 \text{ s}} \approx 10.87 \text{ m/s}$$

$$= 10.87 \times 3.6 = 39.132 \text{ km/h}$$

**Example 6.2**

Calculate the time mean speed and space mean speed of three vehicles travelling over a 1-km length in 1.2 minutes, 2.0 minutes and 2.9 minutes, respectively.

**Answer:** The speeds of vehicles are calculated as

$$V_1 = 1 \div (1.2 \div 60) = (1 \times 60 \div 1.2) = 50.00 \text{ km/hr}$$

$$V_2 = 1 \div (2.0 \div 60) = (1 \times 60 \div 2.0) = 30.00 \text{ km/hr}$$

$$V_3 = 1 \div (2.9 \div 60) = (1\times 60 \div 2.9) = 20.68 \text{ km/hr}$$

$$\text{Time mean speed} = \text{Arithmetic average of speeds}$$

$$= (50.00 + 30.00 + 20.68)/3 = 33.56 \text{ km/hr}$$

$$\text{Average time} = (1.2 + 2.0 + 2.9) \div 3 = 2.03 \text{ minutes}$$

$$\text{Space mean speed} = (1 \text{ km}) \div (2.03 \div 60) \text{ hr} = (1 \times 60 \div 2.03)$$

$$= 29.55 \text{ km/hr}$$

or

$$\text{Space mean speed} = \frac{3}{\left(\frac{1}{50}+\frac{1}{30}+\frac{1}{20.68}\right)} = 29.5\text{km/hr}$$

**Example 6.3**

Determine the running speed and journey speed of a vehicle travelling from A to G. Use the following field data:

Average total travel time including delays due to traffic at junctions = 36 minutes

Average delay at junction B = 2 minutes, C = 3 minutes, D = 6 minutes, E = 4 minutes, F = 5 minutes.

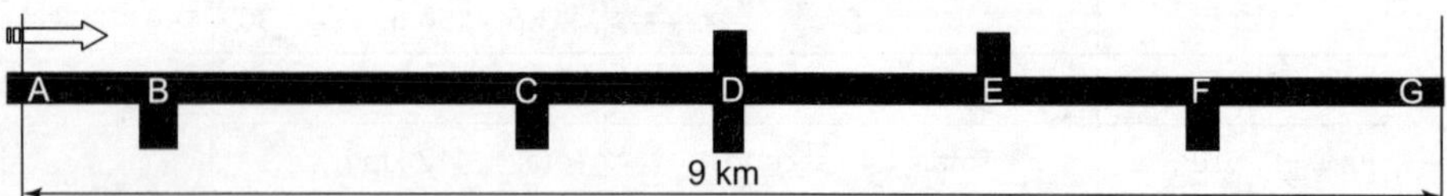


**Answer:** Average journey speed between A and G $= \left(\frac{9 \text{ km}}{\left(\frac{36}{60}\right)\text{hr}}\right) = \left(\frac{9 \text{ km}}{0.6 \text{ hr}}\right)$

$$= 15 \text{ km/hr}$$

$$\text{Total delay} = (2 + 3 + 6 + 4 + 5) = 20 \text{ minutes}$$

$$\text{Average running speed between A and G} = \left(\frac{9 \text{ km}}{\left(\frac{36-20}{60}\right)\text{hr}}\right) = \left(\frac{9 \text{ km}}{\left(\frac{16}{60}\right)\text{hr}}\right)$$

$$= \left(\frac{9 \text{ km}}{0.267 \text{ hr}}\right) \approx 33.7 \text{ km/hr}$$

$$\text{Per cent difference in speed} = \left(\frac{33.7-15}{15}\right)\times 100 = 124.76\%$$

As the journey speed is 124.67% less than the running speed, this indicates that an inefficient traffic control system exists at the junctions, and hence, the system requires improvement.

**Example 6.4**

The speed data collected for a segment of a national highway is given below.

| *S. No.* | *Speed class* | *Vehicle type* | | | |
|---|---|---|---|---|---|
| | | *Cars* | *Trucks/Buses* | *Others* | *Total* |
| 1 | 30–40 | 卌 卌 | 卌 卌 | IIII | 24 |
| 2 | 40–50 | 卌 卌 III | 卌 卌 IIII | 卌 卌 I | 38 |
| 3 | 50–60 | 卌 卌 卌 卌 I | 卌 卌 卌 III | 卌 卌 卌 IIII | 58 |
| 4 | 60–70 | 卌 卌 卌 卌 卌 II | 卌 卌 卌 IIII | 卌 卌 卌 卌 卌 卌 | 76 |
| 5 | 70–80 | 卌 卌 IIII | 卌 卌 IIII | 卌 卌 IIII | 42 |
| 6 | 80–90 | 卌 卌 | 卌 卌 I | 卌 卌 I | 32 |
| 7 | 90–100 | 卌 卌 | 卌 卌 | II | 22 |
| 8 | 100–110 | 卌 卌 | IIII | I | 15 |
| 9 | 110–120 | 卌 | III | I | 9 |
| 10 | 120–130 | II | I | I | 4 |

(a) Draw a histogram of the speeds.

(b) Draw the cumulative frequency curve and find the values of (i) median speed, (ii) modal speed, and (iii) 15$^{th}$, 85$^{th}$ and 98$^{th}$ percentile speeds.

Also determine the (c) mean speed, (d) standard deviation of the speed values, (e) coefficient of variation (COV) of the speeds, (f) standard error (SE) of speed values, (g) range of speed at which the mean speed lies within 95% confidence, and (h) the size of the sample at which the SE is restricted to 0.5 km/hr.

**Answer:** The following calculations are performed on the given data for the initial process of analysis.

| *S. No.* | *Speed class* | *Mid class ($V_i$)* | *Observations or Frequency ($f_i$)* | *% of $f_i$* | *% Cum. of $f_i$* | $V_i \times f_i$ | $(V_i - V_m)^2$ | $f_i \times (V_i - V_m)^2$ |
|---|---|---|---|---|---|---|---|---|
| [0] | [1] | [2] | [3] | [4] | [5] | [6] = 2× 3 | [7] | [8] |
| 1 | 30–40 | 35 | 24 | 7.50 | 7.5 | 840 | 1086.938477 | 26086.52 |
| 2 | 40–50 | 45 | 38 | 11.88 | 19.375 | 1710 | 527.5634766 | 20047.41 |

| S. No. | Speed class | Mid class ($V_i$) | Observations or Frequency ($f_i$) | % of $f_i$ | % Cum. of $f_i$ | $V_i \times f_i$ | $(V_i - V_m)^2$ | $f_i \times (V_i - V_m)^2$ |
|---|---|---|---|---|---|---|---|---|
| 3 | 50–60 | 55 | 58 | 18.13 | 37.5 | 3190 | 168.1884766 | 9754.932 |
| 4 | 60–70 | 65 | 76 | 23.75 | 61.25 | 4940 | 8.813476563 | 669.8242 |
| 5 | 70–80 | 75 | 42 | 13.13 | 74.375 | 3150 | 49.43847656 | 2076.416 |
| 6 | 80–90 | 85 | 32 | 10.00 | 84.375 | 2720 | 290.0634766 | 9282.031 |
| 7 | 90–100 | 95 | 22 | 6.88 | 91.25 | 2090 | 730.6884766 | 16075.15 |
| 8 | 100–110 | 105 | 15 | 4.69 | 95.9375 | 1575 | 1371.313477 | 20569.7 |
| 9 | 110–120 | 115 | 9 | 2.81 | 98.75 | 1035 | 2211.938477 | 19907.45 |
| 10 | 120–130 | 125 | 4 | 1.25 | 100 | 500 | 3252.563477 | 13010.25 |
| | | $\Sigma V_i =$ 802 | $\Sigma f_i = 320$ | | | $\Sigma = 21750$ | | $\Sigma =$ 137487.7 |

(a) The histogram of the speeds is drawn (Fig. 6.3) by plotting the mid-speed values on the *y* axis and frequency values on the *x* axis.

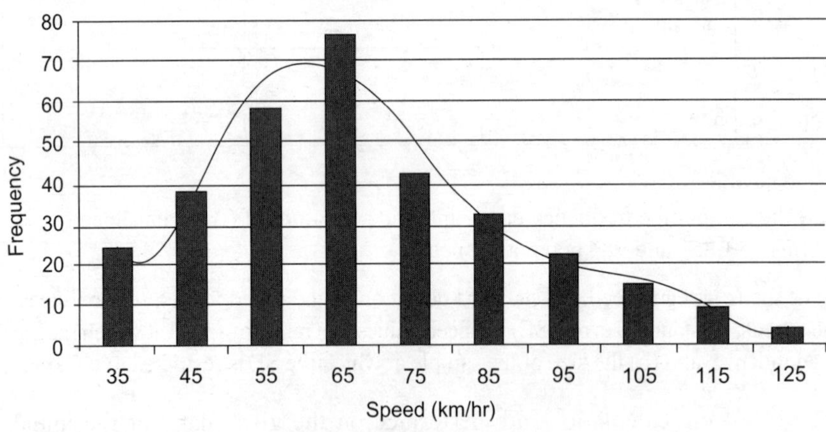


**Fig. 6.3** Histogram of speeds

(b) The cumulative frequency curve is drawn (Fig. 6.4) by plotting the % of cumulative frequency values on the *y* axis and the mid-speed values on the *x* axis.
The following values are obtained from Fig. 6.4:
- (i) Median speed (i.e., 50th percentile speed) = 60 km/hr
- (ii) Modal speed (i.e., peak point on frequency distribution curve, Fig. 6.3) = 65 km/hr
- (iii) 15th percentile speed = 42 km/hr
- (iv) 85th percentile speed (used for traffic regulation) = 86 km/hr
- (v) 98th percentile speed (used for geometric design) = 110.2 km/hr

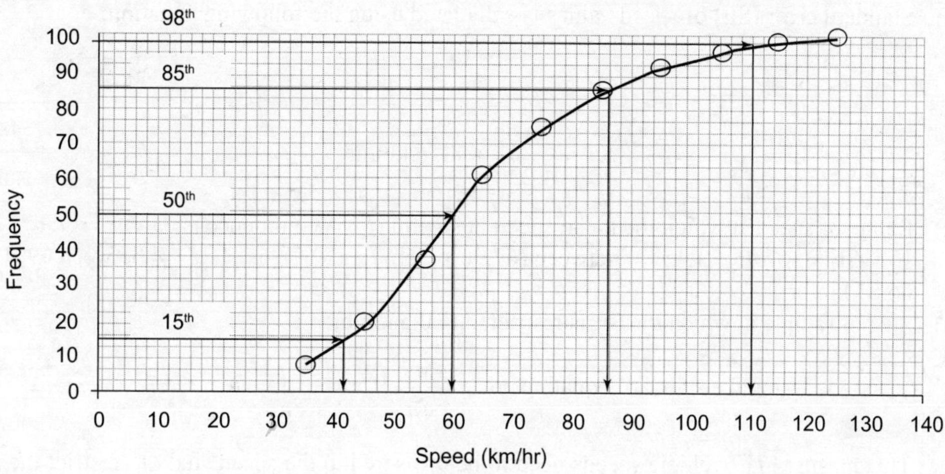


**Fig. 6.4** Cumulative frequency curve

(c) Mean speed is calculated using the following equation:

$$\text{Mean value of individual speeds}, V_m = \frac{\sum_{i=1}^{n} f_i \times V_i}{\sum_{i=1}^{n} f_i} = \frac{21750}{320} \approx 67.96 \text{ km/hr}$$

where $V_i$ = individual deflections corrected for temperature and seasonal variations and $n$ = number of deflections measured in a test section.

(d) The standard deviation value of speeds of grouped vehicles is calculated using the following equation:

$$\text{Standard deviation of speed values} = s = \sqrt{\frac{\sum_{i=1}^{n} f_i \times \left(V_i - V_m\right)^2}{\sum_{i=1}^{n} f_i}} \tag{6.4}$$

$$s = \sqrt{\frac{137487.7}{320}} = 20.72 \text{ km/hr}$$

(e)

$$\text{CoV} = \frac{\text{Standard deviation of the speed values } (s)}{\text{Mean speed value } (V_m)} \times 100 \tag{6.5}$$

$$\text{CoV} = \frac{20.72}{67.96} \times 100 = 30.49\,\%$$

(f) Standard error (SE) of speed values is calculated using the following equation:

$$SE = \frac{s}{\sqrt{n}} \tag{6.6}$$

$$SE = \frac{20.72}{\sqrt{320}} \approx 1.15 \text{ km/hr}$$

(g) Range of speed at which the mean speed lies within 95% confidence is $V_m \pm$ SE = (67.96 ± 1.15) = between (67.96 + 1.15) and (67.96 − 1.15) = between 69, 11 and 66.81 km/hr.

(h) Size of sample at which the SE is restricted to 0.5 km/hr:

Using the standard equation, $SE = \frac{s}{\sqrt{n}}, 0.5 = \frac{20.72}{\sqrt{n}}, \ n \approx 1717$

This means, 1717 vehicle speeds need to be measured in the speed study to restrict the SE to 0.5 km/hr. Then, the range of the speed calculated above will be narrowed to (67.96 ± 0.5) km/hr.

**Example 6.5**

A spot speed study was conducted on a stretch of a highway and the data collected was consolidated as given below:

| Speed range (km/hr) | 0–10 | 10–20 | 20–30 | 30–40 | 40–50 | 50–60 | 60–70 | 70–80 | 80–90 | 90–100 |
|---|---|---|---|---|---|---|---|---|---|---|
| No. of vehicles observed | 15 | 35 | 40 | 60 | 150 | 260 | 290 | 60 | 70 | 20 |

Determine (i) The upper and lower limits of speed for speed regulation of mixed traffic
(ii) The design speed for checking the geometric design of the highway
(iii) The median speed observed

**Answer:**

| *Speed range (km/hr)* | *Mid-speed, V (values plotted on x axis)* | *No. of vehicles observed, f* | *Percentage frequency %f* | *Cumulative frequency, % (values plotted on y axis)* |
|---|---|---|---|---|
| 0–10 | 5 | 15 | 1.5 | 1.5 |
| 10–20 | 15 | 35 | 3.5 | 5 |
| 20–30 | 25 | 40 | 4 | 9 |
| 30–40 | 35 | 60 | 6 | 15 |
| 40–50 | 45 | 150 | 15 | 30 |

| *Speed range (km/hr)* | *Mid-speed, V (values plotted on x axis)* | *No. of vehicles observed, f* | *Percentage frequency %f* | *Cumulative frequency, % (values plotted on y axis)* |
|---|---|---|---|---|
| 50–60 | 55 | 260 | 26 | 56 |
| 60–70 | 65 | 290 | 29 | 85 |
| 70–80 | 75 | 60 | 6 | 91 |
| 80–90 | 85 | 70 | 7 | 98 |
| 90–100 | 95 | 20 | 2 | 100 |
| | Sum = | 1000 | 100 | |

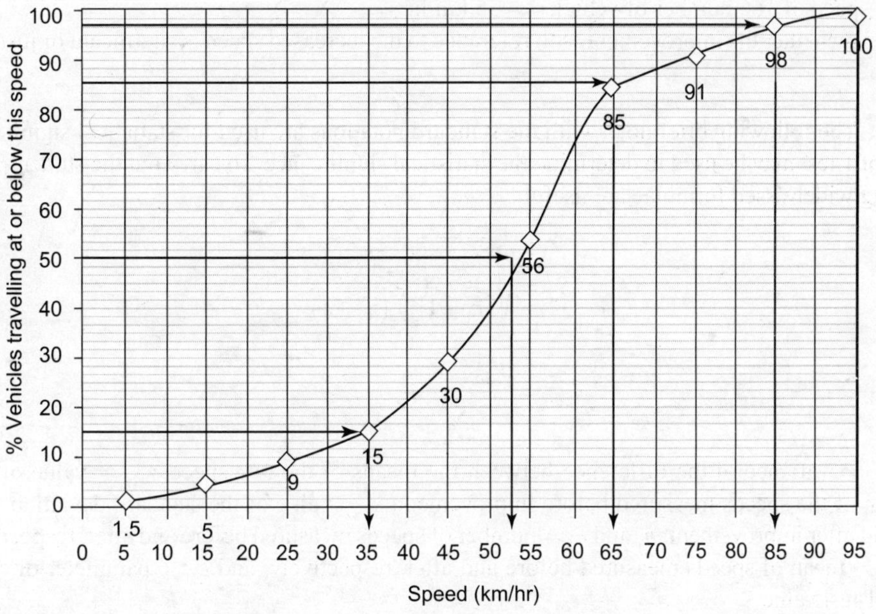


**Fig. 6.5** Cumulative speed distribution

Using the mid-speed and cumulative frequency values, the cumulative frequency distribution curve is plotted (Fig. 6.5) and the following results are obtained.

(i) The upper and lower speed limits for speed regulation of mixed traffic:
The upper speed limit for speed regulation = 85th percentile speed = 65 km/hr.
The lower speed limit for speed regulation = 15th percentile speed = 35 km/hr.

(ii) The design speed for checking the geometric design of the highway:
The speed to check for design elements = 98th percentile speed = 85 km/hr.

(iii) The median speed observed = 53 km/hr.

**Example 6.6**

A segment of road is improved by widening the carriageway and clearing the trees that are obstructing the vision at certain points. The summary of speed studies conducted at the site is given below.

*Before improvement*:
Average speed of the vehicles = 50 km/hr
Standard deviation of the speeds measured = 6.1 km/hr
Number of vehicles observed = 120

*After improvement*:
Average speed of the vehicles = 55 km/hr
Standard deviation of the speeds measured = 4.4 km/hr
Number of vehicles observed = 150
Average speed of 166 number of vehicles = 55 km/hr
Determine whether the improvement with reference to the increased speed is significant or just a change variation?

**Answer:** The following brief notes with the standard notations are used for statistical significance. A student T-test may be used to determine the statistical significance. To carry out the student T-test, the following widely used formulae are useful.

$$S_d = \sqrt{\frac{(s_1)^2}{(n_1-1)} + \frac{(s_2)^2}{(n_2-1)}} \tag{6.7}$$

$$t_{cal} = \frac{|V_{m1} - V_{m2}|}{S_d} \tag{6.8}$$

where $S_d$ = variance of the difference between the means of the two speeds; $s_1$ = value of standard deviation of the speeds measured before improvement; $s_2$ = value of the standard deviation of speeds measured after improvement; $n_1$ and $n_2$ = number of speeds measured before and after, respectively; $V_{m1}$ and $V_{m2}$ = mean of speeds measured before and after, respectively; and $t_{cal}$ = parameter of calculated student T-test value.

The $t_{cal}$ value should be compared with the standard value of the student T-test ($T_{stan}$). The $T_{stan}$ values are determined based on the degree of probability ($P$). Table 6.1 lists the $T_{stan}$ values when $P$ is assumed to be 0.05 (i.e., 95% confidence) for a given degree of freedom (DoF).

**Table 6.1** A few values of $T_{stan}$

| *DoF** | 20 | 21 | 22 | 23 | 24 | 26 | 27 | 29 | 30 | 40 | 60 | 120 | ∞ |
|---|---|---|---|---|---|---|---|---|---|---|---|---|---|
| *P*(0.05) | 2.09 | 2.08 | 2.07 | 2.07 | 2.06 | 2.06 | 2.05 | 2.05 | 2.04 | 2.02 | 2.00 | 1.98 | 1.96 |

* DoF = $(n_1 + n_2 - 2)$

The following procedure may be adopted for checking the statistical significance of the improved facility.

1. When $t_{cal} \leq T_{stan}$, the two means considered are not statistically significant. The speeds measured before and after improvement are *not* different and the two cases should be considered as similar.

2. When $t_{cal} > T_{stan}$, the two means considered are statistically significant. The speeds measured before and after improvement are *different* and the two speeds should be treated as different, and hence, the improvement may be considered as justified.

Substituting the given values in the above equations, we get

$$S_d = \sqrt{\frac{(s_1)^2}{(n_1-1)} + \frac{(s_2)^2}{(n_2-1)}} = \sqrt{\frac{(6.1)^2}{(120-1)} + \frac{(4.4)^2}{(150-1)}} \approx 0.66$$

$$t_{cal} = \frac{|V_{m1} - V_{m2}|}{S_d} = \frac{|50-55|}{0.66} = \frac{5}{0.66} = 7.57$$

$$\text{Degree of freedom (DoF)} = n_1 + n_2 - 2 = (120 + 150 - 2) = 268$$

From Table 6.1, for a DOF of 268, the value of $T_{stan} = 1.96$. In this case, it is found that, $t_{cal}$ (7.57) > $T_{stan}$, (1.96). Hence, the two mean speeds are statistically significant and the improvement may be considered as justified.

*Note*: The above parameters of statistics are commonly used to test the significance of traffic facilities as studies before and after improvements.

## EXERCISES

### Multiple Choice Questions

1. The total perception time taken for an average driver in India is __________ seconds.
   (a) 0.5 (b) 1.5 (c) 2.5 (d) 3.5
2. Regulating speed on high-speed roads is decided based on
   (a) 15th percentile cumulative frequency (b) 50th percentile cumulative frequency
   (c) 85th percentile cumulative frequency (d) 98th percentile cumulative frequency
3. Minimum permissible speed on high-speed roads is decided based on
   (a) 10th percentile cumulative frequency (b) 15th percentile cumulative frequency
   (c) 50th percentile cumulative frequency (d) 85th percentile cumulative frequency
4. An enoscope is used for conducting
   (a) Traffic volume study (b) Spot speed study
   (c) Speed and delay study (d) Intersection delay study
5. Methods of speed and delay studies are
   1. Floating car method 2. Videographic method 3. Spot speed method 4. Non-intrusive techniques
   (a) 1 and 2 (b) 3 and 4 (c) 3 and 2 (d) 1 and 4
6. It is the value of speed that is found to have the highest frequency of occurrence:
   (a) Modal peed (b) Median speed
   (c) Running speed (d) Spot speed
7. The geometric design of highways is carried out in India based on
   (a) 30th percentile cumulative frequency (b) 50th percentile cumulative frequency
   (c) 85th percentile cumulative frequency (d) 98th percentile cumulative frequency

8. In a spot speed study, 100 vehicles were observed and the 85th percentile and 15th percentile speeds were obtained as 50 km/hr and 16 km/hr, respectively. The number of vehicles travelling at speeds between 50 km/hr and 16 km/hr would be
   (a) 66 (b) 70 (c) 44 (d) 50
   (*Hint*: Total number of vehicles observed = 100
   Number of vehicles corresponding to 85th percentile speed of 50 km/hr = 85
   Number of vehicles corresponding to 15th percentile speed 16 km/hr = 15
   The number of vehicles travelling between speeds 50 and 16 km/hr = (85 − 15) = 70)
9. In a spot speed study, 250 vehicles were observed and the 85th percentile and 15th percentile speeds were obtained as 55 km/hr and 10 km/hr, respectively. The number of vehicles travelling at speeds between 55 km/hr and 10 km/hr would be
   (a) 175 (b) 140 (c) 70 (d) 200
10. Speed regulation and control of traffic flow along a street is imposed based on which study?
    (a) Traffic volume studies (b) Spot speed studies
    (c) Speed and delay studies (d) Saturation flow studies
11. Match List I (Traffic surveys) with List II (Method) and select the correct answer using the codes given below the lists:

| *List I* | *List II* |
|---|---|
| A. Spot speed | 1. By video tape |
| B. Traffic volume | 2. By carousel method |
| C. O-D survey | 3. By Doppler radar |
| D. Parking survey | 4. By pneumatic tube |

| | A | B | C | D |
|---|---|---|---|---|
| (a) | 4 | 1 | 3 | 2 |
| (b) | 4 | 3 | 1 | 2 |
| (c) | 3 | 4 | 2 | 4 |
| (d) | 3 | 1 | 2 | 4 |

## Answers

1. (c) 2. (c) 3. (b) 4. (b) 5. (a) 6. (a) 7. (d) 8. (b) 9. (a) 10. (b) 11. (c)

## Questions

1. Define spot speed and explain how spot speed studies are carried out. Also write the most commonly used distances to be considered with the average speed of traffic.
2. State the factors affecting spot speed.
3. What are the applications of spot speed data?
4. Define the following terms: (a) Operating speed (b) Design speed (c) Journey speed (d) Running speed.
5. Define and distinguish time mean speed and space mean speed.
6. Define and distinguish median speed and modal speed.

7. Define pace.
8. Draw a typical sketch plotted with sample spot speed observations and indicate the following parameters: (a) Pace (b) Mode (c) Median (d) $85^{th}$ percentile speed.
9. Determine the time mean speed and space mean speed from the following data:

| *Vehicle number* | *Time (in seconds) taken to travel the distance L = 60 m* |
|---|---|
| 1 | 4 |
| 2 | 3 |
| 3 | 4 |
| 4 | 4 |
| 5 | 5 |
| 6 | 6 |
| 7 | 3 |
| 8 | 5 |
| 9 | 6 |
| 10 | 5 |

10. State why the time mean speed will be greater than the space mean speed on a section of a road.
11. Calculate the time mean speed and space mean speed of three vehicles travelling over a 2-km length in 2.1 minutes, 2.1 minutes and 2.5 minutes, respectively.
12. A spot speed study was conducted on a stretch of a highway and the data collected was consolidated as given below:

| *Speed range (km/hr)* | *0–10* | *10–20* | *20–30* | *30–40* | *40–50* | *50–60* | *60–70* | *70–80* | *80–90* | *90–100* |
|---|---|---|---|---|---|---|---|---|---|---|
| No. of vehicles observed | 15 | 33 | 44 | 105 | 120 | 210 | 270 | 160 | 50 | 26 |

Determine (a) The upper and lower limits of speed for speed regulation of mixed traffic

(b) The design speed for checking the geometric design of the highway

(c) The median speed observed

CHAPTER

# 7 Origin and Destination Studies

## 7.0 INTRODUCTION

Origin and destination (OD) studies are conducted to understand traffic movements across and within a selected study area in a period of time. The survey may be related to specific trip studies pertaining to vehicles, pedestrians, cyclists and goods in a small residential area or a large urban area or any geographical region or crossing key screen lines within a study area.

In these studies, traffic movements are determined, recorded and analysed as trips. Each trip is defined as a one-way movement, from the point a person starts at (i.e., the origin) to the point the person wants to reach (i.e., destination). The greater the number of samples, the more accurate will be the study results.

Information concerning OD studies should be disseminated to the public to obtain their cooperation during collection of data.

## 7.1 OD DATA

The following points may be considered before collecting OD data:

- The purpose and objectives of the OD study must be clearly defined.
- The target geographic area and population of interest whose travel patterns need to be understood must be specified.
- The type and extent of information to be collected should provide an adequate degree of support to transportation planners and decision makers.
- The survey questionnaire should accomplish the goals by incorporating all relevant questions, without seeking unambiguous answers.
- The questionnaire should be short enough and relevant for the development of the intended transportation demand model.
- The quality and quantity of the data collected should be based on the nature and funding needs of the transportation project.

While compiling the above data, the percentage of the sample interviewed should be converted to represent a 24-hour day and consolidated under the following heads:

- Date and time
- Place
- Trip origin (specify the place on the street)
- Trip destination (specify the place on the street)
- Purpose of the trip (business or non-business)
- Trip Frequency
- Type of vehicle used, with occupancy
- Vehicle ownership
- Type of fuel used (for air quality planning)
- Vehicle registration number (for private transport vehicles)
- Total average time of travel, from origin to destination
- Weather conditions during the journey
- Details of the delays that occurred (with place and duration)

## 7.2 THE OD STUDY AREA

Initially, the transportation engineer must identify a study area with its boundary based on an objective. An imaginary line, considered the boundary line of the study area, is termed the **cordon line**. The political borders of an area such as a town, district, mandal, village or a group of adjacent colonies with a similar traffic flow pattern are often considered as cordon lines for OD surveys. Direction-wise (i.e., entry and exit) vehicle flow data across the cordon line is recorded using any one or a combination of methods, discussed in the next section.

The study area is further divided into a suitable number of zones based on similar traffic flow patterns. These zones are often termed traffic analysis zones (TAZs). Each TAZ may have one main and other minor internal traffic centres. The main traffic movements across the cordon line (i.e., boundary line) are identified as Entry/Exit nodes. A sample demarcation of a study area with 14 TAZs and 4 Entry/Exit nodes are presented in Fig. 7.1.

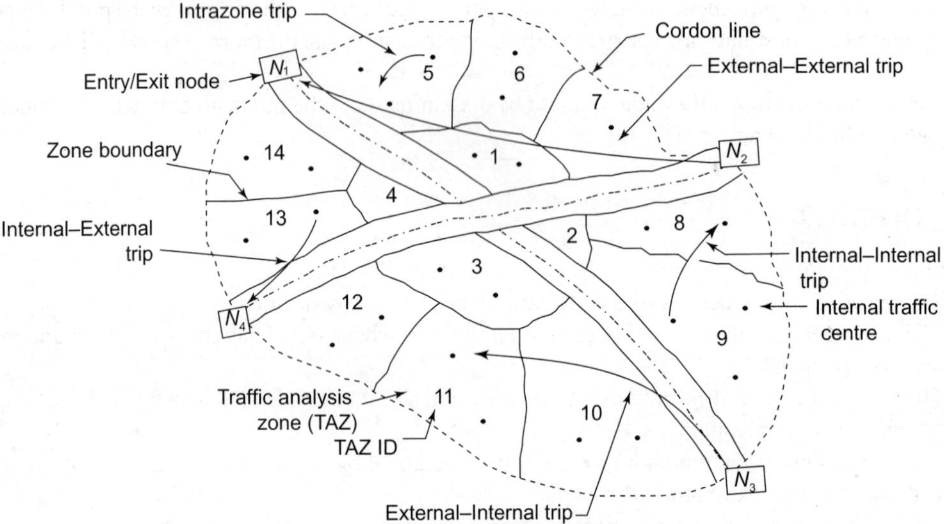


**Fig. 7.1** A sample OD study area with traffic analysis zones (TAZ) and types of trips (modified and elaborated, based on Guy and Fricker, 2005)

After conducting the OD survey and collecting the data, the trip distribution in a study area is presented as an OD matrix. The OD matrix is represented as an $n \times n$ matrix; where $n$ = (Number of TAZs + Number of Entry/Exit nodes) considered in a study area. Based on the origins and destinations in the study area, the trips can be classified into four types:

- I–I = Internal–Internal trips
- I–E = Internal–External trips
- E–I = External–Internal trips
- E–E = External–External (through) trips

The location of these trips on a sample OD study area is presented in Fig. 7.1, and the OD matrix is presented in Table 7.1 by shaded cells and labels. The sum of each row represents the total number of trips originating in each zone and the sum of each column represents the number of trips that have destinations in each zone (Guy and Fricker, 2005).

In reality, OD matrices are much larger as they need to represent hundreds of TAZs and Entry/Exit nodal stations in large cities or metropolitan areas.

**Table 7.1** Sample OD matrix 18 × 18, with reference to Fig. 7.1

| Origin zone | Destination zones | | | | | | | | | | | | | | | | | |
|---|---|---|---|---|---|---|---|---|---|---|---|---|---|---|---|---|---|---|
| | 1 | 2 | 3 | 4 | 5 | 6 | 7 | 8 | 9 | 10 | 11 | 12 | 13 | 14 | N1 | N2 | N3 | N4 |
| 1 | | | | | | | | | | | | | | | | | | |
| 2 | | | | | | | | | | | | | | | | | | |
| 3 | | | | | | | | | | | | | | | | | | |
| 4 | | | | | | | | | | | | | | | | | | |
| 5 | | | | | | | | | | | | | | | | | | |
| 6 | | | | | | | | | | Internal–Internal trips | | | | | | Internal–External trips | | |
| 7 | | | | | | | | | | | | | | | | | | |
| 8 | | | | | | | | | | | | | | | | | | |
| 9 | | | | | | | | | | | | | | | | | | |
| 10 | | | | | | | | | | | | Intra-zonal trips | | | | | | |
| 11 | | | | | | | | | | | | | | | | | | |
| 12 | | | | Internal–Internal trips | | | | | | | | | | | | | | |
| 13 | | | | | | | | | | | | | | | | | | |
| 14 | | | | | | | | | | | | | | | | | | |
| N1 | | | | | | | | | | | | | | | | | | |
| N2 | | | | | | | | | | | | | | | | | | |
| N3 | | | | | | External–Internal trips | | | | | | | | | | External–External trips | | |
| N4 | | | | | | | | | | | | | | | | | | |

## 7.3 OD DATA COLLECTION METHODS

The methods of collecting OD data may be categorised as three types: questionnaire mail-back–based methods, (b) interview-based methods and (c) camera-based detection methods. The camera-based detection methods are suitable in situations where the majority of the respondents are unwilling to share information because of concerns about privacy. The details of the different types of methods widely used for conducting OD surveys are discussed below (Bradshaw, 1999).

### 7.3.1 Registration Questionnaire Survey

A list of the addresses of vehicle owners, arranged according to their licence plate number within a study area, is obtained from the concerned Motor Vehicle Registration Department. Each questionnaire is printed with the full licence plate number. Each vehicle owner is sent the questionnaire to complete and is requested to mail it back within a certain date. Low response rates are usually expected, generally 15–30%. The cost of printing and mailing the survey questionnaire forms will be additional (Quiroga et al., 2000). Alternatively, internet-based online forms may be explored, if possible.

These questionnaires can be more detailed than other methods of data collection, particularly on socioeconomic details and household data. This method is preferred for surveying high-volume locations where stoppage and disturbance of traffic can be avoided. Hence, this method has been found to be safer for both drivers and observers.

The main challenge with this method is that there may be a relatively low response or non-response from the vehicle owners with respect to (timely) return of the completed questionnaire; this may lead to bias in the results as a representation of the sample (Bonsall and McKimm, 1993). Another problem is that the raw licence data needs to be converted into standard OD data in a short period of time.

### 7.3.2 Mail-back Questionnaire or Self-completion Postcard Survey (or Roadside Handout Survey)

A prepaid return postcard printed with the questionnaire is distributed at a given time to all the drivers entering roadways at the boundary of a study area. The cards may also be distributed to drivers who have already stopped/halted at rest areas, toll booths and at specific border crossings. This method is less expensive than the licence plate survey. Traffic disruption is inevitable during distribution of the postcards. This method of surveying is less disruptive than the roadside interview method. However, the challenges associated with the registration questionnaire method hold good for this as well.

Traffic volume count is also made during this period to estimate the proportion of drivers/vehicles with OD data collected for the time period. The response from the vehicle owners in general was found to be poor for several reasons, such as as failure to compete the questionnaire and failure to send it back. For these reasons, a return rate of 25–50% of the distributed postcards is desirable.

### 7.3.3 Roadside Interview Survey

OD data is collected by conducting personal interviews with vehicle drivers at roadside stations; information about their travel characteristics through the study area are also obtained. Some or all passing vehicles are stopped for the interviews. The vehicle drivers are requested to park at the side safely and the information is manually recorded at each road segment within the boundary of a study area. This process of data collection is termed the **vehicle intercept survey technique**.

A minimum of 25% of the sampling should be made during peak hours and the remainder may be suitably distributed to obtain representative information about the locality. Approximately, 100 drivers can be interviewed per hour. The interview stations need not be operated at the same time. Hourly traffic volume counts in each direction are augmented by this survey so that the sample data can be expanded into a 24-hour analysis or to obtain a response or capture rate for expansion of the data to the population.

Advance information about roadside interviews may be given to the public concerned. This method requires a greater number of personnel for conducting interviews and stopping the vehicles. In addition, all the vehicle drivers may not be interested in stopping and being interviewed, during busy trips. This method is often combined with a postcard survey, when drivers are in a hurry.

### 7.3.4 Home Interview Survey

In this survey, the data pertaining to the OD study is collected by selecting sample households within a study area with varying levels of income, age of family members, different types of professions and vehicle types. Trained observers personally go to individual houses in the selected area and record household information with their travel activity by conducting home interviews with them. Household surveys can also be conducted by telephone or mail. Such surveys are conducted for a certain time period, generally 24 or 48 hours. The information is aggregated with other similar household travel trips and summarised as average trip rates and trip lengths for that type of household. By combining this information with trip production and attraction analysis of travel demand modelling, an OD matrix for the study area is defined (Guy and Fricker, 2005).

This survey provides the most detailed data including the social and economic characteristics of transport users that can be used for deriving general travel patterns from a large sample of urban population for all the roadways in a study area. However, it does not provide detailed information on any particular roadway within a particular study area. Accuracy of the OD data depends on the sample interviews in the selected dwelling units. This method involves a greater number of interviewers, is time consuming and costly, faces the challenge of non-availability of the persons at home and sometimes it is difficult to obtain permission for the home interview. However, such difficulties due to non-response in travel surveys may be reduced by proper designing of an easy and simple questionnaire and by announcing small prepaid monetary incentives (Simowski et al., 1997).

### 7.3.5 Tag-on-Car Survey

This method has been found to be suitable for studying through trips. The vehicles tagged with specific colour ribbons at each entry station are tallied at the exit stations, and the resulting origin and destination data of the tagged vehicles is summarised for a study area. This method can be used to answer specific problems only. Matching the colour ribbons with the specific stations is very difficult, but this will dictate the accuracy of the data collected. In addition, some drivers may not like to attach tags to their vehicle; they may remove them before reaching their destination. In other cases, motorists may leave the tag attached even after reaching their destination, which may cause false matches if the vehicles are spotted later in the study (Guy and Fricker, 2005).

### 7.3.6 Carousel Method

A team of trained observers with a driver travel in a survey vehicle on the roadway segment being studied, more slowly than the general traffic stream. The observers enter the licence plate numbers of the passing vehicles in a palmtop. The survey vehicle may travel at speeds of 10–15 miles an hour below the prevailing traffic speed. This method is also known as Baltimore OD survey (Harrold, 1998). This method may cost relatively less than the roadside methods and is suitable when the general traffic speed is at or exceeds legal speed limits. The survey team and the general traffic will be safe, with minimum disturbance to the normal flow of traffic. However, this method may not work well in heavily congested and underutilised road segments (Bradshaw, 1999).

### 7.3.7 Licence Plate Survey

The licence plate numbers of vehicles at each entry and exit station on a road in a study area are recorded and tallied to summarise the OD data. The locations of such roadside stations are identified on the border (i.e., cordon) line of a study area. Traffic volume counts in each direction along with vehicle classification is usually carried out with this survey to obtain a capture rate for expansion of the data to the population.

The following methods are used to record the licence plate numbers of vehicles in this survey (Bradshaw, 1999):

- An observer at a road side station observes the vehicles crossing across the cordon line and calls out their licence plate numbers, as they pass. These numbers can be initially recorded on paper using a pencil, or may be directly entered into an Excel spreadsheet using a computer device by a field helper. This is the conventional method, also termed the clipboard method. Studies have shown that an observer can record about 300 full licence plates per hour or 800 partial licence plate per hour (with three or four digits) (Martin and Bell, 1993).
- An observer at a roadside station observes the vehicles crossing the cordon line and calls out their licence plate numbers; they are then recorded in a tape recorder or palmtop. The audio files are later manually transcribed. and finally, the data is entered into an Excel spreadsheet. Studies have shown that an observer can record between 1000 and 1200 licence plates an hour (Martin and Bell, 1993).
- An observer at a roadside station observes the vehicles crossing the cordon line and dictates the licence plate numbers into a microphone; they are then recorded automatically in an Excel spreadsheet using computer-based voice recognition technology. The data thus entered is time-stamped. One observer is adequate with this system of data collection (Washburn and Nihan, 1997). Background noise from the traffic and letters that sound similar are the main problems in this type of method.
- Roadside stations with mounted video recordable cameras are used to capture images of the moving vehicles with their licence plate numbers. Later, such video tapes are manually transcribed (i.e., someone sits in an office, plays the recorded video files, watches the licence plate numbers of the vehicles, and manually enters the number of each vehicle in a computer).

### 7.3.8 Comments and Remarks

The licence plate numbers obtained from each roadside station are compared to records from other stations within the study area. Matches are noted and counted as a trip, if a licence plate is observed at two stations within a study area in a certain time period. The time period considered for the matches is based on the average speed of the vehicles observed to reach those two stations. The search for each licence plate number from a station record is carried out using suitable software programs. Each licence plate is evaluated for a match with the other stations and the trip data is arranged in the form of E–E, E–I, I–E and I–I types to prepare an OD matrix, as shown in Table 7.1.

The match percentage is calculated as the ratio of the number of matched licence plates divided by the number of collected licence plates, and its value may range from 5% to 20% due to the following factors (Shuldiner et al., 1996; Turner et al., 1998):

- Distance between observation stations and the number of intersecting streets
- Non-uniform speeds, sharp turns, dust/fog, mixed traffic and congested traffic flow
- Percentage of through and internal traffic
- Location of observation stations and the technology adopted for recording/capturing the data/images
- Licence plate syntax used by a state/government
- Type of search and match algorithms used

### 7.3.9 Other Methods Relating to ITS

In addition to the above widely used methods, the following are also used as vehicle tracing methods with advanced satellite and information technology. Vehicle tracing is a technique that utilises electronic satellites or radio-based wireless technology to trace vehicles unobtrusively through a study area (Guy and Fricker, 2005).

- **GPS tracing:** The vehicles equipped with GPS are traced and their trip data such as latitude and longitude with time are recorded for analysis of OD trips. This information is automatically recorded and analysed by a software package, without their knowledge. One of the main limitations of this technique is that only a small proportion of vehicles may be equipped with GPS. And hence, the sample of vehicles traced by this method is not likely to represent the sample of the total population.
- **Wireless phone tracing:** The following two methods are used to obtain the location of vehicles (i.e., indirectly) and their moving paths. In both cases, the coordinates of the wireless phones with time in a study area are analysed to determine the OD trip data.
  - In the *handset method*, the coordinates of wireless phones equipped with a GPS receiver are obtained in a study time period and the data is analysed to determine the OD trip matrix.
  - In the *network-based method*, signals from the wireless phone in the study area with their time of occurrence are received from the nearest cell towers and the data is analysed based on triangulation principles to pinpoint the location of the wireless phone, either by measuring the angles of signal arrival or the difference in time that the signal reaches the respective towers. Finally, the paths traced by those vehicles are plotted to a GIS package to reveal the travel patterns of persons carrying cell phones.

Currently, these methods are being used in the USA. However, further technological improvements are required in the collection of travel information in a study area and their processing techniques to minimise errors.

*Note:* The OD data collected using the methods described above may be augmented with additional studies such as telephone survey, land-use study and classified population growth trend study for new routes or changes in existing routes by the local authorities. The data collected is geo-coded and presented using Geographic Information System (GIS) software. GIS is a standard approach used for storing, managing, classifying, manipulating and displaying all forms of travel survey data (Bradshaw, 1999).

## 7.4 SELECTING A METHOD

The objectives of conducting an OD survey determine the method selected. It must be noted that different methods of OD survey yield different types of information (number of trips between OD pairs, exact origins and destinations, socioeconomic data). Each method has its merits and limitations. Hence, the surveyor must select a suitable method so as to obtain the required information with minimum errors, time and cost. The traffic conditions (i.e., vehicle type, speed and flow rate) on the road for which OD data is being collected is also an important factor.

## 7.5 SAMPLE SIZE

The sample size requirements for each type of survey method depend on the number of completed surveys obtained in a study area. As the response rate increases, the value of the number of completed surveys also increases, and vice versa. Non-response/Poor response by the completed questionnaire method, refusal to do roadside interviews and incorrect detection of licence plate numbers by video/camera techniques dictates the minimum required sample size. In other words, the response rate or failure rate dictates the minimum value of the sample size. Ultimately, the representativeness of the OD survey data relating to the entire vehicle population in a study area is determined based on the sample size, and it can be ascertained using statistical formulae. Some of the widely used formulae considered as general guidelines for determining the minimum statistical sample size for collecting OD data by specific survey methods are presented below.

- The Highways Agency, United Kingdom (2005) recommends the following equation for estimating the sample size for roadside interviews:

$$n = \frac{p(1-p)N^3}{(N-1)\left(\frac{E}{Z}\right)^2 + p(1-p)N^2} \tag{7.1}$$

  where $n$ = sample size = number of completed surveys required; $p$ = estimated proportion of total traffic at the station with a particular destination; $E$ = (percent error × $N$) = absolute error or accuracy desired; $N$ = traffic volume at the survey station; and $Z$ = normal variate for a specified level of statistical confidence (for example, 1.96 for 95% confidence).

- The Ontario Ministry of Transportation and Communication, Canada (Hajek, 1977) recommended the following equation for determining the optimal sample size ($n$) for roadside interview OD surveys, and the same was used by Stokes (1989) for the development of an E–E trip OD matrix.

$$Sample\,rate(r) = \left(\frac{n}{N}\right) = \frac{p(1-p)Z^2}{(N-1)E^2 + p(1-p)Z^2} \tag{7.2}$$

  It is concluded that the sample rate ($r$) increases as the estimated proportion of total traffic ($p$) decreases, for a specific confidence interval ($z$) and error ($E$).

- Eq. 8.4 may be used with reference to the approximate travel time variability. In this, the CoV value may be suitably assumed to be between 10% and 35% for representing low-to-moderate traffic and congested traffic, respectively.

  The sample size ($n$) needs to be increased to account for refusal to participate in the survey, non-response of surveys and errors in data collection (Guy and Fricker, 2005). For example, if the value of drivers who refuse to stop for roadside surveys is 12%, the total number of vehicles to be stopped should be $n/0.88$. In the case of the licence plate follow-up survey technique, if the value of the response rate of the number of licence plates recorded is 20% and the errors assumed in recording the licence plates in the field that does not match with the RTO database is 12%, the total number of licence plates recorded should be = $n/(0.20 \times 0.88)$.

- A study conducted by NCHRP (Lomax et al., 1997) using video and character recognition software found that the minimum sample size of 50 licences with correct matches was adequate for conducting a study with a wide range of travel time variability. Such a sample size may be considered as the target sample for the purpose of planning as well.

- The sampling size may not be a serious problem in the case of *automatic* photo/video-based capturing techniques because, in general, the number of licence plates correctly identified by these methods will mostly be more than 50, and therefore, would be adequate for meeting the statistical sample size requirements.

## 7.6 USES OF OD DATA

The information collected from OD studies can be used for the following purposes:

- To locate expressways and plan for major routes on the desire lines
- To establish preferential routes and their associated needs for budgetary fund allocation
- To anticipate present and future trends of travel patterns, trip purposes, trip frequency and traveller demographics, based on existing traffic demand on the road network, and subsequently, for development of transport models
- To find out the existing number of trips made within, through and across the study area or a specified space boundary, with reference to time of day, mode of travel and occupants per vehicle. Multimodal transport studies may also be considered as part of this study.
- To determine current travel patterns in terms of the areas that generate and/or attract more traffic. Such data, with traffic count, will be used to improve the efficiency of traffic flow per lane with respect to maximum allowable capacity, level of service and safety.
- To evaluate general road plans relating to current or future problems or needs based on analysis of travel characteristics such as travel time, delay and pollution
- To determine the need for revised traffic-flow patterns, alternate routes, bypass roads, ring roads and parking areas
- To support traveller market analysis and demand forecast for high-speed ground transportation alternatives in a state (WSDOT, 1992)
- To anticipate traffic-flow patterns or problems based on future projects, new residential colonies, changes in population, service facilities and zoning limitations

## 7.7 PRESENTATION OF OD DATA

Several methods are available to present the information obtained from OD studies. These include tabular forms and graphical forms. The graphical form is widely used and comprises desire line diagrams and route volume maps.

In the case of desire line diagrams, various locations representing the origins and destinations in a study area are connected with straight lines. The thickness of these straight lines varies according to the thickness scale representing traffic volume in both the directions from one key area to the other. The desire line diagram represents the major traffic-flow patterns in a study area, and are used more often for transport planning purposes.

The route volume map is prepared by drawing lines of varying thickness on an actual topographic road map. The lines bend in the directions indicating the origin and destination of traffic-flow pattern between any two locations on the map.

It is not easy to determine which method or combination of methods is best for obtaining accurate data at minimum cost, time and effort. Each method has a unique set of features that should be considered for planning activities, data collection and data analysis.

The data collected by traditional survey methods tend to be simple, costly, laborious and time consuming. In addition, these methods are generally suitable for very small sample sizes. For these reasons, various advanced techniques like automatic number plate recognition (ANPR) are being used worldwide.

The information collected in OD surveys should not be made available to people outside the project team. The OD information should be treated as confidential business information to the maintain the privacy of the respondents.

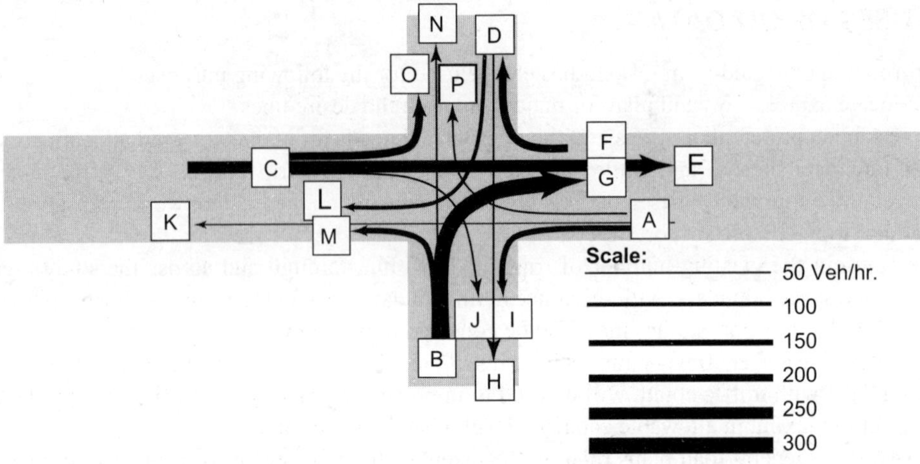


Fig. 7.2 Typical presentation of traffic flow at intersections with traffic volume scaled lines

## 7.8 SOLVED PROBLEM

**Example 7.1**

A survey was conducted in a residential colony to establish the relationship between the number of trips generated per household per day (N) and the number of persons per household (P) and the number of vehicles per house hold (V). The relationship showed that N = 1.5P + 2.0V + 0.35.

Based on this condition, how many trips is a neighbouring colony with 400 households expected to generate if there is an average of 3.8 persons and 2.3 vehicles per household?

Also determine the number of car trips if the modal split of V is assumed to be 40% car, 45% bus and local train and 5% by other modes.

**Answer:** Substituting the given values, we get P = 3.8 persons/household and V = 2.3 vehicle/household. The number of trips generated by 400 households per day (N) = 400 × (1.5P + 2.0 V + 0.35)

Number of trips generated per household per day = N = 400 × ((1.5 × 3.8) + (2.0 × 2.3) + 0.35)

$$N = 400 \times 10.65 = 4260$$

$$\text{Number of car trips per day} = 0.40 \times 4260 = 1704$$

## EXERCISES

### Multiple Choice Questions

1. The following OD survey technique does not give important details such as delay causes, duration and the number of delays within a test section:

   (a) Roadside windshield method (b) Photographic surveillance

   (c) Licence plate survey (d) Carousel method

2. The carousel method is effective on ________.

   (a) Single lanes (b) Multilanes (c) Two lanes (d) Gravel roads

3. Desire lines are drawn based on

   (a) Traffic volume studies (b) Moving car method
   (c) Origin destination studies (d) Elevated observer method

4. The OD survey technique is conducted for the collection of data related to the number of vehicles and passengers in each vehicle, selection of route, intermediate stops and their reason:

   (a) Roadside interview (b) Home interview
   (c) Workplace interview (d) Carousel method

5. In desire line diagrams

   (a) The thickness of the line represents the number of trips in one direction only
   (b) The thickness of the line represents the number of trips in both directions
   (c) Both thickness and length of the line represent the number of trips in one direction only
   (d) Both thickness and length of the line represent the number of trips in both directions

6. A journey from home to work is made by travelling by bike to the railway station and then travelling to another place by train. Finally, the destination is reached on foot. This travel trip is regarded as:

   (a) 4 trips (b) 3 trips (c) 2 trips (d) 1 trip

7. Which of the following traffic survey schemes is most relevant for deciding the location of major routes in a city?

   (a) Traffic volume studies (b) Origin and destination studies
   (c) Speed and delay studies (d) Traffic capacity studies

## Answers

1. (c) 2. (b) 3. (c) 4. (a) 5. (b) 6. (b) 7. (b)

## Questions

1. What are the points to be considered before collecting OD data?
2. Define cordon line and explain its uses in conducting OD surveys in a study area.
3. Define traffic analysis zones. State their functionality relating to OD matrix data collection and types of trips with a typical sketch.
4. Explain the various methods of OD data collection with their suitability and limitations.
5. Explain the registration questionnaire survey method of OD data collection with its advantages and limitations.
6. Explain the mail-back questionnaire survey method of OD data collection with its advantages and limitations.
7. Explain the roadside interview survey method of OD data collection with its advantages and limitations.
8. Explain the home interview survey method of OD data collection with its advantages and limitations.
9. Explain the tag-on-car survey method of OD data collection with its advantages and limitations.

10. Explain the carousel method of OD data collection with its advantages and limitations.
11. Explain the licence plate survey of OD data collection with its advantages and limitations.
12. Explain the methods used for recording the licence plate numbers of vehicles during OD surveys.
13. Write a short note on the applicability of ITS-related techniques in OD data collection.
14. Write the most commonly used equation for estimating the sample size for conducting traffic surveys by roadside interviews.
15. Write an overview of the equations used for estimating the sample size for conducting OD surveys.
16. What are the applications and uses of OD data?
17. Write an overview and explain the usability of various methods of presentation of OD data.

CHAPTER

# 8 Travel Time and Delay Studies

## 8.0 INTRODUCTION

Travel time management is an important aspect of traffic flow control with respect to any improvement in highway facilities. The specific purpose of these surveys is to identify the locations of delays (time spent in any form of stoppage and also due to slow speed) on various stretches of roads in urban and rural areas. In this process, the road facility is evaluated to determine the factors that cause delays and stoppages, and to make recommendations for improving the flow conditions of traffic. In addition, the following specific points may be considered as the objectives of travel time and delay studies:

- To indicate the level of service
- To identify problem locations and the causes of traffic delays on a route
- To ascertain the impact of improved highway facilities, in terms of speed, on a route
- To predict traffic flow diversions at various access points in a study locality
- To analyse the resulting road user benefits due to proposed improvements (Cribbins et al., 1961).

## 8.1 FACTORS AFFECTING SPEED AND DELAY

The travel speed of uninterrupted flow depends on the type of geometry of the road, type of locality of the road and its situation – in commercial, rural or urban areas – and the influence of traffic in mixed flow conditions. Delays at signalised intersections largely depend on signal timings, volume of traffic, saturation flow, queue length and the number of stoppages/halts due to road encroachments.

The following suggestions may be considered as improvements for reducing delays:

- **Short-range improvements**: Establishing controlled access/junction points and channelising the access points. For example, commercial truck traffic is given limited access to the entry points of urban localities, to accommodate free flow conditions, particularly in urban arterial streets. Bypass roads are constructed around urban localities and commercial truck traffic on one side of the highway is diverted to the other side or to other parts of the urban road network. Restricting turning movements is another option.
- **Long-range improvements**: Improving the geometric design of intersection points and road facilities, widening carriageways by providing additional lanes and capacity enhancement by segregating traffic based on vehicular characteristics (i.e., slow speed, high speed, animal drawn, pedal driven and size).

Specific methods are applicable for uninterrupted traffic flow and interrupted traffic flow conditions.

## 8.2 METHODS OF MEASUREMENT OF TRAVEL TIME AND DELAYS ON ANY SELECTED LENGTH OF A ROUTE

Several methods/techniques are available to measure travel time and delays on a selected length of route (Fig. 8.1). Each method is based on a specific type of technique of field measurement and possesses its own advantages and limitations. Any one of the methods discussed below may be selected based on the nature and objectives of the study (Treadway, 1965).

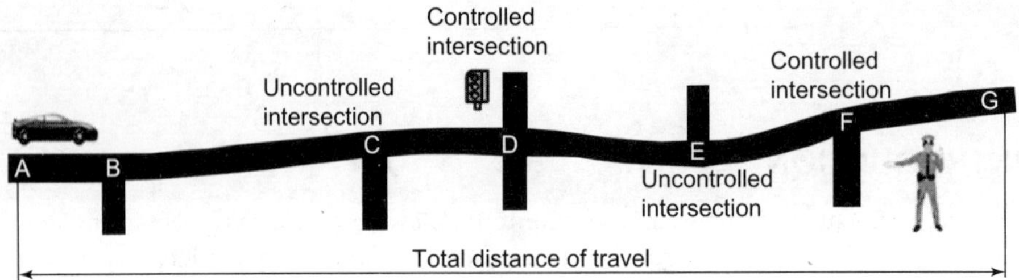


**Fig. 8.1** Schematic diagram of a route selected for conducting speed and delay surveys

## 8.2.1 Floating Car Method (Moving Observer Method)

In this method, the driver is instructed to maintain the speed of the test vehicle at the representative mean speed of all the traffic in the stream (i.e., floating in the traffic stream), and pass approximately the same number of vehicles that pass the test vehicle. However, the driver does not try to overtake fast-moving vehicles (Bone, 1952). Therefore, the test vehicle approximates the behaviour of an average vehicle in the traffic stream. This method is generally applied only on two-lane highways.

At least 12 trips need to be made in each direction to obtain reliable data about average speed and delays in that direction of traffic flow (for example, A→G and G→A, see Fig. 8.1). As part of data collection, three enumerators may be employed in the test vehicle to count the number of vehicles travelling in the opposite direction, the number of overtaking vehicles and overtaken vehicles and note the running speed, distance travelled, delays at each location and their causes. Stopwatches, survey sheets, note pads and voice recording devices may be used to record the speed and delay survey data.

The following equations may be used to analyse the collected data.

Average travel time in a direction,

$$ATT_{\text{Direction}} = t_{a\_\text{Direction}} - \left(\frac{N_{\text{diff}}}{Q_{\text{Direction}}}\right). \tag{8.1}$$

Traffic volume in a direction,

$$Q_{\text{Direction}} = (Veh_{\text{Opp}} + N_{\text{diff}}) / (t_{a\_\text{Direction}} + t_{a\_\text{Opposite Direction}}) \tag{8.2}$$

$$\text{Space mean speed in a direction} = \text{Distance/Average time}$$

$$= \text{Distance} / ATT_{\text{Direction}} \tag{8.3}$$

where $ATT_{Direction}$ = average travel or journey time for all the vehicles in a direction of flow, in minutes; $t_{a\_Direction}$ = average travel time (in minutes) of the test vehicle travelling in the direction of $Q_{Direction}$; $t_{a\_Opposite\ Direction}$ = average travel time (in minutes) of the test vehicle while travelling in the opposite direction; $N_{diff}$ = difference between the number of vehicles overtaking and the number of vehicles overtaken by the test vehicle, while travelling in the same direction; $Q_{Direction}$ = volume of vehicles per minute, in the direction of flow; and $Veh_{Opp}$ = average number of vehicles travelling in the direction opposite that of the test vehicle.

This procedure has been found to be reliable for two-lane two-way highways for low-to-moderate volumes and over long test runs. A major advantage of this method is that the locations of delays and their causes can be readily identified. Measurement of travel time for each individual section length over a long test distance is also possible (Berry, 1954; Cribbins et al., 1961). A major disadvantage of this method is that unreliable results may be obtained in the case of high-volume traffic conditions and multi-lane highways because the average speed of the test vehicle operated by the driver directly depends on their individual behaviour.

### 8.2.2 Test Car Procedure

The time consumed by a test car driven in a traffic stream between selected control points of a study roadway is measured (Bone, 1952). The speed of the test car depends on the judgement of the driver and should be as close as possible to the speed of the traffic in the stream. The test car should be equipped with a voice recorder, stopwatch and tachometer/odometer so that the driver can collect data such as location delays, delay duration and their causes (Table 8.1). This method is suitable for studying road segments with very low traffic. The accuracy of the test results depends on the number of trips repeated over the study road segment; 8–12 such trips in each direction may be recommended to obtain reasonably good results.

**Table 8.1** Sample speed and delay field sheet

| *Site Name: Cotton Highway* | | *Surveyor's Names: Ahmed & Sony* | *Weather: Sunny* | | *Route No: 133* |
|---|---|---|---|---|---|
| *Date: 06.08.15* | | *Start Time: 10:00 am* | *Start Location: Mile Stone 2.0* | | *Direction: East* |
| *Check Point* | *Cumulative Distance along Route (km)* | *Cumulative Travel Time (minutes:seconds)* | *Per Section* | | |
| | | | *No. of Stops* | *Stopped Delay (Sec)* | *Reasons/Remarks* |
| Library | 1.22 | 1:45 | 0 | 0 | - |
| Post Off. | 3.23 | 3:36 | 0 | 0 | - |
| Tea Point | 4.65 | 5:55 | 2 | 48 | Stops due to parked vehicles |
| Sale Mall | 6.78 | 9:14 | 1 | 15 | Stop due to school van |
| Park Rd. | 7.21 | 10:18 | 0 | 0 | - |
| Pharma | 8.21 | 11.34 | 1 | 11 | Stop due to pedestrian |
| Bus Stn. | 9.54 | 13.45 | 3 | 58 | Stops due to signals |

### 8.2.3 Licence Plate Matching Method

Licence plate number data can be collected in two ways: by the manual method or by the automatic number plate recognition (ANRP) technique.

In the case of the manual method, observers at the start and end points of a test length of a road record the licence numbers of vehicles along with their entry and exit times. Each observer located at such strategic points records the last three or four digits of the licence plates. The difference in the time values of a particular licence number of a vehicle is reported as their travel time. The travel time measured by this technique provides a *true measure* of the time that reflects the individual driver's characteristics (Cribbins et al., 1961). The limitations of this method are that only total travel time is measured, irrespective of the number of delays, with the locations and their causes. Also, it is time consuming to record the licence plate numbers of vehicles at each entry and exit point, and matching the same during calculation of travel time is not easy (Walker, 1957).

The licence plate data acquired is analysed to compute the average travel time of each candidate vehicle with reference to survey points, and the results obtained can be used to study intersection performance (in terms of level of service).

### 8.2.4 Fixed Time Interval Photographs

Elevated roadside cameras are mounted to cover a short test length of a road and photographs are taken at regular intervals. The fixed time interval photographs can provide data such as travel/delay time of individual vehicles along the test section, spacing between vehicles, lane usage, travel/turning/crossing/weaving patterns on carriageways, formation of queues, types of vehicles and their relationship to travel time. Daylight and clear weather conditions are essential to obtain quality data. Manual measurement is inevitable though it is cumbersome to quantify each parametric value of the vehicle on photographs.

### 8.2.5 Elevated Observer Method

The travel information of any randomly selected vehicle movement along a specific short test length of road is obtained using elevated roadside video cameras or with a couple of observers stationed at elevated vantage points. The videographs are reviewed later in the office and the pertinent information along the control sections of the road is determined. This method is suitable for the study of delay at intersections and over short test sections.

### 8.2.6 Field Interviews (Survey)

Randomly selected vehicle drivers on a selected test section of road are interviewed on site and the approximate travel time, delay data and their causes are ascertained appropriately. These interviews are conducted preferably at the origin and destination points. A large test area can be covered within a short duration (Walker, 1957). The effort and time consumed by this method is less. However, the reliability of the results largely depends on the quality of the data summarised from the roadside interviews.

### 8.2.7 GIS and GPS Method

Global positioning system (GPS) is used to determine the exact location of a user on earth at any time and in any weather condition. There are 24 GPS satellites in orbit 11,000 nautical miles above the earth. They are continuously monitored by ground stations located worldwide. The radio signals transmitted by these satellites can be detected by any GPS receiver and can be used to determine the user's location with great precision. One limitation, however, is that traffic flow through tunnels and under dense trees and tall buildings may disturb signals and distort the user's current position (Gallagher, 1996).

Geographic information system (GIS) is a computer software package capable of storing, categorising, manipulating and displaying the collected data on a defined geographically referenced map. The geographic maps and data are stored as layers of information and the GIS algorithms enable the user to perform complex analysis for producing reports in a predefined format. In addition, GIS is used as a software platform that is capable of easily integrating/referencing travel time data with a spatial model database.

GPS and GIS are the latest techniques by which travel time data collection is made easy. This method can eliminate the errors and inaccuracies that arise from the use of conventional study methods of travel time and delay. The GPS integrated signal data on a study area of a GIS map can depict the current location and movement of the test vehicle. The surveyor need not record the time elapsed between road junctions. The test vehicle location can be automatically traced through a GPS receiver mounted on the vehicle. The GPS signal data recorded on a digital map of GIS is used to analyse and summarise the travel time, distance travelled and delays in that location. As the total delay, total travel time and total distance covered are known, the average travel and running speeds over the test segment of the road can be easily computed (Mauricio et al., 2003).

### 8.2.8 Bluetooth Technologies

Devices such as smart phones, headsets and in-vehicle navigation systems which are Bluetooth-enabled can be tracked by a Media Access Control (MAC) address. The MAC addresses of each device are considered as their licence plate numbers. Several software programs are available that can detect and record these addresses with timestamps in real time. The recorded real-time data is used to determine space mean speed, travel time and delay along the route (MnDOT, 2013).

## 8.3 SAMPLE SIZE

Determining the minimum number of travel time 'runs' for a study roadway in a given time period is essential for ensuring that the average travel time obtained for a study is within a specified error range of the true average travel time of vehicles, as representing the entire vehicle population. The sample size value may be calculated using the following equation (used in a study conducted by FHWA and Texas Transportation Institute) (Turner et al., 1998).

$$\text{Sample size} = n = \left(\frac{CoV \times T_{\text{stan}}}{Err}\right)^2. \tag{8.4}$$

where $CoV$ = coefficient of variation $= \dfrac{\text{Standard deviation of travel time}}{\text{Mean value of travel time}\,(t_m)} \times 100$

In the absence of the above data, the $CoV$ value may be assumed to be 9–17 (Lomax et al., 1997).

$t_m$ = mean travel time $= \dfrac{\sum_{i=1}^{n} t_i}{n}$; $t_i$ = individual travel time recorded for a test run; $n$ = number of sample runs; $s$ = Standard deviation of travel time $= \sqrt{\dfrac{1}{(n-1)} \times \sum_{i=1}^{n} (t_i - t_m)^2}$

$$Err = \frac{\text{Maximum permissible error value}}{\text{Mean value of travel time}\,(t_{\text{m}})}$$

$Err$ = Relative error permissible in the estimate of travel time (%)

= Tolerence in error

In general, this value is assumed as ±5% for operations and evaluations or ±10% for planning and policy studies (Lomax et al., 1997); and $T_{\text{stan}}$ = student's T-distribution for $n$ – 1 degree of freedom, where the proposed $n$ value is $\leq 30$. If the $n$ value exceeds 30, the z-statistic should be used instead.

The standard value of the Student's T-test ($T_{stan}$) can be obtained based on the degree of freedom. A sample $T_{stan}$ value for the probability of error occurrence in average travel time is listed in Table 6.1 (*Chapter 6*). Take degree of freedom = ($n$ – 1). Its value may be assumed as 1.96 and 3.0 when the confidence level is 95% and 99.7%, respectively.

*Note:* The minimum required runs computed above should be evenly distributed over a period of time, in terms of time of day, weekdays and headways along the study roadway (Turner et al., 1998).

## 8.4 PRESENTATION OF TRAVEL TIME AND DELAY DATA

The data obtained from a travel time delay study can be summarised and presented in any one of the following forms.

### 8.4.1 Tabular Form

Distribution of travel time and delays are summarised in tabular form (Table 8.2), and the required parameters such as total delay, individual values of travel and delay, average travel time and running speed can be easily computed.

**Table 8.2** Typical summary of travel time and delay distribution over a test length

| *Section name* | *Distance (km)* | *Mean travel time (minutes:seconds)* | *Mean delay (seconds)* | *Mean running time (minutes:seconds)* |
|---|---|---|---|---|
| [1] | [2] | [3] | [4] | [5] = [3] – [4] |
| Star lane | 1.1 | 1:58 | 8 | 1:50 |
| Race road | 0.6 | 1:30 | 6 | 1:24 |
| College centre | 1.2 | 1:45 | 3 | 1:42 |
| Hospital road | 1.5 | 1:12 | 2 | 1:10 |
| Park lane | 2.1 | 1:59 | 9 | 1:50 |
| Church road | 1.3 | 1:26 | 8 | 1:18 |
| Market place | 1.4 | 1:28 | 4 | 1:24 |
| Palace road | 1.5 | 1:46 | 7 | 1:39 |
| Total = | 10.7 | 14:04 =14.07 minutes | 47 seconds | 12:17 =12.28 minutes |
| Average Travel Speed = 10.7 /(14.07/60) = 45.64 km/hr | | | Average Running Speed = 10.7 /(12.28/60) = 52.27 km/hr | |

### 8.4.2 Time Contour Map

Contour lines are concentric circles drawn on a map to represent the minutes of travel time from the central business district (CBD) (Fig. 8.2). Average travel times on respective routes between selected points on roads can be easily obtained using these maps.

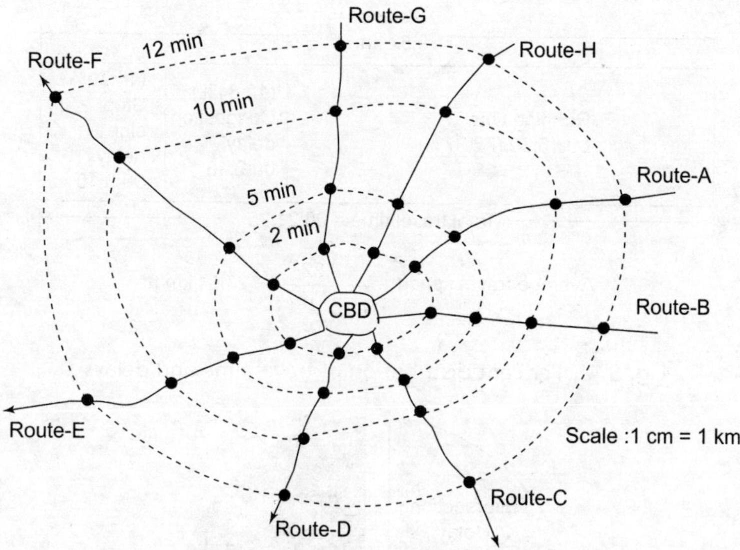


**Fig. 8.2** A typical travel time contour map

### 8.4.3 Time Bar Chart

The delays caused by different factors and their parameters can be represented in a time delay bar chart (Fig. 8.3). The length of each bar indicates the respective parameter per cent distribution. In other cases, travel time and the details of delays with per cent distribution are presented as a single bar (Fig. 8.4) or as a pie chart (Fig. 8.5). Such bar charts can also be prepared to compare travel time and delay data before and after improvements.

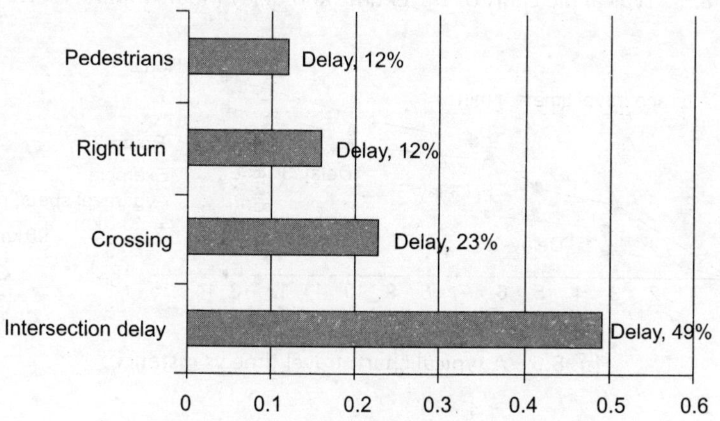


**Fig. 8.3** Typical time delay bar chart

### 8.4.4 Speed Profiles

Speed profile charts can indicate variations in travel time (Fig. 8.6) and travel speed (Fig. 8.7) along the test length of a road. The slope of the lines indicates the rate of change in travel time along the distance.

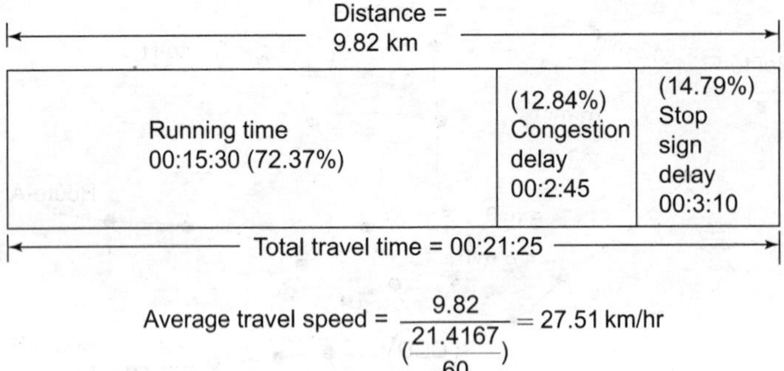


$$\text{Average travel speed} = \frac{9.82}{\left(\frac{21.4167}{60}\right)} = 27.51 \text{ km/hr}$$

**Fig. 8.4** Per cent distribution of travel time and delays

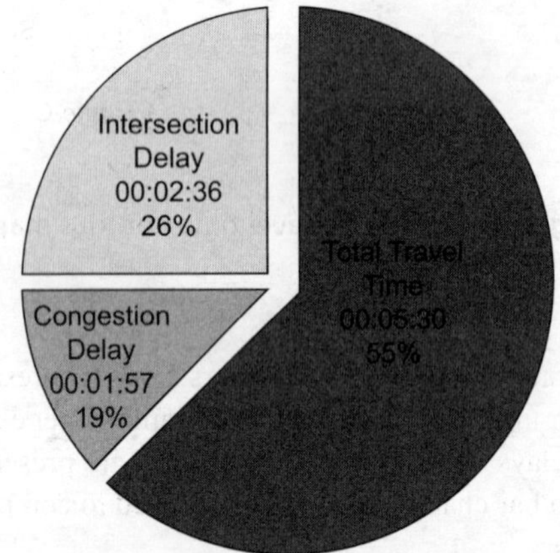


**Fig. 8.5** Typical pie chart of travel time and delay (hours:minutes:seconds)

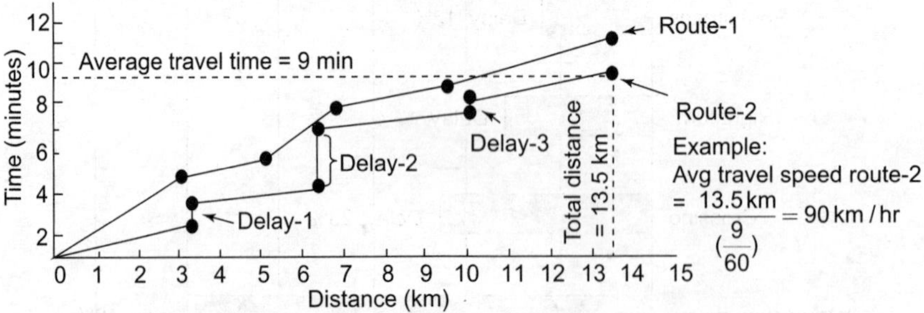


**Fig. 8.6** A typical chart: travel time vs distance

Such charts graphically present the mean travel speed on individual segments of a road along the test section and the overall travel speed and running speeds.

### 8.4.5 Frequency Distribution Curve

The cumulative frequency values of the number of vehicles versus travel time (or delays) indicate the average speed or (delay) of such vehicles that belong to their respective per cent distribution (see *Chapter 6*).

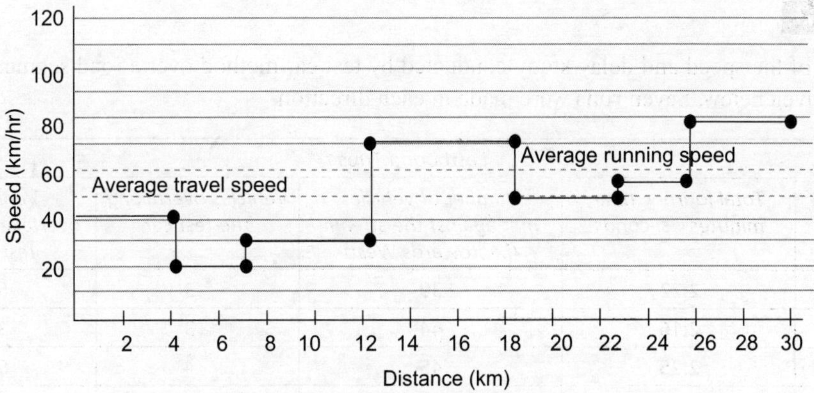


**Fig. 8.7** A typical chart: speed vs distance

## 8.5 SOLVED PROBLEMS

**Example 8.1**

A summary of the values obtained from a spot speed study is given below:

Random sample size = $n$ = 50; Average speed of vehicles = $V_m$ = 44 km/hr; Standard deviation of the speeds measured = $s$ = 10 km/hr.

Calculate the approximate range within which the true (population) average speed lies with a 95% confidence limit?

**Answer:** Let us first calculate the standard error ($SE_m$) of the sample mean = Standard deviation ÷ Square root of sample size

$$SE_m = \frac{s}{\sqrt{n}} = \frac{10}{\sqrt{50}} = \frac{10}{7.07} = 1.41 \text{ km/hr}$$

True mean of the speed lies within a range of ($V_m \pm 1.96 \times SE_m$) km/hr with 95% confidence.

True mean of the speed lies within a range of ($44 \pm 2.76$) km/hr with 95% confidence.

**Example 8.2**

Determine the minimum sample size required if the average spot speed of vehicles = $V_m$ = 44 km/hr, standard deviation of the speeds measured = $s$ = 10 km/hr to an accuracy of standard error of mean ($SE_m$) = 3 km/hr.

**Answer:** We know $$SE_m = \frac{s}{\sqrt{n}}$$

Rewriting the above equation, we get $\sqrt{n} = \frac{10}{3}$; minimum sample size = $n = 11.11 \approx 12$

It is noted that as the value of standard deviation increases with decreased value of standard error of mean, the minimum required sample size increases.

**Example 8.3**

A summary of the speed and delay study conducted by test car method over a road segment length of 2.14 km is given below. Seven runs were made in each direction.

| *Eastbound Trips* | | | | |
|---|---|---|---|---|
| *Trip number* | *Total journey time minutes : seconds* | *Number of vehicles met against the stream (i.e.,towards West)* | *Vehicles overtaking the test car* | *Vehicles overtaken by the test car* |
| 1 | 2:22 | 39 | 3 | 5 |
| 2 | 2:16 | 44 | 3 | 3 |
| 3 | 2:25 | 45 | 4 | 5 |
| 4 | 2:43 | 43 | 3 | 5 |
| 5 | 2:19 | 41 | 2 | 3 |
| 6 | 2:28 | 39 | 5 | 3 |
| 7 | 2:13 | 44 | 3 | 3 |
| Total | 16.77 Min. | 295 | 23 | 27 |
| Average | $t_{a\_East}$ = 2.40 Min. | $Veh_{Opp}$ = (295/7) = 42.14 | $O_{diff}$ = (23-27)/7 = −0.57 | |

| *Westbound Trips* | | | | |
|---|---|---|---|---|
| *Trip number* | *Total journey time minutes : seconds* | *Number of vehicles met against stream (i.e., towards East)* | *Vehicles overtaking the test car* | *Vehicles overtaken by the test car* |
| 1 | 2:32 | 46 | 4 | 3 |
| 2 | 2:12 | 49 | 5 | 3 |
| 3 | 2:46 | 30 | 3 | 4 |
| 4 | 2:29 | 37 | 3 | 5 |
| 5 | 2:17 | 43 | 2 | 4 |
| 6 | 2:32 | 38 | 4 | 3 |
| 7 | 2:35 | 43 | 2 | 4 |
| Total | 17.38 Min. | 286 | 23 | 26 |
| Average | $t_{a\_West}$ = 2.48 Min. | $Veh_{Opp}$ = (286/7) = 40.86 | $N_{diff}$ = (23 – 26)/7 ≈ −0.43 | |

Calculate

(a) Traffic volume in the Eastbound direction; (b) Travel (average journey) time in the Eastbound direction; (c) Space mean speed in the Eastbound direction; (d) Traffic volume in the Westbound direction; (e) Travel (average journey) time in the Westbound direction; (f) Space mean speed in the Westbound direction.

**Answer:** *Speed and delay parameters of Eastbound traffic*:

(a) By Eq. 8.2

Traffic volume in the Eastbound direction $(Q_{East}) = (Veh_{Opp} + N_{diff}) / (t_{a\_East} + t_{a\_West})$

Substituting, we get $Q_{East} = (42.14 + (-0.57)) / (2.40 + 2.48) = (42.14 - 0.57)) / (4.88)$

$Q_{East} = (41.57) / (4.88) \approx 8.52$ vehicles per minute

Or $Q_{East} = 8.52 \times 60 = 511.10$ vehicles per hour

(b) Average travel (average journey) time in the Eastbound direction ($ATT_{East}$):

By Eq. 8.1 $$ATT_{East} = t_{a\_East} - \left(\frac{N_{diff}}{Q_{East}}\right) = 2.40 - \left(\frac{-0.57}{8.52}\right) = 2.46 \text{ minutes}$$

(c) By Eq. 8.3

Space mean speed in the Eastbound direction = Distance/Average travel time
= 2.14/(2.46/60) = 52.19 km/hr

*Speed and delay parameters of Westbound traffic:*

(d) By Eq. 8.2

Traffic volume in the Westbound direction ($Q_{West}$) = $(Veh_{Opp} + N_{diff}) / (t_{a\_East} + t_{a\_West})$

Substituting, we get $Q_{West} = (40.86 + (-0.43)) / (2.40 + 2.48) = (40.86 - 0.43)) / (4.88)$

$Q_{West} = (40.43) / (4.88) = 8.28$ vehicles per minute

Or $Q_{West} = 8.28 \times 60 = 496.8 \approx 497$ vehicles per hour

(e) Average travel (average journey) time in the Westbound direction ($ATT_{West}$):

By Eq. 8.1 $$ATT_{West} = t_{a\_West} - \left(\frac{N_{diff}}{Q_{West}}\right) = 2.48 - \left(\frac{-0.43}{8.28}\right) = 2.53 \text{ minutes}$$

(f) By Eq. 8.3

Space mean speed in the Westbound direction = Distance/Average travel time

= 2.14/(2.53/60) = 50.75 km/hr

**Example 8.4**

A cyclist rode a stretch of 4 km in a period of 40 minutes. He stopped for a period of 10 minutes during the trip. It was observed that 54 vehicles overtook the cyclist during travel time and 30 vehicles passed him during the stop period. Determine the speed of the vehicle stream on the road. Assume that no vehicles were overtaken by the cyclist.

**Answer:** Running speed of the cyclist = 5 km/(40 − 15) minutes = 8 km/hr
Let the velocity of the vehicle stream = $V$

$$\left(\frac{\left(\frac{\text{Vehicles overtaken during the travel}}{\text{Duration in munites}}\right)}{\text{Relative speed of vehicles w.r.t. cyclist while travelling}}\right)$$

$$= \left(\frac{\left(\frac{\text{Vehicles overtaken during the stop}}{\text{Duration in munites}}\right)}{\text{Relative speed of vehicles w.r.t. cyclist at stop}}\right)$$

$$\left(\frac{\left(\frac{54}{40-10}\right)}{V-8\text{ km/hr}}\right)=\left(\frac{\left(\frac{30}{10}\right)}{V-0}\right)$$

Solving, we get $V = 20$ km/hr

**Example 8.5**

A total of six test car travel time and delay runs were conducted on a 3-km route. Determine the average travel speed and average running speed for the test route based on the data given below.

| *Test run #* | *Travel time* | | *Delay (seconds)* |
|---|---|---|---|
| | *Minutes* | *Seconds* | |
| 1 | 6 | 55 | 55 |
| 2 | 7 | 20 | 95 |
| 3 | 8 | 15 | 115 |
| 4 | 6 | 10 | 5 |
| 5 | 8 | 35 | 165 |
| 6 | 6 | 50 | 25 |

**Answer:** The given data is arranged as follows:

| Test run # | Travel time (seconds) | Delay (seconds) |
|---|---|---|
| 1 | 6 × 60 + 55 = 415 | 55 |
| 2 | 7 × 60 + 20 = 440 | 95 |
| 3 | 8 × 60 + 15 = 495 | 115 |
| 4 | 6 × 60 + 10 = 370 | 5 |
| 5 | 8 × 60 + 35 = 515 | 165 |
| 6 | 6 × 60 + 50 = 410 | 25 |
| Total = 2645 | | Total = 460 |

$$\text{Average travel speed} = \frac{\text{Distance travelled}}{\text{Average of total travel time}}$$

$$= \frac{3000\text{ m}}{\left(\frac{2645\text{ sec}}{6}\right)}$$

$$= \frac{3000\text{ m}}{440.83\text{ sec}} = 6.80\frac{\text{m}}{\text{sec}} = 24.48\text{ km/hr}$$

$$\text{Running time} = \text{Total travel time} - \text{Total delay}$$

$$= 2645 - 460 = 2185\text{ sec}$$

$$\text{Average running speed} = \frac{\text{Distance travelled}}{\text{Average of total running time}} = \frac{3000\,\text{m}}{\left(\frac{2185\,\text{sec}}{6}\right)}$$

$$= \frac{3000\text{ m}}{364.16\text{ sec}} = 8.23\text{ m/sec} = 29.628\text{ km/hr}$$

## EXERCISES

### Multiple Choice Questions

1. During a speed and delay study, it was recorded that the average journey time taken to travel along a test stretch of 3 km was 10 minutes and the average stoppage delay observed was 2 minutes. The running speed would be ________ km/hr
(a) 20.5 (b) 33.5 (c) 21.5 (d) 22.5
(*Hint*: Length of the test section = 3 km; Average journey time = 10 minutes; Average stopped delay = 2 minutes; Average running time = 10 – 2 = 8 minutes = (8/60) = hr; Average running speed =3/(8/60) = 22.5 km/hr)
2. In a moving car study conducted over a road stretch of 10 km at a constant speed of 80 km/hr, it was observed that the test car was passed by 20 vehicles more than the number it passed. The number of vehicles met against the stream was 300. The traffic flow rate is ________ veh/hr
(a) 1280 (b) 2240 (c) 2120 (d) 1160
(*Hint*: By Eq. 8.2, Traffic volume in the Eastbound direction ($Q_{\text{East}}$) = ($Veh_{\text{Opp}} + N_{\text{diff}}$) / ($t_{a\_\text{East}}$ + $t_{a\_\text{West}}$)
Substituting, we get $Q_{\text{East}}$ = (300 + 20) / (10/80 + 10/80) = 1280 veh/hr)
3. What would be the value of the average travel time of all the vehicles in the direction of travel of the test car if the traffic flow rate = 1200 veh/hr. The test car is passed by 20 vehicles more that it passes. Average travel time of test car travelling the direction of traffic flow = 30 minutes.
(a) 27 (b) 28 (c) 29 (d) 30
(*Hint*: Average travel (average journey) time in the Eastbound direct ($ATT_{\text{East}}$), by Eq. 8.1,
$$ATT_{East} = t_{a\_\text{East}} - \left(\frac{N_{\text{diff}}}{Q_{\text{East}}}\right) == \frac{1}{2}\text{ hr} - \left(\frac{20}{1200}\right) = 0.48333\text{ hr} = 29\text{ minutes})$$
4. The moving observer method is a technique adopted for the measurement of
(a) Traffic flow of the stream (b) Capacity of the stream
(c) Origin and destination of traffic (d) Spot speed of vehicles
5. What is the minimum sample size required if the average spot speed of vehicles = 44 km/hr, standard deviation of the speeds measured is 10 km/hr to an accuracy of standard error of mean 3 km/hr?
(a) 12 (b) 10 (c) 8 (d) 15
(*Hint*: $\sqrt{n} = \frac{10}{3} = 11.11 = 12$)

## Answers

1. (d) 2. (a) 3. (c) 4. (a) 5. (a)

## Questions

1. State the objectives of conducting travel time and delay studies.
2. What are the improvements most commonly adopted for reducing delays?
3. Explain the methods of measurement of travel time and delays on any selected length of a route.
4. Explain the licence plate matching method of conducting travel time and delay studies on any selected length of a route.
5. Explain the floating car method of conducting travel time and delay studies on any selected length of a route.
6. Differentiate the floating car method and test car procedures of conducting travel time and delay studies on any selected length of a route.
7. Briefly explain the following methods of conducting travel time and delay studies on any selected length of a route:
   (a) Fixed time interval photographs method (b) Elevated observer method
   (c) Field interviews survey (d) GIS and GPS method (e) Bluetooth technologies
8. Explain how minimum sample size is determined for conducting speed and delay studies along a route based on statistical considerations.
9. Write the details of presentation of speed and delay data collected along a route, with appropriate illustrations.
10. Describe the speed profile method of presentation of speed and delay data collected along a route, with appropriate illustrations.

CHAPTER

# 9 Intersection Delay Studies

## 9.0 INTRODUCTION

Intersections are formed when two or more roads meet. Delay at intersections is one of the major problems of most urban road networks. Delay studies are conducted at individual intersections and the data obtained is analysed to evaluate the efficiency or effectiveness of the existing traffic control method. The outcome of the analysis is used to determine the congestion level and level of service of the intersection being studied.

Intersection delay studies should be conducted during periods of congestion. Peak hour delay studies are commonly conducted at isolated intersections that have congestion problems. Average control delay studies at intersections are conducted to evaluate the delay conditions and to explore the option of signalisation.

The following factors affect delays at intersections:

- **Geometric parameters**: Lane width, number of lanes, grades, visibility at turnings, proximity to bus stops, drainage provisions and access control
- **Traffic factors**: Directional traffic volume on each approach, parking provision, predominant vehicles, vehicle composition in the traffic mix, approach speeds and driver characteristics

## 9.1 TYPES OF DELAYS AT INTERSECTIONS

The terminology that is commonly used in intersection delay surveys (MnDOT, 2013) is given below:

- **Delay:** The period of time lost due to causes beyond the control of the driver. The causes may be traffic friction and traffic control devices.
- **Operational delay:** The period of time lost due to delays caused by any impedance of other components of the traffic stream. For example, time lost while waiting for a gap during conflicting traffic streams, congestion, pedestrian flow, parked vehicles or turning movements.
- **Fixed delay:** The period of time lost due to delays caused regardless of the traffic volume and interference present. It may be due to control devices such as traffic signals.
- **Deceleration/Acceleration delay:** The delay incurred by a vehicle by decelerating or accelerating at a signalised intersection. Both these types of delays are often termed signal delays, which can be determined from signal cycle length, volume and the geometric conditions of an intersection.
- **Stopped delay:** The delay that occurs during queuing of vehicles at the entry to an intersection area, or the period of time that a vehicle is not moving. If this delay occurs at the stop line of an intersection, it is termed queue delay (for more information, see *Chapter 16*).
- **Travel time delay**: The difference in time between the actual time required to traverse a section of a street and the time corresponding to the average speed of traffic in uncongested/ideal conditions. This delay includes stopped delay and acceleration and deceleration delays. In other words,

$$\text{Travel time delay} = (\text{Actual travel time} - \text{Ideal travel time}) \text{ through an intersection}$$

Ideal travel time through an intersection is determined as the average speed during uncongested traffic flow.

- **Approach delay**: The travel time delay encountered at the approach to an intersection.
- **Average controlled delay:** The total delay at an intersection caused by a signal, including both stop delay and delay due to deceleration/acceleration to ambient speed.
- **Total vehicle delay**: The total time lost by vehicles in a traffic stream because of non-standard design/operating conditions of the street. It is expressed as vehicle-minutes per km length and obtained by multiplying the peak-hour one-direction volume by the respective delay rate.

## 9.2 METHODS OF MEASURING DELAY THROUGH INTERSECTIONS

A major part of any delay occurs in urban localities at intersections, particularly at signalised intersections (considered as interrupted flow at controlled intersections). Two survey techniques are available to determine the delay or performance of intersections: stopped vehicle count method and elevated observer method. A brief description of each of these techniques is given below.

### 9.2.1 Stopped Vehicle Method

The number of stationary queuing vehicles at the respective approaches of an intersection is counted at fixed time intervals (say, 15 or 30 seconds, for a duration of 5 or 10 minutes). The observation interval must not match the regular signal cycle time. The same fixed observation time is considered for all the approach observations of an intersection. The selected duration (say, 10 or 5 minutes) of observation of delays at an instance may be used for different approaches surveyed in rotation. All the approaches at an intersection should be surveyed simultaneously for the same duration if the delays on those approaches are required to correlate. Classified vehicle volume with turning movements is also counted at the same time. If the longest queues are observed during peak hours that extend back to the upstream intersection, then the floating car method or licence plate matching methods are adopted (TRL, 1993).

A minimum of two observers may be employed at each approach for data collection.

This method provides only the total delay (expressed as vehicle-sec = total number of queuing vehicles counted $\times$ the time interval in sec) and average delay in queue per approach vehicle (i.e., total delay $\div$ number of approaching vehicles during that period) for each approach or for all approaches, for the period of time observed. The value obtained may be multiplied with an empirical adjustment factor ($\approx$0.9) to account for errors due to sampling that tend to overestimate the delay in field studies.

Acceleration/Deceleration delay correction is applied to average delay in queue per approach vehicle with the TRB (2000) recommended values (Table 9.1).

**Table 9.1** Partial table of acceleration–deceleration delay correction factors (*Source*: TRB, 2000)

| *Free flow speed (km/hr)* | *Number of vehicles stopping per lane, per cycle* | |
|---|---|---|
| | *≤7 vehicles* | *8–19 vehicles* |
| ≤60 | +5 | +2 |
| >60–71 | +7 | +4 |
| >71 | +9 | +7 |

A sample of such survey data is presented in Table 9.2. It does not provide details of the delays that occurred with turning movements. In addition, this method is not suitable for studying delays on roundabouts because a rolling queue of vehicles may be present. This method is recommended by the Highway Capacity Manual (TRB, 2000).

**Table 9.2** A partial sample sheet of a completed delay by the stopped vehicle method

| *Site Name: Church Road* | | | *Surveyor's Names: Sira & John* | | | *Weather: Sunny* | | | *Location Sketch:* 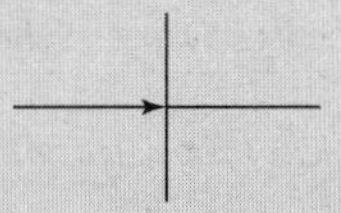 | | |
|---|---|---|---|---|---|---|---|---|---|---|---|
| *Date: 06.06.15* | | | *Time: 9:00 to 9:30* | | | *Interval Time: 30 sec* | | | | | |
| *Approach Name: Park Lane West* | | | *Direction: East* | | | *Number of Lanes: 2 (One way)* | | | *Notes: A road show held 1 km away* | | |
| Time [1] | | | Number of stopped vehicles, by class [2] | | | Time [1] | | | Number of stopped vehicles, by class [2] | | |
| Hr | Min. | Sec | 2W | 3W | 4W | Hr | Min. | Sec | 2W | 3W | 4W |
| 9 | 00 | 00 | 1 | 3 | 2 | | - | - | - | - | - |
| | | 30 | 9 | 0 | 4 | | - | - | - | - | - |
| | 01 | 00 | 6 | 4 | 6 | | - | - | - | - | - |
| | | 30 | 4 | 5 | 0 | | - | - | - | - | - |
| | 02 | 00 | 6 | 4 | 5 | | 29 | 00 | 7 | 3 | 5 |
| | - | - | - | - | - | | | 30 | 9 | 6 | 2 |
| | - | - | - | - | - | 9 | 30 | 00 | 6 | 0 | 6 |

## 9.2.2 Elevated Observer Method

In this method, an observer or a stationary elevated video camera is installed over a study road intersection and the number of vehicles that are stopped at each approach to the intersection is recorded in the study period. Such video cameras are capable of continuously recording traffic flow through an intersection. This data is later transcribed from the video tapes, for each 15- or 30-second interval, over the period of the study. The classified vehicle volume count along with their turning movements is also recorded at the same time. A sample of such survey data is presented in Table 9.3.

This survey method is suitable for determining the delays caused by turning movements at all types of controlled and uncontrolled intersections, including roundabouts, where high-vantage video camera installation points are available. Care should be taken while mounting the video camera over an intersection so that three points are clearly visible from the shooting point, as described below:

- The *start timing point* on an approach, which is at least 10 m beyond the longest anticipated length of the vehicle queue, with an easily identifiable object on the ground
- The *sampling point,* which is about 20 m in advance of the start timing point
- The *finish timing point,* as the stop line representing the exit for the intersection

In addition, the cause of the delay can also be recorded by correlating with signal controls, vehicles parked on kerbs, stopped vehicles, cattle and pedestrians.

**Table 9.3** A sample form for conducting an intersection delay survey by the elevated observer method

<table>
<tr><td colspan="2">Site Name:<br>Beach Road Jn.</td><td colspan="3">Surveyor's Names:<br>Sophia & Yen</td><td colspan="2">Weather: Cloudy</td><td colspan="2" rowspan="2">Location Sketch:</td></tr>
<tr><td colspan="2">Date:12-12-2015</td><td colspan="3">Time: From<br>8:30 to 9:30</td><td colspan="2">Carriageway Widths:<br>7.5m each</td></tr>
<tr><td colspan="2">Approach Name: School lane side</td><td colspan="3">Direction: South</td><td colspan="2">Number of Lanes: 2</td><td colspan="2">Notes: Police help taken</td></tr>
<tr><td rowspan="2">Vehicle Type</td><td rowspan="2">Turning Movement</td><td colspan="3">Time of Passing</td><td rowspan="2">No. of Stops</td><td colspan="3">Total Stopped Time</td></tr>
<tr><td>Hr.</td><td>Min.</td><td>Sec.</td><td>Min.</td><td>Sec.</td><td>Reason due to</td></tr>
<tr><td>Car</td><td>RT</td><td>9</td><td>00</td><td>10</td><td>3</td><td>00</td><td>6</td><td>Pedestrian</td></tr>
<tr><td>Truck</td><td>Thr</td><td>9</td><td>03</td><td>18</td><td>2</td><td>00</td><td>8</td><td>Bus</td></tr>
<tr><td>Van</td><td>LT</td><td>9</td><td>06</td><td>14</td><td>1</td><td>00</td><td>5</td><td>Police</td></tr>
<tr><td>2W</td><td>RT</td><td>9</td><td>08</td><td>32</td><td>1</td><td>00</td><td>4</td><td>Veh. Stopped</td></tr>
<tr><td>-</td><td>-</td><td>-</td><td>-</td><td>-</td><td>-</td><td>-</td><td>-</td><td>-</td></tr>
<tr><td>-</td><td>-</td><td>-</td><td>-</td><td>-</td><td>-</td><td>-</td><td>-</td><td>-</td></tr>
</table>

The total and average delays are derived for individual vehicles based on the difference in travel times observed from the video file and summarised for each approach or for all approaches to a junction.

Intersection delay is a measure of intersection performance, usually presented in the form of average delay per vehicle. The total stopped time of the vehicles (total stopped vehicle-time in units of total vehicle-seconds of delay) is computed by multiplying the total number of stopped vehicles by the time interval between the observations. It must be noted that the selected time interval should not be an exact multiple of the signal cycle length. This criterion enables sampling of data during different parts of the signal phase (Berry and Van Till, 1954).

### 9.2.3 Sample Size

A minimum sample of 60 measurements of the number of vehicles in each approach is a commonly adopted method of measuring intersection delay in field studies (MnDOT, 2013).

## 9.3 SPEED AND DELAY FORMULAE

The following formulae are commonly used to compute the travel speeds and delays (TRL, 1993; Camsys, 2011):

Individual vehicle delay across an intersection for an approach = (Sample vehicle travel time observed across the intersection) – (Average delayed travel time for that class of vehicle) (9.1)

Alternatively, let travel delay across a signalised intersection = $D$ (sec)

$$D\,(\text{sec}) = \text{Travel duration }\;(\text{sec}) - \left[\frac{\text{Length of test section, km}}{\left(\dfrac{\left(V_{\text{before}} + V_{\text{after}}\right)}{2}\right)} \times 3600\right] \qquad (9.2)$$

where $V_{before}$ = average overall speed of traffic on an adjacent section before the intersection (km/hr); and $V_{after}$ = average overall speed of traffic on an adjacent section after the intersection (km/hr).

$$\text{Average delay per approach } (Di) = (\text{Total time of delay of all sample vehicles}) \div (\text{No. of vehicles}) \quad (9.3)$$

$$\text{Total delay per approach} = (\text{Average delay per approach}) \times (\text{Traffic volume of the approach})$$

$$= \sum_{i=1}^{n} (VOL_i \times D_i) \quad (9.4)$$

Average delay at an intersection ($D_{Int}$) is calculated using the following equation

$$D_{Int} = \frac{\sum_{i=1}^{n} (VOL_i \times D_i)}{\sum_{i=1}^{n} VOL_i} \quad (9.4)$$

where $Di$ = average delay for an approach $i$; $VOL_i$ = traffic flow rate (veh/hr); and $n$ = number of approaches to the intersection.

Total delay at an intersection during the survey period = Sum of values of total delays calculated for all approaches to that intersection (9.5)

These parametric values can be used to compare intersections and indicate the necessary improvements relating to geometric design or flow control. They can also be used to compare intersection delay before and after an improvement. It must be noted that when comparing the before and after studies, it is essential to adopt the same survey technique (TRL, 1993).

*Note:* In the case of manual observation, each surveyor may be assigned to count up to three classes of vehicles, and the same should be used for delay surveys. Non-motorised vehicles should not be included in the delay surveys (TRL, 1993).

## 9.4 SOLVED PROBLEMS

**Example 9.1**

The following details pertain to a vehicle delay study conducted on one approach to an intersection:

The stopped vehicle method was used for the sampling procedure. Vehicles occupying the intersection approach at successive intervals of 15 seconds for a study period of 5 minutes were counted (i.e., over 5 cycles). Take the signal cycle length as 116 seconds and the number of lanes as 2.

Vehicles occupying the intersection approach in each successive count of 15 seconds were counted by Observer-1. The number of vehicles passing through the intersection approach in the time period of 5 minutes was counted by Observer-2.

Compute (a) Total delay at the approach to the intersection; (b) Average delay per stopped vehicle; (c) Average delay in the queue per approach vehicle; (d) Per cent of vehicles stopped at the intersection; (d) Per cent of vehicles stopped at the approach; (e) Number of cycles surveyed; (f) Number of vehicles stopping per lane each cycle; and (g) Control delay per vehicle.

| *Observer* | *Observer-1 (Instantaneous Density)* | | | | *Observer-2 (Traffic Volume)* | |
|---|---|---|---|---|---|---|
| | *Number of vehicles in the queue/stationary in the approach, at time* | | | | *Approach volume category* | |
| *Starting time* | *0 Sec* | *15 Sec* | *30 Sec* | *45 Sec* | *Number of vehicles that stopped in more than one observation* | *Number of non-stopped vehicles* |
| 17:30 | 0 | 6 | 8 | 7 | 11 | 7 |
| 17:31 | 4 | 7 | 4 | 6 | 7 | 13 |
| 17:32 | 6 | 9 | 10 | 7 | 12 | 7 |
| 17:33 | 5 | 11 | 6 | 11 | 14 | 6 |
| 17:34 | 7 | 3 | 1 | 5 | 6 | 12 |
| Sub-totals | 22 | 36 | 29 | 36 | 50 | 45 |
| Total | **123** | | | | **95** | |

**Answer:**

(a) Total delay at the approach to the intersection:

Total delay = Total number of vehicles in queue × Time interval
= 123 × 15 = 1845 vehicle seconds

(b) Average delay per stopped vehicle = Total delay ÷ Number of stopped vehicles
= 1845 ÷ 50 = 36.9 seconds

(c) Average delay in the queue per approach vehicle = Total delay ÷ Number of approaching vehicles
= 1845 ÷ 95 = 19.42 seconds

*Note*: This value may be multiplied by an empirical adjustment factor (≈0.9) to account for errors due to sampling that tend to overestimate the delay in field studies.

Average delay in queue vehicle per approach = 19.42 × 0.9 = 17.47 seconds

(d) Per cent of vehicles stopped at the approach = Stopped vehicles ÷ Approaching vehicles
= (50 ÷ 95) × 100 = 52.63%

Fraction of vehicles stopped (FVS) = 0.5263

(e) Number of cycles surveyed = Cycle length/15 = 116 /15 = 7.73

(f) Number of vehicles stopping per lane per cycle = 50/(7.73 × 2) = 3.23 ≤7 vehicles

From Table 9.1 (TRB, 2000), the given value of free flow speed = 55 km/hr (i.e, ≤60 km/hr); so take acceleration/deceleration delay correction factor = +5 seconds

Therefore, acceleration/deceleration correction delay = 0.52 × (5) = +2.6 seconds

(g) Control delay per vehicle = (Average delay in queue per approach vehicle + Acceleration/Deceleration correction delay)

Control delay per vehicle = (17.47 + 2.6) = 20.07 seconds

**Example 9.2**

A vehicle wanted to cross an un-signalised intersection and make a right turn. In this process, the vehicle stops at the stop line of the intersection at 12:00:00 pm and starts entering the intersection at 12:00:06 pm. While making the right turn, it encountered 7 opposing vehicles (i.e., conflicting vehicles) and finally crossed the intersection by 12:00:13 pm. Calculate the instantaneous service delay and conflicting rate at the intersection.

**Answer:**

Service delay at the stop line of the intersection = 12:00:06 − 12:00:00 = 6 seconds

$$\text{Conflicting flow rate} = \left(\frac{\text{Number of conflicting vehicles}}{\text{Total delay}}\right)$$

Number of conflicting vehicles = 7

$$\text{Conflicting flow rate} = \left(\frac{7}{\text{12:00:13-12:00:00}}\right) = \left(\frac{7\text{ veh}}{13\text{ sec}}\right) = 0.538\text{ veh/sec}$$

## EXERCISES

### Multiple Choice Questions

1. Match List-I (speed) with List-II (application) and select the correct answer using the codes given below the lists:

| List-I | List-II |
|---|---|
| A. Space mean speed | 1. Road roughness studies |
| B. Journey speed | 2. Speed regulatory measures |
| C. Running speed | 3. Traffic flow studies |
| D. Spot speed | 4. Congestion studies |

| | A | B | C | D |
|---|---|---|---|---|
| (a) | 1 | 2 | 3 | 4 |
| (b) | 4 | 2 | 3 | 1 |
| (c) | 2 | 3 | 1 | 4 |
| (d) | 3 | 4 | 1 | 2 |

2. Match List-I with List-II and select the correct answer using the codes given below the lists:

List-I

A. Traffic flow rate

B. Traffic density

C. Space mean speed

D. Time mean speed

List-II

1. Number of vehicles occupying a unit length of a study road at a given instant of time

2. Number of vehicles passing a particular section of a road over a period of time in a direction

3. The speed of vehicles traversing a specified section-length of a road, over a specified time period
4. Arithmetic mean speed of all instantaneous vehicle speeds at a specified location on a road length

| | A | B | C | D |
|---|---|---|---|---|
| (a) | 2 | 1 | 3 | 4 |
| (b) | 1 | 2 | 4 | 3 |
| (c) | 2 | 1 | 4 | 3 |
| (d) | 1 | 2 | 3 | 4 |

## Answers

1. (d) 2. (a)

## Questions

1. State the factors affecting delays at intersections.
2. Define (a) Operational delay; (b) Fixed delay; (c) Travel time delay; (d) Approach delay; (e) Average control delay; and (f) Total vehicle delay.
3. Explain the methods of measuring delay through intersections.
4. Explain the stopped vehicle method of measuring delay through intersections.
5. Explain the elevated observer method of measuring delay through intersections.
6. Write the formula used to determine
   (a) Individual vehicle delay across an intersection for an approach
   (b) Travel delay across a signalised intersection (c) Average delay per approach
   (d) Total delay per approach (e) Average delay at an intersection
7. The following details pertain to a vehicle delay study conducted on one approach to an intersection using the stopped vehicle method. The vehicles occupy the intersection approach at successive interval of 15 seconds. Take the signal cycle length as 118 seconds and the number of lanes as 2. Vehicles occupying the intersection approach for each successive count of 15 seconds are counted by Observer-1. The number of vehicles passing through the intersection approach in the time period of 5 minutes is counted by Observer-2.

| *Observer* | *Observer-1 (Instantaneous Density)* | | | | *Observer-2 (Traffic Volume)* | |
|---|---|---|---|---|---|---|
| *Starting time* | *Number of vehicles in the queue/ stationary in the approach, at time* | | | | *Approach volume category* | |
| | *0 Sec* | *15 Sec* | *30 Sec* | *45 Sec* | *Number of vehicles that stopped during more than one observation* | *Number of non-stopped vehicles* |
| 18:31 | 1 | 4 | 2 | 7 | 10 | 8 |
| 18:32 | 4 | 7 | 4 | 6 | 7 | 12 |
| 18:33 | 2 | 9 | 0 | 8 | 12 | 8 |

(Continued)

**Table** (Continued)

| *Observer* | *Observer-1 (Instantaneous Density)* | | | | *Observer-2 (Traffic Volume)* | |
|---|---|---|---|---|---|---|
| *Starting time* | *Number of vehicles in the queue/ stationary in the approach, at time* | | | | *Approach volume category* | |
| | *0 Sec* | *15 Sec* | *30 Sec* | *45 Sec* | *Number of vehicles that stopped during more than one observation* | *Number of non-stopped vehicles* |
| 18:34 | 5 | 10 | 9 | 11 | 12 | 7 |
| 18:35 | 1 | 3 | 3 | 6 | 7 | 11 |

Compute (a) Total delay at the approach of the intersection; (b) Average delay per stopped vehicle; (c) Average delay in the queue per approach vehicle; (d) Per cent of vehicles stopped at the intersection; (d) Per cent of vehicles stopped at the approach; (e) Number of cycles surveyed; (f) Number of vehicles stopping per lane each cycle; and (g) Control delay per vehicle.

CHAPTER
# 10 Traffic Flow Characteristics

## 10.0 INTRODUCTION

To design, evaluate and upgrade the geometrics of a road and the associated traffic control devices and procedures, it is mandatory to collect various types of data related to traffic flow characteristics. The following sections describe the basic parameters that are widely used for this purpose.

## 10.1 TRAFFIC STREAM PARAMETERS

Based on the type of vehicular flow along a selected length of a road, the traffic stream parameters may be categorised as follows (Fig. 10.1):

- Macroscopic parameters
- Microscopic parameters

Parameters such as rate of flow (Q), average speed (V) of vehicles in a stream of traffic and their density (K) are considered to belong to the macroscopic category as they describe the traffic stream as a whole (i.e., considered as a group of vehicles). Parameters such as speed (V), spacing and time headway are considered to belong to the microscopic category and they describe the individual behaviour of vehicles or pairs of vehicles within a stream of traffic.

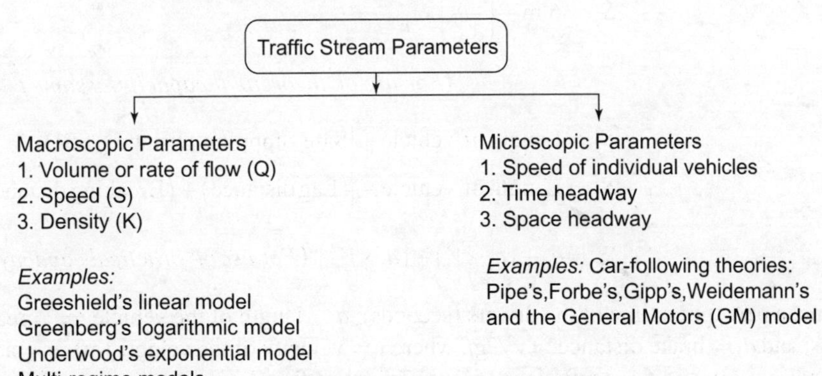


**Fig. 10.1** Classification of traffic stream parameters

### 10.1.1 Space Headway and Time Headway

Spacing is defined as the distance (m) between successive vehicles in a traffic lane, at a given point (Fig. 10.2). Time headway is defined as the time gap (seconds) to pass the space gap between successive vehicles travelling along a lane.

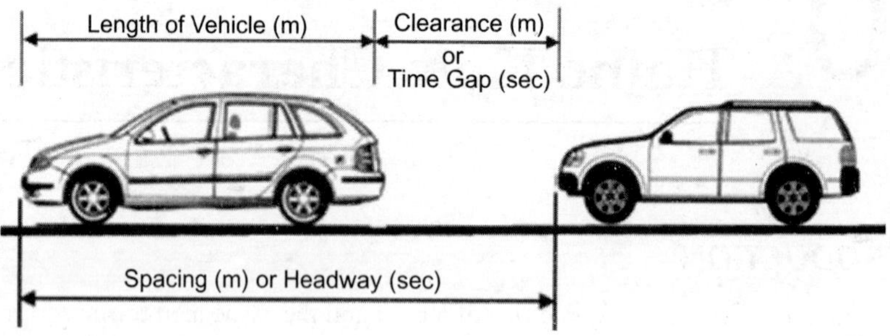


**Fig. 10.2** Measurement of spacing between a pair of vehicles travelling at a point

Space headway and time headway are considered microscopic measures of traffic flow as they apply to pairs of vehicles travelling in a traffic stream. These parameters are useful for determining the number of gaps in a traffic stream or with pedestrians to cross a roadway, and for the design of pedestrian signals at junctions.

The following additional parameters can be determined from field measurements:

Clearance (m) = Spacing (m) – Average vehicle length (m)

Gap (sec) = Time headway (sec) – Time equivalence of the average vehicle length (sec)

Space headway ($S_s$) can be calculated using any one of the following equations:

Space headway = Length of rear vehicle + Space gap between the vehicles

$$S_s = 6\text{ m} + \left(\frac{V}{3.6}\right) \times t_p \tag{10.1}$$

(*For use of theoretical capacity, assume* $t_p = 0.7$ sec)

$$S_s = \text{Length of vehicle} + [\text{Safe Stopping Distance, SSD}]$$

$$S_s = \text{Length of vehicle} + [(\text{Lag distance}) + (\text{Breaking distance})]$$

$$S_s = a_1 + (a_2 \times V) + (a_3 \times V^2) \ (\textit{For use of practical capacity}) \tag{10.2}$$

where $t_p$ = mean perception time of the drivers (seconds); $a_1$= length of the vehicle; $a_2$ = reaction time of the drivers; and $a_3$ = brake distance = $v^2/2gf$, where $v$ = velocity of the vehicle (m); $g$ = acceleration due to gravity (m/sec$^2$); and $f$ = coefficient of longitudinal surface friction.

If the $S_s$ value is expressed in terms of time gap in seconds, it is termed the *time headway* (T) and it can be calculated as follows:

$$\text{Time headway } (T) = \left(\frac{\text{Space gap between the front of one vehicle and the front of the other vehicle considered}}{\text{Average velocity of the vehicles}}\right) \tag{10.3}$$

With reference to Fig. 10.2, Time headway ($T$) = ($t_2$ – $t_1$) sec (10.4)

Average time headway over a length of road segment can be calculated using the following relationship:

$$\text{Average time headway (sec)} = \left( \frac{\text{Average spacing between vehicles moving in a segment (m)}}{\text{Space mean speed (m/sec)}} \right)$$

Average values of spacing and time headway can be used to relate with macroscopic parameters:

$$\text{Density} = K \text{ (veh/km per lane)} = \frac{1000 \text{ m}}{\text{Average spacing in m}} \tag{10.5}$$

$$\text{Volume} = Q \text{ (veh/hr per lane)} = \frac{3600 \text{ sec}}{\text{Average time headway in sec}} \tag{10.6}$$

$$\text{Speed} = V \text{ (m/sec)} = \frac{\text{Average spacing in m}}{\text{Average time headway in sec}} \tag{10.7}$$

Headway data is useful for the design, operation, control and evaluation of road sections, particularly the overtaking zones, lane changing transitions, intersection operations and acceleration/deceleration lanes.

If vehicle movements are not influenced, each driver operates independently. Under these conditions, uniform headway will be maintained by such vehicles. The resulting distribution can be a random occurrence and the probability of such an event occurring can be determined by a Poisson distribution, as shown below:

$$P(k) = \frac{e^{-\mu} \mu^k}{k!}$$

where $P(k)$ = probability that $k$ vehicles will arrive in $t$ interval of time; $t$ = given interval of time (sec); $e$ = base of Naperian logarithm; $\mu$ = average number of vehicles arriving in $t$ interval of time = $\frac{t \times Q}{3600}$; and $Q$ = traffic volume (veh/hr).

If no vehicles will arrive in an interval of time $t$ at traffic volume $Q$, the above expression can be written as:

$$P(0) = \frac{e^{-\mu} \mu^0}{0!} = e^{-\mu}$$

### 10.1.2 Lane Occupancy

It is the ratio of the time in which vehicles travel over a detection loop station in a traffic lane to the total time of observation. Vehicle length significantly affects lane occupancy. As the average vehicle length increases, the lane capacity decreases, and vice versa. Lane occupancy data is used as a surrogate for density in evaluating operational performance. Induction loop detectors are generally used to measure the lane occupancy (Fig. 10.3). Measurement of lane occupancy is easier than measurement of density.

$$\text{Lane occupancy} = \frac{\text{Sum of individual vehicle times when the detector is occupied}}{\text{Total observation time}} = \frac{\sum_{i=1}^{n} t_i}{T}$$

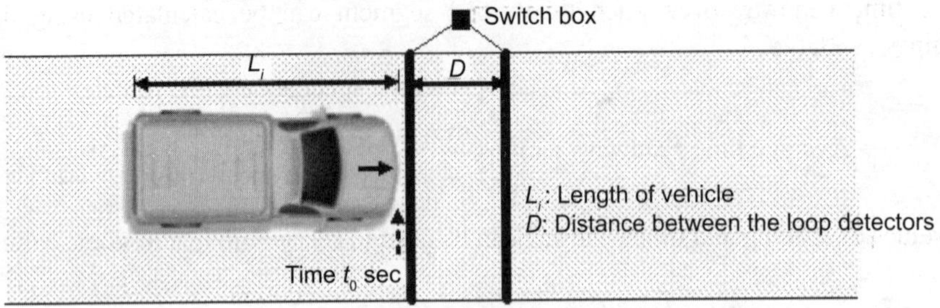


(a) Position of a vehicle before crossing the induction loop detector

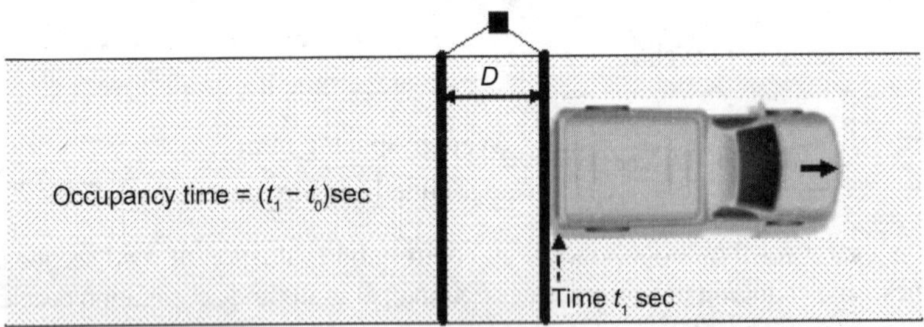


(b) Position of the vehicle after crossing the induction loop detector

**Fig. 10.3** The basic principle of measurement of vehicle occupancy time across a loop detector

where $t_i$ = time taken by a vehicle to travel across the loop detector $= \frac{Li + D}{V_i}$; $Li + D$ = length of vehicle $i$ + distance between loops of detector; and $V_i$ = speed of vehicle $i$.

### 10.1.3 Density

It is defined as the number of vehicles ($n$) travelling on a unit length ($\ell$) of a road, per lane, at an instance (Fig. 10.4). The density or concentration ($K$) value is measured in vehicles/kilometre (veh/km). If there is no space between the vehicles at an instance, the value of density will be maximum. Density is a function of mean vehicle dimension, and this parameter can be used to describe road performance.

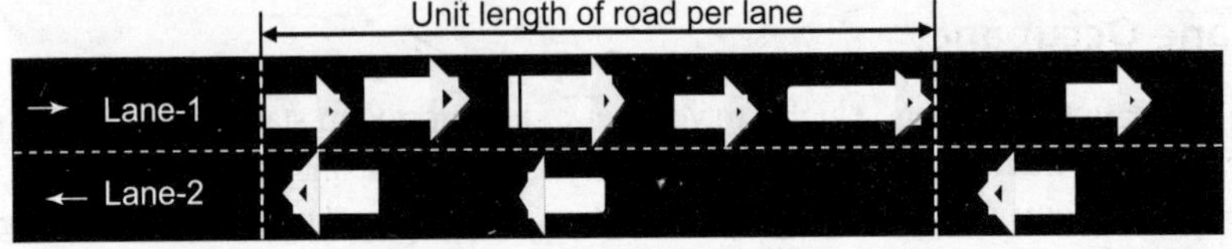


**Fig. 10.4** Density of vehicles

$$K = \frac{\text{Number of vehicles}}{\text{Distance}} = \frac{n}{\ell} = \frac{Q}{V} \tag{10.8}$$

This equation may be re-written as given below, to assess the impact of a development project by determining the LOS of a road segment (SCVTA, 2003):

$$\text{Density} = K \text{ (veh/hr per lane)} = \frac{Q}{N \times V} \tag{10.9}$$

where $Q$ = peak hour traffic volume (i.e., flow rate in veh/hr); $V$ = average speed of vehicles (km/hr); and $N$ = number of travel lanes.

### 10.1.4 Lane Capacity

Capacity is a measure that represents the ability of a roadway to accommodate traffic volume. The theoretical maximum number of vehicles that can *pass* a given point on a lane per unit time under ideal roadway, traffic and control conditions is often termed the *basic capacity of a lane* ($C$). The capacity of a lane can be calculated from the following equation.

$$C = \frac{V}{S_s} \times 1000 \tag{10.10}$$

where $C$ = basic capacity of a single lane (veh/hr per lane); $V$ = average speed or velocity of the vehicles (km/hr); and $S_s$ = safety mean spacing (m) between the front of two consecutive vehicles that are travelling on the same lane at a speed (Fig. 10.5). This parameter is also termed *space headway* or *spacing* (by Eq. 10.1).

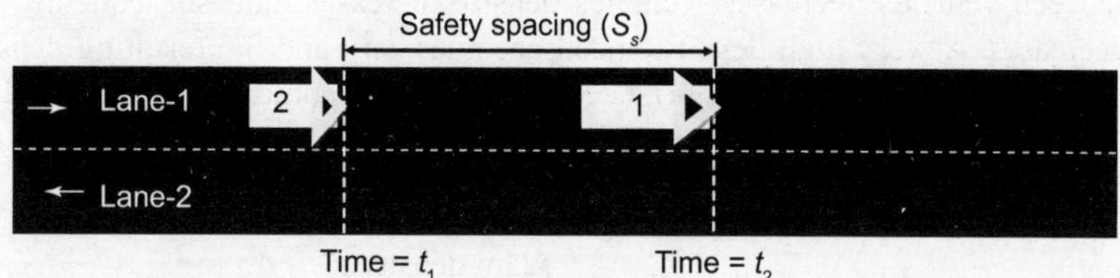


**Fig. 10.5** Concept of determination of safety spacing

**Types of traffic capacity**

- **Basic capacity:** It is the maximum possible number of vehicles that can *pass* a given point on a lane in an hour, under ideal conditions of traffic flow on the roadway.
- **Possible capacity:** It is the maximum number of vehicles that can *pass* a given point on a lane in an hour, under prevailing conditions of traffic flow on the roadway. It must be noted that the possible capacity of a roadway will generally be *less* than the basic capacity, as the prevailing traffic conditions are seldom ideal. This means that possible capacity will vary from zero to basic capacity. Zero density represents jam density ($K_j$).
- **Practical/Design capacity:** It is the maximum number of vehicles that can *pass* a given point on a lane in an hour, without causing any unreasonable delay or restrictions to the driver's freedom to manoeuvre under the prevailing conditions of traffic flow on the roadway. It must be noted that roadway width and the number lanes is designed in such a way that measured traffic volume is less than or equal to its lane design capacity. For this reason, practical capacity is also termed design capacity.

**Factors affecting practical capacity:** The capacity of a roadway depends on several factors relating to road geometrics and mixed traffic conditions (i.e., with the proportion of vehicle types).

- **Space headway:** As the space gap between vehicles increases, there will be a decrease in the value of capacity. If the vehicles are in a jammed condition, the density will be maximum, and it is termed jamming density. In addition, all the factors affecting space headway may also be considered under this factor.
- **Lane width:** As the lane width increases (i.e., the number of lanes increases), the capacity to pass vehicles per hour also increases, and vice versa.
- **Shoulder width**: As the width increases, the capacity value also increases, and vice versa.
- **Proportion of commercial vehicles:** As the proportion of commercial vehicles (large dimensions) increases, the capacity value decreases.
- **Alignment:** As there is no adequate sight distance, there may be a reduction in speed, and hence, the capacity decreases.

## 10.2 FUNDAMENTAL RELATIONSHIPS BETWEEN TRAFFIC FLOW PARAMETERS

The fundamental traffic flow characteristics and their relationships which pertain to uninterrupted traffic flow condition are discussed below.

### 10.2.1 Speed–Density (V–K) Relationship

As spacing between vehicles decreases, vehicles density increases, and subsequently, the speed of the vehicles decreases. At very high density, speed becomes zero and the resulting density is termed *jam density* ($K_j$). With decrease in density (i.e., $K = 0$), speed becomes maximum, and this speed is termed *free speed* ($V_f$). A linear relationship exists between speed and density, and the correlation is represented in Fig. 10.6.

$$\text{Critical density } (K_c) = \left(\frac{\text{Jam density}}{2}\right) = \left(\frac{K_j}{2}\right) \tag{10.11}$$

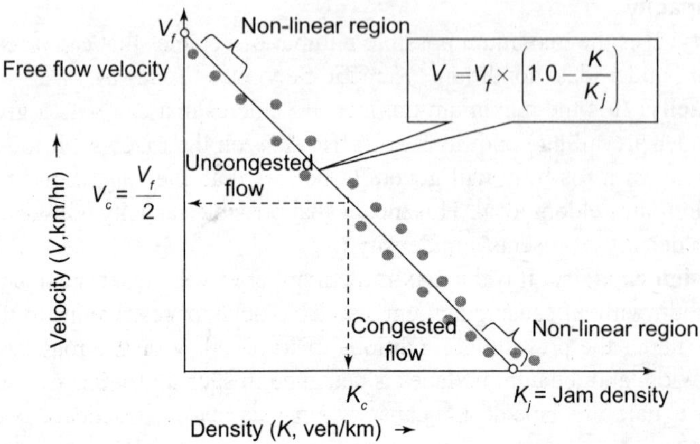


**Fig. 10.6** Relationship between velocity and density

### 10.2.2 Speed–Flow (V–Q) Relationship

As vehicle speed decreases (from free flow condition), the flow rate reduces. As vehicle speed increases, the initial flow rate also increases, up to a certain maximum limit, with decreased space headway. However, as speed increases further, the space headway increases and the flow rate of the vehicle decreases (Fig. 10.7). The speed at which the flow rate is maximum ($Q_m$) is termed optimum speed or *critical speed* ($V_c$).

$$\text{Critical speed } (V_c) = \left(\frac{\text{Free flow velocity}}{2}\right) = \left(\frac{V_f}{2}\right) \tag{10.12}$$

$$\text{Capacity} = \text{Maximum flow } (Q_m) = V_c \times K_c$$

$$= \left(\frac{V_f}{2}\right) \times \left(\frac{K_j}{2}\right) = \left(\frac{V_f \times K_j}{4}\right) \tag{10.13}$$

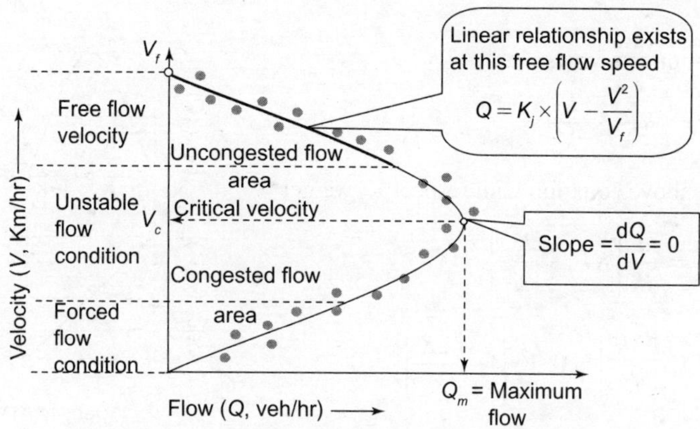


**Fig. 10.7** Relationship between velocity and flow

### 10.2.3 Flow–Density (Q–K) Relationship

As density increases from zero (Fig. 10.8), the vehicle flow rate increases, up to the point of critical density ($K_c$), which corresponds to the maximum flow rate ($Q_m$). Further, as the density reaches maximum – termed jam density ($K_j$) – the vehicles are stopped and flow rate becomes zero.

The following relationship may be used to determine the density or any unknown parameter from the other known values using the standard trend curves (Fig. 10.9).

$$\text{Traffic flow } (Q) = \text{Space mean speed } (V_s) \times \text{Traffic density } (K)$$

$$Q = V_s \times K \tag{10.14}$$

**Derivation of the parabolic flow–density model:** The fundamental relationships of traffic flow models are derived as shown below:

With reference to Fig. 10.9, at maximum flow ($Q_m$), the slope of the curve $= \dfrac{dQ}{dK} = 0$

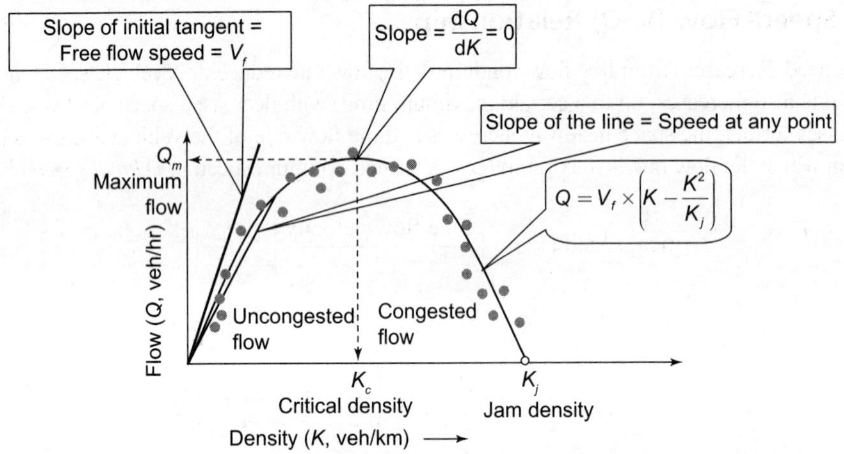


**Fig. 10.8** Relationship between flow and density

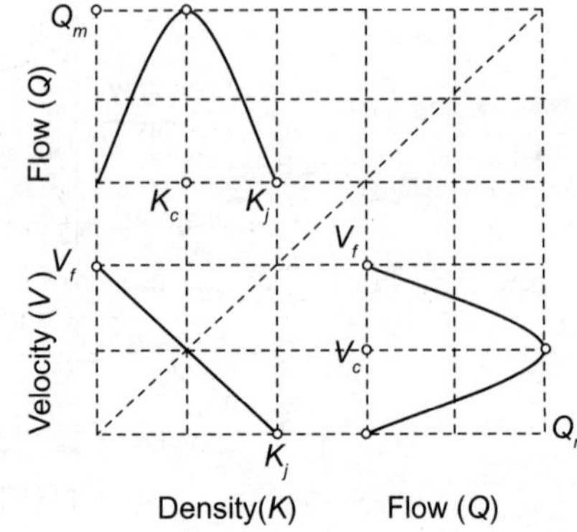


**Fig. 10.9** Velocity–Flow–Density relationship

Greenshield's equation is rewritten as

$$Q = KV_f - \frac{K^2}{K_j}$$

Differentiating the above equation with respect K, we get

$$\frac{dQ}{dK} = \frac{d}{dK}\left(KV_f - \frac{K^2}{K_j}\right) = 0$$

$$V_f - V_f\left(\frac{2K}{K_j}\right) = 0;\ V_f\left(1 - \left(\frac{2K}{K_j}\right)\right) = 0$$

$$\left(\frac{2K}{K_j}\right) = 1;\ K = \left(\frac{K_j}{2}\right) \tag{10.15}$$

Maximum flow or volume occurs when the density is half the jam density. In other words, the density at maximum flow = critical density ($K_c$) = $\frac{K_j}{2}$.

Substituting Eq. (10.15) in Eq. (10.2), we get

$$\frac{K}{K_j} + \frac{V_s}{V_f} = 1.0;\ \frac{\left(\frac{K_j}{2}\right)}{K_j} + \frac{V_s}{V_f} = 1.0$$

$$\frac{1}{2} + \frac{V_s}{V_f} = 1.0;\ \frac{V_s}{V_f} = \frac{1}{2};\ V_s = \frac{V_f}{2} \tag{10.16}$$

The critical speed (i.e., space mean speed, $V_s$) is half the free flow speed at maximum flow (Fig. 10.7).

i.e., $$V_c = \frac{V_f}{2}$$

Using Eq. (10.14) $$Q = V_s \times K; Q_m = \frac{V_f}{2} \times \frac{K_j}{2}$$

$$Q_m = \frac{V_f \times K_j}{4} \tag{10.17}$$

(The above relationships pertain to Greenshield's model, which is explained later in this chapter).
Rewriting Eq. (10.14)

$$Q = K \times V_s$$

To determine the maximum value of $Q$, i.e., $Q_m$, differentiate the above equation with respect to K and equate to zero.

$$\frac{dQ}{dK} = K_c \times \frac{dV_s}{dK} + V_s \times 1 = 0$$

$$K_c \times \frac{dV_s}{dK} + V_s = 0; V_s = -K_c \times \frac{dV_s}{dK}$$

Integrating, we get $$\int \frac{1}{V}.dV_s = -\int \frac{1}{K_c}.dK$$

$$\log V_s = -\frac{K}{K_c} + C \tag{10.18}$$

where $C$ = constant of integration. When $K = 0$, $V = V_f$; $\log V_f = -0 + C$

Therefore, $$C = \log V_f$$

Now, Eq. (10.18) can be written as

$$\log V_s = -\frac{K}{K_c} + \log V_f; \frac{K}{K_c} = \log V_f - \log V_s$$

$$K = K_c . \log \frac{V_f}{V_s} \tag{10.19}$$

*Note*: Numerous other models are also available in literature. Please refer to the appropriate texts.

### 10.2.4 Density–Spacing (K–$S_s$) Relationship

As the spacing between the vehicles travelling on a lane decreases, their density increases. The density value can be determined by measuring the flow rate and average speed of the vehicles travelling on a length of road.

## 10.3 MACROSCOPIC TRAFFIC STREAM MODELS

### 10.3.1 Greenshield's Linear Model

The macroscopic stream models represent the interdependence of one parameter of traffic flow with respect to the other. In other words, as the speed of the traffic stream decreases, the density of the traffic increases.

The speed and density relationship is modelled as a straight line, which is known as Greenshield's linear speed–density model (Fig. 10.6). Greenshield's model is mathematically expressed as follows (Greenshields, 1935):

$$\frac{K}{K_j}+\frac{V_s}{V_f}=1.0 \tag{10.20}$$

The above equation can be written as

$$\frac{V_s}{V_f}=1.0-\frac{K}{K_j};\; V_s=V_f\times\left(1.0-\frac{K}{K_j}\right)$$

We know, $Q=V_s\times K$

After substituting $V_s$ in the above equation, we get

$$Q=V_f\times\left(1.0-\frac{K}{K_j}\right)\times K \tag{10.21}$$

The above model is termed Greenshield's linear model and the following parameters can be derived from it:

- When density is zero, flow will be zero because there will be no vehicles on the roadway.
- When density increases, flow will also increase, up to maximum flow. Further continual increase in density will cause flow to decrease to jam density.
- When density reaches a maximum value (i.e., jam density), the flow value reaches zero because the vehicles line up, as in end-to-end parking.

**Calibration of Greenshield's linear model:** In order to use Greenshield's linear model for any traffic stream of a road, the field representative boundary values of free flow speed ($V_f$) and jam density ($K_j$) must be determined by conducting traffic surveys. This process of determination of such field representative values is termed calibration of the model.

Approximate values of $V_f$ and the $K_j$ can be determined from a number of speed and density observations conducted on a study road section. The data is used to fit a linear regression equation between them, in the form of $v = A + B \times k$, where $v$ = speed and $k$ = density. By linear regression analysis, the coefficients $A$ and $B$ can be solved as

$$B=\frac{\sum_{i=1}^{n}\left(k_i-\bar{k}\right)\left(v_i-\bar{v}\right)}{\sum_{i=1}^{n}\left(k_i-\bar{k}\right)^2} \tag{10.22}$$

$$A=\left(\bar{v}-B\bar{k}\right) \tag{10.23}$$

where $v_i$ and $k_i$ are the samples, $n$ is the sample size, and $\bar{v}$ and $\bar{k}$ are the average values of $v_i$ and $k_i$, respectively.

*Note:* One of the limitations of Greenshield's model is that the critical density does not equal half the jam density.

**Example 10.1**

Determine the regression coefficients of Greenshield's linear model from the following sample field observations. Also determine the maximum flow and density values corresponding to a speed of 25 km/hr.

| Space mean speed, *v* (km/hr) | 8 | 18 | 42 | 56 |
|---|---|---|---|---|
| Density, *k* (veh/km) | 168 | 116 | 50 | 42 |

**Answer**: The given values are substituted and calculated, as shown below:

| | $k_i$ | $v_i$ | $(k_i - \bar{k})$ | $(v_i - \bar{v})$ | $(k_i - \bar{k}) \times (v_i - \bar{v})$ | $(k_i - \bar{k})^2$ |
|---|---|---|---|---|---|---|
| | 168 | 8 | 74 | −23 | −1702.0 | 5476.0 |
| | 116 | 18 | 22 | −13 | −286.0 | 484.0 |
| | 50 | 42 | −44 | 11 | −484.0 | 1936.0 |
| | 42 | 56 | −52 | 25 | −1300.0 | 2704.0 |
| Sum | 376 | 124 | | | −3772.0 | 10600.0 |
| Average | 94 | 31 | | | | |

$$B = \frac{\sum_{i=1}^{n}(k_i - \bar{k})(v_i - \bar{v})}{\sum_{i=1}^{n}(k_i - \bar{k})^2} = \frac{-3772.0}{10600.0} = -0.355$$

$$A = (\bar{v} - B\bar{k}) = (31 - (-0.355 \times 94)) = (31 + (0.355 \times 94)) = 64.37$$

The above coefficients are substituted in the equation $v = A + B \times k$

$$v = 64.37 - 0.355 \times k$$

The observed values of velocity and density are plotted and the calibrated Greenshield's linear model is represented in Fig. 10.10.

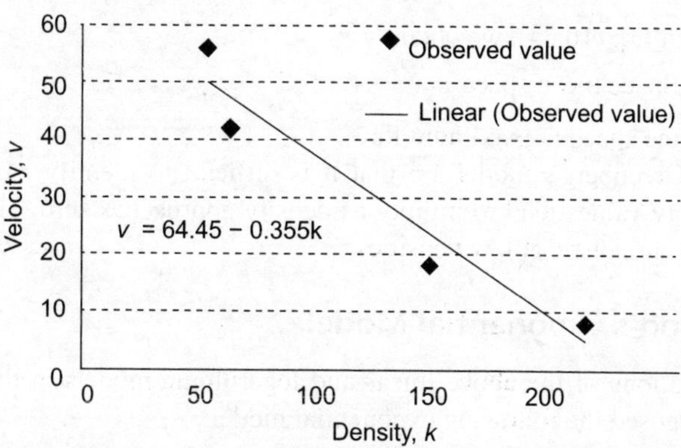


**Fig. 10.10** Calibrated Greenshield's linear model with observed values

Here, free flow velocity = $v_f$ = 64.37 km/hr and the ratio $\frac{v_f}{k_j} = 0.355$.

Hence, jam density $= k_j = 64.37/0.355 = 181.32$ veh/km

By Eq. (10.17),

$$\text{maximum flow} = Q_m = \frac{V_f \times K_j}{4} = (64.37 \times 181.32)/4$$

$$= 2917.89 \text{ veh/hr.}$$

Density corresponding to the speed 25 km/hr can be determined by substituting $v = 25$ in the equation,

$$v = 64.37 - 0.355 \times k;\ 25 = 64.37 - 0.355 \times k$$

Therefore, density at the speed of 25 km/hr $= k = (64.37 - 25)/0.355 = 110.90$ veh/km.

### 10.3.2 Greenberg's Logarithmic Model

Based on the fluid flow analogy concept, Greenberg (1959) developed a logarithmic relation between speed and density and proposed the equation given below:

$$\text{Space mean speed} = v = A\,ln\left(\frac{k_j}{k}\right) \tag{10.24}$$

As we know, $Q = vk$. Substituting $Q/k$ for $v$, we get

$$\frac{Q}{k} = A\,ln\left(\frac{k_j}{k}\right);\ Q = A\,ln\left(\frac{k_j}{k}\right)$$

Differentiating $Q$ with respect to $k$, we obtain

$$\frac{dQ}{dk} = A\left[k\left(\frac{-1}{k}\right) + \left(lnk_j - lnk\right)\right]$$

For maximum flow,

$$\frac{dQ}{dk} = 0$$

$$A\left[-1 + \left(lnk_j - lnk\right)\right] = 0;\ \left(lnk_j - lnk\right) = 1;\ ln\left(\frac{k_j}{k}\right) = 1$$

Substituting the above in Eq. (10.24), we obtain

$$\text{Space mean speed} = v = A$$

Thus, the value of $A$ is the speed at maximum flow.

The limitations of Greenberg's model are that it is difficult to identify jam density and optimal velocity, and the velocity values tend to infinity as density approaches zero. Therefore, the predicted values of speed will be unrealistic at low densities.

### 10.3.3 Underwood's Exponential Model

To overcome the limitations of the above linear and logarithmic models in the free flow condition, Underwood (1961) proposed the following exponential model:

$$v = v_f \times e^{\frac{-k}{k_c}}$$

where $v_f$ = free speed and $k_c$ = critical density corresponding to maximum traffic capacity. The calibration constants to be obtained from field observations are $v_f$ and $k_c$.

One of the main limitations of Underwood's exponential model is that speed becomes zero only when density reaches infinity. Hence, this model cannot be used for predicting speeds at high densities.

### 10.3.4 Multi-regime Models

The above models of speed–density are often referred to as single-regime models as they are valid for the entire range of densities observed in a study traffic stream. As human driving behaviour changes in a low density to high density condition, this single-regime condition doed not hold good. Hence, separate speed–density relations must be determined for different regimes of density (i.e., zones along the *x* axis representing the densities).

For example, Edie (1961) proposed a two-regime model (Fig. 10.11) and used an exponential model in the free flow regime and a logarithmic model in the forced flow regime. Other researchers have proposed a three-regime model, in which the middle potion of the speed–density relationship is taken to be another type of model (Fig. 10.11).

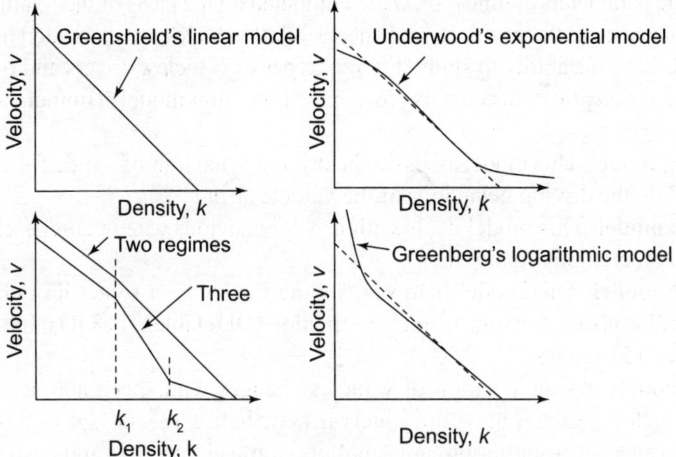


**Fig. 10.11** Typical trend lines of speed–density models

A few of the other widely used speed–density models are listed in Table 10.1.

**Table 10.1** A few widely used speed–density models (from Ardekani et al., 2011)

| *Model* | *Equation* | *Calibration parameters* | *Remarks* |
|---|---|---|---|
| Modified Greenberg | $v = v_c ln\left(\frac{k_j + k_o}{k + k_o}\right)$ | $v_c$, $k_j$ and $k_o$ | $k_o$ is the average minimum density value, which represents a non-zero value |
| Polynomial | $v = A + Bk + Ck^2$ | $A$, $B$ and $C$ | Regression coefficients |
| Quadratic | $v = v_f\left(1 - \frac{k^2}{k_j^2}\right)$ | $v_f$, and $k_j$ | Same as described previously |
| Drake | $v = v_f e^{\left[\frac{-1}{2}\left(\frac{k}{k_c}\right)^2\right]}$ | $v_f$, and $k_c$ | |

**Advantages and disadvantages of macroscopic models:** The main advantage of macroscopic models is that they need fewer parameters than their microscopic counterparts. The macroscopic model uses density, velocity and flow. The main disadvantage of this model is its inability to account for small details or dynamics relating to driving behaviour in a traffic stream and with roadway geometric components. The effect of these parameters can be modelled using microscopic models.

## 10.4 MICROSCOPIC TRAFFIC STREAM MODELS

Microscopic traffic models do not use aggregate variables such as average speed, density or flow rate. They deal with the interactions between/among individual drivers, vehicles and the roadway infrastructure.

Microscopic traffic models describe the interactions between the various vehicles in uninterrupted traffic flow conditions. As it is impossible to precisely predict driver behaviour in a traffic stream under geometric conditions, computer-based simulation software packages are used for this purpose. Driver and vehicle characteristic parameters at time $t + \Delta t$ are computed on the basis of their values at instantaneous time $t$. Based on this concept, the positions and speeds of all vehicles are computed at an interval of $\Delta t$. Microscopic models have the ability to study different types of vehicles, drivers and roadway conditions.

The main component features of most microscopic simulation models (Immers and Logghe, 2003) are as follows:

- **Car-following model**: This model assesses the driving behaviour of a specific vehicle that follows another based on the driving behaviour of the vehicle ahead.
- **Lane change model**: This model deals with driver behaviour when vehicles change lanes based on the vehicles travelling around them.
- **Route choice model**: This model involves finding the shortest route through an infrastructure network on the basis of a dynamic origin–destination (OD) table. The OD table is instantaneously specified every 15 minutes.
- **Additional models**: As the position of vehicles changes with speed and acceleration and each individual vehicle is known at a time interval (say, half a second), it is possible to calculate derived effects such as air pollution, noise pollution, travel time loss and costs.

*Note:* In addition to prediction of driver behaviour and associated vehicle characteristics, it is possible to model a number of varying parameters relating to roadway infrastructure components such as curves, traffic lights, signs, weather and accidents.

### 10.4.1 The Car-Following Theory

This model describes the acceleration or deceleration of a vehicle following another vehicle, in uninterrupted traffic flow conditions. It describes how the driver of the vehicle that follows reacts to changes in the position of the vehicle in front of him.

The car-following theories were developed during the 1950s and 1960s. Reuschel and Pipes were the pioneers in the early 1950s. Parallel efforts were undertaken in the late 1950s and mid-1960s. Kometani and Sasaki in Japan, Forbes at the Institute for Research and Michigan State University and a group of researchers associated with General Motors made significant contributions to the car-following theory. Later, a series of car-following theories were proposed by Adolph May, one of which was published in 1990 with the title *Traffic Flow Theory*. The basic form of these theories is as follows (Koppula, 2002):

$$\text{Response} = f\{\text{Sensitivity, Stimuli}\} \tag{10.25}$$

where response = acceleration or deceleration of the vehicle under consideration, which depends on the sensitivity of the automobile and the driver himself; stimuli = visual and auditory inputs that influence the driver's decision; and sensitivity = ability of the driver to perceive and react to the stimuli.

Most of the car-following models use the equation given below, which defines the interaction between the following vehicle (FV) and the leading vehicle (LV).

$$a(t+T)\propto\frac{v(t)^{\mathrm{m}}}{\left[\Delta x(t)\right]^{\ell}}\Delta v(t) \tag{10.26}$$

where $a$ = acceleration of the FV; $v$ = velocity of the FV; $t$ = time $t$; $\Delta \mathrm{x}(t)$ = distance between the two vehicles at time $t$; $\Delta \mathrm{v}(t)$ = velocity difference between the two vehicles at time $t$; $T$ = delay time between stimulus and response time; and m and $\ell$ = constants.

This formula assumes that the acceleration or deceleration of the FV is proportional to the speed difference with the LV. When both vehicles travel at the same speed, acceleration will be zero. It is assumed that FV acceleration is inversely proportional to the square of the distance to LV. For this reason, it is said that the LV influence increases as the distance between them decreases.

This concept of the car-following theory is described as follows (Koppula, 2002): Let us consider two vehicles travelling from the left to the right direction (Fig. 10.12). Let one vehicle under consideration be denoted by $n + 1$; i.e., the following vehicle (FV), and the vehicle travelling in front of it be denoted by $n$; i.e., the lead vehicle (LV). The lengths of FV and LV are denoted as $Ln + 1$ and $Ln$, respectively. Assume that they are moving at the same time with the gap between them at any time being $t$. As the travelling vehicles change in distance positions, speeds and acceleration rates (or deceleration) over time, the subscript $t$ is used to specify the time.

The acceleration rate of the FV is specified as occurring at time $t + \Delta t$. $\Delta t$ represents an interval of time between the time a unique car following situation occurs ($t$) and the time the driver of the FV decides to apply a specified acceleration rate (or deceleration rate) at ($t + \Delta t$). This interval of time is often referred to as the reaction time. A few details are given below:

The relative velocity of LV and FV = $\dot{x}_n(t) - \dot{x}_{n+1}(t)$

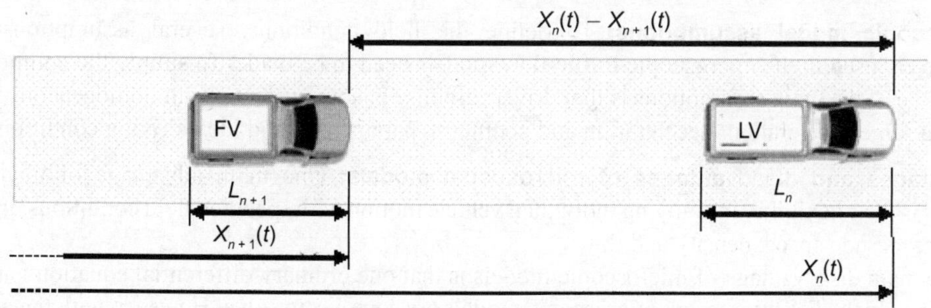


**Fig. 10.12** Car-following theory

*Conditions*:

- The relative velocity will be positive if the speed of LV is greater than FV.
- The relative velocity will be negative if the speed of LV is less than FV.
- The relative velocity will be zero if the speed of LV is equal to FV.

The acceleration rate (or deceleration rate) = $\ddot{x}_{n+1}(t + \Delta t)$

where $n$ = the LV; $n + 1$ = the FV; $\dot{x}_n(t)$ = velocity $n$ at time $t$; $\dot{x}_{n+1}(t)$ = velocity $n + 1$ at time $t$; $\ddot{x}_n(t)$ = acceleration of vehicle $n$ at time $t$; and $\ddot{x}_{n+1}(t)$ = acceleration of vehicle $n + 1$ at time $t$.

This theory has been found to be accurate for single-lane situations where, in reality, every driver reacts only to the vehicle in front of him (Koppula, 2002).

### 10.4.2 GM Model

The General Motors (GM) group and associated researchers conducted field experiments on the GM test track to quantify the parameters relating to the car-following theory. The research team developed five generations of car-following models, which are all in the form of Eq. (10.25). The response was always represented either by acceleration or by deceleration of the FV, while the stimuli was represented by the relative velocity of LV and FV. The final form of the GM model was a generalised approach and the equation is given below:

$$\ddot{x}_{n+1}(t+\Delta t)=\frac{\alpha_{\ell,m}\left[\dot{x}_{n+1}(t+\Delta t)\right]^{m}}{\left[x_n(t)-x_{n+1}(t)\right]^{\ell}}\left[\dot{x}_n(t)-\dot{x}_{n+1}(t)\right] \tag{10.27}$$

where $n$ = the LV; $n + 1$ = the FV; $x_n(t)$ = the position of vehicle $n$ at time $t$; $x_{n+1}(t)$ = the position of vehicle $n + 1$ at time $t$; $\dot{x}_n(t)$ = velocity $n$ at time $t$; $\dot{x}_{n+1}(t)$ = velocity $n + 1$ at time $t$; $\ddot{x}_n(t)$ = acceleration of vehicle $n$ at time $t$; $\ddot{x}_{n+1}(t)$ = acceleration of vehicle $n + 1$ at time $t$; $m$ = speed exponent, with the range of −2 to 2; $\ell$ = distance exponent, with the range of −1 to 4; and $\alpha_{\ell,m}$ = sensitivity coefficient. This factor models the intensity of the reaction of a driver on the basis of changes in the behaviour of the vehicle moving in front.

As proposed by various researchers, changing the values of $m$ and $l$, and their different combinations, can produce different shapes of various macroscopic models, which may be calibrated for the specific traffic flow conditions. For this reason, the GM model can be used as a generalised car-following model so that the calibrated parameters produce results in agreement with the field observations.

In addition to these models, the readers may learn more about Pipe's model, Forb's theory, Gipp's model, the LWR model, the intelligent driver model and others (TRB, 2015).

**Microscopic model assumptions:** To define the field conditions, several assumptions of the parameters used in the microscopic traffic flow models need to be made. To simply the computational work, one of the main assumptions is that driver response is homogeneous with homogeneous vehicles (having same acceleration, deceleration and geometric parameters) and ideal driving conditions.

**Advantages and disadvantages of microscopic models:** One main advantage of microscopic models is their flexibility in studying individual vehicle motion with specified field conditions, including parameters such speed, density and flow.

The main disadvantage of microscopic models is that one ordinary differential equation is required for each vehicle. For this reason, microscopic models are complex to solve. However, with the advent of high-speed computers, it has been made simpler and easy. Towards this necessity, numerous computer packages have been made available in the market to simulate user-defined driver behaviour with a variety of boundary conditions (Scott, 2011).

## 10.5 SHOCK WAVES IN TRAFFIC STREAMS

The phenomenon of shock wave formation in a traffic stream is described below.

Let us assume that a platoon of uniform vehicles is travelling at uniform velocity with uniform space headway on a road. In this traffic flow condition, a truck with a wide body slows down suddenly and compels all the vehicles travelling behind it to decelerate and follow for a certain distance, at that slow speed. As soon as the truck exits or speeds up, the first vehicle following it and all the subsequent vehicles will drive at safe acceleration, try to keep a safe headway, and finally return to their normal speed. This effect of drop in speed in the traffic stream is termed the shock wave.

Shock waves occur when there is a change in the travel condition along the roadway that affects the normal flow of the stream. The situation of sudden change in the speed of a traffic stream may occur due to accidents, reduction in the number of lanes, restricted width of bridge or culvert, signalised intersections, work zones and other such bottleneck conditions.

*Note*: A bottleneck is a fixed location along a road where the capacity is lower than the upstream capacity. Bottlenecks affect traffic flow capacity and can potentially cause flow break down (Austroads, 2008).

### 10.5.1 Shock Wave Speed

The speed of a traffic stream during a shock wave is represented by the slope of the chord connecting the two stream conditions that define the shock wave on a Q–K diagram (see points 1 and 2, in Fig. 10.13).

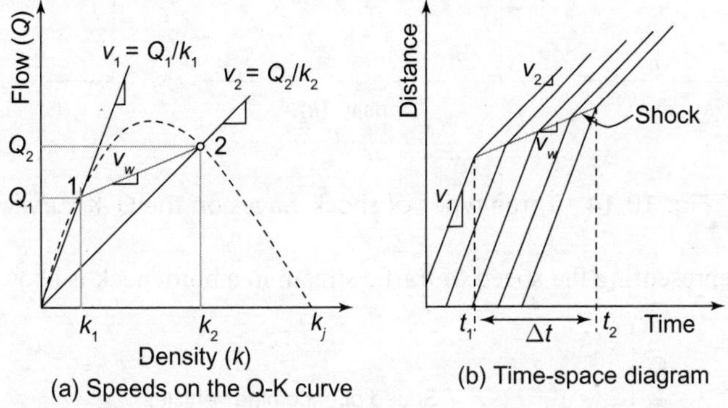


**Fig. 10.13** Illustration of shock wave speed on a flow and density curve

Let us label the first stream condition as '1' and the second stream condition as '2'. The space mean speed, concentration and flow values at Condition-1 = $v_1$, $k_1$ and $Q_1$. The space mean speed, concentration and flow values at Condition-2 = $v_2$, $k_2$ and $Q_2$. Assume, the resulting speed of shock wave = $v_w$. The relative velocity of shock wave vehicles moving at Condition-1 ($v_{r1}$) = $v_1 - v_w$. The relative velocity of shock wave vehicles moving at Condition-2 ($v_{r2}$) = $v_2 - v_w$. The number of vehicles moving at the juncture of Condition-1 in the time duration of $\Delta t = N_1$.

$$N_1 = v_{r1} \times \Delta t \times k_1 = (v_1 - v_w) \times \Delta t \times k_1$$

The number of vehicles moving at the juncture of Condition-2 in the time duration of $\Delta t = N_2$.

$$N_2 = v_{r2} \times \Delta t \times k_2 = (v_2 - v_w) \times \Delta t \times k_2$$

Based on the law of conservation of the traffic stream, let us assume

$$N_1 = N_2$$

$$(v_1 - v_w) \times \Delta t \times k_1 = (v_2 - v_w) \times \Delta t \times k_2$$

$$v_1 \times k_1 - v_w \times k_1 = v_2 \times k_2 - v_w \times k_2$$

$$v_w \times k_2 - v_w \times k_1 = v_2 \times k_2 - v_1 \times k_1$$

$$v_w (k_2 - k_1) = v_2 \times k_2 - v_1 \times k_1$$

$$v_w (k_2 - k_1) = Q_2 - Q_1 \quad \text{(We know that, } Q = vk\text{)}$$

$$v_w = \frac{Q_2 - Q_1}{k_2 - k_1} \text{(km/hr)} \tag{10.28}$$

If the shock wave speed ($v_w$) is positive, it indicates that the shock wave is moving in the same direction as the flow (Fig. 10.14 (a)); if $v_w$ is negative, it indicates that the shock wave is moving in the backward direction (Fig. 10.14 (c)); and if its value is zero, it indicates that the shock wave is stationary (Fig. 10.14 (b)).

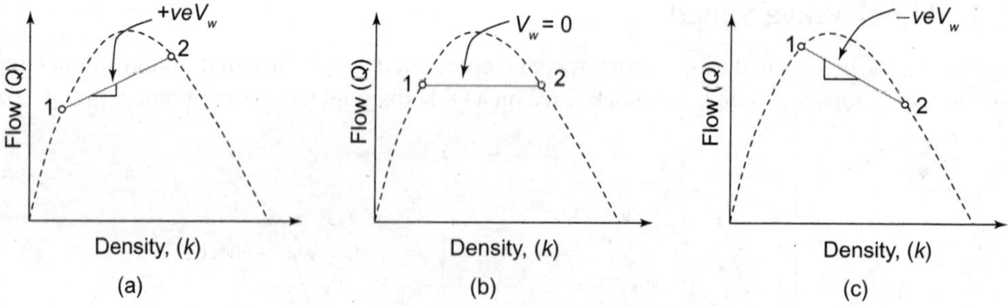


**Fig. 10.14** Three types of shock waves on the Q–K curve

The Q–K curve representing the speed of traffic stream in a bottleneck is shown below:

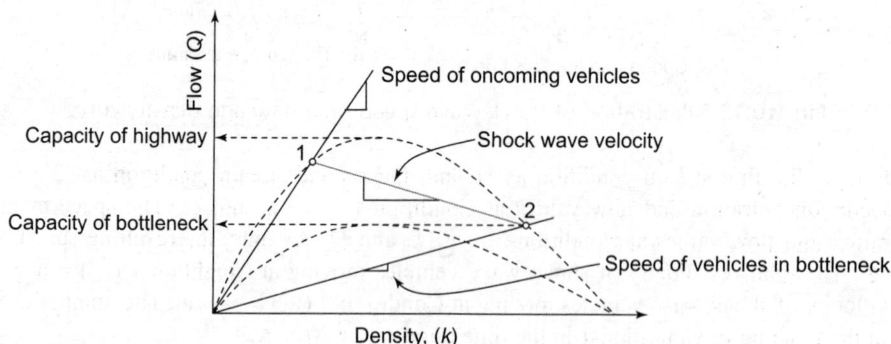


**Fig. 10.15** Capacity of a bottleneck with a shock wave on the Q–K curve

## Example 10.2

It was observed that a traffic stream instantaneously changed from Condition 2 to Condition 1. Take Condition 2 as upstream of the traffic stream, moving at a speed of 40 km/hr with a volume of 1100 veh/hr; Condition 1 is the rear side of the traffic, moving at a speed of 20 km/hr with a volume of 1800 veh/hr. Determine the speed and direction of the resulting shock wave.

**Answer**: $k_1 = (1800\text{ veh/hr}) / (20\text{ km/hr}) = 90\text{ veh/km}$

$k_2 = (1100\text{ veh/hr}) / (40\text{km/hr}) = 27.5\text{ veh/km}$

$$\text{Shock wave speed } = v_w = \frac{Q_2 - Q_1}{k_2 - k_1} = \frac{1100 - 1800}{27.5 - 90}$$

$$= \frac{-700}{-62.5} = 11.2\text{ km/hr}$$

The +ve sign indicates that the shockwave is moving in the forward direction.

**Example 10.3**

A traffic stream was found to be approaching a signalised intersection with an approach speed of 25 km/hr at a density of 50 veh/km. Take the value of jam density as 150 veh/km. Find the speed and direction of the shock wave.

**Answer**: Substitute the given as, $k = 50$ veh/km; $k_j = 150$ veh/km; $v = 25$ km/hr.
Free from speed = $Vf$ = ?

We know that

$$V = V_f \times \left(1.0 - \frac{K}{K_j}\right)$$

Substituting the above values, we get

$$25 = V_f \times \left(1.0 - \frac{50}{150}\right). \; V_f = 37.5\text{ km/hr}$$

We know that

$$Q = VK = (25 \times 50) = 1250\text{ veh/hr}$$

$$\text{Shock wave speed} = v_w = \frac{Q_2 - Q_1}{k_2 - k_1} = \frac{0 - 1250}{150 - 50} = \frac{-1250}{100}$$

$$= -12.5\text{ km/hr}$$

The −ve sign indicates that the shockwave was moving in the reverse direction.

**Example 10.4**

A 20-km/hr office street zone stays critical at 10:00–10:15 am and causes a slow down of traffic stream flow from conditions 1 to 2. Subsequently, the traffic stream is released to traffic flow as Condition 3. The field-observed values along the street are given below:
Approaching zone: Traffic speed = 42 km/hr and flowing at the rate of 900 veh/hr
Middle zone: Traffic speed = 20 km/hr and flowing at the rate of 1050 veh/hr
Downstream zone: Traffic speed = 32 km/hr and flowing at the rate of 1160 veh/hr
Determine the resulting (i) speeds and directions of the two shock waves, (ii) the length of the platoon being created; and (iii) the time required to dissipate it.

**Answer**: The traffic stream flowing conditions 1, 2 amd 3 can be represented as shown in Fig. 10.16.

$$k_1 = \frac{Q_1}{v_1} = \frac{900}{42} = 21.42\text{ veh/km}$$

$$k_2 = \frac{Q_2}{v_2} = \frac{1050}{20} = 52.5 \text{ veh/km}$$

$$k_3 = \frac{Q_3}{v_3} = \frac{1160}{32} = 36.25 \text{ veh/km}$$

$$\text{Shock wave}_{1\to 2} \text{ speed} = v_{w1-2} = \frac{1050-900}{52.5-21.42}$$

$$= \frac{150}{31.08} = +4.82 \text{ km/hr}$$

$$\text{Shock wave}_{2\to 3} \text{ speed} = v_{w2-3} = \frac{1160-1050}{36.25-52.5}$$

$$= \frac{110}{-16.25} = -6.76 \text{ km/hr}$$

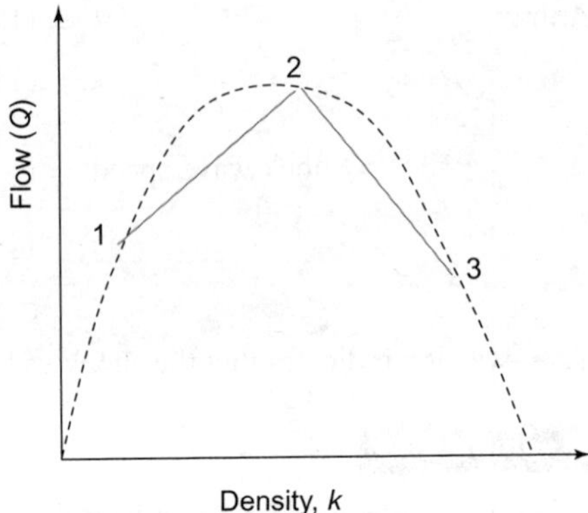


**Fig. 10.16** Shock waves on the Q–K curve

$$\text{Growth of platoon} = 20 - 4.82 = 15.18 \text{ km/hr}$$

Length of the platoon by 10:15 am (i.e., at the end of the 15-minute period)

$$L = 15.18 \text{ km/hr} \times 1/4^{\text{th}} \text{ hour} = 3.795 \text{ km}$$

Number of vehicles involved, $N = 3.795 \times 52.5 \text{ veh/km} = 199.2375$ vehicles

Relative speed of the two shock waves = Front − Rear

Relative speed of the two shock waves ($v_{wr}$) = −6.76 − (+4.82) = −11.58 km/hr

The time required to disperse = $L/v_{wr}$ = 3.795/11.58 = 0.32772 hr = 19.66 minutes.

*Note:* The queue will not dissipate if the relative speed of the two shock waves ≥ 0.

## 10.6 LEVEL OF SERVICE

Level of service (LOS) is a qualitative measure of traffic flow conditions which describes the operational quality within a traffic stream, as experienced by drivers/passengers. It represents the quality of service of a roadway in the prevailing traffic and roadway geometric conditions. The qualitative performance condition is measured on an A–F scale (Fig. 10.17), with LOS-A representing the *best* operating conditions and LOS-F representing the *worst* operating conditions, from the traveller's perspective (TRB, 2000).

LOS is commonly determined using travel projections twenty to thirty years in the future. The general operating conditions for the six levels of service are defined in Table 10.2.

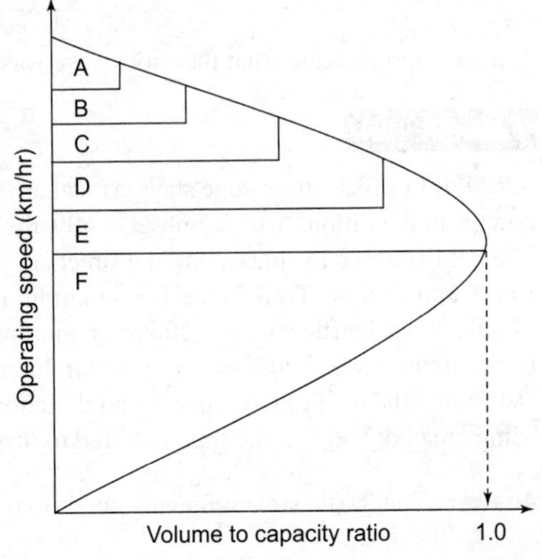


**Fig. 10.17** Level of Service A to F

**Table 10.2** General operating conditions for levels A–F

| *LOS* | *Volume-to-capacity ratio limit* | *Condition* | *Remarks* | *Typical illustration* |
|---|---|---|---|---|
| A | 0.35 | Free flow: Average speed = 90% of free flow speed | With low volume and density, and high speed. Speed is controlled as driver desires. No delay. | |
| B | 0.54 | Stable flow: Average speed = 70% of free flow speed; flow <55% of capacity | Speeds beginning to be restricted due to traffic conditions | |
| C | 0.77 | Stable flow: Average speed = 50% of free flow speed; flow <75% of capacity | Speeds are restricted | |
| D | 0.93 | High density: Average speed = 40% of free flow speed; flow up to 89% of capacity | Drivers have little freedom to manoeuvre | |
| E | 1.0 | Unstable flow: Average speed = 33% of free flow speed; flow at 100% of capacity | With short stoppages | |
| F | >1.0 | Breakdown: Forced flow: Average speed at 25–33% of free flow speed | Congested flow | |

The LOS of a roadway along a selected length measures the effect of the following factors (IRC: 106-1990):

- Speed and travel time
- Road user freedom to manoeuvre
- Traffic interruptions, delays and hindrances
- Driving comfort and convenience
- Safety and operating costs

The following factors affect the LOS on urban roads (IRC: 106-1990):

- Heterogeneity of traffic: The proportion of slow-moving vehicles and fast-moving vehicles, and the size and tractive energy of vehicles. Traffic service quality or traffic flow on the road may vary with time and day.
- Speed regulation/control: One-way, two-way, signalisation, un-signalised intersections
- Frequency of intersections and spacing
- Presence of on-street bus stoppages
- On-street parking
- Intensity of roadside commercial activities and land use
- Speed breaker

- Pedestrian volume
- Roadway geometrics: Number of lanes, lane width, sight distance, curvature
- Access road alignment (suburban and rural areas)
- Weaving length, weaving angle and space headway

To determine the performance of a road, LOS is an important factor. The following six factors are used to determine LOS and such studies will be helpful in the future development of a road facility:

- Volume-to-capacity ratio
- Density
- Speed
- Travel time
- Delay
- Control delay (sec/veh)

The Highway Capacity Manual (HCM) recommends the LOS of intersections based on measured delays (Table 10.3).

**Table 10.3** HCM-recommended criteria for LOS of an intersection

| *LOS* | *Signalised intersection* | *Un-signalised intersection* |
|---|---|---|
| | *Average control delay (sec/veh)* | *Delay (sec/veh)* |
| A | ≤10 | ≤10 |
| B | 10–20 | 10–15 |
| C | 20–35 | 15–25 |
| D | 35–55 | 25–35 |
| E | 55–80 | 35–50 |
| F | >80 | >50 |

## 10.7 EQUIVALENCY FACTORS

The *interaction* and *influence* of slow-moving vehicles on fast-moving vehicles and the *effect* of commercial vehicles (i.e., larger dimensions) on other types of vehicular traffic (in a heterogeneous mix) is one of the important considerations in a study of the performance characteristics of non-uniformity in the traffic stream. Due to the heterogeneous traffic mix, one class of vehicles cannot be considered equal to any other class. Interactions among various classes of vehicles travelling in a stream are accounted for in prevailing roadway and traffic conditions using a common factor, termed passenger car unit (PCU). PCU is a multiplication factor (Table 10.4) used to convert different types of vehicles into the *equivalent* passenger car units, based on their relative interference.

PCU, also called passenger car equivalent (PCE), was initially introduced in the 1965 Highway Capacity Manual to account for the effect of heavy vehicles in the traffic stream (HRB, 1965). Of several methods (Walker's, headway, multiple linear regression, simulation and density), the most suitable one for non-uniform mixed traffic flow conditions in India is the one proposed by Chandra et al. (1995). According to this method, the PCU value is directly proportional to the speed ratio and inversely proportional to the space occupancy ratio with respect to a standard design vehicle, i.e., car. The equation used for estimating the PCU values for a study area is given below (Chandra and Kumar, 2003).

$$PCU_i = \frac{\left(\text{Speed ratio of the car to the } i^{\text{th}} \text{ vehicle}\right)}{\left(\text{Space ratio of the car to the } i^{\text{th}} \text{ vehicle}\right)}$$

$$= \frac{\left(\dfrac{\text{Velocity of car}}{\text{Velocity of the } i^{\text{th}} \text{ vehicle}}\right)}{\left(\dfrac{\text{Space occupied by the car}}{\text{Space occupied by the } i^{\text{th}} \text{ vehicle}}\right)} \tag{10.29}$$

### 10.7.1 Factors Affecting Equivalency Factors (PCU Values)

The PCU conversion factors are subject to *variation,* which depends on the following factors:

- **Width of the pavement:** It was observed that the mean free speed of cars increases linearly with the width of the carriageway over a certain range of width – 5.2–11.3 m (Leong, 1968; Farouki and Nixon, 1976). As the freedom of movement increases with the width, drivers will have more opportunities to increase their speed on the roads. This results in higher PCU values for buses and trucks.
- **Condition of the shoulder:** The PCU value of a vehicle on a road increases with increase in the quality of the shoulder. A better shoulder increases the effective width of the carriageway and results in higher PCU values. This is because of the higher speed differential between cars and heavy vehicles (Sachdeva, 2004).
- **Surface roughness/unevenness characteristics:** It is widely known that the free flow speed of vehicle decreases with increased roughness of the road surface. As speed decreases, the PCU value also decreases.
- **Directional split of traffic:** In two-lane roads, the PCU value decreases as the directional split of traffic deviates from the 50/50 proportion. If the traffic is not balanced in both directions, overtaking opportunities will reduce and the vehicles in the stream will be forced to travel at low speed. Hence, the PCU values for heavy vehicles will be reduced (Sachdeva, 2004).
- **Composition of traffic:** Traffic stream characteristics such as the proportion of slow- and fast-moving vehicles in a heterogeneous traffic mix is an important parameter that influences the performance of a traffic facility. It is well known that capacity decreases with increase in the proportion of slow-moving vehicles in the traffic stream. With large differences in speed between heavy vehicles and cars, the PCU value for heavy vehicles will increase with increase in the proportion of slow-moving vehicles (Sachdeva, 2004).
- **Road geometrics:** Geometrics such as terrain, gradient, curvature and sight distance influence the operating speeds at different locations in an area, and hence, there will be variations in the PCU values.
- **Vehicular characteristics**: Vehicle size and speeding capabilities contribute to change in the value of the PCU.
- **Traffic regulation and control**: As with the prevailing operating speed.
- **Gap/Spacing between vehicles**: The ratio between the time headway of a candidate vehicle and the car used to calculate the PCU of the candidate vehicle is the basic headway method. The following equation is used to estimate the PCU of vehicles (Greenshields et al., 1947):

$$PCU_i = \frac{\text{Average time headway of vehicle type } i \text{ (sec)}}{\text{Average time headway of passenger car (sec)}} \tag{10.30}$$

Several road user cost studies have been conducted in India, and separate sets of PCU values have been recommended for use in India *for specific traffic conditions*:

- Roads in rural areas (IRC: 64-1990)
- Urban roads (IRC: 106-1990)

**Table 10.4** Graphical presentation of PCU factors applicable on rural roads (not to be applied to road intersections), a typical example (Modified from IRC: 64-1990)

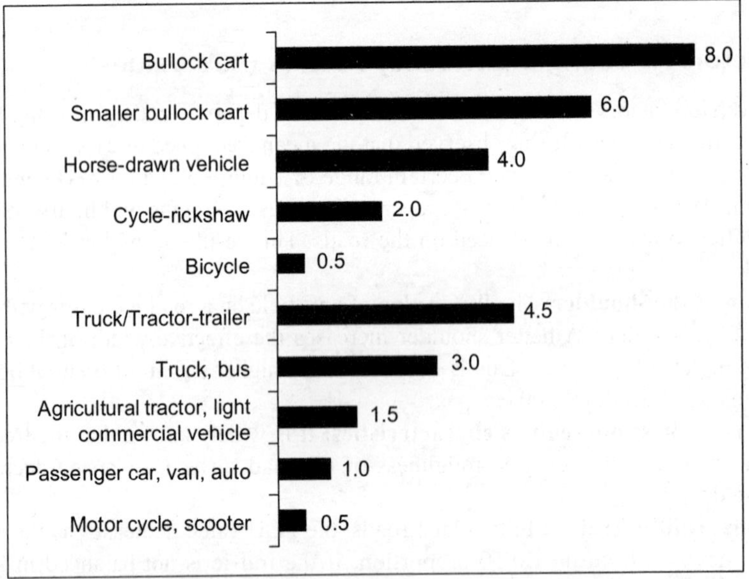


**Table 10.5** Graphical presentation of PCU factors applicable on urban roads in plains, a typical example (Modified from IRC: 106-1990)

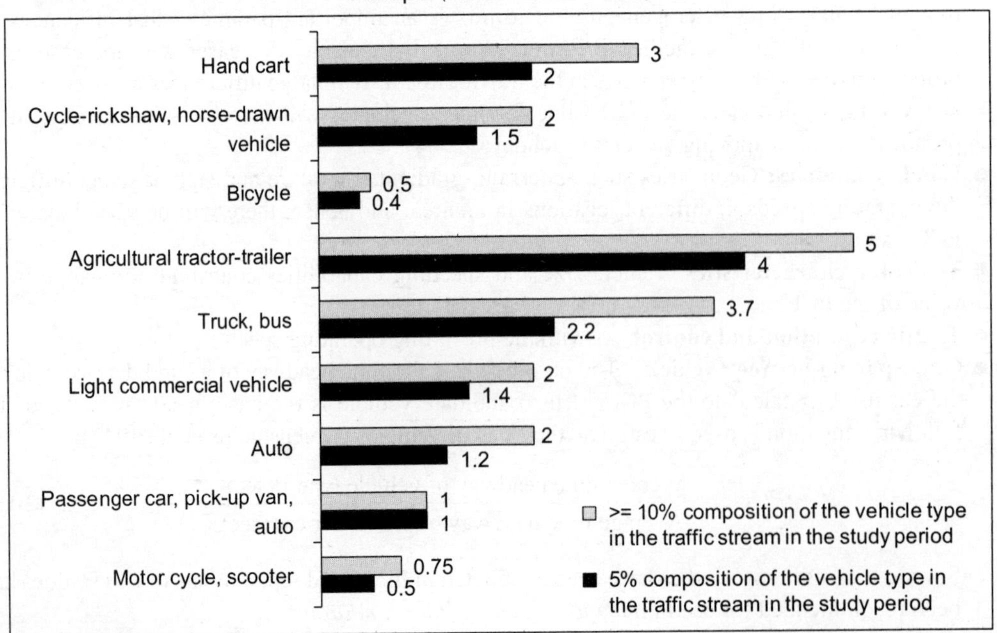

## 10.8 DESIGN SERVICE VOLUME

Design service volume is the maximum hourly traffic volume that is expected to traverse a point or section of a roadway or lane in a given period of time under prevailing geometric, control and traffic flow conditions while maintaining a designated level of service (IRC: 64-1990). As a common practice, it is not advisable to design the width of road pavement for a traffic volume equal to its capacity at LOS E. At this condition, the speeds will be extremely low and restricted. As the volume increases, with design volume close to LOS C or D, there may be a chance of a lower level of LOS. Hence, as a balanced design concept, LOS B and LOS C are assumed as stable flow zones for the design of roads in rural and urban areas, respectively. At these LOS levels, the volume of traffic will be around 0.5 times and 0.7 times the maximum capacity adopted for the design of rural and urban roads, respectively.

The standard recommended service volumes for different types of roads in rural (single lane, intermediate lane) and urban areas (arterial, sub-arterial and collector streets) are available in IRC: 64 and IRC: 106, respectively. The recommended design service volumes are shown in Tables 10.6, 10.7 and 10.8.

**Table 10.6** Design service volumes (PCU/day) for a single lane (3.75 m carriageway with minimum shoulder width of 1.0 m on either side), a typical example (Modified from IRC: 64-1990)

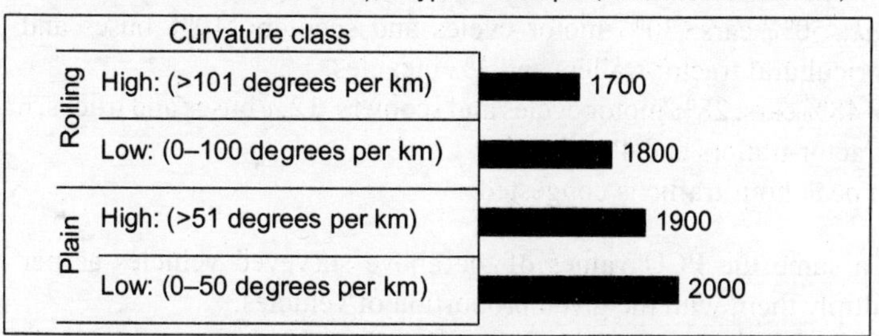


* The capacity values relate to peak hour traffic in the range of 8–10% at LOS-B. Appropriate reduction factors are multiplied with these values if the carriage width is less than the specified. Adopt a value of 35,000 PCU/day for a four-lane divided carriageway with good shoulder on either side and with a minimum 3-m-wide central verge in plain terrain (IRC: 64-1990).

**Table 10.7** Design service volumes (PCU/day) for two lanes (7.0 m carriageway with minimum shoulder width of 1.5 m on either side), a typical example (Modified from IRC: 64-1990)

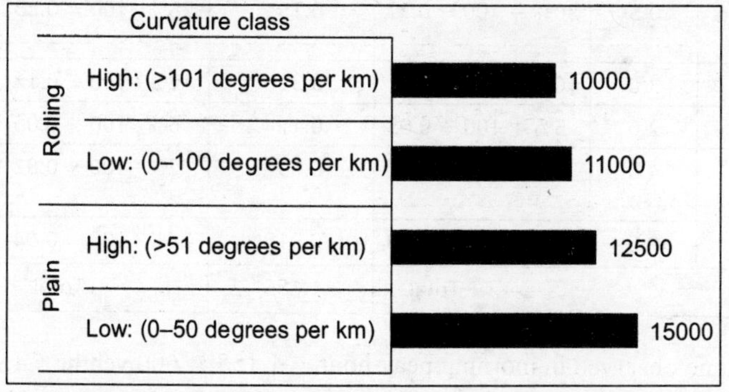


* The design hour volumes may be reduced by about 20–30% when the pavement is not black-topped

**Table 10.8** Recommended total design service volumes (PCU/hr)* for roads in urban areas, a typical example (Modified from IRC: 106-1990)

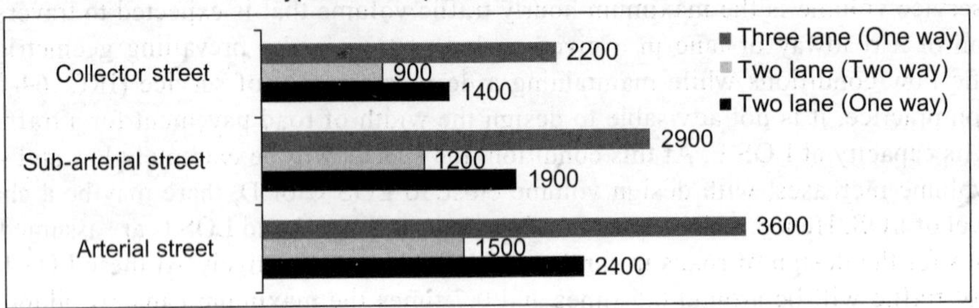


* These values are subject to roadside fringe conditions, as specified by IRC: 106-1990.

## 10.9 SOLVED PROBLEMS

**Example 10.5**

The traffic volume on a section of a rural road in morning peak hour is 600 veh/hr and in the evening peak hour is 600 veh/hr. The details of traffic composition are given below:

*For morning peak*: 50% cars, 30% motor cycles and scooters, 10% buses and trucks, 5% cycle-rickshaws, 2% agricultural tractor-trailers and 3% bicycles.

*For evening peak*: 48% cars, 28% motor cycles and scooters, 12% buses and trucks, 6% cycle-rickshaws, 2% agricultural tractor-trailers and 4% bicycles.

Determine which peak hour traffic is congested.

**Answer**: Let us assume the PCU values of the above surveyed vehicles as per the standard IRC: 64-1990, and multiply them with the given proportion of vehicles.

| *Vehicle type* | *PCU factor* | *Morning peak* | | *Evening peak* | |
|---|---|---|---|---|---|
| | | *Proportion (decimal value)* | *PCU product* | *Proportion (decimal value)* | *PCU product* |
| [1] | [2] | [3] | [4] = [2]×[3] | [5] | [6] = [2] × [5] |
| Cars | 1.0 | 50% ÷ 100 = 0.5 | 0.5 | 48%÷100 = 0.48 | 0.48 |
| Motor cycles and scooters | 0.5 | 30% ÷ 100 = 0.3 | 0.15 | 28%÷100 = 0.28 | 0.14 |
| Buses and trucks | 3.0 | 10% ÷ 100 = 0.1 | 0.3 | 12%÷100 = 0.12 | 0.36 |
| Cycle-rickshaws | 2.0 | 5% ÷ 100 = 0.05 | 0.1 | 6%÷100 = 0.06 | 0.12 |
| Agricultural tractor-trailers | 4.5 | 2% ÷ 100 = 0.02 | 0.09 | 2%÷100 = 0.02 | 0.09 |
| Bicycles | 0.5 | 3% ÷ 100 = 0.03 | 0.015 | 4%÷100 = 0.04 | 0.02 |
| | | Total = | 1.155 | Total = | 1.21 |

$$\text{Traffic volume observed in morning peak hour} = 1.155 \times 600 \text{ veh/hr} = 693 \text{ PCU/hr}$$

$$\text{Traffic volume observed in evening peak hour} = 1.21 \times 600 \text{ veh/hr} = 726 \text{ PCU/hr}$$

This result shows that evening peak hour traffic is more congested than morning peak hour traffic.

**Example 10.6**

Calculate the flow of traffic if the average velocity of vehicles on a lane is 42 km/hr and the stopping sight distance (SSD) is 22 m. Assume the length of rigid wheel base is 6.0 m.

**Answer:** $S_s = \text{Length of vehicle} + SSD = 6.0 + 22 = 28\,\text{m}$

Practical capacity can be calculated using the following equation

$$C = \frac{V}{S_s} \times 1000 = \frac{42}{28} \times 1000 = 1500 \text{ veh/hr/lane}$$

**Example 10.7**

A study was conducted on a selected road section and the one-way traffic stream speed was found to be 42 km/hr. The average length of a vehicle is 6.0 m. Determine (i) theoretical capacity and (ii) practical capacity of the study road in terms of veh/hr/lane.

Assume that reaction time ($t_p$) of the driver is 0.7 seconds for determining the theoretical capacity; average length of the vehicle ($a_1$=) 6 m, reaction time of the driver ($a_2$) = 2.5 seconds and the coefficient of longitudinal friction ($f$) = 0.4.

**Answer:** (i) *Theoretical capacity*: The following equation is used to determine the space headway.

$$\text{Space headway} = \text{Length of rear vehicle} + \text{Space gap between vehicles}$$

$$S_s = 6\,\text{m} + \left(\frac{V}{3.6}\right) \times t_p = \left(6 + \left(\frac{42}{3.6}\right) \times 0.7\right)$$

$$= (6 + (11.67) \times 0.7) = (6 + 8.167) = 14.16\,\text{m}$$

$$\text{Theoretical capcity} = C = \left(\frac{V}{S_s} \times 1000\right) = \left(\frac{42}{14.16} \times 1000\right)$$

$$= 2964.7 \text{ veh/hr/lane}$$

(ii) *Practical capacity*: The following equation is used to determine the space headway.

$$S_s = \text{Length of vehicle} + (\text{Lag distance}) + (\text{Breaking distance})$$

Rewriting the above relationship, we get

$$S_s = a_1 + (a_2 \times V) + \left(\frac{V^2}{2gf}\right)$$

where $a_1$= length of vehicle = 6.0 m (given) and $a_2$ = reaction time of the driver = 2.5 sec (given).

$$\text{Breaking distance} = a_3 \times V^2 = \left(\frac{V^2}{2gf}\right).$$

On substitution, we get

$$S_s = a_1 + (a_2 \times V) + \left(\frac{V^2}{2gf}\right) = 6.0 + \left(2.5 \times \frac{42}{3.6}\right) + \left(\frac{\left(\frac{42}{3.6}\right)^2}{2 \times 9.81 \times 0.4}\right)$$

$$S_s = 6.0 + (2.5 \times 11.67) + \left( \frac{(11.67)^2}{2 \times 9.81 \times 0.4} \right) = 52.52\,\text{m}$$

$$\text{Practical capacity} = C = \left( \frac{V}{S_s} \times 1000 \right) = \left( \frac{42}{52.52} \times 1000 \right) = 799.56\ \text{veh/hr/lane}$$

**Example 10.8**

Determine the mean value of time headway ($T$) of vehicles if the traffic volume ($Q$) in a lane is 2400 veh/hr, assuming there are no mopeds and motor cycles in the traffic stream.

**Answer**: We know that the time-headway ($T$) is equal to the reciprocal of traffic volume, and it can be written as

$$T\,(\text{sec/veh}) = \frac{1}{(Q)} = \frac{1}{\left( \dfrac{2400\ \text{vehicles}}{3600\ \text{seconds}} \right)}$$

$$= 1.5\ \text{seconds per vehicle.}$$

**Example 10.9**

Determine the mean value of space headway ($S$) of vehicles if the traffic density ($K$) in a lane is 2400 veh/km, assuming there are no mopeds and motor cycles in the traffic stream.

**Answer**: We know that the space headway ($S$) is equal to the reciprocal of traffic density, and it can be written as

$$S\,(\text{metre/veh}) = \frac{1}{(K)} = \frac{1}{\left( \dfrac{2400\ \text{vehicles}}{1000\ \text{metres}} \right)}$$

$$\approx 0.416\ \text{metre per vehicle.}$$

**Example 10.10**

Vehicles in a traffic stream move at an average time headway ($T$) of 2 sec. The average speed of the vehicles was found to be 75 km/h. Estimate the flow rate and density of the traffic stream.

**Answer**: We know that the flow rate ($Q$) is equal to the reciprocal of time headway ($T$), and it can be written as

$$Q\,(\text{veh/hr}) = \frac{1}{(T)} = \frac{1}{\left( \dfrac{2\ \text{seconds}}{3600} \right)}$$

$$= 1800\ \text{veh/hr.}$$

By Eq. (10.14), $Q = KV$, can be rewritten as

$$\text{Density} = K = Q/V = 1800/75 = 24\ \text{veh/km.}$$

**Example 10.11**

During a 60-sec period, a loop-detector is occupied by vehicles for the following times: 0.28, 0.32, 0.42, 0.50 and 0.55 sec. Take the distance between the loop detectors as 3 m and the average length of the vehicle as 6.2 m. Calculate the values of lane occupancy, density, speed and flow rate.

**Answer:**

$$\text{Lane occupancy} = \frac{\sum_{i=1}^{n} t_i}{T} = \frac{(0.28+0.32+0.42+0.5+0.55)\,\text{sec}}{60\,\text{sec}}$$

$$= \frac{2.07\,\text{sec}}{60\,\text{sec}} = 0.0345$$

$$\text{Lane occupancy} = 0.0345 \times 100 = 3.45\%$$

$$\text{Density per 1 km length} = 0.0345 \times 1000\text{ m}/(3 + 6.2)\text{ m}$$

$$= (0.0345 \times 1000\text{ m})/9.2\text{ m} = 3.75\text{ veh/km}$$

$$\text{Speed} = (5 \times 9.2\text{ m})/(2.07)\text{ sec} = 22.22\text{ m/sec} = 79.9\text{ km/hr}$$

$$\text{Flow rate} = 3.75\text{ veh/km} \times 79.9\text{ km/hr} = 299.625\text{ veh/hr}$$

**Example 10.12**

The free flow speed of vehicles on a road section is considered as 80 km/hr and the jammed density as 240 veh/km. Estimate the (a) traffic density on the road-section in maximum traffic flow conditions, (b) possible maximum flow rate for the road-section; and (c) average speed of vehicles at that maximum traffic volume.

**Answer:** Density when traffic flow reaches maximum volume

$$K_c = K_j/2 = 240/2 = 120\text{ veh/km.}$$

$$\text{Expected maximum flow} = Q_m = \left(\frac{V_f \times K_j}{4}\right)$$

$$= \left(\frac{80 \times 240}{4}\right) = 4800\text{ veh/hr}$$

Average velocity at $Q_m$,

$$V_c = V_f/2 = 80/2 = 40\text{ km/hr.}$$

**Example 10.13**

Free mean speed measured on an urban road was 60 km/hr and average spacing between vehicles in jam conditions was 6.9 m. Determine the maximum flow for the above condition per lane.

**Answer:**

$$\text{Jam density} = \frac{1}{6.9} = \text{veh/m}$$

Therefore,

$$\text{Jam density} = \frac{1}{6.9} \times 1000\ \text{m} = 144.92\ \text{veh/km}$$

$$\text{Critical density} = 144.92/2 = 72.46\ \text{veh/hr}$$

$$\text{Critical speed} = 60/2 = 30\ \text{km/hr}$$

$$\text{Maximum flow } (Q_m) = V_c \times K_c = 30 \times 72.46$$
$$= 2173.91\ \text{veh/hr}$$

**Example 10.14**

The speed–density relationship for a study stretch of highway was found to be $V = 40 - 0.5\,K$, where $V$ = speed in km/hr and $K$ = density of vehicles in veh/km. Determine the value of maximum flow (i.e., capacity) of the highway, mean density and mean speed. Sketch three plots with these traffic parameters and label the obtained values on these curves. Also write your comments.

**Answer**: We know that $Q = KV$. After substituting the given equation, we get

$Q = K\,(40 - 0.5\,K)$; $Q = 40\,K - 0.5\,K^2$

If $Q = 0$, then $40\,K - 0.5\,K^2 = 0$

$K\,(40 - 0.5\,K) = 0$

$K = 0$ and $K = (40/0.5) = 80$ veh/km

Jam density $= K_j = 80$ veh/km

Now, for maximum value of flow $Q_{max}$, $\dfrac{dQ}{dK} = 0$

$40 - 1K = 0$

Therefore, mean density = critical density

$$= K_c = 40\ \text{veh/km}$$

As maximum flow exists at critical density, we can rewrite the expression as

$$Q_{\max} = 40K_c - 0.5K_c^2$$

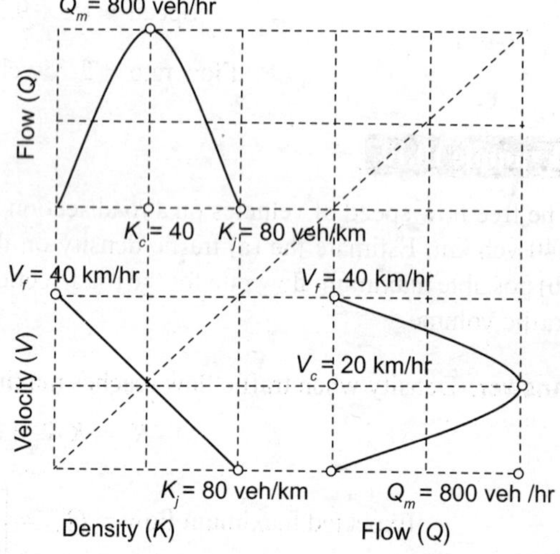


**Fig. 10.18** Velocity–Flow–Density relationship

Substituting the $K_c$ value in the above equation, we get $Q_{\max} = 40 \times 40 - 0.5 \times 40^2 = 800$ veh / hr

Mean velocity occurs at mean density; on substitution, we get

$$\text{Mean velocity} = \text{Critical velocity} = V_c = 40 - 0.5K$$

$$= 40 - 0.5 \times 40 = 20\ \text{km/hr}$$

As density becomes zero, velocity becomes maximum; i.e., free flow exists.

Therefore, $\quad V_f = 40 - 0.5K = 40 - 0.5 \times 0 = 40$ km/hr.

*Summary of results:*

The maximum flow rate is 800 veh/hr; mean speed is 20 km/hr and mean density is 40 veh/km.

*Comments on the results*: As flow becomes zero, density also becomes zero. The flow rate ($Q$) increases with density and attains the maximum value of 800 veh/hr at the critical density value of 40 veh/km. For further increase in density, the flow becomes zero and results in jam density at the value of 80 veh/km.

**Example 10.14**

Calculate the jam density, maximum flow rate, critical density and the corresponding speed at maximum flow, from the following traffic survey data:
Average length of vehicles in a stream of traffic = 6.5 m; Average gap between them = 1.8 m
Average time headway ($T$) at maximum flow ($Q_m$) = 1.2 sec.

**Answer**: Average space headway = Average length of vehicle + Average gap between them

$$S = (6.5 + 1.8) = 8.3 \text{ m}$$

Jam density ($K_j$) per km = (1000 m/ 8.3 m) = 120.48 veh/km.

$$\text{Critical density } (K_c) = \left(\frac{\text{Jam density}}{2}\right) = \left(\frac{K_j}{2}\right)$$

$$= \left(\frac{120.48}{2}\right) = 60.24 \text{ veh/km}$$

Maximum flow rate ($Q_m$) = 1 vehicle per every 1.2 sec (given)

As we know that 1 hr = 3600 sec. Therefore, maximum flow rate ($Q_m$) per 1 hr = 3600/1.2 = 3000 veh/hr. The corresponding speed (i.e., critical speed or maximum speed, $V_c$) and the maximum flow ($Q_m$): $V_c = Q_m/K_c$ = 3000/60.24 = 49.8 km/hr.

**Example 10.15**

Based on a traffic survey conducted on a National Highway (NH), it was found that free speed was 110 km/hr and jam density was 75 veh/km. Calculate the maximum flow rate ($Q_m$) that could be expected on this NH and the corresponding speed ($V_c$).

**Answer**:

$$\text{Critical speed } (V_c) = \left(\frac{\text{Free flow velocity}}{2}\right) = \left(\frac{V_f}{2}\right) = \left(\frac{110}{2}\right) = 55 \text{ km/hr.}$$

$$\text{Capacity} = \text{Maximum flow } (Q_m) = V_c \times K_c = \left(\frac{V_f \times K_j}{4}\right) = \left(\frac{110 \times 75}{4}\right)$$

$$= 2062.5 \text{ veh/hr.}$$

**Example 10.16**

The details of traffic data obtained on a road segment are given below.

| Speed (km/hr) | 59 | 51 | 46 | 41 | 34 | 29 | 22 | 15 | 10 | 4 |
|---|---|---|---|---|---|---|---|---|---|---|
| Density (veh/km) | 11 | 21 | 32 | 42 | 52 | 63 | 70 | 81 | 91 | 100 |

A plot between density and speed was drawn using MS-Excel. A best fit line as a linear relationship was obtained from Fig. 10.19, where speed = 65.539 – (0.6117 × density). Determine the (a) average decrease in speed in case of a 15-vehicles–per-km increase in density, (b) the corresponding volume data, (c) the equation for the speed–volume curve and (d) the optimum or critical speed, jam density and free speed.

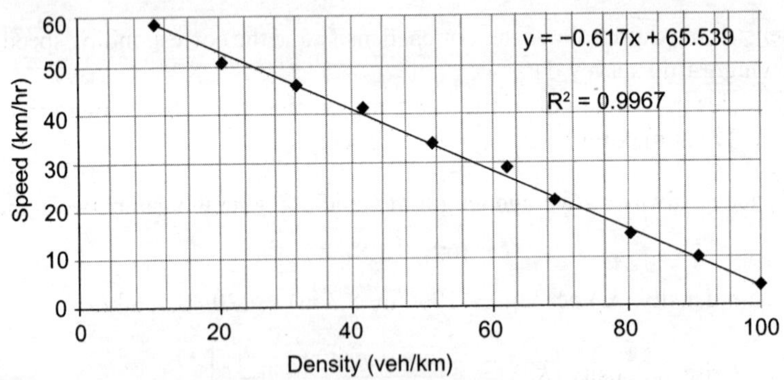


**Fig. 10.19** Speed–Density relationship

**Answer**: The given equation is

$$\text{Speed} = 65.539 - (0.6117 \times \text{Density})$$

(a) Average decrease in speed, in case of a 15-vehicles-per-km increase in density

$$= -0.6117 \times 15 \text{ veh/km} = -9.1755 \text{ km/hr.}$$

(b) Corresponding volume data: We know that
Traffic flow ($Q$) = Traffic density ($K$) × Space mean speed ($V$); i.e., $Q = K \times V$. The details of volume are as follows:

| *Density (veh/km)* | *Speed (km/hr)* | *Volume (veh/hr)* |
|---|---|---|
| *K* | *V* | *Q* = (*K* × *V*) |
| 11 | 59 | 649 |
| 21 | 51 | 1071 |
| 32 | 46 | 1472 |
| 42 | 41 | 1722 |
| 52 | 34 | 1768 |
| 63 | 29 | 1827 |
| 70 | 22 | 1540 |
| 81 | 15 | 1215 |
| 91 | 10 | 910 |
| 100 | 4 | 400 |

The speed–flow or volume curve is drawn in MS Excel (Fig. 10.20).

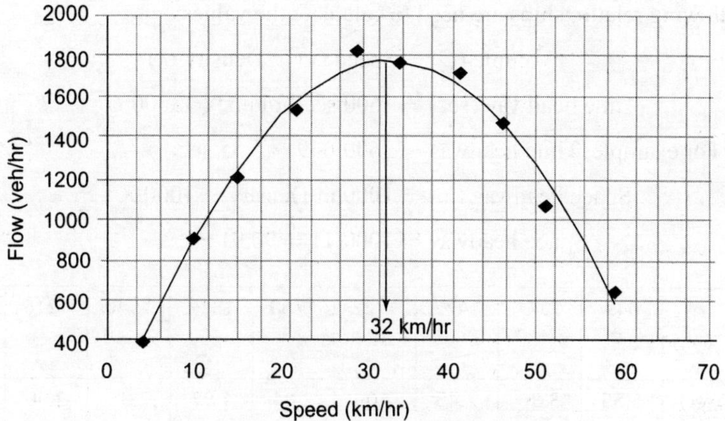


**Fig. 10.20** Speed–Flow relationship

(c) Equation for the speed–volume curve:

$$\text{Speed} = 65.539 - (0.6117 \times \text{Density}) \text{ (given)}$$

Rewriting the above using standard notations

$$V = 65.539 - (0.6117 \times K)$$

Multiplying the above by $V$ on both sides, we get

$$V \times V = \{65.539 - (0.6117 \times K)\} \times V$$

$$V^2 = 65.539 \times V - (0.6117 \times K \times V)$$

$$V^2 = 65.539 \times V - (0.6117 \times Q)$$

(d) Optimum or critical speed, jam density and free speed: As flow increases to the maximum value, speed is considered critical, and its value is read approximately as 32 km/hr (Fig. 10.20). Jam density ($K_j$) occurs when speed becomes zero, and it is obtained from the given equation

$$\text{Speed} = 65.539 - (0.6117 \times \text{Density})$$

$$\text{Speed} = 0 = 65.539 - (0.6117 \times \text{Density})$$

$$\text{Jam density } (K_j) = 65.539/0.6117 = 107.14 \text{ veh/km.}$$

Free flow speed occurs when density becomes zero, and it is obtained from the given equation

$$\text{Speed} = 65.539 - (0.6117 \times \text{Density})$$

$$\text{Free flow speed } (V_f) = 65.539 - (0.6117 \times 0)\ V_f = 65.539 \text{ km/hr.}$$

**Example 10.17**

The details of traffic data obtained on a road segment are given below.

| Speed, *V* (km/hr) | 59 | 51 | 46 | 41 | 34 | 29 | 22 | 15 | 10 | 4 |
|---|---|---|---|---|---|---|---|---|---|---|
| Density, *K* (veh/km) | 11 | 21 | 32 | 42 | 52 | 63 | 70 | 81 | 91 | 100 |

Calculate the volume data and the corresponding time headway and space headway values.

**Answer**: The following relationships are used to calculate the values:

$$\text{Volume } (Q) = \text{Speed (v)} \times \text{Density } (k)$$

$$\text{Time headway (sec)} = 3600 \text{ sec/Volume} = 3600/Q$$

$$\text{For example, Time headway} = 3600/649 = 5.55 \text{ sec}$$

$$\text{Space headway (m)} = 1000 \text{ m/Density} = 1000/K$$

For example,

$$\text{Space headway} = 1000/11 = 90.91$$

| By Eq. (10.14), $Q = V \times K$ Volume (veh/hr) | 649 | 1071 | 1472 | 1722 | 1768 | 1827 | 1540 | 1215 | 910 | 400 |
|---|---|---|---|---|---|---|---|---|---|---|
| Time headway, $T$ (sec) = (3600/$Q$) | 5.55 | 3.36 | 2.45 | 2.09 | 2.04 | 1.97 | 2.34 | 2.96 | 3.96 | 9.00 |
| Space headway, $S$ (m) = (1000/$K$) | 90.91 | 47.62 | 31.25 | 23.81 | 19.23 | 15.87 | 14.29 | 12.35 | 10.99 | 10.00 |

**Example 10.18**

The free flow speed found in a study section of road is 82 km/hr and the jam density is 98 vehicles per km length. Determine the maximum flow expected in the study section and the speed at maximum flow. Assume a linear relationship between speed and density in the study sections.

**Answer**: By Eq. (10.13)

$$\text{Capacity} = \text{Maximum flow } (Q_m) = V_c \times K_c = \left(\frac{V_f}{2}\right) \times \left(\frac{K_j}{2}\right) = \left(\frac{V_f \times K_j}{4}\right)$$

$$\text{Maximum flow } (Q_m) = V_c \times K_c = \left(\frac{82}{2}\right) \times \left(\frac{98}{2}\right) = \left(\frac{8036}{4}\right) = 2009 \text{ veh/hr}$$

Speed at the above maximum flow = Critical speed = $V_c$ = 82/2 = 41 km/hr

**Example 10.19**

Based on a traffic survey conducted on a National Highway (NH), it was found that free speed was 110 km/hr and jam density was 75 veh/km. Calculate (a) the speed at maximum flow rate and the value of the maximum flow rate expected on this segment of the NH; (b) the possible flow speeds if the observed flow is 80% of the capacity (i.e., max. flow); and (c) the corresponding time headway and space headway values. Assume a linear relationship exists between speed and density on the above segment of NH.

**Answer**: (a) Speed at maximum flow rate and value of maximum flow rate:

$$\text{Critical speed } (V_c) = \left(\frac{\text{Free flow velocity}}{2}\right) = \left(\frac{V_f}{2}\right) = \left(\frac{110}{2}\right)$$

$$= 55 \text{ km/hr.}$$

$$\text{Capacity} = \text{Maximum flow } (Q_m) = V_c \times K_c$$

$$= \left(\frac{V_f \times K_j}{4}\right) = \left(\frac{110 \times 75}{4}\right) = 2062.5 \text{ veh/hr}$$

(b) The possible flow speeds if the observed flow is 80% of capacity:

$$\text{The observed flow} = 80\% \text{ of maximum flow}$$

$$= (80/100) \times 2062.5 = 1650 \text{ veh/hr}$$

Assuming a linear relationship between speed and density, (y = c − mx), expressed as follows:

$$\text{Speed} = V_f - (\text{m} \times \textit{Density})$$

At jam density ($K_j$) = 0 veh/km, the corresponding free flow speed ($V_f$) = 110 km/hr.
At jam density = 75 veh/km, the corresponding speed becomes zero; after substituting these values in the assumed linear relationship, we get

$$0 = 110 - (\text{m} \times 75)$$

$$\text{m} = 110/75 = 1.467$$

Therefore, we can now write this as $V = 110 - 1.467 \times K_j$

Multiplying the above equation by V on both sides, we get

$$V \times V = \{110 - 1.467 \times K_j\} \times V$$

$$V^2 = 110 \times V - (1.467 \times K_j \times V)$$

$$V^2 = 110 \times V - (1.467 \times Q_m)$$

Substituting $Q_m$ = 1650 veh/hr in the above equation, we get

$$V^2 = 110 \times V - (1.467 \times 1650)$$

$$V^2 = 110 \times V - (2420.55) = 0$$

After solving

$$\text{Speed} = V = \frac{110 \pm \sqrt{110^2 - 4 \times 2420.55}}{2}$$

$$\text{Speed} = V = \frac{110 \pm \sqrt{2417.8}}{2} = \frac{110}{2} \pm \sqrt{2417.8}$$

$$= \frac{110}{2} \pm 49.17$$

$$V = 55 \pm 49.17$$

Therefore, $V_{non\text{-}congested} = 55 + 49.17 = 104.17$ km/hr

(considered as free flow and non-congested condition)

Or $V_{congested} = 55 - 49.17 = 5.83$ km/hr

(considered as forced flow and congested condition)

$$\text{Density at } V_{non\text{-}congested} = 1650/104.17 = 15.83 \text{ veh/km}$$

$$\text{Density at } V_{congested} = 1650/5.83 = 283.01 \text{ veh/km}$$

$$\text{Space headway at } V_{non\text{-}congested} = 1000/15.83 = 63.17 \text{ m}$$

$$\text{Space headway at } V_{congested} = 1000/283.01 = 3.53 \text{ m}$$

## Example 10.20

A two-lane urban road with one-way traffic flow control was designed for a maximum capacity of 2250 veh/hr. Under jam conditions, the average length of a vehicle is 6.0 m. Assume the speed versus density relationship is linear. What is the density (veh/km) value, if the traffic volume is found to be 1200 veh/hr?

**Answer:**

$$S_s = 6\,\text{m} + \left(\frac{V}{3.6}\right) \times t_p \qquad (\textit{For use of theoretical capacity, assume } t_p = 0.7\,\text{sec})$$

$$S_s = 6\,\text{m} + \left(\frac{V}{3.6}\right) \times 0.7 = 6 + \frac{0.7 \times V}{3.6}$$

We know,

$$\text{Capacity} = C = \frac{V}{S_s} \times 1000$$

Substituting, we get

$$2250 = \frac{V}{\left(6 + \dfrac{0.7 \times V}{3.6}\right)} \times 1000$$

Solving,

$$V = 24\,\text{m/sec}$$

$$\text{Density} = \frac{\text{Observed capacity}}{V} = \frac{1200}{24} = 50 \text{ veh/km}$$

## Example 10.21

Traffic volume count was carried out at a location on a two-lane two-way state highway and found to be 1200 veh/hr on one of the lanes with a directional distribution of 58%. A major district road intersects this state highway. Assume the vehicle arrival rate at any point along the road follows Poisson's distribution and calculate the probability of the number of vehicles travelling along the major lane at a headway of 5 seconds or more.

**Answer**: Poisson's distribution is represented by $P(k) = \dfrac{e^{-\mu}\mu^k}{k!}$

$Q$ = Volume of traffic = 1200 veh/hr in both directions

$Q$ = Volume of traffic on a major lane = 58% of 1200

= 0.58 × 1200 = 696 veh/hr

$\mu$ = Average number of vehicles arriving in the given time gap period, $t$ = 5 sec

$$\mu = \frac{Qt}{3600} = \frac{696 \times 5\,\text{sec}}{3600\ \text{sec}} = 0.967 \ \text{vehicles in 5 sec}$$

The probability that no vehicle (i.e., $x = 0$) arrives in a given time interval $= P(0) = \dfrac{e^{-\mu}\mu^0}{0!}$

$$P(0) = e^{-\mu} = e^{-0.967} = 0.38$$

That means only 38% of vehicles (i.e., 0.38 × 696 = 264.48 vehicles) are travelling at a headway of 5 sec or more.

## EXERCISES

### Multiple Choice Questions

1. The traffic flow equation for a study road is $V = 88 - 0.4\ K$, where $V$ = Speed (km/hr) and $K$ = Density (veh/km). What is the maximum flow expected in veh/hr?
   (a) 2286 (b) 3740 (c) 3500 (d) 2730
   (*Hint*: We know that $Q = KV$. Substituting, we get Flow $= Q = K \times (88 - 0.4\ K) = 88\ K - 0.4\ K^2$; $Q = 88\ K - 0.4\ K^2$. Differentiating and equating to zero, to get maximum flow $\frac{dQ}{dK} = (88 - 0.8K) =$ Zero. Therefore, $K = 88/0.8 = 110$. The maximum flow is expected $= Q =$ $88 \times K - 0.4 \times K^2 = 88 \times 110 - 0.4 \times 110^2 = 3740$ veh/hr.)
2. What is the acceleration due to retardation of a vehicle (in m/sec$^2$) when on a pavement surface having a longitudinal coefficient of friction of 0.38
   (a) 3.7278 (b) 3.800 (c) 3.9812 (d) 3.0808
   (*Hint*: Acceleration due to retardation of a vehicle = 0.38 × 9.81 m/sec$^2$ = 3.7278 m/sec$^2$)
3. What is the basic capacity of a lane if the operating speed is 65 km/hr with a safe stopping sight distance of 90 m? Assume average length of vehicle = 6 m.
   (a) 771 veh/hr (b) 654 veh/hr (c) 677 veh/hr (d) 560 veh/hr
   (*Hint*: Formula for calculation of basic capacity $= C = \dfrac{V}{S_s} \times 1000 = \dfrac{65}{(6+90)} \times 1000 = 677$ veh/hr)
4. Free flow speed on a lane was 60 km/hr and jam density was 90 veh/km. The maximum flow in veh/hr that could be expected on this lane is
   (a) 5400 (b) 2700 (c) 2750 (d) 1350
   (*Hint*: Maximum flow $(Q_m) = V_c \times K_c = \left(\dfrac{V_f}{2}\right) \times \left(\dfrac{K_j}{2}\right) = \left(\dfrac{V_f \times K_j}{4}\right) = \left(\dfrac{60 \times 90}{4}\right) = 1350$ veh/hr)
5. On an urban road, free mean speed was 60 km/hr and the average spacing headway under jam condition was 8 m. The value of maximum flow (veh/hr) per lane for this road would be equal to
   (a) 1875 (b) 2000 (c) 1000 (d) None of these
   (*Hint*: Jam density $K_j = 1000/8 = 125$ veh/km per lane. We know, Maximum flow $(Q_m) = V_c \times K_c$ $= \left(\dfrac{V_f}{2}\right) \times \left(\dfrac{K_j}{2}\right) = \left(\dfrac{V_f \times K_j}{4}\right) = \left(\dfrac{60 \times 125}{4}\right) = 1875$ veh/hr)

6. A linear relationship exists between speed and density on a study road. The free flow observed was 60 km/hr and the jam density found was 120 veh/km per lane. The values of maximum flow expected and the corresponding speed would be.
   (a) 7200 veh/hr and 60 km/hr (b) 3600 veh/hr and 60 km/hr
   (c) 1800 veh/hr and 30 km/hr (d) 900 veh/hr and 20 km/hr

   (*Hint*: We know, Maximum flow $(Q_m) = V_c \times K_c = \left(\frac{V_f}{2}\right) \times \left(\frac{K_j}{2}\right)$

   $= \left(\frac{V_f \times K_j}{4}\right) = \left(\frac{60 \times 120}{4}\right) = 1800$ veh/hr; $V_c = \left(\frac{V_f}{2}\right) = \left(\frac{6.0}{2}\right) = 30$ km/hr)

7. Design capacity is often provided as
   (a) Basic capacity (b) Practical capacity (c) Ideal capacity (d) Possible capacity
8. As per IRC: 64, for smooth flow of traffic, it is not advisable to design the width of road pavement for a traffic volume equal to its capacity, which is available at LOS ________.
   (a) C (b) D (c) E (d) F
9. As per IRC: 64, under normal circumstances, which of the LOS is considered adequate for the design of rural highways at which traffic volume will be around 0.5 times the maximum capacity?
   (a) A (b) B (c) C (d) D
10. As per IRC: 64, what is the value of design service volume that can be tentatively adopted for the design of a four-lane divided carriageway located in plain terrain?
    (a) 15000 PCU/day (b) 25000PCU/day (c) 35000 PCU/day (d) 45000 PCU/day
11. The traffic flow parameter which is not used to measure effectiveness is
    (a) Flow (b) Delay (c) Density (d) Speed
12. As per IRC: 106, in order to capture all the peak traffic, it is usual practice to conduct traffic counts for ________ hours.
    (a) 3 (b) 10 (c) 16 (d) 24
13. As per IRC: 106, the design period adopted for arterial and sub-arterial roads is
    (a) 0–5 years (b) 10–12 years (c) 10–15 years (d) 15–20 years
14. As per IRC: 106, the design period adopted for collector and local streets is
    (a) 0–5 years (b) 5–10 years (c) 10–15 years (d) 15–20 years
15. As per IRC: 106, it is advisable to design road cross-sections for traffic volumes equal to the maximum capacity at LOS ________
    (a) B (b) C (c) D (d) E
16. As per IRC: 106, at LOS C design service volume, the volume of traffic will be around how many times the maximum capacity adopted for the design of urban roads?
    (a) 0.7 (b) 0.3 (c) 10 (d) 30
17. As per IRC: 64, at LOS B design service volume, the volume of traffic will be around how many times the maximum capacity adopted for the design of rural roads?
    (a) 0.7 (b) 0.5 (c) 10 (d) 30

18. Select the correct statement:
    (a) Volume of traffic should always be greater than capacity of traffic
    (b) Spot is the mean speed of a vehicle over a section length
    (c) 85th percentile speed may be more than 95th percentile
    (d) Traffic capacity should always be greater than its volume
19. As per IRC: 106, the recommended PCU (passenger car unit) value for motor cycle and scooter, if $\geq$ 10% composition of these vehicles is found in a traffic stream is
    (a) 0.5 (b) 0.75 (c) 1.0 (d) 2.0
20. As per IRC: 106, the recommended PCU (passenger car unit) value for truck and bus, if $\geq$10% composition of these vehicles is found in a traffic stream during a study period is
    (a) 2.0 (b) 3.0 (c) 2.2 (d) 3.7
21. A roadway had a capacity of 2849 veh/hr when there were only passenger cars plying the road. The same roadway under similar conditions had a capacity of 770 veh/hr of a particular class of vehicle. What is the PCU value of that particular vehicle as per IRC: 106?
    (a) 3.0 (b) 3.7 (c) 0.27 (d) 6.0
22. As per IRC: 106, the recommended PCU (passenger car unit) value for light commercial vehicles, if 5% composition of these vehicles is found in the traffic stream during a study period, is
    (a) 0.75 (b) 1.4 (c) 2.0 (d) 1.2
23. As per IRC: 106, the recommended PCU (passenger car unit) value for auto rickshaw, if 5% composition of these vehicles is found in traffic stream during a study period, is
    (a) 0.75 (b) 1.4 (c) 2.0 (d) 1.2
24. When the speed of the traffic becomes maximum (i.e., free flow condition)
    (a) Traffic density and volume become zero (b) Traffic density and volume become maximum
    (c) Traffic density becomes maximum and volume becomes zero
    (d) Traffic density becomes zero and volume becomes maximum
25. When the flow of the traffic becomes maximum
    (a) Traffic velocity becomes maximum (b) Traffic velocity becomes zero
    (c) Traffic velocity neither become zero or maximum (d) Density and velocity values become half
26. If J is the length of the vehicle in metres, B is the clearance stopping sight distance between two consecutive vehicles, and X is the speed of vehicles in km per hour, the maximum number (N) of vehicles per hour is equal to

    (a) $N = \dfrac{1000B}{J+X}$ (b) $N = \dfrac{1000J}{B+X}$ (c) $N = \dfrac{1000X}{J+B}$ (d) $N = \dfrac{J+X}{1000B}$

27. Which of the following diagrams illustrates the relationship between Q and K?

(a) 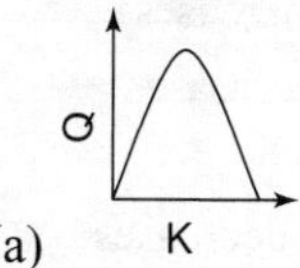


(b) 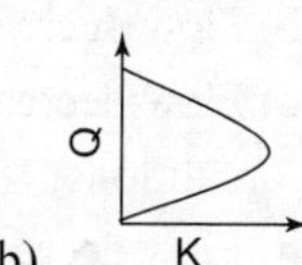


(c) 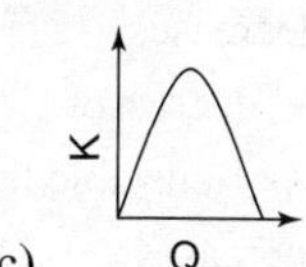


(d) 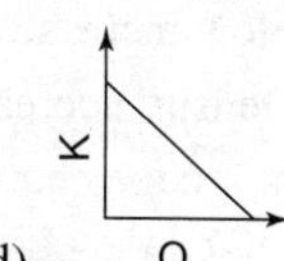

28. With increase in traffic density, traffic flow
    (a) Increases (b) Decreases
    (c) First increases and then decreases after reaching a maximum value at optimum speed
    (d) First decreases and then increases after reaching a maximum value at optimum speed
29. With increase in traffic speed, traffic density
    (a) Increases (b) Decreases
    (c) First increases and then decreases after reaching a maximum value at optimum speed
    (d) First decreases and then increases after reaching a maximum value at optimum speed
30. The space headway in a traffic stream is 40 m; then the density of the stream (veh/km) will be
    (a) 960 (b) 25 (c) 0.04 (d) 100
    (*Hint*: $K = 1000/40 = 25$ veh/km)
31. In a traffic volume study, an observer counts 180 veh/hr at a study location of a road. It is assumed that the arrival of vehicles in the location is Poisson distributed. The probability of having one vehicle arriving over a 10-sec time interval is
    (a) 10% (b) 0.25 (c) 0.040 (d) 0.303
    (*Hint*: $p(n,t) = \frac{e^{-\lambda t}(\lambda t)^n}{n!} = \frac{e^{-180\times10/3600}(180\times10/3600)^1}{1!} = 0.30320$)
32. The average number of cars passing a point on a National Highway is 2000 PCU/hr per lane. The cars travel at an average speed of 50 km/hr. What is the clear distance between the successive cars if the average length of a car is 5.5 m?
    (a) 30.5 m (b) 34.5 m (c) 14.5 m (d) 19.5 m
    (*Hint*: Density ($K$) = Space headway between the cars = Clear distance between the cars = Space headway – Average length of cars = 25 m – 5.5 m = 19.5 m)
33. Space headway is defined as
    (a) Length of vehicle from front to back bumper (b) Space gap between two successive vehicles
    (c) Length from the centre of one vehicle to the centre of another
    (d) Distance between common points of successive vehicles
34. Which of the following is not true under jam density conditions?
    (a) Flow is zero (b) Speed is average
    (c) Density is maximum (d) Speed is zero
35. Which of the following is true under critical density conditions?
    (a) Speed is maximum (b) Density is maximum
    (c) Headway is maximum (d) Flow is maximum
36. What are the characteristics of unstable traffic flow conditions?
    (a) Flow increases, speed decreases (b) Flow decreases, speed increases
    (c) Density decreases, speed decreases (d) Flow decreases, speed decreases
37. What are the characteristics of unstable traffic flow conditions?
    (a) Flow increases, speed increases (b) Flow decreases, density decreases
    (c) Density decreases, speed decreases (d) Density increases, flow decreases

38. A racing car has been designed to attain a speed of 4 sec from a standing position to 54 km/hr. What is the rate of acceleration required from the racing car to achieve this speed?

(a) 3.00 m/sec$^2$ (b) 13.50 m/sec$^2$ (c) 2.50 m/sec$^2$ (d) 3.75 m/sec$^2$

(*Hint*: Let, u = 0 km/hr = 0 m/sec; v = 54 km/hr = 15 m/sec; t = 4 sec; a = v – u/t = (15 – 0/4 = 3.75 m/sec$^2$)

39. At a time, in front of the Jawaharlal Nehru Port in Mumbai, a long queue of trucks were waiting for inspection and permission to enter. The trucks have an average length of 17.0 m and the average space between the front and rear bumpers of successive trucks is 3 m. What is the jam density (in truck/km) in a lane?

(a) 70 (b) 20 (c) 54 (d) 50

(*Hint*: Jam density = (1000 m/km) x (Vehicles/[17 + 3]) = 50 trucks/km)

40. Which one of the following is taken into consideration for computing traffic capacity per lane of a study road?

(a) Passenger cars and two-wheelers (b) Truck and buses

(c) All types of vehicles including pedestrian traffic (d) Passenger car equivalents

41. Traffic capacity is the:

(a) Ability of a roadway to accommodate traffic volume in terms of vehicles per hour

(b) Ability of a roadway to accommodate number of vehicles in terms of vehicles per km

(c) Ability of a roadway to accommodate number of vehicles across its width in terms of vehicles per km

(d) Maximum attainable speed of vehicles

42. What is the theoretical capacity of a lane per hour, if the design speed is 64 km/hr? Take total reaction time of the driver as 2.5 sec and average length of design vehicle as 6.6 m.

(a) 771 veh/hr (b) 554 veh/hr (c) 817 veh/hr (d) 700 veh/hr

(*Hint*: Assume the coefficient of longitudinal friction along the road = 0.4. Formula used for calculation of basic capacity = $S_s = \{(0.278 \times 64 \times 2.5) + [(0.278 \times 64)^2/(2 \times 9.81 \times 0.4)]\} + 8\text{ m} = 91.41579\text{ m}$

$$C = \frac{V}{S_s} \times 1000 = \frac{64}{(91.41579)} \times 1000 = 700.09 \text{ veh/hr})$$

43. Two shock waves were observed in a traffic stream and their speeds calculated as −2 km/hr and −6.67 km/hr, respectively. The queue will dissipate at the rate of

(a) 4.66 km/hr (b) −4.66 km/hr (c) −8.66 km/hr (c) 3.66 km/hr

(*Hint*: Relative speed of the two shock waves = Front − Rear = −6.67 – (−2) = −4.67 km/hr)

44. A shock wave was observed in a traffic stream and its speed was calculated as −2 km/hr. What is the length of the queue after 3 minutes?

(a) 0.1 km (b) 1.0 km (c) 0.66 km (c) 6.0 km

(*Hint*: Length = L = u × t = 2 km/hr × 3/60 hr = 0.1 km)

45. A shock wave was observed in a traffic stream and its speed was calculated as −2 km/hr. How many vehicles are in the queue after 3 minutes? Take the density after the formation of shock wave = 260 veh/km.

(a) 26 (b) 130 (c) 82.5 (c) 13

(*Hint*: Length of queue after 3 minutes = L = u × t = 2 km/hr × 3/60 hr = 0.1 km. No. of vehicles = k × L = 260 × 0.1 = 26 vehicles)

46. What is the value of time gap between two vehicles, if the time headway between the vehicles is 6 sec, average length of a vehicle is 5 m and average space mean speed of the traffic stream is 10 m/sec?
(a) 6 sec (b) 1 sec (c) 5.5 sec (c) 4 sec
(*Hint*: Time headway = 6 sec – (5/10) = 5.5 m)

## Answers

| | | | | | | | | | |
|---|---|---|---|---|---|---|---|---|---|
| 1. (b) | 2. (a) | 3. (c) | 4. (d) | 5. (a) | 6. (c) | 7. (b) | 8. (c) | 9. (b) | 10. (c) |
| 11. (a) | 12. (c) | 13. (a) | 14. (c) | 15. (b) | 16. (a) | 17. (b) | 18. (d) | 19. (b) | 20. (d) |
| 21. (b) | 22. (b) | 23. (d) | 24. (a) | 25. (d) | 26. (c) | 27. (a) | 28. (c) | 29. (b) | 30. (b) |
| 31. (d) | 32. (d) | 33. (d) | 34. (b) | 35. (d) | 36. (d) | 37. (d) | 38. (d) | 39. (d) | 40. (d) |
| 41. (a) | 42. (d) | 43. (b) | 44. (a) | 45. (a) | 46. (c) | | | | |

## Questions

1. Define space headway and time headway. Explain how they are computed from the observations obtained by conducting field surveys.
2. Write the equation used to represent Poisson's distribution for uniform headway between moving vehicles.
3. Define lane occupancy and explain how it is computed from the observations obtained by conducting field surveys.
4. Define traffic density and state how it is represented by an equation.
5. Define lane capacity and explain the different types of traffic capacities.
6. Define (a) Basic capacity (b) Possible capacity, and (c) Design capacity.
7. State the factors affecting practical capacity.
8. Derive and describe an expression that represents the standard relationship between speed and density.
9. Derive and describe an expression that represents the standard relationship between speed and flow.
10. Derive and describe an expression that represents the standard relationship between traffic density and flow.
11. Draw the typical trend of curves showing the fundamental relationships between speed, flow and density of traffic flow characteristics.
12. Explain Greenshield's model representing the speed–density relationship.
13. Derive the parabolic flow–density model.
14. What is meant by level of service? Describe with operating speed and volume-to-capacity ratio.
15. Write the details of general operating conditions of the six LOS with volume to capacity ratio and the speed condition.
16. State the factors affecting the LOS of a roadway.
17. What are the factors affecting LOS on urban roads?
18. State the six factors that are used to determine LOS.
19. Write the HCM-recommended criteria for LOS for signalised and un-signalised intersections based on control delay and delay, respectively.

20. Define passenger car equivalent and explain any method of estimation of PCU values for a study area.
21. Explain the factors affecting PCU values.
22. Write the IRC: 106-recommended PCU factors applicable on urban roads in plain areas.
23. Write the IRC: 64-recommended PCU factors that are applicable on rural roads.
24. Write the significance of design service volume and state a few standard specifications as per IRC: 64.
25. Traffic volume on a section of a rural road during morning peak hour is 800 veh/hr and during evening peak hour is 800 veh/hr. The details of traffic composition are given below:

    For morning peak: 52% cars, 28% motor cycles and scooters, 9% buses and trucks, 6% cycle rickshaws, 3% agricultural tractor-trailers and 2% bicycles.

    For evening peak: 49% cars, 27% motor cycles and scooters, 13% buses and trucks, 5% cycle rickshaws, 3% agricultural tractor-trailers and 3% bicycles.

    Determine which peak hour traffic is the congested condition.
26. Calculate the capacity of traffic flow if the average velocity on the lane is 40 km/hr and the stopping sight distance (SSD) is 25 m. Assume length of rigid wheel base as 6.0 m.
27. A study was conducted on a road section and found that one-way traffic stream speed was 42 km/hr. Average length of the vehicle wheel base = 6.0 m.

    Determine (i) theoretical capacity and (ii) practical capacity of the study road in terms of veh/hr/lane. Assume reaction time ($t_p$) of the driver as 0.8 sec for determination of theoretical capacity; average length of vehicle ($a_1$) = 6 m, reaction time of the driver ($a_2$) = 2.5 sec and coefficient of longitudinal friction (f) = 0.36.
28. Determine the mean values of time headway and space headway of vehicles if the traffic volume in a lane is 1200 veh/hr, assuming there are no mopeds and motor cycles in the traffic stream.
29. Vehicles in a traffic stream moved at an average time headway of 2.5 sec. The average speed of the vehicles was found as 60 km/h. Estimate the flow rate and density of the traffic stream.
30. In a 60-sec period, a loop-detector is occupied by vehicles for the following times: 0.30, 0.32, 0.43, 0.55 and 0.51 sec. Take the distance between the loop-detectors = 3 m and average length of vehicle = 6.0 m. Calculate the values of lane occupancy, density, speed and flow rate.
31. Free flow speed of vehicles on a selected road-section is 75 km/hr and the jam density is 220 veh/km. Estimate the (a) traffic density in maximum traffic flow condition, (b) possible maximum flow rate and (c) average speed of vehicles at maximum traffic volume. Determine the maximum flow for the above condition per lane if the free mean speed = 50 km/hr and average spacing between the vehicles under jam condition was 5.4 m.
32. The speed–density relationship for a study stretch of highway was found as $V = 42 - 0.4K$, where $V$ = speed (km/hr) and $K$ = density (veh/km). Determine the value of maximum flow (i.e., capacity) of the highway, mean density and mean speed. Also sketch three plots with these traffic parameters and label the values obtained on these curves.
33. Determine the values of jam density, maximum flow rate, critical density and the corresponding speed at maximum flow, from the following traffic survey data:

    Average length of vehicles in a stream of traffic = 6.6 m; Average gap between them = 2.0 m

    Average time headway at maximum flow = 1.5 sec.
34. Calculate the volume data and the corresponding time headway and space headway values for the following traffic data obtained on a road segment.

| Speed, $V$ (km/hr) | 60 | 55 | 47 | 42 | 35 | 30 | 22 | 15 | 11 | 5 |
|---|---|---|---|---|---|---|---|---|---|---|
| Density, $K$ (veh/km) | 12 | 21 | 32 | 42 | 52 | 63 | 70 | 80 | 90 | 100 |

35. Define bottleneck and describe how the capacity of a highway and the bottleneck is determined, with a sketch.
36. Calculate the net savings in waiting time between toll collecting booths with two and three toll booths which are proposed to be installed at an entry point of a bridge. The average arrival of vehicles as 1500 veh/hr and the service rate assumed is 10 seconds each to pay the toll. Assume both the arrivals and the departures are exponentially distributed.

CHAPTER 11

# Introduction to Intersections and Design Guidelines for At-Grade Intersections

## 11.0 INTRODUCTION

**Intersections** are the connecting segments between road links. The quality of any road network does not depend solely on the characteristics of the links, but also on the way the links are connected. As different types of road links exist in a network, connections such as intersections and interchanges should be designed to provide efficient, consistent and safe operation of traffic (Rietveld, 1995).

The general design and geometry of at-grade intersections on major interurban and urban roads are described in this chapter. This chapter also covers the general features and design details of the components that form the intersections. The principles and guidelines discussed in this chapter may provide the basis for the design of intersections and also assist in the overall approach to designing any route of a road network.

## 11.1 ACCESS MANAGEMENT

This consists of an integrated approach to controlling access to land developments while preserving the free flow of traffic on the road system, in terms of safety, travel capacity and speed. Land development consists of public highways of all classes and activity centres in urban and rural areas at all hierarchy levels. It is a rational approach to coordinating the transportation facilities and geometric elements with the surrounding land developments. The objectives and elements of access management include (Koepke and Levinson, 1992):

- Defining the appropriate accessibility and relevant spacing standards for various classes of roadways, with reference to road geometric standards and traffic operating characteristics
- Providing alternative treatment options and retrofitting designs/alignment wherever reasonable access cannot be provided
- Establishing standard roadway geometric specifications and associated traffic operations/controls so as to enforce safety and consistency
- Developing guidelines for access denial where reasonable alternative access exists and variations to the access permit criteria

These aspects are achieved by the integrated design of roadway geometrics with the associated traffic operations by adopting standard specifications and design philosophy. The following sections describe the most commonly used geometric design elements with reference to access control and safety.

**Basic principles of design:** Operational integrity, with reference to the access design of roadway intersections with the surrounding activity centres, is the main factor affecting the design of intersections and associated geometric elements. The following basic principles must be considered at all stages in the design of intersections relating to general geometrics and design parameters (Setra, 1998; Chery, 2010):

- The design of intersections should be compatible with the type of road, locality and behaviour of all road users, including cyclists and pedestrians.

- Provision of adequate manoeuvring space is essential for design vehicles.
- The design must fit in with the rationale of the route in terms of consistency in operating standards, uniformity in operating speed and provide smooth continuity or division of roads.
- Conflicts between competing movements in the intersection area should be resolved by providing adequate visibility and space.
- The design of intersections and access points should limit disruption and speed loss caused by exchange movements between the intersecting roads.
- The drivers must be provided facilities to recognise the intersections well in advance, without any difficulty. Suitable advance warning of all possible hazards is required.
- Safety must be prioritised to optimise free flow of traffic on intersecting roads.
- The design must take into account intended groups of driver behaviour relating to pedestrians, trucks, exceptional vehicles, two-wheelers, three-wheelers and animal-drawn vehicles.

Thus, the design principles commonly adopted for rural and urban areas are the same. The basic difference lies in their input values such as design speed, sight distance availability, restricted available land, pedestrian volume, and volume of cyclists, truck traffic and animal-drawn vehicles.

In the case of intersection performance evaluation and redesign, the number of stops observed over a period of peak traffic time on average – the number of times a driver has to stop or experiences delay while travelling through the intersection – is measured. Further studies may be undertaken to explore the effect of balanced and unbalanced volume conditions at major–minor road intersections and left-turn and right-turn volumes of unbalanced traffic through approaches to intersections (Chery, 2010).

## 11.2 CLASSIFICATION OF INTERSECTIONS

**Based on the grade of the intersecting surface:** Road intersections can be broadly classified, based on the grade of the intersecting surface level, into two types (Fig. 11.1).

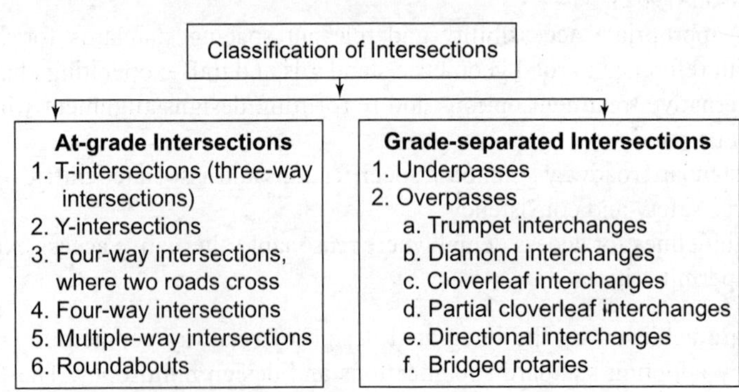


**Fig. 11.1** Classification of intersections based on the intersecting surface level

- **At-grade intersections**: These intersections are formed when the intersecting roads meet at the same level (i.e., on the same plane). Traffic flow manoeuvres such as merging, diverging, crossing and weaving will occur at these intersections. Some of the factors to be considered in the design of at-grade intersections are given below:
  - The area of conflict should be as small as possible.
  - The relative approach speeds should be low.

- The angle of approach of vehicles entering the intersection area should be small.
- Adequate visibility (with reference to the sight triangle) should be available to the vehicles approaching the intersection.
- Sudden change in lane/path should be avoided by providing adequate flared legs.

- **Grade-separated intersections or interchanges**: These intersections are formed when intersecting roads meet at different levels. Traffic flow manoeuvres such as merging, diverging, crossing and weaving will be eliminated at these intersections. For more information, see *Chapter 15*.

**Based on the control methods used**: Intersections can also be classified based the control methods used (Fig. 11.2).

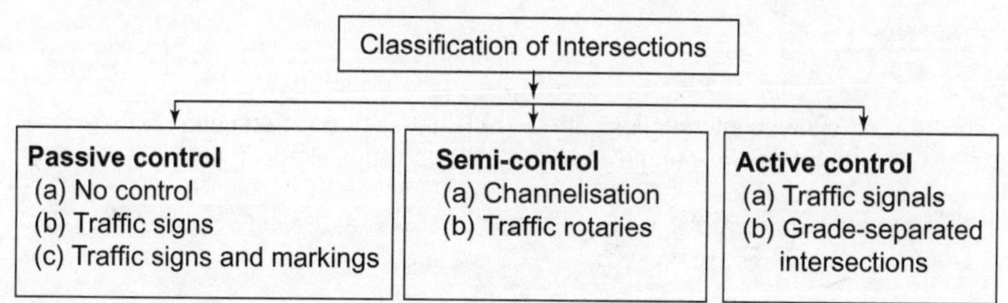


**Fig. 11.2** Operational classification of intersections (based on control type)

## 11.3 CLASSIFICATION OF AT-GRADE INTERSECTIONS

At-grade intersections can be classified based on their geometric form and functionality.

**Based on their geometric form**: The basic configurations of roads intersecting at grade, based on their geometry, are shown in Fig. 11.3. The typical range of the angle between the intersecting legs (the angle between the centre lines of the intersection roads) of different intersections may be as follows:

- T-intersection: 75–120°
- Y-intersection or three-legged intersection: 60–75°
- Right-angled intersection: 75–120°
- Oblique four-legged intersection: 75–120°

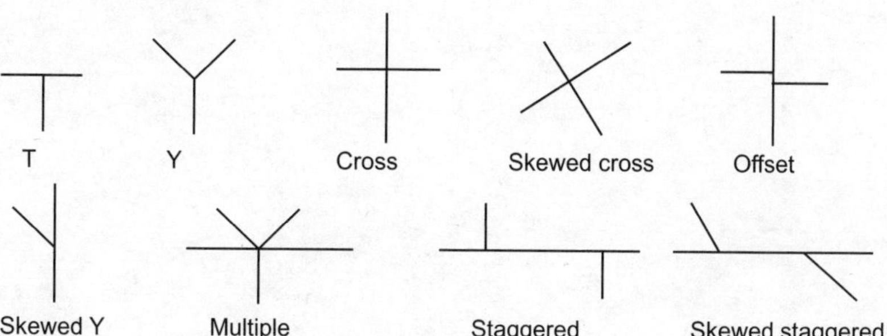


**Fig. 11.3** Basic configurations of intersections

*Skewed intersections*: If the intersecting roads are not exactly perpendicular, then the intersection is said to be skewed (Kaluva, 2013). The angular deviation from the perpendicular is measured in the clockwise or anticlockwise directions (whichever is closer) and it is termed the skew angle (Θ). The method of measurement of the skew angle for a typical three-legged intersection is presented in Fig. 11.4 and that for a four-legged intersection in Fig. 11.5. It must be noted that the area of any skewed intersection will be larger than the area of that intersection with no skew.

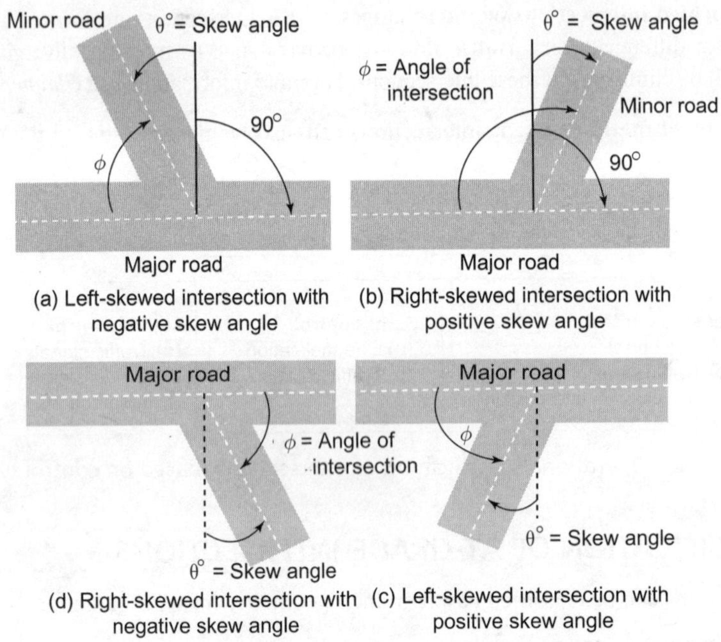


Fig. 11.4 Measurement of the skew angle of intersecting roads (T-intersections)

*Note:* The crossing angle of intersection (Φ) between the major road and the minor road should preferably be in the range 75–120°.

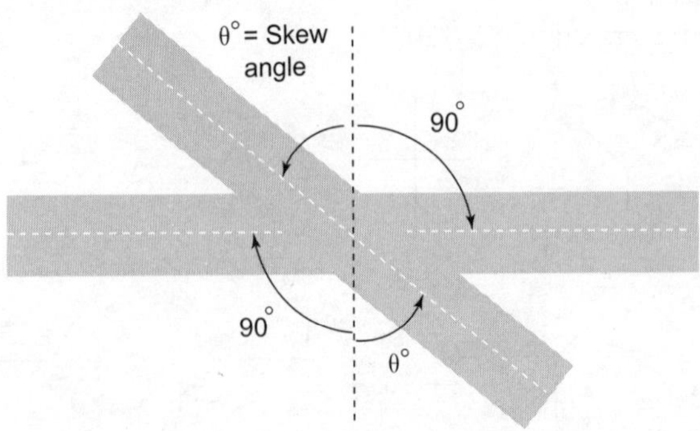


**Fig. 11.5** Measurement of the skew angle of intersecting roads (four legs)

*Note:* Right angle, cross and T-intersections have the smallest area of conflicts as compared to skewed, multi-legged or offset intersections (Delphi et al., 2006).

**Based on channelisation of traffic flow:** At-grade intersections can be classified based on channelisation, as shown in Fig. 11.6.

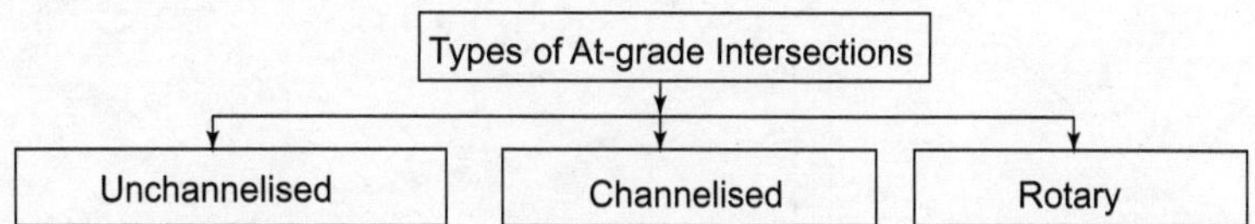


**Fig. 11.6** Classification of intersections based on channelisation of traffic flow

*Unchannelised intersections*: One of the most commonly used types of intersection is the unchannelised intersection. This consists of two or more roadways that intersect at the same elevation and which are connected by radius returns to accommodate the wheel paths of turning vehicles. The most commonly observed characteristics of unchannelised intersections are low turning movements and low overall traffic volume (NDOR, 2012).

The intersection area is paved and the paths of the vehicles are not restricted to traversing along their intended direction of travel. When no additional pavement width is provided to facilitate turning movements, such intersections are termed *plain intersections* (Fig. 11.7). When the pavement width is widened at the intersection area by a lane width or more, such intersections are termed *flared intersections* (Fig. 11.8). As the conflict area increases in flared intersections, vehicular flow paths are not restricted. However, while crossing, the vehicles will have to stop while those in the other direction proceed. Flared approaches to intersections can be provided with channelised and rotary intersections as well.

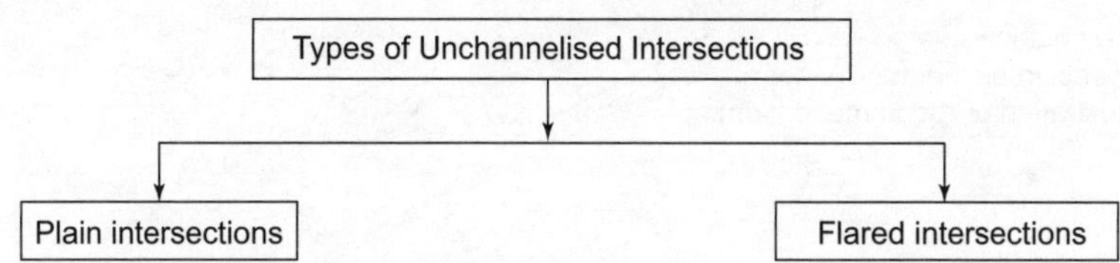


**Fig. 11.7** Classification of unchannelised intersections

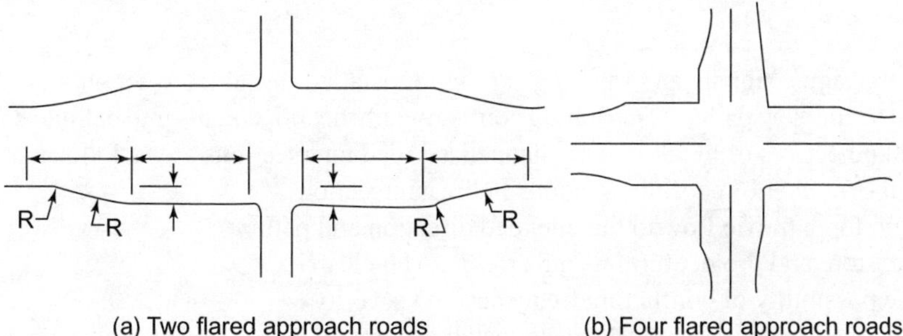


**Fig. 11.8** Flared intersections

*Channelised intersections*: The fundamental concept of channelisation is to divide and direct traffic movement. A variety of components and arrangements such as traffic lanes, raised (kerbed) islands,

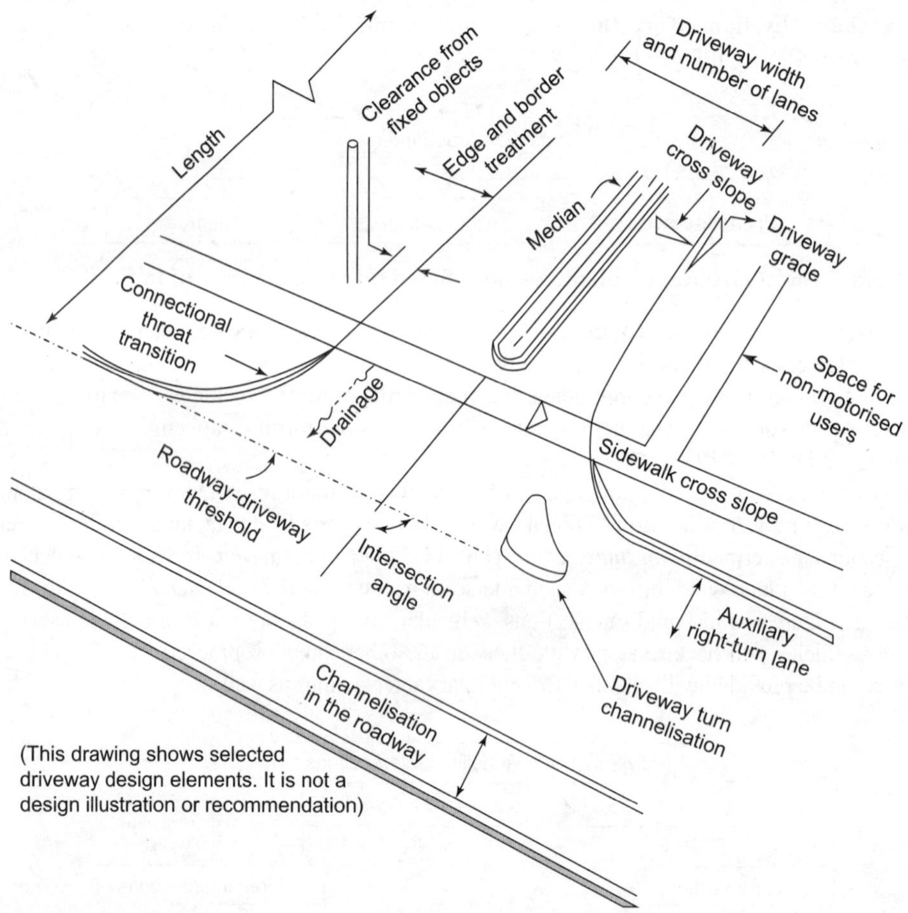


**Fig. 11.9** Typical layout of a three-legged intersection with channelisation design elements (*Source*: Gattis et al., 2010; *Courtesy*: Transportation Research Board, Washington DC, USA)

median dividers, signs, signals and markings are used/installed in the intersection area (Fig. 11.9) to channelise the traffic flow paths by separating conflicting traffic movements into definite paths of travel. The functional objectives of the design of channelised road intersections are as follows:

- Reduce the number of conflicting points in the intersection
- Guide and force traffic flow to the intended direction and path
- Separate, store and protect turning and crossing vehicles
- Limit the possibility of conflicting frequency and severity
- Provide space in medians and islands to install different types of traffic control devices

*Rotary intersections*: This is an enlarged intersection area in which a large, centralised, raised island is installed. Vehicles from the various approach roads converge at the entry to the intersection and are moved out to their respective directions in the clockwise direction by weaving around the island. For more details about rotary intersections, see *Chapter 12*.

**Based on traffic control:** Based on the movement of traffic on a selected length of a route, the flow in at-grade intersections may be classified as uninterrupted flow through uncontrolled intersections or interrupted flow through controlled intersections. In controlled intersections, traffic flow is controlled manually or using signals (Fig. 11.10).

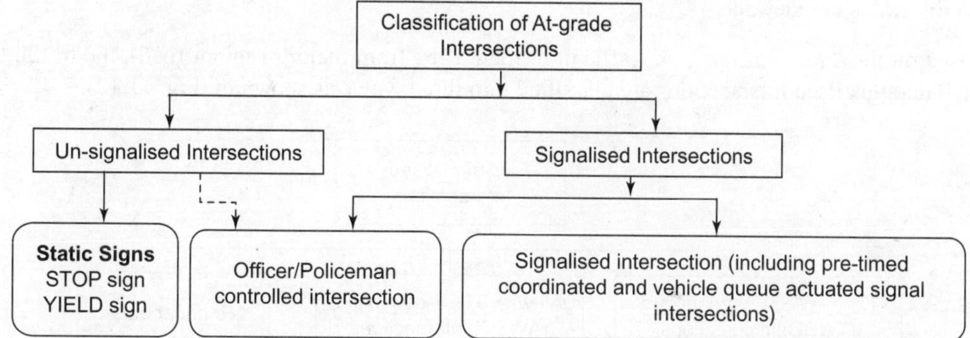


**Fig. 11.10** Classification of at-grade intersections based on traffic flow controls

**Fig. 11.11** A view of an un-signalised three-legged intersection

*Un-signalised intersections:* Intersections that carry lower traffic flow are usually operated as un-signalised intersections. The vehicles passing through un-signalised intersections (including roundabouts) are controlled by two types of *static signs*: STOP-controlled intersections and YIELD-controlled intersections.

- The **STOP sign** at an intersection requires the driver to stop the vehicle before entering the intersection, and to give way (i.e., yield) to any traffic in the intersection area or approaching the intersection; and then proceed when it is safe to pass. In another words, the word *control*, which is used in un-signalised intersection studies, represents the traffic flow in an approach in which a STOP sign is installed. This indicates that vehicles from that approach should wait before entering the intersection and then proceed when it is safe to pass.

- The **YIELD sign** at an intersection requires the vehicles approaching the sign to yield the right of way (i.e., give way) to any oncoming traffic in priority roads (i.e., major roads) and stop if necessary. Installation of this sign is also suitable for pedestrian crossings on the approach to crosswalks or sidewalks.

Based on the *relative priority* of traffic movement (i.e., from major or minor road), the traffic flow through un-signalised intersections are classified into three types, as shown in (Fig. 11.12).

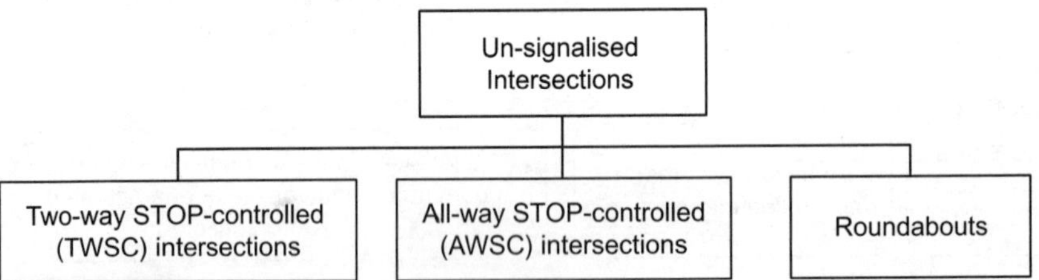


Fig. **11.12** Classification of un-signalised intersections based on relative priority of traffic flow

The classification of un-signalised intersections is explained as follows (TRB, 2000):

- **Two-way STOP control (TWSC)**: In the USA, STOP-controlled approaches are referred to as minor road approaches, while the major road approaches are not controlled by a STOP sign. The details of TWSC with reference to the number of approaches to an intersection are given below:
  - **Two approaches**: The two intersecting roads are one-way streets (Fig. 11.13 (a)). Traffic flow in one street (i.e., the Minor street) is controlled while traffic flow in the other street (i.e., the Major street) is prioritised.
  - **Three approaches**: Traffic flow in the stem-leg approach (i.e., the Minor street) of a T-intersection is controlled, while traffic flow in the other two approaches (i.e., the Major streets) is prioritised (Fig. 11.13 (b)).
  - **Four approaches**: Traffic flow in the minor approaches of a four-legged intersection is controlled, while traffic flow in the other two approaches (i.e., the Major streets) is prioritised (Fig. 11.13 (c)).
- **All-way STOP control (AWSC)**: All vehicles from all the approaches should stop at the intersection before proceeding (Fig. 11.13 (d)). Movement of such vehicles to pass through the intersection depends on the traffic conditions in the other approaches. The vehicle can proceed immediately after stopping if there is no traffic in the other approaches.
- **Roundabouts or traffic rotary**: The traffic moves in one direction around a central island (Fig. 11.14). Vehicles enter in an orderly fashion into the circulating width and move in the clockwise direction. Major conflicts, as may be encountered in un-signalised intersections – between vehicles turning right and those moving through traffic, are eased by converting the movements into merging and diverging movements. The end result is division of a single stream of traffic into separate streams. While moving, vehicles weave over a convenient distance with reduced speed and exit in their desired direction.

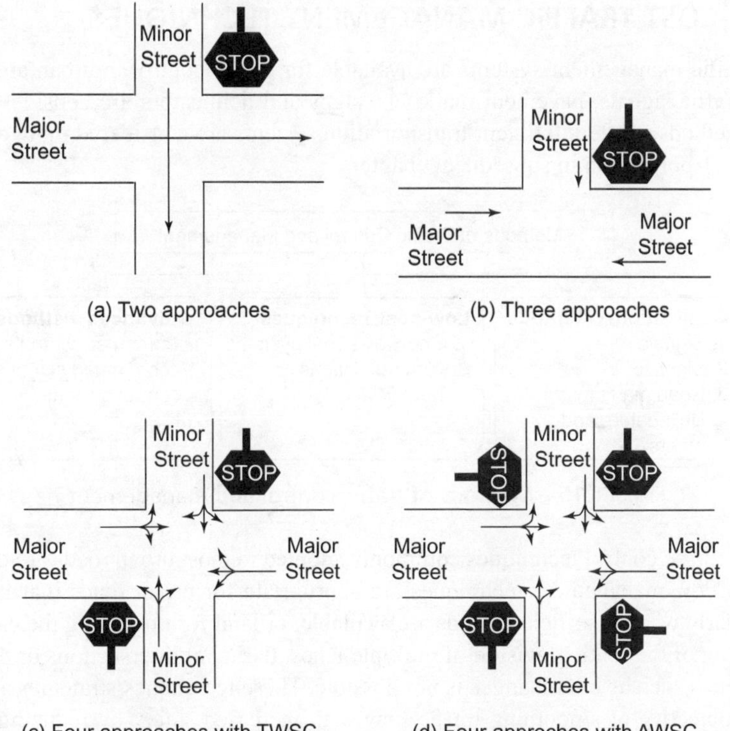


Fig. **11.13** Types of un-signalised STOP-controlled intersections at grade

(a) Three-legged rotary intersection

(b) Four-legged rotary intersection

**Fig. 11.14** Typical aerial view of roundabouts

(*Source*: www.jamartech.com; *Courtesy*: JAMAR Technologies Inc.)

*Signalised intersections*: Please see *Chapter 13* for detailed information.

## 11.4 LOW-COST TRAFFIC MANAGEMENT TECHNIQUES

A number of traffic management systems are available for traffic control in urban areas (Fig. 11.15). Among them, traffic signals, pavement markings, signs and delineators are considered conventional traffic control methods, while intelligent transportation systems are considered more advanced. These methods will be elaborated on in subsequent chapters.

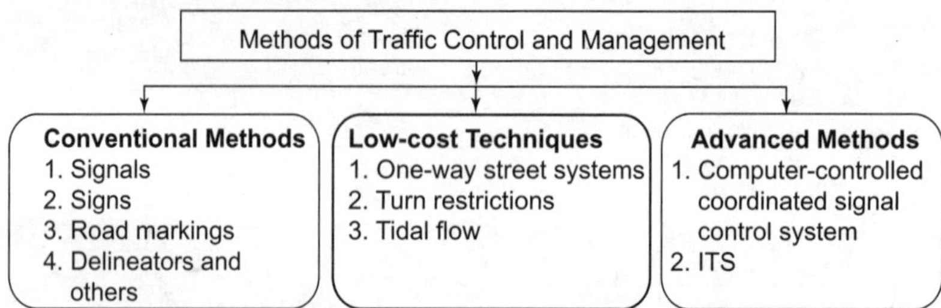


**Fig. 11.15** Methods of traffic control and management

The low-cost traffic control techniques commonly adopted in most urban roads are described below. Low-cost traffic flow management techniques are appropriate for urban roads that have high traffic volume, particularly when insufficient funds are available, or land for upgrading the capacity is scarce or where widening of the road, provision of multiple lanes, flaring at intersections or providing grade-separated structures such as interchanges is not possible. These techniques/strategies are implemented with the broad objective of smoothing traffic flow with increased safety by reducing the number of accidents and also to ensure optimal use of the existing infrastructure.

- **One-way street systems**: Traffic flow is permitted in one direction only. There will be no opposing traffic on these streets. This technique is applicable when demand exceeds lane capacity or when traffic flow needs to be diverted to a particular direction to avoid traffic delays.
- **Turn restrictions**: This strategy is applicable to traffic that turns across the opposing traffic stream by banning left or right or both turns at a signalised intersection. As a common practice, right-turn crossings are banned at signalised intersections so that crossing movements can be eliminated and the number of signal phases can be reduced. To avoid diverging and merging conflict, restrictions on U-turns at intersections or median openings is another traffic control measure.
- **Dedicated bus, taxi and cycle lanes**: Each lane is exclusively assigned for use by a specific type of vehicle. This technique is adopted so that slow-moving, fast-moving and high-occupancy vehicles are segregated from mixed traffic flow conditions.
- **Shared lanes**: At a few locations on urban roads, the left lane of a multi-lane road is given over exclusively for loading and unloading goods during a specified period of a day; the remaining lanes will be operated as usual for through-traffic flow. Variable or changeable message signs may be used for this purpose.
- **Staggered timings**: To avoid excessive traffic-density during peak periods, school timings and office timings in a locality are staggered. The same lanes can be used effectively at the assigned times for the specific trip. The strategy of variable work-hours is a variant of this technique.
- **Odd or even vehicle registration numbers**: Restricting vehicle-use based on the odd or even number of the registration of vehicles is one techniques of avoiding traffic volume and environmental pollution. This technique encourages commuters to car-pool or van-pool or to use mass transit systems such as buses and local trains.

- **Contra flow lanes**: On high-speed arterial roads, only buses (high-occupancy vehicles) are permitted to travel in the reverse direction on specified lanes. This is a common practice in some cities in India, as a part of the Bus Rapid Transit (BRT) system.
- **Tidal flow technique**: In multilane urban arterial roads, the centre lanes are sometimes assigned for moving traffic in one direction in the morning peak period and in the opposite direction during evening peak period. This technique is adopted on a daily basis during peak periods when the recurrent lane capacity is exceeded in one direction and is minimum in the opposite direction.

  Gantry-mounted overhead signals are used for indicating and controlling the traffic flow in the desired directions. Variable-lane assignment signs may also be used for this purpose, as a system of dynamic reversal of lane directions. As an alternative, special equipment may be used to shift the central safety barriers across the lane, either to the left or to the right side lanes. As a standard practice, the lane direction change is implemented on a fixed-time basis, for a predetermined number of hours.
- **Congestion charging or congestion pricing**: It is a system of surcharging road users during peak periods so that traffic congestion is reduced by discouraging traffic flow on certain busy lanes in urban roads, particularly on week days. The road pricing strategy regulates traffic demand and manages traffic without enhancing the capacity of lanes. This practice is common in some busy roads in London.
- **Traffic calming measures**: These are intended to reduce the speed of vehicles, thereby reducing the number and severity of accidents. Calming refers to the calming of residential roads. These measures restrict traffic movement, often by erecting barriers across the road. Some of the traffic calming options commonly used are
  - Installing speed limit signs
  - Installing and operating speed cameras which enable self-enforcement and which are visible as evidence
  - Channelisation at intersections including wider central island refuge for pedestrian crossings at mid-block locations. The wider central medians restrict speed in narrow carriageways
  - Mini-roundabouts, which can reduce the speed of through-traffic
  - Speed breakers/humps
- **Traffic control in the work zone**: Traffic separation from the site of construction adjacent to the roadway is a mandatory requirement of work executors and road administrators. Adequate separation in the horizontal and vertical directions needs to be provided to avoid conflict with traffic flow and to avoid accidents. Traffic cones, arrow boards, flag trees, barriers, barricades, traffic barrels, flashing chevrons and delineators are installed; when work is carried out on the roadside, transition tapers, buffer space, work space and terminal areas are provided. Standard templates of these installation plans are available at IRC: SP: 55 (2014), MUTCD (2009, 2012) and ODOT (2016).

These techniques/strategies may be integrated with the devices/detectors/sensors used in ITS to provide comprehensive traffic control and management for any large road network.

## 11.5 FACTORS CONSIDERED IN THE DESIGN OF INTERSECTIONS

The safe and efficient movement of traffic is the main concern in the design of intersections. The other factors that are considered in the design of intersections are as follows (IRC: SP: 41-1994):

- **Driver/Pedestrian factors** (Table 11.1)
  - The performance of the drivers/cyclists in using the road facility at the intended design speed in terms of skill, perception time and reaction time in understanding how to traverse the intersection

- Pedestrian characteristics such as age and their physical ability to cross the intersection
- Pedestrian habits such as talking on cell phones and listening to music while walking/crossing the road
- Visibility is better at right-angled intersections than at skewed intersections. At skewed intersections, the ability of drivers to recognise conflicting vehicles diminishes when compared to the same at right-angled intersections (Kaluva, 2013).

**Table 11.1** Values of driver/pedestrian factors appropriate for the design of intersections

| *Human factors* | *Design values* | *Design elements affected* |
|---|---|---|
| Perception/Reaction time | 2.5 seconds | Intersection sight distance (ISD) |
| Gap acceptance | 5.5–8.0 seconds | SSD, SD |
| Eye height of driver | 1.2 m | Intersection sight triangle |
| Walking speed of pedestrian | 1.2 m/sec | Pedestrian facilities |

- **Road geometric factors**
  - The intersecting pavement area should increase with the degree of skew of the intersecting roads (Fig. 11.4). As the skew angle ($\theta°$) increases, so does the difficulty in detecting potentially conflicting vehicles crossing the roadways, and the time required to manoeuvre through the intersection, for both vehicles and pedestrians.
  - The sight triangle and the dimensions along the centre line of the intersecting roads. Sight distance at an intersection should be adequate to provide an unobstructed view of the entire intersection area, and the view length should permit control of the vehicle to avoid collision/conflict. The sources of deficiency may be physical obstructions close to the intersection, severe grades and poor horizontal alignment of the intersecting roads. The physical obstructions in the sight triangles may consist of the vehicle stop-line setback for pedestrian crosswalks, trees, fencing, signage, traffic signal control boxes, flyover pillars and other developments (NDOR, 2012).
  - Adequate turning radius must be provided on intersections. For example, in skewed intersections, larger vehicles or semi-trailers are not recommended for use when the skew angle is greater than 200° (Son et al., 2002).
  - The skew angle and the resulting larger area makes turns more difficult, increases pedestrian exposure, and thus, reduces intersection capacity.
  - Inappropriate large kerb radii should redesigned.
- **Traffic factors**
  - Design speed, physical and operating characteristics of vehicles (Table 11.2)
  - Composition of traffic, with the proportion of heavy vehicles
  - Capacity of vehicles to stop, accelerate and decelerate
  - Height of the driver's eye above the carriageway
  - Traffic flow is adversely affected with the degree of skew in terms of distance to be travelled and the sharpness of turns. As a result, drivers are required to travel longer distances through the intersection and take longer time to cover the longer distance (Walker, 1981)
  - Average queue delay increases with the skew angle at STOP-controlled intersections, and a typical trend curve is shown in Fig. 11.16.
- **Safety issues**
  - Long pedestrian crossing distance should be resolved by constructing raised intermediate medians.

- As the intersection area increases with increased value of skew angle, a large conflicting area will form. As a result, the vehicles will be exposed to conflicting traffic for a longer period (Walker, 1981). In other words, large skews increase the pavement intersection area, and thus, crossing vehicles and pedestrians are exposed for longer periods.
- At skewed intersections, the right turns may become wider and the left turns sharper. For example, at left-skewed intersections (Fig. 11.4 (c)), drivers make a sharp left turn, which brings them into conflict with the oncoming traffic. In right-skewed intersection (Fig. 11.4 (d)), drivers making right turns may speed up (due to larger turning radius than the left turn) and maintain unsafe gaps while turning (Gattis and Low, 1997).

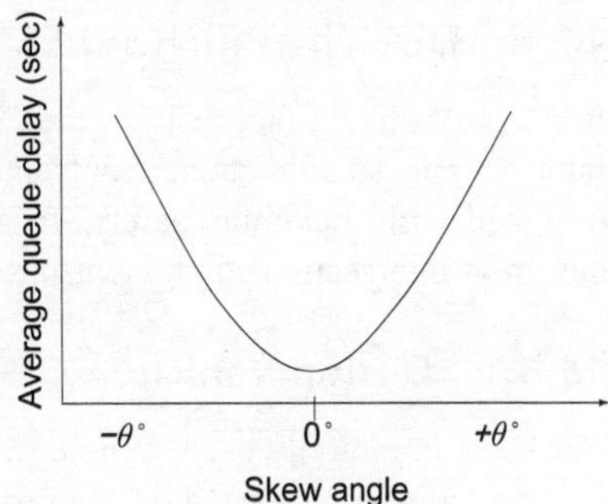


**Fig. 11.16** Effect of skew angle on average queue delay (Redrawn from Bonneson, 1993)

- An investigation revealed that skewed intersections had higher crash and fatality rates than normal intersections (Burchett and Maze, 2006).
- The impact of skew angle on safety is more severe at four-legged STOP-controlled multilane intersections than in three-legged and signalised intersections. This is because the traffic flow configuration in four-legged intersections is more complex than in three-legged intersections.
- At signalised intersections, skew has less effect on safety because vehicle movement through the intersection area is controlled by signals (Oh et al., 2004).
- Proximity to on-street parking spaces/transit stops

- **Environmental factors**
  - Drainage facilities such as inclination of the camber, run-off and the location of side drains, proximity to farmland/agricultural areas, mist that affects SSD, etc.
- **Economic factors**
  - Land acquisition issues and cost, and the cost of constructing/reconstructing all the required installations
  - The use of the adjacent land governs the selection of the type of intersection. For example, in urban areas, pedestrian crossings and on-street parking are some common components. In residential areas, bicycle lanes, pedestrian and school zone crossings are seen. In highways in rural areas, truck storage lanes and parking are essential components (NASRAL).

**Table 11.2** Vehicular characteristics applicable for the design of channelised intersections (NASRAL)

| *Physical characteristics of vehicles* | *Related design elements of intersections* |
|---|---|
| Length | Length of storage lane |
| Width | Width and number of lanes, widening on curved paths |
| Height | Placement of overhead signals and signboards |
| Wheelbase | Turning radius, swept path width |
| *Operational characteristics of vehicles* | *Related design elements of intersections* |
| Acceleration capacity | Length of acceleration lanes and tapers and headway, SSD, ISD |
| Deceleration and brake efficiency | Length of deceleration lanes and tapers |

## 11.6 DESIGN PRINCIPLES OF AT-GRADE INTERSECTIONS

It is a well-known fact that at-grade intersections provide a lower level of safety than other types of intersections, such as roundabouts and interchanges. For this reason, at-grade intersections should be designed with maximum safety. To achieve this objective, the following parameters relating to the design of intersection components are considered important.

### 11.6.1 Design Vehicle

The most commonly used vehicle ($85^{th}$ or $15^{th}$ percentile use in any particular vehicle class) which is intended to be operated on the proposed roadway intersection is designated as the design vehicle (NASRAL). The wheelbase (WB) and width of the design vehicle are the most important dimensions which determine the minimum turning radius, kerb radii and channelisation design. In India, the most commonly used vehicles (along with their WB) are: passenger car (3.34 m), single unit truck (6 m), semi-trailer truck (12.2 m and 15.2 m) and truck trailer (18.2 m). The maximum width of vehicles permitted in India is 2.5 m (IRC: 3-1983). Adequate widening of the intersection area at turnings is provided to accommodate the wider turning paths traced between the outer front tyre and the inner rear tyre of the design vehicle. AASHTO (2004)-recommended turning templates may be adopted for the geometry of study intersection area and for the selected design vehicle WB. The kerb radii should be appropriately adjusted to this requirement. A few sample turning templates are presented in Fig. 11.17.

### 11.6.2 Stopping Sight Distance

Stopping sight distance (SSD) may be defined as the safe minimum visible straight distance of a driver along the centre line of a highway when the height of the driver's eye and an object in the opposite direction above the road surface is 1.2 m and 0.15 m, respectively (Fig. 11.18). With this condition, the SSD is measured along the centre line of the far lane of the carriageway at an intersection, in the plan. The details of calculating minimum SSD are given below (IRC: 66-1976).

Minimum sight distance should be provided for the given conditions, such as design speed, gradient and coefficient of later friction. The sum of lag distance and braking distance is termed the stopping sight distance (SSD), for the assumed driver perception time of 2.5 seconds. The following formula is used to calculate SSD.

Stopping sight distance (SSD) = Lag distance (LD) + Braking distance (BD)

$$\text{SSD} = \text{v t} + \frac{v^2}{2g(f_s \pm 0.01\,n)} \tag{11.1}$$

where $t$ = mean total reaction time of the drivers (seconds); $v$ = velocity of vehicle (m); $g$ = acceleration due to gravity (m/sec$^2$); $n$ = ascending or descending gradient (%); and $f_s$ = coefficient of longitudinal surface friction (0.35–0.4 for a speed range of 100–20 km/hr).

The SSD should be provided continuously on all roadways including at the approaches to intersections.

At un-signalised intersections, provision of clear and adequate length of sight distance on the corners of intersections, on turnings, is very important. Obstruction-free line of sight between a driver waiting at a cross road and the driver of the approach vehicle should be established with the design speed on the approach road. Adequate time must be provided to the waiting driver to pass from the intersection.

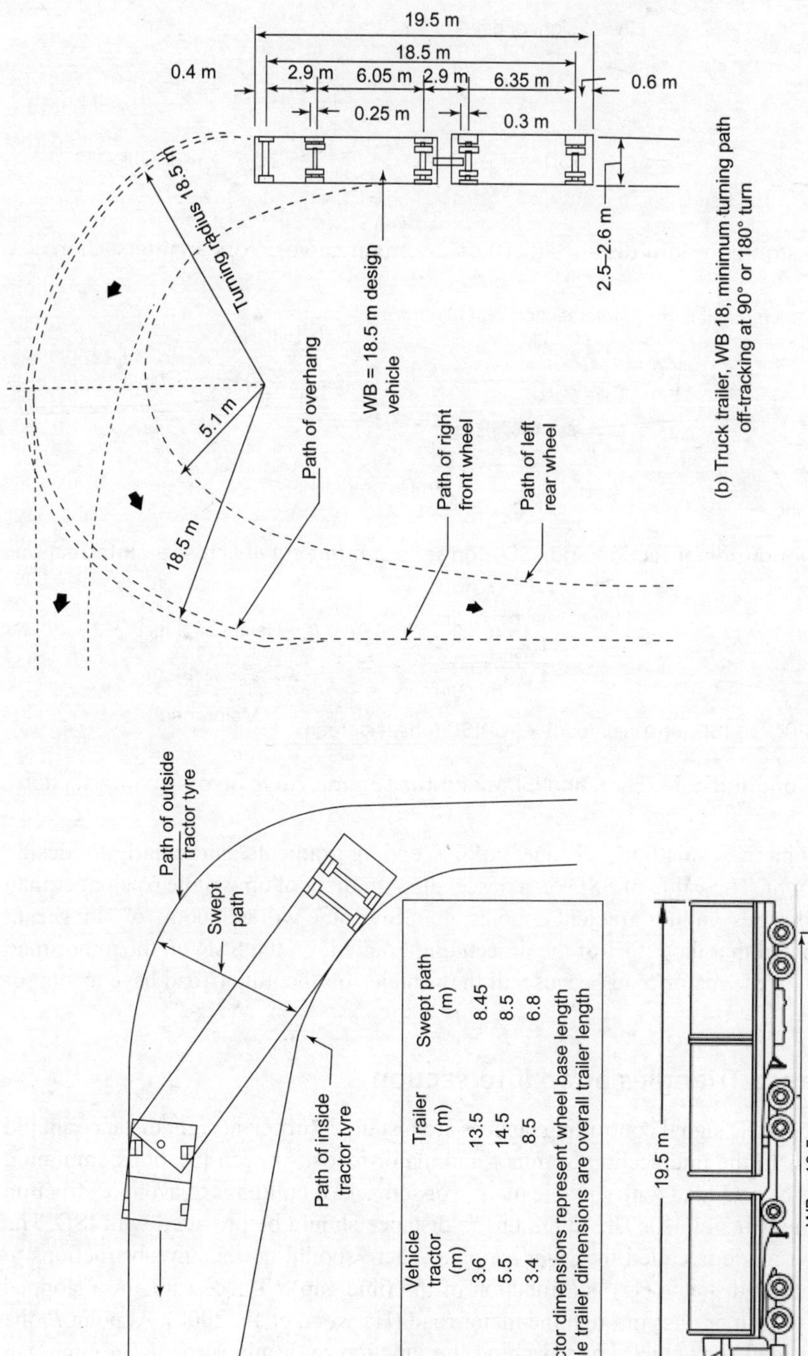


**Fig. 11.17** A sample turning template (Modified from IRC: SP: 41-1994 and NASRAL)

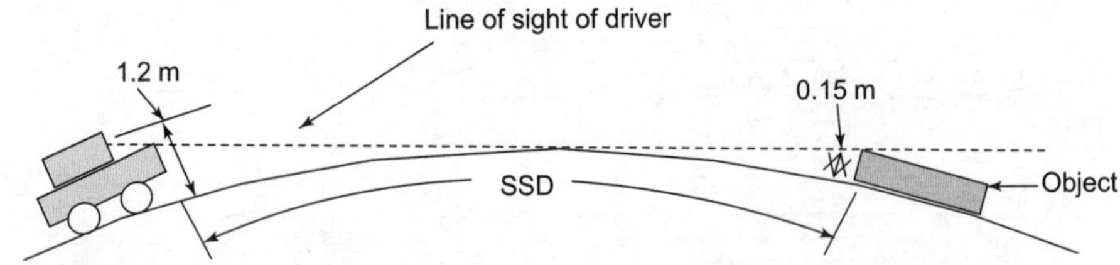


Fig. 11.18 (a) Safe stopping sight distance (SSD) on a summit curve across an intersection

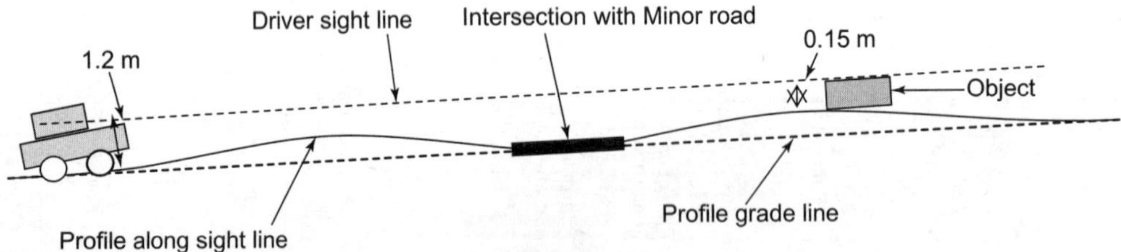


Fig. 11.18 (b) Typical longitudinal section and SSD along the summit curve across an intersection

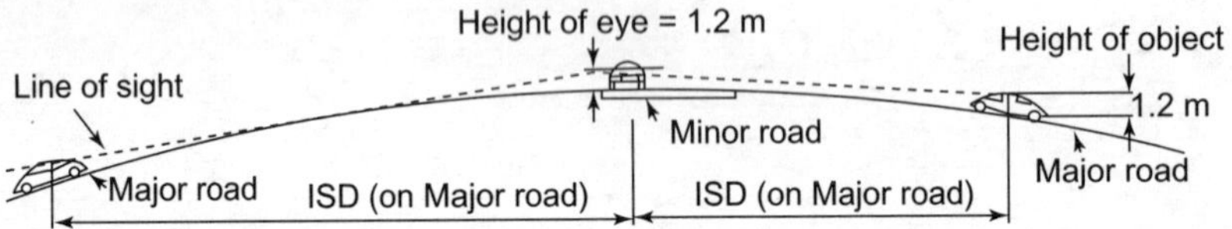


Fig. 11.18 (c) Typical longitudinal section and ISD along the summit curve across an intersection

*Note:* The length of SSD increases with the steepness of descending gradients, particularly for design speeds in excess of 80 km/hr. The value of SSD on a descending gradient of 35 will be 6% longer than that on a level grade. If this descending gradient doubles, the SSD value will be about 16% longer. In this context, it must be noted that the effect of the descending gradient on the SSD on the minor road is more critical than that on the major road because all the vehicles on the minor road have to stop or yield (NASRAL).

### 11.6.3 Sight Distance Triangles at an Intersection

At intersections where no traffic signal controlling devices are installed, sufficient sight distance should be provided at approaches to the intersection of minor and major roads. At such locations, minimum sight distance (*xz*) should be provided with suitable margin or setback of buildings to avoid obstruction of vision along the *xy line* (Fig. 11.19). The minimum *xz* distance should be provided with ISD. The area covered by the *xyz* triangle is called the *sight triangle*. There should not be any obstructions to vision in this area. The sight distance (*xz*) is a function of the time gap ($t_g$) necessary for a stopped vehicle on a minor road to turn or pass through the major road (Harwood et al., 2000). At point *B*, the total STOP board is provided for vehicles approaching the junction on a minor road. The time gap acceptance method is specified by AASHTO (2001) for determining the ISD based on assumptions that (i) the time gap is independent of the approach speed on the major road, (ii) the intersection angle is

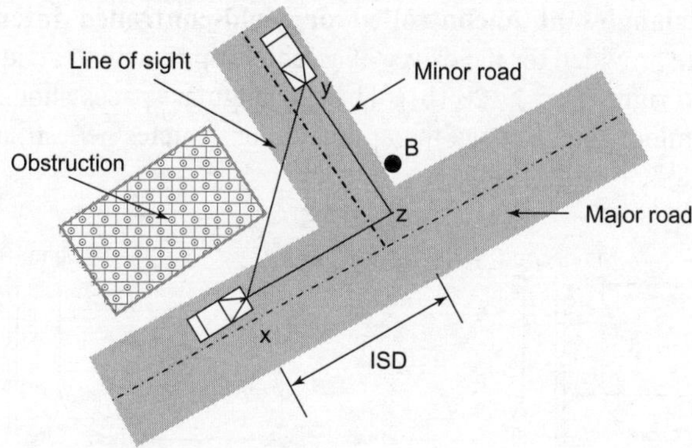


**Fig. 11.19** Details of intermediate sight distance (ISD)

90° and (iii) both the major and minor roads are straight and have no curvature in the horizontal and vertical planes. The required sight distance (*xz*) is calculated using the following relationship.

$$\text{ISD} = \text{Sight distance } (xz) = v \times t_g \tag{11.2}$$

where $v$ is the design speed on the major road (m/s) and $t_g$ is the time gap (sec) required for a vehicle on a minor road to proceed. For different types of vehicles, the value of $t_g$ may be assumed as 7.5 sec. If the approach gradient to the minor road is on an upgrade greater than 3%, increase the ISD value by 10%. More such details may be obtained from AASHTO (2004).

Adequate sight distance should be provided at intersections to view vehicles that are approaching or to stop or reduce speed or cross the main road smoothly without collisions (in the case of uncontrolled intersections). As per Indian practice (IRC: 66-1976), the minimum visibility distance (*yz*) along the minor road is taken to be 15 m and the average time gap to travel the distance (*xz*) along the major road is assumed to be 8 seconds. For example, for a vehicle standing on a minor road 15 m before approaching the major road, the minimum visibility distance required along the major road is $v \times 8$, where $v$ is the design speed in m/s on the major road.

## 11.6.4 Types of Sight Triangles at an Intersection

The driver of a vehicle approaching an intersection or departing from an intersection should have an unobstructed view of the intersection from the left and right approaches. A sufficient length of clear view along the left and right sides of the intersecting legs of an intersection should be provided to the driver entering the intersection so as to enable him to anticipate and avoid potential collisions. Such unobstructed views form triangular areas known as sight triangles.

The area between each adjacent road intersection may be termed a quadrant. Three and four quadrants will be formed for a typical three-legged and four-legged intersection, respectively. The same number of sight triangles will be formed with these configuration intersections. The area enclosed in these triangles should be free from any type of obstruction, including traffic controlling installations. If not, any obstruction might block the driver's view of conflicting vehicles, pedestrians and animals. Based on the position of the driver on the minor road and the intended turning movement, the sight triangles are categorised into two types (AASHTO, 2011):

- **Approach sight triangles at uncontrolled or yield-controlled intersections**: These clear triangular spaces are provided for the driver of a vehicle approaching an intersection from the left (Fig. 11.20 (a)) and right (Fig. 11.20 (b)). These triangular spaces should be large enough that vehicle drivers on minor road can see the approaching vehicles/pedestrians on the main road in sufficient time to stop or avoid collision.

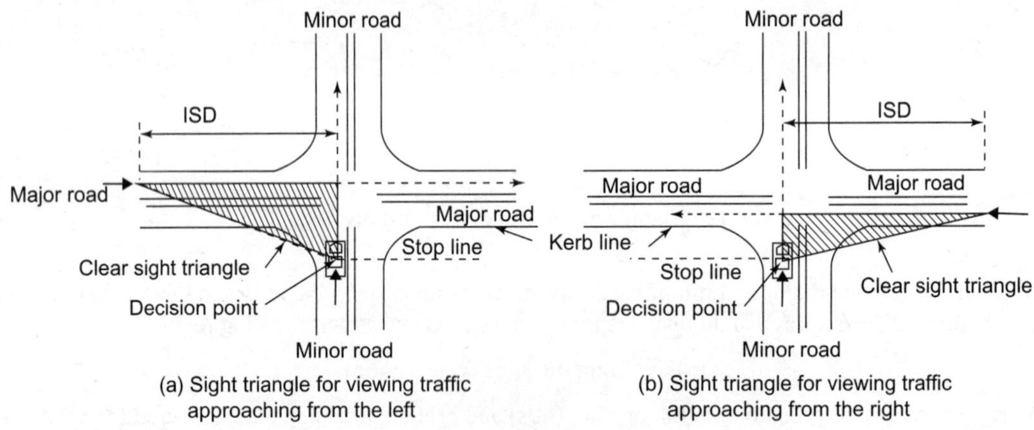


**Fig. 11.20** Approach clear sight triangles at uncontrolled or yield-controlled intersections on a minor road (Redrawn based on AASHTO, 2011)

- **Departure sight triangles at STOP-controlled intersections**: These triangular spaces are provided at intersections for viewing traffic on the main road approaching the minor road from the left (Fig. 11.21 (a)) and right (Fig. 11.21 (b)), when the driver of the vehicle stops at the stop line on the minor road. These triangular spaces are smaller than approach triangles because the driver on the minor road stops at the stop line and waits for a favourable chance to cross or enter the major roadway.

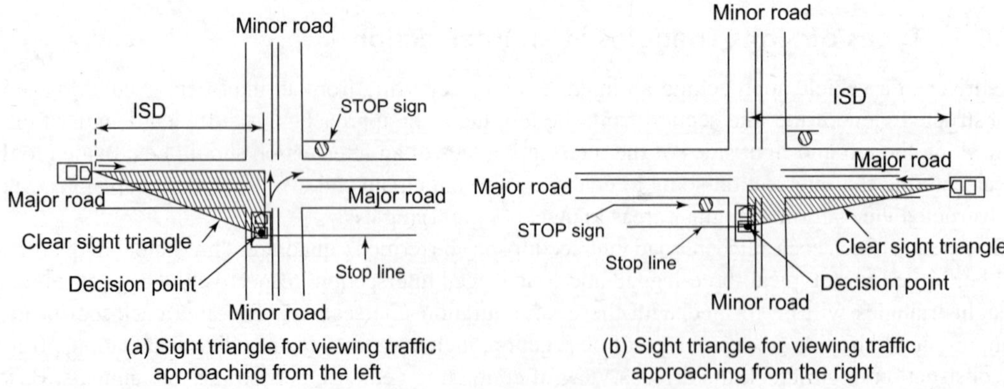


**Fig. 11.21** Departure clear sight triangles at STOP-controlled intersections on a minor road (Redrawn based on AASHTO, 2011)

### 11.6.5 Sight Distance at Median

In the case of a median crossing, vehicles generally stop adjacent to the median and cross or make a U-turn when sufficient distance is available from the oncoming vehicle (B) (Fig. 11.22). To cross the median, the minimum required sight distance (SD) may be determined by the equation

$$SD = v \times T$$

where $v$ = design speed of the vehicle on the opposite lane (m/s).

$$T = t_p + t_c$$

where $t_p$ = perception time (sec) and $t_c$ = time (sec) required by a stopped vehicle adjacent to the median to cross or make a U-turn.

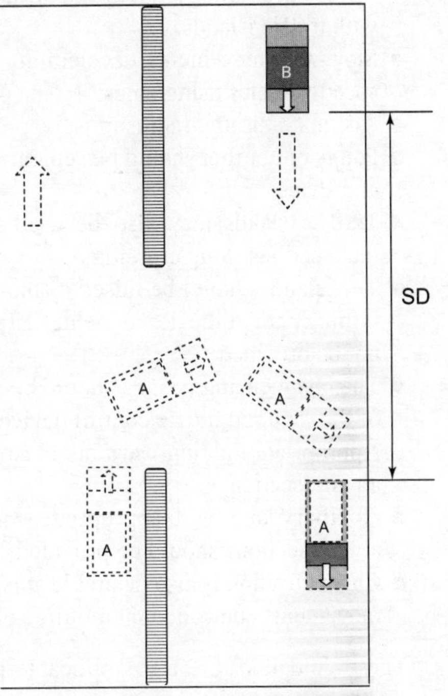


**Fig. 11.22** Sight distance (SD) at a median turning

## 11.7 PRINCIPLES OF CHANNELISATION

The purpose of channelisation is to *reduce* areas of conflict by *separating* and *regulating* traffic movement into predefined paths of travel using raised islands and pavement markings. Traffic islands may be composed of paint, kerbs, pavement edges, raised pavement markers, splitter islands or other devices. These islands can be installed at signalised and un-signalised intersection areas, particularly in urban areas, where the traffic speed is not high. Traffic islands serve the following functions:

- Channelisation confines traffic flow to specific, definite, predefined directions and paths
- Traffic flow can be divided into separate groups based on lanes and/or the direction of flow
- Traffic islands can be used as pedestrian refuges when they cross roadways
- Depressed medians (with flat slopes and a gently rounded bottom) may be used in rural areas as a separator, and for drainage control
- The shape and size of the islands eliminates potential conflicts when pedestrians cross or approach traffic
- Channelisation should not result in the creation of an archipelago of small islands, which would tend to confuse drivers and lead them to selecting the wrong path through the intersection area.

The degree of channelisation required for a roadway midblock/intersection is based on the configuration of the road intersection, and vehicular and pedestrian traffic volume in peak periods. Traffic analysis by directional flow, conflicting movements, relative importance of traffic flow by direction and the nature of the conflict can help in the design of the required channelisation. The following guidelines may be considered while designing a channelised intersection (NDOR, 2012):

- Preference alignment should be given for free flowing major road traffic movements. In this regard, appropriate control should be imposed on minor movements such as stopping, converging and diverging. The paths traced by traffic flow should conform to the natural path of vehicles, thereby promoting smooth and efficient operation. Further, undesirable moves should be restricted with traffic islands.

- Slow-moving vehicles, decelerating vehicles and stopped vehicles should be separated from high-speed vehicles using lanes.
- Adequate turning radius must be provided for the intended use of the design vehicle.
- Points of conflict should be separated and reduced by installing islands/markings of suitable size and shape.
- Traffic islands may also be used as pedestrian refuges, particularly on multilane undivided intersections (long crossings).
- The islands should be raised as mountable kerbs and the curve should be parallel to the natural turning path of the design vehicle. Turns greater than 90° and sudden sharp/reverse curves should not be provided.
- The channelising islands should be designed to provide appropriate locations for the installation of the required traffic control devices, such as signals and signs.
- Appropriate lighting warrants must be checked in the case of raised channelisation designed at an intersection.
- Auxiliary lanes with appropriate storage areas for right-turning vehicles at median openings and at intersections should be provided clear of the through traffic lane (Neuman, 1985).
- Channelisation is also desirable at isolated signal control intersections, to separate various traffic movements controlled by multiple phase signals.

These guidelines are also applicable in the case of signalised intersections, so that coordination exists between signal control and the installations relating to channelisation. The collision records at a study intersection can be viewed as a valuable guide while selecting the type of channelisation required.

## 11.8 ELEMENTS OF CHANNELISATION

The following components/tools are widely used for channelisation.

### 11.8.1 Alignment of Lanes

Traffic lanes should be properly defined and arranged, with the appropriate width, direction, intersection, weaving length, sight distance and location.

### 11.8.2 Islands

An island is a defined area between traffic lanes in an intersection used for separating vehicular movements and for pedestrian refuge. Islands may be designed with different sizes and shapes; and they may be flush with the kerb or raised. The shape of the islands varies to facilitate the predefined driving path. In general, their shape may be triangular and elongated. Elongated divisional islands are normally provided with a width of at least 1.2 m and a length of 6 m. In rural areas with high-speed traffic, the minimum and desirable area of triangular islands is 7 m$^2$ and 9.3 m$^2$, respectively. Smaller islands ($< 9.3$ m$^2$) are acceptable in urban areas with speeds of less than 36 km/hr. However, islands should be least 18.6 m$^2$ when they contain traffic control devices/luminaires or are used for pedestrian refuge (WSDOT, 2014).

Raised islands provide effective delineation and can also control/restrict undesirable turning movements, as compared to pavement markings. Channelisation with raised islands is preferred in urban areas, particularly where higher pedestrian volume is observed. Island height is usually delineated by pavement markings. The allowable height of the kerb or a raised island is a function of the design speed and may be categorised with reference to the classification of roads in rural/urban areas. The following kerb heights may be adopted (NDOR, 2012):

- On high-speed facilities ($\geq$ 80 km/hr): 7.5–10.0 cm height of concrete slope kerb
- On low-speed facilities ($\leq$ 72 km/hr): 15 cm height of integral concrete kerb

## 11.8.3 Kerbs

Kerbs are raised (or flat sloping face) elements located adjacent to the extreme sides of the travel way. Kerbs commonly adjoin walkways along roadways, including at intersections. Kerbs are also used as semi-barriers with storm-water drainage channels (or gutters). Kerbing is a widely used practice of channelisation in urban areas, particularly where vehicle speeds are relatively low. In addition, kerbing is also used for the following purposes:

- To delineate the edges of carriageways and restrict driveways by width and direction
- To prevent unwanted entry onto walkways adjacent to the roadway
- For drainage control
- To protect the edges of surfacing along the carriageway (buried edge blocks)

In addition, vehicles may be allowed to drive off the travel way to park on the verge of flat sloping kerbs, in residential areas.

Different types of road-kerbs are available, such as natural stone, pre-cast concrete and steel. Pre-cast concrete kerbs are the most widely used. Steel kerbs may sometimes be used in restricted urban areas that experience constant over-running of vehicles. Based on the type of installation, kerbs are commonly installed in the raised position and sometimes submerged. Based on their purpose and cross-section, kerbs are classified as follows (Table 11.3).

**Table 11.3** Types of kerbs

| *Shape (in mm)* | *Typical dimensions/Purpose/Suitability/Remarks* |
|---|---|
| Square/Rectangle | Length: 600 cm; Height: 300 cm; Width at bottom: 150 cm<br>The most commonly used kerb |
| Bull-nosed/Round<br>131, R19, 300, 600, 150 | Length: 600 cm; Height: 300 cm; Width at bottom: 150 cm; Width on top: 131 cm; Radius of top edge: 19 cm<br>Provides access to vehicles from the carriageway to private driveways |

(Continued)

**Table 11.3** (Continued)

| *Shape (in mm)* | *Typical dimensions/Purpose/Suitability/Remarks* |
|---|---|
| Half-battered (HB) 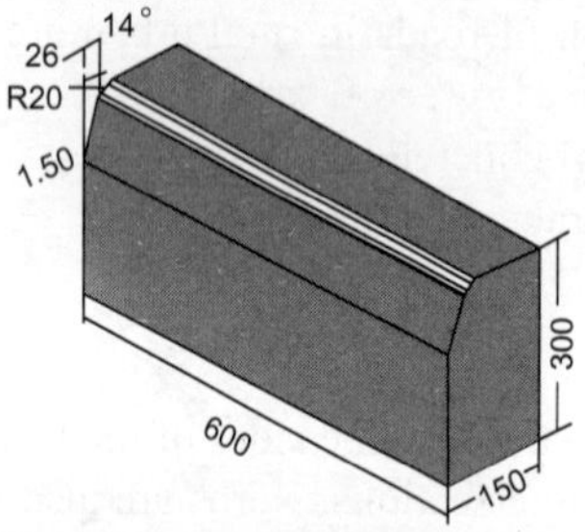<br>14°, 26, R20, 1.50, 300, 600, 150 | Length: 600 cm; Height: 300 cm; Width: 150 cm; Batter: 14° with the vertical<br><br>Most commonly used shape; its purpose is to deflect traffic back to the carriageway, thus protecting pedestrians |
| Splayed (45°) 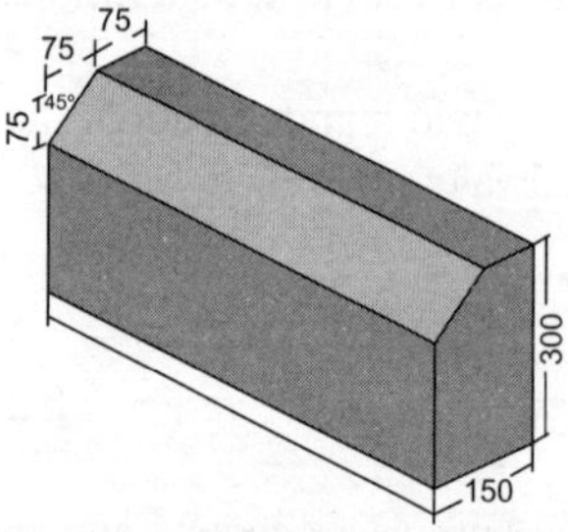<br>75, 75, 45°, 75, 300, 150 | Length: 600 cm; Height: 300 cm; Width: 150 cm; Splay: 75×75 cm<br><br>To allow access to vehicular traffic over itself at parking spaces; these are not used at pedestrian bays, for safety reasons |
| **Specific Purpose Kerbs** | |
| Dropped (45°) 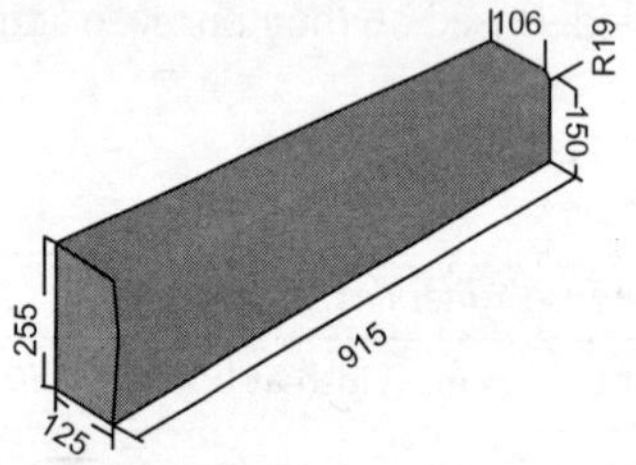<br>106, R19, 150, 255, 915, 125 | Used to link half-battered kerbs to bull-nosed kerbs <br>Half-batter kerb<br>Right-hand dropper kerb HB to BN<br>Bull-nose kerb |
| Lowered<br>Dropper left kerb<br>Dropper right kerb<br>Centre Bull-nose kerb 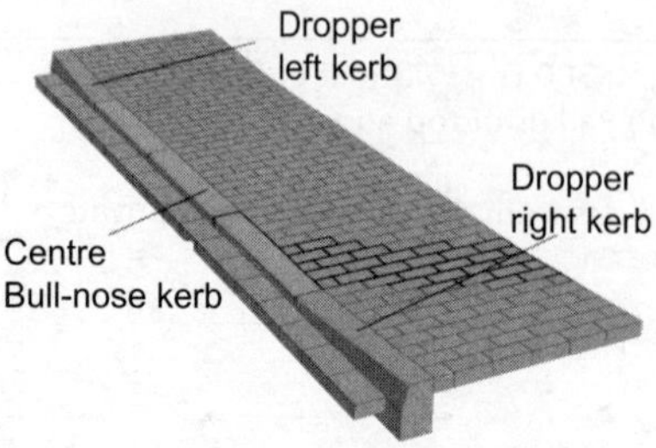 | Easily allow vehicles from the road to cross over onto private property/parking spaces |

(Continued)

**Table 11.3** (Continued)

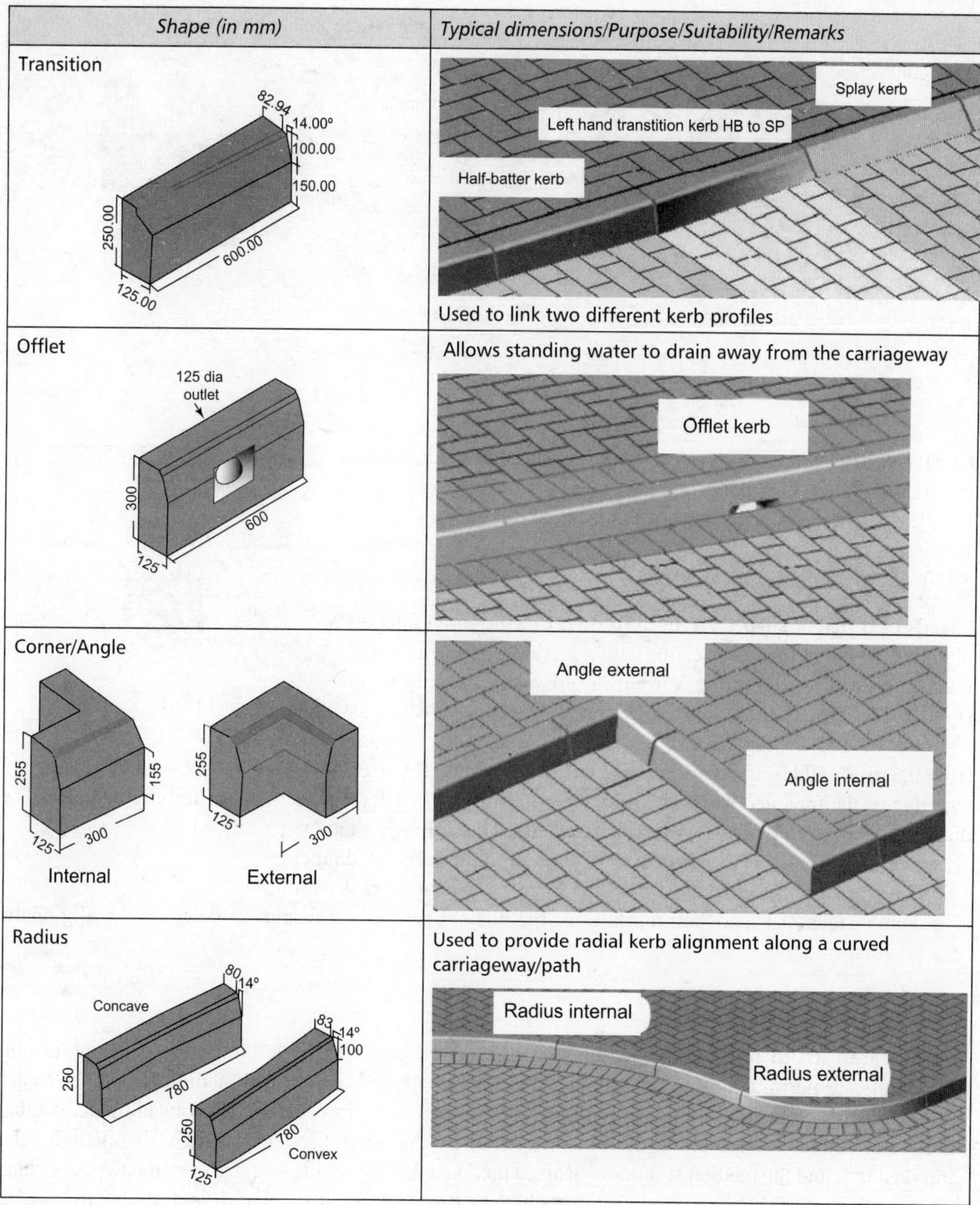


| *Shape (in mm)* | *Typical dimensions/Purpose/Suitability/Remarks* |
|---|---|
| Transition | Used to link two different kerb profiles |
| Offlet | Allows standing water to drain away from the carriageway |
| Corner/Angle<br>Internal<br>External | |
| Radius | Used to provide radial kerb alignment along a curved carriageway/path |

(Continued)

**Table 11.3** (Continued)

| *Shape (in mm)* | *Typical dimensions/Purpose/Suitability/Remarks* |
|---|---|
| Quadrant 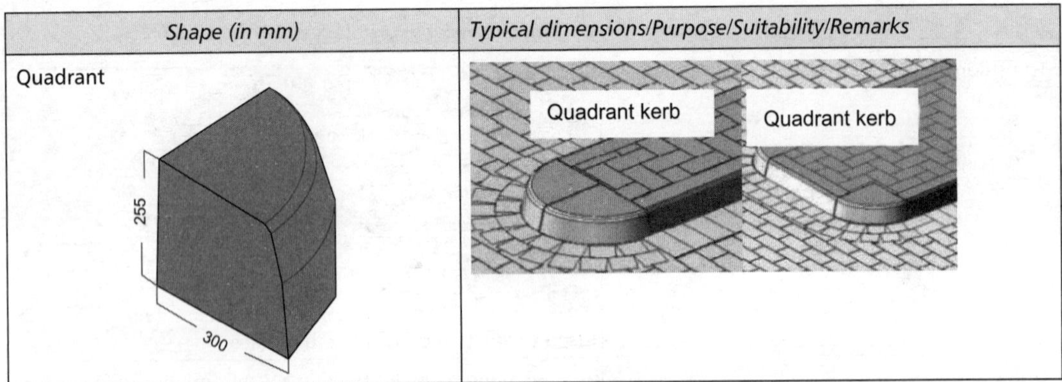  | |

http://vyaratiles.in/presentations/VWP_Kerbs_Presentation_from_VYARA.pdf

**Corner radii:** Corner radii at intersections should be constructed by aligning the kerb as an arc corner having a radius (R) of not less than 1.8–3.0 m and a maximum of 6.0 m (Fig. 11.23).

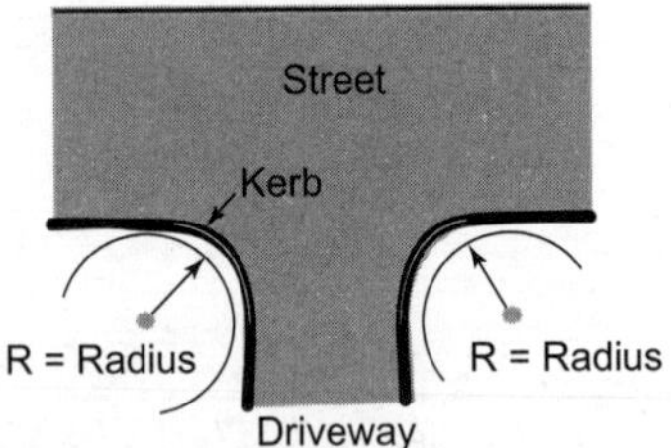


**Fig. 11.23** Kerb radii

## 11.8.4 Directional Islands

Islands of different shapes are used to divert traffic to an intended direction through the intersection. In general, triangular islands with an area of not less than 5 m$^2$ are used with rounded corners (Fig. 11.24). This minimum area criterion is adopted to ensure easy visibility to approaching drivers. The corner radius at the approach (R = 0.6 m) should be greater than the exit corner radius (R = 0.3 m). The offset between the island and the edge of the travelled way should be 0.6–1.0 m to allow lateral placement of kerbing vehicles. The dimensions and shape of triangular islands dictates the following factors (NASRAL):

- The corner radii of the directional islands and the associated tapers
- Alignment of the islands with reference to the intersection skew angle
- Dimensions of islands appropriate for the turning path of the design vehicle and the alignment of adjacent lanes

## 11.8.5 Median Islands

Medians are used to fully separate opposing traffic flow and clearly define where crossing/turning movements are permitted. Medians may be raised or depressed. Depressed medians (Fig. 11.25) are normally provided in rural areas while raised medians are seen in urban areas. Raised medians must have wider islands kerbed along the inner edges of the inside traffic lanes, which also includes the central islands and the median shoulders. Raised median islands provide better protection to pedestrians than painted islands. Median widths commonly vary from 1.2 m to over 9 m. It is desirable to provide medians (width 4.2–6.0 m) on multilane highways to facilitate storage for separate right turns.

Urban median islands are usually narrower than their rural counterparts. To accommodate single right-turn lanes, a median width of M = 5.4 m may be used (W = 3.6 m lane width + M = 1.8 m

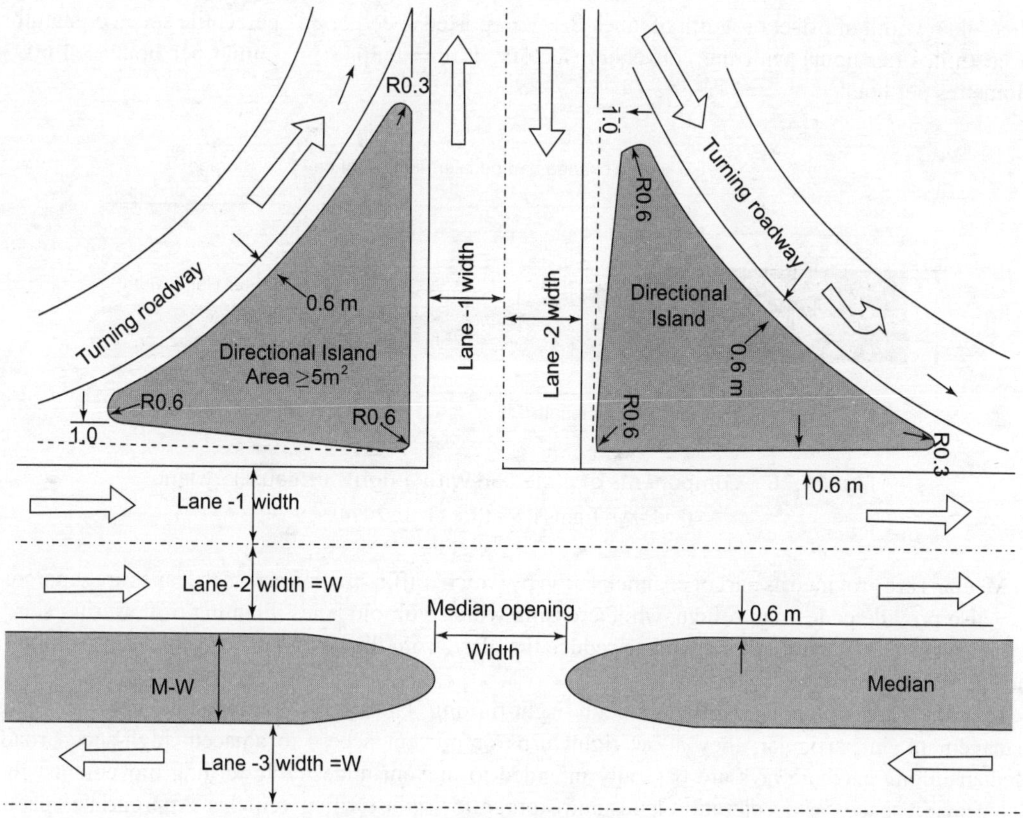

**Fig. 11.24** Typical layout of a three-legged multilane intersection with directional islands and a median opening

width of divider, Fig. 11.25). The 1.8-m divider (M − W = 5.4 − 3.6 = 1.8) provides pedestrian refuge, depending on its design (Fitzpatrick et al., 2005).

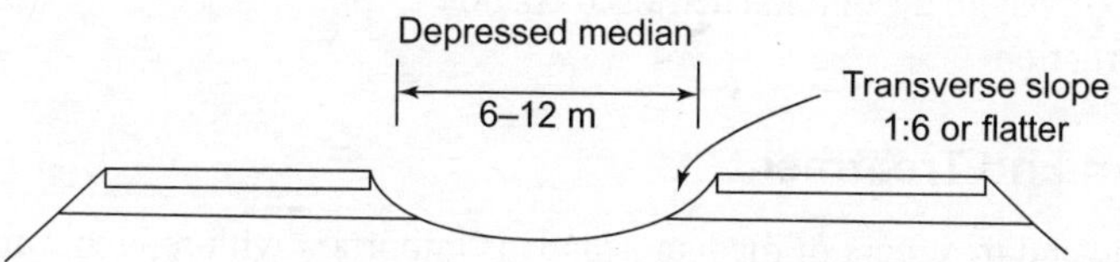


**Fig. 11.25** Sketch of a depressed median

The transition taper or taper length (Fig. 11.26) values can be calculated as recommended by the FHWA (MUTCD, 2009):

$$\text{For speeds of 45 miles per hour or more : Taper length (ft)} = W \times V \tag{11.3}$$

$$\text{For speeds under 45 miles per hour : Taper length (ft)} = \frac{WV^2}{60} \tag{11.4}$$

where $W$ = width of offset or width of lanes (ft); $V$ = posted speed or 85th percentile speed or statutory speed (miles per hour) with the conversion factors: 1 ft = 0.3048 m; 1 mile per hour = 1.60934 kilometres per hour.

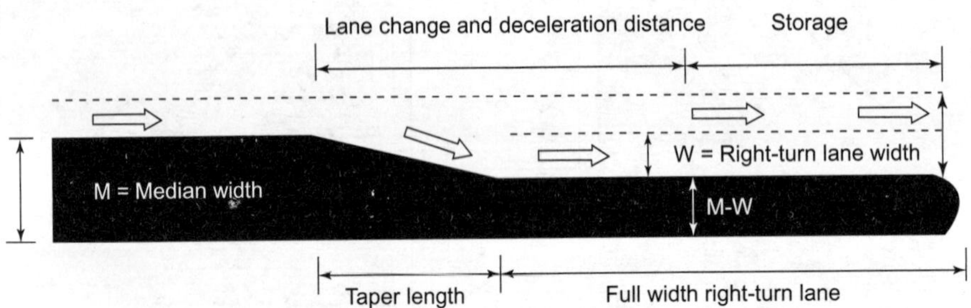


**Fig. 11.26** Components of a median with a right-turn auxiliary lane
(Redrawn from Fitzpatrick et al., 2014)

Medians are provided as part of channelisation by which traffic streams across the lanes are separated; they also provide pedestrian refuge, while crossing wider/multiple lanes. Planting of flowering shrubs is the usual practice on these medians to reduce the glare from the headlights of vehicles travelling on the opposite lane.

In urban areas, medians often contain right-turning lanes. Such alignments help in access management; in particular, they allow right turns or prevent access to adjacent high-speed roads. Medians along carriageways are basically intended to prevent undesirable turning movements; they also provide space for deceleration lanes and storage for right-turning vehicles. Appropriate spacing should be provided for right turns (as U-turns) along a street or roadway.

*Note:* In the case of major haulage routes, where larger trailers and combinations of trucks are prevalent, channelisation with raised islands are typically *omitted* for the following reasons (NASRAL):

- Low volumes of pedestrians
- To facilitate larger turning/manoeuvring area
- Limited right of way to accommodate raised islands
- Cost of construction

### 11.8.6 Median End Treatment

At intersections, the end treatment of median islands is important with reference to directing the turns allowed, skew angle of the intersecting roads and the width of the minor road. Different shapes of end treatments at median openings are used, which are determined by the width of the median (Fig. 11.27). Semi-circular end treatments are used up to a width of 3 m, while bullet-nosed ones are used when the median width is between 3 m and 5 m.

Bullet nose and parabolic curve are the most commonly found shapes of median end treatments; they provide better guidance for the manoeuvre of turning vehicles. The arcs radii of bullet-nosed edges are dictated by the wheel paths of the turning radius of the vehicles (Fig. 11.28). As a common practice, the radius of the nose may be assumed to be between 0.6 m and 1.0 m to fit within the prevailing roadway-turning geometrics.

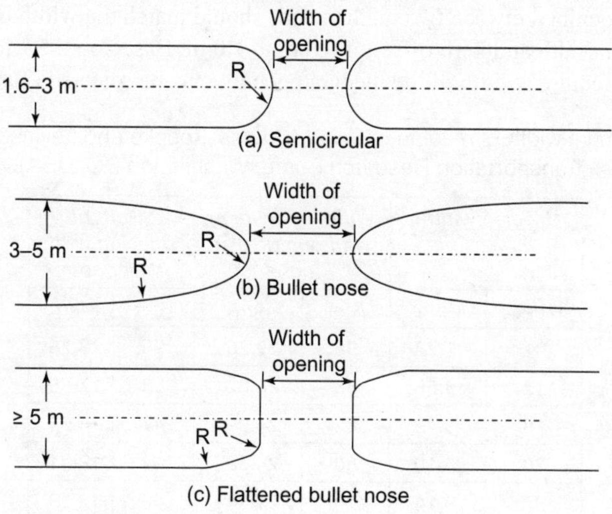


**Fig. 11.27** Commonly used median end treatments

| *Aspect* | *Suggested Design* | *Rationale* |
|---|---|---|
| Length | Minimum 40 ft, preferable 50 ft or more | Need adequate length for conspicuity, effectiveness. |
| Width | Absolute minimum: 4 ft<br>Minimum to provide pedestrian refuge: 6 ft<br>Width for visibility of landscaping: 8 to 10 ft<br>Maximum for a driveway divisional island (width of the part that is unavailable for travel, i.e., not including turn lane widths): 12 to 16 ft. | Absolute minimum based on the Green Book.<br>Maximum based on potential for drivers to mistake one driveway with a median for two separate driveways. |
| End treatment | The 2004 AASHTO Green Book states that for a median island less than 10 ft wide, a semicircular end shape is adequate. For median island widths of 10 ft or more, a bullet nose end shape is suggested.<br>From observations, a bullet nose shape may be desirable for widths of less than 10 ft. | To fit the wheel path of a turning vehicle, per the 2004 Green Book (p.697). |

Geometry of a bullet-nosed median end shape
Radius of nose
Median width / 2
Half-bullet nosed median-end shape
Stop line
Radius of nose

**Fig. 11.28** Median design guidelines

(*Source*: Gattis et al., 2010; *Courtesy*: Transportation Research Board, Washington DC, USA)

The width of the opening between two median ends should match the width of the minor roadway , including the shoulder width and kerb offset of 0.6 m to 1.0 m. Based on the width of the median and its end treatment, the required width of the median opening may be selected (Table 11.4).

**Table 11.4** Minimum width of median opening (*Source*: Koepke and Levinson, 1992; *Courtesy*: Transportation Research Board, Washington DC, USA)

| *Median width* | | *Width of opening with semicircular nose* | | *Width of opening with bullet nose* | |
|---|---|---|---|---|---|
| *Feet* | *Metre* | *Feet* | *Metre* | *Feet* | *Metre* |
| 4 | 1.22 | 96 | 29.26 | 96 | 29.26 |
| 12 | 3.66 | 88 | 26.82 | 58 | 17.68 |
| 18 | 5.49 | 82 | 24.99 | 47 | 14.33 |
| 20 | 6.10 | 80 | 24.38 | 44 | 13.41 |
| 30 | 9.14 | 70 | 21.34 | 40 | 12.19 |
| 40 | 12.19 | 60 | 18.29 | 40 | 12.19 |
| 50 | 15.24 | 50 | 15.24 | | |
| 60 | 18.29 | 40 | 12.19 | | |

*Note:* These values are applicable for single unit and semi-trailer trucks with a wheelbase of 6 m and 12.2 m, respectively, and a control radius of 12.m and 15.2 m, respectively.

These shapes of median end treatments can be used in combination to meet the needs, space restrictions and geometry of the intersection (Fig. 11.29).

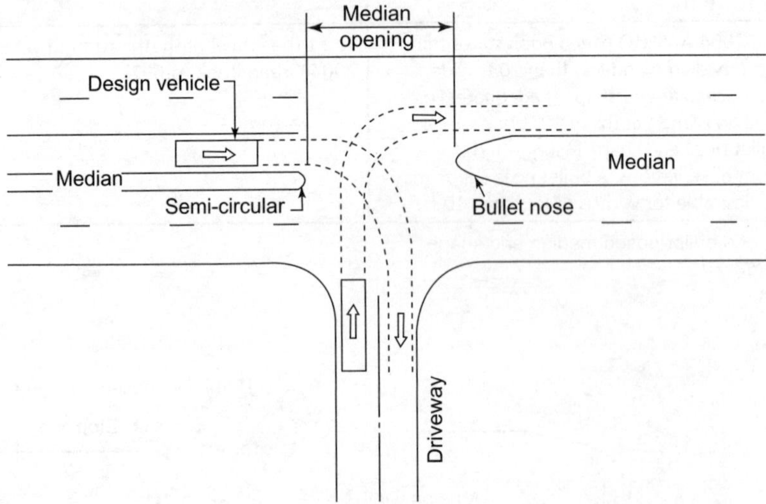


**Fig. 11.29** Elements of median openings
(*Source*: Koepke and Levinson, 1992;
*Courtesy*: Transportation Research Board, Washington DC, USA)

### 11.8.7 Spacing of Median Opening or Median Break Spacing

The spacing of median openings on divided roadways should be determined in light of the roadway geometrics and provision of sufficient length of median to accommodate deceleration taper and storage length. The desirable spacing applicable in urban and rural areas may range between 100 m and 800 m. Minimum desirable spacing of un-signalised median openings may be provided as a function of speed (Table 11.5). The spacing may be appropriately adjusted to the requirements of signal coordination, availability of storage space for right turns and interference of vehicle queues at intersections at peak times (Koepke and Levinson, 1992).

**Table 11.5** Spacing criteria between (un-signalised) median openings on divided highways

| *Speed* | | *Spacing recommendations – desirable minimum* | |
|---|---|---|---|
| *Miles/hr* | *Km/hr* | *Feet* | *Metre* |
| 30 | 44.00 | 370 | 112.77 |
| 35 | 51.33 | 460 | 140.20 |
| 40 | 58.66 | 530 | 161.54 |
| 45 | 66.00 | 670 | 204.21 |
| 50 | 73.33 | 780 | 237.74 |
| 55 | 80.66 | 910 | 277.36 |

*Note:* Provide openings in all arterial roads, all collectors and in some local streets. Desirable minimum spacing value based on 6.5 ft/sec$^2$ (1.8288 m/sec$^2$) average deceleration rate and no deceleration in through-traffic lane. (*Source*: Adapted from Stover, VG, Adkins, WG and Goodknight, JC. *Guidelines for Medial and Marginal Access Control on Major Roadways*, NCHRP Report 93, Transportation Research Board, Washington, DC, pp. 19, 1970).(*Source*: Koepke and Levinson, 1992; *Courtesy*: Transportation Research Board, Washington DC, USA)

A few sample layouts of right-turn median break spacing are described in Figs 11.30 and 11.31.

### 11.8.8 Dividers

Dividers are raised row-like kerb elements provided as part of channelisation along the inner edges of the inside traffic lanes. They are intended to divide/separate the traffic streams across the lanes. However, dividers are never considered as pedestrian refuge. Edge blocks or standing iron bars or a combination of both are constructed along the intended alignment of the divider, and painted with inclined hatched lines. Pedestrians crossing over dividers is discouraged/restricted.

### 11.8.9 Splitter Islands

These usually have a teardrop shape. Their main functions are to divide the traffic streams by direction of flow and warn the driver that an intersection is ahead (Fig. 11.32). The total length of the splitter islands depends on the design speed.

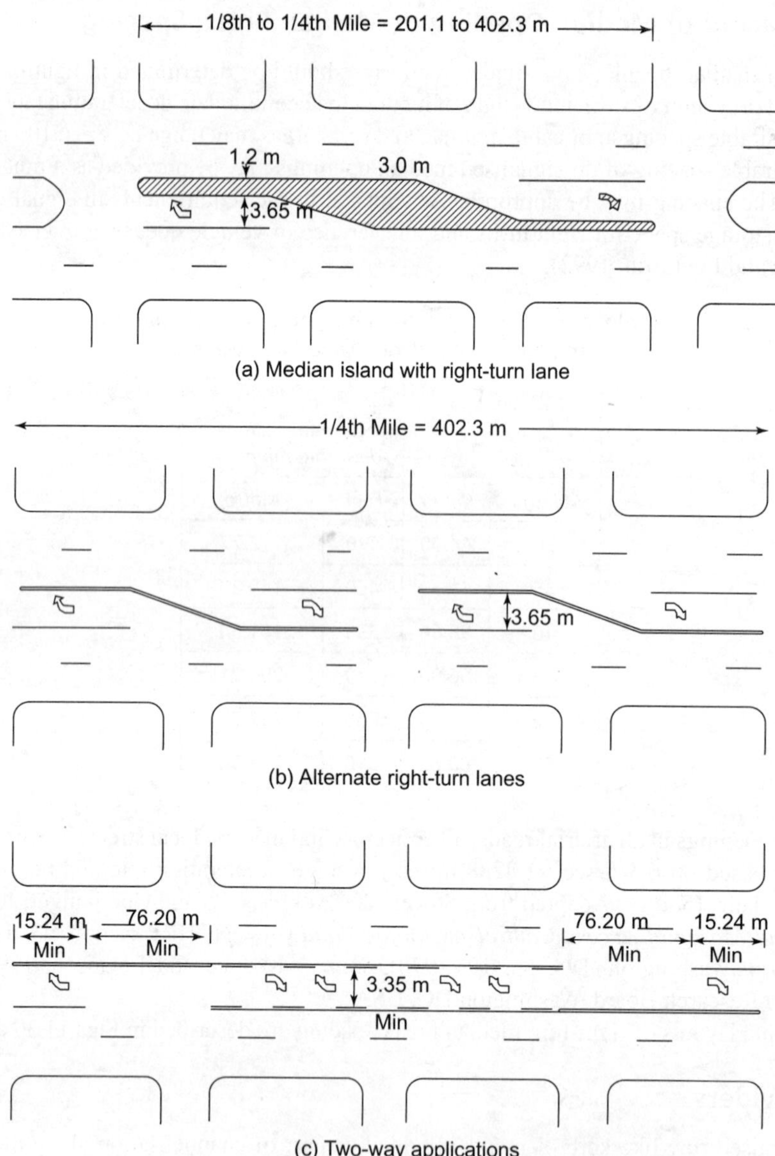


**Fig.11.30** Sample layouts of right-turn lane designs
(Modified based on Koepke and Levinson, 1992; *Courtesy*: Transportation Research Board, Washington DC, USA)

## 11.8.10 Tapers and Transitions

Tapers and transitions are used to shift the path of through-traffic to an intended lane and direction. The tapers should be short enough to provide sufficient visual distance to the driver so that through-traffic can shift over this distance at the prevailing operating speed of the roadway.

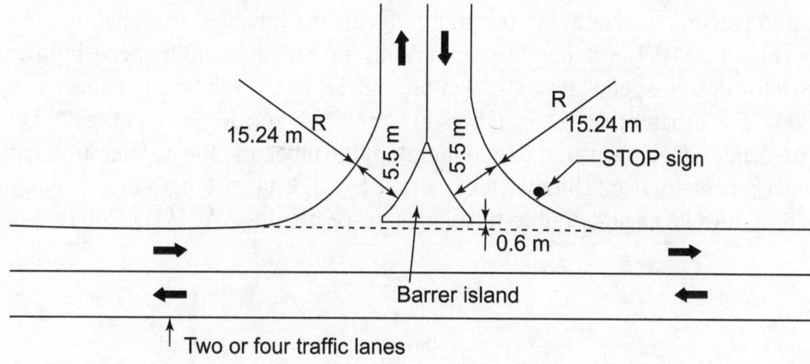


**Fig. 11.31** Right-turn restrictions – undivided roadway

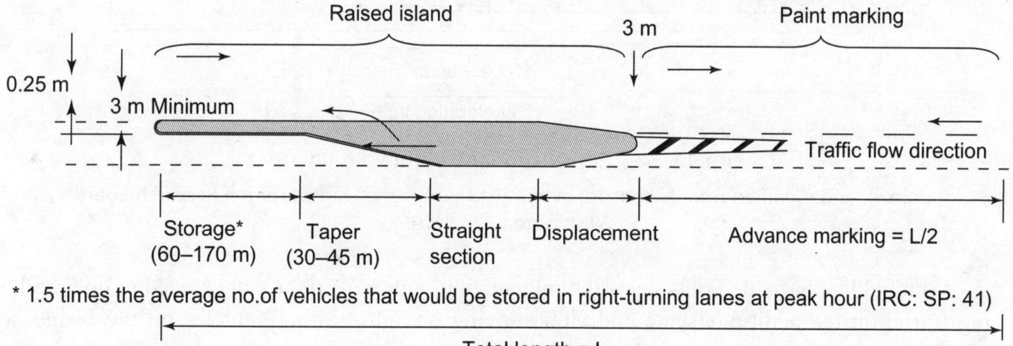


**Fig. 11.32** Elements of splitter islands used in three-legged and four-legged intersections
(Elaborated based on Setra, 1998)

Tapers may be categorised as approach tapers and bay tapers. An approach taper provides storage space for a right-turn lane by moving traffic laterally to the left on a road without a median (Fig. 11.33). Such tapered channelisation is often used in rural conditions where it is beneficial to provide added protection and/or guidance to turning vehicles, particularly at isolated T-intersections (Fitzpatrick et al., 2014).

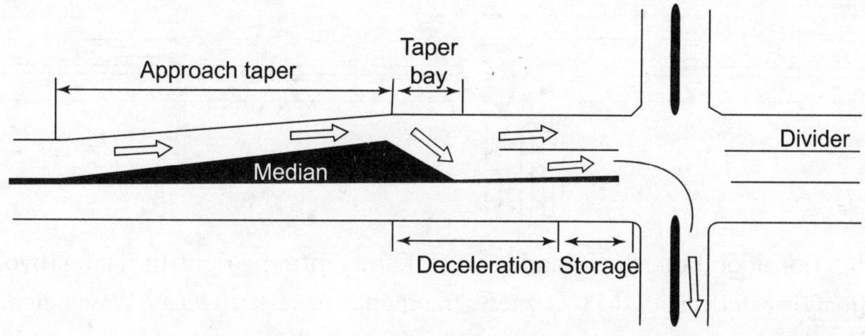


**Fig. 11.33** Right-turn lane added to an undivided two-lane highway (fully shadowed left-turn lane)
(Redrawn from Fitzpatrick et al., 2013; *Courtesy*: Transportation Research Board, Washington, DC, USA)

Bay tapers are reversing curves along the right edge of the travelled way that direct traffic into the right-turn lane (Fig. 11.34). Bay tapers are commonly provided on high-speed highways at a taper rate between 8:1 for design speeds up to 48 km/hr, and 15:1 (L:T) for design speeds up to 80 km/hr (AASHTO, 2004). For urban areas, short tapers are preferred for deceleration lanes because of slow speeds during peak periods. However, the total length of the taper and the deceleration length should be the same if longer tapers are used (Fitzpatrick et al, 2014). The taper length maybe adopted as 30.5 m and 46 m for single-turn lanes and double-turn lanes, respectively (AASHTO, 2004).

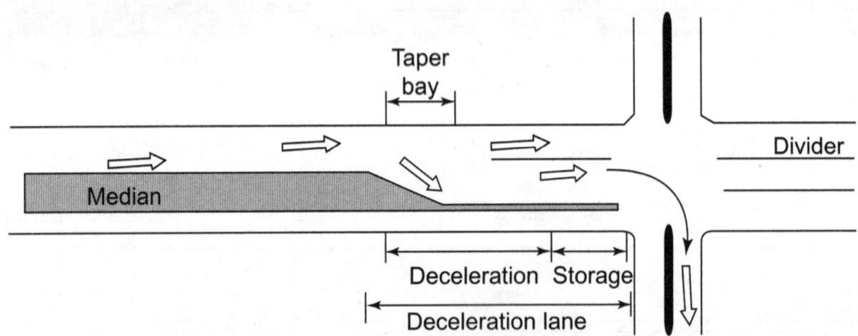


**Fig. 11.34** Right-turn lane within a median

(Redrawn and modified from Fitzpatrick et al., 2013; *Courtesy*: Transportation Research Board, Washington, DC, USA)

An additional lane provided outside of the intersection with transitions can be used to operate fast-moving traffic on the additional lane and slow-moving or right-turning vehicles on the inside lane (Fig. 11.35).

*Note:* The length of the taper and the transition depends on the design speed and the intersection geometrics. As design speed increases, the required taper length also increases, and vice versa. Designers are advised to adopt standard values which may be obtained from standard literature (AASHTO, 2004).

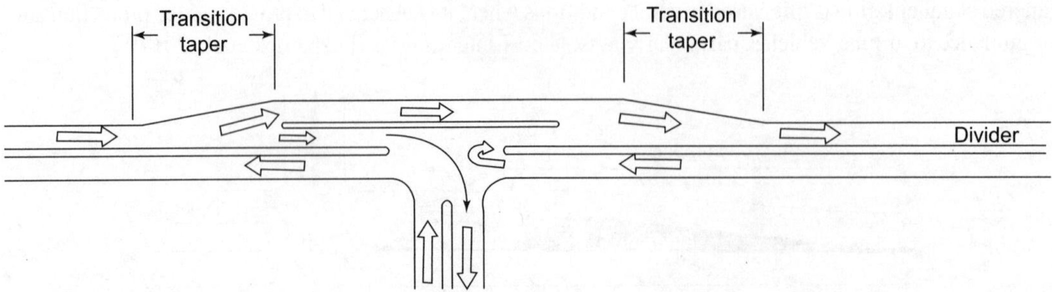


**Fig. 11.35** Transition tapers provided for direct entry into the right-turn lane (bypass lane)

(Modified from Fitzpatrick et al., 2013; *Courtesy*: Transportation Research Board, Washington, DC, USA)

A few other models of channelisation are presented in Figs 11.36, 11.37 and 11.38.

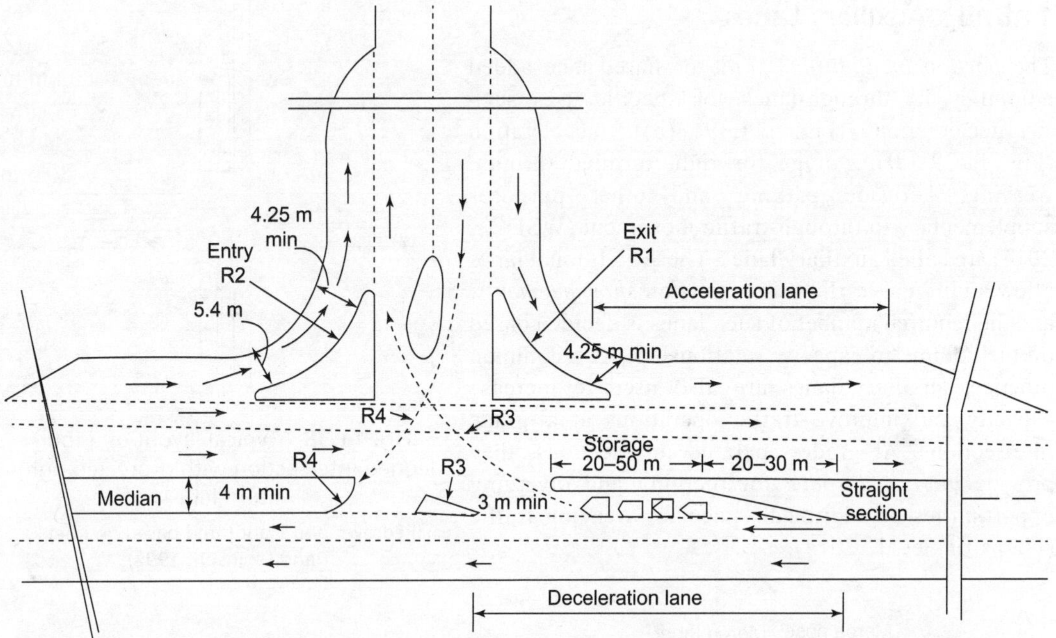


**Fig. 11.36** Typical three-legged intersection with appropriate channelisation (Redrawn and elaborated based on IRC: SP: 41-1994)

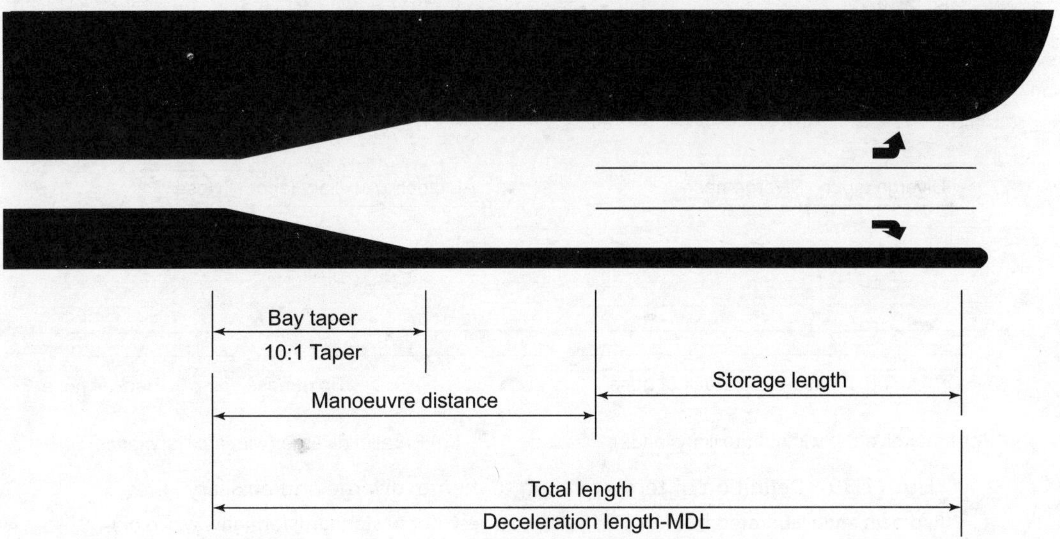


**Fig.11.37** Separating turn lanes

(Redrawn based on Koepke and Levinson, 1992; *Courtesy*: Transportation Research Board, Washington DC, USA)

### 11.8.11 Auxiliary Lanes

The *portion of additional lane* or flared lane added adjoining the through-lanes for specific uses such as deceleration (Fig. 11.39 (b)), acceleration (Fig. 11.39 (d)), storage for right turning, turning, weaving, kerbside parking and other purposes supplementary to through-traffic movement (WSDOT, 2014) are called auxiliary lanes. These additional lanes allow additional traffic movements on *short segments,* and the required number of such lanes is decided based on the volume-to-capacity relationship. As a common practice, auxiliary lanes are also used to increase capacity and improve traffic operations at at-grade intersections. At grade-separated intersections, they provide *increased space* for merging and diverging opportunities or additional space for weaving traffic (Fitzpatrick et al., 2014).

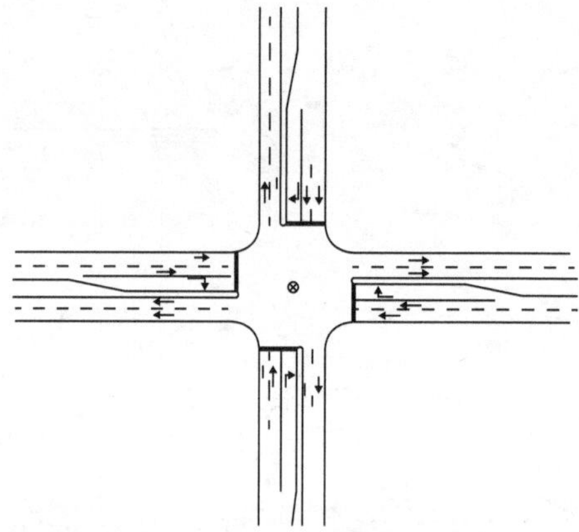

**Fig. 11.38** Typical layout of a four-legged intersection with protected right-turn lane

(Redrawn and elaborated based on Koepke and Levinson, 1992)

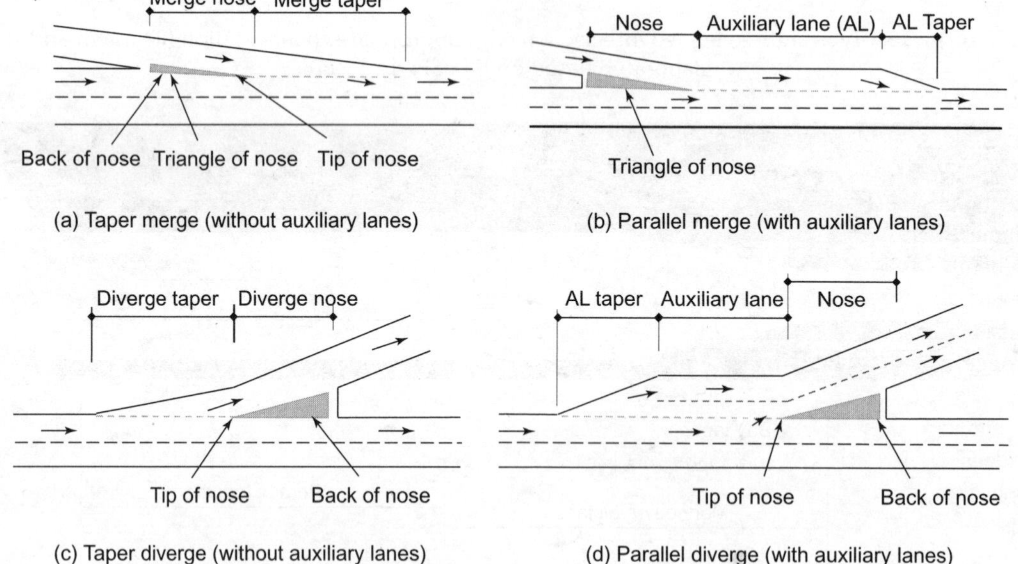


**Fig. 11.39** Definition of terms relating to merge, diverge and auxiliary lanes

(Redrawn and elaborated from DMRB, 2006; *Courtesy*: www.standardsforhighways.co.uk)

Adequate diverge area at the back of the merge nose and the length of nose to be provided must not cause queues and must enable deceleration in reasonable comfort (DMRB, 2006).

The different types of auxiliary lanes provided for specific purposes at an intersection are given below:

- *Acceleration lanes*: Vehicle speeds increase in this lane segment. This lane is advantageous for picking up the speed of the vehicles travelling away from the intersection. However, it is not advantageous when the entering drivers have to wait for an opportunity to merge, without disrupting the through-traffic.
- *Deceleration lanes*: Gradual reduction in the speed of vehicles takes place in this lane segment. A dedicated deceleration lane is useful for separating/diverging slow-moving, weaving and turning vehicles from the through-traffic lane. Its length comprises the deceleration length, storage length and entering taper. For example, at a gradient not exceeding 3%, the deceleration lengths required for right-turn lanes with design speed of 44 km/hr and 73 km/hr range between 52 m and 125 m, respectively (AASHTO, 2004). The minimum storage length required to accommodate at least two cars with a speed of <40 m/hr is 15.5 m; for high speeds, it is 30.5 m.
- *Climbing lanes*: This is used to divert slow traffic from the through-lane.
- *Passing lanes*: An overtaking zone is provided at mid-block sections of a two-lane highway.
- *Drop lanes*: A lane that may be dropped at an intersection with a turn-only lane or extended beyond the intersection with sufficient length to allow smooth merging/diversion. Its length should be not less than the acceleration lane length and deceleration lane length for merging and diversion, respectively.
- *Left-turn lanes*: Additional lanes provided as dedicated left-turn lanes at an intersection.
- *Right-turn lanes*: Additional lanes provided as dedicated right-turn lanes at an intersection (Fig. 11.34). These lanes can be provided at all median cross-overs on divided, high-speed highways and on unstopped approaches of major approaches to intersections. The NCHRP Report 279 (Fitzpatrick et al., 2014) states that exclusive right-turn lanes are required when the right-turn volume is greater than 20% of the total approach volume or when right-turn volume is greater than 100 veh/hr in peak periods (Neuman, 1985). Double right-turn lanes are provided if the volume exceeds 300 veh/hr (TRB, 2000).

Auxiliary lanes are commonly provided close to at-grade intersections to accommodate high traffic volumes for a short distance. Appropriate changes in the geometrics of existing at-grade intersections can be made to reduce the potential speed differential effects on safety through the provision of free left-turn auxiliary lanes, right-turn auxiliary lanes, acceleration lanes onto major roads from minor roads, and vice versa in the case of deceleration lanes (Delphi et al., 2006).

Acceleration lanes and deceleration lanes are termed speed change lanes (Fig. 11.40). Deceleration lanes are provided before entering the intersection and acceleration lanes are provided to pass through the intersection. Speed change lanes are normally provided for at-grade intersections on multilane divided highways with access control. The justification for providing a speed change lane is based on several factors such as approach speed, traffic volume, capacity, highway type, distance between consecutive intersections and collision history (WSDOT, 2014).

## 11.9 RETROFIT ACTIONS

Retrofit actions are inevitable due to increased congestion, accidents, new major construction plans and street improvements. Access management is essential for dealing with such undesirable conditions. Most retrofit actions/treatments involve the application of traffic engineering techniques that limit the number of conflict points, provide reorientation, deal with installing or closing medians, separating traffic flow by direction and regulate the speed limit.

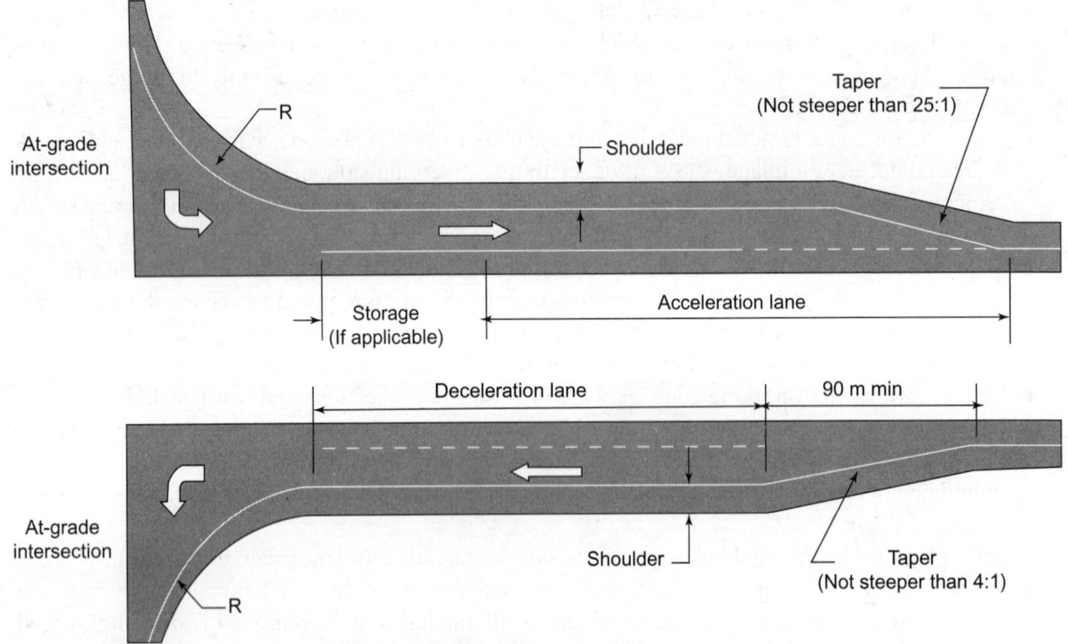


**Fig. 11.40** Typical layouts of speed change lanes
(Redrawn based on WSDOT, 2004)

## 11.9.1 Retrofit Techniques

The guidelines and techniques of access management applicable to retrofit situations, such as new or expanded developments including intended operating changes (Flora and Keitt, 1982) are given below:

- Limit the number of conflict points by installing channelising islands and by controlling turning movements and closing/opening medians.
- Separate basic conflict areas by regulating/providing adequate clearance with abutting/adjacent properties/roadways, and widening roadways.
- Limit deceleration requirements by installing traffic signals and channelising islands, replace kerb parking with off-street parking, and improve sight distances on horizontal and vertical alignments.
- Remove turning vehicles from the through-lanes by installing appropriate turning lanes with storage lanes and limiting speed, additional lanes/bypass roads for facilitating the desired (right or left) turning movements.
- Improve intersection alignment: To cross an intersection or merge with traffic on a major road, drivers on minor roads need adequate sight distance on the approaches to major roads and clear sight triangles. If these criteria are not satisfied, the existing alignment of minor or major roads or both must be modified to ensure adequate sight distance. The following options may be explored and appropriately implemented.
    - *Change the type of traffic control*: For example, provide all-way STOP control or signal control.
    - *Relocate the intersection*: For example, this option may be suitable when heavy earth work is beyond the normal acceptable limit and land acquisition is costly in urban areas. Relocate the intersection when it is located on the curved portion of a major road. The traffic from the minor road turning into

the major road against an adverse superelevation may lose stability. For this reason, intersections should not be located on a curve with a superelevation greater than 6% (NASRAL).

- *Re-align the approach roads to an intersection*: For example, skew intersections are corrected to obtain the adequate clear sight triangles on either side (Figs 11.41, 11.42, 11.43, 11.44). A few such improvements are discussed in the following section.
- *Upgrade the at-grade intersection to a grade-separated intersection*: For example, provide jug handles or quarter link interchanges.

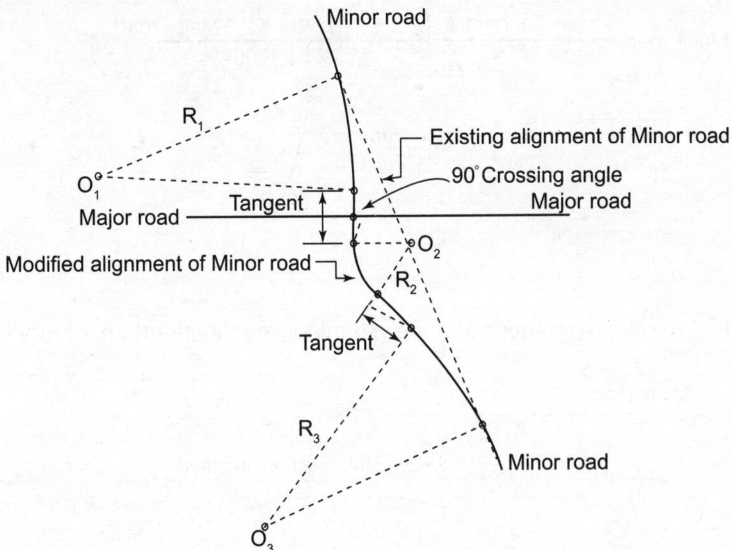


**Fig. 11.41** Typical re-alignment of a skewed minor road (option 1)

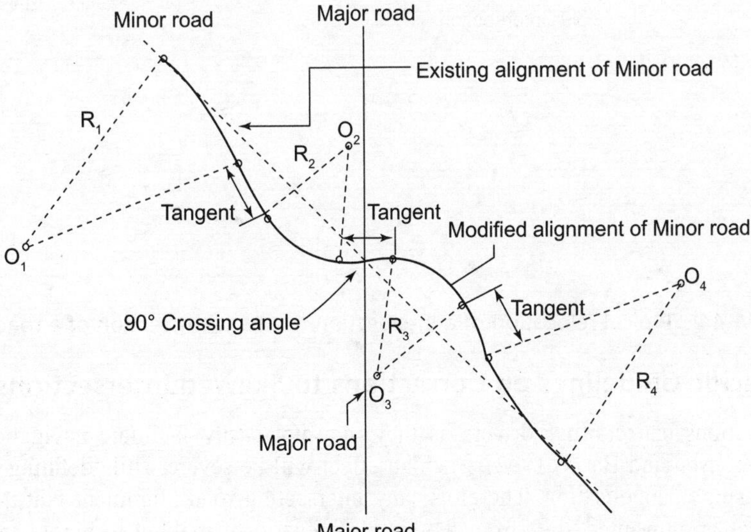


**Fig. 11.42** Typical re-alignment of a skewed minor road (option 2)

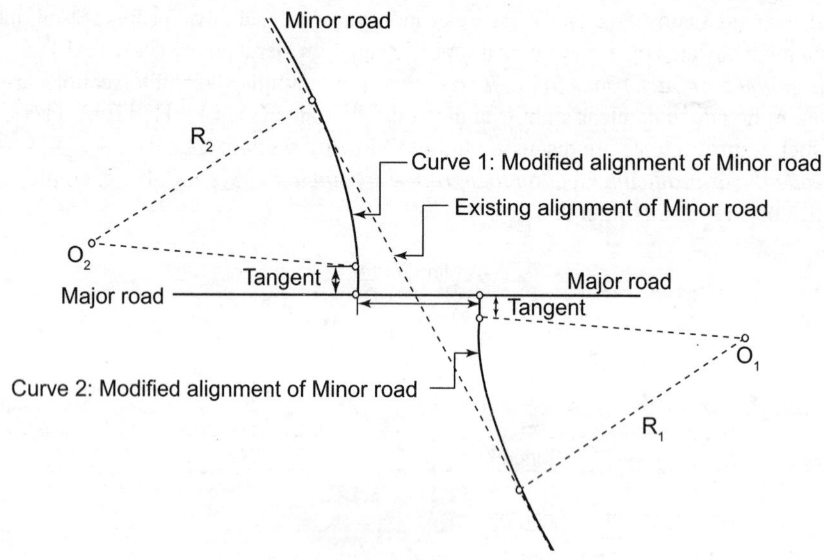


**Fig. 11.43** Typical re-alignment of a skewed minor road in an urban locality (option 3)

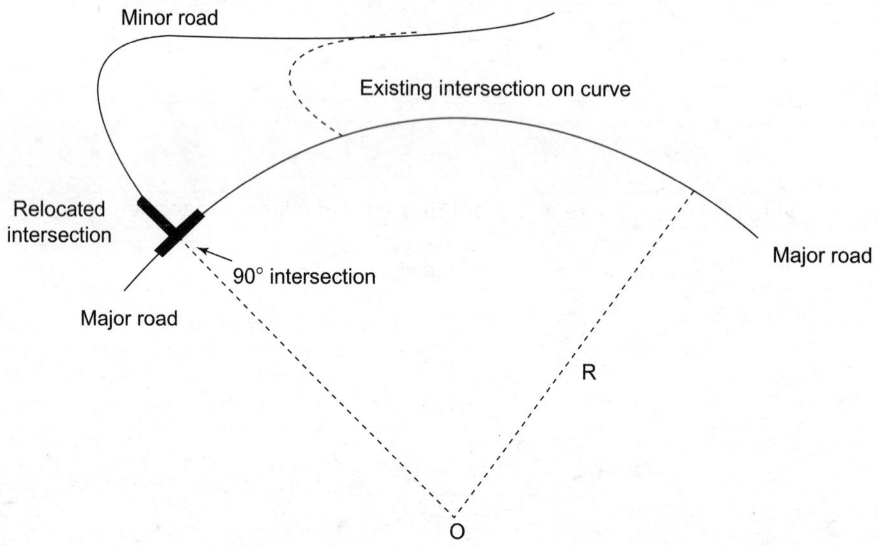


**Fig. 11.44** Typical relocation of a intersection to a tangent section of a roadway

## 11.9.2 Specific Guidelines on Corrections to Skewed Intersections

Skewed intersections can confuse drivers as they may not easily facilitate navigation through an intersection (Woodson and Barry, 1992). Such an effect will be severe while defining longer vehicle turning paths using channelisation. Therefore, any intersecting road alignment that *deviates* from a 90° angle causes an unwanted shift from the simple scenario of making turnings, and hence, such intersections become *less safe*. In contrast, a perpendicular (non-skewed) intersection provides the *best* and *most favourable* operating conditions for drivers as they can easily perceive the direction in which they need to travel, estimate the speeds of the opposing vehicles and smoothly complete turning manoeuvres in a shorter time.

It is desirable that all approaching legs of an intersection be on a tangent alignment (90° angle of intersection). However, minor deviations from right angles are considered acceptable, provided there will be no potentially detrimental impact on the visibility and turning movements of large trucks.

In this connection, AASHTO states that right-angled crossing are preferred. However, skewed intersections should be avoided or re-aligned (*reconstructed*), and in no case should the skew angle be greater than 60° as per AASHTO (1994), 75° as per ITE (1984) and 30° as per the state of Nebraska (NDOR, 2015). However, for designing new and reconstructed projects, a skew of 15° or less is preferred (NDOR, 2015).

Appropriate geometric re-alignment is explored with reference to the existing degree of skewed alignment of minor roads, wetlands, surrounding buildings, land availability, right-turning movements, design speed, minimum sight distance and intersecting traffic volumes, in either direction. Some typical samples and widely used re-alignment options of skewed road intersections are shown in Figs 11.41, 11.42, 11.42 and 11.44. The possibility of adopting re-alignment as shown in Fig. 11.43 may be explored for very low volume conditions or in urban areas where land acquisition is costly and difficult. The minimum length of the tangent may be taken as 100 m (Figs 11.41 and 11.42). It must be noted that as the skew angle increases (in Fig. 11.43) from 20° to 60°, the common tangent length along the main road range from 110 m to 1325 m (NDOR, 2015).

In intersections on curved alignment, the intersecting approaches should be relocated to the nearest intersection to a tangent section of a roadway (Fig. 11.44) or re-aligned as close to 90°, with the approaches having lengths greater than 120 m. The superelevation on the main road may be restricted to 4% on the curve, and the speed on the intersection and approaches should be recalculated for the modified geometric condition.

The following alternative options may be suitably adopted for conditions in general (Setra, 1998).

| *General conditions* | *Alternatives for use* |
|---|---|
| Low traffic on minor roads | • Removal or elimination of intersection and transfer to an adjacent intersection<br>• Modification of existing intersection as roundabout, as a terminal intersection and possibly at a central point |
| Predominant through-traffic on minor roads | Grade separation with a flyover or road overbridge |
| Heavy traffic on both intersecting roads | Suitable grade-separated intersection or interchange; selection of full cloverleaf/half cloverleaf or trumpet/partial interchange may be decided based on the directional distribution of traffic flow (left, right or through) and the site/available space geometrics at the intersection |

Road infrastructure facilities including provision of interchanges along a route must be designed in accordance with a standard code of specification. Access points and approach roads should be aligned or re-designed (in case of existing intersections) to channelise smooth movement of traffic, and thus, prevent detrimental journey times.

The choice is made based on an *examination* of the study location and the route. Scientific analysis may be carried out for multi-criteria design. In this process, several factors are considered essential for any decision-making regarding a new construction or upgradation of an existing intersection. Some important factors may be listed for this purpose, such as, user costs, benefit due to reduction in the number of accidents, benefit due to reduction in delay, financial analysis, and cost–benefit analysis with the intended safety features of the facility.

## EXERCISES

### Multiple Choice Questions

1. Which of the following shows skewed Yconfiguration type of intersection?

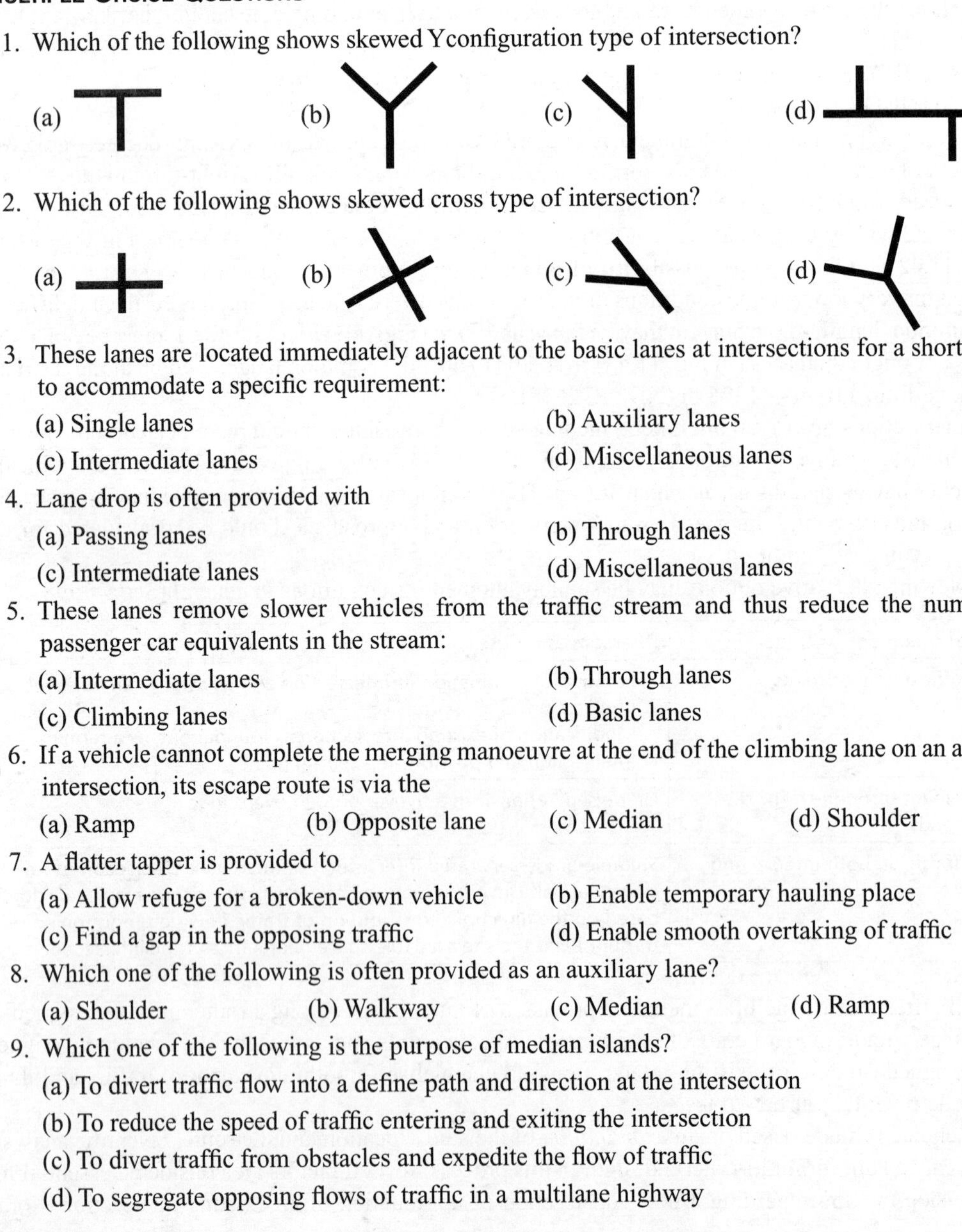


2. Which of the following shows skewed cross type of intersection?

(a) (b) (c) (d)

3. These lanes are located immediately adjacent to the basic lanes at intersections for a short length to accommodate a specific requirement:
   (a) Single lanes (b) Auxiliary lanes
   (c) Intermediate lanes (d) Miscellaneous lanes
4. Lane drop is often provided with
   (a) Passing lanes (b) Through lanes
   (c) Intermediate lanes (d) Miscellaneous lanes
5. These lanes remove slower vehicles from the traffic stream and thus reduce the number of passenger car equivalents in the stream:
   (a) Intermediate lanes (b) Through lanes
   (c) Climbing lanes (d) Basic lanes
6. If a vehicle cannot complete the merging manoeuvre at the end of the climbing lane on an at-grade intersection, its escape route is via the
   (a) Ramp (b) Opposite lane (c) Median (d) Shoulder
7. A flatter tapper is provided to
   (a) Allow refuge for a broken-down vehicle (b) Enable temporary hauling place
   (c) Find a gap in the opposing traffic (d) Enable smooth overtaking of traffic
8. Which one of the following is often provided as an auxiliary lane?
   (a) Shoulder (b) Walkway (c) Median (d) Ramp
9. Which one of the following is the purpose of median islands?
   (a) To divert traffic flow into a define path and direction at the intersection
   (b) To reduce the speed of traffic entering and exiting the intersection
   (c) To divert traffic from obstacles and expedite the flow of traffic
   (d) To segregate opposing flows of traffic in a multilane highway

10. On a four-lane undivided highway in the North–South direction, the left-most lane in the northbound direction has a pothole that needs to be filled for a 1-km-long stretch. The 85$^{th}$ percentile speed on this highway is 43.2 km/hr. All the lanes are 3.5 m wide and the shoulders are 1.8 m wide. What is the most appropriate merge taper length to place the traffic cones for the repair work in the northbound left-most lane?

    (a) 100 m (b) 40 m (c) 50 m (d) 37.5 m

    (*Hint*: As the speed is less than 45 miles/hr, L = WS = 3.5 m x (43.2 x 0.278 m/sec) = 3.5 m x 12 m/sec = 42 m).

11. Match List-I (unit) with List-II (purpose) and select the correct answer using the codes given below the lists:

| List-I | List-II |
|---|---|
| A. Trumpets | 1. Information signs |
| B. Traffic studies | 2. Traffic regulatory signs |
| C. Rotary islands | 3. Traffic flow characteristics |
| D. Road intersection approach signs | 4. Grade separation |

| | A | B | C | D |
|---|---|---|---|---|
| (a) | 1 | 4 | 3 | 2 |
| (b) | 2 | 3 | 1 | 2 |
| (c) | 3 | 1 | 2 | 4 |
| (d) | 4 | 3 | 2 | 1 |

## ANSWERS

1. (c) 2. (b) 3. (b) 4. (a) 5. (c) 6. (d) 7. (c) 8. (a) 9. (d) 10. (b) 11. (d)

## QUESTIONS

1. State the objectives and elements of access management.
2. What are the basic principles to be considered during the design of intersections relating to general geometrics and design parameters?
3. Classify intersections based on the intersecting road surface level.
4. Write the operational classification of intersections based on control type.
5. List the basic requirements of the design of at-grade intersections.
6. Describe the basic geometric configurations of roads intersecting at grade.
7. What is a skew intersection? Draw typical sketches showing left- and right-skewed intersections with negative and positive skew angles.
8. Define flared intersections and classify flared at-grade intersections.
9. Write the classification of at-grade Intersections based on channelisation of traffic flow.
10. What is a channelised intersection? Write the functional objectives of the design of channelisation of road intersections.
11. Draw a typical sketch layout of a three-legged intersection with channelisation design elements.

12. What are the factors considered for the design of intersections relating to (a) Driver/Pedestrian factors, (b) Road geometric factors, (c) Traffic and vehicular factors, (d) Safety issues , (e) Environmental factors, (f) Economic factors?
13. Explain the design principles of at-grade intersections relating to (a) Design vehicle, (b) Stopping sight distance, (c) Sight distance triangles at intersections.
14. What is a sight triangle? Describe the different types of sight triangles at an intersection.
15. Draw typical layouts showing the approach clear sight triangles at intersections with uncontrolled or yield-controlled minor roads.
16. Draw typical layouts showing the departure clear sight triangles at intersections with STOP-controlled minor roads.
17. Explain how sight distance at medians is determined, with a neat sketch.
18. Write the basic functionalities of providing islands as channelisation.
19. Write the principles of channelisation.
20. Write the guidelines considered while designing a channelised intersection.
21. What are the elements of channelisation? State the standards that are most widely adopted.
22. What is an island? Describe the most widely adopted standard dimensions with reference to their geometrical area and kerb height.
23. What is a kerb? Explain the objectives/purpose of providing kerbs as a component of channelisation.
24. Describe the classification of kerbs based on their purpose and cross-section. Also list their most commonly adopted sizes.
25. What is a directional island? State the factors affecting the selection of triangular islands based on their dimension and shape.
26. Draw a typical layout of a three-legged multilane intersection with directional islands and a median opening.
27. Write a short note on median islands, detailing with their standard dimensions and applicability.
28. Draw a typical layout sketch showing the components of a median with a right-turn auxiliary lane. Also state the formulae used to determine the transition taper length.
29. State the conditions in which channelisation with raised islands is typically omitted.
30. List and draw sketches showing the most commonly used median end treatments, indicating their applicable widths and usage.
31. Describe the median design guidelines with reference to parameters: length, width, end treatment, turning radius and radius of nose.
32. Draw a layout of an intersection showing the elements of median openings.
33. Write a short note on the general guidelines applicable for spacing of median opening with reference to speed.
34. Draw a few sample layouts of right-turn lane design.
35. What is a splitter island and where is it used? Also draw typical layouts showing the elements of splitter islands used in three-legged and four-legged intersections.
36. What are tapers and transitions and where are they used? Also draw a typical layout sketch showing the right-turn lane added to an undivided two-lane highway with fully shadowed left-turn lane.

37. Draw a typical layout of an at-grade intersection showing the right-turn lane within a median depicted with a taper bay, deceleration and storage lanes.
38. Draw a three-legged intersection showing the transition tapers provided for direct entry into the right-turn lane (bypass lane).
39. Draw a layout showing a separating turn lane with storage length, manoeuvre distance and bay taper.
40. Draw a typical layout of a four-legged intersection with protected right-turn lane.
41. What are auxiliary lanes? State their functionality with reference to traffic flow and channelisation.
42. Draw typical sketches showing taper merge, taper diverge, parallel merge, parallel diverge with and without auxiliary lanes. Indicate elements such as merge/diverge nose, merge/diverge taper, tip nose, triangle nose and back nose.
43. Explain the different types of auxiliary lanes provided for specific purposes at an intersection.
44. Draw typical layouts showing speed change lanes with (a) Acceleration lane and (b) Deceleration lane.
45. What are the guidelines and techniques of access management applicable to retrofit situations adopted for new or expanded developments of at-grade intersecting road alignment?
46. Write the guidelines specifically used to correct skewed intersections. Draw typical sketches showing any four re-alignments.
47. State the values of human factors that are appropriate for the design of intersections. Explain how the effect of a descending gradient on the SSD of a minor road is more critical than that on a major road intersection.
48. State and explain the three main functions of channelising islands. (*Hint*: Direction, Division and Refuge)
49. Draw the general types and shapes of islands. Label their names and list their functionality with reference to traffic flow and safety.
50. Define traffic island and state the functions of traffic islands.
51. State the situations where grade-separated interchanges are appropriate.

CHAPTER

# 12 Design and Analysis of Rotary Intersections

## 12.0 THE CONCEPT OF TRAFFIC ROTARY OPERATIONS

The **traffic rotary** is a form of circular intersection in which vehicles circulate in the clockwise direction around a central island and pass to the right of it. The entering traffic must give way to the circulating traffic (Fig. 12.1). The rotary is characterised by a central island with a large diameter ($\phi$) (often $>100$ m) and high circulating speeds. **Roundabouts** ($\phi \geq 30$ m) are a similar type of intersection specifically designed with improved safety standards (often termed modern roundabouts) and operated on a yield-control basis. Roundabouts may be categorised as single lane, multilane and mini ($\phi \leq 25$ m). Mini-roundabouts are a form of circulating intersection designed for areas where insufficient right of way is available at the intersection to accommodate a wider central island with the usual high speeds. Such intersections are categorised as **circular intersections**.

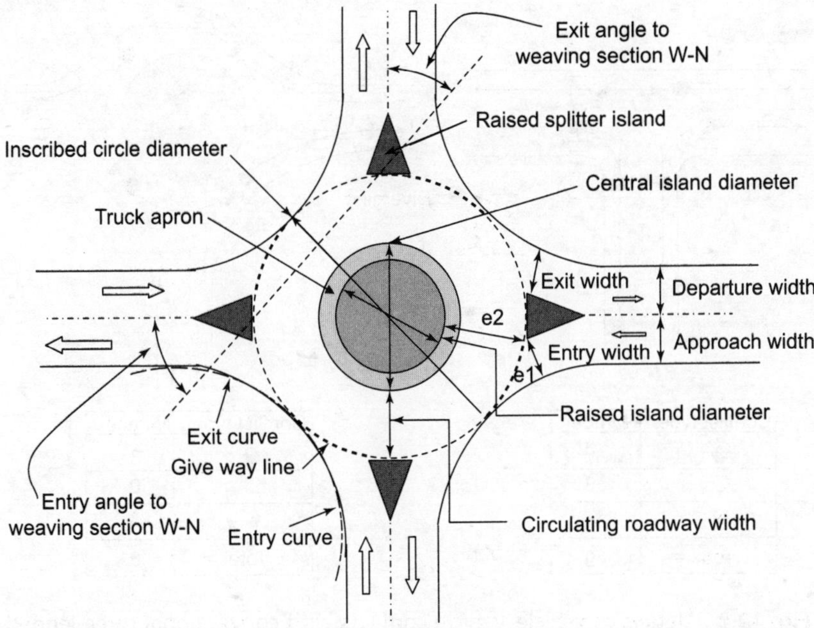


**Fig. 12.1** Basic geometrics and components of a roundabout (Redrawn based on TRB, 2000)

The diameter of the central island and the size of the raised splitter islands must be designed based on the design speed and the area available at the intersection to accommodate the adequate horizontal curvature of vehicle paths. The raised central island and splitter islands on each approach road to the intersection have specific purposes. Each approach entry and exit flow is *channelised* by the splitter islands and the *entry movements* are configured to force vehicles to *slow* down. Wide curvature at each exit facilitates acceleration of traffic. The central island provides a curved kerb that channelises the circulating traffic and facilitates a single stream of traffic to split by converging and diverging movements. This phenomenon is termed **weaving** – the lane changing required is the basic principle of the functionality of a traffic rotary. In addition, the orderly movement of traffic is maintained provided the entering traffic gives way or yields to the circulating traffic. This rule prevents blocking of the rotary intersection by not allowing vehicles to enter the intersection until there are sufficient gaps in the circulating traffic (NCHRP, 2010).

## 12.1 ADVANTAGES OF TRAFFIC ROTARIES

A well-designed circulatory intersection can improve safety, operations and aesthetics, as described below.

**Safety:** Traffic rotaries reduce the frequency of accidents with a wide range of settings (urban, suburban and rural) and traffic control facilities such as two-way stop control and signalised intersections. The safety benefits may be listed as accident reduction with fewer points of conflict between vehicles and between pedestrians and vehicles than in conventional intersections. They reduce the potential for severe conflicts such as right-angle and right-turn head-on collisions. Examples of the benefits in terms of reduced number of conflicts with single-lane intersections are summarised below (Figs 12.2, 12.3 and 12.4).

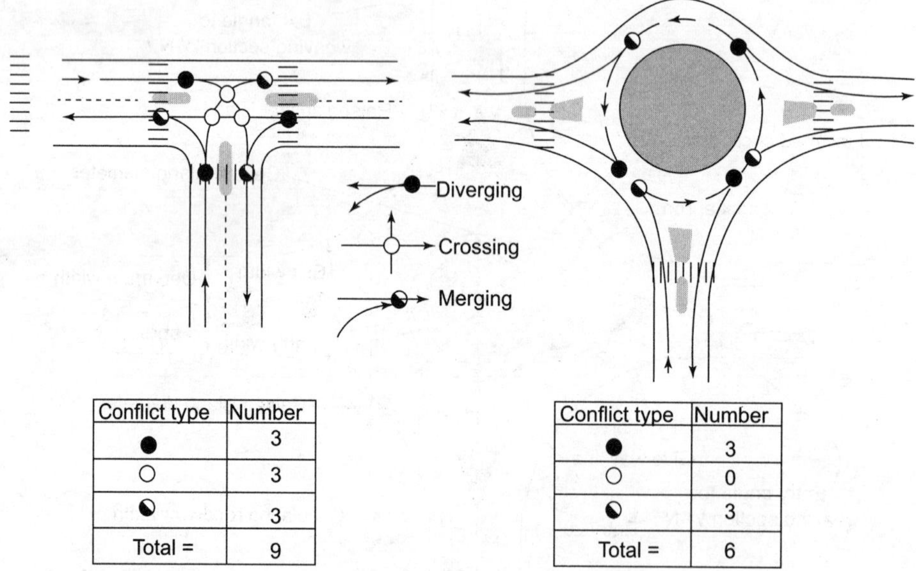


| Conflict type | Number |
|---|---|
| ● | 3 |
| ○ | 3 |
| ◑ | 3 |
| Total = | 9 |

| Conflict type | Number |
|---|---|
| ● | 3 |
| ○ | 0 |
| ◑ | 3 |
| Total = | 6 |

**Fig. 12.2** Details of vehicle–vehicle conflicts with conventional three-legged and rotary intersections

(Modified and elaborated based on NCHRP, 2010; *Courtesy*: Transportation Research Board, Washington DC, USA)

- Eliminate vehicle–vehicle crossing conflicts and separate vehicle–vehicle conflicts from pedestrian conflicts by using raised splitter islands.
- Benefit of vehicle–vehicle conflicts with T-intersection $= \frac{(9-6)}{9} \times 100 = 33.33\%$

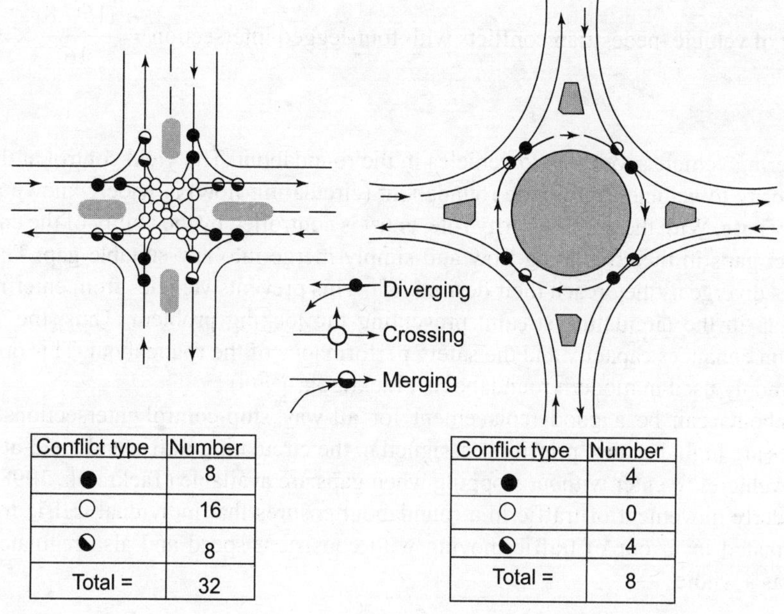


| Conflict type | Number |
|---|---|
| ● | 8 |
| ○ | 16 |
| ◐ | 8 |
| Total = | 32 |

| Conflict type | Number |
|---|---|
| ● | 4 |
| ○ | 0 |
| ◐ | 4 |
| Total = | 8 |

**Fig. 12.3** Details of vehicle–vehicle conflicts with conventional four-legged and rotary intersections

(Modified and elaborated based on NCHRP, 2010; *Courtesy*: Transportation Research Board, Washington DC, USA)

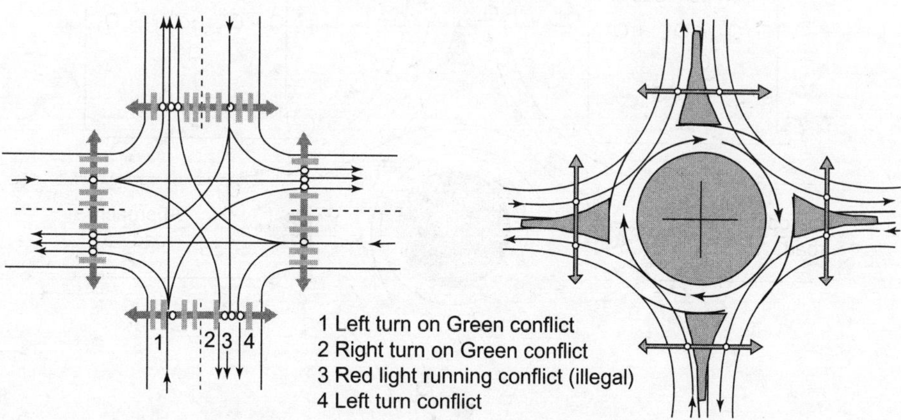


**Fig. 12.4** Details of vehicle–pedestrian conflicts with conventional four-legged and rotary intersections

(Modified and elaborated based on NCHRP, 2010, and ODOT, 2011; *Courtesy*: Oregon Department of Transportation (ODOT) and Transportation Research Board, Washington DC, USA)

- Benefit of vehicle–vehicle conflicts with four-legged intersection $= \dfrac{(32-8)}{32} \times 100 = 75\%$

- Benefit of vehicle–pedestrian conflicts with four-legged intersection $= \dfrac{(16-8)}{16} \times 100 = 50\%$

**Operations**

- An entering vehicle gives way to vehicles in the roundabout. The yield control at the entry point and priority to vehicles within the roundabout (circulating flow) is widely known as the **offside priority rule.** With the above priority rule, entry is controlled by the ability of the entering drivers to detect gaps in the circulating flow and simply merge into any suitable gap. The **circulating vehicles** diverge as they reach their desired exit. This prevents vehicles from entering when there is no gap in the circulating stream, preventing the locking problem. Thus, the yield at entry operation enhances capacity and the safety performance of the roundabout. This operational rule is commonly used in modern roundabouts (Taekratok, 1998).
- Roundabouts can be a good replacement for all-way stop-control intersections when traffic volumes are high, because priority is assigned to the circulating traffic and yield-at-entry control allows vehicles to enter without stopping when gaps are available (Taekratok, 1998).
- The orderly movement of traffic in a roundabout ensures that individual traffic movements are subordinated in favour of traffic moving with consistent speed and also regulates the moving traffic as a whole.

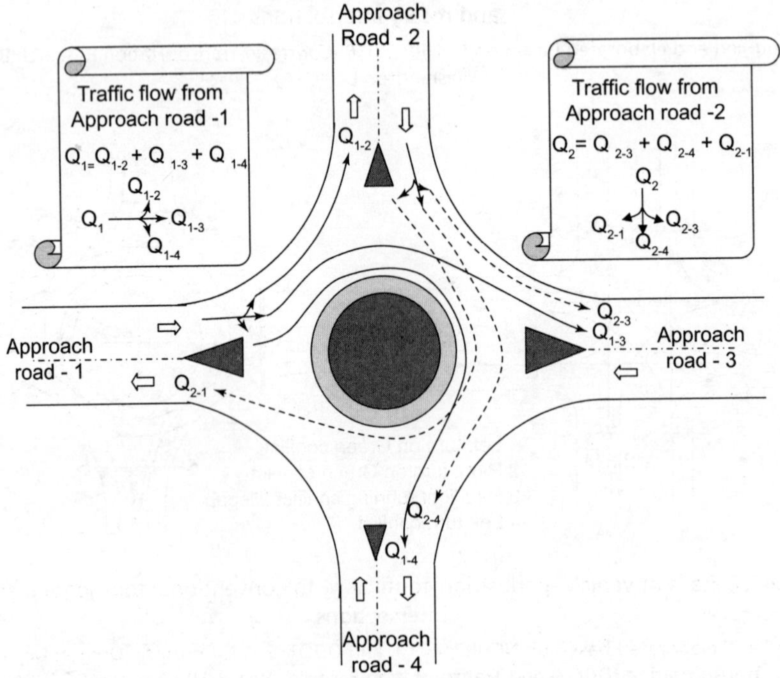


**Fig. 12.5** Flow stream definitions: Example of designation and distribution details of flow streams from Approach Roads 1 and 2

- Traffic operation is improved by the smooth flow of traffic, with uniform speed and fewer conflicts and delays.
- Rotary intersections can be constructed with more than four intersecting road legs.
- Self-control of vehicular flow is possible with a moderate volume of traffic when traffic signals are warranted.

## 12.2 DISADVANTAGES OF TRAFFIC ROTARIES

- Larger area of requirement, initial cost of construction of roundabout with landscape enhancement of the area and its maintenance are some of the primary disadvantages. Non-availability of adequate space may be another limitation in designing such intersections in urban localities.
- Locations within a coordinated signal network where provision of roundabout would disrupt the platoon (NCHRP, 1998).
- No pedestrian signals, to help pedestrians cross the road.
- Bicyclists must merge with motor vehicles to enter roundabouts.
- The right-turning traffic must travel an extra distance around the central island.
- The number of vehicular conflicts may increase with *dual* and *multilane* roundabouts due to increased number of lane changes/crossings. This is particularly predominant in the circulating area when vehicles weave to exit by changing lanes and when vehicles enter next to another vehicle perceived to be circulating in front of a corner of a splitter island due to improper left turn by the lane change. For these reasons, an increased number of accidents may be seen with multilane roundabouts. Hence, specialised geometric design is necessary if two-lane/multilane roundabouts are used.
- Acute angles of intersection may result from very closely spaced adjacent approach roads, which may result in the design of inadequate weaving section length.
- If the number of vehicles in a rotary exceeds capacity, the resulting vehicle queue will force entry vehicles to a complete stop and jam the weaving section. Intervention of the traffic police is essential in such cases.
- Multiple rotary intersections placed at close intervals and frequent passing through such intersections may increase journey time.

## 12.3 SUITABILITY OF TRAFFIC ROTARIES/ROUNDABOUTS

The following conditions are favourable for designing roads that intersect at grade as roundabouts/rotaries.

- The layout of the site consists of plain space with three or more roads intersecting.
- Traffic flow in the intersecting road legs is almost equal in proportion.
- Right-turning traffic at the intersection is predominant and is about equal to or greater than 30% of through and left-turning traffic.
- The traffic volume is low to moderate (500–3000 veh/hr) and enters from all the intersecting roads.
- Intersections with high delays.
- Intersections where traffic signals are not warranted.
- Intersections with unusual geometry, such as Y-junctions and which cross at acute angles.
- Longer queues are created by signalised intersections that cause operational problems.

Different shapes of rotary intersections (at grade) are used based on the site conditions and the number of approach roads with their alignment of intersection. Typical shapes are described in Table 12.1.

**Table 12.1** Some shapes of roundabouts/rotaries

| | | |
|---|---|---|
| Circular (four legs) | Circular (three legs) | Elliptical |
| Square with rounded edges | Small turbo roundabout* | Large turbo roundabout* (Google map) |
| Irregular shape/Longabout | Irregular shape | Square with rounded edges |

* Turbo roundabouts are preferred to double-lane/multilane conventional roundabouts.

## 12.4 DESIGN CRITERIA AND GUIDELINES

The geometrics of a rotary must be appropriately designed based on standard design criteria (IRC: 65-1976) that fit the conditions of the selected site. In this connection, the following guidelines may be adopted.

### 12.4.1 Geometric Standards

Geometric design involves trade-offs between safety, space availability, cost, capacity, operations and accommodation of design traffic and pedestrian volume. The following guidelines may be appropriately adopted based on the site conditions (NCHRP, 2010).

- **Width, position and alignment of approach** roads to the roundabout are one of the primary design considerations to allow entry of traffic with reduced speeds. As the entry radius reduces, the entry speed also reduces, and this type of alignment is desirable (Fig. 12.6 (a) and (b)). In the case of increase in the entry radius (Fig. 12.6 (c)), the entry speeds of vehicles will increase and there is a potential risk of crashes in the weaving section. The path traced by the proposed streams of traffic may be considered to provide adequate sight distance and intersection sight distance with the design speed.

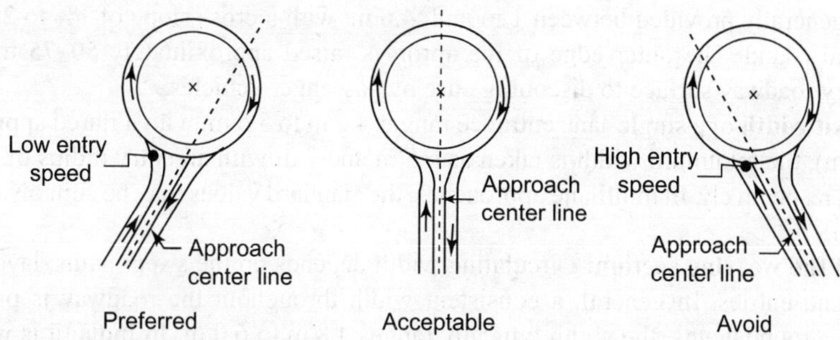


**Fig. 12.6** Typical alignment of approach roads to a roundabout

(Elaborated based on NDOR, 2012; *Courtesy*: Nebraska Department of Roads, USA)

- **Central island:** Larger central islands are preferred to improve the drivers' recognition of the form of intersection treatment. Larger central islands are commonly used for roundabouts in high-speed areas and at intersections with more than four legs. These larger islands can be non-circular to suit the specific intersecting legs, with the long side of the island oriented for major flow. However, islands of less than the nominal 5 m radius should be circular. According to the Florida guidelines, the central island for the state highway system should have a minimum radius of 7.5 m. Single-lane roundabouts designed for high-speed rural areas where two-way roads intersect would typically require central islands with radii in the range of 10–15 m (Taekratok, 1998).
- **Inscribed circle diameter**: It is the sum of the diameter of the central island and twice the width of the circulatory roadway. The dimension must be large enough accommodate the turning radius of the longest vehicle (design vehicle) and the design speed. In general, the diameter should be at least 32 m to accommodate bus or single-unit trucks as design vehicles. The diameter is typically in the range of 40–46 m, if the design vehicle is longer than a bus. Typical ranges of inscribed circle diameters for two-lane and three-lane roundabouts require at least 50–67 m and 61–91 m, respectively.
- **Entry approach angle:** It is selected to provide adequate sight distance and visibility, and its value ranges between 20° and 40°.
- **Entry and exit angles:** The entry angle should be larger than the exit angle. Ideally, the entry angle should be 60° with an exit angle of 30°. Staggering of the approach legs is recommended to achieve an entry angle that is close to 60°.
- **Exit curb radii** should not be less than 15 m (commonly between 30 m and 60 m). Further, larger exit radii may be desirable in areas with high truck volumes to improve ease of exit and reduce the risk of trailers tracking over the outside kerb.
- **Entry radii:** Larger entry radii are often not provided in order to discourage high entry speeds. A low-speed value is commonly recommended in urban areas. Entry radii at the kerb line of

single-lane roundabouts typically range 15–30 m. With reference to design speeds of 40 km/hr and 30 km/hr, the recommended radius of entry may be in the range of 20–35 m and 15–25 m, respectively (IRC: 65-1976).

- **Truck aprons** provide additional paved width around the central island which allows over-tracking of large semi-trailers without compromising the deflection for smaller vehicles. The width of the truck apron is decided based on the swept path of the longest vehicle permitted. Its width is generally provided between 1 m and 4.6 m, with a cross slope of 1% to 2% away from the central island. The outer edge of the apron is raised approximately 50–75 mm above the circulatory roadway surface to discourage use by passenger vehicles.
- **Entry/Exit width** of a single-lane entrance ranges 4.2 m to 5.5 m, with a flared approach. In dual lanes (7 m), the minimum width is taken as 6.5 m and 7 m with an entry radius of 25–35 m and 15–25 m, respectively. In multilane approaches, the standard values may be suitably selected from IRC: 65-1976.
- **Width of the weaving section:** Circulating width depends on the swept paths, layout and width of exits and entries. In general, a consistent width throughout the roadway is preferable. For single-lane roundabouts, the width typically ranges 4.8 m to 6.0 m. In India, it is wider than the mean entry width by 3.5 m. California DOT recommends that the width should not exceed 15 m (Taekratok, 1998).
- **Weaving length:** Adequate weaving length is provided to accommodate the design volume of traffic with reference to the number of lanes, weaving width and proportion of weaving traffic to total traffic. As the design speed increases, the required weaving length increases. As a thumb rule, the length of the weaving section must be equal to four times its width. With reference to design speeds of 40 km/hr and 30 km/hr, the recommended minimum weaving length may be taken as 45 m and 30 m, respectively. As a common guideline, weaving section length is longer by 33–50% to avoid decrease in design speed while weaving. However, over-speeding of traffic in the weaving section is checked by restricting its length to twice the prescribed values (IRC: 65-1976).
- **Grade:** According to French guidelines, any grade lower than 3% is generally considered acceptable, but on roadways, any grade greater than 6% is generally considered unsuitable for the installation of roundabouts (Taekratok, 1998).
- **Superelevation**: Superelevation on curved circulating width through the roundabout is generally not necessary as the operating speeds are constrained (up to 50 km/hr) and the drivers can tolerate higher values of the coefficient of sideways friction within the intersection (MDOT, 1995).

The following observations may be considered during the design of the circulating area of rotary intersections (NCHRP, 2010):

- Entering/Circulating collisions decrease with increase in central island diameter, decreased entry width and increased angle between two intersecting legs.
- Exiting/Circulating collisions decrease with increase in central island diameter, decreasing circulating width and increased angle between two intersecting legs.
- Approach crashes increase with increasing lane width when approach speeds are high.

*Note:* There are no standard designs that can fit all site conditions, and as such, each design must be examined carefully by considering the above or any other standard guidelines or by using design manuals.

### 12.4.2 Operating Standards

Achieving appropriate operating speeds for entering, circulating and exiting the roundabout is a critical design objective which impacts on the safety and comfort of the road users. Hence, operating speeds at entry and through the roundabout are managed with geometric design (Tian et al., 2007).

Approach alignment and slow entry speeds are important considerations while designing the appropriate width of the entry and the size of the raised splitter island. In this connection, a study revealed that reducing the vehicle path radius at the entry decreased the relative speed of the entering-circulating vehicles, and thus, resulted in lowering the number of entering-circulating vehicle accidents (NCHRP, 2010). The circulating speeds of vehicles should be *consistent* so as to enable smooth negotiation of conflicting traffic streams. There is also a need to provide adequate sight distance and visibility through the roundabout to recognise objects and conflicting users ahead. On account of this objective, the following *principles* may be considered essential for attaining consistency in operating speeds of a roundabout which is a fundamental aspect for safety requirements (Fig. 12.7).

- Minimise the difference in relative speeds of traffic streams between consecutive geometric elements of a roundabout
- Minimise the difference in speeds of conflicting traffic streams (entry-circulating and exit-circulating traffic)

The maximum permissible design speeds at *entry* may be considered in the range 32–40 km/hr and 40–48 km/hr, for single-lane and multilane roundabouts (NCHRP, 2010).

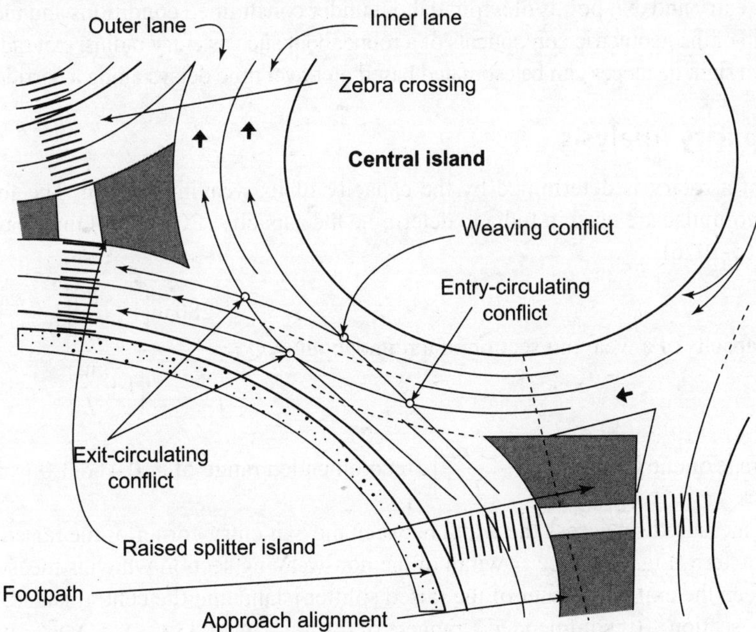


**Fig. 12.7** Illustration of weaving conflict in a dual-lane roundabout
(Elaborated based on Tian et al., 2007; *Courtesy*: California Department of Transportation (CDOT), USA)

## 12.5 DATA COLLECTION AND ANALYSIS

### 12.5.1 Methods of Data Collection

- *Manual recording*: Live recording of turning movements of traffic is carried out by field observers who position themselves at a place from which the entire roundabout is visible. This method is suitable for low-volume conditions. Field observers at each exit may be used to record moderate to low volumes of traffic approaching the exit.
- *Elevated observer method*: Video recordings of the entire circulating roundabout followed by manual extraction of turning movement data video is one of the commonly adopted methods of data collection. Multiple video mountings can also be made; the turning movements must be carefully synchronised to obtain precise vehicular turning movement. This method is suitable for collecting low or high volume traffic data provided the video-mounted position is convenient for covering and capturing the entire study area.
- *Automatic counting* techniques using intrusive or non-intrusive methods may be installed at entries, exits and across the weaving section. The traffic volume data is collected with their turning pattern.
- *Origin-Destination surveys* may be conducted to collect traffic data when multiple intersections are being studied simultaneously, within a selected study area.

### 12.5.2 Operational Performance Evaluation

- The operational performance of existing roundabouts can be evaluated by measuring the control delay. This can be estimated by measuring the average time taken by vehicles to travel between any study entry and exit points of a roundabout under constrained conditions and ideal conditions.
- The design of the geometric components of a roundabout such as entry radius, exit radius, circulating radius with sight distances can be estimated based on travel time delays along a corridor.

### 12.5.3 Capacity Analysis

The capacity of a rotary is determined by the capacity of its weaving section. The following TRL-recommended formulae are used in India to determine the capacity (PCU/hr) of the weaving section of a rotary (IRC: 65-1976).

$$\text{Capacity of a weaving section of a roundabout} = Q = \frac{280\,w\left(1+\dfrac{e}{w}\right)\left(1-\dfrac{p}{3}\right)}{\left(1+\dfrac{w}{L}\right)} \tag{12.1}$$

where $e =$ average of entry widths $= \left(\dfrac{e1+e2}{2}\right)$ (recommended range of $e$: 0.4 $w$–1.0 $w$ m); $e1$ = entry width which is measured/set across the width between the exit entry corner of the raised splitter island and the adjacent kerb (Fig. 12.8); $e2$ = width of the non-weaving section which is measured/set across the width between the exit end corner of the raised splitter island and the central island; $L$ = length of the weaving section (recommended range of L: 2.5w–8.33w m); $w$ = weaving width $= \left\{\left(\dfrac{e1+e2}{2}\right)+3.5\,\text{m}\right\} = \{e+3.5\,\text{m}\}$ (for recommended range of $w$: 6–18 m), $p$ = proportion of weaving traffic to total traffic in the weaving section, (range: 0.4–1.0).

*Adjustments:* The calculated capacity values are adjusted for the following site conditions (IRC: 65-1976):

- Deduct 5% from the calculated capacity of the weaving section if the entry angle is between 0° and 15°.
- Deduct 2.5% from the calculated capacity of the weaving section if the entry angle is between 15° and 30°.
- Deduct 2.5% from the calculated capacity of the weaving section if the exit angle is between 60° and 75°.
- Deduct 5% from the calculated capacity of the weaving section if the exit angle is greater than 75°.
- Deduct 5% from the calculated capacity of the weaving section if the internal angle of the approach roads is greater than 95°.
- Deduct 1/6th of the calculated capacity of the weaving section if pedestrian flow at the exit from the roundabout exceeds 300 per hour.

***Remarks:*** The PCU values considered for the calculation of capacity of the weaving section is 2.8 for buses and heavy commercial vehicles and 4–6 for animal-drawn vehicles. The necessary adjustment factors may be derived from the prevailing site conditions if these values of PCU differ.

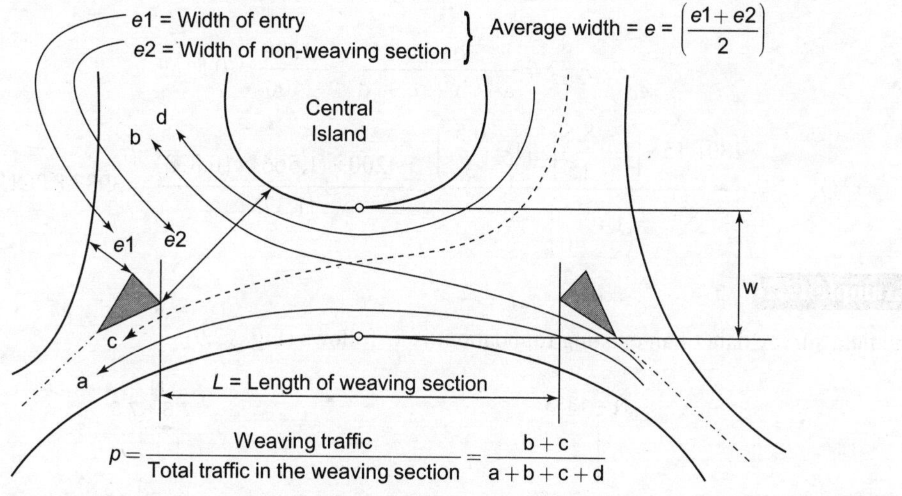


**Fig. 12.8** Definition of average entry width, weaving traffic and total traffic in a weaving section

*Note:*

- The capacity of a rotary is determined based on the highest value of *p*, which is obtained by considering the directions of traffic flow. This criterion holds good only if the rotary geometrics (*w*, *e* and *L*) are the same for all the intersecting approach roads.
- If the rotary geometrics are not the same for all the intersecting approaches, the lowest value of the capacity calculated is said to be the capacity of the rotary intersection.
- If the calculated capacity value is greater than the standard limiting value of a single lane, then an additional lane or a multilane rotary may be designed.

## 12.6 SOLVED PROBLEMS

**Example 12.1**

Determine the capacity of a roundabout for the following values:
Average entry width = 8.5 m; Width of the weaving section = 15 m; Length of the weaving section between channelising islands = 40 m; Conflicting traffic volume observed in the weaving section = 1200 CPU/hr; Total traffic volume observed in the weaving section = 2500 CPU/hr.

**Answer:** The given values are rewritten as
$e$ = Average entry width = 8.5 m; $w$ = Width of the weaving section = 15 m
$L$ = Length of the weaving section between channelising islands = 40 m
b + c = Conflicting traffic volume observed in the weaving section = 1200 CPU/hr
a + b + c + d = Total traffic volume observed in the weaving section = 2400 CPU/hr
The following equation is used to determine the capacity of the weaving section. Substituting, we get

$$\textbf{Capacity of weaving section} = \frac{280\,w\left(1+\dfrac{e}{w}\right)\left(1-\dfrac{p}{3}\right)}{\left(1+\dfrac{w}{L}\right)}$$

$$p_{E\to S} = \frac{\mathrm{b+c}}{\mathrm{a+b+c+d}} = \frac{1200}{2400} = 0.5$$

$$\mathrm{Q_{N\to E}} = \frac{280\times 15\times\left(1+\dfrac{8.5}{15}\right)\times\left(1-\dfrac{0.5}{3}.\right)}{\left(1+\dfrac{15}{40}\right)} = \frac{4200\times(1.566)\times(0.833)}{(1.375)} = 3987.8\,\text{PCU/hr}$$

**Example 12.2**

The field survey data of an existing roundabout is as follows (Fig. 12.9):

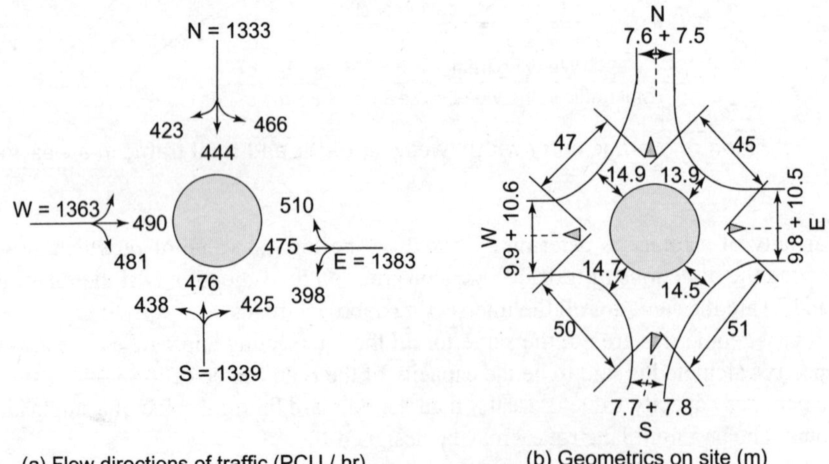


**Fig. 12.9** Rotary details (schematic diagram)

Average entry width of the N approach = 10.4 m; Average entry width of the E approach = 11.0 m; Average entry width of the S approach = 11.2 m; Average entry width of the W approach = 11.4 m.

Draw a sketch showing the inflow and outflow of traffic and determine the individual capacities of the weaving sections and the capacity of the rotary. Write your comments on the results obtained.

**Answer:** Inflow and outflow values:

Towards North: 510 + 476 + 392 = 1378; From North: 423 + 444 + 466 = 1333
Towards East: 490 + 425 + 466 = 1381; From East: 510 + 475 + 398 = 1383
Towards South: 444 + 398 + 481 = 1323; From South: 438 + 476 + 425 = 1339
Towards West: 475 + 238 + 423 = 1336; From West: 481 + 490 + 392 = 1363

*Note:* The maximum two-way flow in the intersection leg (East) is 2764 PCU/hr and the maximum in one direction is 1383 PCU/hr.

North: 2711
1378 ↑ 1333 ↓
West: 2699 { 1363 → ; 1336 ← }
East: 2764 { → 1381 ; ← 1383 }
1339 ↑ 1323 ↓
South: 2662

Directional distribution of traffic across the rotary

Let us determine the capacity of the individual weaving sections.

(i) Weaving section in the direction N→E:
$L$ = Length of weaving section = 45 m; $w$ = Width of weaving section = 13.9 m; $e$ = Average entry width at $N$ = 10.4 m

$a = Q_{N\to E} = 466$ PCU/hr

$b = (Q_{N\to w} + Q_{N\to S}) = (423 + 444) = 867$ PCU/hr

$c = (Q_{S\to E} + Q_{W\to E}) = 425 + 490 = 915$ PCU/hr

$d = Q_{W\to S} = 481$ PCU/hr

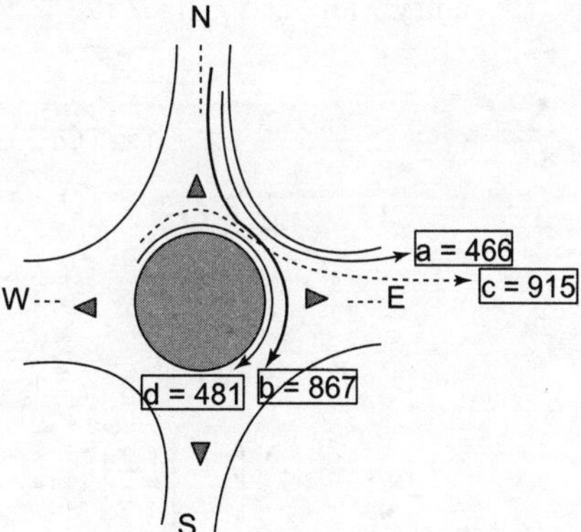


**Fig. 12.10** Flow stream definitions of a, b, c and d

Flow streams in the weaving section N→E

$$p_{N\to E} = \frac{b+c}{a+b+c+d} = \frac{867+915}{466+867+915+481} = 0.653$$

$$\text{Capacity of weaving section N} \to \text{E} = Q_{N\to E} = \frac{280w\left(1+\frac{e}{w}\right)\left(1-\frac{p}{3}\right)}{\left(1+\frac{w}{L}\right)}$$

$$Q_{N\to E} = \frac{280\times 13.9\times\left(1+\dfrac{10.4}{13.9}\right)\times\left(1-\dfrac{0.653}{3}\right)}{\left(1+\dfrac{13.9}{45}\right)} = \frac{3892\times(1.748)\times(0.782)}{(1.308)} = 4067.3\,\text{PCU/hr}$$

(ii) Weaving section in the direction E→S:
$L$ = Length of weaving section = 51 m; $w$ = Width of weaving section = 14.5 m; $e$ = Average entry width at $E$ = 11.0 m

$$\text{a} = Q_{E\to S} = 398\ \text{PCU/hr}$$

$$\text{b} = (Q_{E\to N} + Q_{E\to W}) = (510 + 475) = 985\ \text{PCU/hr}$$

$$\text{c} = (Q_{W\to S} + Q_{N\to S}) = (481 + 444) = 925\ \text{PCU/hr}$$

$$\text{d} = Q_{N\to W} = 423\ \text{PCU/hr}$$

$$p_{E\to S} = \frac{\text{b}+\text{c}}{\text{a}+\text{b}+\text{c}+\text{d}} = \frac{985+925}{398+985+925+423} = 0.699$$

$$Q_{E\to S} = \frac{280\times 14.5\times\left(1+\dfrac{11.0}{14.5}\right)\times\left(1-\dfrac{0.699}{3}\right)}{\left(1+\dfrac{14.5}{51}\right)} = \frac{4060\times(1.758)\times(0.767)}{(1.284)} = 4263.5\,\text{PCU/hr}$$

(iii) Weaving section in the direction S→W:
$L$ = Length of weaving section = 50 m; $w$ = Width of weaving section = 14.7 m; $e$ = Average entry width at $S$ = 11.2 m

$$\text{a} = Q_{S\to W} = 438\ \text{PCU/hr}$$

$$\text{b} = (Q_{S\to N} + Q_{S\to E}) = (476 + 425) = 901\ \text{PCU/hr}$$

$$\text{c} = (Q_{N\to W} + Q_{E\to W}) = (423+475) = 898\ \text{PCU/hr}$$

$$\text{d} = Q_{E\to N} = 510\ \text{PCU/hr}$$

$$p_{S\to W} = \frac{\text{b}+\text{c}}{\text{a}+\text{b}+\text{c}+\text{d}} = \frac{901+898}{438+901+898+510} = 0.654$$

$$Q_{S\to W} = \frac{280\times 14.7\times\left(1+\dfrac{11.2}{14.7}\right)\times\left(1-\dfrac{0.654}{3}\right)}{\left(1+\dfrac{14.7}{50}\right)} = \frac{4116\times(1.761)\times(0.782)}{(1.294)} = 4380.3\,\text{PCU/hr}$$

(iv) Weaving section in the direction W→N:
$L$ = Length of weaving section = 47 m; $w$ = Width of weaving section = 14.9 m; $e$ = Average entry width at $W$ = 11.4 m

$$\text{a} = Q_{W\to N} = 392\ \text{PCU/hr}$$

$$\text{b} = (Q_{W\to S} + Q_{W\to E}) = (481 + 490) = 971\ \text{PCU/hr}$$

$$c = (Q_{E \to N} + Q_{S \to N}) = (510 + 476) = 986 \text{ PCU/hr}$$

$$d = Q_{S \to E} = 425 \text{ PCU/hr}$$

$$p_{W \to N} = \frac{b + c}{a + b + c + d} = \frac{971 + 986}{392 + 971 + 986 + 425} = 0.705.$$

$$Q_{W \to N} = \frac{280 \times 14.9 \times \left(1 + \frac{11.4}{14.9}\right) \times \left(1 - \frac{0.705}{3}\right)}{\left(1 + \frac{14.9}{47}\right)} = \frac{4172 \times (1.765) \times (0.765)}{(1.317)} = 4277.2 \text{ PCU/hr}$$

*Comments on the result*:

1. As the rotary geometrics (values of $w$, $e$ and $L$) are not the same for all the intersecting approaches, the *lowest value* of the capacity calculated ($Q_{N \to E} = 4067.3$ PCU/hr) is considered to be the *capacity* of this rotary intersection.
2. $Q_{N \to E} = 4067.3$ PCU/hr $> 2764$ (E). Hence, this is acceptable.
3. If the traffic volume entering from all the intersection legs of the rotary exceeds 3000 veh/hr, the intersection may be redesigned with widened geometrics or as a signalised intersection (IRC: 65-1976).

**Example 12.3**

The design of a four-legged at-grade intersection (E–W and N–S directions) was proposed as a rotary intersection. The geometrics of the rotary are given below. Determine the capacity of the proposed rotary intersection.

Width of the carriageway approaching the intersection = 15.0 m; Entry width = 10.0 m; Non-weaving section width = 10.2 m

The traffic approaching the intersection from four sides is given below:

| | | | | | | | |
|---|---|---|---|---|---|---|---|
| $Q_{W \to N}$ | 287 | $Q_{N \to E}$ | 338 | $Q_{E \to S}$ | 266 | $Q_{S \to W}$ | 364 |
| $Q_{W \to E}$ | 290 | $Q_{N \to S}$ | 303 | $Q_{E \to W}$ | 377 | $Q_{S \to N}$ | 210 |
| $Q_{W \to S}$ | 255 | $Q_{N \to W}$ | 209 | $Q_{E \to N}$ | 211 | $Q_{S \to E}$ | 300 |
| Total | 832 PCU/hr | Total | 850 PCU/hr | Total | 854 PCU/hr | Total | 874 PCU/hr |

Assume any missing/required data as per the guidelines of IRC: 65-1976.

**Answer:** Average value of entry and non-weaving section width, $e = \left(\frac{e1 + 2}{2}\right) = \left(\frac{10 + 10.2}{2}\right) = 10.1$ m;

Weaving section width, $w = (e + 3.5\text{m}) = 10.1 + 3.5 = 13.6$ m (as per IRC: 65); Weaving length, $L = 4 \times w = 4 \times 13.6 = 54.4$ m (assumed as per IRC: 65)

The proportion of weaving traffic to total traffic is calculated for the four approaches as follows:

(i) Weaving section in the direction N→E:

$$a = Q_{N \to E} = 338 \text{ PCU/hr}$$

$$b = (Q_{N \to w} + Q_{N \to S}) = (209 + 303) = 512 \text{ PCU/hr}$$

$$c = (Q_{S\to E} + Q_{W\to E}) = (300 + 290) = 590 \text{ PCU/hr}$$

$$d = Q_{W\to S} = 255 \text{ PCU/hr}$$

$$p_{N\to E} = \frac{b+c}{a+b+c+d} = \frac{512+590}{338+512+590+255} = 0.650$$

(ii) Weaving section in the direction E→S:

$$a = Q_{E\to S} = 266 \text{ PCU/hr}$$

$$b = (Q_{E\to N} + Q_{E\to W}) = (211+377) = 588 \text{ PCU/hr}$$

$$c = (Q_{W\to S} + Q_{N\to S}) = (255+303) = 558 \text{ PCU/hr}$$

$$d = Q_{N\to W} = 209 \text{ PCU/hr}$$

$$p_{E\to S} = \frac{b+c}{a+b+c+d} = \frac{588+558}{266+588+558+209} = 0.707$$

(iii) Weaving section in the direction S→W:

$$a = Q_{S\to W} = 364 \text{ PCU/hr}$$

$$b = (Q_{S\to N} + Q_{S\to E}) = (210 + 300) = 510 \text{ PCU/hr}$$

$$c = (Q_{N\to W} + Q_{E\to W}) = (209+377) = 586 \text{ PCU/hr}$$

$$d = Q_{E\to N} = 211 \text{ PCU/hr}$$

$$p_{S\to W} = \frac{b+c}{a+b+c+d} = \frac{510+586}{364+510+586+211} = 0.656$$

(iv) Weaving section in the direction W→N:

$$a = Q_{W\to N} = 287 \text{ PCU/hr}$$

$$b = (Q_{W\to S} + Q_{W\to E}) = (255+290) = 545 \text{ PCU/hr}$$

$$c = (Q_{E\to N} + Q_{S\to N}) = (211+210) = 421 \text{ PCU/hr}$$

$$d = Q_{S\to E} = 300 \text{ PCU/hr}$$

$$p_{W\to N} = \frac{b+c}{a+b+c+d}$$

$$= \frac{545+421}{287+545+421+300} = 0.622$$

The highest proportion of weaving traffic to total traffic is 0.707, which is in the direction, E→S. Hence, the minimum traffic flow rate, which is considered as the rotary capacity, is:

$$Q_{E\to S} = \frac{280\times 13.6\times\left(1+\dfrac{10.1}{13.6}\right)\times\left(1-\dfrac{0.707}{3}\right)}{\left(1+\dfrac{13.6}{54.4}\right)}$$

$$= \frac{3808\times(1.742)\times(0.764)}{(1.25)} = 4054.4 \text{ PCU/hr}$$

**Example 12.4**

Calculate the adjusted values of the calculated capacity of a weaving section of 2000 PCU/hr for the following individual site conditions, as per IRC: 65.
(a) If the entry angle is 10°; (b) If the entry angle is 20°; (c) If the exit angle is 62.8°; (d) If the exit angle is 85°; (d) If the internal angle of the approach road is 105°.

Pedestrian flow in the exit from the roundabout is 550 per hour.

**Answer:** The details of the adjustments applied to the calculated capacity of a weaving section of 2000 CPU/hr are given below.

(a) Deduct 5% from the calculated capacity of the weaving section as the given value of the entry angle is 10° which is between 0° and 15°

$$\text{Adjusted } capacity = Q = 2000 - \left(2000 \times \frac{5}{100}\right) = 2000 - 100 = 1900 \text{ CPU/hr}$$

(b) Deduct 2.5% from the calculated capacity of the weaving section as the given value of the entry angle is 20° which is between 15° and 30°

$$\text{Adjusted capacity} = Q = 2000 - \left(2000 \times \frac{2.5}{100}\right) = 2000 - 50 = 1950 \text{ CPU/hr}$$

(c) Deduct 2.5% from the calculated capacity of the weaving section as the given value of the exit angle is 62.8° which is between 60° and 75°

$$\text{Adjusted capacity} = Q = 2000 - \left(2000 \times \frac{2.5}{100}\right) = 2000 - 50 = 1950 \text{ CPU/hr}$$

(d) Deduct 5% from the calculated capacity of the weaving section as the given value of the exit angle is 85° which is greater than 75°

$$\text{Adjusted capacity} = Q = 2000 - \left(2000 \times \frac{5}{100}\right) = 2000 - 100 = 1900 \text{ CPU/hr}$$

(e) Deduct 5% from the calculated capacity of the weaving section as the given value of the internal angle of the approach roads is 105° which is greater than 95°

$$\text{Adjusted capacity} = Q = 2000 - \left(2000 \times \frac{5}{100}\right) = 2000 - 100 = 1900 \text{ CPU/hr}$$

(f) Deduct about 1/6th of the calculated capacity of the weaving section as the given value of pedestrian flow at the exit from the roundabout is 550 per hour, which exceeds 300 per hour.

$$\text{Adjusted capacity} = Q = 2000 - \left(2000 \times \frac{1}{6}\right) = 2000 - 333.33 = 1666.67 \text{ CPU/hr}$$

**Example 12.5**

The details of turning movement volumes obtained at an intersection are given in Fig. 12.11.

Convert the turning movement volumes to roundabout volumes by calculating (i) Entry flow rates (ii) Circulating flow rates and (iii) Exiting flow rates from each approach road.

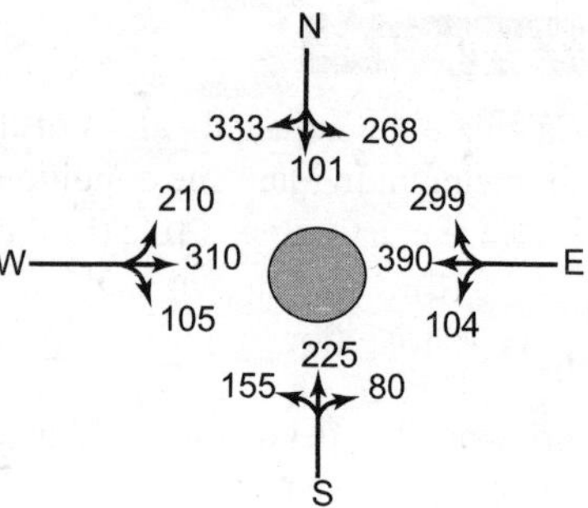


**Fig. 12.11** Turning movement volumes

**Answer:** The turning movement volumes are tabulated as given below:

| $Q_{W\to N}$ | 210 | $Q_{N\to E}$ | 268 | $Q_{E\to S}$ | 104 | $Q_{S\to W}$ | 155 |
|---|---|---|---|---|---|---|---|
| $Q_{W\to E}$ | 310 | $Q_{N\to S}$ | 101 | $Q_{E\to W}$ | 390 | $Q_{S\to N}$ | 225 |
| $Q_{W\to S}$ | 105 | $Q_{N\to W}$ | 333 | $Q_{E\to N}$ | 299 | $Q_{S\to E}$ | 80 |
| Total | 625 PCU/hr | Total | 702 PCU/hr | Total | 793 PCU/hr | Total | 460 PCU/hr |

(i) The entry flow rate = Sum of movement flow rates that enter the roundabout
From West: $Q_W = Q_{W\to N} + Q_{W\to E} + Q_{W\to S} = 625$ pcu/ hr
From North: $Q_N = Q_{N\to E} + Q_{N\to S} + Q_{N\to W} = 702$ pcu/ hr
From East: $Q_E = Q_{E\to S} + Q_{E\to W} + Q_{E\to N} = 793$ pcu/hr
From South: $Q_S = Q_{S\to W} + Q_{S\to N} + Q_{S\to E} = 460$ pcu/hr

(ii) Circulating flow rate for each leg = Sum of all volumes that are conflicting with entering vehicles on the subject approach leg
From West: $Q_W = Q_{E\to N} + Q_{S\to E} + Q_{S\to N} = 299 + 80 + 225 = 604$ pcu/hr
From North: $Q_N = Q_{S\to E} + Q_{W\to S} + Q_{W\to E} = 80 + 105 + 310 = 495$ pcu/hr
From East: $Q_E = Q_{W\to S} + Q_{N\to W} + Q_{N\to S} = 105 + 333 + 101 = 539$ pcu/hr
From South: $Q_S = Q_{N\to W} + Q_{E\to N} + Q_{E\to W} = 333 + 299 + 390 = 1022$ pcu/hr

(iii) Exiting flow rate for each leg = Sum of all flows that exit the roundabout on the subject leg
Towards West: $Q_W = Q_{N\to W} + Q_{E\to W} + Q_{S\to W} = 333 + 390 + 155 = 878$ pcu/hr
Towards North: $Q_N = Q_{E\to N} + Q_{S\to N} + Q_{W\to N} = 299 + 225 + 210 = 734$ pcu/hr
Towards East: $Q_E = Q_{S\to E} + Q_{W\to E} + Q_{N\to E} = 80 + 310 + 268 = 658$ pcu/hr
Towards South: $Q_S = Q_{W\to S} + Q_{N\to S} + Q_{E\to S} = 105 + 101 + 104 = 310$ pcu/hr

(iv) Figure 12.12 illustrates the final volumes converted into roundabout volumes for entering, exiting and circulating flow rates.

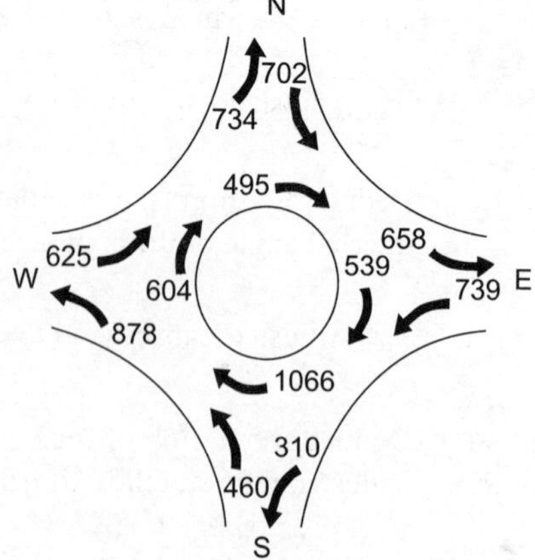


**Fig. 12.12** Entering, exiting and circulating flow rates (PCU/hr)

**Example 12.6**

Estimate how many lanes are required to serve the design-year traffic volumes (PCU/hr) shown in Fig. 12.13. Use NCHRP 672 (2010)-recommended guidelines, as given below:

| Sum of entering + Conflicting volume | Number of lanes required |
|---|---|
| 0–1000 | Single lane |
| 1000–1800 | Double lane |
| >1800 | More than three lanes |

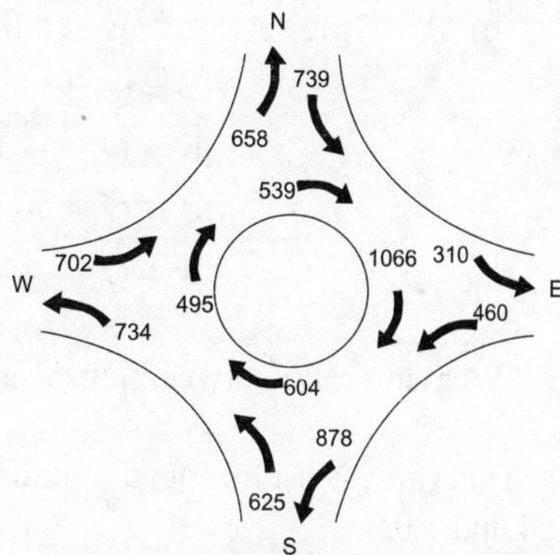


Fig. 12.13 Design year traffic volume

**Answer:**

From South: 625 + 604 = 1229 (double lane required)
From West: 702 + 495 = 1197 (double lane required)
From North: 793 + 539 = 1332 (double lane required)
From East: 460 + 1066 = 1526 (double lane required)

A double lane roundabout is adequate for handling the design year traffic volumes for each individual approach.

**Example 12.7**

A rotary intersection has five legs designated as 1, 2, 3, 4 and 5. Leg 1 is in the North–South direction and the remaining legs are marked in the clockwise direction. The traffic volumes observed during peak hour in the respective directions are given below in terms of PCU/hr.

| $Q_{1\to2}$ | 40 | $Q_{2\to1}$ | 0 | $Q_{3\to1}$ | 450 | $Q_{4\to2}$ | 181 | $Q_{5\to1}$ | 44 |
|---|---|---|---|---|---|---|---|---|---|
| $Q_{1\to3}$ | 299 | $Q_{2\to3}$ | 0 | $Q_{3\to2}$ | 112 | $Q_{4\to2}$ | 50 | $Q_{5\to2}$ | 133 |
| $Q_{1\to4}$ | 65 | $Q_{2\to4}$ | 0 | $Q_{3\to4}$ | 49 | $Q_{4\to3}$ | 21 | $Q_{5\to3}$ | 60 |
| $Q_{1\to5}$ | 55 | $Q_{2\to5}$ | 0 | $Q_{3\to5}$ | 666 | $Q_{4\to5}$ | 116 | $Q_{5\to4}$ | 19 |

(a) Calculate the volume of vehicles leaving and approaching each leg; (b) Calculate the weaving ratio between the legs 1 and 2; (c) State the use of this value; (d) Draw a sketch showing the traffic volumes between legs 1 and 2.

Fig. 12.14 Sketch showing the traffic volumes between legs 1 and 2

**Answer:**

(a) Volume of vehicles leaving and approaching each leg:

Vehicles leaving leg 1 = $Q_{1\to2} + Q_{1\to3} ++ Q_{1\to4} + Q_{1\to5} = 459$

Vehicles leaving leg 2 = $Q_{2\to1} + Q_{2\to3} ++ Q_{2\to4} + Q_{2\to5} = 0$

Vehicles leaving leg 3 = $Q_{3\to1} + Q_{3\to2} ++ Q_{3\to4} + Q_{3\to5} = 1277$

Vehicles leaving leg 4 = $Q_{4\to1} + Q_{4\to2} ++ Q_{4\to3} + Q_{4\to5} = 368$

Vehicles leaving leg 5= $Q_{5\to1} + Q_{5\to2} ++ Q_{5\to3} + Q_{1\to4} = 256$

Vehicle approaching leg 1 = = $Q_{2\to1} + Q_{3\to1} + Q_{4\to1} + Q_{5\to1} = 0 + 450 + 181 + 44 = 675$

Vehicle approaching leg 2 = = $Q_{1\to2} + Q_{3\to2} + Q_{4\to2} + Q_{5\to2} = 40 + 112 + 50 + 133 = 335$

Vehicle approaching leg 3 = = $Q_{1\to3} + Q_{2\to3} + Q_{4\to3} + Q_{5\to3} = 299 + 0 + 21 + 60 = 380$

Vehicle approaching leg 4 = = $Q_{1\to4} + Q_{2\to4} + Q_{3\to4} + Q_{5\to4} = 65 + 0 + 49 + 19 = 133$

Vehicle approaching leg 5 = = $Q_{1\to5} + Q_{2\to5} + Q_{3\to5} + Q_{4\to5} = 55 + 0 + 666 + 116 = 837$

(b) Weaving section between legs 1 and 2:

$$a = Q_{1\to2} = 40 \text{ PCU/hr}$$
$$b = Q_{1\to3} + Q_{1\to4} + Q_{1\to5} = 299 + 65 + 55 = 419 \text{ PCU/hr}$$
$$c = Q_{3\to2} + Q_{4\to2} + Q_{5\to2} = 112 + 50 = 133 = 295 \text{ PCU/hr}$$
$$d = Q_{5\to3} + Q_{5\to4} + Q_{4\to3} = 60 + 19 + 21 = 100 \text{ PCU/hr}$$

$$\text{Weaving ratio between legs 1 and 2} = p_{1\to2} = \frac{b+c}{a+b+c+d} = \frac{419+295}{40+419+295+100} = 0.836$$

This value is useful for calculating the practical capacity of rotary intersections. It should lie between 0.4 and 1.0.

**Example 12.8**

A rotary intersection has five legs designated as 1, 2, 3, 4 and 5. Leg 1 is in the North–South direction and the remaining legs are marked in the clockwise direction. Identify the leg through which traffic flow leaving does not affect the weaving ratio calculated between legs 1 and 2.
(a) 1 (b) 2 (c) 3 (d) 4

**Answer:** (b)

**Example 12.9**

A rotary intersection has five legs designated as 1, 2, 3, 4 and 5. Leg 1 is in the North–South direction and the remaining legs are marked in the clockwise direction. The traffic volumes observed during peak hour in the respective directions are given below in terms of PCU/hr.

| $Q_{1\to2}$ | 40 | $Q_{2\to1}$ | 37 | $Q_{3\to1}$ | 450 | $Q_{4\to2}$ | 181 | $Q_{5\to1}$ | 44 |
|---|---|---|---|---|---|---|---|---|---|
| $Q_{1\to3}$ | 299 | $Q_{2\to3}$ | 52 | $Q_{3\to2}$ | 112 | $Q_{4\to2}$ | 50 | $Q_{5\to2}$ | 133 |
| $Q_{1\to4}$ | 65 | $Q_{2\to4}$ | 11 | $Q_{3\to4}$ | 49 | $Q_{4\to3}$ | 21 | $Q_{5\to3}$ | 60 |
| $Q_{1\to5}$ | 55 | $Q_{2\to5}$ | 29 | $Q_{3\to5}$ | 666 | $Q_{4\to5}$ | 116 | $Q_{5\to4}$ | 19 |

Calculate the weaving ratio between legs 1 and 2, 2 and 3, 3 and 4, 4 and 5 and 5 and 1. Determine the weaving ratio used to calculate the rotary capacity.

**Answer:**

(i) Weaving section between legs 1 and 2:

$$a = Q_{1\to2} = 40 \text{ PCU/hr}$$
$$b = Q_{1\to3} + Q_{1\to4} + Q_{1\to5} = 299 + 65 + 55 = 419 \text{ PCU/hr}$$
$$c = Q_{3\to2} + Q_{4\to2} + Q_{5\to2} = 112 + 50 = 133 = 295 \text{ PCU/hr}$$
$$d = Q_{5\to3} + Q_{5\to4} + Q_{4\to3} = 60 + 19 + 21 = 100 \text{ PCU/hr}$$

$$\text{Weaving ratio between legs 1 and 2} = p_{1\to2} = \frac{b+c}{a+b+c+d} = \frac{419+295}{40+419+295+100} = 0.836$$

(ii) Weaving section between legs 2 and 3:

$$a = Q_{2\to3} = 52 \text{ PCU/hr}$$
$$b = Q_{2\to4} + Q_{2\to5} + Q_{2\to1} = 11 + 29 + 37 = 77 \text{ PCU/hr}$$
$$c = Q_{4\to3} + Q_{5\to3} + Q_{1\to3} = 21 + 60 + 299 = 380 \text{ PCU/hr}$$
$$d = Q_{1\to4} + Q_{1\to5} + Q_{5\to4} = 65 + 55 + 19 = 139 \text{ PCU/hr}$$

$$\text{Weaving ratio between legs 2 and 3} = p_{2\to3} = \frac{b+c}{a+b+c+d} = \frac{77+380}{52+77+380+139} = 0.705$$

(iii) Weaving section between legs 3 and 4:

$$a = Q_{3\to4} = 49 \text{ PCU/hr}$$
$$b = Q_{3\to5} + Q_{3\to1} + Q_{3\to2} = 666 + 450 + 112 = 1228 \text{ PCU/hr}$$
$$c = Q_{5\to4} + Q_{1\to4} + Q_{2\to4} = 19 + 65 + 11 = 95 \text{ PCU/hr}$$
$$d = Q_{2\to5} + Q_{2\to1} + Q_{1\to5} = 29 + 37 + 55 = 121 \text{ PCU/hr}$$

$$\text{Weaving ratio between legs 3 and 4} = p_{3\to4} = \frac{b+c}{a+b+c+d} = \frac{1228+95}{49+1228+95} = 0.886$$

(iv) Weaving section between legs 4 and 5:

$$a = Q_{4\to5} = 116 \text{ PCU/hr}$$
$$b = Q_{4\to1} + Q_{4\to2} + Q_{4\to3} = 181 + 50 + 21 = 252 \text{ PCU/hr}$$
$$c = Q_{1\to5} + Q_{2\to5} + Q_{3\to5} = 55 + 29 + 666 = 750 \text{ PCU/hr}$$
$$d = Q_{3\to1} + Q_{3\to2} + Q_{2\to1} = 450 + 112 + 37 = 599 \text{ PCU/hr}$$

$$\text{Weaving ratio between legs 4 and 5} = p_{4\to5} = \frac{b+c}{a+b+c+d} = \frac{252+750}{116+252+750+599} = 0.583$$

(v) Weaving section between legs 5 and 1:

$$a = Q_{5\to1} = 44 \text{ PCU/hr}$$
$$b = Q_{5\to2} + Q_{5\to3} + Q_{5\to4} = 133 + 60 + 19 = 212 \text{ PCU/hr}$$
$$c = Q_{2\to1} + Q_{3\to1} + Q_{4\to1} = 37 + 450 + 181 = 668 \text{ PCU/hr}$$
$$d = Q_{4\to2} + Q_{4\to3} + Q_{3\to2} = 50 + 21 + 112 = 183 \text{ PCU/hr}$$

$$\text{Weaving ratio between legs 5 and 1} = p_{5\to1} = \frac{b+c}{a+b+c+d} = \frac{212+668}{44+212+668+183} = 0.794$$

The highest proportion of weaving traffic to total traffic is found to be 0.886, in the direction 3→4. Hence, the above value may be taken for the rotary capacity.

## EXERCISES

### Multiple Choice Questions

1. What would be the possible number of total conflict points with a conventional four-legged intersection having a two-lane two-way operation?

   (a) 24 (b) 32 (c) 8 (d) 16

2. What would be the possible number of total conflict points with a conventional three-legged intersection having a two-lane two-way operation?
   (a) 9 (b) 6 (c) 24 (d) 32
3. What would be the possible number of crossing conflict points with a conventional four-legged intersection having a two-lane two-way operation?
   (a) 24 (b) 32 (c) 8 (d) 16
4. What would be the possible number of diverging conflict points with a conventional four-legged intersection having a two-lane two-way operation?
   (a) 24 (b) 32 (c) 8 (d) 16
5. What would be the possible number of merging conflict points with a conventional four-legged intersection having a two-lane two-way operation?
   (a) 24 (b) 32 (c) 8 (d) 16
6. What would be the possible number of diverging conflict points with a conventional three-legged intersection having a two-lane two-way operation?
   (a) 3 (b) 6 (c) 9 (d) 12
7. What would be the possible number of merging conflict points with a conventional three-legged intersection having a two-lane two-way operation?
   (a) 9 (b) 12 (c) 3 (d) 6
8. What would be the possible number of crossing conflict points with a rotary three-legged intersection having a two-lane two-way operation?
   (a) 3 (b) 0 (c) 9 (d) 12
9. What would be the possible number of total conflict points with a rotary three-legged intersection having a two-lane two-way operation?
   (a) 3 (b) 6 (c) 9 (d) 12
10. What would be the possible number of crossing conflict points with a rotary four-legged intersection having a two-lane two-way operation?
    (a) 3 (b) 0 (c) 9 (d) 12
11. What would be the possible number of merging conflict points with a rotary four-legged intersection having a two-lane two-way operation?
    (a) 3 (b) 0 (c) 4 (d) 12
12. What would be the possible number of total conflict points with a rotary four-legged intersection having a two-lane two-way operation?
    (a) 3 (b) 6 (c) 9 (d) 8
13. What is the reduction in the total number of vehicle–vehicle conflict points observed in a rotary compared with four-legged intersections, when both have a two-lane two-way operation?
    (a) 33.33% (b) 100% (c) 75% (d) 50%
14. What is the reduction in the total number of vehicle–vehicle conflict points observed in a rotary compared with a three-legged intersection, when both have a two-lane two-way operation?
    (a) 33.33% (b) 100% (c) 75% (d) 50%

15. What is the reduction in the total number of vehicle–pedestrian conflict points observed in a rotary compared with four-legged intersections, when both have a two-lane two-way operation?
(a) 33.33% (b) 100% (c) 75% (d) 50%
16. What is the total number of vehicle–pedestrian conflict points observed in a rotary with four-legged intersections, having a two-lane two-way operation?
(a) 4 (b) 8 (c) 16 (d) 9
17. What is the appropriate entry approach angle that should be provided to a rotary intersection to ensure adequate sight distance and visibility?
(a) Between 20° and 40° (b) < 20°
(c) >100° (d) Between 60° and 90°
18. As per IRC: 65, the recommended radius of entry may be selected for the design of a rotary intersection in the range of 20–35 m and 15–25 m, respectively, for design speeds of ________ and ________.
(a) 20 and 10 km/hr (b) 10 and 20 km/hr (c) 30 and 40 km/hr (d) 40 and 30 km/hr
19. As per IRC: 65, the recommended minimum width of the weaving section of a rotary intersection should be wider than the mean entry width by ________
(a) 1.5 m (b) 2.5 m (c) 3.5 m (d) 5.5 m
20. As per IRC: 65, with reference to design speeds of 40 km/hr and 30 km/hr, the recommended minimum weaving length for a rotary may be taken as ________ and ________, respectively.
(a) 45 m and 30 m (b) 40 m and 35 m (c) 25 m and 30 m (d) 30 m and 35 m
21. As per IRC: 65, from the calculated capacity of a weaving section of a rotary intersection is 5% is deducted if the entry angle is between ________ and ________.
(a) 75° and 90° (b) 60° and 75° (c) 15° and 30° (d) 0° and 15°
22. As per IRC: 65, from the calculated capacity of the weaving section of a rotary intersection 2.5% is deducted if the exit angle is between ________ and ________.
(a) 75° and 90° (b) 60° and 75° (c) 15° and 30° (d) 0° and 15°
23. As per IRC: 65, deduct about $1/6^{th}$ of the calculated capacity of weaving section if the pedestrian flow at the exit from the roundabout exceeds ________ per hour.
(a) 200 (b) 300 (c) 33 (d) 15
24. As per IRC: 65, the proportion of weaving traffic to total traffic in the weaving section of a rotary intersection should be in the range ________
(a) 0 to 1.0 (b) 0 to 0.4 (c) 1 to 2 (d) 0.4 to 1.0
25. The following condition is favourable for designing roads intersecting at-grade as a roundabout/rotary:
(a) Intersections where traffic signals are warranted
(b) Traffic volume in the range 500–3000 veh/hr
(c) At intersections with fewer delays
(d) Traffic volume is greater than 3000 veh/hr
26. The following condition is favourable for designing roads intersecting at-grade as a roundabout/rotary:
(a) The number of intersecting roads is between 4 and 7
(b) Traffic volume in the range 4000–5000 veh/hr
(c) At intersections with fewer delays
(d) Traffic volume is less than 400 veh/hr

27. When all the roads intersect at the same point and one of the roads is carrying considerably more traffic than the remaining roads, then the suitable alignment of the approach road to the rotary is
    (a) Radial alignment (b) Tangent alignment (c) Elliptical (d) Turbine
28. What is the deduction value if the calculated capacity of the weaving section of a rotary is 2000 CPU/hr. The approach roads internal angle is 105°.
    (a) 1900 PCU/hr (b) 200 PCU/hr (c) 100 PCU/hr (d) 1850 PCU/hr
    (*Hint*: As the internal angle of approach roads is 105° which is greater than 95°

    $$\text{Deduct value}\,(5\%) = \left(2000 \times \frac{5}{100}\right) = 100\,\text{CPU/hr})$$

29. What would be the capacity of the weaving section of a rotary where the length of the weaving section = 45 m; width of weaving section = 13.9 m; average entry width = 10.4 m and the proportion of weaving traffic to total traffic in the weaving section = 0.653.
    (a) 4067.3 PCU/hr (b) 4263.5 PCU/hr (c) 4380.3 PCU/hr (d) 4277.2 PCU/hr

    $$(\textit{Hint: } Q_{N\to E} = \frac{280w\left(1+\frac{e}{w}\right)\left(1-\frac{p}{3}\right)}{\left(1+\frac{w}{L}\right)} = \frac{280\times 13.9\times\left(1+\frac{10.4}{13.9}\right)\times\left(1-\frac{0.653}{3}\right)}{\left(1+\frac{13.9}{45}\right)} = 4067.3\,\text{PCU/hr})$$

30. The entrance and exit curves of a roundabout/rotary should have
    (a) Equal radii
    (b) Entrance radius should be less than exit radius
    (c) Exit radius should be less than entrance radius
    (d) Equal pavement widths but entrance radius should be more than exit radius
31. The types of conflicts that occur in rotary intersections are
    (a) Crossing and diverging
    (b) Crossing and merging
    (c) Crossing, merging and diverging
    (d) Weaving
32. Consider the following parameters relating to a roundabout intersection:
    A. Width of the weaving section B. Length of the weaving section
    C. Proportion of the weaving traffic D. Weaving angle
    E. Width of the carriageway at entry
    Which of the above is not considered while calculating the capacity of a roundabout?
    (a) A (b) B (c) D (d) E
33. Which of the following parameters are not related to roundabout intersections?
    A. Raised splitter islands B. Entry circulating conflict
    C. Proportion of weaving traffic D. Truck apron
    E. Ramp F. Deceleration lanes
    (a) A and B (b) E and F (c) D and E (d) D and A

34. Which of the following geometric features requires the magnitudes of weaving angle and long weaving distance for its design?
    (a) Rotary (b) Roundabout
    (c) Mini-roundabout (d) Grade-separated intersection
35. The practical capacity and required capacity of a roundabout is 2000 and 1500 PCU/hr, respectively. The reserved capacity of the roundabout is ________
    (a) 75.00% (b) 133.33% (c) 25.00% (d) 33.33%
    (*Hint*: Reserve capacity = ((2000 − 1500)/1500) × 100 = 33.33%)

## Answers

| | | | | | | | | | |
|---|---|---|---|---|---|---|---|---|---|
| 1. (b) | 2. (b) | 3. (d) | 4. (c) | 5. (c) | 6. (a) | 7. (c) | 8. (b) | 9. (b) | 10. (b) |
| 11. (c) | 12. (d) | 13. (c) | 14. (a) | 15. (d) | 16. (c) | 17. (a) | 18. (d) | 19. (c) | 20. (a) |
| 21. (b) | 22. (b) | 23. (b) | 24. (d) | 25. (b) | 26. (a) | 27. (b) | 28. (c) | 29. (a) | 30. (b) |
| 31. (d) | 32. (c) | 33. (b) | 34. (a) | 35. (d) | | | | | |

## Questions

1. Explain the concept of traffic rotary operation with a neat sketch, and label all its components.
2. Explain the advantages of traffic rotary with reference to (a) safety and (b) operations.
3. Draw a layout showing details of vehicle–vehicle conflicts with conventional three-legged at-grade intersections and three-legged rotary intersections. Indicate the diverging, crossing and merging conflicting points for these two intersections.
4. Draw a layout showing details of vehicle–vehicle conflicts with conventional four-legged at-grade and rotary intersections. Indicate the diverging, crossing and merging conflicting points for these two intersections.
5. Draw a layout showing details of vehicle–pedestrian conflicts with conventional four-legged at-grade and rotary intersections. Indicate the left turn on Green signal conflicts, right turn on Green signal conflicts, Red light running conflicts and left turn conflict points for these two intersections.
6. State the disadvantages of traffic rotaries.
7. Specify the conditions favourable for designing roads intersecting at-grade as a roundabout/rotary.
8. List the design criteria and standard guidelines for the geometric parameters relating to rotary intersections as per IRC: 65.
9. Draw a layout of the typical alignment of approach roads to a roundabout showing (a) alignment offset right, (b) radial alignment and (c) alignment offset left.
10. State the design criteria and standard guidelines for rotary intersections for the following parameters: (a) Width, position and alignment of approach roads (b) Central islands (c) Inscribed circle.
11. State the design criteria and standard guidelines for rotary intersections for the following parameters: (a) Entry approach angle (b) Entry exit angle (c) Exit kerb radii (d) Entry radii (e) Truck apron (f) Entry/Exit width (g) Width of weaving section (h) Weaving length (i) Grade (j) Superelevation.
12. What aspects need to be considered during the design of the circulating area of rotary intersections?
13. Explain the operating standards with weaving conflicts in a dual-lane roundabout with a neat sketch. Also state the principles considered to be essential for attaining consistency in the operating speeds of a roundabout as a safety requirement.

14. Explain the different methods of data collection with their advantages and limitations.
15. What are the data collection methods used for operational performance evaluation? Specify the details.
16. Explain in detail how rotary capacity analysis is determined, including the adjustments for specific conditions.
17. Draw a rotary layout illustrated with average entry width, weaving traffic, width of non-weaving section, length of weaving section and total traffic components in a weaving section.
18. Determine the capacity of a roundabout for the following data:
    Average entry width = 8.6 m; Width of weaving section = 16 m; Length of weaving section between channelising islands = 42 m; Conflicting traffic volume observed in the weaving section = 1250 CPU/hr; Total traffic volume observed in the weaving section = 2419 CPU/hr.
19. The following details pertain to a four-legged rotary intersection (E–W and N–S directions). Determine its capacity.
    Width of the carriageway approaching the intersection = 14.0 m; Entry width = 12.0 m; Non-weaving section width = 9.8 m.
    The traffic approaching an intersection from four sides is as follows:

| | | | | | | | |
|---|---|---|---|---|---|---|---|
| $Q_{W \to N}$ | 250 | $Q_{N \to E}$ | 308 | $Q_{E \to S}$ | 206 | $Q_{S \to W}$ | 314 |
| $Q_{W \to E}$ | 280 | $Q_{N \to S}$ | 363 | $Q_{E \to W}$ | 397 | $Q_{S \to N}$ | 190 |
| $Q_{W \to S}$ | 202 | $Q_{N \to W}$ | 220 | $Q_{E \to N}$ | 261 | $Q_{S \to E}$ | 330 |

    Assume any missing/required data as per IRC: 65.

20. Calculate the adjusted values of calculated capacity of a weaving section of 2000 PCU/hr for the following site conditions, as per IRC-65:
    If the entry angle is 12°; (b) If the entry angle is 18°; (c) If the exit angle is 64°; (d) If the exit angle is 89°; (e) If the internal angle of the approach roads is 109°. Pedestrian flow at the exit from the roundabout is 590 per hour.
21. A rotary intersection has five legs, designated as 1, 2, 3, 4 and 5. Leg 1 is in the North–South direction and the remaining legs are marked in the clockwise direction. The traffic volumes observed during peak hour in the respective directions are given below in terms of PCU/hr.

| | | | | | | | | | |
|---|---|---|---|---|---|---|---|---|---|
| $Q_{1 \to 2}$ | 40 | $Q_{2 \to 1}$ | 37 | $Q_{3 \to 1}$ | 450 | $Q_{4 \to 2}$ | 181 | $Q_{5 \to 1}$ | 44 |
| $Q_{1 \to 3}$ | 299 | $Q_{2 \to 3}$ | 152 | $Q_{3 \to 2}$ | 162 | $Q_{4 \to 2}$ | 150 | $Q_{5 \to 2}$ | 163 |
| $Q_{1 \to 4}$ | 165 | $Q_{2 \to 4}$ | 111 | $Q_{3 \to 4}$ | 149 | $Q_{4 \to 3}$ | 121 | $Q_{5 \to 3}$ | 160 |
| $Q_{1 \to 5}$ | 55 | $Q_{2 \to 5}$ | 129 | $Q_{3 \to 5}$ | 360 | $Q_{4 \to 5}$ | 116 | $Q_{5 \to 4}$ | 19 |

    Calculate the weaving ratio between these adjacent legs and also determine the value of the weaving ratio used to calculate the rotary capacity.

CHAPTER

# 13 Design of Signalised Intersections

## 13.0 INTRODUCTION

An at-grade intersection is a junction of two or more roads meeting or crossing on the same level. Based on the intersection geometry, they may be categorised as three-way (╦ or Y) intersections, four-way (crossroads X or ╬) intersections or intersections formed with five or more arms. As traffic flow through urban intersections increases – inclusive of a sizable proportion of pedestrians and cyclists – traffic operation becomes critical, with increased accident rates. Hence, intersections are often controlled using traffic lights or signals (Red, Green and Amber).

The objective of signalised intersections with appropriate signal timing is to assign the right of way to alternating traffic movements in order to minimise the average delay to any group of vehicles or pedestrians and to reduce the probability of accident-producing conflicts.

## 13.1 ADVANTAGES OF SIGNAL CONTROL

- Provide orderly movement of traffic in the intersection
- Enhance the traffic handling capacity of the intersection
- Make it possible to alter the natural priority of traffic flow from approach roads so as to provide more balanced delays
- Reduce the frequency of occurrence and the severity of certain types of accidents in intersections (right-angle and pedestrian)
- Enable coordination of intersections in series to provide continuous movement of traffic at a definite speed over a predetermined length of road
- Permit pedestrian flow or bicycle traffic to cross at intersections by interrupting the continuous flow of heavy traffic at intervals
- Promote driver confidence by assigning right of way
- Help monitor queue length and control the Green signal to reduce critical queues on approach roads
- Regulate traffic flow patterns and reduce the need for merging and weaving
- Manage/Regulate unbalanced traffic flow from different approaches in the case of high traffic flow (say, 7000 PCU/hr), especially when volumes are predominant in one direction (Hallworth, 1992).

## 13.2 DISADVANTAGES OF SIGNAL CONTROL

- Excessive delay and fuel consumption may cause drivers/pedestrians to disobey the signal indications during peak periods
- May increase certain types of crashes (rear end collisions)
- Increase the use of less adequate alternative routes
- Significant increase in frequent rear end collisions during signal changes from Green to Amber/Red
- Formation of longer queues during peak hours – sometimes, the queue length may extend into an intersection or a residual queue may be formed in a grid network due to inadequate signal time planning.

## 13.3 WARRANTS FOR TRAFFIC SIGNAL INSTALLATION

A **warrant** is a threshold condition used to determine/evaluate whether a traffic signal installation is justified at an intersection, based on satisfaction of criteria in an engineering study. If any one of the warrant requirements is met, there will be a need to install or operate traffic signals at an intersection. Eight warrants are provided in the MUTCD and five in IRC: 93-1985. Brief details of warrants as per IRC: 93 are given below.

- ***Warrant I:* Minimum vehicular volume**: Intersecting traffic volume is considered as the criterion under this category. This warrant is satisfied if the eight-hour average day traffic volume equals or exceeds the values specified in Table 13.1.

**Table 13.1** Graphical presentation of eight-hour vehicular volume criteria for Warrant I, a typical example (Modified from IRC: 93-1985)

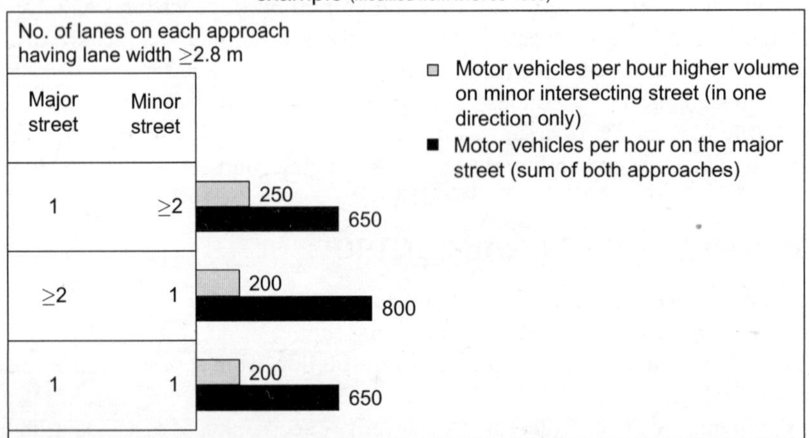


*Note:* The minimum vehicular warrant should be considered as 70% of the requirements specified, subject to the following conditions: (a) if the 85th percentile of the average approach speed exceeds by 50 km/hr or (b) if the proposed signalised intersection lies within the area of an isolated community having a population of less than 2.5 lakhs

- ***Warrant II:* Interruption of continuous traffic**: This warrant is applicable if the traffic volume from the major street is so heavy that it causes excessive delay or is a hazard to the minor street traffic entering or crossing the major street. The warrant is satisfied if the eight-hour average day traffic volume is equal to or exceeds the values specified in Table 13.2.

- ***Warrant III:* Minimum pedestrian volume**: The warrant is satisfied if the eight-hour average day traffic volume is as specified below:
  - If 600 or more motor vehicles enter the intersection from the major street every hour (sum of both approaches) or if there is a raised island of width ≥ 1.5 m and 1000 or more vehicles enter the intersection from the major street every hour (sum of both approaches), and
  - If 150 or more pedestrians cross the major street every hour in the same eight-hour period

  The minimum vehicular warrant should be considered as 70% of the requirements specified above, subject to the same conditions as prescribed for Warrant II.

  The signal installed under this warrant at isolated intersections or mid-block locations should be designed as a push-button traffic-actuated signal to facilitate safe pedestrian crossing. If tampering is suspected, the signal should be operated with a fixed time cycle and phase.

  If signals need to be provided at mid-block sections under this warrant, the location of such facilities should be a minimum of 300 m away from another crosswalk. Kerb-side parking should

**Table 13.2** Graphical presentation of eight-hour vehicular volume criteria for Warrant II, a typical example (Modified from IRC: 93-1985)

No. of lanes on each approach having lane width ≥2.8 m

□ Motor vehicles per hour higher volume on minor intersecting street (in one direction only)
■ Motor vehicles per hour on the major street (sum of both approaches)

| Major street | Minor street | Minor street volume (□) | Major street volume (■) |
|---|---|---|---|
| 1 | ≥2 | 150 | 1000 |
| ≥2 | 1 | 100 | 1200 |
| 1 | 1 | 100 | 1000 |

*Note:* The minimum vehicular warrant should be considered as 70% of the requirements specified, subject to the following conditions: (a) if the 85th percentile of the average approach speed exceeds by 60 km/hr or (b) if the proposed signalised intersection lies within the area of an isolated community having a population of less than 2.5 lakhs

be prohibited 75 m before and after such mid-block crossings. The related standards should be maintained with reference to such installations.

- ***Warrant IV:* Accident experience**: This warrant is satisfied
  - If the frequency of occurrence of accidents at an intersection *does not reduce* even after satisfactory road geometric improvements were made and adequate enforcement was provided.
  - If five or more accidents (each with damage worth > Rs. 2000) were reported within a period of 12 months that are suspected to be due to un-signalised control of traffic
  - Where signalised control will not seriously disrupt progressive flow of traffic
- ***Warrant V:* Combination of warrants**: Sometimes, signal installation may be justified when no signal warrant is fully satisfied but two or more of warrants I, II and III are satisfied with 80% or more of the traffic volume stated in these criteria. However, signal installation is only permitted provided there will be no delay or inconvenience to current traffic flow.

**Preliminary data collection:** With reference to warrant verification at an intersection, traffic volume with directional distribution studies need to be conducted at the proposed intersection on each approach road during 10 consecutive hours of a representative day. At the least, a 16-hour count should be carried out during a representative day (24 hours) during which the greatest percentage of traffic volume is observed. Classified vehicular volume data during AM peak and PM peak are useful for justifying signal installation at an intersection.

Pedestrian volume count and spot speeds at uncontrolled approach roads of the proposed intersection must also be studied during the peak periods mentioned. The collected data is summarised as peak pedestrian flow per hour and the 85th percentile speed of all the vehicles.

## 13.4 TRAFFIC SIGNAL DESIGN ELEMENTS

The design of a traffic signal system is a comprehensive process that involves balancing the basic requirements of traffic flow across an intersection with the prevailing conditions. Before designing the traffic signal control system, it must be determined if the signal is warranted and if there is such a need based on traffic volume, turning movements, accident experience, geometry, sight distances and inputs from local officials. The factors affecting the design of signal systems are categorised below (CTDOT, 2014).

**Geometric parameters:** The number of approach roads to an intersection, the number of lanes, the width of each lane, sight distance triangle, gradient, the intersecting angle of the approach roads and proximity to signals and the possible location/installation of signals, and underground utilities at intersections.

**Operational characteristics of vehicles:** Traffic flow rate during peak hour, peak hour factor, composition of traffic, saturation flow, design speed, 85$^{th}$ percentile speed, possible phasing, degree of need for platoon cohesion, level of service and volume to capacity ratios.

**Operational characteristics of pedestrians:** Pedestrian traffic volume, average walking speed of pedestrians, locality type such as school zone or CBD and crossing and sidewalk facilities.

**Design volumes:** The current traffic is projected to some design year in the future, usually a 20-year projection. The future projected traffic volume should be used to determine signal phasing, cycle length, signal timings, lane arrangements and queue storage requirements. Signal timings and cycle length can be determined based on the current needs of operational traffic volume and the volume expected approximately 2–3 years after installation. Therefore, the future projected 20-year traffic volume should be adjusted to operational volume by any appropriate method to provide the current needs of signal design for AM peak, PM peak and off-peak periods.

**Air quality:** Assessment of air quality based on localised carbon monoxide (CO) levels are also made with reference to the above projected traffic volumes, particularly if the level of service is D or worse.

**Constraints of construction:** Need-based funding is one of the main constraints of any intersection development project. Space constraints due to land acquisition, accommodation of components of intersections, preparation of utility conflicts prior to actual construction and staging plan need to be considered.

## 13.5 DATA COLLECTION

The following data is collected for the design of signal timings at proposed intersections:

- Geometrics of the intersection, with lane usage and their link lengths
- Classified traffic volume count at the proposed site during AM and PM peaks with 15-minute periods
- Details of turning movement counts
- Pedestrian volume count during peak hours with 15-minute periods
- Saturation flow rate or discharge volumes with reference to the highest flow rate during Green time
- Number of vehicles in each overflow queue during cycle-by-cycle counts
- Seasonal traffic/pedestrian counts, if required
- Approach delay data
- Existing signal timing, phasing and type of control

Total station surveys are conducted at study intersections and the geometric data is plotted using Auto-CAD software with reference to Survey of India/NRSC maps and Google Maps.

The data collection procedures and analysis relating to traffic volume counts and pedestrian surveys recommended in *Chapters 2* and *3* should be used.

## 13.6 TRAFFIC CONTROL SIGNAL LIGHTS

Three coloured lights – Red, Amber (or Yellow) and Green – glow facing each direction of traffic flow at an intersection. Red indicates that the vehicles must stop before the stop line, Green indicates permission to move in the specified direction(s), and Amber indicates that traffic should clear the intersection by the end of Green or Red. Typical signal head configurations are shown in Fig. 13.1.

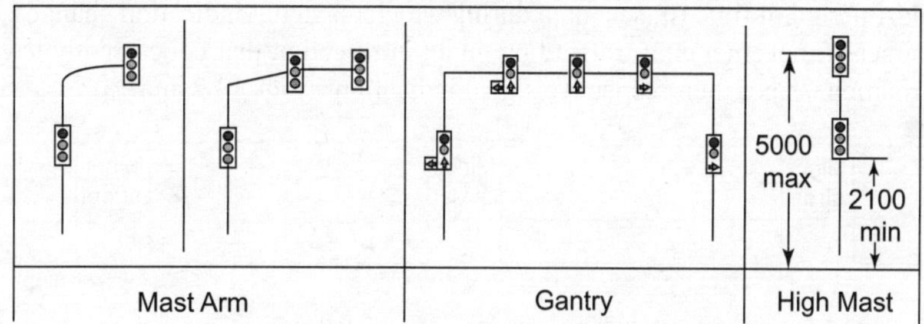


**Fig. 13.1** Signal head configurations
(*Courtesy*: National Roads Authority; *Source*: www.dttas.ie)

Some of the indications relating to traffic control across intersections or mid-block lanes are given below:

- **Flashing Red:** The illuminated Red lens of a signal face rapidly flashes and indicates that the vehicles should stop before entering the stop line or crosswalk.
- **Flashing Yellow:** The illuminated Yellow lens of a signal face rapidly flashes and indicates that the vehicles should proceed with caution.

A digital timer-unit is optionally installed on the signal mast, by which countdown (in seconds) of time intervals during Red and Green indications are displayed to the intersection users.

## 13.7 COMMONLY USED TERMINOLOGY

The following terms are commonly used in the design of signal timings of intersections (MnDOT, 2013; CTDOT, 2014; NIATT, 2015):

- **Traffic signal:** The lights – Red, Amber and Green – used to control movement of traffic to indicate right of way at intersecting roadways.
- **Movement:** A specific direction of movement of traffic flow such as left turn, right turn and through.
- **Platoon**: A group of vehicles travelling together along a study street at almost the same speed.
- **Queue:** A line of vehicles that builds up in front of a stop line when a phase has a Red indication. Queues form and grow when demand exceeds capacity.
- **Protected turn**: A turn that can be made *without* conflicting traffic and which is indicated by a steady burning 300-mm-diameter Green arrow. For example, right turn is adopted as a protected turn in most intersections in India, provided it does not conflict with other traffic flow directions.

The protected right-turn phase may reduce delays for turning vehicles but will likely increase overall intersection delay.

- **Cycle**: It represents one complete sequence of phases or intervals. It is a period of time (in seconds) in which a complete rotation of signal indications takes place (Red→ Green→ Amber) → (Red→ Green→ Amber). The Amber indication that appears between Green and Red is termed clearance Amber. As the approach speed increases (from 50 km/hr to 80 km/hr), Amber time is also suitably increased (from 3.0 seconds to 5.0 seconds).
- **Cycle length**: The time taken to complete one sequence of signal phases (a cycle).
- **Cycle failure**: A condition during which the provided Green time is not sufficient to clear the queued vehicles and a residual queue may form. Cycle failure often causes residual queues.
- **Interval**: A period of time (in seconds) during which no signal indication changes.
- **Phase**: A set of movements of traffic flow in an intersection that concurrently have the right of way on rotation, as per their respective signal indications (Figs 13.2 and 13.3).

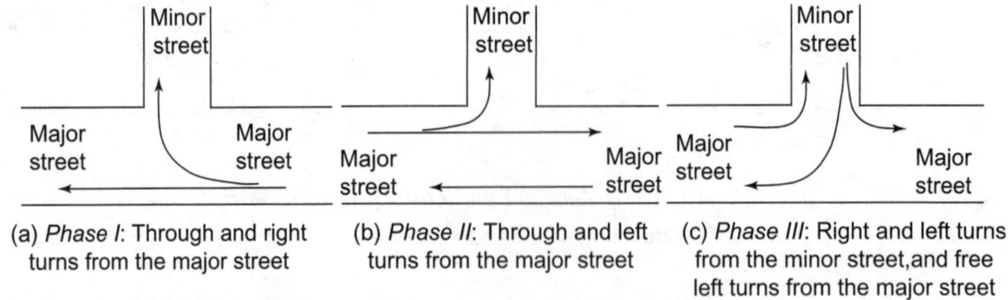


**Fig. 13.2** Typical phasing diagram for a three-legged intersection

- **Change interval**: The time gap allocated to clear the traffic that has already entered the intersection, and alert the drivers that their right of way is about to end, using an Amber indication. This time duration is also termed Amber change interval.
- **Lost time:** The duration (in seconds) in which there are no vehicles in the intersection. It occurs in clearance intervals and at the beginning of Green time. It is counted as the sum of clearance lost time and startup lost time, respectively (Fig. 13.4).
- **Clearance lost time:** The time between the last vehicle entering the intersection and the initiation of Green on the next phase.
- **Startup lost time:** The time lost due to increased space gap between vehicles that generally occurs among the first four vehicles entering the intersection as soon as the Green indication turns on.
- **Clearance interval:** The interval during which *no phase is given the right of way as all indications are Red*. The last vehicles, if any, that entered the intersection earlier, may be cleared before initiation of the next phase. This interval is also termed Red clearance interval (RCI). RCI is considered as the *safety clearance interval* at the end of a phase, when Red is displayed for all traffic movements. All-Red should be in the range of 1–5 seconds, depending on the speed, design vehicle length, visibility, approach grades and traffic composition.
- **Intergreen:** The period between the end of Green in one phase and the start of Green in another phase that is gaining right of way (Fig. 13.5).
- **Offset:** The interval of time between the occurrence of Green timings of successive signals

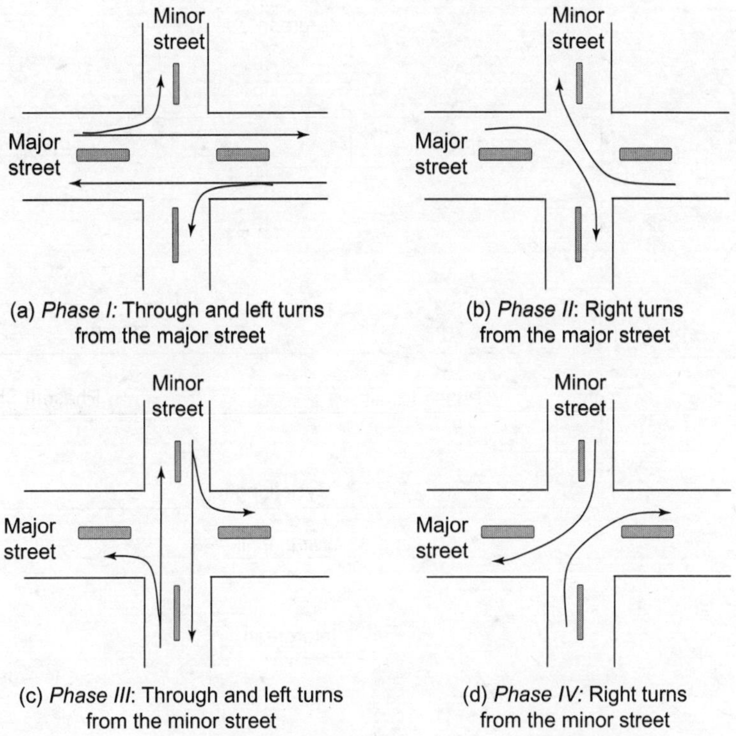


**Fig. 13.3** Typical phasing diagram for a four-legged intersection

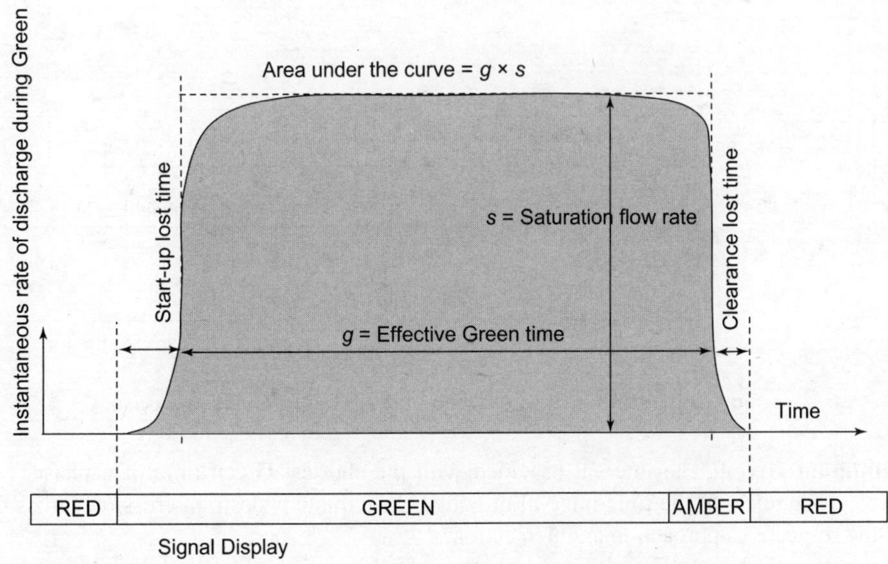


**Fig. 13.4** Details of startup lost time, effective Green time, saturation flow and clearance lost time

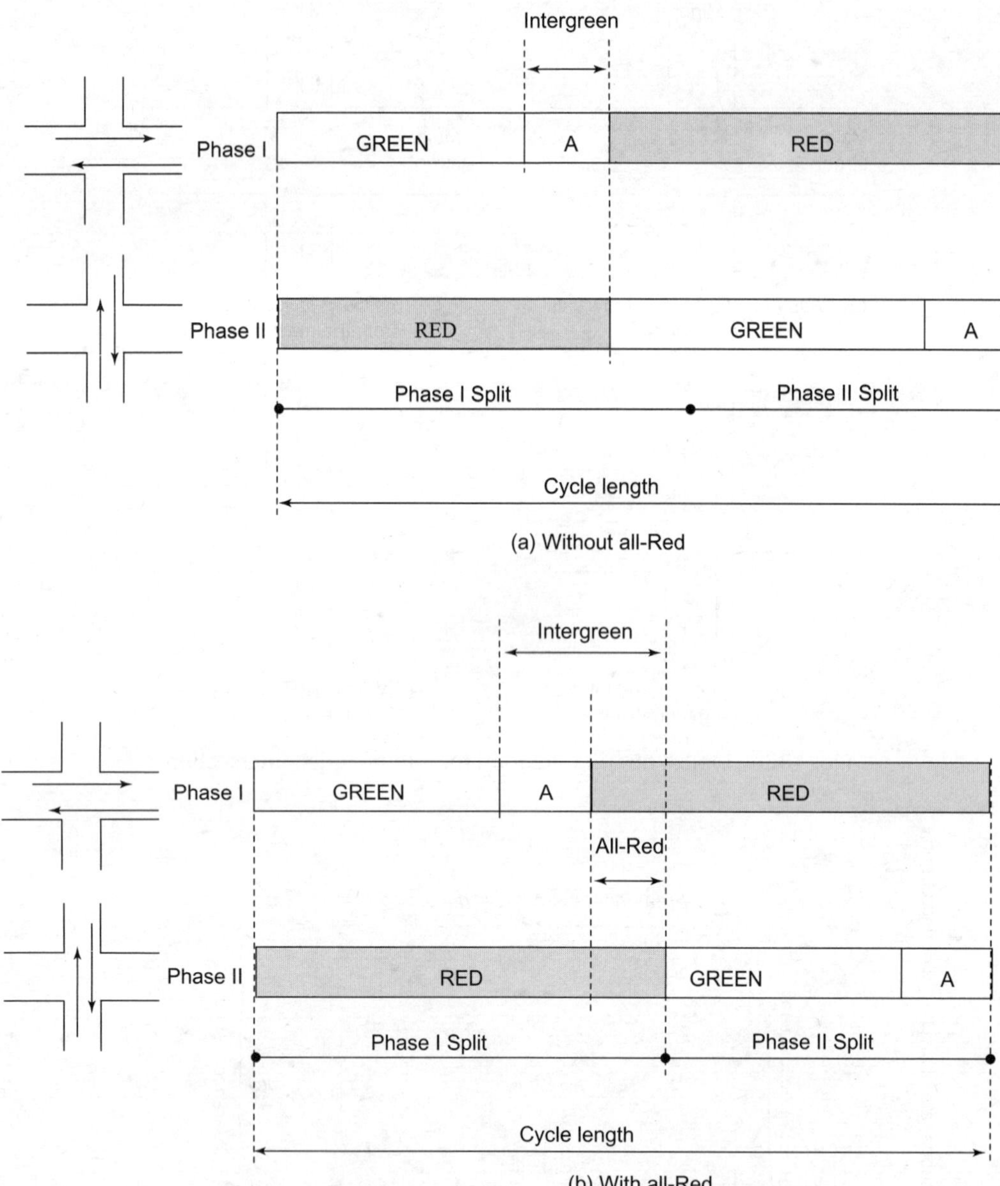


**Fig. 13.5** Details of intergreen and all-Red intervals in a cycle

- **Minimum Green**: The interval provided with the shortest Green time of a phase. Minimum Green time refers to the time interval provided to facilitate pedestrian crossing across the width of the respective approach to an intersection.
- **Saturation flow rate:** The maximum number of vehicles served (passed) by a lane during Green time, expressed as veh/hr (Fig. 13.4).

- **Saturation headway**: The clear gap (distance) between vehicles in a stable flow while moving in a platoon. It can be represented as the vertical distance between successive inclined line trajectories of vehicle flows during a Green period, as shown in Fig. 13.6.

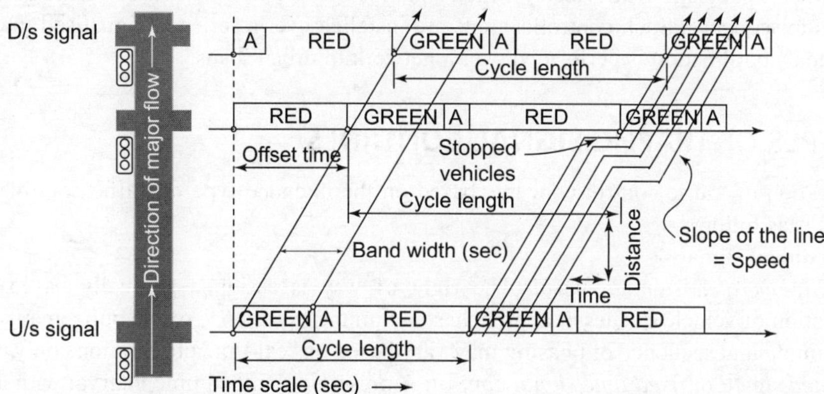


**Fig. 13.6** A typical coordinated signal plan with offsets (time–space diagram)

- **Effective Green time (EGT)**: The period during Green time that a movement/phase of traffic is actually using the intersection (Fig. 13.4). EGT = Cycle length – Effective Red time. As cycle length (seconds) increases, the effective Green time also increases, in terms of seconds per hour of operation.
- **Effective Red time (ERT)**: The period during Red time that a movement/phase of traffic is directed to stop. ERT = Cycle length – Effective Green time.
- **Protected movement**: A movement made without conflicts with others. For example, a right-turn movement in India. In contrast, the left-turn movement is permitted.
- **Permitted movement**: A movement that is always permitted without conflicts with other movements at an intersection. For example, the left-turn movement in India.
- **Signal coordination**: A definite timing relationship established between adjacent intersections of traffic signals as a system.
- **Time–Space diagram**: A chart showing the location of signalised intersections along the vertical axis and signal timing along the horizontal axis, drawn to suitable scale. Such diagrams are used to illustrate the coordination relationships between intersections (Fig. 13.6).
- **Progression**: The order in which the signal of a system turn Green.
- **Flexible progression**: The order of indication of Green time change to accommodate changing traffic flow in a day.
- **Progression speed**: The ratio of the distance between the first and last intersections to the travel time of a platoon of vehicles. Progression speed is represented by inclination of vehicle trajectory lines on a time–space diagram (Fig. 13.6). Progression speed is also termed band speed.
- **Band width**: The duration of the Green time (in seconds) through-band on a time–space diagram (Fig. 13.6)
- **Traffic actuated signals**: Traffic signals designed to vary their intervals based on current traffic demand, as measured by detectors. Queue length is commonly measured to change the signal interval by using intrusive induction-loop detectors.
- **Pre-timed signals**: The individual signal durations and cycle lengths that are preset and which do not change with traffic flow and time of day.

- **Traffic controller**: A device used for controlling the operations of traffic signals.
- **Volume-to-capacity ratio**: The ratio of the volume of traffic in a facility to its capacity. This parameter is widely used to measure/represent congestion of the facility.
- **Controller**: An electronic device by which signal lamps are switched based on the given interval timings in sequence. Such controller units are installed over a series of signalised intersections, and signal coordination is performed over a network of urban roads.

## 13.8 TYPES OF TRAFFIC SIGNAL CONTROLS

The classification of traffic control concepts based on the intended type of traffic control at isolated intersections is as follows:

**(a) Traffic control signals**

- *Manually-operated signal control*: The right of way is assigned manually based on visual observation of vehicle queues at approaches and on the experience of the policeman. The signal cycle timing and sequence of phasing may vary based on real-time observations only.
- *Pre-timed signal* or *fixed time signal control*: A preset/determined time interval with a sequence of phases is assigned for using the right of way at an intersection. The time interval for each signal indication in the cycle is fixed, irrespective of changes in traffic conditions. The main design elements of operation of pre-timed control consist of fixed number of phases with preset sequence, fixed phase length and fixed cycle length. This system is very easy and simple to operate but it does not take into account the fluctuations in traffic demand in real time.
- *Traffic-actuated signal control*: Vehicle queue length detectors installed at each approach to the intersection attempt to adjust Green time continuously, including sequencing and the length of phasing. The vehicle queue length detectors installed with isolated signal control are capable of measuring real-time demand at approaches to an interaction and also work very well with low volumes of traffic flow (Hani et al., 2014). This system of control reduces delay, is effective with multiple phases and takes into account short-term fluctuations in traffic demand in real time. However, the quality of intersection operation depends on the relationship between detection layout and the signal controller settings. For optimum performance, the detector layout and signal settings should be properly designed to suit the geometry of the intersection, its traffic volume and the approach speed of vehicles. In this connection, there is a need to establish a balance between detector location (relative to the stop line), detector length, passage time and minimum Green time for the prevailing conditions (Koonce et al., 2008). The disadvantages of this system are higher cost of detectors and their inspection and maintenance. This type of signal control is often based on non-coordinated phases. Based on the type of control, actuated signal control can be operated as a semi-actuated control or a full-actuated control (Fig. 13.7).

**(b) Pedestrian control signal**: Pedestrian movements across each approach to an intersection or mid-block lane should be controlled with separate pedestrian signal heads. The pedestrian signal face lens consists of a human figure. The following three types of indications are used for pedestrian control (IRC: 93-1985):

(i) The steady Green of the standing-human figure of the signal indicates that pedestrians can move across the intended roadway.

(ii) The steady Red of the standing-human figure of the signal indicates that pedestrians should stop and that they do not have permission to cross the roadway.

(iii) The flashing Red of the standing-human figure of the signal indicates that pedestrians on the roadway must quickly move on to the nearest refuge island/footpath; pedestrians who are not already on the roadway are not allowed to enter it.

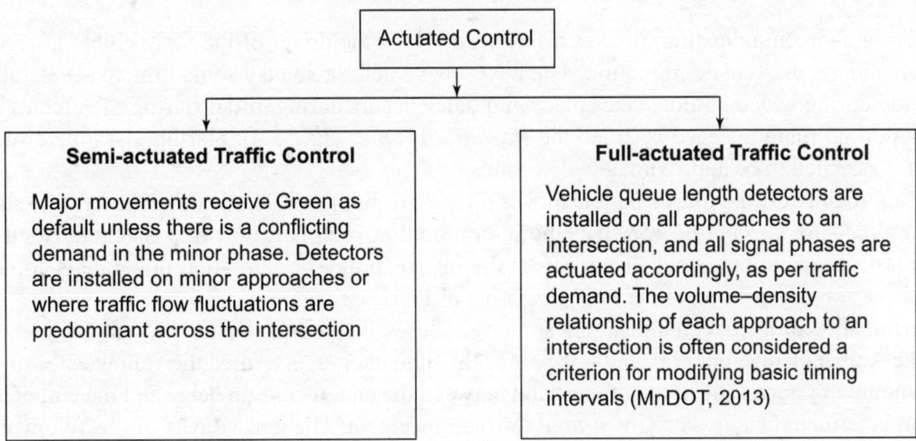


**Fig. 13.7** Categorisation of traffic-actuated control

**(c) Special traffic signal control**: Traffic flow across intersections can be controlled for special purposes such as for emergency purposes or assigning right of way for specific vehicle types. Two examples are given below (MnDOT, 2013):

- *Emergency vehicle pre-emption (EVP)*: This is a system of signalisation at an intersection by which authorised emergency vehicles are allowed to travel through signalised intersections in a safe and timely manner. The authorised vehicles are equipped with activated emitters having infrared LED or a visible strobe light oscillating at a specified frequency. The vehicle approaching an intersection is detected by an EVP detector mounted on the signal mast arm. The controller of the intersection brings up the through phase to allow authorised vehicles to pass through the intersection by terminating conflicting phases.
- *Transit signal priority (TSP)*: Bus priority or TSP is a technique used to improve the service and reduce the delay associated with mass transit vehicles. As the vehicle approaches an intersection, the controller installed at the intersection detects the vehicle and performs the required timing adjustments dynamically to bring up the through phase by terminating the conflicting phases. Electronic devices with detection system and controllers and mass-transit vehicles equipped with special transmitters are used in a special coordination program like EVP.

## 13.9 CONCEPTS OF TRAFFIC SIGNAL TIME DESIGN

### 13.9.1 Phasing

Phasing defines one or a set of movements of traffic used to permit right of way for a predefined time period in a cycle. For example, a phase may control both through movement and left-turn movement on an approach. The typical phasing adopted for three-legged and four-legged intersections is presented in Figs 13.2 and 13.3, respectively.

The predetermined/preset order of phases is the sequence of operation. This order is fixed in a pre-timed controller and the sequence of operation may vary in the case of an actuated controller.

### 13.9.2 Saturation Flow Rate

On receiving a Green indication, it takes a few seconds for the driver of the first vehicle to recognise the Green and get the vehicle in motion. The next few vehicles also take some time to accelerate. The time period during which no flow takes place and delay occurs during initial startup of vehicles before attaining normal running speed is called the *startup lost time* (Fig. 13.4). Startup lost time or delay is commonly assumed to be approximately 2 seconds.

After approximately the fourth vehicle moves forward from the queue, the flow rate tends to stabilise at the *maximum discharge rate*, known as the saturation flow rate. This flow rate is generally sustained until the last vehicle in the queue departs from the intersection. The saturation flow rate is commonly expressed as passenger car units per hour (PCU/hr) or PCU/sec.

On termination of the Green indication, some vehicles will continue to pass through the intersection during the Amber change interval, and this part of the time interval is termed the Yellow extension. The usable amount of Green time in the time period between the end of start up delay and the end of Yellow extension is termed the *effective Green time* for the movement. The unused part of the Yellow change interval and the following Red clearance interval is called the clearance lost time (Koonce et al., 2008).

*Factors affecting saturation flow:* The saturation flow rate depends on the prevailing roadway geometrics and traffic conditions. Roadway geometrics include the layout of the intersection, width of the approach, number of lanes and their widths, approach gradient and site conditions. The traffic conditions include composition of traffic, proportion of traffic flow in through, left and right turns, presence of parked vehicles, crossing pedestrian volume and the surrounding encroachments.

*Measurement of saturation flow:* The value of saturation flow is an important component that is required for the design of optimum cycle time. The information obtained from videos recorded at each approach during peak hour is used to estimate the saturation flow.

### 13.9.3 Estimation of Saturation Flow

In the absence of measured saturation flow rates at an intersection, the following equations are recommended by the Indian Roads Congress (IRC) based on geometric and traffic factors (IRC: SP: 41-1994; TRL, 1996).

*Effect of approach width:* Saturation flow increases with increased approach width, subject to no parked vehicles being present on the approach road. Saturation flow can be estimated using the following equation:

$$s = 525 \times w \text{ PCU/hr} \tag{13.1}$$

where $w$ = width of the approach road (m), measured from the kerb to the nearest median or divider or centre line, whichever is closer (Fig. 13.8).

Flow with previous line expression is valid when the $w$ value is in the range 5.5–18.0 m. If $w <$ 5.5 m, the values of saturation flow are estimated as given in Table 13.3 (TRL, 1996).

**Table 13.3** Saturation flow values for $w < 5.5$ m

| $w$ (m): | 2.70 | 3.0 | 3.5 | 4.0 | 4.5 | 5.0 | 5.5 |
|---|---|---|---|---|---|---|---|
| $s$ (PCU/hr): | 1790 | 1850 | 1890 | 1950 | 2250 | 2550 | 2990 |

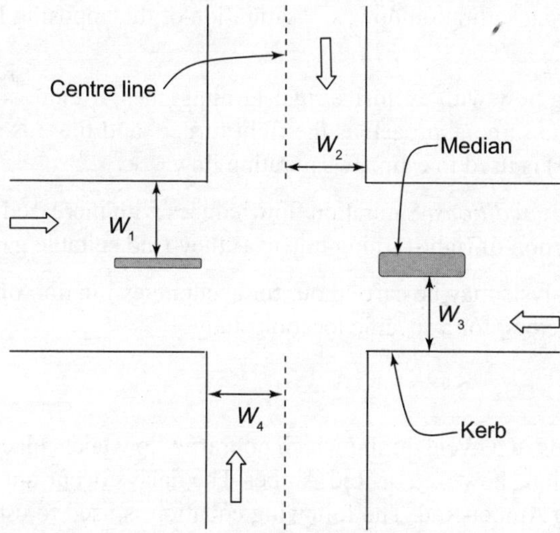


**Fig. 13.8** Measurement of approach width, $w_i$

- *Effect of gradient*: Saturation flow increases at the rate of 3% for each 1% of descending (downhill) gradient. Saturation flow reduces at the rate of 3% for each 1% of ascending (uphill) gradient. The maximum correction value applicable to saturation flow due to uphill gradient is a 30% reduction and to a downhill gradient is a 15% increase (TRL, 1996). Gradient is measured in the longitudinal direction along the approach road. The gradient along each approach road is measured from the stop line to 60 m before it (IRC: SP: 41-1994).
- *Effect of right-turning traffic*: If the right-turning traffic movements from the opposite direction do not cause severe conflicts that may lock the intersection, the effects of right-turning traffic depend on four possible conditions. These conditions are applicable under this category of estimation of saturation flow (IRC: SP: 41-1994):

  **Case 1:** If there is no opposing flow and no exclusive right turn, estimate the saturation flow using Eq. (13.1).

  **Case 2:** For exclusive right-turning lanes with no opposing flow, saturation flow depends on the radius of curvature ($r$, in metres) of the right-turning stream through a right-angled intersection. The equation is given below (TRL, 1996).

$$s = \frac{J}{1+\left(\frac{1.515}{r}\right)} \text{ PCU/hr} \tag{13.2}$$

  where $J = 1800$ for single lanes and 3000 for double lanes.

  **Case 3:** For opposing flow with no exclusive right-turning lanes, the maximum number of right turning veh/cycle that can take advantage of gaps in the flow of the opposing traffic stream is estimated using the equation given below.

$$\text{Right-turning saturation flow, veh/cycle} = s_r \times \left(\frac{(g \times s)-(g \times c)}{(s-g)}\right) \tag{13.3}$$

where $s_r$ = right-turning saturation flow; $s$ = saturation of the opposing leg; $g$ = Green time; and $c$ = cycle time.

**Case 4:** For opposing flow with exclusive right-turning lanes, if there is no delay to the straight-ahead traffic using the same approach as the right turners and there is an effect on cross-phase traffic flow, Eq. (13.3) is used to estimate saturation flow.

- *Effect of mixed traffic conditions*: Saturation flow reduces with increased number of slow-moving traffic and the proportion of right-turning traffic as they find suitable gaps to cross.

  *Note:* Regression analysis may be carried out for accurate estimation of left-turn, right-turn and U-turn adjustment factors, for a specific location study.

### 13.9.4 Lost Time

It is the sum of time periods in a cycle during which no traffic flow takes place in each phase when all the signals in the intersection show Red or Red-Amber. The delays occur during startup of Green and falling of discharge during Amber-Red. The following equation is used to estimate total lost time per cycle ($L$):

$$L = (n \times \ell + R) \tag{13.4}$$

where $n$ = number of phases in a cycle; $\ell$ = average time lost per phase, excluding the time during all-Red or sequent Ambers; and $R$ = all-Red period during each cycle when all the signals in an intersection show Red or Red with Amber.

Average lost time during startup delay in an Amber period may be assumed as 2 seconds per phase. The Highway Capacity Manual (TRB, 2000) defines a default value of 4 seconds per phase as total lost time (sum of startup lost time and clearance lost time).

### 13.9.5 Cycle Time by Webster's Method

Cycle length is the total time period during which one complete sequence of signalisation takes place around an intersection. A complete cycle comprises all the phases with their respective signal indications.

The cycle length with minimum delay is often termed *optimum* cycle length, and it can be determined by Webster's equation:

$$C_o = \frac{1.5L_i + 5}{1 - Y_i}$$

where $C_o$ = optimum cycle length (sec); and $L_i$ = sum of the lost time (sec) for all phases = $\ell_1 + \ell_2 + \ldots + \ell_i$, up to $i$ number of phases.

Lost time per phase = Startup time (usually 1–2 sec) + Clearance lost time

Clearance lost time = A portion of the Amber interval (usually 1–2 sec) + All-Red interval

$Y_i$ = Critical lane volume (in veh/hr) divided by saturation flow, summed over the $i$ phases

$$Y_i = \sum_{i=1}^{n} (y_1 + y_2 + y_3 + y_4 + \ldots\ldots\ldots y_i)$$

$$y_i = \text{Flow ratio}$$
$$= \frac{\text{Maximum flow through a lane during the } i^{\text{th}} \text{ phase}}{\text{Saturation flow for phase } i} \quad (13.6)$$
$$= \frac{Q_i \max}{s_i}$$

*Note:*

- Eq. (13.6) is used for isolated pre-timed signal locations only.
- Webster's formula is applicable if the volume/capacity ratio $\leq 1.0$.
- Eq. (13.6) indicates that delay is not significantly increased if cycle length variations are in the range 0.75–1.5$C_o$.
- A detailed network analysis should be performed (using any software program) to determine cycle length in a coordinated system. The software programs (with their inbuilt models) are capable of performing iterative procedures of varying cycle combinations across a series of signalised intersections, and ultimately finding the optimum signal timing parameters (MnDOT, 2013).

Minimum cycle length ($C_m$) allows just sufficient passing time to the traffic, and it is calculated using the following equation (TRL, 1996):

$$C_m = \frac{L_i}{1 - Y_i} \text{ sec} \quad (13.7)$$

$C_m$ may lead to excessively long delays, but this cycle length may be selected to coordinate signal control when an adequate time margin needs to be maintained for key intersections. However, it is appropriate to use a load of up to 90% of its capacity, and the resulting cycle length is termed practical capacity time ($C_p$). $C_p$ is calculated as follows (TRL, 1996):

$$C_p = \frac{0.9 L_i}{0.9 - Y_i} \text{ sec} \quad (13.8)$$

This value of $C_p$ may be adopted to facilitate minimum delays to pedestrian crossings where the use of short cycle length is preferred with high pedestrian volume.

### 13.9.6 Effective Green Time

It is the net time period of Green with Amber after excluding the total lost time per phase, calculated as follows:

$$g_i = G_i - L_i \quad (13.9)$$

where $g_i$ = effective green time for the $i^{th}$ phase; $G_i$ = (Green period + Amber period) per phase; and $L_i$ = total lost time per phase.

Webster (1958) suggested dividing Green time ($g$) in the proportion of flow ratio ($y$) for each phase and recommended the following equation to establish the relationship between the ratios of effective Green time ($g_i$) and the corresponding $y_i$ values of each phase. The relationship can be extended to three or more phases as well.

$$\frac{g_1}{g_2} = \frac{y_1}{y_2} \tag{13.10}$$

where $g_1$ and $g_2$ are the effective Green times of phases 1 and 2, respectively. $y_1$ and $y_2$ are the ratios of maximum flow to saturation flow of phases 1 and 2, respectively, as explained in Webster's method of cycle time (Eq. 13.5). This is a simple rule useful for Green time allocation to obtain the least overall delay.

### 13.9.7 Green Time

Effective Green time ($g_i$) for the respective $i^{th}$ phase can be calculated from the following expression:

$$g_i = \frac{y_i}{Y_i}(C_o - L_i) \tag{13.11}$$

where $g_i$ = effective Green period of phase $i$; $y_i$ = flow factor of phase $i$; $C_o$ = cycle length; $L_i$ = total lost time; $Y_i$ = sum of $y_i$ factors; and $C_o - L_i$ = total effective Green time.

## 13.10 SIGNAL PERFORMANCE MEASUREMENT

Certain measurements are used to judge the performance of signal timings and operations. The performance measurements typically considered are saturation flow rate, capacity of the intersection, degree of saturation, average delay, travel time, stops and level of service. A few of them are explained below.

### 13.10.1 Capacity of an Intersection

The volume of traffic flow for a particular movement through an intersection depends on the maximum rate of traffic flow during a Green period (saturation flow rate). The capacity of a signalised intersection through an approach road may be estimated from the following expression (IRC: SP: 41-1994).

$$Cap = \frac{(g)}{C_o} \times s \tag{13.12}$$

where $Cap$ = capacity of a signalised intersection from an approach road (veh/hr); $s$ = saturation flow rate of the lane group (veh/hr); $g$ = effective Green time for the movement (sec); and $C_o$ = cycle time or cycle length (sec).

In Fig. 13.9, capacity is graphically represented as the shaded area bounded by the saturation flow rate and effective Green time, and the current volume is represented as a hatched area under the instantaneous flow rate curve.

### 13.10.2 Degree of Saturation

The degree of saturation (DS) is used as a measure of the effectiveness of the signal control system, and it is calculated using the following equation (MnDOT, 2013):

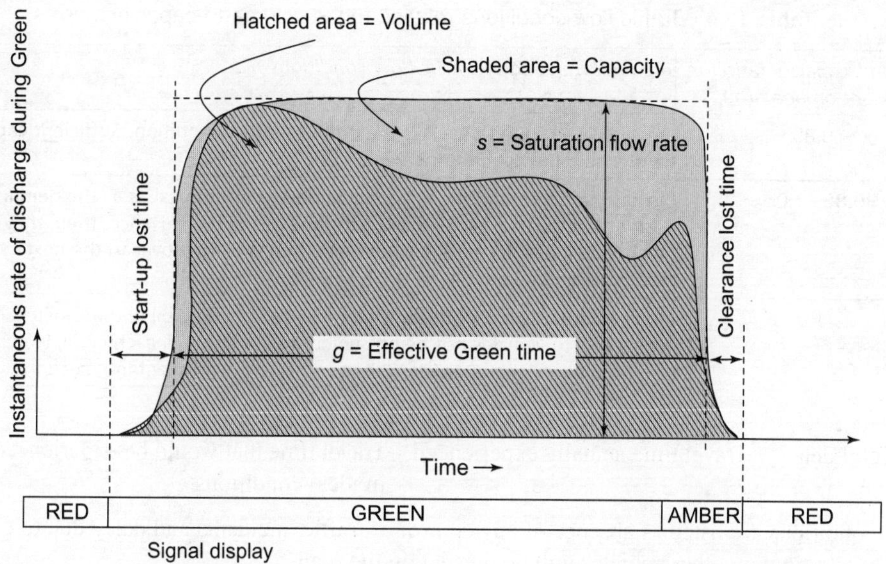


**Fig. 13.9** Typical illustration of volume and capacity of signalised traffic movement
(Redrawn based on Koonce et al., 2008)

$$DS = \frac{Q}{Cap} \tag{13.13}$$

where $Q$ = volume (veh/hr) and $Cap$ = capacity, as defined by Eq. (13.12).

In addition to the above parameter, the performance of a signalised control can also be estimated based on the volume-to-capacity ratio (Eq. 13.14), and flow ratio (Eq. 13.6). The volume-to-capacity ratio is a widely accepted measure of congestion at an intersection (or a facility).

$$\text{Volume-to-Capacity ratio} = \frac{Q}{Cap} \tag{13.14}$$

where $Q$ = demand volume of subject movement (veh/hr).

It must be noted that the volume-to-capacity ratio represents the *proportion* of the area (shaded) that defines the capacity that is occupied by volume (hatched area) (Fig. 13.9). A few remarks on traffic flow conditions relating to the volume-to-capacity ratio are presented in Table 13.4.

Critical movements at an intersection are generally based on the flow ratio. As the $y_i^{th}$ value increases for a particular approach, it indicates the significance and usability of the approach road.

## 13.10.3 Level of Service (LOS)

The delays that occur at signalised intersections are considered a measure of the intersection LOS. Delay indicates the degree of driver discomfort, frustration, fuel consumption and lost travel time. The following equations explain the basic parameters considered for the calculation of delay at an intersection.

**Table 13.4** Traffic flow conditions relating to the volume-to-capacity ratio

| *Volume-to-capacity ratio for movement or lane groups* | *Remarks relating to traffic flow conditions* |
|---|---|
| <0.85 | In saturated conditions, it is the preferred flow condition. Sufficient capacity and stable flow conditions exist. |
| 0.85–1.0 | In saturation condition, a less stable flow condition exists as the demand during the cycle exceeds the Green time of the cycle. Hence, there may be a change in the formation of a queue that is carried over to the next cycle (residual queue). |
| >1.0 | In oversaturation conditions, there is a definite possibility that queues may form due to vehicles not being served by the signal in each cycle, leading to accumulation; the queue length may affect the adjacent intersections. |

$$\text{Total delay} = \text{Travel time actually experienced} - \text{Travel time that would be experienced in ideal conditions} \quad (13.15)$$

Ideal conditions are listed as absence of any control of traffic, incidents and delay due to geometric factors. In addition, no other vehicles will be present on the road.

$$\text{Control delay at a signalised intersection} = \text{Delay due to (Initial deceleration + Queue move-up time + Stoppages + Final acceleration)} \quad (13.16)$$

Details about determination of control delay at signalised intersections are presented in *Chapter 14*. The criteria adopted for the determination of the LOS of a signalised intersection is based on the average control delay per vehicle, typically during a peak 15-minute analysis period. The assessment of the LOS of a signalised intersection can be made using Table 13.5.

**Table 13.5** LOS criteria for signalised intersections (Source: Highway Capacity Manual (TRB, 2000))

<table>
<tr><th rowspan="2">Control delay per vehicle (sec)</th><th colspan="2">LOS by volume/capacity ratio[a]</th></tr>
<tr><th>Volume/Capacity ratio ≤ 1.0</th><th>Volume/Capacity ratio > 1.0</th></tr>
<tr><td>≤ 10</td><td>A</td><td rowspan="6">F</td></tr>
<tr><td>> 10 and ≤ 20</td><td>B</td></tr>
<tr><td>> 20 and ≤ 35</td><td>C</td></tr>
<tr><td>> 35 and ≤ 55</td><td>D</td></tr>
<tr><td>> 55 and ≤ 80</td><td>E</td></tr>
<tr><td>> 80</td><td>F</td></tr>
</table>

[a]For approach-based and intersection-wide assessment based on control delay only.

### 13.10.4 Other Parameters

In addition to the parameters discussed above, travel time, queue length, emissions, fuel consumption, crash data review and performance index (calculated based on cycle length, offsets and splits that achieve optimum value) are some measures that can also be used as signalised control performance indicators (Koonce et al., 2008).

## 13.11 DESIGN OF ISOLATED FIXED TIME SIGNAL

Once lane arrangements and their proposed phasing movements are decided, the next step is to analyse the detailed operational capacity of the intersection with signal timings and cycle length. The following are widely used methods for determining optimum cycle length (with minimum delay) and signal timings. It must noted that a short cycle length should be used as the process of evaluation and design for any intersection signal system. The main details about the design of isolated fixed time signal is as per IRC: 93.

### 13.11.1 IRC: 93 Method

The salient steps/aspects of signal timing design as per IRC: 93 are given below.

*Inputs required from field surveys*

- Number of intersecting approach roads
- Width and number of lanes of each approach road
- Turning radius
- Peak hour volumes on the respective approaches

*Assumptions*

- Average walking speed of pedestrians is 1.2 m/sec.
- Maximum number of phases to be considered for the design is restricted to four.
- The reaction time of drivers to start up from the stop position and for the first vehicle to move is 6 sec and the subsequent vehicles follow the first vehicle with a uniform time headway of 2 sec each.
- Minimum Green time for any vehicular phase should not be less than 16 sec, to ensure built-in safety in the operation of the signal at an intersection.
- The vehicles distribute themselves equally across the lanes in their intended direction of flow.
- The number of vehicles arriving on an approach is equal during all cycles of fixed cycle length.
- It is recommended that signals at an intersection are controlled for a day with three cycle lengths – morning peak period, afternoon off-peak period and evening peak period.

*Overview of the design procedure:* The minimum pedestrian clearance time provided on the major street is calculated by dividing the width of the major street by the average walking speed of pedestrians. For the reaction time of pedestrians, a value of 7 sec must be added to this pedestrian clearance time, and the sum obtained must be allotted for the pedestrian phase on the major street (as minimum Green time for the minor street). Similarly, the pedestrian phase across the minor street should also be calculated.

After adding initial and clearance Amber intervals to the obtained value of pedestrian clearance time on the major street, Green time should be apportioned to the major and minor streets in proportion to the number of vehicles/hr/lane, for the respective streets. The cycle length calculated above should be rounded off to the immediate higher multiple of 5 sec and the total cycle length must be restricted to 120 sec.

*Check for major approach:* The Green time provided on the major street, which is based on the pedestrian clearance period, should be greater than the sum of reaction time taken to start the first and subsequent vehicles that are passing on the major street.

*Check for minor approach:* Similarly, the Green time provided on the minor street, which is based on the pedestrian clearance period, should be greater than the sum of reaction time taken to start the first and subsequent vehicles that are passing on the minor street.

*Optimisation of signal timings:* The cycle length and effective Green times on the major and minor streets are verified against the optimum cycle length calculated by Webster's method.

### 13.11.2 Guiding Standards

Several parameters of signal timing affect the efficiency of intersections, including cycle length, Green time and duration of clearance intervals. Increasing the Green time may reduce delay and the number of vehicles that stop. However, an increase in one approach Green time is generally derived at the expense of increased delay and stops to another movement. Therefore, a good signal timing plan is determined by allocating Green times appropriately, based on the traffic demand at the intersection and keeping cycle lengths to a minimum (Koonce et al., 2008).

The relationship between signal timing and the safety of traffic flow through an intersection with reference to signal timing parameters and design of the intersection is a very important aspect. For example, the Amber change interval should be adequate to facilitate the safe transfer of right of way from one movement to another. The safety benefit of the Amber or all-Red interval is most likely to be realised when its duration is consistent with the needs of drivers approaching the intersection at the onset of the Amber indication. This time interval needs to be related to the local driver's ability to perceive the Amber indication and assess their ability to stop before the stop line or to pass through the intersection safely. This time duration depends on the approach speed. A study also revealed that appropriately timed Amber change intervals can reduce intersection crashes. Signal timing plans with the sequence of rotation of phase can also reduce the number of stops and minimise delays (Retting et al., 2002).

In view of these two concepts, the following guidelines may be considered standard for achieving safe and efficient operation of signalised intersections (MnDOT, 2013).

- Minimum number of phases are used to decrease the amount of lost time due to starting delays and clearance intervals, in each phase.
- Short cycle lengths typically yield the best performance in terms of lowest overall average delay, provided the capacity of the cycle to pass vehicles is not exceeded. The cycle length determined corresponding to minimum delay is often termed optimum cycle length (Fig. 13.10). Delays increase with increased cycle length beyond the optimum cycle length. A thumb rule is that double the cycle time may result in double the delay.

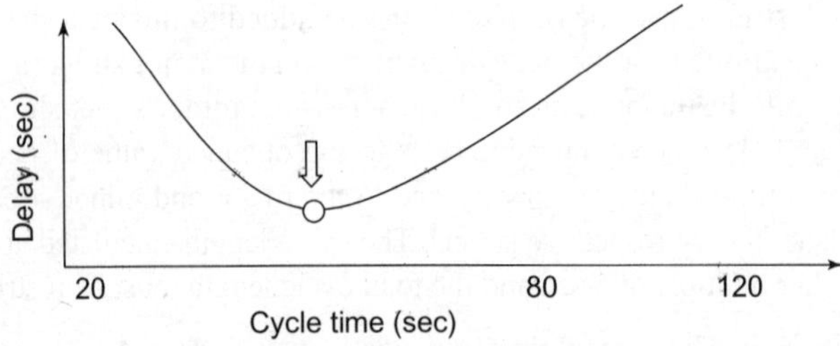


**Fig. 13.10** Determination of optimum cycle length (CCE, 2003)

- Longer cycle lengths may be used during peak periods to provide more Green time for major urban streets, so that longer platoons can be permitted in the peak direction with reduced number of starting delays. Based on the signal data obtained from field studies, separate signal plans need to be prepared for 24-hour signal plans during weekdays, Saturdays and

Sundays. During weekdays, traffic signals in series may be programmed to operate in an isolated mode as traffic-actuated control in the night and in a coordinated way during the day (Hani et al., 2014).

## 13.12 INTRODUCTION TO COORDINATED SIGNAL CONTROL SYSTEMS

A coordinated signal control system is a strategic approach to synchronising the start of Green indications along two or more consecutive closely spaced intersections along a major street/corridor, designed with definite interrelationship of their Green intervals.

**Coordination objectives:** The intent of coordination is that a group of vehicles travelling together (as a platoon) be allowed to pass smoothly along multiple intersections with reduced travel time, minimum delay and stops (NTOC, 2005). As a result, air quality along the corridor can be considerably improved with the advantage of reduced fuel consumption (Homburger et al., 2001). To achieve this objective, the following requirements must be met:

- Coordination among the signal controllers with reference to a master clock
- Compatibility of the equipment with the signal controllers
- The signal timing plan
- Equal cycle times, preferred
- Suitable and adequate spacing between signalised intersections

### 13.12.1 Need for Coordinated Control

When there is major flow of traffic in one direction, the coordinated mode usually works better to allow progression of traffic along the road. Linking of adjacent signals with interrelated Green intervals using a coordinated signal system can be adopted along the main street/corridor.

Coordination of signal control along a selected street is desirable for avoiding frequent stopping and queuing or for extended queue length of vehicles standing at every intersection on Red indications, and for reducing delays due to the reaction time taken to start up at the stop line, time headway of the first and subsequent vehicles, and the acceleration time taken by those vehicles to attain the approach speed, at each intersection.

Consecutive intersections, each within 800 m along a corridor, are suitable for the operation of a coordinated signal control system (Fig. 13.11), unless operated on different cycle lengths (MUTCD, 2003; CCE, 2003).

With reference to another source of information (Henry, 2005), it was recommended that coordinated signal systems be used when intersections are close together (within 1.2 km of each other). The traffic volumes and potential for platoons should be reviewed to determine if the possibility of coordination is beneficial to system operations if the distance between them is greater than 1.2 km. However, in both cases, the traffic conditions and government policies should be considered as part of the decision.

### 13.12.2 Basic Concepts of Signal Coordination

A time–space diagram is drawn, with Time on the x axis and Distance from a reference intersection on the y axis (Fig. 13.6). Time is counted with reference to a single common clock which is termed the master clock. The vehicle trajectory along the length of the major corridor is represented by an inclined line. The distance between successive intersections over the travel time (representing speed) is indicated by the slope of the inclined line, and it is represented by operating speed. It is very important

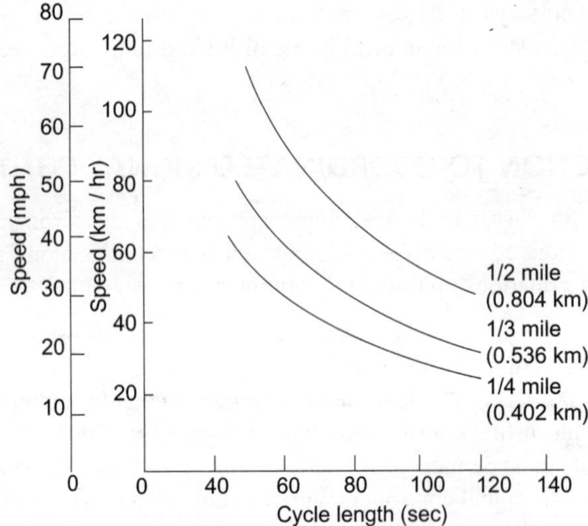


**Fig. 13.11** Relationship between speed, cycle length and signal spacing
(*Source*: CCE, 2003; *Courtesy*: Civil and Construction Engineering, http://cce.oregonstate.edu)

that these plots on both the axes are on scale so that consistency between the Distance and Time units can be maintained.

The vertical distance between two trajectories is the space headway between vehicles and the horizontal distance between the lines is the time headway between the vehicles (see the successive inclined lines in Fig. 13.6).

The directions of the trajectories (representing inclined lines) always move from left to right on the x axis, representing Time, and the distance traversed by the vehicles along the proposed intersections can be either northbound (bottom to top with upward slope) or southbound (top to bottom with downward slope). The upward and downward sloped lines indicate the dual (two-way street) movements on the main corridor. However, stopped vehicles (no change in their position on the y axis) are shown as horizontal lines. The slope of the inclined lines is set with reference to an assumed speed or intended coordination control speed or 85$^{th}$ percentile speed. However, the speed may be corrected with reference to the observed field-measured speed relating to factors such as mixed traffic conditions, signal timing settings, acceleration and deceleration rates of vehicles, progression speed, carriageway width variations and change in gradient along the intended corridor.

### 13.12.3 Factors Affecting Coordination

The following fundamental factors are considered as basic input parameters that need to be balanced and which influence the performance of any system of signal coordination.

Efficiency in the operation of a coordinated signal control system depends on three key timing parameters: cycle length, offset and split (individual phase timings at the intersections).These three parameters are described below:

(a) *Cycle length*: It is the total time taken to complete one sequence of all movements (all phases in sequence around an intersection). For example, one cycle length is the time required to complete any one of the following:

- Phases I through III, in the case of three-legged intersections (Fig. 13.2)
- Phases I through IV in the case of four-legged intersections (Fig 13.3)
- Phases I and II in the case of four-legged intersections (Fig. 13.12)
- Phases I through III in the case of four-legged intersections (Fig. 13.13).

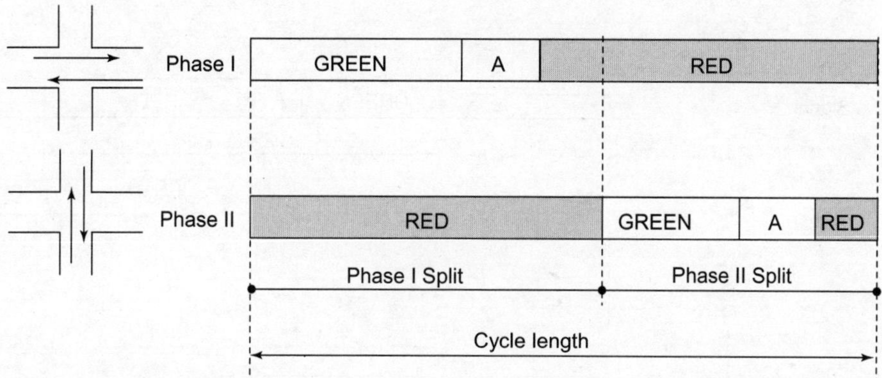


**Fig. 13.12** Timing operation of a basic two-phase pre-timed isolated signal control at the respective intersections (Redrawn based on MnDOT, 2013)

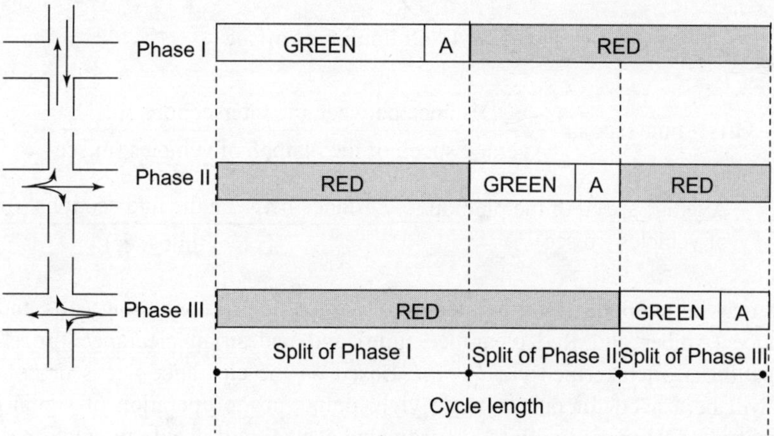


**Fig. 13.13** Timing operation of a three-phase pre-timed isolated signal control at the respective intersections (Redrawn based on MnDOT, 2013)

In order to maintain consistent time relationships between the coordinated signals along a corridor, the cycle lengths must be the *same* (or multiples of each other) for all the intersections. In India, cycle length typically ranges between 50 sec and 120 sec, which depends on the intersection design, spacing between intersections and traffic flow characteristics. Optimum cycle length for a system of several intersections in series is a challenging task for the traffic engineer, and often requires a suitable software model to help balance the coordinated flow of traffic along the major street and minimise delays on the intersecting minor streets.

(b) *Offset*: It is the time gap scheduled between the start of Green indications at successive upstream and downstream signals (Fig. 13.14). It is an important parameter used in the design of coordinated controls. Offset is determined based on the average travel speed of vehicles along the corridor, the number of vehicles in a platoon, time headway between vehicles in the platoon and the distance between successive upstream and downstream signalised intersections.

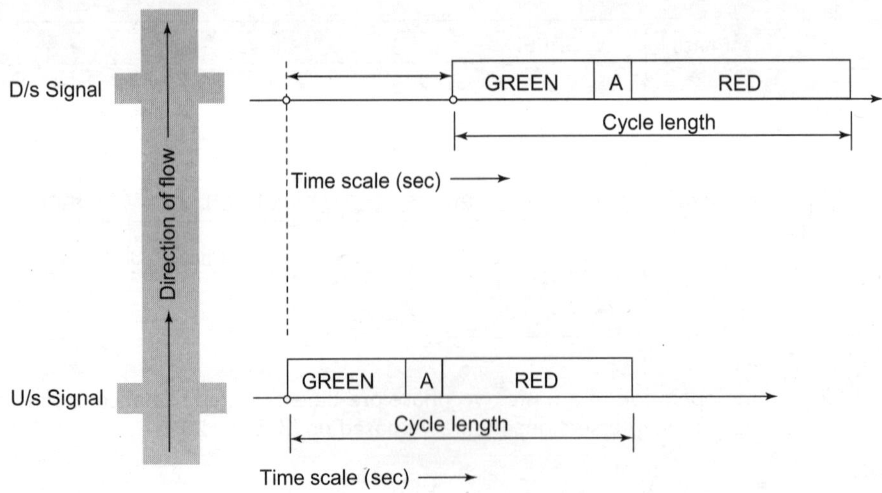


**Fig. 13.14** Definition of offset

$$\text{Offset time (sec)} = \frac{\text{Distance between the intersections (m)}}{\text{Average speed of the platoon of vehicles (m/sec)}} \tag{13.17}$$

where $\begin{array}{l}\text{Average speed of the platoon}\\ \text{of vehicles (m/sec)}\end{array} = \frac{\text{Distance between the intersections (m)}}{\text{Travel time (sec)}}$

Offset may be temporarily terminated or discarded due to conflicting calls and time during clearance (Amber and Red clearance time) and pedestrian clearance time. This type of termination is often termed *yield point*. Based on the clearance needs or conflicting calls, there will be a need for one or more yield points in the operation of signal coordination. The controller associated with the coordinated system will decide the yield point and which phase(s) will be served next.

(c) *Individual phase timings* or *split*: A split is a portion of time within a cycle length during which each phase of an intersection is assigned to use the right of way. Splits are expressed in percentages of the cycle or in seconds. The split percentages also include the Amber and all-Red clearance with the phase. For example, the split consists of time during Green + Amber + Red clearance of each phase. The split lengths are calculated based on phasing at an intersection and the expected demand.

(d) *Relationship between speed, cycle length and spacing between signals* (Fig. 13.11): The demand for increase in cycle length will occur with increase in traffic flow during peak hours. As a result, the speed of operation will reduce. The operating speed increases with increased spacing between signals. In other words, as spacing between signals increases, facility speeds obviously increase. However, appropriate operating speeds may be determined from the following typical values of cycle lengths (CCE, 2003):

- 120–150 sec: Urban major arterials with speeds of 77–116 km/hr
- 90 sec: Suburban arterials with speeds of 103–128 km/hr
- 60 sec: Rural arterials with speeds of 116–141 km/hr

The above values may be suitably adjusted for Indian traffic speed conditions.

*Note:* Maximum efficiency in the operation of a coordinated control system is achieved when the distance between adjacent intersections is not too high and more than 60% of the vehicles in a platoon can clear the intersections along the corridor (IRC: 93-1985). The cycle length must be carefully selected with respect to signal spacing and the anticipated or desired speed in the system (Koonce et al., 2008).

## 13.13 TYPES OF COORDINATED TRAFFIC SIGNAL CONTROL SYSTEMS

Details of the different types of coordinated traffic signal control systems are explained below and illustrated with time–space diagrams (Gordon and Tighe, 2005; CCE, 2003). Any one of these systems may be selected with modifications/refinements in signal timings with regard to the spacing between signals, offset, rectangular grid pattern, block spacing, signal phasing, pedestrian crossing requirements and the anticipated or desired speed in the system (Koonce et al., 2008).

### 13.13.1 Simultaneous Signal Control System

All the signals along a street display the same indication at the same time (Fig. 13.15). This system is suitable when traffic on the major street requires more Green time and the distance between intersections is very short (say, 90–160 m).

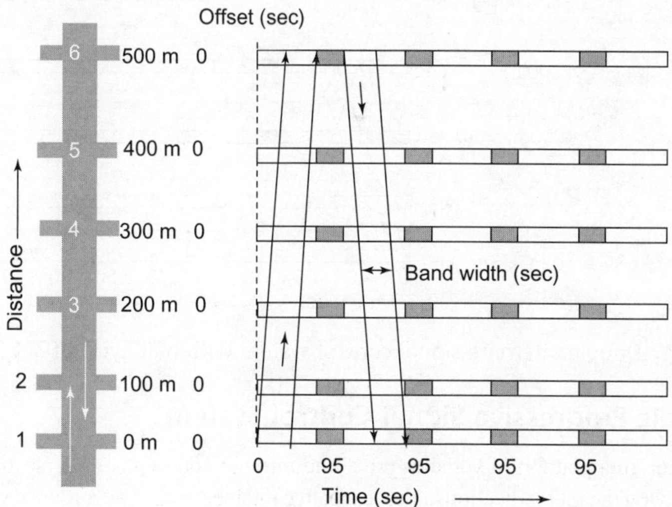


**Fig. 13.15** Simultaneous signal control system with zero offset (schematic)

### 13.13.2 Alternate Signal Control System

Alternate signals or groups of signals along a series of intersections in a street display opposite indications at the same time. This means that every other signal shows the same indication – termed single alternate (Fig. 13.16) – or every other pair of signals shows the same indication – termed double alternate (Fig. 13.17). This system is suitable for road networks having square blocks operating at low speeds.

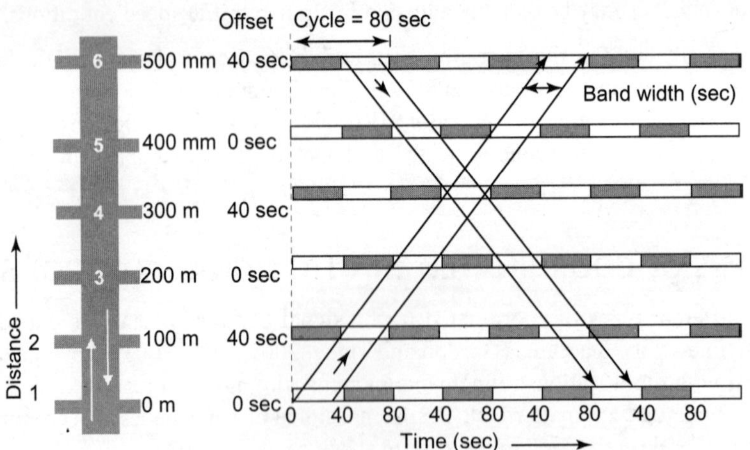


**Fig. 13.16** Single alternate signal control system with half-cycle offset (schematic)

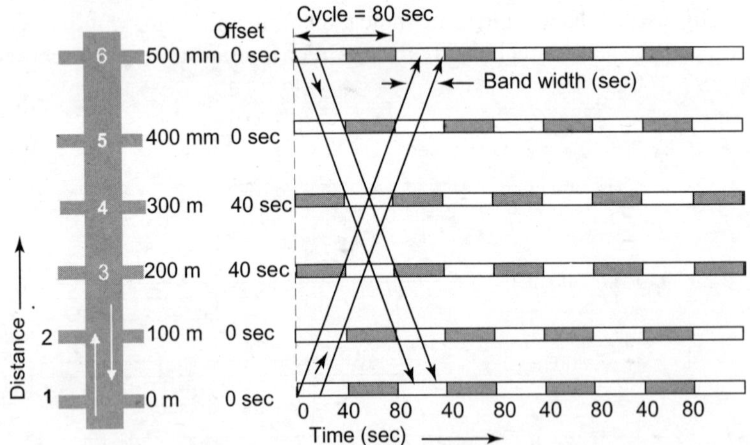


**Fig. 13.17** Double alternate signal control system with half-cycle offset (schematic)

### 13.13.3 Simple Progressive Signal Control System

Continuous operation of a platoon of vehicles travelling through consecutive intersections along a street is facilitated by displaying the GO indication at the respective intersections in accordance with a time schedule, at a predetermined rate of speed. The rate of speed of the platoon travelling between any two consecutive intersections may vary, and the same is scheduled with their respective GO indications, across the street.

### 13.13.4 Flexible Progressive Signal Control System

Signal intervals at each intersection along a street may be independently adjusted to the traffic requirements during which GO indications at the respective intersections may start independently to obtain maximum efficiency in traffic operation along the street (Figs 13.18 and 13.19).

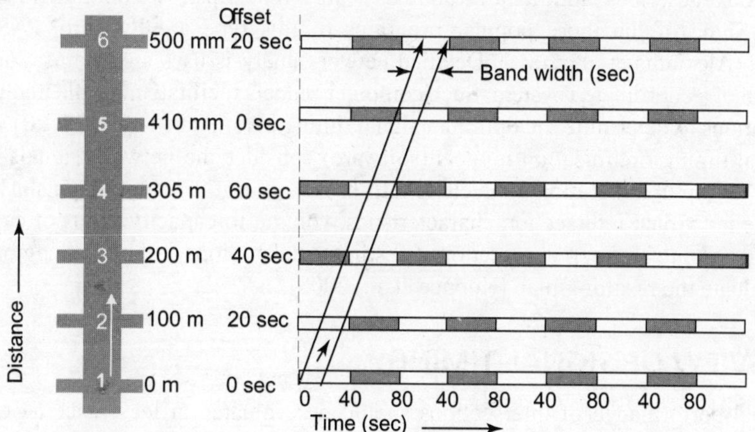


**Fig. 13.18** Typical example of flexible progressive signal control system on a one-way street: Quarter cycle system (schematic)

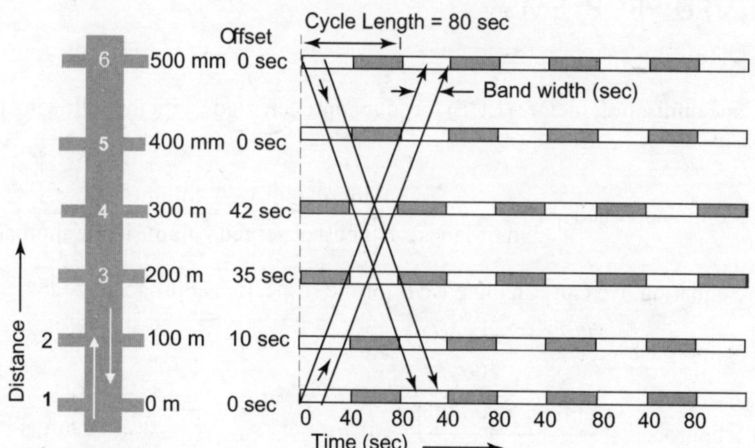


**Fig. 13.19** Typical example of flexible progressive signal control system with different offsets (schematic)

**Area Traffic Control System:** This is a method of controlling and coordinating traffic signals over a complete network of road intersections in an urban area where homogeneous traffic flow operation is required. For this purpose, a centralised control room is equipped with computers loaded with the appropriate software to interrelate and control the traffic signals in the selected urban area. It is a complex operation of traffic control across several routes of common intersections which need Information Technology (IT)-based tools/installations to obtain real-time traffic volume data at each intersection in the network. Using this facility, any one of the above coordinated systems can be implemented in a selected urban area.

To achieve the objective of minimum delay in operation, optimum cycle length must be determined by considering multiple intersections. Most network approaches use signal timing optimisation models that assist in the selection of a cycle length. The Federal Highway Administration's (FHWA's) Traffic Analysis Toolbox describes additional resources (http://www.ops.fhwa.dot.gov/trafficanalysistools/toolbox.htm). Some of the more popular programs of this type are Synchro, PASSER™ II and TRANSYT-7F (Alexiadis et al., 2004). Detailed network analysis uses an optimisation tool to assess the cycle length of a coordinated system. Such computer models facilitate multiple iterations of varying cycle combinations to determine the optimum signal timing parameters (MnDOT, 2013).

These signal timing optimisation models (software) consider the network signals being analysed and determine an optimal solution as cycle length based on a given set of inputs. The optimisation models use the individual intersection characteristics, volume-to-capacity ratios of each intersection, link speeds and spacing between intersections to estimate delay, stops and queue lengths for each cycle length and evaluate the resulting plan (Koonce et al., 2008).

## 13.14 REVIEW OF SIGNAL TIMING

The operational performance of intersections should be evaluated at least once a year. A complete analysis of signal timings and their operational performance should be evaluated every 3–5 years (MnDOT, 2013).

## 13.15 SOLVED PROBLEMS

**Example 13.1**

Determine the lane unitisation factors (LUF) for four approach roads with the following traffic volumes.

**Answer:**

$$\text{LUF of an approach} = \frac{\text{Sum of approach volume}}{\text{No of lanes} \times \text{Highest observed volume in the approach}}$$

Using the above equation, we can calculate LUF for the respective approaches:

$$\text{LUF of N approach} = \frac{(123+244+266)}{3\times 266} = 0.79$$

$$\text{LUF of E approach} = \frac{(110+275+298)}{3\times 298} = 0.76$$

$$\text{LUF of S approach} = \frac{(238+276+125)}{3\times 276} = 0.77$$

$$\text{LUF of W approach} = \frac{(292+290+181)}{3\times 292} = 0.87$$

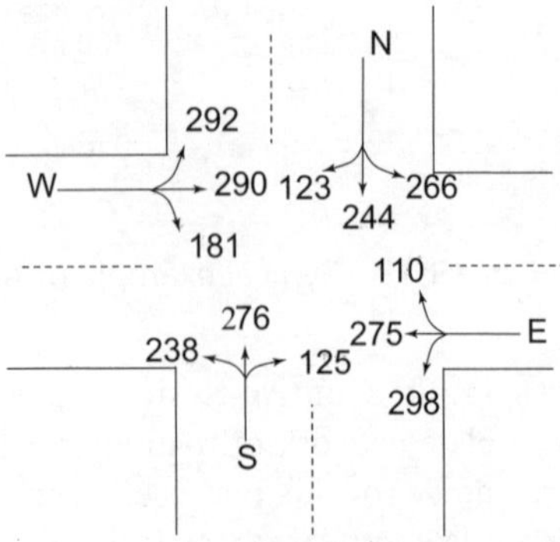

*Note:* Where there is more than one lane, the traffic will not strictly use all the lanes equally. Hence, saturation flow may be affected at signalised intersections.

Fig. 13.20 Directional distribution of traffic volume (veh/hr)

**Example 13.2**

A saturation survey was conducted at an intersection on different days by employing two enumerators with stopwatches. The summary of the field measurements is presented below.

| *Parameter measured* | *Run number* | | | | | | | | | | | | | | |
|---|---|---|---|---|---|---|---|---|---|---|---|---|---|---|---|
| | *1* | *2* | *3* | *4* | *5* | *6* | *7* | *8* | *9* | *10* | *11* | *12* | *13* | *14* | *15* |
| Time duration between the fourth and last vehicles in the queue (sec) | 31 | 29 | 33 | 36 | 22 | 22 | 48 | 45 | 20 | 24 | 27 | 28 | 28 | 44 | 30 |
| Number of vehicles that passed between the fourth and the last vehicles in the queue | 19 | 16 | 20 | 21 | 9 | 10 | 25 | 25 | 10 | 13 | 10 | 16 | 15 | 27 | 16 |

Determine (i) saturation flow rate through an approach to the intersection, (ii) saturation flow rate if the recorded time was 1 sec higher than the measured values, (iii) saturation flow rate if the recorded time was 1 sec lower than the measured values and (iv) average time headway for the three cases.

**Answer:**

| *Parameter measured* | *Run number* | | | | | | | | | | | | | | |
|---|---|---|---|---|---|---|---|---|---|---|---|---|---|---|---|
| | *1* | *2* | *3* | *4* | *5* | *6* | *7* | *8* | *9* | *10* | *11* | *12* | *13* | *14* | *15* |
| Time duration between the fourth and the last vehicles in the queue (sec) | 31 | 29 | 33 | 36 | 22 | 22 | 48 | 45 | 20 | 24 | 27 | 28 | 28 | 44 | 30 |
| Number of vehicles that passed between the fourth and the last vehicles in the queue | 19 | 16 | 20 | 21 | 9 | 10 | 25 | 25 | 10 | 13 | 10 | 16 | 15 | 27 | 16 |
| Time headway (sec/veh) | 1.63 | 1.81 | 1.65 | 1.71 | 2.44 | 2.20 | 1.92 | 1.80 | 2.00 | 1.85 | 2.70 | 1.75 | 1.87 | 1.63 | 1.88 |

Sample calculation for Run Number 1:

$$\text{Time headway for Run Number 1} = \frac{\text{Time duration b/w 4}^{\text{th}}\text{ and last vehicles}}{\text{Number of vehicles during the above period}}$$

$$= \frac{31}{19} = 1.63\,\text{sec/veh}.$$

Similarly, time headways for the remaining 14 run numbers are calculated and shown in the table.

$$\text{Average time headway} = \frac{\begin{pmatrix}1.63+1.81+1.65+1.71+2.44+2.20+1.92+1.80\\+2.00+1.85+2.70+1.75+1.87+1.63+1.88\end{pmatrix}}{15}$$

$$= \frac{28.84}{15} = 1.92 \text{ sec/veh}$$

$$\text{Saturation flow rate} = \frac{1}{(1.92)} = 0.5208 \text{ veh/sec}$$

$$\text{Saturation flow rate} = \frac{1}{(1.92)} \times 3600 = 1875 \text{ veh/sec}$$

(i) *If the recorded time was 1 sec higher*: Each measured duration is increased by 1 sec.

| *Parameter measured* | *Run number* | | | | | | | | | | | | | | |
|---|---|---|---|---|---|---|---|---|---|---|---|---|---|---|---|
| | *1* | *2* | *3* | *4* | *5* | *6* | *7* | *8* | *9* | *10* | *11* | *12* | *13* | *14* | *15* |
| Time duration between the fourth and the last vehicles in the queue (sec) | 32 | 30 | 34 | 37 | 23 | 23 | 49 | 46 | 21 | 25 | 28 | 29 | 29 | 45 | 31 |
| Number of vehicles that passed between the fourth and the last vehicles in the queue | 19 | 16 | 20 | 21 | 9 | 10 | 25 | 25 | 10 | 13 | 10 | 16 | 15 | 27 | 16 |
| Time headway (sec/veh) | 1.68 | 1.88 | 1.70 | 1.76 | 2.56 | 2.30 | 1.96 | 1.84 | 2.10 | 1.92 | 2.80 | 1.81 | 1.93 | 1.67 | 1.94 |

Sample calculation for Run Number 1:

$$\text{Time headway for Run Number 1} = \frac{\text{Time duration b/w } 4^{\text{th}} \text{ and last vehicles}}{\text{Number of vehicles during the above period}}$$

$$= \frac{32}{19} = 1.68 \text{ sec/veh.}$$

Similarly, time headways for the remaining 14 run numbers are calculated and shown in the table.

$$\text{Average time headway} = \frac{\begin{pmatrix}1.68+1.88+1.70+1.76+2.56+2.30+1.96+1.84\\+2.10+1.92+2.80+1.81+1.93+1.67+1.94\end{pmatrix}}{15}$$

$$= \frac{29.85}{15} = 1.99 \text{ sec/veh}$$

$$\text{Saturation flow rate} = \frac{1}{(1.99)} \times 3600 = 1809 \text{ veh/sec}$$

(ii) *If the recorded time was 1 sec lower:* Each measured duration is decreased by 1 sec.

| *Parameter measured* | *Run number* | | | | | | | | | | | | | | |
|---|---|---|---|---|---|---|---|---|---|---|---|---|---|---|---|
| | *1* | *2* | *3* | *4* | *5* | *6* | *7* | *8* | *9* | *10* | *11* | *12* | *13* | *14* | *15* |
| Time duration between the fourth and the last vehicles in the queue (sec) | 30 | 28 | 32 | 35 | 21 | 21 | 47 | 44 | 19 | 23 | 26 | 27 | 27 | 43 | 29 |
| Number of vehicles that passed between the fourth and the last vehicles in the queue | 19 | 16 | 20 | 21 | 9 | 10 | 25 | 25 | 10 | 13 | 10 | 16 | 15 | 27 | 16 |
| Time headway (sec/veh) | 1.58 | 1.75 | 1.60 | 1.67 | 2.33 | 2.10 | 1.88 | 1.76 | 1.90 | 1.77 | 2.60 | 1.69 | 1.80 | 1.59 | 1.81 |

Sample calculation for Run Number 1:

$$\text{Time headway for Run Number } 1 = \frac{\text{Time duration b/w } 4^{\text{th}} \text{ and last vehicles}}{\text{Number of vehicles during the above period}}$$

$$= \frac{30}{19} = 1.58 \text{ sec/veh.}$$

Similarly, time headways for the remaining 14 run numbers are calculated and shown in the table.

$$\text{Average Time headway} = \frac{\begin{pmatrix} 1.58 + 1.75 + 1.60 + 1.67 + 2.33 + 2.10 + 1.88 + 1.76 \\ +1.90 + 1.77 + 2.60 + 1.69 + 1.80 + 1.59 + 1.81 \end{pmatrix}}{15}$$

$$= \frac{27.83}{15} = 1.86 \text{ sec/veh}$$

$$\text{Saturation flow rate} = \frac{1}{(1.86)} \times 3600 = 1940 \text{ veh/sec}$$

Percent difference in saturation flow rate compared between actual record and 1 sec high

$$= \frac{1872 - 1809}{1872} \times 100 = 3.65\% \text{ (Reduced)}$$

Percent difference in saturation flow rate compared between actual actual record and 1 sec low

$$= \frac{1940 - 1872}{1872} \times 100 = 3.63\% \text{ (Increased)}$$

**Example 13.3**

Find the Yellow time based on the fundamental principles for the given dimensions of the intersection and approach speed: Average length of vehicle = 6.5 m; Longest width of the intersection area = 20 m; Approach speed at the intersection, S =50 km/hr.

Assume perception time of driver = 2.5 sec.

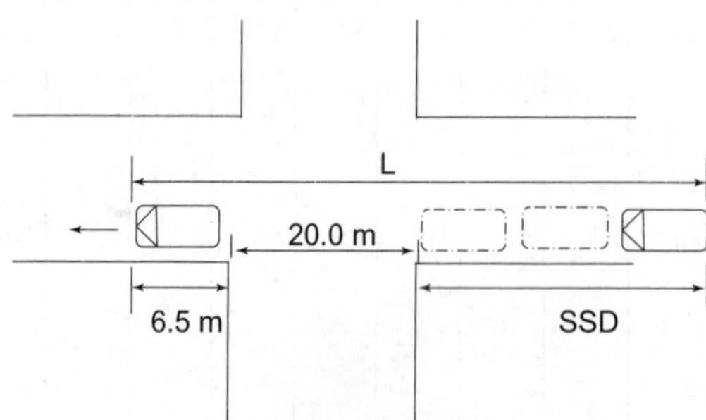


**Fig. 13.21** Minimum Amber time

**Answer:** With reference to Fig. 13.21

$$\text{Minimum Amber time (sec)} = \frac{L}{S} = \frac{6.5\text{ m} + 20.0\text{ m} + \text{SSD}}{S}$$

Speed, $S = 50 \text{ km/hr} = 50 \times 1000/(60 \times 60) = 50 \times 0.278 = 13.9 \text{ m/sec}$

$$\text{Safe Stopping Sight Distance (SSD)} = \text{Perception time} + \text{Breaking Distance} = S_t + \frac{S^2}{2g f_s}$$

Assume acceleration due to gravity = 9.81 m/sec² and coefficient of surface friction = 0.35

$$\text{SSD} = (13.9\text{ m/sec} \times 2.5\text{ sec}) + \frac{13.9\text{ m}^2}{2 \times 9.81\text{ m/sec}^2 \times 0.35}$$

$$\text{SSD} = (34.75\text{ m}) + 28.13\text{ m} = 62.88\text{ m}$$

$$\text{Minimum Amber Time (sec)} = \frac{L}{S} = \frac{6.5\text{ m} + 20.0\text{ m} + 62.88\text{ m}}{13.9\text{ m/sec}} = \frac{89.38\text{ m}}{13.9\text{ m/sec}} = 6.43\text{ sec}$$

**Example 13.4**

Determine the optimum cycle time based on Webster's formula. Take the sum of ratios of normal flow to saturation flow of two directional traffic flow = 0.6. The total lost time per cycle =10.0 sec.

**Answer:** The given input values are expressed as standard Webster notations.

$L = 10.0$ sec and $Y = 0.65$,

Optimum cycle length as per Webster's formula, $C_o = \dfrac{1.5L+5}{1-Y} = \dfrac{1.5\times10.0+5}{1-0.65} = 57.12$ sec.

Rounded to 60.0 sec. As the cycle time is commonly expressed in multiples of 5 sec.

*Note:* The delay is not so sensitive for a certain range of cycle length, and hence, the cycle length may be rounded to a multiple of 5 seconds.

**Example 13.5**

A pre-timed four-phase signal has the following critical lane flow rates observed for the first three-phases: 180, 192 and 198 veh/hr. Take saturation flow for all the phases as 1500 veh/hr. The time lost per phase is 4 sec.

**Answer:** If the optimum cycle length obtained by Webster's formula is 67 sec, the effective Green time (sec) of the fourth phase would be

$$Y_1 = \frac{\text{Normal flow phase 1}}{\text{Saturation flow for pase 1}} = \frac{180}{1500} = 0.12$$

$$Y_2 = \frac{\text{Normal flow phase 2}}{\text{Saturation flow for pase 2}} = \frac{192}{1500} = 0.128$$

$$Y_3 = \frac{\text{Normal flow phase 3}}{\text{Saturation flow for pase 3}} = \frac{198}{1500} = 0.132$$

$$Y_1 + Y_2 + Y_3 = 0.12 + 0.128 + 0.132 = 0.38$$

Total time lost during 4 phases = 4 × 4 sec = 16 sec

Optimum cycle length as per Webster formula $Co = \dfrac{1.5L+5}{1-Y}$ $67 = \dfrac{1.5\times16.0+5}{1-Y}$

$$Y = 0.567164$$

Therefore, $Y_4 = Y - (Y_1 + Y_2 + Y_3)$

$$= 0.567164 - (0.38) = 0.187164$$

Effective green time for 4th pahse $= \dfrac{Y_4}{Y}(C_o - L) = \dfrac{0.187164}{0.567164}(67-16) = 16.83$ sec

**Example 13.6**

The data given below is collected for designing a two-phase fixed signal at an intersection having North–South and East–West approach roads. Straight ahead traffic is permitted to operate in these approach roads.

| *Parameter* | *Approach road* | | | |
|---|---|---|---|---|
| | *North* | *South* | *East* | *West* |
| Design hr. flow (PCU/hr) | 845 | 750 | 850 | 490 |
| Saturation flow (PCU/hr) | 2500 | 2500 | 2100 | 2100 |

The time lost per cycle = 12.0 sec
Determine the optimum cycle length as per Webster's formula. Also calculate (a) effective Green time per phase North–South and (b) effective Green time per phase East–West.

**Answer:** Standard notations of Webster are used to represent the input values as follows

$$y_{North\text{-}South} = \frac{\text{Critical value of normal flowphase North-South}}{\text{Saturation flow for pase North-South}}$$

$$= \frac{845}{2500} = 0.338$$

$$y_{East\text{-}West} = \frac{\text{Critical value of normal flow phase East-West}}{\text{Saturation flow for phase East-West}}$$

$$= \frac{850}{2100} = 0.4048$$

$$Y = (y_{North\text{-}South} + y_{East\text{-}West}) = (0.4048{+}0.4048) = 0.7428$$

Optimum cycle length as per Webster's formula, $C_o = \dfrac{1.5L+5}{1-Y} = \dfrac{1.5\times 12.0+5}{1-0.7428} = 89.41$ sec.

Rounded to 90.0 sec. As the cycle time is commonly expressed in multiples of 5.

Effective Green time per cycle = $C_o - L = 90.0 - 12.0 = 78.0$ sec

Let effective Green time per phase North–South = $g_1$ sec
Let effective Green time per phase East–West = $g_2$ sec

The cycle length is distributed as two Green times to the two phases on the basis of the following proportion:

$$\frac{g_1}{g_2} = \frac{Y_1}{Y_2} = \frac{0.338}{0.4048} = 0.835$$

We know, $g_1 + g_2 = C_o - L$

From the above proportion, we can write $g_1 = 0.835 \times g_2$

Substituting, we get $(0.835 \times g_2) + g_2 = 78.0$ sec

$$g_2 = (78.0/1.835) = 42.5 \text{ sec}$$

Therefore, $g_1 = (78.0 - g_2) = (78 - 42.5) = 35.5$ sec

Cycle = Phase 1 + Phase 2

= 35.5 (45.5%) + 42.5 (54.5%) = 78 sec

## Example 13.7

Calculate the capacity and degree of saturation on an approach road of a signalised intersection for the following data:
Eastbound through traffic volume, $Q = 850$ PCU/hr; Cycle length, $C_o = 90$ sec; Effective Green time, $g = 42.5$ sec; Approach road width, $w = 6.0$ m

**Answer:** Saturation flow can be estimated using Eq. (13.1)

$$s = 525 \times w = 525 \times 6.0 = 3150 \text{ PCU/hr.}$$

The degree of saturation can be calculated using Eq. (13.12):

$$Cap = \frac{(g \times s)}{C_o} = \frac{(42.5 \times 3150)}{90} = 1487.5 \text{ veh/hr}$$

The degree of saturation can be calculated using Eq. (13.13):

$$DS = \frac{Q}{Cap} = \frac{850}{1487.5} = 0.571 = 57.1\%$$

**Example 13.8**

Determine the net Green time for a two-phase pre-timed intersection for the following signal timings: Optimum cycle length = 60 sec; Number of phases = 2; Amber time on Phase I = 4 sec; Amber time on Phase II = 4 sec; Lost time per cycle = (4 sec reaction time for 1st veh. on Phase I) + (4 sec reaction time for 1st veh. on Phase II) = 8 sec.

**Answer:** The net Green time is calculated as follows:
As per IRC: SP: 41 (1994)

$$\text{Total lost time per cycle } (L) = (n \times \ell) + R$$

Substituting, we get

$$L = (n \times \ell) + R = (2 \times 4) + 8 = 16.0 \text{ sec}$$

$$\text{Net Green time} = (C_o - L) = 60.0 - 16.0 = 44.0 \text{ sec}$$

**Example 13.9**

Design signal timing as per IRC: 93 (1985) for the following conditions:
Type of intersection: At-grade right angle (both right and left turns prohibited); Width of major street = 14.1 m (four-lane divided, two-way); Width of minor street = 7.5 m (two-lane, two-way); Peak hour volumes on the four approaches are shown in Fig. 13.22.
Assume: Average walking speed of pedestrians = 1.2 m/sec; Startup time of pedestrian to crosswalk = 7 sec; Reaction time of 1st vehicle to start after stop = 6 sec; Reaction time of subsequent vehicles in platoon to follow the 1st vehicle = 2 seconds each.
Apply necessary checks.

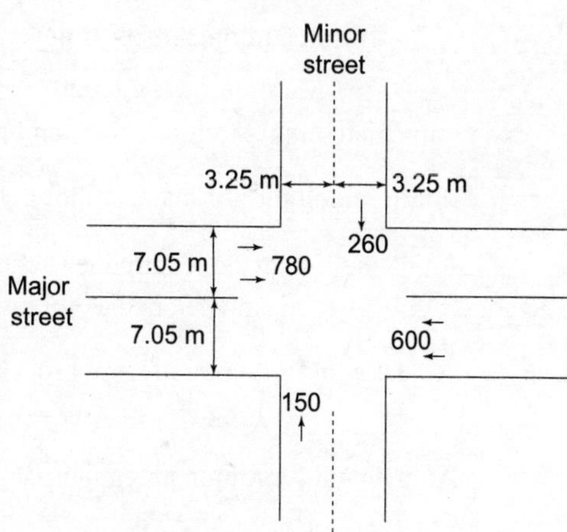


**Fig. 13.22** Design of signal timing

**Answer:** Two phases are considered in the analysis, as show in Fig. 13.23.

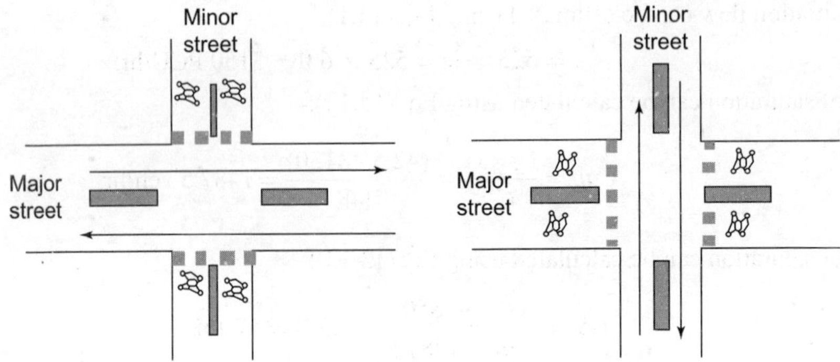


**Fig. 13.23** Details of phases considered for the design

$$\text{Pedestrian clearance time for major street} = \frac{14.1\,\text{m}}{1.2\,\text{m/sec}} = 11.75\,\text{sec}$$

Pedestrian Green time for crossing major street = Pedestrian clearance time + Startup time of pedestrian

Pedestrian reaction time on major street = 11.75 + 7 = 18.75 sec, Rounded to 19.0 sec.

Therefore, minimum Green time required on minor street = 19.0 sec.

$$\text{Pedestrian clearance time for minor street} = \frac{7.5\,\text{m}}{1.2\,\text{m/sec}} = 6.25\,\text{sec}$$

Pedestrian Green time for major street = Pedestrian clearance time + Startup time of pedestrian

Pedestrian Green time for major street = 6.25 + 7 = 13.25 sec, Rounded to 14.0 m.

Critical lane volume on major street = 780/2 = 390 veh/hr per lane

Critical lane volume on minor street = 260/1 = 260 veh/hr per lane

$$\text{Green time for major street} = \frac{390}{260} \times 19\,\text{sec} = 28.5\,\text{sec}$$, Rounded as to 29.0 sec.

Assume initial Amber time = 2.0 sec, and clearance Amber = 2.0 sec.

Minimum cycle length = (2 + 29 + 2) + (2 + 19 + 2) = 56.0 sec

Take the immediate higher multiple of 5, and round tfor minimum cycle length = 60.0 sec

$$\text{Ratio of volumes of major/minor} = \frac{390}{260} = \frac{3}{2}$$

Adjustment of cycle length due to the rounded value = 60.0 – 56.0 = 4 sec.

The above adjustment is apportioned as follows:

$$\text{Adjustment for major street approach} = \frac{3}{3+2} \times 4\,\text{sec} = 2.4\,\text{sec}$$

$$\text{Adjustment for minor street approach} = \frac{2}{3+2} \times 4\,\text{sec} = 1.6\,\text{sec}$$

Let us take the adjustment value for major street = 3 sec and for minor street = 1 sec.

Therefore, add 3.0 sec to Green major street, as = 29.0 + 3.0 = 32.0 sec

Add 1.0 sec to Green minor street, as = 19.0 + 1.0 = 20.0 sec

Summary of calculated signal timing (sec):

| *Signal timing* | *Initial Amber* | *Green time* | *Clearance Amber* | *Red time* | *Cycle length* |
|---|---|---|---|---|---|
| Major street | 2.0 | 32.0 | 2.0 | 2+20+2 = 24 | 60.0 |
| Minor street | 2.0 | 20.0 | 2.0 | 2+32+2 = 36 | 60.0 |

I. *Check for design of signal cycle timings on the basis of vehicular volume:*

(a) Check for major Aapproach:

$$\text{No. vehicles passing per cycle/lane on major street} = \frac{\text{Number of Veh/hr/lane in major street}}{\text{Minimum cycle length}}$$

$$\text{No. vehicles passing per cycle/lane on major street} = \frac{390}{60\,\text{sec}} = 6.5 \text{ No., Rounded to 7.}$$

Green time required on major street based on vehicle flow assumption of reaction time of 1st vehicle to start after stop as 6 sec and reaction time of the subsequent vehicles in platoon to follow the 1st vehicle as 2 sec. = (1 × 6 sec) + ((7 – 1) × 2 sec) = 18.0 sec < 32.0 sec. Hence, this is acceptable.

(b) Check for minor approach:

$$\text{No. vehicles passing per cycle/lane on minor street} = \frac{\text{Number of Veh/hr/lane in minor street}}{\text{Minimum cycle length}}$$

$$\text{No. vehicles passing per cycle/lane on major street} = \frac{260}{60\,\text{sec}} = 4.33 \text{ no., Rounded to 5 no.}$$

Green time required on minor street based on vehicle flow assumption of reaction time of 1st vehicle to start after stop as 6 sec and reaction time of the subsequent vehicles in platoon to follow the 1st vehicle as 2 sec. = (1 × 6 sec) + ((5 – 1) × 2 sec) = 14.0 sec < 20.0 sec; Hence, this is acceptable.

II. *Check for optimisation of signal timings*:

As per IRC: SP: 41 (1994); Total lost time per cycle $(L) = (n \times \ell) + R$; $n$ = Number of phases = 2 ; $\ell$ = Average lost time per phase due to starting delays and reduction in flow rate during Amber (*excluding* all-Red periods or sequent ambers) = 2 + 2 = 4.0 sec per phase

(By assuming, starting delays and reduced flow during Amber period = 2 sec + 2 sec per phase)

$R$ = All-Red period time during each cycle where all signals display Red or Red + Amber

$R = (2 + 2) + (2 + 2) = 8.0$ sec per cycle, i.e.. all-Red time intervals against the Ambers (see Fig. 13.24)

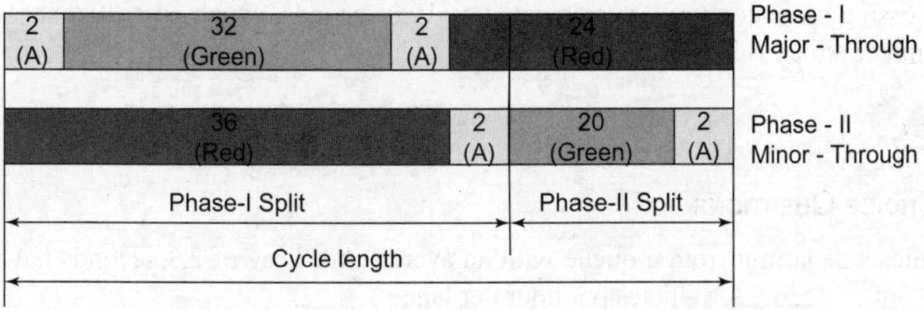


**Fig. 13.24** Signal phasing and band diagrams

Substituting, we get

$$L = (n \times \ell) + R = (2 \times 4) + 8 = 16.0 \text{ sec}$$

As per IRC: SP: 41 (1994), Saturation flow, $s = 525\,w$
where $w$ = width of approach road, measured from the kerb to the inside of the central median or centre line of approach, whichever is less.
On major street, $s = 525 \times 7.01$ m = 3701.25 CPU/hr
On minor street, $s$ (from Table 13.3) = 1870 CPU/hr (by interpolation between 3.0 m and 3.5 m)

$$Y_1 = \frac{\text{Normal flow phase 1}}{\text{Saturation flow for pase 1}} = \frac{780}{3701.25} = 0.210$$

$$Y_2 = \frac{\text{Normal flow phase 2}}{\text{Saturation flow for pase 2}} = \frac{260}{1870} = 0.139$$

$$Y = (Y_1 + Y_2) = (0.210 + 0.139) = 0.349$$

Optimum cycle length as per Webster's formula, $C_o = \dfrac{1.5L+5}{1-Y} = \dfrac{1.5\times 16.0+5}{1-0.139} = 44.54$ sec.

Rounded to 45.0 sec. As the cycle time is commonly expressed in multiples of 5.

Cycle length is distributed to $g_1$ and $g_2$ (respective effective Green times) to each phase, in the following proportion

$$\frac{g_1}{g_2} = \frac{Y_1}{Y_2} = \frac{0.210}{0.139} = 1.51$$

We know, $g_1 + g_2 = C_o$

From the above proportion, we can write $g_1 = 1.51 \times g_2$

Substituting, we get $(1.51 \times g_2) + g_2 = 45$ sec; $g_2 = (45/2.51) = 17.92$ sec

Therefore, $g_1 = (C_o - g_2) = (45 - 17.92) = 27.08$ sec

Effective Green time after deducting initial and clearance Ambers:

$g_1 = 27.08 - 2.0 - 2.0 = 23.08$ sec $< 32.0$ sec, acceptable.

$g_2 = 17.92 - 2.0 - 2.0 = 13.92$ sec $< 20.0$ sec, acceptable.

However, as per pedestrian requirements, effective Green time minor street to accommodate pedestrian crossing on major street is calculated as 19.0 seconds, which is < 20.0 seconds of Green time for the minor street. Hence, this is acceptable.

## EXERCISES

### Multiple Choice Questions

1. The vehicles departing from a queue with an average headway of 2.3 seconds have a saturation flow rate of __________ vehicles per hour per lane.
   (a) 1065 (b) 1265 (c) 1465 (d) 1565
   (*Hint:* Saturation flow rate = 3600 sec/hr ÷ 2.3 sec = 1565).

2. Normal flows on two approach roads at an intersection were observed as 622 and 485 PCU/hr and the saturation flow was 1890 PCU/hr on each road. The total lost time per phase = 16 sec. What would be the optimum cycle time as per Webster's formula?

(a) 60 (b) 70 (c) 80 (d) 90

(*Hint*: $Y_i = (622/1890) + (485/1890) = 0.5857$; $L_i = 16$ sec; $C_o = \frac{1.5L_i + 5}{1 - Y_i} = \frac{1.5 \times 16 + 5}{1 - 0.5857} = 69.99 \approx 70$ sec)

3. The IRC-recommended value of initial walking time of pedestrians is ________ sec.

(a) 2.0 (b) 1.2 (c) 5.0 (d) 7.0

4. As per IRC: 93, the minimum Green time required for a vehicular phase to ensure safety in operation is ________ sec.

(a) 60 (b) 50 (c) 16 (d) 17

5. As per IRC: 93, the reaction time of vehicle drivers to start up from the stop position for the first vehicle is ________ sec.

(a) 2.0 (b) 2.5 (c) 6.0 (d) 7.0

6. As per IRC: 93, the reaction time of drivers to start up from the stop position after the first vehicle is ________ sec.

(a) 2.0 (b) 1.2 (c) 2.5 (d) 1.0

7. With fixed cycle length, the number of vehicles arriving on an approach during all cycles is

(a) Equal (b) Unequal (c) Variable (d) Unknows

8. An isolated intersection is controlled by a three-phase signal system. The critical flow ratios for the three phases are 0.25, 0.3 and 0.15, respectively. The time lost per phase is 4 sec. The optimum cycle length (sec) by Webster's method is

(a) 77 (b) 60 (c) 90 (d) 100

(*Hint*: $C_o = \frac{1.5L_i + 5}{1 - Y_i} = \frac{1.5(3 \times 4) + 5}{1 - (0.25 + 0.3 + 0.15)} = \frac{18 + 5}{1 - (0.7)} = 76.66$ sec, rounded to 77 sec)

9. The total lost time on a traffic signal is noted to be 6 sec, the actual Green time is 32 sec and the Amber time is 3 sec. How much is the effective Green time?

(a) 35 (b) 21 (c) 33 (d) 29

(*Hint:* Effective Green time = Actual Green time + Amber time – Total lost time = 32 + 3 – 6 = 29 sec)

10. The signal intervals at each intersection along a street may be independently adjusted to the traffic requirement in which system of traffic control?

(a) Simple progressive system (b) Flexible progressive system

(c) Alternate system (d) Simultaneous system

11. All the signals along a street display the same indication at the same time in which of the following systems of traffic control?

(a) Simple progressive system (b) Flexible progressive system

(c) Alternate system (d) Simultaneous system

12. The rate of speed of a platoon of vehicles travelling between any two consecutive intersections may vary and the same is scheduled with their respective GO indications, across the street, in which of the following systems?
    (a) Simple progressive system
    (b) Flexible progressive system
    (c) Alternate system
    (d) Simultaneous system
13. The most efficient system of signal coordination along a street with intersecting roads is
    (a) Simple progressive system
    (b) Flexible progressive system
    (c) Alternate system
    (d) Simultaneous system
14. A three-phase traffic signal at a four-legged intersection is designed for flows as shown below: There are six groups of flows identified by the alphabets, A through F. Among these, A, C, D and F are through flows; and B and E are right turnings. Identify which phasing scheme is not recommended.
    (a) W (B) X (c) Y (d) Z

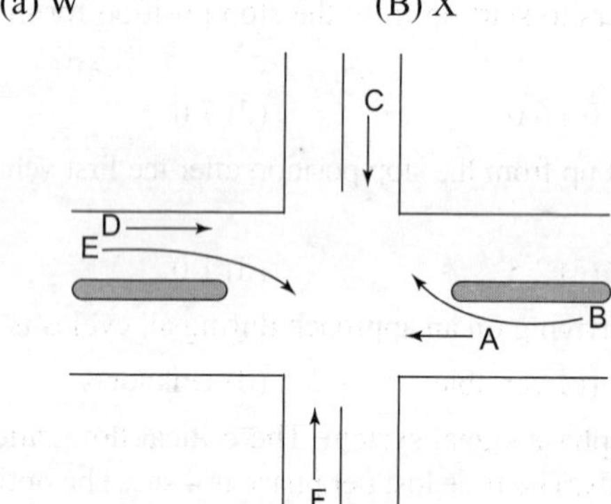


| Combination choice | Phase-I | Phase-II | Phase-III |
|---|---|---|---|
| W | A, B | D, E | C, F |
| D | A, D | C, F | E, B |
| Y | A, D | C, E | B, F |
| Z | A, B | E, F | C, D |

15. Consider the following factors:
    A: Length of vehicle; B: Width of vehicle; C: Approach speed; D: Stopping time for approaching vehicle; E: Intermediate sight distance
    Which of these factors is taken into consideration for determining the Amber time of a traffic signal at an intersection?
    (a) A, B and E (b) B, C and D (c) A, C, E (d) A, C and D
16. Match List-I (type of traffic signal) with List-II (advantages) and select the correct answer using the codes given below the lists:
    List-I
    A. Fixed time signals
    B. Vehicle actuated signals
    C. Semi-vehicle actuated signals
    D. Coordinated traffic signals
    List-II
    1. Suitable for junctions of minor street having lower traffic volume than its main street traffic
    2. Overall optimisation of traffic flow

3. Delay is held to a minimum and the capacity achieved will be maximum
4. Most successfully used in linked system

| | A | B | C | D |
|---|---|---|---|---|
| (a) | 4 | 3 | 2 | 1 |
| (b) | 4 | 3 | 1 | 2 |
| (c) | 3 | 4 | 2 | 1 |
| (d) | 3 | 4 | 1 | 2 |

17. Which one of the following traffic signal systems is useful when there is continuous operation of a group of vehicles along the main road?
(a) Simultaneous signal control system (b) Simple progressive signal control system
(c) Alternate signal control system (d) Flexible progressive signal control system

## ANSWERS

1. (d) 2. (b) 3. (d) 4. (c) 5. (c) 6. (a) 7. (a) 8. (a) 9. (d) 10. (b)
11. (d) 12. (a) 13. (b) 14. (d) 15. (d) 16. (d) 17. (b)

## QUESTIONS

1. What are the advantages and disadvantages of signal control of road intersections?
2. Specify the details of five warrants of traffic signals as per IRC: 93. Also state what data is collected for this purpose of warrant verification at an intersection.
3. Explain the measures for effectiveness of signal control systems in general.
4. Define bandwidth of signal time space. State the factors affecting bandwidth.
5. Explain how the bandwidth can be increased with reference to the offset provided in a coordinated control system.
6. What are the factors affecting the design of a traffic signal system?
7. What data is collected for the design of signal timings at proposed intersections?
8. What are the different types of signal head configurations? Explain their suitability of installation.
9. Define the terms relating to traffic signal systems: (a) Platoon (b) Protected turn (c) Cycle (d) Cycle failure.
10. Define the terms relating to traffic signal systems: (a) Interval (b) Phase (c) Change interval (d) Lost time (e) Clearance lost time (f) Startup lost time.
11. Define the terms relating to traffic signal systems: (a) Clearance interval (b) Intergreen (c) Offset (d) Minimum Green.
12. Define the terms relating to traffic signal systems: (a) Saturation flow rate (b) Saturation headway (c) Effective Green time (d) Protected movement.
13. Define the terms relating to traffic signal systems: (a) Progression speed (b) Bandwidth (c) Traffic-actuated signals.
14. Draw a sketch showing details of intergreen without and with all-Red intervals in a cycle.
15. Explain the various types of traffic signal controls including their merits and demerits.
16. Categorise and write the details of the types of traffic-actuated control.

17. Write a short note on pedestrian control signal systems including the three types of indications used for pedestrian control.
18. Write a short note on special traffic-control signal control with examples: (a) Emergency vehicle preemption (b) Transit-signal priority.
19. Draw typical sketches showing the possible phasing adopted for (a) three-legged intersections and (b) four-legged intersections.
20. Draw a typical sketch showing the signal time–space diagram indicating details of startup lost time, effective Green time, saturation flow and clearance lost time.
21. Define and state the factors affecting saturation flow.
22. Write the details of estimation of saturation flow based on geometric and traffic factors related to (a) Effect of approach width (b) Effect of gradient (c) Effect of right-turning traffic for all four cases (d) Effect of mixed traffic conditions.
23. What is lost time? What is the formula used for estimation of total lost time per cycle?
24. Explain Webster's method of determining optimum cycle length. Also write the formulas used for estimation of minimum cycle length and practical capacity time as per TRL practice.
25. What is effective Green time? How is it determined?
26. What are the parameters considered for traffic signal performance measurement? Explain any two in detail.
27. Explain how the capacity of a signalised intersection through an approach road may be estimated. Also draw a typical sketch showing the volume and capacity of signalised traffic movement presented on a time–space diagram.
28. What is degree of saturation and how it is determined? Also write remarks on flow conditions related to volume-to-capacity ratio.
29. Define control delay and total delay with formulas used to estimate them.
30. Write the details of inputs required from field surveys and the assumptions made for design of signal timing design as per IRC: 93.
31. Write an overview of the design of signal timing as per IRC: 93.
32. What are the guidelines considered as standard to accomplish safe and efficient operation of signalised intersections?
33. State the objectives and requirements for providing coordinated traffic signal system.
34. State the need for coordinated control of traffic signals.
35. Explain the basic concept of coordinated traffic control system with a typical sketch drawn with signal plan and offsets.
36. Explain the factors affecting coordinated traffic signal control system including (a) Cycle length (b) Offset (c) Individual phase timing or split (d) Relationship between.
37. Explain the relationship between speed, cycle length and spacing between the signals and their effect on the coordinated signal control system.
38. Explain the types of coordinated traffic signal control systems, with neat sketches.
39. Explain the simultaneous and alternate types of coordinated traffic signal control systems, with neat sketches.
40. Explain the simple progressive and flexible progressive types of coordinated traffic signal control systems with neat sketches.

41. Explain the three key timing parameters used to evaluate efficiency in the operation of coordinated signal control system.
42. Draw typical sketches showing a simultaneous signal control system with zero offset .
43. Draw typical sketches showing a single alternate signal control system with half-cycle offset.
44. Draw typical sketches showing single and double alternate signal control systems with half-cycle offset.
45. Draw a typical sketch showing a flexible progressive signal control system with different offsets.
46. Write a short note on area traffic control system and its applicability in current conventional traffic control practices in India.
47. Determine the Yellow time based on the fundamental principles for the given dimensions of an intersection and approach speed:

    Average length of vehicle = 6.1 m; Longest width of the intersection area = 25 m; Approach speed at the intersection, S = 46 km/hr. Assume perception time of driver = 2.5 sec.
48. Determine the optimum cycle time based on Webster's formula. Take the sum of ratios of normal flow to saturation flow of two directional traffic flows = 0.55. The total lost time per cycle =11.0 sec.
49. The following data is collected for designing a two-phase fixed signal at an intersection having North–South and East–West approach roads. Straight ahead traffic is permitted to operate in these approach roads.

| *Parameter* | *Approach road* | | | |
|---|---|---|---|---|
| | *North* | *South* | *East* | *West* |
| Design hr. flow (PCU/hr) | 815 | 750 | 850 | 495 |
| Saturation flow (PCU/hr) | 2100 | 2100 | 2500 | 2500 |

    Take the time lost per cycle = 10.0 sec. Determine the optimum cycle length as per Webster's formula. Also calculate effective Green time per phase North–South and East–West.
50. Design signal timing as per IRC: 93 for the following conditions:

    Type of intersection: At-grade right angle (both right and left turns prohibited); Width of major street = 15 m (four-lane divided, two-way); Width of minor street = 8 m (two-lane, two-way).

    Peak hour volumes on the four approaches are shown in the figure.

    Assume: Average walking speed of pedestrians = 1.2 m/sec; Startup time of pedestrian to crosswalk = 7 sec; Reaction time of 1st vehicle to start after stop = 6 sec; Reaction time of the subsequent vehicles in platoon to follow the 1st vehicle = 2 seconds each.

    Apply necessary checks.

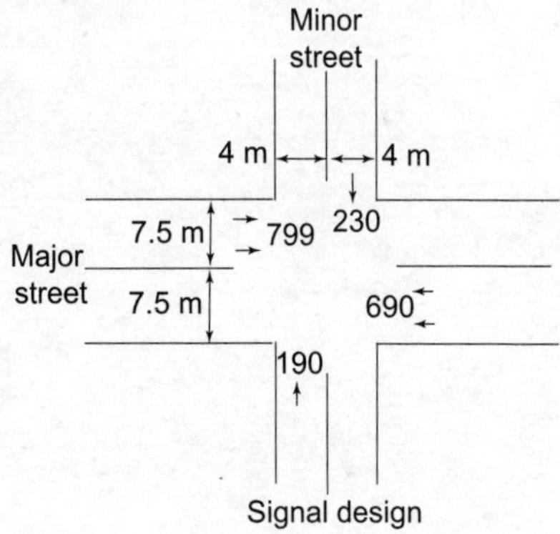

Signal design

CHAPTER

# 14 Basics of Queuing Theory and Delay Analysis

## 14.0 INTRODUCTION TO QUEUING THEORY

Queues occur whenever instantaneous traffic demand exceeds the capacity to provide a service (Fig. 14.1) (Vandaele et al., 2000). Queues may form because of a temporary increase in demand or a transitory decrease in the capacity of a road facility, with uninterrupted and interrupted flow conditions.

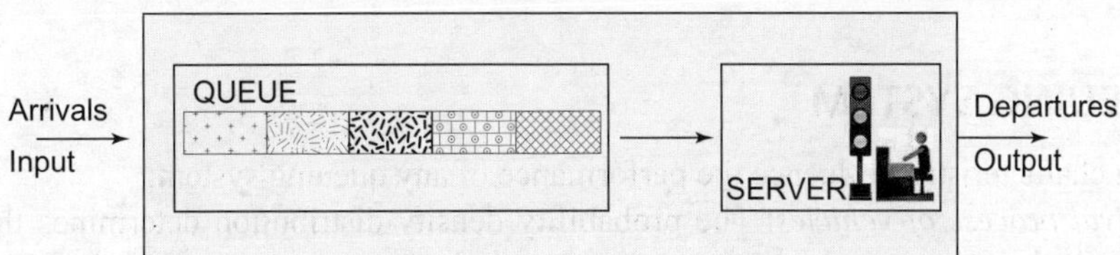


**Fig. 14.1** Queue system

Any temporary increase in demand resulting in queue formation can be controlled by taking appropriate traffic control measures. For example, a road accident or any incident on a roadway that interrupts traffic flow.

Let the assumed uniform arrival and departure of vehicles be represented on a plot with Time on the x axis and Cumulative Number of Vehicles on the y axis (Fig. 14.2). The arrival rate of vehicles is measured at a faraway place from a point of study (or at a bottleneck), and the departure rate of vehicles is measured at the bottleneck or at any point of accident. The departure rate may be assumed to be a constant with a value equal to the capacity at the bottleneck or any observed rate. The slope of the departure curve should represent this value.

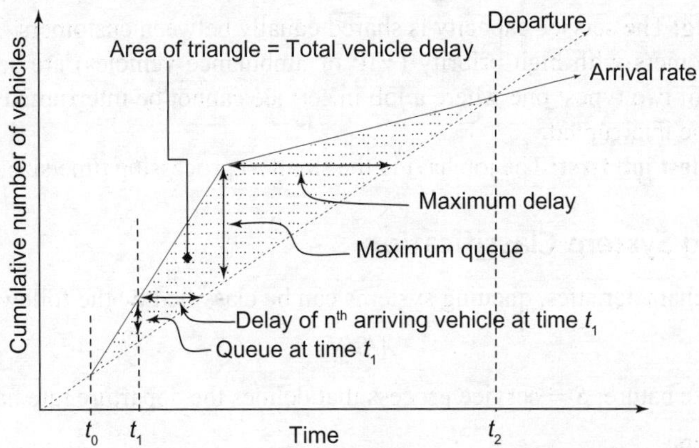


**Fig. 14.2** Queuing diagram for uninterrupted flow (assumed to be a D/D/1 queue)

As shown in Fig. 14.2, there will be no deviation between arrivals and departures up to time $t_0$ (before an accident or any obstruction to flow). When traffic flow is obstructed, there will be a rise in the arrival curve till the obstruction is removed and normal traffic flow is restored. As soon as the queue of vehicles is cleared (at $t_2$), the two curves merge. The following parameters can be derived from this plot:

- Maximum queue, with the number of vehicles
- Maximum delay that occurred to the queuing vehicles
- Total vehicle delay = Area between the arrival and departure curves

Queuing theory involves the mathematical study of these waiting queuing lines. Using appropriate mathematical models, queuing theory can be used to determine various characteristics of the waiting queuing line, like waiting time, length of the queue and the number of vehicles involved in the system (Yang and Yang, 2014).

## 14.1 QUEUING SYSTEM

The following characteristics influence the performance of any queuing system:

- *The arrival process of vehicles*: The probability density distribution determines the customer/vehicle arrivals in the system.
- *The service process that facilitates departure*: The probability density distribution determines the service times for customers/vehicles in the system.
- *The number of servers*: The number of servers available to provide service to the customers/vehicles.

The process involved between the arrival and the departure of a customer/vehicle is one of the main features of any queuing system. The process options based on scheduling policies can be broadly categorised as follows:

- **First in first out (FIFO):** The first customer will be served first and will be the first to depart, and customers are served one at a time.
- **Last in first out (LIFO):** The last customer will be served first and will be the first to depart, i.e., the customer with the shortest waiting time will be served first. Here also, customers are served one at a time.
- **Process sharing:** The service capacity is shared equally between customers.
- **Priority:** Customers with high priority (VIP or ambulance vehicles) are served first. Priority queues can be of two types: one where a job in service cannot be interrupted or one where a job in service can be interrupted.
- **Shortest/Smallest job first:** The job having the smallest processing time/size will be served first.

### 14.1.1 Queuing System Classification

Based on the above characteristics, queuing systems can be classified by the following convention:

$$A/S/N$$

where $A$ = arrival rate nature; $S$ = service process that defines the departure rate nature; and

$N$ = number of servers.

The nature of $A$ and $S$ can be defined by any of the following:

- **M** (Markovian, i.e., random): The inter arrivals of vehicles assume Poisson's distribution (use exponential probability density function)
- **D** (Deterministic): All customers arriving at the queue require the same/fixed amount of service
- **G** (General): Any general/arbitrary probability distribution (with known mean and variance)

### 14.1.2 Little's Theorem

Little's theorem states that the average number of customers ($N$) can be determined using the following equation (Little, 1961):

$$L = \lambda W \tag{14.1}$$

where $L$ = average number of tasks in the system; $\lambda$ = average arrival rate of customers; and $W$ = the average service or response time for a customer (Fig. 14.3).

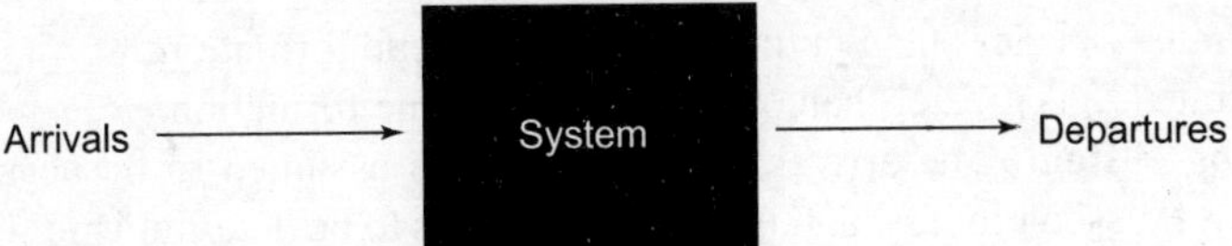


**Fig. 14.3** Little's theorem

The above equation can be generalised as:

$$\text{Average number of customers in the system} = L_s = \lambda W_s \tag{14.1a}$$

$$\text{Average number of customers in the queue} = L_q = \lambda W_q \tag{14.1b}$$

where $W_s$ = average waiting time of a customer or vehicle in the system and $W_q$ = average waiting time of a customer or vehicle in the queue.

By definition, average waiting time of a customer or vehicle in the system:

$$W_s = W_q + \frac{1}{\mu}$$

where $\mu$ = average output or service rate.

If the above equation is multiplied by $\lambda$ on either side, (where $\lambda$ = average arrival rate)

$$\lambda \times W_s = \lambda \times W_q + \frac{\lambda}{\mu}$$

$$L_s = L_q + \frac{\lambda}{\mu} \tag{14.1c}$$

**Example 14.1**

It was observed that 90 cars were travelling in front of a tunnel, and the departure rate of the vehicles from the tunnel was 30 cars/minute. Assume that no other vehicles were travelling in the tunnel during this observation. Determine the average waiting time of vehicles before and in the tunnel.

**Answer:** By Little's theorem, $L = \lambda W$

$$W = L/\lambda = 90/30 = 3 \text{ minutes.}$$

### 14.1.3 Queuing Models

A few of the widely used queuing systems with their conventions are explained below (Kendall, 1953; Kingman 1960; Abate and Whitt, 1987; Brun and Garcia 2000; Yingshi, 2006):

- **D/D/1 queuing system:** Both the inter-arrival time and the service time of the vehicles are assumed to be deterministic. This means that the inter-arrival times are all equal to each other and the service times are all also equal to each other. The system is assumed to have infinite queue length, with one server and FIFO discipline. Important parameters of this model are as follows: $\lambda$ = average arrival rate; $\mu$ = average output or service rate; Average inter arrival time = $1/\lambda$; Average service time = $1/\mu$; Traffic intensity or utilisation coefficient = $\rho = (\lambda/\mu)$
  *Conditions*:
  $\rho < 1$: arrival rate ($\lambda$) is less than output rate ($\mu$): no queue formation.
  $\rho \geq 1$: arrival rate ($\lambda$) is greater than output rate ($\mu$):queue formation.
- **M/D/1 queuing system:** The arrival of the vehicles is assumed to be negative exponentially distributed (say, Poisson process) and the departure times to be deterministic. The system consists of only one server. This queue model assumes the value of traffic intensity ($\rho$) < 1.
  $\lambda$ = average arrival rate; $\mu$ = average output or service rate
  Traffic intensity or utilisation coefficient = $\rho = (\lambda/\mu)$; Probability of no vehicle in the system: $P_o = 1 - \rho$; Average number of vehicles in the queue: $L_q = \dfrac{\rho^2}{2(1-\rho)}$; Average number of vehicles in the system: $L = L_q + \rho$; Average waiting time of vehicles in the queue: $W_q = \dfrac{L_q}{\lambda}$; Average staying time of vehicles in the system: $W = W_q + \dfrac{1}{\mu}$
- **M/M/1 queuing system:** The arrival and service times of vehicles are assumed to be negative exponentially distributed (say, Poisson process). The system consists of one server and is termed the single channel queuing system (Baoqun et al., 2004). This queue model assumes that the value of traffic intensity ($\rho$) < 1. This model assumes that customers arrive randomly and it follows Poisson distribution with FIFO service order (Fig. 14.4).
  The M/M/1 model queuing parameters are determined as follows:

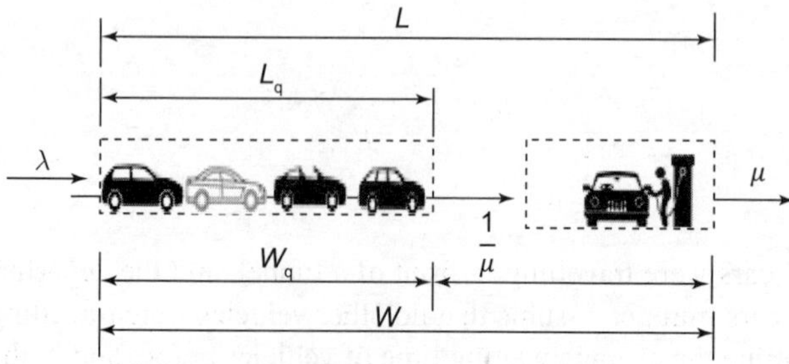


Fig. 14.4 M/M/1 queue model

$\lambda$ = average arrival rate (veh/hr or ped/hr or customer/hr) $\approx$ (density $\times$ speed); $\mu$ = average output or departure or service rate (veh/hr or ped/hr or customer/hr); $\rho = (\lambda/\mu)$ = traffic intensity or utilisation coefficient

Combined with Little's formula, the relevant quantity indexes can be obtained as follows: Probability of no vehicle in the system: $P_o = 1 - \rho$; Probability of $n$ vehicles in the system: $P_n = \rho^n(1-\rho)$; Average number of vehicles in the system: $L = \frac{\lambda}{\mu-\lambda} = \frac{\rho}{1-\rho}$; Average staying time of vehicles in the system: $W = \frac{L}{\lambda} = \frac{1}{\mu-\lambda}$; Average waiting time of vehicles in the queue: $W_q = W + \frac{1}{\mu} = \frac{\lambda}{\mu(\mu-\lambda)}$; Average number of vehicles in the queue: $L_q = \lambda W_q = \frac{\lambda^2}{\mu(\mu-\lambda)} = \frac{\rho^2}{1-\rho}$

- **M/M/N queuing system**: This assumes the same characteristics as in M/M/1 except that it has multiple servers and is termed multichannel queuing system (Chuanlai, 2009). For example, multiple servers ($N$) are present for security checks in airports and at boarding pass issue counters.

Traffic intensity = $\frac{\rho}{N}$. This model assume the value of $\frac{\rho}{N} < 1.0$.

Probability of no vehicles in the system: $P_o = \dfrac{1}{\sum_{k=0}^{N-1}\frac{\rho^k}{k!} + \frac{\rho^N}{N!\left(1-\frac{\rho}{N}\right)}}$

Probability of having $k$ vehicles in the system:

If $k < N$: $\quad P_k = \frac{\rho^k}{k!} \times P_o$

If $k \geq N$: $\quad P_k = \frac{\rho^k}{N!\,N^{k-N}} \times P_o$

Average number of vehicles in queue: $L_q = \frac{\rho^{N+1}}{N!\,N} \times \frac{P_o}{\left(1-\frac{\rho}{N}\right)^2}$

Average number of vehicles in the system: $L = L_q + \rho$

Average staying time of vehicles in the system: $W = \frac{L_q}{\lambda} + \frac{1}{\mu}$

Average waiting time of vehicles in the queue: $W_q = \frac{L_q}{\lambda}$

$$\text{Probability of having to wait in a queue: } P_{k>N} = \frac{\rho^{N+1}}{N!N} \times \frac{P_o}{\left(1-\frac{\rho}{N}\right)}$$

Derivations of the above equations can be obtained from Rao, 2011.

**Example 14.2**

Vehicles arrive at an entrance to an office parking space which has a single entrance. The parking gate opens at 9:30 AM, at which time the vehicles begin to arrive at a rate of 300 veh/hr. After 30 minutes, the vehicle arrival flow rate declines to 100 veh/hr, and it remains the same for the rest of the working hours of the day. If the time required to serve each vehicle is 20 seconds, describe the operational characteristics of the queue by assuming the D/D/1 model.

**Answer:** The given values of arrivals and departures of vehicles are converted to per minute values as:
Arrival rate-1: $\lambda 1 = 300$ vehicles per 60 minutes; $\lambda 1 = 300/60 = 5$ veh/minute $\rightarrow$ for $t \leq 30$ min.
Arrival rate-2: $\lambda 2 = 90$ vehicles per 60 minutes; $\lambda 1 = 90/60 = 1.5$ veh/minute $\rightarrow$ for $t > 30$ min
Departure rate: $\mu = 20$ s per 1 vehicle; $\mu = (60\text{s}/20\text{s}) \times 1 = 3$ veh/ minute $\rightarrow$ for all $t$.

The total number of vehicle arrivals from 9:30 AM to 10:00 AM ($t \leq 30$ minutes) $= 5\ t$, and after 10:00 AM as ($t > 30$ minutes), $150 + 1.5\ (t - 30)$.

Similarly, the number of vehicle departures is $3\ t$ for all times of $t$.

The details of calculation of departure and arrival curve values for the two time durations are given below:

| ***t*** (Minutes) | 0 | 10 | 20 | 30 | 40 | 50 | 60 | 70 | 80 | 90 | 100 |
|---|---|---|---|---|---|---|---|---|---|---|---|
| Departures $= 3 \times t$ | 0 | 30 | 60 | 90 | 120 | 150 | 180 | 210 | 240 | 270 | 300 |
| Arrivals at time | 0 | 50 | 100 | 150 | 165 | 180 | 195 | 210 | 225 | 240 | 255 |
| | $= 5 \times t$ | | | | $= 150 + 1.5\ (t - 30)$ | | | | | | |

The point of queue dissipation can be determined by equating arrivals with departures

$$150 + 1.5\ (t - 30) = 3\ t$$

Solving, we get $t = 70$ minutes.

The queue that started at 9:30 AM dissipated 70 minutes later, after 10:10 AM. By that time, 210 vehicles (3 veh/minute $\times$ 70 minutes) departed.

Based on the assumption of FIFO queuing discipline, the first vehicle that arrived was in the queue till the entire queue cleared – 70 minutes of delay occurred to the first vehicle.

The plot showing the arrival and departure values are presented in Fig. 14.5. The following values can be obtained from Fig. 14.5:

Maximum number of vehicles in the queue = 60; Maximum time spent in the queue = 20 seconds

$$\text{Total vehicle delay} = \text{Summation of the delays of individual vehicles}$$

$$= \text{Area of the triangle}$$

$$\text{Total vehicle delay} = \tfrac{1}{2}\,(60\times30) + \tfrac{1}{2}\,(60\times40) = 2100 \text{ vehicle-minutes}$$

$$\text{Average delay} = \frac{\text{Total vehicle delay}}{\text{No. of vehicles wating up to the point of dissipation}}$$

$$= \frac{2100}{210} = 10 \text{ minutes/vehicle}$$

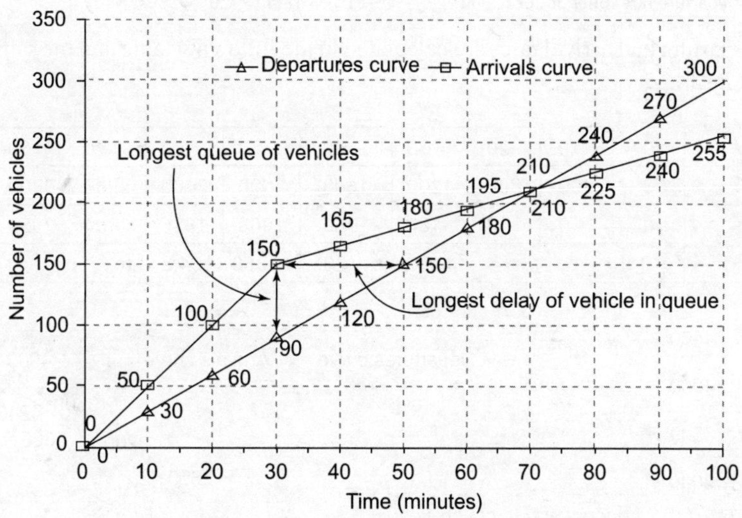


**Fig. 14.5** D/D/1 queue diagram

**Example 14.3**

A State Highway (SH) has a capacity of 3888 veh/hr and operates at a constant average flow rate of 2400 veh/hr. An accident occurred on the SH at 5:00 AM, and the road was completely blocked. At 5:20 AM, the road was partially opened with a capacity of 1080 veh/hr. Finally, all the wreckage was completely cleared by the traffic police and traffic flow was restored to its full capacity at 6:00 AM. Determine the time taken to dissipate the queue, the longest queue length, the longest waiting time, total delay and average delay per vehicle by assuming the D/D/1 model of queue with FIFO discipline.

**Answer:** Let the given values of arrival and departure of vehicles be converted to per minute:

Departure rate with full capacity up to 5:00 AM:

$\mu_{\text{Full}} = 3888$ vehicles per 60 minutes; $\mu_{\text{Full}} = 3888/60 = 64.8$ veh/minute $\rightarrow$ for $t \leq$ 5:00 AM.

Departure rate with partial capacity from 5:00 to 5:20 AM:

$\mu_{\text{Partial}} = 1080$ vehicles per 60 minutes; $\mu_{\text{Partial}} = 1080/60 = 18$ veh/minute for 20 minutes

Arrival rate: $\lambda = 2400$ vehicles per 60 minutes; $\lambda = (2400/60) = 40$ veh/minute $\rightarrow$ for all $t$.

Based on Little's law, the total number of vehicles that arrived after 5:00 AM will be $= \lambda t$, where $t$ is the time in minutes after 5:00 AM. The total number of vehicles departing during these three phases of traffic can be stated as:

| *Number of vehicles departures* | *Time* | *Period* |
|---|---|---|
| 0 | For $t \leq 20$ minutes | 5:00–5:20 AM |
| $\mu_{\text{Partial}}\ (t - 20)$ | For 20 minutes $< t \leq 40$ minutes | 5:20–6:00 AM |
| $360^{*} + \mu_{\text{Full}}\ (t - 40)$ | For $t > 40$ minutes | 6:00 AM onwards |

*Departure rate function for $t > 40 = \mu_{\text{Partial}}\ (t - 20) = 18(40 - 20) = 360$

The above departure and arrival rates are calculated in the following table and the same is graphically represented in Fig. 14.6:

| *t* (Minutes) | 0 | 10 | 20 | 30 | 40 | 50 | 60 | 70 | 80 | 90 | 100 |
|---|---|---|---|---|---|---|---|---|---|---|---|
| Departures = 40 × *t* | 0 | 400 | 800 | 1200 | 1600 | 2000 | 2400 | 2800 | 3200 | 3600 | 4000 |
| Arrivals at time | 0 | 0 | 0 | 180 | 360 | 1008 | 1656 | 2304 | 2952 | 3600 | 4248 |
| | = 0 × t | | | =18 × (*t* – 20) | | = 360 + 64.8 × (*t* – 30) | | | | | |

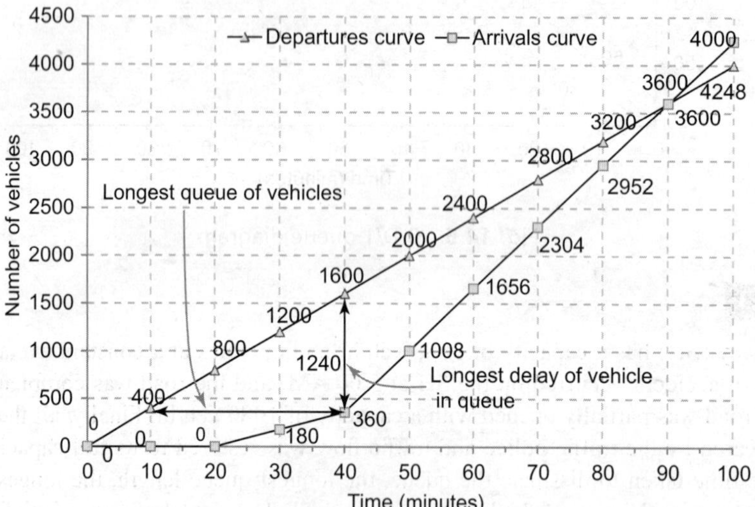


**Fig. 14.6** D/D/1 queue diagram

The point of queue dissipation can be determined by equating arrivals with departures

$$\lambda \times t = 360 + \mu_{\text{Full}} \times (t - 40)$$

Substituting the value of $\mu_{\text{Full}} = 64.8$

$$40 \times t = 360 + 64.8 \times (t - 40)$$

Solving, we get $t = 90$ minutes.

The queue that started at 5:00 AM dissipated after 90 minutes, after 6:30 AM.

The longest queue occurred at 5:40 AM (longest vertical distance between the arrival and departure curves (Fig. 14.6)):

$$Q_{max} = \lambda \times t - \mu_{\text{Partial}} \times (t - 20)$$

Substituting t = 40 minutes, we get

$$Q_{max} = 40 \times 40 - 18 \times (40 - 20)$$

$$Q_{max} = 1240 \text{ vehicles}$$

Longest vehicle delay by FIFO = 30 minutes (Fig. 14.6)

Total vehicle delay = Area between the arrival and departures curves

$$\begin{aligned}\text{Total vehicle delay} &= \{½ \times 3600 \times 90\} - \{(½ \times 20 \times 360) + (50 \times 360) + (½ \times 50 \times 3240)\}\\ &= \{162000\} - \{(3600) + (18000) + (81000)\}\\ &= \{162000\} - \{102600\} = 59400 \text{ veh-minutes.}\end{aligned}$$

$$\text{Average delay} = \frac{\text{Total vehicle delay}}{\text{No. of vehicles wating up to the point of dissipation}}$$

$$= \frac{59400}{1240} = 47.9 \text{ minutes/vehicle}$$

**Example 14.4**

Visitors arrive at an entrance to a ticket issuing desk of a show-mall. The desk opens at 9:30 AM and remains open till the evening. The visitors begin to arrive at a rate of 150 persons/hr and it assumed to be a Poisson distribution with exponential times between the arrivals. *T*, the time required to serve each person, is 15 seconds. Compute the probability of no persons in the system, average value of the number of persons in the queue, number of vehicles in the system, waiting time in the queue and the time spent in the system. Assume the M/D/1 queue model.

**Answer:** Let the given values of arrival and departure of persons be converted to per minute:
Arrival rate: $\lambda$ = 150 persons per 60 minutes; $\lambda$ = 150/60 = 2.5 persons/minute
Departure rate: $\mu$ = 15 sec per 1 person; $\mu$ = (60s/15s) × 1 = 4 persons/ minute
Traffic intensity or utilisation coefficient: $\rho = (\lambda/\mu) = 2.5/4 = 0.625$
Probability of no persons in the system: $P_o = 1 - \rho = 1 - 0.625 = 0.375$

$$\text{Average number of persons in the queue} = L_q = \frac{\rho^2}{2(1-\rho)} = \frac{0.625^2}{2(1-0.625)} = 0.520 \text{ persons}$$

$$\text{Average number of persons in the system: } L = L_q + \rho = 0.520 + 0.625 = 1.145 \text{ persons}$$

$$\text{Average waiting time of persons in the queue: } W_q = \frac{L_q}{\lambda} = \frac{0.52}{2.5}$$

$$= 0.208 \text{ minutes per person}$$

$$\text{Average staying time of persons in the system: } W = W_q + \frac{1}{\mu} = 0.208 + \frac{1}{4} = 0.458 \text{ minutes per person.}$$

**Example 14.5**

A circus show ticket booth has an average arrival rate of 2 persons per minute and service is provided at the rate of 4 persons per minute. Calculate the following characteristics of queuing by the M/M/1 model.

Probability of 1 person in the system; Average number of persons in the system; Average staying time of persons in the system; Average waiting time of persons in the queue; Average number of persons in the queue.

**Answer:**

Arrival rate: $\lambda = 2$ persons/ minute; Departure rate: $\mu = 4$ persons/ minute.
Traffic intensity or utilisation coefficient: $\rho = (\lambda/\mu) = 2/4 = 0.5$
Probability of no persons in the system: $P_o = 1 - \rho = 1 - 0.5 = 0.5$
Probability of 1 person in the system: $P_n = \rho^n (1 - \rho) = \rho^1 (1 - \rho) = 0.5 (1 - 0.5) = 0.25$

Average number of persons in the system: $L = \dfrac{\rho}{1-\rho} = \dfrac{0.5}{1-0.5} = 1$ person

Average staying time of persons in the system: $W = \dfrac{L}{\lambda} = \dfrac{1}{2} = 0.5$ minutes

Average waiting time of persons in the queue: $W_q = W + \dfrac{1}{\mu} = 0.5 + \dfrac{1}{4} = 0.75$ minutes

Average number of persons in the queue: $L_q = \lambda W_q = 2 \times 0.75 = 1.5$ persons

**Example 14.6**

Five toll collecting booths are installed at an entry point of a highway. The average arrival of vehicles observed was 1500 veh/hr and the service rate was assumed as 10 seconds each to pay the toll. Assume both the arrivals and the departures are exponentially distributed and calculate (i) the probability of no vehicle in the system; (ii) average queue length; (iii) time spent in the system and (iv) the probability of waiting in a queue.

**Answer:** Based on the given inputs, the system is considered as an M/M/N model of queueing.

Let the given values of arrivals and departures of the persons be converted to per minute values:
Arrival rate: $\lambda = 1500$ vehicles per 60 minutes; $\lambda = 1500/60 = 25$ veh/minute
Departure rate: $\mu = 10$ sec per 1 vehicle; $\mu = (60\text{s}/10\text{s}) \times 1 = 6$ veh/ minute
Traffic intensity or utilisation coefficient: $\rho = (\lambda/\mu) = 25/6 = 4.166$

$$\rho/N = 4.166/5 = 0.833 \text{ which is less than } 1.0.$$

$$(1 - \rho/N) = 1 - 0.833 = 0.167$$

Probability of no vehicles in the system: $P_o = \dfrac{1}{\sum_{k=0}^{N-1} \dfrac{\rho^k}{k!} + \dfrac{\rho^N}{N!\left(1 - \dfrac{\rho}{N}\right)}}$

$$P_o = \frac{1}{\frac{4.166^0}{0!} + \frac{4.166^1}{1!} + \frac{4.166^2}{2!} + \frac{4.166^3}{3!} + \frac{4.166^4}{4!} + \frac{4.166^5}{5!(0.167)}}$$

$$P_o = \frac{1}{1 + 4.166 + 8.677 + 12.05 + 12.55 + 62.61} = \frac{1}{101.053} = 0.009896$$

Average number of vehicles in the queue: $L_q = \frac{\rho^{N+1}}{N!\,N} \times \frac{P_o}{\left(1 - \frac{\rho}{N}\right)^2}$

$$L_q = \frac{4.166^{5+1}}{5! \times 5} \times \frac{0.009896}{(0.167)^2} = 3.09 \text{ vehicles}$$

Average number of vehicles in the system: $L = L_q + \rho = 3.09 + 4.166 = 7.256$ vehicles

Average staying time of vehicles in the system: $W = \frac{L_q}{\lambda} + \frac{1}{\mu} = \frac{3.09}{25} + \frac{1}{6} = 0.290266$ minutes

Average waiting time of vehicles in the queue: $W_q = \frac{L_q}{\lambda} = \frac{3.09}{25} = 0.1236$ minutes

Probability of having to wait in a queue: $P_{k>N} = \frac{\rho^{N+1}}{N!N} \times \frac{P_o}{\left(1 - \frac{\rho}{N}\right)}$

$$P_{k>N} = \frac{4.166^{5+1}}{5! \times 5} \times \frac{0.009896}{(0.167)} = 0.516306 \text{ minutes}$$

## 14.2 ANALYSIS OF DELAY AND QUEUING AT SIGNALISED INTERSECTIONS

A sudden drop in the capacity of a road may take place due to interrupted traffic flow at signalised intersections. The delay caused at signalised intersections can be termed **control delay,** which comprises the total delay that occurs during deceleration, stopping in a queue before the stop line and acceleration just after the signal is released (Fig. 14.7). Stopped delay is termed **time-in-queue delay** and the summation of delay due to deceleration and acceleration is termed **approach delay**. The slope of the trajectories of vehicles represents the speed of each vehicle at any point of time. The gap in the horizontal direction between these successive line trajectories indicates the time headway between the vehicles.

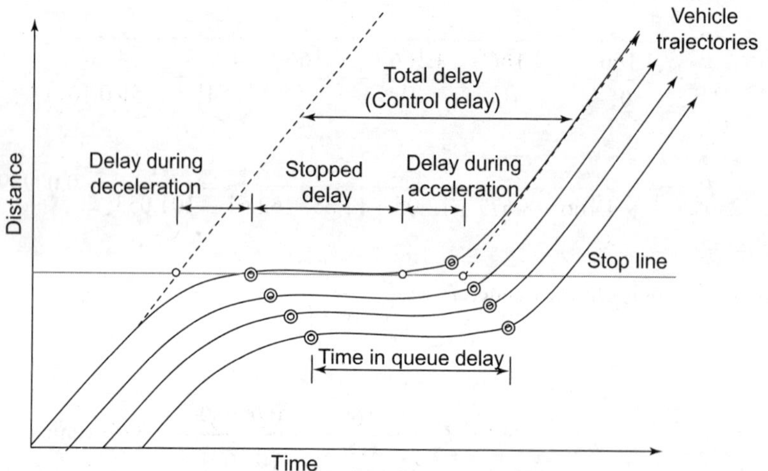


**Fig. 14.7** Traffic delays at an isolated signal-controlled intersection

### 14.2.1 Methods of Delay Analysis at Isolated Signalised Intersections

The traffic flow at an isolated signalised intersection can be represented and the control delay determined using two methods:

- Deterministic (uniform) arrivals (uses D/D/1 model)
- Probabilistic (random) arrivals (uses empirical equations, M/D/1 model)

**Deterministic queuing and delay analysis:** The D/D/1 model assumes a deterministic traffic arrival pattern, a deterministic service pattern and one service channel. The assumption of deterministic pattern means that the pattern is known, uniform and does not change over time. This indicates that there will be no random arrival and departure of vehicles. One service channel implies that only one lane of the approach road is used. This model also assumes that the approach arrivals are less than the departure capacity, and hence, there is no chance for the formation of a backlog queue which is waiting for the next Green. This means that no queue exists at the beginning/end of a cycle (Kyte and Tribelhorn, 2014).

Before discussing this method of analysis in detail, a few of the fundamental definitions and concepts relating to it are described below.

*Vehicle flow profiles:* The uniform arrival and departure pattern of vehicles through a signalised intersection can be represented by a flow profile diagram. The arrival flow rate is assumed to be uniform over the period of study while the service flow rate is assumed to vary during the cycle, as described below (Fig. 14.8).

Departure flow will be zero during the Red interval, and a queue of vehicles will form in front of the stop line. In the initial period of Green, the departure flow rate ($\mu$) will be equal to its saturation flow rate ($s$), while the queue is clearing. As soon as the queue is cleared, the departure flow rate will be equal to the arrival flow rate ($\lambda$), until the end of the Green interval. As the queue clears before the end of every Green, the assumption of the D/D/1 model is satisfied.

The discrete events of vehicle arrivals and departures in the field can be theoretically represented by stair step lines or by continuous lines. However, both discrete and continuous presentation produce the same results (Kyte and Tribelhorn, 2014).

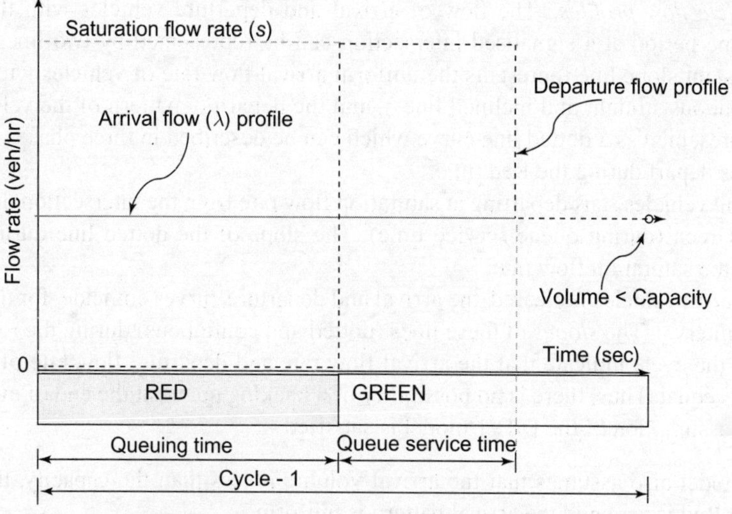


**Fig. 14.8** A typical flow profile diagram

## Example 14.7

Draw a flow profile that represents the following field observations:
Arrival flow rate = 500 veh/hr; Saturation flow rate = 1000 veh/hr; Cycle length = 110 sec; Red time = 50 sec; Green time = 60 sec; Queue service time = 32 sec.

**Answer:** The flow profile arrival and departure rates are spread over the given cycle time, as shown below:

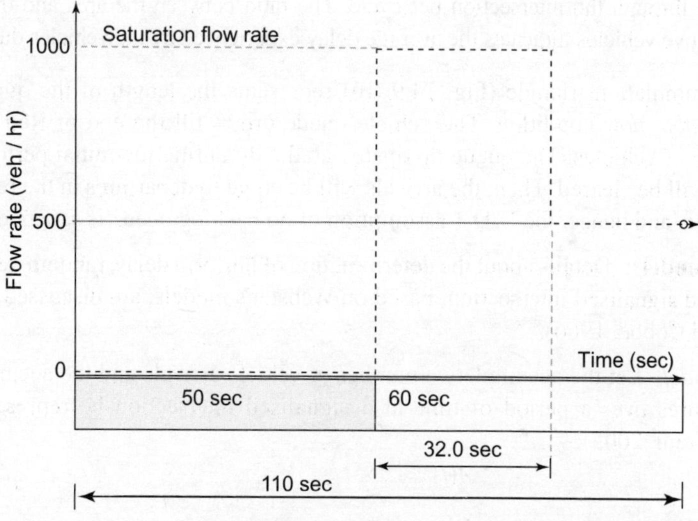


**Flow profile diagram with discrete events of vehicle departures**

*Cumulative vehicle flow profiles:* The flow of arrival and departure vehicles with their cumulative values over a time period at a signalised intersection can be represented by two lines (Fig. 14.9). A straight and constant slope line represents the uniform arrival flow rate of vehicles with constant time headways (continuous straight and inclined lines), and the departure pattern of the vehicles over each cycle time is represented as a dotted line curve which can be described in three phases (Fig. 14.9 (a)):

- No vehicles depart during the Red time.
- The queuing vehicles start departing at saturation flow rate from the intersection during the initial period of Green (during queue service time). The slope of the dotted line during this interval represents the saturation flow rate.
- As the queue of vehicles is cleared, the arrival and departure curves coincide, for the remainder of the Green interval. The *slopes* of these lines (dotted and continuous) during the remaining Green interval of the cycle indicate that the arrival flow rate and departure flow rate of vehicles is the same and is equal. Thus, there is no possibility of a backlog queue at the end of every Green, and hence, the assumption of the D/D/1 model is satisfied.

The D/D/1 model also assumes that the arrival volume is less than the capacity, the queue at the beginning of the Red is zero and the arrival pattern is uniform.

The following observations can be made to **evaluate the performance** of traffic flow at a signalised intersection.

- The vertical distance between the arrival line and the departure curve line indicates the number of vehicles in the queue at any point of time ($t$).
- The horizontal distance between the arrival line and the departure curve line indicates the delay experienced by the $n^{th}$ vehicle in the queue at any point of time ($t$). The left end point of the arrow on the arrival line indicates the vehicle's entry time in the system and the right point of the arrow indicates the time at which the vehicle exited from the system. The time difference between the entry and exit points represents the delay experienced by the vehicle under observation.
- The areas of the individual triangles indicate the total delay experienced by all the vehicles that arrive at and depart through the intersection per cycle. The ratio between the area and the corresponding cycle cumulative vehicles indicates the average delay experienced by the vehicles during that cycle.

The queue accumulation triangle (Fig. 14.9 (b)) represents the length of the queue at any point in time, $t$, with *stable flow* condition. The vehicle queue grows till the end of Red and reaches the maximum number of vehicles. The queue dissipates gradually during the initial period of Green, and finally the queue will be cleared. Then, the arrivals will be equal to departures in the remaining portion of the Green period, and hence, the D/D/1 assumption of no backlog queue is satisfied.

**Webster's delay models:** Details about the determination of uniform delay, random delay and overflow delay at an isolated signalised intersection, based on Webster's models, are discussed below (Webster, 1958; Webster and Cobbe, 1966).

*Uniform delay model*: Let the accumulative number of vehicle arrivals and the accumulative number of vehicle departures over a period of time at a signalised intersection be represented as follows (Teodorovic and Trani, 2005):

$$A(t) = \lambda \times t \tag{14.2}$$

$$D(t) = \mu \times t \tag{14.3}$$

where $\lambda$ = arrival rate (veh/hr); $\mu$ = departure rate (veh/hr); and $t$ = time period (sec).

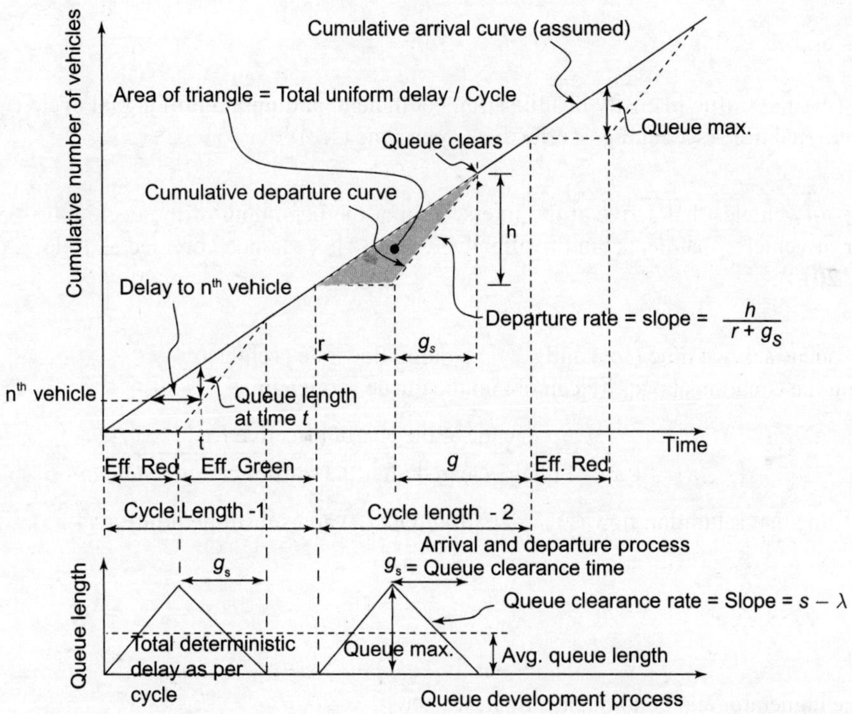


**Fig. 14.9** Components of delay due to queuing at signalised intersections (based on FHWA, 2016)

The uniform delay is incurred in the same signal cycle in which vehicles arrive, and there will not be any backlog queue at the end of the Green time (Olszewsk, 1993). Hence, stable traffic flow exists at signalised intersections based on the D/D/1 model, in the following conditions.

*Condition 1*

Dissipation of the queue in every cycle occurs when the total number of vehicle arrivals during a cycle is less than or equal to the total number of vehicle departures in its effective Green. This condition is represented by $g_s \leq g$

These two durations can be integrated as follows

$$\int_0^{C_o} \lambda\, dt \leq \int_0^{g} \mu dt$$

$$[\lambda \times t]_o^{C_o} \leq [\mu \times t]_0^{g}$$

$$\lambda \times C_o \leq \mu \times g$$

$$\left(\frac{\lambda}{\mu}\right) \leq \left(\frac{g}{C_o}\right)$$

$$\rho \leq \left(\frac{g}{C_o}\right) \tag{14.4}$$

where $\rho = (\lambda/\mu)$ = traffic intensity or utilisation coefficient, and duration of signal cycle $C_o = r + g$ ($r$ = effective Red time (sec) and $g$ = effective Green time (sec)).

*Condition 2*

The number of vehicles that arrive at the intersection at the beginning of the cycle must be equal to the number of vehicles that depart in the effective Green. This can be expressed as follows (Kyte and Tribelhorn, 2014):

$$\lambda \times (r + g_s) = s \times g_s$$

where $g_s$ = queue service time (sec) and $s$ = saturation flow rate (veh/hr).

Rewriting the equation above, we can obtain the queue service time

$$g_s = \frac{\lambda \times r}{(s-\lambda)} = \frac{\text{Queue at the end of effective Red}}{\text{Rate of queue cleared after the start of effective Green}} \tag{14.5}$$

If we assume that saturation flow ($s$) = departure rate ($\mu$) and substitute it in Eq. (14.5), we get

$$g_s = \frac{\lambda \times r}{(\mu - \lambda)}$$

Dividing the numerator and denominator by $\mu$, we get $g_s = \dfrac{\dfrac{\lambda}{\mu} \times r}{\left(\dfrac{\mu}{\mu} - \dfrac{\lambda}{\mu}\right)} = \dfrac{\rho \times r}{(1-\rho)}$ (14.6)

*Condition 3*

The total delay of all the vehicles that arrived during a cycle must be equal to the area of the triangle (Fig. 14.10)

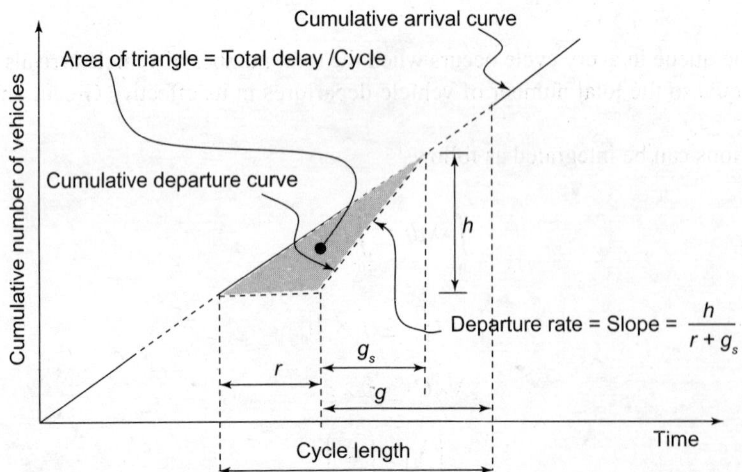


**Fig. 14.10** Cumulative flow of vehicles at a signalised intersection during a cycle

$$\text{Area of a triangle} = \tfrac{1}{2} \times \text{base} \times \text{height}$$

$$\text{Area of a triangle} = \tfrac{1}{2} \times r \times h$$

We know the departure rate = slope $\lambda = \dfrac{h}{(r+g_s)}$ (14.7)

By rewriting the above, $h = \lambda(r + g_s)$ (14.8)

$$\text{Area of triangle} = \tfrac{1}{2} \times r \times \lambda(r + g_s)$$

Total uniform delay of the vehicles arriving in a cycle

$$= d_T = 0.5 \times r \times \lambda(r + g_s)$$

By Eq. (14.6), substituting $g_s$ in the above, we get

$$d_T = 0.5 \times r \times \lambda\left(r + \frac{\rho \times r}{(1-\rho)}\right)$$

$$d_T = 0.5 \times r^2 \times \lambda\left(1 + \frac{\rho}{(1-\rho)}\right)$$

$$d_T = 0.5 \times r^2 \times \lambda\left(\frac{1}{(1-\rho)}\right) \tag{14.9}$$

The average uniform delay per vehicle can be obtained by dividing the total uniform delay by the number of vehicles arriving and departing during the cycle (Newell, 1965):

$$\text{Average uniform delay per vehicle } \left(d_{avg}\right) = \frac{\text{Total delay per cycle}}{\text{Total number of vehicles per cycle}}$$

$$= \frac{d_T}{\lambda \times C_o}$$

$$d_{Avg\text{-}Uniform} = \frac{0.5 \times r^2 \times \lambda\left(\frac{1}{(1-\rho)}\right)}{\lambda \times C_o}$$

$$d_{Avg\text{-}Uniform} = \frac{0.5 \times r^2 \times \lambda}{\lambda \times C_o(1-\rho)} = \frac{0.5 \times r^2}{C_o(1-\rho)} \tag{14.10}$$

*Alternatively,*

If we take $g_s = \dfrac{\lambda \times r}{(s-\lambda)}$ and substitute it below

$$\text{Total delay of the vehicles arriving in a cycle} = d_T = 0.5 \times r \times \lambda(r + g_s)$$

$$d_T = 0.5 \times r \times \lambda \left( r + \frac{\lambda \times r}{(s-\lambda)} \right) \tag{14.11}$$

And

$$d_{Avg\text{-}Uniform} = \frac{0.5 \times r \times \lambda \left( r + \frac{\lambda \times r}{(s-\lambda)} \right)}{\lambda \times C_o} = \frac{0.5 \times r \left( r + \frac{\lambda \times r}{(s-\lambda)} \right)}{C_o} \tag{14.12}$$

Other formulae:

$$\text{Proportion of the cycle with a queue} = P_q = \frac{r + g_s}{C_o} \tag{14.3}$$

$$\text{Proportion of vehicles stopped} = \frac{r + g_s}{C_o} \tag{14.4}$$

**Random arrivals analysis:** The above derivations are based on the assumption that traffic flow is uniform, which may not occur all the time in reality. Traffic flows at signalised intersections are characterised by random fluctuations, and the individual vehicle arrivals depend on the probability density function of the inter-arrival time, signal timings and the time of day. For these reasons, individual vehicles experience various types of delay at signalised approaches. The random arrival of vehicles is considered and the delay at signalised intersections is determined, as discussed below.

**Random delay:** The delay due to variations in the random arrival of vehicles causes short-term backlog queues which will be dissipated in a couple of cycles (Fig. 14.11). Such conditions of flow may occur during peak period, and the queue that forms will be dissipated within a few cycles. This condition is defined as under-saturation when $v/c \leq 1.0$, which represents the overall traffic flow condition as stable over the period of analysis and where demand does not exceed the total capacity. This delay can be calculated using deterministic queuing theory techniques. The area between the arrival curve and the dotted line represents this additional delay and is termed random delay (RD).

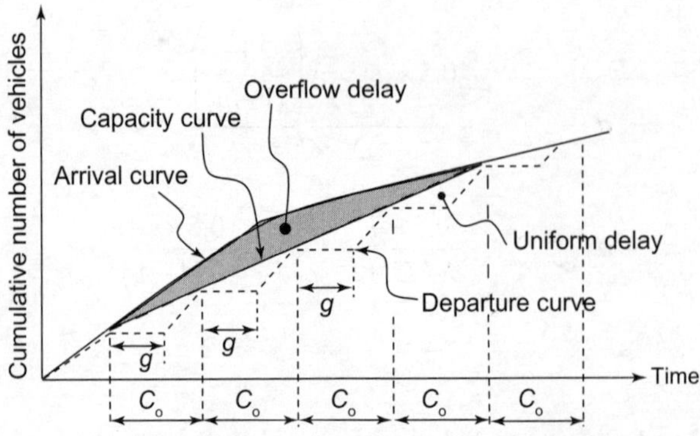


**Fig. 14.11** Random delay and uniform delay due to random variations in arrival flow (for example, in a peak period, it may be acceptable if overall $v/c < 1.0$)

Several random or stochastic delay models are available in the literature and they assume that the inter-arrival times between successive vehicles are distributed exponentially and the service rate is deterministic. For example, the most widely used Webster's model is presented below:

Let the departure rate of $\mu_r$ take place from the artificial queue into a single signal server and let us assume this artificial queue system is of the type M/D/1, which represents Poisson arrivals and deterministic service times with a single server (Kendall, 1951).

The average delay per vehicle as in the M/D/1 queuing system $= d_{Random} = \dfrac{0.5 \times X^2}{\lambda(1-X)}$ (14.15)

where $X$ = utilisation ratio in the M/D/1 queuing system $= \left(\dfrac{\lambda}{\mu_r}\right)$.

The artificial departure rate ($\mu_r$) of the whole cycle (during Red, it is zero, and during Green, it is $\mu \times g$) can be expressed as (Teodorovic and Trani, 2005)

$$\mu_r = \frac{0 \times r + \mu \times g}{C_o} = \mu \times \frac{g}{C_o}$$

Now, substituting $\mu_r$ for $X$, we get

$$X = \left(\frac{\lambda}{\mu}\right) \times \left(\frac{C_o}{g}\right) = \frac{\left(\frac{\lambda}{\mu}\right)}{\left(\frac{g}{C_o}\right)} = \frac{\rho}{\left(\frac{g}{C_o}\right)} = \rho \times \left(\frac{C_o}{g}\right) \tag{14.16}$$

where $X = \left(\dfrac{\lambda}{\mu_r}\right) = \left(\dfrac{\lambda}{\mu \times \frac{g}{C_o}}\right)$.

The quantity of $X$ is also known as the volume-to-capacity ratio in the UK and in Australian literature (Hurdle, 1984). Alternatively, the $X$ value can be written as

$$X = \left(\frac{\text{Arrival volume}}{\text{Capacity}}\right) = \left(\frac{\lambda}{s \times \left(\frac{g}{C_o}\right)}\right) = \left(\frac{\lambda}{s}\right) \times \left(\frac{C_o}{g}\right) \tag{14.17}$$

where $s$ may be replaced by $\mu$, and the ratio of $\dfrac{\lambda}{s} = \rho$.

Therefore, the overall average delay $= d_{Avg\text{-}Uniform} + d_{Random}$ (14.18)

where by Eq. (14.10), $d_{Avg\text{-}Uniform}$ = Average uniform delay $= \dfrac{0.5 \times r^2}{C_o(1-\rho)}$ and by Eq. (14.15), $d_{Random}$ = Random delay $= \dfrac{0.5 \times X^2}{\lambda(1-X)}$

*Note:* The term overall delay can also be called total delay or approach delay. It is the time difference between the actual time taken to traverse through a signalised intersection and the time taken to traverse the same road segment at the desired cruising speed (Olszewski, 1993).

Based on simulation studies and field observations pertaining to unsaturated signalised intersections and deterministic queuing analysis, Eq. 14.18 was found to overestimate the values to some extent, and hence, the following corrected formulae were recommended.

Webster's (1958) formula:

$$\text{Overall delay} = (d_{Avg\text{-}Uniform} + d_{Random}) - AF \tag{14.19}$$

where $AF$ = adjustment factor for overestimation $= 0.65 \times \left(\frac{C_o}{\lambda^2}\right)^{\frac{1}{3}} \times X^{\left(2+\frac{5\times g}{C_o}\right)}$.

AF is frequently assumed to be around 10% of the first two terms ($d_{Avg} + RD$), and thus, the above formula is often simplified as follows (Allsop, 1972):

$$\text{Overall delay} = 0.9 \times (d_{Avg\text{-}Uniform} + d_{Random}) \tag{14.20}$$

**Overflow delay:** The condition of further excessive random arrival of vehicles with accumulated backlog queue is used to describe *prolonged* time periods during which vehicle *arrivals exceeds the capacity* of the intersection approach to discharge vehicles, termed *oversaturation* (Fig. 14.12). This means that the vehicles in queue are not cleared during the first available Green, and this condition is termed *cycle failure*. This condition represents the non-equilibrium state.

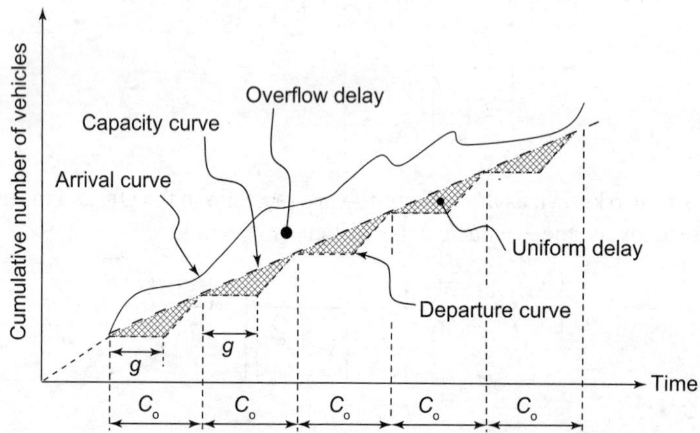


**Fig. 14.12** Intersection with random arrivals in oversaturation condition

As the arrival rate exceeds the capacity, the queue length increases in each cycle and thus causes *overflow delay* to the vehicle in the queue – an additional delay above and beyond the uniform delay. Overflow delay accounts for failure of clearance of the queue over an extended period of cycles, with a portion including random delay as well. When demand exceeds capacity ($v/c > 1.0$), the total delay depends on the length of time during which the oversaturation condition exists.

To simplify the analysis of total delay due to oversaturation condition, the rate of oversaturated arrivals of vehicles is assumed to be uniform and is represented by a straight inclined line (Fig. 14.13).

During oversaturation ($X \geq 1.0$), total delay can be calculated as the summation of two components, uniform delay ($d_{Avg\text{-}Uniform}$) and overflow delay ($d_{Overflow}$).

$$\text{Total overflow delay (TOD)} = d_{Avg\text{-}Uniform} + d_{Overflow} \tag{14.21}$$

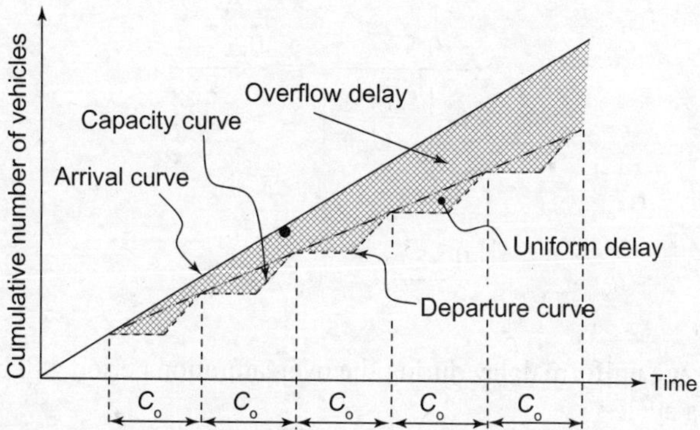


**Fig. 14.13** Overflow delay and uniform delay with uniform arrivals in oversaturated condition (prolonged shortage of capacity when overall $v/c > 1.0$)

These two component delays are derived as follows:

1. $d_{Avg\text{-}Uniform}$: During oversaturation, average delay consists of uniform delay, which is represented by the triangular area between the capacity and departure curves, when the volume-to-capacity ratio ($X$) = 1.0. Hence, let us substitute $X = 1$ in Eq. (14.10). We can rewrite the equation of $d_{Avg\text{-}Uniform}$ for the oversaturation condition as

$$d_{Avg\text{-}Uniform} = \text{Average uniform delay} = \frac{0.5 \times r^2}{C_o(1-\rho)}$$

We know by Eq. (14.17) that

$$X = \left(\frac{\text{Arrival volume}}{\text{Capacity}}\right) = \left(\frac{\lambda}{s \times \left(\dfrac{g}{C_o}\right)}\right) = \left(\frac{\lambda}{s}\right) \times \left(\frac{C_o}{g}\right) = (x)$$

Rewriting,
$$X = \rho \times \left(\frac{C_o}{g}\right)$$

$$\rho = \left(\frac{g}{C_o}\right) \times X$$

Substituting the above values in Eq. (14.10),

$$d_{Avg\text{-}Uniform} = \frac{0.5 \times r^2}{C_o\left(1 - \left(\dfrac{g}{C_o}\right) \times X\right)}$$

Now, substituting $X = 1.0$ due to oversaturation, we get

$$d_{Avg\text{-}Uniform} = \frac{0.5 \times r^2}{C_o\left(1 - \left(\frac{g}{C_o}\right) \times 1.0\right)} = \frac{0.5 \times r^2}{C_o\left(1 - \left(\frac{g}{C_o}\right)\right)}$$

$$= \frac{0.5 \times r^2}{C_o\left(\frac{C_o - g}{C_o}\right)} = \frac{0.5 \times r^2}{C_o - g}$$

We know that $r = (C_o - g)$

$$d_{Avg\text{-}Uniform} = \frac{0.5 \times r^2}{r} = 0.5 \times r \quad (14.22)$$

*Note:* The average uniform delay during the oversaturation period will be equal to half the Red period during a cycle.

2. $d_{Overflow}$: The area between the capacity curve and the arrival curve represents the overflow delay ($d_{Overflow}$). The following is a time-dependent travel time function with deterministic expression used to determine the overflow delay, $d_{Overflow}$ (Hurdle, 1984; Kimber and Hollis, 1979).

*Case 1*: To determine the component of overflow delay with uniform arrival rate, let us only consider the arrival rate and the capacity rate curves (above the departure curve) up to the analysis period, $T$ (sec) (Fig. 14.14).

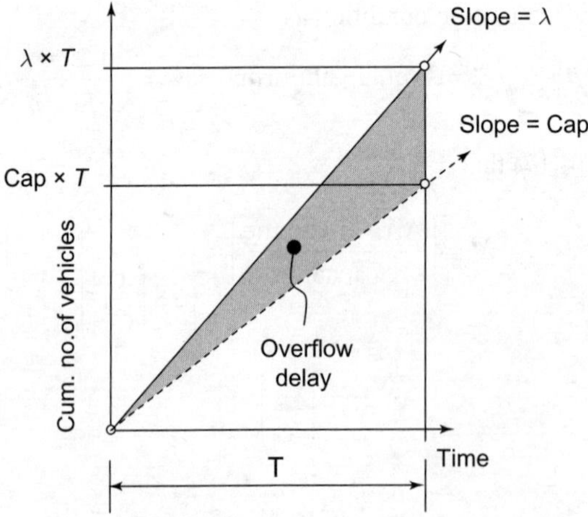


**Fig. 14.14** Overflow delay up to a time in the oversaturated condition

The total vehicles discharged up to time $T$ (sec) = Cap × $T$

where $$Cap = \text{Capacity} = s \times \left(\frac{g}{C_o}\right)$$

Total uniform overflow delay up to $T$ = Triangle area between arrival and capacity curves

Total uniform overflow delay = ½ × base × height

Total uniform overflow delay = ½ × $T$ × [($\lambda$ × $T$) – (Cap × $T$)]

Total uniform overflow delay = 0.5 × $T^2$ × ($\lambda$ – Cap)

Average overflow delay = $d_{Overflow}$

$$= \frac{\text{Total uniform overflow delay up to } T \text{ sec}}{\text{The total vehicles discharged up to the time } T \text{ sec}}$$

$$d_{overflow} = \frac{0.5 \times T^2 \times (\lambda - \text{Cap})}{\text{Cap} \times T} = \frac{0.5 \times T \times (\lambda - \text{Cap})}{\text{Cap}} = 0.5T \times \left(\frac{\lambda}{\text{Cap}} - 1\right)$$

$$d_{Overflow} = 0.5T(X-1) \tag{14.23}$$

where $X = \frac{\text{Arrival volume}}{\text{Capacity}} = \frac{\lambda}{Cap}$.

*Note:* The overflow delay is directly proportional to the length of time T during which the arrival flow exceeds the capacity (Hurdle, 1984).

Therefore, Eq. (14.21) can be written with the parameters derived above as

$$\text{Total overflow delay (TOD)} = d_{Avg\text{-}Uniform} + d_{Overflow} = 0.5 \times r + 0.5T(X-1) \tag{14.24}$$

*Case 2*: To determine the component of overflow delay with uniform arrival rate, let us consider the arrival rate and the capacity rate curves between a study period of $T_1$ and $T_2$ (sec) (Fig. 14.15).

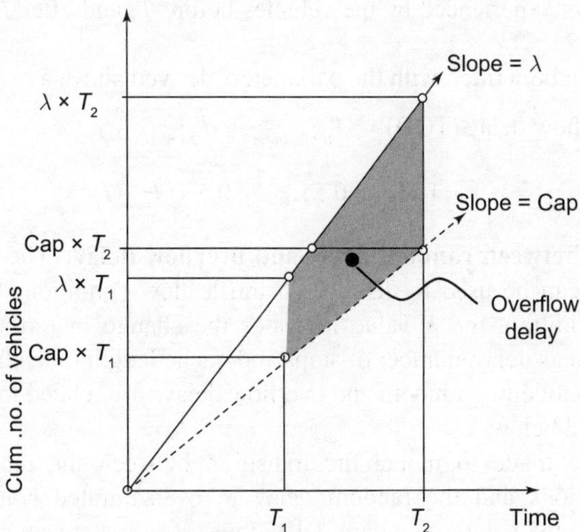


**Fig.14.15** Overflow delay during a time period in the oversaturated condition

The total vehicles discharged in the time period between $T_1$ and $T_2$ = Cap × ($T_2 - T_1$)

Total uniform overflow delay between $T_1$ and $T_2$ = Trapezoid area between arrival and capacity curves

Total uniform overflow delay = Base × Mean height of trapezoid

$$\text{Total uniform overflow delay} = (T_2 - T_1) \times \left( \frac{[(\lambda \times T_2) - (Cap \times T_2)] + [(\lambda \times T_1) - (Cap \times T_1)]}{2} \right)$$

$$\text{Total uniform overflow delay} = (T_2 - T_1) \times \left( \frac{[T_2 \times (\lambda - Cap)] + [T_1 \times (\lambda - Cap)]}{2} \right)$$

$$\text{Total uniform overflow delay} = (T_2 - T_1) \times \left( \frac{T_2 + T_1}{2} \right) \times (\lambda - Cap)$$

$$\text{Total uniform overflow delay} = 0.5 \times (T_2 - T_1) \times (T_2 + T_1) \times (\lambda - Cap)$$

Average overflow delay = $d_{Overflow}$

$$= \frac{\text{Total uniform overflow delay during } T_1 \text{ and } T_2}{\text{The total vehicles discharged during the time period } T_1 \text{ and } T_2 \text{ sec}}$$

$$d_{Overflow} = \frac{0.5 \times (T_2 - T_1) \times (T_2 + T_1) \times (\lambda - \text{Cap})}{\text{Cap} \times (T_2 - T_1)}$$

$$d_{Overflow} = \frac{0.5 \times (T_2 + T_1) \times (\lambda - \text{Cap})}{\text{Cap}} = 0.5 \times (T_1 + T_2) \times \left( \frac{\lambda}{\text{Cap}} - 1 \right) = 0.5 \times (T_1 + T_2) \times (X - 1) \quad 14.25)$$

*Note:* The overflow delays experienced by the vehicles before $T_1$ and after $T_2$ are not included in the above equation.

Therefore, Eq. (14.21) can be written with the parameters derived above as

$$\text{Total overflow delay (TOD)} = d_{Avg\text{-}Uniform} + d_{Overflow}$$

$$\text{TOD} = 0.5 \times r + 0.5 \times (T_1 + T_2) \times (X - 1) \quad (14.26)$$

**Inconsistent modelling between random delay and overflow delay:** The degree of saturation ($X$) is the single factor that can be used to distinguish traffic flow conditions between undersaturation and oversaturation. In addition, the $X$ value indicates the change in traffic flow patterns in other performance indices, such as delay, number of stops and queue length (Akçelik 1981). For this reason, inconsistencies found in modelling random and overflow delays are related to the degree of saturation ($X$ value); this is discussed below.

Several attempts were made to model the transition between the random delay experienced in undersaturated conditions and the random delay in oversaturated conditions. The theoretical determination of the expected value of random delay under steady-state conditions (Webster's model) assumes that delays tend asymptotically to infinity when the demand approaches capacity, i.e., when the degree of saturation $X = 1.0$. On the other hand, the overflow delay model is valid when $X \geq 1.0$

(Akçelik, 1980; Dion et al., 2004). Random and overflow delay models do not predict exact true values, if the $X$ value is on either side of 1.0. This problem can be illustrated (Fig. 14.16). The observations and field studies conducted by researchers confirm that the estimated uniform delay parameter with $X$ value $\leq 0.85$ can be used as a reliable performance indicator of an isolated signalised intersection, except in the case of platooned arrival of vehicles. In this range ($0 < X \leq 0.80$), the value of random delay is very small and there will be no chance for the formation of overflow delay.

Similarly, the theoretical overflow delay, which is presented above, is a reasonable predictor when $X \geq 1.1$. Hence, it is noted that there is a difficulty in exact estimation of true values of delays using steady-state stochastic models and deterministic models, when $0.80 \leq X \leq 1.1$. In this connection, recent research attempts are underway to model the delay between this gap (Chen et al., 2013). Some of the overflow and overall delay models widely accepted in the USA and in Australia are presented below.

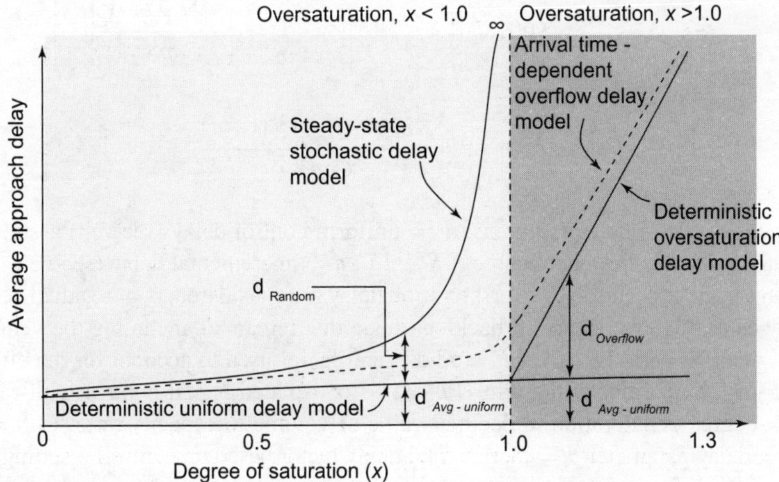


**Fig. 14.16** Relationship between delay and degree of saturation in different conditions of traffic flow at an isolated intersection (Drawn based on Akcelik, 1980; Hurdle, 1984; Dion et al, 2004)

*Akçelik delay model:* To overcome the difficulty in estimation of delay, Akçelik proposed the following model to estimate overflow delay at an isolated signalised intersection; this has been adopted by the Australian Road Research Board (ARRB) (Akçelik, 1980):

$$d_{Overflow} = \frac{\text{Cap} \times T}{4}\left[(X-1) + \sqrt{(X-1)^2 + \frac{12(X - X_o)}{\text{Cap} \times T}}\right] \qquad (14.27)$$

where $T$ = time period of analysis (say, 1 hour); $X$ = volume-to-capacity ratio; $Cap$ = capacity (veh/hr); $s$ = saturation flow rate (veh/sec during the Green); $g$ = effective Green (sec); and $X_o = 0.67 + \frac{s \times g}{600}$ ($X \geq X_o$; if $X \leq X_o$, the overflow delay = 0).

In addition to this model, separate model equations were developed for estimating delay in variable traffic conditions, such as the post-peak period, peak period and non-peak period. A detailed discussion on these models can be obtained from Akçelik and Rouphail (1993).

*HCM delay model:* This model is composed of three components: uniform control delay ($d_1$), incremental delay ($d_2$) and initial queue delay ($d_3$). The average delay per vehicle for a lane group can be estimated using the following equation, which is recommended in the Highway Capacity Manual (TRB, 2000).

$$d = (d_1 \times PF) + d_2 + d_3 \tag{14.28}$$

$$d_1 = \frac{0.5 \times C_o \times \left(1 - \frac{g}{C_o}\right)^2}{\left[1 - \left(\text{min. of } (1, X)\right) \times \frac{g}{C_o}\right]}$$

$$d_2 = 900 \times T \left\{ (X-1) + \left[ (X-1)^2 + \left( \frac{8 \times k \times l \times X}{Cap \times T} \right) \right]^{0.5} \right\}$$

$$d_3 = \text{Zero, or } = \frac{1800 \times Q_b \times t \times (1+u)}{Cap \times T}$$

where $d$ = average control delay (sec/veh); $d_1$ = uniform control delay (sec/veh) assuming perfectly uniform arrivals and stable flow, applicable if $X < 1.0$; $d_2$ = incremental delay (sec/veh), non-uniform arrivals and random delays caused by oversaturation delay are considered; $d_3$ = residual delay (sec/veh), overflow delay caused by oversaturated backlog queue that remained, including the vehicles added at the start of the analysis period $T$ (hr); $PF$ = adjustment factor used to account for the effect of quality of progression in the coordinated system ($PF$ =1.0 for isolated signal control with random arrival vehicles); $X$ = degree of saturation in decimal value of volume-to-capacity ratio; $Cap$ = capacity of lane group; $u$ = delay parameter; $k$ = incremental delay factor based on controller settings ($k$ = 0.5 for pre-timed signals and may vary between 0.04 and 0.50 for actuated controllers); $l$ = upstream filtering/metering adjustment factor ($l$ = 1.0 for an isolated intersection); $T$ = duration of the analysis period (hr); and $Q_b$ = initial unmet demand of traffic at the start period $T$ (veh).

The delay value calculated for each approach of an intersection is averaged and reported as the average control delay per vehicle at a signalised intersection.

*Note:* Eq. (14.28) was revised in 2010 as, $d = d_1 + d_2 + d_3$ with similar definitions. The values of $d_1$ and $d_2$ are the same as in HCM 2000, but a revised formula is recommended for $d_3$. These details can be obtained from the revised Highway Capacity Manual (2010).

**Example 14.8**

Draw the cumulative vehicle diagram for the following field-observed values on one approach to a signalised intersection:

Vehicles arrive uniformly every 5 sec; Cycle length = 50 sec; Red time interval = 23 sec; Green time interval = 25 sec; Vehicles depart every 2 sec after Green begins; Queue service time = 13 sec.

Determine the (i) delay experienced by each vehicle until the queue is cleared and (ii) the total delay for all the vehicles in the system.

**Answer:** The cumulative vehicle flow and delay experienced by the vehicles are presented in the following figure. Details of the computation of total delay are presented in the table.
Vehicle arrival and departure with delay time:

| *Vehicle number* | *Arrival time (sec)* | *Departure time (sec)* | *Delay (time in system) sec/veh* |
|---|---|---|---|
| 1 | 5 | (23 + 2) = 25 | (25 – 5) = 20 |
| 2 | 10 | (25 + 2) = 27 | (27 – 10) = 17 |
| 3 | 15 | (27 + 2) = 29 | (29 – 15) = 14 |
| 4 | 20 | (29 + 2) = 31 | (31 – 20) = 11 |
| 5 | 25 | (31 + 2) = 33 | (33 – 25) = 8 |
| 6 | 30 | (33 + 2) = 35 | (35 – 30) = 5 |
| 7 | 35 | (35 + 2) = 37 | (37 – 35) = 2 |
| Total = | | | 77 sec |

Average uniform delay per vehicle = 77/7 = 10 sec

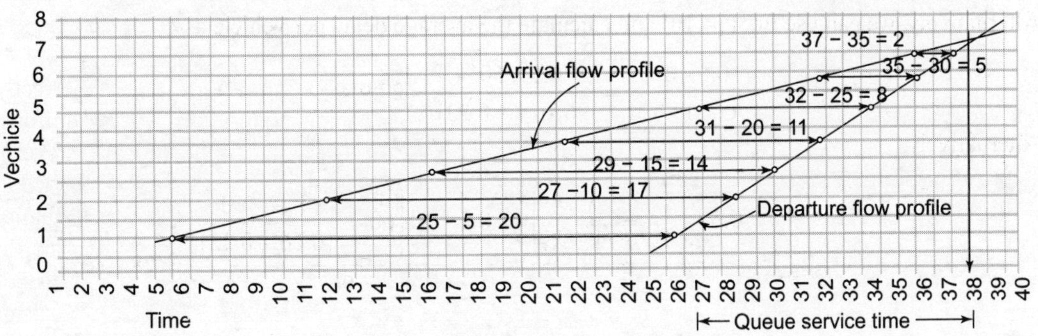


**Cumulative vehicle diagram**

**Example 14.9**

Determine the total delay and average delay experienced by the vehicles for the following field traffic condition at a signalised intersection:

Cycle length = 96 sec; Approach saturation flow rate = 2400 veh/hr; Green time duration = 36 sec; Flow rate = 800 veh/hr. Assume that the D/D/1 model of queuing adequately describes the approach of the intersection.

**Answer:** The given values are taken as follows
$C_o = 96$ sec; $g = 36$ sec; $r = C_o - g = (96 - 36) = 60$ sec

$$\text{Arrival flow rate} = \lambda = 800 \text{ veh/hr} = 800/3600 = 0.22 \text{ veh/sec}$$

$$\text{Departure flow rate} = \mu = 2400 \text{ veh/hr} = 2400/3600 = 0.66 \text{ veh/sec}$$

$$\rho = \left(\frac{\lambda}{\mu}\right) = \left(\frac{0.22}{0.66}\right) = 0.33$$

Check for the assumption about the D/D/1 model: $\rho \le \left(\frac{g}{C_o}\right) = 0.33 \le \left(\frac{36}{96}\right)$

$0.33 \le 0.375$. Hence, it is acceptable. This indicates that undersaturated traffic flow exists here.

Using Eq. (14.9),

$$\text{Total uniform delay per cycle} = d_T = 0.5 \times r^2 \times \lambda\left(\frac{1}{(1-\rho)}\right)$$

$$d_T = 0.5 \times 60^2 \times 0.22\left(\frac{1}{(1-0.33)}\right) = 591.0 \text{ sec}$$

$$\text{Number of vehicles in queue (N)} = \lambda \times C_o = 0.22 \times 96 = 21.12$$

$$\text{Average delay per vehicle} = \left(\frac{d_T}{N}\right) = \left(\frac{591.0}{21.12}\right) = 27.9 \text{ sec}$$

Alternatively, we can use Eq. (14.10) and calculate the average delay per vehicle as

$$d_{Avg\text{-}Uniform} = \frac{0.5 \times r^2}{C_o(1-\rho)} = \frac{0.5 \times 60^2}{96 \times (1-0.33)} = 27.9 \text{ sec}$$

Alternatively,

$$d_{Avg\text{-}Uniform} = \frac{0.5 \times r\left(r + \frac{\lambda \times r}{(s-\lambda)}\right)}{C_o} = \frac{0.5 \times 60\left(60 + \frac{0.22 \times 60}{(0.66-0.22)}\right)}{96} = 28.1 \text{ sec}$$

**Example 14.10**

An intersection approach has an arrival rate of 600 veh/hr and a saturation flow rate of 1800 veh/hr. Take cycle time as 98 sec, effective Red time as 62 sec and effective Green time as 36 sec. Check whether the approach capacity is less than the arrival flow and determine queue service time, total delay time per cycle and average delay experienced by the vehicles.

**Answer:** Let us convert the flow rates from veh/hr to veh/sec.

$$\text{Arrival flow rate} = \lambda = 600 \text{ veh/hr} = \frac{600}{3600} = 0.167 \text{ veh/sec}$$

$$\text{Departure flow rate} = \mu = 1800 \text{ veh/hr}$$

$$= \frac{1800}{3600} = 0.5 \text{ veh/sec}$$

$$\rho = \left(\frac{\lambda}{\mu}\right) = \left(\frac{0.167}{0.5}\right) = 0.334$$

$C_o = 98$ sec; $r = 62$ sec; $g = 36$ sec

$$Cap = s \times \left(\frac{g}{C_o}\right) = 1800 \times \left(\frac{36}{98}\right) = 661 \text{ veh/hr}$$

As arrival volume (600 veh.hr) is less than the capacity (661/ veh/hr), the assumption of D/D/1 model is satisfied.

Using Eq. (14.6) Queue service time $= g_s = \dfrac{\rho \times r}{(1-\rho)} = \dfrac{0.334 \times 62}{(1-0.334)} = 31.09$ sec

Alternatively, queue service time can be calculated as, $g_s = \dfrac{\lambda \times r}{(\mu - \lambda)} = \dfrac{600 \times 62}{(1800-600)} = 31$ sec

Using Eq. (14.9) Total uniform delay per cycle $= d_T = 0.5 \times r^2 \times \lambda \left(\dfrac{1}{(1-\rho)}\right)$

$$d_T = 0.5 \times 62^2 \times 0.167 \left(\frac{1}{(1-0.334)}\right) = 481.9 \text{ sec}$$

Number of vehicles in queue ($N$) $= \lambda \times C_o = 0.167 \times 98 = 16.366$

Average uniform delay per vehicle $= d_{Avg\text{-}Uniform} = \left(\dfrac{d_T}{N}\right) = \left(\dfrac{481.9}{16.366}\right) = 29.4$ sec

**Example 14.11**

A T-intersection leg (Fig. 14.17) has an evening peak-hour volume of 180 veh/hr from the westbound right-turn movement. Take the saturation flow rate as 1800 veh/hr. The signal cycle length is 90 sec and effective Green for the right-turn approach is 14 sec. Assume that each vehicle has a length of 7.5 m including the space gap between vehicles while in queue. At a Red signal indication, calculate the maximum number of vehicles in queue from the STOP line to the back of queue until the queue clears. Also determine whether the right-turn storage bay of length 40 m is sufficient to accommodate such spillback from the queue.

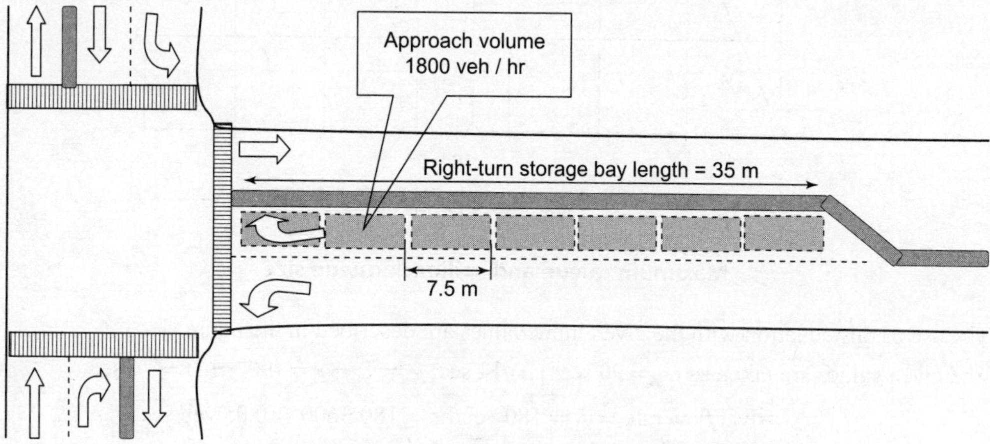


Fig. 14.17 Intersection layout with approach volume

**Answer:** The spillback queue size can be determined by computing the product of the arrival rate ($\lambda$) and the sum of effective Red time ($r$) and the queue service time ($g_s$), as given below (Kyte and Tribelhorn, 2014):

Spillback queue size $$Q_{Spillback\text{-}queue} = \lambda \times (r + g_s) \quad (14.29)$$

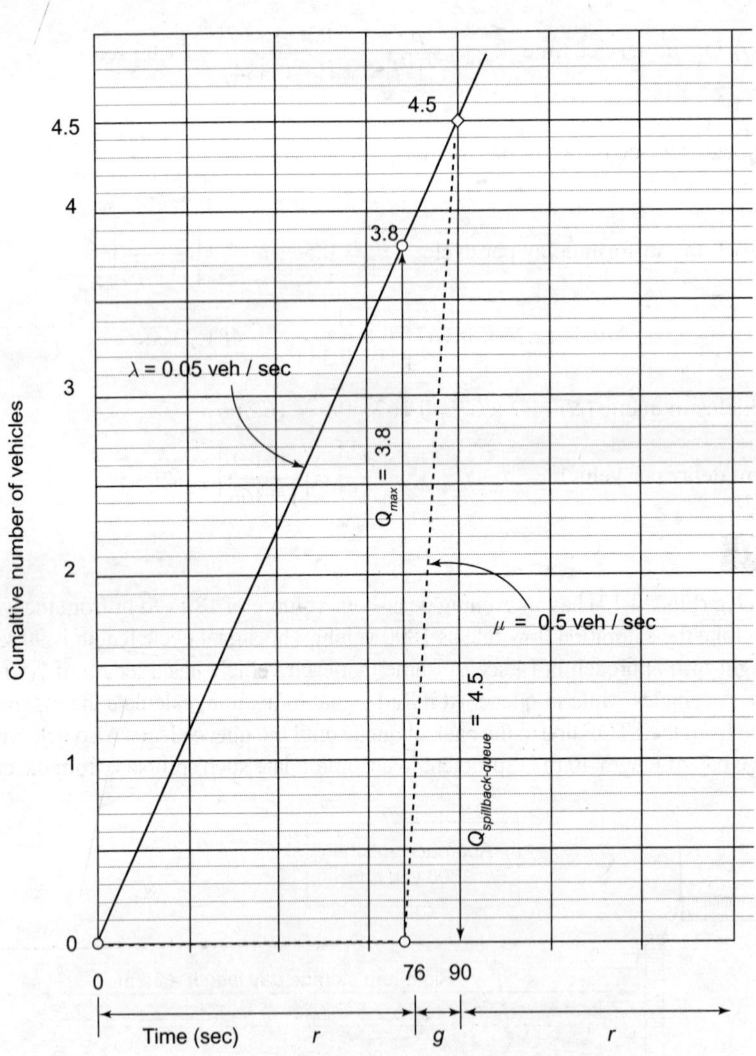


**Maximum queue and spillback queue size**

The use of this equation with the given input values are described in the figure.

The given values are taken as $C_o = 90$ sec; $r = 14$ sec; $g = C_o - r = (90 - 14) = 76$ sec

$$\text{Arrival flow rate} = \lambda = 180 \text{ veh/hr} = 180/3600 = 0.05 \text{ veh/sec}$$

$$\text{Departure flow rate} = \mu = 1800 \text{ veh/hr} = 1800/3600 = 0.5 \text{ veh/sec}$$

$$\text{Queue service time} = g_s = \frac{\lambda \times r}{(\mu - \lambda)} = \frac{0.05 \times 76}{(0.5 - 0.05)} = 8.44 \text{ sec}$$

Using Eq. (14.29), spillback queue size $Q_{Spillback\text{-}queue} = \lambda \times (r + g_s) = 0.05 \times (76 + 8.44) = 4.22$ vehicles

Let us round the value as $Q_{Spillback\text{-}queue} = 5$ vehicles

The distance upstream from the stop line to the position of the farthest vehicle stopped:

$$= 5 \text{ veh} \times 7.5 \text{ m} = 37.5 \text{ m}$$

The distance occupied by the queue is 37.5 m upstream of the stop-line against the available storage lane length of 35 m. Hence, the available storage-lane length provided for the right turn in evening peak hour is not sufficient for accommodating the spillback queue size, and this may cause capacity problems and pose safety issues. The traffic engineer may take appropriate action to increase the storage length to a minimum value of 40 m.

**Example 14.12**

Determine the percentage of stopped vehicles for the following traffic flow conditions of an approach to a signalised intersection:

Cycle length = 100 sec; Saturation flow rate = 1800 veh/hr; Green time duration = 25 sec; Arrival flow rate = 900 veh/hr.

**Answer:** The formula is used to determine the percentage of stopped vehicles during a cycle

$$= \left( \frac{Q_{\text{Spillback-queue}}}{\text{Arrived vehicles during the cycle}} \right) \times 100$$

$$= \left( \frac{\lambda \times (r + gs)}{\lambda \times C_o} \right) \times 100$$

The given values are assigned and calculated as follows:

$C_o = 100$ sec; $\lambda = 600$ veh/hr $= 900/3600 = 0.25$ veh/sec; $\mu = 1800$ veh/hr $= 1800/3600 = 0.5$ veh/sec

$$\rho = \left( \frac{\lambda}{\mu} \right) = \left( \frac{0.25}{0.5} \right) = 0.5 < 1.0$$

Hence, the traffic flow can be modelled as D/D/1.

$$g = 25 \text{ sec}$$

$$r = C_o - g = 100 - 25 = 75 \text{ sec}$$

$$\text{Queue service time} = g_s = \frac{\lambda \times r}{(\mu - \lambda)} = \frac{0.25 \times 75}{(0.5 - 0.25)} = 75 \text{ sec}$$

Spillback queue size

$$Q_{Spillback\text{-}queue} = \lambda \times (r + g_s)$$

$$= 0.25 \times (75 + 75) = 37.5 \text{ vehicles}$$

$$\text{Percentage of stopped vehicles} = \left(\frac{\lambda \times (r + g_s)}{\lambda \times C_o}\right) \times 100 = \left(\frac{37.5}{0.25 \times 100}\right) \times 100 = 150\%$$

**Example 14.13**

A signalised intersection has a cycle length of 100 sec, with the effective Green and Red times as 40 and 60 sec, respectively. Field observations were made and it was found that the vehicle arrival rate varied over three successive cycles as 990 veh/hr, 666 veh/hr and 450 veh/hr, respectively. Determine the average delay for the approach over the three cycles. Take the saturation flow rate as 1800 veh/hr.

**Answer:** The given input values are assigned to the standard notations.

Cycle length $= C_o = 100$ sec; Red $= r = 60$ sec; Green $= g = 40$ sec

Arrival flow rate during the 1st cycle $= \lambda 1 = 990$ veh/hr $= 990/3600 = 0.275$ veh/sec

Arrival flow rate during the 2nd cycle $= \lambda 2 = 666$ veh/hr $= 666/3600 = 0.185$ veh/sec

Arrival flow rate during the 3rd cycle $= \lambda 3 = 450$ veh/hr $= 450/3600 = 0.125$ veh/sec

Saturation flow rate $= s = \mu = 1800$ veh/hr $= 1800/3600 = 0.5$ veh/sec

$$\text{Capacity of the approach intersection } Cap = s \times \left(\frac{g}{C_o}\right) = 1800 \times \left(\frac{40}{100}\right) = 720 \text{ veh/hr}$$

Let us convert the flow rates from veh/hr to veh/sec, and check whether the volume exceeds capacity.

| *Cycle* | *Arrival flow rate (veh/hr)* | *Whether arrival volume exceeds capacity?* | *Remark* |
|---|---|---|---|
| 1st | 990 | 990 > 720 | Exceeded |
| 2nd | 666 | 666 < 720 | Not exceeded |
| 3rd | 450 | 450 < 720 | Not exceeded |

As the arrival flow rate exceeds the capacity, the assumption of the D/D/1 model is not applicable and the above equations derived for the determination of delay cannot be used.

Hence, a graphical approach is adopted by drawing a cumulative vehicle flow diagram.

Using Eq. (14.6), the queue service time for the 1st cycle

$$= g_{s1} = \frac{\rho \times r}{(1-\rho)} = \frac{(0.275/0.5) \times 60}{\left(1 - \left(\frac{0.275}{0.5}\right)\right)} = 73.33 \text{ sec}$$

As the queue service time of the 1st cycle (73.33 sec) > Green (40 sec), it is again confirmed that the arrival volume exceeded the departure and a backlog queue will be formed in the 2nd cycle.

The cumulative vehicle flow diagram and a few main sample calculations are presented below.

| Cycle | Time t | Cumulative number of arriving vehicles | Cumulative number of departing vehicles | Number of vehicles in the queue at t |
|---|---|---|---|---|
| [1] | [2] | [3] | [4] | [3]–[4] |
| 1st | 0 | 0 | 0 | 0 |
| | 60 | 0.275 veh/sec × 60 sec = 16.5 | 0 | 16.5 |
| | 100 | 0.275 veh/sec × 100 sec =27.5 | 0.5veh/sec × 40 sec = 20 | 7.5 |
| 2nd | 160 | 27.5 + (0.185 veh/sec × 60 sec) = 38.6 | 20 | 18.6 |
| | 200 | 27.5 + (0.185 veh/sec × 100 sec) = 46 | 20 + 0.5 veh/sec × 40 sec = 40 | 6 |
| 3rd | 260 | 46 + (0.125 veh/sec × 60 sec) = 53.5 | 40 | 13.5 |
| | 300 | 46 + (0.125 veh/sec × 100 sec) = 58.5 | 40+ 0.5 veh/sec × 40 sec = 60 | −1.5 |

The backlog queue remained after first Green and the second Green but cleared before the third Green was completed. The queue service time during the third Green period can be calculated as:

$$g_{s3} = \frac{\text{Backlog queue} + \lambda \times r}{(\mu - \lambda 3)} = \frac{6 + 0.125 \times 60}{(0.5 - 0.125)} = 36 \text{ sec}$$

The value of $gs_3$ (36 sec) < 3rd Green (40). It indicates that the queue cleared 36 sec after the beginning of the third Green.

Cumulative number of vehicles at which arrival and departure curves meet:

On arrival flow curve = 46 veh+ (0.125 veh.sec × (60+36) sec) = 58

On departure flow curve = 40 veh+ (0.5 veh.sec × 36 sec) = 58

The total delay can be calculated by summing the area between the arrival and departure flow curves. The details of area calculation are as follows:

| Cycle | Shape | Calculation details | Area (veh-sec) |
|---|---|---|---|
| 1st | Triangle ABC | ½ × 60 × 16.5 | 495 |
| | Polygon BCDE | $\frac{16.5+7.5}{2} \times 40$ | 480 |
| 2nd | Polygon EDGF | $\frac{7.5+18.6}{2} \times 60$ | 783 |
| | Polygon FGHI | $\frac{18.6+6.0}{2} \times 40$ | 492 |
| 3rd | Polygon IHJK | $\frac{6.0+13.5}{2} \times 60$ | 585 |
| | Triangle KJL | ½ × 36 × 13.5 | 243 |
| | Total Area = | | 3078 veh-sec |

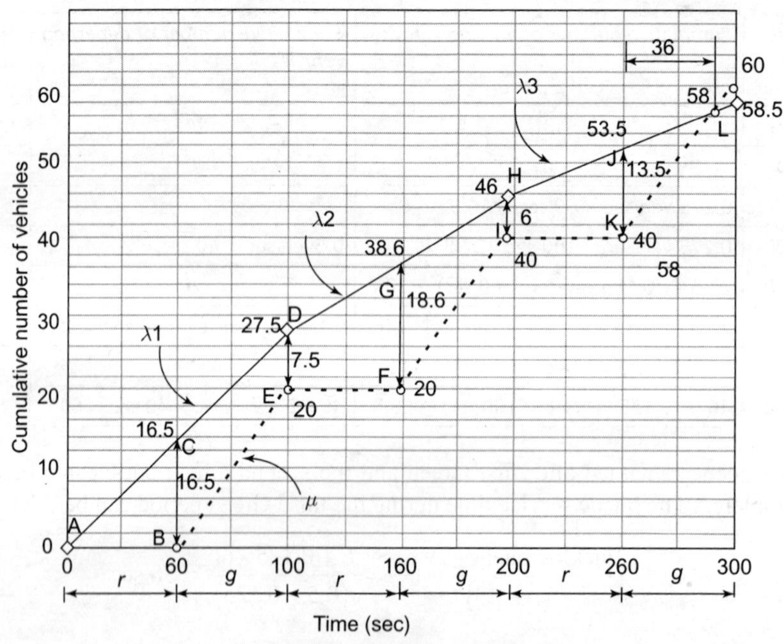


**Cumulative vehicle flow diagram**

The total delay $d_T = 3078$ veh-sec.

No. of vehicles that arrived during the three cycles are as follows:

| *Cycle* | *Calculation details: λ × Co* | *No. of vehicles* |
|---|---|---|
| 1st | 0.275 veh/sec × 100sec | 27.5 |
| 2nd | 0.185 veh/sec × 100sec | 18.5 |
| 3rd | 0.125 veh/sec × 100sec | 12.5 |
| Total number of vehicles during the 3 cycles = | | 58.5 |

$$\text{The average delay}, d_{Avg\text{-}Uniform} = \frac{dT}{\text{No. of vehicles arrived}} = \frac{3078 \text{ veh-sec}}{58.5 \text{ veh}} = 52.6 \text{ sec.}$$

**Example 14.14**

Determine the uniform average delay and overflow delay experienced by vehicles for the following traffic condition at an isolated signalised intersection.

Cycle length = 96 sec; Approach saturation flow rate = 2400 veh/hr; Green time duration = 36 sec; Flow rate = 800 veh/hr.

Also determine the (i) overall delay by Webster's model and Allsop's model and (ii) duration of the Green and Red times necessary to achieve an average delay per vehicle of 26 seconds.

**Answer:** The given values are taken as follows: $C_o = 96$ sec; g = 36 sec.

$$r = C_o - g = (96 - 36) = 60 \text{ sec}$$

$$\text{Arrival flow rate} = \lambda = 800 \text{ veh/hr} = 800/3600 = 0.22 \text{ veh/sec}$$

$$\text{Departure flow rate} = \mu = 2400 \text{ veh/hr} = 2400/3600 = 0.66 \text{ veh/sec}$$

$$\rho = \left(\frac{\lambda}{\mu}\right) = \left(\frac{0.22}{0.66}\right) = 0.33$$

Using Eq. (14.10), $\text{Uniform average delay} = d_{Avg\text{-}Uniform} = \dfrac{0.5\times r^2}{C_o(1-\rho)}$

$$= \frac{0.5\times 60^2}{96\times(1-0.33)} = 27.9 \textit{ sec}$$

Using Eq. (14.16), $X = \rho\times\left(\dfrac{C_o}{g}\right) = \dfrac{0.33\times 96}{36} = 0.88$

Using Eq. (14.15), $\text{Random delay} = d_{Random} = \dfrac{0.5\times X^2}{(1-X)}$

$$= \frac{0.5\times X^2}{\lambda(1-X)} = \frac{0.5\times 0.88^2}{0.22\times(1-0.88)} \approx 14.67 \text{ sec}$$

(i) The overall delay per vehicle by Webster's model

$$AF = 0.65 \times\left(\frac{C_o}{\lambda^2}\right)^{\frac{1}{3}} \times X^{\left(2+\frac{5\times g}{C_o}\right)}$$

$$= 0.65 \times\left(\frac{96}{0.22^2}\right)^{\frac{1}{3}} \times 0.88^{\left(2+\frac{5\times 36}{96}\right)} = 4.97 \text{ sec}$$

Using Eq. (14.19), $\text{Overall delay} = (d_{Avg\text{-}Uniform} + d_{Random}) - AF$

$$= (27.9 + 14.67) - 4.97 = 37.6 \text{ sec}$$

Overall delay per vehicle by Allsop's formula: Using Eq. (14.20), Overall delay $= 0.9 \times (d_{Avg\text{-}Uniform} + d_{Random})$

$$= 0.9 \times (27.9 + 14.67) \approx 38.31 \text{ sec}$$

(ii) Duration of the Green and Red times to achieve an average delay per vehicle of 26 seconds. Using Allsop's formula: Overall delay $= 0.9 \times (d_{Avg} + RD)$

$$26 \text{ sec} = 0.9 \times \left(\frac{0.5\times r^2}{C_o(1-\rho)} + \frac{0.5\times X^2}{\lambda(1-X)}\right)$$

$$= 0.9 \times \left( \frac{0.5 \times r^2}{96 \times (1 - 0.33)} + \frac{0.5 \times 0.88^2}{0.22 \times (1 - 0.88)} \right)$$

$$\frac{26}{0.9} = \left( \frac{0.5 \times r^2}{64.32} + 14.66 \right) \quad 28.88 = \left( \frac{0.5 \times r^2}{64.32} + 14.66 \right)$$

$$= \frac{0.5 \times r^2}{64.32} = 14.22$$

$r^2 = 1829.26$; $r = 42.76$ sec

$$g = C_o - r = 96 - 42.76 = 53.24 \text{ sec}$$

Therefore, 53.24 sec of Green time is required to achieve a delay of 25 seconds per vehicle.

**Example 14.15**

An intersection approach road has an arrival flow rate of 1800 veh/hr with a saturation flow rate of 2500 veh/hr. Take cycle length as 100 sec and effective Green ratio for the approach as 0.58. Calculate the average delay per vehicle experienced under this condition of traffic flow.

**Answer:** The given input values are assigned to the standard notations.
$\lambda = 1800$ veh/hr; $s = 2500$ veh/hr; $C_o = 100$ sec

$$\frac{g}{C_o} = 0.58, \; g = 0.58 \times 100 = 58 \text{ sec}$$

$$r = C_o - g = (100 - 58) = 42 \text{ sec}$$

Let us first determine whether the given traffic flow belongs to the undersaturated or oversaturated condition.

$$Cap = s \times \left( \frac{g}{C_o} \right) = 2500 \times (0.58) = 1450 \text{ veh/hr}$$

$$\text{Degree of saturation} = X = \left( \frac{\text{Arrival volume}}{\text{Capacity}} \right)$$

$$= \left( \frac{\lambda}{Cap} \right) = \left( \frac{1800}{1450} \right) \approx 1.24$$

As $X$ exceeds 1.0 by 0.24, the given condition of traffic belongs to the oversaturated condition of *Case 1*, and hence, Eq. (14.24) is used to determine the total overflow delay per cycle.

$$\text{Total overflow delay (TOD)} = (d_{Avg\text{-}Uniform} \text{ for } X = 1.0) + (d_{Overflow} \text{ for } X = 1.0 \text{ to } 1.24)$$

$$\text{TOD} = 0.5 \times r + 0.5T(X - 1)$$

Assume the study period is 1 cycle and take $T = 100$ sec.

Substituting the values, we get

$$\text{TOD} = (0.5\times 42) + (0.5\times 100(1.24-1))$$

$$= (21 \text{ sec}) + (12 \text{ sec}) = 33 \text{ sec/veh}$$

$$\text{TOD per cycle} = 33 \text{ sec/veh}$$

**Example 14.16**

An intersection approach road has an arrival flow rate of 1550 veh/hr with a saturation flow rate of 2750 veh/hr. Take cycle length as 100 sec and effective Green ratio for the approach as 0.54. Calculate the average delay per vehicle experienced under this condition of traffic flow.

**Answer:** The given input values are assigned to the standard notations.
$\lambda = 1550$ veh/hr; $s = 2750$ veh/hr; $C_o = 100$ sec

$$\frac{g}{C_o} = 0.54,\; g = 0.54 \times 100 = 54 \text{ sec}$$

$$r = C_o - g = (100 - 54) = 46 \text{ sec}$$

Let us first determine whether the given traffic flow belongs to the undersaturated or oversaturated condition or is in the transition gap.

$$Cap = s\times\left(\frac{g}{C_o}\right) = 2750\times(0.54) = 1485 \text{ veh/hr}$$

$$\text{Degree of saturation} = X = \left(\frac{\text{Arrival volume}}{\text{Capacity}}\right) = \left(\frac{\lambda}{Cap}\right) = \left(\frac{1550}{1485}\right) \approx 1.043$$

As $X$ exceeds 1.0 by 0.03, the given condition of traffic belongs to the transition gap, i.e, $0.80 \le X \le 1.1$. Hence, $d_{Avg\text{-}Uniform}$ is calculated up to $X = 1.0$, and the overflow portion (between $X$ 1.0 to 1.043) of the transition gap delay is calculated by Akçelik's equation.
Using Eq. (14.24),

$$d_{Avg\text{-}Uniform} = (0.5\times r) = (0.5\times 46) = 23 \text{ sec}$$

Using Akçelik's Eq. (14.27)

$$d_{Overflow} = \frac{Cap\times T}{4}\left[(X-1)+\sqrt{(X-1)^2+\frac{12(X-X_o)}{\text{Cap}\times T}}\right]$$

where $T$ = time period of analysis (say, 1 hour) = 1 hour = 3600 sec; $X$ = volume–to-capacity ratio = 1.043; Cap = capacity (veh/hr) = 1485; $s$ = saturation flow rate (veh/sec) during Green = 2750/3600 = 0.763; $g$ = effective Green (sec) = 54 sec.

$$X_o = \left(0.67+\frac{s\times g}{600}\right) = \left(0.67+\frac{0.763\times 54}{600}\right) = 0.73867$$

$$d_{Overflow} = \frac{1485 \times 1\text{hr}}{4}\left[(1.043-1)+\sqrt{(1.043-1)^2+\frac{12(1.043-0.73867)}{1485\times 1\text{hr}}}\right]$$

$$d_{Overflow} = 40.33\,\text{sec}$$

Total delay during average uniform and overflow conditions = 23.0 + 40.33 = 63.33 sec/veh

Let us also check the value of total delay using Webster's overflow Eq. (14.24) and determine the total overflow delay per vehicle.

$$\text{Total overflow delay (TOD)} = (d_{Avg\text{-}Uniform} \text{ for X} = 1.0) + (d_{Overflow} \text{ for } X = 1.0 \text{ to } 1.043)$$

$$\text{TOD} = 0.5\times r \ + \ 0.5T(X-1)$$

Assume the study period is 1 hr and take $T = 3600$ sec.

Substituting the values, we get $\quad TOD = (0.5\times 46) + (0.5\times 3600(1.043-1))$

$$= (23\text{ sec}) + (77.4\text{ sec}) = 100.4\text{ sec/veh}$$

$$\text{TOD per cycle} = 100.4\text{ sec/veh}$$

The overflow delay estimated by the Akçelik model is 40.33 sec/veh, against Webster's overflow delay model, which is 77.4 sec. This large difference is because of inconsistencies in the modelling of overflow delay in the transition gap, when $0.80 \leq X \leq 1.1$.

**Example 14.17**

An isolated signalised intersection with a cycle length of 100 sec has a saturation flow rate of 2789 veh/hr. The effective Green ratio is 0.34, and the degree of saturation is 0.92. Assume the control delay is measured over a 15-minute interval and determine the level of service based on the HCM 2000 manual. Take the value of the initial queue delay during overflow delay caused due to backlog as zero.

**Answer:** The given values are assigned to the standard notations. $s$ = 2789 veh/hr; $C_o$ = 100 sec; $\frac{g}{C_o} = 0.34$

$$g = 0.34 \times 100 = 34\text{ sec}$$

$$Cap = s\times\left(\frac{g}{C_o}\right) = 2789\times(0.34) = 948.26\text{ veh/hr}$$

$X = 0.92$; $k = 0.5$ (assumed as pre-timed signal); $l = 1.0$ for an isolated intersection; $T = 0.25$ hr; $PF = 1.0$ (assumed as an isolated signal control with random arrival of vehicles)

Using Eq. (14.28), $\quad$ Total delay $= d = (d_1 \times PF) + d_2 + d_3$

$$d_1 = \frac{0.5\times C_o \times\left(1-\frac{g}{C_o}\right)^2}{\left[1-(\text{min. of }(1,X))\times\frac{g}{C_o}\right]}$$

$$= \frac{0.5 \times 100 \times (1 - 0.34)^2}{[1 - (\text{min. of } (1, 0.92) \times 0.34]} = 31.69 \text{ sec}$$

$$d_2 = 900 \times T \left\{ (X - 1) + \left[ (X - 1)^2 + \left( \frac{8 \times k \times l \times X}{Cap \times T} \right) \right]^{0.5} \right\}$$

$$d_2 = 900 \times 0.25 \left\{ (0.92 - 1) + \left[ (0.92 - 1)^2 + \left( \frac{8 \times 0.5 \times 1.0 \times 0.92}{948.26 \times 0.25} \right) \right]^{0.5} \right\} = 15.31 \text{ sec}$$

$$d_3 = \text{zero sec. (given)}$$

$$d = (d_1 \times \text{PF}) + d_2 + d_3 = 31.69 + 15.31 + 0.00 = 47.00$$

35 < 47 < 55: Level of Service = D.

**Example 14.18**

Determine the average queue length at the end of the Red interval from the following field data from the four approaches to an intersection. Take the average length of a vehicle as 6.0 m, and cycle length as 80 sec. Assume the D/D/1 queue model.

| *Lane #* | *Direction* | *Arrival flow (PCU/hr)* | *Effective Green (sec)* |
|---|---|---|---|
| 1 | Northbound | 810 | 40 |
| 2 | Southbound | 900 | 40 |
| 3 | Eastbound | 450 | 35 |
| 4 | Westbound | 650 | 35 |

**Answer:** The average number of vehicles at the end of the Red interval $= \lambda \times (C_o - g)$

$$= \lambda_i \times (80 - g_i)$$

| *Lane #* | *Arrival flow (PCU/sec)* $\lambda_i$ | *Effective Green (sec)* $g_i$ | *Average number of vehicles in queue at the end of Red (PCU)* $\lambda_i(80 - g_i)$ | *Average length of queue at the end of Red (m)* |
|---|---|---|---|---|
| 1 | 810/3600 = 0.225 | 40 | 0.225 × (80 − 40) = 9 | 9 × 6 = 54 |
| 2 | 900/3600 = 0.25 | 40 | 0.25 × (80 − 40) = 10 | 10 × 6 = 60 |
| 3 | 450/3600 = 0.125 | 35 | 0.125 × (80 − 35) = 5.625 | 5.625 × 6 = 33.75 |
| 4 | 650/3600 = 0.180 | 35 | 0.180 × (80 − 35) = 8.1 | 8.1 × 6 = 48.6 |

**Example 14.19**

Determine the level of service for four approaches and an intersection using the following summary of field observations and delay computations:

| *Approach road* | *Volume (veh/hr)* | *Average control delay (sec/veh)* |
|---|---|---|
| Northbound | 845 | 55 |
| Southbound | 750 | 60 |
| Eastbound | 850 | 65 |
| Westbound | 490 | 18 |

**Answer:** Based on the given delay value, the LOS of the individual approaches is presented below:

| *Approach road* | *Volume (veh/hr)* | *Volume (veh/sec)* | *Delay (sec)* | *LOS as per HCM* |
|---|---|---|---|---|
| Northbound | 845 | 845/3600 = 0.234 | 55 | D |
| Southbound | 750 | 750/3600 = 0.208 | 29 | C |
| Eastbound | 850 | 850/3600 = 0.236 | 65 | E |
| Westbound | 490 | 490/3600 = 0.136 | 18 | B |

The LOS of the intersection as a whole is determined below:

| *Approach road* | *Delay (sec), $d_i$* | *Volume (veh/sec)* $\lambda_i$ | *Vehicles in queue* $d_i \times \lambda_i$ |
|---|---|---|---|
| Northbound | 55 | 0.234 | 55 × 0.234 = 12.87 |
| Southbound | 29 | 0.208 | 29 × 0.208 = 6.032 |
| Eastbound | 65 | 0.236 | 65 × 0.236 = 15.34 |
| Westbound | 18 | 0.136 | 18 × 0.136 = 2.448 |
| | | *Total* = 0.814 veh/sec | *Total* = 36.69 veh |

$$\text{Average delay for the intersection} = \frac{d_i \times \lambda_i}{d_i} = \frac{36.69}{0.814} = 45.07 \text{ sec.}$$

An average delay of 45.07 sec corresponds to intersection level of service = D.

It must be noted that any category beyond LOS C is unacceptable, and hence, the signal timings of the whole intersection must be reviewed and the approaches widened, if required, to increase the saturation flow rates.

**Example 14.20**

Determine the average overall intersection delay using the following summary of field observations:

| Lane # | Direction | Phase | Arrival flow (PCU/hr) | Average overall delay (sec/PCU) |
|---|---|---|---|---|
| 1 | Northbound | 1 | 845 | 25.4 |
| 2 | Southbound | 1 | 750 | 20.2 |

(Continued)

**Table** (Continued)

| Lane # | Direction | Phase | Arrival flow (PCU/hr) | Average overall delay (sec/PCU) |
|---|---|---|---|---|
| 3 | Eastbound | 2 | 850 | 30.3 |
| 4 | Westbound | 2 | 490 | 28.4 |

**Answer:** The given values are calculated as:

| Lane # | Direction | Phase | Arrival flow (PCU/hr) $\lambda_i$ | Average overall delay (sec/PCU) $d_i$ | Weighted delay $\lambda_i \times d_i$ |
|---|---|---|---|---|---|
| 1 | Northbound | 1 | 845 | 25.4 | 21463 |
| 2 | Southbound | 1 | 750 | 20.2 | 15150 |
| 3 | Eastbound | 2 | 850 | 30.3 | 25755 |
| 4 | Westbound | 2 | 490 | 28.4 | 13965 |
| | | | Total = 2935 | | Total = 76333 |

Weighted average of the average overall lane delay for all intersection lanes

$$= (\lambda_i \times d_i)/\ \lambda_i$$

$$= 76333/2935$$

$$= 26.0 \text{ sec/PCU}$$

# EXERCISES

## MULTIPLE CHOICE QUESTIONS

1. What is the effective Green time, if Green is displayed for 15 sec, the sum of the Yellow and Red clearance is 5 sec and the lost time is 4 sec?
   (a) 15 sec (b) 16 sec (c) 19sec (d) 6 sec
   (*Hint:* $g = G + (Y + R_c) - \text{Lost} = 15 + 5 - 4 = 16$ sec)
2. For one approach to a signalised intersection, the saturation flow rate is 2000 veh/hr, effective Green = 15 sec and cycle duration = 60 sec. What is the capacity of the approach?
   (a) 7500 veh/hr (b) 533.33 veh/hr (c) 540 veh/hr (d) 500 veh.hr
   (*Hint:* Capacity = Saturation flow x (*g*/*C*); Capacity = 2000 x (15/60) = 2000 x 0.25 = 500 veh/hr)
3. What is the length of the cycle time, if there are 60 cycles in one hour?
   (a) 60 sec (b) 120 sec (c) 90 sec (d) 30 sec
   (*Hint:* (3600 sec/hr)/(60 cycle/hr) = 60 sec)
4. What is the volume-to-capacity ratio of an approach road, if the volume of the intersection approach is 500 veh/hr, saturation flow rate is 2000 veh/hr and the effective Green ratio is 0.5?
   (a) 0.5 (b) 1.0 (c) 0.75 (d) 0.125

(*Hint:* $Cap = s\ (g/c) = 2000\ (0.5) = 1000$ veh/hr; Volume-to-capacity ratio, $X = v/cap = 500/1000 = 0.5$)

5. What is the time taken to dissipate a queue at a signalised intersection if the queue size at the end of the red interval is 12 vehicles, saturation flow rate is 900 vehicles/hr and vehicle arrival rate is 600 vehicles/hr?
   (a) 144 sec (b) 72 sec (c) 25sec (d) 48 sec
   (*Hint:* $t = q/(\mu - \lambda) = 12/(900 - 600) = 12/300 = 0.04$ hr $= (0.04 \times 3600)$ sec $= 144$ sec)
6. What per cent of vehicles stopped, if the northbound through volume = 600 veh/hr, cycle length = 60 sec, effective Red = 30 sec and saturation flow = 1800 veh/hr?
   (a) 70% (b) 75% (c) 77% (d) 50%
   (*Hint:* $P_{stopped} = \dfrac{r \times s}{C_o \times (s - v)} \times 100 = \dfrac{30 \times 1800}{60 \times (1800 - 600)} \times 100 = 75\%$
7. A signalised four-leg intersection has adjusted flow rates and total delays at each approach road. What is the average control delay of the intersection?

| Approach | EB | WB | NB | SB |
|---|---|---|---|---|
| Adjusted flow rate (veh/hr) | 100 | 200 | 300 | 400 |
| Average control delay by approach (Sec/veh) | 5 | 10 | 15 | 20 |

   (a) 300 sec (b) 15 sec (c) 12.5sec (d) 30 sec
   (*Hint:* Intersection delay $= d_I = \dfrac{\sum d \times \lambda}{\sum \lambda} = \dfrac{\sum[(100 \times 5) + (200 \times 10) + (300 \times 15) + (400 \times 20)]}{\sum[100 + 200 + 300 + 400]}$

$$= \frac{15000}{1000} = 15\,\text{sec}$$

8. A signalised four-leg intersection has adjusted flow rates and total delays at each approach road. What is the LOS of the intersection?

| Approach | EB | WB | NB | SB |
|---|---|---|---|---|
| Adjusted flow rate (veh/hr) | 100 | 200 | 300 | 400 |
| Average control delay by approach (Sec/veh) | 8 | 16 | 24 | 48 |

   (a) LOS-A (b) LOS-B (c) LOS-C (d) LOS-D
   (*Hint:* Intersection delay $= d_I = \dfrac{\sum d \times \lambda}{\sum \lambda} = \dfrac{30400}{1000} = 30.4$ sec ;As per HCM, $> 20$ and $\leq 35$: LOS-C)
9. What is the average waiting time of vehicles before and on a bridge if 120 cars are found travelling in front of the bridge, when the departure rate of these vehicles from the bridge is 40 cars/minute? Assume no other vehicles were travelling on the bridge during this observation.
   (a) 3 sec (b) 60 sec (c) 20 minutes (d) 180 sec
   (*Hint:* By Little's law, $L = \lambda W$, $W = L/\lambda = 120$ cars/(40 car/minute) = 3 minutes = 180 sec)

10. The queuing system in which the arrivals of the vehicles are assumed as negative exponentially distributed and the departure times as deterministic is
(a) D/D/1 (b) M/D/1 (c) M/M/1 (d) D/M/1
11. The queuing system in which the arrival and service times of vehicles are assumed as negative exponentially distributed is
(a) D/D/1 (b) M/D/1 (c) M/M/1 (d) D/M/1
12. What is the probability of no person in the queue system when the arrival rate is 120 persons/hr and departure rate is 15 sec/person. Assume the M/D/1 queue system.
(a) 0.375 (b) 0.5 (c) 0.75 (d) zero
(*Hint:* $\lambda = 120$ persons per 60 minutes; $\lambda = 120/60 = 2.0$ persons/minute; $\mu = 15$ sec per 1 person; $\mu = (60s/15s) \times 1 = 4$ persons/ minute. Traffic intensity or utilisation coefficient: $\rho = (\lambda/\mu) = 2.0/4 = 0.5$; The probability of no persons in the system: $P_o = 1 - \rho = 1 - 0.5 = 0.5$)
13. What is the average number of persons in a queue when their utilisation coefficient is 0.9. Assume the M/D/1 queue system.
(a) 4.05 (b) 90.00 (c) 1.11 (d) 0.10
(*Hint:* Utilisation coefficient: $\rho = 0.9$; Average number of persons in queue $= L_q = \dfrac{\rho^2}{2(1-\rho)} = \dfrac{0.9^2}{2(1-0.9)} = 4.05$ persons)
14. What is the average number of persons in the system when their utilisation coefficient is 0.5. Assume the M/M/1 queue system.
(a) 1.00 (b) 5.00 (c) 1.11 (d) 0.10
(*Hint:* Utilisation coefficient: $\rho = 0.5$. The average number of persons in the system: $L = \dfrac{\rho}{1-\rho} = \dfrac{0.5}{1-0.5} = 1$ person)
15. What is the average number of vehicles in the system if the average number of vehicles in the queue is 3 vehicles and the utilisation coefficient is 4.0. Assume the M/M/N model of queuing.
(a) 1.33 (b) 0.75 (c) 1 (d) 7
(*Hint:* The average number of vehicles in the system: $L = L_q + \rho = 3 + 4 = 7$ vehicles)
16. Overflow delay is directly proportional to the length of time T during which the arrival flow ________ the capacity.
(a) is equal to (b) is less than (c) exceeds (d) does not exceed
17. The range of degree of saturation for which the deterministic and steady-state delay models fail is
(a) 0–1.0 (b) 0.5–1.5 (c) 1.0–1.1 (d) 0.8–1.1
18. What is the value of total uniform delay if the cycle length = 96 sec, effective Green = 36 sec, arrival rate of vehicles = 800 veh/hr and the departure rate = 2400 veh/hr from an approach leg of an isolated signalised intersection?
(a) 300 sec (b) 689 sec (c) 473 sec (d) 591 sec

(*Hint:* By Eq. (14.9), uniform total delay per cycle $= d_T = 0.5 \times r^2 \times \lambda \left(\dfrac{1}{(1-\rho)}\right)$

$C_o = 96$ sec; $g = 36$ sec; $r = C_o - g = (96 - 36) = 60$ sec; Arrival flow rate $= \lambda = 800$ veh/hr $=$ $800/3600 = 0.22$ veh/sec; Departure flow rate $= \mu = 2400$ veh/hr $= 2400/3600 = 0.66$ veh/sec; $\rho = \left(\frac{\lambda}{\mu}\right) = \left(\frac{0.22}{0.66}\right) = 0.33$; $d_T = 0.5 \times 60^2 \times 0.22\left(\frac{1}{(1-0.33)}\right) = 591.0$ sec) )

19. What is the queue service time at an isolated signalised intersection when the utilisation ratio = 0.5 and the Red interval = 60 sec?
(a) 120 sec (b) 30 sec (c) 60 sec (d) 59.5 sec

(*Hint:* By Eq. (14.6), the queue service time $= g_s = \frac{\rho \times r}{(1-\rho)} = \frac{0.5 \times 60}{(1-0.5)} = 60$ sec

20. At a Red signal indication, what is the maximum number of vehicles in queue from the stop-line to the back of queue until the queue clears, when the arrival flow rate = 180 veh/hr, Red interval = 60 sec and queue service time = 8.00 sec?
(a) 24 sec (b) 3.4 sec (c) 2.6 sec (d) 0.13 sec
(*Hint:* Arrival flow rate $= \lambda = 180$ veh/hr $= 180/3600 = 0.05$ veh/sec; $r = 60$ sec; $g_s = 8$ sec. By Eq. (14.29), Spillback queue size $Q_{Spillback\text{-}queue} = \lambda \times (r + g_s) = 0.05\ (60+8) = 3.4$ sec)

21. What is the value of random delay by Webster's formula, when the arrival rate of traffic flow at an approach to an isolated signal controlled intersection = 900 veh/hr, Red interval = 50 sec and degree of saturation = 0.8?
(a) 14.4.0 sec (b) 25.0 sec (c) 10.0sec (d) 6.4 sec
(*Hint:* Arrival flow rate $= \lambda = 900$ veh/hr $= 900/3600 = 0.25$ veh/sec; By Eq. (14.15), Random delay $= d_{Random} = \frac{0.5 \times X^2}{\lambda(1-X)}; = \frac{0.5 \times 0.8^2}{0.25 \times (1-0.8)} = 6.4$ sec)

22. What is the value of average uniform delay by Webster's formula, when the arrival rate of traffic flow at an approach to an isolated signal controlled intersection = 1800 veh/hr, Red interval = 60 sec, cycle length = 96 sec, coefficient of traffic intensity 0.33 and degree of saturation = 1.24?
(a) 30 sec (b) 27.9 sec (c) 14.67 sec (d) 34.92 sec

(*Hint:* When $X > 1.0$, $d_{Avg\text{-}Uniform} = 0.5 \times 60 = 30\,\text{sec}$ )

23. What is the total overflow delay per cycle by Webster's formula, when the arrival rate of traffic flow at an approach to an isolated signal controlled intersection = 1800 veh/hr, Red interval = 42 sec, cycle length = 100 sec, coefficient of traffic intensity 0.33 and degree of saturation = 1.24?
(a) 21.0 sec (b) 12.0 sec (c) 40.92 sec (d) 33.0 sec
(*Hint:* Take $T$ = Cycle length = 100 sec; $TOD = 0.5 \times r + 0.5T(X-1) = 0.5 \times 42 + 0.5 \times 100(1.24-1) = 33$ sec/veh/cycle)

24. A one-way road is provided with a maximum departure rate of 0.6 veh/sec. The arrival rate of vehicles on this road was observed as 0.4 veh/sec. The road was blocked for 30 minutes due to an accident. After the 30 minutes, the road was reopened to full flow of traffic. What is the time required for the queue to dissipate?
(a) 0.2 hour (b) 0.5 hour (c) 1.0 hour (d) 2hours

(*Hint:* Let T = Time required to dissipate the queue (sec); (Total No. of Veh accumulation during 30 min.) + (Arrival rate x $T$) = (Departure rate × $T$) = (0.4 veh/sec × (30 × 60)sec) + (0.4 × $T$) = 0.6 × $T$. By solving $T$ = 3600 sec = 1 hour)

25. A two-lane one-way road has a directional capacity of 2800 veh/hr for each lane. The vehicle arrival rate is 1800 veh/hr. An accident occurred at a point on the road and one of the lanes was blocked for 30 minutes. After 30 minutes, the wreckage was cleared and full flow of traffic was allowed on these two lanes. Assume that the road blockage had no effect on the traffic flowing on the undisturbed lane. How long does it take to dissipate this queue?

    (a) 60 minutes (b) 30 minutes (c) 20 minutes (d) 40 minutes

    (*Hint:*Let $T$ = Time takes to dissipate the queue in hours. (1800 veh/hr × 0.5 hr) + (1800 veh/hr × $T$) – (2800 veh/hr × 0.5 hr) – (2800 veh/hr × $T$) = 0. By solving, $T$ = 0.5 hr = 30 minutes)

## ANSWERS

1. (b) 2. (d) 3. (c) 4. (a) 5. (a) 6. (d) 7. (b) 8. (c) 9. (d) 10. (b)
11. (c) 12. (b) 13. (a) 14. (a) 15. (d) 16. (c) 17. (d) 18. (d) 19. (c) 20. (b)
21. (d) 22. (a) 23. (d) 24. (c) 25. (b)

## QUESTIONS

1. State the causes of queue formation with reference to traffic engineering.
2. Describe the concept of queue formation at an uninterrupted transportation facility with reference to the arrival and departure rates of flow and the cumulative number of vehicle/person data available. Also state the parameters that can be derived from such a plot.
3. State Little's theorem and write its application.
4. What is meant by service discipline? Explain the details of process options based on the scheduling policies and their categorisation.
5. State the characteristics that influence the performance of any queuing system.
6. State any four methods of commonly adopted queuing systems and explain their fundamental characteristics.
7. At a service centre, customers arrive at the rate of 20 an hour and are served at the rate of 25 per hour. Find the average length and average waiting time in the system by assuming Poisson distribution; the service is exponentially distributed.
8. Write short notes on the following queue systems with their basic assumptions of analysis:

    (a) D/D/1 queuing system; (b) M/D/1 queuing system; (c) M/M/1 queuing system; (d) M/M/N queuing system
9. Compare the M/M/1 and M/D/1 queuing models.
10. Vehicles arrive at a fuel-filling station having one serving pump unit at an average of 15 vehicles per hour. The service rate is 3 minutes for each. Assume the arrival time and the service time are distributed as Poisson and exponential distributions, respectively. Determine, (a) average number of vehicles in the system, (b) average waiting time in the queue, (c) average queue length and (d) the probability that the number of vehicles in the system is 3.

11. Pedestrians arrive at an entrance to a park that has only one entrance. The office gate opens at 10:30 AM, at which time, people begin to arrive at the rate of 27 persons/ minute. After 45 minutes, the arrival flow rate decreases to an average of 10 persons/minute, and it continues for the remaining working hours of the day. If the time required to serve each person is 6 seconds, describe the operational characteristics of the queue by assuming the D/D/1 model.
12. A National Highway (NH) has a capacity of 3600 veh/hr and operates at a constant average flow rate of 2200 veh/hr. An accident occurred at 4:15 AM and the road was completely blocked. At 4:56 AM, the road was partially opened with a capacity of 1000 veh/hr. Finally, all the wreckage was cleared by the traffic police and traffic flow was restored to its full capacity at 5:20 AM. Determine the time taken to dissipate the queue, the longest queue length, the longest waiting time, total delay and average delay per vehicle by assuming the D/D/1 model.
13. Students arrive at an entrance to a store. The door opens at 10:00 AM and remains open till 1:00 PM. The visitors begin arriving at a rate of 90 persons/hr. Take the time required to serve each student as 16 seconds. Compute the probability of no persons in the system, average number of persons in the queue, number of vehicles in the system, waiting time in the queue and the time spent in the system. Assume the M/D/1 queue model.
14. Calculate the net savings in waiting time between toll collecting booths with two and three toll booths which are proposed to be installed at an entry point of a bridge. The average arrival rate of vehicles is 1500 veh/hr and the service rate is assumed as 10 seconds each (to pay the toll). Assume that arrivals and departures are exponentially distributed.
15. A mobile repair centre ticket booth has an average arrival rate of 3 persons per minute and service is provided at a rate of 1 person per minute. Calculate the following characteristics of queuing by the M/M/1 model:

    (i) Probability of no person in the system; (ii) Average number of persons in the system; (iii) Average staying time of persons in the system; (iv) Average waiting time of persons in the queue; (v) Average number of persons in the queue
16. Three toll collecting booths were installed at an entry point of an NH. The average arrival of vehicles observed was 1600 veh/hr and their service rate assumed was 12 seconds. Assume both the arrivals and the departures are exponentially distributed and calculate (i) the probability of no vehicle in the system, (ii) average queue length, (ii) the time spent in the system and (iii) the probability of waiting in a queue.
17. Describe how uniform delay and queue pattern of vehicles are determined using a cumulative vehicle diagram.
18. Define the types of delays that occur at isolated signalised intersections with a typical time–distance diagram.
19. What are the methods of delay analysis at isolated signalised intersections? Explain them briefly.
20. Determine the percentage of stopped vehicles for the following traffic flow condition of an approach to a signalised intersection:

    Cycle length = 90 sec; Saturation flow rate = 30 veh/minute; Red time duration = 72 sec; Arrival flow rate = 12 veh/minute
21. A T-intersection is signalised and operated with two phases, as indicated below. The cycle length at this intersection is 60 sec. The Major street has a saturation flow of 2400 veh/hr, effective Green time of 40 sec and flow rate of 700 veh/hr. The Minor street has a saturation flow of 1800 veh/hr, effective Green time of 16 sec and flow rate of 480 veh/hr. Assume the D/D/1 model of queuing

describes the traffic flow operations at this intersection. Calculate (i) the average delay per vehicle for the Major and Minor streets and (ii) allocate effective Red and Green time among these two legs in such a way as to minimise the total delay of this intersection.

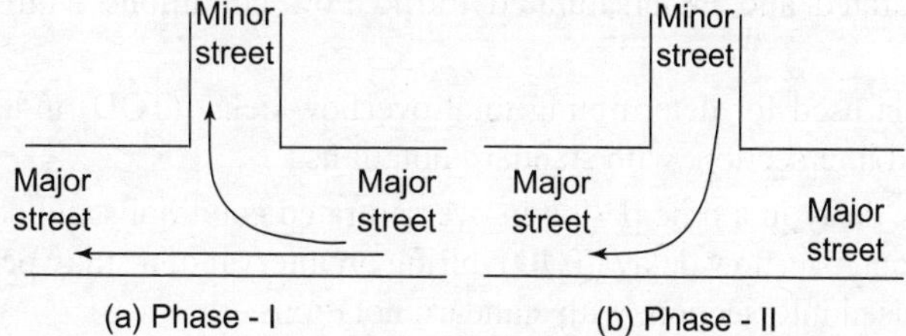


(a) Phase - I (b) Phase - II

(*Hint:* Total delay at the intersection will be minimum when, $d(TD)/d_{r1} = d(d_{T1} + d_{T2})/d_{r1} = 0$

$$d_{T1} = 0.5 \times r_1^2 \times \lambda_1\left(\frac{1}{(1-\rho_1)}\right);\ d_{T2} = 0.5 \times r_2^2 \times \lambda_2\left(\frac{1}{(1-\rho_2)}\right);= 0.5 \times (C_o\text{-}r_1)^2 \times \lambda_2\left(\frac{1}{(1-\rho_2)}\right)$$

We know that, $r_1 + r_2 = C_o$

After differentiation, we get

$$r_1 \times \lambda_1\left(\frac{1}{(1-\rho_1)}\right) = (C_o - r_1) \times \lambda_2\left(\frac{1}{(1-\rho_2)}\right)$$

Finally, determine the $r_1$ value by substituting the respective values. After obtaining $r_1$, obtain the values of $r_2$, $g_1$ and $g_2$.)

22. What are the measures of signalised intersection performance? Explain why and how signalised performance is described by the curves related to various types of delays.
23. Calculate the capacity of an approach road for the following conditions:
Cycle length = 68 sec; Green time = 28 sec; Yellow plus all Red time = 4 sec; Startup lost time = 2 sec; Clearance lost time = 2 sec; Saturation headway = 2 sec/veh
(*Hint*: $s = 3600/2$ sec/veh = 1800 veh/hr; Effective Green = $g$ = Green time + Yellow plus all Red time – Lost time at startup and clearance. $g = 28+4 - 2 - 2 = 28$ sec; $Cap = s\ (g/C_o) = 1800\ (28/68) = 741.1$ veh/hr/lane)
24. State the assumptions made for the analysis of average uniform delay and random delay.
25. Draw a flow profile that represents the following field observations:
Arrival flow rate = 400 veh/hr; Saturation flow rate = 8000 veh/hr; Cycle length = 100 sec; Red time = 60 sec; Green time = 40 sec; Queue service time = 30 sec
26. Draw a typical sketch showing the flow of arrival and departure vehicles with their cumulative values over a time period of three cycles at a signalised intersection. Label the lines and slopes of the triangles.
27. Explain the basic concept of analysis of delay due to queuing at a signalised intersection, with a neat sketch.
28. Derive Webster's model equation used for determining the uniform average delay at an isolated signalised intersection with standard notations, and a typical sketch.
29. Describe uniform delay, random delay and overflow delay, with sketches.

30. Distinguish between random delay and overflow delay with a neat sketch.
31. Write an overview of Webster's delay models including the corrections applicable for the determination of overall delay at an isolated signalised intersection.
32. Describe oversaturated and undersaturated traffic flow conditions with arrival and departure curves.
33. Derive the formula used for determining total overflow delay (TOD) in a time period ($T$) at an isolated signalised intersection, with standard notations.
34. Describe overflow delay in a time period in oversaturated conditions and derive the formula used for determining total overflow delay (TOD) during an interval of a study period (say, $T_1$ and $T_2$) at an isolated signalised intersection, with standard notations.
35. What are the difficulties and inconsistencies found between the random delay and overflow delay models? Draw a typical plot of curves between the parameters degree of saturation and average approach delay. Also indicate the uniform delay, random delay and overflow delay on these curves.
36. Write an short note on the Akçelik delay model and explain its suitability.
37. Write an short note on the HCM delay model and explain its suitability.
38. Draw the cumulative vehicle diagram of the following field observed values on one approach to a signalised intersection:

    Vehicles arrive uniformly every 4 sec; Cycle length = 60 sec; Red time interval = 24 sec; Vehicles depart every 2 sec after the beginning of Green; Queue service time = 14 sec

    Determine the (i) delay experienced by each vehicle until the queue is cleared and (ii) the total delay for all the vehicles in the system.
39. Consider the following field traffic conditions at a signalised intersection:

    Cycle length = 96 sec; Approach saturation flow rate = 35 veh/minute; Green time duration = 36 sec; Flow rate = 10 veh/minute

    Assume the D/D/1 model of queuing system and determine total delay and average delay experienced by the vehicle.
40. An intersection approach has an arrival rate of 20 veh/minute and a saturation flow rate of 24 veh/minute. Take the cycle time as 90 sec, and the effective Red time as 60 sec. Check if the approach capacity is less than the arrival flow and determine the queue service time, total delay time per cycle and the average delay experienced by the vehicles.
41. A lane of an intersection leg has a peak-hour volume of 18 veh/minute for right turns. Take the lane saturation flow rate as 28 veh/minute. Take the signal cycle length as 96 sec and the effective Green for the right turn as 16 sec. Assume that each vehicle occupies 7.0 m of length including the space gap between vehicles in queue. At a Red signal indication, calculate the maximum number of vehicles in the queue from the stop-line to the back of queue until the queue clears. Also check whether the right-turn lane with a straight length of 50 m is sufficient to accommodate such spillback of the queue.
42. An isolated signalised intersection has a cycle length of 90 sec with the effective Green and Red times as 35 and 55 sec, respectively. The vehicles arrival rate varies over two successive cycles as 16 veh/minute and 8 veh/minute, respectively. Determine the average delay for the approach over the two cycles. Take the saturation flow rate as 28 veh/minute.

43. Consider the the following traffic conditions at an isolated signalised intersection:

    Cycle length = 98 sec; Approach saturation flow rate = 40 veh/minute; Red time duration = 62 sec; Flow rate = 14 veh/minute

    (i) Compare the uniform average delay and overflow delay experienced by the vehicles based on Webster's model and Allsop's model and (ii) determine the duration of Green and Red times necessary to achieve an average delay per vehicle of 28 seconds.
44. An intersection approach road has an arrival flow rate of 39 veh/minute and saturation flow rate of 42 veh/minute. Take cycle length as 98 sec and effective Green ratio for the approach as 0.58. Calculate the average delay per vehicle.
45. An intersection approach road has an arrival flow rate of 0.4 veh/sec with a saturation flow rate of 0.75 veh/sec. Take cycle length as 90 sec and effective Green ratio for the approach as 0.55. Calculate the average delay per vehicle.
46. An isolated signalised intersection has a cycle length of 96 sec with a saturation flow rate of 2700 veh/hr. The effective Green ratio is 0.33 and the degree of saturation is 0.96. Assume the control delay is measured over an interval of 1/4$^{th}$ hour interval and determine the level of service based on the HCM 2000 manual. Take the value of initial queue delay during overflow delay as 8 sec and assume any missing data suitably.
47. Determine the level of service for the four approaches and the intersection using the following data:

| *Approach road* | *Volume (veh/hr)* | *Average control delay (sec/veh)* |
|---|---|---|
| Northbound | 445 | 09 |
| Southbound | 790 | 65 |
| Eastbound | 750 | 60 |
| Westbound | 490 | 22 |

# CHAPTER 15 Design Guidelines for Grade-Separated Intersections and Interchanges

## 15.0 INTRODUCTION

An interchange is a system of interconnecting roadways provided with *ramps* and *bridges* that facilitate uninterrupted movement/free flow of traffic between two or more roadways of intersecting highways on different grade-separated levels.

Interchanges are replaced by at-grade intersections, particularly on expressways/freeways, national highways, state highways and also on arterial urban road intersections (Fig. 15.1), where access control is essential.

**Fig. 15.1** Scope of intersecting roadways considered for interchanges

The introduction of an interchange/grade-separated structure is a valid solution that helps avoid many of the problems experienced with at-grade intersections. For example, grade-separated intersections have the advantage of eliminating crossing conflicts and minimising turning conflicts, depending on the configuration of the particular interchange.

The design of an interchange should not be considered in isolation, but with reference to the adjacent intersections, land-use patterns and community activities in that locality, to ensure consistent and safe traffic flow that meets the needs of the community.

## 15.1 GRADE-SEPARATED STRUCTURES

Grade-separated structures (i.e., interchanges) may be used in the following situations (Koepke and Levinson, 1992):

- When two expressways cross or when an expressway crosses any class of road, such as NH, SH, MDR, VR or an urban arterial road.
- Intersection traffic volumes nearing 10,000 PCU/hr result very inefficient traffic flow, and hence, it is beneficial to provide flyovers (IRC: SP: 90-2010).
- Grade separation can be economically justified for design flows of 20,000 AADT on the mainline, depending on turning traffic. However, it could be more economical to service 12,500 AADT on through routes, in case of low traffic flow situations on minor roads (DMRB, 1994).
- Existing at-grade signalised arterial intersection roadways operating at LOS-F and with inadequate capacity.
- A significant reduction in accidents at an at-grade intersection can be realised by constructing an interchange.

- When a new at-grade signalised intersection is facing the possibility of reduction in LOS to E in urban/suburban and D in rural areas.
- At a signalised intersection, where the access point would hamper progressive flow along the roadway and there is a need to provide a direct access point to a major activity centre.
- A major public street at-grade intersection is located adjacent to/close to/passes through a major traffic generator that affects the operation of traffic flow.
- An activity centre is located along a major highway/expressway, where direct access or right turns would be prohibited/undesirable.

**Basic requirements:** The basic requirements of grade-separated intersections are as follows:

- The design speeds should be higher.
- Suitable gradient, i.e., longitudinal slope, should be provided along the interchange ramp.
- Adequate visibility along the ramp path should be provided for the required design speed.
- Adequate superelevation, turning radius and road markings/signs should be provided.
- Sudden change in lane/path should be prevented by providing dividers.
- Temporary halting areas should be provided in case of any vehicle breakdown.
- Adequate drainage facilities should be provided to drain the run-off.

**Advantages of grade-separated intersections**

- Vehicle conflict points can be eliminated, thus preventing unwanted stopping and occurrence of accidents.
- Right-turning movements can be completely avoided by providing indirect interchange ramps.
- Grade-separated intersections allow the operation of higher capacities of traffic flow.
- The alignment and layout of interchange ramps (i.e., separation angles of the ramps) can be easily accommodated/adopted as prescribed by any standard code of practice within most of the intersecting area.

**Disadvantages of grade-separated intersections**

- High cost of land acquisition (in urban areas) and cost construction of interchange ramp.
- It is very difficult to provide such facilities in unfavourable ground conditions and topography, particularly in the case of hilly areas.
- Unwanted construction of ramps with crest and sag curves may occur in the case of plain and rolling terrain.
- Some of the other challenges are (Taleb and Majumder, 2012):
  - Traffic congestion may increase on the exit ramps and surrounding areas
  - Noise and air pollution may increase in the locality
  - Displacement of people and loss of social-cohesion in the affected areas
  - Loss of economic activities in the affected areas
  - Significant damage and relocation of public and private properties
  - Environmental impact

## 15.2 FACTORS AFFECTING THE SELECTION OF AN INTERCHANGE

The following factors/criteria are considered when selecting an appropriate *configuration* for an interchange (Mulinazzi, 1973).

- Terrain and soil support conditions
- Right-of-way controls
- Crossing layout with the number of intersecting legs

- Grades on highways or cross roads
- Design speed
- Traffic and pedestrian volumes
- Directional distribution of traffic (through and turning volumes)
- Character and composition of traffic
- Capacity requirements
- Proximity of adjacent interchanges
- Present and future land use – CBD/urban/rural, and the land value
- Ramp grades
- Ramp junctions
- Arrangement of local streets
- Type of access control
- Ramp intersection volumes and capacity
- Number and length of weaving sections
- Route continuity
- Lane balance
- Sight distance and safety issues
- Accident experience
- LOS continuity through the interchange area and ramps
- Overall traffic operations on cross roads, if signals are required
- Drainage
- Applicable standards
- Stage construction
- Dislocation aspects
- Detrimental impact on community and environment
- Aesthetics
- Cost of construction, maintenance and operation

This list is open-ended; all the factors and sub-factors pertaining to a specific site should be considered.

Literature is available on numerous combinations and types of interchanges in use in various parts of the world. However, site-specific conditions are the main factors affecting the selection of an interchange type. In this context, the following remarks may also be considered as important.

- To determine the environmental impact, an assessment must be made of the anticipated significant beneficial and detrimental effects on the quality of the human-environment which depends on the configuration of the interchange.
- The guidelines provide general standards with conceptual discussion. However, the designers are free to choose site-specific and unique dimensions of the ramp/interchange components that are suitable for the locations relating to operational, signage and safety considerations.

## 15.3 CLASSIFICATION OF GRADE-SEPARATED STRUCTURES

Grade–separated structures (without ramps) are broadly classified into two types: underpass and overpass (Fig. 15.2). The choice of these structures depends on several factors such as topography, slope of intersecting grades, vertical alignment, drainage, aesthetic features, preferential aspects relating to the habits of local people/users and the cost of construction. A brief description of these structures is as follows.

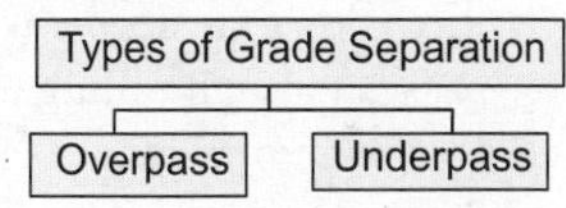


**Fig.15.2** Types of grade–separated structures

### 15.3.1 Overpass or Bridge

An overpass or flyover is an elevated road structure on which major traffic travels *over a point* of an intersecting road, river or railway line. Both the terms are *synonymous* and commonly referred to as a bridge structure in the UK and most commonwealth countries (Fig. 15.3). These structures are also constructed to facilitate crossing of pedestrians and bicyclists.

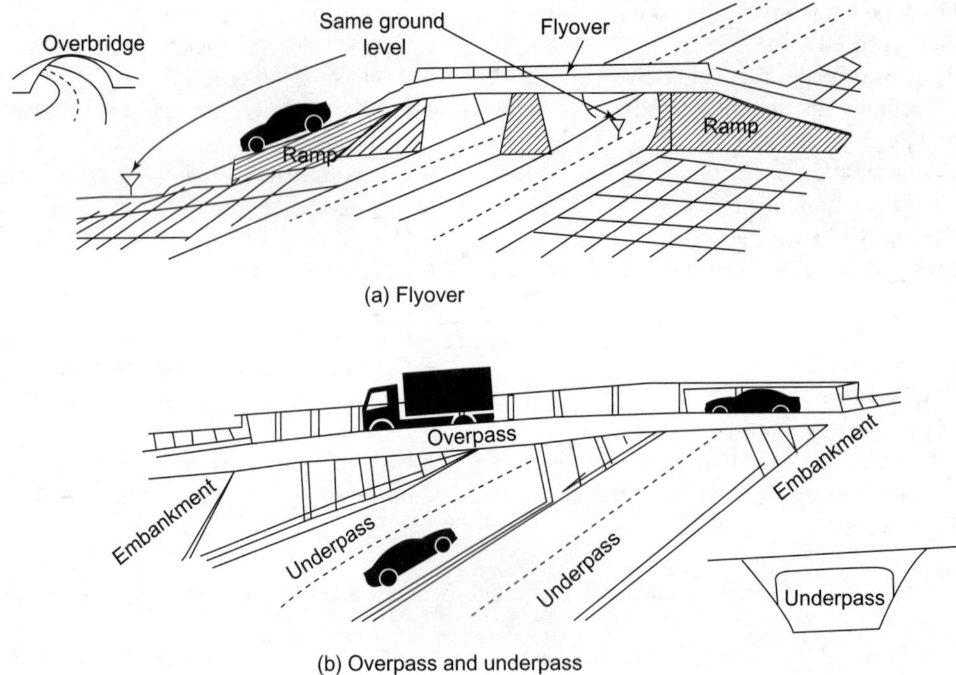


(a) Flyover

(b) Overpass and underpass

**Fig. 15.3** Synonymous meaning of flyover, overpass and underpass

**Advantages**

- Drainage-related problems can be reduced by aligning the major highway above the cross road, instead of providing it as an underpass.
- Overpass of major highways is aesthetically preferred over an underpass as it is associated with a lower sense of restriction and confinement and greater visibility.
- Future widening of existing overpasses is simpler than that of underpasses because sub-structure disturbances may occur during widening.

**Disadvantages**

- Provision of a steeper (or rolling) gradient of the major highway ramp may be required in case it passes through rolling terrain.
- Provision of a longer embankment may become necessary to attain adequate sight distance for the design speed. This may increase the total cost of construction (land cost + construction cost).
- There may be a reduction in design speed due to the steeper gradient of the interchange ramp. Heavy vehicles need more tractive energy to overcome this steeper gradient.

## 15.3.2 Underpass

An underpass is a grade crossing where the traffic flows *under* another structure at a point such as a road, river or railway line. In some cases, the underpass may be constructed as a tunnel through mountains. The underpass created to facilitate the safe crossing of pedestrians, wildlife and bicyclists under a railway line or roadway is often termed a *subway*.

**Advantages**

- Underpasses are useful when the main highway needs to be constructed on the existing grade, without any change in its vertical alignment.
- Underpasses are constructed when the crossing site conditions demand such alignment of the main highway.

**Disadvantages**

- Rainwater may stagnate during the monsoon season and there may also be a rise in ground water levels in agricultural areas nearby.
- The sight distance through the underpass may be inadequate if a sag curve passes under an overhead structure. The structure may restrict the sight distance.
- There may be a feeling of restriction to traffic at the sides while passing along the underpass and this may affect the capacity of the intersection.
- No changes can be made during the construction of underpasses when there is a paucity of funds.

## 15.4 CLASSIFICATION OF RAMPS

A ramp is a short section of a road length that connects intersecting roads, from one level to another. Ramps are classified based on their function and intended use.

### 15.4.1 Classification of Ramps based on Direction of Use

Grade separation may take place using an underpass or an overpass (Fig. 15.4). Transfer of routes from one grade to another is provided by interchange ramps. Based on the applicable type of merging and diverging routes, interchange ramps may be classified as described in Table 15.1. Further classification of this type of intersection is explained later in this chapter.

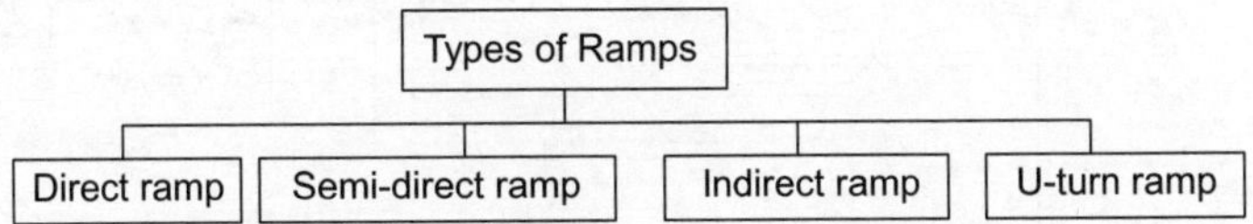


**Fig. 15.4** Classification of ramps based on directional use

**Table 15.1** Types of interchange ramps

| *Type of ramp* | *Illustration of diverging and merging ramp* | *Remarks* |
|---|---|---|
| Direct/Directional ramp (where vehicles turn right to go right and left to go left) | Ramp | Traffic flow is diverted to the right by diverging the vehicles to the right and merging them from the right. As per Indian traffic rules, this may be unsafe and hazardous because both the diverging and the merging of traffic flow takes place on the right. However, the distance to be traversed on direct interchange ramps will be less. The capacity and speed on these ramps will be high. |

(Continued)

**Table 15.1** (Continued)

| *Type of ramp* | *Illustration of diverging and merging ramp* | *Remarks* |
|---|---|---|
| Semi-direct/Semi-directional/ Jughandle ramp (where vehicles must first manoeuvre left to make a right turn and vice versa for a left turn) | 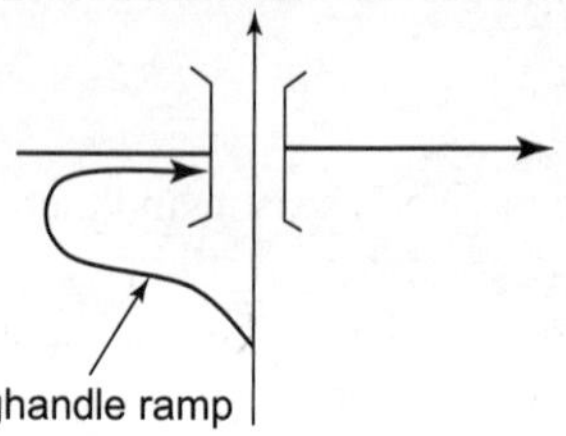 Jughandle ramp | Traffic flow is diverted to the left by diverging the vehicles to the left and merging them from the right. This is a ramp provided on the left-hand side of the road, and used for making right turns. |
| Indirect/Non-directional ramp | 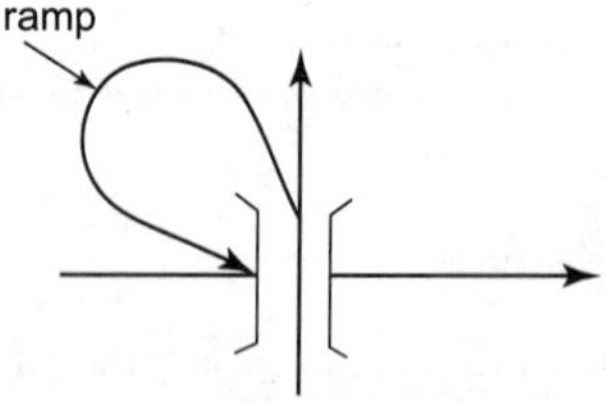 Loop ramp | Traffic is diverted to the left by diverging the vehicles to the left and merging them from the left. As per Indian traffic flow rules, this is considered safe and less hazardous because the diverging and merging of traffic flow takes place on the left. However, the distance to be traversed will be greater. |
| U-turn ramp (where a vehicle must drive on a 360° curve to make a U turn) | 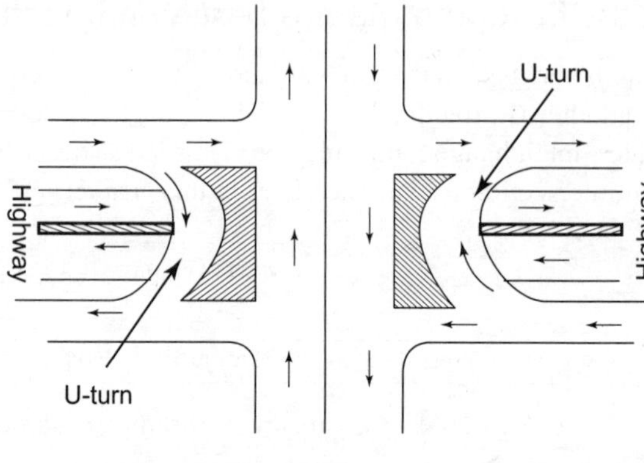 U-turn<br>Highway<br>Highway<br>U-turn<br>This ramp leaves the road in one direction and turns under or over it and re-joins it in the opposite direction (known as Texas U-turn). | |

These types of ramps can be used in combination for a specific flow direction (Figs 15.5 and 15.6). These combinations are useful when designers need to sketch a suitable configuration for an interchange in a study area, during the preliminary design process.

## 15.4.2 Classification of Ramps Based on Access Control

A highway ramp or slip road is a short section of roadway that allows vehicles to enter and exit a highway interchange, particularly at controlled access points. Based on the type of access control of vehicular movement on the highway interchange, these ramps are classified as entry and exit ramps, i.e., on and off ramps (Fig. 15.7).

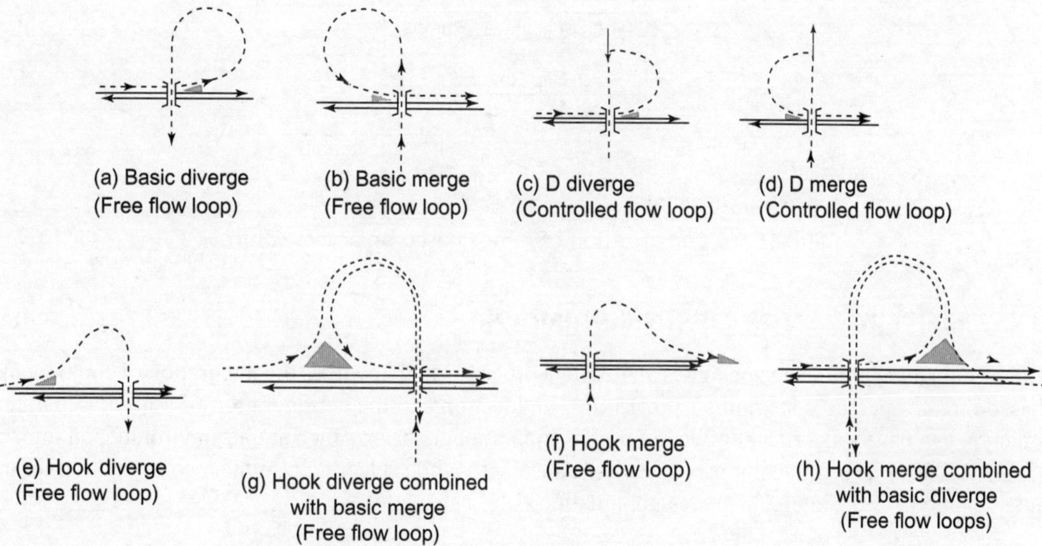


**Fig. 15.5** Loop ramps used for merging and diverging and a combination of both for traffic flow
(Redrawn based on DMRB, 2006 and Harwood et al., 1982)

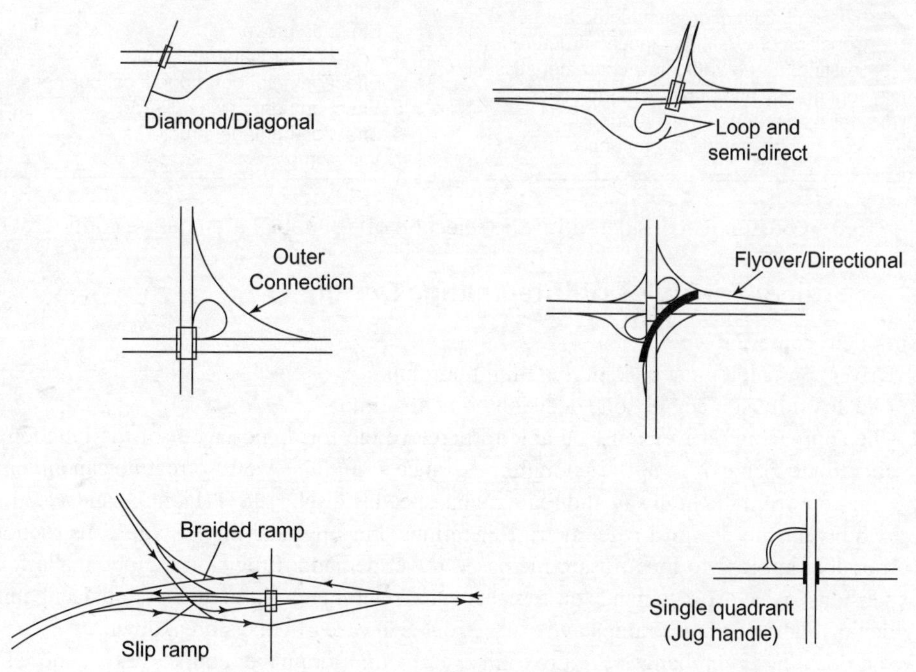


**Fig. 15.6** Loop ramp types (for right-hand traffic)
(*Source*: ODOT, 2012; *Courtesy*: www.oregon.gov/odot/hwy/engservices/)

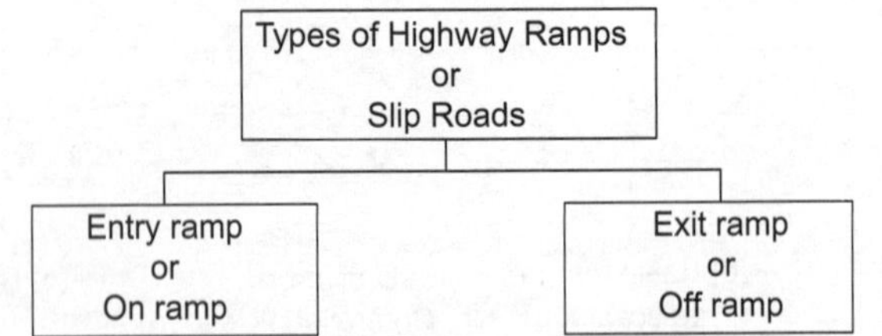


Fig. 15.7 Classification of ramps based on access control

## 15.5 INTERCHANGE DESIGN PRINCIPLES

Intersection and interchange design is an iterative process which involves the integration of components based on site-specific constraints (Fig. 15.8). The location and the configuration of a grade-separated junction can have a significant effect on both its operational performance and its environmental impact (DMRB, 2006). The following principles are considered while selecting a suitable configuration for an interchange in every interchange design situation (Mulinazzi, 1973).

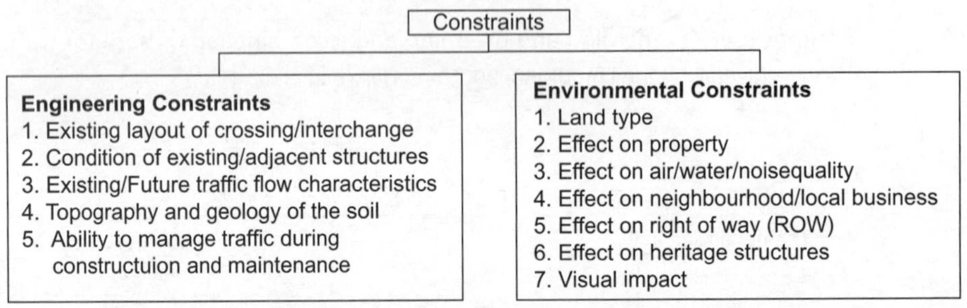


Fig. 15.8 Constraints considered for the selection of a suitable interchange configuration

### 15.5.1 Cardinal Principles of Interchange Design

- Simplicity in design
    - Driver confusion must be kept to a minimum/null.
    - The driver must face only one decision point at a time.
    - The approaching driver must be able to perceive the form and layout of the junction and be able to view an exit ramp far enough in advance at night as well, so that he can anticipate his departure from the highway and regulate his speed (Leisch, 1963) (Figs 3.9 and 11.20).
- Flexibility in the design of geometric operations and upgradability: Stage construction must be possible in order to meet future/increased travel demand. Provision may be made for future expansions or re-arrangements or replacements relating to right of way, adjacent junctions, widening/adding of lanes/ramps and escape routes in case of delays/breakdown.
- Avoid locating interchanges in proximity to natural drainage courses, so as to eliminate unnecessary, complex and costly hydraulic structures (WSDOT, 2014)
- The configuration selected from among the various alternatives should be *justified* based entirely on traffic operational characteristics/safety, existing land use, topography, artificial obstructions such as railroad crossings and roadway crossings and layout requirements, but *not based* on costs.

One of the fundamental principles is that the configuration of interchanges should be studied with reference to the initial route location process.

- The safety concerns include adequate capacity of roadway, sufficient sight distance, setback distance on curves, one decision at a time, no surprise lane drops, channelisation to deter motorists travelling the wrong way, traffic attenuation devices and safe gore area design (Fig. 11.35).
- Overlapping of lane sections will create weaving sections (Fig. 15.9 (a)). High-volume high-speed weaving sections lead to high accident rates and congestion. The number of weaving section(s) should be minimised as this severely reduces the speed and capacity. Therefore, the fewer the weaving sections, the more attractive the interchange design. To reduce the potential weaving problem, the following features may be implemented:
  - Exits should *precede* (come before) entrances in interchanges (Fig. 15.10 (a)).
  - The weaving sections should be lengthened (Fig. 15.9 (b)).
  - The weaving section should be widened by adding one or more lanes. This forces traffic flow to spread out and create gaps.
  - A combination of lengthening and widening of the weaving section should be used (Fig. 15.9 (c)).
  - Any overlapping traffic flow on high-speed lanes can be separated from the mainline by providing a collector–distributer road (Fig. 15.9 (d)).
  - Ramp *braids* that physically separate weaving manoeuvres by placing one roadway over the other may be used (Fig. 15.9 (e)). Such treatments are considered operationally effective approaches (Fig. 15.9 (f)).
- Cross-roads at median openings on interchanges should be properly channelised to minimise wrong way movements.
- Right-hand entrances and exits should be avoided. The creation of ramps on the right-hand side should be avoided because they do not meet driver expectations, create signage issues and confuse drivers attempting to exit the highway (NCHRP 687, 2011).
- Uniform route continuity with *lane balancing* and basic number of lanes must be maintained. Auxiliary lanes should *precede* exit ramps and *succeed* entry ramps. Successive ramps should be connected using auxiliary lanes to maintain continuity in traffic flow along the entry and exit ramps.
- A *balanced design* should also eliminate weaving and maintain speed continuity in all sections, particularly on turnings and ramp terminals.
- Route continuity is often expressed as speed continuity. The design speed differential between the mainline and the interchange ramp should be smaller than 0.3 times the mainline design speed. The larger the differential speed, the greater will be the potential for an accident (CGRA, 1963).
- Adequate spacing requirements relating to consecutive interchanges and ramp spacing should be provided. As a general practice, if the interchanges are closely spaced, auxiliary lanes are used to connect the successive ramps. One mile is usually considered appropriate as the minimum spacing between successive interchanges in urban areas. This spacing requirement is greater than in rural areas, which may be in the range of 1½ to 6 miles. Greater spacing may be provided, if possible, to obtain longer distances between ramps.
- Uniformity of on ramp and off ramp configurations along the highway should be maintained.
- All ramp terminals must have access control.
- All full cloverleaf interchanges must be provided with or have provision for collector–distributor roads (Fig. 15.20). These roads are used to eliminate the weaving problem inherent in the through

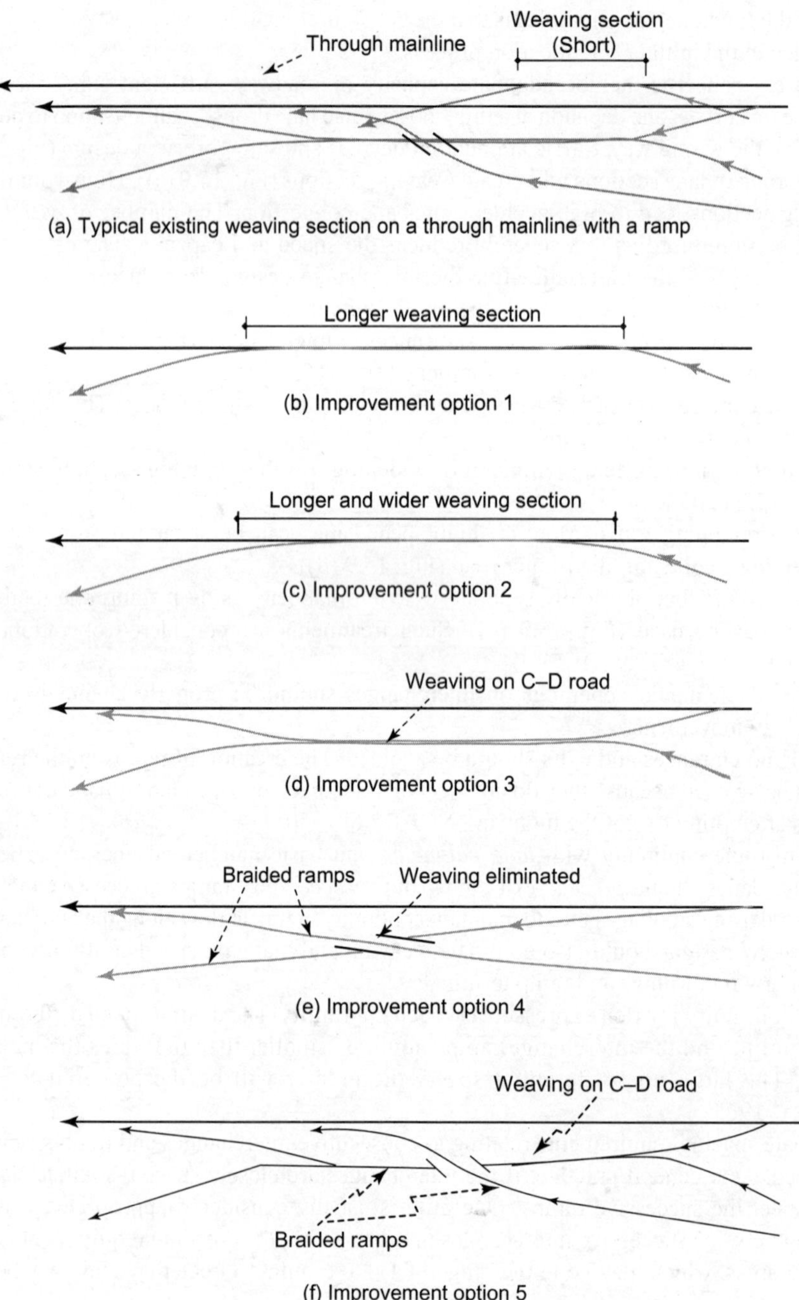


**Fig. 15.9** Weaving improvements: Alternative treatment options

lanes of cloverleaf interchanges. Hence, they should be considered as a basic design element of a cloverleaf interchange.

- Adequate signage must be provided at cross roads, median openings and at all ramp terminals. Providing too much information and signboards that are too close may lead to confusion. Any standard signage templates may be suitably adopted (FHWA, 2012).
- The impact of the interchange on the community, environment and aesthetics must be considered.

## 15.5.2 Secondary Principles of Interchange Design

- The number of junction access points or direct link roads should be reduced. Such unwanted layout planning will lead to turbulent merging and sub-standard weaving length, which may pose potential problems and reduce the capacity of the mainline. In this connection, it is strongly recommended that auxiliary lanes be provided with merge taper and diverge taper. This phenomenon is commonly known as lane gain and lane drop, respectively (Fig. 11.35) (DMRB, 2006).
- Left turning movements are normally handled on direct ramps (Fig. 15.10 (a)). These left-hand off ramps should be provided as exits prior to any cross structures. The offset distance between the off ramp and the crossing eliminates operational problems on the through road.
- Appropriate placement of ramps with interchange configurations may be selected for right turning movements.
- Right-hand ramps with points of major convergence or divergence should be avoided (Fig. 15.10 (b)).
- Loop ramps generally have a worse safety record than other types of ramps and should be avoided where possible (Garber and Fontaine, 1999). However, if they are essential, it is operationally better to have a loop ramp as an on ramp to the highway (Fig. 15.10 (c)), instead of providing

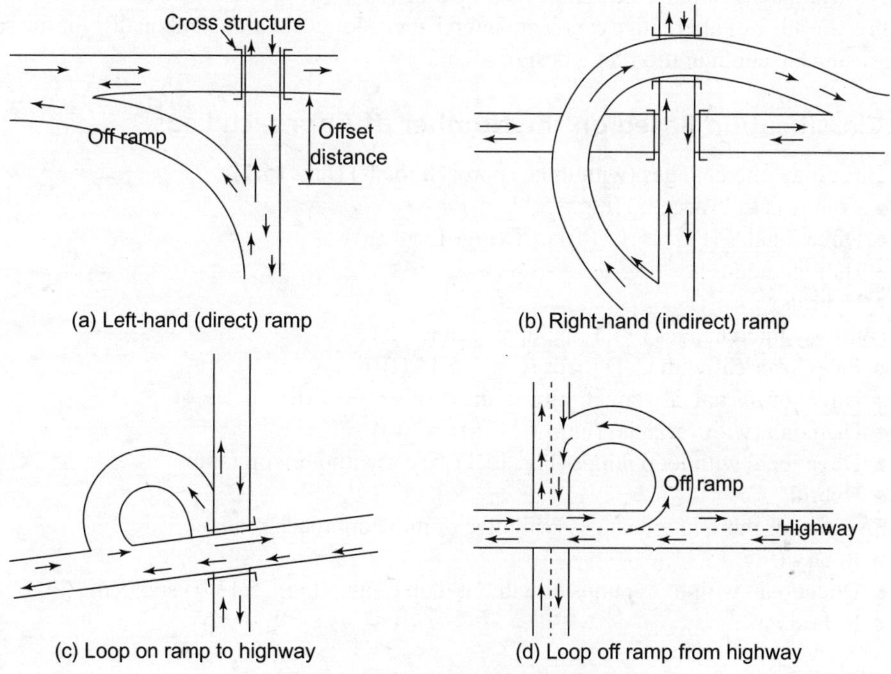


**Fig. 15.10** Commonly used ramps

a loop off ramp from the highway (Fig. 15.10 (d)). This is because the speed of vehicles while traversing on the up-grade loop ramp will be gradually reduced and the vehicles will merge cautiously with the through traffic on the highway (Fig. 15.10 (c)).

- The use of trumpets may be avoided, except where the roadway crossing is discontinued.
- Directional interchanges are preferred for high-volume traffic facilities.
- As per the usual practice of grade separation at crossings, the minor roadway must be elevated over the main through roadway. This may facilitate better sight distance in the off ramp terminal area. This condition holds good when there is no non-motorised traffic on the minor roadway.
- There should not be any lane drop in the basic lanes. Provide lane drops in auxiliary lanes only, as required at speed change, storage for turning, weaving, truck climbing transition lane, acceleration lane, on ramp terminal and off ramp terminal (CDOT, 2005).
- Basic number of lanes may be reduced with caution where there is significant reduction in through traffic.
- Lane drops should preferably be provided on the left-hand side, with a provision for future widening of the lane. This is a good practice when traffic flows at slow speed on the left. Lane drops may be placed at least 750 m past an interchange with an appropriate taper of 50:1 to 70:1.
- All the components must be designed in such a way as to maintain the same LOS throughout the interchange area.

## 15.6 CLASSIFICATION OF INTERCHANGES

Interchanges can be classified based on different parameters. These classifications may be useful as a guide to prepare a list of alternative candidate interchanges for evaluation, and finally, for the selection of the most suitable configuration for a design situation.

### 15.6.1 Classification Based on the Number of Approach Legs

(a) Three-way interchanges (with three approach roads) (Fig. 15.11)
   - Trumpets (all variants) (Fig. 15.11 (d))
   - Directional Y (Fig. 15.11 (b)) or T (Fig. 15.11 (g))
   - Half diamond
   - Hybrids*

(b) Four-way interchanges (with four approach roads)
   - Full cloverleaf with C–D roads ((Fig. 15.11 (f)))
   - Parclo or *par*tial *clo*verleaf (all variants) ((Fig. 15.11 (h))
   - Diamond (with variants) (Fig.15.11 (a) and (c))
   - Directional with loop ramps (Fig. 15.11 (j)) or without loop ramps
   - Hybrids*

(c) Five-way or more interchanges (with five or more approach roads)
   - Rotary (Fig. 15.11 (e))
   - Directional with loop ramps or without loop ramps (Fig. 15.11 (j) and (k))
   - Hybrids*

---

* Hybrid interchange configurations do not exactly fit into any one of the standard geometrical configurations. They are interchanges modified to meet site constraints.

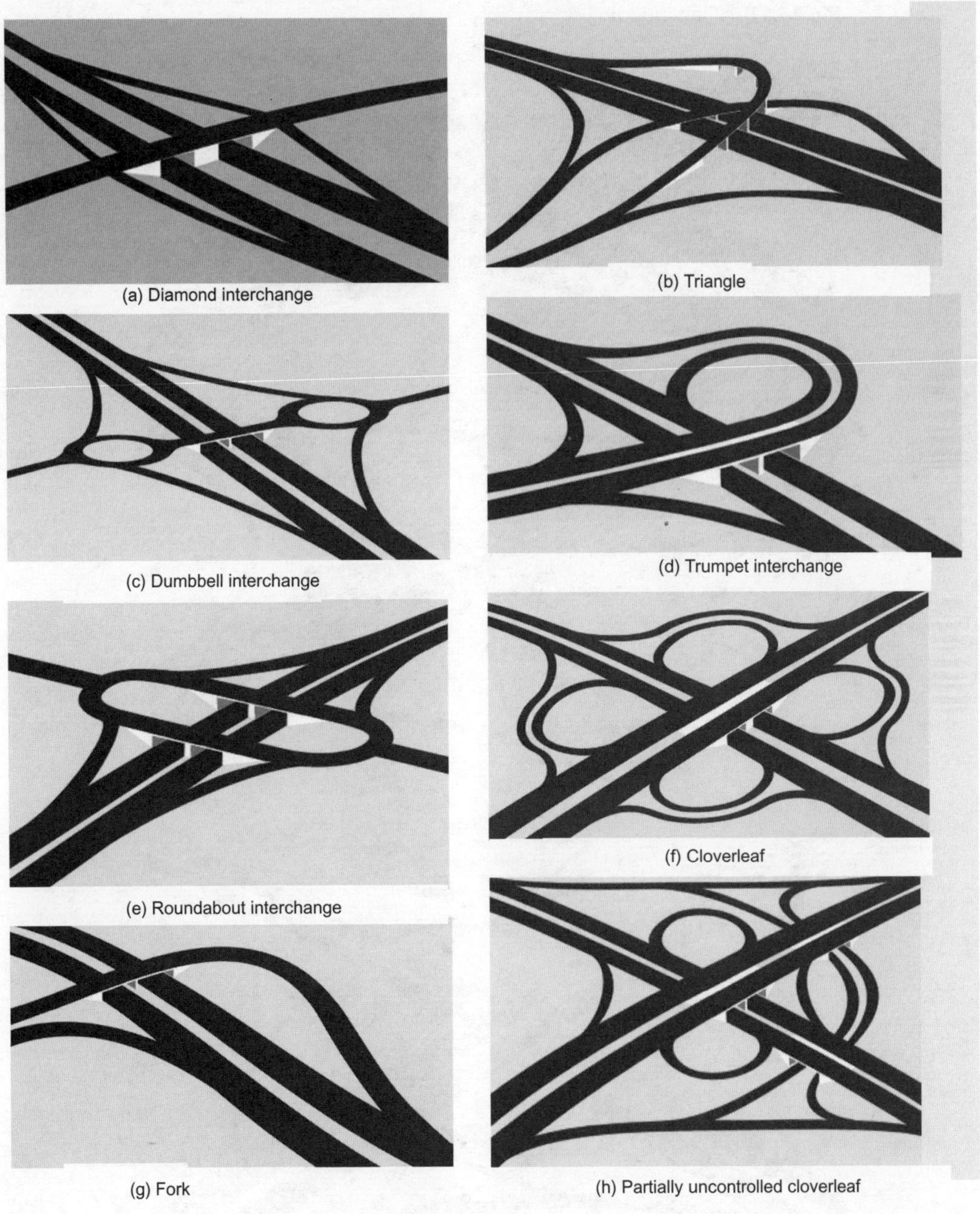

(a) Diamond interchange

(b) Triangle

(c) Dumbbell interchange

(d) Trumpet interchange

(e) Roundabout interchange

(f) Cloverleaf

(g) Fork

(h) Partially uncontrolled cloverleaf

(Continued)

Fig 15.11 (Continued)

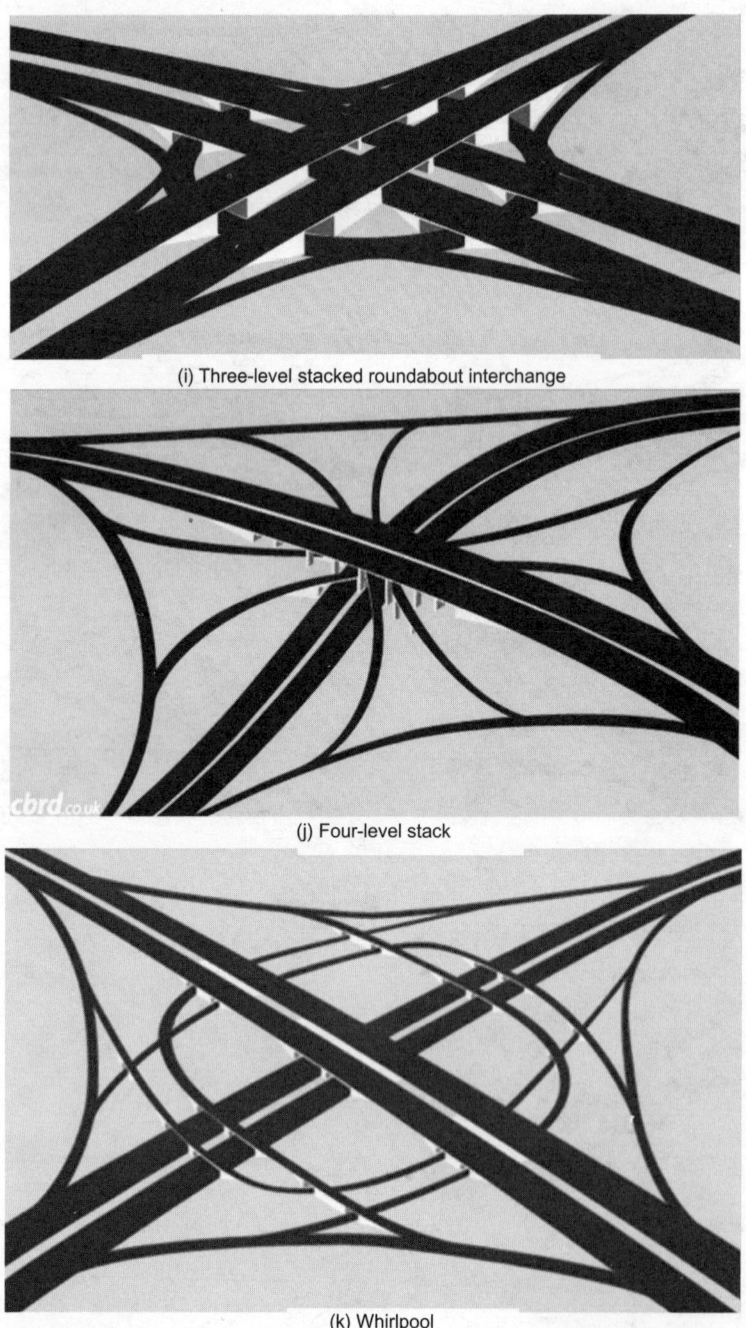


(i) Three-level stacked roundabout interchange

(j) Four-level stack

(k) Whirlpool

Fig. 15.11 Perspective views of a few interchange configurations
(*Source*: www.cbrd.co.uk; *Courtesy*: Mr. Chris Marshall)

### 15.6.2 Classification Based on Directional Use

Left turns are simple and they can be directly configured in the case of left-hand traffic, as is in force in India. However, right-turn movements are considered critical in left-hand traffic. Details of the classification of interchanges based on the direction of the ramp that turns to the right are as follows (Fig. 15.12).

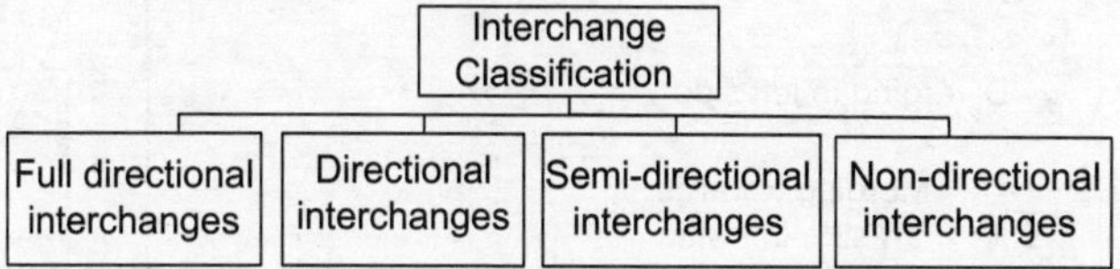


**Fig. 15.12** Classification of interchanges based on the directions of diversion and merge

- **Fully directional interchanges or high interchanges:** All right-turn movements in such interchanges are provided by directional ramps. Both diverging and merging of vehicular movements on a ramp/roadway takes place towards the intended right direction from the *left* (or the *right*) of the through-crossing traffic flow. They permit vehicles to move from major highways to another highway at relatively fast speed with safety. Fully directional interchanges are often preferred at points of right-turn, high-speed traffic flows such as at crossings of expressways or highways or arterials (Garber and Fontaine, 1999). Directional interchanges are the most expensive due to the high standard of the geometrics required to provide the highest capacity and LOS, without weaving sections. Directional interchanges can be configured in many ways (Figs 15.13, 15.14, 15.15 and 15.16). The choice configuration is based on the best fit for the given constraints or situations (Mulinazzi, 1973).
- **Directional interchanges:** An interchange in which *one* or *more* right-turn movements are provided by directional ramps that do not deviate greatly from the intended direction of travel. This means that if a driver wanted to make a right turn, he would have to divert towards the left and then travel on a curved ramp towards the right and finally merge with the through traffic on the left. For example, an interchange configuration comprising one or two ramps of a fully directional interchange.

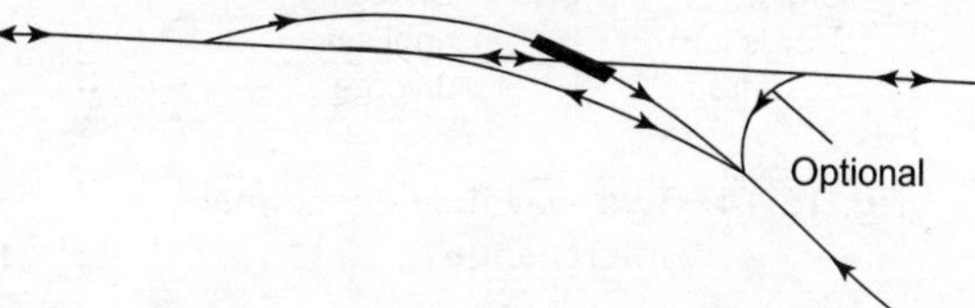


(a) Fully directional Y: Two levels, adaptable to non–high speed with appropriate access control

(*Source*: ODOT, 2012; *Courtesy*: www.oregon.gov/odot/hwy/engservices/)

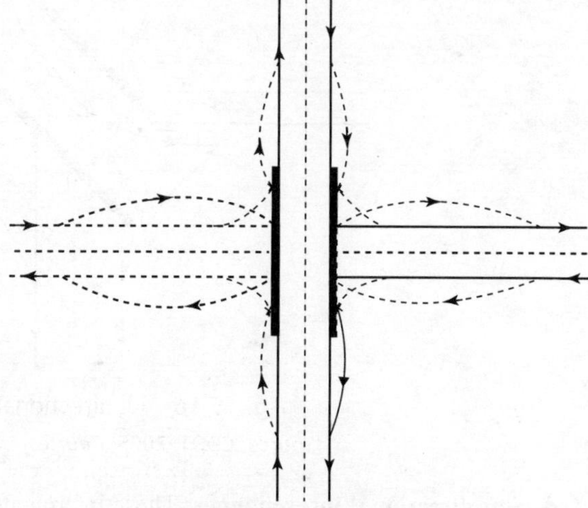

(b) All directional with two entrances and exits

(*Source*: CDOT, 2005; *Courtesy*: www.codot.gov)

**Fig. 15.13** Directional interchanges (altered for left-hand traffic)

- **Semi-directional interchanges:** An interchange in which any one or more *right-turn* movements are provided by *semi-directional* ramps or *jughandle ramps*. The interchanges configured with semi-direct ramps are aligned in a direction opposite to the desired direction of travel (they initially diverge to the left and then turn to the right) (Fig. 15.17).

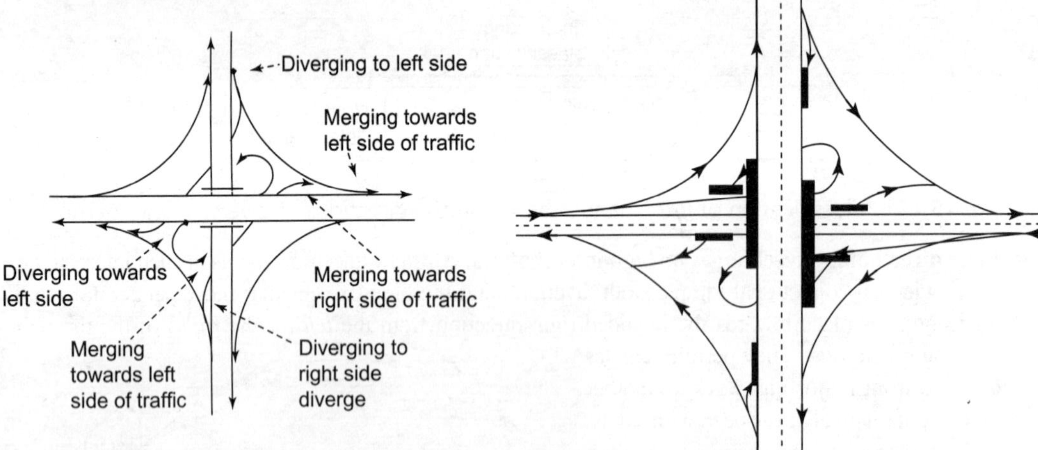


**Fig. 15.14** Two-way fully directional interchange
(*Courtesy*: AASHTO)

**Fig. 15.15** A variant of directional with loops
(*Source*: CDOT, 2005; *Courtesy*: www.codot.gov)

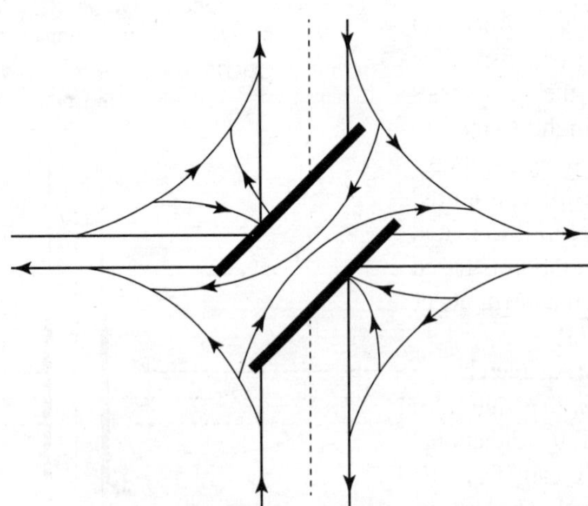

**Fig. 15.16** All directional – four legs
(*Source*: CDOT, 2005; *Courtesy*: www.codot.gov)

- **Non-directional interchanges:** These interchanges comprise *indirect ramp loops* aligned in the direction opposite to the desired direction of travel (Fig. 15.18). One of the best examples is the cloverleaf, in which the loop ramp diverts *left* and then turns towards the desired direction of the *right*. Other examples of non-directional interchanges are the conventional two-level diamond

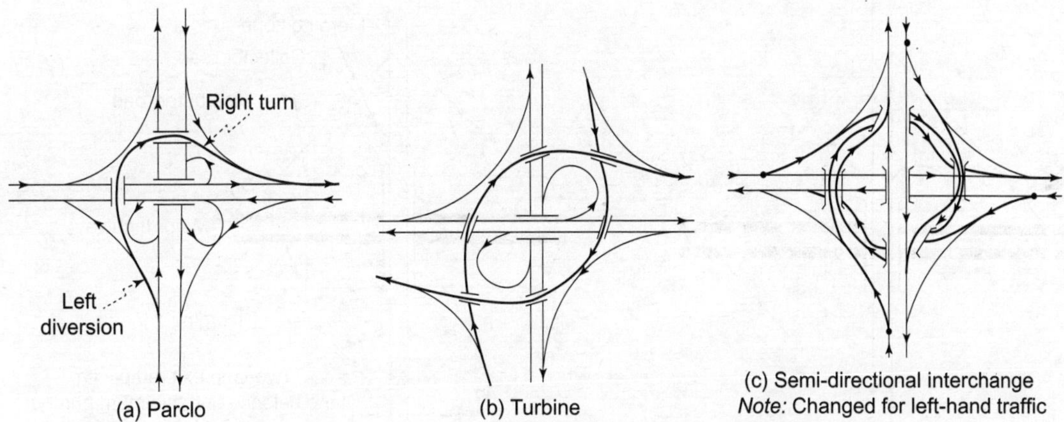


**Fig. 15.17** Examples of semi-directional interchanges
(*Source:* WSDOT, 2014; *Courtesy:* www.wsdot.wa.gov)

(Fig. 15.19), cloverleaf with (Fig. 15.20) and without collector–distributor roads (Fig. 15.21). A few variations of these interchanges may be listed as split diamond, three-level diamond and partial cloverleaf.

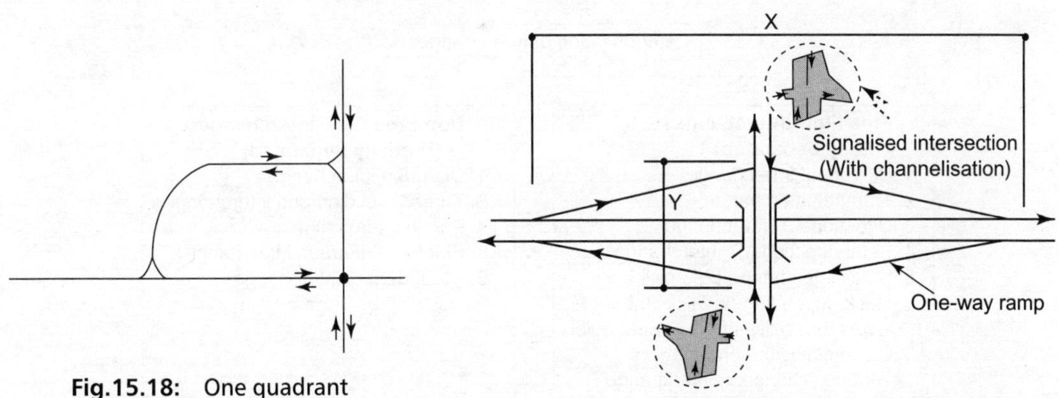


**Fig.15.18:** One quadrant
(*Courtesy*: AASHTO)

**Fig. 15.19** Conventional diamond interchange

These types of interchanges with their respective ramps are used based on the directional distribution of traffic flow at a site, particularly with reference to heavy right-turn movements. The ramps used should reduce travel time, maintain the design speed and capacity and eliminate weaving.

## 15.6.3 Classification Based on Access Control

Access controls in the terminal area of a ramp of any interchange or along a frontage road in the vicinity of an intersection can interfere/affect/hamper the free flow of traffic at the entry and exit to an intersection. Interchanges can be categorised based on the type of access control. Such information

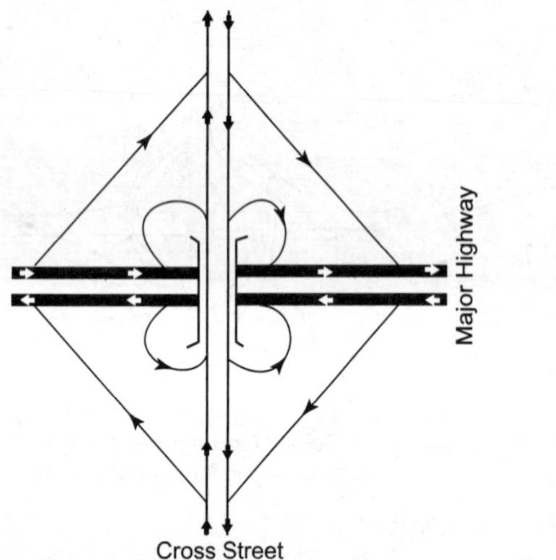


**Fig. 15.20** Cloverleaf or full cloverleaf
(Modified based on Pinnell et al., 1963)

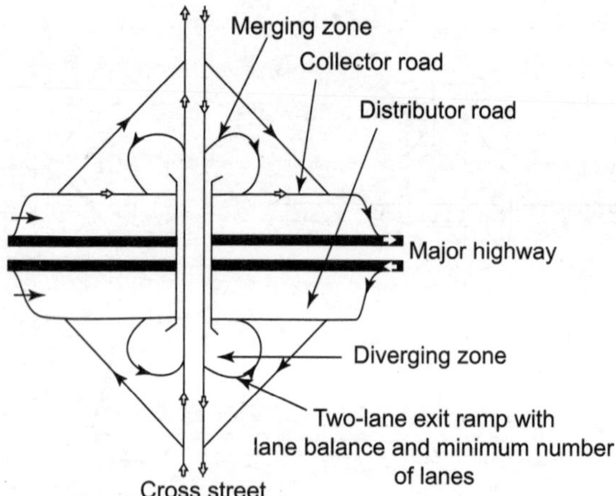


**Fig. 15.21** Full cloverleaf with collector–distributor roads
(Modified and elaborated based on Pinnell et al., 1963)

(Fig. 15.22) can be used by the designer to ensure the degree of free flow or absence/presence of conflicts that prevail with any type of interchange.

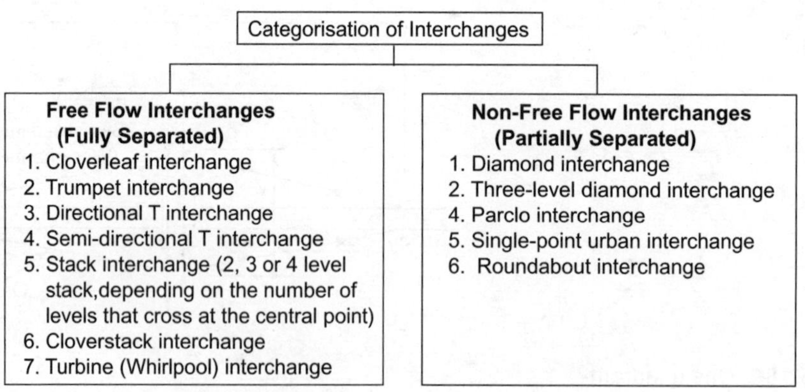


**Fig. 15.22** Categorisation of interchanges based on the type of access control

## 15.7 INTERCHANGE CONFIGURATIONS

The types of interchange configuration based on the number of intersecting roadways are presented in the following sections.

### 15.7.1 Three-leg Interchanges

Three-leg intersections can be configured in different ways. The type of configuration fit for a particular location of an intersection is determined mainly by the alignment of the intersecting roads, the topography and the directional distribution of traffic. It is important to take into consideration that it is very difficult to include a fourth leg in any three-leg intersection, as part of future expansion. To plan for such expansion, any suitable configuration of four-leg intersections which can fit into the current requirement for a three-leg intersection must be selected.

*Trumpets* are the most widely used configuration for three-leg intersections (Fig. 15.11 (d) Fig. 15.23 (a) and (b)). As a thumb rule, preference is given to major turning movements on directional/semi-directional ramps and the lowest traffic movements should be facilitated by a loop ramp.

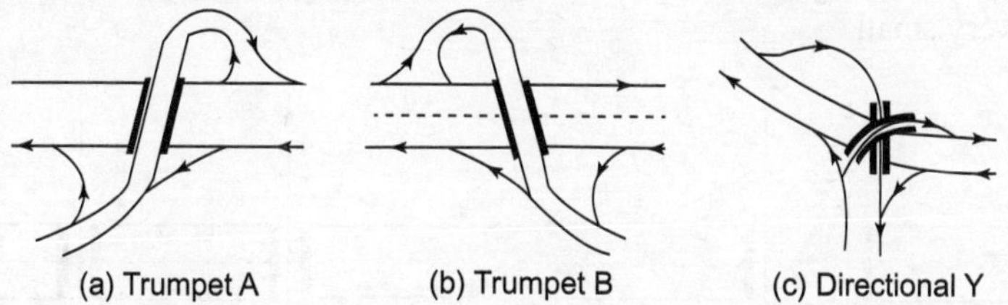


**Fig. 15.23** Three-leg interchanges (*Note*: Changed for left-hand traffic)
(*Source*: CDOT, 2005; *Courtesy*: www.codot.gov)

A composite listing of the advantages and disadvantages of trumpet interchanges is presented in Table 15.2.

**Table 15.2** Advantages and disadvantages of trumpet interchanges

| *Advantages* | *Disadvantages* |
|---|---|
| ✓ Loop serves as off ramp to the highway<br>✓ Fits in T-intersection traffic flow pattern<br>✓ Used for highway-to-arterial connection<br>✓ Can handle large volumes | × Difficult to extend truncated approach road if demand increases<br>× Indirection in movement<br>× Not used for highway-to-highway connections |

*Directional* three-leg interchanges use no loops (Fig. 15.11 (b) and (g), Fig. 15.13 (a), Fig. 15.23 (c), Fig. 15.24). All direct or a few indirect ramps (Fig. 15.25) are also used with more than one grade-separated structure. These configurations are used only when all the movements on the three legs are high.

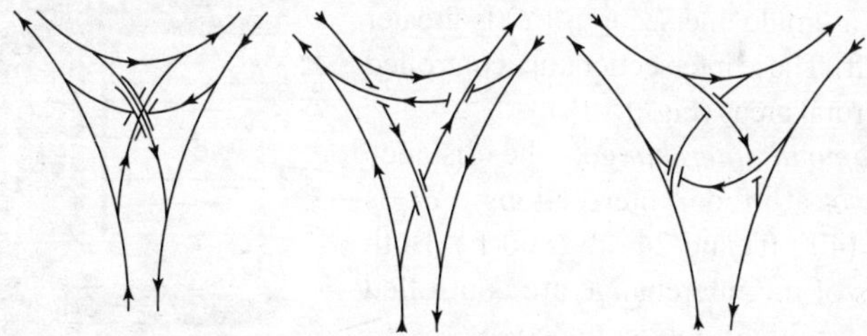

**Fig.15.24** Variants of Y interchanges with directional ramps

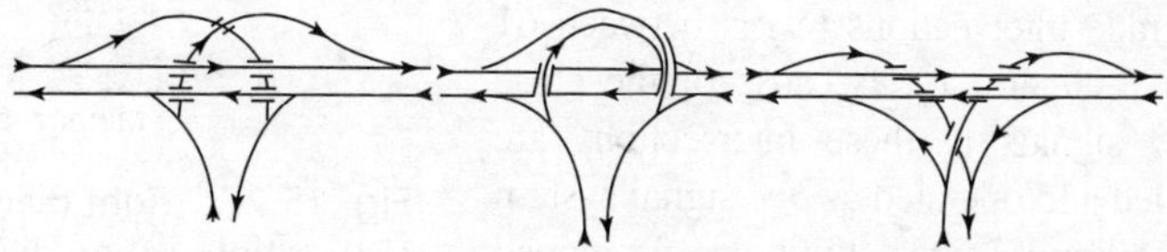

**Fig. 15.25** Variants of T interchanges with semi-directional and directional ramps

### 15.7.2 Diamond Interchanges

These are one of the most commonly used interchanges in urban and rural areas, where highways (major road) and arterial roads (minor road) intersect (Fig. 15.19). Usually, the minor road is provided as an overpass and the turning movements are facilitated by one-way diagonal ramps as interchanges, in each quadrant (Fig. 15.26 (a)). The design allows free flow of operation on the major highway but creates two at-grade intersections on the minor road with ramps. These two at-grade intersections are often signalised if the traffic volume is at medium level or just operated with a Stop sign, when the turning traffic is very small.

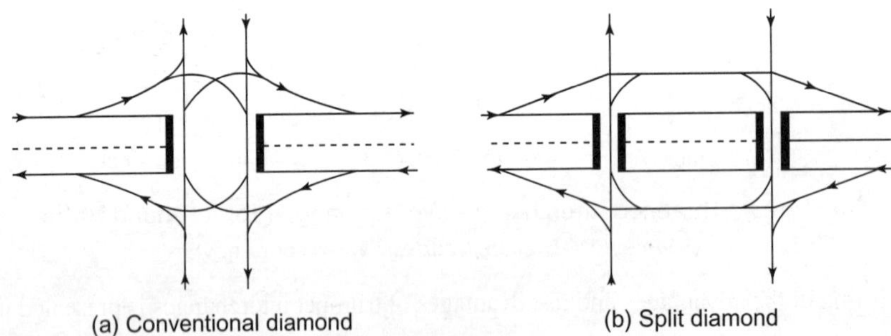


**Fig. 15.26** Typical directions of traffic flow permitted in diamond interchange variants (*Note*: Changed for left-hand traffic)

(*Source*: CDOT, 2005; *Courtesy*: www.codot.gov)

An un-signalised system of diamond interchanges can be operated with an acceptable LOS when the entering volume is up to 1500 veh/hr. The diamonds can operate at LOS D or higher till volume 4500 veh/hr (Garber and Fontaine, 1999).

Depending on the distance between these intersections, diamond interchanges can be classified into three types (Chaudhary and Chu, 2000):

- *Conventional diamond interchange*: The distance between the two at-grade intersections (Y) is greater than 244 m (800 ft). These intersections are controlled by Stop signs in rural areas (Fig. 15.19).
- *Compressed diamond interchange*: The distance between the two at-grade intersections (Y) is between 122 m (400 ft) and 244 m (800 ft). Both the intersections of the interchange are controlled by signals with or without interconnection.
- *Tight diamond interchange* (TDI*)* or *tight urban diamond interchange (TUDI):* The distance between these two at-grade intersections (Y) is between 61 m and 122 m (200–400 ft). Because of the close proximity, *two* signals at these intersections are always controlled and operated as *one* signal system (Fig. 15.27). Optimum signal timing is developed to minimise storage of vehicles between the ramp

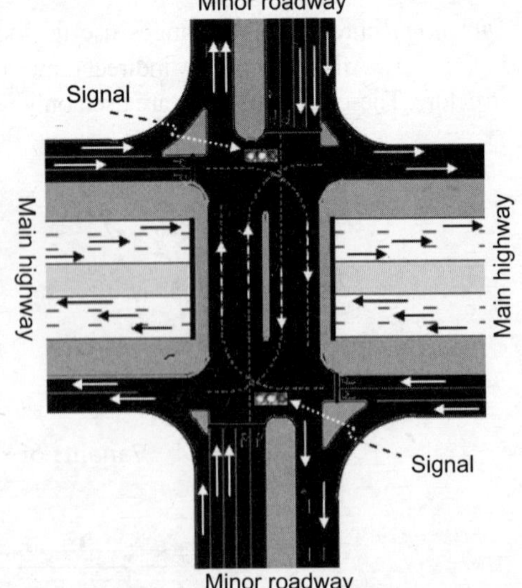


**Fig. 15.27** Tight diamond interchange (TDI) or tight urban diamond interchange (TUDI)

intersections and also to provide coordinated movement of traffic along the street and ramps. TDIs are commonly designed for developed urban areas, where land acquisition is difficult and costly.

A composite listing of a few advantages and disadvantages of TDI is presented in Table 15.3.

**Table 15.3** Advantages and disadvantages of TDI

| *Advantages* | *Disadvantages* |
|---|---|
| ✓ Minimum property required outside of the usual right of way (RoW) along the major highway<br>✓ No weaving on the through road<br>✓ Economical, with one grade-separated structure<br>✓ Flared ramp terminals can improve capacity<br>✓ Single exit from highway at high speed<br>✓ Suitable in rural and urban areas as local road intersection treatment<br>✓ Fits service road pattern in case of urban areas<br>✓ Pedestrian and bicycle movements can be better accommodated than in full cloverleaf<br>✓ Diamonds are preferred over SPUI when frontage roads are present | × The operating capacity is limited due to at-grade intersections on the minor road<br>× There would be inadequate sight distance at ramp terminals on the at-grade intersections<br>× Need to stop on crossroads for all turn movements on the minor road<br>× Possibility of wrong way movement at the at-grade intersections<br>× Difficult to accommodate high right-turn volumes<br>× Three-phase signal control with signal coordination is often required to operate the traffic on the minor road of at-grade intersections |

## 15.7.3 Single Point Urban Interchange (SPUI)

SPUI is a variation of TDI and TUDI where all the approaches converge in *one* intersection, above or below the mainline roadway (Fig. 15.28). This intersection is controlled by a single traffic signal. The

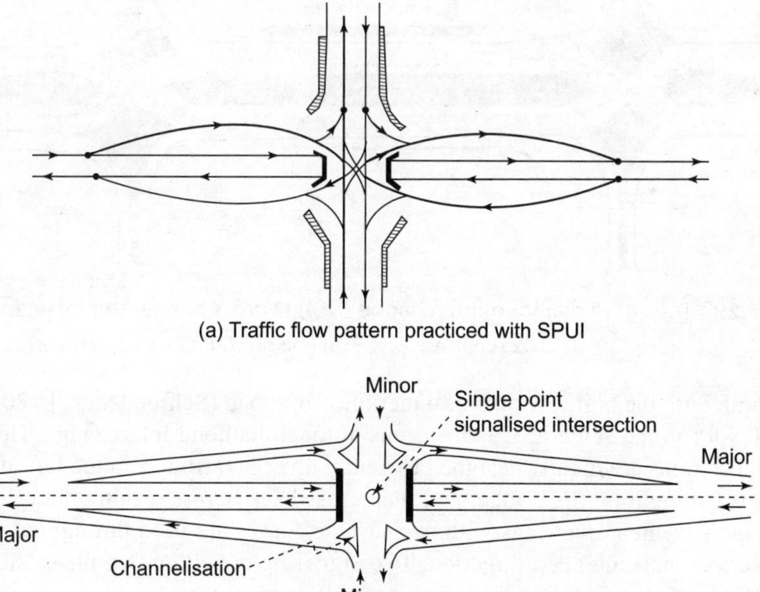


(a) Traffic flow pattern practiced with SPUI

(b) Channelisation and signal position on SPUI

**Fig. 15.28** Schematic representations of single point urban interchange (SPUI)

most important operation of SPUI is that all right-turn movements are aligned/channelised to travel to the inside of one another (i.e., close to the centre point), which reduces the right-turn conflicts that exist in conventional intersections. The construction cost of SPUI is, in general, 10–20% more than that of other interchange types (Garber and Fontaine, 1999).

SPUI is recommended when the AADT on the main through highway is between 20,000 and 35,000 vehicles, or the traffic on the minor road is between 15,000 and 30,000 vehicles (Amer, 2009). A TTI report states that SPUI should be used when the total entering AADT is 40,000 vehicles (Bonneson et al., 2004)

A study tested the equivalent characteristics related to geometric conditions, volume conditions, travel speeds, phase failures and percentage of stops; it revealed that SPUI provides superior traffic operational performance than TUDI. Another finding concludes that TUDI would reach capacity conditions when SPUI was operating at average conditions (Jones and Selinger, 2002). In contrast, pedestrians are not easily accommodated by SPUI, unlike diamonds. In addition, SPUI is not recommended when two roads intersect at a large skew angle because of reduced sight distance and high construction cost (Garber and Fontaine, 1999).

## 15.7.4 Diverging Diamond Interchange (DDI)

DDI works differently from the traditional diamond interchange, with crossover locations at each end of the bridge with two entrance ramps and two exit ramps (Fig. 15.29).

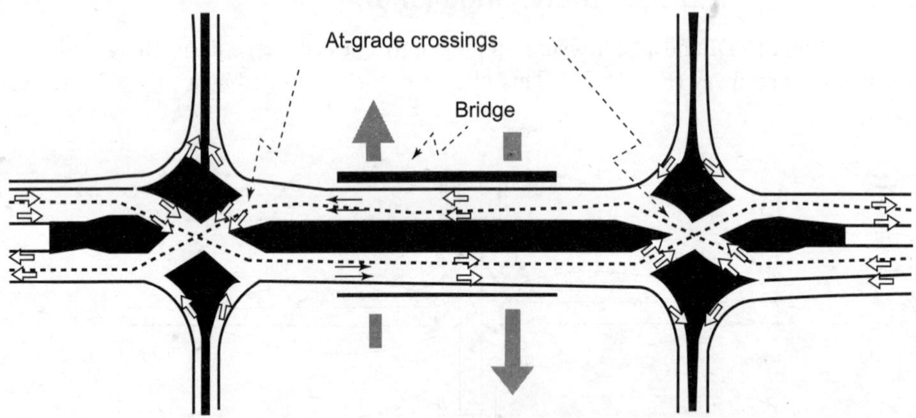


**Fig. 15.29** Diverging diamond interchange (DDI) (*Note*: Changed for left-hand traffic) (*Source*: ODOT, 2012; *Courtesy*: www.oregon.gov/odot/hwy/engservices)

DDI can be built with the highway over or under the cross road (Schroeder et al., 2014). All the left turns in DDI work in the same way as in a conventional diamond interchange. However, the two crossover intersections are mainly designed to gently diverge traffic from the left of the road to the right of the road straight on the bridge, and then cross back before leaving the bride (Fig. 15.30). Guide signs and pavement markings with two reverse curves are used throughout the interchange to direct drivers to their intended direction. Two signal lights with a two-phase signal system are commonly adopted for controlling the opposing traffic movements at these two crossover points. This type of traffic flow makes DDI different from conventional diamond interchanges. This may seem unusual, but these two crossover signal points are actually the reason why DDI functions so smoothly. They allow for better traffic flow with safe and simpler right turns between the interchange and the highway.

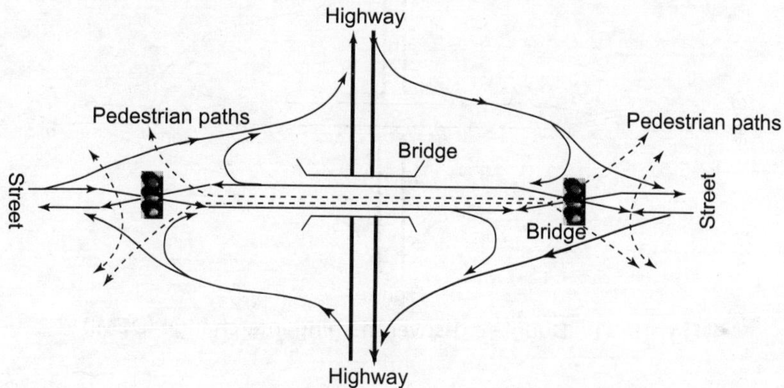


Fig. 15.30 Directions of traffic flow in DDI

The factors affecting the performance of DDI are design speed, curve radii, crossing angle, tangent length, RoW, central median width and vehicle storage for turning lanes of adjacent intersections. One of the main factors, a large acute crossing angle (>40°), must be used while balancing the smooth turning radius against the desirable shorter pedestrian crossing (MoDOT, 2010).

**Table 15.4** Advantages and disadvantages of DDI

| *Advantages* | *Disadvantages* |
|---|---|
| ✓ Reduces the number of potential vehicle conflict points to 14, as against 26 in conventional diamond interchanges<br>✓ Beneficial to pedestrians and cyclists. They can cross and be accommodated on the central long median on the bridge or along the side of the road<br>✓ DDI can be built on the existing bridge deck when the conventional diamond interchange is retrofitted<br>✓ For new DDIs, fewer additional lanes are required with a single bridge structure, and hence, less RoW is required<br>✓ Can eliminate right-angle crashes for right turns<br>✓ Can accommodate right turns with heavy traffic volumes, shorter queue length, less delay with fewer stops and is efficient compared to conventional diamond interchanges | × As through volumes are high, the drivers may be inconvenienced as they must crossover twice to travel through the interchange<br>× There is no U-turn facility<br>× Fewer crossing angles will increase the potential for wrong way movements at crossings by confused drivers<br>× DDI signalling system may not work well in closely spaced adjacent intersections<br>× Public information campaign may be required to educate the local people |

## 15.7.5 Double Crossover Merging Interchange (DCMI)

DCMI is an upgraded version of DDI, and is provided with two additional bridges on either side (Fig. 15.31). Signals control is completely eliminated and uninterrupted traffic flow is established using these two crossovers. The area of DCMI spread over the land is the same as for DDI. Capacity enhancement with minimised queuing and delays across DCMI are a few major advantages. The cost of construction will be greater than with DDI.

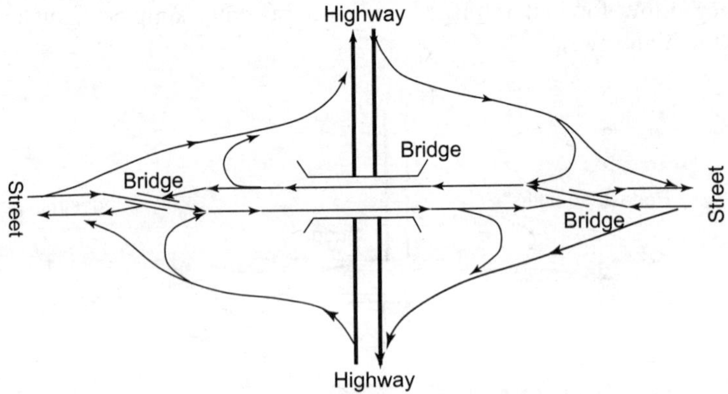


Fig. 15.31 Double crossover merging interchange (DCMI)

### 15.7.6 Three-level Diamond Interchange

Both the through roads are separated on two levels and they are operated as free flow conditions without any stop. All the turning movements are separated from these two through roads, say on a third level, usually in the middle. A square of one-way roads circulates in the clockwise direction. Each corner of the elevated square road is used as an on/off ramp terminal (Fig. 15.32). The overall configuration may be imagined as two diamond interchanges fused together. It is appropriate to use three-level diamond interchanges when the cross streets carry high volumes of traffic in urban areas. It is more economical to use than directional interchanges. However, the construction cost will be more with these three levels of structure and a complex coordinated signal may need to be installed.

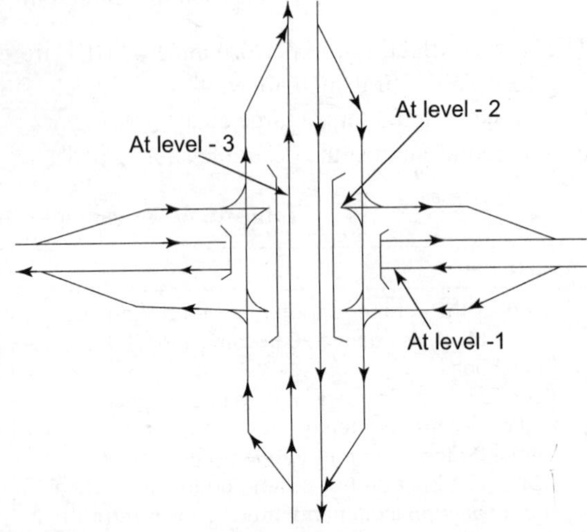


Fig. 15.32 Three-level diamond interchange (*Courtesy*: ODOT)

### 15.7.7 Variants of Diamond Interchanges

- *Single point urban interchange* (SPUI) (Fig. 15.28): Compresses the two intersections into one single signalised intersection. SPUI is preferred over TDI because it can operate with less delay and handle heavier traffic volumes as vehicles need to cross only one signalised intersection.
- *Spread diamond interchange* (SDI) (Fig. 15.33 (c)): The ramps of the diamond interchange are spread wider away from the through mainline. There will be flexibility to provide good sight distance at the ramp terminals at the at-grade intersections. The disadvantages are the same as listed for TDI.
- *Split diamond interchange*: This (Fig. 15.33 (a)) provides good bypass in case of accidents as its configuration provides alternatives for entering and leaving the highway (Fig. 15.33 (b)). There would

be a capacity problem due to queuing which is because of closely spaced at-grade intersections. The split diamond (Fig. 15.33 (d)) can be operated with two-phase signals. It requires two structures. The turning movements are simplified but are not convenient for right turns.

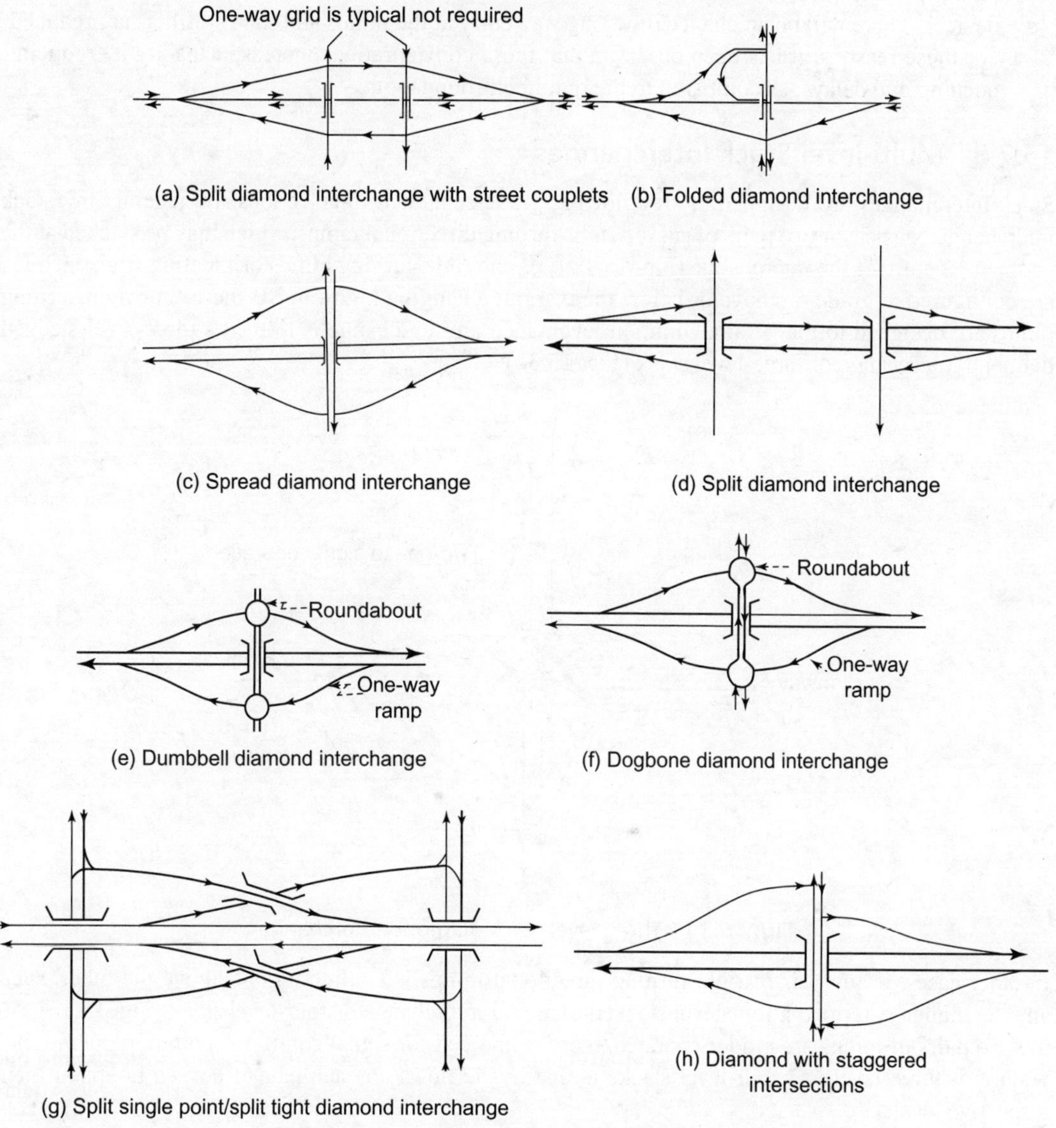


**Fig. 15.33** Variants of diamond interchanges

- *Diamonds with right-turn structures* or *dumbbell diamond* (Fig. 15.33 (e)): This is similar to the diamond interchange, but uses a pair of mini-roundabouts to join the ramps with the non-highway street. Dumbbell diamonds increase the efficiency of the interchange when compared to a conventional diamond interchange.

- *Diamonds with U-turn structures* or *dogbone diamond* (Fig. 15.33 (f)): This is a variation of the dumbbell roundabout and diamond interchange. The central island is shaped like a toy dog bone, in the plan view. The dog bone is formed by two interconnected mini-roundabouts that look like raindrop- or teardrop-shaped islands at either ends. These two roundabouts act as traffic calming devices. There will be no chance of wrong way entry at the ramps and such conflicts are reduced. For these reasons, it has been observed that these configurations increase capacity by reducing queuing and delays, as compared to the dumbbell roundabout.

### 15.7.8 Multi-level Stack Interchanges

Stack interchanges are designed as free-flowing grade-separated intersections. In general, any stack interchange consists of two pairs of curved right-turning directional ramps, which may be stacked at the geometric centre of the interchange (Fig. 15.11 (j)). The right-turning ramps originating from each leg are configured at grade or above or below the two interchanging highways. If these four right-turning ramps are designed to merge at a common level, such an interchange is termed a three-level stacked diamond interchange or three-level stack (Fig. 15.34).

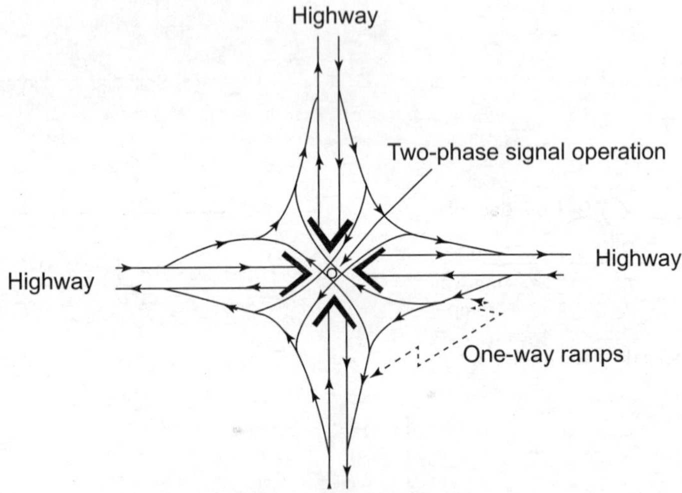


**Fig. 15.34** Three-level stack diamond interchange

In other cases, if any pair of right-turning directional ramps is configured one above the other, such an interchange is termed a four-level stacked diamond interchange or four-level stack (Fig. 15.16). As the two pairs of ramps are grade separated, there is no need for signal control. A common operational feature of three-level and four-level stacks is that traffic flows on each pair of curved ramps at their respective levels are reversed.

Five-level and six-level stacks also have a similar configuration, with additional frontage roads or elevated highways adding the levels. Intersections of such additional frontage roads are typically at grade or nearer to the through road.

All variants of stack interchanges can serve high volumes of traffic at highway speeds, except at signalised intersections of three-level stacks. They occupy smaller land area. Stacks eliminate the problem of weaving and can provide the highest vehicle capacity among similar types of four-way interchanges. Their ramps are directional (or sometimes semi-directional), and hence, their use is very comfortable, with adequate sight distance. Despite these advantages, the construction of raised stack ramps and maintenance costs are higher; they do not usually permit U-turns.

### 15.7.9 Cloverleaf Interchanges

A four-quadrant, full cloverleaf has eight ramps (four direct ramps and four loop ramps) which are used to eliminate right-turning conflicts through the construction of a two-level interchange. Cloverleaf interchanges are better suited to rural areas with low turning volumes (Fig. 15.20).

A full cloverleaf *without* C–D roads is used only when weaving volumes on through roads are small. It was observed in a study that as the number of weaving vehicles approached 1000 veh/hr, a significant reduction in speed took place, and hence, C–D roads (Fig. 15.21) should be used (Garber and Fontaine, 1999).

A few advantages and disadvantages of full cloverleaf are listed in Table 15.5.

**Table 15.5** Advantages and disadvantages of full cloverleaf

| *Advantages* | *Disadvantages* |
|---|---|
| ✓ All movements are free flow<br>✓ No right-turn movements on ramps or cross roads<br>✓ Weaving problem can be solved by providing C–D roads<br>✓ Provide higher capacity than most diamonds since movements at the ramp terminal are usually free flow by indirect loop ramps<br>✓ Considered more appropriate than diamonds where traffic volumes are relatively high | × Large area of right-of-way required<br>× Free flowing exits may pose problems at adjacent limited access control at-grade intersections<br>× All are indirect loop ramps and may offer limited capacity/ restricted capacity in the case of significant truck traffic<br>× Indirection in movement<br>× There will be a change of tight-weaving on the mainline if there are no C–D roads<br>× Additional cost of C–D roads<br>× Bicycle and pedestrian movement along crossings is difficult<br>× Two entrances and two exits<br>× Higher construction cost for right-turning vehicles than diamonds<br>× The travel distance may increase |

### 15.7.10 Parclo (Partial Cloverleaf Interchange)

Cloverleaf interchanges that provide a *limited* number of movements with loops in one, two or three quadrants are referred to as partial interchanges or partially directional interchanges or parclo (Figs 15.35 and 15.36). These types of interchanges are used when unavoidable property/right-of-way

**Fig. 15.35(a)** Four-leaf clover
(*Source*: Cygnus921, https://commons.wikimedia.org/wiki/File:Four_Leaf_Clover_068.jpg)

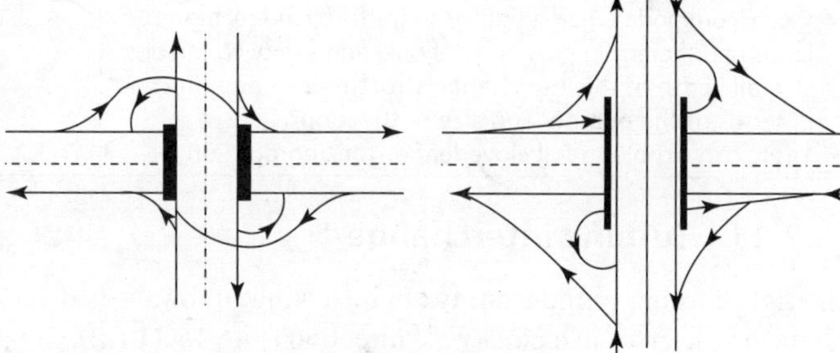

**Fig. 15.35(b)** Variants of partial cloverleaf (*Note*: Changed for left-hand traffic)
(*Source*: CDOT, 2005; *Courtesy*: www.codot.gov)

restrictions preclude ramps in one or more quadrants, and where right-turn/left-turn volumes are *low* or there is a significant impact on the environment due to artificial disturbances. Several examples of parclos are illustrated in Figs 15.15 and Fig. 15.11 (h).

Signalised delays observed with parclo will be less than with diamonds or SPUIs for all entering volumes. This may be because large right-turning traffic flow are handled by ramps and they do not influence phase/signal timings (Garber and Fontaine, 1999).

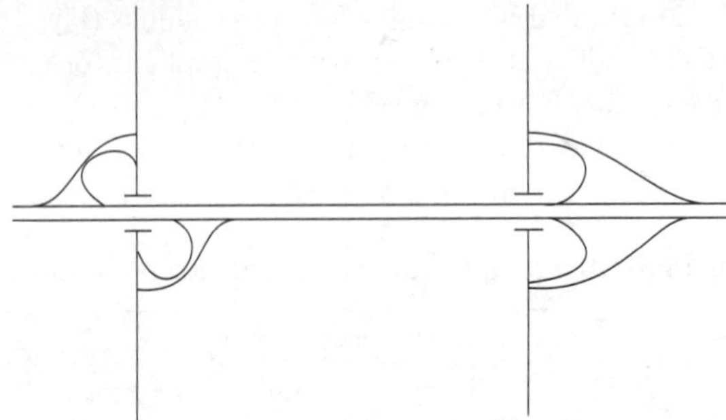

**Fig.15.36** Partial cloverleaf (*Note*: Changed for left-hand traffic)
(*Source*: WSDOT, 2014; *Courtesy*: www.wsdot.wa.gov)

It is preferred that the movements provided in a partial interchange should have the facility to provide the corresponding return movement also, because the driver may expect to *retrace* his route in the return direction on a particular trip.

A composite listing of a few advantages and disadvantages of parclo is presented in Table 15.6.

**Table 15.6** Advantages and disadvantages of parclo

| *Advantages* | *Disadvantages* |
|---|---|
| ✓ Appropriate when one or more quadrants pose adverse terrain/right-of-way problems that preclude the use of partial diamond or full diamond or full cloverleaf interchanges<br>✓ Good design for unbalanced traffic movements<br>✓ Can eliminate or increase weaving distances on major roads, as with full cloverleaf with C–D roads<br>✓ The loops in the opposite quadrants can eliminate the weaving problem associated with full clover<br>✓ Can accommodate heavy right-turn traffic by indirect ramp loops and thereby improve operational and safety conditions<br>✓ Overall width of the interchange structure can be reduced in terms of the number and size of the loops<br>✓ Easily convertible to full cloverleaf or for upgradation | × Right turns are made off the cross road<br>× Channelisation is required at the cross road<br>× At-grade intersections may require three-phase signals<br>× Chances of wrong way movements at these cross roads |

## 15.7.11 Turbine Interchange

The right-turning traffic from two intersecting highways is diverted to a circular directional ramp and the traffic traverses in a clockwise direction (Fig. 15.11 (k)). Smaller, single-span bridges are provided at each crossing location of these ramps. The flatter circular roadway profile provides longer sight distance, even at highway speeds. Construction and maintenance costs of turbine interchanges will be lower than for equivalent stack interchanges.

Details of roundabout interchanges (Fig. 15.11 (i)) and modified configurations of the above interchanges (hybrid) may be obtained from any standard design manuals (AASHTO, 2004).

## 15.8 GUIDELINES FOR THE DESIGN OF INTERCHANGE COMPONENTS

The concepts involved in the design of interchange components, along with their operational and safety considerations, are briefly explained below. The fundamental principles stated in this chapter are applicable for the design of various components of interchanges. For a more detailed discussion on the component design of interchanges, readers are instructed to refer to AASHTO Green Book (2004).

### 15.8.1 Lane Balance

To ensure efficient operation and safety of traffic flow through and beyond an interchange with ramps, there should be balance in the number of lanes, with continuity provided on through roads and ramps. The merging, weaving and diverging points of traffic flow should be smoothly facilitated at and between the entrance and exit terminals of ramps to fully realise the potential capacity (Leisch, 1963). Reduction in the number of lanes (lane drops) and addition of lanes (lane gain) should be properly designed with adequate taper lengths to meet the deceleration and acceleration needs, respectively, of vehicles.

The minimum number of through lanes normally associated with a highway is considered the *basic* number of lanes. This should not be dropped or added to indiscriminately with auxiliary lanes. The basic number of lanes required for a section of highway depends on the volume-to-capacity relationship, exclusive of auxiliary lanes (CDOT, 2005). When the volume-to-capacity ratio is attained at a minimum level of service, another lane is added. This procedure often results in unbalanced lane distribution. In no case should lane reduction below the basic number of lanes be allowed. Hence, there is a need to solve this problem by coordinating the concept of determination of basic number of lanes with lane balance (Mulinazzi, 1973).

In this connection, design traffic volume and capacity analysis determine the minimum (or basic) number of lanes required on the main road and on the ramps. After determination of the basic number of lanes on each roadway, lane balance at and beyond the merging and diverging points of an interchange must be checked against the minimum requirement (Fig. 15.37). This process of maintaining balance between traffic volumes with the level of service is often referred to as *operational flexibility* (Leisch, 1963).

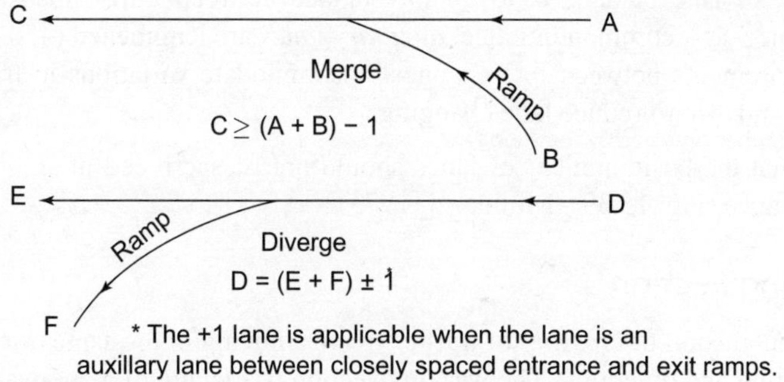


**Fig. 15.37** The fundamental concept of determination of lane balance

The following principles are adopted (SDDOT, 2015; AASHTO, 2004; WSDOT, 2014):

- At entrances, the number of lanes beyond the merging of two traffic streams (C) should be greater than or equal to the sum of *all* traffic lanes on the merging roadways *minus* one, i.e., $C \geq (A + B) - 1$.
- At exits, the number of approach lanes on the main through highway (D) must be equal to *all* the lanes of the highway beyond the exit diverge minus one, i.e., $D = (E + F) - 1$. The following cases are exempted from this rule:
  - At cloverleaf loop ramp exits which follow the loop ramp entrance or at exits between closely spaced interchanges
  - At closely spaced interchanges with an auxiliary lane continuously provided between the entrance and exit (merge and diverge points). Interchanges are considered closely spaced when the distance from the end of the acceleration lane to the beginning of the deceleration lane is less than 640 m.
- It is safe to maintain the basic number of lanes through interchanges. For example, when a ramp with two-lane exit or entrance is used, an auxiliary lane may be provided as a typical treatment for lane balance through the main roadway (Fig. 15.38). The exception to this rule is when the basic number of lanes is changed at an interchange.

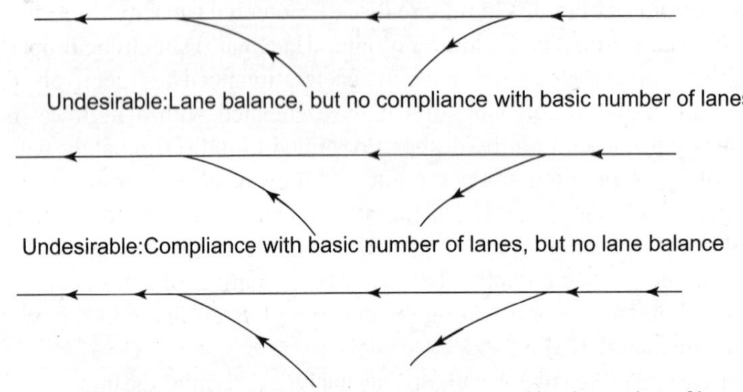


**Fig. 15.38** Coordination of lane balance and basic number of lanes (Redrawn based on AASHTO, 2004 and WSDOT, 2014)

- Not more than one traffic lane should be reduced at a time on any through lane of an interchange.
- The principle of lane balance is to obtain balance between traffic load and the capacity is provided. Hence, as a common practice, *auxiliary lanes* are lengthened (450–900 m) to meet the weaving requirements between interchanges, accommodate variations in traffic pattern across interchanges and also to reduce lane changing.

It must be noted that the basic number of lanes should not be sacrificed in an attempt to follow the concept of lane balance (Fig. 15.39) (Mulinazzi, 1973).

## 15.8.2 Weaving Section

Weaving is the phenomenon of *crossing* of traffic streams moving in the same direction, accompanied by successive merging and diverging. The weaving section is a *length* of a one-way roadway on which vehicles have crossing paths to change lanes or merge or diverge with through traffic. Weaving sections can be classified based on the type of weaving form (Leisch and Leisch, 1984):

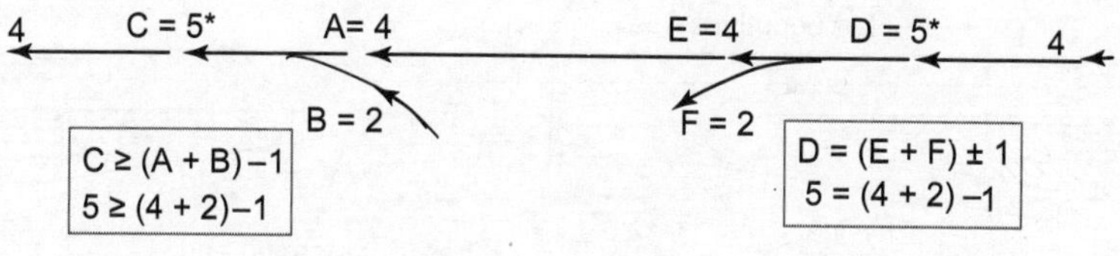


**Fig. 15.39** Another example of lane balancing
(Redrawn from Mulinazzi, 1973)

- **Simple weaving section**: This consists of a single weaving segment in which merging and diverging of traffic takes place across two adjacent undivided roadways, operated as a one-way roadway (Fig. 15.40 (a)).

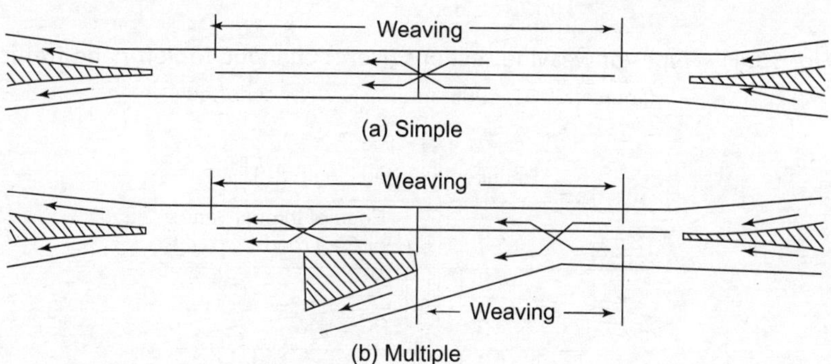


**Fig. 15.40** Types of weaving sections (*Note*: Changed for left-hand traffic)
(*Source*: CDOT, 2005; *Courtesy*: www.codot.gov)

- **Multiple weaving sections**: Multiple weaving sections may be formed in sequence by a series of ramp junctions (Fig. 15.40 (b)). These are considered the most complex operation sections.
- **One-sided weaving section**: A weaving section that is formed by one left-hand entry followed by a left-hand exit is termed a *ramp weave*. This type of weaving section may also be formed on a continuous C–D road when the right-hand entry is followed by a right-hand exit (Fig. 15.41 (a)). **Two-sided weaving section**: A weaving section that is formed by one left-hand entry followed by a right-hand exit, and vice versa (Fig. 15.41 (b)).

As described above, weaving sections may occur within an interchange, between entrance ramps followed by exit ramps and on segments of overlapping roadways. Interchanges should preferably be designed to completely eliminate weaving within the interchange or C–D roads. Otherwise, appropriate channelisation should be designed to minimise/eliminate weaving conflicts.

While selecting the appropriate layout for an interchange for a given topography, the minimum spacing required between the possible combinations of exit–entry ramps and the merging/diverging points of the turning roadways should be considered (Fig. 15.42). However, such minimum weaving distances may not be applicable in the case of cloverleaf loop ramps, where the distance between gores depends on the ramp radii, the roadway and the median width.

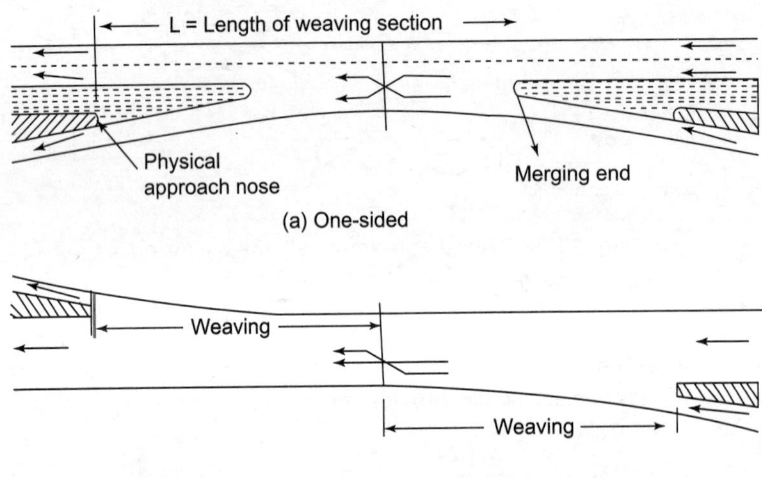


**Fig. 15.41** Types of weaving sections (*Note*: Changed for left-hand traffic)
(*Source*: CDOT, 2005; *Courtesy*: www.codot.gov)

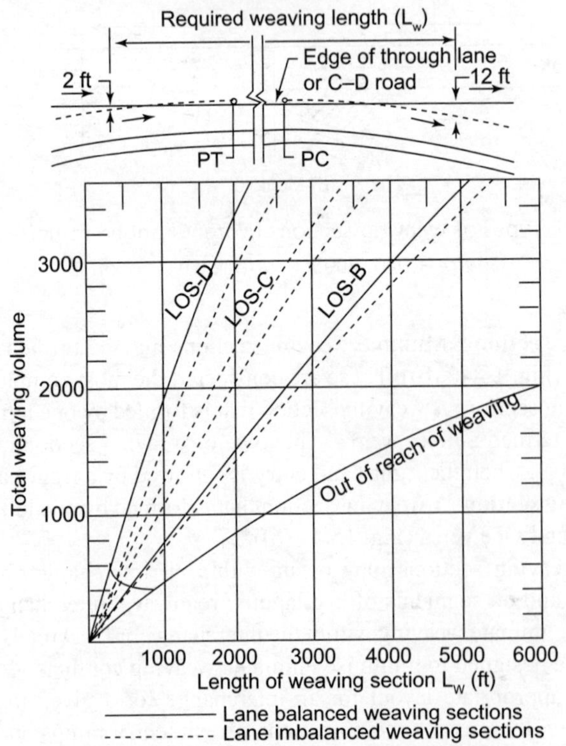


**Fig. 15.42** Length of weaving sections
(*Source*: WSDOT, 2014; *Courtesy*: www.wsdot.wa.gov)

The capacity of a weaving section depends on the provision of adequate length and width and on lane balance. The minimum weaving section length can be obtained from the minimum spacing criteria for ramp pairs (Fig. 15.42). A preliminary design of weaving section length and width must be carried out for a desirable LOS (Fig. 15.43). Other parameters such as proportion of weaving volume and non-weaving traffic volume, service volume, gradients, proportion of truck traffic, design speed, speed of weaving traffic and nearby traffic control devices are also considered in this analysis. The Highway Capacity Manual should be used to finalise the details of a weaving section.

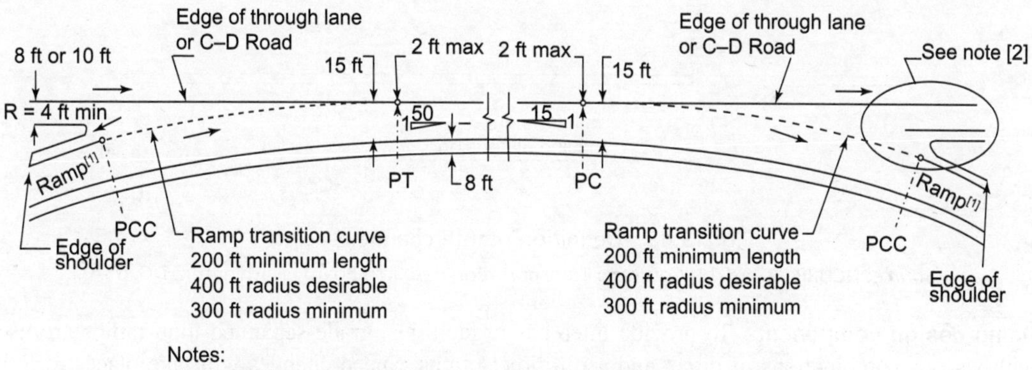


**Fig. 15.43** Typical details of loop ramp connections
(*Source*: WSDOT, 2014; *Courtesy*: www.wsdot.wa.gov)

## 15.8.3 Ramp Design

A ramp at any interchange will terminate or begin with an at-grade intersection, and will be access-controlled with a Stop sign or a traffic signal. Ramp design involves geometrical elements, spacing of ramps and operational parameters (Fig. 15.44) that particularly depend on the class of road and location. Some of the factors affecting ramp design are explained below (NCHRP 687, 2011).

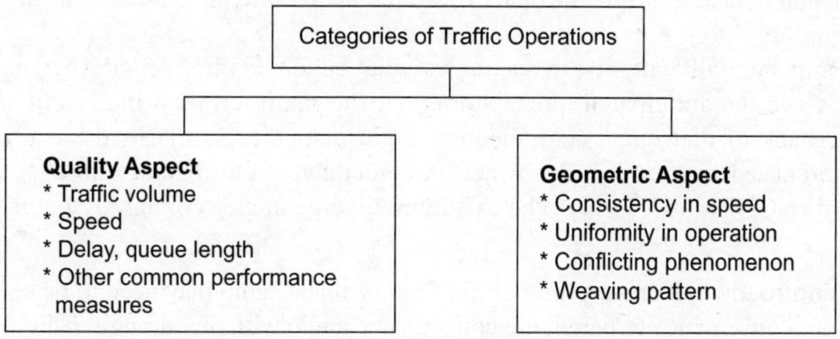


**Fig. 15.44** Categories of traffic operations

**Interchange spacing** This is the distance between the respective centrelines of highway cross-streets that include ramp configurations (Fig. 15.45). The AASHTO (2004) Green Book suggests that the minimum spacing between adjacent interchanges be 1 mile on interstate highways in urban areas and 2 miles on non-interstate highways in rural areas. The 1-mile minimum spacing criterion may be relaxed for designs developed with grade-separated ramps or C–D roads in urban areas. According to the AASHTO (2004) Policy on Design Standards, the minimum recommended spacing is 3 miles in rural areas. However, some states in the USA recommend a spacing of 2 or 6 miles in rural areas. The spacing between interchanges may be finally decided based on the spacing between ramp connections, particularly in urban areas.

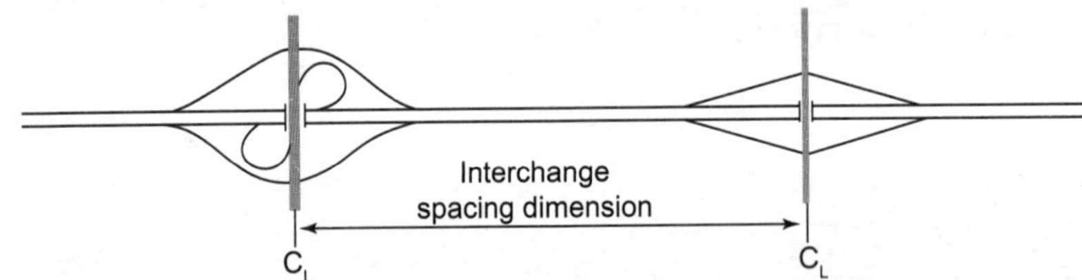


**Fig. 15.45** Definition of interchange spacing

(*Source*: NCHRP 687, 2011; *Courtesy*: Transportation Research Board, Washington, DC, USA)

**Ramp design components** To provide interchange facilities, grade-separated loop ramps are used with various combinations of direct and semi-direct ramps. Speed changes will take place from the main road to the ramp, and vice versa. The directional distribution of traffic volume, traffic intensity and design speed on the main road will determine the choice of interchange and its design components (Fig. 15.46) such as ramp configuration, tangent length of ramp for acceleration/deceleration, speed on ramp, sight distance, queue length, weaving length and ramp pair spacing. An important design aspect is that speed transition principles are applied to determine the length of the ramp required to facilitate decelerating and accelerating traffic, in conjunction with sight distance.

**Ramp entrance–exit configuration** Different configurations of ramp can be aligned to design the ramp entrance and spacing (Fig. 15.47). If the exit ramp terminal downstream is fixed, the turning ramp configuration of the single entrance yields a unique value of ramp spacing (Fig. 15.47 (a)). The spacing could be reduced in the case of multiple successive entrance ramp configuration (Fig. 15.47 (b)). Therefore, upstream interchange ramp configuration influences the spacing of the exit ramp, and hence, this specific spacing value may dictate the choice of upstream side ramp interchange configuration.

The exit ramp should be aligned in such a way that adequate ramp length (beyond the physical gore) should serve the anticipated queue storage at the ramp terminal intersection, deceleration lengths to the back of that queue and adequate sight distance. Similarly, the entry ramp should have adequate curved tangent length to facilitate adequate sight distance and acceleration after the entry point (NCHRP 687, 2011). The exit ramp diverge angles commonly range from 2° to 5° (NCHRP 687, 2011).

**Effect of turning roadway convergence angle** A curvilinear ramp may need to be extended beyond the point of gore convergence to flatten the convergence angle with the through main road. The flatter convergence angle on the ramp shifts the entrance ramp terminal downstream when compared to the location of the ramp terminal with a greater convergence angle (Fig. 15.48). In addition, the flatter convergence angle also facilitates the introduction of superelevation transition and satisfies the sight distance design requirement.

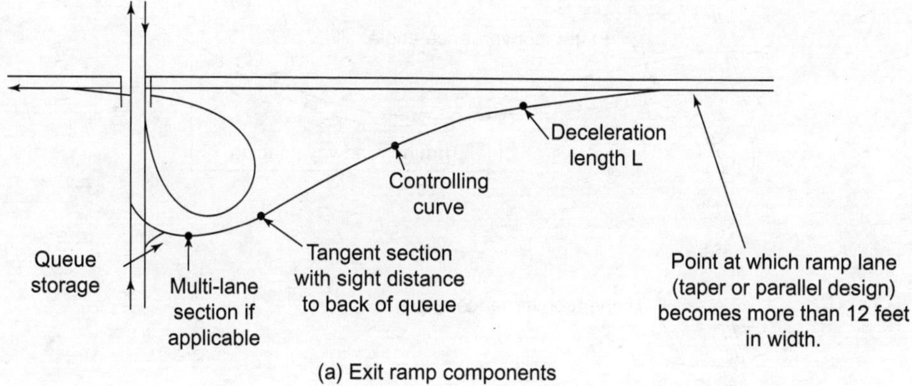


(a) Exit ramp components

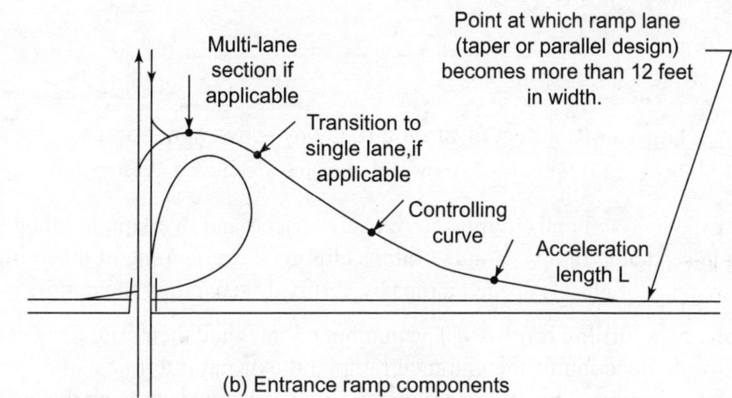


(b) Entrance ramp components

**Fig. 15.46** Ramp design components (*Note*: Changed for left-hand traffic)
(*Source*: NCHRP 687, 2011; *Courtesy*: Transportation Research Board, Washington, DC, USA)

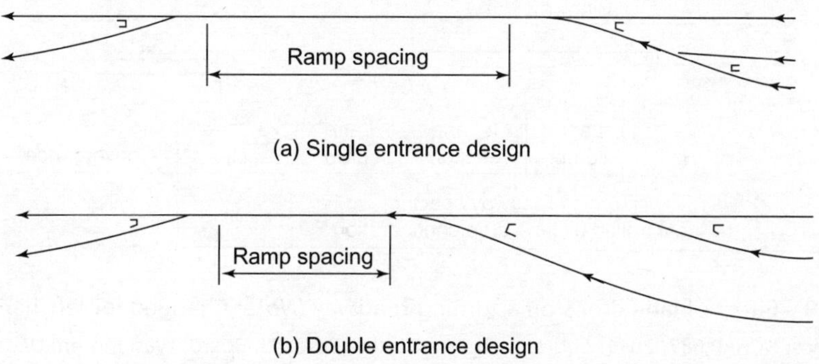


(a) Single entrance design

(b) Double entrance design

**Fig. 15.47** Configuration of single and multiple entrance ramps (*Note*: Changed for left-hand traffic)
(*Source*: NCHRP 687, 2011; *Courtesy*: Transportation Research Board, Washington, DC, USA)

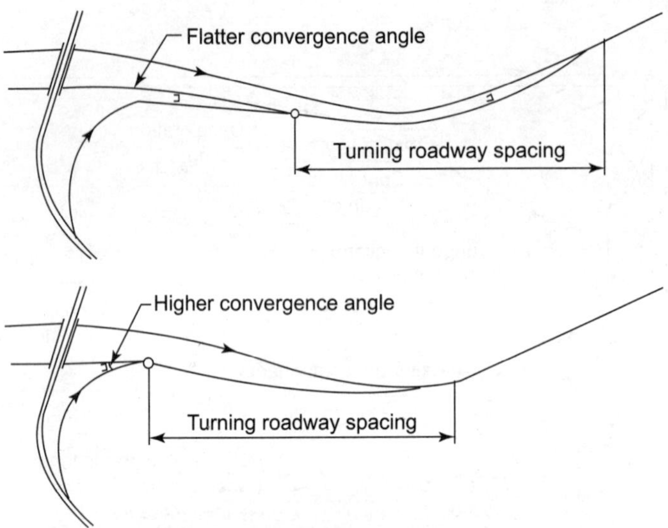


**Fig. 15.48** Effect of turning roadway convergence angle

(*Source*: NCHRP 687, 2011; *Courtesy*: Transportation Research Board, Washington, DC, USA)

**Braided ramps** If the exit and entry ramps are closely spaced and overlapping, their intersection is provided with grade-separated ramps. Braided ramps eliminate overlapping of diverging and merging and the resulting weaving of closely spaced ramps by vertically separating the ramps.

**Effect of lane drops on a turning roadway** The number of lanes and their arrangement for the turning radius will influence the location of the entrance ramp and exit ramp terminals on the highway. The locations of the merging points along the turning ramps are determined based on the need to drop lanes before the ramp–highway junction (Fig. 15.49). This is the basis for determining the length of a ramp, spacing between ramp terminals and spacing between successive interchanges. The lane drop distance may be provided in the range of 500 m to 1000 m (WSDOT, 2014).

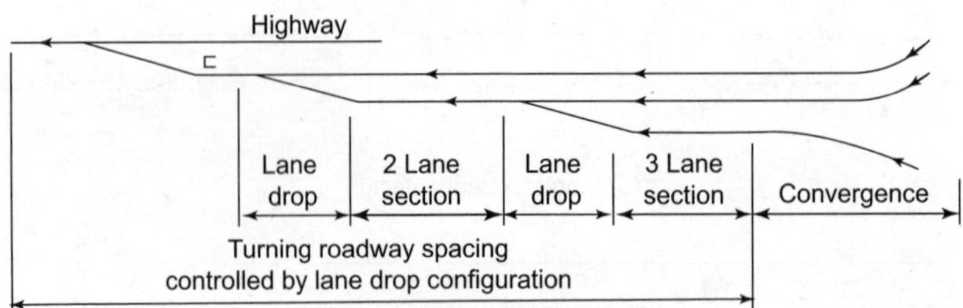


**Fig. 15.49** Effect of lane drops on a turning roadway (*Note*: Changed for left-hand traffic)

(*Source*: NCHRP 687, 2011; *Courtesy*: Transportation Research Board, Washington, DC, USA)

**Cross-section** Conventionally, a ramp cross-section comprises single or double lanes adjoined with shoulders on either side provided with camber (Fig. 15.50). The standard single-lane highway ramp width is 8 m (26 ft). A shoulder with additional width of 0.6 m (2 ft) is provided on the left to accommodate roadside barriers. Similarly, a two-lane ramp is 12.2 m (40 ft) wide, and comprises two lanes in the middle, a 3-m-wide (10 ft) shoulder on the left and a 1.8-m-wide (6 ft) shoulder on the right (ODOT, 2012). Camber (1.7–2.0%) and superelevation (4%) criteria for ramps with shoulders are suitably used based on urban roadway geometric design standards.

The loop ramp shoulder width should accommodate off-tracking of trucks and articulated vehicles. The standard is that loop ramps should be as large as practically possible and with a minimum curve of 36°. For exit ramp curves, the maximum is 30° with a longer spiral (ODOT, 2012).

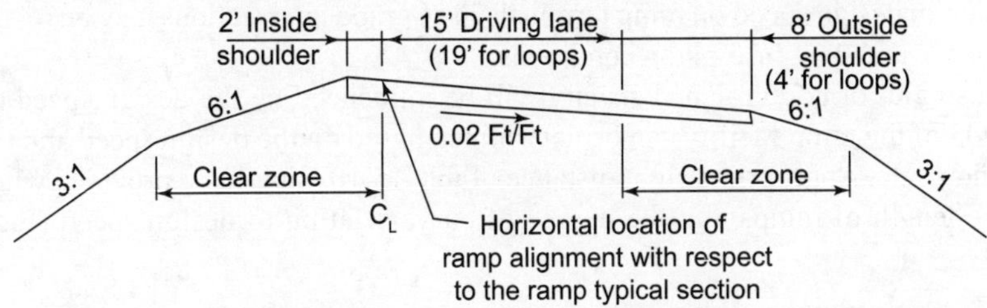


**Fig. 15.50** Typical section of ramp

(*Source*: SDDOT, 2015; *Courtesy*: Mr. Mark A. Leiferman of www.sddot.com)

The minimum ramp width is determined based on the number of lanes and their width (Table 15.7). The shoulder width and medians are appropriately selected with the minimum ramp width.

**Table 15.7** Ramp widths (Source: WSDOT, 2014; Courtesy: www.wsdot.wa.gov)

| | *Number of lanes* | | *1* | *2* |
|---|---|---|---|---|
| | Traveled Way | | 15 [1] | 25 [2] |
| **Ramp Width (ft)** | Shoulders | Right | 8 | 8 |
| | | Left | 2 | 4 |
| | Medians[3] | 6 | 8 | 8 |

**Notes:**

[1] May be reduced to 12 ft on tangents.

[2] Add 12 ft for each additional lane.

[3] The minimum two-way ramp median width (incluuing shoulders) is given. Wider medians may be required for signs or other traffic control devices and their respective clearances. When either the on- or off-ramp is single-lane. use the one-lane column. If both directions are two lanes, use the two-lane column.

1 ft = 0.3048 m. *Note*: Right-hand traffic

**Design speed on ramps** The design speed of ramps varies from 40 km/hr to 120 km/hr and mainly depends on the design speed on the main road. In addition, other factors may influence the design speed on a ramp, such as ramp type (directional or non-directional), configuration (curve alignment), gradient, weaving characteristics, spacing between adjacent ramps and spacing between successive interchanges.

Smooth transitional speed should preferably be provided on connections to ramps and on the through line. The minimum desirable speed on ramps and loop ramps is not less than 40 km/hr. However, upper range values are preferred when relating to mainline design speed (Table 15.8).

**Table 15.8** Ramp design speed (Source: WSDOT, 2014; Courtesy: www.wsdot.wa.gov)

| *Mainline design speed (mph)* | | *50* | *55* | *60* | *65* | *70* | *80* |
|---|---|---|---|---|---|---|---|
| **Ramp Design Speed (mph)** | Upper Range | 45 | 50 | 50 | 55 | 60 | 70 |
| | Midrange | 30 | 40 | 45 | 45 | 50 | 60 |
| | Lower Range | 25 | 25 | 25 | 25 | 25 | 25 |

The above minimum design speed on ramp terminals in at-grade intersections may be relaxed to 15 mi/hr for turning traffic and 0 mi/hr for stop conditions.

Ramps with single- or double-lane capacity will be influenced by the design speed (Table 15.9). Appropriate LOS of the ramp facility can be determined based on the design speed and the proposed/exiting traffic flow rate. Appropriate sight distance (Table 15.10) should be provided while designing the spiral curve length of ramps and the degree of curve relating to design speed (Figs 15.51 and 15.52).

**Table 15.9** Capacity of single-lane ramp (*Source:* ODOT, 2012; Courtesy; www.oregon.gov/odot/hwy/engservices)

| *Approximate Service Flow Rates for Single Lane Ramps (Passenger Cars Per Hour)* | | | | |
|---|---|---|---|---|
| *LOS* | *Ramp Design Speed (mph)* | | | |
| | *21–30* | *31–40* | *41–50* | *Over 50* |
| A | • | * | • | 600 |
| B | • | * | 900 | 900 |
| C | * | 1100 | 1250 | 1300 |
| I) | 12(H) | 1350 | 1550 | 1600 |
| E | 1450 | 1600 | 1650 | 1700 |
| F | Variable | Variable | Variable | Variable |

* Level of service not obtainable due to restricted design speed.

*Note*: For two lane ramps, multiply the values in the table by:

• 1.8 for 21–30 mph • 1.9 for 31–40 mph • 2.0 for 41 mph or over

**Table 15.10** A partial table of sight distance on ramps (Source: ODOT, 2012; Courtesy; www.oregon.gov/odot/hwy/engservices)

| Speed (mph) | 25 | 30 | 35 | 40 | 45 | 50 | 55 | 60 |
|---|---|---|---|---|---|---|---|---|
| Sight Distance (ft) | 155 | 200 | 250 | 305 | 360 | 425 | 495 | 570 |

1 ft = 0.3048 m; 1 mph = 1.60934 km/hr

**Ramp grade** The maximum and desirable gradients applicable to ramps for various design speeds are selected suitably (Table 15.11). In India, a gradient greater than 3.5% should not be provided (IRC: SP: 90-2010).

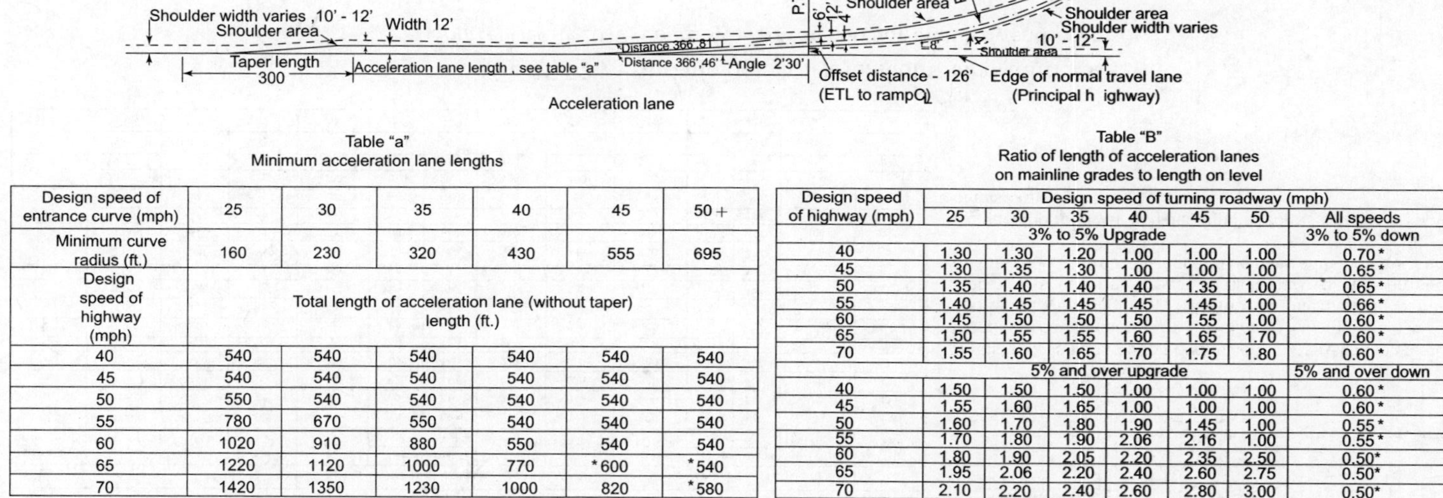


Table "a"
Minimum acceleration lane lengths

| Design speed of entrance curve (mph) | 25 | 30 | 35 | 40 | 45 | 50 + |
|---|---|---|---|---|---|---|
| Minimum curve radius (ft.) | 160 | 230 | 320 | 430 | 555 | 695 |
| Design speed of highway (mph) | Total length of acceleration lane (without taper) length (ft.) | | | | | |
| 40 | 540 | 540 | 540 | 540 | 540 | 540 |
| 45 | 540 | 540 | 540 | 540 | 540 | 540 |
| 50 | 550 | 540 | 540 | 540 | 540 | 540 |
| 55 | 780 | 670 | 550 | 540 | 540 | 540 |
| 60 | 1020 | 910 | 880 | 550 | 540 | 540 |
| 65 | 1220 | 1120 | 1000 | 770 | *600 | *540 |
| 70 | 1420 | 1350 | 1230 | 1000 | 820 | *580 |

*Desirable length is 750'. This allows additional decision time for picking gaps

Table "B"
Ratio of length of acceleration lanes on mainline grades to length on level

| Design speed of highway (mph) | Design speed of turning roadway (mph) | | | | | | |
|---|---|---|---|---|---|---|---|
| | 25 | 30 | 35 | 40 | 45 | 50 | All speeds |
| | 3% to 5% Upgrade | | | | | | 3% to 5% down |
| 40 | 1.30 | 1.30 | 1.20 | 1.00 | 1.00 | 1.00 | 0.70* |
| 45 | 1.30 | 1.35 | 1.30 | 1.00 | 1.00 | 1.00 | 0.65* |
| 50 | 1.35 | 1.40 | 1.40 | 1.40 | 1.35 | 1.00 | 0.65* |
| 55 | 1.40 | 1.45 | 1.45 | 1.45 | 1.45 | 1.00 | 0.66* |
| 60 | 1.45 | 1.50 | 1.50 | 1.50 | 1.55 | 1.00 | 0.60* |
| 65 | 1.50 | 1.55 | 1.55 | 1.60 | 1.65 | 1.70 | 0.60* |
| 70 | 1.55 | 1.60 | 1.65 | 1.70 | 1.75 | 1.80 | 0.60* |
| | 5% and over upgrade | | | | | | 5% and over down |
| 40 | 1.50 | 1.50 | 1.50 | 1.00 | 1.00 | 1.00 | 0.60* |
| 45 | 1.55 | 1.60 | 1.65 | 1.00 | 1.00 | 1.00 | 0.60* |
| 50 | 1.60 | 1.70 | 1.80 | 1.90 | 1.45 | 1.00 | 0.55* |
| 55 | 1.70 | 1.80 | 1.90 | 2.06 | 2.16 | 1.00 | 0.55* |
| 60 | 1.80 | 1.90 | 2.05 | 2.20 | 2.35 | 2.50 | 0.50* |
| 65 | 1.95 | 2.06 | 2.20 | 2.40 | 2.60 | 2.75 | 0.50* |
| 70 | 2.10 | 2.20 | 2.40 | 2.60 | 2.80 | 3.00 | 0.50* |

*If the ratio Table "B" yields a length less than 540', use 540' as the minimum acceleration lane length. The desirable length is 750', this allows additional decision time for picking gaps.

Entrance curves and spirals

| Ramp design speed (mph) | Degree of curve | Radius (ft.) | 0 (%) | Spiral length (ft.) | Spiral parts 'a' value | Spiral parts 'S' value |
|---|---|---|---|---|---|---|
| 70 | 1°00' | 5729.58 | 3.5 | 200 | 0.5 | 1.0 |
| | 2°00' | 2884.79 | 6.5 | 200 | 1.0 | 2.0 |
| | 3°00' | 1909.06 | 8 | 240 | 1.25 | 3.6 |
| 65 | 4°00' | 1432.39 | 9 | 240 | 1.667 | 4.8 |
| 60 | 5°00' | 1145.92 | 9.5 | 250 | 2.0 | 6.25 |
| 55 | 6°00' | 954.93 | 10 | 240 | 2.5 | 7.2 |
| 50 | 7°00' | 818.51 | 10.5 | 240 | 2.917 | 8.4 |
| | 8°00' | 716.20 | 10.5 | 250 | 3.2 | 10.0 |
| 45 | 10°00' | 572.96 | 11 | 250 | 4.0 | 12.5 |
| 40 | 12°00' | 477.46 | 11.5 | 240 | 5.0 | 14.4 |
| | 14°00' | 409.26 | 11.5 | 200 | 7.0 | 14.0 |
| 35 | 16°00' | 358.10 | 12 | 200 | 8.0 | 16.0 |
| | 18°00' | 318.31 | 12 | 200 | 9.0 | 18.0 |
| 30 | 20°00' | 286.48 | 12 | 200 | 10.0 | 20.0 |
| | 24°00' | 238.73 | 12 | 200 | 12.0 | 24.0 |
| 25 | 30°00' | 190.99 | 12 | 200 | 15.0 | 30.0 |
| | 36°00' | 159.15 | 12 | 200 | 18.0 | 36.0 |

Example:

Find: Adjusted acceleration lane length
Given: Principal highway design speed 70 mph
Turning highway design speed 30 mph
Meeting on a 5% upgrade

Solution: From Table "A" acceleration lane length =1350
From Table "B" ratio on upgrade = 2.20
Adjusted acceleration lane length = 1350 × 2.20 = 2970'

Note: If adjusted acceleration lane length appears impractical, a merge analysis should be performed. Lengthening of acceleration lane length may be neccessary due to high truck volumes.

**Fig.15.51** Entrance ramp details (Note: Right-hand traffic)
(*Source*: ODOT, 2012; *Courtesy*: www.oregon.gov/odot/hwy/engservices)

Example 1:
Find: Deceleration lane length
Given: One lane ramp, 24' curve
Design speed -30 mph (from table)
Less than 3% downangle
Solution: T = 420', if sprial length = 100' (from table)

T = 420'
+ 1/2 Sprial length = 100'
Deceleration lane length = 520'

Example 2:
Find adjusted 'T' distance
Given: Data from example 1
Deceleration lane length (calculated in example 1) = 1
0% minimum downgrade
Solution: Adjusted deceleration lane length
520' ×1.30' (from table) = 702' (rounded 705')

Adjusted Length = 700'
± Sprial Length= 100'
Adjusted 'T' distance = 800'

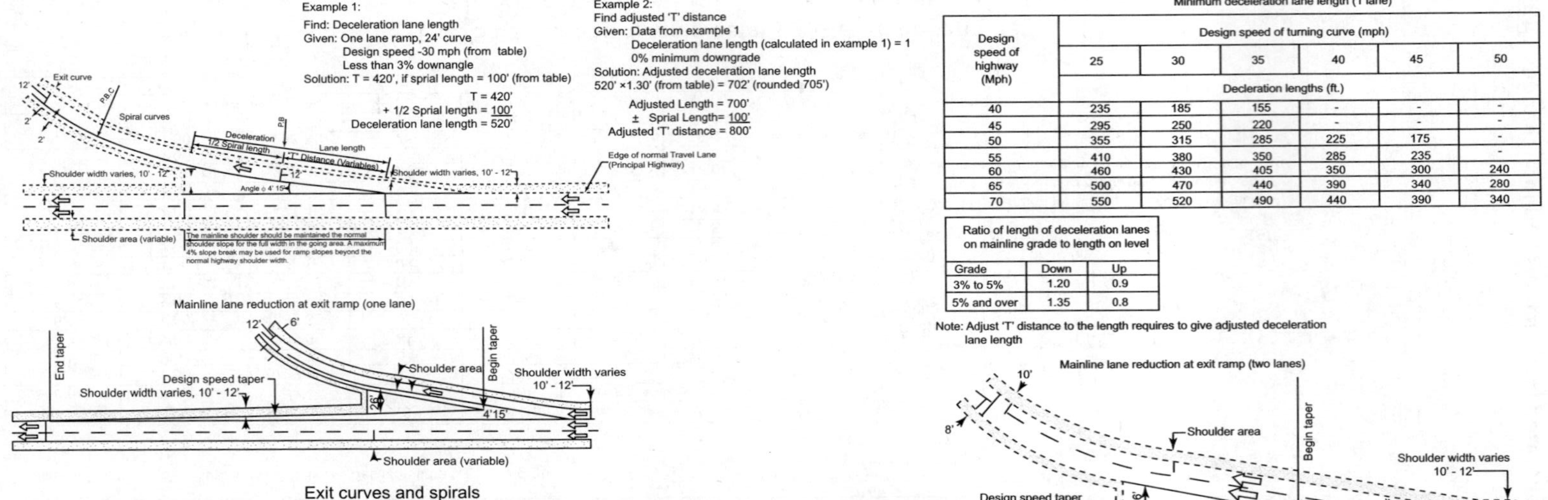


Minimum deceleration lane length (1 lane)

| Design speed of highway (Mph) | Design speed of turning curve (mph) | | | | | |
|---|---|---|---|---|---|---|
| | 25 | 30 | 35 | 40 | 45 | 50 |
| | Decleration lengths (ft.) | | | | | |
| 40 | 235 | 185 | 155 | - | - | - |
| 45 | 295 | 250 | 220 | - | - | - |
| 50 | 355 | 315 | 285 | 225 | 175 | - |
| 55 | 410 | 380 | 350 | 285 | 235 | - |
| 60 | 460 | 430 | 405 | 350 | 300 | 240 |
| 65 | 500 | 470 | 440 | 390 | 340 | 280 |
| 70 | 550 | 520 | 490 | 440 | 390 | 340 |

| Ratio of length of deceleration lanes on mainline grade to length on level | | |
|---|---|---|
| Grade | Down | Up |
| 3% to 5% | 1.20 | 0.9 |
| 5% and over | 1.35 | 0.8 |

Note: Adjust 'T' distance to the length requires to give adjusted deceleration lane length

Mainline lane reduction at exit ramp (two lanes)



'T' distance is based upon normal length. Variation in spiral length will cause change in 'T' distance. If spiral length is shortened then lengthen 'T' distance for minimum deceleration length. 'T' should be increased if high truck volumes are present on the ramp. 'T' truck distance to be used with high truck volumes.

Contact the roadway engineering unit for assistance on any portion of this drawing.

Exit curves and spirals

| | | | | 1 lane (highway design speed 70 mph) | | | | | 2 lane (highway design speed 70 mph) | | | | |
|---|---|---|---|---|---|---|---|---|---|---|---|---|---|
| Ramp design Speed (mph) | Degree of curve | Radius (ft) | (%) | Spiral length ft) | Spiral parts 'a' Value | 'S' Value | 'T' Distance (ft) | 'T' truck Distance (ft) | Spiral length (Tl) | Spiral parts 'a' Value | 'S' Value | 'T' Distance (ft) | 'T' TRUCK Distance (ft) |
| 70 | 1'00' | 5729.58 | 3.5 | 200 | 0.5 | 1.0 | 240 | 340 | 300 | 0.333 | 1.0 | 300 | 400 |
| | 2'00' | 2884.79 | 6.6 | 200 | 1.0 | 2.0 | 240 | 340 | 300 | 0.667 | 3.0 | 300 | 400 |
| | 3'00' | 1909.66 | 8 | 240 | 1.25 | 3.6 | 220 | 320 | 360 | 0.833 | 5.4 | 300 | 370 |
| 65 | 4'00' | 1432.39 | 9 | 240 | 1.667 | 4.8 | 220 | 320 | 360 | 1.111 | 7.2 | 300 | 370 |
| 60 | 5'00' | 1145.92 | 9.5 | 250 | 2.0 | 6.25 | 215 | 316 | 375 | 1.333 | 9.375 | 300 | 385 |
| 55 | 6'00' | 954.90 | 10 | 240 | 2.5 | 7.2 | 220 | 320 | 360 | 1.667 | 10.8 | 300 | 370 |
| 50 | 7'00' | 818.51 | 10.5 | 240 | 2.917 | 8.4 | 220 | 400 | 360 | 1.944 | 12.8 | 300 | 450 |
| | 8'00' | 718.20 | 10.5 | 260 | 3.2 | 10. | 215 | 395 | 376 | 2.133 | 15.0 | 300 | 445 |
| 45 | 10'00' | 572.98 | 11 | 260 | 4.0 | 12.5 | 255 | 470 | 375 | 2.667 | 18.75 | 315 | 420 |
| 40 | 12'00' | 477.48 | 11.5 | 240 | 5.0 | 14.4 | 320 | 545 | 360 | 3.333 | 21.6 | 370 | 595 |
| | 14'00' | 409.28 | 11.5 | 200 | 7.0 | 14.0 | 340 | 565 | 300 | 4.667 | 21.0 | 400 | 625 |
| 35 | 16'00' | 350.10 | 12 | 200 | 8.0 | 16.0 | 360 | 660 | 300 | 5.333 | 24.0 | 450 | 710 |
| | 18'00' | 318.31 | 12 | 200 | 9.0 | 18.0 | 380 | 660 | 300 | 6.0 | 27.0 | 450 | 710 |
| 30 | 20'00' | 286.48 | 12 | 200 | 10.0 | 20.0 | 420 | 700 | 260 | 8.0 | 26.0 | 505 | 785 |
| | 24'00' | 238.73 | 12 | 200 | 12.0 | 24.0 | 420 | 700 | 240 | 10.0 | 28.8 | 510 | 790 |
| 25 | 30'00' | 190.99 | 12 | 200 | 15.0 | 30.0 | 450 | 745 | 240 | 12.5 | 36.0 | 540 | 835 |
| | 36'00' | 159.15 | 12 | 200 | 18.0 | 36.0 | 450 | 745 | 240 | 15.0 | 43.2 | 540 | 835 |

**Fig. 15.52** Exit ramp details (Note: Right-hand traffic)
(*Source*: ODOT, 2012; *Courtesy*: www.oregon.gov/odot/hwy/engservices/)

**Table 15.11** Maximum ramp grade (Source: WSDOT, 2014; Courtesy: www.wsdot.wa.gov)

| *Ramp Design Speed (mph)* | | *25–30* | *35–40* | *45 and above* |
|---|---|---|---|---|
| **Ramp Grade** (%) | Desirable | 5 | 4 | 3 |
| | Maximum′ | 7 | 6 | 5 |
| • On one-way ramps, downgrades may be 2% greater. | | | | |

**Ramp spacing** It is the distance from painted tip to painted tip of gore stripes of successive ramps (Fig. 15.53). The AASHTO (2004) Green Book has specified the minimum spacing between successive ramps for five possible ramp pair (entrance/exit) combinations and weaving length standards which are widely accepted around the world.

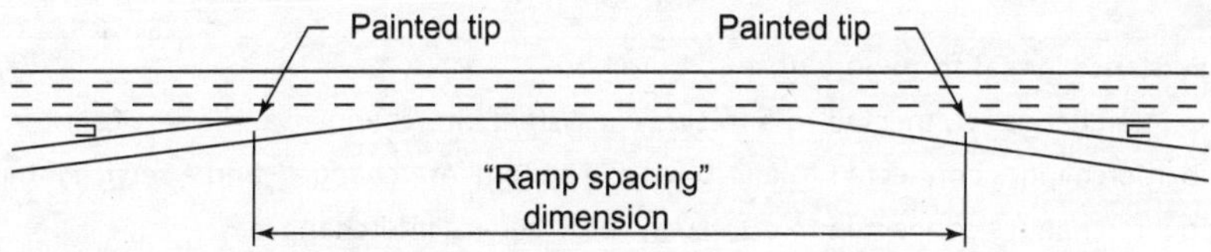


**Fig. 15.53** Definition of ramp spacing

(*Source*: NCHRP 687, 2011; *Courtesy*: Transportation Research Board, Washington, DC, USA)

**Ramp terminal** It is the portion adjacent to the through roadway which comprises speed change lanes, tapers and channelising islands. Ramp terminals may be designed at-grade, such as at the crossroad terminal or as free-flow ramps where traffic merges with or diverges from through traffic at flat angles (SDDOT). The design principle that the terminal should be close to the 90° crossing/re-align skew angle intersection at-grade is applicable here.

## 15.8.4 Ramp Spacing

In urban areas, entrance/exit ramps may need to be provided in close succession. Recent research has indicated that the majority of accidents occur at interchanges and in the weaving sections between closely spaced entrance and exit ramps.

Geometric design principles and site-specific conditions dictate that the minimum spacing between the gore noses of adjacent ramps be considered with the associated components of interchanges (Fig. 15.54). However, intensity of traffic volumes and their weaving characteristics can necessitate increased spacing beyond the need from dimensions based on geometrics. Design/Operating speeds and human factors also need to be considered for signage and interchange spacing.

## 15.8.5 Entrance Ramp Taper Design

The entrance ramp connects with the mainline of the highway. It comprises two components: an acceleration lane and a taper (Fig. 15.51). The acceleration lane allows speeding up and maintains acceptable gaps between successive vehicles on the highway while the taper facilitates entry of vehicles to the through lane.

The minimum acceleration lane length required depends on the design speed of the ramp. If the average gradient is 3% or greater, the obtained length is further increased by multiplying with the factor relating to the average grade of the acceleration lane (Fig. 15.51).

| On–On or Off–Off | | Off–On | | Turning Roadways | | On–Off (Weaving) | | | |
|---|---|---|---|---|---|---|---|---|---|
| L | | L | | L | | L[1] | | | |
| Freeway | C–D Road | Freeway | C–D Road | System[2] Inter-change | Service[3] Inter-change | A | B | C | D |
| 1000 | 800 | 500 | 400 | 800 | 600 | 2000 | 1600 | 1600 | 1000 |

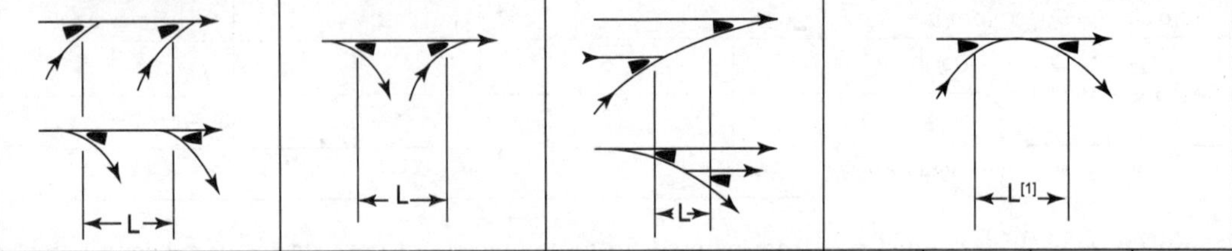

Gore nose

L = Minimum distance in feet from gore nose to gore nose.

A Between two interchanges connected to a freeway: a system interchange[2] and a service interchange.[3]

B Between two interchanges connected to a C–D road: a system interchange[2] and a service interchange.[3]

C Between two interchanges connected to a freeway: both service interchanges.[3]

D Between two interchanges connected to a C–D road: both service interchanges.[3]

*Notes:* These values are based on operational experience, need for flexibility, and signing. Check them in accordance with Exhibit 1360-12 and the procedures outlined in the Highway Capacity Manual and use the larger value.

[1] With justification, these values may be reduced for cloverleaf ramps.

[2] A system interchange is a freeway-to-freeway interchange

[3] A service interchange is a freeway-to-local road interchange.

**Fig.15.54** Minimum ramp connection spacing. *Note*: Right-hand traffic
(*Source*: WSDOT, 2014; *Courtesy*: www.wsdot.wa.gov)

## 15.8.6 Exit Ramp Taper Design

The exit ramp also connects with the mainline of the highway. It comprises two components: a deceleration lane and a taper (Fig. 15.52).

The minimum acceleration lane length is determined based on the design speed of the ramp, speed of the highway and degree of the curve. Deceleration length comprises tangent length (T) and half-spiral length. For the given conditions, the T length and half-spiral length are selected from the table (see top-left of Fig. 15.52). If the down/up gradient is 3% or greater, the obtained length is further increased by multiplying with the adjustment factor for grades (Fig. 15.51).

## 15.8.7 Gores

Gores are the decision points at on ramp/off ramp terminal areas. The gore area comprises the pavement area and an unpaved area. Standard dimensional features of gores are illustrated in Figs 15.51 and 15.52.

The paved area between the physical nose and the front of the gore nose is termed the reserve area (hatched triangle in Fig. 15.55). It is reserved for the installation of a delineator, and the attenuators/barriers should preferably be placed just after the gore nose. Minimum length of the reserve area is determined based on the design speed of the mainline.

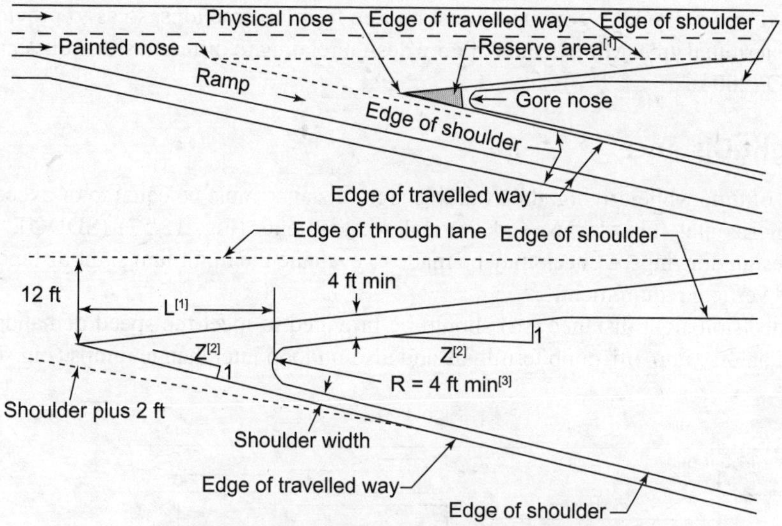

(a) Single-lane off connections: No lane reduction

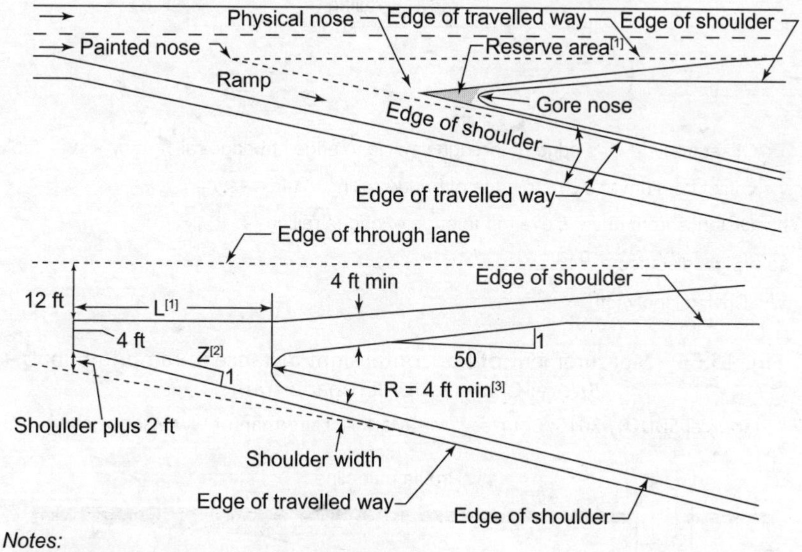

*Notes:*

(1) The reserve area length (L) is not less than:

| Mainline Design Speed (mph) | 40 | 45 | 50 | 55 | 60 | 65 | 70 | 80 |
|---|---|---|---|---|---|---|---|---|
| L (ft) | 25 | 30 | 35 | 40 | 45 | 50 | 55 | 70 |

(2) $Z = \dfrac{\text{Design Speed}}{2}$ design speed is for the mainline

(3) Radius may be reduced, when protected by an impact attenuator

(b) Single-lane, one-lane reduction off-ramp connections and all two-lane off connections

**Fig. 15.55** Gore area characteristics with reference to design speed on the mainline (Note: Right-hand traffic)

(*Source*: WSDOT, 2014; *Courtesy*: www.wsdot.wa.gov)

The unpaved area behind the gore nose must be free from obstructions, so as to provide clear sight distance in the terminal area. More details of gore design relating to different situations can be obtained from AASHTO (2004).

### 15.8.8 Sight Distance

At any instant of time while driving, the available sight distance must be equal to or exceed the largest value of the horizontal (Fig. 15.56) and vertical requirements (Fig. 15.57) (SDDOT, 2015). After preliminary design drawings, a check may be made to evaluate the minimum sight distance against the horizontal and vertical requirement.

Adequate decision sight distance (sd) should be provided to meet the speed demands of traffic at decision spots on on-ramp, off-ramp terminals and also on loop interchange ramps (Fig. 15.56).

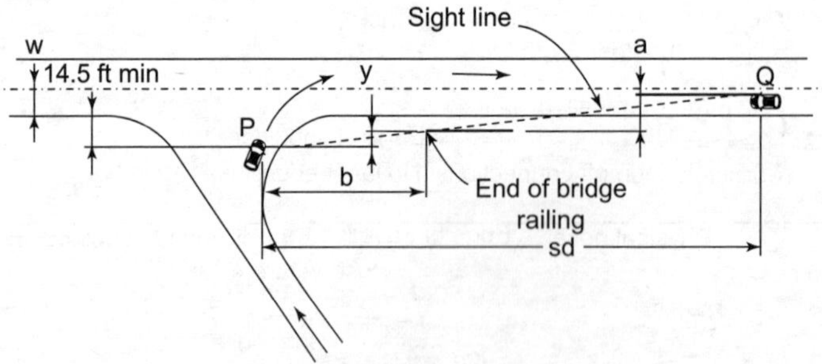


a = Offset from driver's edge of through vehicle to edge of bridge railing = w + sw – 2.5'

y = Offset from driver's eye to edge of bridge railing = 14.5' – sw

b = Distance from driver's eye to the edge of bridge railing

sd = Sight distance = $\frac{b\,(a + y)}{y}$

w = Driving lane width

**Fig. 15.56** Measurement of horizontal sight distance at ramp terminals (*Note*: Changed for left-hand traffic)

(*Source*: SDDOT, 2015; *Courtesy*: Mr. Mark A. Leiferman of www.sddot.com)

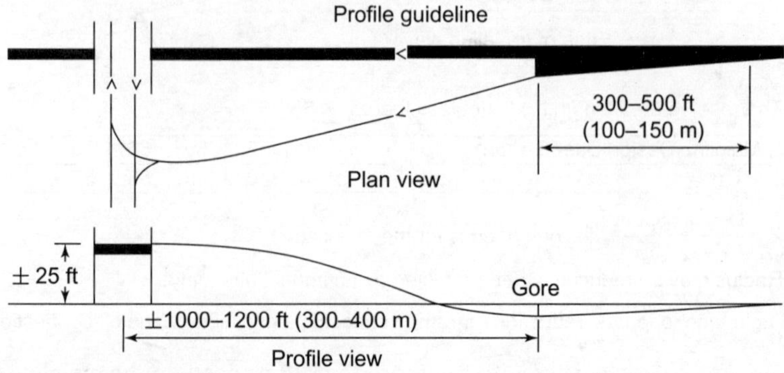


**Fig. 15.57** Conceptual plan and profile (example) (*Note*: Changed for left-hand traffic)

(*Source*: NCHRP 687, 2011; *Courtesy*: Transportation Research Board, Washington, DC, USA)

### 15.8.9 Other Considerations

A few other design aspects relating to ramp components are described below.

- A sample profile view of a right running crossroad passing over a highway is depicted in Fig. 15.57. The profile length of the ramp is dictated by site conditions and the sight distance along and at the exit terminal.
- An exit ramp with adequate deceleration length is provided before the queue to accommodate additional storage of vehicles approaching the crossroad (Fig. 15.58).
- With reference to vertical alignment, adequate decision/stopping sight distance should be provided on entry/exit ramps (Fig. 15.59). The ramp is extended in case of downhill gradient, which requires additional deceleration length before the cross-street because of the increased speed due to gravitation effect.
- The Highway Capacity Manual (TRB) states that a default value of 2% grade is provided on state highways, 4% for extended grade on rolling terrain, and 6% as maximum extended grade on highways in mountainous terrain.

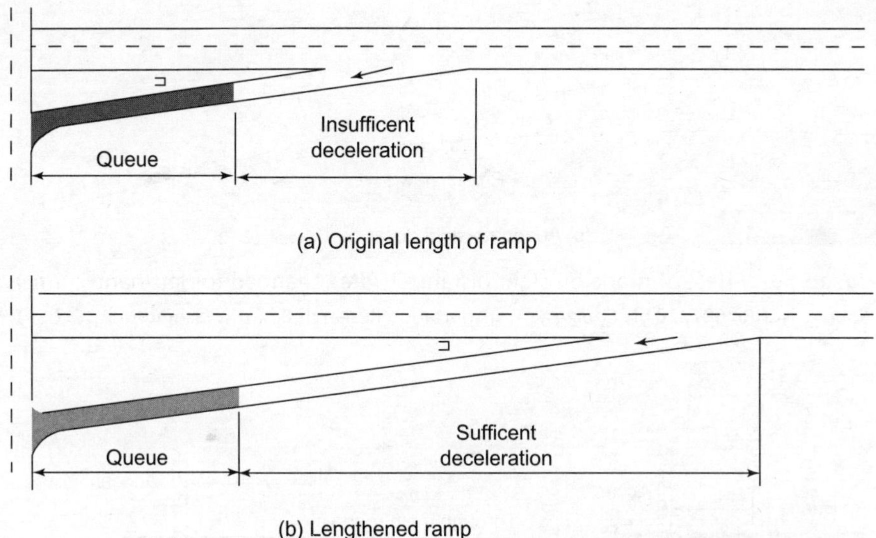


**Fig.15.58** Effect of queue storage (*Note*: Changed for left-hand traffic)
(*Source*: NCHRP 687, 2011; *Courtesy*: Transportation Research Board, Washington, DC, USA)

### 15.8.10 Access Control

Access control is of two types: full control and limited control at the entry to an interchange or grade -separated locations. Limited access control at at-grade intersections may be restricted by the type of turn (right-in/right-out). Full control of access is designed with appropriate channelisation and restricted crossing movements at the ramp terminal or at at-grade intersections. Full access control is preferred to reduce the number and frequency of accidents. The AASHTO Policy on Design Standards – Interstate System (1991) recommends that access control be extended beyond the ramp terminal for a minimum distance (L) of 30.5 m (100 ft) and 91.5 m (300 ft), respectively, in urban and rural areas (Fig. 15.60).

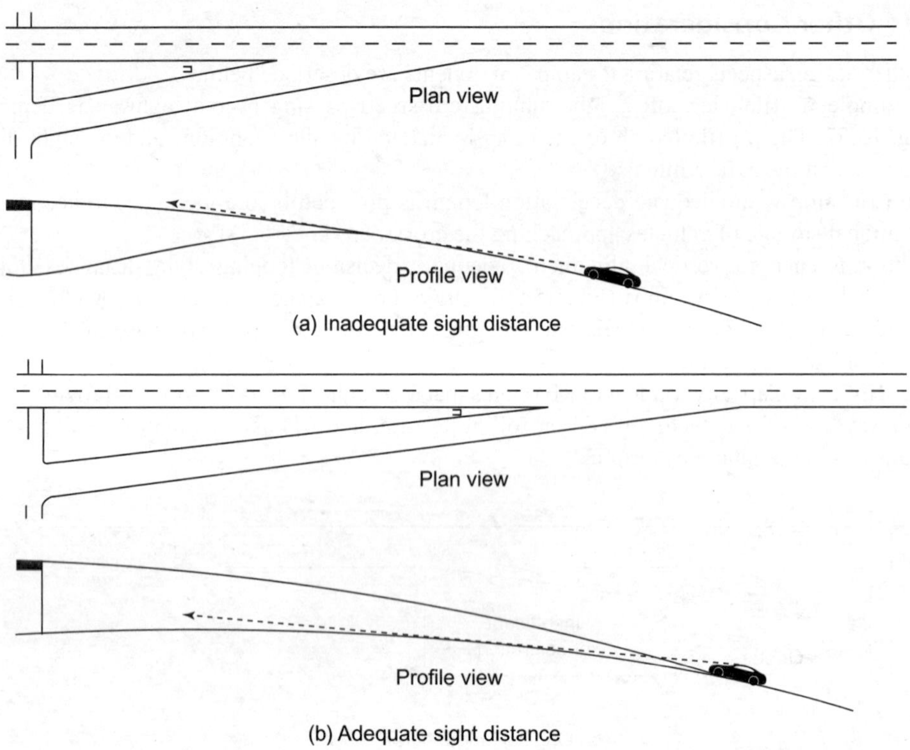


**Fig. 15.59** Effect of grade on sight distance (*Note*: Changed for left-hand traffic)
(*Source*: NCHRP 687, 2011; *Courtesy*: Transportation Research Board, Washington, DC, USA)

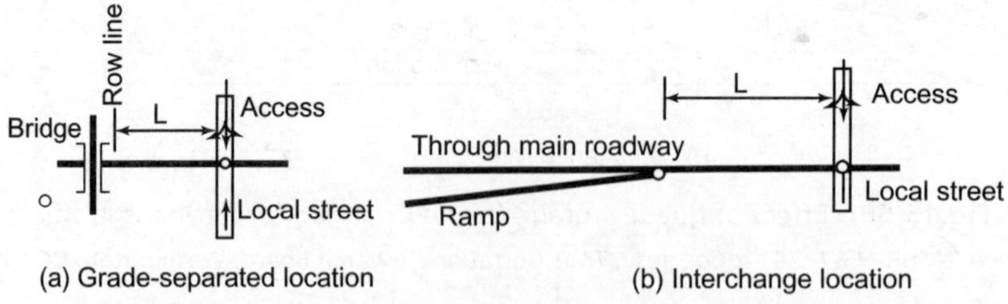


**Fig. 15.60** Access control and measurement of access distance

In addition, entry of traffic flow to a mainline can be controlled by installing signals at the entry or at the merge-mouth of a ramp terminal or link road. This signal permits the entry of vehicles on to the mainline at a 'metered rate'. The signal is linked to a system of sensors measuring speed and lane occupancy rates of traffic using the mainline. The signal is automatically actuated with the appropriate metering rate for the current traffic conditions. This system of limiting excess entry of vehicles to the mainline is termed *ramp metering* (DMRB, 2006).

### 15.8.11 Comparative and Evaluation Studies

Comparative studies are conducted to select the best interchange layout based on several factors. The main factors may be considered as capacity, efficiency (in terms of delay) and cost. Some of these factors are discussed below.

- **Capacity:** If traffic is restricted by signals, the interchange capacity is defined in terms of total interchange volume. The capacity value may vary depending on the number of lanes available on each approach to an interchange. Therefore, any standard procedure may be adopted for evaluating the capacity of a given type of interchange. If traffic is not restricted by signals, the interchanges can be evaluated based on change in loop volume with the connecting roadway volume and the specific design volume.
- **Efficiency:** The delay encountered by traffic in passing through a given type of interchange is used to define the efficiency of an interchange. Delay studies are conducted on specific movements in the interchange. The traffic movements may be categorised based on the direction of travel from one road to another or any specific interchanging movement, and the delays are compared. For example, the categories of movement in a study interchange may be split as through traffic, traffic turning right and traffic turning left. The delays observed in all candidate types of interchange layouts are compared and the interchange with the least delay observed is selected.
- **Travel time:** Travel time to total travel time (including delays) observed with respect to the above set of movements may also be considered for evaluation of existing interchanges, comparison with signalised interaction and proposed interchange types and comparison with the proposed interchange types by *simulation*.
- **Geometrics:** The interchanging loop with its ramp geometrics can be evaluated with various combinations of entrance/exit ramps with assumed speeds.
- **Cost comparison:** To facilitate a comparison of cost between the selected types of interchanges, it is necessary to develop standard design proposals and then apply a unit cost figure to each cost item for the types of interchanges considered. In order to include all of the cost items, the major inputs are listed as area required, amount of earthwork, number of lanes for both AM and PM peaks, loops, acceleration lanes, deceleration lanes, ramps and maintenance costs. Reduction in the number of accidents with and without improvements may also be considered, if required.

## EXERCISES

### Multiple Choice Questions

1. As per IRC: SP: 90, it is appropriate to provide flyovers, when the intersection traffic volumes near __________ PCU/hr.

   (a) 5000 (b) 7000 (c) 8000 (d) 10000

2. The jughandle ramp is an example of

   (a) Direct ramp (b) Semi-direct ramp (c) Indirect ramp (d) D merge ramp

3. The ramp on which the vehicles must first manoeuvre left to make a right turn is the

   (a) Direct ramp (b) Semi-direct ramp (c) Indirect ramp (d) U-turn ramp

4. As per Indian traffic flow rules, this ramp is safe and less hazardous:
   (a) Semi-direct ramp (b) Direct ramp (c) Indirect ramp (d) U-turn ramp
5. As a common guideline of route continuity, the design speed differential between the mainline and the interchange ramp should be smaller than ________ of the mainline design speed.
   (a) 0.3 times (b) 0.1 times (c) 0.01 times (d) 1.0 times
6. As a general practice, if the interchanges are closely spaced, ________ lanes are used to connect the successive ramps.
   (a) Auxiliary (b) Slip (c) Ramp (d) Loop
7. Usually, the minimum spacing between successive interchanges in an urban area that is considered appropriate is
   (a) 0.5 mile (b) 2/3rd mile (c) 1/3rd mile (d) 1 mile
8. What roads are used to eliminate the inherent weaving problem in the through lanes of a cloverleaf interchange?
   (a) Slip (b) Collector–Distributor
   (c) Loop lane (d) Auxiliary lane
9. To reduce the number of junction-access points as a part of interchange design, the phenomenon commonly known as lane gain and lane drop is providing auxiliary lanes with ________, respectively.
   (a) Merge taper and diverge taper (b) Diverge taper and merge taper
   (c) D diverge and D merge (d) Indirect ramp with direct ramp
10. How many lane drops should be provided in basic lanes?
    (a) 1 (b) 2 (c) 0 (d) Depends on road condition
11. Lane drops should preferably be provided on
    (a) Left-hand side (b) Right-hand side
    (c) Any side (d) Depends on the field condition
12. The lane drops may be placed at least how many metres past an interchange with an appropriate taper of 50:1 to 70:1.
    (a) 100 m (b) 250 m (c) 500 m (d) 750 m
13. All right-turn movements in an interchange are provided by
    (a) Directional ramps (b) Non-directional ramps
    (c) Semi-directional ramps (d) Auxiliary lanes
14. These interchanges comprise indirect ramp loops which are aligned in a direction opposite to the desired direction of travel
    (a) Jughandle ramp (b) Turbine (c) One quadrant (d) Parclo
15. The partially separated or not free flow interchange is
    (a) Cloverleaf (b) Trumpet (c) Turbine (d) Diamond
16. The fully separated or free flow interchange is
    (a) Parclo (b) Roundabout (c) Diamond (d) Turbine

17. An un-signalised diamond interchange can be operated with an acceptable LOS when the entering volume is up to _________ veh/hr
    (a) 1500 (b) 2000 (c) 2500 (d) 3000
18. In compressed diamond interchange, the distance between the two at-grade intersections is _________
    (a) More than 244 m (b) Between 122 m and 244 m
    (c) Between 61 m and 122 m (d) More than 500 m
19. In case of tight diamond interchange, the distance between the two at-grade intersections is
    (a) More than 244 m (b) Between 122 m and 244 m
    (c) Between 61 m and 122 m (d) Less than 44 m
20. The type of diamond interchange with crossover locations at each end of the bridge with two entrance ramps and two exit ramps is
    (a) Conventional diamond (b) Split diamond
    (c) Tight diamond (d) Diverging diamond
21. Its use reduces the number of potential vehicle conflict points to 14, as against 26 in conventional diamond interchanges
    (a) Single point urban interchange (b) Split diamond interchange
    (c) Tight diamond Interchange (d) Diverging diamond interchange
22. This type of weaving section is formed by one left-hand entry followed by a right-hand exit, and vice versa:
    (a) Simple weaving section (b) Multiple weaving section
    (c) One-sided weaving section (d) Two-sided weaving section
23. Exit ramp diverge angles commonly range from
    (a) 2° to 5° (b) 1° to 2° (c) 45° to 65° (d) 65° to 90°
24. If the exit and entry ramps are closely spaced and overlapping, their intersection is provided with _________ ramps.
    (a) Braided (b) Loop (c) Auxiliary (d) Slip lane
25. The lane drop distance may be provided in the range of _________.
    (a) 250 m to 500 m (b) 500 m to 1000 m
    (c) 2000 m to 2500 m (d) 4000 m to 5000 m
26. As a standard, loop ramps should be as large as practically possible and with a minimum curve of
    (a) 10° (b) 36° (c) 15° (d) 26°
27. As per IRC: SP: 90, the maximum and desirable gradients applicable to ramps for various design speeds should not exceed _________
    (a) 3.5% (b) 4.5% (c) 5.5% (d) 7.5%
28. The portion adjacent to the through roadway which comprises speed change lanes, tapers and channelising islands is
    (a) Ramp terminal (b) Loop lane (c) Weaving section (d) Braided ramp

## Answers

1. (d) 2. (b) 3. (b) 4. (c) 5. (a) 6. (a) 7. (d) 8. (b) 9. (a) 10. (c) 11. (a)
12. (d) 13. (a) 14. (c) 15. (d) 16. (d) 17. (a) 18. (b) 19. (c) 20. (d) 21. (d) 22. (d)
23. (a) 24. (a) 25. (b) 26. (b) 27. (a) 28. (a)

## Questions

1. State the differences between grade separation and interchange. (*Hint:* Traffic may be interrupted in the case of grade separation but never with an interchange)
2. State in what situations it is appropriate to provide interchanges.
3. State the basic requirements of grade-separated intersections.
4. Write the advantages and disadvantages of grade-separated intersection.
5. What are the factors affecting the selection of an appropriate configuration of interchange?
6. Write the advantages and disadvantages of overpass structures.
7. Write the advantages and disadvantages of underpass structures.
8. Classify and detail the constraints considered for the selection of a suitable interchange configuration.
9. Explain the cardinal principles of interchange design. Write with suitable sketches.
10. Explain the secondary principles of interchange design. Write with suitable sketches.
11. Explain with sketches how alternative treatment options can be developed for weaving improvement.
12. Describe the classification of interchanges based on the directional type. Draw a few examples of directional and non-directional interchange layouts.
13. Classify the ramps based on their configuration with sketches for the each case. Also explain their suitability and limitations with reference to the Indian practice of driving on the left.
14. Classify ramps based on the directional used and explain their applicability with sketches.
15. Describe the operational characteristics of these interchanges with example sketches:
    (a) fully directional interchanges. (b) directional interchanges,
    (c) semi-directional interchanges and (d) non-directional interchanges.
16. Draw sketches showing various forms of loop ramps used for merging and diverging and combination of traffic flows.
17. Classify ramps based on access control type and draw a few example sketches for each case.
18. What is a trumpet? Draw sketches showing configurations of three-leg intersections. Also state the advantages and advantages of trumpet interchanges.
19. What is meant by directional three-leg interchanges? Draw any three configuration variants of Y interchanges with directional ramps and T interchanges with semi-directional and directional ramps.
20. Write the advantages and disadvantages of diamond interchanges.
21. When is single point urban interchange (SPUI) used? Draw a typical configuration and write its operating features including the applicable AADT values.
22. Describe the diverging diamond interchange (DDI). Explain its operational features and its suitability with a sample configuration.

23. Write the advantages and disadvantages of diverging diamond interchange (DDI).
24. Describe the double crossover merging interchange (DCMI). Explain its operational features and its suitability with an example configuration.
25. What is a three-level diamond interchange? Describe it with a configuration.
26. List any five variants of diamond interchanges and describe their operational features with an appropriate configuration.
27. Draw the configurations of these diamond interchange variants:
    (a) Split Single Point/ Split Tight Diamond Interchange (b) Folded Diamond Interchange
    (c) Split Diamond Interchange (d) Dogbone Diamond Interchange
    (e) Diamond with Staggered Intersections
28. Draw a sketch showing the configuration of a three-level stack diamond interchange and write its operational advantage.
29. Define and describe directional, non-directional and semi-directional interchanges.
30. Describe with sketches the two types of lane drops: a basic lane drop and an auxiliary lane drop.
31. Write the classification of interchanges based on the number of approach legs such as for (a) three-way interchanges (b) four-way interchanges and (c) five-way or more interchanges.
32. What are the criteria used determine the type of interchange?
33. What are the relationships between accidents and the interchange configuration?
34. Write the classification of interchanges based on directional use. State the advantages and disadvantages.
35. Draw line diagrams of any four types of directional interchange configurations.
36. Differentiate between fully directional and semi-directional interchanges. Draw any two types of sketches for each case.
37. Define and draw typical sketches showing non-directional, semi-directional and directional type of interchanges.
38. Describe various forms of three-leg interchanges and state the advantages and disadvantages of trumpet interchange.
39. Describe various forms of Y interchanges and T interchanges that are commonly in use.
40. Classify diamond interchanges based on the distance between the intersections. Draw three sketches and explain their operating principles.
41. Describe with a sketch how C–D roads can be considered as a basic design element of a cloverleaf interchange.
42. State the operational characteristics of traffic that are used for the selection of suitable interchanges among the various alternatives.
43. Write the advantages and disadvantages of each basic type of the following interchanges:
    (a) Diamond (b) Parclo (c) Cloverleaf
    (d) Trumpet (e) Directional
44. What are the criteria used to determine the type of interchange to design? Briefly explain them.
45. Draw a configuration showing a full cloverleaf with C–D roads and label their component road parts.
46. Draw the configuration of any two variants of partial clover leafs and state their uses.

47. Explain how topography affects the interchange type. Draw any three sketches and explain.
48. What is a turbine interchange? Draw its configuration.
49. How do you design lane drops?
50. Explain how you take a cross road into consideration in the selection an interchange type. Draw any three sketches and explain.
51. How do you try to eliminate wrong way manoeuvres on interchange ramps?
52. How do you select the interchange type related to the functional use of the crossroad?
53. When would you recommend the use of C–D roads?
54. Explain how all stack interchanges are directional interchanges. Draw a few sketches showing their configurations with traffic flow directions.
55. Describe the basic differences between three-level and four-level stack diamond interchanges. Write the merits and demerits of each case.
56. Explain why many highways are constructed with on ramps before exit ramps.
57. Describe any four types of possible alternative treatments that can be proposed to a weaving section on a through mainline with a ramp so as to make the weaving section an operationally effective approach. Draw sketches for each case.
58. Define lane balance. Explain how operational flexibility is attained by lane balancing.
59. Describe the fundamental concept of determination of lane balance by drawing a sketch.
60. What are the principles adopted as standard for lane balance purposes?
61. Define the concept of lane balancing used for the design of interchanges. Also explain the process adopted for lane balancing with reference to basic number of lanes. [*Hint:* First determine the basic number of lanes and then consider lane balancing]
62. Describe the concept of coordination of lane balance with the basic number of lanes. Explain with an example.
63. Define weaving section. Explain the classification of weaving sections based on the weaving forms. Draw sketches for each case.
64. State the categories of traffic operational parameters considered for the design of ramps.
65. Define interchange spacing. What are the details of standard spacing recommended for rural and urban areas?
66. Draw sketches showing the typical components of exit ramp and entry ramp.
67. State the significance of entry–exit ramp configurations with reference to single and double entrance ramps.
68. State the factors/design elements to be considered for determination of adequate length of entrance and exit ramps.
69. Describe the effect of turning roadway convergence angle for determination of adequate length of entrance and exit ramps.
70. What are braided ramps? Draw a few sketches showing examples of braided ramp configurations.
71. Describe the effect of lane drops on turning roadway angle for determination of appropriate location of entrance and exit ramp terminals.
72. Draw a typical cross-section of a single-lane ramp and shoulders with standard geometric standards.

73. State the factors affecting design speed on ramps.
74. Distinguish parallel versus tapered ramp terminal designs.
75. Define ramp spacing. What are four possible ramp configurations recommended for determination of minimum ramp connection spacing?
76. Draw sketches showing typical component features of entrance ramp and exit ramp terminals.
77. What is a gore? Draw typical sketches showing gore area characteristics of (a) single-lane off connections: no lane reduction (b) single-lane, one-lane reduction off-ramp connections and all two-lane off connections.
78. Write a short note on how sight distance is provided with reference to horizontal and vertical controls.
79. Draw a typical sketch showing how horizontal sight distance is measured at ramp terminals.
80. Define access control. Draw a sketch showing how access distance is measured for (a) grade-separated locations and (b) interchange locations.
81. Write an overview of comparative studies and evaluation studies conducted on the performance of an interchange in a locality.

CHAPTER

# 16 Parking Studies

## 16.0 INTRODUCTION

Parking is an important part of any transportation system. Parking is the act of stopping a vehicle and leaving it unoccupied in a space for a specified duration, when it is not in use. Parking is one of the basic functional requirements of any road traffic facility; vehicles are usually permitted to park on the roadside or off the roadside. Parking studies are used to evaluate the road user facilities relating to the needs of parking demand, parking volume and parking space inventory with parking characteristics.

**Parking policy:** The policy goals of the governing agency concerning parking are as follows:

- To provide adequate parking facility for vehicles travelling to an attraction, service, business or residential community area
- To separate/design parking facilities, away from regular traffic flow
- To limit the number of public parking spaces to reduce traffic congestion/pollution (in the CBD) by discouraging the proportion/share of private vehicle modes of travel; mass transit facilities are planned for this purpose
- To recommend the use of intelligent transportation system (ITS) technologies to provide smart parking operations at competitive market prices

## 16.1 BASIC TERMINOLOGY

**Parking space/stall:** An area provided to accommodate one vehicle to be parked, without any access restrictions due to adjacent parked vehicles. A parking space may be larger than a stall (3 × 6 m) as additional space is provided for loading/un-loading operations. According to IRC: SP: 12 (1973), the standard minimum space recommended for individual cars is 3 × 6 m, car parking lots is 2.5 × 5.0 m and trucks is 3.75 × 7.50 m. Parking spaces may be categorised as free, paid, for differently abled persons and illegal.

**Loading and unloading berths:** A berth space of 3.75 × 7.50 m is provided for all kinds of developments, except residential warehouses and godowns, having an initial floor area of 500–1500 $m^2$. Additional berths are provided at the rate of one for every subsequent 1000 $m^2$ of floor area. In the case of regular warehouses, two such berths are provided for an initial floor area of 500–1500 $m^2$. Additional berths are provided at the rate of one for every subsequent 500 $m^2$ floor area (IRC: SP: 12-1973).

**Stall length:** The longitudinal dimensions of a stall, usually taken as 5.6–6.0 m.

**Stall width:** The width of a stall measured crosswise to the parked vehicle; usually taken as 2.6–3.0 m.

**Headroom:** The vertical clearance provided at the entrance and inside a parking plaza. The minimum headroom is 2.4 m for private cars and taxis.

**Other vehicle types:** The space requirement details for other vehicles (length × width × headroom) are as follows:

- Light goods vehicles: 7.0 × 3.5 × 3.6 m
- Medium/Heavy vehicles: 11.0 × 3.5 × 4.7 m
- Big buses:12.0 × 3.5 × 3.8 m
- Mini-buses: 8.0 × 3.0 × 3.3 m
- Container vehicles: 16.0 × 3.5 × 4.7 m
- Motorcycles: 2.5 × 1.0 m

**Turning radii:** The largest turning radii of the design vehicle or any specified vehicle that is the largest is provided as a function of parking angle and end island design. The turning radius should not be less than 5.5 m.

**Study period:** The time during which a study is conducted, usually during peak parking hours. The time period may be between 9:00 AM and 6:00 PM for offices and between 2:00 AM and 4:00 AM in residential areas. This time period may change based on the type of the location.

**Aisle:** The portion of the driveway provided in a parking lot which is intended to allow immediate access to the parking spaces (Fig. 16.1). Adequate aisle width is provided to accommodate turning movements of vehicles during parking in the stall and reverse movement. The following aisle width (m) may be suitably provided (CoP, 2011):

| | $X_1$ | $X_2$ | $X_3$ |
|---|---|---|---|
| 30° parking | 3.6 | 4.2 | 6.3 |
| 45° parking | 4.2 | 4.8 | 6.3 |
| 60° parking | 4.8 | 4.8 | 6.6 |

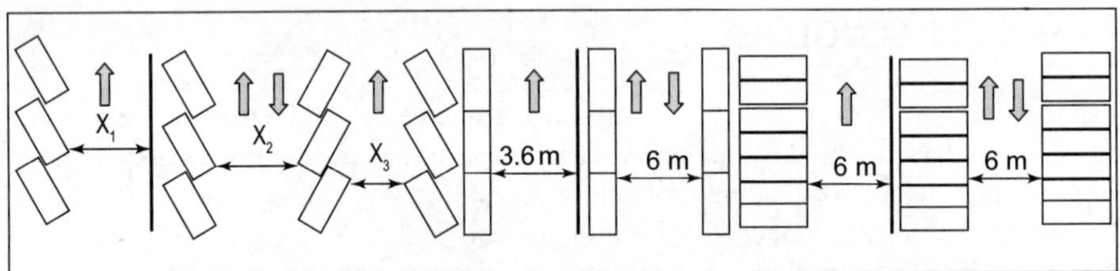


**Fig. 16.1** Typical parking aisle
(Redrawn from CoP, 2011; *Courtesy*: www.lta.gov.sg)

**Module:** A portion of the parking lot covering the width of one aisle and serving a row of parking spaces, on each side of the aisle.

**Parking volume:** The total number of vehicles parked in a study area in a specified period of time (say, per hour) (Fig. 16.2).

**Parking duration:** The time period during which a vehicle is parked in a parking space (Fig. 16.3). If the duration exceeds three continuous hours, it is termed long-term parking.

$$\text{Parking duration} = \frac{\text{Total vehicle hours}}{\text{Number of vehicles parked}}$$

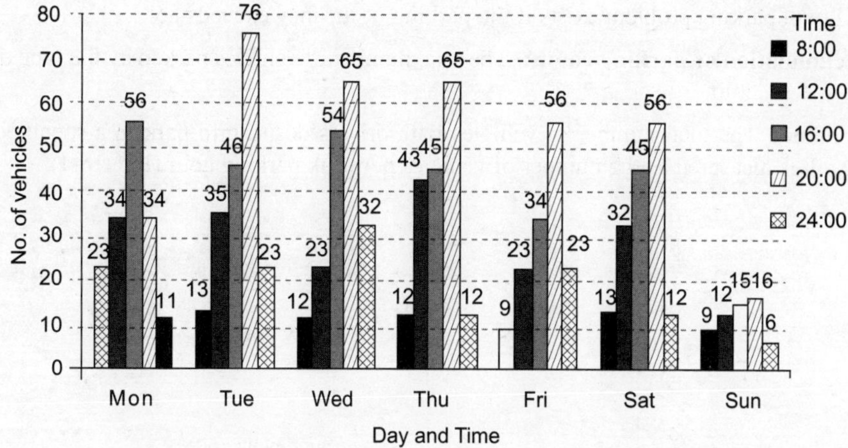


**Fig. 16.2** Typical presentation of total traffic entering and leaving a parking plaza

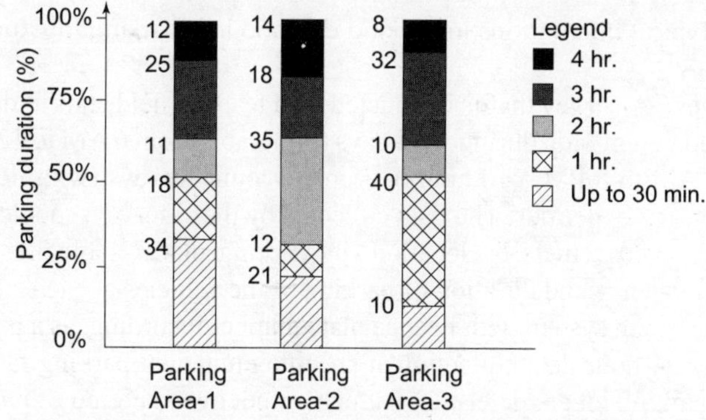


**Fig. 16.3** Typical parking duration distribution

**Average parking duration:** The average duration of time spent by a car in a parking space.

**Space hour:** The use of a single parking space for a period of one hour.

**Parking load:** The area under the accumulation curve, between two specific times. It is expressed as vehicle-hours.

**Parking capacity:** The total number of parking spaces for each parking category.

**Parking index:** $\dfrac{\text{Total vehicle hours}}{\text{Total number of spaces available}} \times 100$

**Turnover:** The total number of vehicles parked (occupied or utilised) in a parking plaza, over a time period (say, per day); this is also expressed as the occupancy rate by the per cent of spaces occupied. This parameter is considered as a measure of the utilisation of a parking plaza.

$$\text{Parking turnover} = \text{Occupancy} = \frac{\text{All parked vehicles during a specified period}}{\text{Number of parking spaces}}.$$

The possibility of parking problem exists if the parking occupancy exceeds 85%.

**Parking accumulation or parking count:** The total number of vehicles parked by the type of parking facility, at a specified time.

**Parking demand:** The total number of vehicles with drivers desiring to park in a specified parking plaza or area. It is measured as the number of vehicles per peak parking hour (Fig. 16.4).

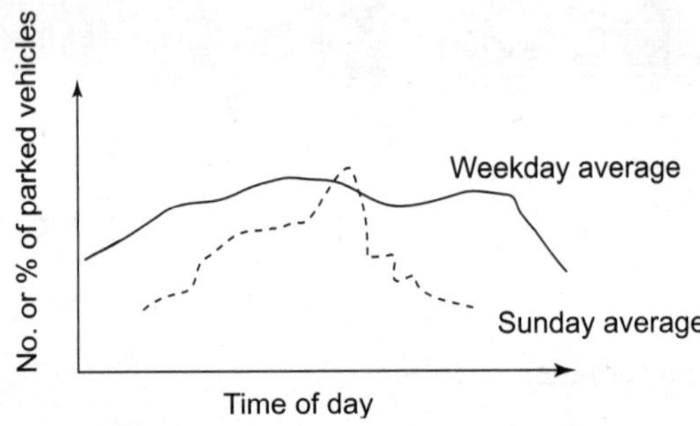


**Fig. 16.4** Typical fluctuations in parking demand in a parking lot/study locality

**Parking demand survey:** A survey that is conducted by a team of field staff to determine the parking occupancy data every hour, non-stop during weekdays (preferably from 6 AM to 12 AM – 18 hours) and on Saturdays (from 10 AM to 8 PM – 10 hours). Vehicle count surveys (with licence plate numbers) are conducted during the study periods. The data collected by these surveys includes the partial licence plate number of each parked vehicle by location (by parking plaza or block). This data is used to determine the nature, frequency and duration of parking of the vehicles at each study location.

Automatic traffic recording system with licence plate number recording, as a part of a smart parking system, can be used for 24-hour data collection at specific municipal parking facilities, by which the busiest times for each parking lot are determined. Such output over a period of months can be used for future provision against parking supply.

**Parking supply:** The number of parking stalls/spaces available for use, usually classified based on the type of parking facility, namely, on-street kerb parking and off-street lot/garage/terminal building. Parking supply may be categorised based on public-use stalls and private-use spaces.

**Parking supply inventory survey:** A survey carried out by a team of field staff to record the number and type of parking spaces provided in each parking facility/area within a municipal/regional street.

## 16.2 CLASSIFICATION OF PARKING FACILITIES

The classification of parking facilities, based on the location and associated facilities, is shown in Fig. 16.5. The importance/attractiveness of these parking facilities is determined based on the delay, difficulty/ease and cost of parking.

### 16.2.1 On-street/ Kerb/Roadside Parking

Parking bays are located alongside the kerb of the road. It is usually controlled by the local municipality. On-street parking can be categorised based on the *angle* at which the vehicles are parked in relation to the kerb (Fig. 16.6).

Types of Parking
On-street parking
Parallel parking
30°parking
45°parking
60°parking
90°or right-angled parking
Off-street parking
Manual car parking
Mechanical car parking
Multi-level car parking
Roof car parking
Surface car parking

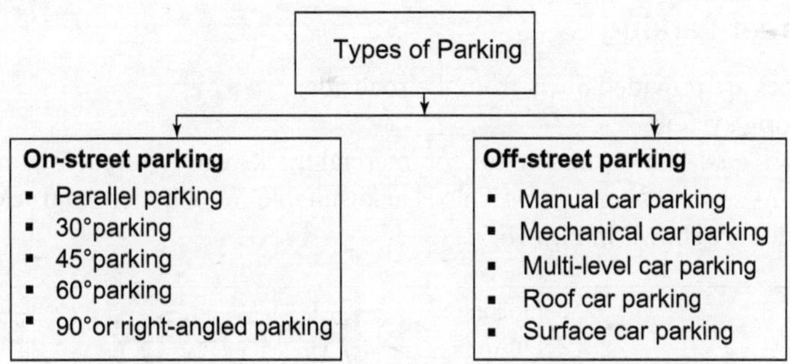

**Fig. 16.5** Classification of parking facilities

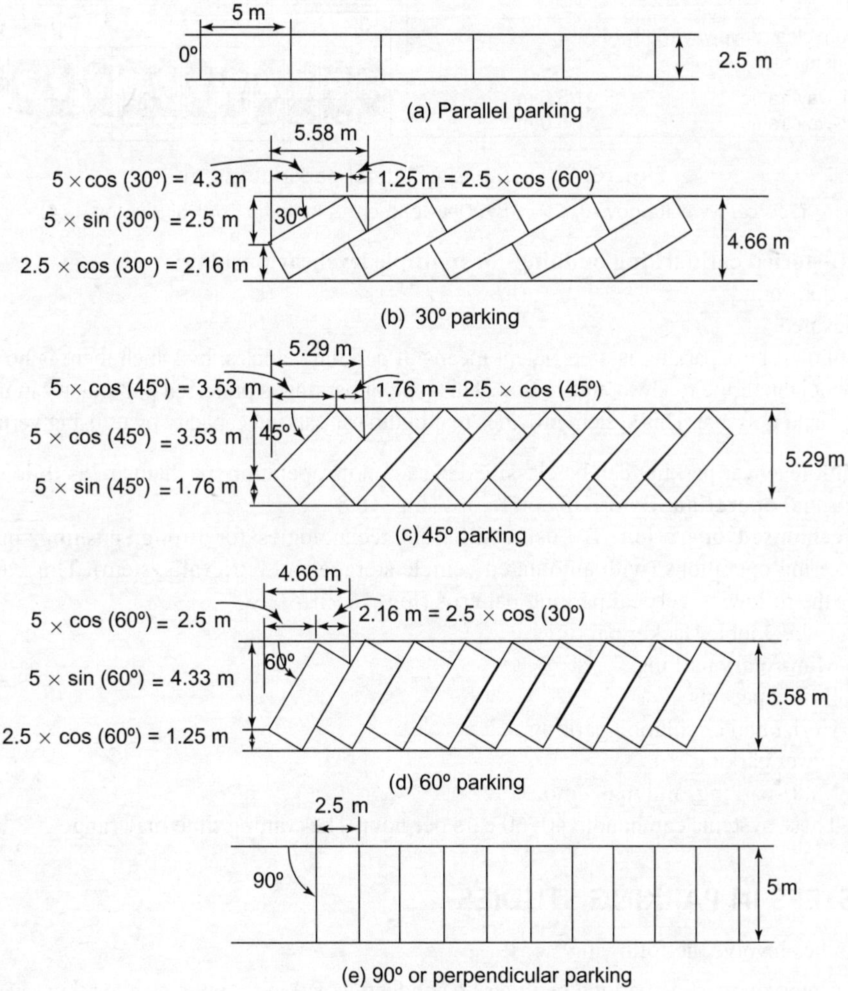

5 m
0°
2.5 m
(a) Parallel parking
5.58 m
5 × cos (30°) = 4.3 m
1.25 m = 2.5 × cos (60°)
5 × sin (30°) = 2.5 m
30°
4.66 m
2.5 × cos (30°) = 2.16 m
(b) 30° parking
5.29 m
5 × cos (45°) = 3.53 m
1.76 m = 2.5 × cos (45°)
5 × cos (45°) = 3.53 m
45°
5.29 m
5 × sin (45°) = 1.76 m
(c) 45° parking
4.66 m
5 × cos (60°) = 2.5 m
2.16 m = 2.5 × cos (30°)
60°
5 × sin (60°) = 4.33 m
5.58 m
2.5 × cos (60°) = 1.25 m
(d) 60° parking
2.5 m
90°
5 m
(e) 90° or perpendicular parking

**Fig. 16.6** Types of on-street parking

### 16.2.2 Off-street Parking

These parking spaces are provided *away* from the roadside.

- **Lateral parking system**
  - *Open surface lots*: This is suitable for commercial trucks with/without loading/unloading bays.
  - *Closed garages or parking plazas*: This is also suitable for commercial trucks.

  A few examples are shown in Fig. 16.7.

| Platform size (min) | 5.4 m long x 2.4 m wide |
|---|---|
| Minimum manoeuvring space (sideways) | 3.0 m |
| Minimum manoeuvring space (lengthways) | 7.2 m |
| Minimum clear driveway width (after installation) | |
| - 1-way traffic | 3.6 m |
| - 2-way traffic | 6.0 m |

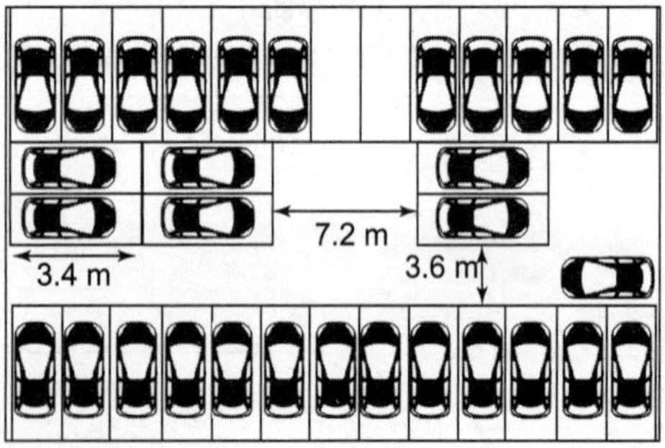

7.2 m
3.4 m
3.6 m

**Fig. 16.7** Examples of lateral parking system

(*Source*: www.lta.gov.sg; *Courtesy*: Development & Building Control Division, LTA)

- **Multi-storied car parking buildings or multiple level car parking**
  - Underground
  - Elevated

  Multilevel car parking is an efficient means of parking vehicles, by which there is no disruption to the neighbouring roadway traffic flow. This type of parking is preferred particularly in urban areas where land is scarce. This system provides maximum car parking capacity by utilising vertical space.

  Multilevel car parking can be classified based on its operation mechanism, as follows:

  - **Manual operation**: By driving on ramps (Fig. 16.8)
  - **Mechanised operation**: By using different technologies for lifting, pushing, pulling and lowering operations (with automated vehicle storage and retrieval system). This can be done for the following vertical parking patterns (Fig. 16.9):
    - Twin/Double stacker parking
    - Mini-rotary parking
    - Puzzle parking
    - Horizontal circulation parking
    - Tower parking
    - Multi-parking, and many more

    These systems can handle 40–60 cars per hour. The waiting time may range 2–3 minutes.

## 16.3 STEPS IN PARKING STUDIES

Parking studies involve the following steps:

1. Inventory of existing parking facilities by conducting parking supply inventory surveys in a study locality. The information collected comprises the type and number of parking spaces, time of operation, time limit of parking duration, type of facility, owner, parking fee, type of control:

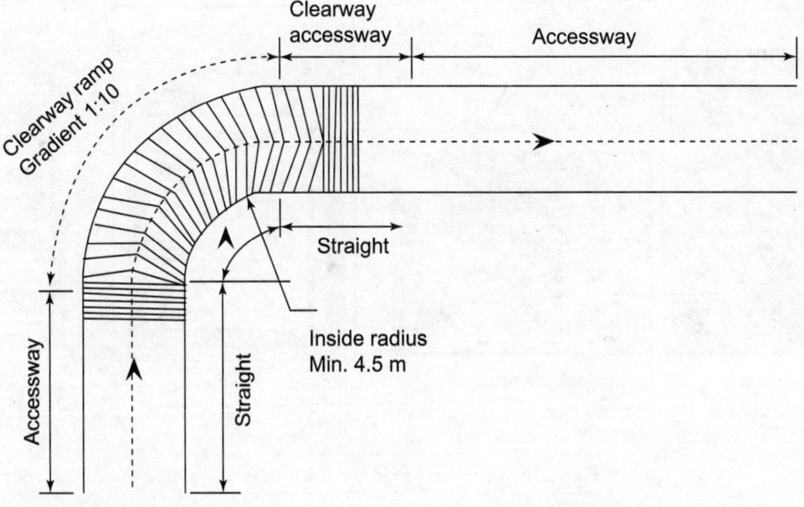


| | Single lane | Multi-lane |
|---|---|---|
| Width of straight clearway ramp and accessway | 3600 mm | 3000 mm per lane |
| Width of inside lane of curved clearway ramp and accessway | 4200 mm | 3600 mm per lane |
| Width of outside lane of curved clearway ramp and accessway | 4200 mm | 3300 mm per lane |
| Inside radius of curved clearway ramp and accessway | 4500 mm | |
| Gradient of clearway ramp and accessway | 1:10 (10%) preferred<br>1:8.3 (12%) maximum | |

**Fig. 16.8** Example of clearway dimensions recommended for off-street parking
(*Source*: www.lta.gov.sg; *Courtesy*: Development & Building Control Division, LTA)

manual/semi-automatic/automatic, and other restrictions such as one-way route, proximity to public transit terminals and bus stops. This data may also be collected by distributing and collecting questionnaires. The questionnaires are designed for a parking facility in a specific locality.

2. Collection of parked vehicle data to determine parking accumulation, parking turnover and parking duration.
3. Identification of parking generators (trip attraction facilities) within and near the study locality.
4. Collection of information on parking demand in the locality by day, week and month. For example, trip purpose (Fig. 16.10).

## 16.4 PARKING SURVEYS

Parking surveys are conducted to collect data about parking inventory, characteristics and distribution details. The following three types of parking surveys are commonly used.

Tower-type vertical system

Stack-type vertical system

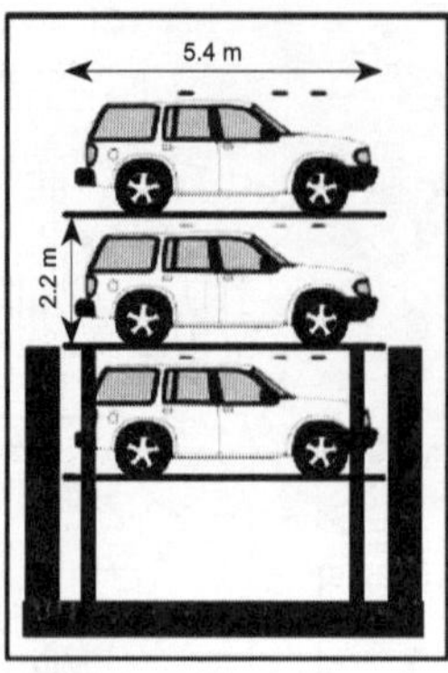


Critcal dimensions for vertical parking systems

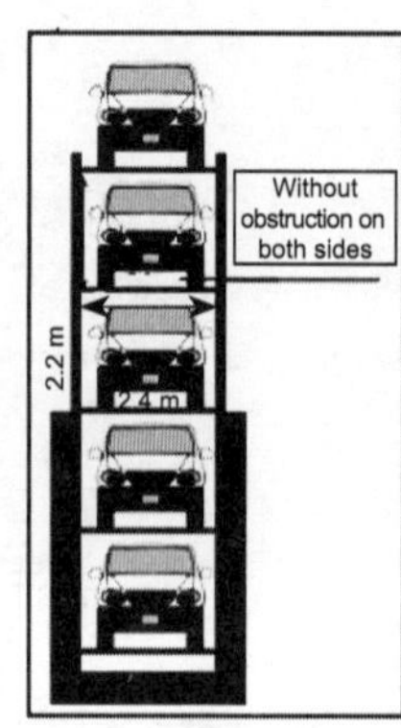


Critical dimensions for vertical parking system

| Platform size | 5.4 m long × 2.4 m wide |
|---|---|
| Holding bay | At entrance and exit |
| Height limit | 2.2 m clear |

Requirements for vertical parking system

**Fig. 16.9** Typical methods and details of a vertical parking system
(*Source*: www.lta.gov.sg; *Courtesy*: Development & Building Control Division, LTA)

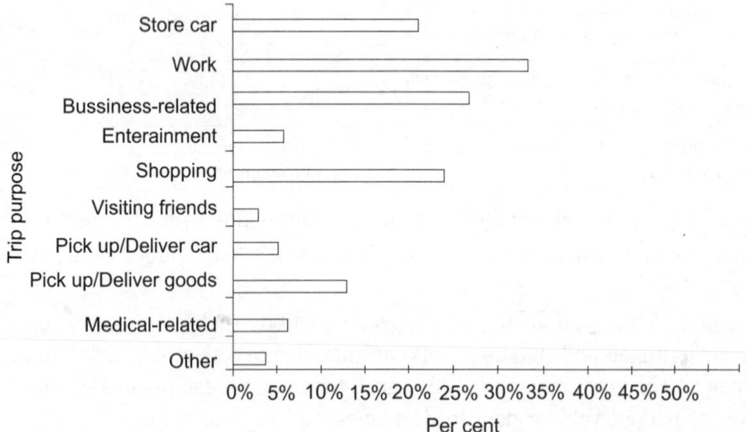


**Fig. 16.10** Typical presentation of parking spaces based on trip purpose

### 16.4.1 In–Out Survey

The occupancy count of a selected parking plaza is taken at the beginning of the survey. Then, the vehicles entering and leaving the parking plaza are recorded for a suitable interval of time. Finally, the current occupancy count is recorded and the survey count is stopped. One surveyor is sufficient for collecting the vehicle count data. Despite this advantage, the following limitations should also be considered:

- Parking duration of individual vehicles and turnover cannot be obtained.
- Parking fares cannot be estimated .

### 16.4.2 Fixed Period Sampling

This survey method is similar to the in–out survey method. The initial and final counts of all vehicles that occupy a parking lot during a predetermined fixed time period (1 hour or 3 hours) is recorded. No intermediate counts are taken during any interval of time. Such data collection may be repeated, if required. Hence, the data collection is very simple and quick. However, the number of vehicles left in the parking lot in any short period within the fixed study period may be missed.

### 16.4.3 Licence Plate Method of Survey

The occupancy duration of each parking space, with reference to the licence plate number of each vehicle entering and leaving it, is continuously recorded for an interval of 15 minutes over a study period (say, a day, week, month or year). This data provides information on the duration of vacancy of parking spaces and the occupancy by each vehicle, including the fares collected. The time interval may be reduced to avoid missing any vehicle that is parked for a very short duration. This method is more accurate than the ones discussed above. However, this requires more surveyors, and it is categorised as a labour-intensive method.

## 16.5 CHARACTERISTICS OF PARKING FACILITIES

Data about the following characteristics are collected in parking studies:

- **Parker characteristics:** When, where, why and how many vehicles are parked
- **Parking supply characteristics:** Location, number, cost of a space/hour and the type of control over the parking spaces
- **Parking needs:** Space requirements based on existing and future developments in the locality
- **Traffic factors:** Rate of growth of motor vehicle registration, population density and travel patterns
- **Walking distance:** As the walking distance to and from the parking space increases, the parking demand for that parking space will decrease
- **Economic factors:** Competitive price changes in the vicinity and free spaces
- **Changes/Developments in public transit facility** or change in accessibility to the CBD
- **Public policies** influencing private and public parking facilities with transit terminals
- **Technological factors:** Regulation/Use of non-renewable energy sources such as battery powered vehicles

## 16.6 ANALYSIS OF PARKING DATA

The data collected from these surveys is used to determine the following output parameters that are, in general, used as measures of a parking facility:

- Number and duration of vehicles parked legally and illegally in a parking plaza.
- Estimation of parking demand, including space hour calculation.

The following equation is used to estimate the parking demand for the period that a parking facility is open:

$$D = \sum_{i=1}^{N} (n_i \times t_i)(1) \tag{16.1}$$

where $D$ = space vehicle-hours demand when the parking plaza is open; $n$ = number of vehicles of $i^{th}$ class parked; $N$ = total number of vehicle classes considered (total of $i$); and $t$ = average duration of each class of vehicle.

Determination of parking supply: The following formula may be used to determine the parking supply.

$$S = f \times \sum_{k=1}^{P} (t_k)(2)$$

Or $$S = f \times (P \times \text{average occupied time per space}) \tag{16.2}$$

where $S$ = parking supply during a specific period of time defined in terms of space hours; $P$ = number of parking spaces available in a parking plaza; $f$ = factor that takes account of the time lost during turnovers; and $t_k$ = total duration of time (hr) when the $k^{th}$ space is legally parked.

**Example 16.1**

Approximately how many car parking spaces can be accommodated in a total paved area of 2392 sq. yards for the parking slot configuration given below:
Note that 5% of the paved area must be provided for drainage and other purposes. Take the conversion factor as 1 sq. yard = 0.836127 sq.m.

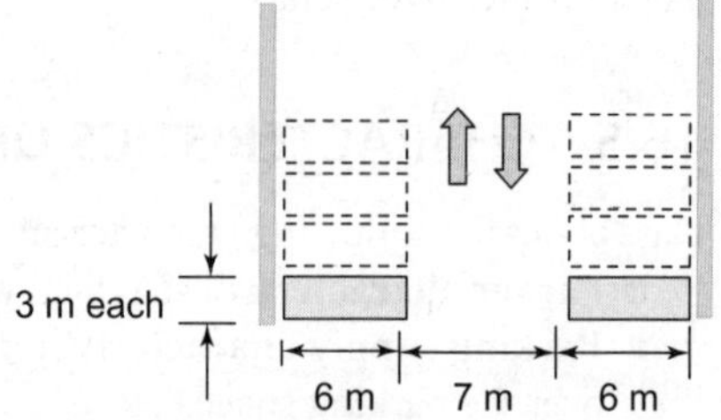


**Answer:** Area required for one car parking space including the aisle area = [(6 + 7 + 6) × 3] = 57 m²
Area required for one car parking space including the aisle area = ½ × 57 = 28.5 m²

The paved area = 2392 sq. yards × 0.836127 sq.m = 2000 m²

Net paved area available after deducting 5% for other purposes = (1 − 0.05) × 2000 = 1900 m²

Number of car spaces = 1900/28.5 = 66.66

Therefore, provide 66 car parking spaces in the given area.

*Note:* This method of determining the number of parking spaces may be used based on the size of vehicle, angle of parking and aisle width with or without handicapped van spaces.

**Example 16.2**

A commercial building is proposed to be constructed with a total carpet area of 10,000 m². This area is allotted for different uses: 50% for office space, 15% for commercial shops space, 5% for installation of air conditioners and other equipment space and the remaining for warehouse and goods storage. Determine the approximate number of parking spaces that can be provided based on the minimum space requirement criteria given below:

| *Use of building carpet area* | *Parking space requirement* |
|---|---|
| Offices | 1 space/ 70 $m^2$ |
| Commercial | 1 space/ 30 $m^2$ |
| Warehouse and goods storage | 1 space/ 100 $m^2$ |

**Answer:** The break-up of the space allotted is as follows

$$\text{Area assigned for office use} = A_o = 0.5\times 10000 = 5000 \text{ m}^2$$

$$\text{Area assigned for commercial use} = A_c = 0.15\times 10000 = 1500 \text{ m}^2$$

$$\text{Area assigned for equipment use} = A_e = 0.05\times 10000 = 500 \text{ m}^2$$

$$\text{Sub-total} = 7000 \text{ m}^2$$

Therefore, area assigned for warehouse and goods storage = $A_w$ = 10000 − 7000 = 3000 $m^2$
Let the total number of parking spaces and its break-up details can be written as

$$N_T = N_o + N_c + N_w$$

$$\text{Number of spaces required for office use} = N_o = 5000/70 \approx 72$$

$$\text{Number of spaces required for commercial use} = N_c = 1500/30 = 50$$

$$\text{Number of spaces required for warehouse and goods storage} = N_w = 3000/100 = 30$$

$$N_T = 152$$

*Note:* The $N_e$ value will be zero because equipment does not require any parking space.

If 1 car parking space with an average aisle area = 3 × 9 m = 27 $m^2$

$$\text{Total parking space required to accommodate} = 152\times 27 \text{ m}^2 = 4104 \text{ m}^2$$

Therefore, the builder has to accommodate the required parking space and design it either as an underground or as a rooftop facility.

**Example 16.3**

Determine the space vehicle-hours demand of a parking plaza using the following data:

| *Class of vehicle* | *No. of vehicles* | *Average duration of parking (hr.)* |
|---|---|---|
| Visitors | 12 | 1.5 |
| Commuters | 134 | 8 |
| Shoppers | 76 | 2.0 |

**Answer:** Substituting the given values in Eq.16.1, we get

$$D = \sum_{i=1}^{N} (n_i \times t_i)$$

$$= \sum_{i=1}^{3} (12\times 1.5) + (134\times 8) + (76\times 2.0)$$

$$= \sum_{i=1}^{3} (18) + (1072) + (152) = 1272 \text{ veh-hrs}$$

**Example 16.4**

An in–out survey was conducted for a parking plaza which comprised 110 spaces. The initial count was found to be 52. The number of vehicles entering and leaving the parking plaza for a time interval of 15 minutes are tabulated below. Calculate (a) parking accumulation, (b) total parking load, (c) average occupancy and (d) efficiency of the parking plaza.

| *Time (Minutes)* | *0–15* | *15–30* | *30–45* | *45–60* | *60–75* | *75–90* | *90–105* | *105–120* | *120–135* | *135–150* | *150–165* | *165–180* |
|---|---|---|---|---|---|---|---|---|---|---|---|---|
| In vehicles | 11 | 7 | 9 | 13 | 11 | 13 | 9 | 10 | 10 | 10 | 12 | 14 |
| Out vehicles | 7 | 9 | 6 | 14 | 9 | 9 | 11 | 9 | 12 | 5 | 9 | 13 |

**Answer:** The details and sequence of calculations are given below:

| *Time* | *In* | *Out* | *Difference* | *Accumulation (number)* | *Occupancy or Parking index (%)* | *Parking load (minutes)* |
|---|---|---|---|---|---|---|
| [1] | [2] | [3] | [4] = [2]–[3] | [5] | [6] = ([5]/110) × 100 | [7] = [5] × 15min. |
| 15 | 11 | 7 | 4 | 52 + 4 = 56 | (56/110) × 100 = 50.91 | 56 × 15 = 840 |
| 30 | 7 | 9 | −2 | 56 − 2 = 54 | (54/110) × 100 = 49.09 | 54 × 15 = 810 |
| 45 | 9 | 6 | 3 | 54 + 3 = 57 | (57/110) × 100 = 51.82 | 57 × 15 = 855 |
| 60 | 13 | 14 | −1 | 57 − 1 = 56 | (56/110) × 100 = 50.91 | 56 × 15 = 840 |
| 75 | 11 | 9 | 2 | 56 + 2 = 58 | (58/110) × 100 = 52.73 | 58× 15 = 870 |
| 90 | 13 | 9 | 4 | 58 + 4 = 62 | (62/110) × 100 = 56.36 | 62 × 15 = 930 |
| 105 | 9 | 11 | −2 | 62 − 2 = 60 | (60/110) × 100 = 54.55 | 60 × 15 = 900 |
| 120 | 10 | 9 | 1 | 60 + 1 = 61 | (61/110) × 100 = 55.45 | 61 × 15 = 915 |
| 135 | 10 | 12 | −2 | 61 − 2 = 59 | (59/110) ×100 = 53.64 | 59 × 15 = 885 |
| 150 | 10 | 5 | 5 | 59 + 5 = 64 | (64/110) × 100 = 58.18 | 64 × 15 = 960 |
| 165 | 12 | 9 | 3 | 64 + 3 = 67 | (67/110) × 100 = 60.91 | 67 × 15 = 1005 |
| 180 | 14 | 13 | 1 | 67 + 1 = 68 | (68/110) × 100 = 61.82 | 68 × 15 = 1020 |
| | | | Total= | 722.00 | 656.36 | 10830 Veh-min, |

Number of 15-minute intervals = 12

Average occupancy for each time interval = 656.36/12 = 54.69%

Total parking load = 10830 veh-min. = (10830/60) =180.5 veh-hr.

**Example 16.5**

The sample data collected by a licence plate parking survey conducted at a parking lot in Hyderabad City is given below. The licence plate numbers and occupancy on specific spaces are recorded. Calculate (a) average occupancy, (b) average turnover, (c) parking load, (d) parking capacity and (e) efficiency of the parking lot.

| *Space #* | *Licence plate numbers of vehicles with duration of stay in minutes* | | | |
|---|---|---|---|---|
| | *0–15 min.* | *15–30 min.* | *30–45 min.* | *45–60 min.* |
| 1 | 1336 | 1336 | - | - |
| 2 | 3245 | 3245 | 3245 | 3245 |
| 3 | 0635 | - | 7980 | - |
| 4 | 6606 | 7777 | 7777 | 5555 |
| 5 | 8181 | 4400 | 8383 | 0127 |
| 6 | 6745 | 6745 | 6745 | 6745 |
| 7 | 2020 | 1947 | 1947 | 2015 |
| 8 | 2099 | 2099 | 1010 | 1010 |
| 9 | 6611 | 6611 | 2244 | 2244 |
| 10 | 6699 | 7117 | 0909 | 1212 |
| 11 | - | 3333 | - | 2900 |
| 12 | 0404 | - | 3000 | - |
| 13 | 0986 | 0986 | 9999 | 9999 |
| 14 | 1122 | 1234 | 1234 | 1234 |
| 15 | - | - | 1336 | 1336 |

**Answer:** Details of the calculations are given below:

| *Space #* | *Licence plate numbers of vehicles with duration of stay in minutes* | | | | *Occupancy status of each vehicle in time slot* | | | | *Turnover per hour* |
|---|---|---|---|---|---|---|---|---|---|
| | *15* | *30* | *45* | *60* | *15* | *30* | *45* | *60* | *Per each space* |
| 1 | 1336 | 1336 | - | - | 1 | | 0 | | 1 + 0 = 1 |
| 2 | 3245 | 3245 | 3245 | 3245 | 1 | | | | 1 |
| 3 | 0635 | - | 7980 | - | 1 | 0 | 1 | 0 | 1 + 0 + 1 + 0 = 2 |
| 4 | 6606 | 7777 | 7777 | 5555 | 1 | 1 | | 1 | 1 + 1 + 1 = 3 |
| 5 | 8181 | 4400 | 8383 | 0127 | 1 | 1 | 1 | 1 | 1 + 1 + 1 + 1 = 4 |
| 6 | 6745 | 6745 | 6745 | 6745 | 1 | | | | 1 |
| 7 | 2020 | 1947 | 1947 | 2015 | 1 | 1 | | 1 | 1 + 1 + 1 = 3 |
| 8 | 2099 | 2099 | 1010 | 1010 | 1 | | 1 | | 1 + 1 = 2 |
| 9 | 6611 | 6611 | 2244 | 2244 | 1 | | 1 | | 1 + 1 = 2 |
| 10 | 6699 | 7117 | 0909 | 1212 | 1 | 1 | 1 | 1 | 1 + 1 + 1 + 1 = 4 |
| 11 | - | 3333 | - | 2900 | 0 | 1 | 0 | 1 | 0 + 1 + 1 + 0 = 2 |
| 12 | 0404 | - | 3000 | - | 1 | 0 | 1 | 0 | 1 + 0 + 1 + 0 = 2 |
| 13 | 0986 | 0986 | 9999 | 9999 | 1 | | 1 | | 1 + 1 = 2 |
| 14 | 1122 | 1234 | 1234 | 1234 | 1 | 1 | 1 | 1 | 1 + 1 + 1 + 1 = 4 |
| 15 | - | - | - | - | 0 | | | | 0 |
| | | | | Total = | 13 | 12 | 12 | 11 | 33 |

$$\begin{aligned}
\text{Total number of spaces} &= 15 \\
\text{Accumulation between 0 and 15 min.} &= 13 \text{ vehicles} \\
\text{Occupancy between 0 and 15 min.} &= (13/15) \times 100 = 86.66\% \\
\text{Accumulation between 15 and 30 min.} &= 12 \text{ vehicles} \\
\text{Occupancy between 15 and 30 min.} &= (12/15) \times 100 = 80\% \\
\text{Accumulation between 30 and 45 min.} &= 12 \text{ vehicles} \\
\text{Occupancy between 30 and 45 min.} &= (12/15) \times 100 = 80\% \\
\text{Accumulation between 45 and 60 min.} &= 11 \text{ vehicles} \\
\text{Occupancy between 45 and 60 min.} &= (11/15) \times 100 = 73.33\% \\
\text{Accumulation between 0 and 60 min.} &= (13 + 12 + 12 + 11) \\
&= 48 \text{ vehicles during the hour} \\
\text{Average occupancy per 15-min. interval} &= (48/4) \times 100 = 80\% \\
\text{Sum of turnover of the parking lot} &= \text{Parking volume} \\
&= 33 \text{ vehicles per hour} \\
\text{Average turnover of the parking lot} &= (\text{Sum of turnover/Total no. of spaces}) \\
&= (33/15) = 2.2 \text{ vehicles per hour} \\
\text{Average duration of the parked} &= (48 \times 15 \text{ min.})/33 \\
\text{vehicles in that parking lot} &= 21.81 \text{ minutes/vehicle} \\
\text{Parking capacity} &= \text{Total number of spaces} \times \text{Number of hours} \\
&= 15 \times 1 = 15 \text{ vehicle-hours or} = 15 \text{ space hours} \\
\text{Parking load} &= 48 \times (15 \text{ min.}/60) = 12 \text{ vehicle-hours} \\
\text{Efficiency} &= (\text{Parking load/Total number of spaces}) \times 100 \\
&= (12/15) \times 100 = 80\%
\end{aligned}$$

## 16.7 PARKING MANAGEMENT STRATEGIES

Inefficient utilisation of existing parking spaces in a locality may result in non-availability of parking. It does not mean that such non-availability arises due to deficient parking supply. Towards this objective, parking management is essential to efficiently manage existing parking spaces by using the appropriate techniques and strategies. The advantages of adopting parking management strategies are:

- Increased efficiency of existing parking lots
- Improved driver/user convenience
- Cost saving
- Encourage the use of alternative modes of transport
- Indirectly reduces traffic congestion problems
- Supports smart growth

The most widely adopted parking management strategies are as follows:

- **Shared parking:** Parking spaces are shared by multiple users (more than one user is allowed to use a space, rather than assigned use of spaces within a lot). This may be applicable in mixed land use, where the residential parking spaces in an apartment may be used by commercial/store vehicles during office working hours.

- **Regulate parking facility use:** The parking spaces in a facility are managed and regulated to give priority to higher value trips. Priority users may be defined as taxi drivers, customers, residents or differently abled users. The maximum duration of parking may be restricted to 60 minutes, and there will be time regulation (for example, no parking during 10 AM to 5 PM).
- **Remote or overflow parking:** Encouraging longer term parkers (permanent office staff) to use less convenient off-site or fringe parking facilities will free up more convenient spaces for priority users (customers). Parking space sharing agreements are planned with nearby off-site and overflow parking facilities.
- **Improving user information:** Convenient and accurate information must be provided on the parking available within a vicinity, along with price, maps, signs, brochures, electronic/internet-based communication and mobile apps. Appropriate direction signs may be installed for prohibited parking places to divert and guide the visiting parkers.
- **Improved walking conditions**: Pedestrian conditions amy be improved to allow parkers to conveniently walk to parking facilities and to various nearby destinations. This reduces parking demand and traffic congestion in CBD zones.
- **Increasing the capacity of existing parking facilities:** More parking spaces are added to existing parking lots by allocating unused space, re-sizing spaces for smaller vehicles and motorcycles, and by upgrading to mechanised/automatic car parking systems.
- **Parking pricing:** Fees may be fixed for motorists directly using the parking facilities. This strategy may reduce parking demand by about 20% and increase transit ridership. Strategic fee collection methods may be adopted for on-street parking by metered system with exceptions to local residents so that visitors will have to pay to use the spaces. Use for longer periods, day and location-based prices enable the efficient use of existing parking facilities.
- **Multiple payment options:** The fee payment options should be made simple, easy to understand and short. User choice must be respected with regard to payment by credit cards, RFID, debit cards, cell phones and coins.
- **Unbundle parking:** Rentable building units (flats) with multiple car parking facilities may be offered on rent with one free car parking. The remaining parking spaces may be offered on additional rent as shared parking.
- **Parking facility management and enforcement:** The parking facility design may be improved to address occasional peaks, overflow problems, and the safety and security of the users.

## EXERCISES

### MULTIPLE CHOICE QUESTIONS

1. The maximum number of vehicles can be parked with
   (a) Parallel parking (b) Perpendicular parking
   (c) 45° angle parking (d) 75° angle parking
2. As per IRC: SP: 12, the standard minimum space for car parking lots recommended is
   (a) 3 × 6 m (b) 2.5 × 5.0 m (c) 3.75 × 7.75 m (d) 2.5 × 6.0 m
3. As per IRC: SP: 12, the standard minimum space for truck parking recommended is
   (a) 3 × 6 m (b) 2.5 × 5.0 m (c) 3.75 × 7.75 m (d) 2.5 × 6.0 m

4. What is the minimum headroom provided for private cars and taxis?
   (a) 2.1 m (b) 2.2 m (c) 2.3 m (d) 2.4 m
5. What is the minimum space (Length × Width × Headroom) required for parking a container vehicle?
   (a) 7.0 × 3.5 × 3.6 m (b) 11.0 × 3.5 × 4.7 m
   (c) 12.0 × 3.5 × 3.8 m (d) 16.0 × 3.5 × 4.7 m
6. The commonly adopted turning radius provided in parking plazas should not be less than
   (a) 5.5 m (b) 5.4 m (c) 5.0 m (d) 4.8 m
7. As a common observation, there may be a parking problem, if the parking occupancy
   (a) Exceeds 85% (b) Is between 50% and 75%
   (c) Is less than 75% (d) Exceeds 50%
8. As per IRC: SP: 12, the standard minimum space for residential flats having floor area greater than 100 sq.m. is ________ spaces per flat.
   (a) 1 (b) 2 (c) 3 (d) 4
9. As per IRC: SP: 12, 1 space each is recommended as minimum requirement for offices having ________ sq.m of floor area.
   (a) 50 (b) 70 (c) 150 (d) 100

## Answers

1. (b) 2. (a) 3. (c) 4. (d) 5. (d) 6. (a) 7. (a) 8. (a) 9. (b)

## Questions

1. What are the objectives of conducting parking studies?
2. Define the terms and write the standard space used for (a) parking stall, (b) loading and unloading berths, (c) stall length and width, (d) headroom, and (e) turning radii.
3. What is the standard space (Length × Width × Headroom) requirement for
   (a) Light goods vehicles (b) Medium/Heavy vehicles
   (c) Big buses (d) Mini-buses
   (e) Container vehicles (f) Motorcycles
4. Define an aisle. Draw the typical layout of a parking aisle and indicate the aisle space requirements for 30°, 45° and 60° parking.
5. Define these terms used for parking studies: (a) module, (b) parking volume, (c) parking duration, and (d) space hour.
6. Define these terms used for parking studies: (a) parking load, (b) parking capacity, (c) parking index, and (d) parking turnover.
7. Define these terms used for parking studies: (a) parking demand and (b) parking supply.
8. Classify parking facilities and briefly explain their suitability.
9. Draw sketches representing the types of on-street parking layouts and describe the space requirement of each parking unit with trigonometric formulas.

10. Draw a sketch representing the off-street lateral car parking system layout and describe the space requirements of each parking unit with minimum manoeuvring space and minimum clear driveway width for one-way and two-way traffic.
11. Classify and describe multilevel car parking systems based on operation mechanism.
12. Draw a typical sketch showing the clearway dimensions recommended for off-street parking.
13. Write details of the minimum parking space requirements based on the purpose of use, as per IRC: SP: 12.
14. Explain how parking studies are carried out in four steps.
15. List the types of parking surveys and explain their data collection procedure, including their applicability.
16. What are the characteristic data collected in parking studies? Write the formulas used for determination of (a) space vehicle-hour demand when a parking plaza is open and (b) parking supply during a specific period of time defined in terms of space hours.
17. Determine the space vehicle-hours demand of a parking plaza using the following data:

| *Class of vehicle* | *No. of vehicles* | *Average duration of parking (hr.)* |
|---|---|---|
| Visitors | 135 | 1.0 |
| Commuters | 234 | 7.5 |
| Shoppers | 86 | 2.5 |

18. What are the objectives of parking management strategies? Explain the most widely adopted parking management strategies.

CHAPTER

17

# Road Markings

## 17.0 INTRODUCTION

Road markings are an essential component of any roadway system. They provide guidance and advance information, and convey regulations and warnings to road users. Road markings are the coloured paint lines on the road surface, symbolic objects and word messages, which are applied systematically with standard geometrics and intended control of traffic flow on different locations on roads.

Most pavement markings are retroreflective – they redirect rays from headlights back in the direction of the driver. This retroreflective property of pavement markings (and signs) is essential for efficient and safe vehicle operations, particularly at night.

The commonly adopted **objectives and advantages** of road markings are given below:

- To channelise/segregate traffic flow along the intended lanes/locations on carriageways, thereby avoiding conflicts. This is an effective traffic flow control measure used when the construction of raised medians/kerbs is not possible due to narrow carriageway width, particularly on urban roads.
- To guide/instruct and provide specific directions using arrows.
- Markings discourage meandering of vehicles across lanes and eliminate confusion about traffic lanes on a multilane carriageway, thus ensuring maximum capacity.
- To serve as a psychological barrier and indicate the delineation of traffic path with other objects on the roadway; for example, objects, kerbs, bridge abutment walls, edge markings and other obstructing objects.
- To convey information to drivers using word paintings, without distracting their attention from the carriageway.
- To indicate specific use of lanes with word paintings; for example, cycle track only, no overtaking zone and pedestrian pathway only.
- To restrict/warn about parking spaces and speed limit indications, specify pedestrian crossing locations and lane direction conversions.
- To help senior (elderly) drivers better understand and easily adapt to traffic rules and regulations.
- To impart traffic discipline and thereby promote road safety during the day and at night.

## 17.1 CLASSIFICATION OF ROAD MARKINGS

Based on the location (Fig. 17.1), the intended function/control, colour (Fig. 17.2) and thickness (Fig. 17.3), road markings can be categorised as follows.

**Audio tactile line marking (ATLM):** ATLM is created by installing a series of raised paint marking lines along/across a road surface. A typical thermoplastic ATLM profile used in South Australia is presented in Fig. 17.4. The raised markings are usually created by extruding molten thermoplastic material onto the existing lines. The spacing and height of ribs determines the level of audio and tactile performance of the ATLM (TASS, 2009).

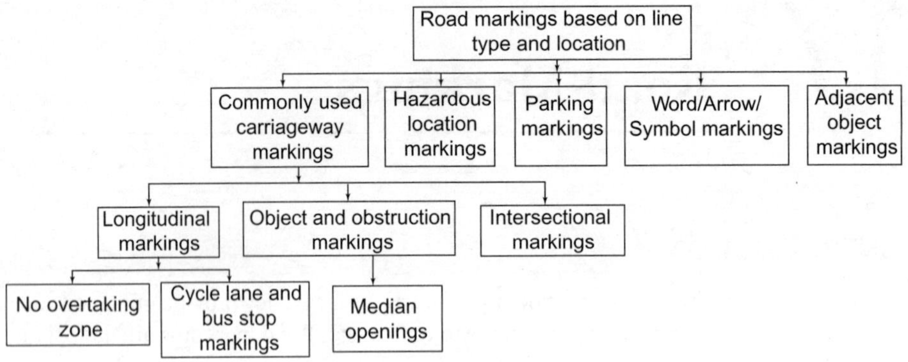


**Fig. 17.1** Classification of road markings based on line type and location

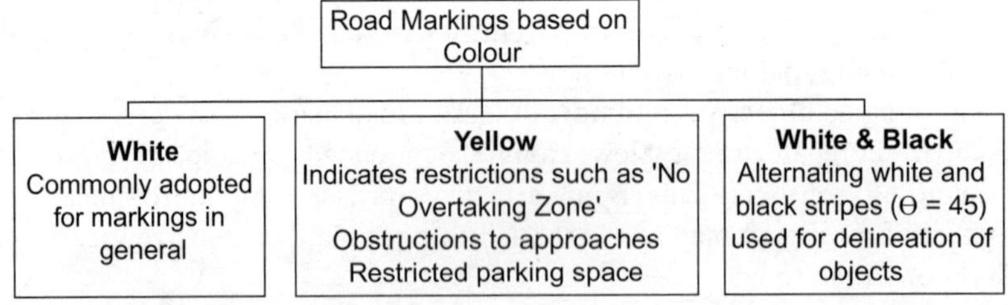


**Fig. 17.2** Classification of road markings based on colour

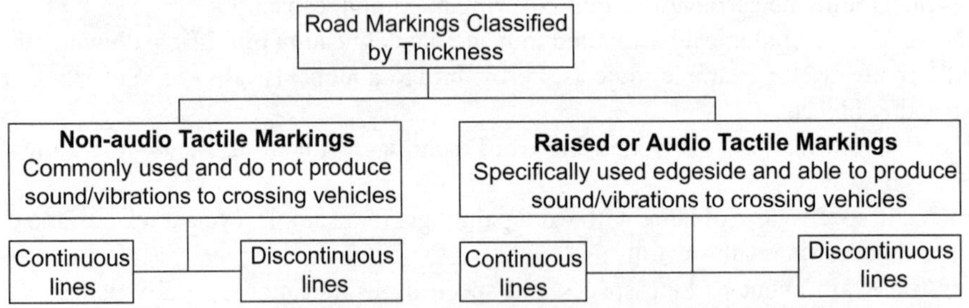


**Fig. 17.3** Classification of road markings based on thickness

ATLM used as an active warning device for fatigued drivers, especially on high-speed roadways. The purpose is to reduce road crashes by providing a noise (audio) and a vibratory (tactile) warning to road users suffering from fatigue or poor visibility, because they have fallen asleep or because of rain or fog. ATLM also provides dazzling delineation during wet weather under headlights.

## 17.2 LINE MARKINGS

Line markings placed along the direction of traffic flow on a roadway belong to this group. These lines are used for separating traffic flow towards a specified lane; they also indicate the driver's position on the roadway. Longitudinal lines are usually in white and yellow. The white lines are used for separating traffic flow in the same direction, while the yellow lines are used for separating traffic flow in the opposite direction and also to warn users of pavement edges. A few such details are described in Table 17.1.

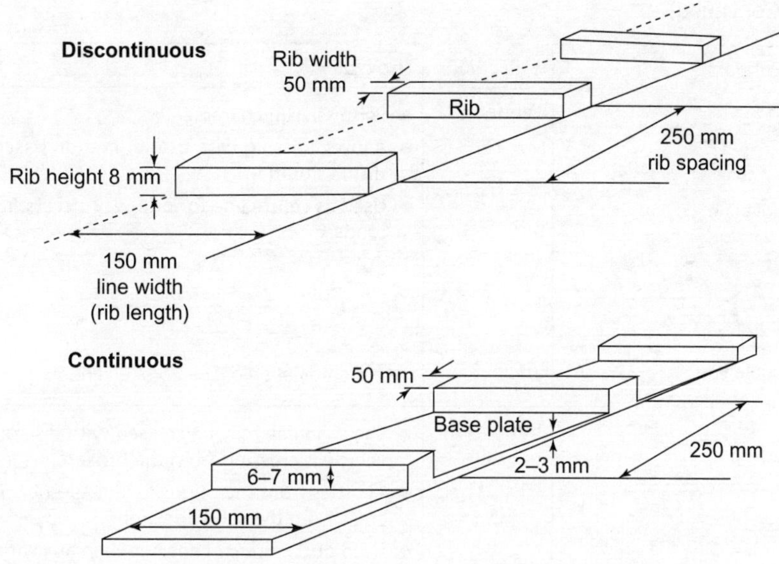


**Fig. 17.4** Example of thermoplastic raised markings (audio tactile edge lines)
(*Source*: Austroads, 2009; *Courtesy*: www.austroads.com.au)

**Table 17.1** Guiding principles of line markings (DDA, 2009)

| *Type with description* | *Line colour* | *Use type* |
|---|---|---|
| Continuous single line | White | • Restrictive in character<br>• Does not allow crossing of vehicles, except for entry or exit to a side road<br>• Used for separating traffic flow in the same direction |
| Continuous single line | Yellow | • Restrictive in character<br>• Does not allow crossing of vehicles, except for entry or exit to a side road<br>• Used for separating traffic flow in the opposite direction. |
| Continuous double line | White | • Seriously restrictive in character<br>• Does not allow crossing of vehicles<br>• Used for separating traffic flow in the same direction |
| Continuous double line | Yellow | • Seriously restrictive in character<br>• Does not allow crossing of vehicles, except in an emergency<br>• Used for separating traffic flow in the opposite direction |

(Continued)

**Table 17.1** (Continued)

| *Type with description* | *Line colour* | *Use type* |
|---|---|---|
| Broken single line<br>Broken double line | White | • Permissive in character<br>• Allows crossing with user discretion based on the traffic situation<br>• Used as centre line for one-way streets, ramps and slip roads<br>Combination of solid/broken lines |
| Continuous and broken line<br>Continuous and broken line (Segment =1000; gap =1000 mm; width = 100 mm, separated by 100 mm) | Both white or white with yellow | • The solid line may be crossed with discretion, if the vehicle is crossing from the broken line lane<br>• Vehicles from the opposite lane are not permitted to cross from the solid line lane<br>• Used at continuity lines between auxiliary lanes and through lanes |
| Solid single edge line | White/ Yellow | • Pavement edge/border is indicated<br>• Warns traffic to keep away from the edge<br>• Desirable on narrow brides/culverts, pavement width transitions, fog/mist-covered pavement localities<br>• Marked 200 mm and 150 mm from the edge of expressways and other roads, respectively. |

The choice of line depends on the width, length and spacing of road segments as well as on the speed, location and category of roads. However, the following situations may allow the use of wider (width greater than 100 mm) markings to ensure the greatest benefit (Gates and Hawkings, 2002):

- Locations where a higher degree of lane or roadway definition is perceived, such as on horizontal curves, narrow roadways with/without shoulders and at construction work zones
- Locations where low luminance contrast of road markings is common
- Locations where older drivers are prevalent, and thus, require additional visibility of roadway, in all conditions

## 17.2.1 Centre Line

Longitudinal markings such as centre line and edge line are provided along the length of traffic flow. The centre line demarcates the centre of the carriageway, along its length. It streamlines traffic flow and separates the same and opposite directions of traffic movement. The centre line is provided on carriageways in the following locations:

- To demarcate the restricted/narrow width of the carriageway. For example, at carriageway width transition locations on culverts and bridges where the width between the kerbs is $\geq$6.0 m.
- To indicate auxiliary lanes at unflared and flared intersections.

- To demarcate the portion of carriageway width that is reserved for on-street parking.
- On extra-widened curved portions of a carriageway.

The centre line may be provided as a solid single/double line, broken single/double line and a combination solid–broken line, for the intended traffic control requirements. Variations of these line configurations in white are prescribed by IRC: 35 (2015).

The broken line variants with the length (of the broken line segment), gap length and width as per IRC: 35 are given below (Figs 17.4 and 17.5):

- 3000–6000–100 or 150 or 200 mm
- 6000–3000–100 or 150 or 200 mm
- 1500–3000–100 or 150 mm
- 3000–1500–100 or150 mm
- 3000–3000–100 or 150 or 200 mm
- 1500–1500–100 or 150 or 200 mm
- 600–300–100 or 150: continuity lines for medians and islands; stop and give way lines
- 500–500–100 or 150: continuity lines for right-turn markings
- 1000–1000–100 or 150: continuity lines for left-turn markings; continuity lane
- 15000–3000–200

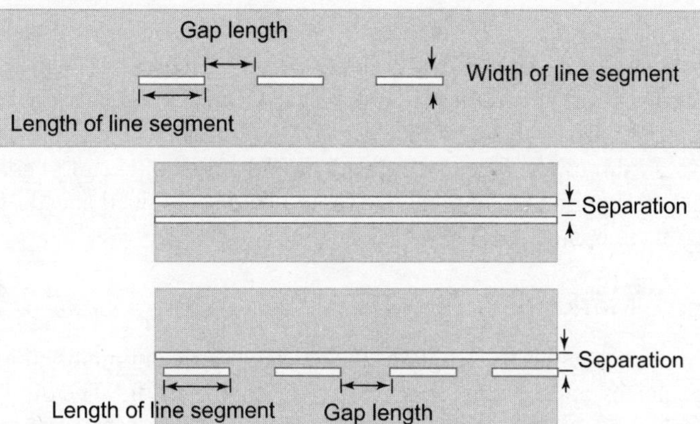


**Fig. 17.5** Configuration and measurement of broken and continuous line characteristics

Continuous solid lines have the following variants:

- Single line width: 100 or 150 or 200 mm
- Double continuous lines, each having width 100 mm, separated by 100 or 150 or 200 mm
- Continuous and broken lines (1000–1000), each line width = 100 mm, separated by 100 mm

The continuous single solid line in yellow is provided to strictly prohibit straddling by vehicles. This line is generally provided as an edge or verge line (Fig. 17.6).

**Warrants to use:** As a common practice, dividing lines and barrier lines (at prohibited overtaking) are warranted on two-lane roads, including on bridges with a width of 5.5 m or more, if the AADT is in excess of 300 and 2500 vehicles on rural and urban roads, respectively (Austroads, 2008). Despite these volume warrants, continuous or isolated sections of separation barrier lines may be provided for special conditions such as sub-standard curves, fog locations, approaches to the main road or pedestrian crossings and accident-prone locations.

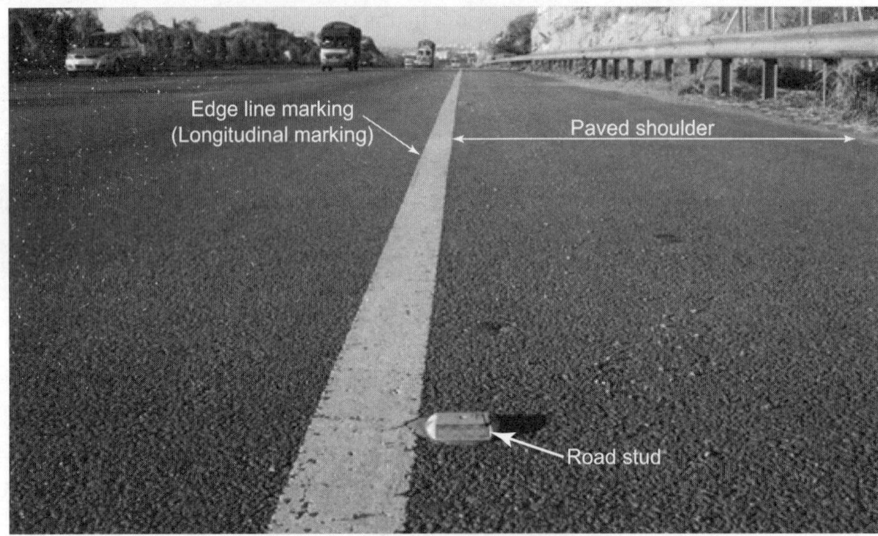


Fig. 17.6 Edge line marking with a stud

**Restrictions of use:** Usually, no centre line is provided on carriageways with a width of less than 5.5 m, as these entail discomfort and hazard (IRC: 35-2015), and also for a road width of less than 5 m and having more than four lanes with a central verge (DDA, 2009).

In case of restricted width of carriageway on bridges and culverts (kerb-to-kerb distance less than 6.0 m), the centre line marking on the approaches should be discontinued ahead of them by 30–35 m along the abutments, on either side (IRC: 35-2015).

### 17.2.2 Transverse Markings

These markings are provided across the length of the carriageway. Continuous single/double lines and broken lines accompanied by appropriate signs are used to control traffic. The Stop line and Give Way markings are commonly used.

Transverse line variants with broken and continuous solid lines are provided as per IRC: 35 (Fig. 17.5):

- Continuous single solid line (white): 200 or 300 mm width
- Continuous double solid line (white): 200 mm width; separated by 300 mm

These two patterns of lines are used as Stop lines (Fig. 17.7). The Stop lines are used in the vicinity ($\geq$2 m and $<$3 m in advance) of pedestrian zebra crossings; 1 m in advance of the nearest traffic signals; and 1.25–9 m in advance of minor road approaches merging with the major road. Normally, the Stop word message encryption (IRC: 35-2015) and a Stop sign will be supplemented at the prescribed locations (IRC: 67-2012).

*Note*: The Stop markings and Stop signs should not be used in conjunction with the Give Way markings/signs at un-signalised intersections as this may cause confusion about which vehicle would have priority (IRC: 35-2015).

- Broken single line (white) length–gap–width: 600–300–100 or 150 or 200 mm
- Broken double line (white) length–gap–width–separation: 600–300–200–300 mm

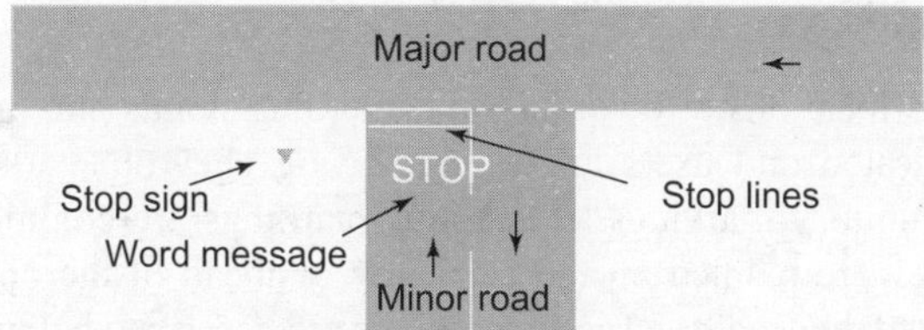


**Fig. 17.7** Stop line markings (schematic)

- These two patterns of lines are used as Give Way lines (Fig. 17.8). The Give Way lines are marked across minor roads at uncontrolled intersections (no Stop sign or signal or police). As a common practice, the Give Way lines are supplemented with hollow triangular Give Way approach markings and a Give Way roadside sign (Fig. 17.9).

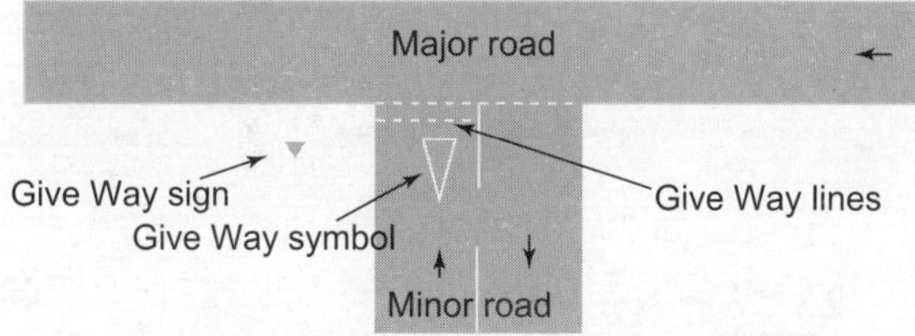


**Fig. 17.8** Give Way markings (schematic)

**Fig. 17.9** Two views of Give Way markings on a road

- A set of bar markings across the full width of the carriageway (yellow): Six stripes of 300 mm each separated by 600 mm with a depth of 5 mm (Fig. 17.10). The 5-mm depth is useful for guiding visually challenged pedestrians when they cross the road, while the crossing vehicles will experience vibrations and therefore slow down while passing over these markings.

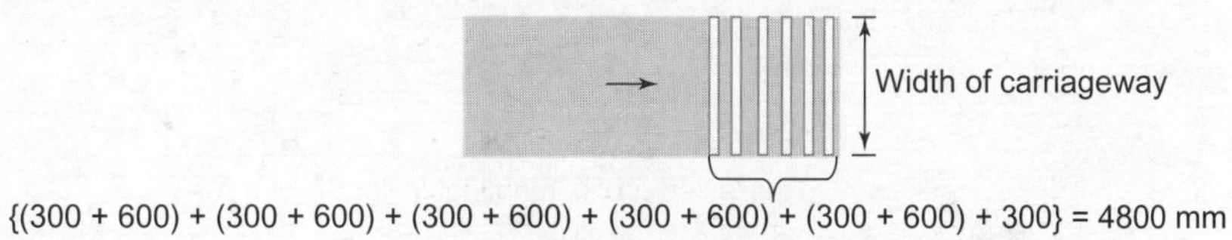


**Fig. 17.10** A set of bar markings across the width of a carriageway
(Redrawn based on IRC: 35-2015)

### 17.2.3 Hazard Markings

Prohibited and hazardous locations at diverging and merging points and any object obstructions on the carriageway are indicated and users are warned using chevron markings, diagonal markings, hatch markings and prohibitory markings. Hazardous markings are commonly accompanied by the appropriate signs and they are painted on the carriageway ahead of the obstruction or merging and diverging points. A few widely used hazardous markings are described below.

- Chevron-diverging (Fig. 17.11 (a)): The area covered by chevron markings at the nose of a channelising island is demarcated as a neutral area. It diverts the entering traffic at the proper angle so that it acts as a transition to the acceleration lane (Fig. 17.12).
- Chevron-converging (Fig. 17.11 (b)): It smoothly converges the exiting traffic at the proper angle so that it acts as a transition to the deceleration lane.
- Chevron-diagonal (Fig. 17.11 (c)): It separates and diverts traffic flow in the opposite direction.

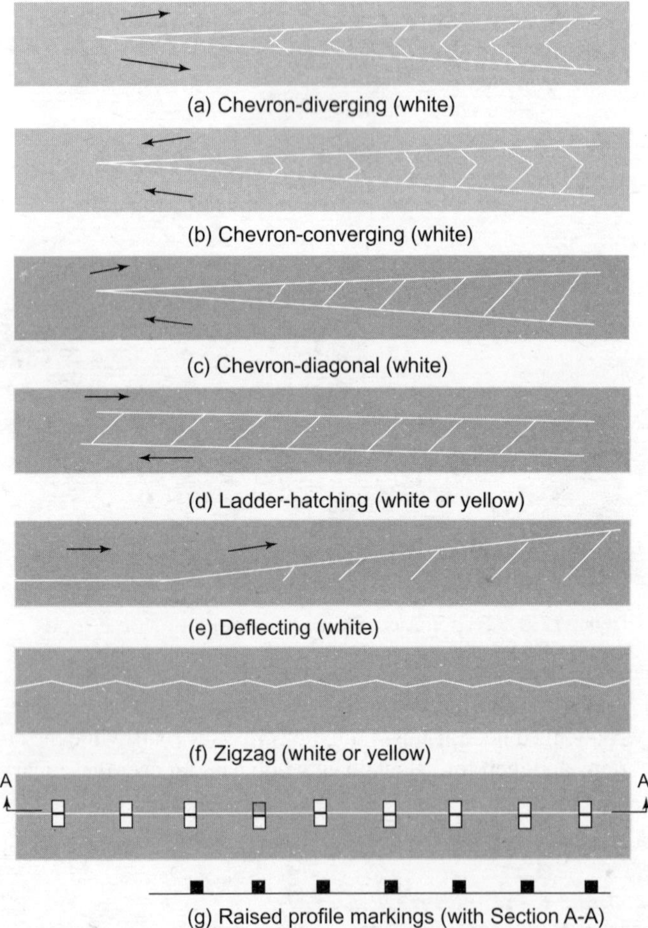


**Fig. 17.11** Types of hazardous markings (schematic)

- Ladder-hatching: It is used to separate opposing traffic flow on undivided two-lane carriageways in the straight segments (Fig. 17.11 (d)) and on curves (Fig. 17.13). This marking is also effective on acute curved portions that provide adequate gap between opposite adjacent traffic flows and on No Overtaking Zones such as at summit curves, horizontal curves and elsewhere on undivided, two-lane, bi-directional and three-lane highways.
- It is preferred that the two solid lines of the ladder be either 100 mm or 150 mm in length and separated by a minimum of 600 mm. A diagonal of length 4000 mm should be provided for speeds of up to 60 km/hr, and 6000 mm for speeds exceeding 65 km/hr.
- Deflecting (Fig. 17.11 (e)): This is used to deflect traffic flow away from an edge or any portion of a carriageway. The deflecting hatchline marking is provided when the edge line marking is ineffective. Such markings are also used to create refuge islands and for achieving lane reduction.
- Zigzag (Fig. 17.11 (f)): These are used as centre line and edge line markings in vulnerable road sections, such as in areas where human activities and pedestrian presence necessitate speed reduction, such as near speed breakers. They also help keep vehicles at a distance from extremities.
- Raised profile marking (Fig. 17.11 (g)): This is used as an alternative to edge markings. The vertical dimension (height raised to 4–8 mm) facilitates clear visibility above water sheets in the wet season. In addition, it also provides audible and vibratory warnings when vehicles cross over them.
- Single continuous solid line: 100 or 150 mm width; yellow/Double continuous solid line: 100 or 150 or 200 mm width; separated by 100 mm; painted yellow: A No Overtaking stretch is provided by double-line demarcation (Fig. 17.14).
- Bar markings of width 300 mm spread across the carriageway, leaving a gap of 150 m on either end to facilitate run-off; height = 4–5 mm. This is often provided in series at railway level crossings.

More details of line width and layout can be obtained from IRC: 35-2015.

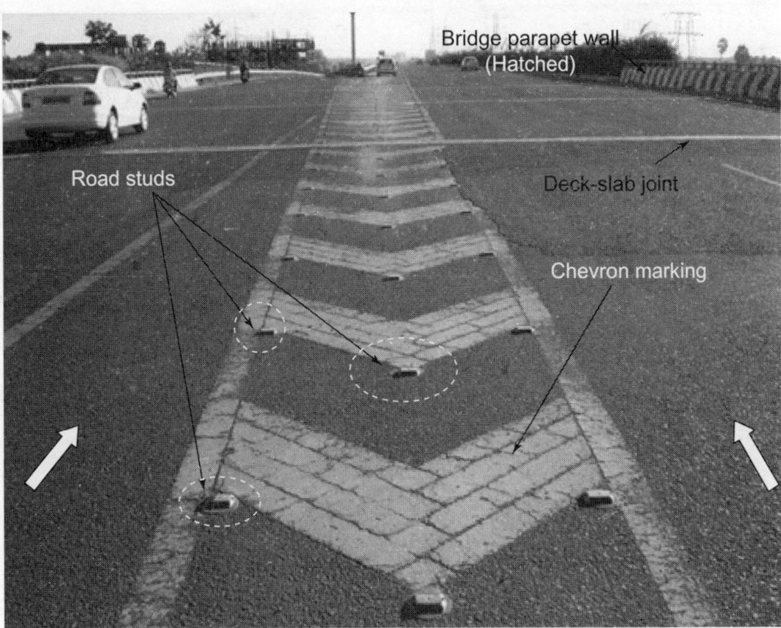


**Fig. 17.12** Chevron-diverging with road studs

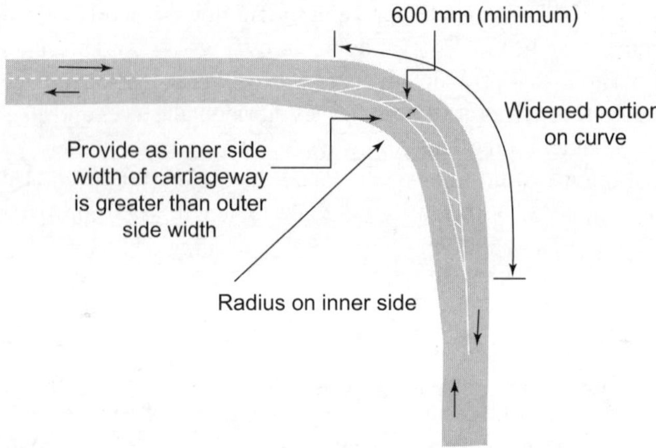


**Fig. 17.13** Schematic view of ladder-hatching markings on the horizontal curve
(Redrawn based on IRC: 35-2015)

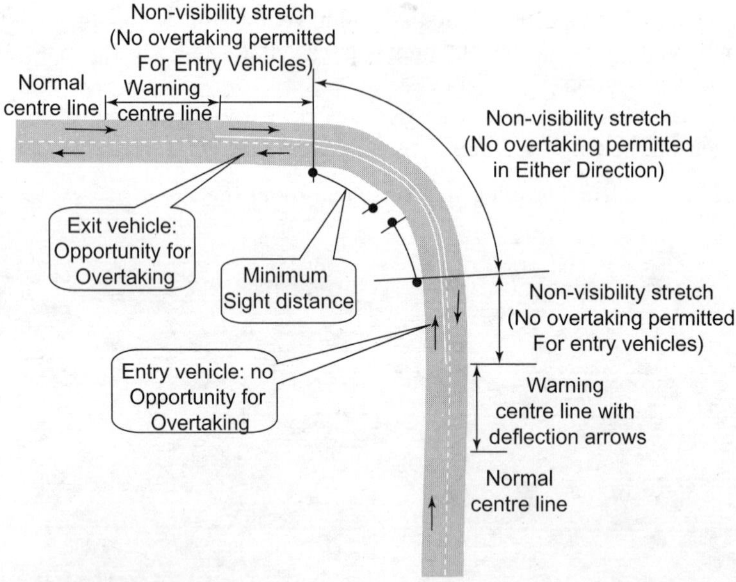


**Fig. 17.14** Schematic view of demarcation of No Overtaking stretch and minimum sight distance on a horizontal curve using road markings
(Redrawn based on IRC: 35-2015)

### 17.2.4 Block Markings

These markings resemble a block of paint and are used for different locations depending on the intended control function. Details of different types of block markings are as follows (IRC: 35-2015):

- Rectangular block (Fig. 17.15 (a)): Used as zebra crossing
- Triangular block (Fig. 17.15 (b)): Used for speed breakers/road humps
- Chequer block (Fig. 17.15 (c)): Used for speed breakers/road humps

- Yellow box marking (Fig. 17.15 (d)): Indicates prohibitory intersection area in which no vehicle is permitted to stop/halt
- Give Way symbol (Fig. 17.15 (e)): Provides priority of flow
- Cycle symbol (Fig. 17.15 (f)): Used to indicate dedicated cycle lanes

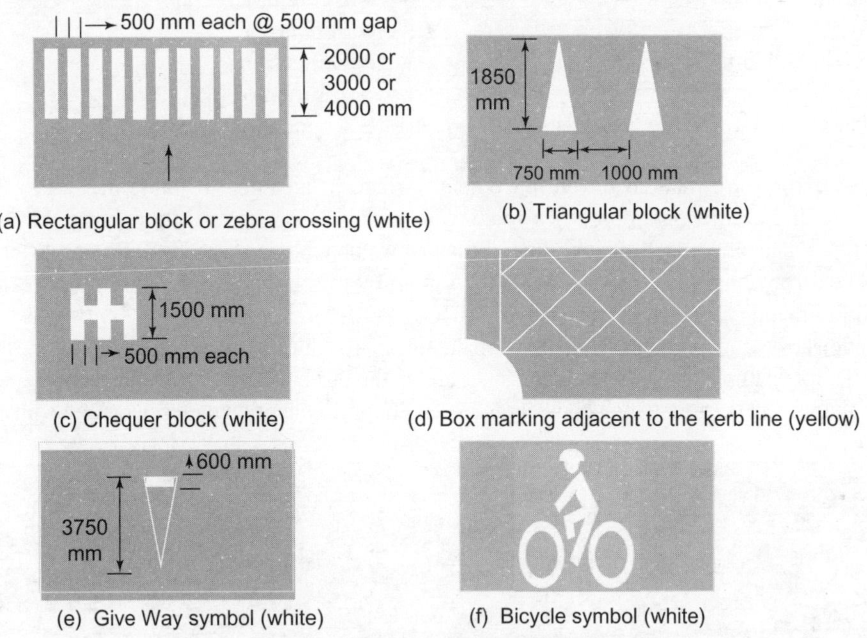


**Fig. 17.15** Different types of block markings
(Redrawn based on IRC: 35-2015)

The visibility of block markings is very important (Fig. 17.16). Hence, a special-quality thermoplastic paint is used for these markings.

**Fig. 17.16** Rectangular block markings used at different locations

### 17.2.5 Arrow Markings

Painted arrows (all in white) indicate the mandatory direction of travel and are categorised as follows (Fig. 17.17):

- Straight
- Left
- Straight and left
- Straight and right
- Right and left
- Straight, left and right
- Forward
- Deflection
- Bifurcation
- Roundabout rotation

Directional arrows are marked on the pavement surface in advance to guide drivers to the correct lane by avoiding confusion/ambiguities relating to entry restrictions at busy intersections and lane restrictions such as one-way flow. The size of the arrows depends on clear visibility at different design speeds. The arrow length is provided as 3.5 m, 5 m and 9 m for speeds $\leq$50 km/hr, between 50 km/hr and 100 km/hr, and $>$100 km/hr, respectively (IRC: 35-2015).

Deflection arrows are used in advance to warn drivers approaching a restriction and direct the traffic to the correct lane; for example, at No Overtaking zones and in the bus bay lane. Bifurcation arrows are painted at the beginning of tapers, from which point commencement of deceleration or acceleration begins.

Fig. 17.17 Arrow road markings: left, straight, and straight and left

### 17.2.6 Directional Markings

Word messages of a directional nature are categorised under this type of marking:

- STOP
- SLOW
- SPEED LIMIT
- KEEP CLEAR
- SCHOOL KEEP CLEAR
- NO PARKING
- NO ENTRY
- TURN LEFT
- TURN RIGHT
- KEEP LANE

Arrows supplemented with directional markings are termed mandatory turn arrows. For example, a left arrow followed by a TURN LEFT word message marking.

### 17.2.7 Facility Markings

The facilities provided for different purposes and their intended uses are indicated by these symbols and word messages. A few examples are listed below:

- Parking markings: Mid-block, Last Block
- Word message markings: BUS, BUS STOP, BUS LANE, CYCLIST, TAXIS and DISABLED

The dimensional details of these numerals and letters used for design speeds of up to 50 km/hr, 51–100 km/hr and >100 km/hr are available in IRC: 35-2015.

### 17.2.8 Object Markings

Any physical obstruction on or adjacent to a carriageway which is considered hazardous to traffic flow is delineated by being painted with dual-coloured hatched stripes. The main goal of painting alternating white-black or yellow-black stripes on these objects is to provide more visibility at night to avoid collision of vehicles with these objects.

The objects may be listed as any type, for example, piers, abutments, restrictive vertical clearance beams, electric poles, sign/signal posts, trees guard rails/stones, central islands/medians, culvert walls and subways/ramps supporting structural columns. The objects within the carriageway are marked with yellow-black stripes sloping downward at an angle of 45° towards traffic (Fig. 17.18). The diagonal stripes should be uniform in width which shall not be less than 300 m. Appropriate chevron markings are painted on the carriageway to cover these objects.

The height restricting beam/object within a carriageway is marked with vertically aligned yellow-black stripes spread over the full width of the beam. The diagonal stripes should be of uniform width which shall not be less than 300 m.

The objects adjacent to the carriageway, such as abutments and any side walls, are marked with white-black stripes sloping down at an angle of 45° towards traffic. Similarly, the width of the diagonal stripes shall not be less than 300 m and they should be supplemented with appropriate chevron markings.

Electric poles are marked with alternating horizontal black-white stripes, not less than 100 mm wide, up to a height of 1.25 above the road surface. Guard rails/stones/drums/trees are painted a solid white colour up to a height of 1.25 m above the road surface. A black band 300 mm wide is marked in the middle of the 1.25 m height of white paint.

Fig. 17.18 Views of object markings at a toll booth

Kerbs and concrete barriers (Fig. 17.19) on bridges/flyovers shall be painted with vertical black-white stripes of 500 mm width. All objects located within 2.4 m of the shoulders/kerbs shall be painted, in addition to the object markings. The object marker shall be at least 1.2 m above the nearest traffic lane.

Fig. 17.19 A few road markings

### 17.2.9 Colour Pavement Markings

Appropriate admixtures are used during construction of the pavement surface to provide the prescribed colours on the bituminous and cement concrete pavements.

## 17.3 ROAD STUDS

These are surface-mounted devices used to delineate road edges, centre lines and pedestrian crossing areas (Fig. 17.20). Based on their intended use, they are classified as follows:

- **Non-reflecting road studs:** These are used on pedestrian crossings as a tactile demarcation. Non-reflective road studs are cast using hard-grade aluminium alloys or stainless steel. They are available as a round or square (100 mm) top profile in aluminium or stainless steel and installed in small holes drilled on the pavement surface using hot poured bitumen.
- **Reflective road studs:** These are used to supplement line markings, longitudinal and transverse. Based on the direction of reflectivity, they can be uni-directional or bi-directional. The uni-directional studs are used on one-directional carriageways of undivided roads, operated as one-way routes or slip roads. The bi-directional studs can be used for two-directional roads to delineate the centre line and edge line of pavements. Reflective road studs are called raised reflective pavement markers (RRPM).

The most commonly used stud in India is the bi-directional road stud made of abrasive resistant plastic and fixed directly onto the pavement surface using hot melt bituminous/epoxy material. They provide guidance at night, including conditions of inclement/adverse weather conditions, when other traffic markings have been obliterated. In addition, the studs fixed across the carriageway in advance of junction/crossing/end of flyover ramp acts as a speed arrestor and as warning by creating a rumbling sensation to the driver through tyre interaction.

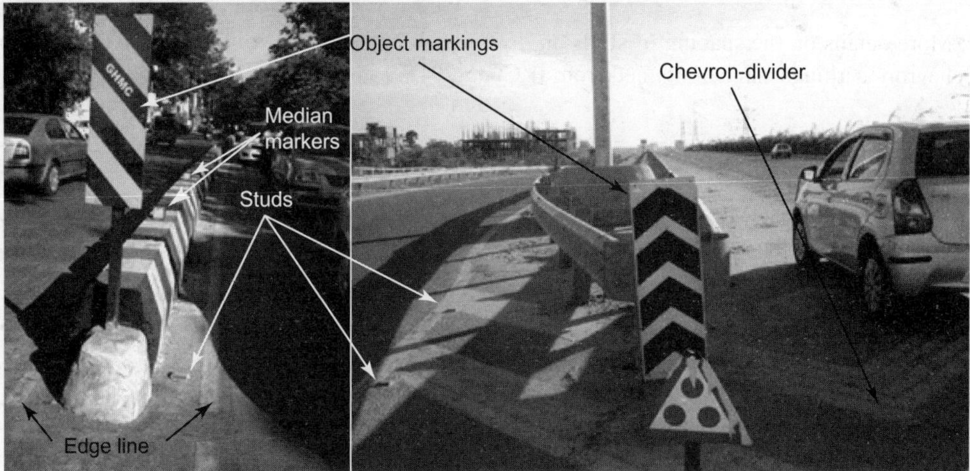


**Fig. 17.20** Road studs with other markings

The factors affecting the choice of stud and guidelines for installation are as follows (IRC: 35-2015):

- The studs come in three colours:
  - *White*: Indicates the traffic lane line and the centre of the carriageway
  - *Red*: Indicates prohibited crossings, prohibited left shoulder edge line
  - *Yellow*: Indicates prohibited crossings, prohibited right edge, i.e., median edge line of a multi-lane divided carriageway
  - *Green*: Indicates crossable edge line such as lay-byes and the boundary of the acceleration or deceleration line on the left of a multi-lane divided carriageway
  - Bi-directional carriageways use two coloured reflector faces (red-white, white-white or yellow-yellow)
- Solar-powered LED road studs may be used when the performance of normal road studs is not effective due to street lighting and other roadside activities. The critical locations may be listed as crash-prone/fog locations and highly hazardous locations. The LED studs are visible at a distance greater than 800 m in favourable conditions.
- Placement of road Studs: The studs are to be placed:
  - In line on broken longitudinal markings at the centre gaps
  - In the case of a paved shoulder, outside the shoulder side edge line, at a setback distance of 50 mm from the edge line
  - In the case of a median side edge, on the median line marking, at at least 50 mm setback distance from the median side edge line by ensuring a minimum clearance of 10 mm from the vertical face of the raised median
  - On a firm surface, where no dust and mud accumulation takes place over the stud

- Longitudinal placement: The desirable spacing along the length of the road and colour usage is summarised below:
  - Normal section: at an interval of 18 m
  - Warning section: at an interval of 9 m
  - No Overtaking zone: at an interval of 6 m
  - Color: White for traffic lane line, Red for shoulder edge line, and Yellow for the median side edge line

*Note*: More details on the spacing of studs on divided, undivided and bi-directional roads, and ramps with chevron markings can be obtained from IRC: 35-2015.

## 17.4 PREVIEW DISTANCE REQUIREMENTS

Road markings must be clearly visible during the day and at night. Drivers travelling at different speeds must be able to detect the guidance markings at a distance equivalent to a minimum of 2.0 sec of travel time (IRC: 35-2015). This is known as the minimum preview distance criterion.

**Example 17.1**

What is the minimum preview distance requirement for a speed of 40 km/hr. Assume IRC: 35 recommended criterion.

**Answer:** Speed = 40 km/hr = 11.11 km/sec. The distance travelled during 2.0 sec = 2 × 11.11 = 22.22 ≈ 22 m. With reference to IRC: 35 (2015), the minimum preview distance for the given speed = 22 m

## 17.5 PHENOMENON OF DIFFUSION, REFLECTION, REFRACTION AND RETROREFLECTIVITY

Surface brightness is measured in terms of luminance, which indicates the brightness intensity of light diffused reflected from an object towards the driver's eye in proportion to the apparent surface area ($m^2$). The luminous intensity of light is measured in units of candela (cd) and over a surface area ($m^2$), which is represented as $cd/m^2$.

In the case of mixed reflection (combination of diffusion and reflection) from the surface, it is described in units of luminance coefficient ($cd/m^2$ per lx) and is determined as follows:

$$\text{Coefficient of luminance} = \frac{\text{(Luminous intensity of the light towards the driver's eye in proportion to the surface } (cd/m^2))}{\text{(The illuminance at the surface created by light source (lx))}}$$

The commonly used measurement unit of coefficient of luminance as observed in the day time (Fig. 17.21 (a)) diffused condition ($Q_d$) and retroreflectivity in the night (Fig. 17.21 (b)) driving condition ($R_L$) is Milli–Candela/$m^2$ per lux ($mcd/m^2/lx$). It indicates the intensity of reflected light per lux (one metre-candle, or illumination provided by one standard candle at a distance of one meter) of light source (Fu and Wilmot, 2008). Pavement markings with a higher value of reflectivity indicate higher levels of visibility/brightness of the marking.

In the case of night driving, the driver can use his headlights to see objects with improved visibility because of retroreflected rays from the object's surface (Figs 17.21 (b) and 17.22). The glass beads embedded in the road marking will enhance the brightness and visibility of the pavement markings and also greatly improve the readability of retroreflective road signs. The retroreflection process in a glass bead occurs in three phases: refraction, reflection and refraction (Fig. 17.21 (c)) (DELTA, 2004; Babic et al., 2014).

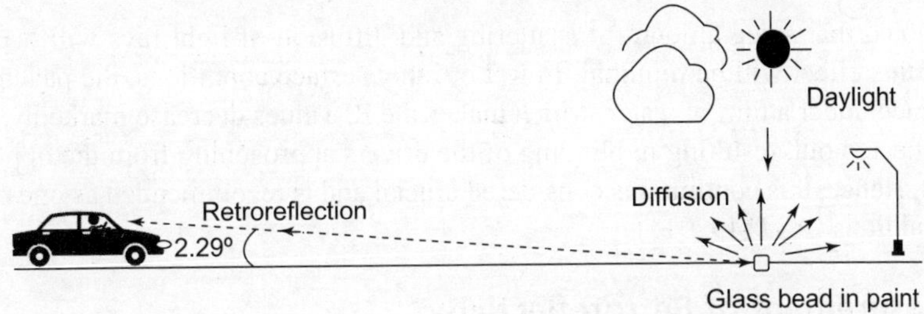


(a) Retroreflection from glass beads embedded in paint marking during the day

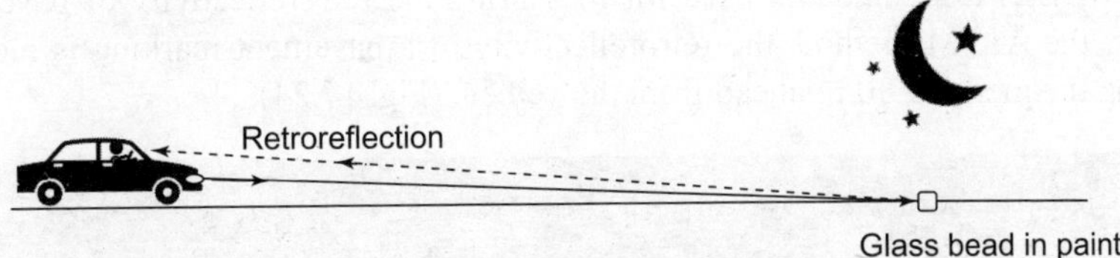


(b) Retroreflection from glass beads embedded in paint marking at night

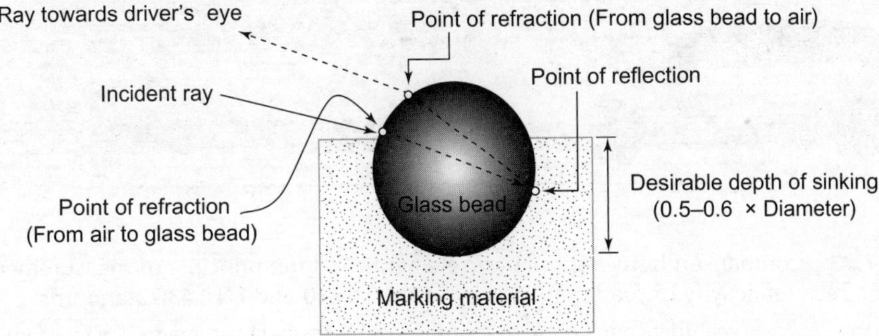


(c) Phenomenon of refraction and reflection of light rays from a glass bead embedded in a paint material

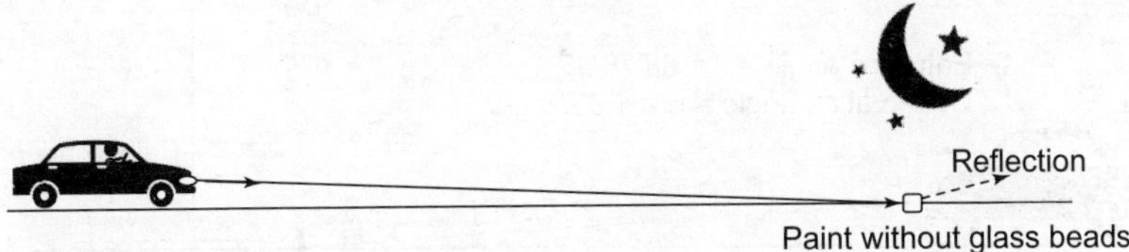


(d) Reflection from paint marking at night and also on a wet surface

**Fig. 17.21** Schematic representation of diffusion, reflection, refraction and retroreflection of light rays

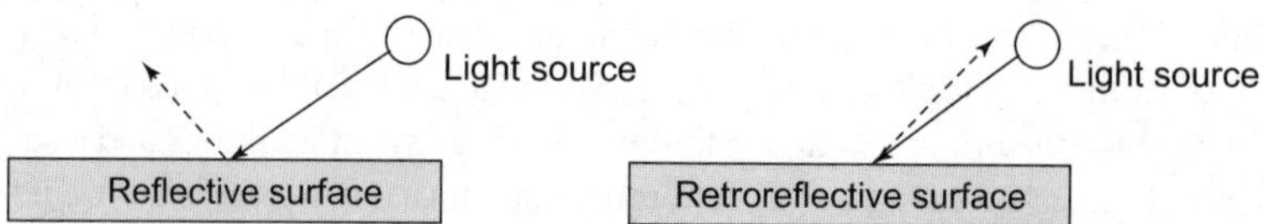


**Fig. 17.22** Schematic diagram showing the difference between reflection and retroreflection

It must be noted that some amount of scattering and diffusion of light rays will occur at night as well. However, this effect will be minimal. In wet or rainy surface conditions, the pavement markings may be submerged under a film of water, which makes the $R_L$ values decrease markedly, down to zero. This may be reflected out, resulting in blinding of the drivers approaching from the opposite direction (Fig. 17.21 (d)). Hence, this condition is considered crucial and is recommended as one of the standard surface test conditions (ASTM, 2011).

## 17.5.1 Measurement of Retroreflectivity

The American Society for Testing of Materials (ASTM) adopted the European Committee for Standardization (CEN) test standards used for measuring the retroreflectivity of pavement marking (Fig. 17.23). By the ASTM method, the retroreflectivity of a pavement marking is measured as if a driver is viewing the marking 30 m ahead from the vehicle (Fig. 17.24).

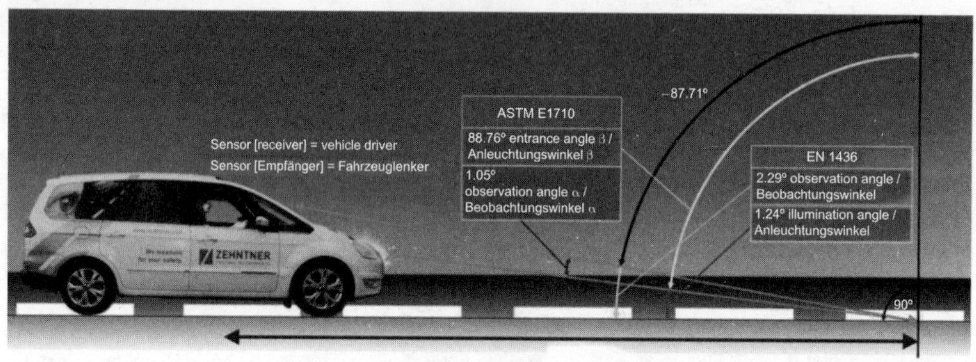


**Fig.17.23** Comparison between the basic elements and the principle of measurement of retroreflectivity of road markings by ASTM E 1710 and EN 1436 standards

(*Source*: www.zehntner.com; *Courtesy*: Zehntner GmbH Testing Instruments, Switzerland)

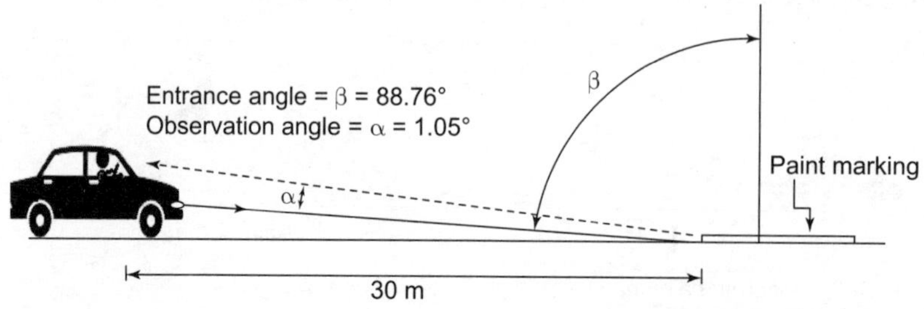


**Fig. 17.24** Standard 30-m geometry of measurement of retroreflectivity

The headlight mounting height is 0.65 m and the eye height of the driver is 1.2 m, directly over the stripe. This standard defines the requirements of retroreflectometers to simulate night-time visibility for the average driver in a car.

The measured value of the coefficient of luminance depends mainly on the conditions of the test. The following criteria are adopted to determine the performance of pavement markings:

- Reflectivity in diffuse lighting (daylight or street lighting) (Qd)
- Retroreflectivity in dry night-time conditions ($R_L$)
- Retroreflectivity when the road is wet ($R_L$)
- Retroreflectivity in rain conditions ($R_L$)

In addition, the colour of the road markings is defined by the luminance factor and chromacity vertices (x, y). Separate chromacity axes are defined by EN 1436 and it can be measured using a spectrophotometer. This instrument is used to evaluate the reflectivity/brightness of surfaces, including small and irregular shapes.

Appropriate test methods are adopted for measuring the performance of markings under the above reflectivity conditions based on any standard (ASTM E 1710 or EN 1436).

## 17.5.2 Instruments Used to Measure Retroreflectivity

The retroreflectivity of road markings, studs and signs can be measured using retroreflectometers (Fig. 17.25). Such instruments are available as handheld (Fig. 17.26) or mobile units (Fig. 17.26(a)). Handheld instruments are less costly but require more crew members to cover the same distance per day and also require traffic control while testing. In contrast, mobile instruments require only a single operator and the readings are taken at driving speed, without disturbing the traffic flow.

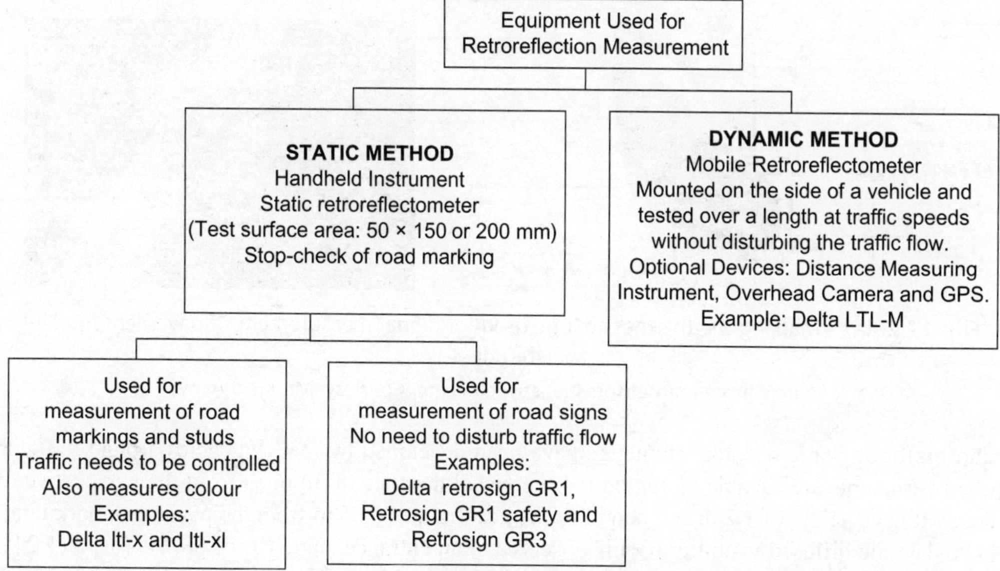


**Fig. 17.25** Types of equipment used to measure the retroreflectivity of road markings, studs and signs

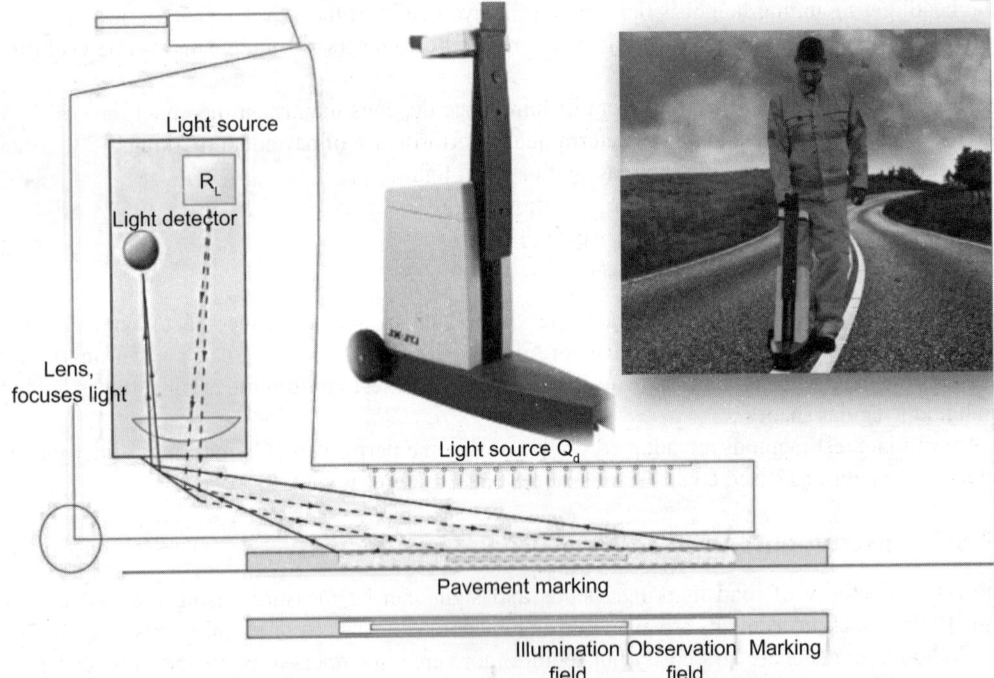


**Fig. 17.26** A handheld retroreflometer, Model: Delta LTL-XL
(*Source*: http://roadsensors.madebydelta.com; *Courtesy*: Mr. Kjeld Aabye)

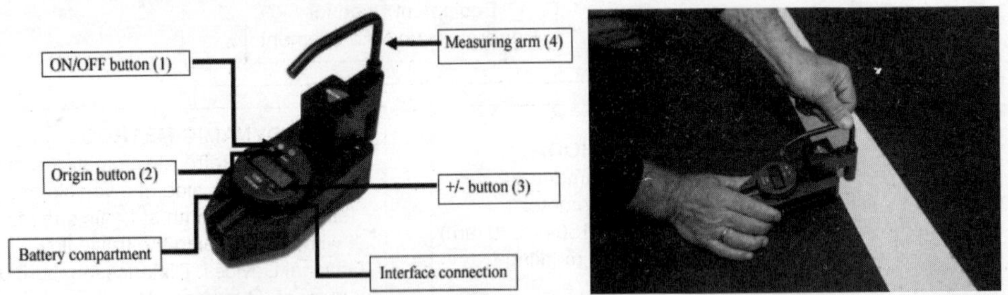


**Fig. 17.26(a)** Marking the thickness gauge used for digital measurement of pavement marking thickness
(Source: http://roadsensors.madebydelta.com, Courtesy: Mr. Kjeld Aabye)

**Static method:** Static retroreflectometers have been developed by many agencies around the globe. These instruments are capable of *simulating* the visual distance of 30 m as from the perspective of a driver, with eye height of 1.2 m and height of the headlight as 0.65 m from the pavement marking.

The daytime diffused visibility ($Q_d$) is observed at an entrance angle ($\beta$) of 88.76° (or = (1.24° + 1.05°) = 2.29° from the horizontal plane, where (90° – 88.76°) = 1.24°), at a distance of 30 m. This represents the value of the diffuse scattered light received by the observer.

Night-time visibility ($R_L$) is observed at a reflected angle of 2.29° with an input angle of 1.24°, at a distance 30 m. This represents the value of the retroreflectivity of light received by the observer.

**Dynamic method:** Night-time visibility ($R_L$) by the dynamic method is the same as that in the static method (simulating night visibility conditions). A dynamic retroreflectometer is mounted outside a vehicle and the test vehicle is driven along the road at highway speeds such that the retroreflectometer runs over the road markings (Figs 17.27 and 17.28) and takes continuous measurements of the retroreflectivity of the road marking, at a length interval of 50 or 100 m. Dynamic retroreflectometers are capable of measuring night and daytime visibility in dry and wet conditions, including profiled markings.

The above test procedures are adopted as per the standard (EN 1436).

## 17.6 MATERIALS USED FOR ROAD MARKINGS

Road markings are made of different types of material. To select the right kind of material, various factors such as traffic intensity, type of pavement surface, climate, rainfall intensity in the locality, type of road marking task and the type of equipment available need to be considered.

In general, the road marking material consists of binders, pigments, fillers and glass beads ($\Phi$ 60–850 μm). Binders with plastics and resins increase the thickness of the marking material and help them adhere to the road surface. Pigment imparts colour to the material while fillers impart durability. The quality and quantity of pigments used in markings (titanium dioxide induces brightness in white markings) affect the retroflection (Babic et al., 2014).

Based on the type of materials used, road markings can be classified into five categories (Fu and Wilmot, 2008; Asdrubali et al., 2013):

- **Paint**: Paints are of two types: waterborne and solvent-based. Waterborne paint is the most common material. Solvent-based paints are being replaced by waterborne paints because the former contain volatile organic compounds (VOCs) (440 g/l against 150 g/l) that are harmful to the environment. In any case, the VOC should be kept to a minimum of 65% (ASTM D 1644) and the liquid paint density should have a minimum of 1.3 gm/cc (ASTM D 1475). Paint is applied in the warm seasons. It does not require heating and can be applied quickly at speeds of up to 15 km/hr. Tiny glass beads can be dropped on the paint while it is wet. The service life of paint is short (range 3–36 months). Their initial retroreflectivity is 275 mcd/m$^2$/lx for white and 180 mcd/m$^2$/lx for yellow. Paints are suitable for low-volume roads, where wear-and-tear is less.
- **Epoxy (cold plastic):** This comprises a mix of a couple of compounds that together create a chemical reaction and form a solid adhesive for road markings. The compound consists of epoxy resins, pigments, extenders, fillers and a catalyst to accelerate the setting time. Epoxy resins may be replaced by polyester, polyurea or methacrylate. All these material have a smaller content of VOCs. Methacrylate-based materials produce very high retroreflectivity but may require a longer cure time (30 minutes), and they applied at slow speed (5 km/hr). Polyurea-based materials cure/dry very fast (3–8 minutes) and are not affected by humidity. They are more expensive than conventional paints but can provide a higher level of retroreflectivity (300 mcd/m$^2$/lx for white and 180 mcd/m$^2$/lx for yellow). They also have a longer service life (range 24–48 months), depending on traffic volume.
- **Thermoplastic**: This is a blend of hot molten ingredients comprising resin, pigments and fillers. Thermoplastics can be applied on the pavement surface by spraying/screeding or extrusion. It is a hot ($\geq$200°C) material applied at a speed of 5 km/hr with a thickness of 1.5–3 mm. Glass beads can be applied at 250 gm/m$^2$ area. Thermoplastics are of two types: hydrocarbon-based plastics derived from petroleum and alkyd, a naturally occurring resin that is suitable for high traffic and adverse climate areas. It has tiny glass beads embedded in the material, as premix beads. Thermoplastics do not contain

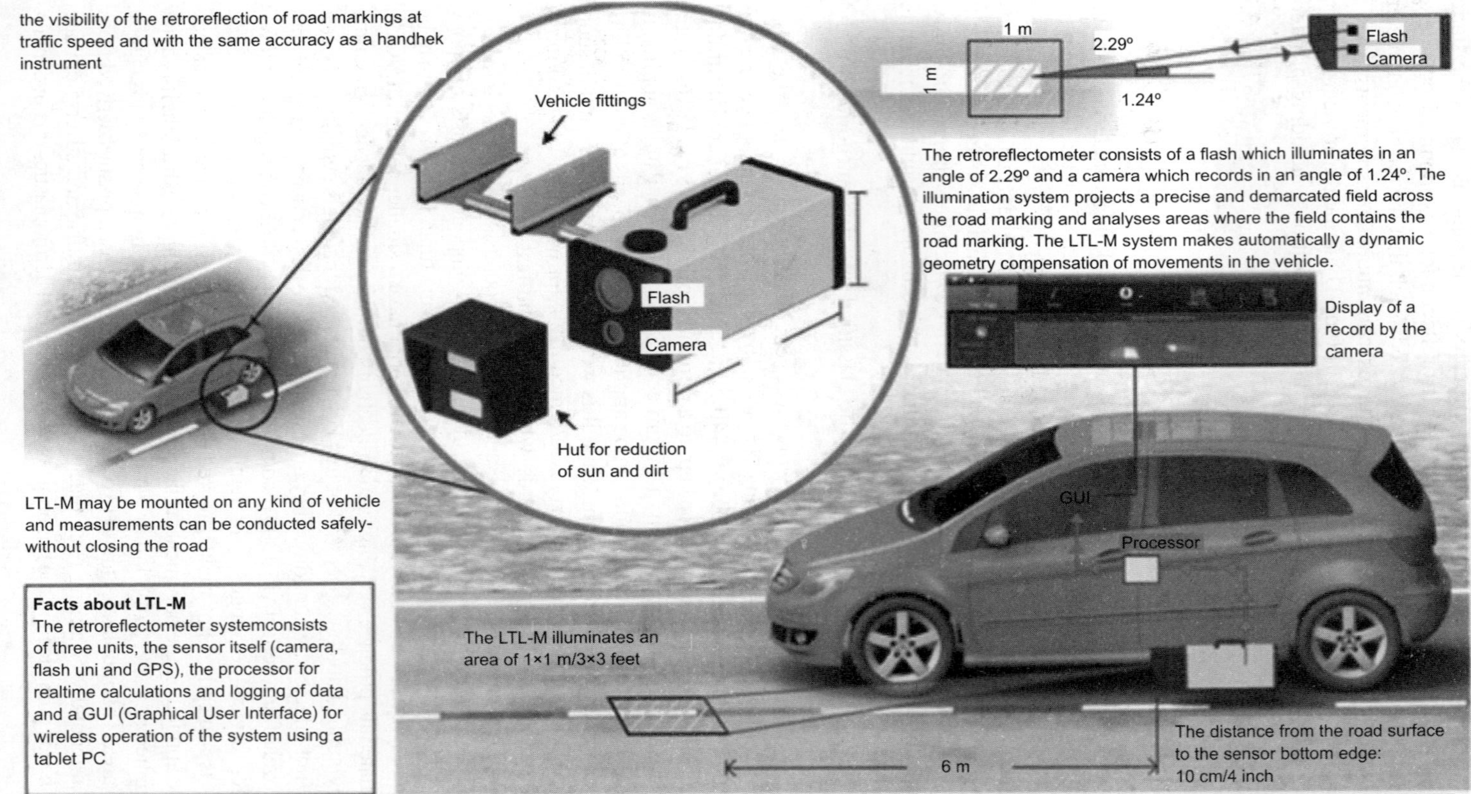


**Fig 17.27** Working principle and basic components of mobile retroreflometers (Model: Delta LTL-M)

(*Source*: http://roadsensors.madebydelta.com; *Courtesy*: Mr. Kjeld Aabye)

**Fig. 17.28** A vehicle-mounted mobile retroreflometer with GUI tablet PC (Model: Delta LTL-M)
(*Source*: http://roadsensors.madebydelta.com; *Courtesy*: Mr. Kjeld Aabye)

solvents. They are more expensive than conventional paints but can provide a higher service life (48–84 months), depending on the traffic volume, application time and equipment used.

- **Preformed thermoplastic:** This is also made of thermoplastic material but its shape is preformed in moulds in factories according to standard profiled pavement markings (PPMs), signs, symbols and text words. For example, bus, arrow, bicycle symbol, disabled symbol and pedestrian crossings. PPMs are used to enhance the visibility under wet and night conditions, where snowploughs are not used. Preformed markings are applied on the pavement surface using a gas-burner.
- **Tape**: This is a preformed flexible polymer (thermoplastic) tape with pre-applied colour and beads. Tape markings can be applied on the existing surface by hot-rolling or gluing into the prepared grooves. Tape is used as a fast marking paint. It is costlier than thermoplastic. Its service life is longer (up to 96 months). Its initial retroreflectivity value is 1100 mcd/m$^2$/lx for white and 800 mcd/m$^2$/lx for yellow. Tapes are suitable for high traffic volume roads. They may also be used as temporary marking as they are easy to remove.

These materials have been used in a variety of conditions, including at different traffic levels, on different pavement surfaces and with different line types and colours (Fu and Wilmot, 2008).

**Glass beads:** Glass beads usually have a diameter in the range 60–850 μm. They are categorised as two types, based on their application and diameter (Fig. 17.29). They should be transparent, colourless, pure and have a minimum true sphere of 70%. Glass beads should have a minimum refractive index of 1.50 and pass the free flow test (BS: 6088 and BS: 3262 (Part 1)).

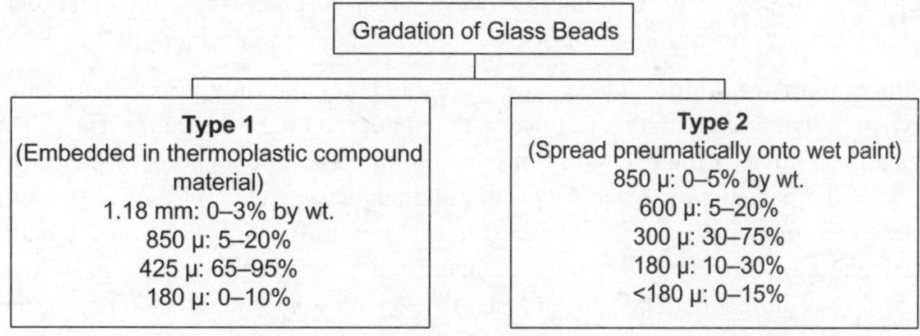


**Fig. 17.29** Types of glass beads based on gradation requirements
(*Source*: MORTH, 2013; *Courtesy*: Ministry of Road Transport & Highways, Govt. of India)

The tiny glass beads embedded in the road markings reflect the headlights back to the driver and enable visibility of the road markings, particularly at night (Fig. 17.30). If no glass beads are embedded in the markings, the headlight rays may be diffused and reflected in all directions. In contrast, too many glass beads in the marking cannot help improve visibility at night because the extra glass beads, behind those in front and which reflect the headlight beam, will be covered in shadow, and hence, no retroreflection can take place from those extra glass beads.

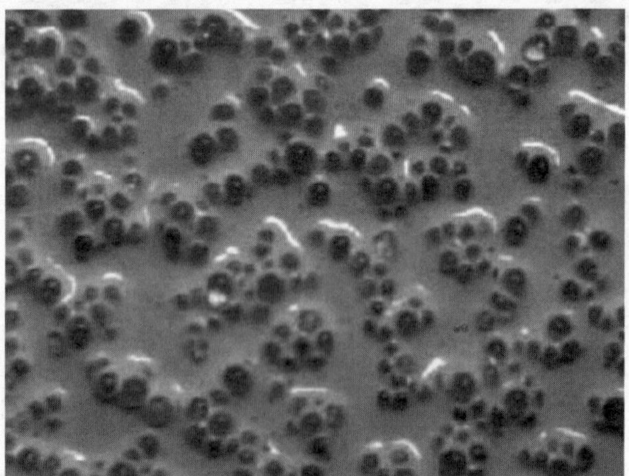

**Fig. 17.30** A view of glass beads embedded in a road marking
(*Source*: www.zehntner.com; *Courtesy*: Zehntner GmbH Testing Instruments, Switzerland)

Glass beads have a refraction index between 1.5 and 1.55 and can induce high retrorefraction if their degree of sinking into the marking is between 55% and 60% (Fig. 17. 31). As the sinking reduces 50% of their diameter, their resistance to abrasive forces under tyre load also reduces (Asdrubali et al., 2013).

## 17.7 PERFORMANCE ASSESSMENT OF ROAD MARKINGS

The retroreflectivity of road marking have a significant effect on road safety, Therefore, the performance of the retroreflection needs to be assessed and tests run to determine the appropriate restoration action to be taken.

The performance of pavement markings is measured based on various functional characteristics such as daytime visibility, night-time visibility, wet reflectivity and wear durability.

**Wear durability of paint markings:** The degree of wear of paint markings is assessed as the percentage of the coloured area that remains intact (leftover painted area) on the pavement surface. The degree of wear (%) depends on the wearing resistance of the paint material, the location of the traffic operation and the quality of the pavement surface supporting the painted area. As pavement distress increases, the wearing rate also increases. Similarly, wear increases as traffic operation on the painted surface increases, particularly on left-hand edge lines and intersection approaches.

IRC: 35-2015 has recommended that the threshold level of wear by visual observation should be markings with less than 70% remaining.

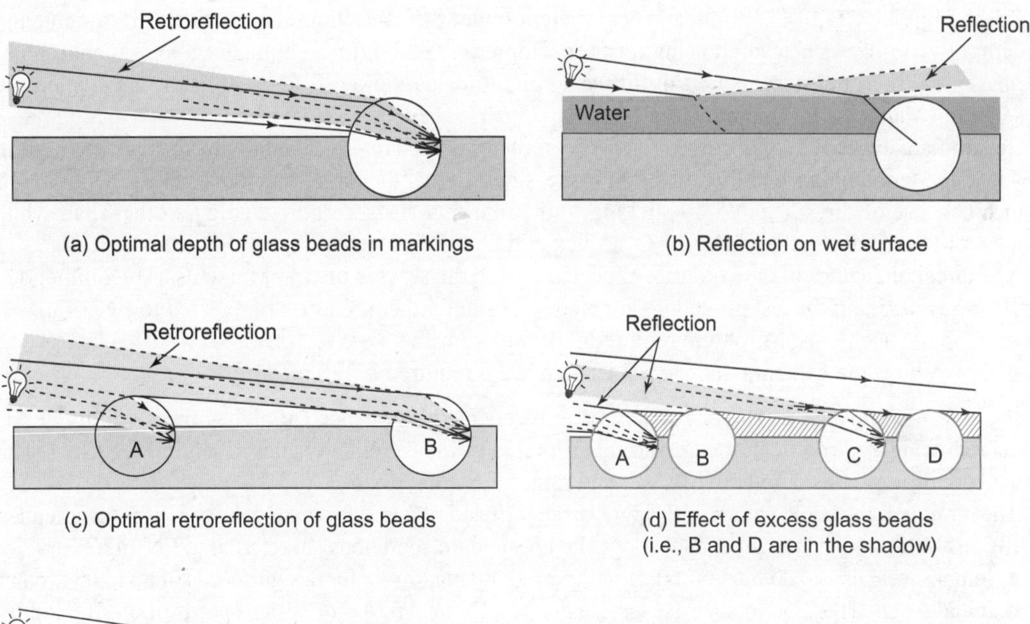


**Fig. 17.31** Phenomenon of retroreflectivity from glass beads embedded in road paint
(*Source*: www.zehntner.com: *Courtesy*: Zehntner GmbH Testing Instruments, Switzerland)

**Fig. 17.32** More than 90% of the paint area is intact

**Fig. 17.33** Less than 30% of the paint area is intact

**Daytime visibility ($Q_d$):** The luminance coefficient under diffused illumination ($Q_d$), for daytime and streetlight visibility, is measured using retroreflectometers. The diffused luminance coefficient ($Q_d$) of the pavement marking is measured in diffused conditions as viewed from a 30-m distance when the observation angle is 2.29° from the road surface (Fig. 17.21 (d)).

To simulate the above conditions, a retroreflectometer is used to project a light beam at an inclination of 45° to the target plane, and the reflection is measured perpendicular to that plane. The reflection will reduce because of dirt, dust and oil sticking to the marking surface and for colours other than white (yellow or faded white), therefore, the $Q_d$ values will be lower.

The threshold values of $Q_d$ which are expected during the service life of pavements as recommended by IRC: 35-2015 and the test procedure adopted with standard conditions are referred to EN 1436.

- Asphalt pavements: Minimum required is 100 mcd/m²/lx
- For yellow and cement concrete roads: Minimum required is 130 mcd/m²/lx

**Night-time visibility ($R_L$):** The coefficient of retroreflected luminance (in night-time visibility) $R_L$ is measured using a retroreflectometer, and the principle of measurement is described in Fig. 17.24. The critical conditions considered are dry, wet and rain.

The minimum threshold values of $R_L$ and warranty period required for up to two years as recommended by IRC: 35-2015 and the test procedure adopted with standard conditions are referred to EN 1436.

- Initial seven days: 200 mcd/m²/lx; service period: 80 mcd/m²/lx for design speeds of up to 65 km/hr
- Initial seven days: 250 mcd/m²/lx; service period: 120 mcd/m²/lx for design speeds of 65–100 km/hr
- Initial seven days: 350 mcd/m²/lx; service period: 150 mcd/m²/lx for design speeds above 100 km/hr
- Initial seven days: 300 mcd/m²/lx; service period: 50 mcd/m²/lx for optional 'wet night' assessment for all speeds

The European Union Road Federation (ERF) recommends a minimum preview time of approximately 3 sec for testing a road marking with a minimum width of 15 cm, and the $R_L$ values should not be less than 150 mcd//m²/lx in dry conditions and 35 mcd//m²/lx in wet conditions.

For more details, readers may refer to the standard codes of practices adopted in different countries (European EN 1436, ASTM E 2177, ASTM E 1710 and ASTM E 2832).

## 17.8 ANTI-SKID TESTS ON ROAD MARKINGS

Skid resistance on road markings are evaluated using several instruments (Fig. 17.34). British Pendulum Apparatus (Fig. 17.35) is used, in general, for measurement of spot skid resistance and of road markings. Alternatively, the Micro Grip Tester (Fig. 17.36) can be used for this purpose.

Standard criteria for skid resistance measurement on markings by British Pendulum Tester can be referred to in EN 1436. The threshold values of skid resistance on road markings, as recommended by IRC: 35-2015, are as follows:

- Normal locations in general: Not less than 45 British Pendulum Number
- Critical locations such as large markings, zebra crossings, letters, numbers, arrows and transverse yellow markings: Not less than 55 British Pendulum Number

**Maintenance inspection frequency:** The following guidelines are recommended by IRC: 35 (2015) for maintenance inspection of the performance of road markings:

- Road markings shall be routinely inspected once in six months for all characteristic requirements.
- Longitudinal markings on National Highways and Expressways at network level should be surveyed on a routine annual basis. The measurements made by mobile retroreflectometers should

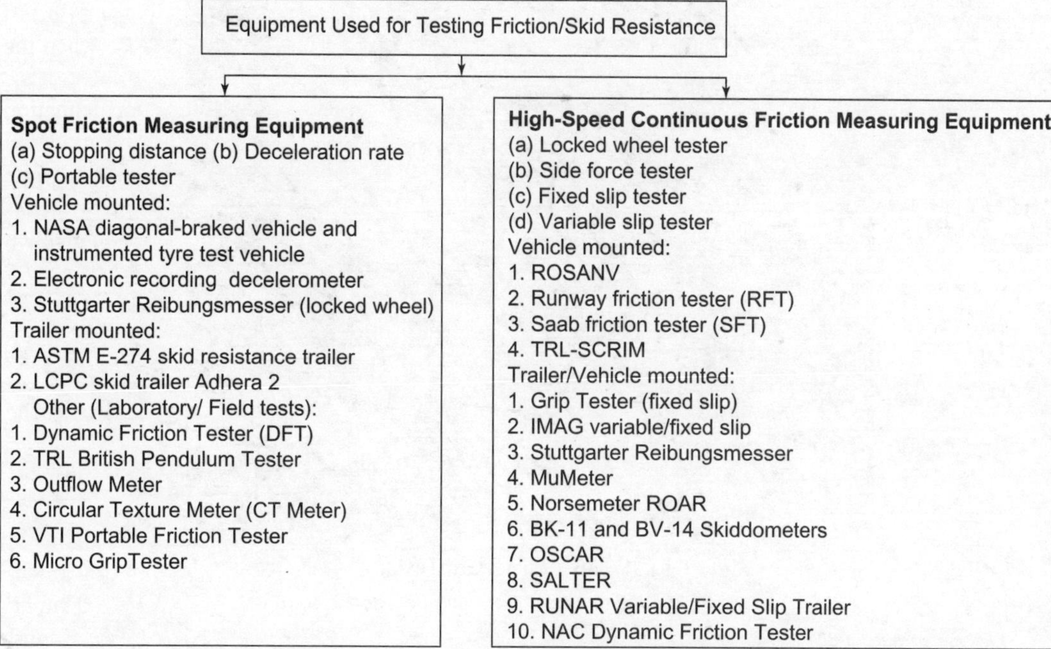


**Fig. 17.34** List of equipment used for measuring the skid resistance of a pavement surface (Srinivasa, 2014)

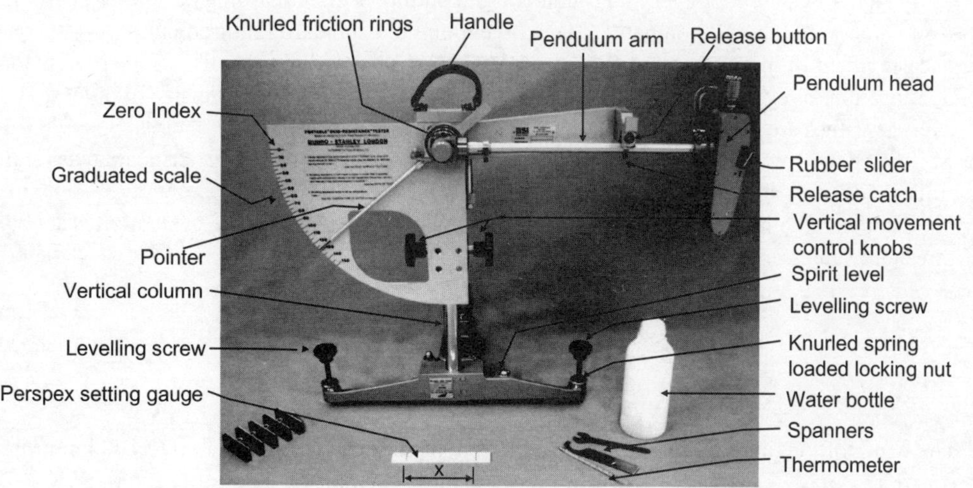


**Fig. 17.35** TRL portable skid resistance tester
(*Courtesy*: Mr. B.S. Pressland of www.munroinstruments.co.uk)

**Fig. 17.36** Micro GripTester
(*Courtesy*: Mr. Alister Miller of www.findlayirvine.com)

be averaged for every 100 m and spot locations such as arrows and STOP location markings should be tested with static/handheld retroreflectometers.

- Wear assessment must be carried out on at least 50% of the road markings in an area, or on a 5-m length of an area for every 20 m interval of continuous road markings. An average of the test results reported must be compared against the minimum standard requirements.
- Measurement of luminance coefficient survey must be carried out on 50% of the road markings at a location, or every 20 m for continuous road markings. Take a sample size of at least 20. The average values are compared.
- Skid resistance measurements are carried out annually on a quarter of the critical areas (at most-trafficked areas) and averaged.

More details on the inspection of new and repaired road markings and the evaluation criteria, with their threshold values, may be obtained from IRC: 35 (2015).

## EXERCISES

### Multiple Choice Questions

1. The width of longitudinal markings in rural sections for design speeds of up to 100 km/hr is
(a) 50 mm (b) 100 mm (c) 150 mm (d) 200 mm
2. The width of longitudinal markings in rural sections for design speeds greater than 100 km/hr is
(a) 50 mm (b) 100 mm (c) 150 mm (d) 200 mm
3. The drying time for ordinary road marking material hot applied thermoplastic compound is set to bear traffic in not more than ________ minutes.
(a) 15 (b) 30 (c) 45 (d) 60

4. In general locations, skid resistance should be _________ (by British Pendulum Tester).
   (a) Not less than 45 (b) Greater than 30 (c) More than 10 (d) Equal to 55
5. IRC recommends that the softening point of ordinary road marking material hot applied thermoplastic compound be at _________°C.
   (a) $100.0 \pm 10.0$ (b) $102.5 \pm 9.5$ (c) $120 \pm 10.5$ (d) $90 \pm 10.1$
6. IRC recommends that the yellow index for white thermoplastic paint shall be
   (a) Not more than 0.10 (b) Not more than 0.20
   (c) Not more than 0.30 (d) Not more than 0.12
7. The glass beads used in road markings should have a minimum true sphere of _________
   (a) 50% (b) 60% (c) 70% (d) 80%
8. The minimum value of refractive index of glass beads used in road markings is
   (a) 1.0 (b) 1.5 (c) 2.0 (d) 2.5
9. The pavement temperature shall not be less than _________°C during application of thermoplastic road marking material.
   (a) 0 (b) 10 (c) 20 (d) 4
10. Thermoplastic paint should be applied in intermittent or continuous line of uniform thickness of at least _________ mm.
    (a) 2.5 (b) 5.0 (c) 7.5 (d) 4.0
11. The rate of spread of glass beads of type 2 in a mono-layer on the hot paint line in quick succession of the paint spraying operation is
    (a) 200 gm/m$^2$ (b) 250 gm/m$^2$ (c) 100 gm/m$^2$ (d) 300 gm/m$^2$
12. The width of lines and other road markings shall not deviate from the specified width by more than _________ %
    (a) 2 (b) 3 (c) 4 (d) 5
13. The position of lines, letters and symbols of road markings shall not deviate from the specified position by more than _________ %
    (a) 5 (b) 10 (c) 15 (d) 20
14. The alignment of any edge of a longitudinal line shall not deviate from the specified alignment by more than _________ in 15 m.
    (a) 15 mm (b) 15 mm (c) 10 mm (d) 20 mm
15. The length of a segment of broken longitudinal lines shall not deviate from the specified length by more than_________ mm.
    (a) 100 (b) 150 (c) 200 (d) 250
16. The paint shall not be applied when the surface temperature of the pavement is greater than _________ °C
    (a) 40 (b) 50 (c) 60 (d) 100
17. The minimum proportion of titanium dioxide in cold applied white paint (by % weight) should be
    (a) 25 (b) 20 (c) 16 (d) 10

18. Skid resistance for a beaded dry film of cold applied reflective paint shall not be less than __________ as per BS6044.
    (a) 30 (b) 40 (c) 45 (d) 50
19. The minimum thickness of a wet unbeaded coat of paint shall not be less than __________ microns.
    (a) 200 (b) 300 (c) 400 (d) 500
20. The minimum thickness of dry unbeaded coat of paint shall not be less than __________ microns.
    (a) 200 (b) 300 (c) 400 (d) 500
21. The thickness of the raised transverse bars on the shoulders and centre line shall be
    (a) 10–15 mm (b) 11–14 mm (c) 10–20 mm (d) 5–10 mm
22. White audible and vibratory markings shall attain an initial retroreflectance of not __________ mcd/m²/lx
    (a) <300 (b) >300 (c) <250 (d) >80
23. Yellow audible and vibratory markings shall attain an initial retroreflectance of
    (a) <300 (b) >300 (c) <250 (d) >80
24. What area the number of days the pavement markings shall be subject to observation under normal traffic? If the marking do not perform satisfactorily in this period, the contractor has to replace the same at his own cost.
    (a) 30 days (b) 60 days (c) 180 days (d) 360 days
25. The dimensions of solar-powered stud road markings shall not be less than
    (a) 100 × 100 × 10 mm (b) 200 × 100 × 10 mm
    (c) 150 × 100 × 10 mm (d) 100 × 150 × 5 mm
26. The minimum threshold distance of visibility requirement when using solar-powered LED stud road markings is
    (a) 500 m (b) 600 m (c) 700 m (d) 800 m
27. When solar-powered LED stud road markings are used, the flashing rate shall not be less than
    (a) 0.1 Hz (b) 0.5 Hz (c) 0.75 Hz (d) 1.0 Hz
28. The road stud marker height shall not be less than __________ mm.
    (a) 5 (b) 10 (c) 15 (d) 20
29. The road stud marker height shall not exceed __________ mm.
    (a) 5 (b) 10 (c) 15 (d) 20
30. The free-flowing property of glass beads shall be assessed by conducting free-flow test using a funnel having an orifice of
    (a) 1 mm (b) 1.18 mm (c) 6 mm (d) 850 micron
31. As per IRC: 35, what would be the minimum preview distance required for a design speed of 80 km/hr?
    (a) 44 m (b) 50 m (c) 56 m (d) 61 m

## Answers

1. (b) 2. (c) 3. (a) 4. (a) 5. (b) 6. (d) 7. (c) 8. (b) 9. (b) 10. (a) 11. (b)
12. (d) 13. (d) 14. (c) 15. (b) 16. (a) 17. (b) 18. (c) 19. (c) 20. (a) 21. (b) 22. (a)
23. (c) 24. (c) 25. (a) 26. (d) 27. (d) 28. (b) 29. (d) 30. (c) 31. (a)

## Questions

1. State the objectives and functions of road markings.
2. Classify road markings based on function/location and the colours used.
3. State the intended use of road marking colours for specific control types. Give a few examples for each case.
4. Describe the guiding principles of line markings.
5. Describe the types of longitudinal markings, with their intended control type.
6. State the locations where centre line markings are provided on carriageways.
7. Explain how configuration and measurement of broken and continuous line characteristics are measured. Give a few examples for each case.
8. Briefly write the details of continuous single- and double-line variants marked; broken line segment, gap length and width may be provided as per IRC: 35.
9. Briefly write the details of broken-line variants marked, with broken line segment, gap length and width provided as per IRC: 35.
10. State where no centre line is marked along the carriageway. Also state the marking details in case of restricted width of carriageway on bridges and culverts.
11. Describe the transverse line variants with broken and continuous solid lines as recommended by IRC: 35.
12. Describe and write the salient details of markings used as hazardous marking variants, as recommended by IRC: 35.
13. Draw a schematic view of a ladder-hatching marking on a horizontal curve, labelled with the main components.
14. Draw a schematic diagram labelled with a No Overtaking stretch and minimum sight distance on a horizontal curve by road markings.
15. List and give details of the different types of block markings used for different locations.
16. List and give details of the different types of arrow markings used for different locations.
17. List the different types of directional markings used for different locations.
18. Write the details and standard specifications of object markings.
19. What are functions of road studs? Classify them and explain their details.
20. Write the factors affecting the selection and a few guidelines of installation/placement of road studs.
21. Define the phenomenon of diffusion, reflection, refraction and retroreflectivity. Draw sketches.
22. Explain the measurement of retroreflectivity of road markings with standard 30-metre geometry.
23. State the criteria adopted for determination of performance of pavement markings.
24. List the characteristics considered for maintenance inspection of pavement markings. State their threshold values, method of inspection frequency and sampling, as recommended by IRC: 35.
25. Classify and write the details of the different types of equipment used for measuring the retroreflectivity of road markings, studs and signs. State the merits and demerits of each equipment.
26. Write a short note on the static method and dynamic method of measurement.
27. Classify and give details of the various types of materials used for road markings. Also state their suitability with reference to chemical compounds, longevity and retroreflectivity.

28. State the functions, types and gradation requirements of glass beads used in road markings.
29. Write a short note on glass beads used in road markings. Explain the characteristic features required for road markings.
30. Explain how the performance of road markings is evaluated based on wear durability of paint markings. Also state the threshold values considered as per IRC: 35.
31. Explain how the performance of road markings is evaluated for day-time visibility criterion. Also state the threshold values considered as per IRC: 35.
32. Explain how the performance of road markings is evaluated for night-time visibility criterion. Also state the threshold values considered as per IRC: 35.
33. Explain how the performance of road markings is evaluated for anti-skidding criterion. Also state the threshold values considered as per IRC: 35.
34. Write a short note on maintenance inspection frequency of road markings.
35. Write the limitations of road markings.

CHAPTER

# 18 Road Traffic Signs

## 18.0 INTRODUCTION

Road signs form a vital and integral part of the road transportation system. They are used for traffic operation and control, and provide information and safety to vehicular and pedestrian traffic. Traffic signs are categorised under traffic control devices. They consist of easily recognisable, readable and understandable symbols, pictures, number and words. Vehicular and pedestrian traffic must know the meaning of all the signs to be able to drive/use the road safely.

## 18.1 CLASSIFICATION OF SIGNS

Traffic signs can be classified based on their purpose (Fig. 18.1).

- **Regulatory signs:** Traffic movements are regulated by assigning priority and indicating the regulations that are legally in force. Any failure in complying with these mandatory traffic regulations is treated as a constitutional offence. The regulatory signs may be categorised into three groups (MOTSAM, 2010):
  - *Prohibitory:* These signs indicate that road users must not take a particular action. For example, No Parking, Right Turn Prohibited, Bullock Carts Prohibited.
  - *Mandatory:* These signs indicate that road users must take a particular action. For example, Minimum Speed Limit, Compulsory Turn Right, Pedestrian Only.
  - *Operational:* These signs indicate that road users must abide by a particular restriction. For example, Axle Load Limit, Length Limit, Maximum Speed Limit.

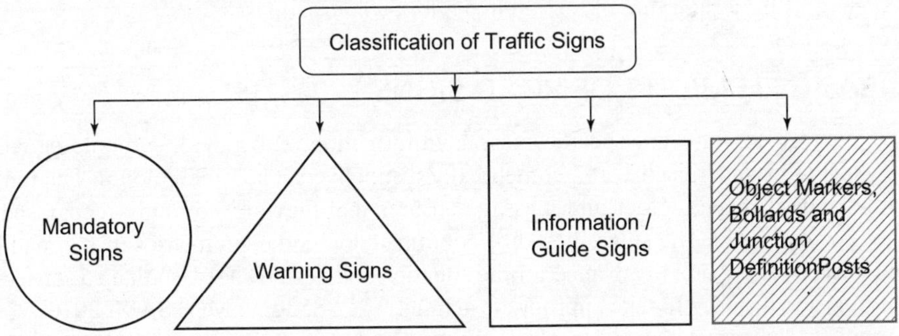


**Fig. 18.1** Classification of traffic signs

- **Warning signs:** Road users are warned of hazards on or adjacent to the road, which may be temporary, unexpected or unusual.

- **Information/Guide signs:** Road users are guided along their journey with information about route directions, distances, destinations, location services and other important/related information along the roadways. These signs may be categorised into three groups:
- **Direction signs**
  - *Advance direction signs*: These are located at approaches to intersections and indicate the destinations ahead.
  - *Intersection direction signs*: These are located at intersections and provide directional information.
  - *Confirmation direction signs*: These are located beyond intersections and reassure the drivers about the route on which they are travelling.
- **Motor services and tourist signs**
  - *Motorist services signs*: These indicate the services available ahead relating to motorist needs, such as repair services and fuel stations.
  - Tourist information signs
- **General information signs**
  - Local place/feature names
  - Local authority boundaries
  - Facilities provided for persons with disabilities
  - Route marker signs
  - Other miscellaneous information
- **Hazard markers:** Road users are informed about the presence of obstructions or change in directions, if any, ahead on the route. These marker signs may be categorised as:
  - Chevron markers
  - Object hazard markers
  - Two-way hazard markers
  - Delineation posts
  - Guide posts
  - Miscellaneous devices

More details of these signs are discussed in the following sections.

## 18.2 BASIC ELEMENTS OF ROAD SIGNS

Signs are intended to convey a specific message/information, either with symbols or with legends or both, which are standardised with geometrically distinctive shapes and colours. The uniformity of signs with standard shape, colour and message ensures that they are promptly recognised, read and understood by road users. The signs should be coordinated with the geometrics of the road layout and should be located far enough in advance for the situation and design speed, so that road users will have sufficient time and distance to take appropriate action (DTTAS, 2010; MOTSAM, 2010).

Uniformity in the application of signs of different classes and similarity in the conditions of road geometrics installed with the same/consistent signs in a country are essential features by which road users readily anticipate the course of action required. In other words, uniformity means treating similar traffic situations in the same way and consistently. Signs should be erected to optimally meet the needs of road users. Unnecessary, incorrect and inconsistent signs lead to confusion and non-compliance. Any variance in the uniform and standard use of signs will confuse and potentially create hazardous situations/road locations.

The factors discussed below affect the application of signs (MUTCD, 2003; IcRC: 67-2012).

### 18.2.1 Shape

Signs come in seven basic shapes, and each shape has a specific meaning. The standardised shapes of signs and their intended use are described in Table 18.1

**Table 18.1** Shapes of signs and their significance

| *Shape* | *Shape geometry* | *Use* |
|---|---|---|
| Octagon | 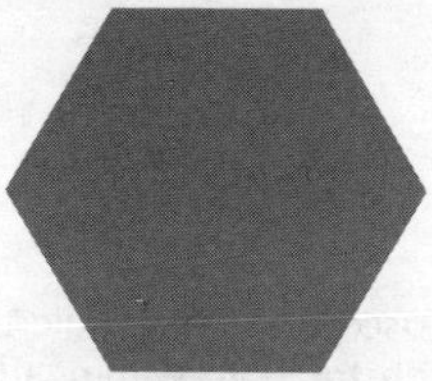 | Exclusively reserved for the Stop sign |
| Circle | | Used for regulatory/mandatory/compulsory direction signs, operational control signs, prohibition signs, speed/weight/width/length limit signs, restriction end signs, and signs relating to pedestrian safety and railway level crossings |
| Equilateral triangle (with one end pointing downwards) |  | Exclusively reserved for the Give Way sign |
| Equilateral triangle (with one end pointing upwards) | 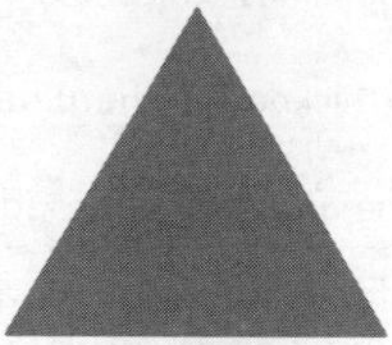 | Used for cautionary/warning signs |
| Rectangle (with long vertical axis) | | Information signs – facility information and direction information, route marker signs, stack advance direction signs, reassurance signs (gantry mounted/shoulder mounted) |
| Rectangle (with long horizontal axis) | | Information signs – map and advance direction signs, flag direction signs, place identification signs (gantry mounted/shoulder mounted) |

(Continued)

**Table 18.1** (Continued)

| *Shape* | *Shape geometry* | *Use* |
|---|---|---|
| Rectangle (with a triangular edge) | | Information signs – flag direction signs, park and ride signs |
| Diamond (with yellow or orange background) | | Used to warn and alert road users about obstructions within, adjacent to and at the end of the roadway; specific animal crossings, also used in work zones |

## 18.2.2 Colour

Different colours are used for road signs. Each colour has a specific meaning. The standard colours commonly used for signs are presented in Table 18.2.

**Table 18.2** Colours of signs and their significance

| *Colour* | *Use* |
|---|---|
| Red | • Used as background colour for the Stop sign and the No Entry sign<br>• Used for most circular border lines (thickness 70 mm) for prohibition/vehicle control/speed control signs<br>• Used as a triangular border (thickness 70 mm) for cautionary warning signs and Give Way signs<br>• Used as the legend depth-line of flood gauges (depth above 600 mm) |
| White | • Used as background colour for Restriction Ends and Give Way signs<br>• Background colour for most prohibitive circular signs<br>• Background colour for all triangular cautionary warning signs/vehicle control/speed/weight/length limit and other regulatory signs<br>• Used as legend/text/arrow symbol colour for circular compulsory control and other circular signs, parking signs, signs for persons with disabilities and some information/guide signs[c, d]<br>• Used as thin border line colour for information/guide signs[c, d] |
| Yellow | • Background colour for specific warning signs and body colour of mile posts (partial depth) |
| Orange | • Used as background colour for National Highway route marker signs |
| Green | • Used as background colour for State Highway route marker signs, and most direction and information/guide signs[a, b, c] |

(Continued)

**Table 18.2** (Continued)

| *Colour* | *Use* |
|---|---|
| Blue | • Used as background colour for circular compulsory/speed/direction control signs, No Parking signs, community facility information signs and persons with disabilities rectangular[a] signs[d] |
| Brown | • Used as background colour for Asian Highway route marker signs[a] and tourist route/information signs |
| Black | • Used as symbols/text/legend colour for circular prohibition regulation/ operational control signs/width marker signs, No U-Turn signs, triangular cautionary/warning signs and some facility information signs<br>• Used for the legend depth-line of flood gauges (depth below 600 mm)<br>• Used as background colour for object (left/right)/two-way hazard marker signs[a] and double/multiple chevron signs[b] |
| Fluorescent Yellow-Green | • Used as background colour for facility information signs |
| Fluorescent Yellow | • Used as background colour |
| Fluorescent Orange | • Used as background colour for rectangular signs, parking restriction signs[b], and road work signs<br>• Used as background colour for single/double/multiple chevron signs<br>• Body colour of traffic cones and Children Crossing flags |

[a] Rectangular signs with long vertical axis; [b] Rectangular signs with long horizontal axis; [c] Used at approaches to at-grade intersections; [d] Used at approaches to flyovers/interchanges

### 18.2.3 Lettering

Letters and numerals of different height are used on various highway signs, such as information/guide/route marker signs and kilometre stones. In India (IRC: 30-1968), the size range of the letters and numerals is 50–300 mm.

Sentence casing or all-capitalised letters are used for the text/words on direction information signs. The numbers on the circular vehicular speed limit signs represent the maximum permitted speed in km/hr. Lowercase letters are used for abbreviations such as pm (post meridiem), am (ante meridiem), m (metres) and km (kilometres). Capital letters are also used for some abbreviations, such as T for tonnes.

The font size and legibility distance are fixed so as to enable drivers to read the signs at design speeds. The legibility distance requirement is discussed later in this chapter.

### 18.2.4 Symbols

Symbols are artwork with a specific meaning. They are arranged in such a way that they can be seen, read and interpreted at a greater distance ahead of the decision-making point.

## 18.3 FUNCTIONAL ASPECTS OF ROAD SIGNS

The functional aspects of road signs are discussed in this section.

### 18.3.1 Regulatory or Mandatory Signs

Regulatory signs regulate and control vehicular and pedestrian traffic. These signs are used to prohibit, restrict or command road users to compulsorily perform certain preset actions and to grant certain permissions. Failure to obey these mandatory signs is considered an offence and is punishable with a fine or imprisonment or both.

All regulatory/mandatory signs are circular in shape. They have a black legend on a white background with a red circular border (Fig. 18.2). In addition, prohibition to use certain vehicles, perform certain actions or park is indicated by a diagonal red bar, marked over the legend/symbol. These signs may be categorised as follows, based on the purpose of application:

- Mandatory signs, such as Keep Left or Bus Only
- Stop and Yield signs
- Restrictive signs, such as Speed Limit or Weight Limit
- Prohibitory signs, such as No Parking, No Right Turn or No U-Turn

For a direction change that must be performed compulsorily, compulsory/direction control signs are used (Fig. 18.3). These signs are also circular in shape and have white arrows on a blue background, without a border. Additional symbols used include Pedestrian Only, Compulsory Sound Horn and Minimum Speed Limit.

Exceptions in the shape of this sign category are the Stop sign and the Give Way or Yield sign, which use octagonal and triangular shapes, respectively (Fig. 18.4). These two signs respectively indicate prohibition of use and permission for right-of-way to drivers.

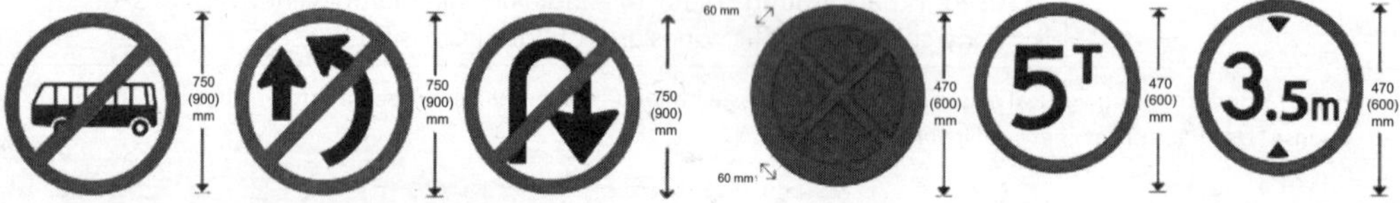


**Fig. 18.2** Sample regulatory signs: Bus Prohibited, Overtaking Prohibited, U-Turn Prohibited, No Parking & Standing, Weight Limit and Height Limit

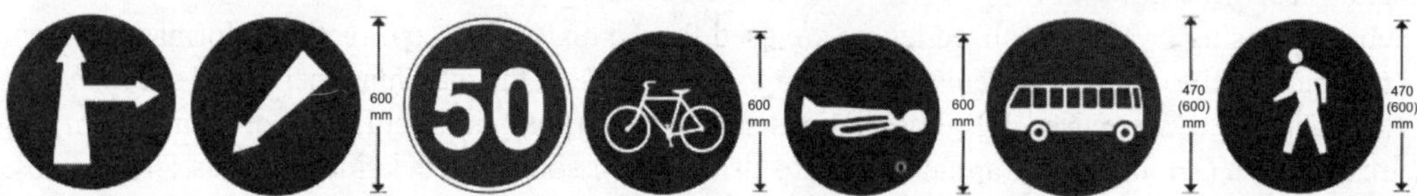


**Fig. 18.3** Compulsory signs: Ahead or Right Turn, Keep Left, Minimum Speed Limit, Cycle Track, Sound Horn, Buses Only and Pedestrians Only

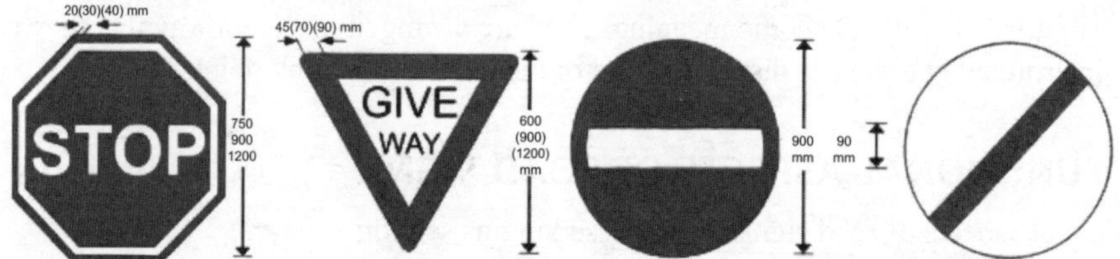


**Fig. 18.4** Mandatory signs: Stop, Give Way, No Entry and Restriction Ends

## 18.3.2 Warning, Cautionary or Precautionary Signs

Warning signs alert road users in advance about the dangerous conditions ahead. All warning signs are triangular in shape. They are composed of a black symbol on a white background with a triangular red border (Fig. 18.5). These can be categorised as follows, based on the purpose of application:

- Direction ahead indications
- Road geometric features ahead
- Roadwork signs

**Fig. 18.5** Warning signs: Right-Hand Curve, Side Road Left, Roundabout, Merging Traffic from Left, Narrow Bridge, Gap in Median and Unguarded Railway Crossing

## 18.3.3 Information/Guide Signs

These signs are used to aid road users and provide route guidance and information about local facilities along the route. The purpose of guidance signs is to help navigation by providing advance information when the road user is travelling to a particular place. Information facility signs provide general information to the road user about the roadway ahead. These signs may be categorised as follows, based on the location and application (Fig. 18.6):

- Route markers
- Direction signs
- Diagrammatic signs
- Overhead signs
- Local facility information signs

All information facility/guidance signs are rectangular in shape, with a blue background.

The direction/guide signs indicate the names of places, with arrows pointing towards the corresponding directions. Direction signs are rectangular in shape, with white legend/text/arrows on a green background for at-grade intersections and a blue background for grade-separated intersections/expressways/urban-city roads.

The exception is for signs in rural roads (other district roads (ODR) and village roads (VR)), which use black legend/symbols on a white background (IRC: 67-2012).

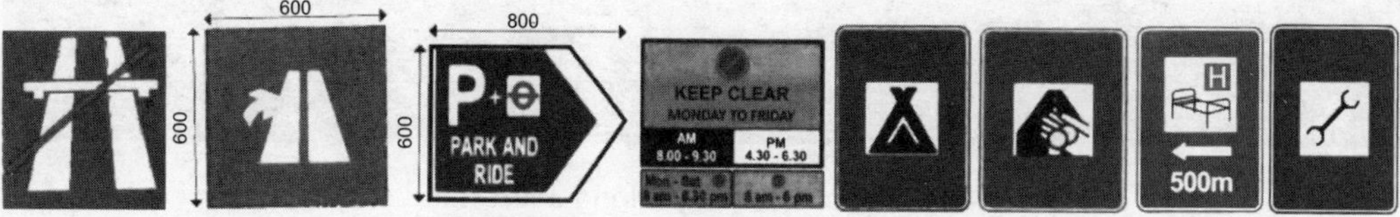


**Fig. 18.6** Sample facility signs: End of Expressway, Entry Ramp to Expressway, Park and Ride, Parking Restriction, Camp Site, Toll Road Ahead, Hospital and Repair Facility

### 18.3.4 Variable Message Signs

Variable message signs (VMSs) show different messages at different times and may also be switched off if they are not essential at a particular time. VMSs should be used along the road for traffic management and regulation, but not for advertisements. The message may be displayed as regulatory information or warning or information. VMSs are discussed in detail in *Chapter 22*.

### 18.3.5 Object/Curve/Hazard Markers, Bollards and Junction Definition Posts

Traffic signs may also take the form of chevrons, object markings, posts and bollards. These reflectors are installed on roadways at appropriate locations to delineate the objects on the roadside, indicate a particular direction or provide awareness about hazards to road users.

Chevron arrows are mounted on rectangular signs (Fig. 18.7). They are used to indicate a sharp change in the direction of travel at night. These signs may be posted in various locations to warn drivers to start slowing down. A series of single chevrons at suitable intervals are provided along road curves/bends (Fig. 18.8). Double and multiple chevron markings are mounted on the objects on the roadside (around roundabout islands, at the narrowing of roads or any other sharp changes). Chevron signs are also used temporarily during road construction and maintenance work. In such cases, they are used to direct drivers to merge/diverge with other lanes to navigate around/away from the work zones.

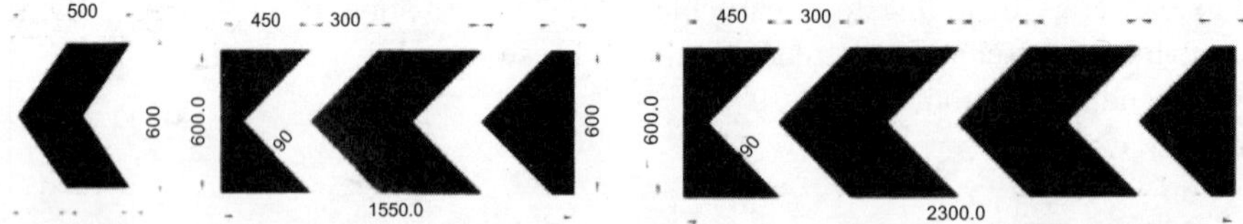


**Fig. 18.7** Chevron markings (black and yellow): single, double and multiple

**Fig. 18.8** Chevron markings along a left-hand curve

Object markers (Fig. 18.9) are used to mark obstructions within or adjacent to the roadway (at narrow shoulders, drop-offs, gores, small islands and abrupt changes in roadway alignment). They are also used to mark the end of a roadway and the front nose of a median opening (within the roadway). The objects may be bridges/culvert abutments (left and right), piers, wall poles, columns or any obstruction on the roadway.

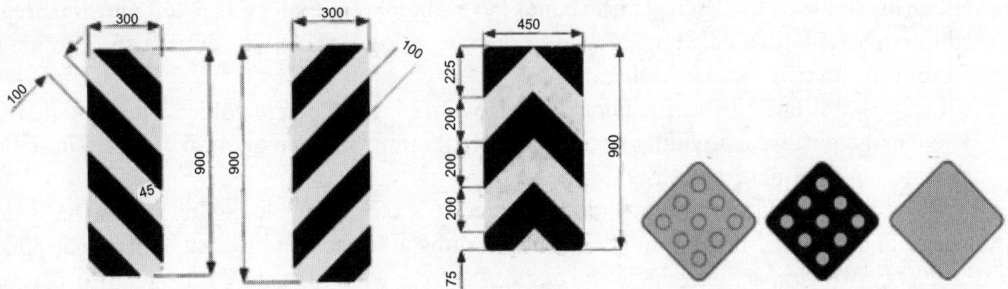


**Fig. 18.9** Object markings: Obstruction on left, obstruction on right, Two-way Hazard and Hazard within the Roadway

(*Courtesy*: http://mutcd.fhwa.dot.gov)

A bollard is a short post marked with fluorescent red or orange reflective tape, having a height of 1500 mm. Bollards are designed to identify or section-off a location demarcated as an extended boundary, to guide traffic (Fig. 18. 10) or to protect pedestrians from vehicular traffic in sensitive areas such as police stations. Bollards can provide an effective visual barrier for vehicles and pedestrians. They may also be used for aesthetic improvement as ornamentation to enhance a building or landscape in a residential locality. Numerous types of bollards are available for specific purposes and locations.

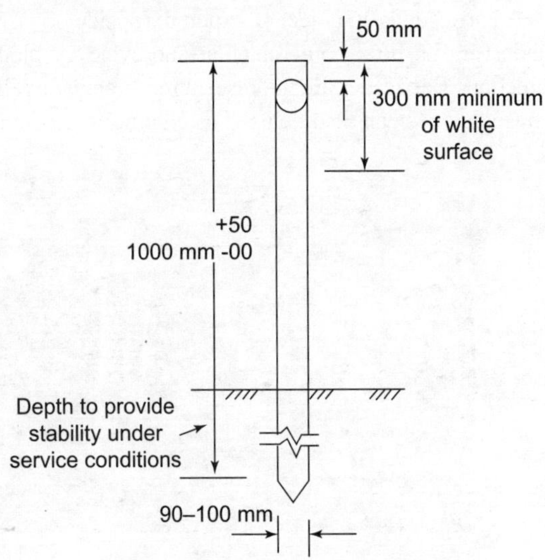


**Fig. 18.10** A typical guide post

(*Source*: Austroads, 2009; *Courtesy:* www.austroads.com.au)

Junction definition posts are high-visibility posts. They are designed to delineate junctions in rural areas (on T-junctions or in a school/factory). They are mostly used to alert motorists to the presence and position of a junction, especially at night. Junction definition posts are made of LDPE (low density polyethylene) or precast cement or galvanised iron with a diameter of 200 mm and height of 1000 mm. The outer surface of the post is marked with alternating white and green hoops using reflective materials.

Delineator posts or tubular markers are installed adjacent to a roadway and are used to delineate/indicate the verge, speed breaker or pedestrian crossing. Reflective strips (red or white on any other background colour such as green, red or white) are marked on the outer surface to ensure high visibility to both motorists and pedestrians. Delineator posts (hollow or solid) are made of polyethylene and have a diameter of 70–95 mm and height of 1380–1800 mm. In this series, similarly shaped posts are used as markers with other colours to indicate various objects on the roadside, such as water mains, OFC cables and electric cables.

Details of various conventional forms of road delineators may be obtained from IRC: 79-1981.

### 18.3.6 Miscellaneous Devices/Signs

Traffic cones are used for channelising purposes, such as to restrict traffic movement to a specific area or to temporarily divide traffic lanes. Traffic cones have a height of not less than 460 mm and are made from bright orange–coloured polyethylene-based materials (Fig. 18.11). The cones used at night must have bands of reflective material near the top.

Vertical panels are reflective traffic control devices of a bright orange colour with diagonal white stripes. The panels must have a width of not less than 200 mm and a height of 610 mm. These panels are also used to guide vehicle/pedestrian traffic.

Construction barrels or drums are cylindrical devices used for channelising traffic flow or diverging traffic at work zones or in critical gore areas. These drums are used as impact attenuators designed to absorb the energy of an impacting vehicle and reduce the force exerted on the passenger to an acceptable level. Drums are made of industrial-grade orange plastic and hooped with alternating orange-white retroreflective stripes. These drums are usually filled with ballast or sand to maintain stability against wind/thrust. Truck-mounted attenuators or trailer-mounted attenuators with flashing arrow panels are also used on a temporary basis.

A barricade is an object or structure (upside down V-shaped foldable plastic frame) that creates a barrier against traffic flow, both vehicular and cycle, or blocks the passage of pedestrians/crowds in the desired direction. Barricades are fluorescent red-orange or yellow-orange stripes with 45° inclinations marked on it. These devices can also be used as channelising devices.

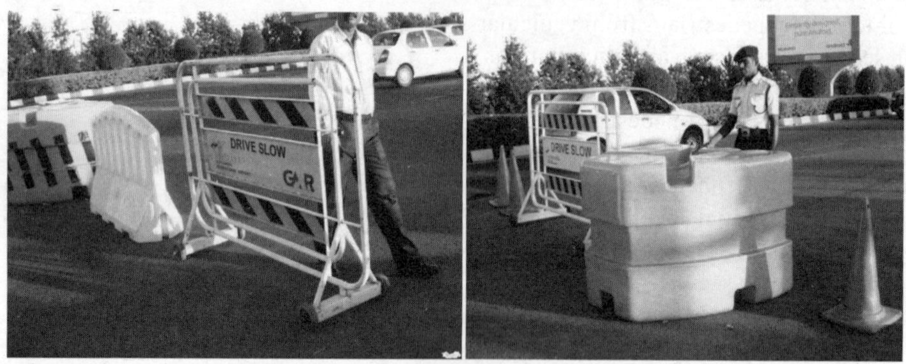


Fig. 18.11 Some traffic barricades/barriers and cones commonly used in India

Rumble strips are roughened paved/loose-rubble surfaces that are embossed or recessed. When a vehicle drives over the strip, a loud rumbling sound is heard and a vibration is produced. Rumble strips may be placed across/adjacent to traffic lanes to alert drivers about potentially hazardous locations, such as a work zone or an isolated intersection.

Raised bars, grooved bars and overlays with exposed coarse aggregate forms of transverse rumble strips are commonly used in Australia for longitudinal delineation (to delineate lanes or the edge of the road) (Austroads, 2009).

## 18.4 SIGN LEGIBILITY DISTANCE

The levels of retroreflectivity, letter height and the number of letters/legends all have a significant effect on the legibility of a sign. Sign legibility increases with letter size. In contrast, the narrower the letter/

spacing, the less legible the sign. The following factors are considered while determining the correct size of letters/legends used in signs (NZTA, 2010):

- Speed of the vehicles approaching the sign
- Number of lines of legend on the sign
- Font used on the sign
- Width of the road and lateral offset to the sign
- Vertical mounting height/position of the sign, in relation to the driver

The size of the sign depends on the letter height (100–300 mm) and the number of words/legends/symbols on the sign. The minimum size requirements for different types of roads and localities can be obtained from any standard code of practices.

### 18.4.1 Australian Practice

The legibility distance of a new legend/sign along a roadway may be determined from first principles. The procedure adopted for determining the legibility distance of a sign on a roadway can be shown schematically, as in Fig. 18.12. One-fourth of the minimum legible distance ($L$) is required for a driver to observe/identify a sign ahead. He starts reading and reacts while he travels along the remaining one-third distance from the sign.

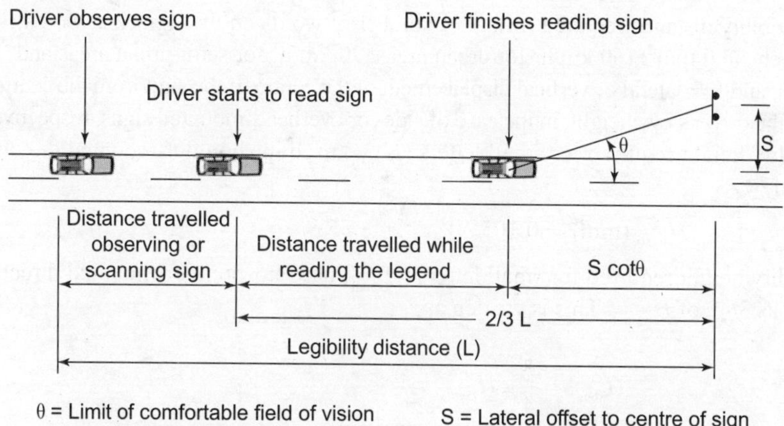


**Fig. 18.12** Sign legibility distance – Australian practice
(*Source*: Austroads, 2009; *Courtesy*: www.austroads.com.au)

The distance travelled while reading the legend or any place name in a graphically represented legend depends mainly on the *number of words* to be read. The average time ($t$ (sec.)) taken by an average driver to correctly read a sign with two to eight words can be estimated using the formula developed at Melbourne University (Jacobs and Cole, 1978):

$$t = 0.32\,N - 0.21 \text{ sec.} \tag{18.1}$$

where $N$ = number of words on the sign. The time taken to locate a place name (seven letters) on a sign of a vertical stack of up to eight words is estimated by

$$t = 0.25\,N \text{ sec.} \tag{18.2}$$

Research indicates that there is a marginal difference between these two estimated times, for up to $N = 5$, and hence, Eq. (18.2) is recommended for estimating the time (sec.) taken to read a sign of up to five words.

Therefore, the distance travelled while reading the five words on the sign

$$t \times v_{m/sec} = (0.25\ N) \times v_{m/sec}$$

where $v_{m/sec}$ is the speed of the approaching vehicle (m/sec).

It is generally accepted that the angular range of lateral ($\theta_H$) and vertical ($\theta_V$) vision of a driver (within 2/3rd $L$) while approaching the sign is limited to 10° horizontally leftward and 5° vertically upward (Austroads, 2009; IRC: 67-2012). These angles give the values

$$\cot(\theta_H) = \cot(10°) = 5.7 \text{ and } \cot(\theta_V) = \cot(5°) = 11.4$$

Based on Eqs (18.1) and (18.2) and the (θ°) values, the 2/3rd distance of the sign legibility distance can be derived as follows (Austroads, 2009)

$$\frac{2}{3}L = (0.25N) \times v_{m/sec} + S\cot(\theta) \tag{18.3}$$

$$L = 0.105\ NV_{km/hr} + 8.55\ S \text{ for side-mounted signs} \tag{18.4}$$

$$L = 0.105\ NV_{km/hr} + 17.1\ S \text{ for overhead-mounted signs} \tag{18.5}$$

where $L$ = legibility distance required (m); $N$ = number of words on the sign (2–5 words); $V$ = speed of approaching vehicle (km/hr) (60 km/hr for urban areas, 70 km/hr for semi-urban areas and 80–100 km/hr for rural areas); and $S$ = lateral or vertical displacement of the centre of the sign from the centre of the traffic lane or above the driver's eye height, in the case of side- or overhead-mounted signs, respectively.

The legibility height required for small letters ($H_{small}$) used in shoulder-mounted direction signs can be determined by

$$H_{small}\ (\text{mm}) = 0.105\ NV_{km/hr} + 8.55\ S \tag{18.6}$$

The legibility height required for small letters ($H_{small}$) used in shoulder-mounted direction signs can be calculated as 3/4th of $H_{capital}$. This is written as

$$H_{small} = \frac{3}{4} \times H_{capital} \tag{18.7}$$

Hence, the expression used for determining the initial capital letter height ($H_{capital}$) can be written as

$$H_{capital} = \frac{4}{3} \times H_{small} = \frac{4}{3} \times (0.105\,NV_{km/hr} + 8.55S)$$

Therefore $$H_{capital}\ (\text{mm}) = 0.14\ NV_{km/hr} + 11.4\ S \tag{18.8}$$

Similarly, the $H$ required for overhead signs may be determined based on Eq. (18.5).

In Australian practice (Astroads, 2009), the alphabet series used on direction signs based on the legibility distance of letters are categorised as C, D, E and Modified E. Eqs (18.6) and (18.8) used to calculate the height of letters are categorised under the Modified E alphabet series, having a stroke width of 1/5th of the letter height.

The required height for other series of letters and the corrections applicable to Eqs (18.6) and (18.8) with specific site conditions are as follows (Astroads, 2009):

- Increase the height of the other alphabet series letters as calculated by Eqs (18.6) and (18.8). Increase $H$ (mm) by 7% for the E, 24% for the D and 50% for the C series of alphabet letters.
- To compensate for adverse visual effect due to bright advertising signs and building height in urban areas, increase $H$ by 50%.
- The factor $S$ used in the formula must be multiplied by 2, where the value of $\theta_V$ does not exceed 5°.
- If the overhead sign is mounted higher than 3 m above the edge the pavement, the equivalent value of the $S$ factor ($S_E$) is calculated using the following formula

$$S_E = \sqrt{\left\{(S_L)^2 + 4(S_V)^2\right\}} \tag{18.9}$$

where $S_E$ = equivalent $S$ factor; $S_L$ = lateral offset of the sign from the driver's path; and $S_V$ = vertical height measured between the driver's eye level and the centre of the overhead sign

*Note*: Take $S = S_E$ to determine $H$ for overhead signs, and also for elevated shoulder-mounted signs with height ≥3 m.

### 18.4.2 Indian Practice

IRC: 67 (2012) recommends that the angular lateral vision ($\theta_H$) of a driver while reading the sign be taken as 10°, i.e., horizontally leftward. The reading time ($t$) is determined by

$$t = 2 + \left(\frac{N}{3}\right) \tag{18.10}$$

The distance travelled while reading the legend on the sign = $t \times v_{m/sec}$ where $N$ = number of words or destinations on the sign, normally restricted to or not exceeding 8.

Minimum legibility distance ($L$) = Reading distance + Cut-off distance

$$L = (t \times v_{m/sec}) + (S\cot(\theta)) \tag{18.11}$$

where $S$ and $\theta$ carry the same meaning.

The legible height of the letters should be determined based on the distance of the driver from the sign, when he starts reading it. The minimum required height ($H$) may be taken as 100 mm up to an initial distance of 60 m, and thereafter, increased proportionately every 30 m by 50 mm. However, $H_{max}$ should not exceed 400 mm. More details on the size of alphabets applicable in Indian traffic conditions may be obtained from IRC literature (IRC: 67-2012 and IRC: 30-1968).

**MORTH 5th Revision (2013) Recommendation:** The size of the letters must range between 100 mm and 400 mm for a design speed range of 40–120 km/hr. For advance direction signs on non-urban roads, the minimum letter height should be 150 mm for expressways, national highways and state highways and 100 mm for other roads including urban areas. For overhead-mounted signs, the minimum letter height should be 300 mm.

For definition plates attached to normal-sized signs, the letter size should be 100 mm or 150 mm. The thickness of the letters could be in the range of 1/5th to 1/6th of the letter height ($H$). The size of the initial uppercase letter should be 1–1/3rd. However, the word messages may be written in uppercase letters only, for better visibility and comprehension.

### 18.4.3 USSC Recommended Practice of Thumb Rule

The Unites States Sign Council (USSC) recommends the following conditions for determining the size of a sign (Bertucci, 2006):

- The average reading or message scanning time of a sign in simple typography is 0.5 seconds per word. The number of words on a sign may be in the range of 4–8. It is assumed that the average number of letters in one word is 5, and the scan rate is 1/10[th] of a second per letter.
- The universally recognised symbols, directional arrows, logos and icons displayed on signs are also considered equivalent to one word or five letters, yielding a reading or scan time of 0.5 seconds per symbol.
- The average of total viewer reaction time (VRT) required for a pre-sign manoeuvre is 8, 10 and 11 seconds, respectively, for simple, complex and multilane environments.
- VRT is used to determine the viewer reaction distance (VRD) that a driver travels in the VRT interval. Hence, VRD is used to determine sign legibility and size. VRD indicates the initial relative distance of detection of a sign from the driver's current position, i.e., the detection distance ahead of the sign.

$$\text{VRD (ft)} = \text{Travel speed (ft/sec)} \times \text{VRT (sec)}$$

For example, $\text{Travel speed} = 25 \text{ mi/hr} \times 1.47 = 36.75 \text{ ft/sec}$

$$\text{VRT} = 8 \text{ sec}$$

Substituting, we get $\text{VRD} = 36.75 \text{ ft/sec} \times 8 \text{ sec} = 294 \text{ ft}$

The USSC standard legibility index or letter height is given below:

- The overall legibility of a sign is a function of the height, colour, illumination conditions, font characteristics of the letters, observation distance and its orientation towards road alignment.
- Letter height = VRD/Legibility index

where legibility index = distance measured in feet for every inch of capital letter height at which a sign may be read. Several variants are given. However, the average value = 30 ft/inch.

For example, if VRD = 294 ft and $LI$ = 30 ft/inch, letter height = 294/30 = 9.8 inch

Area of the sign can be determined using the following equation:

$$Area_{sign} = \frac{3 \times \text{No. of letters in the sign}}{80}\left[\frac{\text{VRT in sec} \times \text{Speed in miles per hour}}{\text{Legibility index}}\right]^2 \qquad (18.12)$$

For example, the number of letters in a sign = 5 × 6 words = 30 letters.

VRT = 8 sec; Speed = 25 mi.hr; Legibility index = 30 ft/inch = 30 × 30

Substituting, we get $Area_{sign} = \frac{3 \times 30}{80}\left[\frac{8 \text{ sec} \times 25 \text{ miles per hour}}{30 \text{ ft/inch}}\right]^2 = 50 \text{ sq.ft}$

The minimum height of shoulder-mounted signs is 2.0 m (7 ft)

*Note:* Readers may obtain more details from www.ussc.org.

**Example 18.1**

Ahead of a 100-m distance along a straight and plain road, what is the minimum lateral offset or setback required to the left of the driver's eye position, with the standard cone of vision?

**Answer:** Assuming the standard cone angle $= \theta_H = 10°$. With reference to Fig. 18.13, the minimum value of offset $= S = 100 \times \tan(10°) = 17.63$ m.

*Note:* The minimum lateral distance may be increased further on account of roadside geometry, number of road lanes, width of the shoulder, and width of any utility or future right-of-way easements.

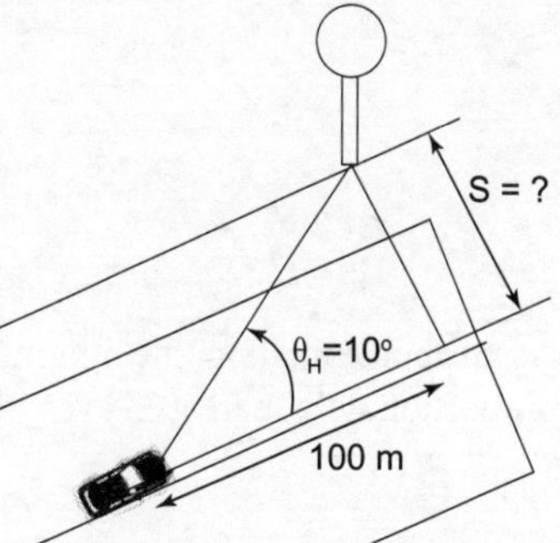


**Fig. 18.13** Lateral offset

**Example 18.2**

Ahead of a 100-m distance along a straight and plain road, what is the minimum vertical offset or sign height required to the left of the driver's eye position, with the standard cone of vision?

**Answer:** Assuming the standard cone angle, in vertical plane $= \theta_v = 5°$. Assume the standard height of the driver's eye level above the road surface = 1.05 m (IRC: 35). With reference to Fig. 18.14

Minimum vertical offset or sign height required $= 1.05 \text{ m} + S_v = 1.05 \text{ m} + 100 \times \tan(5°)$

$= 1.05 \text{ m} + 100 \times 0.0874 = 9.79$ m

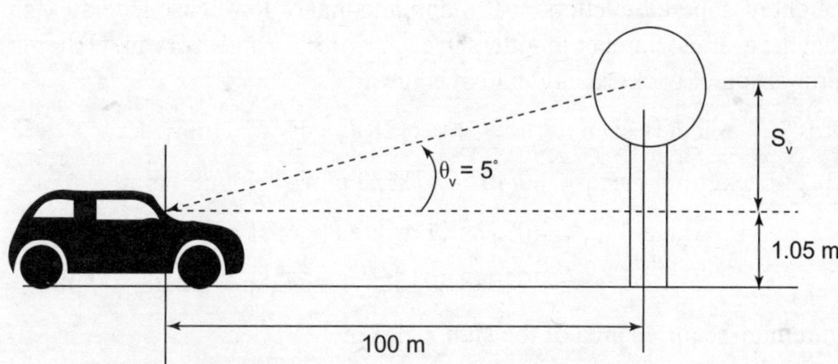


**Fig. 18.14** Minimum vertical offset

*Note:* The minimum $S_v$ maybe changed subject to roadside topography.

**Example 18.3**

Determine the legibility distance and height of the letters of a proposed shoulder-mounted sign for the following requirements based on Indian practice:
Speed = 50 km/hr; Number of words on the proposed sign = 6; Number of lanes in one direction = 2.

**Answer:** As it is a shoulder-mounted sign, $\theta = \theta_H = 10°$ (assumed); $v = 50$ km/hr = 13.9 m/sec (given); $N = 6$ (given); and $S = 3.75$ m, as it is across a single lane.

$$\text{Reading time (t)} = \{2 + \left(\frac{N}{3}\right)\} = \{2 + \left(\frac{6}{3}\right)\} = 4 \text{ sec.}$$

By Eq. (18.11)
$$L = \left(t \times v_{\text{m/sec}}\right) + \left(S\cot(\theta)\right)$$

$$L = (4\,\text{sec} \times 13.9\text{ m/sec}) + (3.75\,\text{m} \times \cot(10^{\circ}))$$

$$L = (4\,\text{sec} \times 13.9\,\text{m/sec}) + (3.75\,\text{m} \times 5.7))$$

$$L = (55.6) + (21.375)) = 76.975 = 77\,\text{m}$$

As an approximation, the minimum height of letter (H) = 100 mm up to an initial distance of 60 m, and proportionately increased every 30 m by 50 mm.

$$H_{\text{capital}} = 100\text{ mm} + \left\{\left(\frac{77 - 60\,\text{m}}{30\,\text{m}}\right) \times 50\,\text{mm}\right\}$$

$$= 100\text{ mm} + 28.33\text{ mm} = 128.33\text{ mm} \approx 128\text{ mm}$$

Let us assume $H_{\text{small}} = 72\%$ of $H_{\text{capital}}$.

Hence, $H_{\text{small}} = 0.72 \times H_{\text{capital}}$

$$H_{\text{small}} = 0.72 \times 128\text{ mm} = 92.16 \approx 92\text{ mm}$$

Therefore, the height of uppercase letters = 128 mm and that of lowercase letters = 92 mm.

Let assume that the average number of letters in each word = 5 and every first letter is a capital letter. Assume 60/40 proportion of background/text coverage area.

Area of 6 words with 5 letters each = $(6 \times 5) \times (128)^2 = 491520\text{ mm}^2$

Area of 60% of background area = $60/40 \times 491520\text{ mm}^2 = 737280\text{ mm}^2$

Grand total = 40% as Copy area + 60% as Negative area

$= 297048\ 491520 + 737280 = 12{,}28{,}800\text{ mm}^2 = 1.2288\text{ m}^2$

Therefore, the minimum required area of the sign = $1.3\text{ m}^2$

## 18.5 LATERAL CLEARANCE AND LOCATION OF SIGNS

Normally, signs are placed on the left side of the road, without any visual obstruction, and can be read from vehicles while travelling at design speeds. In multilane divided roads too, the signs are placed on the left each carriageway. To provide additional emphasis, duplicate signs may be placed on the left of sharp curves where adequate lateral clearance against the column from the shoulder is not provided.

Signs should be located sufficiently well in advance to enable the driver to react appropriately. Similarly, appropriate separation distance should be provided between a series of successive advance signs.

The following seven aspects are considered as general principles of positioning signs (DTTAS, 2010):

- Signs should be installed along a roadway so that they can seen and read at the design speeds.
- The position of the sign should be relevant to the junction, interchange ramp, hazard or other feature of the location and at a minimum of 1000 mm above ground level from the point below the sign.
- The sign should be located with reference to the geometric/cross-sectional features of the road such as edge, curve, railway level crossing, speed/weight/height limit or any such location. For example, the sign should be located laterally away so that it does not limit the sight distance for drivers entering from a side road (Austroads, 2009).

- Its height above the road, lateral distance from the kerb side and longitudinal distance from decision points, if any, should be considered.
- Its orientation with road alignment should be kept in mind.
- It should meet all safety requirements; for example, signs located beyond the clear zone or behind safety barriers (Austroads, 2009)
- Its relationship with other signs in adjacent areas should be considered.

As a common rule, signs should be positioned as far away as practicable from the edge of the roadway, subject to the following (NZTA, 2010):

- Maximum and minimum criteria recommended by a standard code of reference manual
- Constraints due to vision obstructions on the roadside

As a common standard practice, the lateral clearance (offset distance measured in the transverse direction of traffic flow) shall be measured from the edge of the sign nearest the road edge to:

- The kerb line
- The outer/nearest edge of the road-shoulder
- The face of the guard rail or nearest lane, whichever of the above are critical

## 18.5.1 Lateral Placement and Height

Some important details pertaining to siting and lateral placement of signs are given below:

- In rural areas, signs should be erected not less than 1500 mm above the level of the adjacent verge/shoulder and not less than 1000 mm above the adjacent ground level (Fig. 18.15).
- In urban areas or locations where pedestrians are likely walk under the sign, the desirable mounting height is 2300 mm, and when cyclists pass under the sign, the desirable mounting height is 2500 mm (DTTAS, 2010).
- Low-level direction signs used at junctions and roundabouts should be mounted at a height of not less than 750 mm, and care should be taken that they do not obstruct the sight lines.

Figs 18.16–18.21 depict a few important and commonly adopted lateral and vertical clearances of signs. Readers are instructed to refer to any standard code of practice (MUTCD, 2009; NZTA, 2010; IRC: 67-2012).

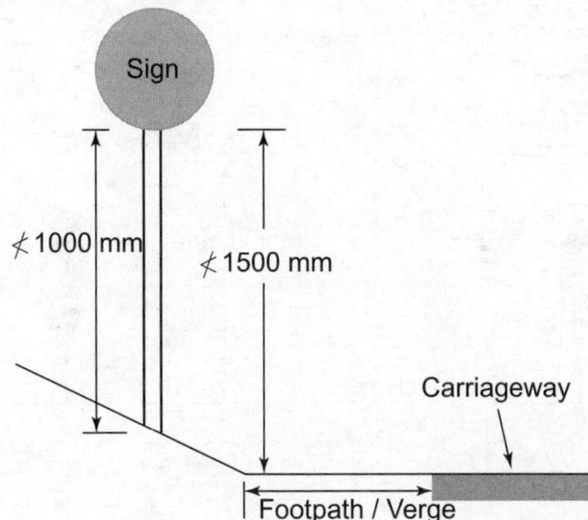


**Fig. 18.15** Height clearance for signs mounted adjacent to footpaths

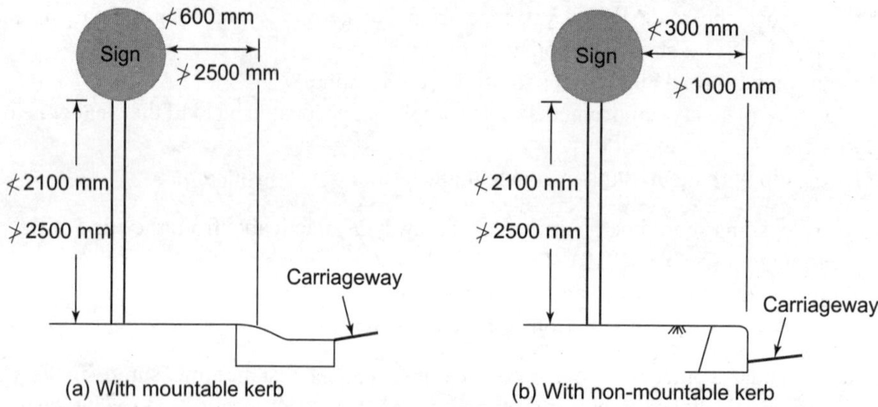


**Fig. 18.16** Lateral and height clearances for signs in kerbed roadways
(Redrawn and elaborated based on IRC: 67-2012 and NZTA, 2008, 2010)

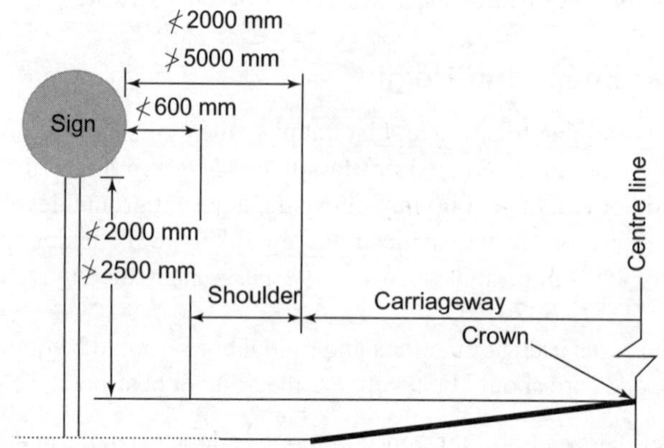


**Fig. 18.17** Lateral and height clearances for signs in unkerbed roadways
(Redrawn and elaborated based on IRC: 67-2012 and NZTA, 2008)

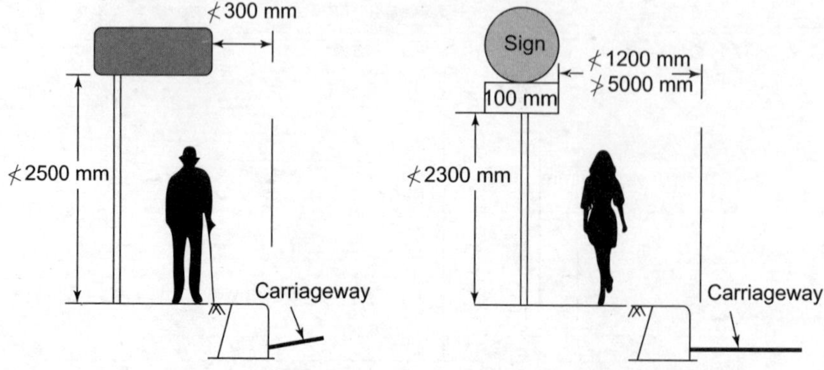


**Fig. 18.18** Lateral and height clearances for signs in footpaths
(Redrawn from NZTA, 2008)

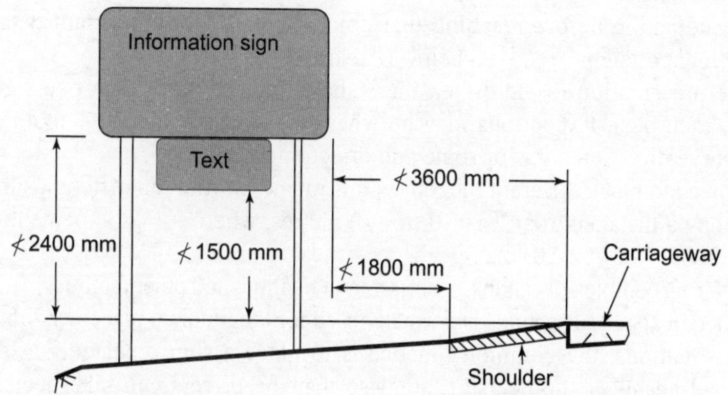


**Fig. 18.19** Lateral and height clearances for signs in kerbed roadways
(Redrawn and elaborated based on IRC: 67-2012; MUTDC, 2009 and NZTA, 2008)

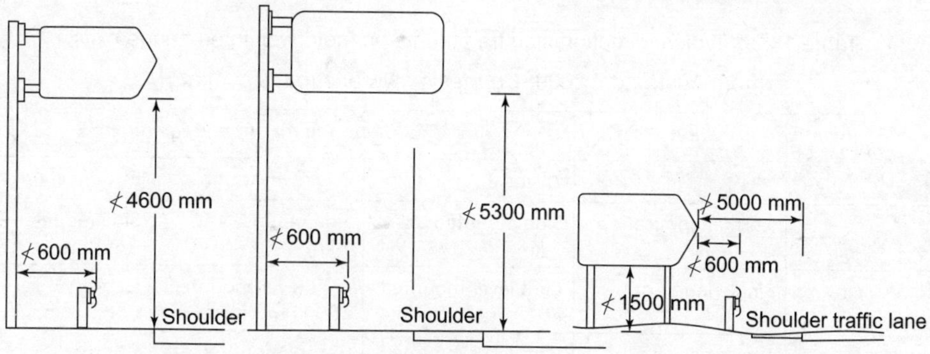


**Fig. 18.20** Lateral and height clearances for signs in rural areas
(Redrawn and elaborated based on NZTA, 2010)

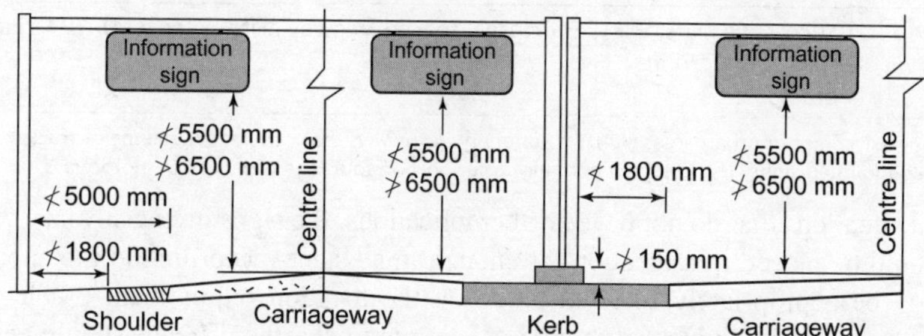


**Fig. 18.21** Lateral and height clearances for gantry-mounted signs in kerbed/unkerbed roadways
(Redrawn and elaborated based on IRC: 67-2012 and MUTDC, 2009)

## 18.5.2 Longitudinal Placement

The road sign must be relevant to the location road geometrics and *far enough* in advance of the object/feature about which it is intended to bring notice. The driver must also be able to see, read, analyse and take appropriate action before reaching that point. Some of the important details of siting and longitudinal placement of signs are given below (Austroads, 2009):

- For effective presentation to the driver, there should not be more than one sign of a particular category on each post. Exceptions may be when one sign supplements another, such as speed signs with curve direction signs or route and direction signs.
- If there is a need to place different message signs in one location on different supports, they must be separated by a distance of at least $(0.6 \times V)$ metres, where $V = 85^{th}$ percentile speed (km/hr) (Austroads, 2009; IRC: 67-2012).
- For signs requiring complex decisions, such as tourist facilities and other facilities, a distance equivalent to that travelled in about 12 seconds is normally used for longitudinal placement of the signs.
- In complex situations, the common practice is to place a sign repeatedly at suitable intervals along the road ahead of the decision point so that the drivers can take decisions and actions progressively before reaching the final location. For example, placement of ramp exit signs at specific locations with appropriate applications (Table 18.3). The road geometrics of the location must be considered with reference to these standards.

**Table 18.3** Typical direction sign treatments on motorway interchange exits (from Austroads, 2009; *Courtesy*: www.austroads.com.au)

| *Motorway location* | *Application* | *Signs and mounting required* | | |
|---|---|---|---|---|
| | | *1st advance sign* | *2nd advance sign* | *Exit direction sign* |
| Rural | Normal application | Side mounted at 2 km | Side mounted at 1 km | Side mounted at exit taper |
| Urban | Single-lane exit | Cantilever mounted at 1 km | Cantilever mounted at 500 m | Gantry |
| All locations | Two-lane exit | Cantilever mounted at 1 km | Gantry mounted at 500 m (generally start of auxiliary lane) | Gantry |
| All locations | Successive exits spaced at 1.2 km or less | For closely spaced freeway exits, refer to AS 1742 Part 15 | | |

*Note:* At low-volume exit ramps in urban areas (AADT$<$ 3000) on motorways that carry a low volume of trucks, consideration may be given to side-mounted signs. This case will normally occur only on four-lane outer urban motorways.

- In general situations that do not have any complications, the signs giving advance warning to the driver should be placed in a location far enough in advance, according to the approach speeds, which may be appropriately selected based on the location requirement (Table 18.4). If any additional complexity is involved or a descending gradient, the distance should be increased by the equivalent of one second of the travel time (Austroads, 2009).
- If the signs require complex decisions, such as tourist facilities and services, a travel distance equivalent to about 12 seconds is normally used.
- These distances may be decreased appropriately when larger-than-normal signs with increased legibility distance are used.
- The maximum distance at which any advance direction sign is located from a hazard to the decision point, or the distance between the second advance sign, which has the same message,

should not exceed the equivalent to travel time of about 15 seconds (a limit due to short-term memory refractory period)

**Table 18.4** Longitudinal location of warning signs (from Austroads, 2009; *Courtesy*: www.astroads.com.au)

| *Road environment* | *A (m)* | | | *B(m)* |
|---|---|---|---|---|
| | *Must or may need to stop* | *Significant speed reduction required* | *No, low or moderate speed reduction required* | |
| $V_{85}$: <75 km/h | 80–120 | 60–80 | 40–60 | 50 |
| $V_{85}$: 75–90 km/h | 120–180 | 80–120 | 60–80 | 60 |
| $V_{85}$: > 90 km/h | 180–250 | 120–180 | 80–120 | 70 |

*Notes:* $V_{85}$ = 85th percentile approach speed measured 1.5 to 2 times 'A' in advance of hazard; A = distance from sign to hazard (or nearest sign to hazard where there are two or more signs); B = minimum distance between successive signs where there are two or more. Source: AS1742.2.

## 18.5.3 Sign Orientation

The common practice in the orientation of signs, except for some parking signs, is for all signs to be erected in a vertical plane. The sign face should be oriented at approximately right angles to the line-of-sight of the approaching driver. On curved alignments, sign orientation should be determined based on the line-of-sight of the approaching driver and not on the alignment of the road at the sign post.

Reflectorised signs should be turned about 5° away from the driver's line of sight (Fig. 18.22). Chevron boards need to be turned about 10° away. This orientation helps reduce undesirable specular reflection from the retroreflected sign surface (NZTA, 2010).

As a good practice, the performance of the signs installed along a roadway must be checked by conducting motor vehicle trial runs during the day and at night.

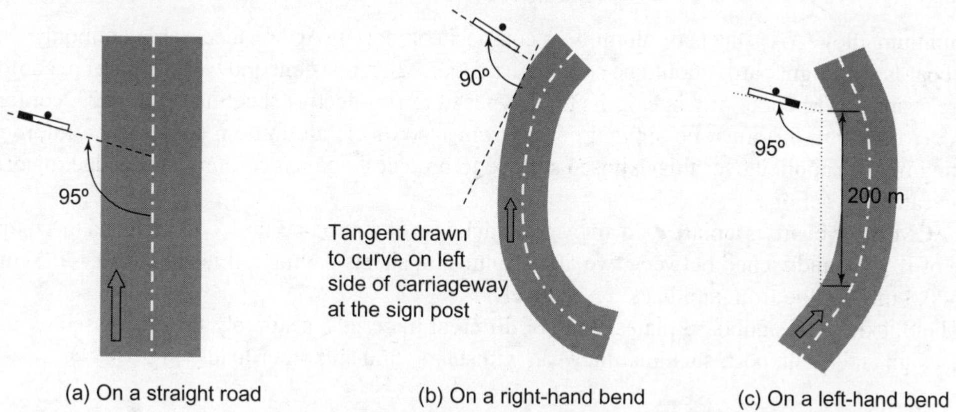


**Fig. 18.22** Desirable orientation of signs helps against specular reflection from retroreflectorised signs – the three critical locations along the road

(Redrawn from IRC: 67-2012 and DTTAS, 2010)

### 18.5.4 Sign Support Standards

Some important mounting details of signs are given below (IRC: 67):

- The support posts should be GI pipes (IS: 1239) or rectangular/square hollow sections (IS: 3589).
- The support pedestal should be used with M25 grade of cement concrete.
- A single supporting post may be sufficient to mount a sign with an area up to 0.9 $m^2$; for a greater area, two supports may be used.

The signs mounted on overhead supporting structures should be properly designed to withstand a wind load of 150 kg/$m^2$ and 30 kg/$m^2$, respectively, in the normal and transverse directions to the face of the sign. In addition to the dead load, a concentrated load of 250 kg/$m^2$ is also considered as walkway load on the overhead sign structure.

### 18.5.5 Warrants

The warrants of sizes, dimensions and font sizes that are applicable in India for the installation of mandatory, warning and information/guide signs with design speeds and related road geometric features have been discussed in the above sections; more details can be obtained from IRC: 67 (2012). It must be noted that, as design speed increases, the size of the sign and font size requirement also increases.

## 18.6 ILLUMINATION AND RETROFLECTION

Signs should be illuminated or reflectorised so that they display their colours, shape, symbols and text legibly at night as well as in the day. Illumination may be required when reflectorisation is ineffective; for example, for overhead signs on which the headlights cannot be focussed. In contrast, reflectorisation may be ineffective in locations with high intensity of street lighting (FHWA, 2003).

A translucent material may be used and an independently mounted light source be designed to provide adequate illumination over the sign face. Appropriate retroreflective materials are used for legend/background/border of signs.

## 18.7 MATERIALS USED FOR SIGNS

Aluminium alloy (AA) sheets or aluminium composite material (ACM) sheets are commonly used for signboards. The signboards should be smooth, hard, corrosion resistant and weatherproof to conform to the relevant ASTM code. Any one of the three types of retroreflective sheeting (Fig. 18.23, conforming to ASTM D 4956-09) must be applied on the front face of the aluminium sheet, and a fluorocarbon coating may be applied over this exposed surface to protect it against colour fastness due to corrosive and weathering actions.

The ACM sheet-board is a plate with an overall thickness of 4 mm or 3 mm, comprising a thermoplastic core of LDPE sandwiched between two aluminium skin sheets having a thickness of 0.4–0.5 mm and 0.25–0.3 mm for the front and back, respectively.

The thickness of signboard plates made of different materials is suitably selected based on the size of the sign and its support, such that they can withstand wind and other loads (Fig. 18.24).

## 18.8 THE CONE OF RETROREFLECTION

Light rays from the vehicle headlights are retroreflected towards the source vehicle in the shape of a narrow cone (Fig. 18.25). The retroreflection cone angle applicable under the headlight visibility of

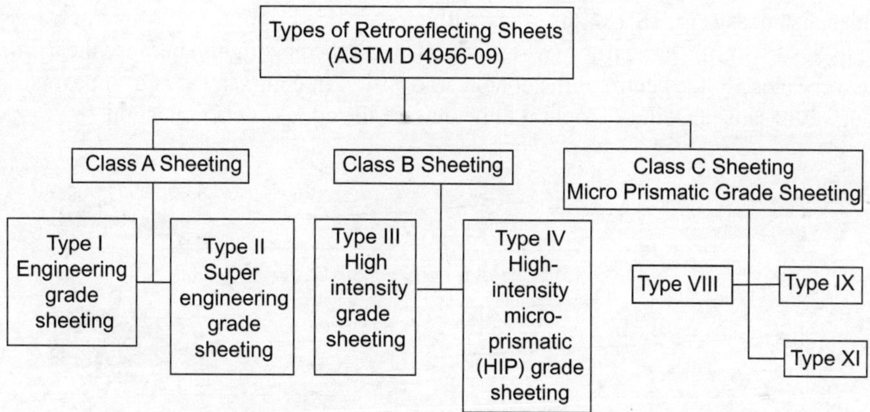


**Fig. 18.23** Classification of retroreflecting sheets used for signs (as recommended by IRC: 67-2012)

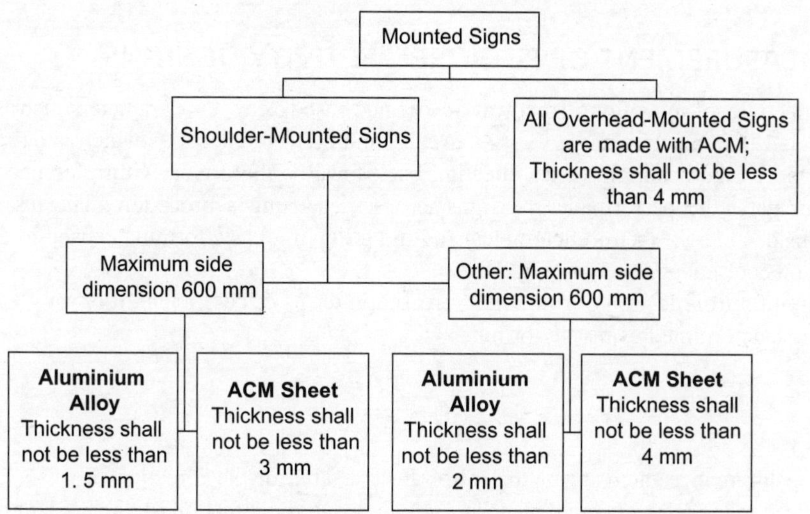


**Fig. 18.24** Plate thickness requirement based on mounted signs (as recommended by IRC: 67-2012)

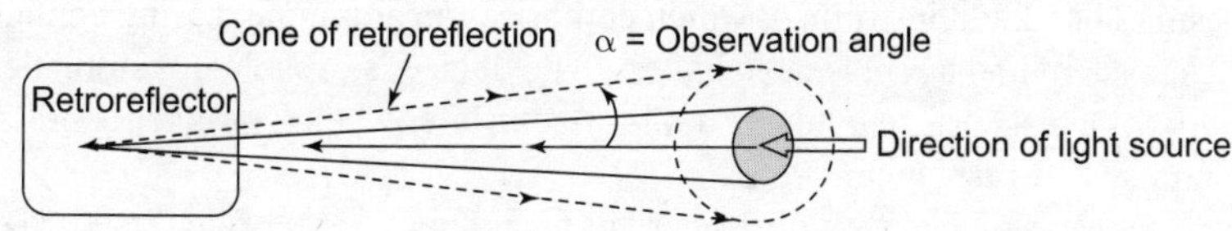


**Fig. 18.25** Schematic representation of cone of retroreflection for road signs

road signs is normally measured as between 0.2° and 2.0°. Half of this cone angle is denoted as the observation angle ($\alpha$).

At longer distances (Fig. 18.26), the signs will appear very bright to the driver of a passenger car compared to the driver of a large truck (who sits well above the headlights) because the car driver's line of sight is very close to the centre of the cone, i.e., $\alpha \leq 1°$. In contrast, the sign appears less bright to the truck driver because his line-of-sight will be outside the cone of retroreflection.

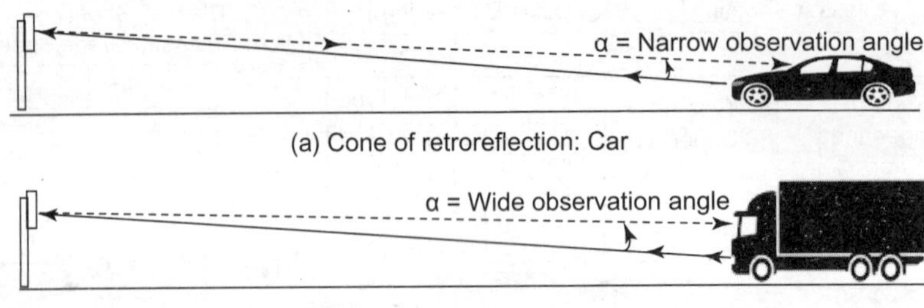


(b) Cone of retroreflection: Truck

**Fig. 18.26** Typical cones of retroreflection for cars and trucks
(Redrawn from www.atssa.com)

## 18.9 MEASUREMENT OF RETROREFLECTIVITY OF SIGNS

The retroreflectivity of signs along a roadway are defined based on a set of parameters such as observation angle ($\alpha$: 0.1–1.0°), entrance angle ($\beta$: −4° to 30°) and the reference axis which is perpenducular to the test surface. Like headlights retroreflecting back towards the driver, with reference to the above set of parameters, the performance of signs is assessed using a procedure. The test condition is simulated inside a device (retroreflctometer) and the following basic formula is used to determine the retroreflection coefficient of a sign (Fig. 18.27).

The intensity of retroreflection of a sign is expressed in terms of coefficient of retroreflection ($R_A$, cd/m²/lx) and it is determined using the formula

$$R_A = \frac{L_{out}}{E \times A}$$

where $L_{out}$ = luminance intensity of the retroreflective area in the direction of observation (cd); $E$ = illumination on the retroreflective plane at right angles to the direction of the incident light (lx); and $A$ = illuminated plane sample sign surface (m²).

$R_A$ is also used to represent the efficiency of the retroreflector, which is the ratio between returning light and source light at that particular geometry of observation and entrance angle. The retroreflectometer (Fig. 18.28) is a small handheld instrument which comprises an aluminum casing containing electronic circuits, an optical system and a replaceable battery. The device is capable of automatically measuring the retroreflectivity of a test surface by using a micro-controller when its trigger is pressed. The output result is shown on its LCD display unit.

The retroreflectometer can simulate a set of standard test conditions as set by any standard code of practice (Fig. 18.29) and the coefficient of retroreflectivity of sign sheeting is determined. These direct means of measurement are useful for monitoring whether the retroreflectivity levels of signs is being maintained.

The measured retroreflectivity of signs data is used to evaluate its performance, and replacement options are selected if they do not meet the minimum retroreflectivity levels.

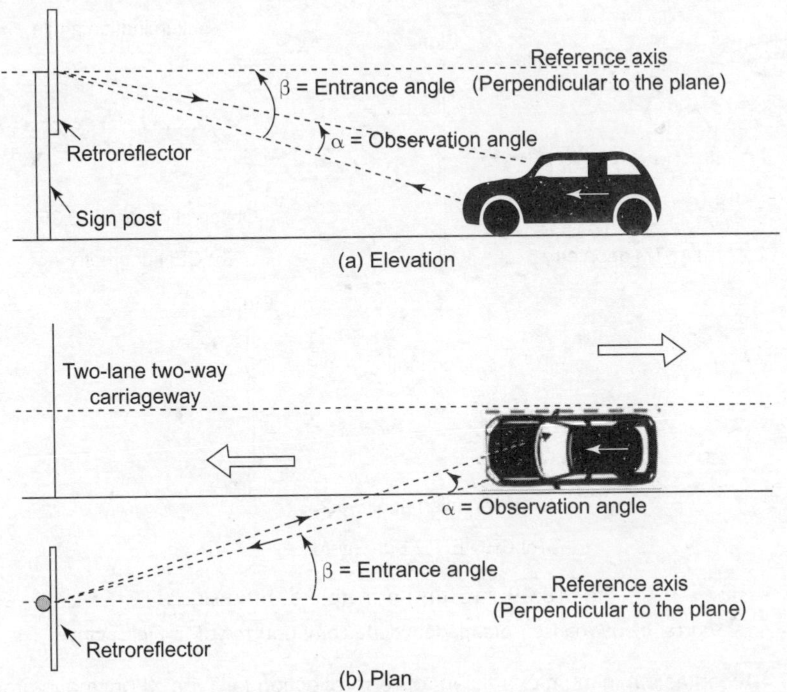


**Fig. 18.27** Schematic representation of entrance angle (β) and observation angle (α)

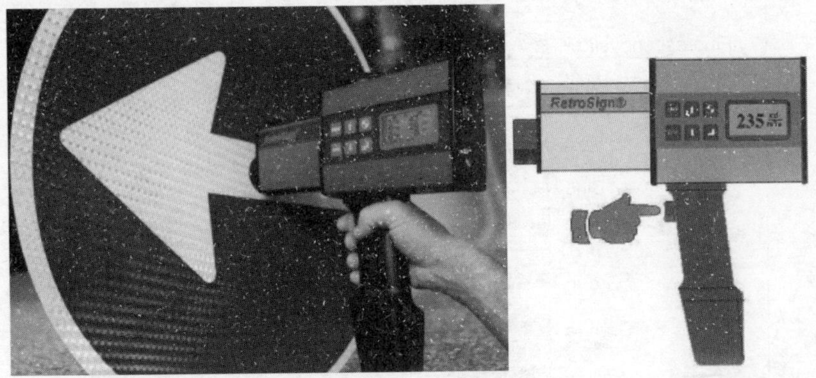


**Fig. 18.28** Handheld Delta retroflectometer used for testing road signs
(*Source*: http://roadsensors.madebydelta.com; *Courtesy*: Mr. Kjeld Aabye

The performance characteristics of each retroreflective sheeting, comprising seven different colours based on chromacity coordinates and daytime luminance factors with observation angle and entrance angle, are available in IRC: 67.

In Indian practice, the minimum acceptable values of coefficient of retroreflection of various colours used for signs with specific observation angles (α: 0.1°, 0.2°, 0.5° and 1.0°) and entrance angles (β: −4.0° and 30°) are prescribed for classes A, B and C of sheeting (Table 18.5).

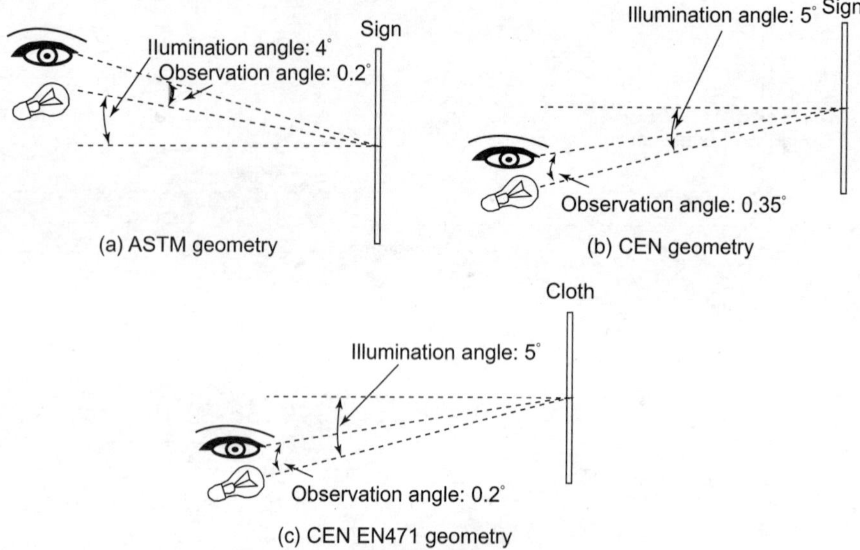


**Fig. 18.29** Testing principle of road signs
(*Source*: http://roadsensors.madebydelta.com; *Courtesy*: Mr. Kjeld Aabye)

**Table 18.5** Acceptable minimum coefficient of retroreflection for Type XI prismatic grade sheeting (cd/lx/m²) colours when the observation angle is $\alpha^\circ = 0.2$, a typical example (Modified from IRC: 67-2012)

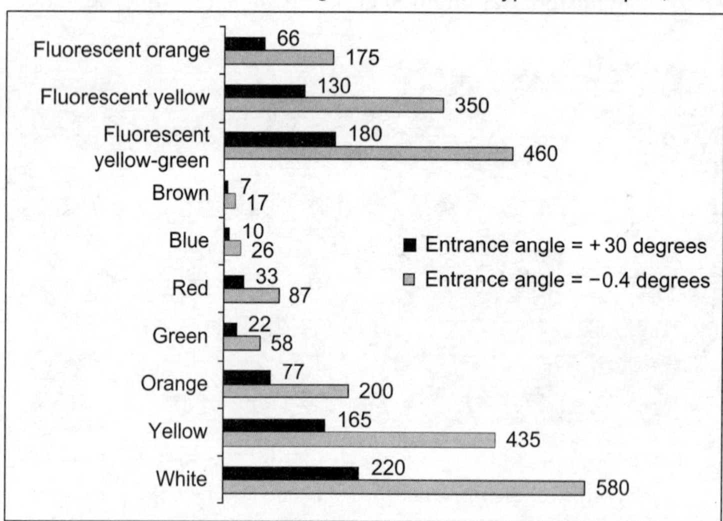


A few guidelines on signs used with retroreflection are given below (IRC: 67):

- It has been noted that sign sheeting with high coefficient of retroreflection at small observation angles gives good performance for drivers viewing from long distances. This means that as the retroreflection cone becomes longer and narrower, the visibility of the sign to the driver will increase under standard conditions.

- The sheeting types A, B and C shall be cleaned with soap-water and their retroreflectance tested when in totally wet condition. All the sheeting shall not show less than 90% of the standard minimum acceptable values when determined in accordance with ASTM D 4956-09.
- The retroreflective sheeting types A, B and C shall cover all types of field-performance compliance under 5, 7 and 10 years of warranty, respectively. At the end of these respective warranty periods, the sheeting shall retain *residual* retroreflectivity of at least 50% for type A, and 80% for types B and C of its original retroreflectance (as recommended in IRC: 67).

# EXERCISES

## Multiple Choice Questions

1. The shape of the Stop sign adopted in Indian practice as per IRC: 67 is
   (a) Circular (b) Triangular (c) Octagonal (d) Rectangular
2. As per IRC: 67, for better visibility, the signs relating to parking should be fixed at an angle to the carriageway of
   (a) 90° (b) 95° (c) 15° (d) 0°
3. The diameter of the disc used for a compulsory control sign which has a blue background is
   (a) 600 mm (b) 470 mm (c) 900 mm (d) 1200 mm
4. The light reflecting devices used to guide the driver along the proper alignment are called
   (a) Traffic signs (b) Attenuators (c) Delineators (d) Road markings
5. As per IRC: 25, the boundary stone shall project above the ground level at
   (a) 500 mm (b) 400 mm (c) 300 mm (d) 250 mm
6. The retroreflective sheeting type A shall be covered under how many years of warranty for field performance of the signs?
   (a) 5 (b) 6 (c) 7 (d) 4
7. The retroreflective sheeting type C shall be covered under how many years of warranty for field performance of the signs?
   (a) 5 (b) 6 (c) 7 (d) 10
8. The retroreflective sheeting type B shall be covered under how many years of warranty for field performance of the signs?
   (a) 5 (b) 6 (c) 7 (d) 4
9. At the end of the standard warranty period, the retroreflective sheeting class A type I shall retain at least how much percent of its original retroreflectance?
   (a) 50 (b) 90 (c) 80 (d) None of these
10. At the end of standard warranty period, the retroreflective sheeting class A type II shall retain at least how much percent of its original retroreflectance?
    (a) 60 (b) 90 (c) 80 (d) None of these
11. At the end of the standard warranty period, the retroreflective sheeting class B type III shall retain at least how much percent of its original retroreflectance?
    (a) 80 (b) 90 (c) 50 (d) None of these

12. At the end of the standard warranty period, the retroreflective sheeting class B type IV shall retain at least how much percent of its original retroreflectance?
    (a) 50 (b) 90 (c) 80 (d) None of these
13. At the end of standard warranty period, the retroreflective sheeting class C type XI shall retain at least how much percent of its original retroreflectance?
    (a) 50 (b) 80 (c) 90 (d) None of these
14. As per IRC: 67, the acceptable minimum co-efficient value of retroreflective sheeting of class C type VIII, when totally wet, shall not show how much value of retroreflection determined in accordance with ASTM D 4956-09?
    (a) More than 90 (b) Less than 90 (c) 100 (d) More than 80
15. The IRC: 67-suggested usage of retroreflective sheeting for urban/city road signs is
    (a) Class A Type I (b) Class A Type II (c) Class B Type IV (d) None of these
16. The IRC: 67-suggested usage of retroreflective sheeting for rural road signs is
    (a) Class C Type VIII (b) Class C Type IX (c) Class C Type XI (d) None of these
17. The IRC: 67 suggested usage of retroreflective sheeting for expressway signs is
    (a) Class A Type II (b) Class B Type IV (c) Class C Type XI (d) Class B Type III
18. The IRC: 67 does not suggest the use of this type of retroreflective sheeting for NH/SH signs:
    (a) Class A Type II (b) Class B Type IV (c) Class C Type XI (d) Class B Type III
19. As per IRC: 67 specification for ACM, the acceptable peeloff strength with retroreflective sheeting by Drum Peel Test in accordance with ASTM D 903 is
    (a) Maximum 4 N/mm$^2$ (b) Minimum 4 N/mm$^2$
    (c) Minimum 3 N/mm$^2$ (d) Between 3 to 4 N/mm$^2$X
20. The commonly used retroreflective sheeting for delineator posts is
    (a) Class A Type II (b) Class B Type IV (c) Class C Type XI (d) Class A Type I
21. The commonly used retroreflective sheeting for gantry-mounted overhead signs is
    (a) Class A Type II (b) Class B Type III (c) Class C Type XI (d) Class A Type I
22. The commonly used retroreflective sheeting for gantry-mounted overhead signs
    (a) Engineering grade sheeting (b) Super engineering grade sheeting
    (c) High intensity grade sheeting (d) Micro prismatic grade sheeting
23. The value of light reflecting efficiency of glass beaded sheeting as compared to the micro prismatic sheeting will be
    (a) Smaller (b) Double (c) Equal (d) Greater
24. MORTH specifies that the white audible and vibratory marking shall attain an initial retroreflectance of not
    (a) $<$ 300 mcd/lx/m$^2$ (b) $>$ 300 mcd/lx/m$^2$ (c) $>$ 250 mcd/lx/m$^2$ (d) $>$ 250 mcd/lx/m$^2$
25. MORTH specifies that the yellow audible and vibratory marking shall attain an initial retroreflectance of not
    (a) $<$ 300 mcd/lx/m$^2$ (b) $>$ 300 mcd/lx/m$^2$ (c) $>$ 250 mcd/lx/m$^2$ (d) $<$ 250 mcd/lx/m$^2$
26. MORTH specifies that the road stud retroreflecting surface slope shall be to the base as
    (a) $35 \pm 5°$ (b) $45 \pm 5°$ (c) $25 \pm 5°$ (d) $90 \pm 5°$

27. MORTH specifies that the area of each retroreflecting surface of road stud shall not be less than
(a) 10 $cm^2$ (b) 11 $cm^2$ (c) 12 $cm^2$ (d) 13 $cm^2$
28. MORTH specifies that each omni-directional stud shall have a C.I.L. of not
(a) < 1 mcd/lx (b) < 2 mcd/lx (c) > 2 mcd/lx (d) > 1 mcd/lx
29. As per IRC :67, a traffic sign indicating 'merging traffic ahead' should be of
(a) Circular shape with white background and red border
(b) Triangular shape with white background and red border
(c) Triangular shape with blue background and red border
(d) Circular shape with blue background
30. Compulsory signs are exhibited in what shape?
(a) Circular shape (b) Triangular shape
(c) Rectangular shape (d) Equilateral triangle (with one point downward)
31. The triangular sign is __________
(a) A warning sign (b) A prohibitive sign
(c) An indicative sign (d) A compulsory sign
32. In Indian practice, this sign indicates Rumble Strip:
(a)  (b)  (c)  (d) 
33. In Indian practice, this sign indicates No Entry:
(a)  (b)  (c)  (d) 
34. The spacing of single chevron signs on a curve having radius 100 m should be
(a) 45 m (b) 15 m (c) 20 m (d) 1 m
35. Which of the following is applicable to Type IV high intensity micro-prismatic grade sheeting of a sign
(a) Fluorescent yellow-green (b) Fluorescent yellow
(c) Fluorescent orange (d) Brown
36. Which of the following is applicable to Type VIII prismatic grade sheeting of a sign
(a) Fluorescent white (b) Fluorescent blue
(c) Fluorescent orange (d) Fluorescent red
37. Which of the following will have the greater value of retroreflectivity (cd/lx/$m^2$) of a new sign
(a) Fluorescent yellow-green (b) Fluorescent yellow
(c) Fluorescent orange (d) White
38. Which of the following will have the least value of retroreflectivity (cd/lx/$m^2$) of a new sign
(a) Brown (b) Blue (c) Green (d) Red

39. Wherever the local language is other than Hindi, the local language shall be used for signs at
    (a) 200 m (b) 300 m (c) 400 m (d) 500 m
40. For toll plazas, advance direction signs shall be provided ahead of the toll plaza, at
    (a) 1 km and 0.5 km (b) 1 km and 2 km (c) 0.5 km and 1 km (d) 1 km and 1.5 km
41. The spacing of single chevron signs before a curve having a radius of 100 m should be
    (a) 40 m (b) 15 m (c) 20 m (d) 30 m
42. The end of dual carriageways may be posted at ________ from the end of the dual carriageway.
    (a) 50–100 m (b)75–100 m (c) 100–150 m (d) 150–200 m
43. The diameter (mm) of regulatory signs used in conjunction with traffic light signals for normal and design speeds of 81–100 km/hr
    (a) 600 (b) 750 (c) 900 (d) 1200
44. The border dimension used for the 'Stop' sign used at approach speeds on minor roads > 65 km/h should be
    (a) 0 mm (b) 40 mm (c) 50 mm (d) 25 mm
45. Identify the warning sign from the list below.
    (a) School zone (b) prohibited
    (c) No parking (d) Speed limit
46. Which of the following is not a valid warrant for installing a traffic control signal?
    (a) 30th highest hour (b) Peak hour
    (c) School zone (d) Accident prone location

## Answers

1. (c) 2. (c) 3. (a) 4. (c) 5. (c) 6. (a) 7. (d) 8. (c) 9. (a) 10. (d) 11. (a)
12. (c) 13. (b) 14. (b) 15. (c) 16. (d) 17. (c) 18. (a) 19. (b) 20. (b) 21. (c) 22. (d)
23. (a) 24. (a) 25. (d) 26. (a) 27. (d) 28. (b) 29. (a) 30. (a) 31. (a) 32. (a) 33. (d)
34. (c) 35. (d) 36. (c) 37. (d) 38. (a) 39. (d) 40. (a) 41. (a) 42. (c) 43. (c) 44. (b)
45. (a) 46. (a)

## Questions

1. Classify road signs under the broad categories and subgroups.
2. Write about the subgroups of regulatory signs with three examples for each case.
3. Write about the four subgroups of information/guide signs with three examples for each case.
4. What are hazard markers? Write about any six types of hazard markers.
5. What are the seven standardised shapes of signs? Write about their intended use.
6. What are the eight standard colours and the three fluorescent colours used in signs? Write about their indications.
7. Write about the details of letters and numerals of different height used on various highway signs.
8. Where are regulatory/mandatory/compulsory signs used? Write about the four subgroups of regulatory/mandatory signs with examples for each case.
9. Where are warning /cautionary/precautionary signs used? Write about the three subgroups of warning /cautionary/precautionary signs with examples for each case.

10. State why illumination and retroreflection become ineffective in different conditions. What are the remedial measures to overcome these effects?
11. Write about the applications of VMS applicable to a range of situations.
12. Write about intended use, types and location of installation and include sketches showing the standard dimensions of (a) Chevron markers (b) Object markers (c) Bollards (d) Junction definition posts (e) Delineator posts (f)Traffic cones (g) Vertical panels (h) Construction barrels (i) Barricades (j) Rumble strips.
13. Define sign legibility distance and write about the factors affecting it.
14. List the factors considered for determination of appropriate size of signs and the letters/legends used in signs.
15. Describe the Australian practice of determination of sign legibility distance, with a schematic diagram.
16. Write the formulas used for determination of height of letters (small and capital) used for shoulder-mounted direction signs as per the Australian practice.
17. Explain the Indian practice of determination of sign legibility distance and the minimum required legible height of letters as per IRC: 67 and IRC: 30.
18. Write a short note on the Indian practice of minimum required legible height and thickness of letters with design speeds as per MORTH recommendations.
19. Write about the details of the Unites States Sign Council (USSC)-recommended conditions of determination of sign size and letter dimensions.
20. At a distance of 120 m before a sign on a straight, plain road, what is the minimum lateral setback and vertical height required to the left of the driver's eye position, with the standard cone of vision for a horizontal of 10° and vertical plane or 5°? Assume any missing value as per IRC: 35.
21. Determine the legibility distance and height of letters of a proposed shoulder-mounted sign for the following conditions based on IRC: 35 recommendation:

    Speed = 80 km/hr; Number of words on the proposed sign = 8; Number of lanes in one direction = 3
22. What are the seven main aspects considered as general principles for positioning a road sign?
23. Describe the important details of siting height and lateral placement of road signs.
24. Draw sketches and indicate the standard dimensions of placement and mounting height of road signs for the following conditions:

    (a) Height clearance for signs mounted adjacent to footpaths

    (b) Lateral and height clearances for signs in kerbed roadways

    (c) Lateral and height clearances for signs in unkerbed roadways

    (d) Lateral and height clearances for signs in footpaths
25. Draw sketches and indicate the standard dimensions of placement and mounting height of road signs for the following conditions:

    (a) Lateral and height clearances for signs in kerbed roadways

    (b) Lateral and height clearances for signs in rural areas

    (c) Lateral and height clearances for gantry-mounted signs in kerbed/unkerbed roadways
26. What are the factors taken into account for siting and longitudinal placement of signs?
27. What is the significance of sign orientation? Describe the desirable orientation of signs located on a (a) straight road (b) right-hand curve (c) left-hand curve.

28. Classify retroreflecting sheets and the material used for signs.
29. Write about the details of plate thickness requirements based on the type of sheet and mounting type of signs as per IRC: 67.
30. Describe the significance of visibility of road signs with a schematic diagram representing the cone of retroreflection from a sign.
31. What will be the effect on visibility from a truck and car with reference to the cone of retroreflection?
32. Describe the basic principle of measurement of retroreflectivity of signs with observation and entrance angles. Also write the basic formula used to determine the retroreflection coefficient of a sign.
33. Describe the testing principle of road signs based on ASTM and CEN geometries.
34. Write about the important guidelines on sign sheets used with retroreflection as per IRC: 67.

CHAPTER

# 19 Roadway Lighting

## 19.0 INTRODUCTION

Good visibility is one of the basic criteria in the design of any roadway facility. It ensures safe and efficient movement for road users. The purpose of lighting on roadways is to provide a comfortable level of visibility to enable motorists and pedestrians to see objects on the roadway quickly, along with the geometric alignment details, including obstructions. Provision of adequate street lighting is essential and needs to be maintained not only at night but also during the day, particularly in tunnels, underpasses and subways.

The design of roadway lighting is a primary responsibility of the Department/Municipality of a region and should be taken seriously, as the safety of the public is directly impacted by any wrong design decision. The designer should apply a logical and systematic approach by adopting standard design specifications. The designer should always remember that the ultimate purpose of the design of any roadway lighting including its maintenance system is that it should be safe and cost-effective.

## 19.1 FACTORS AFFECTING NIGHT-TIME VISIBILITY

- Flux intensity and distribution of luminaire/lamp
- Sides of objects
- Brightness of objects
- Brightness/Glare of the background
- Roadway surface characteristics that influence the reflection of light
- Time of visualisation
- Lighting pollution
- Surface wetness, rainfall and fog

## 19.2 OBJECTIVES OF ROADWAY LIGHTING

The primary purpose of light is to provide comfortable visibility along roadways, streets, underpasses, underbridges, tunnels and for overhead signs (Fig. 19.1). The objectives commonly specified as the safety benefits of roadway lighting are as follows:

- Promotion of safety, efficiency and comfortable visibility at night to drivers and pedestrians.
- Improvement of visibility of traffic safety aids at night so that the driver can read the information placed outside the headlight beams (Pliszczak, 2010). The safety aids may be listed as signs, signal poles, delineators, and other fixed objects and obstructions, especially at level crossings, complex turning lanes and road intersections, and also at zebra crossings and signal-controlled intersections.
- Provision of illumination in underpasses, subways and tunnels.
- Deterring crime on the roadside at night by providing good visibility for pedestrians and other road users. For example, chain-snatching, picking of pockets and theft behind commercial buildings can be controlled by providing security lights (bright lights or floodlights).

- Enhancement of commercial activities at night by providing good visibility and lighting on the roadside. For example, during festivals and temporary melas (street shows).
- Developing tourism in a locality by highlighting the historic areas or landmarks at night.

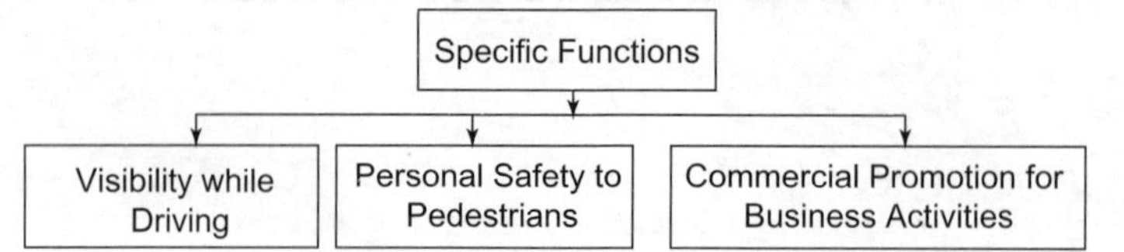


**Fig. 19.1** Broad categories of functions of roadway lighting (EEC, 2013)

## 19.3 GLOSSARY OF TERMS

The following terminology is commonly used in the design of lighting facilities on roadways:

**Luminaire:** A complete electrical device used as a lighting fixture. The luminaire unit comprises lamps, lamp sockets, ballasts, reflective materials, refractors, lenses, louvres, wires and a protective housing/casing. The pole, post or bracket arm is not considered a part of the luminaire (UNAM, 1992).

**Ballast:** An electronic or electromagnetic circuit system using which lamps are illuminated. The ballast regulates the flow of power and current to the lamp based on the design requirement of its luminance. For example, an electronic circuit with a starter unit is used for lighting a fluorescent tubelight.

**Lamp:** An electrically operated light source that has a higher wattage ($\geq$150 W).

**Bulb**: A general purpose lighting source of wattage up to 100 W.

**Lamp wattage:** A measure of the consumption of electrical energy by a light source. The electrical energy consumed will be transformed into lighting and the rest will be released as heat output.

**Luminous flux ($\phi$)** (Fig. 19.2): A measure of the amount of light emitted/radiated from a light source. It indicates the total output of lamps and is measured in units of lumen (lm). Typically, the lumen output for 1000 hours of life is quoted. A few examples of luminous flux are given below (EC, 2011):

- A candle flame may generate about 12 lumens
- A high-pressure sodium (HPS) lamp of 150 W is rated as 17,500 lumens
- A metal halide (MH) lamp of 250 W is rated as 21,000 lumens

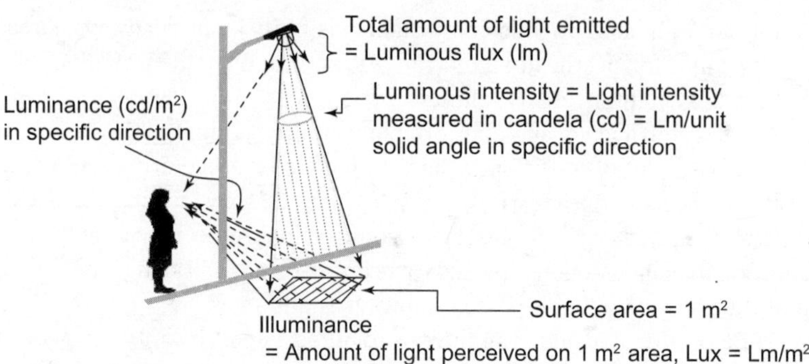


**Fig. 19.2** Illustration of illumination and luminance from a luminaire

**Luminous intensity (I)** (Fig. 19.2): A measure of the brightness of a light source (Lm) projected in a particular direction through a unit solid angle onto a surface (Fig. 19.3). The quantity of the intensity of light from a source is expressed in units of candela (cd for a solid angle).

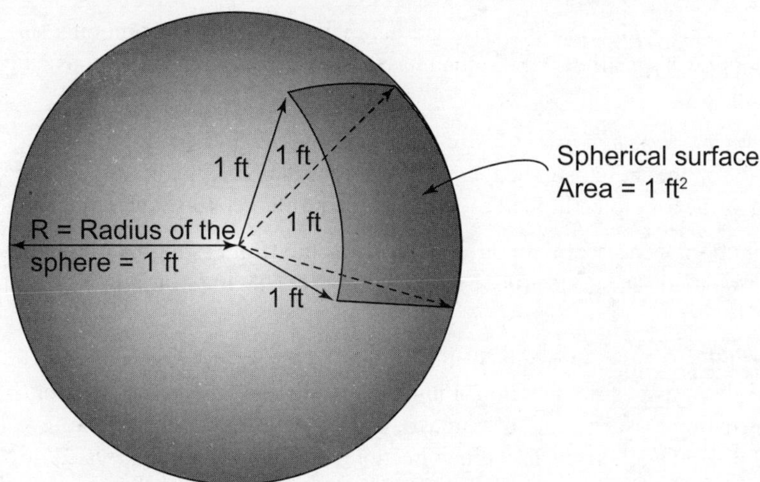


**Fig. 19.3** The solid angle subtended by the spherical area 1 ft$^2$ = Area/R$^2$ = 1 Steradian (Redrawn from IES Lighting Handbook, 1972)

**Illuminance (E):** The count of the amount of luminous flux incident (which falls) on a unit surface. The unit of illuminance is lux ($\approx$lm/m$^2$). When the unit of luminous flux is expressed in lumen and the area is measured in square feet, the unit of illumination is the foot-candela (fc).

$$\text{Illuminanace}(\text{E in lx}) = \frac{\text{Luminous flux}(\text{lm})}{\text{Area}(\text{m}^2)}$$

Or

$$\text{Illuminanace}(\text{E in fc}) = \frac{\text{Luminous flux}(\text{lm})}{\text{Area}(\text{ft}^2)}$$

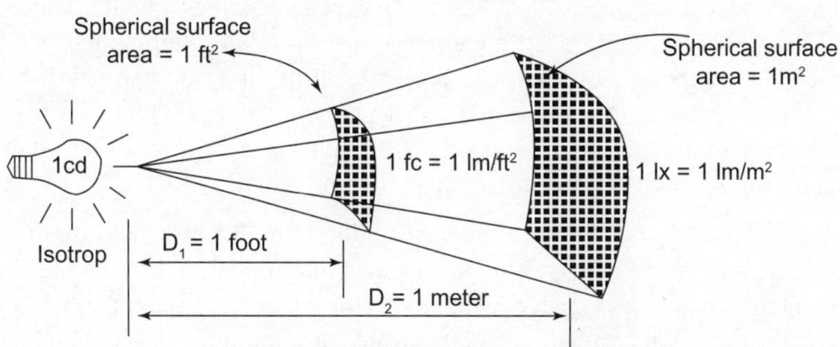


**Fig. 19.4** Differences between fc and lux

**Foot-candle (fc)**: The amount of luminous flux per square foot on a sphere of radius one foot. This quantity is often referred to as density of light flux, and it is expressed as foot-candle (fc). fc = Lumens/

Area. For example, if 3000 lm is emitted on an area of $10 \times 10$ ft$^2$, fc = 30 lm/ft$^2$. It is an important measuring unit used for calculating the desired illumination with the layout of the luminaire (Fig. 19.4).

*Note:* 1 foot-candle = 1fc ≈ 10.764 lux = 10.764 metre-candle. 1 cd/ft$^2$ = 10.764 cd/m$^2$.

The illumination value can be measured using a lux meter. The light incident on a surface may come from multiple sources. Typically, the full moon has 0.1 lux, an emergency light has 1 lux, a winter day has 10,000 lux and a very bright summer day has 100,000 lux.

*Differences between lumens and lux:* 1 lux of light intensity is equivalent to 1 lm spread over an area of 1 m$^2$. If the area of illumination is to be increased but have the same lux, more lamps will be required. This means that as the area to be lit increases, the number of lumens also increases (total light output).

**Average illumination:** For street lighting, the luminous flux from a luminaire will not be uniformly distributed over the roadway surface (not a spherical surface), so the resulting illumination calculated on check points is averaged.

**Luminance (L)** (Fig. 19.2): A measure of the luminous intensity of light reflected from a unit area within a unit angle in a specific direction. Luminance indicates how much luminous intensity (light intensity reflected from a surface as brightness) will be detected by the human eye looking at the surface within the solid angle from a specific direction. The unit of luminance is cd/m$^2$. Luminance on the vertical plane is considered for the design of pedestrian visibility while luminance on the horizontal plane is considered for the design of roadway visibility.

**Luminaire efficacy:** The light output of an entire luminaire divided by the total power consumed by the lamps and its ballasts. It is measured in lm/W. The higher the efficacy value, the greater the energy efficiency of the lighting system.

**Dimming:** The voltage supply is lowered using step transformers, and the light output from a luminaire is regulated using dimmers (in the range of 100% to 50% late at night and early in the morning). These dimmers are mostly coordinated with lamp ballast. For example, a 300-W lamp may be regulated to emit light equal to a 150-W lamp when the expected traffic volume is very low for the time period 1:00 am to 4:00 am. Therefore, the dimming system will result in considerable energy saving and be financially justified. However, dimmers are applicable for street lighting only when the supply voltage is not less than 220 V (ECO-3, 2010).

**Photometry**: The science of measurement of light, which describes the luminaire's light output characteristics. The most commonly used photometric data are lamp-luminaire–specific data charts such as isofoot-candle performance chart, coefficient of utilisation curve, vertical and lateral light distribution data, lumen maintenance curve and dirt depreciation curve. These charts can be obtained from the manufacturer of the selected lamp-luminaire and be used to calculate the intensity of a particular luminaire for a proposed roadway lighting system (MnDOT, 2010).

**Isofoot-candle curve diagram:** A graphical representation of points of equal illumination connected by a continuous line – it resembles the contour lines drawn on a topographical map for level surveying. Each of these contour lines is associated with a specific value representing the foot-candle on a horizontal plane from a selected luminaire or a composite lamp-luminaire at a definite mounting height (Fig. 19.5). In order to make these curves applicable to all conditions of mounting, they are given correction factors for a few proposed mounting heights, which are expressed in ratios of actual lateral roadway distance. Such correction factors for different mounting heights are usually supplied by the manufacturer, along with the isofoot-candle curve diagram for each type of luminaire.

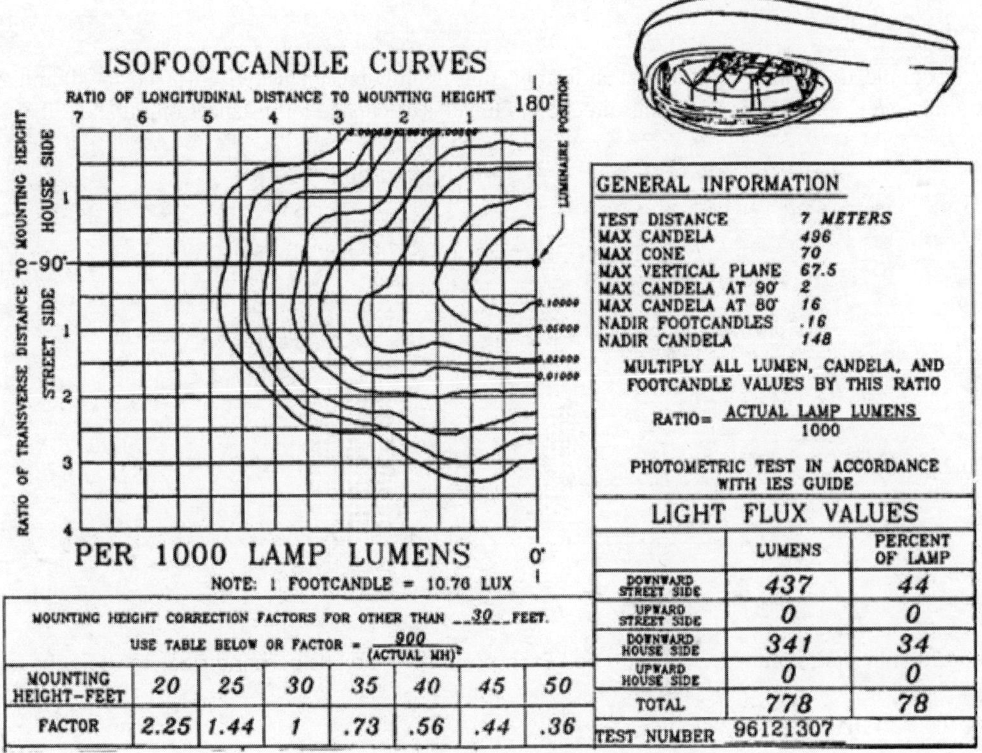


**Fig. 19.5** Isofoot-candle curves diagram (for HPS luminaire)

(*Source*: MnDOT, 2010; *Courtesy*: Office of Traffic, Security and Technology, MnDOT)

*Note:* If the linear measurement used is in metres, these diagrams are termed isolux curve diagrams.

**Colour rendering index (CRI):** A scale from 0 to 100 that represents the ability of a light source to render true colours or colour shift. The value of 100 represents no colour distortion of objects under light and the best representation in sunlight. A light source that has a CRI in the range 100–75 is considered excellent, 75–65 good, 65–55 fair and 55–0 poor (Szary et al., 2005).

**Glare:** A kind of lighting pollution caused by high luminance or excessive luminance differences in the visual field. Disability glare and discomfort glare are the two main types pertaining to civil engineering. Discomfort glare is a major concern in indoor lighting, while disability glare is the main concern in outdoor/street lighting designs.

## 19.4 LIGHTING LAWS

A few basic concepts of lighting relating to roadway lighting design are discussed in this section.

### 19.4.1 Inverse Square Law

The illumination ($E$) on a surface is directly proportional to the luminous intensity (I) of the source and inversely proportional to the square of the distance between the source and the surface normal (Fig. 19.6). The inverse square law refers to point sources and is expressed as

$$E_{normal} = \frac{I}{D^2} \tag{19.1}$$

For example, if a point source of light uniformly emits a luminious intensity $I = 100$ cd, the illumination on a surface ($E_{normal}$) that is normal to its direction of travel is calculated for distances of 1 ft and 2 ft as

$$E_{normal@1ft} = \frac{100}{1^2} = 100 \text{ fc} = 100 \text{ lumen / ft}^2$$

$$E_{normal@2ft} = \frac{100}{2^2} = 25 \text{ fc} = 25 \text{ lumen / ft}^2$$

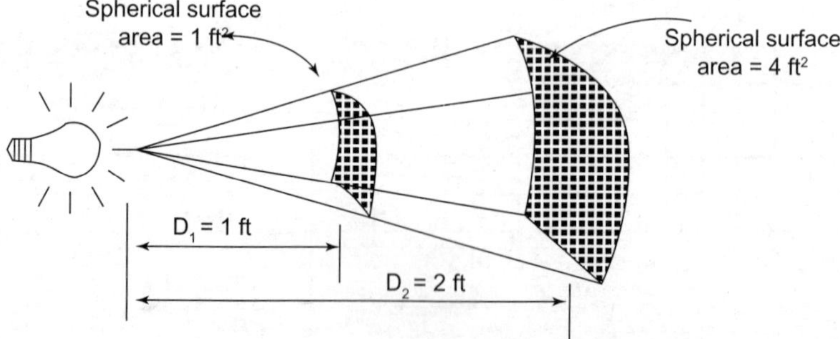


**Fig. 19.6** Inverse square law (schematic): Each point in an area is perpendicular to the light rays

## 19.4.2 Cosine Law

The illumination on a road surface (not perpendicular to the incident ray) varies as the cosine (Fig. 19.7) of the angle of incidence ($\theta$). Therefore, illumination ($E$) is calculated as

$$E_{\theta} = \frac{I}{D^2} \times \cos(\theta) = E_{normal} \times \cos(\theta) \tag{19.2}$$

The total illumination obtained from more than one source $= E_{source1} + E_{source2} + .....$ (19.3)

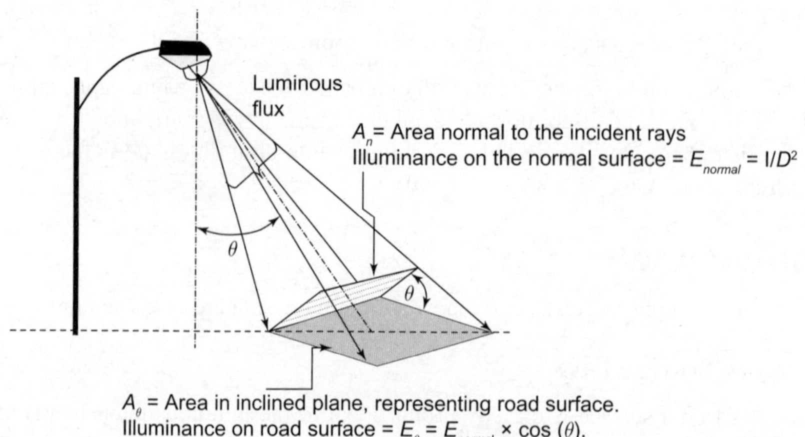


**Fig. 19.7** Concept of cosine law

## 19.5 TYPES OF LAMPS

Based on the technology, different types of lamps are used for road illumination. The most commonly used lamps for street/roadway lighting and other purposes are described in this section (Aalto, 2010; EC, 2011; MnDOT, 2010). The general classification and characteristics of lamps (Table 19.1) that are most widely used for roadway lighting are presented in Fig. 19.8.

**Table 19.1** General characteristics of lamps widely used for roadway lighting

| *Type of light* | *Initial light output lumens × 10³* | *Approximate efficacy lumens/watt* | *Approximate lamp life hours × 10³ʼ²* |
|---|---|---|---|
| Metal Halide | 34–100 | 85–100[1] | 10–15 |
| High Pressure Sodium | 9.5–140 | 95–140[1] | 15–28 |
| Low Pressure Sodium | 1.8–33 | 100–183[1] | 10–18 |
| Induction Lighting | 3.5–12 | 67-74 (based on 100 h) | 100 |
| Light Emitting Diaode (LED)[3] | 16–20 | 71 | 50 |

Notes:

1. These values exclude wattage losses due to ballast.
2. Number of hours for a group of lamps at which 50 percent will remain in operation; based on 10 hours of operation per start.
3. Estimated values. This value is now proposed as 100 ×10³ with a luminous efficacy of 120 lm/W (Zarling, 2016).

(*Source*: MnDOT, 2010; *Courtesy*: Office of Traffic, Security and Technology, MnDOT, USA)

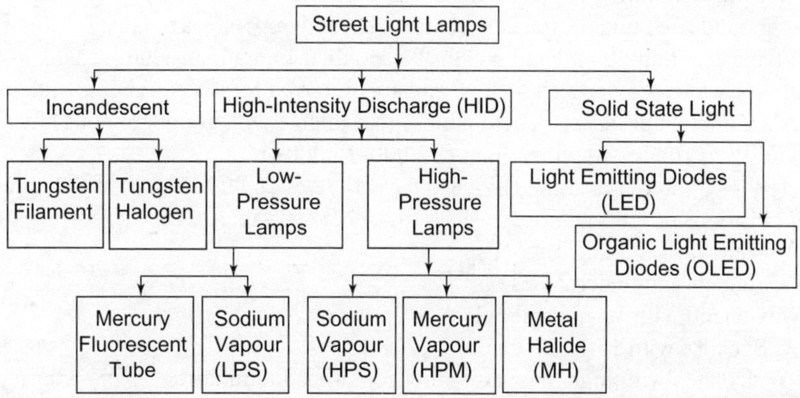


**Fig. 19.8** Family tree of electric lamps commonly used for lighting of streets/roadways

- **Incandescent lamps**: A filament unit is electrically heated and it glows to produce light. Based on the filament, incandescent lamps can be of two types: tungsten and tungsten halogen.
  - Tungsten filament lamps are rarely, if ever, used for street lighting because of their low efficiency and short life (~750–1000 hours) compared to HID lamps. The typical luminous efficacy of these lamps is in the range of 5–15 lm/W.
  - Tungsten halogen lamps are derived from incandescent lamps. They emit a bright white light that contains a small capsule/linear double-ended lamps filled with halogen gas, with or without reflectors. Its life is longer than that of tungsten filament lamps of the same wattage. Its life may range 2000–4000 hours. The typical luminous efficacy of these lamps is in the range of 12–35 lm/W.

- **Fluorescent tubes:** In these tubes, mercury vapour is used as a low-pressure gas to produce light. Less heat is produced and this is more energy efficient than incandescent and halogen bulbs. It is commonly used in homes, workshops and for general use. It has good luminous efficiency (100 lm/W) and a long life (10,000–16,000 hr). However, these lamps are not commonly used for street lighting because the output light is adversely affected by cold weather (<35° C) and it is difficult to obtain good light control with shorter luminaires. It requires ballast and is hence not widely used for street lighting or for sign lighting.
- **Compact fluorescent light (CFL) electrode-less lamp:** It is a compact variant of the fluorescent lamp. CFLs are commonly used for domestic and office lighting applications. They last about 10,000 hours, which is about 10 times that of incandescent bulbs, and their luminous efficacy is four times higher. This is a best replacement for incandescent lamps.
- **High-pressure mercury vapour lamp:** This is the oldest variant of HID lamps which produces light by gas discharged using an electric arc through vaporised mercury, at high pressure. These lamps are available with low luminous efficacy, typically between 30 lm/W and 60 lm/W, and have a life of up to 24,000 hr. As they contain traces of mercury, the disposal of these lamps was found to be difficult, and hence, they have been banned in some developed countries.
- **High-intensity discharge (HID) lamps:** HID lamps emit a large amount of light using relatively little power. They include metal halide with a quartz arc or ceramic arc tube, high-pressure sodium and low or high-pressure mercury vapour lamps. HID lamps are predominately used for street lighting and in parking lots.

  Both metal halide (MH) and high-pressure sodium (HPS) lamps are commonly used in street lighting for different kinds of applications. For example, MH lamps provide clear lighting with true colours and are suitable for city centres. These lamps are available with good luminous efficacy, typically from 50 lm/W to 100 lm/W, and their life ranges 6000–12,000 hr.

  HPS lamps provide a golden-yellow light and are preferred for street lighting, with the advantage of minimal attraction of insects. These lamps are available with good luminous efficacy, typically from 80 lm/W to 100 lm/W, with a life of 12,000–16,000 hr.

  A few varieties of non-crystalline HPS also have a life of 25,000–30,000 hr (Alberta, 2003).

  A few applications of HPS with wattage are given below (Alberta, 2003):

  - 100 W: Walkways and bikeways
  - 150 W: Tunnels and underpasses
  - 250 W: General with 13-m davit pole
  - 400 W: Specific with 15-m davit pole
  - 1000 W: Isolated with high mast excluding cobra-head luminaires

- **Induction (electrode-less fluorescent) lamps:** These are different from the traditional discharge lamps. They emit white light and have good lumen maintenance with a lamp wattage of 55–165 W, luminous efficacy of 60–80 lm/W and life of 100,000 hr. Induction lamps can also be used in inaccessible locations such as in tunnels and on/under bridges.
- **Light emitting diode (LED):** It is a solid state lamp that uses light emitting diodes. LEDs are small point sources with high intensities of light; arrays of such individual sources can form luminaires of different shapes and sizes. It has a long service life (100,000 hr), contains no mercury, is suitable for low-voltage operation, has design flexibility, does not use infrared, has the potential for colour mixing and has high efficacy compared to incandescent lamps. The luminous efficacy of a white-light LED depends on the desired wavelengths and CRI.

The current barriers/challenges to the use of LEDs in mainstream applications are their insufficient efficacy compared to fluorescent and HID lamps, lack of industrial standards and high prices (Pohl, 2007).

*Note:* Certain countries and regions in the USA, Europe and Australia have banned or are on their way to banning, or are considering banning inefficient light sources from their market through legislation and voluntary measures. For example, incandescent lamps, mercury lamps and certain inefficient fluorescent and HID lamps (Aalto, 2010).

## 19.6 LIGHTING REQUIREMENTS

Prior to designing roadway lighting, the extent and type of lighting requirements in a locality is determined based on the following categories:

**Continuous lighting:** The lighting that runs continuously along a roadway segment, between intersections or interchanges, i.e., the area covered by intersections or interchanges are not considered under this category. Continuous lighting is generally used in urban areas, but it may not be compulsory in rural areas.

**Full lighting:** The lighting that covers the entire travel portion of a roadway continuously, including the straight and curved areas, intersections, ramps and interchange areas.

**Partial lighting:** The lighting that is provided only in selected areas or points of interest on roadways. For example, on potential pedestrian–vehicle conflict areas, hazardous locations, approaches to an intersection/interchange, tunnels, underpasses and any isolated traffic conflict areas such as midblock crosswalks, exit areas, merge areas and railway level crossings (Fig. 19.9).

**Delineation lighting:** Special lighting that is provided on intersection/interchange approaches or any hazardous locations to delineate the key components of channelisation, diversion points, merging points, gore areas, warning/information/guide signs, vehicle inspection stations and obstructions on the roadway.

## 19.7 WARRANTS FOR LIGHTING

Warrants indicate the conditions or circumstances that should be satisfied to justify the installation of lighting on roadways. Roadway/Street lighting is commonly adopted with the following conditions of warrants:

- Lighting on expressways, national highways, state highways, other roads/streets, interchanges, flyovers and bridges should be considered where the night-to-day ratio of accidents is greater than 0.5 and the lighting system is cost effective (INDOT, 2013).
- Lighting is provided for locations where there is high potential for accidents to occur due to complex geometric features including inadequate stopping sight distance, channelised islands, speed change lanes, truck weighing stations, toll collection booths, intersections with significant night-time pedestrian and commercial activity or with residential development.
- The deceleration and adjacent lanes at interchanges must be illuminated unless otherwise specified. The deceleration lighting unit should be installed for the safe stopping sight distance from the physical gore area. Only two units shall be installed at the gore area: one in the ramp area and the other along the main line (Fig. 19.10) (NJDOT, 2015).
- The acceleration lanes beginning at the entrance gore area need not be illuminated (NJDOT, 2015).
- Lighting is required in parking areas and rest areas that have pedestrian activity.

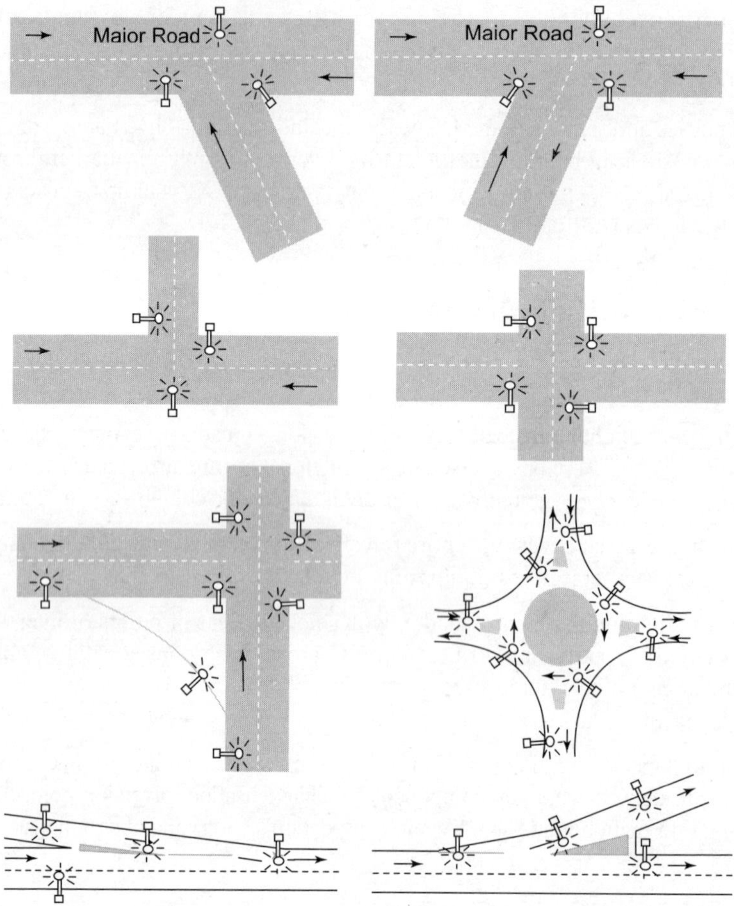


**Fig. 19.9** Some examples of partial street lighting layouts

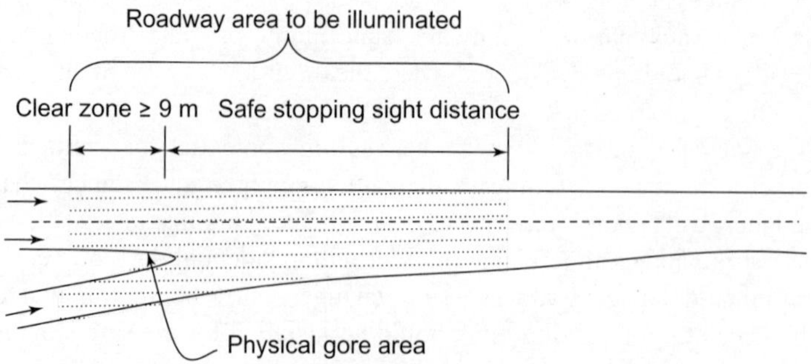


**Fig. 19.10** Exit ramp or deceleration lane – roadway area to be illuminated
(Redrawn from NJDOT, 2015)

- Continuous lighting arrangement at night should be provided for underpasses of length greater than 15 m. Underpasses longer than 60 m should be provided with a lighting system during the day as well (MnDOT, 2010).
- Lighting of underpasses with length less than 22 m is not normally required. Daytime lighting should be considered for tunnels or underpasses with a length-to-height ratio of 10:1 (INDOT, 2013).
- Lighting should be provided in potential vehicle–vehicle and vehicle–pedestrian conflict areas and higher illumination levels should be provided at ramp entry/exit, gore, diverging and merging areas.
- Sign lighting should be considered if an overhead-mounted sign with super-high-intensity retroreflective sheeting is located at a viewing angle such that its retroreflectivity is inadequate at a sight distance of 300 m (Alberta, 2003) or which is not sufficient for night-time visibility (INDOT, 2013).

**Warrant analysis:** Prior to designing the roadway lighting system, the designer must determine whether there is a need for lighting in a particular section, area or location. Evaluating the need for lighting on a roadway is a multi-step process. A warrant analysis should be carried out to evaluate the need for lighting based on data collected from field surveys relating to on-site geometric and traffic conditions.

NJDOT (2015) recommends using the following cases of warrant points analysis for specific road geometric conditions. These warrants must be met to be considered for provision and installation of lighting.

To evaluate the need for:

- Continuous lighting for an existing highway for controlled access facility - Form 1 (Table 19.2)
- Complete or partial lighting for existing interchange - Form 2
- New highways or alignments for controlled access facility with continuous lighting - Form 3. All travelled lanes are required to be illuminated, except express lanes separated from local lanes by concrete islands, barriers or grass.
- Partial or complete lighting for new interchange highway alignments with controlled access facility - Form 4.
- Existing alignments with non-controlled access facility (land service highway) - Form 5
- New alignments with non-controlled access facility - Form 6
- Existing roadway intersection lighting - Form 7
- New roadway intersection lighting - Form 8

**Table 19.2** Warrant evaluation form for providing continuous freeway lighting with controlled access facility

| *Evaluation Form for Controlled Access Facility (Continuous Freeway Lighting)* | | | | | | | | | *Lighting Form 1* |
|---|---|---|---|---|---|---|---|---|---|
| | *Rating* | | | | | | | | |
| *Classification Factor* | *1* | *2* | *3* | *4* | *5* | *Unlit Weight (A)* | *LIghted Weight (B)* | *Diff (A B)* | *Rating X(A-B)* |
| Geometric Factors | | | | | | | | | |
| Number of Lanes | 4 | | 6 | | $\geq$8 | 1.0 | 0.8 | 0.2 | |
| Lane width | >12′ | 12′ | 11′ | 10′ | $\leq$9′ | 3.0 | 2.5 | 0.5 | |
| Median Width | >40′ | 24′–39′ | 12′–23′ | 4′–11′ | 0–3′ | 1.0 | 0.5 | 0.5 | |

(Continued)

**Table 19.2** (Continued)

| *Evaluation Form for Controlled Access Facility (Continuous Freeway Lighting)* | | | | | | | | | *Lighting Form 1* |
|---|---|---|---|---|---|---|---|---|---|
| | *Rating* | | | | | | | | |
| *Classification Factor* | *1* | *2* | *3* | *4* | *5* | *Unlit Weight (A)* | *Lighted Weight (B)* | *Diff (A B)* | *Rating X(A-B)* |
| Shoulders | 10′ | 8′ | 6′ | 4′ | 0 | 1.0 | 0.5 | 0.5 | |
| Slopes | ≥8:1 | 6:1 | 4:1 | 3:1 | 2:1 | 1.0 | 0.5 | 0.5 | |
| Curves | $0-\frac{1}{2}°$ | $\frac{1}{2}-1°$ | 1–2° | 2–3° | 3–4° | 13.0 | 5.0 | 8.0 | |
| Grades | <3% | 3.0–3.9% | 4.0–4.9% | 5.0–6,9% | >7% | 3.2 | 2.8 | 0.4 | |
| Interchange Frequency | 21,000′ | 16,000′ | 10,500′ | 5000′ | <5,000′ | 4.0 | 1.0 | 3.0 | |
| | | | | | | Geometric Total | | | |
| Operational Factors | | | | | | | | | |
| Level of service | A | B | C | D | E | 6.0 | 1.0 | 5.0 | |
| (Any dark hour) | | | | | | | | | |
| | | | | | | Operational Total | | | |
| Environmental | | | | | | | | | |
| % Development | 0% | 25% | 50% | 75% | 100% | 3.5 | 0.5 | 3.0 | |
| Offset to Development | 200′ | 150′ | 100′ | 50′ | <50′ | 3.5 | 0.5 | 3.0 | |
| | | | | | | Environmental Total | | | |
| Accidents | | | | | | | | | |
| Ratio of Night to Day Accidents | 1.0 | 1.0–1.2 | 1.2-1.5 | 1.5–2.0 | 2.0* | 10.0 | 2.0 | 8.0 | |
| | | | | | | Accident Total | | | |
| * Continuous Lighting Warranted | Geometric Total | | | | = ________* | | | | |
| | Operational Total | | | | = ________* | | | | |
| | Environmental Total | | | | = ________* | | | | |
| | Accident Total | | | | = ________* | | | | |
| | Sum | | | | = ____ Points | | | | |
| | Warranting Condition | | | | = 95 Points | | | | |

(*Source*: NJDOT, 2015; *Courtesy*: New Jersey Department of Transportation, USA)

In addition, specific warrants are recommended by (a) AASHTO Roadway Lighting Design Guide, (b) ANSI/IESNA RP-22-11 publication on American National Standard Practice for Tunnel Lighting (c) ANSI/IESNA RP-8 on Transition Lighting (d) NCHRP Report 152 (e) NCHRP Report 672 for

Roundabouts Lighting and (f) Publication No. FHWA-HRT-14-050, June 2014 on Adaptive Lighting (to provide lighting only when and where it is needed).

## 19.8 FACTORS AFFECTING THE SELECTION OF APPROPRIATE LIGHTING SYSTEM

- Highway class/Street type
- Roadway width
- Sidewalk width
- Typical height of buildings
- Desired level of illumination
- Mounting height
- Type of light source
- Luminaire type
- Pedestrian movements
- Pedestrian–vehicle conflict areas
- Location: Intersection/Curve characteristics
- Number, placement and types of trees

## 19.9 LIGHTING CORRECTION FACTORS

The following terminology and correction factors are used in the design of any roadway lighting system.

$$\text{Light output ratio (LOR)} = \frac{\text{Light emitted by the luminaire}}{\text{Light output of the lamps in that unit}}$$

The energy efficiency of a luminaire is characterised by the LOR, and its typical value for street lighting may be between 0.6 and 0.9, i.e., 60–90% (EC, 2011). The LOR depends on the type of materials used (type of lamp) and the design of the luminaire unit (control gear and optical components).

$$\text{Installed test lamp lumen (ITL) factor} = \frac{\text{Installed lamp lumens}}{\text{Test lamp lumens}}$$

For example, if the installed lamp lumen value = 19,667 and the test lamp lumen value = 20,500, then the LLC = (19667/20500) ≈ 0.96, i.e., 96%. This means that 100 − 96 = 4% of its tested value is reduced due to variations in luminaire casing and reflectors.

*Luminous intensity distribution* from a light source indicates the direction and spread of light (say, across the vertical plane and on the horizontal surface).

$$\text{Utilisation ratio} = \frac{\text{Required average luminance on a surface}}{\text{Minimum maintained luminance on the surface, obtained after all corrections}}$$

*Utilisation factor (UF)* is the per cent utilisation of luminous flux received by a road from the full luminous flux produced by the bare lamp. The following formula may be used for determination of utilisation factor of a street lighting lamp. The UF indicates how best a luminaire is used for a specific roadway location.

$$\text{Utilisation factor} = \frac{\text{Luminous flux (Lumens) received on a roadway surface}}{\text{Luminous flux (Lumens) emitted by the bare lamp source}} \times 100$$

The utilisation curve of a luminaire lamp defines the degree of the total lumen output that reaches a roadway surface as a percentage of its initial lamp lumens (Fig. 19.11), but reveals little about the way in which the light is distributed. For this reason, the performance of any street lighting system with a series of luminaires mounted along a roadway area is evaluated by calculating the ratio of the average minimum illumination to the minimum obtained value. This ratio is termed uniformity ratio (UNAM, 1992).

$$\text{Coefficient of utilisation, CU} = \frac{\text{Luminous flux (Lumens) received on a roadway surface}}{\text{Luminous flux (Lumens) emitted by the bare lamp source}}$$

There are two coefficient of utilisation curves for any type of luminaire: street-side utilisation curve, i.e., in the desired area to be lit; and house-side utilisation curve, i.e., the direction away from the primary lit direction. The street-side curve represents the desired area to be lit by the bare lamp in per cent, and increases as the ratio of the transverse distance to mounting height increases. Similarly, the house-side curve is used to represent the lighting level in the direction away from the primary lit direction.

For example (Fig. 19.12), the pavement area CU may be obtained for a ratio between distance and height as (36 + 13)/40 = 1.23, and the obtained utilised lumen value (from Fig. 19.11 street curve) = 37%. Similarly, the shoulder area CU may be obtained for a ratio between distance and height as (13)/40 = 0.33, and the obtained utilised lumen value (from Fig. 19.12 house curve) = 11%. Therefore, the CU for the roadway area = (37 − 11) = 26% = 0.26.

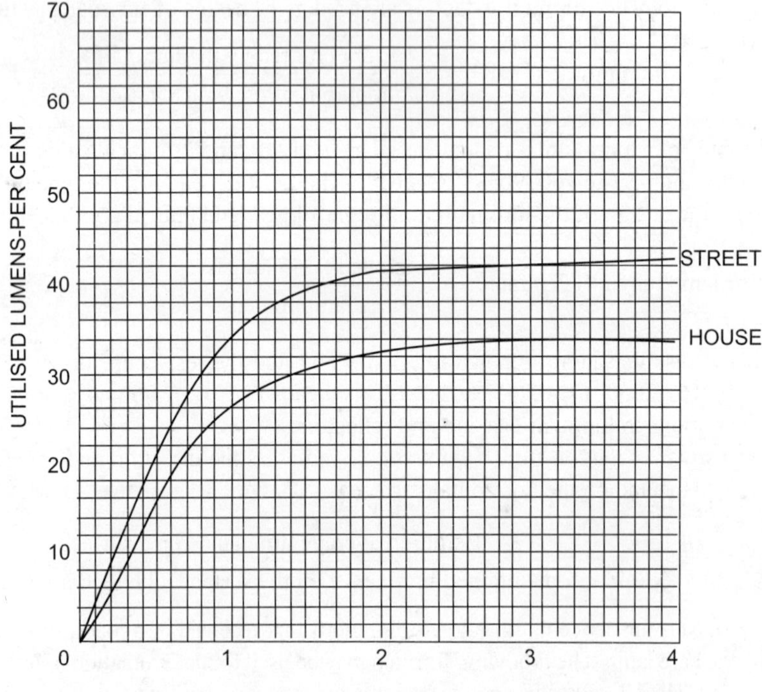


**Fig. 19.11** Coefficient of utilisation curve (for cobra-head high-pressure sodium lamp)
(*Source*: MnDOT, 2010; *Courtesy*: Office of Traffic, Security and Technology, MnDOT, USA)

*Lamp lumen depreciation (LLD) factor:* The lumen output from any lamp reduces as its service life increases. The initial lumen output from a luminaire is multiplied by a depreciation factor to compensate for the anticipated reduction in lumen over its service period. This information is provided

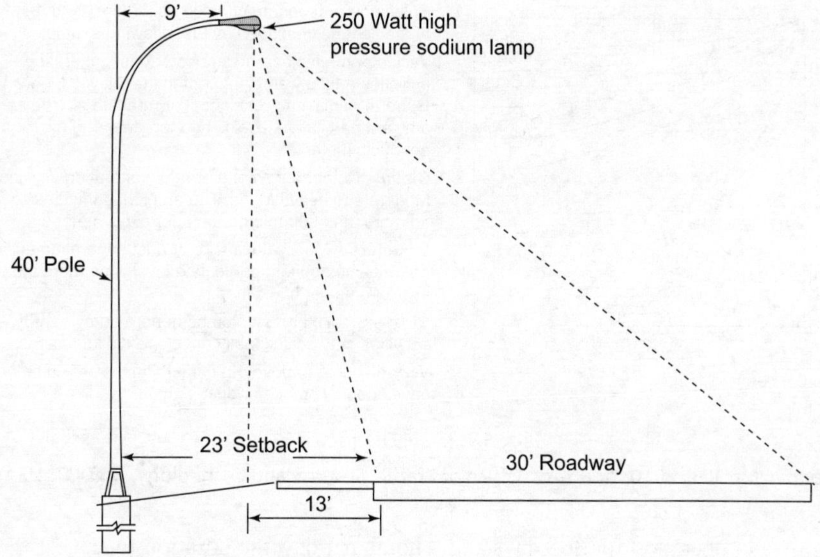


**Fig. 19.12** Coefficient of utilisation calculation example
(*Source*: MnDOT, 2010; *Courtesy*: Office of Traffic, Security and Technology, MnDOT, USA)

by the manufacturer for determining the minimum rated output of any light source (Fig. 19.13). The re-lamping schedule is commonly carried out based on these individual LLD values of luminaires. The most commonly adopted LLD value is 0.88, based on a four-year maintenance cycle (Alberta, 2003). Mn/DOT uses the LLD value of 0.90. LLD can be estimated using the following formula:

$$\text{LLD} = \frac{\text{Mean Lumen at rated operation conditions}}{\text{Initial output of the lamp}}$$

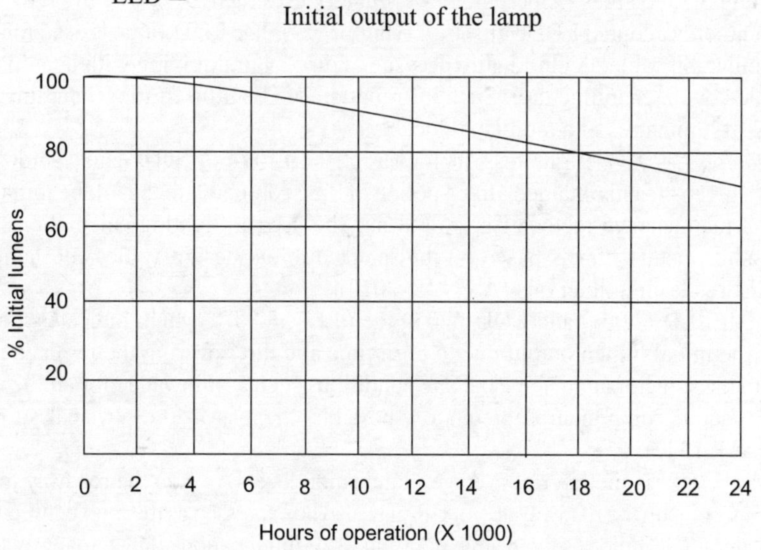


**Fig. 19.13** Typical lamp lumen maintenance curve (LLD% for HPS luminaire)
(*Source*: MnDOT, 2010; *Courtesy*: Office of Traffic, Security and Technology, MnDOT, USA)

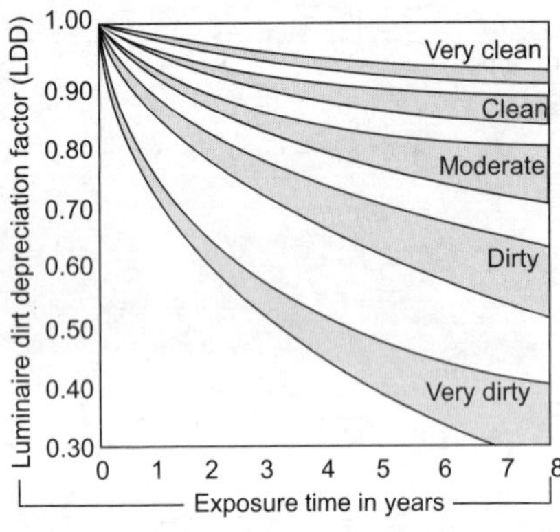


Select the appropriate curve in accordance with the type of ambient as described by the following examples:

**Very clean** – No nearby smoke or dust generating activities and low ambient contaminant level. Light traffic. Generally limited to residential or rural areas. The ambient particulate level is no more that 150 micrograms per cubic meter.

**Clean** – No nearby smoke or dust generating activities. Moderate to heavy traffic. The ambient particulate level is no more than 300 micrograms per cubic meter.

**Moderate** – Moderate smoke or dust generating activities nearby. The ambient particulate level is no more than 600 micrograms per cubic meter.

**Dirty** – Smoke or dust plumes generated by nearby activities may occasionally envelope the luminaires.

**Very dirty** – As above but the luminaires are commonly enveloped by smoke or dust plumes.

**Fig. 19.14** LDD factor

(*Source*: MnDOT, 2010; *Courtesy*: Office of Traffic, Security and Technology, MnDOT, USA)

If the luminaire is in operation for 365 days and 9 hours per day, the duration for 1 year = 365 × 9 = 3285 hours per year. Let the lamp's assumed life = 4 years. The total duration for the four years will be = 4 × 3285 = 13,140 hours. Looking at the curve (Fig. 19.13) at 13,140 hours on the x axis, the lamp depreciation factor on the curve = 89% of initial lumen = 0.89.

*Luminaries dirt depreciation (LDD) factor:* Dirt deposited on any part of the luminaire will reduce the amount of light that reaches the roadway. As the mounting height of the luminaire increases, there will be less accumulation of dirt on it. However, the accumulation of dirt depends on several factors such as dusty roads, dust pollution from different sources and automobile exhausts. Based on a four-year maintenance cycle, the commonly adopted value of LDD is 0.88 for general locations and 0.82 for heavily industrial/polluted areas (Alberta, 2003). MnDOT uses the LDD value of 0.90.

The most commonly adopted LDD value is 0.9 (Fig. 19.14). The LDD value is used for determination of reduced illumination value of a luminaire becasue of dirt, by multiplying with its initial illumination as it was provided by the manufacturer. Such information can be utilised for scheduling the frequency of dirt-cleaning of luminaires in a municipal locality.

*Maintenance factor (MF)* = Light loss factor (LLF) = LLD × LDD. This factor compares the illumination of a given area maintained after a period of time compared to that of the initial illumination. As a standard, a maintenance factor of up to 0.8 may be used in lighting calculations. Flexibility in assuming this adjustment factor is based on the selected initial design values and lamp replacement value and luminaire cleaning intervals (AS/NZS, 2010).

For example, if LLD = 0.90 and LDD = 0.9, then LLF = 0.81, which indicates there would be a loss of 19% of the initial lumen output due to burn time and dirt covering the luminaire. Adjustments are applicable to these factors under special site conditions such as allowance for ambient temperature, voltage, ballast factor, component depreciation, possible changes in the physical surroundings and burn-outs (UNAM, 1992).

*Uniformity ratio (UR)* is the ratio of average illumination level ($E_{avg}$) on the roadway to the minimum illumination at any point on the roadway ($E_{min}$). UR may also be expressed as the ratio between the maximum and minimum values. UR indicates the degree of evenness of illumination along a roadway length. The UR value 1.0 indicates complete uniformity and excellent design of the lighting system along a roadway. The specification requirement for each road class may be obtained from any standard code of practice.

## 19.10 DESIGN ELEMENTS OF LIGHTING SYSTEM

### 19.10.1 Lateral Luminaire Light Distribution

Different types of luminaires are designed to provide lighting to fit many conditions of roadway geometrics. No luminaire is suitable for all conditions, and hence, specific luminaires for specific conditions or a combination of luminaries can be modified to produce more than one pattern of light distribution.

The lateral distribution patterns for streetlights as developed by the Illuminating Engineering Society of North America (IESNA) and the American National Standards Institute (ANSI) may be broadly categorised into five types with two variants (Table 19.3).

**Table 19.3** Road coverage by type of luminaire

| *Type* | *Coverage geometry* | *Applicability* |
|---|---|---|
| Type 1 | 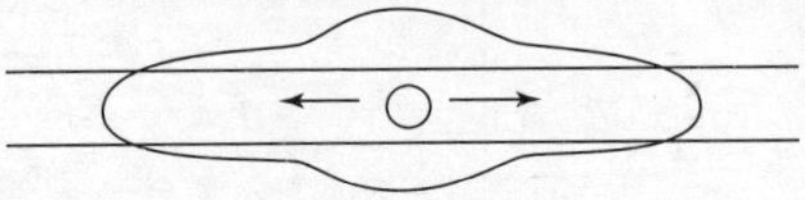 | At median openings, narrow roadway and narrow midblock crossings |
| Type 2 | 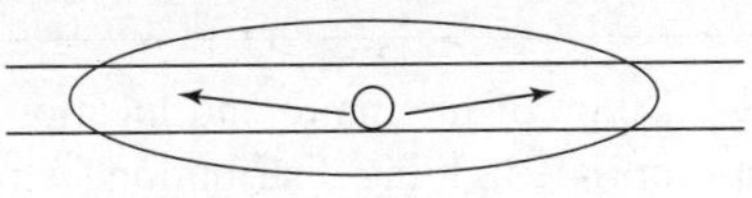 | Wider than Type 1<br>At narrow roadways having a width about 1.6 times its mounting height and the lamp located near one side; walking paths, single-lane highways, single-lane on and off ramps; continuous lighting purpose |
| Type 1 (four-way) | 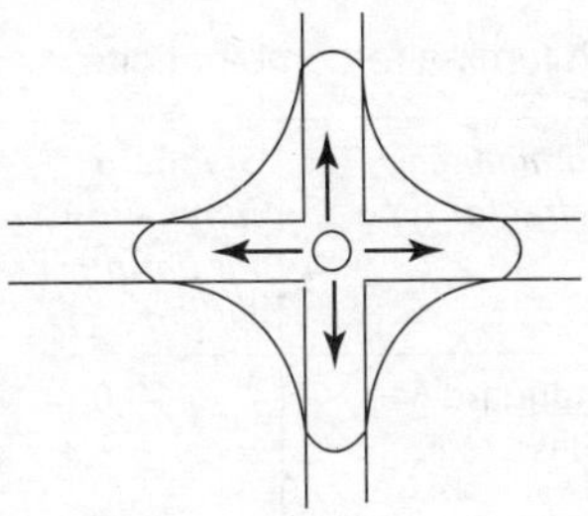 | At narrow roadway intersections; the luminaire is positioned at its centre |
| Type 2 (four-way) | 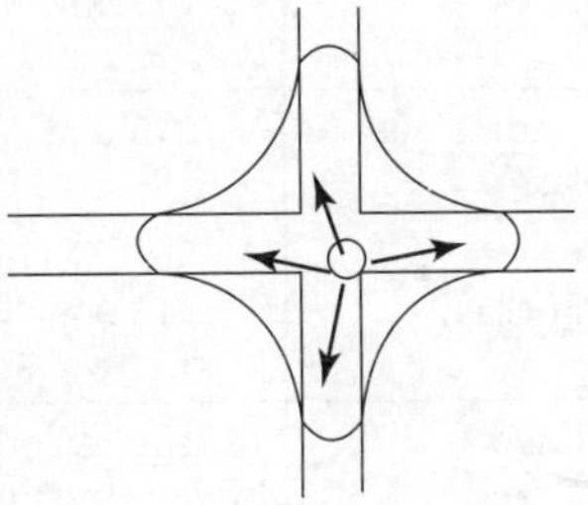 | At small to medium-sized intersections; the lamp is not near the centre of the intersection |
| Type 3 | 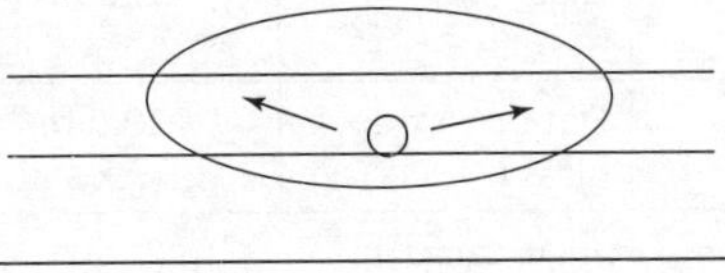 | Shorter but wider than Type 2;<br>at wide streets of width up to 2.7 times the mounting height and multiple-lane highways for continuous lighting purpose; crosswalks, pathway; bikeway |

(Continued)

**Table 19.3** (Continued)

| *Type* | *Coverage geometry* | *Applicability* |
|---|---|---|
| Type 4 | 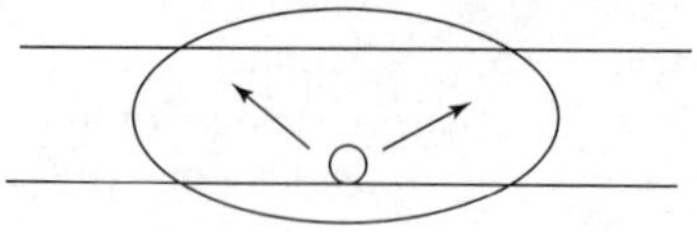 | Shorter but wider than Type 3; at roadway width up to 3.7 times the mounting height; at multi-fixture configurations used with high mast lighting applications, large intersections; better uniformity with staggered–opposite pole configuration like Type 3; isolated lighting |
| Type 5 | 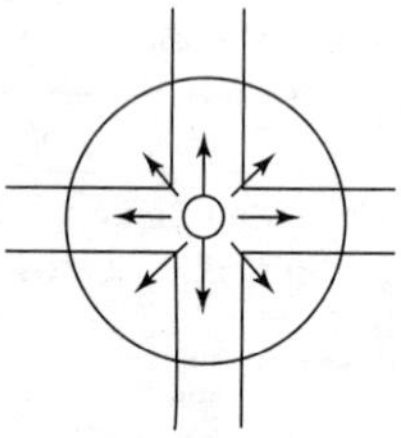 | Circular and wide symmetrical pattern of light distribution; high mast luminaire; in the interior of parking lots, parks and rest areas; isolated lighting |

Pole height, setback, arm length, spacing, cut-off type of luminaire and location of the roadway will influence the selection of luminaire with the appropriate lighting distribution (Alberta, 2003).

As a standard guideline, the applicability of luminaire distribution with the type of lamp-luminaire and its pole height may be suitably selected from Table 19.4.

**Table 19.4** Typical pole and luminaire combinations

| *Lamp wattage and type* | *Line current (A) at 240 V* | *Luminaire type* | *Luminaire refractor type* | *Standard luminaire mounting height* | *IESNA luminaire distribution types* |
|---|---|---|---|---|---|
| 250 W HPS (clear) | 1.3 | Cobra head | Standard Sag Glass | 13.0 m | Type 2 or 3, medium semi-cut-off |
| 250 W HPS (clear) | 1.3 | Cobra head | Flat Glass | 13.0 m | Type 2 or 3, cut-off |
| 400 W HPS (clear) | 1.9 | Cobra head | Standard Sag Glass | 15.0 m | Type 2 or 3, medium semi-cut-off |
| 400 W HPS (clear) | 1.9 | Cobra head | Flat Glass | 15.0 m | Type 2 or 3, cut-off |
| 150 W HPS (coated) | 0.8 | Wall mount | Glass | As required on bridge structure | Type 2 or 3 |
| 100 W HPS (clear) | 0.6 | Post top | Glass | 6.0 m | Type 3 or 5 |
| 1000 W HPS (clear) | 4.8 | High mast | Flat or convex glass | 30.0 m | Type 4 or 5, cut-off |

A = Ampere. (*Source*: Alberta, 2003; *Courtesy*: Alberta Transportation, Canada)

## 19.10.2 Vertical Light Distribution

Light distribution in the vertical plane is characterised by the distance from the luminaire to the point where the beam of maximum candle power strikes the roadway surface. Vertical light distribution is classified into three groups: short, medium and long (Fig. 19.15).

The maximum beam length between adjacent luminaries is joined along the roadway surface and the maximum spacing (S) between the luminaries is determined. In general, medium distribution is often used, as the spacing (S) should not exceed 5–6 times the mounting height (MH). Short distribution is not preferred extensively for reasons of economy; long distribution is also not used because of the glare produced by the high beam angle of maximum cd power (MnDOT).

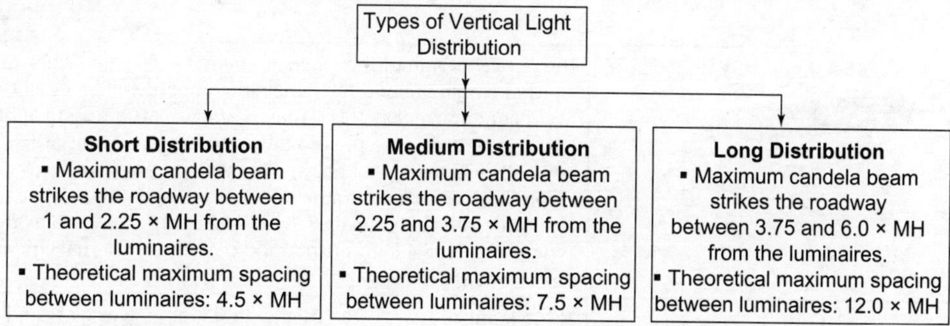


**Fig. 19.15** Classification of vertical light distribution (from MnDOT, 2010)

## 19.10.3 Luminaire Cut-off Classification

Numerous types of luminaire fixtures are available in the market. The performance characteristics of each type and their lamp combinations are unique, and such photometric data can be obtained from the manufacturer. Therefore, it is important to select the optimum luminaire type for the optimum pole.

Luminaires are designed to provide different degrees of control of light distribution, with different photometric criteria, using an appropriate optical arrangement (Fig. 19.16). The Illuminating Engineering Society of North America (IESNA) categorised the limits of light distribution control applicable to outdoor lighting into four types: non–cut-off, semi–cut-off, cut-off and full cut-off. Each cut-off type is described in Table 19.5 (IESNA, 2000):

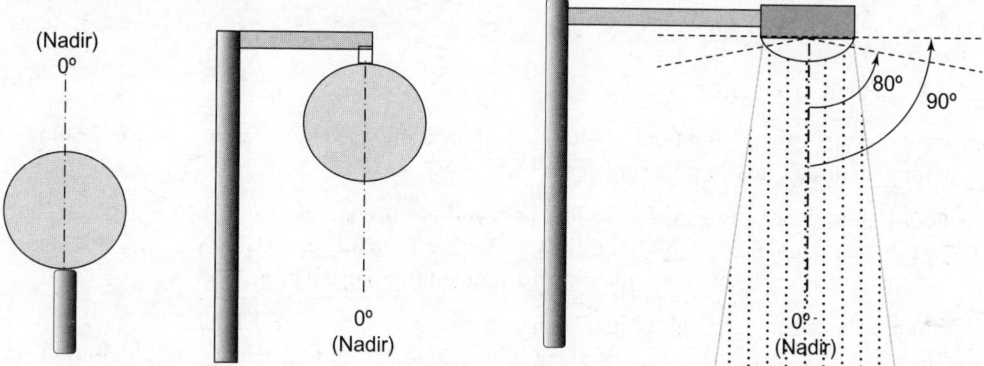


**Fig. 19.16** Details of angles relating to IESNA light control cut-off classification

**Table 19.5** IESNA cut-off classification of control of outdoor light distribution

| *Illustration of light control* | *Description with applications* |
|---|---|
| Non–cut-off | The optical design of the luminaire permits light to emit in all directions around its periphery; it is most commonly used in a Type 5 light distribution pattern. It may create light pollution and cause glare. It is not used on thoroughfare where the glare affects the drivers' vision. Example: Post-top luminaires |
| Semi–cut-off | The optical design of the luminaire permits 95% of its light to be emitted within 90° and the remaining amount will be permitted to up-light above 90°. This design is effective at diffusing light over a large area onto the ground. As these luminaires are generally erected on tall poles, glare will not affect the drivers' vision. It is preferred in tunnels and underpasses and also used for all cobra-head luminaires in rural locations. It is used as for Type 3 and Type 4 light distribution patterns for partial lighting arrangements. |
| Cut-off | These luminaires are often used with flat or convex glass and less than 2.5% of the light is permitted as up-light above 90°. They are also mounted on high masts where pole spacing is preferred. They are preferred in parks and urban areas to limit light pollution, such as sky glow, light trespass and glare. They are used for Type 1 light distribution pattern. |
| Full cut-off | The optical design of the luminaire permits light to flow onto the defined ground space, i.e., only within 90° as a continuous lighting arrangement. With this advantage, more area may be covered on the roadway (as with cut-off/semi–cut-off) by increasing the mounting height of the luminaire. Full cut-offs are preferred near airports. |

*Note:* The standard and general practice is to use semi–cut-off cobra head luminaires with high-pressure sodium lamps on davit poles.

## 19.10.4 Types of Poles

Numerous varieties of streetlight poles with bracket arms are used in different countries. Some of the most commonly used types of poles are described below.

**Mild steel poles:** They are available at low cost and are preferably installed using the anchor bolt system. At higher maintenance cost, their lifespan ranges 15 to 30 years. However, they are susceptible to corrosion and require heavy equipment for handling during installation.

**Aluminium poles:** They are lightweight, non-corrosive and may have a longer lifespan of up to 50 years. However, they have low strength and are suitable only for limited heights for mounting. They can be embedded or anchored to a base plate concrete footing.

**Cement concrete poles:** They are strong, less susceptible to wind vibrations, non-corrosive and may have a long lifespan of up to 50 years. They should be installed as embedded poles. Heavy equipment may be required during handling and installation.

**Wooden poles:** They are lightweight but subject to decay and low strength. Hence, they are not in use widely.

**Galvanised iron poles:** They consist of thin-walled tapered hollow tubes. These are the poles most commonly used in India.

**Steel poles:** I-section or C-section or tubular poles are the common types of poles used for street lighting.

**Fibreglass poles or glass-fibre reinforced plastic poles:** These hollow tubes are made of plastic material reinforced with glass fibres. They are lightweight, easy to install, non-conductive, non-corrosive and rust resistant. In addition, their lifespan ranges 15 to 20 years with low maintenance and they not susceptible to harmonic vibration. Their disadvantages may be a susceptibility to UV damage, delaminating skin, lack of recycle value and expensive disposal. They are generally used with limited mounting heights. They can be embedded or anchored to a base plate fixed on cement concrete footing.

**Carbon fibre reinforced polymer/plastic poles:** These hollow tubes are made of complex material comprising a polymer matrix reinforced with carbon fibres.

**Self-break posts or breakaway posts:** These are specially designed posts. When a car hits, they break and cause no danger to the occupants. These posts meet AASHTO 2001 breakaway requirements. They are used when speeds exceed 40 miles per hr and are located within the roadway. Breakaway posts are installed in a suitable base plate, such as stainless steel progressive shear base with a stainless steel shaft, frangible cast aluminium transformer base with an aluminium pole shaft and arm, a slip base pole, and an aluminium shoe base plate (MnDOT, 2010).

*Note:* The current common practice is to use poles that have round conical or octagonal conical cross-sections.

## 19.10.5 Pole Height

Pole height is an important parameter that determines the performance of any street lighting system in terms of the intensity of illumination, uniformity of brightness along the roadway, coverage area and glare.

Shorter poles are less expensive and are typically used for mounting luminaires of lower wattage; however, uniformity of lighting along the roadway may be affected and the possibility of glare affecting the driver's vision does exist.

As mounting height increases, uniformity of lighting is greater, coverage is wider and so fewer poles need to be installed with adequate setback. This also decreases the glare and dirt, but there may be loss/decrease in the foot-candle level on the roadway surface. Such undesirable decrease in foot candle can be corrected using higher wattage lamps or/and more luminaires per location (UNAM, 1992). As a guideline to preserve longitudinal uniformity, the space–height ratio of luminaires should generally be greater than 3 (ECO-3, 2010).

Typical pole heights are 7.5 m, 9.0 m, 12.0 m and 15.0 m. High-mast poles used for mounting single or multiple 1000-W luminaires have heights varying from 30 m to 45 m. They are also used in wide/complex intersections/interchanges, sports facilities, commercial and other critical locations. Their use is restricted to near power lines, residential neighbourhoods and airports.

As a thumb rule, shorter poles are used for narrower roads and close to residential areas while longer poles are used for wider roads (Gazzar, 2008). However, the mounting height of luminaires along a proposed street layout should be checked by computing the *point of minimum illumination criteria* using the photometric data of the lamp-luminaire characteristics.

### 19.10.6 Pole Arms

Pole arms hold and support luminaires at a lateral distance from the pole (Table 19.6). The arm length is usually 1.8 m, 2.7 m or 3.6 m. The davit and upsweep arm types are found to have a common arm length of 1.8 m. The truss arm is used for longer projection distance. Hover and overhang should not exceed 1/4th of the mounting height, to avoid reduced visibility of kerbs, obstacles and footpaths (CoC, 2003).

**Table 19.6** Some commonly used pole arms

| *Arm type* | *Single* | *Double* |
|---|---|---|
| Davit<br>Pipe arm length = 1.8 m, 2.7 m, 3.65 m.<br>On 12 m, 15 m poles, spacing:73–76 m | 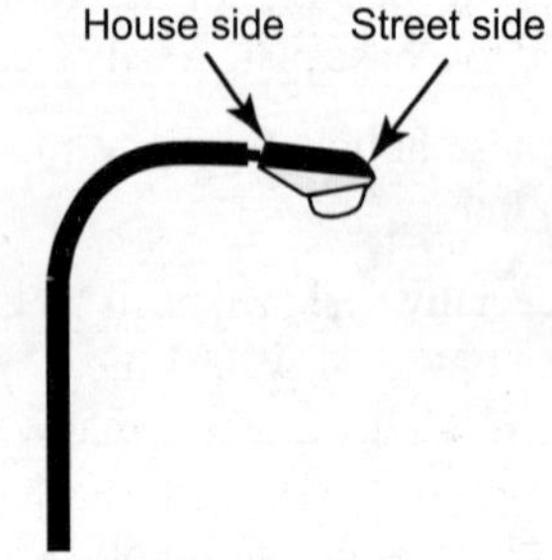 House side Street side | 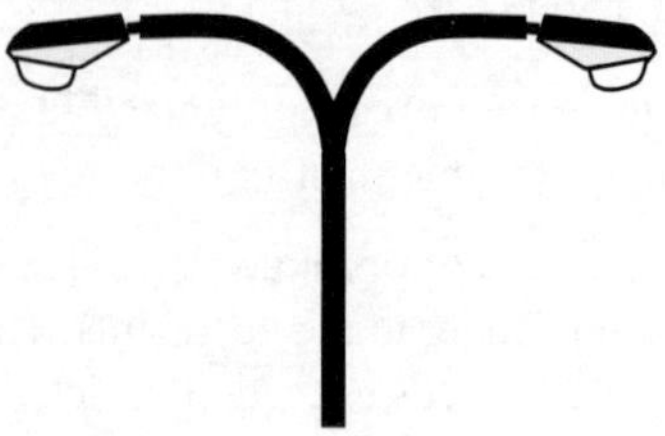 |
| Tapered elliptical | 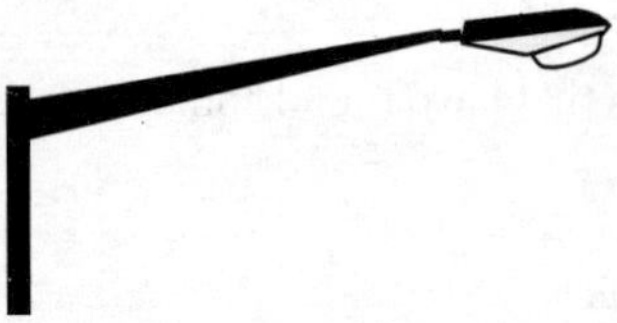 | 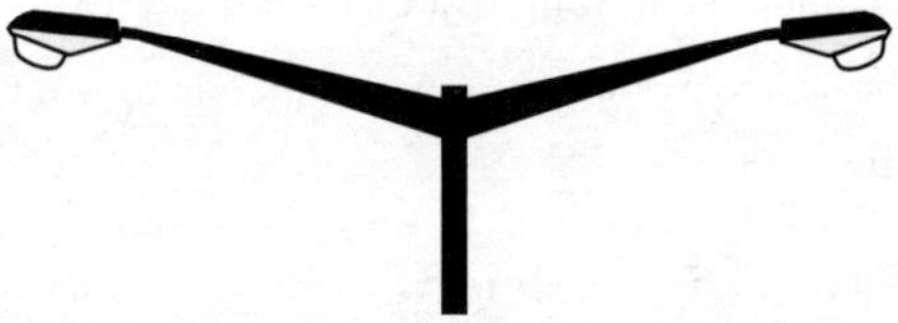 |
| Truss arm (pipe arm length >2.7 m) | 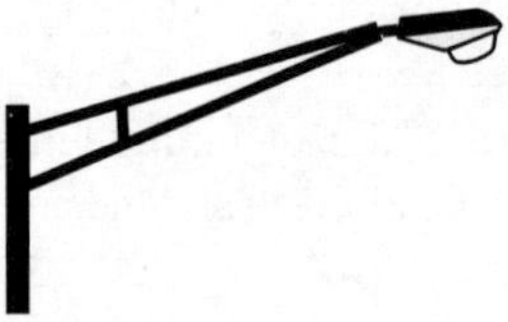 | 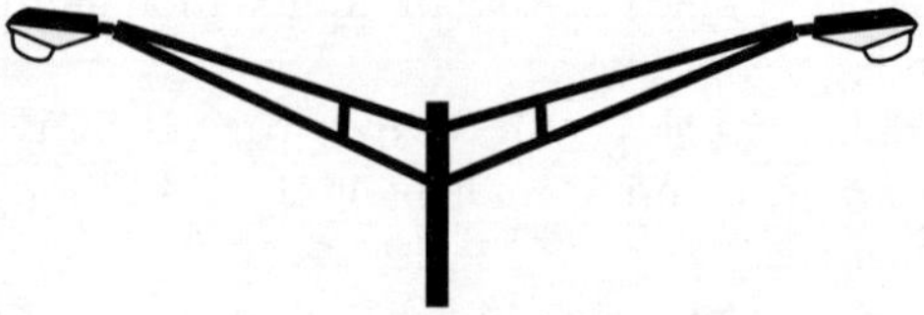 |
| Double guy | 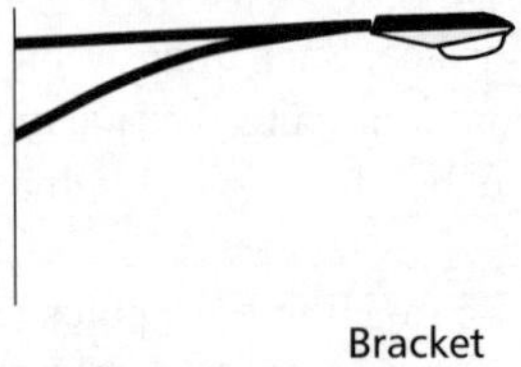 Bracket | 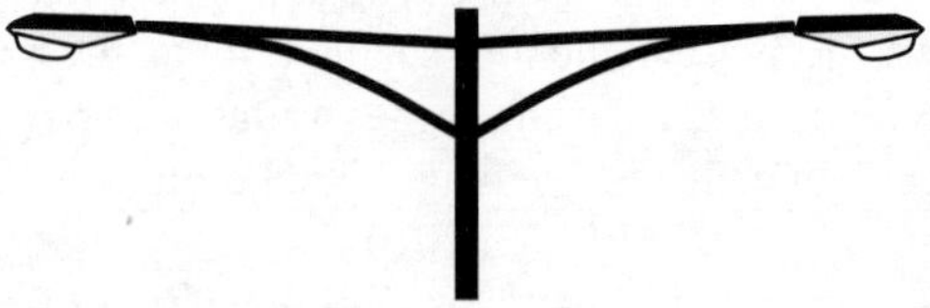 |

(Continued)

**Table 19.6** (Continued)

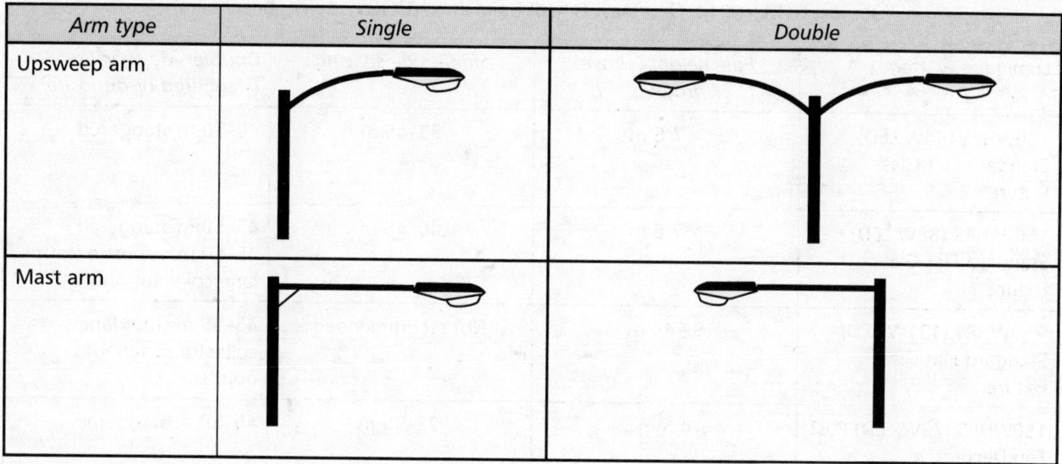

| *Arm type* | *Single* | *Double* |
|---|---|---|
| Upsweep arm | | |
| Mast arm | | |

Other types of pole arms such as ornamental and decorative pole arms are also in use.
*Note:* The vertical mount pole does not have any arms.

## 19.10.7 Mounting Height and Type of Luminaire

Details of the designation of pole height, mounting height, arm projection length, overhang, setback and outreach are illustrated in Fig. 19.17. There can be a difference between the height of the pole and the mounting height of the luminaire. If a 9-m-high pole is placed on a 50-cm-high concrete pedestal, the mounting height is 9.5 m.

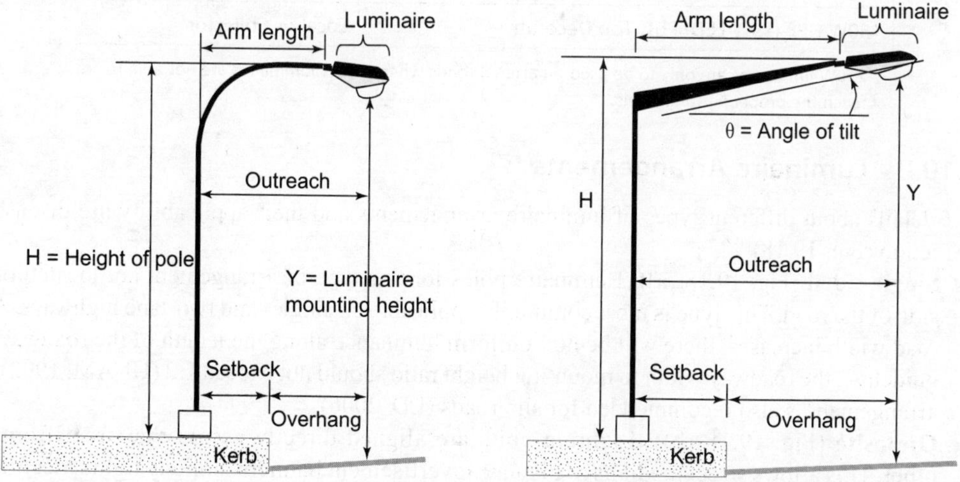


**Fig. 19.17** Component details of electric poles mounted with luminaire (cobra head)

Lighting equipment combinations along with details of lamp wattage, luminaire type, cut-off type, pole height and spacing may be selected from standard guidelines (Tables 19.4, 19.7 and 19.8).

**Table 19.7** Pole height and spacing

(Source: COSJ, 2014; Courtesy: City of Saint John (COSJ), www.saintjohn.ca)

| *Luminaire wattage* | *Pole height (above ground)* | *Single side spacing* | *Double side spacing (staggered or opposite)* |
|---|---|---|---|
| 100WHPS (44W LED) Standard Flat Glass Fixture | ≥ 7.6 m | 35–45 m | 40–50 m staggered |
| 150W HPS (88W LED) Standard Flat Glass Fixture | ≥ 7.6 m | 40–45 m | 45–50 m staggered<br>45–50 m opposite (four lane collector only) |
| *250WHPS (131W LED) Standard Flat Glass Fixture | 9.144 m | Not recommended | 45–55 m (four lane collector /arterial) opposite |
| 150WHPS (75W LED) Post Top Decorative | 4.5 m | 28–35 m | 45–50 m staggered |

*250W lumlnaires are only to be used on artenal roads where 150W luminaires are not able to reach the proper lighting levels.

**Table 19.8** Luminaire wattage

(Source: COSJ, 2014; Courtesy: City of Saint John (COSJ), www.saintjohn.ca)

| *Luminaire wattage* | *Roadway classification* |
|---|---|
| 100W HPS (44W LED) Standard Flat Glass Fixture | Rural or Local |
| 150W HPS (88W LED) Standard Flat Glass Fixture | Rural, Local or Collector |
| *250W HPS (131W LED) Standard Flat Glass Fixture | Arterial |
| 150W HPS (75W LED) Post Top Decorative | Local or Collector |

* 250W lumlnaires are only to be used on arterial roads where 150W lumlnaires are not able to reach the proper lighting levels.

## 19.10.8 Luminaire Arrangements

Brief details about different types of luminaire arrangements and their applicability are discussed in this section (Fig. 19.18):

- **Single-sided** (Fig. 19.18 (a)): Luminaire poles for this type of arrangement are located on one side of the road. This type is most commonly applicable on single- and two-lane highways. As the road width increases, there will be non-uniform luminance along the length of the roadway. As a guideline, the roadway width to mounting height ratio should not exceed 1.2 (UNAM, 1992). This arrangement is also recommended for slip roads (LD, 2006).
- **Opposite** (Fig. 19.18 (b)): Lighting units are aligned directly across the street from each other. This allows street medians to display advertisement banners, holiday lights, decorations and sheet barriers. As a guideline, opposite arrangement is preferred when the width of the road is greater than 1.5 times or starts approaching twice the mounting height (LD, 2006; USNAVY, 1998).

- **Staggered or zigzag** (Fig. 19.18 (c)): Staggering lighting posts across the street produces better theoretical uniformity with greater spacing than the one-side arrangement. This condition holds good when the width of the road is significantly greater than the mounting height (USNAVY, 1998). Opposite and staggered arrangements are most commonly applicable in multilane highways with narrow medians. The designer may explore the possibility of attaining the required level of uniformity using a staggered arrangement, if the roadway width to mounting height ratio is between 1 and 1.5 times, and does not exceed 2.5 (LD, 2006; UNAM, 1992).
- **Twin central** (Fig. 19.18 (e)): This is most commonly applicable for dual carriageways and multilane highways with wider medians. The poles are located on the central reserve and each pole accommodates two luminaires in back-to-back orientation on either side of the carriageway. This type of arrangement is economical but may pose potential hazards during mobile operation or repair.
- This type of luminaire arrangement on a central reserve when used with opposite arrangement is termed combined twin central and opposite arrangement. It is suitable when hard shoulders are provided in expressways, with exceptionally wide carriageways (LD, 2006).

  These lighting arrangements can be provided as a continuous lighting system.
- **Midway suspended or single-central** (Fig. 19.18 (d)): This is suitable for underpasses used as bikeways and walkways. In standard practice, a luminaire spacing of 3.0 m may be provided for pedestrian and bicycle tunnels having walkways that are 3.0 m wide (Alberta, 2003).

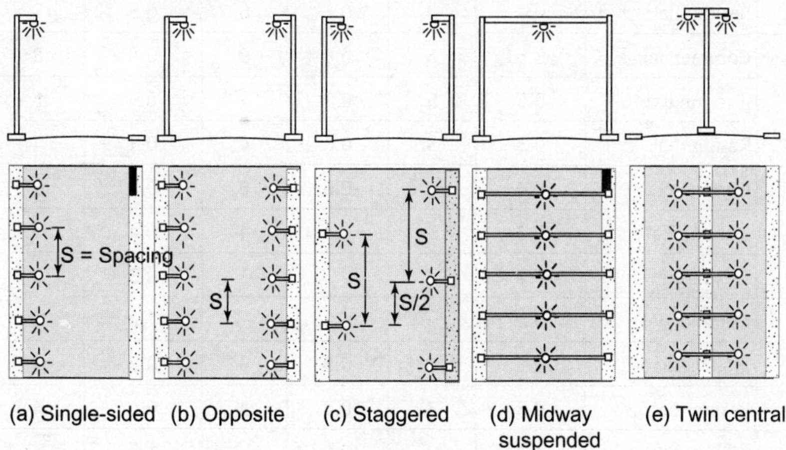


**Fig. 19.18** Common formats of installation of street lights along a roadway segment

## 19.10.9 Lighting Level Requirements

Lighting level criteria are provided with specific conditions relating to roadway class, design speed and type of use. Some commonly adopted lighting levels for different types of roadway class in the USA and Canada are listed in Tables 19.9 and 19.9 (a), but may require adjustments for special situations such as for pedestrian and bicycle tunnels (Table 19.10), walkways and bikeways (Table 19.11), rest areas and vehicle inspection stations (Table 19.12) and for overhead signs (Table 19.13).

**Table 19.9** Minimum average maintained illumination and maximum uniformity ratios by facility classification and pavement classification

| | | *R1* | | *R1 and R3* | | *R4* | | *Max Unif.* |
|---|---|---|---|---|---|---|---|---|
| *Roadway and walkway classification* | | *foot candles* | *lux* | *foot candles* | *lux* | *foot-candles* | *lux* | *avg/mm* |
| Interstate 3 no Uln6f Freeways | Commercial | 0.6–1.1 | 6–12 | 06–1.1 | 6–12 | 06–1.1 | 6–12 | 3:1 or 4:1 |
| | Intermediate | 0 6–0.9 | 6–10 | 06–0.9 | 6–10 | 0.6–0.9 | 6–10 | 3:1 or 4:1 |
| | Residential | 0 6–0 8 | 6–8 | 06–0.8 | 6–8 | 0.6–0.8 | 6–8 | 3:1 or 4:1 |
| Other Principal Artenals | Commercial | 1.1 | 12 | 1.6 | 17 | 1.4 | 15 | 3:1 |
| | Intermediate | 0.8 | 9 | 1.2 | 13 | 1.0 | 11 | 3:1 |
| | Residential | 0.6 | 6 | 0.8 | 9 | 0.8 | 8 | 3:1 |
| Minor Arterial | Commercial | 0.9 | 10 | 1.4 | 15 | 1.0 | 11 | 4:1 |
| | Intermediate | 0.8 | 8 | 1.0 | 11 | 0.9 | 10 | 4:1 |
| | Residential | 0.5 | 5 | 0.7 | 7 | 0.7 | 7 | 4:1 |
| Collectors | Commercial | 0.8 | 8 | 1.1 | 12 | 0.9 | 10 | 4:1 |
| | Intermediate | 0.6 | 6 | 0.8 | 9 | 0.8 | 8 | 4:1 |
| | Residential | 0.4 | 4 | 0.6 | 6 | 0.5 | 5 | 4:1 |
| Local | Commercial | 0.6 | 6 | 0.8 | 9 | 0.8 | 8 | 6:1 |
| | Intermediate | 0.5 | 5 | 0.7 | 7 | 0.6 | 6 | 6:1 |
| | Residential | 0.3 | 3 | 0.4 | 4 | 0.4 | 4 | 6:1 |
| Alleys | Commercial | 0.4 | 4 | 0.6 | 6 | 0.5 | 5 | 6:1 |
| | Intermediate | 0.3 | 3 | 0.4 | 4 | 0.4 | 4 | 6:1 |
| | Residential | 0.2 | 2 | 0.3 | 3 | 0.3 | 3 | 6:1 |
| Sidewaks | Commercial | 0.9 | 10 | 1.3 | 14 | 1.2 | 13 | 3:1 |
| | Intermediate | 0.6 | 6 | 0.8 | 9 | 0.8 | 8 | 4:1 |
| | Residential | 0.3 | 3 | 0.4 | 4 | 0.4 | 4 | 6:1 |
| Pedestrian Ways and Bike Ways | | 1.4 | 15 | 2.0 | 22 | 1.8 | 19 | 3:1 |
| Rest Areas | Roadways | . | . | 0.6–0.8 | 6–9 | . | | 3:1 or 4:1 |
| | Parking Areas | . | . | 1.0 | 11 | . | . | 3:1 or 4:1 |

Notes:
1. R1 = cement/concrete
2. R2 = asphalt/gravel and R3 = asphalt/rough texture (typical highway).
3. R4 = asphalt/smooth texture

(*Source*: MnDOT, 2010; *Courtesy*: Office of Traffic, Security and Technology, MnDOT, USA)

**Table 19.9 (a)** Roadway lighting design selection criteria

| *Road* | *Pedestrian Conflict Area* | *Average Luminance L(avg) Cd/ m^2* | *Uniformity Ratio Lavg/ Lmin Max. allowed* | *Uniformity Ratio Lmax/ Lmin Max allowed* | *Veiling Luminance Ratio Lvmax/ Lavg Max. allowed* | *Min Maintained Average Illuminance on Pavement R1 in fc* | *Min Maintained Average Illuminance on Pavement R2 & R3 in fc* | *Min Maintained Average Illuminance on Pavement R4 in fc* | *Uniformity Ratio Eavg/ Emin* | *STV Criteria Weighting Average VL* | *LavgCd/m ^2 Median <7.3 m* | *Lavg Cd/ m^2 Median >=7.3 m* |
|---|---|---|---|---|---|---|---|---|---|---|---|---|
| Freeway Class A | | 0.6 | 3.5 | 6 | 0.3 | 0.6 | 0.9 | 0.8 | 3 | 3.2 | 0.5 | 0.4 |
| Freeway Class B | | 0.4 | 3.5 | 6 | 0.3 | 0.4 | 0.6 | 0.5 | 3 | 2.6 | 0.4 | 0.3 |
| Expressway | High | 1 | 3 | 5 | 0.3 | 1 | 1.4 | 1.3 | 3 | 3.8 | 0.5 | 0.4 |
| Expressway | Medium | 0.8 | 3 | 5 | 0.3 | 0.8 | 1.2 | 1 | 3 | | | |
| Expressway | Low | 0.6 | 3.5 | 6 | 0.3 | 0.6 | 0.9 | 0.8 | 3 | | | |
| Major | High | 1.2 | 3 | 5 | 0.3 | 1.2 | 1.7 | 1.5 | 3 | 4.9 | 1 | 0.8 |
| Major | Medium | 0.9 | 3 | 5 | 0.3 | 0.9 | 1.1 | 1.1 | 3 | 4 | 0.8 | 0.7 |
| Major | Low | 0.6 | 3.5 | 6 | 0.3 | 0.6 | 0.9 | 0.8 | 3 | 3.2 | 0.6 | 0.6 |
| Collector | High | 0.8 | 3 | 5 | 0.4 | 0.8 | 1.2 | 1 | 4 | 3.8 | 0.6 | 0.5 |
| Collector | Medium | 0.6 | 3.5 | 6 | 0.4 | 0.6 | 0.9 | 0.8 | 4 | 3.2 | 0.5 | 0.4 |
| Collector | Low | 0.4 | 4 | 8 | 0.4 | 0.4 | 0.6 | 0.5 | 4 | 2.7 | 0.4 | 0.4 |
| Local | High | 0.6 | 6 | 10 | 0.4 | 0.6 | 0.9 | 0.8 | 4 | 2.7 | 0.5 | 0.4 |
| Local | Medium | 0.5 | 6 | 10 | 0.4 | 0.5 | 0.7 | 0.6 | 4 | 2.2 | 0.4 | 0.3 |
| Local | Low | 0.3 | 6 | 10 | 0.4 | 0.3 | 0.4 | 0.4 | 4 | 1.6 | 0.3 | 0.3 |

(*Source*: Gazzar; *Courtesy*: www.dewberry.com)

**Table 19.10** Lighting levels for pedestrian and bicycle tunnels

| | $E_H$ (lux) | $E_{vmin}$ | $E_{av}/E_{min}$ |
|---|---|---|---|
| Day | 100 | 50 | 3.0 |
| Night | 40 | 20 | 3.0 |

$E_H$ = Illuminance; $E_{vmin}$ = Minimum vertical illumination 1.5 m above the grade measured in both directions parallel to main pedestrian flow; $E_{av}$ = Average horizontal illumination on the horizontal plane; $E_{min}$ = Minimum illuminance

(*Source*: Alberta, 2003; *Courtesy*: Alberta Transportation, Canada)

**Table 19.11** Lighting level for walkways and bikeways

| *Area class* | $E_H$ (lux) | $E_{vmin}$ | $E_{av}/E_{min}$ |
|---|---|---|---|
| Urban | 20.0 | 10.0 | 4.0 |
| Suburban | 5.0 | 2.0 | 4.0 |
| Rural | 2.0 | 0.6 | 10.0 |

(*Source*: Alberta, 2003; *Courtesy*: Alberta Transportation, Canada)

**Table 19.12** Illumination levels for rest areas and vehicle inspection stations

| | | *Rest areas* | | *Vehicle inspection stations* | |
|---|---|---|---|---|---|
| *Area description* | *Luminaire type* | *Lux (avg)* | *Uniformity (avg:min)* | *Lux (avg)* | *Uniformity (avg:min)* |
| Interior roads (Ingress and Egress roads) | Cobra head (flat glass) | 12.0 | 6:1 | 15.0 | 6:1 |
| Parking Area | Cobra head (flat glass) | 12.0 | 6:1 | 20.0 | 3:1 |
| Public Areas (walkways, washrooms, telephone, info areas) | Post top | 22.0 | 6:1 | N/A | N/A |
| Inspection Area | Cobra head (flat glass) | N/A | N/A | 30.0 | 3:1 |
| Lighting Weigh Scales | Wall mount on buildings or cobra head | N/A | N/A | 50.0 | 3:1 |
| On/Off Ramps | | Refer to TAC for luminance levels. | | | |

Certain site conditions may require modification of lighting levels as defined by Alberta Transportation

(*Source*: Alberta, 2003; *Courtesy*: Alberta Transportation, Canada)

**Table 19.13** Illumination and luminance level for overhead signs

| *Area class* | *Maintained illuminance* | *Maintained luminance (cd/m²)* |
|---|---|---|
| Rural | 13.0 | 20.0 |
| Suburban | 26.0 | 40.0 |

(Continued)

**Table 19.13** (Continued)

| *Area class* | *Maintained illuminance* | *Maintained luminance (cd/m²)* |
|---|---|---|
| Urban | 52.0 | 80.0 |

(*Source*: Alberta, 2003; *Courtesy*: Alberta Transportation, Canada)

## 19.10.10 Methods of Spacing of Lighting Units

The design of any roadway lighting must be site specific and must provide the level and uniformity of illumination required to meet the needs of a road class (Table 19.10). The lighting design can be carried out using different methods, and they may be broadly categorised into three types: luminance method, illuminance method of lighting design and small target visibility study.

The luminance method is based on a measure of the light reflected from the roadway surface (brightness of the pavement surface (cd/m$^2$)), which is visible to the driver. The illuminance method is based on a measure of the amount of light incident on the pavement surface (in lux). Small target visibility is the weighted average luminance of the target and it considers factors such as luminance of the target, luminance of the immediate background, adopted level of the adjacent surroundings and light pollution in terms of glare and light trespass (Gazzar 2008).

**Illuminance methods:** The longitudinal spacing requirement may be calculated using the following formula:

$$S = \frac{LL \times CU \times LLD \times LDD}{E_h \times B} \tag{19.4}$$

where $S$ = spacing between successive poles, measured parallel to the centre line of the roadway; $LL$ = initial lamp output in lumens (obtained from the photometric data of a selected light source as supplied by the luminaire manufacturer); $CU$ = coefficient of utilisation factor, determined using the utilisation curves that belong to the lamp type and roadway class; $LLD$ = lamp lumen depreciation factor; $LDD$ = lamp dirt depreciation factor; $B$ = width of the lighted roadway in feet or metres, measured kerb-to-kerb or edge-to-edge of the pavement, or the edge of pavements for all open roadways; and $E_h$ = average maintained level of horizontal illumination on the roadway (Tables 19.9 and 19.10–19.13). The average required illumination level depends on the roadway classification, surface material characteristics of the roadway and the usage type of the area.

The $E_h$ input may be used in units of lux or fc. If the lux value is used, the $B$ value should be substituted in metres, and the output of the spacing of poles will be in metres; otherwise, if fc is used, the $B$ value should be substituted in feet and the output of spacing will be in feet.

Eq. (19.4) calculates the spacing requirement based on the average level of illumination ($E_h$ in lux = 1 m/m$^2$ or fc = 1 m/ft$^2$) on the area of roadway. This is determined by calculating the illumination levels at all critical locations in a discrete area – a study area of interest, say along a section length of roadway.

## 19.10.11 Lateral Placement of Poles (Luminaire)

Lamp posts can be viewed as obstacles near the roadway which may be a danger to road users. From the safety point of view, the poles are set an adequate distance apart, laterally away from traffic flow.

The pole setback distance, measured from the face of the kerb, is often termed pole setback. It varies depending on the area and design speed of the roadway. The standard pole setback recommended by Alberta Transportation (2003) is given below:

- The pole setback for high-speed roadways (up to 120 km/hr) in rural areas shall be 5.0 m from the edge of the through lane/carriageway to the face of the lighting pole. This distance may be further increased, with appropriate barrier protection, if the design speed exceeds 120 km/hr.
- The pole setback in urban areas shall be 3.0 m from the face of kerb to the centre of the lighting pole. This space, which is adjacent to the carriageway, allows pavement use and placement of landscaping. Based on the space constraints along municipal roadways, the pole setback may be reduced to as low as 0.5 m, where the design or posted speed is up to 60 km/h.
- High-mast lighting assembly setback shall be a minimum of 9 m measured from the face of the kerb or edge of the pavement to the centre line of the pole. A lower setback may be used with approval by providing appropriate protection around the base (NJDOT, 2015).

A few safety-related factors are given below (MnDOT, 2010):

- The pole should not be placed in front of a guide rail. It may be placed at least 0.6 m behind the guard rail or at a distance that will allow the guard rail to deflect safely on impact.
- Poles should not be closer than 6.0 m in any direction from power lines.
- Poles should not be closer than 7.6 m to 15.2 m in any direction from overhead signs.
- Poles should not be placed adjacent to each other or in ditches for maintenance purposes.
- A minimum gap of at least 0.6 m should be maintained adjacent to the pole against any structure/wall/noise barrier, for maintenance purposes.

However, the following site considerations may be used while selecting the appropriate setback distance adjacent to the roadway:

- Existing guard walls
- Narrow shoulders
- Power lines/cables/fibre optic lines
- Nearby airports
- Railway level crossings
- Traffic signals/signs
- Residential neighbourhoods
- Roadside rocks/trees
- Sufficient shoulder width to accommodate a maintenance truck

## 19.11 DESIGN OF ROADWAY/STREET LIGHTING SYSTEM

The design of a lighting system comprises the following steps (MnDOT, 2010):

1. *Objectives*: The lighting needs are assessment based on the roadway class with reference to the minimum requirement of foot-candle (fc) (Tables 19.9 and 19.10–19.13). In this step, the requirements are studied.
2. *Specifications*: The suitable lamp, luminaire, fixtures and poles are selected for the assessed needs. The light source that is selected should meet the site-specific lighting requirements relating to lumen output, efficacy, energy requirements, lamp life, colour, optical controllability, temperature range and environmental effects (Fig. 19.8 and Table 19.1). The mounting height along with the selected light source with minimum illumination and uniformity criteria should be suitably selected from an applicable standard (Tables 19.4, 19.10 and 19.11).
3. *Spacing*: The lateral distance from the roadway edge and longitudinal spacing of the luminaire must be determined. The latter is decided based on the minimum required setback distance and site restrictions/prevailing conditions. The longitudinal spacing between luminaires may be determined by the luminance or illumination method by appropriately selecting the other required input values, such as mounting height, pole arm projection and lamp-luminaire type/efficacy.
4. *Point-by-point analysis and determination of illumination on the roadway*: An appropriate pole configuration should be proposed from a set of alternative layouts of roadway pole

layouts (Fig. 19.18). Approximate foot-candles at selected grid points (say, check-points P1, P2 and P3) on a roadway are determined for each of these proposed alternative layouts of pole configuration, as shown for one-sided (Fig. 19.19), staggered (Fig. 19.20), opposite (Fig. 19.21) and twin-central (Fig. 19.22) configurations. Isofoot-candle curve diagram or inverse-square method is used for calculation of foot-candle values at these check-points from the combined effect of luminaires along the roadway. Isofoot-candle values at these check points on the roadway because of the effect of each luminaire are separately calculated and their respective summation values are reported.

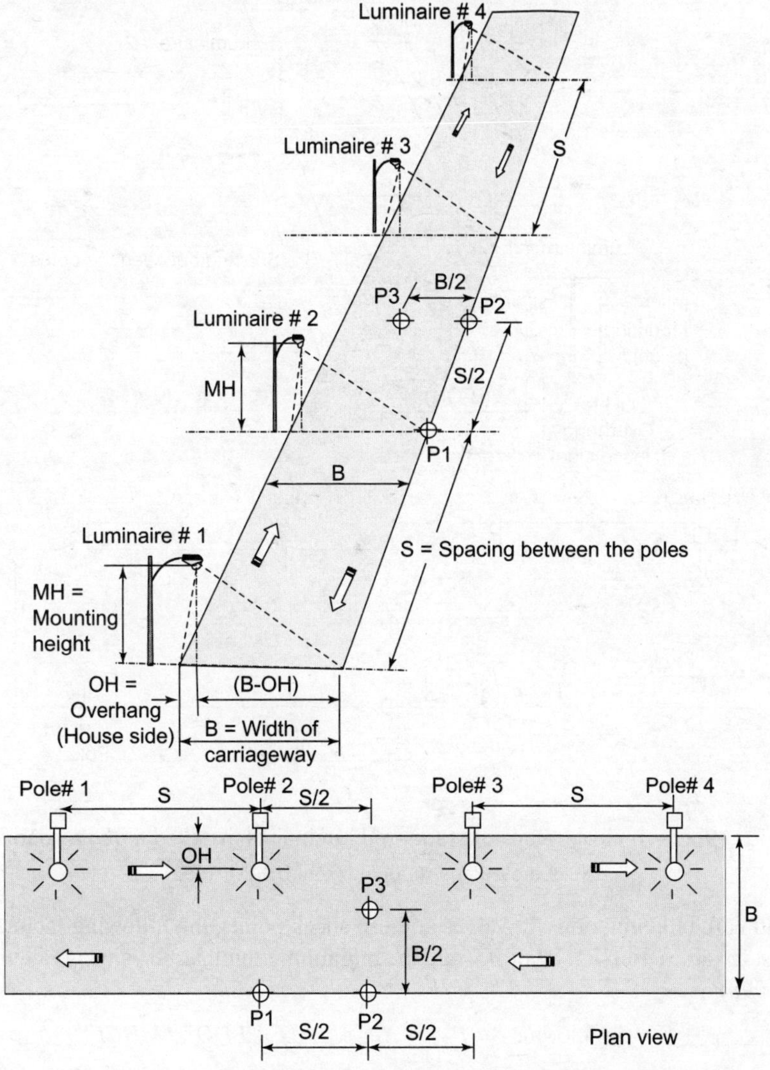


**Fig. 19.19** Typical details of one-sided luminaires installed along a roadway (Elaborated and modified from UNAM, 1992)

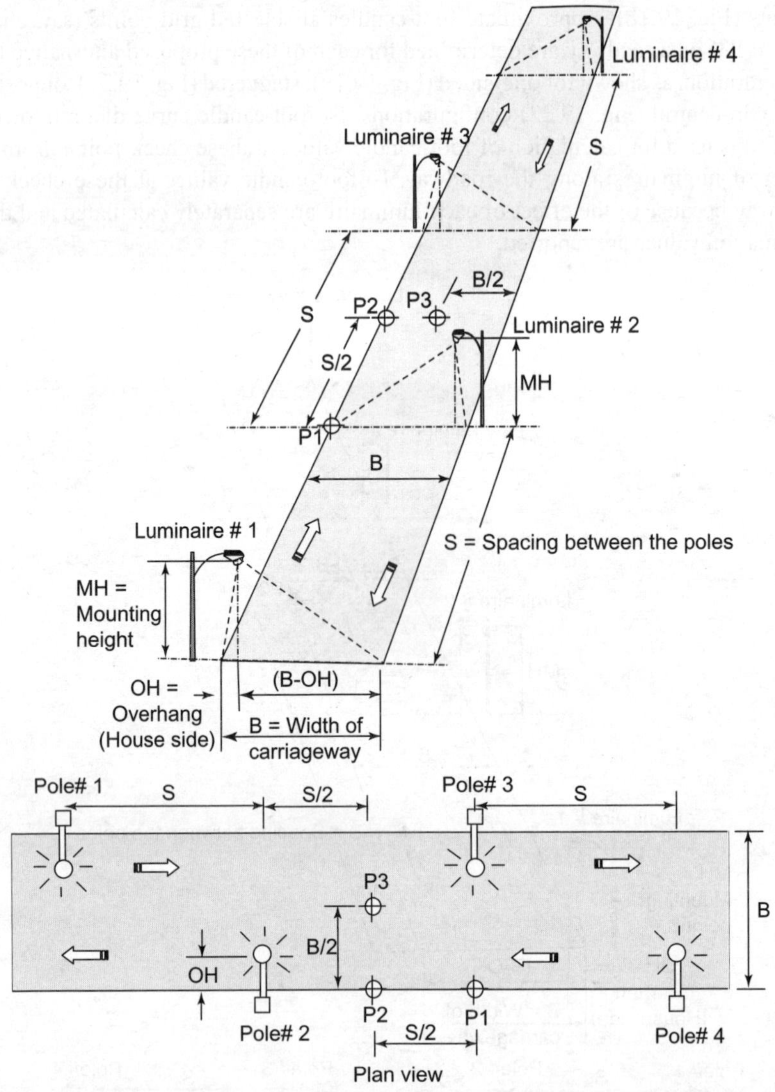


**Fig. 19.20** Typical details of staggered luminaires installed along a roadway
(Elaborated and modified from UNAM, 1992)

After calculating the illumination (fc) values at these check points, the following factors are applied to the *raw* lowest fc value (from P1, P2 or P3) and the minimum illumination value is calculated as follows (USNAVY, 1998):

$$P_{min} \text{ (in fc or lux)} = P_{imin} \times LF \times LLD \times LDD \times MHCF. \tag{19.5}$$

where $P_{imin}$ = raw value of the point of minimum illumination (fc or lux), i.e., the darkest spot, which is selected from P1, P2 or P3; $LF$ = lamp factor; $LLD$ = lamp lumen depreciation factor; $LDD$ = lamp dirt depreciation factor; and $MHCF$ = mounting height correction factor, for other mounting heights of the luminaire.

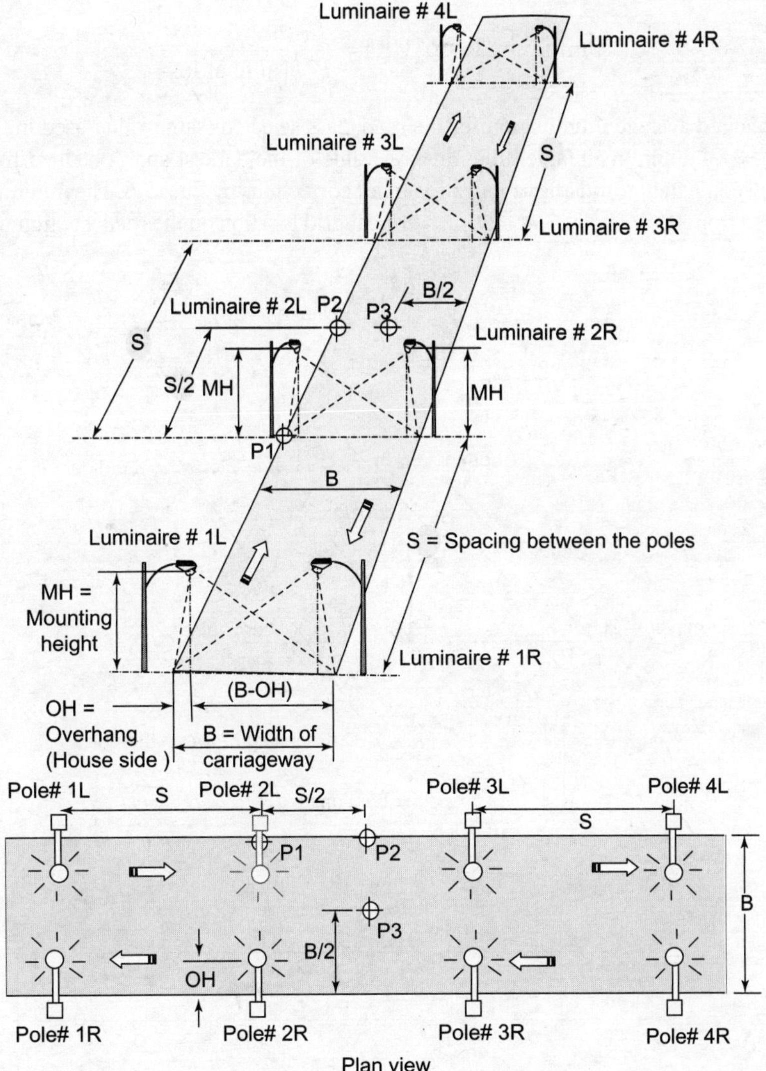


**Fig. 19.21** Typical details of opposite luminaires installed along a roadway (Elaborated and modified from UNAM, 1992)

5. *Design precision based on the uniformity ratio*: The quality of illumination along the roadway is determined based on the minimum illumination criterion by calculating the uniformity ratio (UR).

The AASHTO Roadway Lighting Design Guide defines UR as the ratio of the average maintained horizontal illumination (lx or fc) to the darkest spot (lx or fc) on the roadway at the end of the rated life of the lamp, after all corrections. The UR value at any point on the roadway lighting map is determined using the following equation:

$$\text{Uniformity Ratio}\left(\text{UR}\right)=\frac{E_h\left(\text{in fc or lux}\right)}{P_{min}\left(\text{in fc or lux}\right)} \tag{19.6}$$

where $E_h$ = required average illumination for the road class selected (same value used in the pole spacing formula); $P_{min}$ = minimum maintained illumination value of the darkest spot, obtained from all the grid/check points by computation and after application of all corrections by Eq. 19.6. The minimum illumination value is selected from P1, P2 and P3 or any such critical grid location on the roadway lighting area.

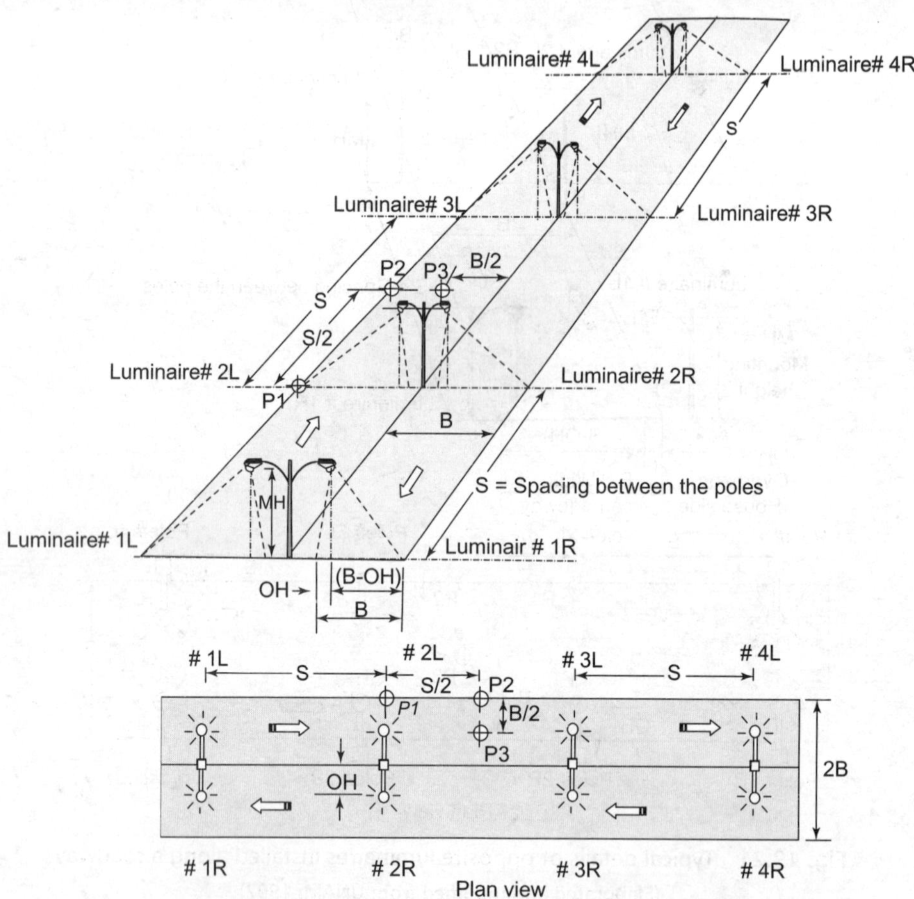


**Fig. 19.22** Typical details of twin central luminaires installed along a roadway (Modified and Elaborated based on UNAM, 1992)

The $E_h$ value (sometimes used in units of illuminance, in lux – E value) is considered the standard requirement of a roadway class and is suitably selected from any applicable specification (Table 19.10).

The calculated minimum ($P_{min}$), maximum ($P_{max}$) and average ($E_h$) values of illuminations within the study roadway grid area are compared to meet the standard criteria of average-to-minimum ratio ($E_{av}/E_{min}$) and maximum-to-minimum illumination ratio ($E_{max}/E_{min}$) values. It must be noted that the $E_{av}/E_{min}$ ratio is important. The calculated values should be equal to or less than the standard recommended criteria. For example, the UR value should be equal to or less than 3 for the criterion of 3:1, or 4 for the criterion of 4:1. For more details, see Table 19.10.

On some occasions, the calculated values of illumination in the study area *exceed* the standard criteria of average-to-minimum ratio ($E_{av}/E_{min}$) and maximum-to-minimum ratio ($E_{max}/E_{min}$). Then, the designer must slightly increase the mounting height, change the projected arm length, decrease the spacing between the luminaire poles based on site conditions and attempt to increase, not only the average level of illumination bu also the minimum level of illumination at all grid/check-points in the roadway.

Alternatively, the designer may replace any proposed luminaire with another lamp of higher wattage, relocate a few luminaire poles or locate additional luminaire poles towards the point of minimum illumination level, or tilt the luminaire without causing glare.

Sometimes, a study area receives more illumination than the maximum specified. This may be due to a concentration of luminaire poles or spill-over light from adjacent/neighbouring sources. Such higher lighting points are identified by calculation and eliminated by relocating the luminaire poles (Gazzar, 2008).

The lighting design process with the sequence of steps, including the design of power distribution systems, are presented in Fig. 19.23.

**Advantage of software tools:** Several software or spreadsheet tools are available to model and determine the optimum spacing of luminaire poles and simulate the street lighting design. Design software use CAD files of the roadway to scale. The photometric characteristics of the proposed luminaire file in the IES format, supplied by the manufacturer, are used. The calculation details of point-by-point illumination levels at various grid locations along the study section length of roadway paved at different grade levels (along ramps) are produced. The pole layout, calculation of illumination levels with applicable corrections and the possible combinations with pole spacing, lateral pole locations, luminaire-lamp type, cut-off type and pole height with arm projection length can be better manipulated accurately and easily within a short period of time. This method is not very different from the manual one, but software makes the design process easier and simpler.

## 19.12 METHODS OF POLE INSTALLATION

Poles can be installed in two ways: embedded or anchored. Poles that are buried in the ground are termed embedded poles, while those that are installed on a base plate fixed in a prepared concrete pedestal foundation using bolts or a welded base plate are anchored. Pole bases can be of three types: tapered high base, anchored base and standard transformer base (MnDOT, 2010).

As a thumb rule, the buried depth of an embedded pole may be about 10% of its length, plus half a metre. The depth of embedding may vary based on the soil type. This is a low cost and time-saving method, which eliminates the use of an expensive anchoring system, as is essential in the case of an anchored base system. Cement concrete poles are commonly mounted as embedded poles. Further, it has been observed that embedded and concrete poles do not experience wind-induced harmonic vibration, and hence, may be considered safe against frequent wind force.

Wind speed maps developed over a period of at least 50 years are used for the design of pole supports. The AASHTO-recommended wind speed with three-second gusts may be considered for pole installations along coastal districts and other wind-intensive zones.

**Ohio DOT-suggested procedure for light tower foundation design:** The following guidelines are useful for determining the caisson depth of tower foundations without the need for extensive soil investigation at each tower location (ODOT, 2014).

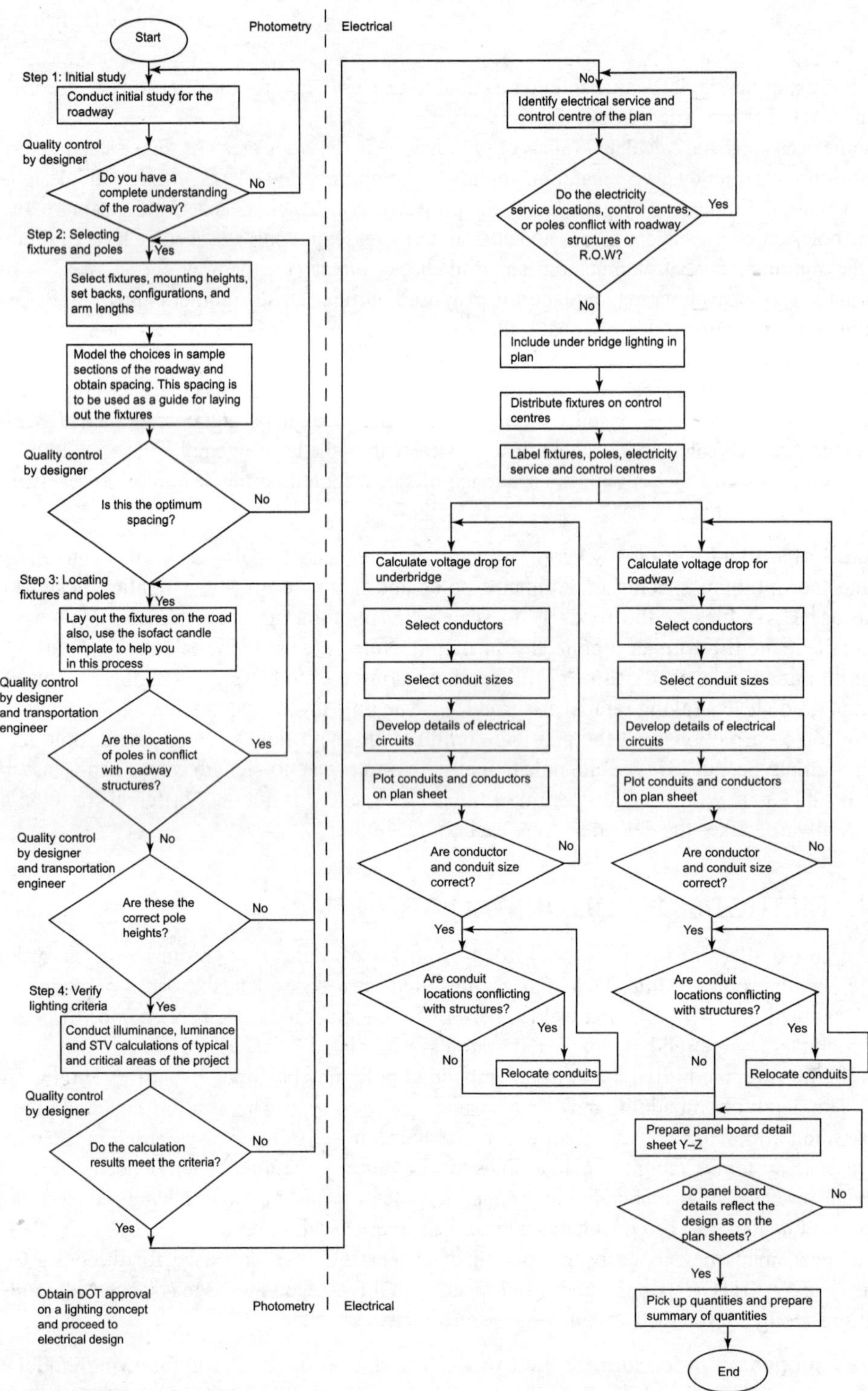


**Fig. 19.23** General procedure adopted for the design of a roadway lighting system
(Source: Gazzar, 2008; Courtesy: www.dewberry.com)

Regardless of the type of foundation, soil pressure is one of the main input parameters of foundation design. Approximate and most probable values of lateral soil pressure can be assumed from Table 19.14. In addition, maximum load reactions on the tower, horizontal shear, and uplift and overturning movements are considered for the design of the foundation for high-mast towers.

**Table 19.14** Recommended lateral soil pressure from foundations

| *Recommended Lateral Soil Pressure (Pounds Per Square Foot Per Foot of Depth)* | |
|---|---|
| *Class of material* | *Value* |
| Rock in Natural Beds - Limited by the Stress in the Pile | |
| Compact Well Graded Gravel | 400 |
| Hard Dense Clay | 400 |
| Compact Coarse Sand | 350 |
| Compact Coarse and Fine Sand | 300 |
| Medium Stiff Clay | 300 |
| Compact Fine Sand | 250 |
| Ordinary Silt | 200 |
| Sandy Clay | 200 |
| Compact Inorganic Sand and Silt Mixtures | 200 |
| Soft Clay | 100 |
| Loose Organic Sand and Silt Mixtures and Muck or Bay Mud | 0 |

(Source: ODOT, 2014; Courtesy: Ohio Department of Transportation, USA)

The design calculations of load reactions and horizontal shear on the tower should comply with the latest AASHTO Standard Specifications for Structural Supports for Highway Signs, Luminaires and Traffic Signals – the design wind load is based on 90 mile/hr, with six luminaires each weighing 35 kg (75 lb) mounted on a horizontal plane with a maximum projected area of 3.5 sq.ft. (ODOT, 2014).

After determining the allowable lateral soil pressure, total lateral force and total moment about the resisting surface of the tower, the nomograph (Fig. 19.24) is used to determine the foundation embedment depth. This nomograph is based on E Czerniak's recommendations for the given lateral soil pressures and classifications. In Ohio, this caisson type foundation is valid where the embedded depth does not exceed 10 times the foundation diameter.

The input values of lateral soil pressure and overturning moments must be rounded-off to a value per foot of caisson width. In general, the foundation diameter is 36 inches, unless the tower anchor base plate and bolt circle require a diameter of 42 inches (ODOT, 2014). The foundation depth obtained from the nomograph should be increased by 1 or 2 ft as soil cover, and the final value must be rounded-off to the next greater depth that is a multiple of a five (when in FPS units).

Alternatively, tower foundation depths for round tapered shafts designed in accordance with the 1975 AASHTO Standard Specifications for Structural Supports for Highway Signs, Luminaires and Traffic Signals can be obtained from Table 19.15, for the following conditions:

- Wind zone category: 90 miles per hour
- Six cylindrical luminaires, each weighing 35 kg (75 lb) mounted on a horizontal plane with a maximum projected area of 3.5 sq.ft: CD = 0.5
- One cylindrical head from the assembly with a projected area of 5.3 sq.ft and 340 lb top-latched lowering device: CD = 1.0

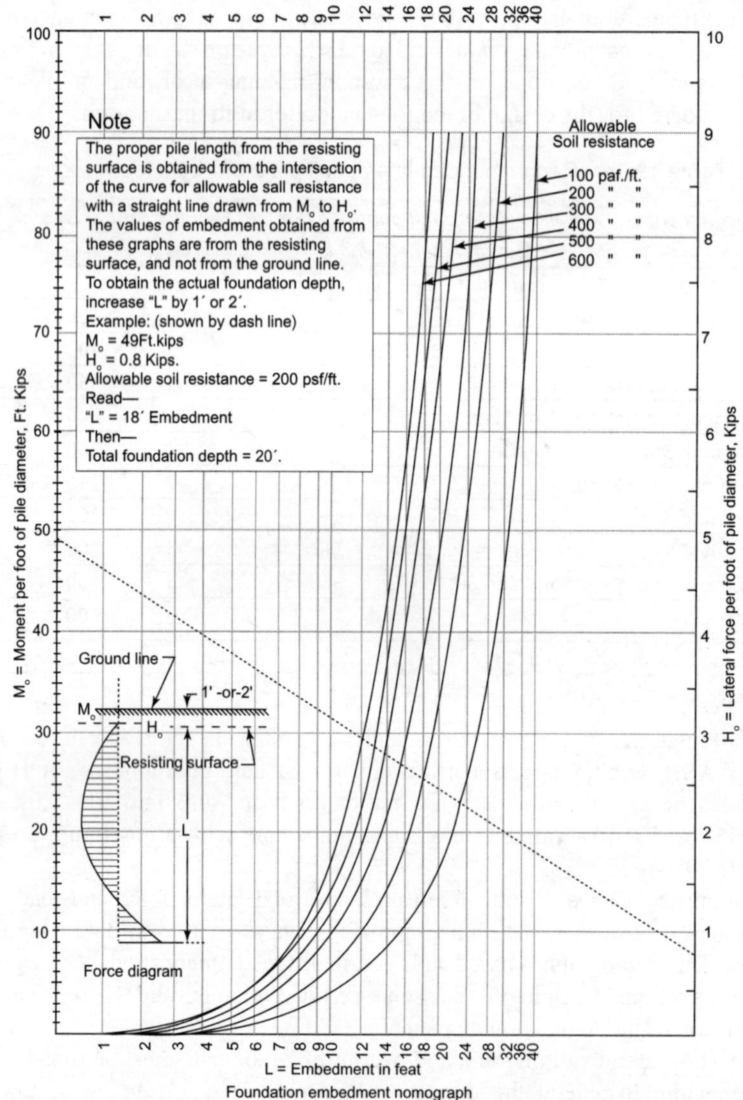


**Fig. 19.24** Foundation embedment nomograph

(Source: ODOT, 2014; Courtesy: Ohio Department of Transportation, USA)

**Table 19.15** Allowable lateral soil resistance

| *Tower Height (Feet)* | *Foundation Diameter (Feet)* | *Allowable Lateral Soil Resistance (psf/ft. of depth)* | | | | | |
|---|---|---|---|---|---|---|---|
| | | *100* | *200* | *300* | *400* | *500* | *600* |
| | | *Foundation Depth (Feet)* | | | | | |
| 70 | 3 | 25 | 20 | 15 | 15 | 15 | 15 |
| 80 | 3 | 25 | 20 | 15 | 15 | 15 | 15 |

(Continued)

**Table 19.15** (Continued)

| *Tower Height (Feet)* | *Foundation Diameter (Feet)* | *Allowable Lateral Soil Resistance (psf/ft. of depth)* | | | | | |
|---|---|---|---|---|---|---|---|
| | | *100* | *200* | *300* | *400* | *500* | *600* |
| | | *Foundation Depth (Feet)* | | | | | |
| 90 | 3 | 25 | 20 | 20 | 15 | 15 | 15 |
| 100 | 3 | 30 | 25 | 20 | 20 | 15 | 15 |
| 120 | 3 | 30 | 25 | 20 | 20 | 20 | 20 |
| 130 | 3.5 | 30 | 25 | 25 | 20 | 20 | 20 |
| 140 | 3.5 | 35 | 25 | 25 | 20 | 20 | 20 |

(Source: ODOT, 2014; Courtesy: Ohio Department of Transportation, USA)

A few useful conversion factors: 1 foot = 0.3048 m, 1 in$^2$ = 6.45 cm$^2$, 100 pounds per cubic foot = 1602 kg/m$^3$, 1 kip = 453.59 kg, 1 pound = 0.453592 kg, 1 lb-ft = 0.1382548 kg-m, 1 kips-ft =138 kg-m.

## 19.13 SMART/INTELLIGENT STREET LIGHTING SYSTEM

To reduce the consumption of electricity and wastage of energy, control of the existing street lighting network must be optimised by adopting the latest technology, widely known as the smart street lighting (SSL) system. This concept will enable street lights to switch on/off and adjust their brightness automatically based on real-time traffic conditions and changing natural conditions; for example, the lights will glow at half of their normal intensity during a full moon night (Jagadeesh et al., 2015).

The SSL system is essentially an intelligent street lighting (ISL) system, in which the street lights located in an area/locality/highway are controlled, regulated, monitored, managed and maintained efficiently by artificial intelligence using non-intrusive sensors, smart ballasts/printed circuit boards (PCBs) fitted with luminaires, microcomputers and computer/tablet software. The control command/data is communicated or exchanged through a wireless data communication system (say, GSM) between the control centre and the sensors mounted on the roadside (Fig. 19.25). Lamps integrated with infrared and passive infrared sensors are controlled using microcontrollers which deal with pre-set levels of brightness, dimming time, power on/off switches with local sunset and sunrise timings.

Based on the sensors used and their applicability, the SSL system may be categorised as follows (Fig. 19.26).

**Intelligent street lighting system:** Pre-programmed microcontrollers analyse traffic movements along the lane and dynamically control the intensity of the street lights. In addition, with reference to the sun illumination level, the street lights can be turned off/on automatically using light-dependent resistors (LDR). LDRs automatically switch off lights when sunlight falls on them. The microcontrollers remain in sleep mode during the day and switch to active mode at night or when it is dark. This operation is automatically implemented based on the witnessed intensity of natural lighting, and also on traffic movements (Jagdeesh et al., 2015).

This system uses internet protocol (IP) connectivity through gateways that enable remote monitoring, control and management of individual lights along a roadway/street. The advantages of using an ISL system are as follows (Rajput et al., 2013):

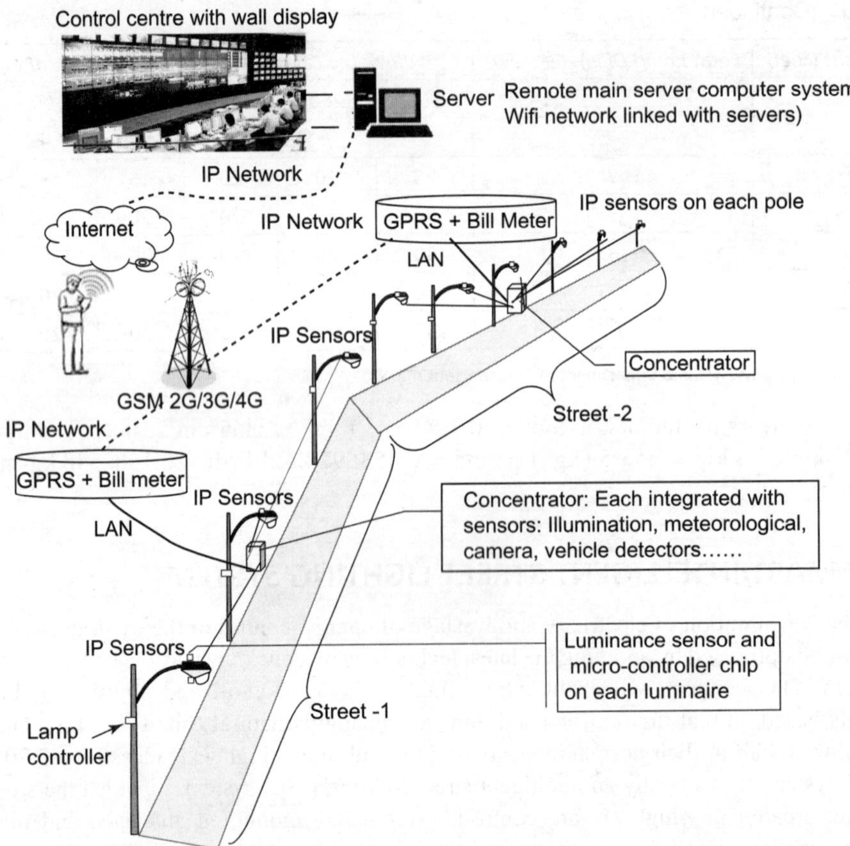


**Fig. 19.25** Typical details of intelligent/smart street lighting system

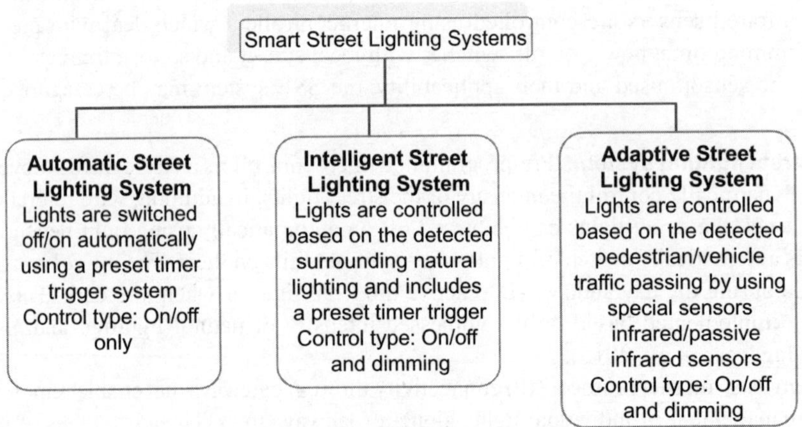


**Fig. 19.26** Classification of SSL systems

- Remote control for on/off
- Remotely controllable dimming and on-site status checks
- System fault detection/alarm
- Anti-theft detection/alarm
- Date-wise management relating to energy consumption report
- 24-hour online monitoring of the lighting system
- Reduced energy use
- Reduced maintenance costs
- Enhanced lamp life

The poles embedded with autonomous special devices with sensors can be also used to monitor environmental parameters along the street, such as sound, fog, temperature, rainfall, dust pollution and carbon monoxide emission.

In addition, any faulty lamp will be automatically detected and energy wastage caused by its ballast can be avoided. Lamp illumination can also be controlled: lamps can be turned on to provide low illumination when there is such a need, particularly on rainy or cloudy days.

**Adaptive street lighting system:** It is a part of the ILS system of street lighting that adapts to movements by pedestrians, cyclists and vehicles. The lamps are dynamically controlled and regulated, by dimming when no activity is detected and brightening as and when traffic movement is detected. Infrared (IR) and passive infrared (PIR) sensors mounted on the poles at specific intervals, on either side of the roadway, are used for precise detection of pedestrian/traffic movements. They communicate with neighbouring sensors to brighten the lights nearby. These sensors are connected to microcontrollers embedded in the luminaires through which data is processed and the lighting levels are adjusted (Jadgeesh et al., 2015). This system of street lighting is different from the fixed-time dimming system, where lamps are dimmed at pre-set times, say between 1:00 am and 4:00 am. The main advantage of this system is that, when a passerby is detected by a camera or sensor, this information is communicated to the neighbouring street lights and they are brightened/dimmed according to the movement along the street (Fig. 19.27). As the pedestrians/vehicles go past, the lights are turned off automatically. This system can be effectively used for enhancement of pedestrian safety by providing a continuous path of lighting at night, with maximum energy efficiency.

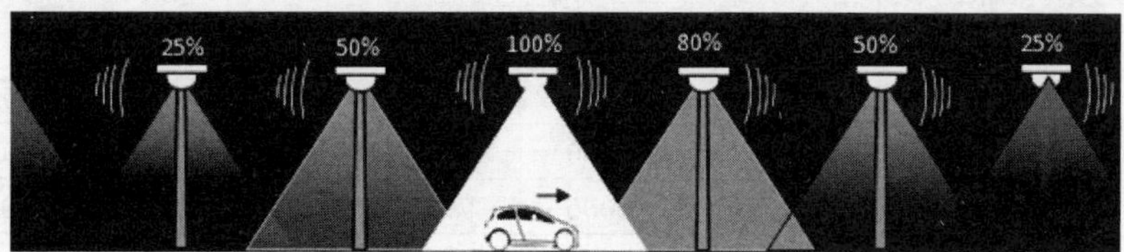


**Fig. 19.27** Typical view of adaptive street lighting along a roadway during vehicle movement

## 19.13.1 Components of ISL

The ISL that is part of an SSL system comprises the following sensors/devices, designed as additional infrastructure:

- Chips consisting of a microcontroller used to control and measure the energy of the ballast
- GSM modules for wireless data communication
- Concentrator
- Computers installed with relevant software, termed main server and slaves

- Sensors:
  - Light/luminous intensity sensor: light dependent resistor (LDR)
  - Non-intrusive traffic monitoring sensor (optional)
  - $CO_2$/CO sensor (optional)
  - Fog sensor (optional)
  - Noise sensor (optional)
  - Optical dust sensor (optional)
  - Humidity and temperature sensor (optional)
  - Air velocity sensor (optional)
- CCTV camera (optional)

## 19.14 LIFE CYCLE COST EVALUATION

The life cycle cost estimate is intended to compare candidate projects with their alternative options. Such candidate design options shall be considered for life cycle cost analysis for a period of 25 years. The proposed cost components of the lighting facility should provide details of issues such as road safety, maintenance, standards and aesthetics. A typical method of analysis used for life cycle cost evaluation of a proposed street lighting project is given below (Alberta, 2003).

*Life Cycle Cost Example:*
Project title: University Road, Hyderabad Urban Development Authority
Description of Design Option: 75 × 250 watt HPS
Pole Type: Single davit; one side

**Part 1**: Design option estimated capital cost: ₹1,000,000/-
(Includes cost of longitudinal wiring, power boxes and other costs)

**Part 2**: Maintenance Estimated Cost:

| | *Proposed number of locations in the project* | *Material average cost (₹ per unit)* | *Estimated labour cost to install* | *Failure rate in 20 years* | *20-year life cycle cost (₹)* |
|---|---|---|---|---|---|
| Lamps | 20 | 1500 | 150 | 4 years | 20 × (1500 + 150) × (20/4) = ₹165,000 |
| Luminaires | 20 | 3000 | 500 | 40% | 20 × (3000 + 500) × (40/100) = ₹28,000 |
| Glass on luminaires | 20 | 300 | 300 | 60% | 20 × (900 + 300) × (60/100) = ₹14,400 |
| Pole + base | 20 | 6000 | 6000 | 12% | 20 × (6000 + 6000) × (12/100) = ₹28,800 |
| | | | | Total of Part 2 = | ₹236,200 |

**Part 3**: Estimated Energy Cost:

| *Proposed no. of luminaires in the project* | *Luminaire wattage (kW)* | *Average operating hours each year* | *Electricity rate (example only, actuals will change)* | *Estimated energy cost in 25 years* |
|---|---|---|---|---|
| 20 | 0.295 | 365 d × 10 hr = 3650 hr | ₹1.0/kWh | 20 × 0.295 × 3650 × 1.0 × 20 = ₹430,700 |

The estimated total 25-year life cycle cost of the street lighting project = (Part 1 + Part 2 + Part 3)

Part 1: 1,000,000
Part 2: 236,200
Part 3: 430,700

---

Total = ₹1,666,900

---

**Example 19.1**

Determine the coefficient of uniformity if the proposed HPS lamp is used at a mounting height of 9 m, pole arm projection of 2.7 m, shoulder width of 3 m, and the width of road is 7.0 m (Fig. 19.28).

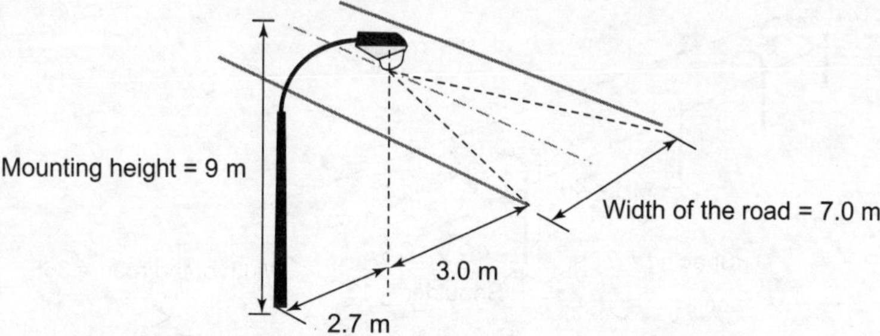


**Fig. 19.28** Details of roadway geometrics with pole height

**Answer:** Total ratio = (Width of road + Width of shoulder)/Mounting height = (7.0 + 3.0)/9 = 1.11
The pavement area CU towards the street obtained for the ratio of 1.11 (from Fig. 19.28 (a) street-curve) = 34.0%

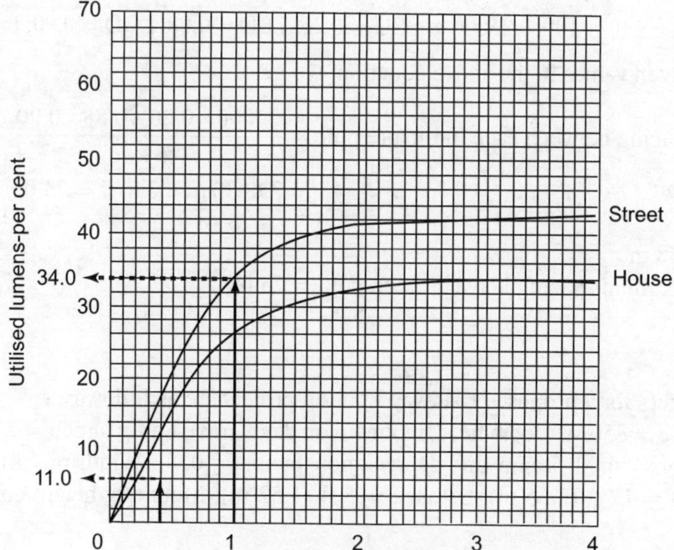


**Fig. 19.28 (a)** Example: Coefficient of utilisation curve of high-pressure sodium luminaire
(Source: MnDOT, 2010; Courtesy: Office of Traffic, Security and Technology, MnDOT, USA)

Ratio of house-side = Width of shoulder/Mounting height = 3.0/9.0 = 0.33

Similarly, the shoulder area CU towards the house obtained for the ratio of 0.44 (from Fig. 19.28 (a)) = 11%.

Therefore, CU for the road area = (34 − 11) = 23.0% = 0.23.

**Example 19.2**

Determine the spacing required between each street light pole for the given input parameters:

Width of carriageway = 11 m (36 ft); Mounting height (MH) = 12 m (40 ft); Luminaire of each pole: 250-watt HPS lamp of 100 lumens/watt; Required lux level = 6 lux ≈ 0.6 foot-candles; Coefficient of utilisation (CU) factor = 0.26; Lamp lumen depreciation factor = 0.88; Luminaire dirt depreciation factor = 0.90.

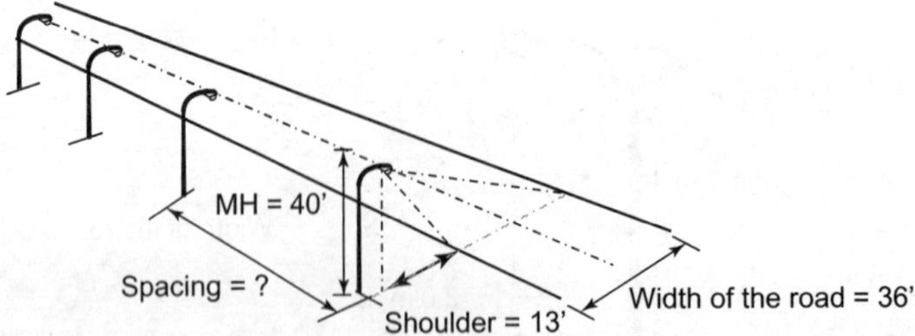


**Fig. 19.29** Proposed layout of luminaires with details of roadway geometrics

**Answer:** The following equation is used

$$\text{Maximum spacing between each luminaire pole} = \frac{\text{LL} \times \text{CU} \times \text{LLD} \times \text{LDD}}{\text{E}_h \times \text{B}}$$

Lamp output = LL = 250 × 100 = 25000 lm; CU = 0.26; LLD = 0.88; LDD = 0.90; $E_h$ = 0.6 fc; B = 36 ft.

Substituting the given values in the above equation, we get

$$\text{Maximum spacing between each luminaire pole} = \frac{25000 \times 0.26 \times 0.88 \times 0.90}{0.6 \times 36}$$
$$= 238.33\,ft \approx 240\,ft = 73.0m$$

*Note:* 1 ft = 0.3048 m.

Therefore, the maximum spacing between each pole = 73 m.

**Example 19.3**

The input parameters for a proposed roadway lighting project are as follows:

Lamp-luminaire = 250 W (cobra head, mounted on davit arm); Lamp lumen = 27,000; Lamp lumen depreciation factor = 0.90; Lamp dirt depreciation factor = 0.90; Width of carriageway = 24.6 ft (7.5 m); Outreach = 12 ft (3.65 m); Setback = 5 ft (1.52 m); Mounting height corrections factors = $900/(MH)^2$

Assume (a) single-sided pole arrangement, (b) roadway classification: cement concrete pavement surface categorised as Other Principal Arterial and used for commercial purposes with required minimum illumination as 1.1 fc (12 lx), (c) maximum uniformity ratio as 3:1.

Design the roadway lighting system by determining (i) spacing between the luminaire poles; (ii) minimum illumination value by conducting a point-by-point analysis of any three critical point-locations on the roadway, and (iii) design precision based on the uniformity ratio.

**Answer:** Let us draw a sketch (Fig. 19.30) and assign the given values to the standard notations as follows.

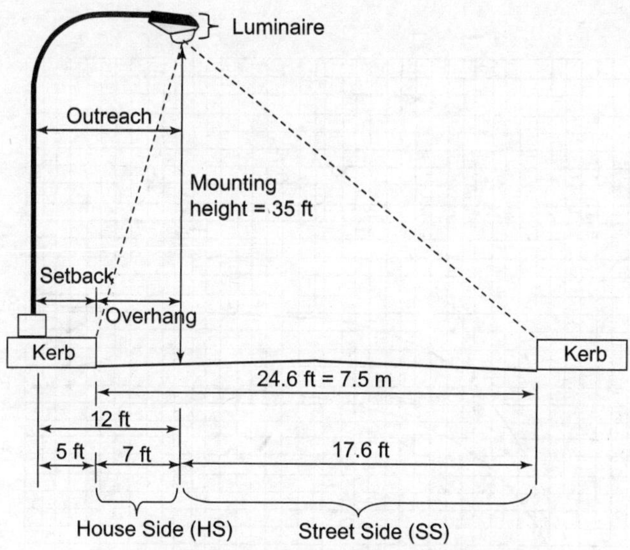


**Fig. 19.30** Roadway geometrics

*Road geometrics*:

Mounting height = MH = 35 ft (10.67 m); Overhang = (Outreach – Setback) = (12 – 5) = 7 ft;

Width of roadway = B = 24.6 ft; Width of road street side (SS) = 17.6 ft; Width of road/shoulder house side (HS) = OH = 7 ft

*Luminaire characteristics*:

$E_h$ = 1.1 fc; LLD = 0.90; LDD = 0.90; LL = 27,000

Let us determine the coefficient of utilisation (CU) factor, using the utilisation curves that belong to the lamp type and roadway class (Fig. 19.31).

$$\text{Ratio of street side} = \frac{17.6\,\text{ft}}{35\,\text{ft}} = 0.503$$

$$\text{Ratio of house side} = \frac{7\,\text{ft}}{35\,\text{ft}} = 0.20$$

The SS area CU obtained for the ratio of 0.53 (from Fig. 19.31 street-curve) = 20.0%
Similarly, the HS area CU obtained for the ratio of 0.2 (from Fig. 19.30) = 6.0%.
Therefore, total CU for the road area = (20.0 + 6.0) = 26.0% = 0.26.

(i) Theoretical maximum longitudinal spacing (S) requirement may be calculated using the following formula: $S = \dfrac{LL \times CU \times LLD \times LDD}{E_h \times B}$

Substituting the given and determined input values, we get

$$S = \frac{27000 \times 0.26 \times 0.90 \times 0.90}{1.1 \times 24.6} = 210.133 \text{ ft} = 64.048 \text{ m}$$

Let us take spacing = S = 210 ft.

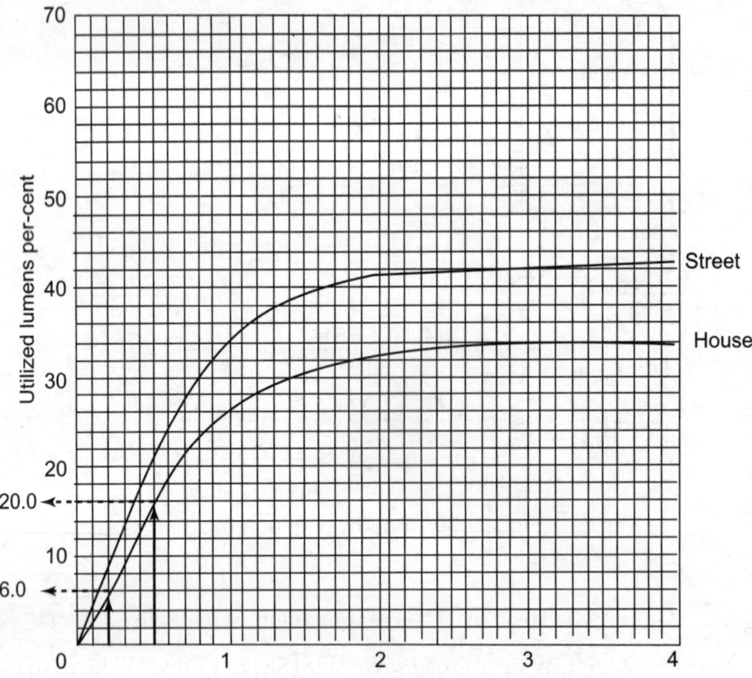


**Fig. 19.31** Coefficient of utilisation curve

(ii) Let us compute the initial illumination (fc) values at the three critical points (see check-points P1, P2 and P3 in Fig. 19.32) using the isofoot-candle curve diagram of the proposed lamp-luminaire (Fig. 19.33).

**Check-point P1:**

Contribution of luminaires #1, #2, #3 and #4 on check-point P1 is computed as follows:

Contribution of luminaires #1 and #3:

Ratio of transverse distance to mounting height $= \dfrac{B - OH}{MH} = \dfrac{17.6}{35} = 0.50$

Ratio of longitudinal distance to mounting height $= \dfrac{S}{MH} = \dfrac{210}{35} = 6.00$

Initial illumination from #1 = 0.000 fc

As the location of luminaire #3 to the point P1 is similar to #1, the initial illumination from #3 = 0.000 fc

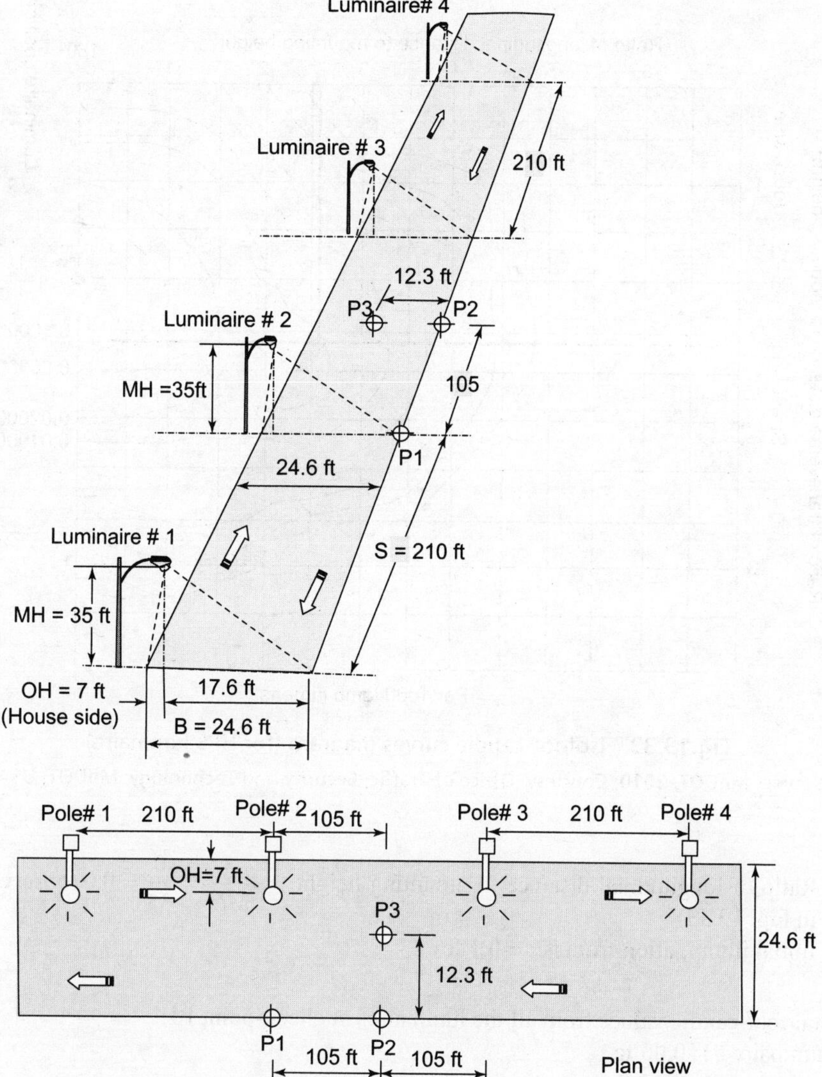


**Fig. 19.32** One-sided luminaires installed along a roadway

Contribution of luminaire #4:

As the contribution from luminaires #1 and #3 are zero, the contribution of luminaire #4 on the check-point P1 will be = 0.000 fc (it is further away than #1 and #3)

Contribution of luminaire #2:

Ratio of transverse distance to mounting height = $\frac{B-OH}{MH} = \frac{17.6}{35} = 0.50$ (observe SS on the $y$ axis in Fig. 19.33)

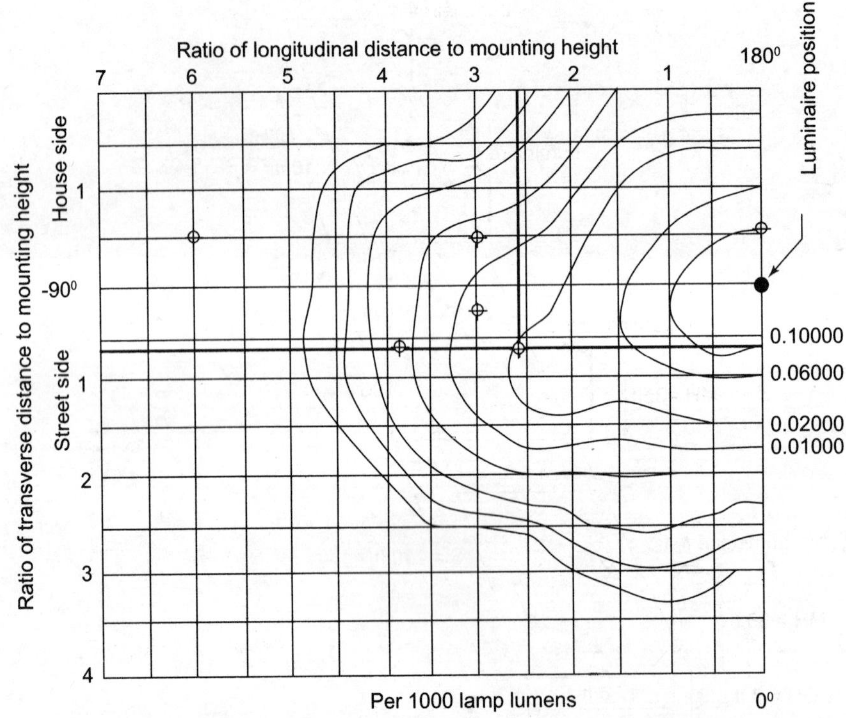


**Fig.19.33** Isofoot-candle curves diagram (for HPS luminaire)
(Source: MnDOT, 2010; Courtesy: Office of Traffic, Security and Technology, MnDOT, USA )

Ratio of longitudinal distance to mounting height $= \dfrac{S}{MH} = \dfrac{0}{35} = 0.0$ (observe the $x$ axis in Fig. 19.33)
Initial illumination from #2 = 0.1 fc

Total foot-candle values from all the luminaires at check-point P1:
Luminaire #1: 0.00 fc

Luminaire #2: 0.10 fc
Luminaire #3: 0.00 fc
Luminaire #4: 0.00 fc

-----------------------------------

Total raw fc at P1 = 0.1 fc

-----------------------------------

**Check-point P2:** Contribution of luminaires #1, #2, #3 and #4 on check-point P2 is computed as follows:

Contribution of luminaires #2 and #3:

Ratio of transverse distance to mounting height $= \dfrac{B-OH}{MH} = \dfrac{24.6-7}{35} = 0.50$

Ratio of longitudinal distance to mounting height $= \dfrac{S/2}{MH} = \dfrac{105}{35} = 3.00$

Initial illumination from #2 = (0.01 + 0.005)/2 = 0.008 fc (it is midway between the contours: 0.01 and 0.005)
As the location of luminaire #3 to point P2 is similar to #2, the initial illumination from #3 = 0.008 fc

Contribution of luminaires #1 and #4:

Ratio of transverse distance to mounting height $= \dfrac{B-OH}{MH} = \dfrac{17.6}{35} = 0.50$ (observe the $y$ axis in Fig. 19.33)

Ratio of longitudinal distance to mounting height $= \dfrac{S+S/2}{MH} = \dfrac{210+105}{35} = 9.00$ (observe the $x$ axis in Fig. 19.33)

Initial illumination from #1 = 0.000 fc
Initial illumination from #4 = 0.000 fc

Total foot-candle value from all the luminaires at check-point P2:

Luminaire #1: 0.000 fc
Luminaire #2: 0.008 fc
Luminaire #3: 0.008 fc
Luminaire #4: 0.000 fc

-----------------------------------

Total raw fc at P2 = 0.016 fc

-----------------------------------

**Check-point P3:** Contribution of luminaires #1, #2, #3 and #4 on check-point P3 is computed as follows:

Contribution of luminaires #2 and #3:

Ratio of transverse distance to mounting height $= \dfrac{\frac{B}{2}-OH}{MH} = \dfrac{12.3-7}{35} = 0.15$ (On SS)

Ratio of longitudinal distance to mounting height $= \dfrac{S/2}{MH} = \dfrac{105}{35} = 3.00$

Initial illumination from #2 = 0.012 fc
As the location of luminaire #3 to the point P2 is similar to #2, the initial illumination from #3 = 0.012 fc

Contribution of luminaires #1 and #4:

$$\text{Ratio of transverse distance to mounting height} = \frac{\frac{B}{2} - OH}{MH} = \frac{12.3 - 7}{35} = 0.15$$

$$\text{Ratio of longitudinal distance to mounting height} = \frac{S + S/2}{MH} = \frac{315}{35} = 9.00$$

Initial illumination from #2 = 0.000 fc

As the location of luminaire #4 to the point P3 is similar to #1, the initial illumination from #4 = 0.000 fc

Total foot-candle values from all the luminaires at check-point P2:

luminaire#1: 0.000 fc
luminaire#2: 0.012 fc
luminaire#3: 0.012 fc
luminaire#4: 0.000 fc

-----------------------------------

Total raw fc at P2 = 0.024 fc

-----------------------------------

Now, select the minimum raw fc value = $P_{raw\text{-}min}$ = 0.016

After calculating the initial illumination (fc) values at these check-points, the following factors are applied to the raw value of the lowest fc value (select P1, P2 or P3) and the minimum illumination value is calculated as follows:

$$P_{min} \text{ (in fc)} = P_{raw\text{-}min} \times LF \times LLD \times LDD \times MHCF$$

$$P_{min} \text{ (in fc)} = 0.016 \times (27000/1000) \times 0.90 \times 0.90 \times (900/35^2)$$

$$P_{min} \text{ (in fc)} = 0.016 \times (27.0) \times 0.90 \times 0.90 \times (0.734) = 0.257$$

(iii) The quality of illumination along the roadway is determined based on the point of minimum illumination criterion by calculating the uniformity ratio (UR)

$$\text{Uniformity ratio (UR)} = \frac{E_h \text{ (in fc)}}{P_{min} \text{ (in fc)}} = \frac{1.1}{0.257} = 4.280$$

As 4.280 is not less than the criterion of 3.0, the design is inadequate (our proposed values do not meet the design UR criterion of 3:1). Hence, the spacing must be reduced; S = 180 ft (≈20% reduction) and the design must be revised.

*Second Attempt*:
Now, take S = 180 ft and retain the same values for the rest of the input parameters.

**Check-point P1:**

Contributions of luminaires #1, #2, #3 and #4 on check-point P1 are computed as follows:

Contribution of luminaires #1 and #3:

Ratio of transverse distance to mounting height $= \dfrac{B-OH}{MH} = \dfrac{17.6}{35} = 0.50$ (On HS)

Ratio of longitudinal distance to mounting height $= \dfrac{S}{MH} = \dfrac{180}{35} = 5.14$

Initial illumination from #1 = 0.000 fc

As the location of luminaire #3 to the point P1 is similar to #1, the initial illumination from #3 = 0.000 fc

Contribution of luminaire #4:
As the contribution of luminaires #1 and #3 is zero, the contribution of luminaire #4 on check-point P1 will be = 0.000 fc (it is further away than #1 and #3)

Contribution of luminaire #2:

Ratio of transverse distance to mounting height $= \dfrac{B-OH}{MH} = \dfrac{17.6}{35} = 0.50$ (on HS, observe the *y* axis in Fig. 19.33)

Ratio of longitudinal distance to mounting height $= \dfrac{S}{MH} = \dfrac{0}{35} = 0.0$ (observe the *x* axis in Fig. 19.33)

Initial illumination from #2 = 0.1 fc

Total foot-candle value from all the luminaries at check-point P1:

Luminaire #1: 0.00 fc
Luminaire #2: 0.10 fc
Luminaire #3: 0.00 fc
Luminaire #4: 0.00 fc

-----------------------------------

Total raw fc at P1 = 0.1 fc

-----------------------------------

**Check-point P2:** Contributions of luminaires #1, #2, #3 and #4 on check-point P2 are computed as follows:

Contribution of luminaires #2 and #3:

Ratio of transverse distance to mounting height $= \dfrac{B-OH}{MH} = \dfrac{24.6-7}{35} = 0.50$ (On HS)

Ratio of longitudinal distance to mounting height $= \dfrac{S/2}{MH} = \dfrac{90}{35} = 2.571$

Initial illumination from #2 = (0.01 + 0.005)/2 = 0.009 fc (it is midway between the contours: 0.01 and 0.005)

As the location of luminaire #3 to the point P2 is similar to #2, the initial illumination from #3 = 0.009 fc

Contribution of luminaires #1 and #4:

Ratio of transverse distance to mounting height = $\frac{B-OH}{MH}=\frac{17.6}{35}=0.50$ (on HS; observe the *y* axis in Fig. 19.33)

Ratio of longitudinal distance to mounting height = $\frac{S+S/2}{MH}=\frac{90+45}{35}=3.857$ (observe the *x* axis in Fig. 19.33)

Initial illumination from #1 = 0.0025 fc
Initial illumination from #4 = 0.0025 fc

Total foot-candle value from all the luminaires at check-point P2:

Luminaire #1: 0.0025 fc
Luminaire #2: 0.0090 fc
Luminaire #3: 0.0090 fc
Luminaire #4: 0.0025 fc

-----------------------------------

Total raw fc at P2 = 0.023 fc

-----------------------------------

**Check-point P3:** Contributions of luminaires #1, #2, #3 and #4 on check-point P3 are computed as follows:

Contribution of luminaires #2 and #3:

Ratio of transverse distance to mounting height = $\frac{\frac{B}{2}-OH}{MH}=\frac{12.3-7}{35}=0.15$ (On SS)

Ratio of longitudinal distance to mounting height = $\frac{S/2}{MH}=\frac{90}{35}=2.571$

Initial illumination from #2 = 0.023 fc

As the location of luminaire #3 to the point P2 is similar to #2, the initial illumination from #3 = 0.023 fc

Contribution of luminaires #1 and #4:

Ratio of transverse distance to mounting height = $\frac{\frac{B}{2}-OH}{MH}=\frac{12.3-7}{35}=0.15$

Ratio of longitudinal distance to mounting height = $\frac{S+S/2}{MH}=\frac{135}{35}=3.875$

Initial illumination from #2 = (0.001+0.005)/2 = 0.003 fc

As the location of luminaire #4 to the point P3 is similar to #1, the initial illumination from #4 = 0.003 fc

Total foot-candle value from all the luminaries at check-point P2:

Luminaire #1: 0.003 fc
Luminaire #2: 0.023 fc
Luminaire #3: 0.023 fc
Luminaire #4: 0.003 fc
-----------------------------------
Total raw fc at P2 = 0.052 fc
-----------------------------------

Now, select the minimum raw fc value = $P_{raw\text{-}min}$ = 0.023

After calculating the initial illumination (fc) values at these check-points, the following factors are applied to the raw value of the lowest fc-value (select P1, P2 or P3) and the minimum illumination value is calculated as follows:

$$P_{min} \text{ (in fc)} = \text{Praw-min} \times \text{LF} \times \text{LLD} \times \text{LDD} \times \text{MHCF}$$

$$P_{min} \text{ (in fc)} = 0.023 \times (27000/1000) \times 0.90 \times 0.90 \times (900/35^2)$$

$$P_{min} \text{ (in fc)} = 0.023 \times (27.0) \times 0.90 \times 0.90 \times (0.734) = 0.369$$

*Design precision based the uniformity ratio*

The quality of illumination along the roadway is determined based on the point of minimum-illumination criterion by calculating uniformity ratio (UR).

$$\text{Uniformity ratio (UR)} = \frac{E_h \text{ (in fc)}}{P_{min} \text{ (in fc)}} = \frac{1.1}{0.369} = 2.981$$

As 2.981 is less than 3.0, the design is adequate (our proposed values meet the design UR criterion of 3:1). Hence, the design is OK.

## EXERCISES

### Multiple Choice Questions

1. The type of lighting distribution pattern that is preferred near astronomical observatories or airports is __________
   (a) Cut-off (b) Semi–cut-off (c) Full cut-off (d) Non–cut-off
2. The type of lighting distribution pattern that is never used near airports is
   (a) Cut-off (b) Semi–cut-off (c) Full cut-off (d) Non–cut-off
3. What is the unit of brightness?
   (a) Lux (b) Lumen (c) Lambert (d) Candela
4. What are the units used to represent a lamp light distribution curve?
   (a) Lux (b) Lumen (c) Lambert (d) Candela

5. What is the unit of measurement of CRI?
   (a) Centigrade (b) Fahrenheit (c) No dimension (d) Armstrong
6. A sphere contains how many steradians?
   (a) 1 (b) 4 (c) π (d) 4 π
7. A point source of 200 cd is 5 ft from a point; the normal illumination at the point is ________
   (a) 2 fc (b) 0.5 fc (c) 4 fc (d) 8 fc
   (*Hint:* $E = I/D^2 = 200/25 = 8$ fc)
8. A point of source of 200 cd is 6 ft above and 8 ft towards the street side. The illumination at the point is ________
   (a) 1 fc (b) 6 fc (c) 1.2 fc (d) 0.6 fc
   $$\left(\textit{Hint}: E_\theta = \frac{I}{D^2} \times \cos(\theta) = \left(\frac{200}{100}\right) \times \left(\frac{6}{10}\right) = 1.2 \text{ fc}\right)$$
9. What is the value of average lux if a lamp initial lamp lumen rating = 20,000, maintenance factor = 0.80, coefficient of utilisation = 0.90, spacing of poles = 200 m and width of the roadway = 10 m?
   (a) 7.2 lux (b) 0.63 lux
   (c) 0.315 foot-candles (d) 0.63 foot-candles
   (*Hint:* Average = (20000 × 0.9 × 0.8)/(200 × 10) = 7.2 lux)
10. What is the value of spacing between luminaire poles when installed in staggered pattern, if the luminaire initial lamp lumen rating = 25,000, maintenance factor = 0.8, coefficient of utilisation = 0.9, average illumination = 10 lux and width of the roadway = 10 m?
    (a) 90 m (b) 180 m (c) 225 m (d) 360 m
    (*Hint:* Spacing = (25000 × 0.9 × 0.8)/(10 lux × 10) =180 m)
11. A luminaire transverse projection of overhang should not exceed what percentage of its mounting height?
    (a) 25% (b) 30% (c) 35% (d) 40%
12. Which of the following measurement of mounting height should not be used for a lighting luminaire?
    (a) 5.5 m (b) 6.0 m (c) 7.6 m (d) 9.0 m
13. What does a luminaire utilisation curve indicate?
    (a) The amount of light emitted within the luminaire
    (b) The distribution of illumination on the roadway surface
    (c) The amount of light that falls on the roadway surface
    (d) The magnitude and direction of light coming from the luminaire
14. What does a luminaire isolux curve indicate?
    (a) The amount of light generated within the luminaire
    (b) The distribution of illumination on the roadway surface in the vicinity of the luminaire
    (c) The amount of light that falls on the roadway surface in the vicinity of the luminaire
    (d) The magnitude and direction of light coming from the lamp

15. For the design of a roadway lighting system, what factor should be included to compensate for gradual loss of illumination due to accumulation of dirt on the surface of luminaire?
    (a) Installed test lamp factor (b) Power factor
    (c) Illumination factor (d) Maintenance factor
16. What value of the maintenance factor should be used in the calculation to compensate for the gradual loss of illumination, if the manufacturer does not supply the required information?
    (a) 0.76 (b) 0.70 (c) 0.65 (d) 0.9
17. What would be the equivalent value of illumination for a mounting height of 10 m, if 12 lux level of illumination is found at 15 m of mounting height?.
    (a) 3 (b) 12 (c) 18 (d) 27
    (*Hint:* (Lux) × $(10)^2$ = (12) × (15)2; Lux = 27 lux)
18. The sensor which automatically switches off lights when sunlight falls on it is called?
    (a) Passive infrared (b) Active infrared
    (c) Ballast (d) Light-dependent resistor
19. Match the following:

| | |
|---|---|
| 1. Luminous flux | A. Candela |
| 2. Illumination | B. Foot-candle |
| 3. Luminous efficacy | C. Lumen/Watt |
| 4. Solid angle | D. Lumen |
| 5. Luminous intensity | E. Degrees |
| | F. Steradian |

## Answers

1. (c) 2. (d) 3. (a) 4. (a) 5. (c) 6. (d) 7. (d) 8. (c) 9. (a) 10. (b) 11. (a)
12. (a) 13. (c) 14. (b) 15. (d) 16. (b) 17. (d) 18. (d) 19. 1(D), 2(B), 3(C), 4(F), 5(A)

## Questions

1. What are the major factors affecting driver visibility?
2. State the objectives and write about the broad categories of functions of roadway lighting.
3. Define the following terms: (a) Luminaire (b) Ballast (c) Glare.
4. Define and illustrate the following terms: (a) Luminous flux (b) Luminous intensity with reference to steradian (c) Illuminance (d) Luminance (e) Luminaire efficacy (f) Colour rendering index.
5. Illustrate the following
   (a) Illumination and luminance from a luminaire
   (b) Solid angle subtended by a steradian
   (c) Difference between foot-candle and lux
   (d) Inverse square law
   (e) Concept of cosine law
   (f) Component details of electric poles mounted with luminaire
6. Illustrate a few typical examples of partial street lighting layouts.
7. Explain the two lighting laws used in the design of street lighting.

8. Sate the differences between foot-candle and lux. Describe with a sketch.
9. What is photometry? Explain how this data is useful in the design of street lighting.
10. What is is isolux curve diagram? Briefly describe that data that can be obtained from these diagrams.
11. Explain the cosine and inverse square law with schematic diagrams.
12. Draw a sketch showing the family tree of electric lamps commonly used for lighting streets/roadways.
13. Give an overview of the different types of lamps used for roadway lighting, with the applicable design specifications.
14. Prior to designing roadway lighting, under what categories should the extent and type of lighting requirement in a locality be determined?
15. What warrants indicate the conditions that should be satisfied to justify the installation of lighting on roadways?
16. What are the factors affecting the selection of appropriate lighting systems along a roadway?
17. Define and briefly explain the following correction factors used for the design of roadway lighting: (a) Light output ratio (b) Installed test lamp lumen (c) Utilisation ratio (d) Utilisation factor.
18. Define and explain the importance of the following factors used for roadway lighting: (a) Lamp lumen depreciation factor (b) Lamp dirt depreciation factor (c) Maintenance factor.
19. Define and explain the significance of uniformity ratio in the design of a roadway lighting system.
20. List and describe the luminaire light distribution types as developed by IESNA and ANSI. Also write the applicability of each type with reference to street lighting.
21. What is meant by luminaire cut-off? Give details of the IESNA cut-off classification of the control of outdoor light distribution.
22. Define and write the applicable standard specifications with reference to current practices on (a) Pole height and mounting height (b) Pole arms.
23. Describe the types of luminaire arrangements and their applicability with reference to roadway geometrics and mounting height.
24. Write an overview of the lighting level requirement of street/roadway lighting.
25. State and explain the three methods of spacing design of lighting units.
26. Will uniform horizontal illumination on a roadway surface produce uniform luminance to the road user?

    [*Hint:* As an ideal surface condition, the uniform illumination on a perfectly diffused surface would produce uniform luminance to the drivers/pedestrians. In most cases, the roadway surface material characteristics never uniformly diffuse the illumination. In addition, factors such as spacing, height and luminaire type will also influence the uniformity of illumination]

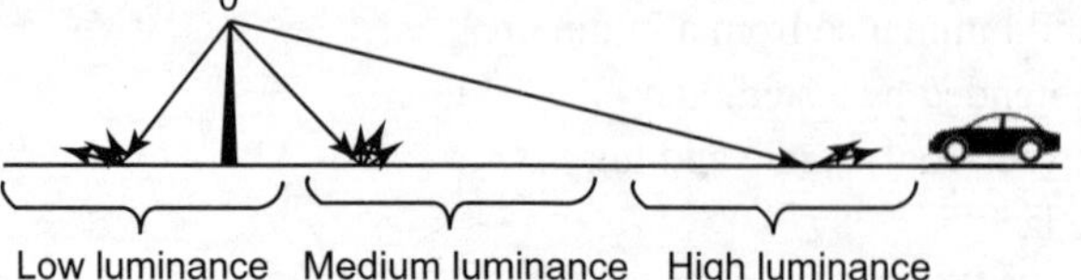


Non-uniform luminance with wide-angle emission of street lighting along the roadway (schematic)

(Redrawn from UNAM, 1992)

27. What is the difference between initial and maintained foot-candela (fc)?
28. Define and write where uniformity coefficient and foot-candela values are used.
29. Explain the possible alternatives of design of roadway lighting if the criterion of uniformity ratio (UR) fails with a proposed set of input variables.
30. Distinguish between LLD and LDD factors.
31. Write an overview of the pole heights used for roadway lighting.
32. Write an overview of the poles used for roadway lighting.
33. List the types of pole arms and their applicability.
34. Write the broad categories of lamps used for roadway lighting. Also write their distinguishing features for the selection of specific lamps/luminaires.
35. Write the factors affecting the selection/choice of luminaire.
36. Explain how often a luminaire fixture needs to be cleaned in accordance with the type of ambient contamination level.
37. What is lateral placement of a pole and a luminaire? Explain a few standard practices.
38. State the factors considered while selecting the appropriate setback distance for installing poles adjacent to roadway.
39. Explain the design of a roadway lighting system with step-wise details.
40. Draw a typical sketch showing the geometric details of one-sided and staggered luminaires installed along a roadway.
41. Draw a typical sketch showing the geometric details of opposite and twin central luminaires installed along a roadway.
42. Draw a flow chart describing the typical lighting design process with the sequence of steps including the design of power distribution systems.
43. State the advantages of using software tools for designing street lighting.
44. Determine the coefficient of uniformity if the proposed HPS lamp is used with a mounting height of 10 m, pole arm projection of 2.5 m, shoulder width of 2 m and the width of carriageway is 7.5 m.
45. Determine the spacing required between each street light pole for the given input parameters: Width of carriageway = 10.5 m; Mounting height (MH) = 10 m; Luminaire of each pole: 300-watt (HPS) lamp capable of light of up to 100 lumen/watt; Required lux level = 11 lux; Coefficient of utilisation (CU) factor = 28%; Lamp lumen depreciation factor = 0.89; Luminaire dirt depreciation factor = 0.90.
46. An area measuring 22 × 16 m is to be lit by a single light mounted on a 10-m pole, as shown the figure. A lamp of 150 W 16,000 lumen is proposed to be used. Take the total lumen loss factor as 0.78. Based on the cosine rule, determine the horizontal illumination at points A, B and C.

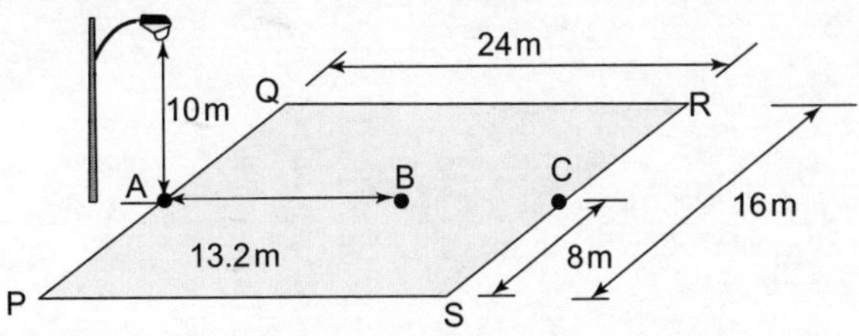

47. The input parameters proposed be used for a project on roadway lighting are as follows:

    Lamp-luminaire = 250 W (cobra head; mounted on davit arm); Lamp lumen = 27,500; Lamp lumen depreciation factor = 0.89; Lamp dirt depreciation factor = 0.90; Width of carriageway = 10.5 m; Outreach = 4.0 m; Setback = 1.8 m; Mounting height correction factor = 900/(MH)2

    Assume (a) single-sided pole arrangement, (b) roadway classification: asphalt concrete pavement surface categorised as Minor Arterial and used for intermediate purpose with required minimum illumination as 10 lx, and (c) maximum uniformity ratio as 4:1.

    Design the roadway lighting system by determining (i) spacing between the luminaire-poles, (ii) minimum illumination value by conducting point-by-point analysis of any three critical point-locations on the roadway and (iii) design precision based on the uniformity ratio.

    *Note:* The relevant photometric data curves can be used. Take 1 fc = 10.764 lx.

48. The design computation findings proposed to be used for a project on roadway lighting are as follows:

    Lamp lumen factor = 27.5; Lamp lumen depreciation factor = 0.89; Lamp dirt depreciation factor = 0.90; Mounting height factor = 0.73.

    Minimum illumination values determined by conducting point-by-point analysis of three critical point-locations on the roadway are P1= 1.5 lx; P2 = 0.21 lx; P3 = 0.26 lx.

    The roadway classification is asphalt concrete pavement surface categorized as Collectors and used for commercial purposes with the required minimum illumination as 12 lx and maximum uniformity ratio as 4:1. Check the design precision based on the uniformity ratio.

49. What do you understand by smart street lighting system? Explain the fundamental concept of intelligent street lighting.
50. Write the classification of smart street lighting systems and explain each of them.
51. What are the advantages of using intelligent street lighting systems?
52. Draw a typical roadway layout of an intelligent/smart street lighting system with basic components.
53. What is the difference between automatic street lighting system and adaptive street lighting system?
54. What are the devices designed as additional infrastructure as part of a smart street lighting system?

CHAPTER

# 20 Road Traffic Accident Studies and Reconstruction

## 20.0 INTRODUCTION

Traffic safety studies are conducted to evaluate, examine, identify and review the causes of road traffic accidents based on the information collected, and recommend necessary improvements to road geometrics with their safety operations and appurtenances. The objective is to implement the necessary corrections to eliminate hazardous conditions in the highway right-of-way.

Traffic safety studies/surveys are conducted to collect information relating to traffic characteristics, vehicular characteristics, driver/pedestrian/user characteristics, road surface condition characteristics, road geometrics, deficient safety appurtenances, weather conditions and accident evidence.

There are two types of road safety studies (Fig. 20.1). The traffic accident study (TAS) uses police accident reports and investigates the causes of accidents based on the information and evidence collected. Reconstruction and investigation of the accidents recorded, with possible improvements in the roadway geometrics including traffic operations, are considered under this category of studies.

A road safety audit (RSA) is a formal study conducted by an independent audit team which evaluates the safety performance of a road or intersection. The safety audit can be carried out for an existing roadway or for one in the planning stage (FHWA, 2004).

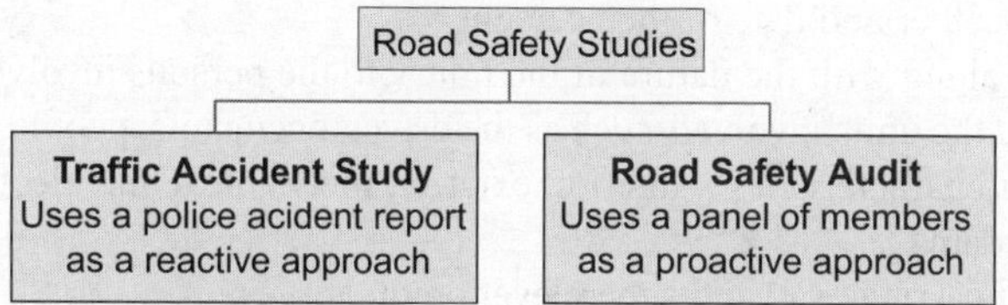


**Fig. 20.1** Types of road safety studies

RSAs have been conducted internationally since the 1980s. However, the first US road safety audit was conducted by the Pennsylvania Department of Transportation in 1997. Since then, several state and local jurisdictions in different countries around the globe have started using this tool to improve safety.

The RSA is a formal examination and review of the safety performance of an existing or future road/intersection by an independent audit team. Based on the RSA report, which consists of identified potential safety issues, assessment and recommendations, the jurisdiction owning the road responds to the problems identified. It is the sole responsibility of the owner of the road to determine the feasible action or to redesign/modify the road geometrics; otherwise, there is a need to document the reason for not acting on such RSA suggestions/recommendations. RSAs of existing roads are sometimes termed RSA reviews (FHWA, 2004; IRC: SP: 88-2010).

## 20.1 OBJECTIVES

Different groups of people are interested in road safety and in finding solutions for related issues. These agencies include the police, lawyers, road safety officers, highway engineers, researchers, politicians, teachers, statisticians, insurance companies and the general public (IRC: 53-2012). The investigations carried out on road accidents and related surveys are conducted for the following objectives:

- To locate/record the exact location, time and day of the accident
- To identify the people/witnesses and objects involved in a type of accident
- To collect on-scene evidence and document the details with accident diagrams describing the collision
- To determine the causes that contributed to the accident
- To find the factors influenced/affected, relating to weather, persons, site conditions, vehicular characteristics, visibility, road geometrics and driver condition
- To estimate the loss/injury/damage caused by the accident and the information required for insurance/judiciary
- To identify the possible precautionary measures or develop site-specific improvements
- To establish post-injury management
- To propose national policies on road safety

## 20.2 BASIC REPORT INFORMATION

The systematic recording and reporting of accident data is a prerequisite for any road accident analysis, as well as for the road safety management system. Relevant and reliable data is recorded to enable identification of the contributing factors for each accident and to offer the best way to explore the prevention of accidents, including ways to implement measures to reduce accident severity (PIARC, 2007).

Information and evidence must be documented as a report in a consistent and uniform manner. The following list includes the basic types of information that should be reported with adequate evidence and a description of the collision (Agent and Pigman, 1993).

- Location, time, weather and date of the accident
- Type of accident, along with the nature of the injury to the persons involved
- Information about the units involved, such as the driver, occupants, people, animals and witnesses
- Vehicles and their characteristics – interior, exterior, mechanical and electrical parts
- Vehicle/Property damage
- Widened width of carriageway and superelevation on curves
- Measurement of sight distance on the horizontal and vertical curves
- Geometrics, signs, markings, safety appurtenances and road surface characteristics of the location
- Vehicle occupant information
- Name of the hospital where the injured persons were admitted
- Contributing factors
- Debris location, if any
- Accident description
- Physical location details of post-collision objects
- Collision/Accident scene measurements sketched as a diagram
- Photographs/Video clips
- Measurement of skid marks, yaw marks, scratches, gouges, drag factor, stop location of the vehicles/objects/pedestrians and dents on collided objects
- Distances measured relating to installation of advance signs, presence of delineators, height of signposts and dimensions of signs and lettering size
- Retroreflectivity of pavement markings and road signs
- Statements from the vehicle occupants/witnesses and investigators

### 20.2.1 Accident Location

Google Maps or Survey of India maps or any such urban roadway maps can be used to locate or pinpoint the location of accidents and thus create accident site maps or index maps. GIS/GPS-based maps can also be used as soft copies for analysis and investigation of accidents using appropriate software tools.

Global positioning system (GPS) is the fastest and most accurate way of obtaining reliable data on accident location registration, subject to the requirement that it is suitable in countries with an existing geographic information system (GIS)-based accident data collection program (PIARC, 2007).

The GPS method is also suitable for resolving safety analysis issues relating to road accident locations and for locating accident black-spots (locations on roadways observed to experience > 4 accidents in a year). Visualisation of road accidents with digital geographic localisation can be performed. Mobile GPS devices with sufficient accuracy to obtain accident location data are used for this purpose. The data collected from these mobile devices can then be transferred to a PC for further analysis and the road safety database management system.

## 20.3 ASSESSMENT OF ACCIDENT DATA

Road safety deficiencies are determined from the available accident data. High frequency and concentration of occurrence of accidents on locations are determined, and such sections of the road or road locations are often termed *black-spots*.

Several methods can be used to detect deficiencies based on the parameters assessed at a location, such as accident frequency, accident rate and accident severity. Accident data is assessed based on the following objectives:

- To determine the common characteristics of accidents and their counter measures
- To identify black-spots on road networks with reference to traffic volume data

## 20.4 COLLISION DIAGRAMS

Collision diagrams are a graphical representation of an accident that occurred on a road location. Each accident location is pictorially drawn, and present and past historical accident data with the date, time and type of collision, along with the objects involved, are drawn to (1:200–1:500) or not to scale.

Direction arrows with symbols are used to represent the nature of the collision, including the path traced by the vehicles and the pedestrians with reference to fixed objects on the roadway (Table 20.1). The fixed objects may be listed as electric poles, signal poles, fencing, kerb lines, compound walls and so on. The representing arrows may not be shown on the exact path and collision spot of the accident; this is because of overlap of accidents that occurred on the same spot. The arrows are shown with details of each accident, including the date and time of day to the nearest hour. A typical collision diagram at an intersection of a road is shown in Fig. 20.2.

**Table 20.1** Symbols used for the development of collision diagrams

| Symbol | Description | Symbol | Description |
|---|---|---|---|
| | Trees | —×—×— | Fence |
| | Guard rail | | Building |

(Continued)

**Table 20.1** (Continued)

| | | | |
|---|---|---|---|
|  | Any fixed object on the roadway |  | Signboard mounted on single post |
|  | Signboard mounted on two posts | 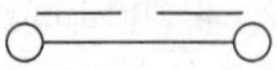 | Overhead sign |
|  | Dotted line arrow indicating pedestrian movement in a direction | 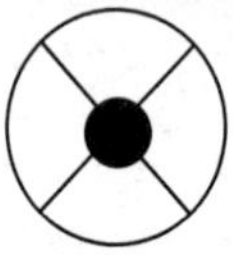 | Pedestrian push button |
|  | Traffic signal |  | Arrow indicating vehicle movement in a direction: The shape of the arrow shows the actual movement, which corresponds to the direction of travel just before the accident |
| 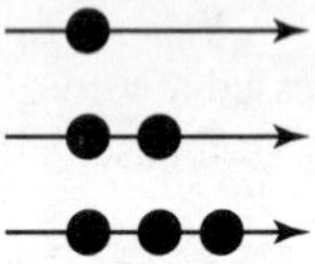 | Arrows with solid round marks indicate the approach speed of vehicles. Each solid round mark indicates the speed of the vehicle with a range of 10 km/hr. For example, an arrow with four solid round marks indicates that the approach vehicle is in the range of 40–50 km/h. | | Head-on collision |
|  | Rear-end collision |  | Side swipe |
| 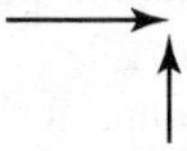 | Angle collision |  | Retarding vehicle movement |
|  | Reversing |  | U-turn |
| 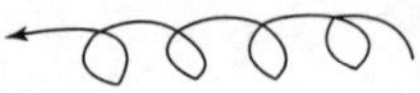 | Uncontrolled vehicle movement/Vehicle out of control |  | Overturned vehicle |
|  | Non-fatal injury | | Fatal injury |
| 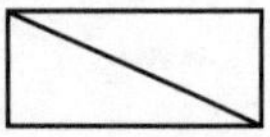 | Parked vehicle | 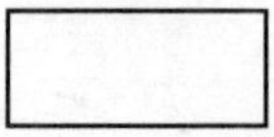 | Stationary vehicle |

(Continued)

**Table 20.1** (Continued)

| | | | |
|---|---|---|---|
| D | Dry road condition | W | Wet road condition |
| I | Ice/Snow road condition | O | Other road condition |
| R | Rain | F | Fog |
| C | Clear | → B | Vehicle collision with a bicycle |
| → P | Vehicle collision with a pedestrian | → A | Vehicle collision with an animal |
| →□ | Vehicle collision with a fixed object | ⩘ | Side swipe in the same direction |
| ⩘ | Side swipe in the opposite direction | | |

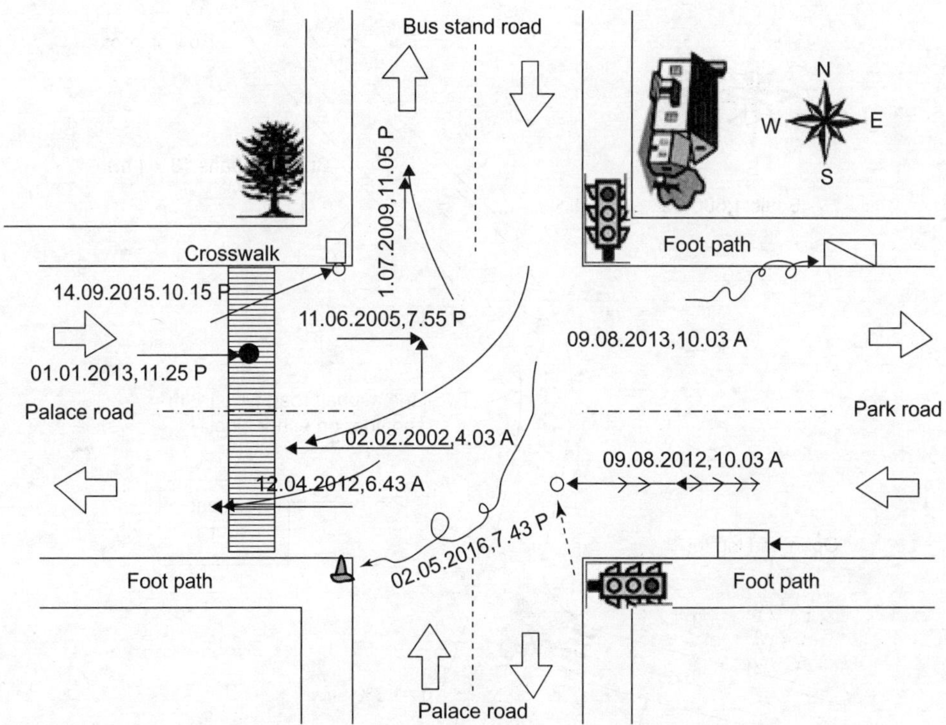


**Fig. 20.2** A typical collision diagram recorded for a location

The main purpose of collision diagrams is to depict the accident type and pattern. They are used for evaluating the possible causes of an accident at a specific site. Such diagrams can be used for analysis within a single intersection (Nyerges and Cihon, 1989).

## 20.5 CONDITION DIAGRAMS

To prepare a collision diagram, the layout of the intersection or road segment, along with its geometric features, must be known. Such a diagram is called a condition diagram. The patterns of accidents with any specific correlation with the features of the site and the existing road geometric conditions, including details pertaining to the approach roads, are represented in condition diagrams drawn to scale (Fig. 20.3). Condition diagrams provide detailed information that may not be easily communicated in a tabular form.

A condition diagram should provide a complete presentation of the physical/geometrical layout of the accident study location as it currently exists, including the appurtenances. Construction drawings done to scale can be used for this purpose. Condition diagrams may be developed to any convenient scale such as in the range 1:200–1:1000.

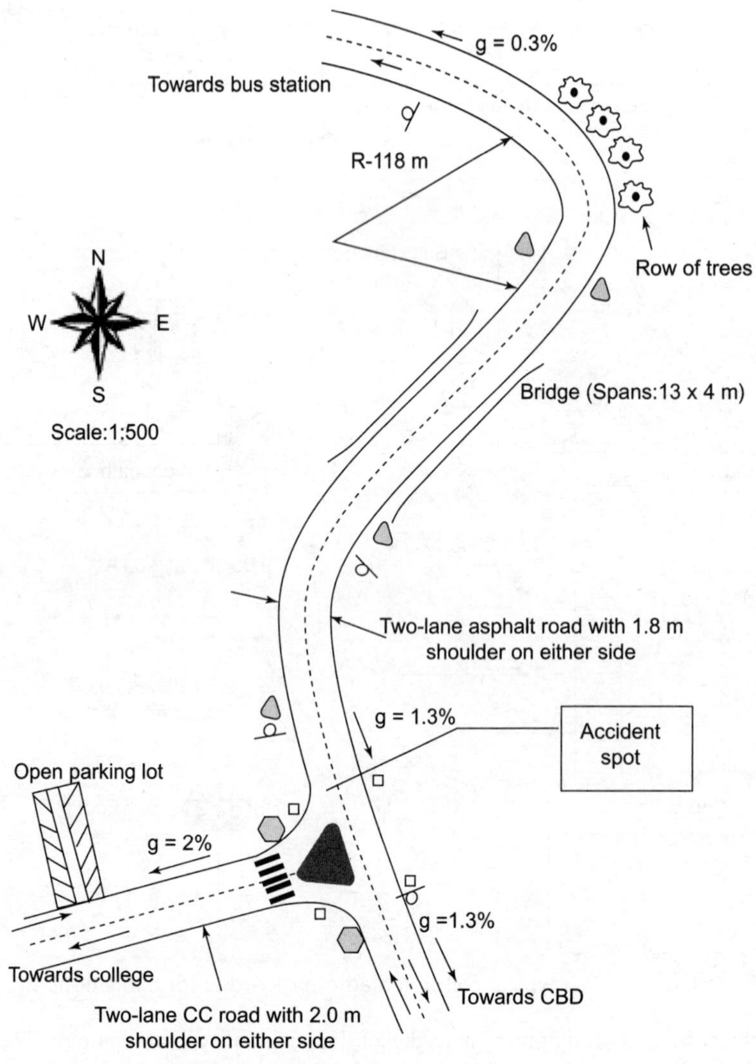


**Fig. 20.3** A typical condition diagram

The following items should be included in the condition diagram which represents the on-site features (TxDOT, 2011):

- A plan drawn to scale with the North direction
- Geometric components of the road site, including
  - Edges of pavement, kerb lines and shoulders with dimensions
  - Roadway limits
  - Widths and grades of approach roads
  - Surface characteristics such as skid resistance, pot holes, ruts and pavement edge drop-offs and other distresses
  - Channelisation details
  - Roadside parking and restriction details
  - Nearby bus stops and pedestrian crossings
  - Illumination details with locations of electric poles
- Roadside developments
  - Obstructions to sight distance due to any object on the roadside such as trees, plants, signs, poles, fences, utility poles and guard rails
  - Adjacent land use
- Position of traffic control devices
  - Traffic flow directions
  - Design speed and 85$^{th}$ percentile speed
  - Signs and markings
  - Traffic signals
  - Distance to adjacent traffic signals within 1.0 km
  - Any other traffic control devices
  - Level crossings with railways and type of control

Local law enforcement agencies use condition diagrams to investigate the current condition in a particular crash location. Condition diagrams are useful for correlating ground features on the road and detecting road deficiencies, for future road geometric design correction.

Collision diagrams and condition diagrams with 'before' and 'after' corrective measures are useful for conducting comparative studies and statistical analysis.

## 20.6 MEASUREMENT SYSTEMS

After locating the spots on the site, the next step is to take measurements of the scene. The main objective of taking measurements of the scene of collision is to produce an after-collision map to scale. The three most widely used methods for measuring scene details are given below (WiDOJ, 2010):

- **The coordinate system**: The distance to vehicles, pedestrians and debris is measured with reference to a nearby straight roadway edge/kerb line gutter line (Fig. 20.4 (a)). In general, the perpendicular offset distances to the points on the road are measured. This method is suitable if an accident has occurred on a straight stretch of a street/road.
- **The triangulation system**: Measurements are taken by using any two permanent landmarks nearby as reference points. These reference points may be electric poles, lamp posts, building corners, manhole corners and so on. The oblique offset distances are measured from each point of interest, thus forming triangle geometry. This method is suitable for accidents that occur on a road curve/bend.

- **Combination method**: This method uses a combination of the coordinate system and the triangulation system (Fig. 20.4 (b)). Both perpendicular and oblique offset distances are measured. This method is suitable for accidents that occur on a road intersection or in any wide-area location. The total station, i.e., the electronic distance measuring instrument may be used for measuring distances and angles.

If required, the objects of interest on the roadway can be marked with spray paint as permanent marking on dry surfaces, or crayon lime on wet surfaces as a temporary marking.

The common practice is to collect evidence and preserve the positions of a few important points in the accident site; go-bags are used for collecting evidence. They usually include a camera, notepad, safety vest, plastic zip-loc bags with labels for collecting evidence, stopwatch, flashlight, calculator, spray paint, marker pen, a piece of chalk, hand level, measuring wheel, 30-m tape measure, string-line, nails, hammer and orange traffic cones (FHWA, 2004).

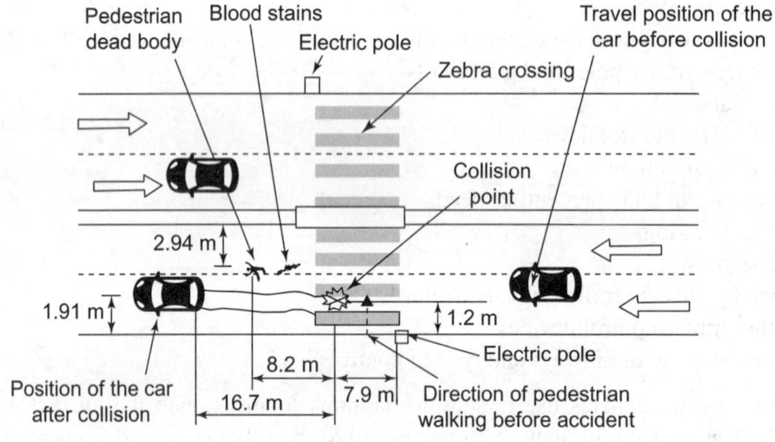


(a) The coordinate system of measurement

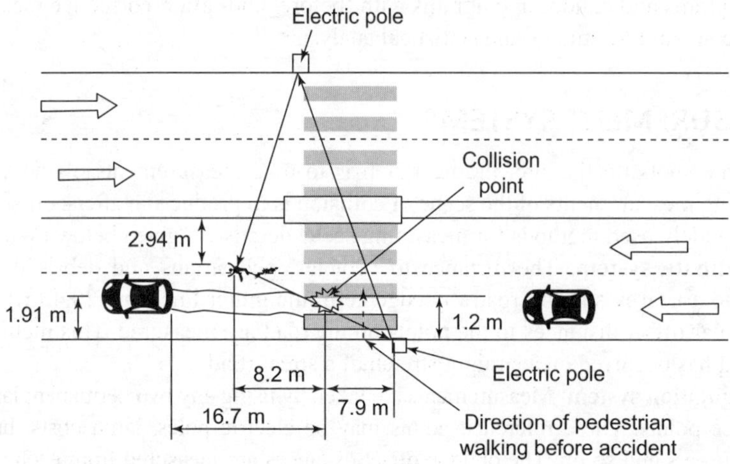


(b) The combination method of measurement

**Fig. 20.4** A typical measurement illustration of an accident

## 20.7 FUNDAMENTAL PHYSICS FORMULAE USED FOR RECONSTRUCTION OF AN ACCIDENT

Some fundamental Physics formulae that are required for understanding and interpreting accident scenes are given below.

$$\text{Momentum} = \text{Mass} \times \text{Velocity}$$

*Momentum equation* (Fig. 20.5): This also known as conservation of linear momentum (COLM) (Momentum In) = (Momentum Out)

$$(W_1 \times V_1 + W_2 \times V_2) = (W_1 \times V_3 + W_2 \times V_4) \tag{20.1}$$

where $W_1$ = weight of the vehicle A; $V_1$ = velocity of the vehicle A before collision; $W_2$ = weight of the vehicle B; $V_2$ = velocity of the vehicle B before collision; $V_3$ = velocity of the vehicle A after collision; and $V_4$ = velocity of the vehicle B after collision.

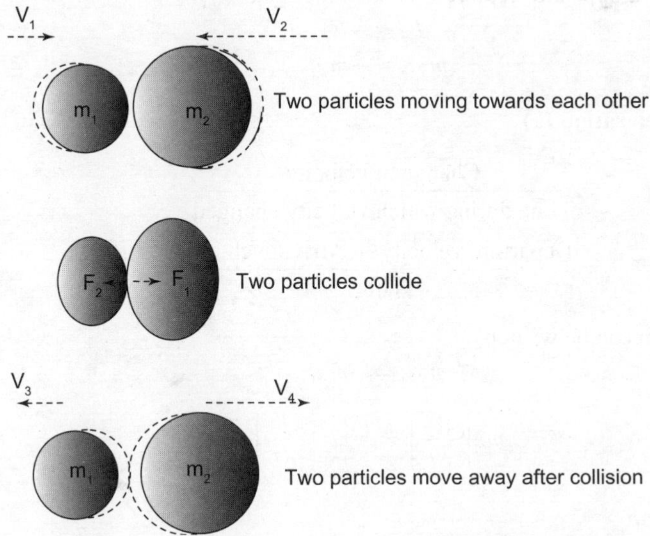


**Fig. 20.5** Phenomenon of linear momentum of two particles

If the two vehicles move together after collision from the rear side, the above equation may be written as:

$$(W_1 \times V_1 + W_2 \times V_2) = (W_1 + W_2) \times V_3 \tag{20.2}$$

where, $V_3 = V_4$

*Note*: We know that mass = weight/acceleration due to gravity ($m = W/g$). Mass of the vehicles is considered as weight here because the acceleration due to gravity ($g \sim 9.81$ m/s$^2$) is the same for all objects on earth.

The pre-impact motion of vehicles is often characterised by kinetic energy (APRI, 2003). The following commonly used form of mathematical description involves the vehicle's weight and speed. As collision commences, the vehicles' kinetic energy and speed reduces by any form, such as stopping (final rest position or FRP), sliding, and damage to vehicles and objects.

$$\text{Kinetic energy } (KE) = \tfrac{1}{2} mV^2 \tag{20.3}$$

where $$m = \text{mass} = \frac{Weight}{g} = \frac{W}{g} \tag{20.4}$$

$V$ = velocity (m/s) and g = acceleration due to gravity = 9.81 m/s$^2$

We can rewrite this as $$KE = \frac{WV^2}{2g} \tag{20.5}$$

**By Newton's 2nd Law**

$$\text{Force} = (\text{Mass}) \times (\text{Acceleration}) \text{ or } F = ma \tag{20.6}$$

With reference to Fig. 20.5, $F_1 = m_1a_1$ and $F_2 = m_2a_2$

**By Newton's 3rd Law**

Forces are equal and opposite. With reference to Fig. 20.5

$$F_2 = -F_1 \tag{20.7}$$
$$m_2a_2 = -m_1a_1$$

**Definition of Acceleration (a)**

$$a = \frac{\text{Change in velocity}}{\text{Time during which velocity changed}} = \frac{\text{Departure velocity} - \text{Arrival velocity}}{\text{Time}} = \frac{\Delta V}{\Delta t} \tag{20.8}$$

The above equation can be written as

$$m_2a_2 = -m_1a_1$$
$$m_2\left(\frac{\Delta V_2}{\Delta t}\right) = -m_1\left(\frac{\Delta V_1}{\Delta t}\right)$$

**Equation of motion**

The distance travelled by a car during acceleration is calculated using the formula

$$(V_{after})^2 - (V_{before})^2 = 2 \times (\text{Acceleration}) \times (\text{Distance travelled}) \tag{20.9}$$

Which is in the form $v^2 - u^2 = 2as$

**Definition of impulse (P)**

Based on the fundamental principle of change in linear momentum, Impulse = Change in momentum ($\Delta$m) This can be expressed as

If $$F = ma, F = m\left(\frac{\Delta V}{\Delta t}\right) \tag{20.10}$$

$$F(\Delta t) = m\,(\Delta V)$$

$$\text{Change in momentum} = m\,(\Delta V) = F(\Delta t) \tag{20.11}$$

The change in momentum will be equal and opposite, when the forces and the time acting on each vehicle are the same.

**Coefficient of restitution**

With reference to linear collision between two bodies (Fig. 20.5 and Eq. 20.1), Newton defined the coefficient of restitution as (20.12)

$$\varepsilon = \frac{\text{Post-collision } \Delta V \text{ of the two vehicles}}{\text{Pre-collision } \Delta V \text{ of the two vehicles}} = \frac{\Delta V \text{ departure}}{\Delta V \text{ arrival}} = \frac{V_4 - V_3}{V_1 - V_2} \tag{20.12}$$

**Drag factor or coefficient of friction**

The drag factor (*f*) depends on the tyre and the road surface characteristics, but is independent of the weight of the vehicle. Drag factor is also termed coefficient of kinetic friction. It is often used in forensic science (in Physics, *f* is termed the coefficient of friction).

$$f = \frac{a}{g} \tag{20.13}$$

Rewriting, we get $a = fg$

Substituting in Newton's 2nd Law,

$$F = ma = \left(\frac{w}{g}\right) \times (fg) = wf \tag{20.14}$$

Drag factor can also be expressed as

$$f = \frac{\text{All parallel forces between the tyres and the road surface}}{\text{Normal weight}} = \frac{F}{W} \tag{20.15}$$

Or

$$f = \frac{F \text{ (Horizontal force)}}{W \text{ (Vertical force)}}$$

Therefore, the friction force can be calculated as

$$F = f \times W$$

In forensic investigations of road vehicle accidents, on-site tests are conducted to determine the drag factor, and the test is termed the drag test (Fig. 20.6). Specialists drag one of the weighted tyres along the roadway to determine the force and find the drag factor.

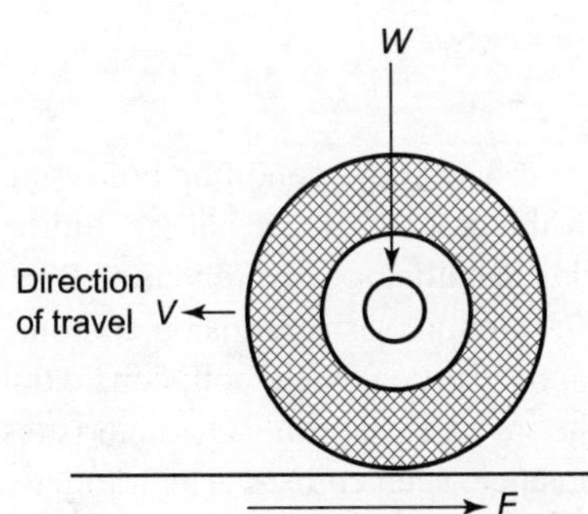


**Fig. 20.6** Free-body diagram of forces acting on a rolling wheel

**Delta-V**

Delta-V = $\Delta V$ = Difference in velocities. $\Delta V$ refers to the change in velocity between the pre-collision and post-collision trajectories of vehicles, and it is written as (Shelby, 2011):

$$\Delta V = V_{after} - V_{before} \tag{20.16}$$

In accident reconstruction analysis, $\Delta V$ is used as the best single predictor of injury and fatality risk in a crash or the crash severity (Joksch et al., 1998). $\Delta V$ can be calculated for the given condition:

Let a vehicle-1 with mass $m_1$ travel at a pre-collision velocity of $V_1$ and a slow-moving vehicle-2 with mass $m_2$ travel at a pre-collision velocity of $V_2$. The post-collision velocities of vehicle-1 and vehicle-2 are represented as $\tilde{V}_1$ and $\tilde{V}_2$, respectively. The change in velocity of the respective vehicles is calculated as:

$$\Delta V_1 = \tilde{V}_1 - V_1 \tag{20.17}$$

$$\Delta V_2 = \tilde{V}_2 - V_2 \tag{20.18}$$

If both the vehicles are travelling in the same linear direction and their collision is assumed to be *inelastic* (like two clay balls colliding), the two vehicles stick together (with the combined mass $m_1 + m_2$) and move together after the collision at velocity $\tilde{V}$ ($\tilde{V}_1 = \tilde{V}_2 = \tilde{V}$).

The equation for this relationship, which conserves momentum, is written as

Momentum of vehicle-1 = $m_1V_1$ Momentum of vehicle-2 = $m_2V_2$

Momentum of the two vehicle stuck together post-collision = $(m_1+m_2)\,\tilde{V}$

$$m_1V_1 + m_2V_2 = (m_1+m_2)\tilde{V}$$

$$\tilde{V} = \frac{m_1V_1 + m_2V_2}{m_1 + m_2} \tag{20.19}$$

If we assume, $\tilde{V}_1 = \tilde{V}_2 = \tilde{V}$, then the above equations (Eq. 20.17 and Eq. 20.18) may be written as follows:

$$\Delta V_1 = \tilde{V} - V_1 = \frac{m_1V_1 + m_2V_2}{m_1 + m_2} - V_1$$

$$= \frac{m_1V_1 + m_2V_2 - m_1V_1 - m_2V_1}{m_1 + m_2} = \frac{m_2}{m_1 + m_2}(V_2 - V_1) \tag{20.20}$$

Similarly,

$$\Delta V_2 = \tilde{V} - V_2 = \frac{m_1}{m_1 + m_2}(V_1 - V_2) \tag{20.21}$$

*Note*: The above condition holds good in the case of inelastic collision.

In the case of rubber balls colliding, they rebound off each other and the collision is termed a pure elastic collision. This rebound effect is modelled using the coefficient of restitution, which equals zero (0.0) for an elastic collision and one (1.0) for an inelastic collision.

In reality, vehicle collisions exhibit neither pure elastic nor pure inelastic collision. It depends on the speed, road profile characteristics, mass and the material characteristics of the colliding vehicles. In practice, such collisions tend to have a coefficient of restitution of about 0.4 at low speed for bumper-to-bumper collisions and may reduce to 0.1 for higher speed collisions, where the vehicle body begins to cave in (dent) on collision (Nordoff, 2005).

## 20.8 ACCIDENT RECONSTRUCTION FORMULAE

The quantification of characteristics that lead/led to an accident can be useful in designing safer roadways and aid the legal system in administrating justice (Reed, 1983). Therefore, a uniform method of investigation and reconstruction of accidents is important to ensure safer roads. A few simple cases of reconstruction of accidents are described, derived and solved in this section (Lofgren,1983).

### 20.8.1 Determination/Measurement of Drag factor of a Road Surface

The drag factor (*f*) of a road surface on which skid marks are found can be measured by any one of the following methods (APRI, 2003).

**Conventional vehicle test:** A vehicle and a measuring tape are used for this test. The investigator drives the vehicle at the documented speed (usually 50 km/hr) and skids to a stop. Later, the skid mark length is measured and the coefficient of friction is calculated using the following equation:

$$f = \frac{V^2}{2gD}$$

where $f$ = average coefficient of friction; $V$ = velocity of the vehicle prior to the skid (m/s); $D$ = length of the skid (m); and g = gravitation acceleration (9.81 m/s$^2$).
Derivations of the above equation are presented in subsequent paragraphs of this chapter. The reliability of the test result depends on the vehicle brake efficiency, tyre characteristics and the temperature.

For example, if an investigator drives at 50 km/h (0.278 × 50 = 13.9 m/s) and skids to a stop and the skid mark length measured was 14 m, the coefficient of friction is calculated as:

$$f = \frac{13.9^2}{2 \times 9.81 \times 14} = 0.703$$

**Drag sled test:** A piece of tyre and a spring scale are used to measure the drag factor of a road surface. The piece of tyre is dragged with a spring scale over the test surface and the pulling force (due to road surface friction) is measured (Fig. 20.7). The tyre is pulled at constant speed until the pull is smooth in the direction travelled by the vehicle and is as close as possible to the actual tyre marks. With the measured weight of the piece of tyre, the drag factor is determined as

$$\text{Drag factor} = \frac{\text{Friction force of the road surface}}{\text{Normal force of the tyre}}$$

Or

$$\text{Drag factor} = \frac{\text{Pulling force measured on the spring scale}}{\text{Weight of the tyre sled}}$$

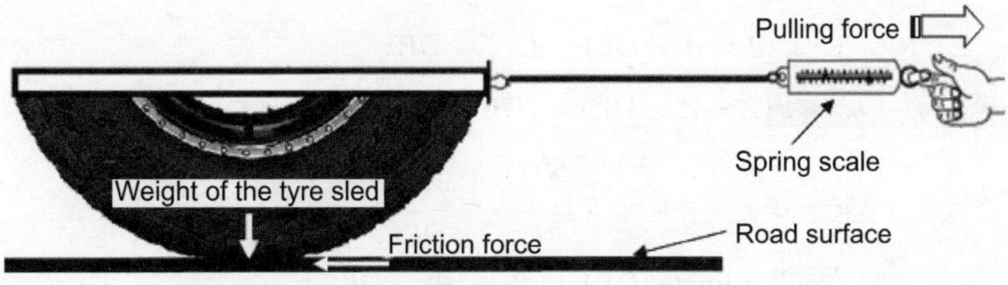


**Fig. 20.7** Basic principle of drag sled test

The investigators may use the lowest measured value of drag factor.
For example, if the weight of the pressure-tyre sled used is 170 N and the pulling force measured by the spring scale is 120 N, the drag factor is calculated as:

$$\text{Drag factor} = \frac{120\,N}{170\,N} = 0.705$$

*Note*: The drag sled test is not recommended for a wet road surface because its weight would squeegee the water out from under the tyre tread. However, the surface condition should be the same when the accident occurred.

The following is a list of possible ranges of drag factor values pertaining to specific compositions of road surfaces:

- Dry bituminous and cement: 0.60–0.80
- Wet bituminous and cement: 0.45–0.70
- Loose ice/snow: 0.10–0.25

### 20.8.2 Determination of the Radius of a Simple Circular Road Curve

Determination of the radius (R) and its mid-ordinate (DE) of a circular curve are the most important parameters (Fig. 20.8) used for calculation of maximum allowable speed and sight distance requirement, respectively. The details of these derivations are given below.

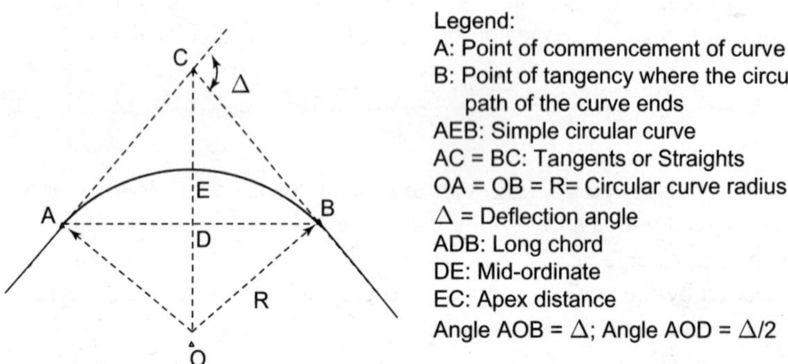


**Fig. 20.8** Elements of a simple circular curve

From angle ADO, based on the Pythagorean theorem, $OB^2 = OD^2 + DB^2$

Substituting, $$R^2 = (R - DE)^2 + DB^2$$

$$R^2 = R^2 + DE^2 - (2 \times R \times DE) + DB^2$$

$$R^2 - R^2 + (2 \times R \times DE) = DE^2 + DB^2$$

$$(2 \times R \times DE) = (DE^2 + DB^2)$$

$$R = \frac{(DE^2 + DB^2)}{2 \times DE} \tag{20.22}$$

where, $$\text{mid-ordinate (DE)} = R - \sqrt{R^2 - DB^2} \tag{20.23}$$

*Note*: The values of DB can be obtained as half the distance measured between the two tangent points A and B.

### 20.8.3 Determination of Critical Speed on Superelevated Road Curves

If a vehicle is travelling along a horizontal curve, there is a chance that it may fall either by overturning or by skidding due to centrifugal force. The intensity of the centrifugal force depends on the radius of the curve, and the weight and velocity of the travelling vehicle. To counteract the centrifugal force acting on the vehicle while travelling on the horizontal curve, the outer edge of the carriageway surface is elevated with respect to the inner side. This elevated outer edge of carriageway or the provision of a slope in the transverse direction of the carriageway along a horizontal curve is called *superelevation* (*e*) or *cant* or *banking*. The different forces acting on a vehicle travelling on a curved path with respect to its centre of gravity (cg) are presented in Fig. 20.9.

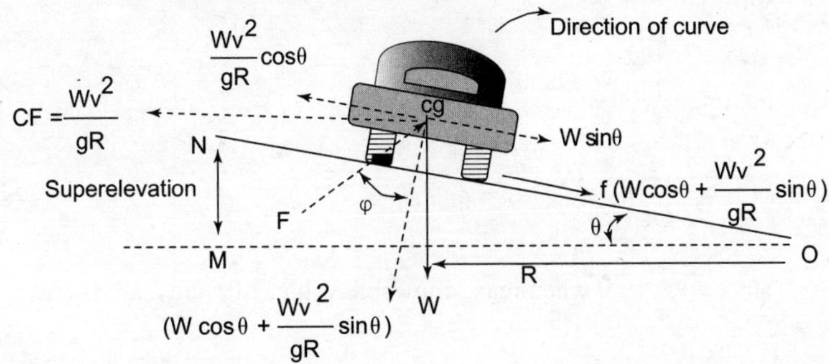


**Fig. 20.9** Forces acting on a vehicle travelling on a circular curve provided with superelevation (*Source*: Srinivasa, 2011).

The superelevation may be expressed as given below.

(i) Rate of superelevation (in decimal form) $= e = \tan\theta = \dfrac{MN}{ON}$

(ii) 1 in $n$ = 1 unit rise for $n$ units of horizontal distance; for example: 1 in 700 = 1 cm rise for 700 cm of horizontal distance.

(iii) $n$ per cent = $n$ unit rise for 100 units of horizontal distance; for example: 7% superelevation for 7-m-wide carriageway means the outer edge of the carriageway is raised by [(7/100) × 700 cm)] = 49 cm above its inner edge.

The relation between the forces that cause friction across the carriageway and the centrifugal force of the vehicle travelling on a circular curve provided with superelevation is given below.

Let $W$ = weight of the vehicle; g = acceleration due to gravity (m/s$^2$); $v$ = velocity of the vehicle (m/s); $R$ = radius of the curve (m); θ = angle of carriageway surface or transverse slope of carriageway; $F$ = resultant of the weight of the vehicle and the centrifugal force; $f = \tan(\phi)$ = coefficient of the side friction between the tyres and the road surface or the side-friction factor (a non-dimensional value); and CF = centrifugal force which pushes the travelling vehicle outward from the centre of the horizontal curve. It is perpendicular to the force of gravity. The centrifugal force is inversely proportional to the curve radius. As the radius of the curve becomes smaller, the CF becomes greater, and vice versa.

In highway design, the CF is counteracted by side friction and the component of the weight of the vehicle caused by superelevation (Awadallah, 2005). Consider the superelevated surface (Fig. 20.9). CF = Horizontal component of the weight of the vehicle + Force caused due to surface friction of the carriageway against sliding

$$\frac{Wv^2}{gR}\cos\theta = W\sin\theta + f\left(W\cos\theta + \frac{Wv^2}{gR}\sin\theta\right)$$

$$\frac{v^2}{gR}\cos\theta = \sin\theta + f\cos\theta + f\frac{v^2}{gR}\sin\theta$$

$$\frac{v^2}{gR}(\cos\theta - f\sin\theta) = \sin\theta + f\cos\theta$$

Divide by $\cos\theta$ on both sides,

$$\frac{v^2}{gR}(1 - f\tan\theta) = \tan\theta + f$$

$$\frac{v^2}{gR}(1 - f \times e) = e + f \qquad \text{(since } \tan\theta = e\text{)}$$

$$\frac{v^2}{gR}(1 - 0) = e + f \text{ (since } f \times e \approx 0 \text{ when max. allowable values of } f \text{ and } e \text{ are taken)}$$

$$e + f = \frac{v^2}{gR}$$

$$v = \sqrt{gR(e + f)} \qquad (20.24)$$

From the above relationship (20.24), the critical (maximum permissible) speed can be determined for the given values of radius of the horizontal curve, the maximum values of superelevation and the side friction factor.

**Example 20.1**

What is the maximum allowable speed on a horizontal road-curve if the available drag factor = 0.164, superelevation = 7% and the radius of the horizontal curve = 215 m?

**Answer:** By Eq. 20.24,

$$v = \sqrt{gR(e + f)}$$

$$v = \sqrt{9.81 \times 215(0.07 + 0.164)} = 21.54 \text{m/s}$$

$$= 77.48 \text{ km/h}$$

### 20.8.4 Determination of Minimum Speed of the Vehicle Before Sliding (From Straight Line Skid Marks)

When a vehicle moves on a road, its tyres will roll and no significant mark will be formed on the road surface. Once the brake is applied, the wheels will be locked against rolling and the locked tyres slip on the road, causing skid marks on the road surface. The skid marks begin when the brake is applied and end when the vehicle stops. The friction between the tyres and the road surface causes the vehicle to decelerate and stop.

Based on the geometry, the skid marks may be categorised as straight line skid marks and yaw marks. The *straight line skid mark* indicates that the vehicle wheels are locked due to braking and the tyres are not rotating while slipping along its direction of travel. The drag factor is measured parallel to the direction of travel.

By measuring the length of the skid marks, it is possible to estimate the vehicle speed just before sliding, i.e., immediately after the wheels are locked, using the following formula:

$$V = \sqrt{2gfD} \tag{20.25}$$

where $V$ = minimum speed of the vehicle just before sliding, i.e., to make skid mark (m/s); $f$ = drag factor; $D$ = length of the skid mark (m), when it is assumed that the braking efficiency of the vehicle is 100%.

The above equation can be derived by equating kinetic energy with the work done by the vehicle before it will slide to a stop (Fig. 20.10).

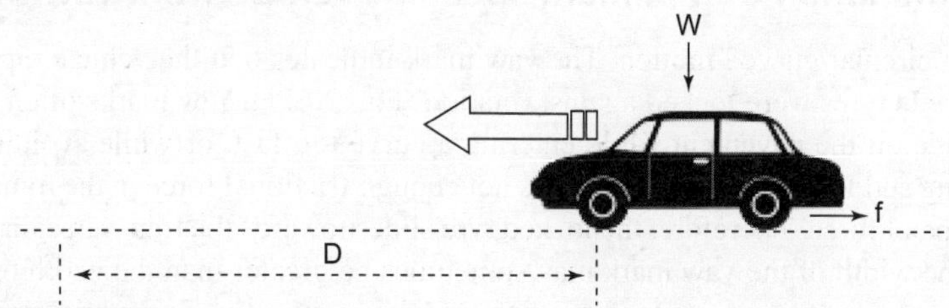


**Fig. 20.10** Visualisation of determination of minimum speed from a skid mark

The workdone can be expressed as follows:

(i)

$$\text{Work done} = \text{Force} \times \text{Distance}$$

Or

$$\text{Work done} = \text{Weight of vehicle} \times \text{Drag factor of the road surface} \times \text{Distance slided to stop}$$

$$= WfD$$

(ii)

$$KE = \textit{Work done}$$

$$\frac{1}{2}mv^2 = Force \times Dis\tan ce$$

$$\frac{1}{2}mv^2 = F \times D$$

$$v^2 = \frac{2FD}{m} = \frac{2(ma)D}{m} = 2(a)D \quad \because F = ma$$

$$v^2 = 2(fg)D \quad \because a = fg$$

$$V = \sqrt{2gfD}$$

where $D$ = skid length (m) and $f$ = drag factor.

*Note*: The above equation may underestimate the speed because the tyre marks start forming after braking commences (APRI, 2003).

**Example 20.2**

A jeep skids to a stop on a road, leaving four skid marks under its four wheels, with an average length of 40 m. Skid tests reveal a drag factor of 0.70. Estimate the speed of the jeep just before sliding.

**Answer:**

$$V = \sqrt{2fgD} = \sqrt{2 \times 9.81 \times 40 \times 0.70} = 23.43\text{m/s}$$

or

$$V = \sqrt{254Df} = \sqrt{254 \times 40 \times 0.70} = 84.33 \text{ km/h}$$

*Note*: The above formula assumes that the vehicle comes to a stop at the end of the skid without hitting any object, such as a vehicle or a tree

## 20.8.5 Determination of Minimum Speed of Vehicle from Yaw Mark

Vehicles skid in a circular/curved motion. The yaw mark indicates that the vehicle rapidly turned in a *curved path* while its tyres were locked against rotation (Fig. 20.11). Yaw marks often occur when the locked wheels slide on the pavement while entering a curve too fast, or while avoiding an object by turning the steering suddenly. As a result, there is not enough frictional force in the transverse direction against the centripetal force. Therefore, the locked tyre slides towards the transverse direction of travel. For this reason, the width of the yaw mark may sometimes be greater than the width of a straight skid line. The drag factor is measured perpendicular to the direction of the yaw mark, since the frictional force will be developed in the traverse direction of its turning path.

The speed at the point of yawing of the vehicle can be derived by balancing the centripetal force and the frictional force that develops while yawing.

$$\text{Centripetal force} = \text{Frictional force}$$

$$\frac{WV^2}{gR} = f \times W; \quad \frac{V^2}{gR} = f$$

$$V^2 = fgR; V = \sqrt{fgR} \tag{20.26}$$

where $W$ = weight of the yawing vehicle; $V$ = vehicle vef the yaw skid mark (m) (see Fig. 20.11); $f$ = drag factor; and g = acceleration due to gravity, 9.81 m/s$^2$.

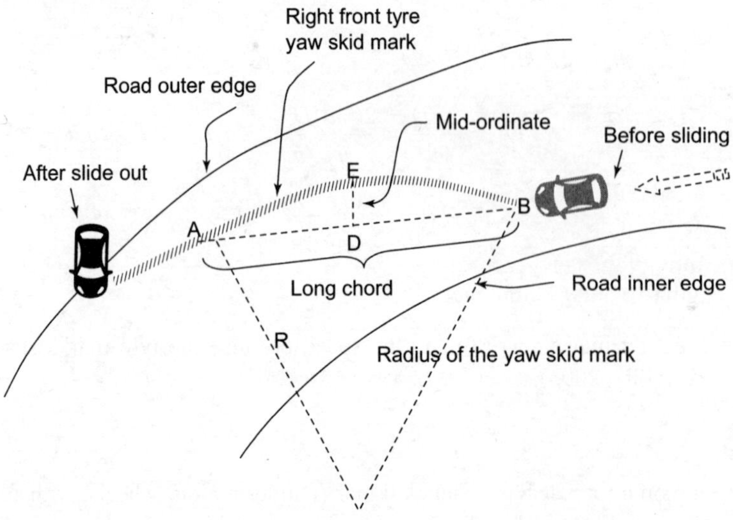


**Fig. 20.11** Details of the measurement of a typical yaw skid mark

**Example 20.3**

Determine the velocity of a vehicle just before sliding, from the following data:
Mid-ordinate of the yaw skid mark = 1.75 m; Length of the long chord = 18 m; Drag factor = 0.72.

**Answer:** With reference to Fig. 20.11, let us calculate the radius of the yaw skid mark using Eq. (20.22).

$$R = \frac{\left(\mathrm{DE}^2 + \mathrm{DB}^2\right)}{2 \times DE}$$

Let AB = Length of the long chord = 18 m (given).

$$\mathrm{AD} = \mathrm{BD} = ½\mathrm{AB} = 18/2 = 9\ \mathrm{m}$$

$$\text{Radius of the yaw skid mark} = \mathrm{R} = \frac{\left(1.75^2 + 9^2\right)}{2 \times 1.75} = 24.018\,\mathrm{m}$$

By Eq. (20.26),

$$V = \sqrt{fgR} = \sqrt{0.72 \times 9.81 \times 24.018}$$
$$= \sqrt{169.644} = 13.025\,\mathrm{m/s} = 46.889\,\mathrm{km/h}$$

## 20.8.6 Determination of Acceleration Factor of a Vehicle from Known Distance or Travel Time

The acceleration factor is defined as the ratio of the rate of acceleration of a vehicle to the acceleration due to gravity.

$$\text{Acceleration factor} = f_a = \frac{\text{Rate of acceleration of the test vehicle}}{\text{Acceleration due to gravity}\,(\mathrm{g})} \tag{20.27}$$

Acceleration factor is determined by conducting investigation test runs on the same road surface and in the same direction travelled by the vehicle that took part in the accident. The test vehicle used for determining the acceleration factor should be similar to that of the vehicle involved in the accident. The test vehicle characteristics may be listed as dimensions, weight, body features, horse power, design speed of the location and brake efficiency.

In this test, the test vehicle is accelerated to the speed known/determined and the time required to reach this speed is recorded. Based on these parameters, we can determine the rate of acceleration of the test vehicle as follows:

$$\text{Rate of acceleration of the test vehicle} = a = \frac{\text{Velocity of the test vehicle}\,(\mathrm{m})}{\text{Tme taken to reach that speed}\,(\mathrm{s})} \tag{20.28}$$

The acceleration factor can also be determined by conducting a test run of the vehicle to accelerate till it reaches the known speed and reaches the known distance. Rewriting Eq. (20.28), we get

$$V = \sqrt{2gfD}$$

$$f_a = \frac{V^2}{2gD} \tag{20.29}$$

If the acceleration factor is known or determined based on the test runs conducted on the site, the distance travelled during the acceleration period or the time taken to travel the distance can also be calculated using the above equation.

**Example 20.4**

What is the rate of acceleration and acceleration factor of a test vehicle when it goes from 0 km/hr to 39.6 km/hr in 7 s?

**Answer:** Let us convert the speed 39.6 km/hr $= 0.278 \times 39.6 = 11$ m/s
By Eq. (20.28),

$$\text{Rate of acceleration of the test vehicle} = a = \frac{\text{Velocity of the test vehicle (m)}}{\text{Tme taken to reach that speed (s)}}$$

$$a = \frac{11\,\text{m/s}}{7\,\text{s}} = 1.57\ \text{m/s}^2$$

By Eq. (20.27) $$\text{Acceleration factor} = f_a = \frac{\text{Rate of acceleration of the test vehicle}}{\text{Acceleration due to gravity (g)}}$$

$$\text{Acceleration factor of the test vehicle on that site} = f_a = \frac{1.57\ \text{m/s}^2}{9.81\ \text{m/s}^2} = 0.16$$

**Example 20.5**

What is the acceleration factor of a test vehicle when it goes from 0 km/hr to 39.6 km/hr in 33 m?

**Answer:** Let us convert the speed 39.6 km/hr $= 0.278 \times 39.6 = 11$ m/s

By Eq. (20.29), $$f_a = \frac{V^2}{2gD} = f_a = \frac{11^2}{2 \times 9.81 \times 33} = 0.18$$

**Example 20.6**

A test vehicle goes from 0 km/hr to 39.6 km/hr with a known acceleration factor of 0.17. What is the distance it travels?

**Answer:** Let us convert the speed 39.6 km/hr $= 0.278 \times 39.6 = 11$ m/s

By Eq. (20.29), $$D = \frac{V^2}{2gf_a} = \frac{11^2}{2 \times 9.81 \times 0.17} = 36.27\,\text{m}$$

*Note*: Sometimes, this distance can be compared with the measured length of the skid mark.

**Example 20.7**

A vehicle running at a speed of 61.2 km/h braked and locked its wheels. The vehicle slid to a stop with a drag factor of 0.70. What is the time required for the vehicle to stop?

**Answer:** Let us convert the speed 61.2 km/hr = 0.278×61.2 = 17 m/s

From the fundamental principles, we know that Time = Velocity/Acceleration, where acceleration (deceleration) = g × Drag factor = 9.81× 0.70 = 6.867

Time taken for the vehicle to stop = 17/6.867 = 2.47 s

**Example 20.8**

A vehicle accelerates for 10 s with an acceleration factor of 0.16. What is the velocity of the vehicle that can be attained?

**Answer:** From the fundamental principles, we know that Velocity = Acceleration × Time

By Eq. (20.13), acceleration of the vehicle = g × Drag factor = 9.81× 0.16 = 1.5696 m/s²

Substituting, we get Velocity = 1.5696 m/s² × 10 s = 15.696 m/s

= 3.6 × 15.696 m/s ≈ 56.5 km/h.20.8.7

## 20.8.7 Combined Speed of a Vehicle Skidding on Road Surface

Suppose, the wheels of a vehicle in motion are locked and it skids on a road surface and comes to a standstill, the following equation holds true

$$\text{Kinetic energy} = \frac{1}{2}mV_{after}^2 - \frac{1}{2}mV_{initial}^2$$

The above expression can also be written as

$$\text{Work done} = \frac{1}{2}mV_{after}^2 - \frac{1}{2}mV_{initial}^2$$

We know, Weight of vehicle × Drag factor of the road surface × Distance slided to stop = $WfD$

$$-\mathrm{W}f\mathrm{D} = \frac{1}{2}mV_{after}^2 - \frac{1}{2}mV_{initial}^2$$

The minus sign is used because the vehicle stops due to deceleration. Rewriting, we get

$$-2\mathrm{W}f\mathrm{D} = \frac{1}{2}mV_{after}^2 - \frac{1}{2}mV_{initial}^2$$

$$mV_{initial}^2 = mV_{after}^2 + 2\mathrm{W}f\mathrm{D}$$

$$V_{initial}^2 = V_{after}^2 + 2\frac{\mathrm{W}}{\mathrm{m}}f\mathrm{D}$$

$$V_{initial}^2 = V_{after}^2 + 2\mathrm{g}f\mathrm{D}$$

$$V_{initial} = \sqrt{V_{after}^2 + 2\mathrm{g}f\mathrm{D}}$$

We know, $\mathrm{V}_{\mathrm{before}}$ = Speed of the vehicle before sliding $= V = \sqrt{2gfD}$

Finally, we can rewrite this as $V_{Combined} = \sqrt{V_{after}^2 + 2\mathrm{g}f\mathrm{D}} = \sqrt{V_{after}^2 + V_{before}^2}$

**Example 20.9**

In a multi-phase accident, the vehicle may have undergone several speed changes. For example, a vehicle loses 50 km/h in sliding before impact, 40 km/h in the collision and 20 km/h in sliding to a stop after the impact. Determine the original speed of the vehicle.

**Answer:** Original speed of the vehicle $= V = \sqrt{V_1^2 + V_2^2 + V_3^2}$

Substituting, $V = \sqrt{50^2 + 40^2 + 20^2} = \sqrt{2500 + 1600 + 400} = \sqrt{4500} = 67.08\,\text{km/h}$

## 20.8.8 Speed of a Vehicle Falling from a Downgrade/Upgrade

A vehicle is assumed to have run off an embankment at uncontrolled speed and travel a certain distance in the air before falling to the ground (Fig. 20.12). The initial velocity of the vehicle can be determined from the fundamental principles, when the free fall height ($H$) and horizontal distance ($D$) measurements are taken from the accident location.

Horizontal distance travelled by a vehicle $= D = V \times t$ (20.30)

where $V$ = velocity of the vehicle (m/s) and $t$ = duration of the travel time when the vehicle is in the air (s). The height of fall can be determined from the kinetic energy principle when the height of the embankment is known.

$$H = \tfrac{1}{2}gt^2 \tag{20.31}$$

$$t = \sqrt{\frac{2H}{g}}$$

Substituting Eq. (20.31) in Eq. (20.30), we get $D = V \times \sqrt{\frac{2H}{g}}$

Rewriting, we get $V = D \times \sqrt{\frac{g}{2H}}$ (20.32)

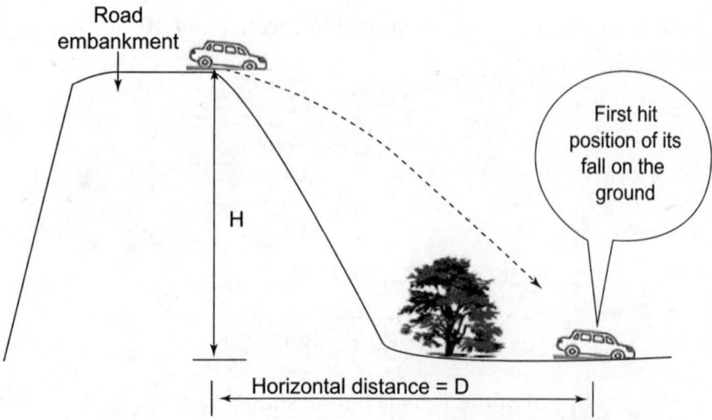


Fig. 20.12 A car falling from an embankment (Elaborated based on Lofgren, 1983)

The above formula changes if the vehicle was moving down a grade and later fell down through the air (Fig. 20.13) (Lofgren, 1983). With reference to Fig. 20.14, the distance travelled in the vertical plane

$$H = (V \times \text{Sin } \theta) \times T + \tfrac{1}{2} g T^2 \tag{20.33}$$

$$\text{Horizontal distance} = D = (V \times \text{Cos } \theta) \times T$$

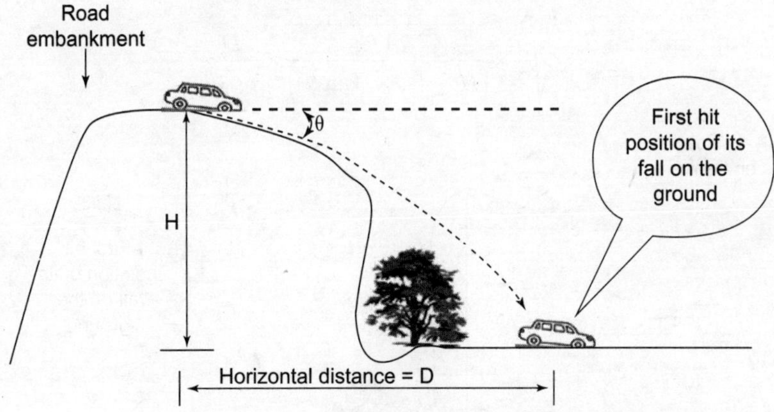


**Fig. 20.13** A car moving down a grade before falling from an embankment
(Modified and elaborated based on Lofgren, 1983)

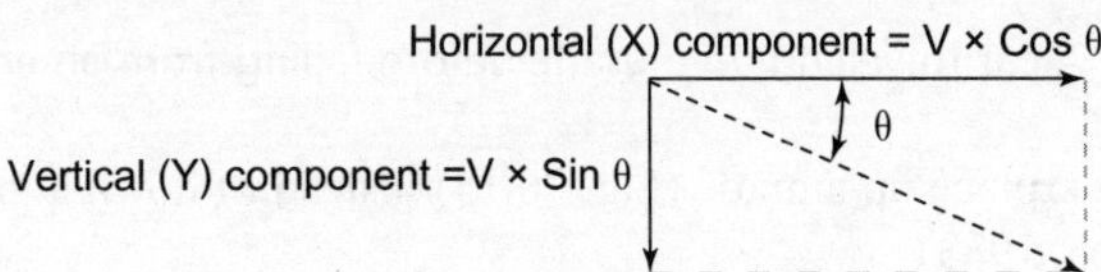


**Fig. 20.14** Components of speed (V)

Rewriting, we get
$$T = \frac{D}{(V \times \text{Cos } \theta)} \tag{20.34}$$

Substituting Eq. (20.34) in Eq. (20.33), we get $H = (V \times \text{Sin } \theta) \times T + \tfrac{1}{2} g T^2$

$$H = (V \times \text{Sin } \theta) \times \frac{D}{(V \times \text{Cos } \theta)} + \tfrac{1}{2} g \left( \frac{D}{(V \times \text{Cos } \theta)} \right)^2$$

$$H = (D \times \text{Tan } \theta) + \tfrac{1}{2} g \left( \frac{D}{(V \times \text{Cos } \theta)} \right)^2$$

$$H = (D \times \text{Tan } \theta) + \tfrac{1}{2} g \times \frac{D^2}{V^2 (\text{Cos } \theta)^2}$$

$$H - (D \times \text{Tan } \theta) = \frac{g}{2} \times \frac{D^2}{V^2 (\text{Cos } \theta)^2}$$

$$V^2 \times (H - (D \times \text{Tan}\,\theta)) = \frac{g}{2} \times \frac{g \times D^2}{(\text{Cos}\,\theta)^2}$$

$$V^2 = \frac{g}{2} \times \frac{1}{(H - (D \times \text{Tan}\,\theta))} \times \frac{g \times D^2}{(\text{Cos}\,\theta)^2}$$

$$V = \sqrt{\frac{g}{2} \times \frac{1}{(H - (D \times \text{Tan}\,\theta))}} \times \frac{D}{\text{Cos}\,\theta} \qquad (20.35)$$

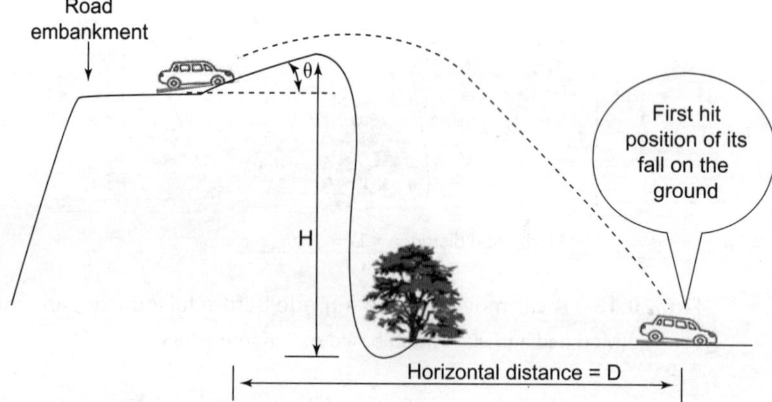


**Fig. 20.15** A car travelling up a grade before falling from an embankment

If the vehicle take-off took place up a grade (Fig. 20.15) instead of down a grade, the above equation can be written as (Lofgren, 1983)

$$V = \sqrt{\frac{g}{2} \times \frac{1}{(H + (D \times \text{Tan}\,\theta))}} \times \frac{D}{\text{Cos}\,\theta} \qquad (20.36)$$

For approximation, as the Cos of zero or small angles is 1.0, it may be removed from the above equations, and Tan θ may be replaced by the grade value in %. For example, if $\theta = 6°$, Tan $(6°) = 0.105$. Hence, the gradient = 10.5%.

**Example 20.10**

A car ran off an embankment and fell in a ditch. The measured horizontal distance from the verge of the embankment to the ditch was 6 m and the height between the two points was 2.1 m. Determine the falling speed of the car from the embankment.

**Answer:** By Eq. (20.32),

$$V = D \times \sqrt{\frac{g}{2H}}$$

Substituting, we get

$$V = 6 \times \sqrt{\frac{9.81}{2 \times 2.1}} = 9.17 \text{ m/s}$$

$$= 3.6 \times 9.17 = 33.0.12 \text{ km/h}.$$

### 20.8.9 Determination of Velocities of Vehicles at Impact/Collision

The law of conservation of momentum is used for this purpose. It states that, for any collision, the vector sums of the momentums (mass times velocity with direction) due to the colliding bodies before and after collision are equal. This principle is useful because once the vector sum of momentum after collision is determined, it is possible to determine the momentum before the impact (see Eqs (20.1) and (20.2)). That means that the velocities of the vehicles at impact can be determined.

The measured length of skid marks on the road surface, direction of skid marks that lead from the point of impact to the point of rest, the approach angle of the colliding vehicles, the approximate weights including payload and the accelerating/decelerating capacity of the vehicles involved in an accident are the parameters commonly used for this type of analysis. These details must be collected/obtained at the collision site. It must be noted that no momentum equation holds good if the collided vehicle positions are altered or if the vehicles were driven off the site without leaving any evidence.

The following construction of the collision of two vehicles may be considered for the application of the momentum equation.

**Case I: Right-angle collision:** Let us resolve the velocity vectors in the X direction (Fig. 20.16)

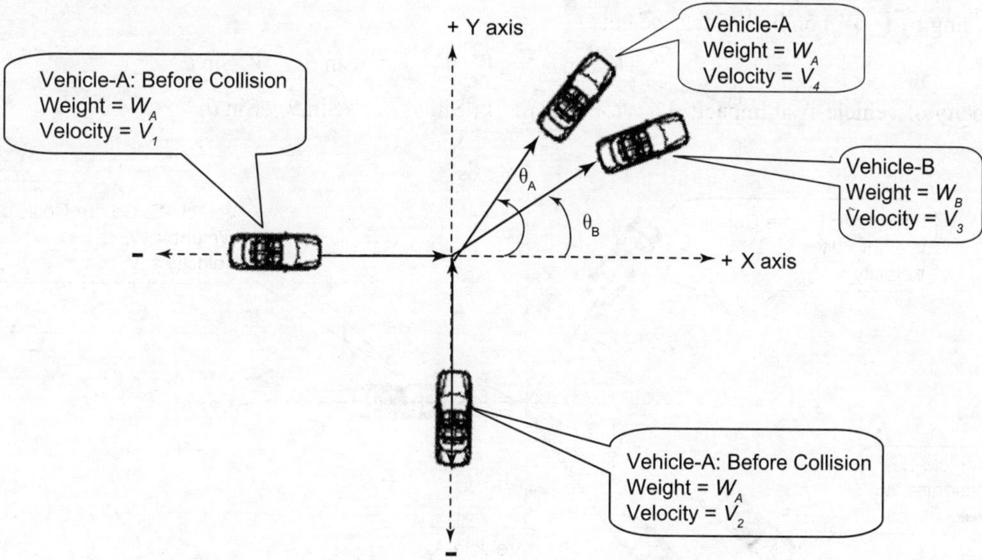


Fig. 20.16 Typical construction of collision of vehicles: right-angle collision

Momentum before collision = Momentum after collision

$$(W_A \times V_1) = (W_A \times V_4 \text{ Cos } \theta_A + W_B \times V_3 \text{ Cos } \theta_B)$$

Velocity of vehicle-A at impact: $$V_1 = V_4 \text{ Cos } \theta_A + \frac{W_B}{W_A} \times V_3 \text{ Cos } \theta_B$$

Let us resolve the velocity vectors in the Y direction

Momentum before collision = Momentum after collision

$$(W_B \times V_2) = (W_A \times V_4 \text{ Sin } \theta_A + W_B \times V_3 \text{ Sin } \theta_B)$$

Velocity of vehicle-A at impact: $V_2 = \frac{W_A}{W_B} \times V_4 \text{ Sin } \theta_A + V_3 \text{ Sin } \theta_B$

where $V_3$ and $V_4$ are the separation vehicle velocities, immediately after impact.

**Case II: Partial head-on collision:** Let us resolve the velocity vectors in the X direction (Fig. 20.17)

$$\text{Momentum before collision} = \text{Momentum after collision}$$

$$(-W_A \times V_1) + W_B \times V_2 \text{ Cos } \theta_1 = (-W_A \times V_4 \text{ Cos } \theta_3 - W_B \times V_3 \text{ Cos } \theta_2)$$

Dividing by $(-W_A)$ on both sides, we get

$$(V_1) - (W_B/W_A) \times V_2 \text{ Cos } \theta_1 = V_4 \text{ Cos } \theta_3 + (W_B/W_A) \times V_3 \text{ Cos } \theta_2$$

Velocity of vehicle-A at impact: $V_1 = (V_4 \text{ Cos } \theta_3 + (W_B/W_A) \times V_3 \text{ Cos } \theta_2) + (W_B/W_A) \times V_2 \text{ Cos } \theta_1)$

Let us resolve the velocity vectors in the Y direction

$$\text{Momentum before collision} = \text{Momentum after collision}$$

$$-W_B \times V_2 \text{ Sin } \theta_1 = (-W_A \times V_4 \text{ Sin } \theta_3 - W_B \times V_3 \text{ Sin } \theta_2)$$

Dividing by $(-W_B)$ on both sides, we get

$$V_2 \text{ Sin } \theta_1 = (W_A/W_B) \times V_4 \text{ Sin } \theta_3 + V_3 \text{ Sin } \theta_2$$

Velocity of vehicle-A at impact: $V_2 = ((W_A/W_B) \times V_4 \text{ Sin } \theta_3 + V_3 \text{ Sin } \theta_2)/ \text{ Sin } \theta_1$

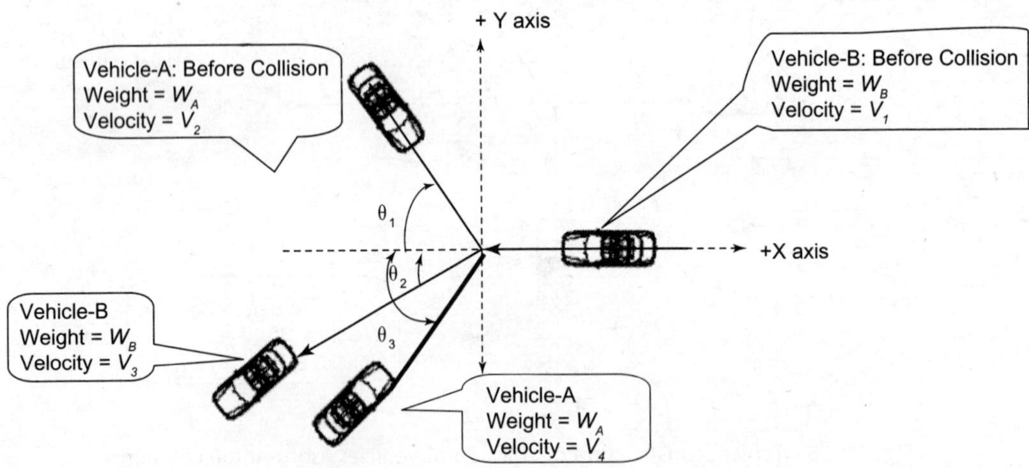


**Fig. 20.17** Typical construction of collision of vehicles: partial head-on collision

**Case III: Collision at an acute angle:** Let us resolve the velocity vectors in the X direction (Fig. 20.18)

Momentum before collision = Momentum after collision

$$(W_A \times V_1) + W_B \times V_2 \text{ Cos } \theta_1 = (W_A \times V_4 \text{ Cos } \theta_3 + W_B \times V_3 \text{ Cos } \theta_2)$$

Dividing by $(W_A)$ on both sides, we get

$$(V_1) + (W_B/W_A) \times V_2 \text{ Cos } \theta_1 = V_4 \text{ Cos } \theta_3 + (W_B/W_A) \times V_3 \text{Cos } \theta_2$$

Velocity of vehicle-A at impact:

$$V_1 = (V_4 \text{ Cos } \theta_3 + (W_B/W_A) \times V_3 \text{ Cos } \theta_2) - ((W_B/W_A) \times V_2 \text{ Cos } \theta_1)$$

Let us resolve the velocity vectors in the Y direction

Momentum before collision = Momentum after collision

$$W_B \times V_2 \text{ Sin } \theta_1 = (W_A \times V_4 \text{ Sin } \theta_3 + W_B \times V_3 \text{ Sin } \theta_2)$$

Dividing by ($W_B$) on both sides, we get

$$V_2 \text{ Sin } \theta_1 = (W_A/W_B) \times V_4 \text{ Sin } \theta_3 + V_3 \text{ Sin } \theta_2$$

Velocity of vehicle-B at impact:

$$V_2 = ((W_A/W_B) \times V_4 \text{ Sin } \theta_3 + V_3 \text{ Sin } \theta_2)/ \text{ Sin } \theta_1$$

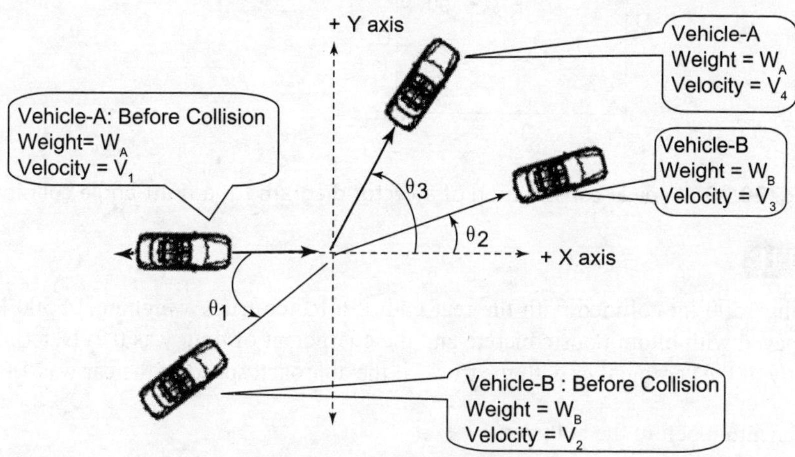


**Fig. 20.18** Typical construction of collision of vehicles: collision at an acute angle

## 20.8.10 Determination of Velocities of Vehicles Before Impact Using Vector Diagram Graphical Method

Let us take the example of the right-angle collision described above to demonstrate how to determine vehicle velocities before impact (Fig. 20.16). A few field investigations need to be carried out at the accident location. The after-impact velocities of the vehicles ($V_3$ and $V_4$) are determined from the skid marks, approach angle (90° to each other) and departure angles ($\theta_A$ and $\theta_B$) of the vehicles measured on the site.

By finding out the approximate weights of each vehicle, it is easy to determine the post-collision momentum values, $W_AV_3$ and $W_BV_4$. A convenient scale of momentum is taken and lines representing the vector quantities $W_1V_3$ and $W_2V_4$ are drawn and labelled A and B, respectively. Based on the field investigation data (Fig. 20.19), let us draw a line A with angle $\theta_A$ and line B with angle $\theta_B$ on the X axis, with the direction of travel of the two vehicles A and B. Next, draw a line B to the end of line A, and vice versa. This now looks like a parallelogram. Now, draw a line R between A and B, representing the resultant of these two vector lines. Extend R downwards with the same length.

With reference to the approach angle between the two vehicles, draw lines of unknown length initially. Next, draw parallel lines (C and D) and construct the second parallelogram by connecting the two ends of the line R. Now, measure the length of C and D and calculate $V_1$ and $V_2$ as follows:

$$V_1 = \text{Length of line C}/W_A$$

$$V_2 = \text{Length of line D}/W_B$$

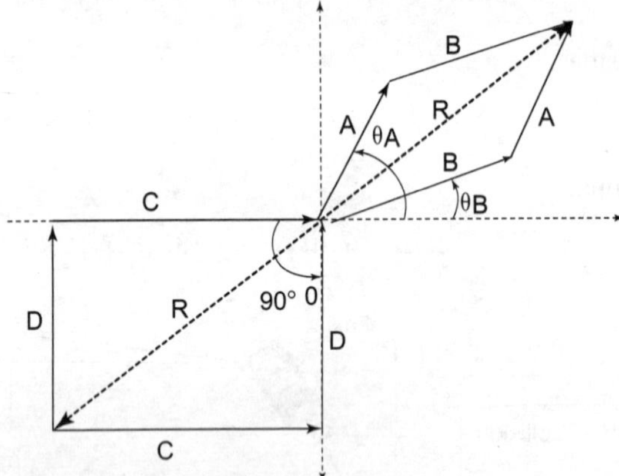


**Fig. 20.19** Typical construction of a vector diagram for a right-angle collision

**Example 20.11**

A car weighing 1500 kg collided with the rear end of a halted truck weighing 14,500 kg. The road surface was paved with bituminous concrete and the coefficient of drag was 0.7. Determine the post-impact velocity of the car together with the truck, if the approach speed of the car was 18 km/h.

**Answer:** Momentum before the collision

$$\text{Momentum} = \text{Mass (car)} \times \text{Velocity (car)} + \text{Mass (truck)} \times \text{Velocity (truck)}$$

$$= 1500 \times 18 + 14500 \times 0 = 27{,}000$$

As the external forces acting during the impact are minimal, the brakes of the truck are not engaged. Hence, momentum is conserved. Therefore, after the collision Momentum before = Momentum after

$$27{,}000 = 1500 \times V + 14500 \times V$$

$$16000 \times V = 27000$$

Post-impact combined speed of the vehicles = $V = 27000/16000 = 1.6875$ km/h = 0.46875 m/s

Thus, the car loses 16.3125 km/h = 4.53 m/s due to the action of the retarding impact force.

**Example 20.12**

Two cars are travelling towards each other, as shown below. They collide, lock together and move forward (to the right) post collision. Determine the speed of the cars together, just after the collision.

Before

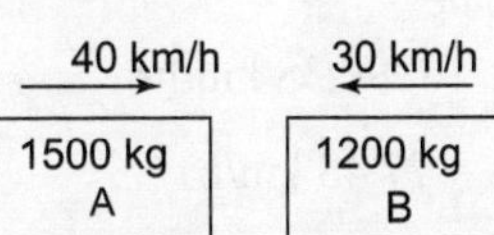


After

?

1500 kg A | 1200 kg B

Take motion towards right as +

Momentum of A = 1500 × 40 = 60,000

Momentum of B = 1200 × –30 = –36,000

Momentum of A and B after collision

$(m_1 + m_2)V = (1500 + 1200)V$

$= 2700\ V$

Total momentum before = Total momentum after

$$60000 - 36000 = 2700\ V$$

$$V = \frac{24000}{2700} = 8.88 \text{ km/h to the right}$$

## EXERCISES

### Multiple Choice Questions

1. On collision diagrams, the symbol shown below indicates:

(a) Overturned vehicle (b) Vehicle out of control
(c) Colliding vehicle (d) Overspeeding vehicle

2. On collision diagrams, the symbol shown below indicates:

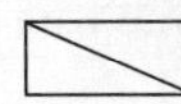

(a) Any fixed object on the roadside (b) Stationary vehicle
(c) Building (d) Parked vehicle

3. As a common practice, condition diagrams may be developed to any convenient scale such as in the range between
(a) 1:1000 and 1:5000 (b) 1:1000 and 1:50000
(c) 1:1000 and 1:10000 (d) 1:200 and 1:1000

4. An investigator drives at 50 km/h and skids to a stop. The skid mark length measured was 14 m. The coefficient of friction would be
(a) 0.304 (b) 0.501 (c) 0.608 (d) 0.703

(*Hint*: $f = \dfrac{V^2}{2gD} = \dfrac{(0.278 \times 50)^2}{2 \times 9.81 \times 14} = 0.703$)

5. The maximum safe speed on a horizontal road-curve if the available drag factor = 0.164, superelevation = 7% and radius of the horizontal curve = 215 m would be
   (a) 55.21 km/hr (b) 77.48 km/hr (c) 89.25 km/hr (d) 90.38 km/hr
   (*Hint*: $v = \sqrt{gR(e + f)}$, $v = \sqrt{9.81 \times 215(0.07 + 0.164)} = 21.54$ m/s = 77.48 km/h)
6. A car skids to a stop on a road, leaving four skid marks under its four wheels, with an average length of 20 m. Skid tests reveal a drag factor of 0.70. The estimated speed (km/hr) of the car just before sliding could be
   (a) 23.43 (b) 59.63 (c) 32.23 (d) 60.12
   (*Hint*: $V = \sqrt{254Df} = \sqrt{254 \times 20 \times 0.70} = 59.63\,\text{km/h}$)
7. The radius of the yaw skid mark if the mid-ordinate of the yaw mark = 2 m and length of the long chord = 20 m is
   (a) 10 m (b) 16 m (c) 26 m (d) 52 m
   (*Hint*: Radius of the yaw skid mark = $R = \dfrac{(2^2 + 10^2)}{2 \times 2} = 26\,\text{m}$)
8. What is the velocity of a vehicle just before sliding, if the measured radius of the yaw skid mark =20 m and the drag factor = 0.7?
   (a) 11.7 km/hr (b) 21.2 km/hr (c) 38.4 km/hr (d) 42.1 km/hr
   (*Hint*: $V = \sqrt{fgR} = \sqrt{0.7 \times 9.81 \times 20} = 11.71\,\text{m/s} = 42.1\,\text{km/h}$)
9. The acceleration factor if a test vehicle goes from 0 km/hr to 50 km/hr in 50 m would be
   (a) 0.16 (b) 0.17 (c) 0.18 (d) 0.19
   (*Hint*: $V = 0.278 \times 50 = 13.9$ m/s, $f_a = \dfrac{V^2}{2gD} = f_a = \dfrac{13.9^2}{2 \times 9.81 \times 50\,\text{m}} = 0.196$)
10. A test vehicle goes from 0 km/hr to 10 m/sec with a known acceleration factor of 0.15. What is the distance it travels? Assume acceleration due to gravity as 10 m/sec².
    (a) 33.33 m (b) 10 m (c) 56.66 m (d) 15 m
    (*Hint*: $D = \dfrac{V^2}{2gf_a} = \dfrac{10^2}{2 \times 10 \times 0.15} = 33.33\,\text{m}$)
11. A vehicle running at a speed 14 m/sec applied its brakes to lock its wheels. The vehicle slides to a stop on a drag factor of 0.70. What is the time required to get the vehicle to stop? Assume acceleration due to gravity as 10 m/sec².
    (a) 2 sec (b) 10 sec (c) 0.98 sec (d) 9.8 sec
    (*Hint*: Acceleration (deceleration) = g × Drag factor = 10 × 0.70 = 7
    Time = Velocity/Acceleration; Time take to the stop = (14 m/sec)/7 = 2 s)
12. A vehicle accelerates for 10 s with an acceleration factor of 0.16. What is the velocity of the vehicle that can be attained? Assume acceleration due to gravity as 10 m/sec².
    (a) 56.5 km/hr (b) 57.6 km/hr (c) 16 km/hr (d) 66.66 km/hr

(*Hint*: Acceleration of the vehicle = g × Drag factor = 10× 0.16 = 1.6 m/s$^2$
Velocity = 1.6 m/s$^2$ × 10 s = 16 m/s = 57.6 km/hr)

13. With a multi-phase accident, a test vehicle undergoes several speed changing events: it loses 38 km/h in sliding before impact, 18 km/h in the collision and 9 km/h in sliding to a stop after the impact. What is the the original speed of the vehicle?

    (a) 21.66 km/h (b) 8.06 km/hr (c) 65 km/h (d) 43

    (*Hint*: $V = \sqrt{38^2 + 18^2 + 9^2} = 43\,\text{km/h}$ )

14. A car ran off an embankment and fell in a ditch. The measured horizontal distance from the verge of the embankment to the ditch was 7 m and the height between the two points was 0.5 m. What would be the falling speed of the car from the embankment? Take acceleration due to gravity = 10 m/sec$^2$.

    (a) 75.6 km/hr (b) 84.6 km/hr (c) 89.6 km/h (d) 21.6 km/hr

    (*Hint*: $V = D \times \sqrt{\dfrac{g}{2H}} = 7 \times \sqrt{\dfrac{10}{2\times 0.5}}$ = 21 m/s = 3.6×21 = 75.6 km/h)

## Answers

1. (b) 2. (d) 3. (d) 4. (d) 5. (b) 6. (b) 7. (c) 8. (d) 9. (d) 10. (a) 11. (a) 12. (b) 13. (d) 14. (a)

## Questions

1. Write an overview of road safety studies.
2. State the types of road safety studies and explain their significance in reduction/prevention of road accidents.
3. State the objectives of conducting investigations and studies on road accidents.
4. List the basic types of information that should be reported with adequate evidence and give a description of the collision as basic accident information.
5. Write the various methods adopted for locating road accidents. Also state their merits and limitations.
6. State the objectives based on which accident data is assessed.
7. What is a collision diagram? Explain how it is useful.
8. Draw any 20 widely used symbols for the development of collision diagrams.
9. Draw a typical collision diagram recorded at a location showing the collision of parked vehicle, swipe collision, uncontrolled vehicle hitting a traffic cone and a fatal accident on a zebra crossing. Indicate using different types of symbols. Provide sample date and time for each case.
10. What is a condition diagram? Explain how it is useful.
11. List the items that should be included in a condition diagram.
12. Draw a typical condition diagram showing a few salient features that are essential.
13. Explain the three widely used methods for measuring on-scene details. Write their merits and limitations.
14. Draw a typical diagram using the coordinate system of measurement and showing the measured distances and angles.

15. Draw a typical diagram using the combination method of measurement and showing the measured distances and angles.
16. Explain the basic concepts of Newton's laws that are useful for accident reconstruction purposes.
17. Describe the phenomenon of linear momentum of two particles as applicable in the reconstruction of a collision between two vehicles on a roadway.
18. Define drag factor and state how it is useful for accident reconstruction purposes.
19. Explain how the drag factor of a road surface is determined and measured. Write the formula with standard notations.
20. List the possible range of drag factor values pertaining to specific compositions of road surface materials.
21. Explain how the radius of a simple circular road curve is determined. Draw a typical sketch and write the formulas with standard notations.
22. Derive an expression used for determination of safe speed on a superelevated road curve. Adopt the usual notations.
23. Explain how the minimum speed of a vehicle before sliding (from straight line skid marks) is determined. Draw a typical sketch and write the formulas with standard notations.
24. Explain how the minimum speed of a vehicle is determined from yaw skid marks. Draw a typical sketch and write the formulas with standard notations.
25. Derive an expression for determination of combined speed of vehicle skids on a road surface.
26. Explain how the speed of a vehicle falling from an upgrade/downgrade is determined. Draw a sketch showing the fall of a car from an wmbankment and explain.
27. Derive an expression for determination of velocity of a vehicle moving up a grade before falling from an embankment.
28. Explain how the velocities of vehicles at impact/collision are determined in right-angle collisions. Draw a sketch showing the details with the usual notations.
29. Explain how the velocities of vehicles at impact/collision are determined in partial head-on collisions. Draw a sketch showing the details with the usual notations.
30. Explain how the velocities of vehicles at impact/collision are determined in acute angle collisions. Draw a sketch showing the details with the usual notations.
31. Explain the graphical concept of determination of velocities of vehicle before impact using a vector diagram.
32. Vehicle-A approaches Vehicle-B from behind. The vehicles are moving at the speeds shown below. After the collision, the front of Vehicle-A travels at 38 km/h. Calculate the speed of Vehicle-B after the collision.

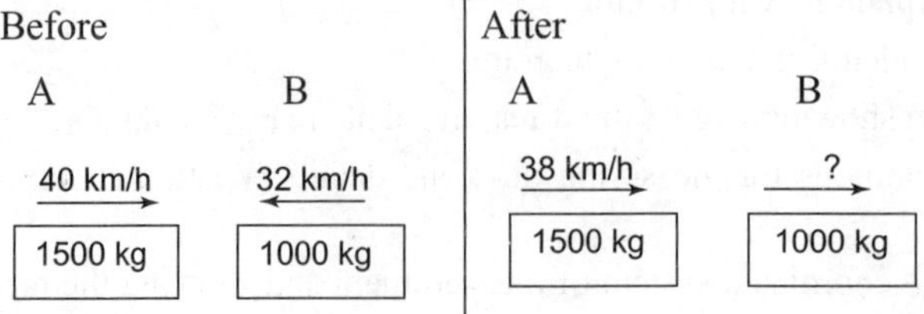

33. Calculate the velocities before the impact of the two vehicles A and B, from the following data: Weight of vehicle A = 2200 kg; Weight of vehicle B = 1800 kg; $\theta_A = 50°$; $\theta_B = 10°$.
The velocities at impact were found to be 50 km/h and 40 km/hr for vehicles A and B.

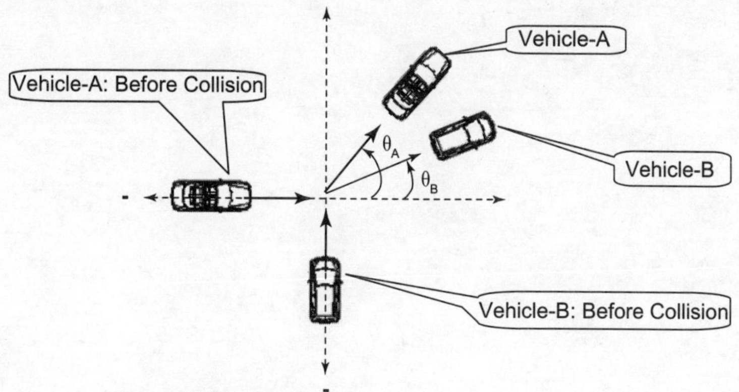


Right-angle collision

34. Vehicle A approaches from the east and vehicle B from the north. After collision, A skids 66° south of west and B skids 33° north of west. The skid distance before collision is 20 m for A and 30 m for B. The skid distances after collision are 31 m and 16 m, respectively. A weighs 5000 kg and B 7000 kg. Take the skid resistance of the pavement as 0.55. Determine the pre-collision speeds of the vehicles.

CHAPTER

# 21 Analysis of Accident Data and Road Safety

## 21.0 ANALYSIS OF ACCIDENT DATA

Analysis of accident data over a specified period of time and along a selected stretch of road is useful for adopting effective road safety initiatives. The analysis results are also useful for identifying the parameters that influence road traffic safety with reference to the type of highway, season, type of driver, type of pedestrian and type of vehicle.

**Rate of accidents per kilometre:** Accident hazard is expressed as the number of accidents of all types per kilometre distance of any selected street or highway.

$$\text{Rate of accidents per kilometre} = \frac{N_a}{L}$$

where $N_a$ = total number of accidents of all types that occurred on a road in a study period – generally one year and $L$ = length of the road (km).

**Rate of accident involvement:** Accident hazard is expressed as the number of accident involvements per 100 million vehicle-kilometres of travel distance of any selected street or highway.

$$\begin{array}{l}\text{Rate of accident involvement per} \\ \text{100 million vehicle-km of travel}\end{array} = \frac{N_d}{VK}\times 10^8$$

where $N_d$ = total number of drivers of vehicles involved in accidents in a study period and $VK$ = vehicle-kilometres of travel on the road section in the study period.

Alternatively, the accident rate for a roadway segment can be calculated using the formula (Shenker et al., 2015):

$$R = \frac{N_a}{(365 \times AADT \times Y \times L)} \times 10^8$$

where $R$ = rate of accidents that occurred on a roadway segment, expressed as the number of accidents per 100 million vehicle-kilometres of travel in a year; $N_a$ = total number of accidents that occurred on a roadway in the study period; $AADT$ = traffic volume, expressed as average annual daily traffic; $Y$ = number of years of data; and $L$ = length of the roadway segment (km).

*Note:* 1 lakh = $10^5$ = 100,000; 1 million = $10^6$ = 10 lakhs; 1 crore = $10^7$ = 10 million; 100 million = $10^8$ = 10 × $10^7$ =10 crores = 1000 × $10^5$ = 1000 lakhs

**Death rate based on population:** Death hazard to life in a community is expressed as the number of fatal accidents that occurred for every 1 lakh population in an area/locality considered.

$$\begin{array}{l}\text{Death rate for every 1 lakh} \\ \text{population in a locality}\end{array} = \frac{D}{P}\times 10^5$$

where $D$ = total number of deaths that occurred due to traffic in a year and $P$ = population of the study area.

**Death rate based on registered vehicles:** Death hazard to life in a community is expressed as the number of fatal accidents that occurred for every 10,000 vehicles registered in an area/locality considered.

$$\begin{array}{l}\text{Death rate for every 10,000 registered} \\ \text{vehicles in a locality}\end{array} = \frac{D}{V} \times 10^4$$

**Rate of accidents that occurred for vehicle-kilometres of travel:** Accident hazard is expressed as the number of accidents that occurred per 100 million vehicle-kilometres of travel distance in any selected street or highway.

$$\begin{array}{l}\text{Rate of accident observed per 100 million} \\ \text{vehicle-km of travel}\end{array} = \frac{N_a}{VK} \times 10^8$$

where $N_a$ = total number of accidents of all types that occurred on a road in a study period and $VK$ = vehicle-kilometres travelled on the road section in the study period.

**Example 21.1**

Calculate the registration death rate, population death rate and accident rate per vehicle-km for the following summary of road accident statistics pertaining to a city in a selected period of study:

Total fuel consumed by the motor vehicles travelling in a city = 6.2 billion litres; Number of fatalities observed = 1111; Major and minor motor vehicle injuries observed = 255,800; Registered motor vehicles in the city = 6,821,000; Estimated population of the city = 20,190,200; Average fuel consumption of vehicles = 11.7 km per litre.

**Answer:**

(i) Registration death rate:

$$\begin{array}{l}\text{Death rate for every 10,000 registered} \\ \text{vehicles in a locality}\end{array} = \frac{D}{V} \times 10^4 = \frac{1111}{6821000} \times 10^4 = 1.62$$

(ii) Population death rate:

$$\begin{array}{l}\text{Death rate for every 1 lakh} \\ \text{population in a locality}\end{array} = \frac{D}{P} \times 10^5 = \frac{1111}{20190200} \times 10^5 = 5.50$$

(iii) Accident rate per vehicle-km of travel:

Vehicle-kilometres of travel in the city = $VK$

$VK$ = Total fuel consumption of the vehicles in litres × Average fuel consumption of vehicles

$$VK = (6.2 \times 10^9) \text{ litres} \times 11.7 \text{ km per litre}$$

$$= 72.54 \times 10^9 \text{ km}$$

Number of fatalities observed = 1111 (given)

$$\text{Rate of accidents per 100 million vehicle-km of travel} = \frac{N_a}{VK} \times 10^8$$

$$\text{Rate of accidents per 100 million vehicle-km of travel} = \frac{(1111+255800)}{72.54\times10^9}\times10^8 = 354.16$$

$$\text{Rate of deaths per 100 million vehicle-km of travel} = \frac{1111}{72.54\times10^9}\times10^8 = 1.53$$

**Example 21.2**

The following accident data pertains to a 25-km section of a national highway:

| *Type of accident* | *2014* | *2015* | *2016* | *2017* |
|---|---|---|---|---|
| Fatal | 1 | 3 | 5 | 1 |
| Personal injury | 45 | 56 | 67 | 32 |
| Property damage only | 132 | 532 | 123 | 102 |
| ADT (vehicles) | 16900 | 17940 | 18974 | 25178 |

Determine the fatal and injury accident rate.

**Answer:** As we need to determine the fatal and injury accident rate, let us rewrite the table to avoid property damage data:

| *Type of accident* | *2014* | *2015* | *2016* | *2017* | *Total* |
|---|---|---|---|---|---|
| Fatal | 1 | 3 | 5 | 1 | 210 |
| Personal injury | 45 | 56 | 67 | 32 | |
| ADT (vehicles) | 16900 | 17940 | 18974 | 25178 | 78992 |

The following formula is used to calculate the rate of accidents per 100 million vehicle kilometres travelled per year:

$$\frac{\text{No.of accidents}\times10^8}{\text{ADT}\times\text{No. of years}\times365\text{ days}\times\text{Length}}$$

$$= \frac{210\times10^8}{78992\times4\times365\text{ days}\times25\text{ km}} = 7.28 \text{ accidents per 100 million vehicle km}$$

**Example 21.3**

A series of interchanges were constructed in a metropolitan city. It was later identified that the corridor locations recorded the highest number of accidents.

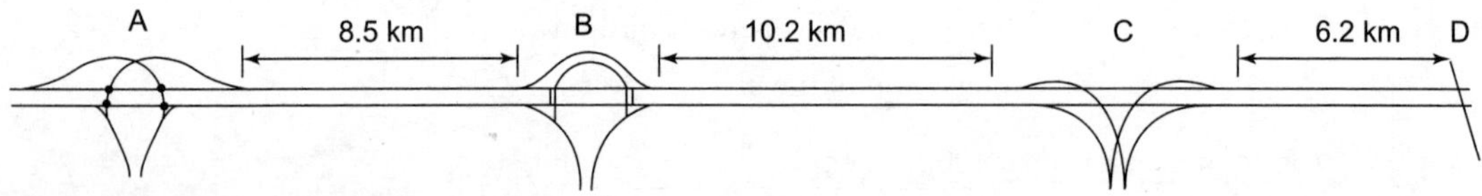


The corridor comprises three interchanges and three expressway segments. Rank the interchanges and segments based on the following accident data:

| *Interchange* | *ADT Veh/day* | *Accidents per year* |
|---|---|---|
| A | 18800 | 34 |
| B | 29560 | 56 |
| C | 20380 | 49 |

| *Expressway segment* | *ADT Veh/day* | *Accidents per year* | *Length (km)* |
|---|---|---|---|
| A–B | 6790 | 23 | 8.5 |
| B–C | 14890 | 35 | 10.2 |
| C–D | 12020 | 52 | 6.2 |

**Answer:** The following formula is used to calculate the rate of accidents per million vehicles per year:

$$R = \frac{\text{No. of accidents} \times 10^6}{\text{ADT} \times \text{No. of years} \times 365 \text{ days}}$$

Ranking of interchanges:

$$R_A = \frac{34 \times 10^6}{18800 \times 1 \times 365} = 4.95 \text{ accidents per million vehicles per year}$$

$$R_B = \frac{56 \times 10^6}{29560 \times 1 \times 365} = 5.19 \text{ accidents per million vehicles per year}$$

$$R_A = \frac{49 \times 10^6}{20380 \times 1 \times 365} = 6.59 \text{ accidents per million vehicles per year}$$

The following formula is used to calculate the rate of accidents per million vehicle kilometres travelled per year:

$$R = \frac{\text{No. of accidents} \times 10^6}{\text{ADT} \times \text{No. of years} \times 365 \text{ days} \times \text{Length}}$$

$$R_{A-B} = \frac{23 \times 10^6}{6790 \times 1 \times 365 \times 8.5} = 1.09 \text{ accidents per million vehicles per year}$$

$$R_{B-C} = \frac{35 \times 10^6}{14890 \times 1 \times 365 \times 10.2} = 0.63 \text{ accidents per million vehicles per year}$$

$$R_{C-D} = \frac{52 \times 10^6}{12020 \times 1 \times 365 \times 6.2} = 1.91 \text{ accidents per million vehicles per year}$$

The R values with their rankings are summarised below:

| *Interchange* | *R* | *Ranking* |
|---|---|---|
| A | 4.95 | 3rd |
| B | 5.19 | 2nd |
| C | 6.59 | 1st |

| *Expressway segment* | *R* | *Ranking* |
|---|---|---|
| A–B | 1.09 | 2nd |
| B–C | 0.63 | 3rd |
| C–D | 1.91 | 1st |

## 21.1 ROAD SAFETY VISION

A road safety vision is a description of a desirable state of road traffic system in the future based on safety principles of interaction among the three basic components: driver/user, vehicle and environment/surroundings. Any road safety vision is formulated as a goal that may be attained through effort over an unspecified period of time. The vision must provide a direction to road safety work and actions and should generate reflection on improvements in order to attain the desirable state based on preset road safety plans and programmes (DG TREN, 2007).

The core concept of road safety vision is to change the road traffic system by eliminating all known opportunities for human error and to reduce possible physical damage due to accidents. In general, the vision is shared by all the stakeholders and the responsibilities are shared by all types of road users, system designers, road authorities, car manufacturers, educationalists and the media.

## 21.2 ROAD SAFETY PROGRAMMES

A road safety programme is more specific and covers a shorter time frame than a road safety vision. A road safety programme systematically describes specific goals and principles for the organisation of road safety work and specifies the target actions for the next 5–10 years. Road safety programmes also define the responsibilities and provides funding with incentives for the best implementation of effective safety measures (DG TREN, 2007).

## 21.3 ROAD SAFETY MEASURES

Several safety measures are in use in different countries. Some are relatively simple and low cost while others are less easy to implement and may require a larger budget. A short list of the road safety measures that are intended to mitigate the circumstances leading to accidents or otherwise decrease the number of accidents is given below. These safety measures are categorised as engineering measures, enforcement measures and educational measures (DG TREN, 2007; Austroads, 2016).

**Engineering measures**

- *Land use and network planning*: The hierarchical integrity of the road network based on their functional objectives is integrated with the adjacent land use.
- Channelisation and directional control/regulation of the traffic streams, including morning and evening peak traffic regulations.
- Changes/Improvements in the geometric components of roadways, including providing safe stopping sight distance, overtaking sight distance, acceleration/deceleration tapers, and access control on high-speed lanes/highways/expressways.
- Provision of low-speed zones in residential areas.
- Provision of off-street parking facilities.

- *Improving/Modifying road intersections*: For example, replacing a junction with a roundabout. As roundabouts aim to lower junction speeds and thus eliminate right-angle and head-on collisions, with greater capacity than normal give-way or signalised junctions.
- Treatments to the centre line and medians for improved delineation/warning purposes; and treatments to other road infrastructure as part of speed management on curves, skid resistance and overtaking zones (Austroads, 2016).
- *Measures against tree collisions*: Decreased severity of road accidents by ensuring protection of rows of trees by providing guard rails wherever possible, or by felling the trees.
- *High-risk/Accident-prone site management*: Site visits should involve the relevant safety engineers, traffic police, emergency services and other stakeholders.
- *Signing and marking*: Maintaining the retroreflectivity of signs and road markings, including installing them at standard recommended locations on the roadway.
- *Rumble strips*: These are milled into the asphalt surface of a road shoulder. Rumble strips vibrate and make a noise when a vehicle passes over them and alert drivers to the potential accident/hazard.
- *Variable message signs*: Speed limits and communication of warnings can be instantaneously provided on-site.
- *Road safety audits*: Audit is a formal proactive procedure by which independent assessment of the potential and likely safety performance of a specific design of a road or traffic scheme is carried out. The recommendations of the audit team are honoured by the client/municipality by incorporating those improvements in the designs drawings (IRC: SP: 88-2010).
- *Road safety inspections*: Safety inspection involves a periodic review or specific inspection of a road site by trained road safety experts. It involves visiting the road network to identify physical defects in the road infrastructure.
- *Vehicles and safety devices*: Safe car/vehicle designs including headlights, acceleration and braking capacity.
- *Daytime running lights* (DRL): Headlights equipped with DRL increases visibility and improves the distance and speed perception of motor vehicles so that the other road users can detect such motor vehicles during the day and adjust their driving behaviour.
- *Bicycle/Two-wheeler side reflection*: This implies that the front and rear wheels of a body frame are equipped with reflective material to increase visibility at night and at twilight.
- *Intelligent speed assistance* (ISA): The ISA systems establish the position of a vehicle and compare its current speed with the posted speed limit there. In case of excess speed, the system gives feedback to the driver and restricts the speed accordingly.
- *Alcohol ignition interlock* (Alcolock): It is an electronic device that prevents a vehicle from being started if the driver has consumed alcohol beyond the accepted level.
- *Event data recorder* (black box): It is an electronic system capable of monitoring a number of vehicle variables such as speed, acceleration and deceleration forces, use of lights, gears, seat belts and others. These data recorders can collect data just before and after an accident or/and for each trip. The data collected by these systems can be used for reconstructing a road accident, presenting evidence of accident contributing parameters and also for research purposes.
- *Promoting the implementation of eCall systems*: eCall is an automatic in-vehicle emergency call service developed in the European Union. An eCall-equipped vehicle has a terminal with a satellite-based wireless communication system by which accidents, rollovers and fire occurrences in the vehicle can be detected. When an accident occurs, the terminal sends information about the vehicle's position and type of accident to the nearest emergency response centre. This system is also capable of opening a voice connection between the vehicle occupants and the emergency response centre.

**Enforcement measures**

- Traffic law enforcement aims to prevent traffic offences by imposing appropriate rules and regulations which ensure safe and efficient movement of traffic.
- *Safety camera programme*: These cameras are mounted on the roadside at clearly marked locations so that road users see them well in advance and control their behaviour.
- *Automatic speed enforcement*: Speeding vehicles are detected by these automatic cameras and these vehicles are photographed with their licence plate number. Such information is immediately passed on to a central processing office for identification of the vehicle owners.
- *Section controls*: Any particular length of road passing through a densely populated suburb is intended to operate as a low-speed zone. Such control sections may be identified to improve air quality along that road. For example, introduction of a reduced speed limit of 60 km/hr where it was previously 80 km/hr.
- Random breath testing (RBT) aims to identify drivers who exceed the legal alcohol limit. With the RBT device, drivers are stopped and tested for alcohol levels by the police.
- *Targetted seat belt enforcement*: Police target drivers who do not wear their seat belts.
- Penalty point system (PPS) aims at prosecuting repeat offenders. When a traffic violation is detected, the violator receives one or more penalty points, and when the accumulated penalty points exceeds the limit, the violator's licence will be suspended temporarily.

A list of **emergency services** is given below:

- *Efficient emergency response*: The core of road safety measures is the agreement between insurance companies, hospitals and the police department that a rescue car/ambulance will be sent to the incident location as soon as an accident is reported.
- *Emergency lanes in congestion*: If there are more than two lanes, all the cars on the outer left lane must move to the left, while the others must move to the right. This free lane enables all emergency vehicles to provide fast and efficient help in congested traffic conditions.
- *Use of a mobile intensive care unit* (MICU): These provide pre-hospital treatment for road accident victims on the road site.
- *Transport of road crash victims by helicopter*: Severely injured road accident victims can be transported to a hospital in a trauma helicopter within a short duration.
- *Post-accident care*: This measure is important in rural areas where emergency services cannot arrive at the accident spot within 5–15 minutes (international standard). The first hour after the accident is the most crucial and important for the accident victim's first aid, and hence is termed 'golden hour'.
- Psychological support for road crash victims should be provided at the first intervention stage, i.e., immediately after the crash and also during hospitalisation.

**Educational measures**

- Driver training is important for learner drivers as well as for regular drivers and to familiarise them with the latest traffic regulations and teach them to assess risks.
- Road safety halls containing educational aids designed to encourage active and accurate use of in-car safety equipment among novice drivers and school children.
- Rehabilitation seminars for novice drivers aim to avoid repeat offences by motivating them to behave more safely and responsibly.
- Newspapers, TV shows, mobile phone SMS alerts, internet-based awareness campaigns and school educational programs are widely used to create awareness among children and road users. Organising Road Safety Weeks is an example of an educational measure in India.

## 21.4 EVALUATION OF EFFECTIVENESS OF ROAD SAFETY IMPLEMENTATION

To judge the merits of road safety improvement and determine its effectiveness in reducing accidents, the following tests are conducted based on statistical significance. All accident data before and after improvements are obtained for this purpose.

### 21.4.1 Chi-Square Test to Find Goodness of Fit

This test is conducted to determine whether the observed frequencies are close to the expected frequencies, or if they are really quite different from each other. The following formula is used for computation of Chi-square: $\chi^2 = \sum_{i=1}^{n}\left[\frac{(O_i - E_i)^2}{E_i}\right]$ (21.1)

Chi-square ($\chi^2$) is a single number summary that represents the discrepancy between obtained frequencies and expected frequencies. The bigger the calculated value of $\chi^2$, the greater the difference between the observed and expected frequencies.

The calculated $\chi^2$ value is compared to the critical $\chi^2$ value (a value from Table 21.1) and conclusions are drawn (Graham, 2000):

1. If calculated $\chi^2$ > critical $\chi^2$, any one of the following possibilities holds good:
   (a) A high value of $\chi^2$ implies a poor fit between the observed and expected frequencies.
   (b) The fit is not acceptable at that level of significance.
   (c) There is a difference between the observed and expected frequencies. The expected frequency (theoretical frequency) distribution is *not* a good description of the population.
   (d) The discrepancy was not simply due to change.
   (e) The discrepancy between the observed and theoretical frequency distribution is not a good description of the population. In other words, the observed sample poorly reflects the characteristics of the population from which it was taken.
2. If calculated $\chi^2$ < critical $\chi^2$, the following possibilities hold good:
   (a) The fit is acceptable at that level of significance.
   (b) There is no evidence to indicate that the true theoretical distribution differs from the observed distribution (Gerlough, 1955).

(c) Some relationship exists between the two variables in question. However, it does not indicate the type of relationship.
(d) The two variables are associated in some way, to be precisely determined by the investigator.

**Table 21.1** A partial table of chi-square distribution for 95% confidence level ($\alpha = 0.05$)

| *Degree of Freedom (DoF)* | 1 | 2 | 3 | 4 | 5 | 6 | 7 | 8 | 9 | 10 | 11 | 12 | 13 | 14 | 15 |
|---|---|---|---|---|---|---|---|---|---|---|---|---|---|---|---|
| 0 $\chi^2$ | 3.841 | 5.991 | 7.815 | 9.488 | 11.070 | 12.592 | 14.067 | 15.507 | 16.919 | 18.307 | 19.647 | 21.026 | 22.362 | 23.685 | 24.996 |

DoF = Number of data sets − 1 = (n − 1)

### 21.4.2 Chi-Square Test to Determine Association between Two Variables

This test is used to determine the association between two independent and continuous variables that may be related to *before* and *after* road safety improvements. The observed frequencies are the recorded data obtained before road improvement. The formula used is exactly the same as above, but the expected frequencies are calculated as given below (which we would expect to see by chance).

$$\text{Expected frequency} = \frac{\text{Row total} \times \text{Column value}}{\text{Grand total}}$$

$$\text{Degree of freedom} = (\text{No. of rows} - 1) \times (\text{No. of columns} - 1)$$

Take the null hypothesis as: "The differences between the two sample variables are only due to chance fluctuations in sampling".

The calculated $\chi^2$ value is compared to the critical $\chi^2$ value and conclusions are drawn:

1. If calculated $\chi^2$ > critical $\chi^2$, reject the null hypothesis (there is a significance difference between the two variables due to improvement; that means that the road safety measures are effective)
2. If calculated $\chi^2$ < critical $\chi^2$, accept the null hypothesis.

### 21.4.3 Chi-Square Test to Determine Effectiveness of Before and After Improvements from One Group of Data

The following formula is used to determine the effectiveness of before and after improvements from one group of data:

$$\chi^2 = \left[ \frac{\left(N_{before} \times t_{after} - N_{after} \times t_{before}\right)^2}{\left(t_{before} \times t_{after}\right) \times \left(N_{before} + N_{after}\right)} \right]$$

where $t_{\text{before}}$ = period of time before the improvement for which accident data is available; $t_{\text{after}}$ = period of time after the improvement for which accident data is available; $N_{\text{before}}$ = number of accidents observed in the $t_{before}$ time period; and $N_{\text{after}}$ = number of accidents observed in the $t_{after}$ time period.

The calculated $\chi^2$ value is compared to the critical $\chi^2$ value.

**Example 21.4**

A list of accidents observed in a city over a period of one year is given below. Find out whether the observed accidents are uniformly distributed over the year using the chi-square test.

| *Month* | *Jan* | *Feb* | *Mar* | *Apr* | *May* | *Jun* | *Jul* | *Aug* | *Sep* | *Oct* | *Nov* | *Dec* |
|---|---|---|---|---|---|---|---|---|---|---|---|---|
| Observed frequency | 34 | 45 | 20 | 22 | 56 | 61 | 56 | 45 | 32 | 31 | 24 | 12 |

**Answer:** The computational details are as follows. Read from left to right.

| *Month* | *Frequency* | | $O_i - E_i$ | $(O_i - E_i)^2$ | $\frac{(O_i - E_i)^2}{E_i}$ |
|---|---|---|---|---|---|
| | *Observed,* $O_i$ | *Expected,* $E_i = 438/12$ | | | |
| January | 34 | 36.5 | 34 – 36.5 = –2.5 | $(-2.5)^2 = 6.25$ | 6.25/36.5 = 0.1712 |
| February | 45 | 36.5 | 45 – 36.5 = 8.5 | $(8.5)^2 = 72.25$ | 72.25/36.5 = 1.9795 |
| March | 20 | 36.5 | 20 – 36.5 = – 16.5 | $(-16.5)^2 = 272.25$ | 272.25/36.5 = 7.4589 |
| April | 22 | 36.5 | 22 – 36.5 = – 14.5 | $(-14.5)^2 = 210.25$ | 210.25/36.5 = 5.7603 |
| May | 56 | 36.5 | 56 – 36.5 = 19.5 | $(19.5)^2 = 380.25$ | 380.25/36.5 = 10.4178 |
| June | 61 | 36.5 | 61 – 36.5 = 24.5 | $(24.5)^2 = 600.25$ | 600.25/36.5 = 16.4452 |
| July | 56 | 36.5 | 56 – 36.5 = 19.5 | $(19.5)^2 = 380.25$ | 380.25/36.5 = 10.4178 |
| August | 45 | 36.5 | 45 – 36.5 = 8.5 | $(8.5)^2 = 722.25$ | 722.25/36.5 = 1.9795 |
| September | 32 | 36.5 | 32 – 36.5 = – 4.5 | $(-4.5)^2 = 20.25$ | 20.25/36.5 = 0.5548 |
| October | 31 | 36.5 | 31 – 36.5 = – 5.5 | $(-5.5)^2 = 30.25$ | 30.25/36.5 = 0.8288 |
| November | 24 | 36.5 | 24 – 36.5 = – 12.5 | $(-12.5)^2 = 156.25$ | 156.25/36.5 = 4.2808 |
| December | 12 | 36.5 | 12 – 36.5 = – 24.5 | $(-24.5)^2 = 600.25$ | 600.25/36.5 = 16.4452 |
| Total | 438 | 438 | | Total = | **76.7397** |

$$\chi^2 = \sum_{i=1}^{n} \left[ \frac{(O_i - E_i)^2}{E_i} \right] = 76.7397$$

Degree of Freedom = (12 – 1) = 11
Assume the significance level as 5% and the critical $\chi^2$ value = 19.647 (from Table 21.1).
Null hypothesis $H_o$: Accidents are uniformly distributed over the year.

As calculated $\chi^2$ (76.7397) > critical $\chi^2$ (19.674), the null hypothesis is *rejected;* the accidents observed in the city are *not* uniformly distributed over the year.

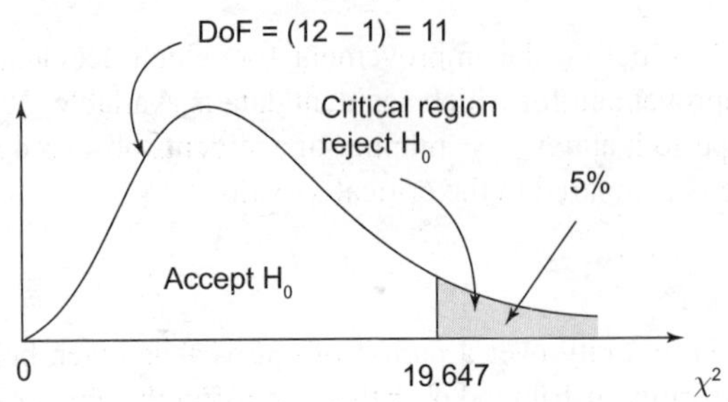


**Fig. 21.1** Typical presentation of critical $\chi^2$ value for 5% significance level

**Example 21.5**

A list of accidents observed in a city over a period of one year is given below. Find out whether the observed accidents are uniformly distributed over the hours using the chi-square test.

| *Hours* | *00:00–03:00* | *03:00–06:00* | *06:00–09:00* | *09:00–12:00* | *12:00–15:00* | *15:00–18:00* | *18:00–21:00* | *21:00–24:00* |
|---|---|---|---|---|---|---|---|---|
| Observed frequency | 21 | 19 | 09 | 10 | 17 | 10 | 11 | 15 |

**Answer:** The computational details are as follows. Read from left to right.

| *Hours* | *Frequency* | | $O_i - E_i$ | $(O_i - E_i)^2$ | $\frac{(O_i - E_i)^2}{E_i}$ |
|---|---|---|---|---|---|
| | *Observed, $O_i$* | *Expected, $E_i = 112/8$* | | | |
| 00:00–03:00 | 21 | 14 | 21 − 14 = 7 | $(7)^2 = 49$ | 49/14 = 3.5000 |
| 03:00–06:00 | 19 | 14 | 19 − 14 = 5 | $(5)^2 = 25$ | 19/14 = 1.7857 |
| 06:00–09:00 | 09 | 14 | 09 − 14 = −5 | $(-5)^2 = 25$ | 09/14 = 1.7857 |
| 09:00–12:00 | 10 | 14 | 10 − 14 = −4 | $(-4)^2 = 16$ | 10/14 = 1.1429 |
| 12:00–15:00 | 17 | 14 | 17 − 14 = 3 | $(3)^2 = 9$ | 17/14 = 0.6429 |
| 15:00–18:00 | 10 | 14 | 10 − 14 = −4 | $(-4)^2 = 16$ | 10/14 = 1.1429 |
| 18:00–21:00 | 11 | 14 | 11 − 14 = −3 | $(-3)^2 = 9$ | 11/14 = 0.6429 |
| 21:00–24:00 | 15 | 14 | 15 − 14 = 1 | $(3)^2 = 1$ | 15/14 = 0.0714 |
| Total = | 112 | 112 | | Total = | **10.7143** |

$$\chi^2 = \sum_{i=1}^{n}\left[\frac{(O_i - E_i)^2}{E_i}\right] = 10.7143$$

Degree of Freedom = (8 − 1) = 7
Assume the significance level as 5% and the critical $\chi^2$ value = 14.067 (from Table 21.1).
Null hypothesis $H_o$: Accidents are uniformly distributed over the hours in a year.

As calculated $\chi^2$ (10.7143) < critical $\chi^2$ (14.067), the null hypothesis is *accepted;* the accidents observed in the city are uniformly distributed over the hours in a year.

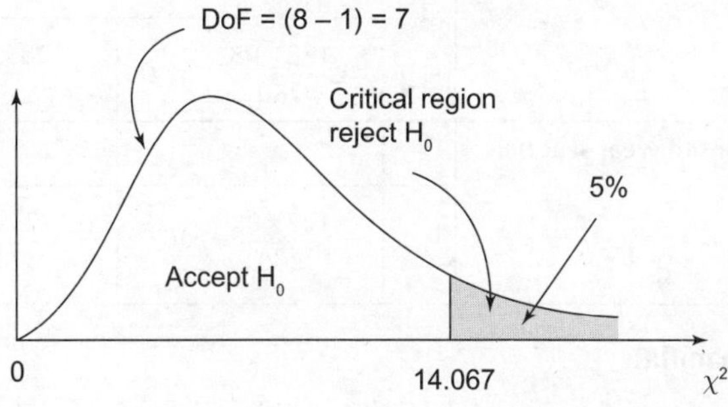


**Fig. 21.2** Typical presentation of critical $\chi^2$ value for 5% significance level

**Example 21.6**

The contingency table of two data sets obtained from a particular database belonging to the state revenue department is given below. Find out whether the distributions of sample variables differ because of chance fluctuations, based on results obtained from a chi-square test.

| *Study* | *Before improvement* | *After improvement* |
|---|---|---|
| Wants to wear seat belt | 55 | 83 |
| Does not want to wear seat belt | 43 | 79 |

**Answer:** Let us calculate the subtotals of the row and column frequencies of the observed frequencies; the grand total is given below:

| Study on | Before | After | Sum of row frequencies |
|---|---|---|---|
| Wants to wear seat belt | 55 | 83 | 138 |
| Does not want to wear seat belt | 43 | 79 | 122 |
| Sum of column frequencies | 98 | 162 | **260** |

Now, calculate the expected frequencies using the following formula:

$$\text{Expected frequency} = \frac{\text{Row total} \times \text{Column value}}{\text{Grand total}}$$

$$\begin{matrix}\text{Expected frequency of those who want} \\ \text{to wear seat belt, before improvement}\end{matrix} = \frac{138 \times 98}{260} = 52.0$$

$$\begin{matrix}\text{Expected frequency of those who want} \\ \text{to wear seat belt, after improvement}\end{matrix} = \frac{138 \times 162}{260} = 86.0$$

The remaining expected frequencies of the observed frequencies are presented below:

| *Study* | | *Before* | *After* |
|---|---|---|---|
| Wants to wear seat belt | $O_i$ | 55 | 83 |
| | $E_i$ | $= \frac{138 \times 98}{260} = 52.0$ | $= \frac{138 \times 162}{260} \approx 86.0$ |
| Does not want to wear seat belt | $O_i$ | 43 | 79 |
| | $E_i$ | $= \frac{122 \times 98}{260} \approx 46.0$ | $= \frac{122 \times 162}{260} = 76.0.$ |

Using the **chi-square formula:**

$$\chi^2 = \sum_{i=1}^{n} \left[ \frac{(O_i - E_i)^2}{E_i} \right]$$

Let us calculate the chi-square for each data set and sum:

$$\chi^2 = \left[\frac{(55-52.0)^2}{52.0}\right] + \left[\frac{(83-86.0)^2}{86.0}\right] + \left[\frac{(43-46.0)^2}{46.0}\right] + \left[\frac{(79-76.0)^2}{76.0}\right]$$

$$\chi^2 = \left[\frac{(3.0)^2}{52.0}\right] + \left[\frac{(-3.0)^2}{86.0}\right] + \left[\frac{(-3.0)^2}{46.0}\right] + \left[\frac{(3.0)^2}{76.0}\right]$$

$$\chi^2 = [0.1730] + [0.1046] + [0.1956] + [0.1184] = 0.5916$$

**Calculated** $\chi^2 = 0.5916$

$$\text{Degree of Freedom} = (\text{No. of rows} - 1) \times (\text{No. of columns} - 1)$$
$$= (2-1) \times (2-1) = 1$$

Assume the significance level as 5% at DoF = 1, and the critical $\chi^2$ value = 3.841 (from Table 21.1).

As calculated $\chi^2$ (0.5916) < critical $\chi^2$ (3.841), the null hypothesis is *accepted;* there is no significant difference between the accidents observed and the expected frequencies.

## Example 21.7

The frequency/contingency table (cross tabulation) for two independent variables is given below. The accident data observed in a year is obtained from a District State Highway Accident Records file. Find out whether the road safety improvements are effective or not based on the results obtained from a chi-square test.

| *Predominance accident types* | *Annual average of the no. of accidents obtained* | |
|---|---|---|
| | *Before improvement* | *After improvement* |
| Rear end | 44.2 | 50.1 |
| Angle collision at intersections | 18.6 | 21.6 |
| Right-turn head-on | 10.9 | 9.9 |
| Sideswipe | 21.7 | 29.0 |
| Other | 51.1 | 20.9 |

**Answer:** Let us rewrite the given frequencies and sum of the column and row frequencies, as shown below:

| *No. of accidents* | *Observed frequencies* | *Rear end* | *Angle collision at intersections* | *Right-turn head-on* | *Side swipe* | *Other* | *Sum of row frequencies* |
|---|---|---|---|---|---|---|---|
| Before | Oi | 44.2 | 18.6 | 10.9 | 21.7 | 51.1 | 146.5 |
| After | Oi | 50.1 | 21.6 | 9.9 | 29.0 | 20.9 | 131.5 |
| Sum of column frequencies | | 94.3 | 40.2 | 20.8 | 50.7 | 72 | 278 |

Details of the calculation of expected frequencies ($E_i$) against observed frequencies are presented below:

| *No. of accidents* | *Freq* | *Rear end* | *Angle collision at intersections* | *Right-turn head-on* | *Side swipe* | *Other* |
|---|---|---|---|---|---|---|
| Before | $O_i$ | 44.2 | 18.6 | 10.9 | 21.7 | 51.1 |
| | $E_i$ | $\frac{146.5 \times 94.3}{278}$ $= 49.69$ | $\frac{146.5 \times 40.2}{278}$ $= 21.18$ | $\frac{146.5 \times 20.8}{278}$ $= 10.96$ | $\frac{146.5 \times 50.7}{278}$ $= 26.71$ | $\frac{146.5 \times 72}{278}$ $= 37.94$ |
| After | $O_i$ | 50.1 | 21.6 | 9.9 | 29.0 | 20.9 |
| | $E_i$ | $\frac{131.5 \times 94.3}{278}$ $= 44.60$ | $\frac{131.5 \times 40.2}{278}$ $= 19.01$ | $\frac{131.5 \times 20.8}{278}$ $= 9.83$ | $\frac{131.5 \times 50.7}{278}$ $= 23.98$ | $\frac{131.5 \times 72}{278}$ $= 34.05$ |

Using the **chi-square formula:**

$$\chi^2 = \sum_{i=1}^{n} \left[ \frac{(O_i - E_i)^2}{E_i} \right]$$

| *No. of accidents* | *Rear end* | *Angle collision at intersections* | *Right-turn head-on* | *Side swipe* | *Other* |
|---|---|---|---|---|---|
| Before | $\left[\frac{(44.2 - 49.69)^2}{49.69}\right]$ | $\left[\frac{(18.6 - 21.18)^2}{21.18}\right]$ | $\left[\frac{(10.9 - 10.96)^2}{10.96}\right]$ | $\left[\frac{(21.7 - 26.71)^2}{26.71}\right]$ | $\left[\frac{(51.1 - 37.94)^2}{37.94}\right]$ |
| After | $\left[\frac{(50.1 - 44.6)^2}{44.6}\right]$ | $\left[\frac{(21.6 - 19.01)^2}{19.01}\right]$ | $\left[\frac{(9.9 - 9.83)^2}{9.83}\right]$ | $\left[\frac{(29 - 23.98)^2}{23.98}\right]$ | $\left[\frac{(20.9 - 34.05)^2}{34.05}\right]$ |

| *No. of accidents* | *Rear end* | *Angle collision at intersections* | *Right-turn head-on* | *Side swipe* | *Other* |
|---|---|---|---|---|---|
| Before | 0.6066 | 0.3142 | 0.0003 | 0.9397 | 4.5647 |
| After | 0.6783 | 0.3528 | 0.0004 | 1.0508 | 5.0784 |

Grand total of the above values = 13.5862

$$\chi^2 = \sum_{i=1}^{n} \left[ \frac{(O_i - E_i)^2}{E_i} \right] = 13.5862$$

$$\text{Degree of Freedom} = (\text{No. of rows} - 1) \times (\text{No. of columns} - 1)$$
$$= (2 - 1) \times (5 - 1) = 5$$

Assume significance level as 5% at DoF = 5, and the critical $\chi^2$ value = 11.070 (from Table 21.1).

As calculated $\chi^2$ (13.5862) > critical $\chi^2$ (11.070), the null hypothesis is *rejected;* there is a significant difference between the accidents before and after improvement. It may be concluded that the improvements and road safety measures are effective.

**Example 21.8**

Before realignment of a road intersection, 11 accidents occurred in 7 years. After the realignment, it was observed that only 5 accidents occurred in 1 year. Determine the effectiveness of realignment of the road intersection using the chi-square test.

**Answer:** The given values are substituted in the following formula:

$t_{before} = 7, t_{after} = 1, N_{before} = 11, N_{after} = 5$

$$\chi^2 = \left[\frac{\left(N_{before} \times t_{after} - N_{after} \times t_{before}\right)^2}{\left(t_{before} \times t_{after}\right) \times \left(N_{before} + N_{after}\right)}\right]$$

$$= \left[\frac{(11 \times 1 - 5 \times 7)^2}{(7 \times 1) \times (11 + 5)}\right] = 5.14$$

The calculated $\chi^2$ value (5.14) > critical $\chi^2$ value (3.841 at 95% confidence level); hence, it is concluded that the realignment improvement was effective.

## 21.4.4 Standard Normal Probability Method

The standard normal probability curve model can be used in hypothesis testing such as for tests on proportions and differences between two means. The area under the whole of a normal curve is 100%, say 1.00 in decimal form (Fig. 21.3) and the area under the curve with reference to a critical Z value at any selected significance level (say $\alpha = 0.05$. (Table 21.2)) is compared to a computed value, say the z score. The critical Z value (table value) indicates the coverage of area under the curve at any particular point (Fig. 21.3).

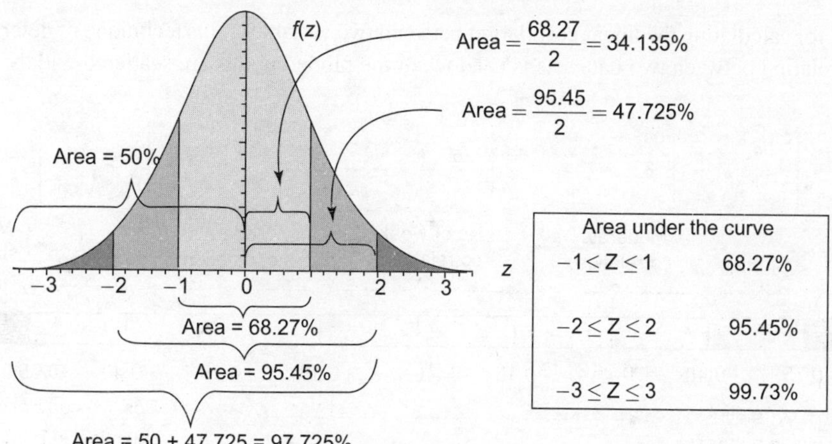


**Fig. 21.3** Area under the normal distribution curve with Z values

**Table 21.2** Critical Z value

| Confidence level (% area under curve) | 99.0 | 98.0 | 95.45 | 95.0 | 90.0 | 68.27 |
|---|---|---|---|---|---|---|
| Significance level (decimal) | 0.01 | 0.02 | 0.0455 | 0.05 | 0.10 | 0.3173 |
| Z value* | 2.58 | 2.33 | 2.00 | 1.96 | 1.65 | 1.00 |

*The other critical Z values can be read off any standard normal curve area table.

The z score may be computed using the following formula (Hald, 1952):

$$\text{z score} = \frac{O_1 - O_2 - 1}{\sqrt{O_1 + O_2}} \tag{21.2}$$

The z score is a measure of by how many standard deviations, below or above the population, mean differs from the observed data. If the computed z score < critical Z value (table value), the two observed values may be considered to have *similar* accident potential.

**Example 21.9**

The number of accidents observed in a year at two road intersections is as follows:

Ram Nagar intersection: 8; College Road intersection: 11

Compare these observed data and determine the accident potential of these two road intersections.

**Answer:** The given values are substituted in the following equation and the z score is computed.

$$\text{z score} = \frac{O_1 - O_2 - 1}{\sqrt{O_1 + O_2}} = \frac{11 - 8 - 1}{\sqrt{11 + 8}} = 0.4588 < 1.96$$

Take the critical Z value at 5% confidence level = 1.96 (from Table 21.2). As the computed z score < critical Z value, there is no evidence of a difference in accident potential between the Ram Nagar intersection and the College Road intersection.

## 21.4.5 Spearman Rank–Order Correlation Coefficient

This is used for calculating the linear correlation between two variables. This technique of determining the type of correlation between two data sets is used when the plotted points are scattered widely (Fig. 21.4)

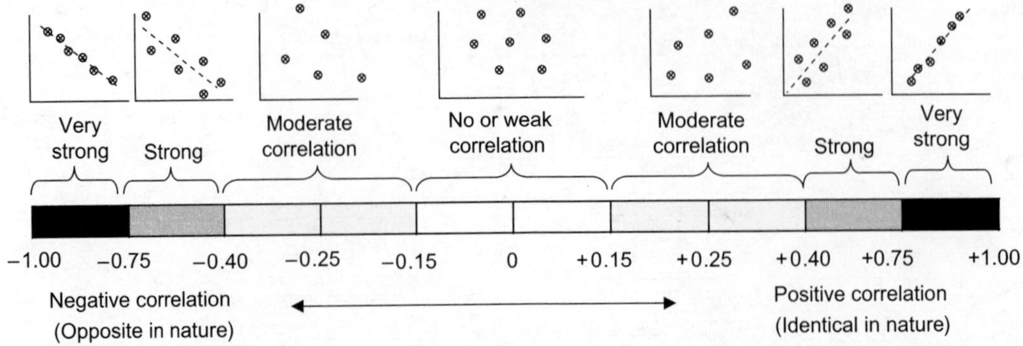


**Fig. 21.4** Interpretation of strength of Spearman rank–order correlation coefficient ($\rho$)

and when it is difficult to make an accurate judgement. This statistical technique is applicable to clarify the level of confidence about whether a correlation exists between any two selected data set variables. The given pair of data sets is arranged stepwise, and the $\rho$ value is calculated:

1. List the scores of the observed data sets, in two columns.
2. Rank each set of data in ascending order, and write them in the third and fourth columns.
3. If a data set has the *same* values, then rank them as *tied*. For example, a data set has two occurrences of 13 in a column, so rank them as third and fourth, average the ranks $(3 + 4)/2 = 3.5$ and enter 3.5 as their rank. Continue to rank the rest as fifth, sixth, and so on. In another example of *tied rank*, a data set has three occurrences of 10 in a column, so rank them as $13^{rd}$, $14^{th}$ and $15^{th}$, average these ranks $(13 + 14 + 15)/3 = 14$ and enter 14 as their ranks. Continue to rank the rest as $16^{th}$, $17^{th}$, etc.
4. After ranking, adopt any appropriate method for calculating the $\rho$ value.

Two methods can be used to calculate Spearman's rank–order correlation coefficient ($\rho$), based on the ranking type: untied and tied. The following formulas are used for calculating $\rho$ (Rho) values based on the tied/untied ranks.

(a) When the data does not have tied ranks:

$$\rho = 1 - \frac{6\sum d_i^2}{n\left(n^2 - 1\right)} \tag{21.3}$$

where $\rho$ = Spearman rank–order correlation coefficient; $d_i$ = difference in paired ranks; and $n$ = number of cases (number of rows in the data set).

(b) When the data has tied ranks (with averaged rank values):

$$\rho = \frac{\sum_i \left(x_i - \ddot{X}\right)\left(y_i - \bar{y}\right)}{\sqrt{\sum_i \left(x_i - \ddot{X}\right)^2 \times \sum_i \left(y_i - \bar{y}\right)^2}} \tag{21.4}$$

where $i$ = number of paired cases; $x_i$ = rank value of data set-1 with the $i^{th}$ case; $\ddot{X}$ = mean value of the ranks of data set-1; $y_i$ = rank value of data set-2 with the $i^{th}$ case; and $\bar{y}$ = mean value of the ranks of data set-2.

*Interpretation of the calculated $\rho$ value:* Always use the default null hypothesis – there is no correlation between the two sets of data. To accept or reject this hypothesis, compare the computed $\rho$ value with critical $\rho$ values obtained from Table 21.3 and conclude as follows:

(a) If the calculated $\rho$ value > critical $\rho$ value, reject the null hypothesis.
(b) If the calculated $\rho$ value < critical $\rho$ value, accept the null hypothesis.

**Table 21.3** A partial table of critical $\rho$ values for 95% confidence level with two-tail curve ($\alpha = 0.05$ or 5% significance level)

| *No. of pairs of data (n)* | *5* | *6* | *7* | *8* | *9* | *10* | *11* | *12* | *13* | *14* | *15* | *16* | *17* | *18* | *19* |
|---|---|---|---|---|---|---|---|---|---|---|---|---|---|---|---|
| Critical $\rho$ value* | 1.000 | 0.886 | 0.786 | 0.738 | 0.700 | 0.643 | 0.613 | 0.587 | 0.560 | 0.538 | 0.521 | 0.503 | 0.488 | 0.472 | 0.460 |

*The other values can be read off any standard table.

*Note:* As the $\rho$ value ranges between +1 and −1, it is usually two-tailed. However, a one-tailed test may be used for specific theoretical reasons, provided the prediction is either positive or negative (TNF, 2008).

**Example 21.10**

The number of accidents observed in a city against cases of drivers booked for being above the legal alcohol limit at 10 road intersections in a year is given below:

| Road\Name | R-1 | R-2 | R-3 | R-4 | R-5 | R-6 | R-7 | R-8 | R-9 | R-10 |
|---|---|---|---|---|---|---|---|---|---|---|
| No. of alcohol cases booked | 20 | 45 | 76 | 34 | 31 | 33 | 67 | 39 | 43 | 29 |
| No. of accidents observed | 19 | 34 | 33 | 32 | 12 | 14 | 46 | 11 | 23 | 21 |

Is there a linear relationship between the rankings produced by these two sets of data? Compare the observed number of accidents with the number of alcohol-related bookings and comment on the test results based on Spearman's rank–order correlation coefficient ($\rho$).

**Answer:** The given values are ranked and listed below. It is observed that the data sets do not have tied ranks; so the calculations are carried out as per the first method.

| Road No. | No. of | | Rank of | | $d_i$ | $di^2$ |
|---|---|---|---|---|---|---|
| | Alcohol cases | Accidents observed | Alcohol cases | Accidents observed | | |
| R-1 | 20 | 19 | 10 | 7 | (10 − 7) = 3 | $3^2 = 9$ |
| R-2 | 45 | 34 | 3 | 2 | (3 − 2) = 1 | $1^2 = 1$ |
| R-3 | 76 | 33 | 1 | 3 | (1 − 3) = −2 | $2^2 = 4$ |
| R-4 | 34 | 32 | 6 | 4 | (6 − 4) = 2 | $3^2 = 4$ |
| R-5 | 31 | 12 | 8 | 9 | (8 − 9) = −1 | $-1^2 = 1$ |
| R-6 | 33 | 14 | 7 | 8 | (7 − 8) = −1 | $-1^2 = 1$ |
| R-7 | 67 | 46 | 2 | 1 | (2 − 1) = 1 | $1^2 = 1$ |
| R-8 | 39 | 11 | 5 | 10 | (2 − 10) = −5 | $5^2 = 25$ |
| R-9 | 43 | 23 | 4 | 5 | (4 − 5) = −1 | $-1^2 = 1$ |
| R-10 | 29 | 21 | 9 | 6 | (9 − 6) = 3 | $3^2 = 9$ |
| No. of pairs of data = $n$ = 10 | | | | | $\sum_d i^2 =$ | 56 |

Substituting in the following equation, Spearman's rank–order correlation coefficient $\rho$ value is computed.

$$\rho = 1 - \frac{6\sum d_i^2}{n(n^2-1)} = 1 - \frac{6 \times 56}{10(10^2-1)} = 0.661$$

The calculated $\rho$ value = 0.661

Take the critical $\rho$ value at 5% confidence level = 0.643 (from Table 21.3 for n = 10).

*Interpretation of the calculated $\rho$ value:* The null hypothesis can be stated as follows: "There is no correlation between the two sets of data". To accept or reject this hypothesis, let us compare the calculated $\rho$ value with the critical $\rho$ values.

The calculated $\rho$ value (0.661) > critical $\rho$ value (0.643), so *reject* the null hypothesis. Moderate correlation exists between the rankings of the two data sets, but it is not strong because the calculated value is not closer to 1.0. Therefore, it is evident that the sets of rankings are positively related and there is moderate association between them.

*Other Observation 1:* Let us also find out whether the calculated $\rho$ value could have arisen by chance and calculate the standard t-test, as shown below:

$$t = \frac{\rho\sqrt{n-2}}{\sqrt{1-\rho^2}} = \frac{0.661\times\sqrt{10-2}}{\sqrt{1-0.661^2}} = \frac{1.8695}{0.7503} = 2.491$$

The critical t-value (at Degree of Freedom 10 – 2 = 8 and significance level 5%) = 2.306 (from Table 21.4)

**Table 21.4** A partial table of student t-value distribution for two-tailed 95% confidence level

| *Degree of Freedom (DoF)* | *1* | *2* | *3* | *4* | *5* | *6* | *7* | *8* | *9* | *10* | *11* | *12* | *13* | *14* | *15* |
|---|---|---|---|---|---|---|---|---|---|---|---|---|---|---|---|
| Critical t-value | 12.710 | 4.303 | 3.182 | 2.776 | 2.571 | 2.447 | 2.365 | 2.306 | 2.262 | 2.228 | 2.201 | 2.179 | 2.160 | 2.145 | 2.131 |

As the calculated t-value (2.491) > critical t-value (2.306), it is concluded that the $\rho$ value (0.661) is significant and the rejected null hypothesis is once again validated.

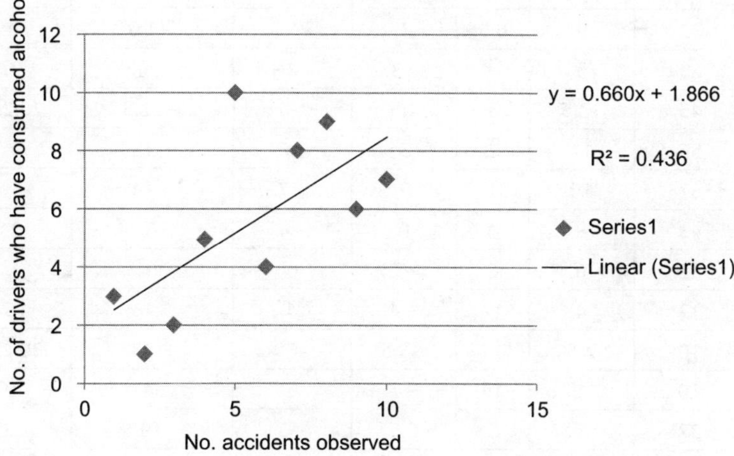


**Fig. 21.5** Coefficient of correlation between the rankings

*Other Observation 2*: Alternatively, the coefficient of correlation between the rankings of the two data sets can be calculated as $R^2 = 0.662^2 = 0.438$.

Pearson's regression coefficient or $R^2$ value of the rankings of the two data sets can be calculated as $R^2 = \rho^2 = 0.660^2 = 0.4356$. Also, the slope of this best fit linear line = $\rho$ value ≈ 0.661 (Fig. 21.5).

## Example 21.11

A summary of the measured values of coefficient of retroreflected luminance (night time visibility) $R_L$ values (mcd/m²/lx) at various locations of different candidate roads and the observed road accidents on road locations are given below.

| Road | R1 | R2 | R3 | R4 | R5 | R6 | R7 | R8 | R9 | R10 | R11 | R12 | R13 | R14 | R15 |
|---|---|---|---|---|---|---|---|---|---|---|---|---|---|---|---|
| $R_L$ values | 150 | 100 | 80 | 120 | 50 | 60 | 70 | 90 | 110 | 130 | 150 | 160 | 36 | 36 | 50 |
| No. of accidents | 10 | 19 | 33 | 23 | 45 | 56 | 67 | 43 | 32 | 12 | 10 | 10 | 37 | 31 | 33 |

Find the possibility of any linear relationship between the rankings produced by the coefficient of retroreflected luminance ($R_L$ values) and the number of road accidents. Compare the observed number of accidents to $R_L$ values and comment on the test results based on Spearman's rank–order correlation coefficient ($\rho$).

**Answer:** As the ranks are tied, let us use the second method of computation of Spearman's rank–order correlation coefficient ($\rho$). The details of the computations are as follows:

| *Road Name* | *$R_L$ (x)* | *Accidents (y)* | *Rank of* | | $x_i - \ddot{X}$ | $Y_i - \bar{y}$ | $(x_i - \ddot{X}) \times (y_i - \bar{y})$ | $(x_i - \ddot{X})^2$ | $(y_i - \bar{y})^2$ |
|---|---|---|---|---|---|---|---|---|---|
| | | | $x_i$ | $y_i$ | | | | | |
| R1 | 150 | 10 | 2.5 | 14 | 2.1 8 = −5.5 | 14 − 8 = 6 | −5.5 × 6 = −33 | 30.25 | 36 |
| R2 | 100 | 19 | 7 | 11 | −1 | 3 | −3 | 1 | 9 |
| R3 | 80 | 33 | 9 | 6.5 | 1 | −1.5 | −1.5 | 1 | 2.25 |
| R4 | 120 | 23 | 5 | 10 | −3 | 2 | −6 | 9 | 4 |
| R5 | 50 | 45 | 12.5 | 3 | 4.5 | −5 | −22.5 | 20.25 | 25 |
| R6 | 60 | 56 | 11 | 2 | 3 | −6 | −18 | 9 | 36 |
| R7 | 70 | 67 | 10 | 1 | 2 | −7 | −14 | 4 | 49 |
| R8 | 90 | 43 | 8 | 4 | 0 | −4 | 0 | 0 | 16 |
| R9 | 110 | 32 | 6 | 8 | −2 | 0 | 0 | 4 | 0 |
| R10 | 130 | 12 | 4 | 12 | −4 | 4 | −16 | 16 | 16 |
| R11 | 150 | 10 | 2.5 | 14 | −5.5 | 6 | −33 | 30.25 | 36 |
| R12 | 160 | 10 | 1 | 14 | −7 | 6 | −42 | 49 | 36 |
| R13 | 36 | 37 | 14.5 | 5 | 6.5 | −3 | −19.5 | 42.25 | 9 |

(Continued)

**Table** (Continued)

| Road Name | $R_L$ (x) | Accidents (y) | Rank of | | $x_i - \ddot{X}$ | $Y_i - \bar{y}$ | $(x_i - \ddot{X}) \times (y_i - \bar{y})$ | $(x_i - \ddot{X})^2$ | $(y_i - \bar{y})^2$ |
|---|---|---|---|---|---|---|---|---|---|
| | | | $x_i$ | $y_i$ | | | | | |
| R14 | 36 | 31 | 14.5 | 9 | 6.5 | 1 | 6.5 | 42.25 | 1 |
| R15 | 50 | 33 | 12.5 | 6.5 | 12.5 − 8 = 4.5 | 6.5 − 8 = −1.5 | −6.75 | 20.25 | 2.25 |
| **n = 15** | | Sum = | **120** | **120** | | | **−208.75** | **278.5** | **277.5** |

No. of pairs of data sets = n = 15

Mean value of the ranks of data set-1 = $\ddot{X}$ = 120/15 = 8

Mean value of the ranks of data set-2 = $\bar{y}$ = 120/15 = 8

The computed values from the above table are summarised as:

$$\sum_i (x_i - \ddot{X})(y_i - \bar{y}) = -208.75$$

$$\sum_i (x_i - \ddot{X})^2 \times \sum_i (y_i - \bar{y})^2 = (278.5) \times (277.5) = 77283.75$$

Substituting these values in the formula recommended by the second method of computation of Spearman's rank–order correlation coefficient ($\rho$), we get

$$\rho = \frac{\sum_i (x_i - \ddot{X})(y_i - \bar{y})}{\sqrt{\sum_i (x_i - \ddot{X})^2 \times \sum_i (y_i - \bar{y})^2}} = \frac{-208.75}{\sqrt{77283.75}} = -0.7509$$

The calculated $\rho$ value = −0.7509

The critical $\rho$ value at 5% confidence level = 0.521 (from Table 21.3, for n = 15)

*Interpretation of the calculated $\rho$ value:* The null hypothesis can be stated as follows: "There is no correlation between the two sets of data". To accept or reject this hypothesis, let us compare the calculated $\rho$ value with critical $\rho$ values.

The calculated mod $\rho$ value (0.7509) > critical $\rho$ value (0.521), so *reject* the null hypothesis.

Moderate to strong negative correlation exists between the rankings of the two data sets. Therefore, it is evident that the sets of rankings are inversely related and good association is found between them.

The negative correlation indicates that as the measured values of coefficient of retroreflected luminance ($R_L$ values, mcd/m²/lx) decrease, there the possibility of an increase in road accidents.

As n >10, let us find out whether the calculated $\rho$ value could have arisen by chance, and calculate the standard t-test, as shown below:

$$t = \frac{\rho\sqrt{n-2}}{\sqrt{1-\rho^2}} = \frac{0.7509 \times \sqrt{15-2}}{\sqrt{1-0.7509^2}} = \frac{2.707}{0.6604} = 4.099$$

The critical t-value (at Degree of Freedom 15 − 2 = 13 and significance level 5%) = 2.160 (from Table 21.4). As the calculated t-value (4.099) > critical t-value (2.160), it is concluded that the $\rho$ value calculated above (0.75.9) is significant and the rejected null hypothesis is once again statistically validated.

## EXERCISES

### MULTIPLE CHOICE QUESTIONS

1. What is the intersection accident rate (i.e., accidents per million vehicles) if 34 accidents were reported for 6 years, with the measured values of ADT on both directions as 9000 for the minor approach and 13,500 for the major approach?
   (a) 1.15 (b) 0.69 (c) 1.72 (d) 3.45
2. The death rate for every 10,000 registered vehicles, if the observed number of deaths = 1000 and the number of registered motor vehicles in a city was 1,000,000 is
   (a) 1.0 (b) 5 (c) 10 (d) 1.62
   (*Hint:* $\frac{\text{Death rate for every } 10{,}000}{\text{registered vehicles in a locality}} = \frac{D}{V}\times 10^4 = \frac{1000}{1000000}\times 10^4 = 10$)
3. Death rate for every 1 lakh population in a locality, if the observed fatal accidents = 10,000 and estimated population =1,000,000,000 is
   (a) 1.0 (b) 5 (c) 10 (d) 1.62
   (*Hint:* Death rate for every 1 lakh pop $= \frac{D}{P}\times 10^5 = \frac{10000}{100000000}\times 10^5 = 1$)
4. What is a blind spot in driving or riding?
   (a) A location not seen due to poor visibility
   (b) An area not reached by headlights
   (c) An area not visible due to physical obstruction along a curved road
   (d) An area not seen through the rear view and wing mirrors
5. Alcohol consumption and driving is prohibited because
   (a) The driver might fall asleep while driving (b) It can raise the energy of the driver
   (c) Violations by the driver increases (d) Reaction time reduces
6. The probability of an event occurring with a certain severity is
   (a) Accident (b) Hazard (c) Risk (d) Fault
7. A large number of accidents were found to occur at a signalised intersection and the reason found was that the carriageway surface was very smooth. What counter measure can be taken to reduce such accidents in that intersection?
   (a) Redesign the cycle timings (b) Install advance signals
   (c) Reduce the approach speeds (d) Install ITS sensors

### ANSWERS

1. (b) 2. (c) 3. (a) 4. (d) 5. (d) 6. (c) 7. (c)

### QUESTIONS

1. Write the formulas with usual notations used for computation of
   (a) Rate of accidents per kilometre (b) Rate of accident involvement

(c) Death rate based on the population (d) Death rate based on the registered vehicles
(e) Rate of accidents for vehicle-kilometres of travel

2. Define road safety vision and state its core concept.
3. Define road safety measures. Explain the (a) Engineering Measures, (b) Enforcement Measures, and (c) Education Measures.
4. Explain how evaluation of the effectiveness of road safety implementation are carried out based on the chi-square test and the same is validated by the student t-test.
5. What are the possible conclusions with reference a comparison between calculated $\chi^2$ and critical $\chi^2$ values?
6. Explain the technique of application of Spearman's rank–order correlation coefficient for analysis of road accident data.
7. Describe the interpretation of strength of Spearman rank–order correlation coefficient ($\rho$) with reference to scatter sketch diagrams.
8. Explain the two methods of calculation of Spearman's rank-order correlation coefficient ($\rho$). Write the formulas with usual notations.
9. The contingency table of two data sets is given below. Find out whether the distribution of frequencies in the sample variables differs only due to chance fluctuations, based on the results obtained from the chi-square test.

| *Employment group* | *Revenue department staff* | *NGO group staff* |
|---|---|---|
| Wants to wear seat belt | 51 | 55 |
| Does not want to wear seat belt | 13 | 45 |

10. Define road safety vision. Also state the differences between road safety vision and road safety programmes.
11. The list of observed accident frequencies in a city in two different years 2016 and 2017 are shown below. Find out whether the observed accidents are consistent with the month using the chi-square test.

| *Month* | *Jan* | *Feb* | *Mar* | *Apr* | *May* | *Jun* | *Jul* | *Aug* | *Sep* | *Oct* | *Nov* | *Dec* |
|---|---|---|---|---|---|---|---|---|---|---|---|---|
| 2016 | 34 | 45 | 20 | 22 | 56 | 61 | 56 | 45 | 32 | 31 | 24 | 12 |
| 2017 | 33 | 57 | 21 | 19 | 61 | 55 | 24 | 43 | 54 | 12 | 22 | 10 |

12. The list of accidents observed in a city over a period of one year is given below. Find out whether the observed accidents are uniformly distributed over the hours using the chi-square test.

| *Hours* | *00:00–03:00* | *03:00–06:00* | *06:00–09:00* | *09:00–12:00* | *12:00–15:00* | *15:00–18:00* | *18:00–21:00* | *21:00–24:00* |
|---|---|---|---|---|---|---|---|---|
| Observed frequency | 210 | 109 | 409 | 109 | 179 | 140 | 131 | 195 |

13. Compare the following accidents observed over a period of one year and determine the accident potential between these two road intersections.

    Road Intersection-A: 8; Road Intersection-B: 11
14. The list of accidents observed for one year before and after improvements arranged by age of respondent is given below. Find out whether the observed accidents are consistent with age group using the chi-square test.

| *Age group* | *20–25* | *25–30* | *30–35* | *35–40* | *40–45* | *45–55* | *55–65* |
|---|---|---|---|---|---|---|---|
| Before | 8 | 54 | 25 | 17 | 29 | 15 | 12 |
| After | 18 | 63 | 39 | 26 | 23 | 12 | 17 |

15. The list of accidents observed for one year before and after improvements arranged by the gender of respondent is given below. Find out whether the observed accidents are consistent with the gender using the chi-square test.

| *Gender* | *Male* | *Female* |
|---|---|---|
| Before | 151 | 31 |
| After | 189 | 42 |

16. The list of observed accidents reported for one year before and after improvements arranged by marital status of the respondent is given below. Find out whether the observed accidents are consistent with marital status using the chi-square test.

| *Marital status* | *Married* | *Divorced* | *Separated* | *Single* |
|---|---|---|---|---|
| Before | 141 | 12 | 1 | 31 |
| After | 187 | 7 | 3 | 32 |

17. Before improvement of a road intersection, there were 14 accidents in 6 years. After improvements, only 4 accidents occurred in 2 years. Determine the effectiveness of improvement of the road intersection using the chi-square test.
18. A road safety engineer observed that 6 consecutive kilometres of a National Highway had the following number of accidents in a year.

| *Km No.* | *1* | *2* | *3* | *4* | *5* | *6* |
|---|---|---|---|---|---|---|
| No. of Accidents | 12 | 23 | 45 | 65 | 32 | 23 |

Assume similar road geometrics and traffic volume along the study length of the road. Conduct a chi-square test and determine whether there is any evidence to show that location makes a difference. Test at the 5% significance level.

19. A summary of fatal accidents recorded for two different years is shown below. Compute Spearman's coefficient of correlation and find the potential of month-wise pattern in fatalities.

| *Month* | *Number of fatal accidents recorded during* | |
|---|---|---|
| | *2015* | *2016* |
| January | 75 | 10 |
| February | 22 | 55 |
| March | 36 | 22 |
| April | 49 | 12 |
| May | 25 | 10 |
| June | 15 | 35 |

(Continued)

**Table** (Continued)

| | | |
|---|---|---|
| July | 105 | 55 |
| August | 12 | 7 |
| September | 25 | 19 |
| October | 4 | 10 |
| November | 19 | 35 |
| December | 75 | 80 |

20. Conduct a chi-square test for the following hourly distribution of accidents in a year.

| *Hours* | *00:00–03:00* | *03:00–06:00* | *06:00–09:00* | *09:00–12:00* | *12:00–15:00* | *15:00–18:00* | *18:00–21:00* | *21:00–24:00* |
|---|---|---|---|---|---|---|---|---|
| Observed frequency | 34 | 44 | 112 | 136 | 129 | 197 | 101 | 93 |

The null hypothesis is that the accidents are uniformly distributed over the hours of the day.

21. A summary of the fatal accidents recorded in a year is given below. Conduct a chi-square test and verify statistically whether the accidents are uniformly distributed over the study year 2016.

| *Month* | *Number of fatal accidents recorded during 2016* |
|---|---|
| January | 7.5 |
| February | 10.5 |
| March | 6 |
| April | 9 |
| May | 2.5 |
| June | 5 |
| July | 10.5 |
| August | 12 |
| September | 2.5 |
| October | 4 |
| November | 1 |
| December | 7.5 |

CHAPTER

# 22 Intelligent Transportation System

## 22.0 INTRODUCTION

It is well known that there has been a significant increase in transportation demand due to the growing needs of urbanisation and rural development. This increased demand has inevitably resulted in congested traffic flow, and it contributes to air and noise pollution, reduces business productivity and degrades the overall performance of the transport network.

Intelligent transportation systems (ITS) are information, electronic and communication technologies applied to road infrastructure, vehicles and users in order to improve efficiency in traffic operation and safety, with reduced traffic congestion (Eq. 22.1), delays and impact on the environment. ITS is installed to regulate traffic based on real-time flow. In general, ITS is used as a primary solution to traffic congestion problems (Fig. 22.1) and for improved operation.

$$\text{Congestion criterion} = \frac{\text{Traffic demand on the road at a point in time}}{\text{Traffic capacity of the road}} > 1.0 \qquad (22.1)$$

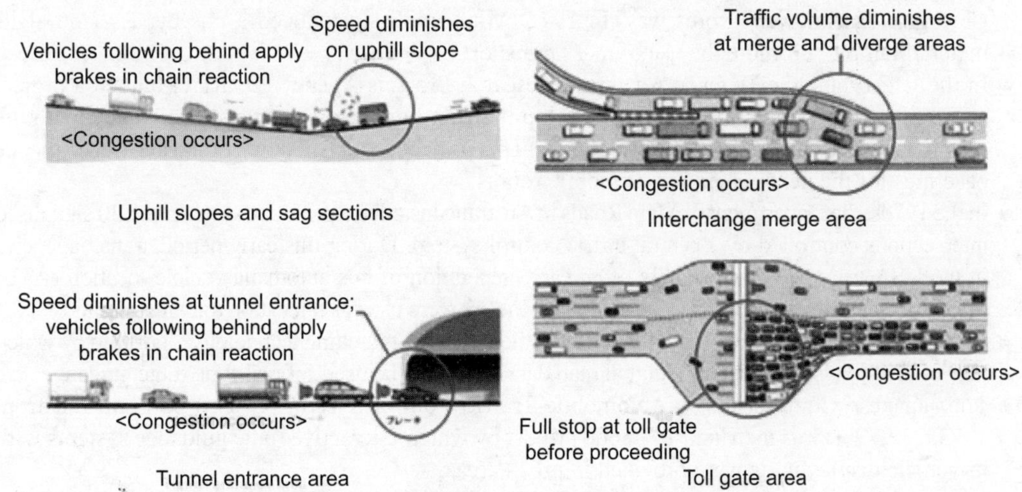


**Fig. 22.1** Main causes of traffic congestion

(*Source*: www.e-nexco.co.jp; *Courtesy*: East Nippon Expressway Company Limited)

It has been noted though, that ITS alone cannot solve all the transportation problems that have arisen due to rapidly increasing traffic demand and changing travel patterns. However, it can be used as a feasible solution to manage transportation challenges by improving the safety and efficiency of the existing infrastructure.

ITS uses information and telecommunications technology to simplify various activities relating to traffic control, current/advance traveller information, incident management, toll transactions, vehicle-to-vehicle and vehicle-to-infrastructure cooperation (DOTMR, 2013).

## 22.1 HISTORICAL DEVELOPMENT OF ITS

With reference to time, the development of ITS may be listed in three phases. The first phase begins in the 1960s and extends to the 1980s; the second phase is between 1980 and 1994; and the third phase starts in 1994 and continues to the present. A brief description of the major developments relating to ITS is given below (Nowacki, 2012; Auer et al., 2016):

*Phase I*

- In the 1960s, several companies in Europe, Japan, China, Australia and USA started developing complex systems for broadcasting a code at the start of a message so that cars equipped with receivers could obtain that information. For example, General Motor's Driver Aided Information and Routing System (DAIR) was tested in the mid-1960s to transmit emergency messages to a service centre, including information on road conditions. In this series, the Experimental Route Guidance System (ERGS) of the Bureau of Public Roads associated with several motor manufacturers tested radio communications between vehicles and roadside units in the late 1960s.

  An electronic traffic sign termed dynamic message sign (DMS) was first used in the USA for a variety of messages such as traffic congestion, upcoming accidents, roadwork zones and changing speed limits. In 1963, the first semi-automated ramp meters were deployed along Chicago's Eisenhower Expressway. In 1972, Minneapolis introduced a bus bypass lane at its metered ramp to promote the use of mass transport.
- In the late 1960s, the Defence Advanced Research Projects Agency (DARPA) funded a project at the Stanford Research Institute to develop the first mobile robot, known as Shakey, with the ability to perceive and reason its own actions. This project established functional and performance baselines for the next generation of mobile robots.
- In the 1970s, the Department of Main Roads in Australia installed a primitive system with 30 signalised intersections controlled by a central traffic control system. During this early period, transit agencies in the USA were the first adopters of an early generation of bus automatic vehicle location (AVL) mapping technology where wayside signpost beacons were used as a location tracking method.
- In 1971, Robert L French of the US developed the first map-matching technology algorithm by which networks of roads modelled as a digital map database could be used for real-time route guidance.
- The Japanese Comprehensive Automobile Traffic Control System (CACS) program ran from 1973 to 1979. It was the first PPP mode project by which interactive route guidance systems with in-vehicle display units were provided in urban areas.
- In 1974, the European Standardisation Committee (CEN) was registered as a non-profit technical association for drafting and disseminating European ITS standards.
- In 1974, a highway radio system (ARI) in Germany used FM technology to alleviate traffic congestion on north-bound autobahns in the summer holidays.
- The US government sponsored an in-vehicle navigation and route guidance system (ERGS). It was an example of a major initial initiative in ITS development in the USA.

The initial phases of development in different countries shared a common emphasis on route guidance and the use of central processing systems with computer software and communications systems.

*Phase II*

- Several technological reforms in memory, processing capacity and communications systems took place in different countries, from 1980 to 1994.
- The DARPA-sponsored artificial intelligence demonstrations that began in the 1960s with Shakey were tested again in the early 1980s with the DARPA Autonomous Land Vehicle (ALV). ALV was built as an all-terrain vehicle equipped with sensors such as video cameras and laser scanners. It contained six racks of computers programmed with algorithms and could steer safely along the road without human assistance, using images from the rooftop camera.
- In 1984, the RACS project began in Japan and formed the basis for their car navigation system.
- In 1984, the Los Angeles Automated Traffic Surveillance and Control System was the first to integrate vehicle detectors, CCTVs and coordinated signal timing.
- In 1984, the Arizona and Oregon DOTs established the Heavy Vehicle Electronic License Plate Program (HELP) to conduct research on commercial vehicle weigh-in-motion and automated vehicle identification (AVI) technologies. This program grew as a consortium of agencies in the USA and Canada to conduct multi-state demonstrations of truck pre-screening and pre-clearance approaches at weigh stations between Texas and British Columbia. This project was known as the Crescent Demonstration Project.
- In 1985, a second-generation traffic management system known as TRACS was installed in Australia.
- In 1986, PROMETHEUS (Program for a European Traffic System with Higher Efficiency and Unprecedented Safety) was developed by auto manufacturers. Another program called Dedicated Road Infrastructure for Vehicle Safety in Europe (DRIVE) was also developed. PROMETHEUS was initiated as part of the EUREKA program, a pan-European initiative to improve the competitive strength of Europe by integrating information technology, telecommunications, robotics and transport technology. The program involved the participation of 18 European automobile companies, state authorities and over 40 research institutions.
- The California Program on Advanced Technology for the Highway (PATH) as a formal research program focused on advanced technology in surface transportation and was founded in 1986 as a collaboration between the University of California at Berkeley and the California Department of Transportation (Caltrans). It is still an active and leading research institution.
- TRANSCOM was established in 1986 in the USA as the Transportation Coordinating Committee for the region (New York, New Jersey and Connecticut) to ensure a common approach for developing solutions, including technology-based solutions for incident notification, regional incident management, and construction coordination using ITS technology.
- In 1988, the European Telecommunications Standards Institute (ETSI) was formed. It is considered equivalent to IEEE. The ETSI objectives include drafting standards for European telecommunications and dissemination of these standards.
- In 1989, the Mobility 2000 group was created in the USA, and led to the formation of Intelligent Vehicle Highway Systems (IVHS) by 1990. The main functions of IVHS were to act as a Federal Advisory Committee for the US DoT, to consider the IVHS program as national policy, and to develop an ITS-based intermodal transport system. It played an important role in bringing together advocates from a variety of institutions, including state and federal governments, industry, consulting and academia to represent the new technology perspective in policy formation.
- In 1989, Operation Greenlight of the USA was started as a joint effort between the Illinois DOT and other regional transportation agencies to reduce vehicle congestion in the region and to reduce demand on the regional highway network so as to increase capacity.

- In 1991, the European Road Transport Telematics Implementation Coordination Organisation (ERTICO) was created with the support of the European Commission (EC) as a private–public partnership with an interest in ITS deployment in Europe. ERTICO provided a platform for defining ITS development, technology usage, deployment and funding needs.
- In 1991, a non-profit organisation known as ITS America was founded. Its main objective was to foster the use of advanced technologies in surface transportation systems. ITS America's members comprise more than 450 private corporations, public agencies, academic institutions and research centres. ITS America has also spread worldwide and has links with various organisations in Europe and Japan, Canada and Brazil. In this series, re-authorisation of the Federal-Aid Highway Program resulted in a significant public commitment in 1991 to institutionalise IVHS. This step established the foundation for the Federal-Aid ITS program as a public–private partnership model.
- Parallel to the founding of IVHS America in 1991, the Intermodal Surface Transportation Efficiency Act of 1991 (ISTEA) was signed by the then President of the USA and was regarded as the first transportation bill of the post-Interstate era. It established policies encouraging the development and application of ITS advanced technologies.
- In 1991, the E-ZPass Interagency Group (IAG) was created to develop an interoperable tolling system that worked for seven independent toll agencies throughout the USA. It was first deployed in August 1993 at the New York State Thruway.
- In 1992, ITS Australia, a non-profit organisation, was established with various members such as ITS suppliers, government authorities, academics and transport businesses and users. Its main objective was to promote the development and deployment of advanced technologies to deliver safe and efficient sustainable transport across all public and private modes by air, sea, road and rail.

*Phase III*

- Since 1994, several applications of ITS have been put into effect and modified with reference to intermodal transportation facilities, including automobile traffic on roads.
- In February 1994, the USDOT established a Joint IVHS Program Office (JPO) to coordinate intermodal policy in the implementation of the IVHS program. In 1994, the IVHS program was renamed intelligent transportation system (ITS) at the first World Congress, held in Paris. It provided for overall management and oversight of the IVHS program, including those of FHWA, the Federal Transit Administration (FTA), NHTSA and the Research and Special Programs Administration (RSPA).
- The Fourth EU Framework Program (1994–1998) envisaged the development of transport telematics and ITS applications. Adopted by the European Council and Parliament, it included telematics as a major topic of research and intended to support the work of standardisation of ITS deployment in traffic management.
- In 1994, ITS Japan was established for promoting research, development and implementation of ITS in cooperation with five national ministries. The policies of ITS Japan include development of national/regional/local architecture, standardisation, and research and development by international cooperation. This Council worked with several national and international organisations such as Vehicle, Road, and Traffic Intelligence Society (VERTIS). As part of these efforts, Vehicle Information and Communication System (VICS) and Advanced Traffic Information System (ATIS) have been in operation in Japan.
- Ford and General Motors formed the Crash Avoidance Metrics Partnership in 1995 to develop crash avoidance countermeasures to improve traffic safety. The CMP has provided a useful way

for vehicle manufacturers and stakeholders to collaborate on pre-competitive crash avoidance research projects. Much of today's intelligent vehicle research has been in this direction.

- VICS was started in April 1996 in Tokyo and Osaka. The VICS Centre was supported by the Japan Ministry of Construction, Ministry of Telecommunications and National Police Agency. The VICS Centre receives real-time traffic information from roadside highway traffic information centres and provides control information using roadside infrared beacons and FM broadcasting. More details about VICS are presented in this chapter.
- The USDOT established the ITS Standards Program in 1996 to encourage the widespread use of ITS technology in surface transportation systems.
- In 1996, the USDOT announced three major ITS deployment initiatives: Metropolitan Model Deployment Initiative (MMDI), Advanced Rural Transportation System and Commercial Vehicle Information Systems and Networks (CVISN). These ITS models addressed the challenges of developing an integrated, multimodal intelligent transportation infrastructure with a vision of providing efficient transportation systems, including better traveller information.
- In 1996, Georgia Navigator, an Advanced Traffic Management System, was operated by the Georgia DoT. This system uses video detection cameras to gather traffic flow data, which is then directed to the Traffic Management Center in Atlanta. The system dispatches the processed data to the highway emergency response operators (HEROs) for further use/communication.
- GPS-based AVL systems were generally adopted by many transit agencies around the globe, in the late 1990s.
- In 1998, the TRAC and South East Freeway systems in Australia merged to form STREAMS Version 1. In subsequent versions, it was integrated and implemented for the management of traffic signals, incident management, highway management, vehicle priority, traveller information and parking guidance.
- In 1998, MIT's Microsystems Technology Laboratories established the Intelligent Transportation Research Center (ITRC). ITRC focuses on key aspects of ITS technologies, including an integrated network of transportation information, automatic crash and incident detection, notification and response, advanced crash avoidance technology and advanced transportation monitoring and management.
- The Intelligent Transportation Systems Society (ITSS) is governed by the Institute of Electrical and Electronics Engineers (IEEE) and began implementation in 1999. The purpose of ITSS is to provide a platform for scientists and engineers for ITS-related advances.
- A key step in 1999 for automated vehicles was when the Federal Communications Commission (FCC) of USA decided to allocate 75 megahertz (MHz) of spectrum to transportation services for improving highway safety and efficiency. The bandwidth of 5.850 to 5.925 GHz was allocated for a variety of dedicated short-range communication (DSRC) uses as a part of the USA national ITS program.
- The EU Common Transport Policy for 2001–2010 and the European Commission have taken initiatives to introduce an e-call emergency system in new cars.
- In 1997, the concept of transport telematics evolved as a branch of knowledge, and ITS activities were integrated with information technology and telecommunications.
- In December 2000, the Ministry of Construction and Transportation (MOCT) of South Korea announced its master plan for Intelligent Transportation Systems 21, with the aim of creating efficient transportation systems in major cities.
- The USDOT launched the Vehicle Infrastructure Integration (VII) program in 2003. The vision was to adopt wireless communications with and between vehicles as a part of ITS applications. In

2005, the USDOT developed and tested a 5.9 GHz-based VII proof of concept (POC). This effort was a continuation of AVCS, as envisioned in Mobility 2000.

- The Safe, Accountable, Flexible, Efficient Transportation Equity Act: A Legacy for Users (SAFETEA-LU) of the USA bill was signed in 2005. SAFETEA-LU affirmed the growing return on ITS investment and encouraged the use of embed ITS in mainstream surface transportation.
- The USDOT established the Integrated Vehicle-Based Safety System (IVBSS) initiative to develop and test integrated safety systems on road vehicles through partnerships with private vehicle industries. Under the IVBSS project, in November 2005, the IVBSS was designed to prevent rear-end, lane change and run-off-road crashes.
- In 2006, the USDOT collaborated with CAMP to develop and test prototype V2V safety applications. The goal was to determine whether this technology would work better with V2V or V2I.
- In 2007, the Polish Association of Transport Telematics (PATT) was registered to provide a common platform for universities, research institutes, and national and private transportation companies. PATT and Intelligent Transportation Systems Slovakia signed a memorandum of agreement for mutual partnership for the development of ITS in those countries.
- In 2007, a meeting of the ITS Association of Poland was held. It envisaged providing ITS solutions to improve transport efficiency and safety in Europe and other parts of the world
- In 2008, PATT, as a member of ERTICO, hosted the Network of National ITS Associations.
- In 2009, the USDOT's ITS-Joint Program Office entered into an agreement with the European Commission (EC) and established the Standards Harmonization Working Group (SHWG). SHWG is intended to execute harmonised standards for the vehicle platform with the EC. Additional agreements have also been reached with Canada and Japan.
- In December 2009, the TCIP Implementation, Requirements and Capabilities Editor (TIRCE-1.2) support tool was released. It is available to any transit agency or vendor, at no charge, to assist in the development of transit communication interface profiles (TCIP)-compliant procurement packages and product specifications.
- The largest real-world test of connected vehicle technology, Connected Vehicle Safety Pilot Model Deployment, occurred from 2012 to 2013 in Ann Arbor, Michigan, USA. The vehicles used wireless safety technology to help everyday drivers avoid crashes as they travelled. The safety applications warn drivers using alerts; for example, braking vehicles ahead, vehicles in blind spots, or impending Red light violations.
- In August 2010, the Washington State Department of Transportation (WSDOT) launched the active traffic and demand management (ATDM) system to reduce traffic collisions due to congestion and blocked lanes. The system uses overhead lane signs to provide advanced warning information based on variable speed limit signs to direct drivers to reduce speeds, and symbols to indicate lane blocks.
- Google's Self-Driving Car project started in 2009. It is one of the most widely known initial efforts in automated vehicles that uses a removable steering wheel.
- The USDOT's ITS Standards program (2011–2014) is a strategic plan designed to ensure the development of the necessary standards for deployment and support of USDOT's overall goals for the ITS program. It also focuses on vehicular connectivity within the context of the current ITS research agenda.
- In September 2015, President Obama announced a new Smart Cities Initiative that led to similar practices in various developing countries like India.
- Adoption and eventual deployment of wireless connected vehicle systems including NHTSA were planned in 2016 for V2V safety messaging and communications based on DSRC technology.

Numerous ITS developments are in progress and are operational in different countries around the world.

## 22.2 OBJECTIVES OF ITS

The following objectives may be considered for any type of ITS deployment and design integration:

- To improve road safety
- To improve the functional efficiency of the road network and infrastructure (reduce congestion/delays)
- To improve the operational efficiency of the transport system (by providing real-time information/guidance)
- To improve energy (fuel) saving
- To reduce the environmental impact of the transport system (reduce air/noise pollution)

## 22.3 BENEFITS OF INTELLIGENT TRANSPORTATION SYSTEM

The following benefits can be obtained by deploying various subsystems of ITS (Nexco, 2008):

- **Area-wide traffic control or traffic management and control** (Fig. 22.2)**:** Traffic signal timings in urban areas are controlled based on the traffic flow demand observed at various intersections using real-time data collection and analysis. In this connection, vehicle detectors, roadside transmitters and variable message sign (VMS) boards operations are integrated.

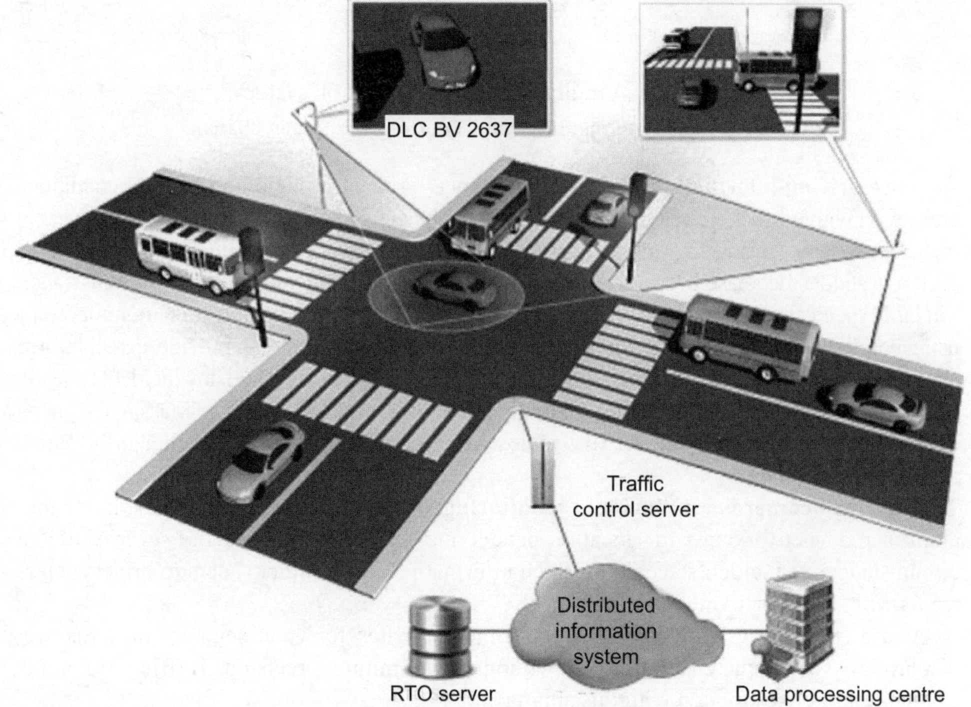


**Fig. 22.2** Traffic control/enforcement system
(*Source*: www.metroinfrasys.com, *Courtesy*: Mr. Sachin Bhatia)

Such an integrated multifunctional enforcement system (Fig. 22.3) can automatically detect signal violations, lane violations and speed violations by a vehicle. The collected violation data (image, violation type, place, time and more) is sent to the Control Centre and violation tickets are automatically issued to the vehicle-owner.

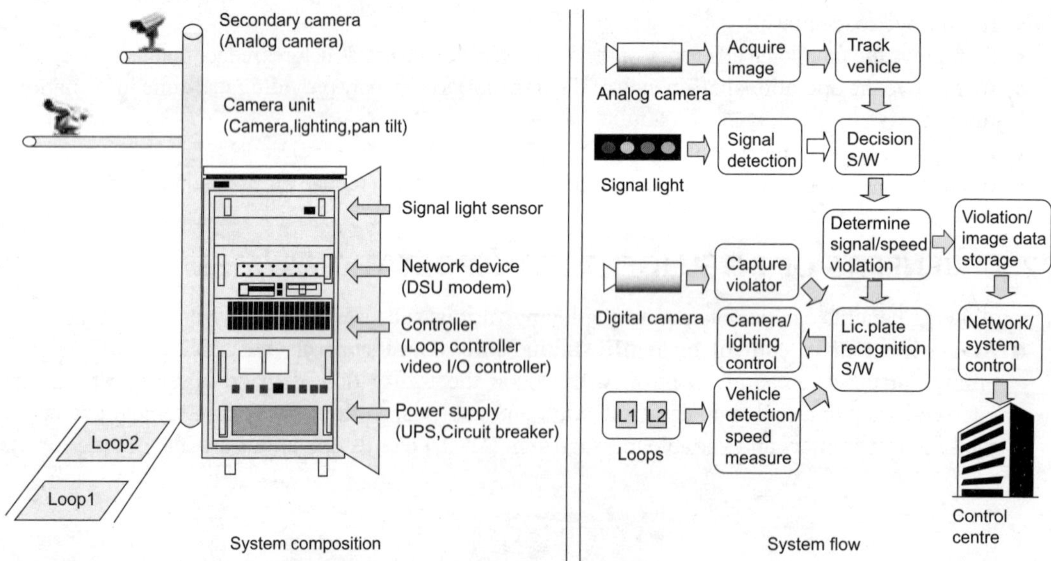


**Fig. 22.3** A multifunctional information system
(*Source*: www.metroinfrasys.com, *Courtesy*: Mr. Sachin Bhatia)

- **Advance driving information and route guidance**: Prior information about road conditions, rest areas with/without fuel/electric recharging facilities, hospitals and restrictions due to congestion and inclement weather makes drivers maintain better control. It also prevents fluctuations in traffic speed such as a sudden decrease in speed at sag curves and turnings, thereby reducing $CO_2$ emissions.
- **Variable speed control and safety information service**: On-board car navigation monitors connected using wireless communication from on-road mounted sensors/devices can provide spatial information about route alignment, traffic volume, merging vehicles, fallen objects and any incidents ahead.
- **Sophisticated graphical message signage:** Multi-colour LED variable message sign (VMS) boards are used to display alternative routes, congestion information, weather information and accurate travel times.
- **Traffic video cameras and incident monitoring:** Video cameras installed at suitable intervals along a road network provide location details and information about the severity of unusual circumstances or incidents/accidents. Such information can be quickly sent to drivers and rescue teams with appropriate medical/equipment.
- **Accurate travel time estimation:** Travel time estimates for each segment of a roadway are displayed based on current traffic data with up-to-the-minute precision. Traffic jams and delays can be accurately estimated with this approach.
- **ETC toll booths:** Stopping, queuing and paying of toll/fee manually at toll booths can be eliminated by using an Electronic Toll Collection (ETC) system. The toll payment is made

using prepaid cards or deducted from bank accounts. The accounts and vehicle database can automatically store this information for future use.

- **Automatic vehicle location:** On-board GPS enables owners to locate their vehicles travelling in a road network.
- **Logistics information:** In-vehicle mounted GPS or OBE-ETC devices can be used to locate the current position of trucks/cargos of distribution companies. Such information can foster business productivity and reduce delays in cargo delivery.
- **Truck traffic consolidation/separation:** Strategic travel planning at interchange ramp meters, rest areas and cargo-distribution centres close to expressways make trucks travel/halt efficiently in groups. Large truck traffic can be separated from the general traffic by establishing freight-only lanes, which provide safety and comfort.
- **Customer/Passenger/Driver inquiries/emergencies:** Dedicated telephone landlines are provided with default connection to the customer care centre.
- **Doctor-heli:** In case of road accidents, medical airlifts can be made by doctors, assisted by nurses.

**Factors to be considered:** ITS comprises electronic surveillance, communications, traffic analysis and control technologies that are intended to benefit users and managers of the transportation system. Road users gain from the advance information and on-path guidance provided by ITS. Transportation managers and agencies profit from improved ability to acquire real-time traffic data at network level, route and control traffic flow and disseminate information. The following points may be considered for ITS deployment.

The variable message, speed limit and lane control signs should be co-located and coordinated to explain the reason for a speed change through the use of *single* variable message sign. Default static speed limit signs must also be placed in case of breakdown of the system.

For maintenance purposes, future traffic needs and the possible widening of road should be considered while deciding the place of installation of any ITS devices, with their supporting structures.

Appropriate locations are optimally selected along the road network for easy accessibility of control cabins and control rooms for troubleshooting. Construction of barriers and protection measures should be taken against vandalism.

ITS relies on different types of traffic flow sensors to provide vehicle detection, incident detection, automatic traveller surveillance, real-time traffic adaptive signal control, archival data, and data for traveller, commercial, technical and emergency information services. The traffic operators/administrators need to select the appropriate combination of subsystems (as a bundle of instrument subsystems) required for a road network/section of road.

The success of ITS largely depends on the technology used, proper design, installation and maintenance of sensors and the associated component system.

*Note:* Numerous ITS systems are being developed and implemented in different countries by different agencies. However, their basic objectives are eliminating congestion, providing safe and comfortable driving, and fuel and time saving, under local road/weather conditions to enable a pollution-free life with sustainable economic growth.

## 22.4 ITS DESIGN

Provision for and location of various components of road infrastructure and geometric elements with the proposed electronic sensor/display devices must be planned in advance. Planning of the following

ITS subsystems, along with their deployment design, is essential for any typical road network in an area (Fig. 22.4):

- Traffic flow and road user data collection systems
- Road condition data collection system
- Weather data collection system
- Road network system with local infrastructure facilities
- Traffic management/speed, height and weight control systems
- Toll collection and enforcement systems
- Incident management systems
- Heavy-vehicle management systems
- Advance travel information system
- Public transport priority/emergency vehicle pre-emption systems

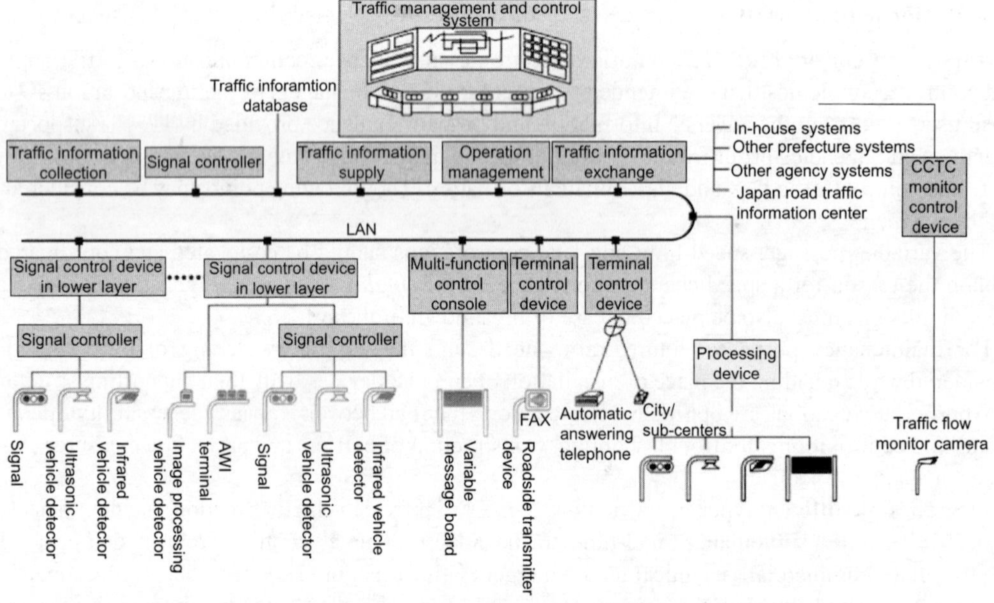


**Fig. 22.4** System structure diagram

(*Source*: IUTR, 1998; *Courtesy*: Japan Traffic Management Technology Association, Japan)

These subsystems should be integrated properly to achieve the objectives of ITS. These systems use different technologies (wired and wireless) and sensors such as vehicle detectors (intrusive/non-intrusive), pedestrian detectors, various types of signboards/display monitors, closed circuit televisions (CCTV) and wired/wireless systems. The data collected is passed on to a nearby traffic control centre (Fig. 22.5). Then, information from the different subsystems is analysed in real time by software and translated into navigational information (volume, speed, congestion length, saturation traffic flows and estimated travel time). This data is finally passed back to the local roadside device – VMS, OBE, in-vehicle navigation unit or traffic signal (Fig. 22.2).

The in-vehicle unit used to receive information may be any one or a combination of the following:

- **Character-based display:** Used to display the speed limit, over-height limit, and distance to specific facilities

- **Diagram-based display:** Used to display route alignment with distance and on-route stations using multi-coloured lines and symbols
- **Map-based display:** Used to display an area map of the road network showing the origin–destination and other related details

As part of the Japanese Universal Traffic Management System (UTMS21), the traffic control centre is termed the integrated traffic control system centre (ITCS). ITCS can collect traffic information and control signals in real time by two-way communication with in-vehicle equipment.

Based on requirements, the traffic operation management of subsystems can be evaluated by traffic supervisors with real-time information through computer interfaces such as dynamic wall maps, GIS map display walls and CRT/LCD displays (Fig. 22.5). The traffic supervisors are authorised to change the system control parameter settings and regulate the entire system so that it is operationally stable and consistent (IUTR, 1998).

**Fig. 22.5** Inside a traffic control centre (TCC), traffic management centre (TMC) or traffic information centre (TIC)

(*Source*: Nexco, 2014; *Courtesy*: Central Nippon Expressway Company Limited, Japan)

Necessary construction-related provisions for laying cables, ladders, antennas, cabinets, ducts and supporting poles/girders are made during preconstruction, maintenance and to meet future needs, especially due to road widening, changed speed limits and increased traffic demand.

## 22.5 TECHNOLOGIES USED IN ITS

ITS relies on a wide range of enabling technologies and software. A broad categorisation of the *sensing system* that uses different technologies/communications for detection, tracing, class and speed data of vehicles are given below:

- **Static sensing**
  - Intrusive sensors such as tubes, electrical induction loops and magnetic sensors that detect passing vehicles

  - Non-intrusive sensors such as video cameras, web cameras, acoustic/passive sensors and radio frequency/radar sensors detect passing vehicles and vehicle/passenger presence
- **Mobile sensing:** GPS on vehicles or public transport, GPS on smart phones used for long-range and short-range detection, tracing and speed of vehicles constitute mobile sensing. Devices with other sensors such as dedicated short-range communications (DSRC) are installed on-board and used as in-vehicle wireless communication systems for applications such as ETC/EFC, VICS, and vehicle-to-vehicle and vehicle-to-roadside equipment communication.
- **Hybrid sensing**
  - Bluetooth is a short-range electronic wireless interconnection technology for mobile phones, computers and other electronic devices. It is used to transfer photographs, text and other types of data through wireless communication from one electronic device to another.
  - Radio frequency identification (RFID) is an automatic short-range wireless one-way identification and data capture technology that uses electromagnetic fields to identify and track vehicles embedded with/carrying RFID tags when they come closer to a reader. The data is passed through broadband wireless devices mounted on roadside poles.

    RFID tags contain electronically stored specific and unique information for a vehicle/object. The information may be the vehicle registration number, vehicle owner or type of product baggage. In addition, RFID cards can be used to control access to public transport facilities.

A brief explanation of the technologies used in ITS deployment is as follows (Ezell, 2010).

## 22.5.1 Geographic Positioning System

Satellite-assisted global positioning system (GPS) is capable of providing precise, global and continuous position fix (with latitude, longitude and altitude) in all weather conditions to a properly equipped vehicle. GPS is a radio-navigation system capable of providing highly precise velocity and time information about the vehicle. GPS radio signals travel at the speed of light ($\approx$300,000 km/sec). This technology is used to find the location of a vehicle, and help with in-vehicle navigation, route guidance, and tracking and tracing of vehicles over a road network, in real time.

## 22.5.2 Communication Technologies

Microwave, short-range radio and infrared electromagnetic wave forms are used for short-range wireless communication. For example, DSRC is widely used for ETC/EFC and commercial vehicle operations. A few examples of wireless communication systems are given below:

- **Dedicated short-range communication (DSRC):** DSRC is a *one-way* or *two-way* short-to-medium-range wireless communication channel specifically designed for automotive use. DSRC enables communication between vehicles through embedded on-board equipment (OBE) and roadside equipment (RSE). This system is often termed V2X. A DSRC beacon is used in many ITS applications such as vehicle-to-vehicle (V2V) communication, vehicle-to-RSE (V2X) communication, ETC/EFC or electronic road pricing, information provision and congestion charging (in the UK). The DSRC channel uses the 5.8 GHz band in Japan and the 5.9 GHz band in the USA. Europe is in the process of upgrading their DSRC band from 5.8 to 5.9 GHz.
- **Mobile telecommunications** can transmit information over 3G or 4G mobile telephone networks. As the network adequately covers a wide area of the road network in a region, it enables users to exchange information. Mobile telecommunication technology can be used for real-time travel information, fleet management and incident management. For example, using the GPS-derived

location of a smart phone; a GPS chip on a vehicle can send information about its location and speed to its owners through satellite ground stations.

- **Broadband-based Wi-Fi internet communication** allows rapid communication between vehicles and roadside devices, within a limited range of a few hundred metres. This limited range can be extended by communicating the information through each successive vehicle or RSU and passing it on to the next vehicle or RSU. Wireless communication can be used for real-time travel information, trip planning, artificial visual surveillance and online issue of challans/penalty payments.
- **FM radio waves** are used for broadcasting information by voice communication.
- **Roadside radio transmitters** are used to supply voice traffic information edited by voice synthesizers at traffic control centres. Radio transmitters and antennae installed on the roadside at suitable intervals emit radio waves. Drivers travelling within the coverage area can access this information on their car-radios (IUTR, 1998).

### 22.5.3 Geographical Information System (GIS)

GIS software is used for the pictorial representation of the transportation network in an area. The location of vehicular and ground facilities database is featured on this GIS platform. GIS provides an interface between sensor data and user services.

### 22.5.4 Artificial Vision System

Closed circuit television (CCTV) and web cameras installed at various locations on a road network are used for surveillance, enforcement, detection and security purposes.

### 22.5.5 Digital Mapping

A city topography map with transportation networks is geo-referenced and a GIS digital map is prepared by marking the infrastructure/facility features, as a layered system. Such digital maps are a key building block for ITS deployment and are used for numerous applications, such as for area traffic management/control, route guidance, vehicle tracking, parking management, incident management and so on.

### 22.5.6 Data Acquisition and Exchange System

Electronics-based devices are capable of acquiring data from different types of sensors using wired and wireless communication systems. The data acquired is stored, processed and exchanged with other instruments using appropriate software.

## 22.6 ITS ARCHITECTURE

ITS architecture describes the *blueprint* of a structured framework of subsystems specifically/strategically located along a road network in a region. The classification of the system is based on their specific performance requirements, conceptual stage-wise development of subsystems, and the role of stakeholders with integrated activity/action plan defined and set as a future course of action/improvement.

ITS architecture does not assume the use of specific technologies or installations such as products, sensors with software and supporting infrastructure. It is *generic* and gives system developers *flexibility* in design to obtain optimised solutions, while meeting the set standards and accommodating the interfaces required for operational efficiency and future extensions.

ITS architecture resembles slices or a GIS layered structure. National/Regional architecture may be pictorially depicted in the layered structure, which comprises three layers: institutional, transportation and communication (Fig. 22.6). The institutional layer is considered the bottom layer, on which the transportation network with the proposed framework of ITS subsystems is envisaged.

The institutional layer is the source of objectives and transport requirements and includes policies, institutions, requirements of user services, funding mechanisms and procedures relating to the effective implementation, operation and maintenance of ITS. The transportation layer is the basic framework of the region/area covered with road networks linked to underlying functional aspects interfaced with vehicle facilities, transport infrastructure components and user services. The communication layer includes the subsystems of ITS with related communication services and technological aspects.

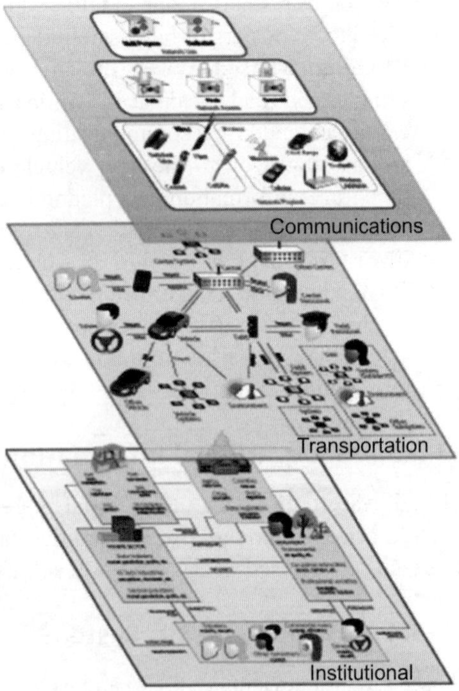

**Fig. 22.6** Overview of typical ITS architecture

(*Courtesy*: Iteris Inc. and US Department of Transportation, 2015)

National/Regional/State/Local government will sketch out the broad framework of ITS architecture based on their transportation policies for safe and efficient use by the general public. The envisaged ITS architecture is deployed as per legislative, funding and resource constraints.

*Use in transportation planning:* The framework of objects and goals set for a specific area of a road network is implemented by the concerned agencies by coordinating and integrating the services, as *channelised* in the National/State ITS architecture. It can be used as a tool to support planning decisions intended to be made in the direction of the *vision* strategy. The steps involved in the implementation of ITS architecture in transportation planning are presented in Fig. 22.7.

## 22.7 V-DEVELOPMENT MODEL FOR ITS PROJECTS

The V-development model is widely used for system development of any ITS project, with various tasks to be carried out over its life cycle phases (Fig. 22.8). The ITS project wholly or in part funds the acquisition of technologies with devices, system tool, supporting infrastructure, work-plans and the related user services. This V-model development combines the main component features of the classical Waterfall model and the Spiral Development model used primarily for software development (Caltrans, 2009).

The scope of the system engineering process includes the tasks to be undertaken in sequence, such as defining project objectives, early development cycle, documentation, design, deployment of technologies, validation, and maintenance and replacement by the end of the project design life.

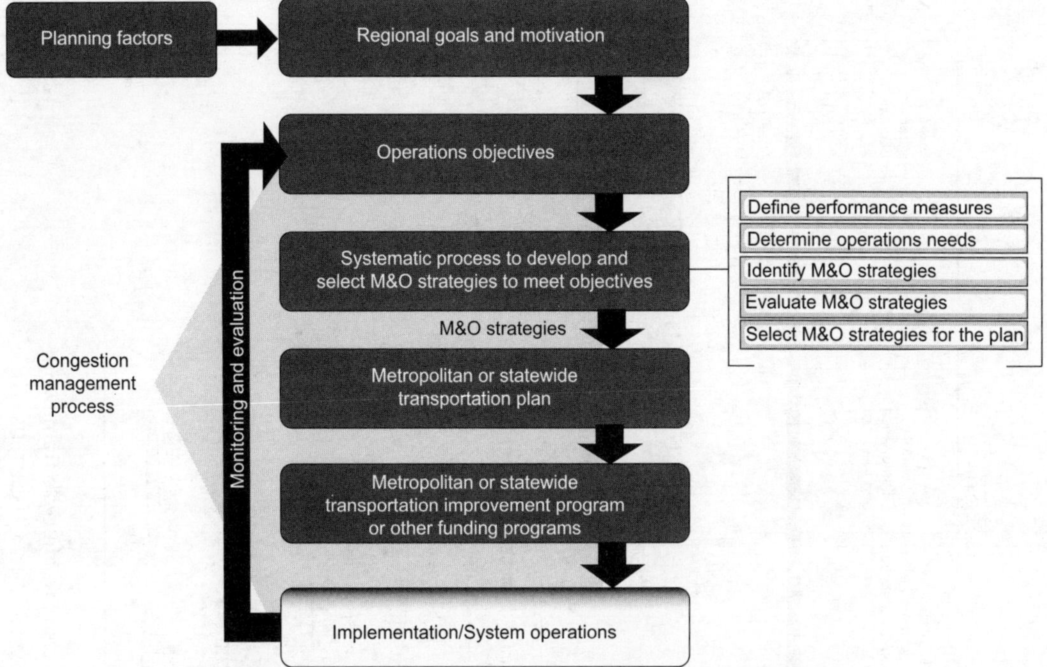


**Fig. 22.7** Use of ITS architecture in transportation planning
(*Courtesy*: Iteris Inc. and US Department of Transportation, 2015)

In short, systems engineering provides a systematic process and tools that directly support project management in all the stages (Ryen, 2008).

The V-development model comprises various stages of the system engineering process of ITS project phases in sequence, ranging from the early stages of project development to the end results, including retirement/replacement plans. Brief details of the V-development model, with inter-related step-by-step (from left to right of Fig. 22.8) descriptions of the systems engineering process for ITS projects are given below:

1. **Regional ITS architecture:** The project scope and objectives are defined and efficient incremental implemental strategies are planned for based on needs assessment. Planning of regional ITS components by integrating with existing and proposed developments provides an overview of the direction of development.
2. **Feasibility study/Concept exploration:** Technical, operational, economic, financial and political feasibility studies are conducted to assess the most viable option for the project, with identification of risks.
3. **Concept of operations:** The technical course of the project with the sequence of proposed activities is well defined, documented, understood and agreed upon by the stakeholders (system owners, developers, operators and maintainers). High-level identification of user needs with system capabilities and validation of the system at the end of project development is essential.
4. **System requirements:** In this step, requirements are elicited based on the problem defined. Needs/Requirements of the proposed operations of the system are assessed, reviewed with constraints, and system acceptance plans with validation are created.

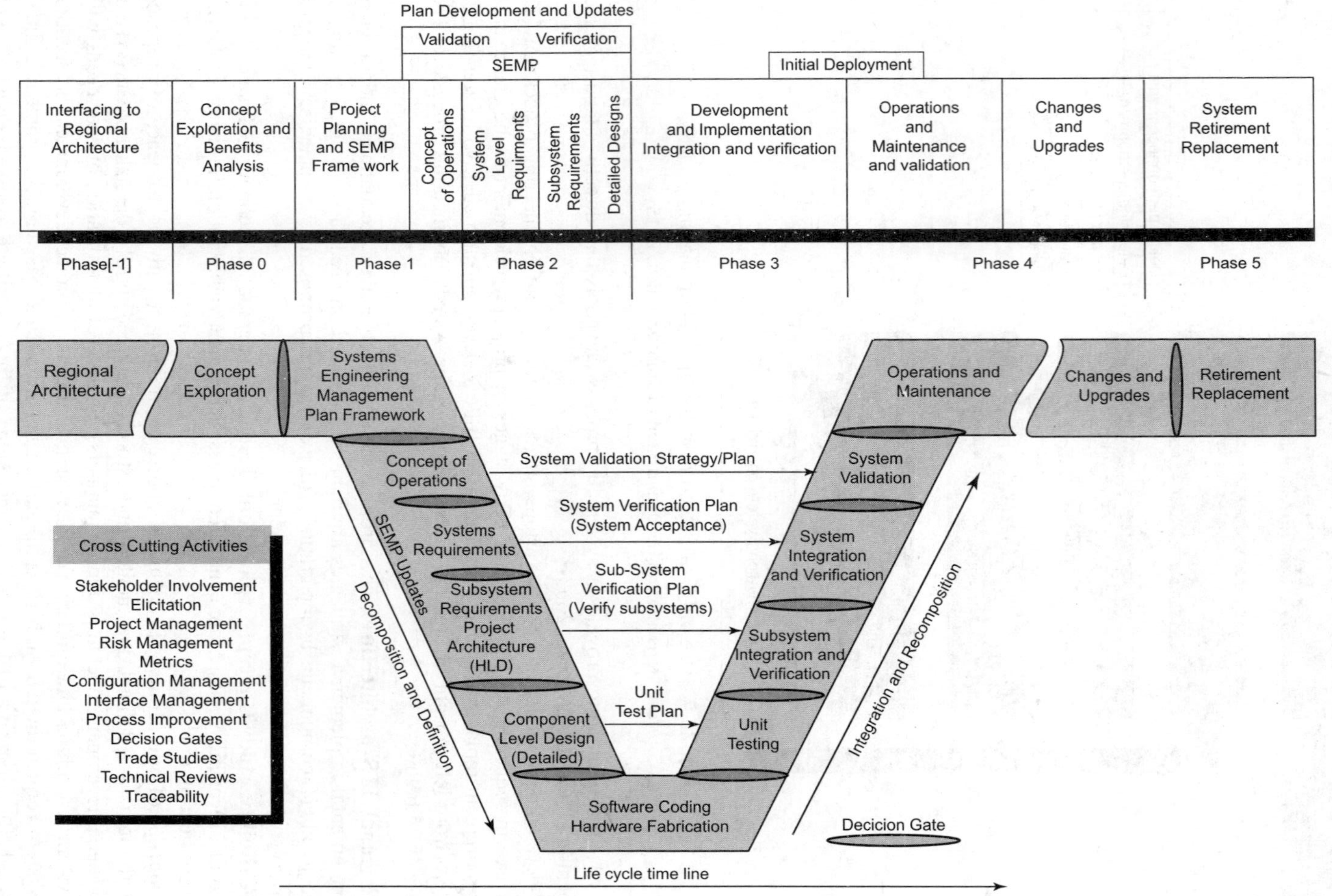


**Fig. 22.8** V-development model of ITS projects with life cycle framework
(*Source*: Caltrans, 2009; *Courtesy*: California Division, FHWA, USDOT, USA)

5. **System design:** The system is divided into suitable divisions of subsystems. Each subsystem is designed, interfaced and integrated to fit into the pre-defined overall framework of the system. Detailed documentation of the designs, specifications and drawings are prepared for the product and software tools selected.
6. **Hardware/Software development and testing:** ITS hardware and software solutions are created for the proposed components in the system design. Testing of these tools with reference to the design specifications is verified. The components are procured and kept ready for integration and field validation.
7. **Integration and validation:** Individual components are examined, verified and integrated to produce higher level assemblies or subsystems. Similarly, all ITS subsystems must be interfaced correctly and the entire system is integrated, verified and confirmed to produce higher level assemblies.
8. **Initial deployment:** System components are installed. In case of paucity of funds at the initial stage, stage-wise development/deployment may be carried out on a priority basis to spread the costs across several fiscal years. The project development team must transfer the support equipment, documentation, operator training and required tools to the operations and maintenance teams.
9. **System validation:** The workability of the installed/deployed components is again tested against the intended use and validated as defined in the concept operations by the concerned, system owner, state DOT, agency or consultant.
10. **Operations and maintenance:** Once the customer accepts the ITS installations with steady-state performance, system evaluation begins with operations and maintenance activities. In this course, difficulties, issues, changes, improvements and required upgradations are implemented based on the latest technological developments and funds available. This process continues till the ITS system reaches the end of its operation life.
11. **Retirement/Replacement:** When the operation and maintenance costs exceed the cost of development of a new ITS system, the retirement plan of a system (or a wing of a subsystem) is generated along with a replacement strategy. Appropriate audits may be carried out for disposal of the devices, and the final documentation is archived for the disposal of the system.

More details of the V-development model and its application to ITS can be obtained from Caltrans (2008).

## 22.8 SUBSYSTEMS OF ITS

The subsystems of ITS applications with the appropriate technologies, communication channels, devices and supporting infrastructure are explained below.

### 22.8.1 Vehicle Detection Station (VDS)

Raw data on traffic volume, occupancy, vehicle class with the number of axles, axle weight, speed, time headway and space headway with time of stamping is collected at these stations. VDSs are located at suitable intervals along the road network so that traffic volume, congestion and incident detection in road sections can be detected on a real-time basis.

The sensors installed on or embedded in the pavement, and mounted on the roadside at VDSs are given in Table 22.1.

**Table 22.1** Sensors at VDSs

| *Sensors installed on or embedded in the pavement* | *Sensors mounted on the roadside* |
|---|---|
| Pneumatic tubes | Photoelectric sensors |
| Inductive loops | Video imaging |
| Piezoelectric sensors | Doppler radar |
| Magnetic sensors | Infrared (optical beacons) |
| | Ultrasonic sensors |
| | Microwave radar |

For more details about the data collected by these sensors, please see *Chapter 5*.

The data collected is stored and sent to a nearby processing centre for various ITS applications. For specific applications, such data may be used for signal design, adaptive signal coordination, ramp meters for interchange access control, variable message signs and variable speed limit signs.

## 22.8.2 Variable Message Signs (VMSs)

Variable message signs with display boards are installed on a column/gantry at locations along a road network and used to provide *advance information* to motorists (Figs 22.9 and 22.10). They are used for displaying a number of messages that may be changed or switched on/off as required. VMSs are designed to change/alter signs at different times in different conditions.

VMSs were officially introduced in the 1968 Vienna Convention on Road Signs and Signals by an amendment, and put into effect on 30 November 1995 (UNECE, 2010). They are *stationary* signs that are permanently mounted on the roadside. They use LED display technology. Instantaneous warning/control/

**Fig. 22.9** Typical variable message sign (VMS)
(*Source*: Hanson et al. 2010; *Courtesy*: International Road Dynamics Inc.)

Fig. 22.10 Graphic information board

(*Source*: Nexco, 2014; *Courtesy*: Central Nippon Expressway Company Limited, Japan)

information messages are sent to a VMS by an operator from a nearby TMC. The information may be about routes and travel modes indicating expected delays, beginning of steep slopes, tunnel entrances, congestion ahead, accidents/incidents, road work, weather information, speed restrictions on curves, school zone speed limits, speed limit changes in peak hour, reversible lane usage and weigh station operations. The size of the characters and the signboard depend on the speed of traffic and the sight distance.

### 22.8.3 Changeable Message Signs (CMS) Systems

A portable trailer-mounted VMS is often termed a changeable message sign (CMS). CMSs are used to warn road users about incidents such as accidents, fire, falling objects and work zones in areas where a permanent VMS is not available or is not close enough to act as a preventive measure. It must be noted that CMSs are capable of displaying different kinds of information, like VMSs. CMSs are usually positioned temporarily and are used to strategically warn users about delay times/alternative routes, till the incident area is cleared.

Trailer-mounted VMSs can be equipped with devices such as radar (speed detection), cameras, graphic user interfaces and wireless communications units. Solar-powered charging options are often used.

### 22.8.4 Variable Speed Limit Sign (VSLS)/Lane Control Sign (LCS)

Congestion in traffic flow due to road work, peak hour demand, school zone, inclement weather and other traffic conditions requires periodic or programmable change in the speed limit. Lane control signs are used to warn approaching traffic by displaying a message indicating that the road is open/closed/closing because of work or some other reason.

VMSs can be used for these purposes as well.

### 22.8.5 Speed Enforcement System or High-speed Driving Deterrence System

Speed enforcement is essential for reducing accident probability and severity on any road section that has been identified as dangerous. Speed enforcement can be made based on spot speed or the average speed of vehicles travelling over a section of road-length. Advanced digital camera technology and vehicle identification software are used for real-time calculation/detection of spot speeds and average speeds of vehicles (Fig. 22.11). Messages for vehicles with speed violations can be sent to the nearest VMS for immediate warning (Slow Down) and their license numbers are reported to a nearby control station for necessary action. In this connection, it is also possible to control signals ahead of the overspeeding vehicles and turn on Red forcibly, so the vehicles stop at the signal and drive safely from there (Fig. 22.12).

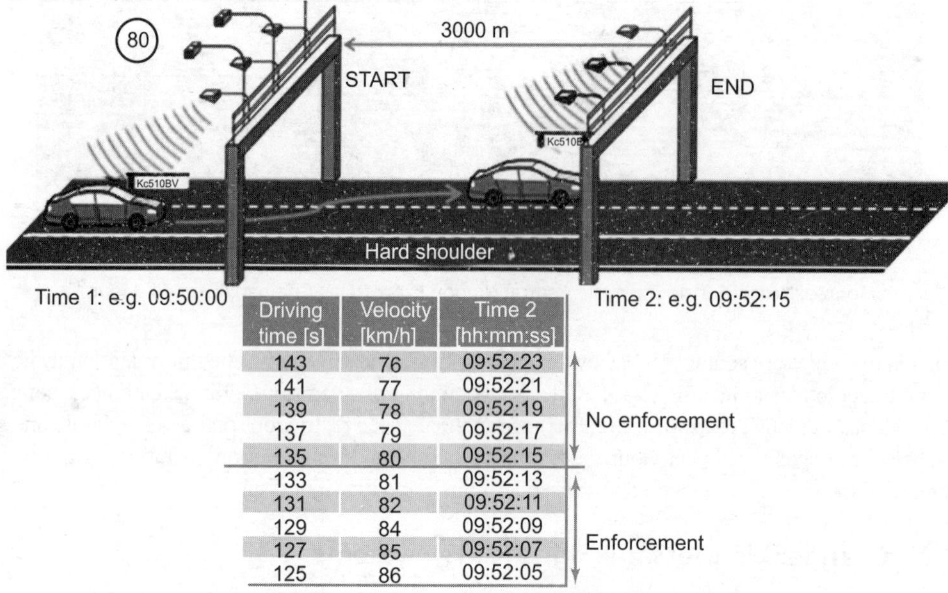


| Driving time [s] | Velocity [km/h] | Time 2 [hh:mm:ss] |
|---|---|---|
| 143 | 76 | 09:52:23 |
| 141 | 77 | 09:52:21 |
| 139 | 78 | 09:52:19 |
| 137 | 79 | 09:52:17 |
| 135 | 80 | 09:52:15 |
| 133 | 81 | 09:52:13 |
| 131 | 82 | 09:52:11 |
| 129 | 84 | 09:52:09 |
| 127 | 85 | 09:52:07 |
| 125 | 86 | 09:52:05 |

**Fig. 22.11** Speed enforcement system installed over a road section
(*Source*: www.metroinfrasys.com, *Courtesy*: Mr. Sachin Bhatia)

The average speed method is typically used on state highways and urban roads where there is a need to control longer distances and stretches of road by restricting the speed limit. This subsystem can harmonise traffic flow over a section length and reduce environmental pollution/noise as an automatic speed enforcement operation.

### 22.8.6 Red Light Violation Detection System

This is an enforcement system capable of automatically detecting signal violations, lane violations and speed violations by vehicles passing on urban roads.

The system comprises loop sensors, digital and analogue cameras, signal light sensors, camera and sensor controller, network device and power supply. Two loop sensors installed on the road detect vehicles and the speed at which they are travelling using high-resolution digital cameras to capture the

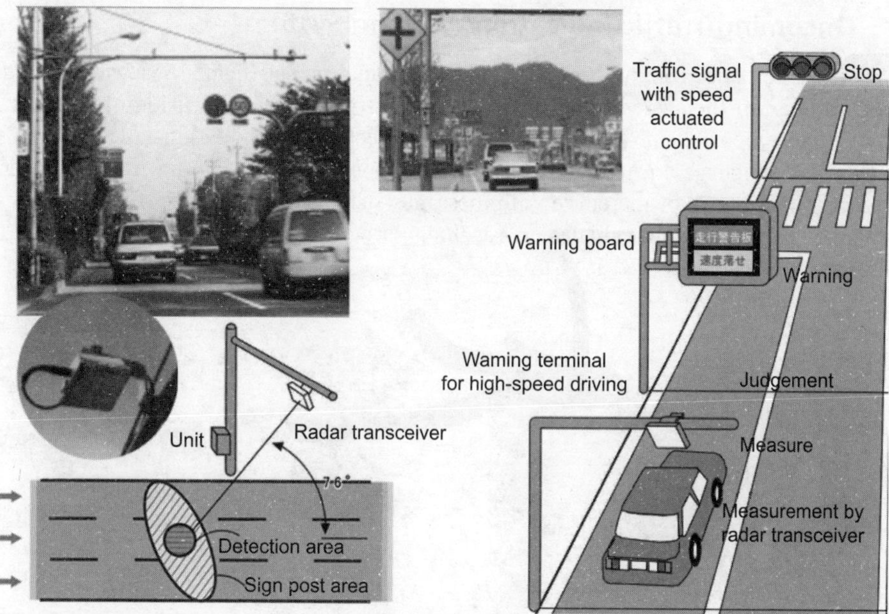


**Fig. 22.12** High-speed driving deterrence system with speed-actuated signal control
(*Source*: IUTR, 1998; *Courtesy*: Japan Traffic Management Technology Association, Japan)

license plates of the violating vehicles (Fig. 22.13). A secondary analogue camera is installed to record the image of the moving vehicle, as evidence.

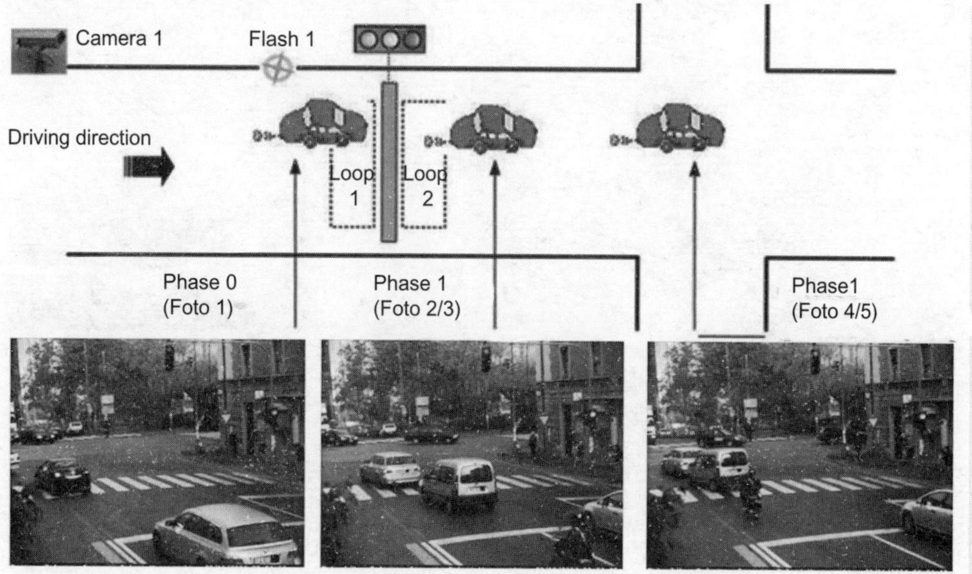


**Fig. 22.13** Red light violation detection system
(*Source*: www.metroinfrasys.com, *Courtesy*: Mr. Sachin Bhatia)

## 22.8.7 Oncoming Traffic Indication/Warning System

The drivers of vehicles need an indication if the sight distance is inadequate to view oncoming vehicles on curves. The aim of the system is to prevent accidents caused by protrusion of vehicles from the centre line into the opposite lane (Fig. 22.14). Vehicle detectors (ultrasonic) installed on opposite lanes at the far end of the curve are used for detection of approaching oncoming traffic, and drivers are warned using VMSs ahead of the curve with messages such as Oncoming Traffic Ahead or Go With Care. In addition, overspeeding vehicles approaching corners/curves can also be warned by using the speed enforcement system (IUTR, 1998).

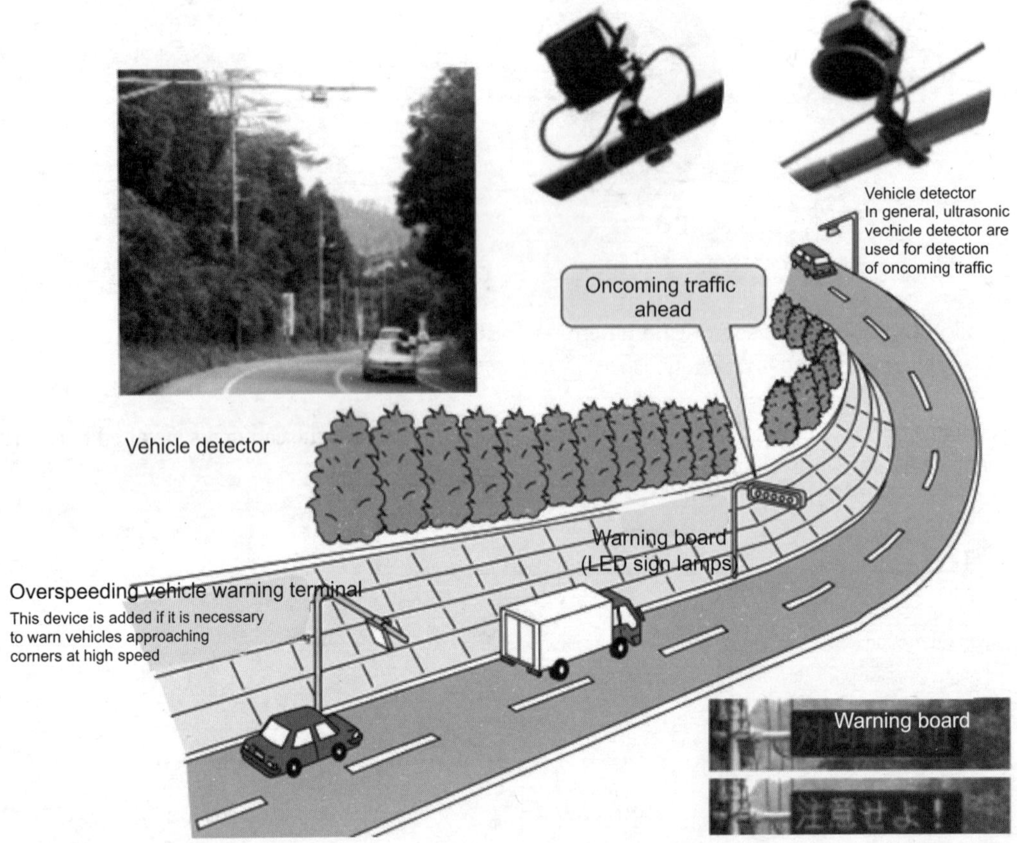


Fig. 22.14 Oncoming traffic indication system
(*Source*: IUTR, 1998; *Courtesy*: Japan Traffic Management Technology Association, Japan)

## 22.8.8 Centre Line Shifting System

Centre line shifting is a traffic flow management technique, where more lanes are allotted in the direction of greater traffic volume, during rush hour. This system deals with predictable variations in traffic demand with time. For example, increase in the number of up-lanes in the morning and down-lanes in the evening, or vice versa, based on the traffic flow observed along a route. This means that as traffic flow increases in a direction, the number of lanes allotted to that direction of flow will be higher, with reference to real-time traffic flow analysis. This technique is useful for deterring or alleviating congestion and improves the efficiency of a limited number of lanes.

The centre line shifting system uses a start point display gate with a signboard to announce the lane reversing, an intermediate display gate with a signboard to announce regulations/control of subordinate roads, and an end point display gate with a signboard to announce lane reversing. In addition, VMSs, illuminated road stud lights, special lights, centre line shifting terminal controllers, operation monitoring boards and other devices set up the centre line shifting sections (Fig. 22.15).

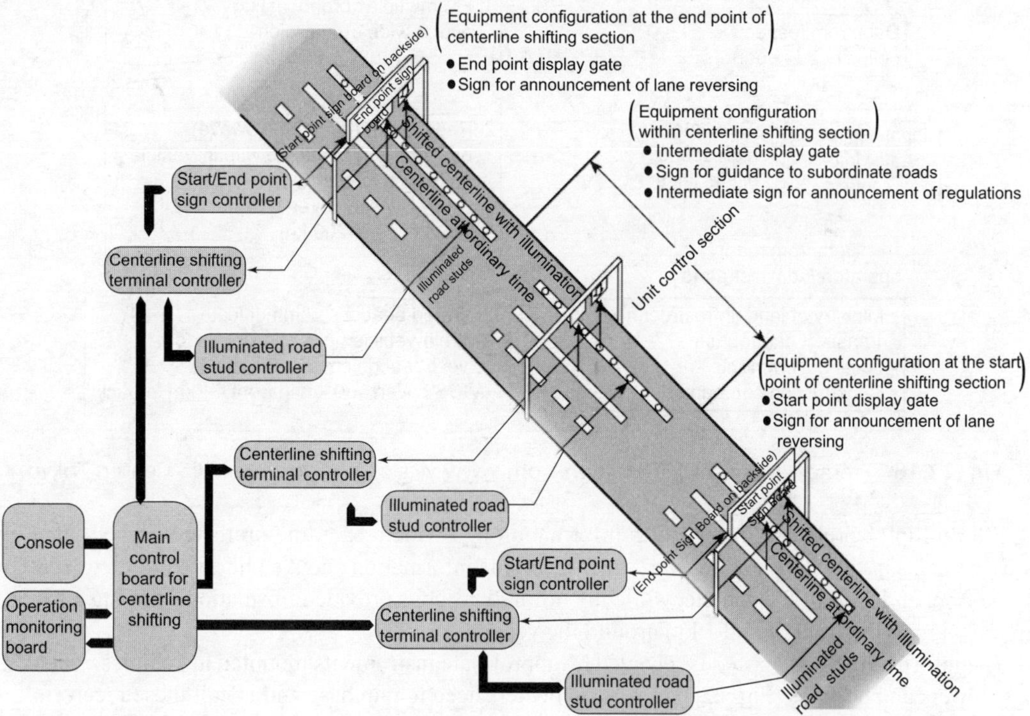


Fig. 22.15 Centre line shifting system

(*Source*: IUTR, 1998; *Courtesy*: Japan Traffic Management Technology Association, Japan)

Necessary publicity about the shift should be provided to drivers before implementing it on a route. Police officers should also be involved at an early stage to help guide and control vehicles (IUTR, 1998).

## 22.8.9 Vehicle Information and Communication (VIC) System

VICS for in-vehicle traffic information has been in use in Japan since 1996. The main objectives of VICS is to avoid traffic congestion en-route, shorten the travel time by suggesting shortest/congestion-free routes and reduce driving-related stress by providing real-time information on traffic congestion, travel time, traffic restrictions due to accidents, construction, disasters and weather conditions, and parking availability/lots, 24 hours a day and 365 days a year.

Data obtained from different police headquarters and from traffic administrators/operators is fed into the Japanese Road Traffic Information Centre, which is connected by dedicated high-speed digital lines to the VICS Centre (Fig. 22.16). The VICS Centre transmits the processed information to

- FM multiplex broadcasting stations spread across a wide area/region
- Radio wave (quasi-microwave) beacons
- Infrared beacons or infrared vehicle detectors, which emit infrared rays in the direction of the road surface. Vehicles passing under these rays are detected by comparing the rays reflected from the vehicles and from the road surface. Communication takes place bi-directionally, using on-board equipment (OBE) and roadside equipment (RSE).

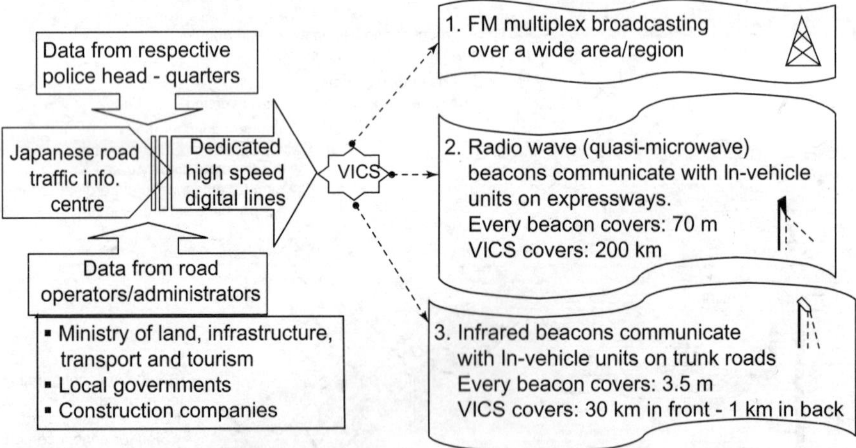


**Fig. 22.16** Framework of VICS (Redrawn from www.vics.or.jp; *Courtesy*: VICS Center, Tokyo,

FM multiplex broadcasting provides information to a wide area at the prefectural level. The radio wave beacons installed mostly on expressways provide information about a wide area of up to 1000 km to the front and back of the vehicle, while the infrared beacons provide information covering an area of up to 30 km to the front of and 1 km around the vehicle.

The information broadcasted is received by appropriate instruments mounted in vehicles, when they pass towers and beacons (Fig. 22.17). Vehicle owners need to purchase and install these receivers, and

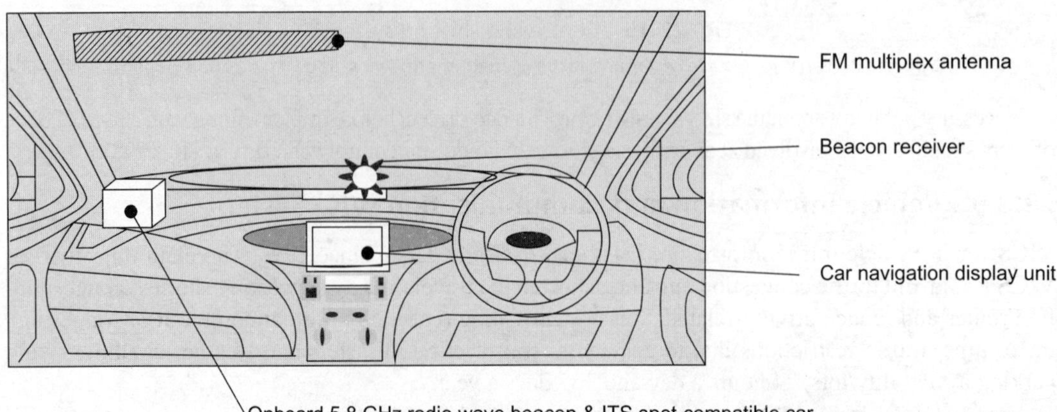


**Fig. 22.17** In-vehicle instrumentation to receive information from VICS
(Redrawn from www.vics.or.jp: *Courtesy*: VICS Centre, Tokyo, Japan)

the information is provided to them free of cost. The information can be in the form of a digital map display, simple graphic display or text display (Fig. 22.18). The driver of the vehicle can make use of the real-time information and take appropriate travel decisions. More details on VICS can be obtained at www.vics.or.jp.

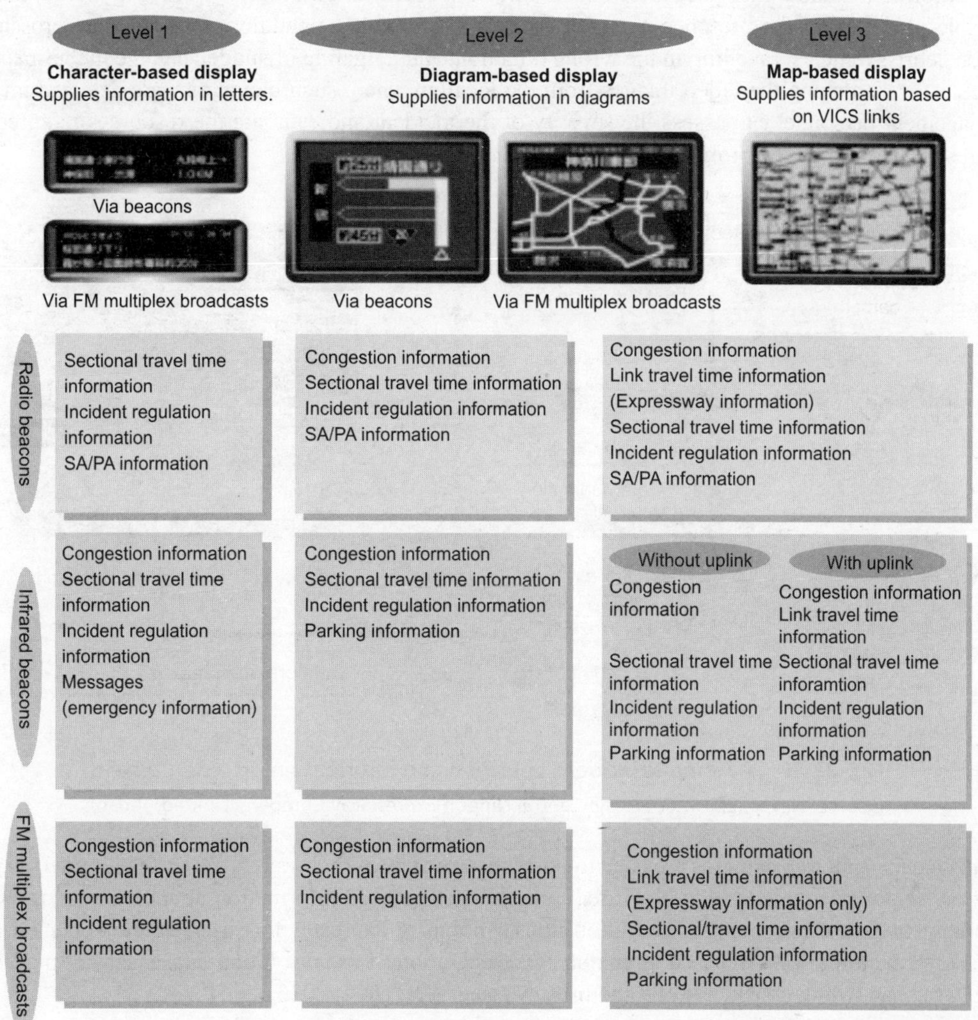


**Fig. 22.18** VICS service menu and information contents

(*Source*: IUTR, 1998; *Courtesy*: Japan Traffic Management Technology Association, Japan

## 22.8.10 Automatic Incident Detection (AID) and Information Provision Systems

High-bandwidth and high-resolution CCTV cameras are installed on the roadside or in tunnels, at suitable intervals along a road network. These CCTV cameras are mounted on any existing structure, such as poles, overpasses or on rooftops. Real-time video images from various spots are used for continuous, round-the-clock *surveillance/monitoring* of the road network (Fig. 22.19). These images are processed and analysed based on speed changes in the objects viewed using appropriate software. In case of violations, the traffic management centre operator is prompted about the detected incident with information about the exact location and direction of travel. This AID system (Fig. 22.20) is also capable of detecting the presence of smoke, stationary vehicles, unauthorised pedestrian crossings, fallen debris, vehicles travelling in the wrong direction, and tailgating or sudden change in the speed of vehicles. By watching the video images from the location where such incidents were detected image, management personnel can assess the severity of the incident and estimate the resources required by the rescue team to fix the problem.

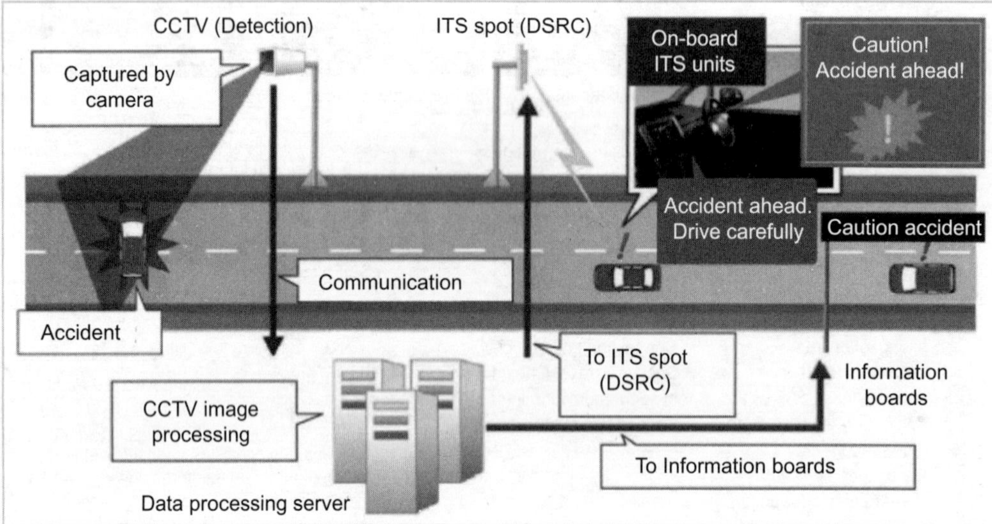


**Fig. 22.19** Automatic incident detection and information provision system
(*Source*: Nexco, 2014; *Courtesy*: Central Nippon Expressway Company Limited, Japan)

In Queensland, web cameras are used to post video images of highway traffic on the internet, as part of *traveller information service*. Such real-time images provide information about the road network condition and the electronic road condition signs at points of interest before users embark on a journey. Such information can also be used for maintenance operations. Low-resolution images may be produced due to the bandwidth considerations and privacy issues (DOTMR, 2013).

## 22.8.11 Road Weather Monitoring Systems

CCTV/Web cameras, rain gauges, weather vanes, anemometers and air and pavement temperature sensors with appropriate data storage and wireless communication system units mounted along a road network are often termed road weather monitoring stations.

These stations provide on-spot real-time information about air temperature, pavement temperature, wind direction, wind speed, precipitation, humidity and cloud cover. This information is transmitted to a traffic management centre and operators send suitable *warning* messages and speed limit information to selected VMSs.

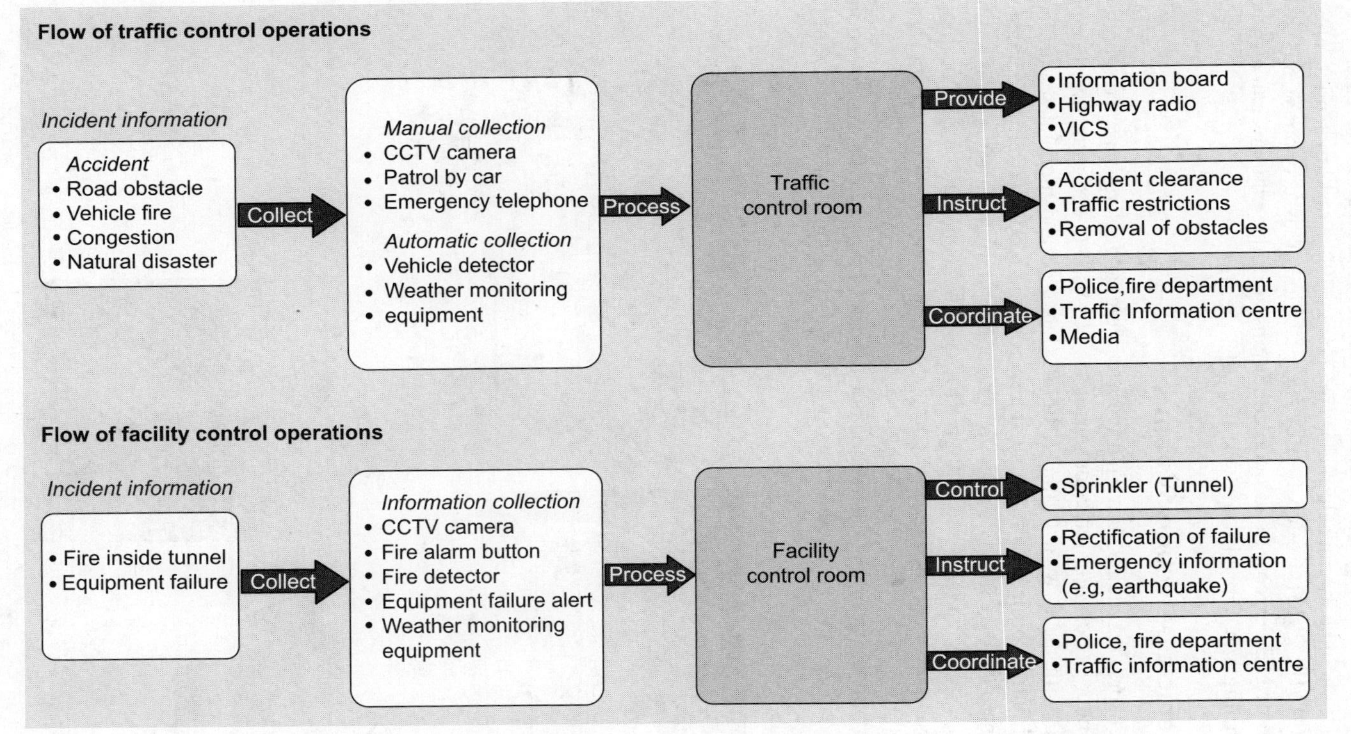


**Fig. 22.20** Typical activities of the incident management system as part of ITS
(*Source*: Nexco, 2010; *Courtesy*: Central Nippon Expressway Company Limited, Japan)

Thus, not only is on-spot real-time weather information made available but it also improves traveller safety by enabling call-box or help-phone communication with the TMC operators for emergency services/additional information. Spacing between help-phones may range 200 m to 1 km in urban areas and 2–5 km in rural areas (DOTMR, 2013). Co-location of such facilities may economise the cost of installation and maintenance of ITS devices. Road flood conditions at bridge columns can also be monitored using sensors with stilling tubes installed downstream of water flow.

### 22.8.12 Over Height Vehicle Detection System

Over height vehicle detection sensors are required to warn drivers at interchange ramp entrances, tunnel entrances, before bridges and under passes. Infrared laser sensors can be used to detect the over height of vehicles, without interfering with the flow of traffic. A tightly focussed and modulated infrared laser beam with narrow band filters is projected from one side across the road, and on the opposite side, a passive reflector is mounted (Fig. 22.21). Based on the preset height of the infrared laser beam reflection, an over height warning message is passed on to a VMS as a warning sign or with flashing lights. All such events are logged in the controller to allow complete documentation of this sensing and activation module (SAM) operation.

This SAM system is used not only for over height detection but also for traffic surveillance, detection of stationary vehicles, moving vehicles, cyclists, pedestrians and any object up to a minimum size of 5 cm, at a speed of up to 100 km/hr.

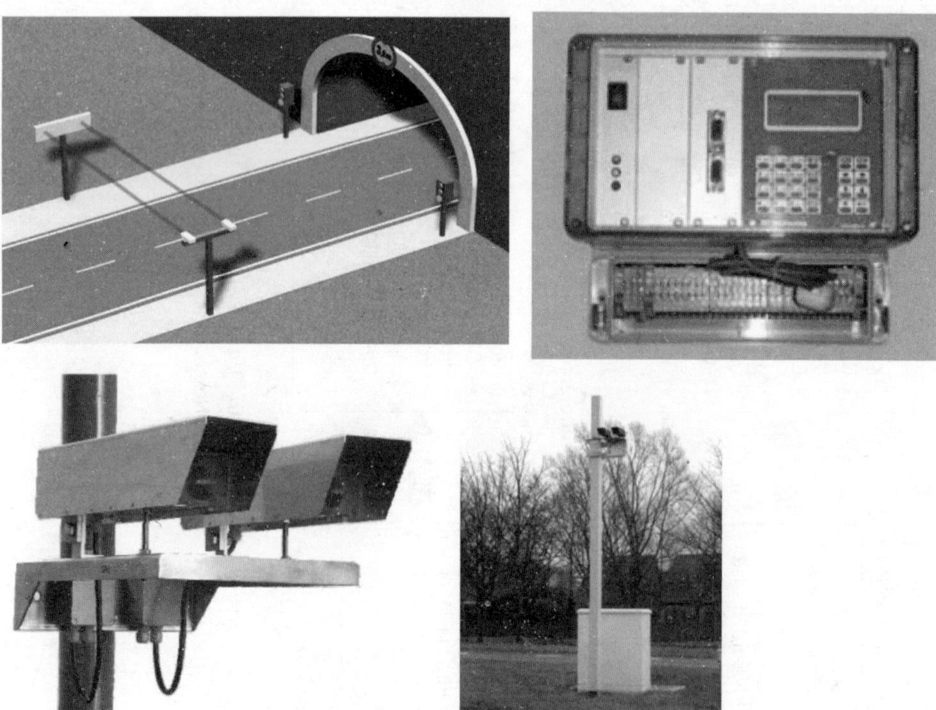

**Fig. 22.21** Over height vehicle detection system: height sensor and controller (*Source*: www.schuhco.com; *Courtesy*: Schuh & Co., Germany)

The laser sensor and SAM system is also used to detect the presence of vehicles, distinguish passenger cars from truck/buses (Fig. 22.22), lane distinction of vehicles based on distance measurements, and drawbar detection between trucks and trailers (Fig. 22.23), and to distinguish close individual car traffic.

**Fig. 22.22** Vehicle detection system

(*Source*: www.schuhco.com; *Courtesy*: Schuh & Co., Germany)

**Fig. 22.23** Draw bar detection system

(*Source*: www.schuhco.com; *Courtesy*: Schuh & Co., Germany)

## 22.8.13 Automatic Number Plate Recognition (ANPR) System or Automatic Vehicle Licence Plate Recognition System

Roadside stations with mounted video cameras are used to capture images of moving vehicles along with their licence plate number (Figs 22.24 and Fig 22.25). Special cameras with data processing

**Fig. 22.24** Camera, over height sensor and electronic cabinet, Deerwood NM

(*Source*: Hanson et al., 2010; *Courtesy*: International Road Dynamics Inc., Canada)\

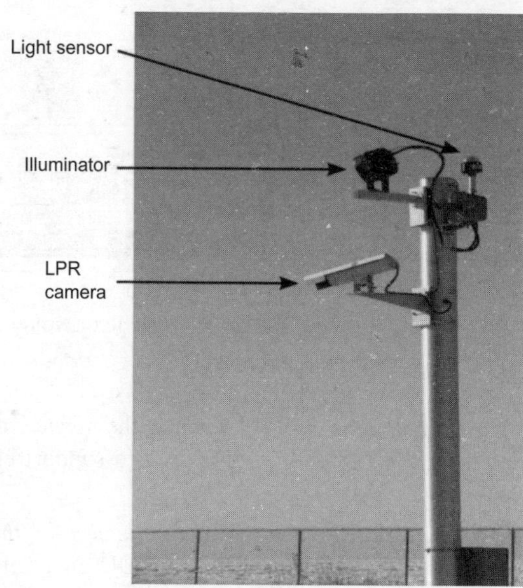


**Fig. 22.25** Typical licence plate reader (LPR) mounted on a pole

(*Source*: Evans and Klashinsky; *Courtesy*: International Road Dynamics Inc., Canada)

units are used to collect and store the licence plate image data of every vehicle with a time stamp across a specific roadway location (Fig. 22.26). Such images or video tapes are electronically transcribed with optical character recognition (OCR) technology. The software associated with these analysis units is capable of generating a licence plate number file by following the steps given below:

1. Acquiring the image
2. Transferring these images to memory
3. Finding the licence plate in the image scene
4. Reading the number
5. Recording the number

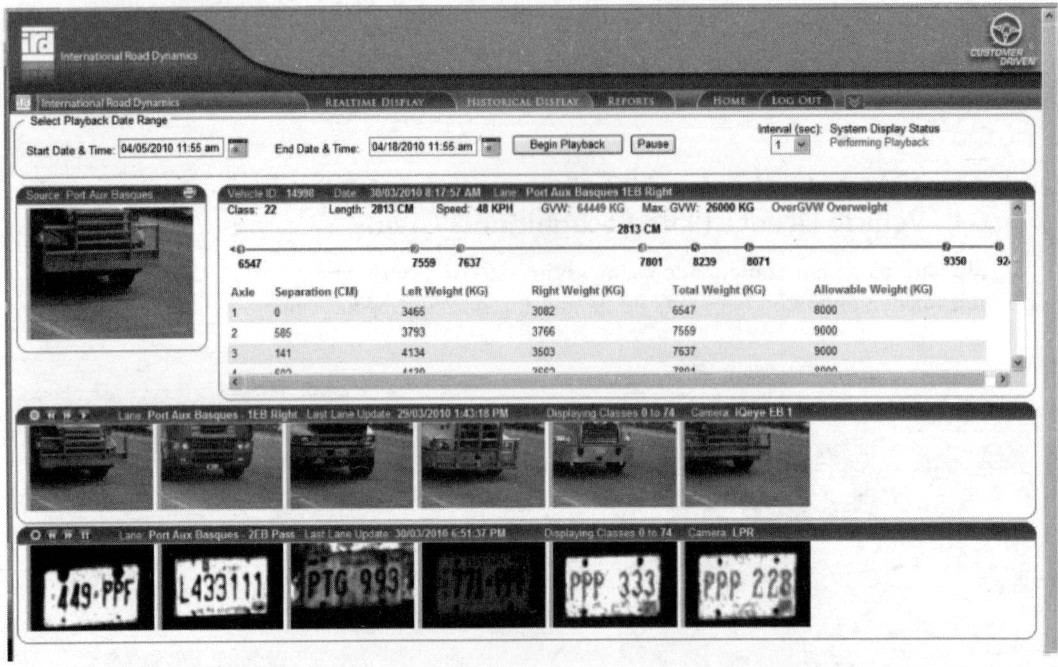


Fig. 22.26 Vehicle list display by licence plate reader method

(*Source*: Hanson et al., 2010; *Courtesy*: International Road Dynamics Inc., Canada)

This technique is widely known as the *licence plate recognition method* and has been in use in the USA since the 1980s. It is also used to support electronic toll collection (ETC) at highway speeds (Bates et al., 1999).

*Performance measurement of data collection by the ANRP system:* The following equations may be used to evaluate the accuracy of the OD data collection system by automatic recognition methods (Turner et al., 1998).

Accuracy of automatic number plate recognition (ANRP) system = (Number of plates correctly interpreted by the system ÷ Number of plates to be read by a person using raw video files on a computer) (22.2)

Accuracy of automatic number plate = (Rate of correctly recognised plates in field of view ×recognition (ANRP) system Number of successfully interpreted plates with its entire content) × 100 (22.3)

(*Note*: Convert these values into decimals)

*Merits and limitations of ANRP:* Manually reading and recording the licence plate numbers (by transcription) of all the survey vehicles is laborious and very difficult and the data entry quality may degrade with time. In a study (WSDOT, 1995), it was realised that more than 10 hours are required to transcribe 1 hour of video footage. However, when OCR technology is used, 1 hour of video footage can be transcribed in 1 hour. For this reason, video recording and the associated methods of ANPR are widely used. Technology can be used to automatically *convert* the licence plate strings from videos/still photographs to digital characters, in a few steps of processing. The effort required for the data reduction process by these methods is simpler than that of the questionnaire methods. Video technology–based automatic processing techniques are cost effective and can also be applied for collecting and evaluating larger data samples, including on high-speed lanes at several roadside stations concurrently (Shuldiner et al., 1996).

Despite these advantages, video camera–based techniques suffer from the following limitations:

- The mounted cameras cannot view outside their fixed field of view.
- The cameras cannot look around obstacles.
- Inclement weather, dirty and light coloured licence plates may be difficult to decipher (conversion of coded text or signals into normal language)
- Software programs may mismatch the licence plate numbers when similar text characters are encountered.
- Large amounts of data can be lost due to equipment/power failure or vandalism.
- Data acquisition may be poor due to improper installation of cameras at locations where vehicles travel freely at high speeds and may pass too quickly or near signals where vehicles travelling in platoons at low speed may block the view of other vehicles' licence plates (Martin and Bell, 1993).
- Camera settings may need to be modified in a day/season due to sunlight glare.

However, ANRP systems today achieve accuracy rates of up to 97–99% with the recent improvements in OCR technology and the use of infrared illuminators at night (Tamagawa et al., 1997). In addition, a study conducted in Japan revealed that still photographs taken using fixed overhead cameras on highways had a 90% capture rate of licence plates with a margin of error of 5% of recognizable characters. The cameras could also record the colour and size of the licence plates (Asakura et al., 2000)

Thus, in general, the licence plate number method does not disrupt or delay traffic flow in any manner. However, with this method, it is difficult to match each licence plate number of each entry vehicle with the numbers listed in the licence plate database, obtained from all exit stations installed on the cordon line.

## 22.8.14 Electronic Toll Collection (ETC) System

The ETC or electronic fee collection (EFC) system comprises two types of components (Fig. 22.27 and 22.28):

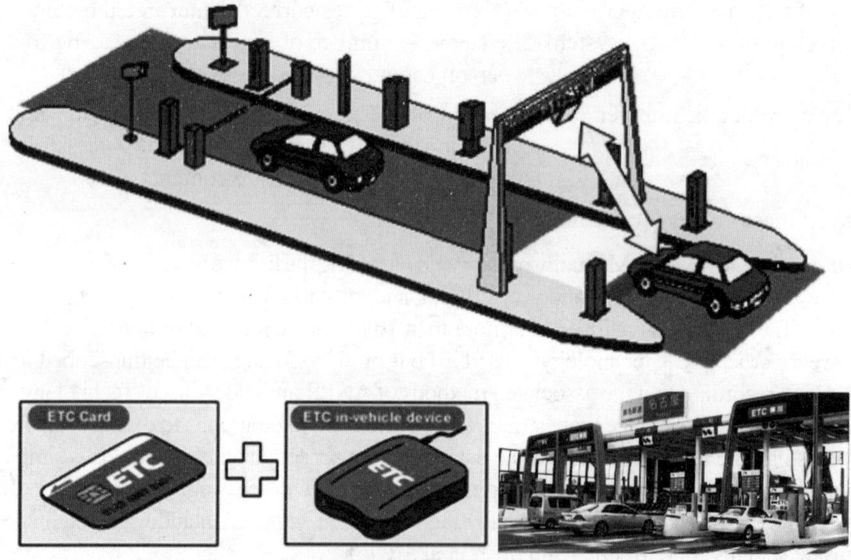


Fig. 22.27 Typical view of ETC plaza with IC card and OBE
(*Source*: http://global.c-nexco.co.jp; *Courtesy*: Central Nippon Expressway Company Limited, Japan)

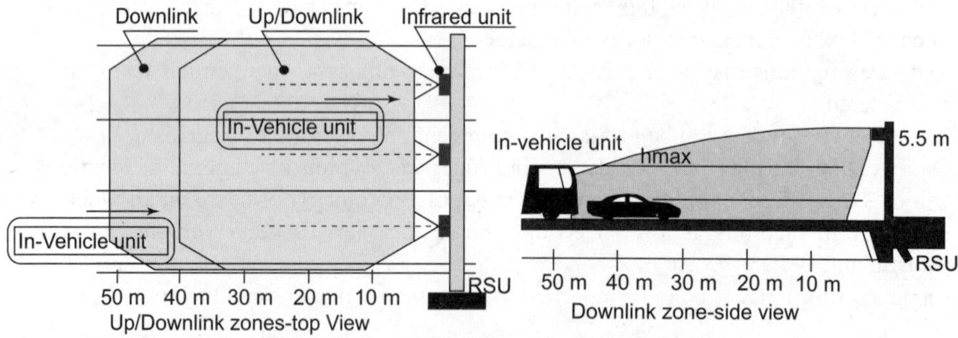


Fig. 22.28 ETC system using infrared beacons
(*Source*: IUTR, 1998; *Courtesy*: Japan Traffic Management Technology Association, Japan)

- **On-board device:** This is mounted on each vehicle. It consists of two parts: the on-board equipment (OBE) and the IC card. The OBE with built-in IC card communicates with the roadside unit.
- **Roadside unit:** This is an electronic unit mounted on the ceiling of the toll collection girder under which the vehicles must pass. The electronic unit is a dedicated short-range communication (DSRC) system operated at the 5.8 GHz band active system.

As soon as the vehicle approaches the toll booth (very close to the ETC girder), the DSRC detects the OBE and debits the pre-set amount from the IC card, based on the type of vehicle and destination.

All these activities are integrated and coordinated with VMS display units, automatic licence plate number recorder cameras, helplines/call boxes and CCTV cameras.

The following advantages can be availed by deploying the ETC system in place of a manual toll plaza:

- Eliminate traffic congestion, queuing and waiting near toll gates. For example, for the congestion charge scheme in London, as the vehicle enters the charging zone, automatic fee collection is done as shown in the ECT system and the vehicle licence plate number is read by the ANPR system, as evidence in the event of a challenge.
- Reduce maintenance/management costs by eliminating manned toll gates and boost the processing capacity of toll gates.
- Eliminate the need to handle cash and improve user convenience.

## 22.8.15 Weigh-in-Motion (WIM) Systems

Truck overloading causes rapid deterioration of highway pavements. Inspection, detection of overloaded trucks, recording of vehicle data and related enforcement including electronic toll collection are carried out in one place using WIM systems.

A basic WIM system consists of a load cell, bending plate or piezoelectric sensor embedded in the pavement surface. The weight of the moving vehicles is measured as they pass over this sensor. These sensors may be installed either temporarily or permanently to check the weight of vehicles (Fig. 22.29).

**Fig. 22.29** WIM sensor

(*Source*: Hanson et al., 2010; *Courtesy*: International Road Dynamics Inc., Canada)

To perform the tasks of inspection and collection of relevant data, WIM sensors with electronic data collection cabinet facility are commonly integrated with other sensors, such as:

- Intrusive loops: Used to detect vehicles, the number of axles and classify vehicles
- CCTV cameras: Capture video and images with pan, tilt and zoom features
- High-speed cameras: Automatically detect and record vehicle registration numbers
- High performance laser: Used as an overheight detector
- Variable message signs (VMSs): Used to display variable instructions to truck drivers at appropriate locations in WIM stations and ETC stations

Some examples of WIM systems integrated with sensors are shown in Figs 22.30, 22.31 and 22.32.

Fig. 22.30 WIM system integrated with CCTV cameras and overheight sensors

(*Source*: Klashinsky and Hanson, 2011; *Courtesy*: International Road Dynamics Inc., Canada)

Fig. 22.31 A typical view of a virtual WIM station integrated with a high-speed camera and licence plate recorder

(*Source*: www.irdinc.com; *Courtesy*: International Road Dynamics Inc., Canada)

Fig. 22.32 WIM scales with optional cameras and licence plate recorder
(*Source*: www.mtwim.com; Used with permission from METTLER TOLEDO, India)

WIM systems can be used for different purposes: virtual WIM station and WIM station with ETC system. These two ITS applications are described below.

**Virtual WIM station:** Commercial vehicle weights are inspected at these stations (Figs 22.33 and 22.34). Advance information of the location of the WIM and the speed limit is displayed to approaching traffic prior to their arrival at the truck inspection station. A VMS is used to limit the speed of trucks entering the inspection station, based on the trucks in queue. Initially, the trucks entering the WIM station are weighed using WIM scales, then the vehicle class is detected using loops, its image is captured and the licence plate number is detected using appropriate cameras. The vehicle is permitted to proceed if its weight is within the legal limit. Otherwise, if it is above the weight limit, the vehicle is diverted to the ramp for further verification (accurate weighing using static weighing scales). Permission to proceed or an order to divert for each vehicle is given using the VMS. Additional VMSs are to display the excess weight, licence plate number and the way to a temporary parking place for unloading of excess weight.

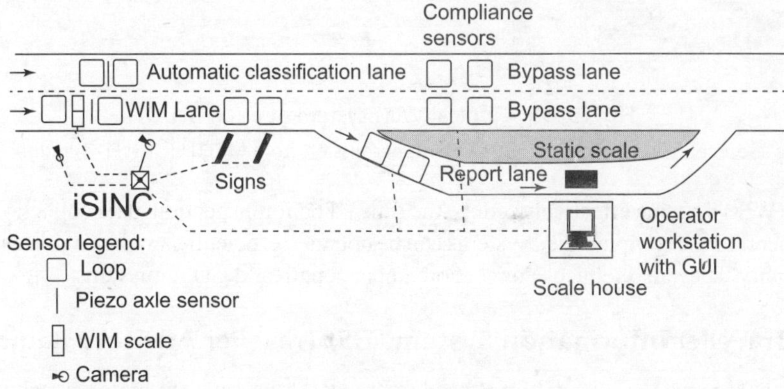


Fig. 22.33 Typical layout of a WIM station sorting system
(*Source*: Hanson et al., 2010; *Courtesy*: International Road Dynamics Inc., Canada)

In case of violations where trucks disobey the VMS message, the same information is immediately communicated through a wireless system to the vehicles patrolling nearby.

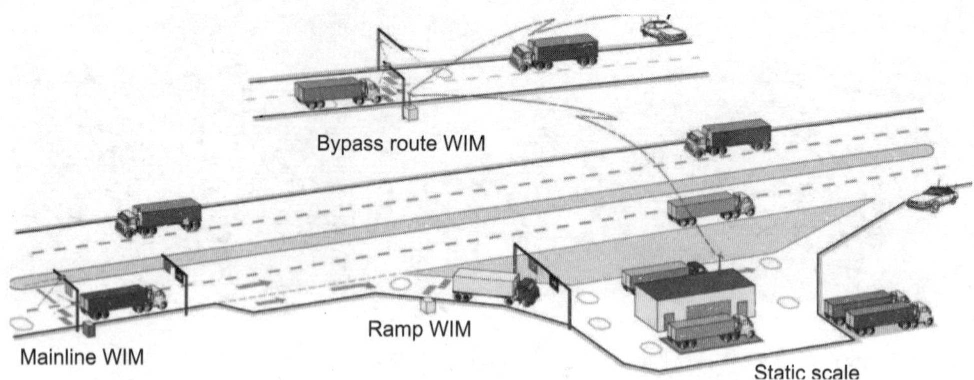


**Fig. 22.34** A typical enforcement-oriented WIM system
(*Source*: www.mtwim.com; Used with permission from METTLER TOLEDO, India)

**WIM station with ETC system:** The concept of virtual WIM station is also applicable for conventional toll collection plazas or electronic toll collection (ETC) plazas (Fig. 22.35). The toll amount can be collected based on the magnitude of excess weight of the vehicles.

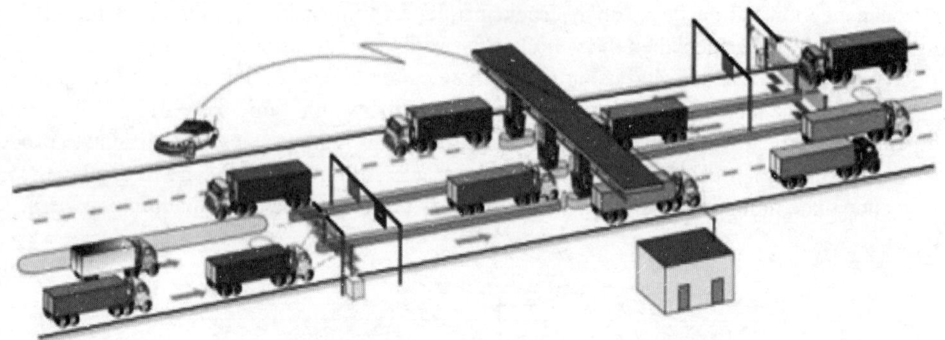

**Fig. 22.35** Typical WIM system at a toll plaza
(*Source*: www.mtwim.com; Used with permission from METTLER TOLEDO, India)

The above WIM systems effectively reduce the time taken for inspection, data collection, monitoring and enforcement. These computerised systems can be operated efficiently any time with less interference to human activity. They also reduces congestion, queue length and $CO_2$ emissions.

## 22.8.16 Traveller Information System (TIS)/Traveller Advisory Radio (TAR)

A low-powered AM or FM radio station is used to broadcast messages to motorists who receive them on their automobile radios. Similar information to that displayed on VMSs, but with more description, can be broadcasted by this system and each message may be up to two minutes long. The broadcast information can be any one of the following types:

- Congestion locations with alternative routes
- Hazardous conditions

- Estimated travel time
- Special event information
- Parking locations
- Rest area locations
- Hospital locations
- Weather reports and road conditions
- Construction/Road closure/ Diversion information.

### 22.8.17 Parking Management and Guidance Systems

The main objectives of this system are to provide real-time information on parking spaces in nearby plazas, automatically disseminate information on the availability of empty parking floors/spaces and guide drivers through the different stages of entry/exit (Fig. 22.36) within parking facilities using VMSs and including EFC.

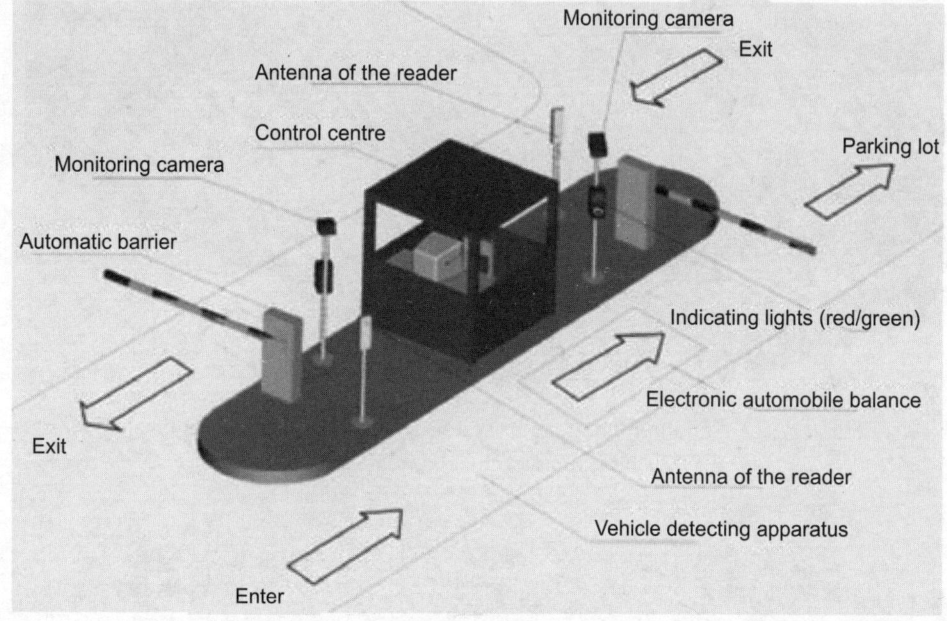


**Fig. 22.36** Component features at an entry/exit booth

Information about the parking spaces available will be automatically transmitted from the central computing/parking management centre to the VMS boards installed at various locations outside and inside a parking plaza and to users on smart phones (if integrated with the system, like VICS in Japan).

The parking fee may be collected using a semi-automatic or automatic method, like with ETC (Fig. 22.37). Fee collection methods may be of two types: pay-in and pay-out. The fee is collected on entry in the pay-in method while it is collected on exiting in the pay-out method. The former is suitable for cinema halls, where the parking time is the same for most customers.

Wireless ultrasonic sensors are embedded in each parking space to detect its occupancy status in real time, and the information is updated at a nearby traffic/parking management centre (Fig. 22.38). Such sensor information can also be used to detect potentially offending vehicles with expired time limits or illegal parking. This information along with parking fee details can be provided by smart phone apps, so that users can easily locate parking lots/spaces near their destination.

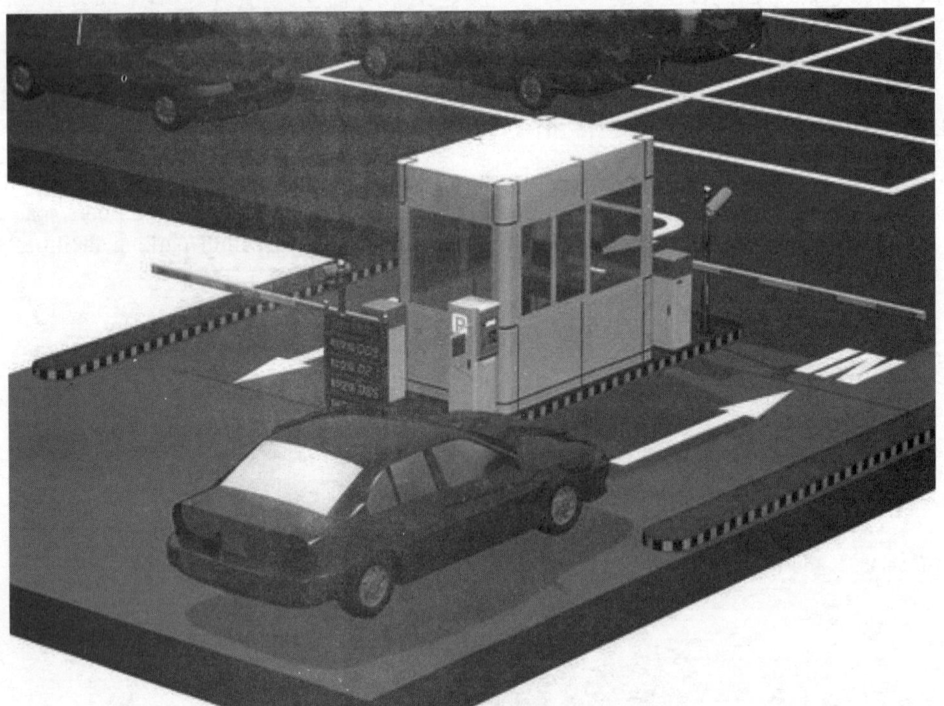


**Fig. 22.37** Parking management system with entry/exit control booth
(*Source*: www.metroinfrasys.com; *Courtesy*: Mr. Sachin Bhatia)

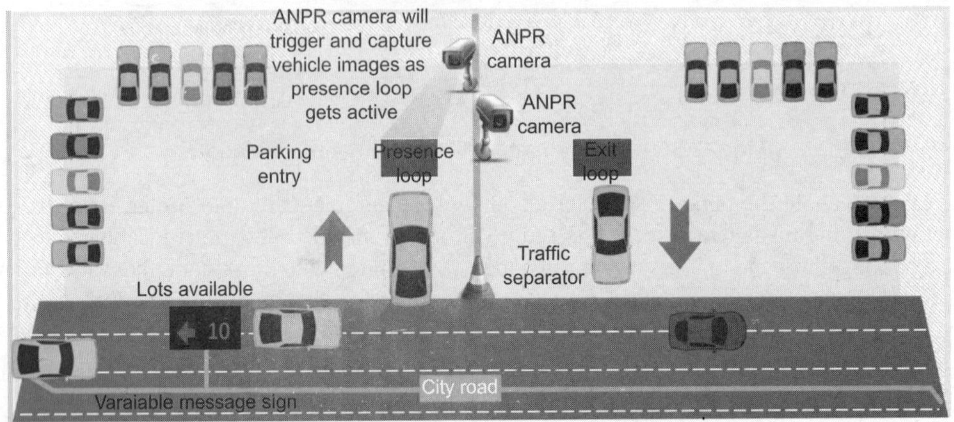


**Fig. 22.38** Vehicle image capturing by an ANPR system in a parking plaza
(*Source*: www.metroinfrasys.com; *Courtesy*: Mr. Sachin Bhatia)

Complete visibility of user activities in different stages of parking outside and inside the parking plaza can be achieved using CCTV cameras, and the images can be used for surveillance and ANPR purposes (Fig. 22.39). This real-time mechanism saves time for enforcement officers and helps them guide the violators or take other necessary action quickly.

The above ITS-based parking system can be adopted for multi-level parking and on-street parking management.

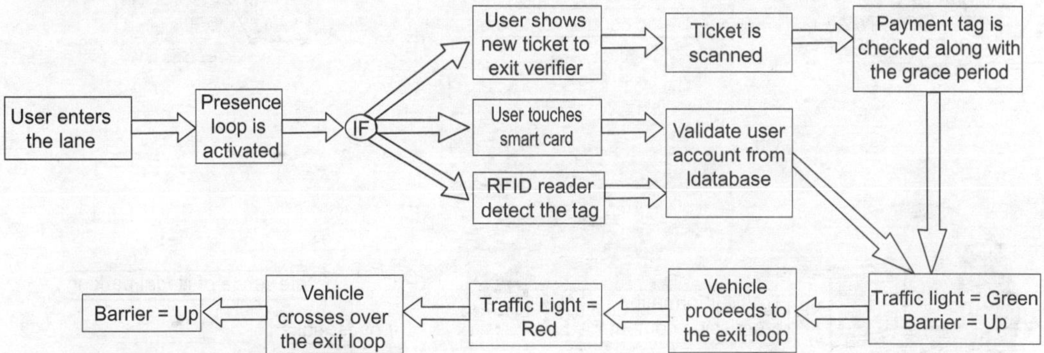


**Fig. 22.39** Typical process flow activities of an automated parking management system
(*Source*: www.metroinfrasys.com; *Courtesy*: Mr. Sachin Bhatia)

**Intelligent illegal parking deterrence system:** Illegal parking on the roadside at intersection areas, pedestrian crossings and bus stops needs to be immediately detected and the vehicles removed forcibly to avoid traffic congestion and accidents.

The Japanese intelligent illegal parking deterrence (IIPD) system comprises CCTV cameras, roadside communication system, loud speakers, microphones and a traffic control centre (Fig. 22.40). The CCTV cameras installed at major intersections, routes and in the vicinity of railway/bus stations are used for monitoring illegally parked vehicles in the traffic control centre. Digital image processing technology is used for detection of stationary vehicles. The detected vehicles are warned by the traffic control centre through loud speakers. In addition, the system is capable of controlling signals based on real-time traffic volume and notifies the police about the incidents detected.

This system can also be used to warn drivers of illegal parking, maintain traffic discipline and guide them to nearby parking lots. In addition, automatic video recording in the traffic control centre can be used as an evidence of before and after accidents.

**Public transportation priority systems (PTPS):** The Japanese PTPS system secures bus lanes and prioritises traffic signals for bus movement through intersections. Priority passage of buses through intersections and along specified routes is facilitated so that high-volume public transportation is efficiently operated. Thus, the general public can enjoy the benefits of reduced travel time while the demand for privately owned vehicles individual cars can be lowered.

Infrared beacons installed at suitable intervals along any specific route can receive bus-specific ID as the bus passes under them (Fig. 22.41). The buses are equipped with in-vehicle units that communicate with the infrared beacons and information from the beacons is sent to a nearby traffic control centre. With reference to the current location and average speed of the buses, the signals ahead of the buses are controlled by the traffic control centre, so that the buses can travel without stopping at intersections. The signal timings are controlled in real time to extend the Green time or shorten the Red time.

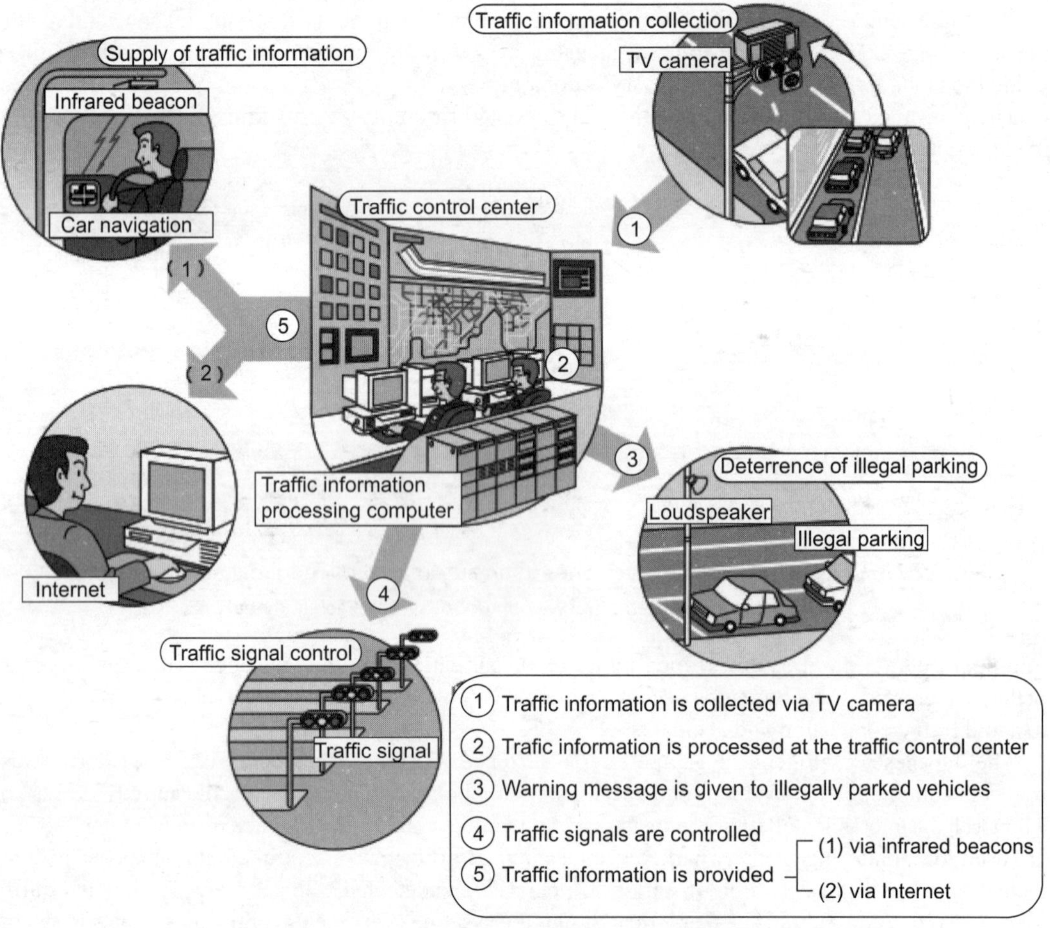


**Fig. 22.40** Intelligent illegal parking deterrence system
(*Source*: IUTR, 1998; *Courtesy*: Japan Traffic Management Technology Association, Japan)

Bus information such as the current location, travel time to destination, and expected arrival of a bus at a specific station can be obtained from the real-time information received from the infrared beacons.

**Environmental protection management systems:** Air pollution due to automobile emissions such as $CO_2$, CO, NOx and suspended particulate matter (SPM) are serious concerns of global warming worldwide. The Japanese Environmental Protection Management Systems (EPMS) is intended to mitigate this problem. Based on the information collected from different sources, such as roadside environmental sensors and other organisations, information is provided to drivers via in-vehicle units and VMSs (Fig. 22.42). With real-time data on pollution and traffic congestion, the EPMS is capable of diverting traffic and/or controlling signals along routes. For guidance, the relationship between vehicle speed and the emitted pollutant concentration can be developed and vehicle speeds reduced accordingly to reduce automobile emissions (IUTR, 1998).

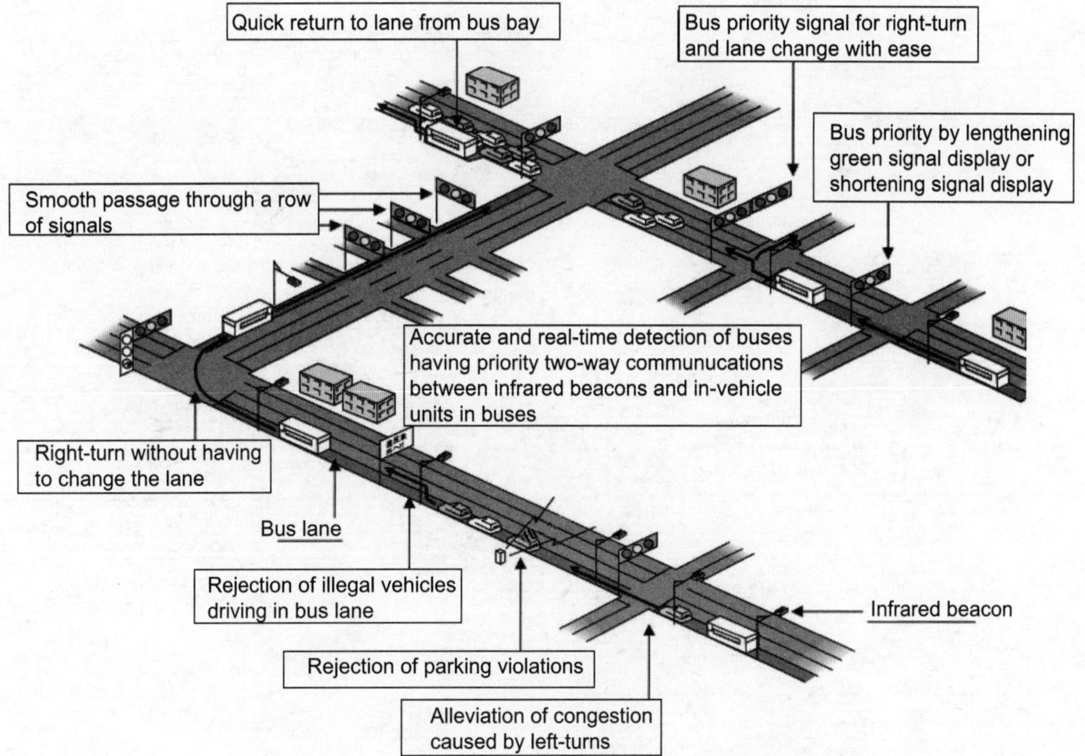


**Fig. 22.41** Outline of public transportation priority systems
(*Source*: IUTR, 1998; *Courtesy*: Japan Traffic Management Technology Association, Japan)

**Driving safety support system (DSSS):** The Japanese DSSS is intended to provide safety and reduce the burden on elderly drivers. The details are given below.

*Crossing collision warning system (Fig. 22.43 (a)):* The infrared beacons installed at un-signalised intersections are used to detect and warn approaching drivers about oncoming vehicles. Information about whether vehicles are coming from the left or right is provided. The warning information is provided to drivers via in-vehicle units (display/audio) or information boards a few metres before the intersection (IUTR, 1998).

*Right-turn accident prevention system (Fig. 22.43 (b)):* Like the warning system, vehicles including motor cycles in opposite lanes are detected and warning information is provided to right-turning vehicles at intersections. It is used to prevent possible collisions of moving motorcycles in the driver's blind spots. The infrared beacons of this system are also capable of providing information/warning to drivers approaching intersections just before Green changes to Amber (IUTR, 1998).

*Pedestrian support and safety system (Fig. 22.43 (c)):* Pedestrians crossing at intersections are detected by the infrared beacons, and the systems alert drivers by displaying boards and voice alerts in in-vehicle units. Two-way communication via the beacons can also supply information (text or voice) to the pedestrians about the surrounding area (IUTR, 1998).

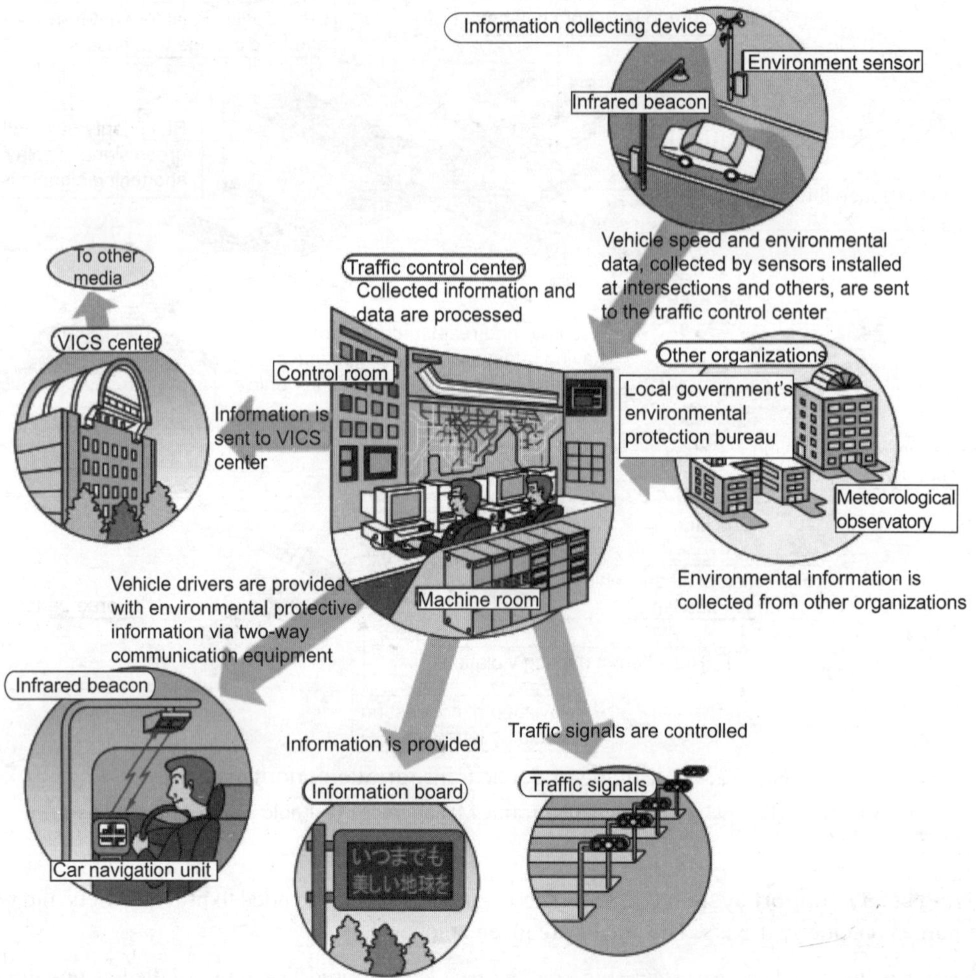


**Fig. 22.42** Configuration of environmental protection management systems
(*Source*: IUTR, 1998; *Courtesy*: Japan Traffic Management Technology Association, Japan)

**Emergency vehicle priority system:** Infrared transmitters mounted on emergency vehicles are capable of sending requests and the ID number of the vehicle to infrared receivers installed on the upper part of traffic signal poles (Fig. 22.44). When such vehicles come to a point about 800 m before an intersection installation, the receiver catches the signal from the vehicle and turns the signal Green. This Japanese system is capable of prioritising the operation of emergency vehicles in signalised intersections. This system can be adopted on routes approaching hospitals in urban areas.

## 22.8.18 Tunnel Systems

Tunnels provide the shortest routes and relieve traffic congestion. Deployment of ITS ensures traveller safety and improves traffic efficiency in tunnels (Fig. 22.45). As part of tunnel systems, emergency

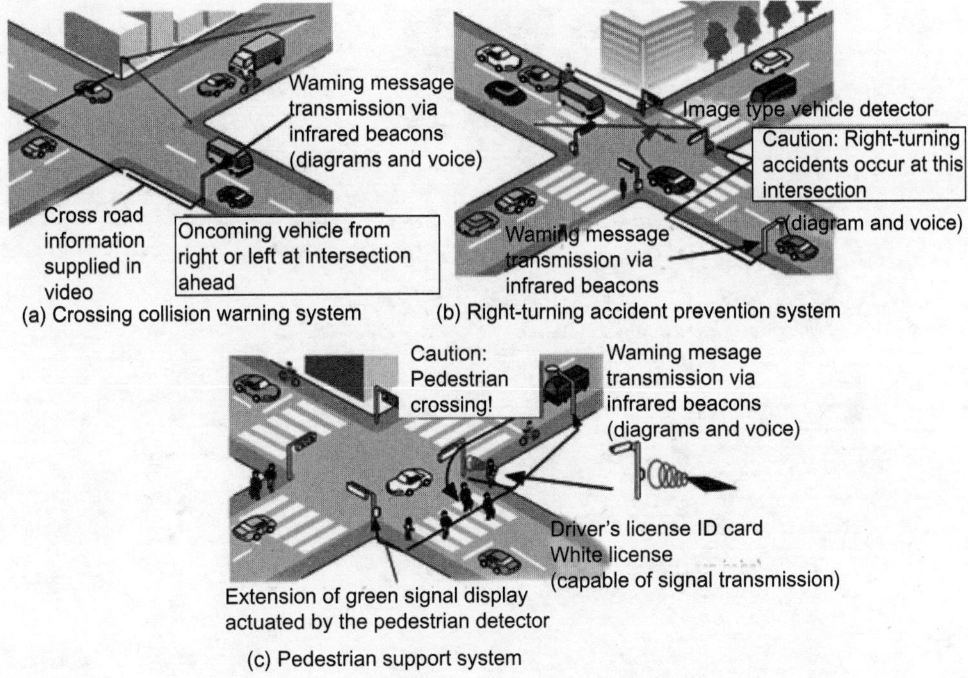


**Fig. 22.43** Examples of driving safety support systems
(*Source*: IUTR, 1998; *Courtesy*: Japan Traffic Management Technology Association, Japan)

response system (ERS) is installed with several sensors in tunnels such as fire detectors, heat detectors, smoke detectors, carbon monoxide detectors, CCTV cameras including with AID, jet fans & emergency lighting system with remote control operation and VMS/CMS. The ERS stations are commonly equipped with fire extinguishers, wireless phones and first aid medical kits. In an event of fire or earthquake or accident, the ERS will be activated for rescue operations, incident clearance and mass evacuation. In addition to this, tunnel radio system will be integrated with nearby radio stations to re-broadcast emergency messages/abnormal conditions in real-time to motorists. The provision of radio re-broadcasting is mandatory for tunnels exceeding 1km in length and carrying at least 2000 veh/lane/day (DOTMR, 2013). The ERS stations can also be used for operation and maintenance activities of the tunnels for continuous traffic management.

## 22.8.19 Road Condition Information Signs (RCIS) Systems

RCIS is a variant of VMS and CMS. These systems are placed at strategic locations on the road network and warn motorists about floods/snow, load restrictions and the pavement surface condition. Operators in the TMS nearby control these messages from time to time.

## 22.8.20 Ramp Metering Systems (RMS)

Excess entry of vehicles and the resulting congestion in the main lanes of an interchange is controlled by installing a traffic signal (with Red, Amber and Green) at the ramp entry. The rate of entry is controlled

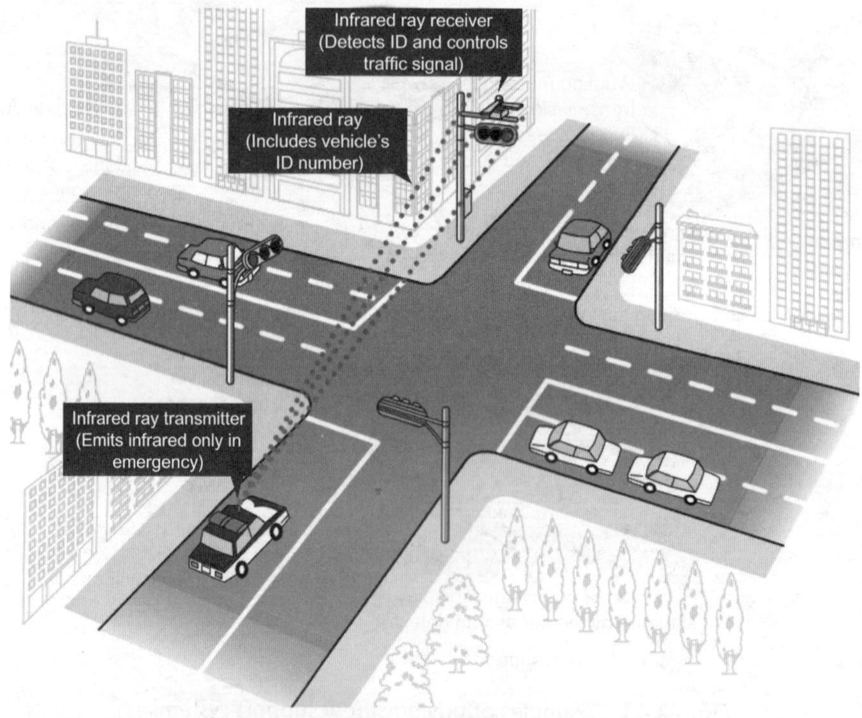


**Fig. 22.44** Emergency vehicle property system
(*Source*: IUTR, 1998; *Courtesy*: Japan Traffic Mangement Technology Association, Japan)

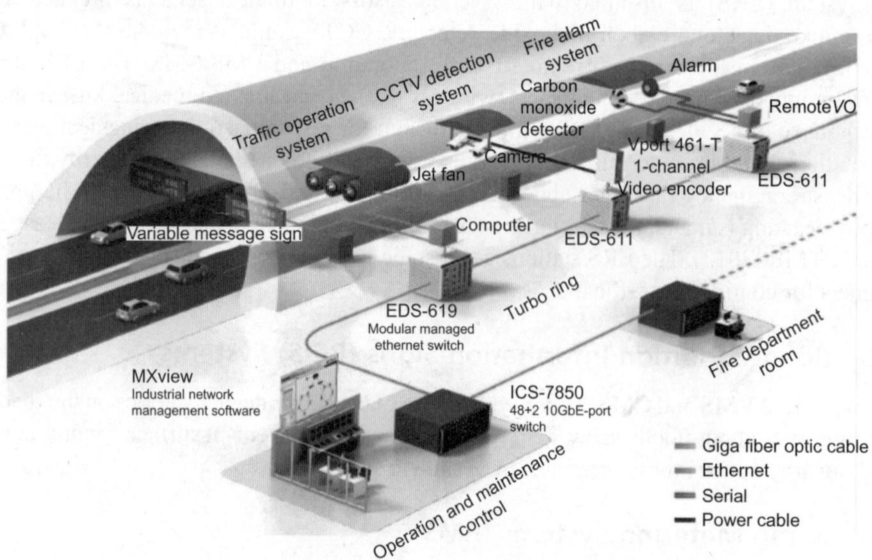


**Fig. 22.45** Typical tunnel emergency response system
(*Source*: www.moxa.com; *Courtesy*: Moxa, Taiwan)

by the signal set for different meter rates to optimise traffic flow and minimise congestion. A queue detector is integrated with an advance warning sign which should also be installed in advance of the RMS to divert traffic to alternate different route, if required.

## 22.8.21 Work Zone System

These sensors are used to prevent accidents and injuries in work zones and work zone areas by alerting workers and errant vehicle drivers of an intrusion into the work zone installation. These ITS monitoring systems detect speed and volume to adapt the length of the work zone and warn about traffic conditions on a real-time basis using VMSs or CMSs. These systems are considered a part of ITS and used for temporary traffic control.

## 22.8.22 Other Systems

Numerous subsystems of ITS are in use:

- **In-vehicle ITS or advanced driver assistance systems (ADAS):** The objective is to provide advance information/warning to drivers on a route using in-vehicle communication and sensor technology. This support can range from simple information presentation to voice communication.
- **Vehicle-to-vehicle (V2V) and vehicle-to-roadside equipment (V2X):** Vehicles in motion can be communicated with for various purposes/applications.
- **Freight management system:** This is a service for truck/commercial/heavy vehicle operation. It is similar to commercial vehicle operation (CVO). The bus rapid transit (BRT) system is an example, in which exclusive bus transit facilities are provided.
- **Adaptive traffic control system:** This provides priority for road-based public transport vehicles by adjusting the signal timings based on traffic demand, in real time. Different kinds of signal controllers are used for the operation of site control by a pedestrian push button/sensor actuated system, coordinated signal system and wide-area control of signals. Examples of this system are the British SCOOT (Split, Cycle and Offset Optimisation Technique), Australian SCATS (Sydney Coordinated Adaptive Traffic System) and the Japanese MODERATO (Management by Origin–Destination-Related Adaptation for Traffic Optimisation) (IUTR, 1998).

# EXERCISES

### Multiple Choice Questions

1. The fixed VMS supported on a concrete pedestal column should be flush with the ground but in no case should it protrude more than __________ m

   (a) 50 (b) 25 (c) 15 (d) 20

2. Electronic Call Box (ECB) shall be identified by a reflective guide sign placed ahead of it by

   (a) 1 m (b) 5 m (c) 8 m (d) 10 m

3. The fixed VMS is mounted on a gantry structure whereby the minimum vertical clearance provided from the road surface is

   (a) 5.5 (b) 6.0 m (c) 6.5 m (d) 7.7 m

4. The minimum distance of the VMS on expressways prior to the decision point should be

   (a) 1250 m (b) 1500 m (c) 750 m (d) 1000 m

5. The minimum distance of the VMS on national highways prior to the decision point should be
   (a) 250 m (b) 500 m (c) 750 m (d) 1000 m
6. The VMS should not be located on curves or on a highway sections having a grade exceeding
   (a) 1% (b) 2% (c) 3% (d) 4%
7. The minimum distance between road signs and VMS on expressways should be ________
   (a) 250 m (b) 500 m (c) 750 m (d) 1000 m
8. The minimum distance of the VMS on national highways prior to the decision point should be
   (a) 150 m (b) 100 m (c) 200 m (d) 250 m
9. The VMS structure should be designed such that it is capable of bearing wind loads up to
   (a) 150 km/hr (b) 100 km/hr (c) 200 km/hr (d) 250
10. The luminous intensity (LED ) of VMS should be provided as
    (a) < 3000 mcd (b) > 3000 mcd (c) < 1000 mcd (d) > 250 mcd
11. The probable range of wheel base considered for classification of truck/buses in India is
    (a) 0–1350 mm (b) 1400–1800 mm (c) 3401–5600 mm (d) 5000–18000 mm
12. The probable range of wheel base considered for classification of three-wheelers (Auto/Tempo) in India is
    (a) 0–1350 mm (b) 1400–1800 mm (c) 3401–5600 mm (d) 5000–18000 mm
13. The probable range of wheel base considered for classification of cars/jeeps in India is
    (a) 0–1350 mm (b) 1400–1800 mm (c) 3401–5600 mm (d) 5000–18000 mm
14. In Indian practice, the CCTV camera shall provide a useable picture at a minimum illumination of ________ lux.
    (a) 0.1 (b) 0.02 (c) 35 (d) 250
15. On national highways, the portable VMS powered by battery showing the sign of 'Men at work' in a construction zone should be placed at ________ prior to the decision points and repeated every 500 m.
    (a) 1250 m (b) 1500 m (c) 1000 m (d) 2000 m

## Answers

1. (c) 2. (d) 3. (a) 4. (b) 5. (d) 6. (d) 7. (a) 8. (a) 9. (c) 10. (b) 11. (c)
12. (b) 13. (a) 14. (b) 15. (d)

## Questions

1. Define ITS and state its need.
2. Write about the historical development of ITS with the main developments in different countries.
3. What are the objectives of the deployment and design integration of ITS?
4. What are the benefits of ITS? Briefly explain and give an example for each case.
5. State and define the concept of ITS design.
6. Draw a typical sketch showing the components involved in a multifunctional enforcement system.
7. Define ITS architecture. Write about the uses of ITS architecture.
8. Draw a sketch showing the uses of ITS architecture in transportation planning.
9. What is the V-development model used for ITS projects?

10. Draw a sketch showing the details of the V-development model with a life cycle framework.
11. Explain the salient features of the V-development model used for ITS projects with life cycle framework.
12. Explain the technologies used in ITS.
13. State the differences between the functionalities of on-board equipment and in-vehicle equipment.
14. Write a short note on(a) GPS,(b) Communication technologies and(c) GIS used for ITS.
15. Explain the uses, functional similarities and differences between RFID and DSRC systems of communications?
16. What is the role of Intelligent Transportation Systems (ITS) in implementing road pricing for congestion management? Discuss any three subsystems of ITS deployments with sketches.
17. Write about the salient features of vehicle detection stations and state how such stations are useful for ITS.
18. Explain the operational features and significance of the speed enforcement system.
19. Explain the operational features and significance of the Red light violation detection system.
20. Explain the components and operational details of the Japanese VIC system. Draw sketches showing the component features.
21. Explain the details of automatic incident detection (AID) and information provision system.
22. Draw a sketch showing the typical activities of the incident management system as a part of ITS.
23. Write a short note on road weather monitoring system.
24. Define VMS. What are the uses of VMS?
25. Define changeable message signs (CMS) systems and state how these signs are different from VMS.
26. Write a short note on(a) Road Condition Information Signs (RCIS) Systems,(b) Variable Speed Limit Sign (VSLS) / Lane Control Sign (LCS) and(c) Ramp Metering Systems (RMS).
27. Explain the concept of over height detection system. Also describe the functionality of vehicle detection system.
28. Explain the details of Automatic Number Plate Recognition (ANPR) system.
29. Write the merits and limitations of the ANRP system.
30. State how the performance measure of data collection by the ANRP system is evaluated.
31. Describe the salient features and components used for Electronic Toll Collection (ETC) system
32. List the equipment used for the WIM system.
33. Explain the concept of virtual WIM station. What is the typical layout of a WIM Station sorting system?
34. Explain the functionality and components involved in a WIM Station with ETC system.
35. Write a short note on traveller information system (TIS).
36. Write the objectives of the parking management and guidance system.
37. Write the salient features and installations required for a parking management and guidance system.
38. Draw a typical sketch showing the component features of an Entry/ Exit booth of a parking plaza.
39. Describe the Japanese proposed Driving Safety Support Systems (DSSS). Draw sketches with the main operational details.
40. Describe the Japanese Environmental Protection Management System.
41. Draw a sketch showing the details of typical process flow activities of the automated parking management system.
42. Write a short note on ITS-based tunnel systems and work zone systems.

# References


Aalto. *Guidebook on Energy Efficient Electric Lighting for Buildings*, International Energy Agency (IEA) – Energy Conservation in Buildings and Community Systems (ECBCS) Annex 45, Liisa Halonen, Eino Tetri and Pramod Bhusal (eds), Aalto University School of Science and Technology, Lighting Unit, Otaniemi, Finland, 2010. *http://lightinglab.fi/IEAAnnex45*

AASHTO. *Policy on Geometric Design of Highways and Streets.* American Association of State Highway and Transportation Officials (AASHTO), Washington, DC, USA. 1984, 1990, 1992, 1994, 2001, 2004 and 2011.

Abate, J and Whitt, W. Transient Behavior of the M/M/1 Queue: Starting at the Origin, *Queuing Systems* (2): 41–65, 1987.

Agent, KR and Pigman, JG. *Traffic Accident Investigation, Research Report KTC-93-10*, Kentucky Transportation Center, University of Kentucky, Lexington, USA, 1993.

Akçelik, R and Rouphail, NM. Estimation of delays at traffic signals for variable demand conditions, *Transportation Research* 27B (2): 109–131, 1993.

Akçelik, R. *Traffic Signals: Capacity and Timing Analysis*, Australian Road Research Board (ARRB), Vermont South, Australia, 1981.

Akçelik, R. *Time-dependent Expressions for Delay, Stop Rate and Queue Length at Traffic Signals. Internal Report AIR 367-1*, Australian Road Research Board (ARRB), Vermont South, Australia, 1980.

Alberta. *Highway Lighting Guide*, Alberta Transportation, August 2003. *www.transportation.alberta.ca/Content/docType233/Production/lighting.pdf*

Alexiadis, V, Jeannotte, K and Chandra, A. *Traffic Analysis Toolbox Volume I: Traffic Analysis Tools Primer*, FHWA-HRT-04-038, June 2004.

Algadhi, SAH, Mahmassani, HS and Herman, R. A speed-concentration relation for bi-directional crowd movements. In: *Pedestrian and Evacuation Dynamics*, M Schreckenberg and S Sharma (eds), Springer, Berlin, pp. 3–20, 2001.

Allsop, RE. Delay at a Fixed Time Traffic Signal – I: Theoretical Analysis, *Transportation Science* 6 (3): 260–285, 1972.

Amer, A. *Tight Diamond Interchange versus Single Point Urban Interchange*, Paper submitted on February 27, 2009 to the Student Paper Competition of the 2nd International Symposium on Freeway and Tollway Operations, 2009.

APRI. *Crash Reconstruction Basics for Prosecutors: Targeting Hardcore Impaired Drivers, Special Topics Series*, American Prosecutors Research Institute, March 2003.

Ardekani, SA, Ghandehari, M and Nepal, SM. *Macroscopic Speed-Flow Models for Characterization of Freeway and Managed Lanes*, Buletinul Institutului Politehnic, DIN IAŞI, Publicat de Universitatea Tehnică, Gheorghe Asachi din Iaşi Tomul LVII (LXI), Fasc. 1, 2011 Secţia Construcţii. Ărhitectură, 2011. *www.bipcons.ce.tuiasi.ro/Archive/222.pdf*

AS/NZS. *Lighting for Roads and Public Spaces, Part 1,2: Vehicular Traffic (Category V) Lighting. Guide to Design, Installation, Operation and Maintenance*, Australian/New Zealand Standard 1158.1.2, 2010.

Asakura, Y, Hato, E and Kashiwadani, M. *Origin-Destination Matrices Estimation Model Using Automatic Vehicle Identification Data and its Application to the Han-Shin Expressway Network Transportation*. Kluwer Academic Publishers, Netherlands, pp. 419–438, 2000.

Asdrubali, F, Buratti, C, Moretti, E, D'Alessandro, F and Schiavoni, S. Assessment of the Performance of Road Markings in Urban Areas: The Outcomes of the CIVITAS RENAISSANCE Project, *The Open Transportation Journal* 7: 7–19, 2013.

ASPO. *The Pedestrian Count, Report No. 199*, American Society of Planning Officials (ASPO), June 1965.

ASTM. *ASTM E 1710 - Standard Test Method for Measurement of Retroreflective Pavement Marking Materials with CEN-Prescribed Geometry Using a Portable Retroreflectometer*, ASTM International, 2011.

Auer, A, Feese, S and Lockwood, S. *History of Intelligent Transportation Systems*, FHWA-JPO-16-329, US Department of Transportation (USDOT), Intelligent Transportation Systems Joint Program Office, Washington DC, USA, 2016.

Austroads. *Austroads Guide to Traffic Management Part 2: Traffic Theory*, Australian Road Research Board (ARRB), Vermont South, Australia, 2008.

Austroads. *Guidance on Median and Centre Line Treatments to Reduce Head-on Casualties*, Prepared by David Beck and Alex Duerden, AP-R519-16, Austroads Incorporated, Sydney, Australia, 2016.

Austroads. *Guide to Traffic Management Part 10: Traffic Control and Communication Devices, Austroads Publication No. NP 1380, Austroads Publication No. AGTM10/09*, Prepared by Gary Veith et al., Australian Road Research Board (ARRB), Austroads Incorporated, Sydney, Australia, 2009. *www.austroads.com.au*

Awadallah, F. Theoretical analysis for horizontal curves based on actual discomfort speed. *ASCE Journal of Transportation Engineering* 131 (11): 843–850, 2005.

Babic, D, Fiolic, M and Prusa, P. Evaluation of Road Markings Retroreflection Measurement Methods, *European Scientific Journal, Special Ed. Vol. 3*, 2014.

Baoqun, Duan, Hongsheng, Xi and Yaping, Zhou. *Queuing System Performance Analysis and Markov Control Process (M)*, The press of the University of Science and Technology of China, Heifei, China, 2004.

Bates, T, Leach, R and Zimmerman, M. *Perfecting the Licence Plate Reader*, Traffic Technology International '99, Dorking, United Kingdom, 1999.

Bennett, CR and Greenwood, ID. Modelling Road User and Environmental Effects, HDM-4, Volume 7. *The Highway Development and Management Series,* 2001.

Benz, GP. *Pedestrian Time-Space Concept: A New Approach to the Planning and Design of Pedestrian Facilities*, Parsons Brinckerhoff Quade & Douglas, Inc., 1986.

Berry, DS and Van Till, CJ. Comparison of Three Methods for Measuring Delay at Intersections, *Traffic Engineering* 25: 93–99, 1954.

Bertucci, A. *Sign Legibility Rules of Thumb*, Unites States Sign Council (USSC), 2006. *www.ussc.org/SignLegibilityLettersize.pdf*

Bloomberg, MR, Burden, AM, et al. *New York City Pedestrian Level of Service Study Phase I*, NYC Department of City Planning and Transportation Division, New York, USA, 2006.

Bone, AJ. *Travel Time and Gasoline Consumption Studies in Boston*, Proceedings of the Highway Research Board 31: 440–466, 1952.

Bonneson, J, et al. *Recommended Ramp Design Procedures for Facilities Without Frontage Roads*, TT Institute, Editor. 2004.

Bonneson, JA. Bridge Size and Clearance Time of Single-Point Urban Interchange, *Journal of Transportation Engineering*, 119 (1), American Society of Civil Engineers (ASCE), Washington, DC, USA, 1993.

Bonneson, JA. Side friction and speed as controls for horizontal curve design. *ASCE Journal of Transportation Engineering* 125 (6): 473–480, 1999.

Bonsall, P and McKimm, J. Non-Response Bias in Roadside Mail-Back Surveys, *Traffic Engineering & Control*, 34 (12): 582–591, 1993.

Božić, M, Kopić, D and Mihoci, F. *Traffic Counting on the Roadways of Croatia in 2011 – Digest*, HRVATSKE CESTE d.o.o., Zagreb, May 2012. *www.hrvatske-ceste.hr*

Bradshaw, C. *Origin and Destination Studies Literature Search Results*, *Final Research Report WA-RD 478.1*, Research Contract T1803, Task 9, I-5 O/D Literature Search, Washington State Transportation Commission, US Department of Transportation (USDOT), Federal Highway Administration (FHWA), October 1999.

Bruce, GD. Reaction time in automobile driving. *Journal of Applied Psychology* 20 (3): 355, 1936.

Brun, O and Garcia, JM. Analytical solution of finite capacity *M*/*D*/1 queues, *Journal of Applied Probability* 37 (4): 1092–1098, 2000.

Burchett, GD and Maze, TH. Rural Expressway Intersection Characteristics as Factors in Reducing Safety Performance, *Transportation Research Record* 1953: 71–80, Transportation Research Board (TRB), National Research Council, Washington, DC, USA, 2006.

Caltrans. *Systems Engineering Guidebook for Intelligent Transportation Systems*, *Version 3.0*, Caltrans Division, Federal Highway Administration (FHWA), US Department of Transportation (USDOT), November 2009.

Cameron, RM. Pedestrian Volume Characteristics, *Journal of Traffic Engineering*, Institute of Transportation Engineers (ITE), Washington DC, USA, pp. 33–37, January 1976.

Camsys. *Los Angeles County Metropolitan Transportation Authority Congestion Mitigation Fee Nexus Analysis Methodology*, Technical Memorandum prepared by Cambridge Systems, Inc., Robert D. Spencer d/b/a Urban Economics, July 2011. *www.camsys.com*

CCE. *Continuing Education Papers: Signal Progression*, Civil and Construction Engineering (CCE), Oregon State University, Corvallis, USA, 2003. *http://cce.oregonstate.edu/continuing-education-papers.*

CDOT. *CDOT Roadway Design Manual: Chapter 10: Grade Separations and Interchanges*, Colorado Department of Transportation (CDOT), 2005. *https://www.codot.gov/business/designsupport/bulletins_manuals/roadway-design-guide*

CGRA. *Manual of Geometric Design Standards for Canadian Roads and Streets Geometric Design Committee, Ottawa*, Canadian Good Roads Association, Ottawa, Canada, 1963.

Chadda, HS and Cross, SE. Speed (road) bumps: Issues and opinions. *ASCE Journal of Transportation Engineering* 111(4): 410–418, 1985.

Chan, CY and Bu, F. *Literature Review of Pedestrian Detection Technologies and Sensor Survey.* Mid-Term Report for Task Order 5200, Experimental Transit Vehicle Platform for Pedestrian Detection, California PATH Institute of Transportation Studies University of California at Berkeley, USA, 2005.

Chandra, S and Kumar, U. Effect of lane width on capacity under mixed traffic conditions in India, *Journal of Transportation Engineering* 129 (2): 155–160, 2003.

Chandra, S, Kumar, V and Sikdar, PK. *Dynamic PCU and Estimation of Capacity of Urban Roads*, Indian Highways 23(4), Indian Roads Congress (IRC), New Delhi, 1995.

Chaudhary, NA and Chu, C-L. *Guidance for Timing and Coordinating Diamond Interchanges with Adjacent Traffic Signals*, *Report No. TX-00/4913-2*, Texas Department of Transportation (TxDOT), Construction Division, Austin, USA, November 2000.

Chen, P, Liu, H, Qi, H-S and Wang, F-J. *Journal of Zhejiang University-SCIENCE A* (Applied Physics & Engineering), 14(10): 691–704, 2013.

Chery, S. *A Comparison of Non-Traditional Intersection Designs Using Microscopic Simulation*, M.Sc. Thesis submitted to the Faculty of College of Engineering and Computer Science, Florida Atlantic University, Florida, USA, May 2010.

Chuanlai, Lu. *The Queuing Theory, 2nd ed. (M)*. The press of Beijing Institute of Posts and Telecommunications, Beijing, China. 2009.

Clement, JP. Letter to Editor. *Journal of ITE* pp. 6–7, March 1983.

CoC. *Spotlight on Public Lighting*, Corporation of Chennai (CoC), Chennai, India, 2003.

CoP. Code of Practice VEHICLE PARKING PROVISION, In: *Development Proposals*, Development & Building Control Division, Land Transport Authority, Edition 2011, Singapore, 2011. *www.lta.gov.sg/content/dam/ltaweb/corp/Industry/files/VPCOP2011.pdf*

CoPR. *Chapter 10: Lighting, Highway Design Manual*, Commonwealth of Puerto Rico, Department of Transportation and Public Health, Highway Authority, Puerto Rico, 1979. *www.dtop.gov.pr/carretera/det_content.asp?cn_id=272*

COSJ. *Municipal Street Lighting - Best Management Practice*, City of Saint John (COSJ), Canada, April 30, 2014. *www.saintjohn.ca/site/media/SaintJohn/COSJ%20Street%20Lighting%20Best%20Mangement%20Practices.pdf*

Craus, J and Polus, A. Aspects of spiral transition curve design. *Transportation Research Record* 631: 1–4. Transportation Research Board (TRB),Washington, DC, USA, 1977.

Cribbins, PD, Horn, JW and Vick, CE. *Development and Use of Maximum-car Technique for Measuring Travel Time, Bulletin 303*, Highway Research Board, pp. 94–102, 1961.

CTDOT. *Traffic Control Signal Design Manual*, Connecticut Department of Transportation (CTDOT), Bureau of Engineering and Construction, Division of Traffic Engineering, Newington, USA, 2009; revised September 2014.

Daamen, W, Hoogendoorn, SP and Bovy, PHL. First-order Pedestrian Traffic Flow Theory, In: *Transportation Research Board Annual Meeting*, Washington DC, National Academy Press, pp. 1–14, 2005.

Daamen, W. *Modelling Passenger Flows in Public Transport Facilities*, Ph.D thesis, Department of Transport & Planning, Trail Thesis Series, T2004/6, The Netherlands TRAIL Research School, Delft University of Technology, The Netherlands, 2004.

Dabbour, E, Easa, SM and Abd El Halim, AO. Radius requirements for reverse horizontal curves on three-dimensional alignments. *ASCE Journal of Transportation Engineering* 130(5): 610–620, 2004.

Daly, PN, McGrath, F and Annesley, TJ. Pedestrian speed/flow relationships for underground stations, *Traffic Engineering and Control* 32(2): 75–78, 1991.

DDA. *Guidelines for Road Markings*, Unified Traffic and Transportation Infrastructure (PLG & ENG) Centre (UTTIPEC), Delhi Development Authority, India, November 2009.

Delphi, MRC, Bulluz, L, Morrall, J and Smith, G. *Rural Intersection Safety Handbook, Final Report*, Transport Canada, Ottawa, Canada, March 2006.

DELTA. *Reflection and Retro-reflection, Technical Note RS 101*, DELTA, Denmark, 7-10- 2004. *http://www.ennisflint.com/getmedia/6d89a9d5-f0fa-461f-a2c2-27a6fb369fc6/OperationsManual_Delta-Retroreflectometers_Technical-note-101*

DELTA. *Marking Thickness Gauge: User Manual-Digital Measurement of Pavement Marking Thickness,* DELTA, Denmark. *http://www.assets.madebydelta.com/docs/roadsensors/Manualer_Quick-guides/Marking-Measurement-Gauge-User-Manual.pdf*

DELTA. *The LTL-M Mobile Retroreflectometer Features,* DELTA, Denmark. *http://www.assets.madebydelta.com/docs/roadsensors/Leaflets/English/LTL-M_Mobile_retroreflectometer_features.pdf*

DELTA. *The LTL-XL and LTL-X Retroreflectometer Features*, DELTA, Denmark. *http://www.assets.madebydelta.com/docs/roadsensors/Leaflets/English/LTL-XL_LTL-X_retroreflectometer_features.pdf*

DELTA. *The Retrosign GR3 and GR1 Rretroreflectometer Features*. DELTA, Denmark. *http://www.assets.madebydelta.com/docs/roadsensors/Leaflets/English/RetroSign_GR3_GR1_Leaflet.pdf*

Despotopoulos, PT. A new easement curve in highway work. *Civil Engineering and Public Works Review* XLII (495): 378–381, September 1947.

Dewar, R. *Chapter 18: Pedestrians and Bicyclists, Human Factors in Traffic Safety*, Lawyers & Judges Publishing Company, Tucson, USA, 2002.

DG TREN. Supreme: Best Practices in Road Safety, *Handbook for Measures at Country Level –Part C*, SER-TREN/E3-2005-SUPREME-S07.53754, Commissioned by DG TREN of the European Commission, 2007.

Dines, NT and Brown, KD. *Landscape Architect's Portable Handbook*, McGraw-Hill Professional, 2001.

Dion, F, Rakha, H and Kang, YS. Comparison of delay estimates at under-saturated and over-saturated pre-timed signalized intersections, *Transportation Research Part B: Methodological* 38 (2): 99–122, 2004.

DMRB. Road Geometry Junctions, Layout of Grade Separated Junctions, Part 1, TD 22/06, *Design Manual for Roads and Bridges (DMRB)*, 2006.

DMRB. The Design of Major Interchanges, Part-4 TD 39/94, Layout of Compact Grade Separated Junctions, Part 5, TD 40/94, *Design Manual for Roads and Bridges (DMRB)* 6(2), 1994. *http://www.standardsforhighways.co.uk/dmrb/vol6/section2/td3994.pdf*

DOTMR. *Road Planning and Design Manual, Queensland Practice, Vol. 5 - Intelligent Transportation Systems, 2nd Ed*, Department of Transport and Main Roads, State of Queensland, Australia, 2013.

DTTAS. *Traffic Signs Manual,* Department of Transport An Roinn Iompair, *www.transport.ie*, National Roads Authority Dublin, Ireland, 2010. *www.dttas.ie/sites/default/files/publications/roads/english/traffic-signs-manual-2010/traffic-signs-manual-2010-chapter-1-introduction-and-sign-location.pdf*

Easa, SM and Hassan, Y. Design requirements of equal-arc unsymmetrical vertical curves. *ASCE Journal of Transportation Engineering* 124 (5): 404–410, 1998.

Easa, SM, Dabbour, E and Ali, MZA. Three-dimensional model for stop control intersection sight distance. *ASCE Journal of Transportation Engineering* 130 (2): 261–270, 2004.

Easa, SM. Distributing super elevation to maximize highway design consistency. *ASCE Journal of Transportation Engineering* 129 (2): 127–133, 2003.

Easa, SM. Exact minimum sight distance on sag curve with centered overpass, *ASCE Journal of Transportation Engineering* 118 (4): 588–592, 1992.

Easa, SM. Lateral clearance needs on compound horizontal curves. *ASCE Journal of Transportation Engineering* 119 (1): 111–123, 1993.

Easa, SM. Lateral clearance to vision obstacles on horizontal curves. *Transportation Research Record* 1303*: 22–32*. Transportation Research Board (TRB), Washington DC, USA, 1991.

EC. *Green Public Procurement: Street Lighting and Traffic Lights*, *Technical Background Report*, European Commission, DG Environment-C1, Brussels, Belgium, 2011. *http://ec.europa.eu/environment/gpp/pdf/tbr/street_lighting_tbr.pdf*

ECO-3. *Guidelines: Energy Efficient Street Lighting*, USAID ECO-II Project, Version 2.0, New Delhi, India, June 2010.

Edie, LC. Following and Steady-State Theory for Non-congested Traffic, *Operations Research* 9: 66–76, 1961.

EEC. *Lighting Design, Section 300, Electrical & Traffic Engineering Design Guidelines*, Electrical Engineering Centre (EEC), British Columbia, Canada, April 2013. *www.th.gov.bc.ca/publications/eng_publications/electrical/electrical_and_traffic_eng/Electrical_Signing_Design_Manual/Section%20300/Section%20300.pdf*

EN 1436. *Road Marking Materials – Road Marking Performance for Road Users*, European Committee for Standardization, CEN/TC 226, EN 1436:2007+A1:2008, 2007–2008.

Evans, D and Klashinsky, R. *Weigh-In-Motion System for the Reduction of Commercial Vehicle Traffic Delays at the Borden-Carleton, PEI Weight Inspection Station*, IRD Inc., Canada. *www.irdinc.com/public/uploads/itssolution_document/24/1387813682_tac_presentation_borden_pei_wim.pdf*

Ezell, S. *Explaining International IT Application Leadership: Intelligent Transportation Systems*, The Information Technology & Innovation Foundation, 2010. *www.itif.org/files/2010-1-27-ITS_Leadership.pdf*

Farouki, OT and Nixon, WJ. The effect of width of suburban roads on the mean-free speeds of cars, *Traffic Engineering Control* 17(12): 518–519, 1976.

FDOT. *Project Traffic Forecasting Handbook*, Florida Department of Transportation (FDOT), USA, 2012.

FHWA. *Manual on Uniform Traffic Control Devices (MUTCD)*, US Department of Transportation (USDOT), Federal Highway Administration (FHWA), Washington DC, USA, 2003.

FHWA. *Evaluation of Design Consistency Methods for Two-lane Rural Highways. Publication No. FHWA-RD-99-173*, Executive Summary. Federal Highway Administration (FHWA), Washington DC, USA, August 2000. *http://www.tfhrc.gov/safety/ihsdm/pdfs/99-173.pdf*

FHWA. *Speed Prediction for Two-lane Rural Highways. Publication No. 99-171*, Federal Highway Administration (FHWA), Washington DC, USA, August 2000. *http://www.tfhrc.gov/safety/ihsdm/pdfs/99-171.pdf*

FHWA. *Revised Monograph on Traffic Flow Theory, A Report of Federal Highway Administration Research and Technology*, Federal Highway Administration (FHWA), Washington DC, USA, 2016. Contact: Dr. Henry Lieu, Henry.Lieu@dot.gov, *www.fhwa.dot.gov/publications/research/operations/tft/*

FHWA. *Manual on Uniform Traffic Control Devices (MUTCD)*, Federal Highway Authority (FHWA), Washington DC, USA, 2009. *http://mutcd.fhwa.dot.gov/*

FHWA. *Manual on Uniform Traffic Control Devices (MUTCD) for Streets and Highways*, US Department of Transportation (USDOT), Federal Highway Administration (FHWA), Washington DC, USA, 2012. *http://mutcd.fhwa.dot.gov/pdfs/2009r1r2/mutcd2009r1r2edition.pdf*

FHWA. *National Park Service: 2004 Traffic Data Report*, prepared for Federal Highway Administration and Eastern Federal Lands Highway Division in association with National Park Service, Park Operation and Education and Facility Management Division, by Applied Research Associates, Inc. June 2005.

FHWA. *Road Safety Fundamentals.* Contract DTFH61-03-D-00105, Federal Highway Administration (FHWA), Washington DC, USA, 2004.

FHWA. *Short Course in Highway Lighting*, Federal Highway Administration (FHWA), Washington DC, USA, 1973. *http://files.eric.ed.gov/fulltext/ED089022.pdf*

FHWA. *Traffic Monitoring Guide*, Federal Highway Administration (FHWA), Washington DC, USA, September 2013.

Fitzpatrick, K, Brewer, M, Dorothy, P and Park, ES. *Design Guidance for Intersection Auxiliary Lanes, Project 03-102, NCHRP Report 780*, National Cooperative Highway Research Program, Transportation Research Board (TRB), Washington DC, USA, 2014.

Fitzpatrick, K, Wooldridge, MD and Blaschke, JD. *Urban Intersection Design Guide: Volume 1—Guidelines and Volume 2—Applications, Report No. FHWA/TX-05/0-4365-P2*, Texas Transportation Institute, The Texas A&M University System, College Station, USA, 2005.

Fitzpatrick K, Brewer, MA, Gluck, JS, Eisele, WL, Zhang, Y, Levinson, HS, von Zharen, W, Lorenz, MR, Iragavarapu, V and Park, ES. Development of Left-Turn Lane Warrants for Unsignalized Intersections, *NCHRP Web-Only Document 193*, Transportation Research Board (TRB), Washington DC, USA. 2013

Flora, JW and Keitt, KM. *Access Management Retrofit Techniques*, *Access Management for Streets and Highways*, IP-82-3, Federal Highway Administration (FHWA), Washington DC, USA, June 1982.

Fruin, JJ and Benz, GP. *Pedestrian Time Space Concept for Analyzing Corners and Crosswalks*, Paper presented at the Annual Meeting of the Transportation Research Board, Washington, DC, USA, January 1984.

Fruin, JJ. Design of Pedestrians: A Level of Service Concept, *Highway Research Record* 355: 1–15, 1971*a*.

Fruin, JJ. *Pedestrian Planning and Design*, Metropolitan Association of Urban Designers and Environmental Planners, Inc., New York, USA, 1971*b*.

Fu, H and Wilmot, CG. *Assessing Performance of Alternative Pavement Marking Materials*, *Technical Assistance Report No. 08-4TA*, Louisiana Department of Transportation and Development, Louisiana Transportation Research Centre, USA, August 2008.

Fwa, TF and Tan, LS. Geometric characterization of road humps for speed-control design. *ASCE Journal of Transportation Engineering* 118(4): 593–598, 1992.

Gallagher, J. *Travel Time Data Collection Using GPS*, National Traffic Data Acquisition Conference, New Mexico, pp.150–154, 1996.

Garber, NJ and Fontaine, MD. *Guidelines for Preliminary Selection of the Optimum Interchange Type for a Specific Location*, *Final Report VTRC 99-R15*, Virginia Transportation Research Council, Virginia, USA, 1999.

García, GA and Conesa, LM. Improving steep bends as hairpin curves on mountainous roads (Discussion on this section). *ASCE Journal of Transportation Engineering* 118 (6): 882–884, 1992.

García, GA and Conesa, LM. Unit curve for design of highway transitions by SK Rao. *ASCE Journal of Transportation Engineering* pp. 244, May/June 1997.

Gårder, PE. The impact of speed and other variables on pedestrian safety in Maine, *Accident Analysis & Prevention* 36 (4): 533–542, 2004.

Gates, TJ and Hawkings, HG. *The Use of Wider Longitudinal Pavement Markings*, *Report No. 02-0024-1*, Texas Transportation Institute, The Texas A&M University System, College Station, USA, 2002.

Gattis, JL and Low, ST. *Intersection Angles and Driver's Field of View*. Mack-Blackwell Transportation Center, University of Arkansas, USA, 1997.

Gattis, JL, Gluck, JS, Barlow, JM, Eck, RW, Hecker, WF and Levinson, HS. *Guide for the Geometric Design of Driveways, NCHRP Report 659*, National Cooperative Highway Research Program, Transportation Research Board (TRB), Washington DC, USA, 2010.

Gattis, JL, Vinsin II, BFI and Duncan, LK. Low-speed horizontal curve friction factors. *ASCE Journal of Transportation Engineering* 131(2): 112–119, 2005.

Gazzar, El MG. *A Logical Approaches to Roadway Lighting Design*, 2008. *www.dewberry.com/docs/default-source/documents/a_logical_approach_to_roadway_lighting_design.pdf*

GDOT. *About Pedestrians*, *Pedestrian Facilities Design Guide*, draft version, Georgia Department of Transportation, Atlanta, USA. *http://www.dot.state.ga.us/dot/plan-prog/planning/projects/bicycle/ped_facilities_guide/about_pedestrians.pdf*

Georgia Department of Transportation. *Pedestrian & Streetscape Guide*, Georgia Department of Transportation, Atlanta, USA, prepared by Otak, Inc., September 2003.

Gerlough, DL. *Use of Poisson Distribution in Highway Traffic*, The Eno Foundation for Highway Traffic Control, Saugatuck, USA, 1955.

Gibreel, GM, Easa, SM, Hassan, Y and El-Dimeery, IA. State of the art of highway geometric design consistency. *ASCE Journal of Transportation Engineering* 125(4): 305–313, 1999.

Glennon, JC. New and improved model of passing sight distance on two-lane rural highways. *Transportation Research Record* 1195: 132–137. Transportation Research Board (TRB), Washington DC, USA, 1988.

Gordon, RL and Tighe, W. *Traffic Control Systems Handbook, Report No. FHWA-HOP-06-006,* Federal Highway Administration (FHWA), Washington DC, USA, 2005.

Graham, H. *Research Methods 1 Handouts: Chi-Square Test*, COGS-Version 1.0, pp.1–8, September 2000. *http://users.sussex.ac.uk/~grahamh/RM1web/sthand1.pdf*

Greenberg, H. An Analysis of Traffic Flow, *Operations Research* 7: 78–85, 1959.

Greenshields, BD, Shaspior, D and Erickson, EL. *Traffic Performances at Urban Intersections*, Bureau of Highway Traffic, Technical Report No. 1, Yale University, Connecticut, USA, 1947.

Greenshields, BD. A Study in Highway Capacity, *Proceedings of Highway Capacity*, 14: 458, 1935.

Greenstein, J. Transition curves for speed changes. *ASCE Journal of Transportation Engineering* 102 (No.TE1): 699–713, 1976.

Guy, BP and Fricker, JD. *Guidelines for Data Collection Techniques and Methods for Roadside Station Origin-Destination Studies, Final Report, FHWA/IN/JTRP-2005/27*, Joint Transportation Research Program, Purdue University, West Lafayette, USA, December 2005.

Hajek, JJ. *Optimal Sample Size of Roadside Interview O-D Surveys*, Ontario Ministry of Transportation and Communication, Ontario, Canada, 1977.

Hald, A. *Statistical Theory with Engineering Applications*, John Wiley and Sons, pp.725–726, 1952.

Hallworth, MS. Signaling Roundabouts, *Traffic Engineering and Control* 33 (6): 354–363, 1992.

Hani, SM, Tian, H, Jiwon, Kim, Ying, C, Zihan, H, Hooram, H and Robert, H. *Implementation of a Weather Responsive Traffic Estimation and Prediction System (TrEPS) for Signal Timing at Utah DOT, Report No. FHWA-JPO-14-140*, US Department of Transportation (USDOT), Federal Highway Administration (FHWA), Washington DC, USA, 2014.

Hanson, R, Klashinsky, R and McGibney, S. *ITS Technologies for Commercial Vehicle Compliance in the Maritime, Paper and Efficiency*, Annual Conference of Transportation Association of Canada, Halifax, Nova Scotia, September 26–29, 2010.

Harrold, C. *The Carousel Data Collection Methodology Comparison with Other Collection Methodology*, Annual Meeting Preprints, Transportation Research Board, Washington DC, USA, 1998.

Harwood, DW, Mason, J and Brydia, R. Sight distance for stop-controlled intersections based on gap acceptance. *Transportation Research Record* 1701: 32–41. Transportation Research Board (TRB), Washington DC, USA, 2000.

Harwood, DW, Glennon, JC and Graham, JL. *Procedures and Guidelines for Rehabilitation of Existing Freeway-Arterial Highway Interchanges Volume II, Design Procedures for Rehabilitation of Freeway-Arterial Interchanges, Report No. FHWA/RD-81/104*, Federal Highway Administration (FHWA), Washington DC, USA, May 1982.

Hawkins, HG Jr. How modern headlamp performance impacts sag vertical curve design. *ASCE Journal of Transportation Engineering* 133(4): 223–231, 2007.

Henry, RD. *Signal Timing on a Shoestring, Report No. FHWA-HOP-07006*, Federal Highway Administration (FHWA), Washington DC, USA, March 2005.

Hickerson, TF. *Route Location and Design*. New York: McGraw-Hill Book Co. Inc. 1964.

Hinkle, AH. The engineer's part in making the highway safe. *Transactions of ASCE* 94: 522–529, 1930.

Hirsh, M, Prashker, JN and Ben-Akiva, M. New approach to geometric design of highways. *Transportation Research Record* 1100: 50–57. Transportation Research Board (TRB), Washington DC, USA, 1986.

HITEC. *Summary of Evaluation Findings for 30-Meter Handheld and Mobile Pavement Marking Retroreflectometers*, Highway Innovative Technology Evaluation Center, CERF #40525, Transportation Research Record, 2001.

Homburger, WS, Hall, JW, Reilly, WR and Sullivan, EC. *Fundamentals of Traffic Engineering, 15th ed.*, University of California, Berkley (UCB-ITS-CN-01-1), USA, January 2001.

HRB. *Highway Capacity Manual*, National Research Council, Department of Traffic and Operations, Special Report 87, Committee on Highway Capacity, Washington, DC, 1965.

Hurdle, VF. Signalized Intersection Delay Models – A Primer for the Uninitiated, *Transportation Research Record* 971, Transportation Research Board (TRB), National Research Council, Washington DC, USA, pp. 96–105, 1984.

IESNA. *IESNA Lighting Handbook*, Illuminating Engineering Society of North America, New York, USA, 2000.

Immers, LH and Logghe, S. *Traffic Flow Theory*, Course H 111, Verkeerskunde Basis, Faculty of Engineering Department of Civil Engineering, Traffic and Infrastructure, Heverlee, Belgium, 2003. *www.kuleuven.be/traffic/dwn/H111part3.pdf*

INDOT. *Indiana Design Manual, Chapter 502*, Indiana Department of Transportation (INDOT), 2013. *www.in.gov/indot/design_manual/files/Ch502_2013.pdf*

IRC: 103-2012. *Guidelines for Pedestrian Facilities*, Indian Roads Congress (IRC), New Delhi, India, 2012.

IRC: 106-1990. *Guidelines for Capacity of Urban Roads in Plain Areas*. Indian Roads Congress (IRC), New Delhi, India. 1990.

IRC: 15-1974. *Ribbon Development along Highways and Its Prevention,* Indian Roads Congress (IRC), New Delhi, India, 1974.

IRC: 30-1968. *Standard Letters and Numerals of Different Heights for Use on Highway Signs*, Indian Roads Congress (IRC), New Delhi, India, 1968.

IRC: 3-1983. *Dimensions and Weights of Road Design Vehicles*. Indian Roads Congress (IRC), New Delhi, India, 1983.

IRC: 35-1970. *Code of Practice for Road Markings (with Paints)*. The Indian Roads Congress (IRC), New Delhi, India, 1970.

IRC: 35-2015. *Code of Practice for Road Markings*, Indian Roads Congress (IRC), New Delhi, India, 2015.

IRC: 37-2012. *Guidelines for Design of Flexible Pavements*, Indian Roads Congress (IRC), New Delhi, India, 2012.

IRC: 39-1986. *Standards for Road–Rail Level Crossings*. Indian Roads Congress (IRC), New Delhi, India, 1986.

IRC: 52-2001. *Recommendations about the Alignment Survey and Geometric Design of Hill Roads*. Indian Roads Congress (IRC), New Delhi, India, 2001.

IRC: 53-2012. *Road Accident Recording Forms A-1 and A-4*, Indian Roads Congress (IRC), New Delhi, India, 2012.

IRC: 54-1974. *Lateral and Vertical Clearances at Underpasses for Vehicular Traffic*, Indian Roads Congress (IRC), New Delhi, India, 1974.

IRC: 58-2015. *Guidelines for the Design of Plain Jointed Rigid Pavements for Highways*, Indian Roads Congress (IRC), New Delhi, India, 2015.

IRC: 64-1990. *Guidelines for Capacity of Roads in Rural Areas. First Revision*, Indian Roads Congress (IRC), New Delhi, India, 1990.

IRC: 65-1976. *Recommended Practice for Traffic Rotaries*, Indian Roads Congress (IRC), New Delhi, India, 1976.

IRC: 66-1976. *Recommended Practice for Sight Distance on Rural Highways*. Indian Roads Congress (IRC), New Delhi, India, 1976.

IRC: 67-1977. *Code of Practice for Road Signs*. Indian Roads Congress (IRC), New Delhi, India, 1977.

IRC: 67-2012. *Code of Practice for Road Signs, 3rd Revision*, Indian Roads Congress (IRC), New Delhi, India, 2012.

IRC: 69-1977. *Space Standards for Roads in Urban Areas*. Indian Roads Congress (IRC), New Delhi, India, 1977.

IRC: 70-1977. *Guidelines on Regulation and Control of Mixed Traffic in Urban Areas*. Indian Roads Congress (IRC), New Delhi, India, 1977.

IRC: 73-1980. *Geometric Design Standards for Rural (Non-Urban) Highways*. Indian Roads Congress (IRC), New Delhi, India, 1980.

IRC: 79-1981. *Recommended Practice for Road Delineators*, Indian Roads Congress (IRC), New Delhi, India, 1981.

IRC: 86-1983. *Geometric Design Standards for Urban Roads in Plains*. Indian Roads Congress (IRC), New Delhi, India, 1983.

IRC: 9-1972. *Traffic Census on Non-Urban Roads*, Indian Road Congress (IRC), New Delhi, India, 1972.

IRC: 93-1985. *Guidelines on Design and Installation of Road Traffic Signals*, Indian Roads Congress (IRC), New Delhi, India, 1985.

IRC: 98-1997. *Guidelines on Accommodation of Utility Services on Roads in Urban Areas, First Revision*. Indian Roads Congress (IRC), New Delhi, India, 1997.

IRC: 99-1988. *Tentative Guidelines on the Provision of Speed Breakers for Control of Vehicular Speeds on Minor Roads*, Indian Roads Congress (IRC), New Delhi, India, 1988.

IRC: SP: 12-1973. *Tentative Recommendations on the Provision of Parking Spaces for Urban Areas*, Indian Roads Congress (IRC), New Delhi, India, 1973.

IRC: SP: 15-1974. *Ribbon development along highways and its prevention*. Indian Roads Congress (IRC), New Delhi, India, 1974.

IRC: SP: 23-1993. *Vertical Curves for Highways*. Indian Roads Congress (IRC), New Delhi, India, 1993.

IRC: SP: 41-1994. *Guidelines for the Design of At-grade Intersections in Rural and Urban Areas, Special Publication No. 41*, Indian Roads Congress (IRC), New Delhi, India, 1994.

IRC: SP: 48-1998. *Hill Road Manual*. Indian Roads Congress (IRC), New Delhi, India, 1998.

IRC: SP: 62-2014. *Guidelines for the Design and Construction of Cement Concrete Pavements for Low Volume Rural Roads*, Indian Roads Congress (IRC), New Delhi, India, 2014.

IRC: SP: 72-2014. *Guidelines for the Design of Flexible Pavements for Low Volume Rural Roads*, Indian Roads Congress (IRC), New Delhi, India, 2014.

IRC: SP: 88-2010. *Manual on Road Safety Audit*, Indian Roads Congress (IRC), New Delhi, India, 2010.

IRC: SP: 90-2010. *Manual for Grade Separators & Elevated Structures*, Indian Roads Congress (IRC), New Delhi, India, 2010.

IRC: SP: 55-2014, *Guidelines on Traffic Management in Work Zones*, Indian Roads Congress (IRC), New Delhi, India, 2014.

IS: 1944. *Code of Practice for Lighting of Public Thoroughfares*, IS 1944-7:1981 (R2003), Bureau of Indian Standards, New Delhi, India, 1981.

IS: 2713. *Specification for Tubular Steel Poles for Overhead Power Lines*, IS 2713: Parts 1 to 3. Bureau of Indian Standards, New Delhi, India, 1980.

ITE, *Traffic Control Devices Handbook*, Institute of Transportation Engineers (ITE), Washington DC, USA, 2001.

IUTR. *ITS Developed by Japanese Police*, Japan Traffic Management Technology Association (JTMTA), Published by Institute of Urban Traffic Research (IUTR), Osaka, Japan, 1998. *http://tmt.or.jp/research/pdf/its-e.pdf*

Jacobs, RJ and Cole, BL. *Acquisition of Information from Alpha-Numeric Signs*, Australian Road Research Board (ARRB) Conference, 9th, Brisbane, Queensland, Australian Road Research Board (ARRB), Vermont South, Australia, 9(5): 390–5, 1978.

Jagadeesh, YM, Akilesh, S and Prasanth. *Intelligent Street Lights*, SMART GRID Technologies, ScienceDirect, Pricedia Technology 21: 547–551, August 6–8, 2015.

Jalihal, SA, Ravinder, K and Reddy, TS. Traffic Characteristics of India, *Proceedings of the Eastern Asia Society for Transportation Studies* 5: 1009–1024, 2005.

Jarvis, JR. The off-road testing of road humps for use under Australian conditions. *Proceedings of the 10th ARRB Conference, Vol. 4.* Traffic Australian Road Research Board (ARRB), Vermont South, Australia, August 1980.

Jia, X, O'Mara, M and Guan, M. Rethinking Geometric Design Standards for Bike Paths, *Journal of Transportation Engineering* 133 (19), 2007.

Jiang, X, Vaziri, SH, Haas, C, Rothenburg, L, Kennepohl, G and Haas, R. Improvements in Piezoelectric Sensors and WIM Data Collection Technology, *Annual Conference of the Transportation Association of Canada (TAC)*, Vancouver, British Columbia, October 18–21, 2009. *http://tac-atc.ca/en/bookstore-and-resources/free-resources-and-tools/conference-proceedings*

Joksch, H, Massie, D and Pichler, R. *Vehicle Aggressivity: Fleet Characterization Using Traffic Collision Data, Report No. DOT-VNTSC-NHTSA-98-1.* US Department of Transportation (USDOT), 1998.

Jones, EG and Selinger, MJ. *A Comparison of the Operations of Single Point and Tight Urban Diamond Interchanges*, Submitted to the Transportation Research Board for presentation and publication on July 27, 2002.

Kaluva, M. *The Effect of Skew Angle on Average Queue Delay at Tee-Intersections: A Simulation Study Using the Texas Model*, M.Sc. Thesis, Civil Engineering, DigitalCommons@University of Nebraska – Lincoln, USA, December 2013.

Kamal, TA. *Roadway Lighting Design Guidelines*, 2007. *http://talalkamal.com/pdf_files/Traffic_Lighting_Design.pdf*

Kanellaidis, G. Aspects of highway super elevation. *ASCE Journal of Transportation Engineering* 117 (6): 624–632, 1991.

Kanellaidis, G. Human factors in highway geometric design. *ASCE Journal of Transportation Engineering* 122(1): 59–66, 1996.

Kendall, DG. Some Problems in the Theory of Queues. *Journal of Royal Statistical Society B (Methodological)* 13(2): 151–183, 1951.

Kendall, DG. Stochastic Processes Occurring in the Theory of Queues and their Analysis by the Method of the Imbedded Markov Chain, *The Annals of Mathematical Statistics* 24 (3): 338–354, 1953.

Kimber, RH and Hollis, M. *Traffic Queues and Delay at Road Junctions*. TRL LR, UK, 1979.

Kingman, JFC. The Single Server Queue in Heavy Traffic, *Mathematical Proceedings of the Cambridge Philosophical Society* 57 (4): 902–904, 1960.

Klashinsky, R and Hanson, R. *Deployment of ITS at Automated Truck Inspection and Weighing Stations in the Maritime*, CITE, International Road Dynamics Inc., Canada, 2011.

Klein, LA, Mills, MK and Gibson, DRP. *Traffic Detector Handbook, Vol. I, 3rd ed.*, FHWA-HRT-06-108, Federal Highway Administration (FHWA), Washington DC, USA, 2006*a*.

Klein, LA, Mills, MK and Gibson, DRP. *Traffic Detector Handbook, Vol. II, 3rd ed.*, FHWA-HRT-06-108, Federal Highway Administration (FHWA), Washington DC, USA, 2006*b*.

Koepke, FJ and Levinson, HS. *Access Management Guidelines for Activity Centers, NCHRP Report 348*, National Cooperatve Highway Research Program, Transportation Research Board (TRB), Washington DC, USA, 1992.

Koonce, P, Rodegerdts, L, Kevin, L, Quayle, S, Beaird, S, Braud, C, Bonneson, J, Tarnoff, P and Urbanik, T. *Traffic Signal Timing Manual, Report No. FHWA-HOP-08-024*, US Department of Transportation (USDOT), Federal Highway Administration (FHWA), Washington DC, USA, June 2008. *http://ops.fhwa.dot.gov/publications/fhwahop08024/fhwa_hop_08_024.pdf*

Koppula, N. *A Comparative Analysis of Weaving Areas in HCM, TRANSIMS, CORSIM, VISSIM and INTEGRATION*, M.Sc. Thesis, Faculty of the Virginia Polytechnic Institute and State University, USA, 2002. *https://theses.lib.vt.edu/theses/available/etd-05242002-011326/unrestricted/thesis.pdf*

Kyte, M and Tribelhorn, M. *Operation, Analysis and Design of Signalized Intersections*, A module for the Introductory Course in Transportation Engineering, University of Idaho, USA, 2014.

Lam, HK, Morrall, JF and Ho, H. Pedestrian Flow Characteristics in Hong Kong, *Transportation Research Record* 1487: 56–62, 1995.

Lam, WHK and Cheung, CY. Pedestrian speed/flow relationships for walking facilities in Hong Kong, *Journal of Transportation Engineering* 126(3), 2000.

Lamm, R and Choueiri, E. Recommendations for evaluating horizontal design consistency based on investigations in the state of New York. *Transportation Research Record* 1122: 68–78, Transportation Research Board (TRB), Washington DC, USA, 1987.

Lamm, R, Choueiri, E and Mailaender, T. Side friction demanded versus side friction assumed for curve design on two-lane rural highways. *Transportation Research Record* 1303: 11–21, Transportation Research Board (TRB), Washington DC, USA, 1991.

Lamm, R, Choueiri, E, Hayward, J and Paluri, A. Possible design procedures to promote design consistency in highway geometric design on two-lane rural roads. *Transportation Research Record* 1195: 111–122, Transportation Research Board (TRB), Washington DC, USA, 1988.

Lamm, R, Hayward, JC and Cargin, JG. Comparison of different procedures for evaluating speed consistencies. *Transportation Research Record* 1100: 10–20, Transportation Research Board (TRB), Washington DC, USA, 1986.

LD. *Public Lighting Design Manual*, Lighting Division, Highways Department, Hong Kong, 2006. *www.oshc.org.hk/others/bookshelf/WB112003E.pdf*

Lee, Yu-Chen. *Pedestrian Walking and Choice Behavior on Stairways and Escalators in Public Transport Facilities*, Master of Science Thesis, Department of Transport & Planning, Faculty of Civil Engineering and Geosciences, Delft University of Technology, The Netherlands, 2005.

Lefloch D. *Real-Time People Counting System Using Video Camera*, Thesis - Master of Computer Science, Department of Computer Science and Media Technology Gjøvik University College, Gjøvik, Norway, 2007.

Leisch, J. *Designing Operational Flexibility into Urban Freeways*, 33rd Annual Meeting, Institute of Transportation Engineers (ITE), Toronto, August 1963.

Leisch, JE and Leisch, JP. New concepts in design speed application. *Transportation Research Record* 631: 4–14, Transportation Research Board (TRB), Washington DC, USA, 1977.

Leong, HJW. Distribution and trend of free speed on two-lane two way rural highway in New South Wales, *Proceedings of the 4th ARRB Conference, Part 1*, Australian Road Research Board (ARRB), Vermont South, Australia, pp. 791–814, 1968.

Lieberman, EB. Model for calculating safe passing distance on two-lane rural roads, *Transportation Research Record (*869), Transportation Research Board (TRB), Washington DC, USA, 1982.

Liggett, L, Burris, M and Turner, S. Analysis of Seasonal and Day-of-Week Traffic Patterns at National Parks, *Journal of Transportation Research Board*, TRR No. 2119, Washington DC, USA, 2009.

Lipičnik, M. New form of road/railway transition curve. *ASCE Journal of Transportation Engineering* 124(6): 546–556, 1998.

Little, JDC. A proof of the queuing formula: L = λW, *Operations Research* 9: 383–387, 1961.

Lofgren, MJ. *Handbook for the Accident Reconstructionist*, Institute of Police Traffic Management, University of North Florida, USA, 1983.

Lomax, T, Turner, S, Shunk, G, Levinson, HS, Pratt, RH, Bay, PN and Douglas, GB. *Quantifying Congestion: User's Guide, NCHRP Report 398: Volume II*, Transportation Research Board (TRB), Washington DC, USA, November 1997.

Losh, AR. *Transactions of the American Society of Civil Engineers (ASCE)*, 1932.

Mahmassani, HS, Hu, T and Jayakrishnan, R. Network Performance Under System Optimal and User Equilibrium Dynamic Assignments: Implications for Advanced Traveller Information Systems, *Transportation Research Record* 1408: 83–93, Transportation Research Board (TRB), National Research Council, Washington DC, USA, 1993.

Maria, G and Braaksma, JP. Application of the Time-Space Concept in Analyzing Metro Station Platforms, *ITE Journal*, pp. 33–37, May 1986.

Martin, PT and Bell, MC. Vehicle Tracking Through Un-sampled Registration Plate Observation, *Traffic Engineering and Control* 34 (1), London, January 1993.

Mauricio, IC, Santos, RC, Regidor, JRF and Tiglao, NCC. Travel Time and Delay Analysis Using GIS and GPS, *Proceedings of the Eastern Asia Society for Transportation Studies* 4: 688– 700, October 2003.

MDOT. *Roundabout Design Guideline*, Maryland Department of Transportation (MDOT), USA, 1995.

Medina, AMF and Tarko, APT. Speed changes in the vicinity of horizontal curves on two-lane rural roads. *ASCE Journal of Transportation Engineering* 133(4): 215–222, 2007.

Michaels, RM and Cozan, LW. Perceptual and Field Factors Causing Lateral Displacement. *Highway Research Record* 25: 1–13, 1963.

Miranda, EAS and Carrasco, JGY. *Pedestrian Volumes Studies: A case study in the city of Gothenburg, Sweden*, Master of Science Thesis in the Master's Programme Geo and Water Engineering, Department of Civil and Environmental Engineering, Chalmers University of Technology, Göteborg, Sweden, 2011.

MnDOT. *MnDOT Roadway Lighting Design Manual*, Minnesota Department of Transportation (MnDOT), Office of Traffic, Security and Technology, USA, May 2010. *www.dot.state.mn.us/trafficeng/lighting/2010_Roadway%20 Lighting_Design_Manual2.pdf*

MnDOT. *MnDOT Traffic Signal Timing and Coordination Manual*, Minnesota Department of Transportation (MnDOT), USA, May 2013.

MoDOT. Missouri's Experience with a Diverging Diamond Interchange – Lessons Learned, *Report No. OR 10-021*, Missouri Department of Transportation (MoDOT), USA, 2010. *http://library.modot.mo.gov/RDT/reports/UnNumbrd/or10021.pdf*

MORTH. *Specifications for Road and Bridge Works, 5th Revision*, Indian Roads Congress on behalf of the Government of India, Ministry of Road Transport & Highways, New Delhi, India, 2013.

MOSRTH. *Manual of Specifications and Standards for four-laning of National Highways through Public Private Partnership*, Ministry of Shipping, Road Transport & Highways, Government of India, published by the Indian Roads Congress, New Delhi, India, 2008.

MOTSAM. *Manual of Traffic Signs and Markings, Part-1: Traffic Signs*, NZ Transport Agency, New Zealand, 2010. *www.nzta.govt.nz/resources/motsam/part-1/motsam-1.html*

Mulinazzi, TE. *Guidelines for the Selection of an Interchange Configuration*, Project No. C-36-68C, No. 23, Joint Highway Research Program, Purdue University and Indiana State Highway Commission, USA, September 1973.

MUTCD. *Manual on Uniform Traffic Control Devices*, Federal Highway Administration (FHWA), US Department of Transportation (USDOT), Washington DC, USA, December 2009. *http://mutcd.fhwa.dot.gov/pdfs/2009r1r2/mutcd2009r1r2edition.pdf*

MUTCD. *Manual on Uniform Traffic Control Devices for Streets and Highways (MUTCD)*, US Department of Transportation (USDOR), Federal Highway Administration (FHWA), USA, 2012. *http://mutcd.fhwa.dot.gov/pdfs/2009r1r2/mutcd2009r1r2edition.pdf*

MUTCD. *Manual of Uniform Traffic Control Devices, Part 1: General Introduction and Sign Illustrations*, Department of Transport and Main Roads, Queensland Government, Australia, 2003. *www.tmr.qld.gov.au/business-industry/Technical-standards-publications/Manual-of-uniform-traffic-control-devices.aspx*

MWT. *Traffic Data Collection and Analysis*, Roads Department, Ministry of Work and Transport (MWT), Gaborone, Botswana, February 2004.

NASRAL, "NRA Geometric Design Guidelines", compiled by CSIR on behalf of the South African National Road Agency, South African National Roads Agency Limited. *www.nra.co.za/content/GDGGeometricDesignGuide.pdf*

NCHRP 687. *Guidelines for Ramp and Interchange Spacing*. Ray BL, Schoen J, Jenior P, Knudsen J, Porter R, Leisch JP, Mason J, Roess R, and Traffic Research & Analysis, Inc., Transportation Research Board (TRB), Washington DC, USA, 2011. *http://onlinepubs.trb.org/onlinepubs/nchrp/nchrp_rpt_687.pdf*

NCHRP. *Modern Roundabout Practice in the United States, Synthesis of Highway Practice 264*, National Cooperative Highway Research Program, Transportation Research Board (TRB), Washington DC, USA, 1998. *http://onlinepubs.trb.org/onlinepubs/nchrp/nchrp_syn_264.pdf*

NCHRP. *Roundabouts: An Informational Guide, NCHRP Report No. 672, 2nd Ed.*, National Cooperative Highway Research Program, Transportation Research Board (TRB), Washington DC, USA, 2010. *http://onlinepubs.trb.org/onlinepubs/nchrp/nchrp_rpt_672.pdf*

NCUT. *Determining Travel Time, Procedure Manual 3B*, National Committee on Urban Transportation (NCUT), USA, 1958.

NDOR. Chapter Four: Intersections, Driveways and Channelization. In: *Roadway Design Manual*, Nebraska Department of Roads, USA, April 2012. *www.transportation.nebraska.gov/traffeng/mutcd.htm*

NDOR. *Roadway Design Manual*, Nebraska Department of Roads, USA, 2015. *http://www.transportation.nebraska.gov/roadway-design/rw-design-man-chapters.htm*

Neuman, TR. *Intersection Channelization Design Guide, NCHRP Report 279*, National Cooperative Highway Research Program, Transportation Research Board (TRB), Washington DC, USA, 1985.

Neuman, TR and Glennon, JC. Cost-effectiveness of improvements to stopping sight distance. *Transportation Research Record* 923: 26–34. Transportation Research Board (TRB), Washington DC, USA, 1984.

Newell, GF. Approximation methods for queues with application to the fixed cycle traffic light, *SIAM Review* 7 (2): 223–240, 1965.

Nexco. *Annual Report 2008*, Central Nippon Expressway Company Limited, Nagoya, Japan, 2008.

Nexco. *Annual Report 2010*, Central Nippon Expressway Company Limited, Nagoya, Japan, 2010.

Nexco. *Annual Report 2014*, Central Nippon Expressway Company Limited, Nagoya, Japan, 2014.

NIATT. *Signal Timing Tutorial*, The National Institute for Advanced Transportation Technology, University of Idaho, USA, 2015. *http://www.webpages.uidaho.edu/niattproject*

Nicholson, A. Super elevation, side friction and roadway consistency. *ASCE Journal of Transportation Engineering* 124 (5): 411–418, 1998.

NJDOT. *Roadway Design Manual-Highway Lighting System, Section 11*, RDM, New Jersey Department of Transportation (NJDOT), USA, 2015. *www.state.nj.us/transportation/eng/documents/RDM/sec11.shtm*

Nordoff, LS. *Motor Vehicle Collision Injuries: Biomechanics, Diagnosis, and Management*, Jones and Bartless Publishers, Inc., 2005.

Nowacki, G. Development and Standardization of Intelligent Transport Systems, *International Journal of Marine Navigation and Safety of Sea Transportation* 6(3), September 2012.

NTOC, *National Traffic Signal Report Card*, National Transportation Operations Coalition, 2005.

Nyerges, TL and Cihon, RF. *Automated Collision Diagram Production*, WA-RD 180.1, Final report, Washington State Transport Center (TRAC), USA,1989.

NZTA. *NZTA TCD Manual: Part 1 General requirements for traffic signs,* Consultation draft, NZ Transport Agency, New Zealand, December 2008. *www.nzta.govt.nz/assets/resources/traffic-control-devices-manual/docs/draft-tcd-traffic-signs.pdf*

NZTA. *Section 1: Signs Introduction.* NZ Transport Agency, Waka Kotahi, New Zealand, August, 2010. *www.nzta.govt.nz/assets/resources/motsam/part-1/docs/motsam-1-section-01.pdf*

O'Flaherty, CA and Parkinson, MH. Movement on a City Centre Footway, *Traffic Engineering and Control* 13: 434–438, 1972.

ODOT. *Traffic Control Plans Design Manual*, Oregon Department of Transportation (ODOT), Technical Services Traffic-Roadway Section Traffic Control Plans Unit, USA, May 1, 2016. *www.oregon.gov/ODOT/HWY/TRAFFIC-ROADWAY/docs/pdf/tcp_manual/TCP_DM_Edition_12.pdf*

ODOT. *1100 Highway Lighting, Traffic Engineering Manual (TEM)*, 11-107, Ohio Department of Transportation (ODOT), October 23, 2002, Revised July 18, 2014. *www.dot.state.oh.us/Divisions/Engineering/Roadway/DesignStandards/traffic/TEM/Documents/Part_11_Complete_011516Revision_122215_bookmarked.pdf*

ODOT. *Bicycle and Pedestrian Design Guide*, Oregon Highway Design Manual, Appendix-L, Oregon Department of Transportation (ODOT), USA, 2011.

ODOT. *Highway Design Manual*, Oregon Department of Transportation (ODOT), USA, 2012.

ODOT. *Modern Roundabouts: A Safer Choice*, a PPT presented in ODOT Traffic Engineering Conference, Wilsonville, Oregon, USA, June 2011.

Oh, J, Washington, S and Choi, K. Development of Accident Prediction Models for Rural Highway Intersections, *Transportation Research Record* 1897: 18–27, Transportation Research Board (TRB), National Research Council, Washington DC, USA, 2004.

Olszewski, P. Overall Delay, Stopped Delay and Stops at Signalized Intersections, *Journal of Transportation Engineering* 119(6): 835–852, November/December 1993.

Ottesen, J and Krammes, R. Speed profile model for a design consistency evaluation procedure in the United States. *Transportation Research Record* 1701: 76–85. Transportation Research Board (TRB), Washington DC, USA, 2000.

Pawar, DS and Patil, GR. Pedestrian temporal and spatial gap acceptance at mid-block street crossing in developing world. *Journal of Safety Research* 52: 39–46, 2015.

PIARC. *ITS Handbook*, Technical Committee on Network Operations. www.road-network-operations.piarc.org

PIARC. *Road Accident Investigation for Road Engineers*, The World Road Association PIARC Technical Committee - 3.1 "Road Safety", Working Group 1, August 2007.

Pinnell, C, Barnett, JD, Haynes, J and Radke, M. *Guide to the Selection of Non-directional Interchanges*, Cooperative Research with the Texas Highway Department and thc Department of Commerce, Bureau of Public

Roads, Final Report on Research Project 2-8-58-6: Ramps and Interchanges, Texas Transportation Institute, Texas A & M University, College Station, Texas, November 1963.

Pliszczak, R. *Street Lighting Management System-Jupiter*, Master's thesis, Faculty of Electrical Engineering, Institute of Electrical Engineering Fundamental, Wroclaw University of Technology, Poland, 2010.

Pohl, W. Trends in Lamps, *E³Light*, Annexe 45, Biannual Newsletter, May 2007.

Prisk, CW. Passing practices on rural highways. *Proceedings of Highway Research Board* 21: 366–378, Washington DC, USA, 1941.

Pushkarev, B and Zupan, J. Capacity of walkways, *Transportation Research Record* 538: 1–15, 1975.

Quiroga, C, Henk, R and Jacobson, M. Innovative Data Collection Techniques for Roadside Origin-Destination Surveys, *Transportation Research Record* 1719: 140–146, Transportation Research Board (TRB), National Research Council, Washington DC, USA, 2000.

Rajput, KY, Khatav, G, Pujari, M and Yadav, P. Intelligent Street Lighting System Using GSM, *International Journal of Engineering Science Invention*, 2 (3), March 2013. *www.ijesi.org/papers/Vol(2)3%20(Version-3)/J236069.pdf*

Rao, SG. *Probability and Statistics for Science and Engineering*, Universities Press, 2011.

Rao, SK. Unit curve for design of highway transitions. *ASCE Journal of Transportation Engineering* 121(2): 169–175, 1995.

Reed, WS. *Automobile Accident Reconstruction by Dynamic Simulation, Research Report 302-1F*, Center for Transportation Research, The University of Texas at Austin, Texas, USA, 1983.

Retting, RA, Chapline, JF and Williams, AF. Changes In Crash Risk Following Re-Timing of Traffic Signal Change Intervals, *Accident Analysis and Prevention* 34:215–20, 2002.

Rietveld, P. Some notes on Interconnectivity in Transport Networks. In: *Overcoming Isolation: Information and Transportation Networks in Development Strategies for Peripheral Regions*, Cocossis, H and Nijkamp, P (eds) Berlin, Springer, 18–28, 1995.

Roebuck, R, Isola, R, Goodrum, W, Cebon, D and Collop, A. Vehicle-Pavement Interaction (VPI) Modelling, HVTT12, *International Symposium on Heavy Vehicle Transport Technology, Setting Future Standards*, Stockholm, Sweden, September 16–19, 2012.

Rouphail, N, Hummer, J, Milazzo II, J and Allen, P. Literature Synthesis for Chapter 13: Pedestrians, *Highway Capacity Manual*, Federal Highway Administration (FHWA), Washington DC, USA, February 1998.

Roy, BK. Geometric design of compound horizontal curves. (Discussion on this section). *ASCE Journal of Transportation Engineering* 120(4): 674, 1994.

Roy, BK. Incorporating transition lengths. (Discussion on this section). *ASCE Journal of Transportation Engineering* 118 (6): 887, 1992.

Ryen (Ed.). *Overview of the System Engineering Process*, North Dakota Department of Transportation (NDDOT), USA, March 2008.

Sachdeva, SN. *Speed-Flow Relationship and Capacity Analysis for an Identified Intercity Road Network*, Ph.D. thesis, Kurukshetra University, Kurukshetra, India, 2004.

Sarkar, AK and Janardhan, KSVS. A Study on Pedestrian Flow Characteristics. In: *CD-ROM with Proceedings*, Transportation Research Board (TRB), Washington DC, USA, 1997.

Schroeder, B, Cunningham, C, Ray, B, Daleiden, A, Jenior, P, Knudsen, J, Kittelson and Associates, Inc. *Diverging Diamond Interchange Informational Guide*, FHWA-SA-14-067, Federal Highway Administration Office of Safety, USA, 2014.

Scott, M. *Traffic Modelling*, December 2011. Note from *http://home2.fvcc.edu/~dhicketh/DiffEqns/Spring11projects/Scott_Miller/Project.pdf*

SCVTA. *Traffic Level of Service Analysis Guidelines, Congestion Management Program*, Santa Clara Valley Transportation Authority (SCVTA), USA, June, 2003.

SDDOT. *SDDOT Road Design Manual: Chapter 13: Interchanges*, South Dakota Department of Transportation (SDDOT), USA, 2015. *http://sddot.com/business/design/forms/roaddesign/default.aspx*

Setra. *The design of interurban intersections on major roads-At-grade intersections*, Produced and disseminated by Service d'Etudes Techniques des Routes et Autoroutes Centre de la Sécurité et des Techniques Routières (Setra), France, 1998. *http://www.infra-transports-materiaux.cerema.fr/IMG/pdf/US_ACI-P_GB.pdf*

Shelby, SG. Delta-V as a Measure of Traffic Conflict Severity, *Proceedings of the 3rd International Conference on Road Safety and Simulation*, Indianapolis, USA, September 14–16, 2011.

Shenker, R, Chowksey, A and Sandhu, AS. Analysis of Relationship between Road Safety and Road Design Parameters of 4-lane National Highway in India, *IOSR Journal of Business and Management* 17 (5) Ver III: 60–70, May 2015.

Shuldiner, PW, D'Agostino, SA and Woodson, JB. Determining Detailed Origin-Destination and Travel Time Patterns Using Video and Machine Vision License Plate Matching, *Transportation Research Record* 1551: 8–17, Transportation Research Board (TRB), National Research Council, Washington DC, USA, 1996.

Simowski, M, Tourangeau, R, Ghadialy, R and Pedlow, S. *Non-Response in Household Travel Surveys*, US Department of Transportation (USDOT), Washington DC, USA, October 1997.

Skszek, SL. *State-of-the-Art Report on Non-Traditional Traffic Counting Methods*, *Report No. FHWA-AZ-01-503*, Arizona Department of Transportation, USA, 2001.

Son, T, Kim, SG and Lee, JK. Methodology to Calculate Sight Distance Available to Drivers at Skewed Intersections, *Transportation Research Record* 1796: 41–47, Transportation Research Board (TRB), National Research Council, Washington DC, USA, 2002.

Sparks, GA, Neudorf, RD, Robinson, JB and Good, D. Effect of vehicle length on passing operations. *ASCE Journal of Transportation Engineering* 119(2): 272–283, 1993.

Srinivasa, RK. *Textbook of Highway Engineering,* Universities Press, India, 2011.

Srinivasa, RK. *Pavement Evaluation and Maintenance Management System,* Universities Press, India, 2014.

Steyn, WJ vd M, Monismith, CL, Nokes, WA, Harvey, JT, Holland, TJ and Burmas, N. Challenges Confronting Road Freight Transport and the Use of Vehicle-Pavement Interaction Analysis in Addressing these Challenges. *Journal of the South African Institution of Civil Engineering* 54(1): 814, 14–21, 2012.

Stokes, RW and Chivala, TC. Design and Implementation of Intercity Origin-Destination Surveys, *Transportation Research Record 1236*, Transportation Research Board (TRB), National Research Council, Washington DC, USA, 1989.

Stucki, P, Gloor, C and Nagel, K. *Obstacles in Pedestrian Simulations*, Department of Computer Sciences, ETH Zurich, Switzerland, September 2003.

Szary, P, Maher, A, Strizki, M and Moini, N. *Use of LED or Other New Technology to Replace Standard Overhead and Sign Lighting (Mercury and/or Sodium)*, FHWA-NJ-2005-029, New Jersey Department of Transportation, Federal Highway Administration (FHWA), US Department of Transportation (USDOT), USA, 2005. *www.state.nj.us/transportation/refdata/research/reports/FHWA-NJ-2005-029.pdf*

Taekratok, T. *Modern Roundabouts for Oregon*, *Report No. OR-RD-98-17*, Oregon Department of Transportation (ODOT), USA, 1998. *http://www.oregon.gov/odot/hwy/engservices/docs/modernroundabouts.pdf*

Taiganidis, J and Kanellaidis, G. Method for grading design of highways. *ASCE Journal of Transportation Engineering* 117(6): 706–712, 1991.

Tamagawa, M, Miyamoto, K and Kayatama, S. Development of Vehicle Licence Plate Recognition Systems, *4th World Congress on Intelligent Transport Systems*, Berlin, Germany, October 1997.

Tanariboon, Y, Hwa, SS and Chor, CH. Pedestrian characteristics study in Singapore, *ASCE Journal of Transportation Engineering* 112(3): 229–235, 1986.

TASS. *Audio-Tactile Line Marking, Operational Instruction 2.13*, Traffic Access Standards Section (TASS), Department of Transport, Energy and Infrastructure, Government of South Australia, Australia, 2009.

Taleb, AM and Majumder, S. Impact of Flyovers in Dhaka City of Bangladesh on the Affected People in the Adjacent Area, *IACSIT International Journal of Engineering and Technology* 4(1), February 2012. *www.ijetch.org/papers/327-TLE16.pdf*

Teodorovic, D and Trani, AA. *Applications of Queueing Theory to Intersection Analysis Level of Service*, Civil and Environmental Engineering Virginia Polytechnic Institute and State University Blacksburg, Virginia Spring, USA, 2005. *http://128.173.204.63/courses/cee3604/cee3604_pub/Intersection_Analysis.pdf*

The Highways Agency. *Traffic Surveys by Roadside Interview*, The Scottish Office Development Department, The Welsh Office and The Department of the Environment for Northern Ireland, United Kingdom, 2005. *www.archive2.officialdocuments.co.uk/document/deps/ha/dmrb/vol5/section1/ta118 1a.pdf*

Tian, ZZ, Xu, F, Rodegerdts, LA, Scarbrough, WE, Ray, BL, Bishop, WE, Ferrara, TC and Mam, S. *Roundabout Geometric Design Guidance, Report No. F/CA/RI-2006/13*. Division of Research and Innovation, California Department of Transportation, USA, June 2007.

TNF. *Fact Sheet: Using Excel for Inferential Statistics.* Advance Applied Science: GCE A2 Units, The Nuffield Foundation, p.4/7, 2008. *www.nuffieldfoundation.org/sites/default/files/excel_inferential_stats.pdf*

Todd, KB and Kulakowski, BT. Simple Computer Models for Predicting Ride Quality and Pavement Loading for Heavy Trucks: In *Pavement Management and Rehabilitation. Transportation Research Record* 1215: 137–150, Transportation Research Board (TRB), Washington DC, USA, 1989.

TranSafety Inc. Making Intersections Safer for Pedestrians, *Road Management and Engineering Journal*, April 1997.

TRB. *Traffic and Transportation Simulation,* Transportation Research Circular Number E-C195, Transportation Research Board (TRB), Washington DC, USA, 2015.

TRB. *Highway Capacity Manual, Special Report No. 209*, National Academy of Sciences, Washington, USA, 2000.

TRB. *Special Report 209: Highway Capacity Manual, 3rd ed.*, Transportation Research Board (TRB), National Research Council, Washington DC, USA, 1998.

TRB. *Traffic and Transportation Simulation*. Transportation Research Circular, Transportation Research Board (TRB), Washington DC, USA, 2015.

Treadway, TB. *An Analysis of Travel Speed and Delay on High-Volume Highway*, Publication FHWA/IN/JHRP-65/10. Joint Highway Research Project, Indiana Department of Transportation and Purdue University, Indiana, USA, doi: 10.5703/1288284313674, 1965.

Treat, JR, Tumbas, NS, McDonald, ST, Shinar, D, Hume, RD, Mayer, RE, Stanisfer, RL and Castellan, NJ. *Tri-level Study of the Causes of Traffic Accidents, Report No. DOT-HS-034-3-535-77 (TAC)*, 1977.

TRL. *The Use of Traffic Signals in Developing Cities, Overseas Road Note 13*, Overseas Centre, Transport Research Laboratory (TRL), United Kingdom, 1996.

TRL. *Urban Road traffic Surveys, Overseas Road Note 11*, Transport Research Laboratory (TRL), United Kingdom, 1993.

TRRL. *A Guide to Geometric Design*. Overseas Road Note No. 6. Transport and Road Research Laboratory (TRRL), United Kingdom, 1988.

Tsapakis, I. *Determination of Seasonal Adjustment Factors and Assignment of Short-Term Counts to Factor Groupings*, Ph.D. Thesis, The Graduate Faculty of The University of Akron, USA, 2009.

Turner, SM, Eisele, WL, Benz, RJ and Holdener, DJ. *Travel Time Data Collection Handbook, Report No. FHWA-PL-98-035*, Federal Highway Administration (FHWA) and Texas Transportation Institute, USA, March 1998.

Turvey, IG, May, AD and Hopkinson, PG. *Counting Methods and Sampling Strategies Determining Pedestrian Numbers, Working Paper 242*, Institute of Transport Studies, University of Leeds, United Kingdom, 1987.

TxDOT. *Traffic Signals Manual*, Texas Department of Transportation (TxDOT), USA, April 01, 2011. *http://onlinemanuals.dot.state.tx.us/txdotmanuals/tff/condition_diagram.htm*

UNAM. *Terminologia, Iluminacion Exterior: Principios, Disero Y Aplicaciones,* Cursos Abiertos, Division de Educacion Continua, Facultad De Ingenieria UNAM, Palacio de Mineria, Mexico, 1992.

Underwood, RT. *Speed, Volume and Density Relationships - Quality and Theory of Traffic Flow*, Yale University 395 Report, Connecticut, USA, 1961.

UNECE. *Consolidated Resolution of Road Signs and Signals*, ECE/TRANS/212, United Nations Economic Commission for Europe, Inland Transport Committee, 2010.

USAB. *Proposed Accessible Guidelines for Pedestrian Facilities in the Public Right-of-Way*, United States Access Board, Washington DC, USA July 26, 2011. *http://www.access-board.gov/attachments/article/743/nprm.pdf*

USNAVY. *Construction Electrician Intermediate*, Non-resident Training Course, NAVEDTRA, 14027, Naval Education and Training Professional Development and Technology Center, USA, 1998. *www.maritime.org/doc/pdf/constelect2.pdf*

UTTIPEC. *Street Design Guidelines*, Unified Traffic and Transportation Infrastructure (Planning & Engineering) Centre, Delhi Development Authority, New Delhi, India, 2009. *http://uttipec.nic.in/StreetGuidelines-R1-Feb2011-UTTPEC-DDA.pdf*

Vandaele, N, Woensel, TV and Verbruggen, A. A Queueing Based Traffic Flow Model, *Transportation Research-D: Transport and Environment* 5(2): 121–135, January 2000.

Virkler, MR and Elayadath, S. Pedestrian speed-flow-density relationships, *Transportation Research Record* 1438: 51–58, 1994.

Walker, RJ. Coordination of Basic Design Elements: Overview, *Transportation Research Record* 1385: 51–59, Transportation Research Board (TRB), National Research Council, Washington DC, USA, 1981.

Walker, WP. *Speed and Travel Time Measurement in Urban Areas*, *Bulletin 156*, Highway Research Board, pp. 27–44, 1957.

Wang, N and Teng, H. VMT Estimation Associated with ITS Data and Maintenance of Loop Detectors, *Research Report UVACTS-15-0-88*, University of Virginia, USA, August 2004.

Wang, T. *Improved Annual Average Daily Traffic (AADT) Estimation for Local Roads using Parcel-Level Travel Demand Modelling*, Ph.D. Thesis, Florida International University, USA, 2012.

Wang, Y and Cartmell, MP. New model for passing sight distance on two-lane highways. *ASCE Journal of Transportation Engineering* 124(6): 536–545, 1998.

Washburn, S and Nihan, N. Using Voice Recognition to Collect Licence Plate Data for Travel Time Studies, *Transportation Research Record* 1593: 41–46, Transportation Research Board (TRB), Washington DC, USA, 1997.

Webster, FV. *Traffic Signal Settings*, *Road Research Technical Paper No. 39*, Road Research Laboratory, Her Majesty's Stationery Office, Berkshire, United Kingdom, 1958.

Webster, FV and Cobbe, BM. *Traffic signals*, *Road Research Technical Paper No: 56,* Her Majesty's Stationery Office, London, United Kingdom, 1966.

Weidmann, U. *Transporttechnik der Fussgaenger*, ETH Zuerich, Schriftenreihe IVT-Berichte 90, Zurich, Switzerland, 1993.

Wen, T. *Traffic Volumes Monthly Report*, NZ Transport Agency, Waka Kotahi, New Zealand, February 2013.

WiDOJ. *Traffic Crash Investigations, A Training Guide for Law Enforcement Officers*, Wisconsin Department of Justice (WiDOJ), Law Enforcement Standards Board, USA, 2010.

Woodson, WE and Barry, T. *Human Factors Design Handbook. Information and Guidelines for Design of Systems, Facilities, Equipment and Products for Human Use*, McGraw Hill, New York, p. 93, 1992.

WSDOT. *Automated Video Based Survey of Travel Times in HOV vs. General-Purpose Lanes*, Washington State Department of Transportation (WSDOT), Olympia, USA, October 1995.

WSDOT. *Field Guide for Accessible Public Rights of Way*, Washington State Department of Transportation (WSDOT), USA, 2012. *www.wsdot.wa.gov/publications/fulltext/Roadside/ADA_Field_Guide.pdf*

WSDOT. *High Speed Ground Transportation Study Final Report*, Washington State Department of Transportation (WSDOT), USA, October 1992.

WSDOT. *Pedestrian Facilities Guidebook*, Washington State Department of Transportation (WSDOT), USA, 1997.

WSDOT. *WSDOT Design Manual: Chapter 1360: Interchanges*, Washington State Department of Transportation (WSDOT), USA, M 22-01.11, July 2014. *www.wsdot.wa.gov/publications/manuals/fulltext/M22-01/1360.pdf*

Yang, S and Yang, X. The Application of the Queuing Theory in the Traffic Flow of Intersection, *International Journal of Mathematical, Computational, Physical, Electrical and Computer Engineering* 8 (6): 986–989, 2014.

Yingshi, Niu, *Operations Research (M)*, Shanxi: the press of the Xian Jiaotong University, China, 2006.

Zarling, S. *Email communication with Ms. Sue Zarling*, Minnesota Department of Transportation (MnDOT), OTST, Traffic Electrical Systems Engineer, USA, 2016.

Zegeer, CV, Seiderman, C, Lagerwey, P, Cynecki, M, Ronkin, M and Schneider, R. *Pedestrian Facilities Users Guide - Providing Safety and Mobility*, FHWA-RD-01-102, Federal Highway Administration (FHWA), US Department of Transportation (USDOT), USA, March 2002.

Zehntner. *Basic Principles of Retroreflection $R_L$ (Night Visibility) of Road Markings Caused by Glass Beads Version 1.6*, Zehntner GmbH Testing Instruments, Switzerland, December 2012. *www.zehntner.com/download/prospekt_retroreflektion_glasperlen_e.pdf; www.zehntner.com/download/prospekt_intro_road_d_e.pdf*

Zukerman, M. *Introduction to Queuing Theory and Stochastic Teletraffic Models, Class Notes*, EE Department, City University of Hong Kong. *www.ee.cityu.edu.hk/~zukerman/classnotes.pdf*

APPENDIX

# Roadway Lighting Evaluation Forms

**Table A.1**

| *Evaluation Form for Interchange Lighting* | | | | | | | | | *Lighting Form 2* |
|---|---|---|---|---|---|---|---|---|---|
| | *Rating* | | | | | | | | |
| *Classification Factor* | *1* | *2* | *3* | *4* | *5* | *Unlit Weight (A)* | *Lighted Weight (B)* | *Diff (A-B)* | *Rating X(A-B)* |
| Geometric Factors | | | | | | | | | |
| Ramp Types | Direct | Diamond | Button Hook Clover Leaf | Trumpet | Scissor & Leftside | 2.0 | 1.0 | 1.0 | |
| Cross Road Channelization | None | | Continuous | | At interchange intersection | 2.0 | 1.0 | 1.0 | |
| Frontage Roads | None | | One Way | | Two Way | 1.5 | 1.0 | 0.5 | |
| Freeway Lane Width | >12′ | 12′ | 11′ | 10′ | <10′ | 3.0 | 2.5 | 0.5 | |
| Freeway Median Width | >40′ | 34′–40′ | 12′–23′ | 4′–11′ | <4′ | 1.0 | 0.5 | 0.5 | |
| Number of Freeway Lanes | ≤4 | | 6 | | ≤8 | 1.0 | 0.8 | 0.2 | |
| Main Lane Curves | $<\frac{1}{2}^{\circ}$ | 1–2° | 2–3° | 3–4° | >4° | 13.0 | 5.0 | 8.0 | |
| Grades | <3% | 3.0–3.9% | 4.0–4.9% | 5.0–6.9% | ≥7% | 3.2 | 2.8 | 0.4 | |
| Sight Distance Cross Road Intersection | >1,000 | 700′ –1,000′ | 500′–700′ | 400′ –500′ | <400′ | 2.0 | 1.8 | 0.2 | |
| | | | | | | Geometric Total | | | |
| Operational Factors | | | | | | | | | |
| Level Of Service (Any Dark Hour) | A | B | C | D | E | 6.0 | 1.0 | 5.0 | |

(Continued)

**Table A.1** (Continued)

| *Evaluation Form for Interchange Lighting* | | | | | | | | | *Lighting Form 2* |
|---|---|---|---|---|---|---|---|---|---|
| | *Rating* | | | | | | | | |
| *Classification Factor* | *1* | *2* | *3* | *4* | *5* | *Unlit Weight (A)* | *Lighted Weight (B)* | *Diff (A-B)* | *Rating X(A-B)* |
| | | | | | | Operational Total | | | |
| Environmental | | | | | | | | | |
| % Development | None | 1 QUAD | 2 Quad | 3 Quad | 4 Quad | 2.0 | 0.5 | 1.5 | |
| Set Back Distance | >200′ | 150′ –200′ | 100′ –150′ | 50′ –100′ | <50′ | 0.5 | 0.3 | 0.2 | |
| Cross Road Approach Lighting | None | | Partial | | Complete | 3.0 | 2.0 | 1.0 | |
| Freeway Lighting | None | | Interchange Only | | Continuous* | 5.0 | 3.0 | 2.0 | |
| | | | | | | Environmental Total | | | |
| Accidents | | | | | | | | | |
| Ratio Of Night-To-Day Accidents | <1.0 | 1.0–1.2 | 1.2–1.5 | 1.5–2.0 | >2.0* | 10.0 | 2.0 | 8.0 | |
| | | | | | | Accident Total | | | |
| * Complete Lighting Warranted | Geometric Total<br>Operational Total<br>Environmental Total<br>Accident Total<br><br>Sum<br>Warranting Condition<br>Partial Lightning<br>Warranting Condition | | | | = __________ *<br>= __________ *<br>= __________ *<br>= __________ *<br><br>= Points<br><br>= 90 Points<br>= 60 Points | | | | |

*Source*: NJDOT, Roadway Design Manual-Highway Lighting System, Section 11, RDM, NJDOT, 2015.1 www.state.ni.us/transportation/eng/documents/RDM/secl 1 .shtml

(*Source*: NJDOT, 2015; *Courtesy*. The State of New Jersey Department of Transportation, USA)

**Table A.2**

| *Evaluation Form for Controlled Access Facility (New Roadway/Alignment)* | | | | | | | | | *Lighting Form 3* |
|---|---|---|---|---|---|---|---|---|---|
| | *Rating* | | | | | | | | |
| *Classification Factor* | *1* | *2* | *3* | *4* | *5* | *Unlit Weight (A)* | *Lighted Weight (B)* | *Diff (A-B)* | *Rating X(A-B)* |
| Geometric Factors | | | | | | | | | |
| Number Of Lanes | 4 | | 6 | | ≥8 | 1.0 | 0.8 | 0.2 | |
| Lane Width | >12′ | 12′ | 11′ | 10′ | ≤.9 | 3.0 | 2.5 | 0.5 | |
| Median Width | >40′ | 24′–39′ | 12′–23′ | 4–11′ | 0–3′ | 1.0 | 0.5 | 0.5 | |
| Shoulders | 10′ | 8′ | 6′ | 4′ | 0 | 1.0 | 0.5 | 0.5 | |
| Slopes | >8:1 | 6:1 | 4:1 | 3:1 | 2:1 | 1.0 | 0.5 | 0.5 | |
| Curves | $0-\frac{1}{2}°$ | $\frac{1}{2}-1°$ | 1–2° | 2–3° | 3–4° | 13.0 | 5.0 | 8.0 | |
| Grades | <3% | 3.0–3.9% | 4.0– 4.9% | 5.0– 6.9% | >7% | 3.2 | 2.8 | 0.4 | |
| Interchange Frequence | 21,000′ | 16,000′ | 10,500′ | 5,000′ | <5,000′ | 4.0 | 1.0 | 3.0 | |
| | | | | | | Geometric Total | | | |
| Operaiional Factors | | | | | | | | | |
| Level Of Service (Any Dark Hour) | A | B | C | D | E | 6.0 | 1.0 | 5.0 | |
| | | | | | | Operational Total | | | |
| Environmental | | | | | | | | | |
| % Development | 0% | 25% | 50% | 75% | 100% | 3.5 | 0.5 | 3.0 | |
| Offset To Development | 200′ | 150′ | 100′ | 50′ | <50′ | 3.5 | 0.5 | 3.0 | |
| | | | | | | Environmental Total | | | |
| | Geometric Total<br>Operational Total<br>Environmental Total<br>Accident Total<br><br>Sum<br><br>Warranting Condition | | | | = ________*<br>= ________*<br>= ________*<br><br><br>= ____ Points<br><br>= 70 Points | | | | |

(*Source*: NJDOT. 2015: *Courtesy*: The State of New Jersey Department of Transportation, USA)

**Table A.3**

| *Evaluation Form for Interchange Lighting (New Alignment)* | | | | | | | | | *Lighting Form 4* |
|---|---|---|---|---|---|---|---|---|---|
| | *Rating* | | | | | | | | |
| *Classification Factor* | *1* | *2* | *3* | *4* | *5* | *Unlit Weight (A)* | *Lighted Weight (B)* | *Diff (A-B)* | *Rating X(A-B)* |
| Geometric Factors | | | | | | | | | |
| Ramp Types | Direct | Diamond | Button Hook Cloverleaf | Trumpet | Scissor & Leftside | 2.0 | 1.0 | 1.0 | |
| Cross Road Channelization | None | | Continuous | | At Interchange Intersection | 2.0 | 1.0 | 1.0 | |
| Frontage Roads | None | | One way | | Two way | 1.5 | 1.0 | 0.5 | |
| Freeway Lane Width | >12′ | 12′ | 11′ | 10′ | <10′ | 3.0 | 2.5 | 0.5 | |
| Freeway Median Width | >40′ | 34′–40′ | 12′–23′ | 4′–11′ | <4′ | 1.0 | 0.5 | 0.5 | |
| Number Of Freeway Lanes | ≤4 | | 6 | | ≥8 | 1.0 | 0.8 | 0.2 | |
| Main Lane Curves | $<\frac{1}{2}^{\circ}$ | 1–2° | 2–3° | 3–4° | >4° | 13.0 | 5.0 | 8.0 | |
| Grades | <3% | 3.0–3.9% | 4.0–4.9°/o | 5.0– 6.9% | ≥7% | 3.2 | 2.8 | 0.4 | |
| Light Distance Cross Road Intersection | >1,000′ | 700′–1,000′ | 500′–700′ | 400′–500′ | <400′ | 2.0 | 1.8 | 0.2 | |
| | | | | | | Geometric Total | | | |
| Operational Factors | | | | | | | | | |
| Level Of Service (Any Dark Hour) | A | B | C | D | E | 6.0 | 1.0 | Level of service (Any dark hour) | A |
| | | | | | | Operational Total | | | |
| Environmental | | | | | | | | | |
| % Development | None | 1 Quad | 2 Quad | 3 Quad | 4 Quad | 2.0 | 0.5 | 1.5 | |
| Set Back Distance | >200′ | 150′–200′ | 100′–150′ | 50′–100′ | <50′ | 0.5 | 0.3 | 0.2 | |
| Cross Road Approach Lighting | None | ---- | Partial | ---- | Complete | 3.0 | 2.0 | 1.0 | |

(Continued)

**Table A.3** (Continued)

| *Evaluation Form for Interchange Lighting (New Alignment)* | | | | | | | | | *Lighting Form 4* |
|---|---|---|---|---|---|---|---|---|---|
| | *Rating* | | | | | | | | |
| *Classification Factor* | *1* | *2* | *3* | *4* | *5* | *Unlit Weight (A)* | *Lighted Weight (B)* | *Diff (A-B)* | *Rating X(A-B)* |
| Freeway Lighting | None | | Interchange Only | | Continuous* | 5.0 | 3.0 | 2.0 | |
| | | | | | | Environmental Total | | | |
| * Complete Lighting Warranted | Geometric Total<br>Operational Total<br>Environmental Total<br><br>Sum<br>Complete Lighting Warranting Condition<br>Partial Lighting<br>Warranting Condition | | | | | = ________*<br>= ________*<br>= ________*<br><br>= Points<br>= 65 Points<br><br>= 35 Points | | | |

(*Source*: NJDOT. 2015: *Courtesy*. The State of New Jersey Department of Transportation, USA)

**Table A.4**

| *Evaluation Form for Non-Controlled Access Facility Lighting* | | | | | | | | | *Lighting Form 5* |
|---|---|---|---|---|---|---|---|---|---|
| | *Rating* | | | | | | | | |
| *Classification Factor* | *1* | *2* | *3* | *4* | *5* | *Unlit Weight (A)* | *Lighted Weight (B)* | *Diff (A-B)* | *Rating X(A-B)* |
| Geometric | | | | | | | | | |
| Number Of Lanes | ≤4 | - | 6 | - | ≥8 | 1.0 | 0.8 | 0.2 | |
| Lane Width | >12′ | 12′ | 11′ | 10′ | <10′ | 3.0 | 2.5 | 0.5 | |
| Medlan Opening Per Mile | <4 or one way | 4.0–8.0 | 8.1–12.0 | 12.1–15.0 | >15 Or No Access | 5.0 | 3.0 | 2.0 | |
| Curb Cuts | <10% | 10–20% | 20–30% | 30–40% | >40% | 5.0 | 3.0 | 2.0 | |
| Curves | <3.0% | 3.1–6.0° | 6.1–8.0° | 8.1–10.0° | >10° | 13.0 | 5.0 | 8.0 | |
| Grades | <3% | 3.0–3.9% | 4.0–4.9% | 5.0–6.9% | >7.0% | 3.2 | 2.8 | 0.4 | |
| Sight Distance | >700′ | 500′–700′ | 300′–500′ | 200′–300′ | <200′ | 2.0 | 1.8 | 0.2 | |
| Parking | Prohibited Both Sides | Loading Zone Only | Off Peak Only | Permitted One Side | Permitted Both Sides | 0.2 | 0.1 | 0.1 | |
| | | | | | | Geometric Total | | | |

(Continued)

**Table A.4** (Continued)

| *Evaluation Form for Non-Controlled Access Facility Lighting* | | | | | | | | | *Lighting Form 5* |
|---|---|---|---|---|---|---|---|---|---|
| | *Rating* | | | | | | | | |
| *Classification Factor* | *1* | *2* | *3* | *4* | *5* | *Unlit Weight (A)* | *Lighted Weight (B)* | *Diff (A-B)* | *Rating X(A-B)* |
| Operational | | | | | | | | | |
| Signals | All Major Intersections | Majority Of Intersections | Most Major Intersections | $\frac{1}{2}$ The Intersections | Frequent Nonsignalized Intersections | 3.0 | 2.8 | 0.2 | |
| Left Turn Lane | All Major Intersections Or One Way | Majority Of Intersections | Most Major Intersections | $\frac{1}{2}$ Major Intersections | Infrequent Or Undivided Street | 5.0 | 4.0 | 1.0 | |
| Median Width | 30′ | 20′–30′ | 10′–20′ | 4′–10′ | 0–4′ | 1.0 | 0.5 | 0.5 | |
| Speed Limit | <25 | 30 | 35 | 40 | >45 | 1.0 | 0.2 | 0.8 | |
| Night Pedestrian Traffic (Peds/Km) | Few Or None | 0–50 | 50–100 | 100–200 | >200 | 1.5 | 0.5 | 1.0 | |
| | | | | | | Operational Total | | | |
| Environmental | | | | | | | | | |
| %Development | 0% | 0–30% | 30–60% | 60–90% | 100% | 0.5 | 0.3 | 0.2 | |
| Type Of Development | Undeveloped | Residential | $\frac{1}{2}$ Residential And/Or Commercial | Industrial Or Commercial | Strip Industry Or Commercial | 0.5 | 0.3 | 0.2 | |
| Set Back Distance | >200′ | 150′–200′ | 100′–150′ | 50′–100′ | <50′ | 0.5 | 0.3 | 0.2 | |
| Advertising/ Area Lighting | None | 0–40% | 41–60% | 61–80% | Continuous | 3.0 | 1.0 | 2.0 | |
| Raised Curb Median | None | Continuous | At All Intersections | All Signallized Intersections | Few Locations | 1.0 | 0.5 | 0.5 | |
| Crime Rate | Extremely Low | Lower Than City Average | City Average | Higher Than City Average | Extremely High | 1.0 | 0.5 | 0.5 | |
| | | | | | | Environmental Total | | | |

(Continued)

**Table A.4** (Continued)

| *Evaluation Form for Non-Controlled Access Facility Lighting* | | | | | | | | | *Lighting Form 5* |
|---|---|---|---|---|---|---|---|---|---|
| | *Rating* | | | | | | | | |
| *Classification Factor* | *1* | *2* | *3* | *4* | *5* | *Unlit Weight (A)* | *Lighted Weight (B)* | *Diff (A-B)* | *Rating X(A-B)* |
| Accidents | | | | | | | | | |
| Ratio Of Night- To-Day Accidents | <1.0 | 1.0–1.2 | 1.2–1.5 | 1.5–2.0 | >2.0* | 10.1 | 2.0 | 8.0 | |
| | | | | | | Accident Total | | | |
| *Continuous Lighting Warranted | Geometric total<br>Operational total<br>Environmental total<br>Accidental total<br><br>Sum<br>Warranting condition | | | | | = ______*<br>= ______*<br>= ______*<br>= ______*<br><br>= ____ Points<br>= 85 Points | | | |

(*Source*-. NJDOT. 2015: *Courtesy*. The State of New Jersey Department of Transportation, USA)

**Table A.5**

| *Evaluation Form For Non-Controlled Access Facility Lighting (New Alignment)* | | | | | | | | | *Lighting Form 6* |
|---|---|---|---|---|---|---|---|---|---|
| | *Rating* | | | | | | | | |
| *Classification Factor* | *1* | *2* | *3* | *4* | *5* | *Unlit Weight (A)* | *Lighted Weight (B)* | *Diff (A-B)* | *Rating X(A-B)* |
| Classification Factor | 1 | 2 | 3 | 4 | 5 | Weight (A) | Weight (B) | Diff (A-B) | Rating X(A-B) |
| Geometric | | | | | | | | | |
| Number Of Lanes | ≤4 | | 6 | | ≥8 | 1.0 | 0.8 | 0.2 | |
| Lane Width | >12′ | 12′ | 11′ | 10′ | <10′ | 3.0 | 2.5 | 0.5 | |
| Median Opening Per Mile | <4 Or One Way | 4.0-8.0 | 8.1-12.0 | 12.1-15.0 | >15 or no access | 5.0 | 3.0 | 2.0 | |
| Curb CuTS | <10% | 10–20% | 20–30% | 30–40% | >40% | 5.0 | 3.0 | 2.0 | |
| Curves | <3.0° | 3.1–6.0° | 6.1–8.0° | 8.1–10.0° | >10° | 13.0 | 5.0 | 8.0 | |
| C-Rades | <3% | 3.0–3.9% | 4.0–4.9% | 5.0–6.9% | >7.0% | 3.2 | 2.8 | 0.4 | |
| Sight Distance | >700′ | 500′–700′ | 300′–500′ | 200′–300′ | <200′ | 2.0 | 1.8 | 0.2 | |

(Continued)

**Table A.5** (Continued)

| *Evaluation Form For Non-Controlled Access Facility Lighting (New Alignment)* | | | | | | | | | *Lighting Form 6* |
|---|---|---|---|---|---|---|---|---|---|
| | *Rating* | | | | | | | | |
| *Classification Factor* | *1* | *2* | *3* | *4* | *5* | *Unlit Weight (A)* | *Lighted Weight (B)* | *Diff (A-B)* | *Rating X(A-B)* |
| Parking | Prohibited both sides | loading zone only | off peak only | permitted one side | permitted both sides | 0.2 | 0.1 | 0.1 | |
| | | | | | | Geometric Total | | | |
| Operational | | | | | | | | | |
| Signals | All Major Intersections | Majority Of Intersections | Most Major Intersections | $\frac{1}{2}$ The Intersections | Frequent Non-Signalized Intersections | 3.0 | 2.8 | 0.2 | |
| Left Turn Lane | All Major Intersections Or One Way | Majority Of Intersections | Most Major : Intersections | $\frac{1}{2}$ Major Intersections | Infrequent Or Undivided Street | 5.0 | 4.0 | 1.0 | |
| Median Width | 30′ | 20′–30′ | 10′–20′ | 4′–10′ | 0–4′ | 1.0 | 0.5 | 0.5 | |
| Speed Limit | <25 | 30 | 35 | 40 | >45 | 1.0 | 0.2 | 0.8 | |
| Night Pedestrian Traffic (Pe D S/ Km) | Few Or None | 0–50 | 50–100 | 100–200 | >200 | 1.5 | 0.5 | 1.0 | |
| | | | | | | Operational Total | | | |
| Environmental | | | | | | | | | |
| % Development | 0% | 0–30% | 30–60% | 60–90% | 100% | 0.5 | 0.3 | 0.2 | |
| Type Of Development | Undeveloped | Residential | $\frac{1}{2}$ Residential And/Or Commercial | Industrial Or Commercial | Strip Industry Cpc Cvvepcial | 0.5 | 0.3 | 0.2 | |
| Set Back Distance | >200′ | 150′–200′ | 100′–150′ | 50′–100′ | <50′ | 0.5 | 0.3 | 0.2 | |
| Advertising/ Area Lighting | None | 0–40% | 41–60% | 61–80% | Continuous | 3.0 | 1.0 | 2.0 | |
| Raised Curb Median | None | Continuous | At All | All Signalized Intersections | Few Locations | 1.0 | 0.5 | 0.5 | |

(Continued)

**Table A.5** (Continued)

| *Evaluation Form For Non-Controlled Access Facility Lighting (New Alignment)* | | | | | | | | | *Lighting Form 6* |
|---|---|---|---|---|---|---|---|---|---|
| | *Rating* | | | | | | | | |
| *Classification Factor* | *1* | *2* | *3* | *4* | *5* | *Unlit Weight (A)* | *Lighted Weight (B)* | *Diff (A-B)* | *Rating X(A-B)* |
| Crime Rate | Extremely Low | Lower Than City Average | City Average | Higher Than City Average | Extremely High | 1.0 | 0.5 | 0.5 | |
| | | | | | | Environmental Total | | | |
| | Geometric Total<br>Operational Total<br>Environmental Total<br><br>Sum<br>Warranting ConditioN | | | | | = ______*<br>= ______*<br>= ______*<br><br>= Points<br>= 60 Points | | | |

(*Source*: NJDOT, 2015; *Courtesy*: The State of New-Jersey Department of Transportation, USA)

**Table A.6**

| *Evaluation Form For Intersection Lighting* | | | | | | | | | *Lighting Form 7* |
|---|---|---|---|---|---|---|---|---|---|
| | *Rating* | | | | | | | | |
| *Classification Factor* | *1* | *2* | *3* | *4* | *5* | *Unlit Weight (A)* | *Lighted Weight (B)* | *Diff (A-B)* | *Rating X(A-B)* |
| Geometric | | | | | | | | | |
| Number Of Legs | | 3 | A | s | >6 Incl Traffic Circle | 3.0 | 2.5 | 0.5 | |
| Approach Lane W1dth | >12′ | 12′ | 11′ | 10′ | <10′ | 3.0 | 2.5 | 0.5 | |
| Channelization | No Turn Lanes | Left Turn On Major Leg | Left Turn All Legs Right Turn On Major Legs | Left And Right Turn On Major. Legs | Left And Right Turn On All Legs | 2.0 | 1.0 | 1.0 | |
| Curvature On Approach Legs | <3.0° | 3.1–6.0° | 6.1–8.0° | 8.1–10.0° | >10° | 13.0 | 5.0 | 8.0 | |
| Graces On Approach | <3% | 3.0–3.9% | 4.0-4.9% | 5.0–6.9% | >7.0% | 3.2 | 2.8 | 0.4 | |
| Approach Sight Distance | >700′ | 500′–700′ | 300′–500′ | 200′–300′ | <200′ | 2.0 | 1.8 | 0.2 | |
| Parking | Prohibited Both Sides | Loading Zone Only | Off Peak Only | Permitted One Side | Permitted Both Sides | 0.2 | 0.1 | 0.1 | |

(Continued)

**Table A.6** (Continued)

| *Evaluation Form For Intersection Lighting* | | | | | | | | | *Lighting Form 7* |
|---|---|---|---|---|---|---|---|---|---|
| | *Rating* | | | | | | | | |
| *Classification Factor* | *1* | *2* | *3* | *4* | *5* | *Unlit Weight (A)* | *Lighted Weight (B)* | *Diff (A-B)* | *Rating X(A-B)* |
| | | | | | | Geometric Total | | | |
| Operational | | | | | | | | | |
| Type Of Control | All Phases Signalized (Incl Turn Lane) | Left Turn Lane Signal Control | Thru Traffic Signal Control Only | 4-Way Stop Control | Stop Control To Minor Leg Or No Control | 3.0 | 2.7 | 03 | |
| Channelization | Left And Right Signal Control | Left And Right Turn Lane Signal Control Major Leg | Left Turn Lane Signal Control All Leg | Left Turn Lane Signal Control Major Leg | No Turn Lane Control | 3.0 | 2.0 | 1.0 | |
| Level Of Service (Load Factor) | A<br>0 | B<br>0–0.1 | C<br>0.1–0.3 | D<br>0.3–0.7 | E<br>0.7–1.0 | 1.0 | 0.2 | 0.8 | |
| Speed Limit On Approach Legs | ≤25 | 30 | 3S | 40 | ≥45 | 1.0 | 0.2 | 0.8 | |
| Night Pedestrian Traffic (Feds/Km) | FEW OR NONE | 0-50 | 50-100 | 100-200 | >200 | 1.5 | 0.5 | 1.0 | |
| | | | | | | Operational Total | | | |
| Environmental | | | | | | | | | |
| % Development | 0% | 0–3% | 30–60% | 60–90% | 100% | 0.5 | 0.3 | 02 | |
| Type Of Development Near Intersection | Undeveloped | Residential | $\frac{1}{2}$ And/Or Commercial | Industrial Or Commercial | Strip Industrial Or Commercial | 0.5 | 0.3 | 0.2 | |
| Lighting In Immediate Vicinity | None | 0–40% | 41–60% | 61–80% | Continuous | 3.0 | 1.5 | 1.5 | |
| Crime Rate | Extremely Low | Lower Thak City Average | City Average | Higher Than Cm Average | Extremely High | 1.0 | 0.5 | 0.5 | |
| | | | | | | Environmental Total | | | |
| Accidents | | | | | | | | | |
| Ratio Of Night-To-Day Accidents | <1.0 | 1.0–1.2 | 1.2–1.5 | 1.5–2.0 | >2.0* | 10.0 | 2.0 | 8.0 | |
| | | | | | Accident total | | | | |

(Continued)

**Table A.6** (Continued)

| *Evaluation Form For Intersection Lighting* | | | | | | | | | *Lighting Form 7* |
|---|---|---|---|---|---|---|---|---|---|
| | *Rating* | | | | | | | | |
| *Classification Factor* | *1* | *2* | *3* | *4* | *5* | *Unlit Weight (A)* | *Lighted Weight (B)* | *Diff (A-B)* | *Rating X(A-B)* |
| *Continuous Lighting Warranted | Geometric Total<br>Operational Total<br>Environmental Total<br>Accident Total<br><br>Sum<br>Warranting condition | | | | = ______*<br>= ______*<br>= ______*<br>= ______*<br><br>= Points<br>= 75 Points | | | | |

(*Source*: NJDOT, 2015; *Courtesy*. The State of New Jersey Department of Transportation, USA)

**Table A.7**

| *Evaluation Form For Intersection Lighting (New Roadway/Alignment)* | | | | | | | | | *Lighting Form 8* |
|---|---|---|---|---|---|---|---|---|---|
| | *Rating* | | | | | | | | |
| *Classification Factor* | *1* | *2* | *3* | *4* | *5* | *Unlit Weight (A)* | *Lighted Weight (B)* | *Diff (A-B)* | *Rating X(A-B)* |
| Geometric | | | | | | | | | |
| Number Of Legs | | 3 | 4 | 5 | ≥6 Incl Traffic Circle | 3.0 | 2.5 | 0.5 | |
| Approach Laws Width | >12′ | 12′ | 11′ | 10′ | <10′ | 3.0 | 2.5 | 0.5 | |
| Channelization | No Turn Lanes | Left Turn On Major Leg | Left Turn All Legs-Right Turn On Major Legs | Left And Right Turn On Major Legs | Left And Right Turn On All Lege | 2.0 | 1.0 | 1.0 | |
| Curvature On Approach Legs | <3.0° | 3.1-6.0° | 6.1-8.0° | 8.1-10.0° | >10° | 13.0 | 5.0 | 8.0 | |
| Grades On Approach | <3% | 3.0-3.9% | 4.0-4.9% | 5.0-6.9% | >7.0% | 3.2 | 2.8 | 0.4 | |
| Approach Sight Distance | >700 ′ | 500′-700′ | 300′-500′ | 200′-300′ | <200′ | 2.0 | 1.8 | 0.2 | |
| Parking | Prohibited Both Sides | Loading Zone Only | Off Peak Only | Permitted One Side | Permitted Both Sides | 0.2 | 0.1 | 0.1 | |
| | | | | | | Geometric Total | | | |
| Operational | | | | | | | | | |

(Continued)

**Table A.7** (Continued)

| *Evaluation Form For Intersection Lighting (New Roadway/Alignment)* | | | | | | | | | *Lighting Form 8* |
|---|---|---|---|---|---|---|---|---|---|
| | *Rating* | | | | | | | | |
| *Classification Factor* | *1* | *2* | *3* | *4* | *5* | *Unlit Weight (A)* | *Lighted Weight (B)* | *Diff (A-B)* | *Rating X(A-B)* |
| Type Of Control | All Phas Es Signalized (Incl Turn Lane) | Left Turn Lane Control | Thru Traffic Signal Control Only | 4-Way Stop Control | Stop Contro To Minor Leg Or No Control | 3.0 | 2.7 | 0.3 | |
| Channelization | Left And Right Signal Control | Left And Right Turn Lane Signal Control Major Leg | Left Turn Lane Signal Control Legs | Left Turn Lane Signal Control Major Leg | No Turn Lane Control | 3.0 | 2.0 | 1.0 | |
| Level Of Service (Load Factor) | A 0 | B 0–0.1 | C 0.1–0.3 | D 0.3–0.7 | E 0.7–1.0 | 1.0 | 0.2 | 0.8 | |
| Speed Umit On Approach Legs | ≤25 | 30 | 35 | 40 | ≥45 | 1.0 | 0.2 | 0.8 | |
| Night Pedestrian Traffic Peds/Km | Fewor None | 0-50 | 50-100 | 100-200 | >200 | 1.5 | 0.5 | 1.0 | |
| | | | | | | Operational Total | | | |
| Environmental | | | | | | | | | |
| %Development | 0% | 0-30% | 30-60% | 60-90% | 100% | 0.5 | 0.3 | 0.2 | |
| Type Of Development Near Intersection | Undeveloped | Residential | $\frac{1}{2}$ Residentialand/ Or Commercial | Industrial Or Com Mercia. | Strip Industry Or. Commercial | 0.5 | 0.3 | 0.2 | |
| Lighting In Immediate Vicinity | None | 0–40% | 41–60% | 61–80% | continuous | 3.0 | 1.5 | 1.5 | |
| Crime Rate | Extremely Low | Lower Than City Average | City Average | Higher Than City Average | Extremely High | 1.0 | 0.5 | 0.5 | |
| | | | | | | Environmental Total | | | |
| | Geometric Total<br>Operational Total<br>Environmental Total<br><br>Sum<br>Warranting Condition | | | | | = _________ *<br>= _________ *<br>= _________ *<br><br>= Points<br>= 50 Points | | | |

*Source*: NJDOT, 2015; *Courtesy*. The State of New Jersey Department of Transportation, USA)

# Index

## B



## C

## D

## Q

## S

## T

## U

## V



## W



## Y